Origins, Worlds, and Life

A Decadal Strategy for Planetary Science and Astrobiology 2023–2032

Committee on the Planetary Science and Astrobiology Decadal Survey

Space Studies Board

Division on Engineering and Physical Sciences

A Consensus Study Report of

The National Academies of
SCIENCES · ENGINEERING · MEDICINE

THE NATIONAL ACADEMIES PRESS
Washington, DC
www.nap.edu

THE NATIONAL ACADEMIES PRESS   **500 Fifth Street, NW**   **Washington, DC 20001**

This study is based on work supported by Contract No. NNH17CB02B/NNH17CB01T with the National Aeronautics and Space Administration and Grant No. 2040016 with the National Science Foundation. Any opinions, findings, conclusions, or recommendations expressed in this publication do not necessarily reflect the views of any agency or organization that provided support for the project.

International Standard Book Number-13: 978-0-309-47578-5
International Standard Book Number-10: 0-309-47578-3
Digital Object Identifier: https://doi.org/10.17226/26522
Library of Congress Control Number: 2023938864

Cover: Design by Paul Byrne.
Images: Foreground waves: Conrad Ziebland; protoplanetary disk: ESO/L. Calçada; Saturn: NASA; Earth: ESA/OSIRIS Team MPS/UPD/LAM/IAA/RSSD/INTA/UPM/DASP/IDA/Gordan Ugarković; magma ocean world: ESA/Hubble, M. Kornmesser; comet: NASA/ESA/Hubble Heritage Team (STScI-AURA); hydrothermal vent: MARUM–Zentrum für Marine Umweltwissenschaften, Universität Bremen; starfield: ESO/S. Brunier.

Copies of this publication are available free of charge from:

Space Studies Board
National Academies of Sciences, Engineering, and Medicine
500 Fifth Street, NW
Washington, DC 20001

This publication is available from the National Academies Press, 500 Fifth Street, NW, Keck 360, Washington, DC 20001; (800) 624-6242 or (202) 334-3313; http://www.nap.edu.

Printed in the United States of America

Suggested citation: National Academies of Sciences, Engineering, and Medicine. 2023. *Origins, Worlds, and Life: A Decadal Strategy for Planetary Science and Astrobiology 2023–2032*. Washington, DC: The National Academies Press. https://doi.org/10.17226/26522.

The National Academies of
SCIENCES · ENGINEERING · MEDICINE

Consensus Study Reports published by the National Academies of Sciences, Engineering, and Medicine document the evidence-based consensus on the study's statement of task by an authoring committee of experts. Reports typically include findings, conclusions, and recommendations based on information gathered by the committee and the committee's deliberations. Each report has been subjected to a rigorous and independent peer-review process and it represents the position of the National Academies on the statement of task.

Proceedings published by the National Academies of Sciences, Engineering, and Medicine chronicle the presentations and discussions at a workshop, symposium, or other event convened by the National Academies. The statements and opinions contained in proceedings are those of the participants and are not endorsed by other participants, the planning committee, or the National Academies.

For information about other products and activities of the National Academies, please visit www.nationalacademies.org/about/whatwedo.

COMMITTEE ON THE PLANETARY SCIENCE AND ASTROBIOLOGY DECADAL SURVEY

Steering Group

ROBIN M. CANUP, NAS,[1] Southwest Research Institute, *Co-Chair*
PHILIP R. CHRISTENSEN, Arizona State University, *Co-Chair*
MAHZARIN R. BANAJI, NAS, Harvard University
STEVEN J. BATTEL, NAE,[2] Battel Engineering
LARS E. BORG, Lawrence Livermore National Laboratory
ATHENA COUSTENIS, National Centre for Scientific Research, Meudon Observatory
JAMES H. CROCKER, NAE, Lockheed Martin Space Systems Company
BRETT W. DENEVI, Johns Hopkins University Applied Physics Laboratory
BETHANY L. EHLMANN, California Institute of Technology
LARRY W. ESPOSITO, University of Colorado Boulder
ORLANDO FIGUEROA, Orlando Leadership Enterprise
JOHN M. GRUNSFELD, Endless Frontier Associates
JULIE HUBER, Woods Hole Oceanographic Institution
KRISHAN KHURANA, University of California, Los Angeles
WILLIAM B. McKINNON, Washington University in St. Louis
FRANCIS NIMMO, NAS, University of California, Santa Cruz
CAROL RAYMOND, Jet Propulsion Laboratory
BARBARA SHERWOOD LOLLAR, NAS/NAE, University of Toronto
AMY SIMON, NASA Goddard Space Flight Center

Panel on Giant Planet Systems

JONATHAN I. LUNINE, NAS, Cornell University, *Chair*
AMY SIMON, NASA Goddard Space Flight Center, *Vice Chair*
FRANCES BAGENAL, NAS, University of Colorado Boulder
RICHARD W. DISSLY, Ball Aerospace and Technologies Corporation
LEIGH N. FLETCHER, University of Leicester
TRISTAN GUILLOT, Nice Observatory
MATTHEW HEDMAN, University of Idaho
RAVIT HELLED, University of Zurich
KATHLEEN E. MANDT, Johns Hopkins University Applied Physics Laboratory
ALYSSA RHODEN, Southwest Research Institute
PAUL M. SCHENK, Lunar and Planetary Institute
MICHAEL H. WONG, SETI Institute

Panel on Mars

VICTORIA E. HAMILTON, Southwest Research Institute, *Chair*
BETHANY L. EHLMANN, California Institute of Technology, *Vice Chair*
WILLIAM B. BRINCKERHOFF, NASA Goddard Space Flight Center
TRACY K.P. GREGG, University of Buffalo
JASPER S. HALEKAS, University of Iowa
JOHN "JACK" W. HOLT, University of Arizona
JOEL HUROWITZ, Stony Brook University

---

[1]Member, National Academy of Sciences.
[2]Member, National Academy of Engineering.

BRUCE M. JAKOSKY, University of Colorado Boulder
MICHAEL MANGA, NAS, University of California, Berkeley
HARRY Y. McSWEEN, NAS, University of Tennessee
CLAIRE E. NEWMAN, Aeolis Research
ALEJANDRO M. SAN MARTIN, NAE, Jet Propulsion Laboratory
KIRSTEN L. SIEBACH, Rice University
AMY WILLIAMS, University of Florida
ROBIN D. WORDSWORTH, Harvard University

Panel on Mercury and the Moon

TIMOTHY L. GROVE, NAS, Massachusetts Institute of Technology, *Chair*
BRETT W. DENEVI, Johns Hopkins University Applied Physics Laboratory, *Vice Chair*
JAMES DAY, University of California, San Diego
ALEXANDER J. EVANS, Brown University
SARAH FAGENTS, University of Hawaii at Manoa
WILLIAM M. FARRELL, NASA Goddard Space Flight Center
CALEB I. FASSETT, NASA Marshall Space Flight Center
JENNIFER L. HELDMANN, NASA Ames Research Center
MASATOSHI HIRABAYASHI, Auburn University
JAMES TUTTLE KEANE, Jet Propulsion Laboratory
FRANCIS McCUBBIN, NASA Johnson Space Center
MIKI NAKAJIMA, University of Rochester
MARK P. SAUNDERS, Independent Consultant
SONIA M. TIKOO-SCHANTZ, Stanford University

Panel on Ocean Worlds and Dwarf Planets

ALEXANDER G. HAYES, Cornell University, *Chair*
FRANCIS NIMMO, NAS, University of California, Santa Cruz, *Vice Chair*
MORGAN L. CABLE, Jet Propulsion Laboratory
ALFONSO DAVILA, NASA Ames Research Center
GLEN FOUNTAIN, Johns Hopkins University Applied Physics Laboratory
CHRISTOPHER R. GERMAN, Woods Hole Oceanographic Institution
CHRISTOPHER R. GLEIN, Southwest Research Institute
CANDICE HANSEN, Planetary Science Institute
EMILY S. MARTIN, National Air and Space Museum, Smithsonian Institution
MARC NEVEU, University of Maryland
CAROL S. PATY, University of Oregon
LYNNAE C. QUICK, NASA Goddard Space Flight Center
JASON M. SODERBLOM, Massachusetts Institute of Technology
KRISTA M. SODERLUND, University of Texas Institute for Geophysics

Panel on Small Solar System Bodies

NANCY L. CHABOT, Johns Hopkins University Applied Physics Laboratory, *Chair*
CAROL RAYMOND, Jet Propulsion Laboratory, *Vice Chair*
PAUL A. ABELL, NASA Johnson Space Center
WILLIAM F. BOTTKE, Southwest Research Institute

HAROLD C. CONNOLLY, JR., Rowan University
THOMAS D. JONES, Association of Space Explorers
STEFANIE N. MILAM, NASA Goddard Space Flight Center
EDGARD G. RIVERA-VALENTÍN, Lunar and Planetary Institute
DANIEL J. SCHEERES, NAE, University of Colorado Boulder
RHONDA STROUD, Naval Research Laboratory
MEGAN BRUCK SYAL, Lawrence Livermore National Laboratory
MYRIAM TELUS, University of California, Santa Cruz
AUDREY THIROUIN, Lowell Observatory
CHAD TRUJILLO, Northern Arizona University
BENJAMIN P. WEISS, Massachusetts Institute of Technology

Panel on Venus

PAUL K. BYRNE, Washington University in St. Louis, *Chair*
LARRY W. ESPOSITO, University of Colorado, *Vice Chair*
GIADA N. ARNEY, NASA Goddard Space Flight Center
AMANDA S. BRECHT, NASA Ames Research Center
THOMAS E. CRAVENS, University of Kansas
KANDIS-LEA JESSUP, Southwest Research Institute
JAMES F. KASTING, NAS, The Pennsylvania State University
SCOTT D. KING, Virginia Polytechnic Institute and State University
BERNARD MARTY, Université de Lorraine
THOMAS NAVARRO, University of California, Los Angeles
JOSEPH G. O'ROURKE, Arizona State University
JENNIFER M. ROCCA, Jet Propulsion Laboratory
ALISON R. SANTOS, Wesleyan University
JENNIFER L. WHITTEN, Tulane University

Staff

DAVID H. SMITH, Senior Program Officer, Space Studies Board, *Study Director*
DWAYNE A. DAY, Senior Program Officer, Aeronautics and Space Engineering Board
DANIEL NAGASAWA, Program Officer, Space Studies Board
JORDYN WHITE, Program Officer, Committee on National Statistics
CARL-GUSTAV ANDERSON, Associate Program Officer, Board on Mathematical Sciences and Analytics
MIA BROWN, Research Associate, Space Studies Board
MEGAN A. CHAMBERLAIN, Senior Program Assistant, Space Studies Board
GAYBRIELLE HOLBERT, Program Assistant, Space Studies Board
DIONNA ALI, Associate Program Officer, Intelligence Community Studies Board
ELI NASS, Research Assistant, Division on Engineering and Physical Sciences
KATHERINE DZURILLA, Temporary Research Assistant, Space Studies Board
LUCIA ILLIARI, Temporary Research Assistant, Space Studies Board
JEAN DE BECDELIEVRE, Christine C. Mirzayan Science and Technology Policy Fellow (2021)
JACOB N.H. ABRAHAMS, Lloyd V. Berkner Space Policy Intern (2021)
TARINI KONCHADY, Lloyd V. Berkner Space Policy Intern (2021)
DYLAN PITTS, Lloyd V. Berkner Space Policy Intern (2023)
COLLEEN N. HARTMAN, Director, Space Studies Board

Preface

The Planetary Science Division (PSD) of NASA's Science Mission Directorate (SMD) is the primary source of funding for planetary science, astrobiology, and planetary defense activities in the United States. In addition, the National Science Foundation (NSF) provides modest, but highly important, support for a variety of supporting ground-based activities, most notably access to world-class, ground-based optical and radio telescopes.

The allocation of resources within and among spacecraft missions, supporting research activities, and technology development is determined to a major extent via a relatively mature strategic planning process that relies heavily on inputs from the scientific community to establish the scientific basis and direction for its space-science flight- and ground-research programs and technology development activities.

The primary sources of this guidance are the independent scientific analyses and recommendations provided by reports of the National Academies of Sciences, Engineering, and Medicine (e.g., by the Space Studies Board [SSB] and its committees) and, to a lesser extent, by parallel inputs coming from community-based, but NASA-organized, analysis/assessment groups (e.g., the Mars Exploration Program Analysis Group and the Outer Planets Assessment Group). The science strategies developed by the SSB and the analysis/assessment groups form input to subsequent program development activities conducted by the FACA-chartered NASA Advisory Council and its associated committees (e.g., NASA's Planetary Science Advisory Committee).

The SSB's primary vehicles for the provision of strategic advice to NASA are the space science decadal surveys. The National Academies decadal surveys are widely recognized among policymakers and program managers as key resources in determining where a field of research is and where it is headed. Indeed, the decadal survey process has proved so useful that Section 1104 of the NASA Authorization Act of 2008 requires that the NASA "Administrator shall enter into agreements on a periodic basis with the National Academies for independent assessments, also known as decadal surveys, to take stock of the status and opportunities for Earth and space science discipline fields and aeronautics research and to recommend priorities for research and programmatic areas over the next decade."

The most recent effort for planetary science and astrobiology resulted in the publication of *Vision and Voyages for Planetary Science in the Decade 2013–2022* in 2011. While it is generally regarded that *Vision and Voyages* was especially successful in its outcomes—as witnessed by the facts that the survey's top two large-class mission priorities are both under development and that PSD's annual budget has doubled over the past decade—a new survey is needed to address the challenges of the coming decade.

Following informal requests in the early months of 2019 from the director of PSD, the SSB and its Committee on Astrobiology and Planetary Science (CAPS) began the task of defining the specific actions and issues that needed to be address in a new decadal survey. CAPS's activities culminated in the convening of a decadal survey organizing meeting, held at the California Institute of Technology's Keck Institute for Space Studies in September 2019. Negotiations between NASA and the SSB continued through the final months of 2019 and eventually settled on a statement of task calling for a decadal survey that provided a clear exposition of the following:[1]

1. An overview of planetary science, astrobiology, and planetary defense: what they are, why they are compelling undertakings, and the relationship between space- and ground-based research.
2. A broad survey of the current state of knowledge of the solar system.
3. The most compelling science questions, goals, and challenges that should motivate future strategy in planetary science, astrobiology, and planetary defense.
4. A coherent and consistent traceability of recommended research and missions to objectives and goals.
5. A comprehensive research strategy to advance the frontiers of planetary science, astrobiology, and planetary defense during the period 2023–2032 that will include identifying, recommending, and ranking the highest priority research activities (research activities include any project, facility, experiment, mission, or research program of sufficient scope to be identified separately in the final report). For each activity, consideration should be given to the scientific case, international and private landscape, timing, cost category and cost risk, as well as technical readiness, technical risk, lifetime, and opportunities for partnerships. The strategy should be balanced by consideration of large, medium, and small research activities for both ground and space.
6. Recommendations for decision rules, where appropriate, for the comprehensive research strategy that can accommodate significant but reasonable deviations in the projected budget or changes in urgency precipitated by new discoveries or technological developments.
7. An awareness of the science and space mission plans and priorities of NASA human space exploration programs and potential foreign and U.S. agency partners reflected in the comprehensive research strategy and identification of opportunities for cooperation, as appropriate.
8. The opportunities for collaborative research that are relevant to science priorities among SMD's four science divisions (for example, comparative planetology approaches to exoplanet or astrobiology research); between NASA SMD and the other NASA mission directorates; between NASA and the NSF; between NASA and other U.S. government entities; between NASA and private sector organizations; and between NASA and its international partners.
9. The state of the profession, including issues of diversity, inclusion, equity, and accessibility; the creation of safe workspaces; and recommended policies and practices to improve the state of the profession. Where possible, provide specific, actionable, and practical recommendations to the agencies and community to address these areas.

In response to this request, the National Academies established the Committee on the Planetary Science and Astrobiology Decadal Survey (hereafter, the "survey committee" or the "committee") consisting of a 19-member steering group and 78 additional experts organized into six topical panels. The co-chairs of the survey committee were appointed in May 2020, and the members of the panels were identified and appointed in the subsequent spring and summer months.

The steering group held its first meeting on September 30, 2020, and held its 22nd and final meeting on November 2, 2021. The six panels each held at least 20 meetings during the period October 2020 to September 2021. Notably, each and every single meeting was held virtually because of the ongoing COVID-19 pandemic. The work of the survey committee can be divided into three distinct phases: the last 3 months of 2020, the first 9 months of 2021, and late summer/early autumn of 2021.

In phase one, the steering group deliberated on and defined the key science questions around which the report would be structured. In parallel, the panels ingested and assessed candidate missions already studied, and identified additional concepts deemed worthy of study. Phase one ended with the development of two key items: first, a cross-survey consensus that the most appropriate key questions had been identified, and second, the prioritization

---

[1] See Appendix A for the letter requesting this study, the full text of the statement of task, and additional, nonbinding guidelines.

by the steering group of 10 new mission concepts worthy of additional study. These 10 new concepts were subsequently forwarded to NASA for detailed study. To ensure that the panels would perform their initial task in an expeditious manner, they were organized and appointed so that each would have responsibility for different portions of the solar system—that is, Mercury and the Moon, Venus, Mars, giant planet systems, ocean worlds and dwarf planets, and small solar system bodies.

During the second phase, the panels worked with mission-design teams at the Jet Propulsion Laboratory, NASA Goddard Space Flight Center, and at the Johns Hopkins University Applied Physics Laboratory to develop the 10 new mission concepts. In parallel, a series of approximately 20 informal, cross-survey writing groups—each consisting of 5–10 members from the steering group and the panels—were established to create the initial drafts of the chapters in this report devoted to the key science questions and to programmatic issues such as the state of the profession, research and analysis, and technology development. Once the additional mission studies were completed, their sponsoring panels performed a comparative assessment of the degree to which the new concepts and other proposed and studied missions could address the survey's key science questions. This phase of the survey ended with the completion of the initial drafts of 20 of this report's 23 chapters and the prioritization by the steering group of 17 mission concepts (some new and some old) for detailed technical risk and cost evaluation (TRACE) by the Aerospace Corporation.

Phase three involved the scheduling of some 20 survey-wide "summit meetings," during which the text produced by each writing group was subjected to intense comment, review, and subsequent revision. In parallel, the steering group assessed the results of the TRACE analyses, selected the most promising ones, prioritized them, and, thus, established the survey's list of recommended mission activities for the coming decade. In addition, the steering group worked with the leaders of each writing group to integrate the draft text of the various chapters into a self-consistent and coherent program of activities for the next decade. The final task performed was the drafting by the steering group of the summary and the chapter describing the recommended program of activities for the period 2023–2032.

Final sections of the report were drafted, assembled, and integrated in October and November 2021. The text was sent to external reviewers in December, was revised between January and February 2022, was formally approved for release by the National Academies on March 24, 2022, and was publicly unveiled on April 19, 2022.

The work of the committee was made easier thanks to the important help given by individuals too numerous to list—indeed, printing just the names of those individuals who made public presentations to the survey committee would require two full pages of text—at a variety of public and private organizations, who made presentations at committee meetings, drafted white papers, and participated in mission studies. Important contributions were also made by the TRACE team at the Aerospace Corporation, led by Russell Persinger, Justin Yoshida, and Mark Barrera. Last, the survey committee thanks Kellie Mendelow for her invaluable record keeping, file management, and editorial assistance.

Acknowledgment of Reviewers

This Consensus Study Report was reviewed in draft form by individuals chosen for their diverse perspectives and technical expertise. The purpose of this independent review is to provide candid and critical comments that will assist the National Academies of Sciences, Engineering, and Medicine in making each published report as sound as possible and to ensure that it meets the institutional standards for quality, objectivity, evidence, and responsiveness to the study charge. The review comments and draft manuscript remain confidential to protect the integrity of the deliberative process.

We thank the following individuals for their review of this report:

Irene Blair, University of Colorado,
Mark Boslough, University of Arizona,
John Chambers, Carnegie Institution for Science,
Leroy Chiao, Independent Consultant,
Gerhard Drolshagen, Carl von Ossietzky University of Oldenburg,
Katherine H. Freeman, NAS,[1] The Pennsylvania State University,
B. Scott Gaudi, The Ohio State University,
Gerald F. Joyce, NAS/NAM,[2] Salk Institute for Biological Studies,
Antonio Lazcano, National Autonomous University of Mexico,
Mark S. Marley, University of Arizona,
Timothy J. McCoy, National Museum of Natural History, Smithsonian Institution,
Melissa A. McGrath, SETI Institute,
Charles Norton, Jet Propulsion Laboratory,
Louise M. Prockter, Johns Hopkins University Applied Physics Laboratory,
Joseph H. Rothenberg, Independent Consultant,
Teresa Segura, Boeing Horizon X Ventures,
Sean C. Solomon, NAS, Lamont-Doherty Earth Observatory, Columbia University,
David J. Stevenson, NAS, California Institute of Technology,

---

[1] Member, National Academy of Sciences.
[2] Member, National Academy of Medicine.

Sarah T. Stewart, University of California, Davis,
Grant H. Stokes, NAE,[3] MIT Lincoln Laboratory,
Theresa A. Sullivan, University of Virginia, and
Orenthal J. Tucker, NASA Goddard Space Flight Center

Although the reviewers listed above have provided many constructive comments and suggestions, they were not asked to endorse the conclusions or recommendations of this report nor did they see the final draft before its release. The review of this report was overseen by Rosaly M. Lopes, Jet Propulsion Laboratory, and Norman H. Sleep, NAS, Stanford University. They were responsible for making certain that an independent examination of this report was carried out in accordance with the standards of the National Academies and that all review comments were carefully considered. Responsibility for the final content of the report rests entirely with the authoring committee and the National Academies.

---

[3]Member, National Academy of Engineering.

Contents

Dedication

*This report is dedicated to the memory of H. Jay Melosh (1947–2020)
who had agreed to play an important role in this study.*

Summary

This report of the Committee on the Planetary Science and Astrobiology Decadal Survey of the National Academies of Sciences, Engineering, and Medicine identifies a research strategy to maximize advancement of planetary science, astrobiology, and planetary defense in the 2023–2032 decade. Federal investment in these activities occurs primarily through NASA's Planetary Science Division (PSD); important activities are also conducted by the National Science Foundation (NSF). The decadal survey committee evaluated potential activities by their capacity to address the priority science questions identified by the committee (Table S-1), cost and technical readiness as assessed through independent evaluation, programmatic balance, and other factors. This summary highlights the committee's top findings and recommendations.

STATE OF THE PROFESSION

The state of the profession (SoP)—including issues of diversity, equity, inclusivity, and accessibility (DEIA)—is central to the success of the planetary science enterprise. Its inclusion here, *for the first time in a planetary science decadal survey*, reflects its importance and urgency. Ensuring broad access and participation is essential to maximizing excellence in an environment of fierce competition for limited human resources, and to ensuring continued American leadership in planetary science and astrobiology (PS&AB). A strong system of equity and accountability is required to recruit, retain, and nurture the best talent into the PS&AB community. The committee applauds the hard-earned progress that has been made—most notably with respect to the entry and prominence of women in the field—as well as the exemplary goals and intentions of NASA science leadership with respect to DEIA. However, much work remains to be done, in particular to address persistent and troubling issues of basic representation by race/ethnicity.

The committee's eight SoP recommendations (see Chapter 16) address:

1. *An evidence gathering imperative.* Equity and accountability require accurate and complete data about the SoP. There is an urgent need for data concerning the size, identity, and demographics of the PS&AB community, and workplace climate. Without such data, it cannot be known if the best available talent is being utilized, nor how involvement may be undermined by adverse experiences.
2. *Education of individuals about the costs of bias and improvement of institutional procedures, practices, and policies.* The committee recommends that the PSD adopt the view that bias can be both unintentional and

TABLE S-1 The Twelve Priority Science Questions

Scientific Themes	Priority Science Question Topics and Descriptions
(A) Origins	Q1. *Evolution of the protoplanetary disk.* What were the initial conditions in the solar system? What processes led to the production of planetary building blocks, and what was the nature and evolution of these materials?
	Q2. *Accretion in the outer solar system.* How and when did the giant planets and their satellite systems originate, and did their orbits migrate early in their history? How and when did dwarf planets and cometary bodies orbiting beyond the giant planets form, and how were they affected by the early evolution of the solar system?
	Q3. *Origin of Earth and inner solar system bodies.* How and when did the terrestrial planets, their moons, and the asteroids accrete, and what processes determined their initial properties? To what extent were outer solar system materials incorporated?
(B) Worlds and Processes	Q4. *Impacts and dynamics.* How has the population of solar system bodies changed through time, and how has bombardment varied across the solar system? How have collisions affected the evolution of planetary bodies?
	Q5. *Solid body interiors and surfaces.* How do the interiors of solid bodies evolve, and how is this evolution recorded in a body's physical and chemical properties? How are solid surfaces shaped by subsurface, surface, and external processes?
	Q6. *Solid body atmospheres, exospheres, magnetospheres, and climate evolution.* What establishes the properties and dynamics of solid body atmospheres and exospheres, and what governs material loss to space and exchange between the atmosphere and the surface and interior? Why did planetary climates evolve to their current varied states?
	Q7. *Giant planet structure and evolution.* What processes influence the structure, evolution, and dynamics of giant planet interiors, atmospheres, and magnetospheres?
	Q8. *Circumplanetary systems.* What processes and interactions establish the diverse properties of satellite and ring systems, and how do these systems interact with the host planet and the external environment?
(C) Life and Habitability	Q9. *Insights from terrestrial life.* What conditions and processes led to the emergence and evolution of life on Earth; what is the range of possible metabolisms in the surface, subsurface, and/or atmosphere; and how can this inform our understanding of the likelihood of life elsewhere?
	Q10. *Dynamic habitability.* Where in the solar system do potentially habitable environments exist, what processes led to their formation, and how do planetary environments and habitable conditions co-evolve over time?
	Q11. *Search for life elsewhere.* Is there evidence of past or present life in the solar system beyond Earth, and how do we detect it?
Crosscutting A–C linkage	Q12. *Exoplanets.* What does our planetary system and its circumplanetary systems of satellites and rings reveal about exoplanetary systems, and what can circumstellar disks and exoplanetary systems teach us about the solar system?

pervasive, and provides actionable steps to assist NASA in identifying where bias exists and in removing it from its processes.

3. *Broadening opportunities to advance the SoP.* Engaging underrepresented communities at secondary and college levels to encourage and retain them along PS&AB career pathways is essential to creating and sustaining a diverse community.

4. *Creating an inclusive and inviting community free of hostility and harassment.* Ensuring that all community members are treated with respect, developing and enforcing codes of conduct, and providing ombudsperson support to address issues are important for maintaining healthy and productive work environments.

Together, the SoP findings and recommendations aim to assist NASA's PSD in boldly addressing issues that concern its most important resource: the people who propel its planetary science and exploration missions.

MISSION CLASSES, BALANCE, AND ONGOING ACTIVITIES

The committee's statement of task (Appendix A) defines missions in three cost classes—small, medium, and large. The Discovery program supports small, principal investigator (PI)-led missions that address focused science objectives with a high launch cadence. Medium-class New Frontiers missions are PI-led and address broader science goals. Large ("Flagship") missions address broad, high-priority science objectives with sophisticated instrument payloads and mission designs. Balance across these classes is important to enable a steady stream of new discoveries and the capability to make major scientific advances.

Currently operating PSD spacecraft include the ongoing Mars orbiter missions, the Curiosity and Perseverance Mars rovers; the Lunar Reconnaissance Orbiter; the InSight and Lucy Discovery missions; and the New Horizons, Juno, and OSIRIS-REx New Frontiers (NF) missions. Missions in development include four small SIMPLEx missions, the Psyche, DAVINCI, and VERITAS Discovery missions, the Dragonfly NF mission, and the Europa Clipper large strategic mission. NASA also contributes to international missions (e.g., ESA's BepiColombo, JUICE, and EnVision and JAXA's MMX). The committee strongly supports (1) continuation of these missions and contributions in their current operational or development phases and (2) the Senior Review process for evaluating the merit of additional extended mission phases.

MARS SAMPLE RETURN

The Perseverance rover on Mars is collecting samples from Jezero crater, a former lake basin carved into >3.7-billion-year-old stratigraphy. This was the highest priority large mission in the prior decadal survey, *Vision and Voyages*. NASA, with ESA partnership, is now undertaking Mars Sample Return (MSR) to return those samples to Earth. Sedimentary, igneous, water-altered, and impact-formed rocks accessible in the Jezero region will provide a geological record crucial for understanding Mars's environmental evolution and, potentially, its prebiotic chemistry and biology, in ways that cannot be addressed in situ or with martian meteorites. MSR will provide an invaluable sample collection to the benefit of future generations.

Recommendation: The highest scientific priority of NASA's robotic exploration efforts this decade should be completion of Mars Sample Return as soon as is practicably possible with no increase or decrease in its current scope. (Chapter 22)

Recommendation: Mars Sample Return (MSR) is of fundamental strategic importance to NASA, U.S. leadership in planetary science, and international cooperation and should be completed as rapidly as possible. However, its cost should not be allowed to undermine the long-term programmatic balance of the planetary portfolio. If the cost of MSR increases substantially (≥20 percent) beyond the $5.3 billion level adopted in this report or goes above ~35 percent of the Planetary Science Division budget in any given year, NASA should work with the Administration and Congress to secure a budget augmentation to ensure the success of this strategic mission. (Chapter 22)

MARS EXPLORATION PROGRAM

The Mars Exploration Program (MEP) has a record of success in advancing our understanding of Mars and the evolution of terrestrial planets, technology development, joint mission implementations, and public enthusiasm for planetary science. The committee strongly supports the continuation of MEP and prioritizes Mars Life Explorer (MLE) as the next medium-class Mars mission (see Appendix C).[1] While ancient biosignatures are a focus of MSR, MLE will seek extant life and assess modern habitability through examination of low-latitude ice. MLE will characterize organics, trace gases, and isotopes at a fidelity suitable for biosignature detection; and assess ice stability and the question of modern liquid water via chemical, thermophysical, and atmospheric measurements.

> **Recommendation: NASA should maintain the Mars Exploration Program, managed within the Planetary Science Division, that is focused on the scientific exploration of Mars. The program should develop and execute a comprehensive architecture of missions, partnerships, and technology development to enable continued scientific discovery at Mars.** (Chapter 22)

> **Recommendation: Subsequent to the peak-spending phase of Mars Sample Return, the next priority medium-class mission for the Mars Exploration Program should be Mars Life Explorer.** (Chapter 22)

LUNAR DISCOVERY AND EXPLORATION PROGRAM

The Lunar Discovery and Exploration Program (LDEP) supports industry partnerships and innovative approaches to accomplishing exploration and science goals, including the Commercial Lunar Payload Services (CLPS) program for lunar landing services. LDEP is funded within PSD, but budgetary responsibility is split between PSD and the Exploration Science Strategy and Integration Office (ESSIO). No single organizational chain has authority for executing lunar science and missions; as a result, LDEP activities are currently not optimized to accomplish high-priority science. A structured, science-led approach to setting goals and measurement objectives for the Moon is needed for LDEP and to provide scientific requirements for Artemis.

> **Recommendation: The Planetary Science Division should execute a strategic program to accomplish planetary science objectives for the Moon, with an organizational structure that aligns responsibility, authority, and accountability.** (Chapter 22)

> **Recommendation: The advancement of high-priority lunar science objectives, as defined by the Planetary Science Division based on inputs from this report and groups representing the scientific community, should be a key requirement of the Artemis human exploration program. Design and implementation of an integrated plan responsive to both NASA's human exploration and science directorates, with separately appropriated funding lines, presents management challenges; however, overcoming these is strongly justified by the value of human-scientific and human-robotic partnerships to the agency and the nation.** (Chapter 22)

The committee prioritizes the medium-class Endurance-A lunar rover mission (see Appendix C).[2] Endurance-A will traverse diverse terrains in the South Pole Aiken (SPA) basin, collect approximately 100 kg of samples, and deliver the samples to a location for return to Earth by astronauts. Endurance-A will address the highest priority lunar science, revolutionizing our understanding of the Moon and the early history of the solar system recorded in its most ancient impact basin. Return of Endurance-A samples by Artemis astronauts is the ideal synergy between NASA's human and scientific exploration of the Moon, producing flagship-level science at a fraction of the cost to PSD through coordination with Artemis.

> **Recommendation: Endurance-A should be implemented as a strategic medium-class mission as the highest priority of the Lunar Discovery and Exploration Program. Endurance-A would utilize Commercial Lunar Payload Services to deliver the rover to the Moon, a long-range traverse to collect a substantial mass of high-value samples, and astronauts to return them to Earth.** (Chapter 22)

---

[1] The full Mars Life Explorer mission study report is available at https://tinyurl.com/2p88fx4f.

[2] The full Endurance-A mission study report is available at https://tinyurl.com/2p88fx4f.

RESEARCH AND ANALYSIS

Robotic solar system exploration is driven by the desire to increase knowledge. Strong, steady investment in research and analysis (R&A) is needed to ensure (1) maximal return from mission data; (2) that data drives improved understanding and novel, testable hypotheses; (3) that advances feed into future mission development; and (4) training a diverse workforce. The fraction of PSD's budget devoted to R&A has decreased from 14 percent in 2010 to a projected 7.7 percent by fiscal year (FY) 2023. It is essential to the nation's planetary science efforts that this trend be reversed. The openly competed R&A programs drive innovation, provide rapid response to new discoveries, identify the most meritorious ideas, and attract new and increasingly diverse investigators.

Recommendation: The Planetary Science Division (PSD) should increase its investment in research and analysis (R&A) activities to achieve a minimum annual funding level of 10 percent of the PSD total annual budget. This increase should be achieved through a progressive ramp-up in funding allocated to the openly competed R&A programs, as defined in this decadal survey. Mid-decade, NASA should work with an appropriately constituted independent group to assess progress in achieving this recommended funding level. (Chapters 17 and 22)

PLANETARY DEFENSE

The Planetary Defense Coordination Office within PSD coordinates and supports activities to protect Earth from impacts by near-Earth objects (NEOs). Congressionally directed NEO detection goals will be ideally advanced by the Near-Earth Object Surveyor (NEO Surveyor)—a dedicated, space-based mid-infrared survey currently pending confirmation. Advancement in planetary defense will require assessment of mitigation techniques, as well as the ability to characterize newly identified hazardous objects. NASA's Double Asteroid Redirection Test (DART) mission, scheduled to impact the moonlet of the binary asteroid 65803 Didymos in September 2022, will demonstrate one approach to asteroid deflection.

Recommendation: NASA should fully support the development, timely launch, and subsequent operation of NEO Surveyor to achieve the highest priority planetary defense near-Earth object survey goals. (Chapters 18 and 22)

Recommendation: The highest priority planetary defense demonstration mission to follow Double Asteroid Redirection Test (DART) and the Near-Earth Object Surveyor should be a rapid-response, flyby reconnaissance mission targeted to a challenging near-Earth object (NEO) population—~50- to 100-m-diameter objects posing the highest probability of a destructive Earth impact. Such a mission should assess the capabilities and limitations of flyby characterization methods to better prepare for a short-warning-time NEO threat. (Chapter 18)

DISCOVERY PROGRAM

The Discovery program supports relatively frequent missions that address any science achievable within a specified cost cap, with a central goal to maximize innovative science per total mission cost. The program has made fundamental contributions to planetary exploration and the committee strongly supports its continuation. The committee assessed the cost cap and structure needed to (1) address decadal-level science[3] questions, (2) more clearly anticipate mission life-cycle cost, and (3) maximize science return per dollar.

Recommendation: The Discovery Phase A through F cost cap should be $800 million in fiscal year 2025 dollars, exclusive of the launch vehicle, and periodically adjusted throughout the decade to account for inflation.

---

[3] Decadal-level science is that which results in significant, unambiguous progress in addressing at least one of the survey's 12 priority science questions.

This cap will enable the Discovery Program to continue to support missions that address high-priority science objectives, including those that can reach the outer solar system. (Chapter 22)

NEW FRONTIERS PROGRAM

New Frontiers (NF) missions address broader and/or more technically challenging scientific questions, with higher costs and less frequent launches. NF missions are managed by a limited number of centers, and extensive resources are required for NF mission proposals. It is thus essential that NF missions be strategically designed to address the most important science. Decadal surveys provide the ideal opportunity for a large, diverse group representing the community to prioritize NF mission themes.

> **Recommendation: New Frontiers (NF) mission themes for the NF-6 and NF-7 calls should continue to be specified by the decadal survey. Additional concepts that may arise mid-decade owing to new discoveries should be evaluated by an appropriately constituted group representing the scientific community and considered for addition to NF-7.** (Chapter 22)

Mission life-cycle costs are the primary factor in determining launch cadence for a cost-bounded program like NF. In evaluating the NF cost structure, the committee prioritized enabling access to all targets across the solar system at the potential expense of launch cadence. New Frontiers missions in development, as well as the most scientifically compelling new concepts considered by the committee, have estimated life-cycle costs substantially greater than the prior NF cost cap. These missions are representative of the nature and breadth of science optimally addressed in the NF program.

> **Recommendation: New Frontiers should have a single cost cap that includes both Phase A–D and the primary mission Phase E–F costs, with a separate, additional cost cap allocation for a mission's quiet cruise phase. This approach will enable the NF Program to optimize mission science, independent of cruise duration.** (Chapter 22)

> **Recommendation: The New Frontiers (NF) Phase A–F cost cap, exclusive of quiet cruise phase and launch vehicle costs, should be increased to $1.65 billion in fiscal year 2025 dollars. A quiet cruise allocation of $30 million per year should be added to this cap, with quiet cruise to include normal cruise instrument checkout and simple flyby measurements, outbound and inbound trajectories for sample return missions, and long transit times between objects for multiple-target missions.** (Chapter 22)

NEW FRONTIERS MISSIONS

The committee considered a broad range of medium-class missions, and from these prioritized the following eight mission themes (in no specific order) for the New Frontiers 6 (NF-6) call:

- Centaur Orbiter and Lander
- Ceres Sample Return
- Comet Surface Sample Return
- Enceladus Multiple Flyby
- Lunar Geophysical Network
- Saturn Probe
- Titan Orbiter
- Venus In Situ Explorer

The themes recommended for New Frontiers 7 (NF-7) include all those not selected from the above list, with the addition of:

- Triton Ocean World Surveyor

Theme descriptions are provided in Chapter 22.

NEW LARGE MISSIONS

The committee prioritizes the Uranus Orbiter and Probe (UOP) as the highest-priority new Flagship mission for initiation in the decade 2023–2032.[4] UOP will deliver an in situ atmospheric probe and conduct a multi-year orbital tour that will transform our knowledge of ice giants in general and the uranian system in particular. Uranus is one of the most intriguing bodies in the solar system. Its low internal energy, active atmospheric dynamics, and complex magnetic field all present major puzzles. A primordial giant impact may have produced the planet's extreme axial tilt and possibly its rings and satellites, although this is uncertain. Uranus's large ice-rock moons displayed surprising evidence of geological activity in limited Voyager 2 flyby data, and are potential ocean worlds. UOP science objectives address Uranus's (1) origin, interior, and atmosphere; (2) magnetosphere; and (3) satellites and rings. UOP will provide ground-truth relevant to the most abundant, similarly sized class of exoplanets. UOP can launch on an existing launch vehicle. Optimal launch opportunities in 2031 and 2032 utilize a Jupiter gravity assist to shorten cruise time; other opportunities from 2032 through 2038 (and beyond) utilize inner solar system gravity assists with an increased cruise time.

The second-highest priority new Flagship mission is the Enceladus Orbilander.[5] Enceladus is an ice-rock world with active plumes of gas and particles that originate from its subsurface ocean. Study of plume material allows direct study of the ocean's habitability, addressing a fundamental question: Is there life beyond Earth and if not, why not? Orbilander will analyze fresh plume material from orbit and during a 2-year landed mission. Its main science objectives are (1) to search for evidence of life; and (2) to obtain geochemical and geophysical context for life detection experiments. Commencing Orbilander late in the decade supports arrival at Enceladus in the early 2050s, when optimal illumination of the south polar region begins. Should budgetary constraints not permit initiation of Orbilander, the committee includes the Enceladus Multiple Flyby (EMF) mission theme in NF. EMF provides an alternative pathway for progress this decade on the crucial question of ocean world habitability, albeit with greatly reduced sample volume, higher velocity of sample acquisition and associated degradation, and a smaller instrument component to support life-detection.

REPRESENTATIVE FLIGHT PROGRAMS

The committee developed two representative programs for the 2023–2032 decade. The *Level Program* assumes currently projected funding for PSD, including inflation at 2 percent/year, while the *Recommended Program* can be achieved with ~17.5 percent higher decade funding. Decision Rules are provided to accommodate significant budgetary deviations (Chapter 22). Both programs continue missions in operation and in development; initiate the Uranus Orbiter and Probe Flagship mission; increase R&A funding to 10 percent or more of the annual PSD budget by mid-decade; incorporate cost realism and cost cap recommendations for Discovery and New Frontiers; and maintain support for planetary defense, including at least one new mission start (Table S-2); support the Lunar Discovery and Exploration Program with a mid-decade start of the Endurance-A rover; and continue the Mars Exploration Program.

The two programs differ in their support for new initiatives. The *Recommended Program* is aspirational and inspirational: it enables robust development of diverse science and engineering communities, drives technology development, and maintains U.S. leadership in solar system exploration. It begins the UOP Flagship in FY 2024 to support a launch in the early 2030s that minimizes cruise length and complexity and initiates the Orbilander Flagship late in the decade to reveal the astrobiological conditions of an ocean world. It also restores the *Vision and Voyages* recommendation, endorsed by the committee, for two NF missions per decade, with NF-5 (which was to be the second NF mission from the prior decade) completed early in the decade, followed by a mid-decade selection of two NF missions in NF-6. The Mars Life Explorer would be initiated late in the decade through the Mars Exploration Program.

---

[4] The full Uranus Orbiter and Probe mission study report is available at https://tinyurl.com/2p88fx4f.

[5] The full Enceladus Orbilander mission study report is available at https://science.nasa.gov/solar-system/documents.

TABLE S-2 Comparison of Representative Programs

Recommended Program	Level Program
Continue Mars Sample Return.	Continue Mars Sample Return.
Five new Discovery selections at recommended cost cap.	Five new Discovery selections at recommended cost cap.
Support LDEP with mid-decade start of Endurance-A.	Support LDEP with mid-decade start of Endurance-A.
R&A increased by $1.25 billion.	R&A increased by $730 million.
Continue Planetary Defense Program with NEO Surveyor and a follow-on NEO characterization mission.	Continue Planetary Defense Program with NEO Surveyor and a follow-on NEO characterization mission.
Gradually restore MEP to pre-MSR level with late decade start of Mars Life Explorer.	Gradually restore MEP to pre-MSR level in late decade with no new start for Mars Life Explorer.
New Frontiers 5 (1 selection).	New Frontiers 5 (1 selection).
New Frontiers 6 (2 selections).	New Frontiers 6 (late, or not included).
Begin Uranus Orbiter and Probe in FY 2024.	Begin Uranus Orbiter and Probe in FY 2028.
Begin Enceladus Orbilander in FY 2029.	No new start for Enceladus Orbilander this decade.

TABLE S-3 Mission Portfolio Assessment Matrix

Mission Name	Priority Science Questions											
	1	2	3	4	5	6	7	8	9	10	11	12
Mars Sample Return												
Uranus Orbiter and Probe												
Enceladus Orbilander												
Endurance-A												
Mars Life Explorer												
Centaur Orbiter and Lander												
Ceres Sample Return												
Comet Surface Sample Return												
Enceladus Multiple Flyby												
Lunar Geophysical Network												
Saturn Probe												
Titan Orbiter												
Triton Ocean World Surveyor												
Venus In Situ Explorer												

NOTES: Assessment of the science questions addressed by MSR and each of the other large- and medium-class missions prioritized in this report. The top rows include MSR and the two new large strategic missions prioritized here. Endurance-A and Mars Life Explorer are highly ranked medium-class missions recommended for the LDEP and MEP programs, respectively. The remaining rows are the prioritized New Frontiers mission themes in alphabetical order. Yellow represents a modest contribution—typically a "substantial" advance in addressing one to a few of a priority science sub-questions—whereas the increasing intensity of green indicates increasing levels of "breakthrough" or "transformative" advances—that is, addressing an increasing number of sub-questions. Note that Q9 focuses on terrestrial life and is therefore not the primary focus of most planetary missions, but rather is supported through astrobiology research programs.

MISSION TRACEABILITY TO SCIENCE GOALS

The large- and medium-class strategic and PI-led missions prioritized and recommended in this report were selected based on their ability to address the priority science questions, as well as programmatic balance, technical risk and readiness, and cost. After these missions had been selected, the committee evaluated this portfolio of new missions to assess how well they covered the breadth of the priority science questions (Q1–Q12) discussed in Chapters 4–15 (see Table S-1). The committee considered whether each mission would likely contribute to a "substantial," "breakthrough," or "transformative" advance for each of the sub-questions in Q1 through Q12. The tabulated and normalized results are displayed in a mission portfolio assessment matrix (Table S-3) on a scale of modest (yellow) to high (dark green) contribution. This matrix illustrates that the collective suite of prioritized missions in the *Recommended Program* does an excellent job of addressing the full breadth of the priority planetary science questions and does so at a diverse set of destinations.

KEY ADDITIONAL RECOMMENDATIONS

Recommendation: NASA should evaluate plutonium-238 production capacity against the mission portfolio recommended in this report and other NASA and national needs and increase it, as necessary, to ensure a sufficient supply to enable a robust exploration program at the recommended launch cadence. (Chapters 20 and 22)

Recommendation: NASA should continue to invest in maturing higher-efficiency radioisotope power system technology to best manage its supply of plutonium-238 fuel. (Chapters 20 and 22)

Recommendation: NASA's Planetary Science Division (PSD) should strive to consistently fund technology advancement at an average of 6 to 8 percent of the PSD budget. (Chapters 21 and 22)

1

Introduction to Planetary Science, Astrobiology, and Planetary Defense

In spring 2011, when the last planetary decadal survey *Vision and Voyages for Planetary Science in the Decade 2013–2022* was released, the New Horizons spacecraft was speeding toward Pluto, Europe's Rosetta was heading toward a rendezvous with a comet, Cassini was still orbiting Saturn, and the GRAIL, Curiosity, InSight, Perseverance, OSIRIS-REx, and Juno missions had not yet launched (see Figures 1-1 through 1-4). Since that time, some of these spacecraft and others have completed their primary missions and dramatically expanded our understanding of the solar system. They have studied the atmosphere and interior of Jupiter, the interior of Saturn, the water plumes of Enceladus, the topography and geochemistry of Pluto and its moon Charon, the seismology and habitability of Mars, the surface of the asteroid Bennu, and the icy chemistry of a comet. They have contributed to planetary science and astrobiology in tremendous ways. The intent of this chapter is to provide general background for the nontechnical reader wishing to know something more about planetary science, astrobiology, and planetary defense.

"Planetary science" is the shorthand definition for an array of scientific disciplines that collectively seek to answer questions about how the solar system formed, what initial conditions and subsequent processes shape how planetary bodies evolve and interact with each other and the environment, and how these factors enabled the conditions for life to form on at least one planet in the solar system. The latter feeds into the growing field of "astrobiology," the study of the origin and evolution of life on planetary bodies. These activities are tightly interlinked, and both have advanced substantially in the decade since *Vision and Voyages* (NRC 2011). Further advances are dependent not only upon new space missions to study the solar system but also on basic research to understand the scientific data and to formulate new hypotheses, as well as on technology development to enable future mission and experimental studies.

PLANETARY SCIENCE AND ASTROBIOLOGY

Planetary science is a multidisciplinary activity involving members of the geology, geophysics, geochemistry, astronomy, atmospheric science, and space physics communities. These communities' study planetary bodies as well as Earth. Astrobiology is, at its most basic, the study of the origin, evolution, and distribution of life in the universe (NASEM 2019a). Astrobiology was recognized as an organized scientific discipline more recently than planetary science and is inherently even more multidisciplinary, encompassing biology, aspects of heliophysics (often referred to as solar and space physics), planetary science, and astronomy. Astrobiology includes laboratory activities as well as field studies in terrestrial surface and marine environments, theoretical work, and sample analyses.

11

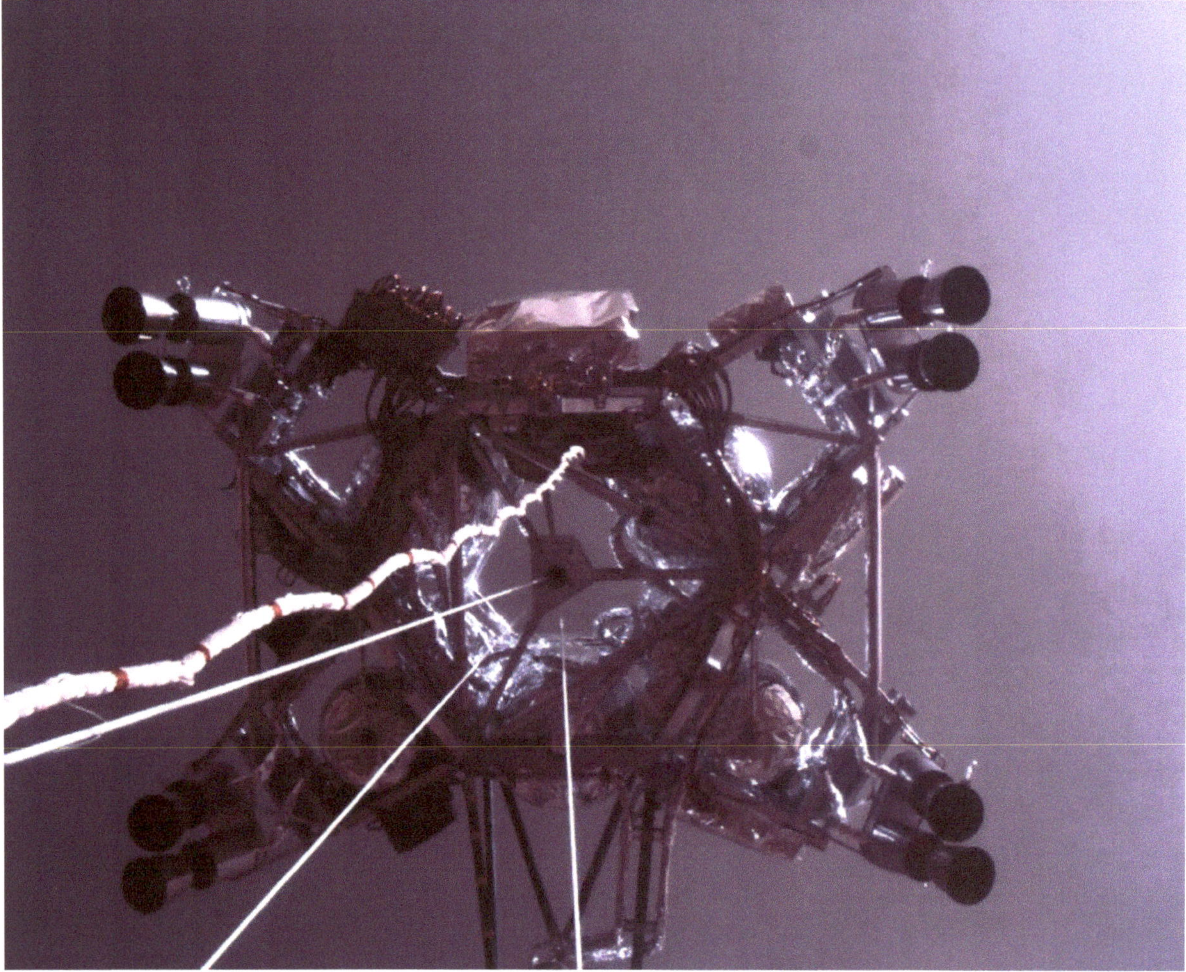

FIGURE 1-1 The descent stage of Mars 2020 hovers as the Perseverance rover (not visible) is lowered to the martian surface in 2021. SOURCE: Courtesy of NASA/JPL-Caltech.

The search for life in the solar system and beyond has been a focus of many current and future spaceflight missions conducted by NASA and other space agencies. A new concept of dynamic habitability has emerged in recent decades that views habitability—the ability of a specific planetary environment to support life—as a continuum. An environment may transition from inhabitable to uninhabitable over time, a function of planetary and environmental evolution. Astrobiology and planetary science take an integrated, systems-level view of the origin and evolution of planetary bodies, seeking to understand how life and its environment may have changed together or co-evolved.

PLANETARY DEFENSE

Planetary defense is an international cooperative effort to detect and track objects that could pose a threat to life on Earth. As such, its motivations are more concerned with human health and safety rather than the advancement of scientific understanding. The threat posed by extraterrestrial bodies to Earth and its inhabitants was amply demonstrated in 2013, when a 20-m-diameter asteroid detonated at an altitude of some 23 km over the Siberian city of Chelyabinsk. The resulting explosion released ~2 petajoules of energy (i.e., the equivalent of some 450 kilotons

FIGURE 1-2 An artist's impression of the Juno spacecraft at Jupiter. Juno is currently studying Jupiter's interior, composition, and atmosphere. SOURCE: Courtesy of NASA/JPL.

of high explosive) and caused nonfatal injuries to more than 1,600 people. This event highlighted the fact that Earth travels around the Sun amid millions of small objects in similar orbits that sometimes cross Earth's path (see, e.g., NASEM 2019b). Planetary science and exploration provide knowledge and tools to detect, track, and characterize such objects, key inputs to developing realistic detection and mitigation strategies against these natural disasters. Starting in the 1990s, Congress and presidential administrations have directed NASA to take a lead role in planetary defense and that role has grown in the past decade. NASA, NSF, and other government agencies collaborate in activities in support of planetary defense.[1] This decadal survey is the first to include planetary defense in its charter.

---

[1]Although assets maintained for national security purposes by the newly established U.S. Space Force are relevant to the detection and tracking of objects in near-Earth space that might pose a hazard, planetary defense is not currently included in the U.S. Space Force mission statement.

FIGURE 1-3 The InSight spacecraft during final assembly. InSight has provided data on the internal structure of Mars via seismology. SOURCE: Courtesy of NASA/JPL-Caltech/Lockheed Martin.

THE RELATIONSHIP BETWEEN GROUND- AND SPACE-BASED RESEARCH

Planetary science is a multidisciplinary endeavor and is conducted by a synergistic combination of ground- and space-based activities. No one type of research approach (e.g., spacecraft missions, telescopic observation, and theoretical studies) is more or less important that the others. All research approaches and techniques have a role to play if progress is to be made in addressing key scientific issues.

The first planetary scientists explored the solar system from the ground, using increasingly powerful telescopes to first discover and then study the planets. Mercury, Venus, Mars, Jupiter, and Saturn were all visible in the night sky with the naked eye. The Galilean satellites were discovered by Galileo Galilei in 1610, Uranus was discovered by William Herschel in 1781, Neptune by Johann Galle and Urbain Le Verrier in 1846, and Pluto by Clyde Tombaugh in 1930. Even as the United States and Soviet Union began sending spacecraft to the Moon and then to Mars and Venus starting in the late 1950s, ground-based astronomy played an important role in understanding the solar system, such as by analyzing radio signals from Jupiter or conducting radar observations of Mercury, Venus, and asteroids.

Today, ground- and space-based telescopic observations continue to provide key support to robotic space missions—for example, by characterizing targets in advance of spacecraft encounters—and ongoing observations provide data between missions as well. Many ground-based telescopes used in planetary science research are supported by the National Science Foundation (NSF), although specific research funding to use them may come from NASA or other sources.

FIGURE 1-4 The New Horizons spacecraft during final preparation for launch at Kennedy Space Center in January 2006. New Horizons flew past Pluto and its large moon Charon in 2015, and then past 486958 Arrokoth, a Kuiper belt object, in 2019. SOURCE: Courtesy of NASA.

Spacecraft and telescopic observations are not the only means researchers use to study planetary bodies. Significant amounts of work are undertaken in laboratories and by field studies in relevant terrestrial and marine environments. Data analysis and theoretical and computational modeling also play fundamental roles. Moreover, the work of a relatively small number of planetary scientists and astrobiologists worldwide would go for naught without the backup and support of a far greater number of engineers, technicians, program managers, and administrators in agencies, organizations, and private companies who keep the space-science enterprise viable.

SUPPORT FOR PLANETARY SCIENCE AND ASTROBIOLOGY

The principal federal organizations that support the nation's programs in planetary science are the Planetary Science Division (PSD) of NASA's Science Mission Directorate, and the Division of Astronomical Sciences (AST) in NSF's Directorate for Mathematical and Physical Sciences Division. Other federal departments, agencies, and organizations provide important financial and other support for aspects of planetary science and astrobiology.

The Department of Energy's (DOE's) national laboratories, for example, support groups whose primary roles are areas such as nuclear forensics and shock physics. Such groups have important secondary roles in the geochemical analysis of extraterrestrial materials and the modeling of asteroid impacts, respectively. DOE's National Nuclear Security Administration, through its Stewardship Science Academic Alliances supports work relevant to the behavior of matter under the conditions found in the interiors of planetary bodies at its Capital/DOE Alliance

Center and the Center for Matter under Extreme Conditions. Similarly, the massive gene-sequencing and computational capabilities of DOE's Genomic Science Program are directly relevant to aspects of astrobiology. Also of relevance to astrobiology are a subset of activities supported by the National Institutes of Health. Examples include the support of individual researchers investigating the chemical and physical processes that facilitated the transition from chemical evolution to biological evolution on the early Earth.

Private research and philanthropic entities are also involved in supporting various aspects of planetary science and astrobiology, but their contributions are beyond the scope of this report. The remainder of this section is devoted to a more detailed look at activities under way at NASA and NSF.

The primary goals of NASA's PSD are to ascertain the origin and history of the solar system, to understand the potential for life beyond Earth, and to characterize hazards and resources present as humans explore space. Spacecraft missions, technology development, research infrastructure, and basic research and analysis programs are supported by PSD to advance these goals. The majority of PSD's budget is devoted to the development, construction, launch, and operation of robotic spacecraft. PSD conducts large strategic (so-called Flagship) missions and smaller Discovery and New Frontiers missions that are proposed and led by principal investigators (PIs). Examples of past and current Flagship missions include Cassini (Figure 1-5), the Curiosity and Perseverance rovers, and Europa Clipper; examples of New Frontiers missions include New Horizons, Juno, and OSIRIS-REx.

The primary purpose of NSF-AST is to support research in ground-based optical, infrared, and radio astronomy. NSF-AST provides access to world-class research facilities and supports the development of new instrumentation and next-generation facilities. NSF-AST also supports basic research in planetary astronomy. However, NSF-AST does not, in general, support activities that are also funded by NASA—for example, the analysis of data returned by planetary spacecraft missions.

FIGURE 1-5 Artist conception of the Cassini spacecraft during its final plunge into Saturn's atmosphere in 2017. SOURCE: Courtesy of NASA/JPL-Caltech.

Relevant Activities in Other NASA Divisions and Directorates

Planetary science activities at NASA are strongly coupled to the agency's other science programs in the Astrophysics, Heliophysics, and to a more limited extent, Earth Science and Biological and Physical Science divisions. Similarly, activities under way in other NASA directorates are of direct relevance to planetary science and astrobiology. Each is addressed below.

Astrophysics Division

The major science goals of the Astrophysics Division (APD) are to discover how the universe works, explore how the universe began and evolved, and to search for planetary environments that may hold keys to life's origins or even might themselves sustain life. APD assets such as the Hubble Space Telescope have played major roles in advancing planetary science through the study of solar system bodies such as the atmospheres of the outer planets. Hubble also was used to study the vicinity of Pluto to plan for New Horizons' 2015 flyby and to identify possible future target for the spacecraft to study. The James Webb Space Telescope is expected to make substantial contributions to planetary science. Another key area where the interests of APD and PSD overlap is in the study of extrasolar planetary systems (see, e.g., NASEM 2018a).

Heliophysics Division

NASA's Heliophysics Division sponsors research in solar and space physics, with particular emphasis on understanding the Sun and its interactions with Earth and other bodies in the solar system. This research also includes study of the particle and field environments of other solar system bodies and includes comparative studies of planetary magnetospheres, ionospheres, and upper atmospheres. Spacecraft such as the Voyagers are taking measurements at the distant edges of the Sun's influence and the beginning of the interstellar medium.

Earth Science Division

NASA's Earth Science Division (ESD) also has important connections to the study of planetary science. The major scientific goal of this division is to advance Earth system science to meet the challenges of climate and environmental change. A better understanding of Earth provides data that enable understanding of the origin and evolution of a terrestrial planetary biosphere. A common interest of both astrobiologists and Earth scientists is how biospheres interact with their host planetary environments. However, the domains of interest to the two communities are somewhat dissimilar. Astrobiologists are mostly interested in the impact of interactions over geological timescales (~100 million to a billion years), whereas Earth science is most interested in changes over much shorter times (~1 to a million years).

The science, technologies, and observational techniques developed for remote sensing of Earth help inform planetary science and astrobiology. However, planetary spacecraft operate in more difficult environments than Earth-orbiting spacecraft and have different design requirements and mass and power limitations, meaning that Earth observation instruments are not typically directly applicable to planetary science needs.

Biological and Physical Science Division

NASA's Biological and Physical Science Division (BPSD) was recently incorporated into the Science Mission Directorate. Much of BPSD's research pertains to how microgravity and partial gravity environments affect contemporary biological processes (e.g., adaption of organisms to the space environment and the health and safety of astronauts). BPSD is also interested in the response of physical systems to low-gravity environments. Familiar and well-understood processes—for example, fluid flow through pipes, combustion, and material effects such as the formation of alloys—behave in a fundamentally different manner when gravitational effects are reduced or eliminated. While BPSD has limited overlap with planetary science and astrobiology, some aspects of biological and physical research are directly relevant to future activities such as the in situ utilization of planetary resources—for example, extraction of oxygen from the martian atmosphere or ice mining on the Moon—or the creation of long-lived life-support systems.

Space Technology Mission Directorate

NASA's Space Technology Mission Directorate develops a wide range of technologies to support agency needs in the mid- to long-term. Some of these technologies, such as communications and in-space propulsion, are of significant value to planetary science missions. Technologies that are more specific to the near-term needs of planetary science and astrobiology, such as spacecraft instrumentation, are supported directly by PSD.

Exploration Systems Development Mission Directorate

In September 2021, as this report was being written, NASA established the Exploration Systems Development Mission Directorate (ESDMD) to oversee the Artemis program to send humans to the Moon. In particular, ESDMD is responsible for developing systems to support human operations on the Moon. PSD collaborates with ESDMD to develop precursor lunar robotic missions and to define those scientific activities that astronauts will conduct on the Moon and, eventually, Mars. A current major area of collaboration between these two parts of NASA is in the development of the Volatiles Investigating Polar Exploration Rover, currently scheduled to land in the Nobile region near the Moon's South Pole in 2024.

Relevant Activities in Other NSF Divisions and Directorates

As already mentioned, the principal source of planetary science funding within NSF is in its Division of Astronomical Sciences. However, other parts of NSF—particularly activities within the Directorate for Geosciences—make important contributions to planetary science and astrobiology. However, much of these planetary science activities are concerned with focused studies of the Earth system and, as such, are beyond the scope of this study (see Appendix A, Scope, item 4). Nevertheless, a small subset of activities funded by the Directorate for Geosciences (e.g., geochemical and cosmochemical of terrestrial and extraterrestrial materials) is very important to the planetary science communities. Similarly, other federal agencies and organizations provide niche support for small subsets of the planetary science and astrobiology communities. While these activities do not support a significant number of planetary scientists or astrobiologists, they are important because they maintain key linkages between space scientists and the very much larger community of researchers studying aspects of, for example, the Earth system, matter under conditions of extreme temperatures and pressures, and fundamental biology. Such linkages provide important means for cross-fertilizing ideas, concepts, and breakthroughs between what might seem disparate research communities. Subsequent sections highlight the important work supported by some of NSF's divisions and directorates.

Office of Polar Programs

The Office of Polar Programs (OPP) provides access to and logistical support for researchers working in Antarctica. One of the key U.S. activities in the southern polar region is the Antarctic Search for Meteorites Program, initiated in 1975 and run as a cooperative activity involving OPP, NASA, and the Smithsonian Institution. The meteorites collected in Antarctica have provided insights into many planetary bodies, including the Moon and Mars. The Smithsonian's National Museum of Natural History is responsible for initial examination and characterization of meteorites collected in Antarctica. The Astromaterials Acquisition and Curation Office at NASA's Johnson Space Center is responsible for long-term curation and distribution of samples to the research community. Antarctic research is also relevant to other aspects of planetary science and astrobiology. Important examples of OPP activities include support for the study of sub-glacial lakes and Mars-analog environments in the Dry Valleys. The former are terrestrial analogs to the oceans known to exist beneath the icy surfaces of objects such as Enceladus (see Figure 1-6). OPP activities have been severely impacted by the COVID-19 pandemic, with most, if not all, activities planned for the 2020–2021, 2021–2022, and 2022–2023 field seasons cancelled.

FIGURE 1-6 Artist impression of the Cassini spacecraft against the backdrop of the ice plumes of Enceladus. The presence of liquid water below Enceladus's icy surface is of particular interest to astrobiologists. Laboratory studies as well as field activities to study the permanently ice-covered lakes in Antarctica inform the study of icy bodies such as Enceladus and Europa. SOURCE: Courtesy of NASA/JPL-Caltech.

Division of Atmospheric and Geospace Sciences

This part of NSF supports fundamental research regarding physical, chemical, and biological processes that impact the composition and physical phenomena and behavior of matter between the Sun and the surface of Earth. Important areas of research synergies with planetary science include the development of atmospheric and general circulation models for other planets and comparative studies of the plasma process.

Division of Earth Sciences

Research in this division focuses on understanding the structure, composition, and evolution of Earth, the life it supports, and the physical and chemical processes governing the formation and behavior of minerals, rocks, and other materials. One area of this division's interest is directly relevant to the study of extraterrestrial materials. The geochemical techniques developed to understand the formation and behavior of terrestrial rocks and minerals are directly applicable to the analysis and study of meteorites, cosmic dust, and samples returned to Earth from other solar system bodies.

Division of Ocean Sciences

This part of NSF is responsible for research, infrastructure, and educational activities designed to improve knowledge and understanding of Earth's oceans and oceanic basins and their interactions with the integrated Earth system. Access to research ships, deep diving submersibles, and core samples from oceanic drilling programs not only inform understanding of the structure and evolution of Earth and its biosphere but also provide a much-needed context for studies of other planetary environments. A particularly relevant example involves contributions to astrobiology made by studies of hot and cold deep-sea vents and their associated biospheres. Most telling is that life found in such systems employs a food-chain whose base is driven by chemical reactions at, for example, water–mineral interfaces and not by solar energy. As such, these marine biospheres may be directly analogous to ones possibly operating in the sub-surface oceans known to exist in the outer solar system.

Division of Physics

Another part of NSF playing an important supporting role in planetary science activities is the Division of Physics. An activity most worthy of mention here is support for theoretical, modeling, and experimental studies of the behavior of matter at the temperatures and pressures found in the interiors of planetary bodies. For example, NSF's Physics Frontiers Center for Matter at Atomic Pressures is largely motivated by major planetary science problems concerning the study the physical properties of matter at extreme pressures.

Other NSF Programs

Niche support for planetary science and astrobiology is also provided by other parts of NSF. While many of these activities are of limited scope and/or duration, they foster interdisciplinary research by bringing together researchers who would not normally interact with each other. A notable recent example was the so-called Ideas Lab on the origin of life, sponsored by NSF's Biological Sciences and Geosciences directorates and NASA's Astrobiology Program. Ideas Labs consist of a series of intensive workshops, whose participants are selected via a competitive process, designed to find innovative approaches to the study of major science questions. The near-term goal of this specific Ideas Lab was to develop a theoretical framework for events transpiring on the early Earth that encompasses the rival metabolism-first versus RNA-first theories for the origin of life.

INTERNATIONAL COOPERATION

Planetary exploration is an increasingly international endeavor, with the United States, Russia, Japan, Canada, China, India, Israel, United Arab Emirates, and many European nations independently or collaboratively mounting major planetary missions. As budgets for space programs come under increasing pressure and the complexity of the missions grows, international cooperation becomes an enabling component. New alliances and mechanisms to cooperate are emerging, enabling partners to improve national capabilities, share costs, build common interests, and eliminate duplication of effort.

NASA's planetary science and astrobiology programs may have prompted other nations, large and small, to undertake similar activities. But that is not all. NASA is an extraordinary soft-power asset in that the results from its missions have changed the scope of courses and textbooks used in schools, colleges, and universities around the world. Moreover, NASA's images of extraterrestrial objects are now commonplace in the national and international media. The extraordinary success of space missions is such that many graduate students and young scientists are willing to bet their careers on the results that can be obtained from space exploration.

The soft-power aspects of NASA's activities aside, international agreements and plans for cooperation need to be crafted with care, for they also can carry risks. The establishment of the NASA Astrobiology Institute (NAI), for example, prompted the establishment of similar organizations in other nations. However, the demise of NAI left many these non-U.S. organizations in limbo because their specific relationship with NASA activities became unclear. The management of international spacecraft missions adds layers of complexity to their technical specification, management, and implementation. Different space agencies use different planning horizons, funding approaches, selection processes, and data dissemination policies. An informed estimate of cost (e.g., the TRACE process described in Appendix C) need not be a "show-stopper" for international collaborations. But attempting to get good estimates of mission costs when there is significant international sharing of costs raises a number of complications. In some cases, for example when an instrument is flown on a foreign spacecraft instead of being a NASA free-flyer, the cost may be low enough that it does not meet the threshold for independent costing in a decadal survey.

With the emergence of a new and highly entrepreneurial commercial space sector in the United States, NASA is fundamentally changing the traditional landscape for implementing space missions. However, such pioneering efforts as NASA's Commercial Lunar Payload Services program have not yet been fully adopted by other countries. Such activities bring new players and stakeholders into the equation and may potentially complicate activities between NASA and other national and international space agencies. For example, a foreign space agency willing to enter into a government-to-government partnership might be less sanguine about a three-way partnership with NASA and a commercial entity. Nonetheless, international cooperation remains a crucial element of the planetary program; it may be the only realistic option to undertake some of the most ambitious and scientifically rewarding missions. Advance planning—through bilateral (or multilateral) agency discussions, scientific community involvement (via workshops and congresses, for example), and informed cost estimates and sharing of tasks—is the most effective way to reap the benefits of such collaborations.

Mechanisms and Recent Examples of Cooperation

Flagship missions afford the greatest potential for NASA and other space agencies to unite resources to meet difficult challenges. The joint NASA-ESA (European Space Agency) Cassini-Huygens mission to explore the saturnian systems was a superb example of international cooperation (see Figures 1-5 and 1-6). Large strategic missions like Cassini-Huygens are complex to manage and implement, as they involve integrating major spacecraft components supplied by different nations (engines, antennas, probes, dual spacecraft) into a single flight system. Still, to minimize the high fractional costs of launch and orbital insertion or landing, this architecture can be the most cost-effective one overall. The Cassini-Huygens mission was composed of two elements separately developed by NASA and ESA and delivered to the saturnian system by the same spacecraft. Such a separation of tasks/responsibilities has proven very effective and successful in the past and will also be beneficial in the future. Indeed, NASA and ESA have been considering undertaking joint missions of this integrated form, such as in the joint Europa Jupiter System Mission (EJSM) and Titan Saturn System Mission (TSSM) concept studies in the late 2000s, that failed to materialize in the end owing to cost issues.

Common collaborative arrangements range in scale from data-sharing arrangements to the provision of resources to foreign partners by NASA. These resources might include, for example, instruments, other key flight elements, and/or science-team members. NASA contributions to foreign missions have been funded by a variety of competitive programs such as the past "Mission of Opportunity" or the present Stand-Alone Missions of Opportunity (SALMON). Examples of foreign missions incorporating NASA-provided instruments include, for example, India's Chandrayaan-1 lunar orbiter; ESA's BepiColombo Mercury and JUICE Ganymede orbiters; and Japan's Hayabusa 1 and 2 asteroid sample return missions and the forthcoming Mars Moons Exploration spacecraft, designed to return samples from the martian moon Phobos in the late 2020s.

International cooperation is a two-way street. NASA has contributed instruments, other items of hardware, and has provided communications and navigational support to a variety of non-U.S. missions. Examples include the following: the Lunar Reconnaissance Orbiter includes a Russian instrument; the Juno Jupiter orbiter carries an Italian auroral experiment; the Mars Exploration Rovers and Phoenix lander included instruments and team members from Germany, Denmark, and Canada; and Russia, Canada, and various European nations contributed elements of the Curiosity and Perseverance Mars rovers. These collaborations dramatically expand mission capabilities and are crucial to developing a strong and effective national and international scientific community.

Guidelines for International Cooperation

Notwithstanding the enormous benefits, both societal and scientific, that international cooperation affords, such agreements are not necessarily of mutual benefit. As such, cooperative ventures require due consideration of all the pluses and minuses. Complicating aspects of cooperative ventures include the following: different goals for the endeavor, misaligned fiscal timelines and commitment schedules; use of mismatched proposal requirements and selection processes; inconsistent technical specifications; management by multiple, sometimes competing interests; agreement on implementation and integration procedures; and the impact of the International Trafficking in Arms Regulations. The time and effort to resolve these and other issues can lead to cost and schedule growth. As NASA pursues opportunities for collaboration with foreign partners, it needs to do so with full understanding of the potential risks and how they can be managed. Clearly articulated and readily understood cooperation guidelines are essential. As a result, the survey committee endorses, as a starting point, the following general principles and guidelines laid out in the joint report of the Space Studies Board and the European Space Science Committee titled *U.S.–European Collaboration in Space Science* (NRC 1998):

1. Support through peer review that affirms the scientific integrity, value, requirements, and benefits of a cooperative mission;
2. Historical foundation built on an existing international community, partnership, and shared scientific experiences;
3. Shared objectives that incorporate the interests of scientists, engineers, and managers in common and communicated goals;
4. Clearly define responsibilities and roles for cooperative partners, including scientists, engineers, mission managers;
5. Agreed-upon processes for data calibration, validation, access, and distribution;
6. Establish a sense of partnership recognizing the unique contributions of each participant;
7. Beneficial characteristics of cooperation; and
8. Reviews for cooperative activities in the conceptual, developmental, active, or extended mission phases—particularly for foreseen and upcoming large-class spacecraft missions.

Despite the negative consequences that may potentially accrue if cooperative activities are not planned and conducted in a manner consistent with the principles listed above, the committee strongly supports international efforts and encourages the expansion of international cooperation on planetary missions to accelerate technology maturation and share costs. From experience in the past decades, it appears that international cooperation generally provides resilience to long-term space programs and allows optimal use of an international workforce and expertise. Multiple international space powers (both traditional national space agencies and the private sector) have now mastered major technological challenges required to explore the solar system. As such, international cooperation will remain a key element of the nation's planetary exploration program. An internationally engaged program of solar system exploration can unite stakeholders worldwide and lay the roadmap for humans to venture into space in the next phases of exploration.

PLANETARY SCIENCE DECADAL SURVEYS AND RELATED REPORTS

In the 1970s and 1980s, science strategies for exploring the solar system were drafted by the National Research Council's (NRC's) Committee on Planetary and Lunar Exploration (COMPLEX), which addressed separately the inner planets, the outer planets, and primitive/small bodies.[2] In the early 1990s, COMPLEX crafted a single solar system strategy that united and updated the several preexisting documents, resulting in the report *An Integrated Strategy for the Planetary Sciences: 1995–2010* (NRC 1994). In 2001, the NRC undertook the first planetary science decadal survey. This produced the 2002 report *New Frontiers in the Solar System: An Integrated Exploration Strategy* (NRC 2003). That report outlined science priorities and identified new initiatives needed to address the scientific priorities established in the decadal survey. The study advocated the creation of a new class of medium-sized missions, named New Frontiers. New Horizons was the first New Frontiers mission, launched in 2006.

In 2010, the NRC undertook the second planetary science decadal survey, which resulted in the spring 2011 delivery of the report *Vision and Voyages for Planetary Science in the Decade 2013–2022*. The 2011 decadal survey's statement of task from NASA called for prioritized missions binned in small, medium, and large categories

---

[2]The NRC is the operating arm of the National Academies of Sciences, Engineering, and Medicine. Until 2017, the NRC name appeared on all National Academies reports. The name of the National Academies of Sciences, Engineering, and Medicine now appears on the covers of publications.

TABLE 1-1 Priority Mission Recommendations for 2013–2022 from the *Vision and Voyages* Decadal Survey

Vision and Voyages Recommendation	Priority and Mission Type	Disposition Prior to This Decadal Survey	Current Status
Mars Astrobiology Explorer-Cacher	First priority large strategic mission.	Implemented by NASA as Mars 2020/Perseverance.	Currently collecting and caching samples for return to Earth.
Jupiter Europa Orbiter	Second priority large strategic mission.	Implemented by NASA as Europa Clipper.	Currently under construction for launch in 2024.
Uranus Orbiter and Probe	Third priority large strategic mission.	NASA initiated a science definition team to examine Uranus and Neptune orbiter. Neptune Orbiter study undertaken via PMCS process.	n/a
Enceladus Orbiter	Joint fourth priority large strategic mission.	Not implemented. Enceladus orbiter/lander study undertaken via the PMCS process.	n/a
Venus Climate Orbiter	Joint fourth priority large strategic mission.	Not implemented. Venus Flagship mission study undertaken via the PMCS process.	n/a
New Frontiers Program	A line of medium-class, PI-led missions. At least two to be selected each decade.	New Frontiers-4, Dragonfly, selected in 2019.	Dragonfly to launch in 2027. NF-5 announcement of opportunity to be released in 2023–2024 and launched in the early 2030s.
Discovery Program	A line of small-class, PI-led missions. At least five to be selected each decade.	Lucy and Psyche selected in 2017. DAVINCI and VERITAS selected in 2021.	Lucy launched in 2021. Psyche to launch in 2023. DAVINCI and VERITAS under development for launch in late 2020s/early 2030s. Next announcement of opportunity scheduled for 2024–2025.

with respective costs of less than $325 million, less than $650 million, and more than $650 million in then-year dollars.[3] In addition, NASA was congressionally mandated to ask the decadal surveys to conduct an independent cost estimation process. In the 2011 decadal survey, this was referred to as the Cost and Technical Evaluation (CATE) process. In 2020, the name of the CATE process was changed to Technical Risk and Cost Evaluation (TRACE) to emphasize the importance of technical risk assessment.

Vision and Voyages produced a range of recommendations across the entire planetary science field, including for research and analysis and technology spending. It also included a set of priority mission recommendations that are summarized in Table 1-1.

In addition to the decadal surveys, the National Academies has also undertaken mid-decade reviews of NASA's planetary science and astrobiology programs. In 2018, the National Academies produced *Visions into Voyages for Planetary Science in the Decade 2013–2022—A Midterm Review* (NASEM 2018b). The report concluded that

---

[3]Then-year dollars being those including the effects of inflation and/or reflect the price levels prevailing during the year at issue.

FIGURE 1-7 The Perseverance rover on Mars. This mission is collecting samples for later return to Earth. SOURCE: Courtesy of NASA/JPL-Caltech/MSSS.

NASA had made substantial progress accomplishing the goals of the decadal survey and recommended additional actions leading to the current decadal survey. As an example, NASA designed, built, launched, and landed the Perseverance rover on Mars, a direct result of the 2011 decadal survey's recommendations. (See Figure 1-7.)

SCIENTIFIC SCOPE OF THIS REPORT

The scientific scope of this report spans two dimensions: first, the principal scientific disciplines that collectively encompass the ground- and space-based elements of planetary science and astrobiology—that is, planetary astronomy, geology, geophysics, atmospheric science, magnetohydrodynamics, celestial mechanics, and relevant aspects of the life sciences—and second, the physical territory within the committee's purview—that is, the solar system's principal constituents and extrasolar planetary systems. This territory includes the following:

- The major rocky bodies in the inner solar system—that is, Mercury, Venus, the Earth-Moon, and Mars;
- The giant planets in the outer solar system—that is, Jupiter, Saturn, Uranus, and Neptune—including their magnetospheres;
- The rings and satellites of the giant planets;

- Dwarf planets in the asteroid and Kuiper belts;
- Primitive solar system bodies (also called small bodies—that is, the comets, asteroids, satellites of Mars, interplanetary dust, meteorites, Centaurs, Trojans, and Kuiper belt objects);
- The abiotic sources of organic compounds in the solar system;
- Origins of life and the coevolution of life and the physical environment;
- Habitable environments in the solar system and, where appropriate, associated biosignatures; plus
- All of the above as they relate to planetary systems around other stars.

Other aspects of astrobiology—such as synthesis and function of macromolecules in the origin of life and early life, and increasing complexity—are beyond the scope of this report.

Previously, planetary defense prioritization was addressed outside the decadal survey process. This decadal survey was charged to address this topic for the first time. The planetary defense findings and recommendations in this report are presented in the framework of the *National NEO Preparedness Strategy and Action Plan* (NSTC 2018), which identifies NASA as the key U.S. government agency to lead such activities and directs NSF to provide support. The plan's five strategic goals underpin the nation's effort to enhance preparedness for dealing with the threat posed by near-Earth objects (NEOs) and other potentially hazardous extraterrestrial impactors (NSTC 2018):

- Enhance NEO detection, tracking, and characterization capabilities.
- Improve NEO modeling, prediction, and information integration.
- Develop technologies for NEO deflection and disruption missions.
- Increase international cooperation on NEO preparation.
- Strengthen and routinely exercise NEO impact emergency procedures and action protocols.

This report concentrates on the first three items above.

The survey's statement of task (see Appendix A) contained a series of nonbinding guidelines designed to ensure that this report contained actionable advice and maintained consistency with other recently provided advice developed by the National Academies.

A GUIDE TO READING THIS REPORT

The survey committee recognizes that this is a long report. Its length derives, in part, from the addition to the statement of task four items not included and/or emphasized in *Vision and Voyages*:

1. Greater emphasis on astrobiology than the two preceding surveys.
2. Inclusion of a discussion of planetary defense and future mission priorities in this area.
3. Addition of considerations of the state of the profession and the provision of specific, actionable, and practical recommendations concerning diversity, inclusion, equity, accessibility, and the creation of safe workspaces.
4. Organizing the report according to priority research questions rather than destinations requiring defining and describing these questions.

A more important reason for the length of this report is that a decadal survey, like other reports of the National Academies, is a multiuser document (Hicks et al. 2022). Different readers are looking for different things. Potential users of this report include the following:

- Policy makers in in the U.S. Congress and their staff (likely to be most interested in the key recommendations).
- Agency officials at NASA and NSF (likely to be most interested in the mission, programmatic, and scientific recommendations).
- Members of the planetary science and astrobiology communities (likely most interested in the programmatic recommendations and priority open questions for the coming decade).
- Graduate students and early-career researchers (likely most interested in key open questions).
- Undergraduate students and public (likely most interested in current state of knowledge and recent discoveries).

TABLE 1-2 A Guide to Reading This Report

Topic	Primary Discussion	Additional Discussion	Recommendations
Issues Related to Nine Priority Topics Identified in the Statement of Task			
1. Overview of planetary science, astrobiology, and planetary defense	Chapter 1	n/a	n/a
2. Broad survey of the current state of knowledge	Chapter 2	n/a	n/a
3a. Compelling questions, goals, and challenges for planetary science	Chapters 3 to 11	n/a	n/a
3b. Compelling questions, goals, and challenges for astrobiology	Chapters 3 and 12–14	Chapter 22	n/a
3c. Compelling questions, goals, and challenges for planetary defense	Chapter 18	Chapter 22	Chapters 18 and 22
4a. Recommended research traceable to objectives and goals	Chapters 4–15	Chapter 22	n/a
4b. Recommended missions traceable to objectives and goals	Chapter 22	Appendix C	Chapter 22
5a. Comprehensive research strategy for planetary science, astrobiology, and planetary defense	Chapter 22	n/a	Chapter 22
5b. Timing, cost, risk, and technical readiness of recommended missions	Chapter 22	Appendix C	n/a
6. Decision rules	Chapter 22	n/a	Chapter 22
7a. Human exploration	Chapter 19	Chapter 22	Chapters 19 and 22
7b. International cooperation	Chapter 1	n/a	Chapter 20
8. Intra- and inter-agency collaboration	Chapters 1 and 18–21	Chapter 22	Chapters 19, 20, 21, and 22
9. State of the profession	Chapter 16	Chapter 22	Chapters 16 and 22
Other Topics Discussed in the Report			
Apophis 2029 encounter	Chapter 18	n/a	Chapter 18
Arecibo	Chapters 18 and 20	Chapter 22	Chapters 18, 20, and 22
Artemis program	Chapter 19	Chapter 22	Chapters 19 and 22
Budgetary projections	Chapter 22	n/a	n/a
Deep Space Network	Chapter 19	n/a	n/a
Discovery program	Chapter 22	n/a	Chapter 22
Europa Clipper	Chapter 22	n/a	n/a
Ground- and space-based telescopes	Chapter 20	Chapter 18 and Appendix E	n/a

continues

TABLE 1-2 Continued

Topic	Primary Discussion	Additional Discussion	Recommendations
International Mars Ice Mapper	Chapter 22	Chapter 19	Chapter 22
Launch vehicles	Chapter 20	Chapter 22	Chapter 22
Lunar Exploration and Discovery program	Chapter 22	Chapter 19	Chapter 22
Mars Exploration program	Chapter 22	Chapter 19	Chapter 22
Mars Sample Return	Chapter 22	n/a	Chapter 22
Mission studies, PMCS, and SDT	Appendix C	Appendix D	n/a
Mission studies, future	Chapter 23	Chapter 22	n/a
Mission studies, decadal survey	Appendix C	Appendixes D and E	n/a
New Frontiers program	Chapter 22	n/a	Chapter 22
NSF facilities and programs	Chapter 20	Chapter 1	n/a
Planetary Data System	Chapter 17	Chapter 20	Chapter 17
Planetary radar facilities	Chapter 18	Chapter 22	Chapters 18 and 22
Plutonium-238	Chapter 20	Chapter 22	Chapters 20 and 22
Research and analysis programs	Chapter 17	Chapter 22	Chapters 17 and 22
Sample receiving and curation facilities	Chapter 20	Chapter 22	Chapters 20 and 22
SIMPLEx program	Chapter 22	n/a	Chapter 22
Technology development	Chapter 21	Chapter 22	Chapters 21 and 22
Technical risk and cost evaluation	Appendix C	Chapter 22	n/a
White papers received	Appendix B	n/a	n/a

Multiple users and their diverse needs make for a long document and some necessary repetition. The survey committee is under no illusion that all readers will start at the beginning and work their way through to the very last page. Nor is it necessary to read chapters sequentially. For example, the chapters devoted to recent discoveries (Chapter 2) and the 12 key-science questions around which the report is structured (Chapters 4 to 15) are designed to stand alone.

Indeed, there are many ways individuals can and will read this report. As such, the survey committee has endeavored to design this report so that it is accessible to readers with varied interests and possessing varying degrees of technical sophistication. Therefore, the desires of most readers will be satisfied by the selective reading of different parts of this report. With the selective reader in mind, Table 1-2 is included as a reader's guide. It is organized according to the topics mentioned in the survey committee's charge and related tasks (see Appendix A). In general, acronyms are spelled out at the first use in each chapter that appear. To help readers who want to delve into the report's more technical aspects, a glossary of acronyms and technical terms can be found in Appendix F.

REFERENCES

Hicks, D., M. Zullo, A. Doshi, and O.I. Asensio. 2022. "Widespread Use of National Academies Consensus Reports by the American Public." *Proceedings of the National Academy of Sciences* 119(9):e2107760119. https://doi.org/10.1073/pnas.2107760119.

NASEM (National Academies of Sciences, Engineering, and Medicine). 2018a. *Exoplanet Science Strategy*. Washington, DC: The National Academies Press. https://doi.org/10.17226/25187.

NASEM. 2018b. *Visions into Voyages for Planetary Science in the Decade 2013–2022—A Midterm Review*. Washington, DC: The National Academies Press. https://doi.org/10.17226/25186.

NASEM. 2019a. *An Astrobiology Strategy for the Search for Life in the Universe*. Washington, DC: The National Academies Press. https://doi.org/10.17226/25252.

NASEM. 2019b. *Finding Hazardous Asteroids Using Infrared and Visible Wavelength Telescopes*. Washington, DC: The National Academies Press. https://doi.org/10.17226/25476.

NRC (National Research Council). 1994. *An Integrated Strategy for the Planetary Sciences: 1995–2010*. Washington, DC: National Academy Press. https://doi.org/10.17226/9264.

NRC and ESF (European Science Foundation). 1998. *U.S.-European Collaboration in Space Science*. Washington, DC: National Academy Press. https://doi.org/10.17226/5981.

NRC. 2003. *New Frontiers in the Solar System: An Integrated Exploration Strategy*. Washington, DC: The National Academies Press. https://doi.org/10.17226/10432.

NRC. 2011. *Vision and Voyages for Planetary Science in the Decade 2013–2022*. Washington, DC: The National Academies Press. https://doi.org/10.17226/13117.

NRC. 2015. *The Space Science Decadal Surveys: Lessons Learned and Best Practices*. Washington, DC: The National Academies Press. https://doi.org/10.17226/21788.

NSTC (National Science and Technology Council) Interagency Working Group for Detecting and Mitigating the Impact of Earth-Bound Near-Earth Objects. 2018. *National Near-Earth Object Preparedness Strategy and Action Plan*. Washington, DC: Executive Office of the President. https://www.nasa.gov/sites/default/files/atoms/files/ostp-neo-strategy-action-plan-jun18.pdf.

2

Tour of the Solar System: A Transformative Decade of Exploration

The past decade has witnessed an explosive growth in the state of knowledge of planetary science and astrobiology through the invaluable combination of new missions and data, supporting theoretical and modeling research, telescopic observations, and laboratory and experimental advances. In this chapter, the committee discusses some of the most exciting advances from the past decade, organized by destination or destination class as represented by the committee panels. Each section ends by enumerating specific highly impactful discoveries.

MERCURY

Prior to the past decade, the innermost planet had been visited only by spacecraft flybys. That changed in March 2011, when NASA's Mercury Surface, Space Environment, Geochemistry, and Ranging (MESSENGER) spacecraft became the first to orbit Mercury, providing a wealth of new observations of the planet's interior, surface, exosphere, magnetosphere, and heliospheric environment until the end of the mission in April 2015.[1] MESSENGER's orbital measurements of Mercury defied premission predictions and transformed our thinking about the formation processes for rocky worlds.

Mercury is unique among the terrestrial planets, starting at its core, which makes up ~70 percent of the planet's mass, more than any other rocky body in the solar system. Mercury's silicate-rich crust and mantle are but a thin veneer (combined thickness 420 km) atop the comparatively enormous metallic core (radius 2020 km). Geodetic and magnetic data indicate that the majority of this core is fluid, and geodetic evidence suggests the presence of a small solid inner core. The fluid outer core presently supports an active magnetic field that is about 100 times weaker at Mercury's surface than Earth's. Mercury's magnetic field is predominantly axisymmetric and dipolar, closely aligned with Mercury's spin axis, although the magnetic equator is offset along the spin axis to the north by ~20 percent of the planet's radius, for reasons that are not fully understood. MESSENGER also mapped crustal magnetic anomalies across varied geologic terranes on 20 percent of Mercury, with measurements limited to areas of the northern hemisphere where the spacecraft was close to the planet. The recognition that ancient volcanic plains in these areas have significant remanent magnetization implies that Mercury's core dynamo was active prior to at least 3.7–3.9 Ga, and may have had a field strength similar to present-day Earth.

---

[1]A glossary of acronyms and technical terms can be found in Appendix F.

Some of the most notable discoveries of the past decade include measurements of Mercury's crustal geochemistry: elevated abundances of sulfur and carbon, low abundances of iron and oxygen, and a nearly chondritic chlorine/potassium ratio. These characteristics have important implications for the thermochemical evolution of Mercury and point to a planet that is surprisingly rich in volatiles and that formed under highly reducing conditions. The origin of Mercury's highly reduced, volatile-rich composition and large core are likely key to understanding its formation (and the formation of other planets and planetesimals at the inner edge of protoplanetary disks). MESSENGER's geochemical measurements ruled out many theories for the origin of Mercury's metal-rich composition, and currently two remain: (1) Mercury was once a large planet with a typical metal/silicate ratio expected from formation from chondrites (~30 percent metal, 70 percent silicate), but an early giant impact stripped away a large proportion of silicates that never reaccreted, leaving behind a metal-rich planet, or (2) Mercury accreted its metal/silicate ratio from highly reduced and Fe-rich material in the protoplanetary disk. At present, these two formation models cannot be distinguished from available data, nor can either model seemingly satisfy all observations.

The surface geochemistry measurements from MESSENGER have also led to new ideas about the variety of outcomes from global-scale magma oceans. The terrestrial planets are all thought to have experienced a phase where they were largely or completely molten with subsequent crystallization and silicate-liquid fractionation resulting in the formation of a crust and mantle of varying thickness and composition, depending on their starting conditions. This process is best known on the Moon, and many researchers originally thought that Mercury also possessed a flotation crust of similar composition (plagioclase). However, the low abundance of iron oxide in Mercury's silicates means that plagioclase would not have been less dense than the melt, and thus could not have risen to the surface of a magma ocean on Mercury. In fact, experimental studies used MESSENGER data to determine that all typical rock-forming minerals are denser than Mercury magmas, except for several atypical low-density mineral candidates, such as graphite and oldhamite (calcium magnesium sulfide). Given that MESSENGER data indicated elevated sulfur and carbon at the surface, both calcium-rich sulfides and graphite have been considered as possible constituents of Mercury's earliest crust. In the case of graphite, its presence at the surface could also reconcile the seeming contradiction of the planet's low average surface reflectance with its bulk composition that would otherwise yield bright surface materials. Most of the regions where Mercury's reflectance is lowest, and thus may have the highest abundance of graphite, are found either in areas with the highest density of large craters, or in the ejecta of some impact craters and basins.

The past decade has also revealed Mercury's geologic history to have been shaped extensively by volcanism. Its vast smooth plains formed from flood-like eruptions that covered approximately a third of the planet. Furthermore, stratigraphic relationships, compositional similarities, and crater size-frequency distributions also suggest that much of the intercrater plains, the most extensive terrain on Mercury, also formed through volcanic eruptions earlier in Mercury's history. Regions with the strongest evidence for volcanism are typically higher in reflectance and superpose the more ancient low-reflectance crustal materials, suggesting that their mantle source regions did not contain substantial abundances of carbon or other possible darkening agents. Explosive volcanic deposits, which are thought to deliver materials to the surface that are the most representative of mantle composition, are also among the highest reflectance, with the largest deposit shown to have low abundances of carbon and sulfur relative to the rest of Mercury's surface. The unexpectedly large number of such explosive eruptions further attests to the high abundance of volatiles native to Mercury, as these pyroclastic eruptions are driven by exsolution of magmatic volatiles.

Vast outpourings of flood basalts may account for a substantial portion of Mercury's crust, even though Mercury is in a state of global contraction owing to secular cooling. MESSENGER observed a surface dominated by long-wavelength folding and contractional tectonic features including lobate scarps, high-relief ridges, and wrinkle ridges (Figure 2-1). Extensional features are comparatively rare and are found only within volcanic plains located inside impact basins. The most abundant tectonic features on Mercury by far are the lobate scarps, thought to be the surface expression of thrust faults, which can result in cliffs several kilometers high. The global distribution of lobate scarps provide insight into how single-plate planets cool and suggest the buckling of the lithosphere is the result of contraction that has resulted in Mercury's radius shrinking by up to 7 km over the history recorded by its tectonic features. Although such contractional stresses could deter volcanism, Mercury's magmas may have been aided in reaching the surface by impact fracturing and thinning of the crust, and the fact that magmas are buoyant at any depth within the mantle and crust. MESSENGER geochemical measurements further suggest that

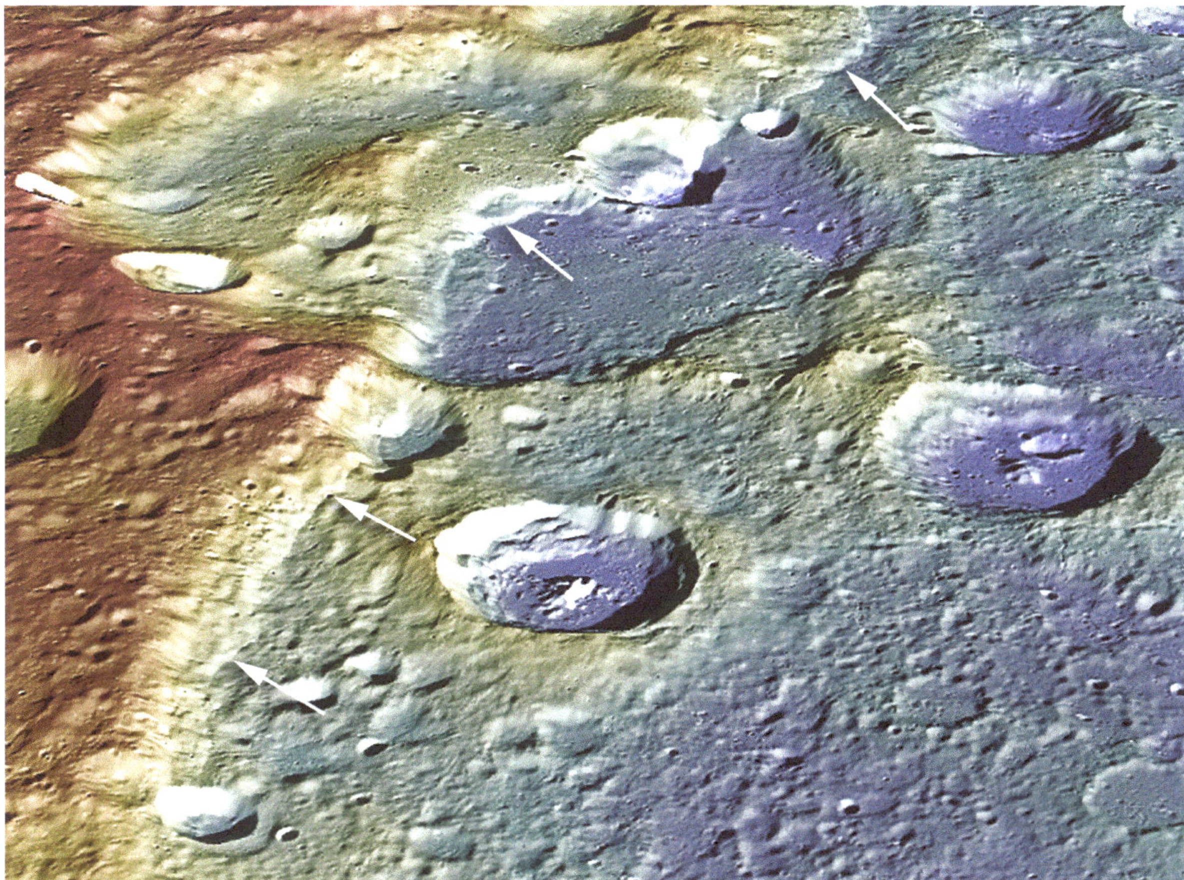

FIGURE 2-1 Giant tectonic landforms like Carnegie Rupes, which is more than 2 km high in places, provide evidence that Mercury shrunk as it cooled, creating a lobate scarp (arrowed). The image has been color-coded to indicate differences in topography: regions in red are higher standing terrain, regions in blue lower. SOURCE: Courtesy of NASA/Johns Hopkins University Applied Physics Laboratory/Carnegie Institution of Washington.

the volcanic deposits were a product of high degrees of partial melting that were likely possible only early in Mercury's history before cooling and global contraction took over.

Because of its orbit close to the Sun, Mercury has experienced more intense impact bombardment than anywhere else in the solar system, and impactors strike the surface at an average velocity at least three times that of the Earth–Moon system. However, in contrast to the expectation of a heavily cratered surface, MESSENGER revealed that Mercury has a lower total population of large impact craters (>20 km diameter) and basins (>300 km diameter) than the Moon. This implies that Mercury's oldest exposed crust is relatively young (<4.1 Ga). Both widespread volcanism and the intense early bombardment of Mercury likely played some role in obscuring older crust. Additionally, the distribution of large impact basins on Mercury is not uniform; more are identified on one hemisphere. This asymmetry in basin distribution may be owing to nonuniform resurfacing by volcanic flows and/or the spin-orbit evolution of Mercury, where the rotational period was originally synchronized with its orbital period, allowing more impact cratering on the western hemisphere than the eastern hemisphere. The impact cratering record also forms the basis for understanding chronology of major landforms on Mercury. These data suggest that most of Mercury's largest impact basins date to the period ~3.8–4.1 Ga, and that the majority of volcanic plains on Mercury are probably older than ~3.0–3.5 Ga.

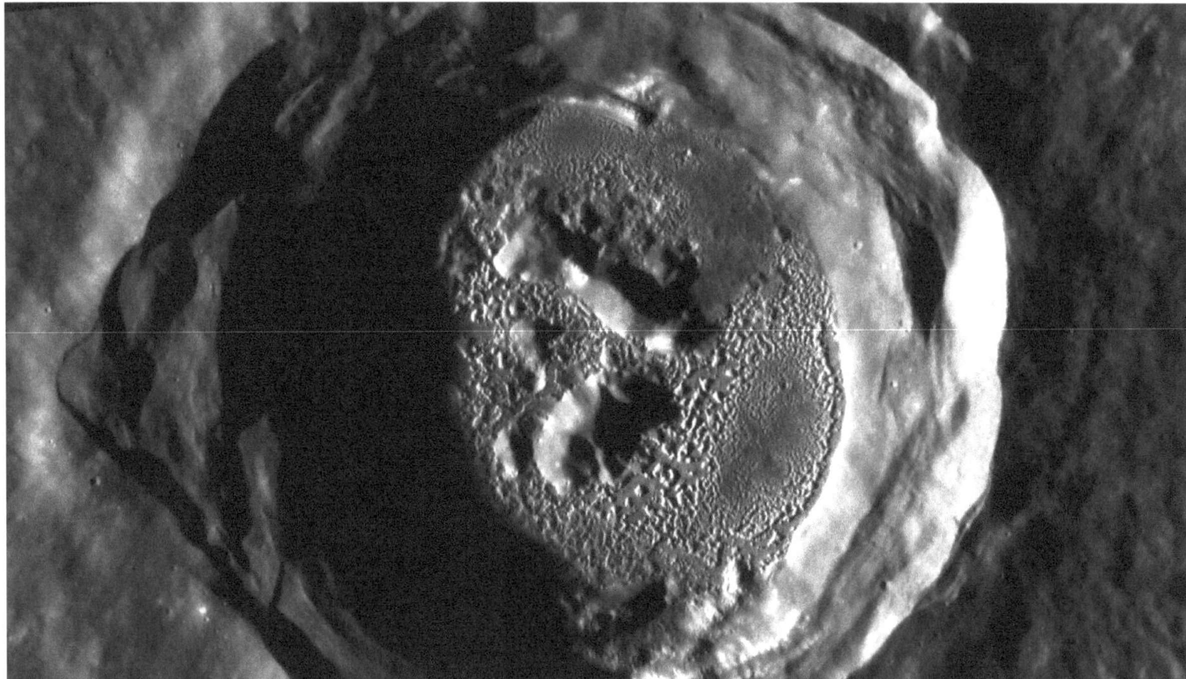

FIGURE 2-2 Volatile-rich rocks on Mercury are thought to sublime when exposed by impact events, leaving behind features known as hollows, such as those that cover floor Kertész crater (31 km in diameter). SOURCE: Courtesy of NASA/Johns Hopkins University Applied Physics Laboratory/Carnegie Institution of Washington.

While most of Mercury's geologic activity occurred early in its history, one type of landform unique to Mercury is thought to be extremely young. High-resolution images from the orbital phase of MESSENGER's mission revealed groups of small (typically less than ~1 km across), irregularly shaped, flat-floored depressions known as hollows (Figure 2-2). These features are found within impact craters, and their extremely crisp morphology suggests their formation is ongoing to this day. Hollows are thought to form owing to sublimation or erosion of a volatile-rich component that is found in Mercury's crust but is not stable on the surface after it has been exhumed by an impact event. Hollows are thus an additional signifier of Mercury's surprisingly volatile-rich nature: some of its lithologies may be too volatile-rich to survive the harsh surface environment for long.

One environment on Mercury is, however, well suited to volatile preservation: regions of permanent shadows in impact craters and topographic lows near Mercury's poles. Deposits of ice were initially suggested to exist near the poles based on Earth-based radar observations, and MESSENGER provided multiple lines of evidence that these deposits are mostly water ice, including hydrogen abundances inferred from neutron spectroscopy, reflectance measurements, and models of the thermal environment (Figure 2-3). Reflectance data and long-exposure images of regions in permanent shadow showed that some deposits have high reflectance, consistent with exposed surface ice, whereas others are lower in reflectance than material anywhere else on the planet. The low-reflectance polar deposits coincide with areas where thermal models indicate that ice is not stable at the surface, and thus they may be covered by an organic-rich lag deposit. The deposits on Mercury appear to be of higher purity and represent a larger total inventory of ice than at the Moon's poles. The source of the water ice in Mercury's polar deposits is unknown. The volatiles may have been delivered relatively recently (10–100 Ga) by comets or asteroids, or water ice may come from the interaction of Mercury's surface with the solar wind or by outgassing of interior volatile species.

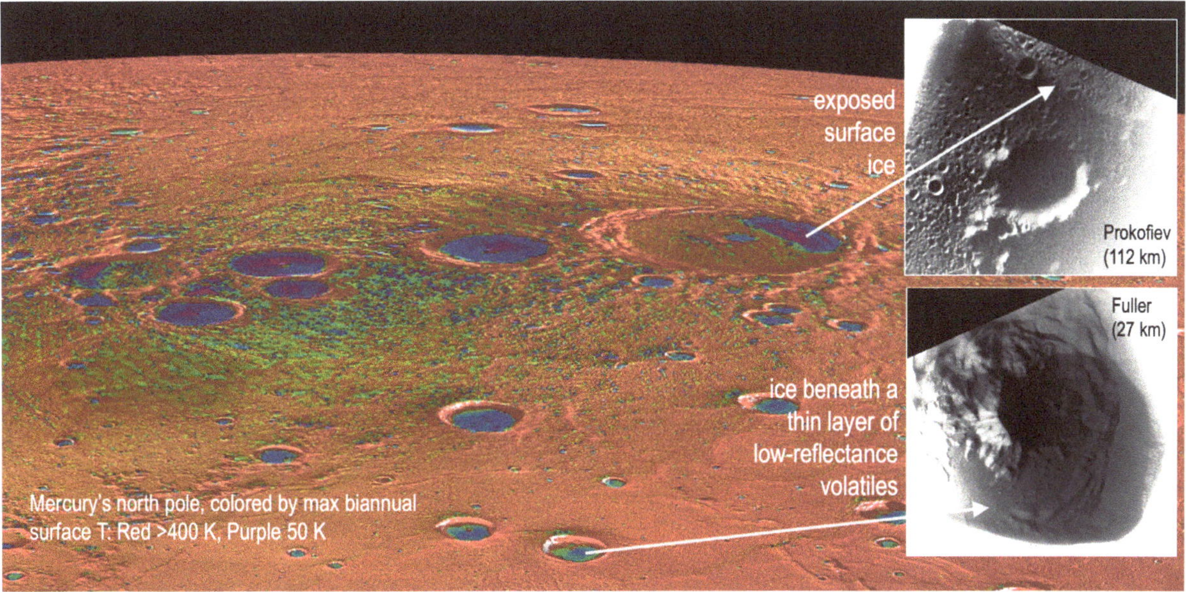

FIGURE 2-3 Despite their hot surroundings, permanently shadowed regions near Mercury's poles stay cold enough to host abundant deposits of water ice. SOURCE: Courtesy of N.L. Chabot with MESSENGER data from NASA/Johns Hopkins University Applied Physics Laboratory/Carnegie Institution of Washington; and Armytage et al. (2018).

Mercury is classically considered an environment inhospitable to life, but the astonishing detection of volatiles such as water ice and complex organic compounds at Mercury's surface suggest that life's essential ingredients may be more abundant throughout the solar system than previously imagined. Evidence of past explosive volcanism in the form of extensive pyroclastic deposits indicates that volatiles are also present in the deep interior of Mercury, although the type and extent of volatiles in the interior (e.g., CO_2, H_2O, and SO_2) remain unclear. Intriguingly, the polar regions of Mercury may provide a unique opportunity to investigate chemical reactions that occur between water, organic materials, and other volatiles at extremely cold temperatures, including how prebiotic molecules such as amino acids and nucleobases abiotically form. Although Mercury likely never sustained life, Mercury may indeed hold precious unadulterated information regarding the distribution and construction of life's building blocks throughout the solar system and inform on the ability of a planetary body to retain volatile-rich material through the catastrophic accretionary events of the early solar system.

Measurements from MESSENGER's orbital mission also revealed the complex and dynamic ways in which Mercury interacts with the extreme space environment, particularly the interplay between the solar wind, magnetosphere, and core. The magnetosphere was found to be strongly affected by events such as coronal mass ejections, allowing solar wind to intrude deeply within Mercury's dayside magnetosphere. Intense solar wind driven magnetospheric variations can induce currents that increase magnetic flux levels around the planet. New results suggest the dayside magnetosphere has competing processes: induced currents to fortify the magnetic field that can also undergo fast erosion by enhanced reconnection to allow deeper solar wind entry.

MESSENGER observations demonstrated that plasma contained within the magnetosphere is predominantly solar wind protons that enter the magnetosphere via dayside magnetic field reconnection. The entering plasma subsequently flows to a high latitude "mantle" region and then into the central anti-sunward magnetic tail. The magnetoplasma circulation times (from dayside entry, tail, and back to dayside) are unexpectedly fast, on timescales of a few minutes—much faster than the ~1 hour circulation time in the analogous terrestrial magnetosphere. Some fraction of the exospheric neutral species also becomes ionized and subsequently incorporated into Mercury's magnetoplasma circulation. Mercury's neutral exosphere contains nine known species: hydrogen, helium, sodium, potassium, calcium, magnesium, aluminum, iron, and manganese. The ultraviolet-visible spectrometer onboard MESSENGER discovered emissions from exospheric magnesium and manganese and also confirmed the presence of highly energetic exospheric calcium. The exospheric calcium energy is consistent with a temperature of 70,000 K—much higher than any known process occurring in the analogous lunar exosphere—and hypothesized to be related to CaO or CaS molecular chemistry.

The MESSENGER mission to Mercury represents a milestone in our understanding of the innermost rocky planet of the solar system. Comparisons with the Moon, based on its outward appearance, have proved incorrect. Instead, the formation of Mercury remains enigmatic but resulted in a thin, volatile-rich but reduced, silicate shell around a massive iron-rich core. Subsequent contractional processes, impacts, and volcanism have shaped the surface of this thin silicate outer shell. MESSENGER showed that Mercury's magnetosphere combined with its geochemical properties make it an important environment for understanding exosphere stability in the solar system. The BepiColombo spacecraft arrives at Mercury in late 2025 to start a 1- to 2-year mission of orbital measurements of the surface composition, geophysics, exosphere and magnetosphere that will substantially augment our knowledge of the planet. After BepiColombo, the next step will be to return to Mercury with a lander. Measurements of the minerals in surface materials will go a long way in helping us understand Mercury's highly reduced, volatile-rich composition and large core. Similarly, surface measurements of the magnetic field and the exosphere will help us understand the way that Mercury interacts with its space environment.

Key Discoveries from the Past Decade

- **Mercury is a volatile-rich world.** Contrary to earlier predictions, and despite being closer to the Sun, concentrations of elements that evaporate at moderate temperatures are more abundant on Mercury than on Venus or Earth, and are comparable with those on Mars. Additionally, high abundances of sulfur and low abundances of iron in the silicate mantle of Mercury indicate that the planet formed with less oxygen than the other bodies of the inner solar system, providing insight into the building blocks and formation of terrestrial worlds.
- **Mercury's offset magnetic field and dynamic magnetosphere were revealed by MESSENGER.** Mercury's large metallic core generates an axially dipolar core dynamo magnetic field that is enigmatic owing to its low intensity relative to Earth's and the offset of the center of the dipole by approximately 20 percent of the planet's radius to the north. Interaction between Mercury's magnetic field and the solar wind results in the generation of currents that induce external magnetic fields. With induced external fields that can be as large as the planetary field, the dynamic magnetosphere at Mercury is a unique natural laboratory for exploring magnetospheric physics and exospheres.
- **Permanently shadowed regions near Mercury's poles hold abundant water ice.** Many regions on Mercury where water ice is predicted to be thermally stable appear to host such ice deposits either at the surface in a relatively pure, thick layer or beneath organic-rich lag deposits. The origin of these deposits is not yet known, but water ice may have been delivered by comets or asteroids or micrometeoroids, from the solar wind, or outgassing from Mercury's interior.
- **Volcanism played a critical role in shaping Mercury's surface.** Mercury's oldest surfaces contain high abundances of carbon that may be remnants of a primary graphite flotation crust from an early magma ocean. Much of the surface, however, was subsequently covered by extensive floods of lava of varying composition indicating geochemically diverse terrains on Mercury's surface that hint at a heterogeneous interior.

Further Reading

Charlier, B., and O. Namur, eds. 2019. "Planet Mercury." *Elements* 15(1):9–45. Mineralogical Society of America. https://pubs.geoscienceworld.org/elements/issue/15/1.

Solomon, S.C., L.R. Nittler, and B.J. Anderson, eds. 2018. *Mercury: The View After MESSENGER*. Cambridge Planetary Science, Cambridge: Cambridge University Press. https://doi.org/10.1017/9781316650684.

THE MOON

A wealth of data from new missions, reinterpretation of old data, and new analyses of Apollo samples and meteorites, have, over the past decade, provided new insight into the evolution of the Moon, Earth, and the early solar system. The Lunar Reconnaissance Orbiter (LRO), in lunar orbit since 2009, continues to return critical data, mapping the lunar surface in unprecedented detail as well as characterizing landing sites for future robotic and human exploration. The Gravity Recovery and Interior Laboratory (GRAIL) mission orbited twin spacecraft in tandem around the Moon in 2012 and mapped the Moon's gravity at higher resolution than available even for Earth, to reveal the Moon's interior and near-surface structure. The Lunar Atmosphere Dust Environment Explorer (LADEE) spacecraft orbited the Moon from 2013–2014 to characterize the natural state of the lunar exosphere and dust environment prior to disruption from future landed robotic and human missions. Missions led by other countries, including China (Chang'e 3, 4, and 5) and India (Chandrayaan-2), have also enhanced our understanding of the state of the lunar surface and interior, duration of volcanism, distribution of volatiles, and impact processes over the past decade.

The most widely accepted hypothesis of lunar formation posits that the Moon formed from the debris of a cataclysmic impact of a Mars-size planetary body with the proto-Earth. Although this giant impact scenario is not new, it has been extensively revised in the past decade. The original model of the Moon forming by an impact between the proto-Earth and a Mars-size impactor successfully explained the size of the lunar core, angular momentum of the Earth-Moon system, and mass of the Moon, but not their isotopic similarities. The problem has been that most of the Moon was thought to form from the impactor, rather than being a mixture of impactor and Earth materials, suggesting Earth and the Moon would be different isotopically, unless the impactor happened to have the same isotopic ratios as the proto-Earth. New models propose different impact speeds, impact sizes, or even multiple impact events as scenarios that could result in more mixing between the materials that formed Earth and Moon, and invoke transfer of angular momentum from the Earth–Moon system to the Sun–Earth system. These models can broadly explain the isotopic similarity, but may not satisfy other constraints.

In the past decade, there has also been debate over the formation age of the Moon. Knowing the timing of this event would enable an understanding of when Earth and the Moon attained their present-day configuration and when the last stages of planetary formation occurred in the solar system. Lunar samples from the Apollo missions and from meteorites give a minimum age for the Moon of 4.4 Ga. To determine the true age, two main strategies have been used with different results. The first provides model ages by using measured isotopic compositions of lunar materials and back-calculating when they were equal to an assumed starting value, yielding a lunar formation age of 4.51–4.52 Ga. This method also gives a model age for the most evolved known reservoir in the Moon (known as KREEP for its potassium, rare earth element, and phosphorous content) of about 4.38 Ga, perhaps dating the end of lunar magma ocean crystallization. These ages suggest the Moon formed about 50–60 million years after solar system formation. The second method uses assumptions of how materials were accreted to planets, including the abundances of highly siderophile elements (HSEs), as a constraint for the Moon formation time. The HSEs currently found in the terrestrial mantle would have been delivered after or during the Moon-forming event because any HSEs accreted prior to the impact would have been partitioned to Earth's core. Using estimates of the impact flux and composition of planetesimals, models of the rate of accretion of materials to Earth tend to predict formation ages for the Moon that are substantially younger than other methods—as much as ~100 million years after the formation of the solar system. Determining the formation time of the Moon is crucial because the impact event defined the initial condition of Earth. Moreover, the Moon-forming impact was likely the last catastrophic event in the inner solar system, and therefore it also defined the end of the accretion phase. In summary, the timing of the Moon's origin remains an open question.

Further insight into the Moon's formation and thermochemical evolution has been gained over the past decade from intense study of its inventory of indigenous volatile elements. Whereas all terrestrial planetary bodies are depleted in volatile elements relative to CI chondrites and the Sun's photosphere, the Moon's depletion is more extreme and results from a combination of the cataclysmic circumstances under which it is thought to have formed, the geochemical signature of the protoplanet impactor, and/or the process of planetary differentiation. However, while the Moon was once thought to contain less than one part per billion H_2O in its interior, new measurements show that at least some portions of the lunar mantle contain hundreds of parts per million H_2O, and remote detection of hydrated materials in some uplifted crustal materials further attest that the Moon is far from "bone dry." The discovery that stable isotope compositions of moderately volatile and volatile elements in lunar samples are fractionated relative to terrestrial values further suggests that the Moon lost a substantial fraction of its volatile elements both during and after it formed. Volatile loss from the lunar interior was likely important for generating transient atmospheres and for distributing some of the volatile species presently on the lunar surface.

Major advances in our understanding of the interior and crustal structure of the Moon and its bulk composition have come from new ultra-high-resolution gravity and topography measurements obtained by GRAIL and LRO, respectively. These data, when combined with Apollo seismic constraints, indicate that the lunar crust has high porosity and a mean thickness of 34–43 km, with an average crustal thickness of ~55 km on the farside and ~30 km on the nearside (Figure 2-4).

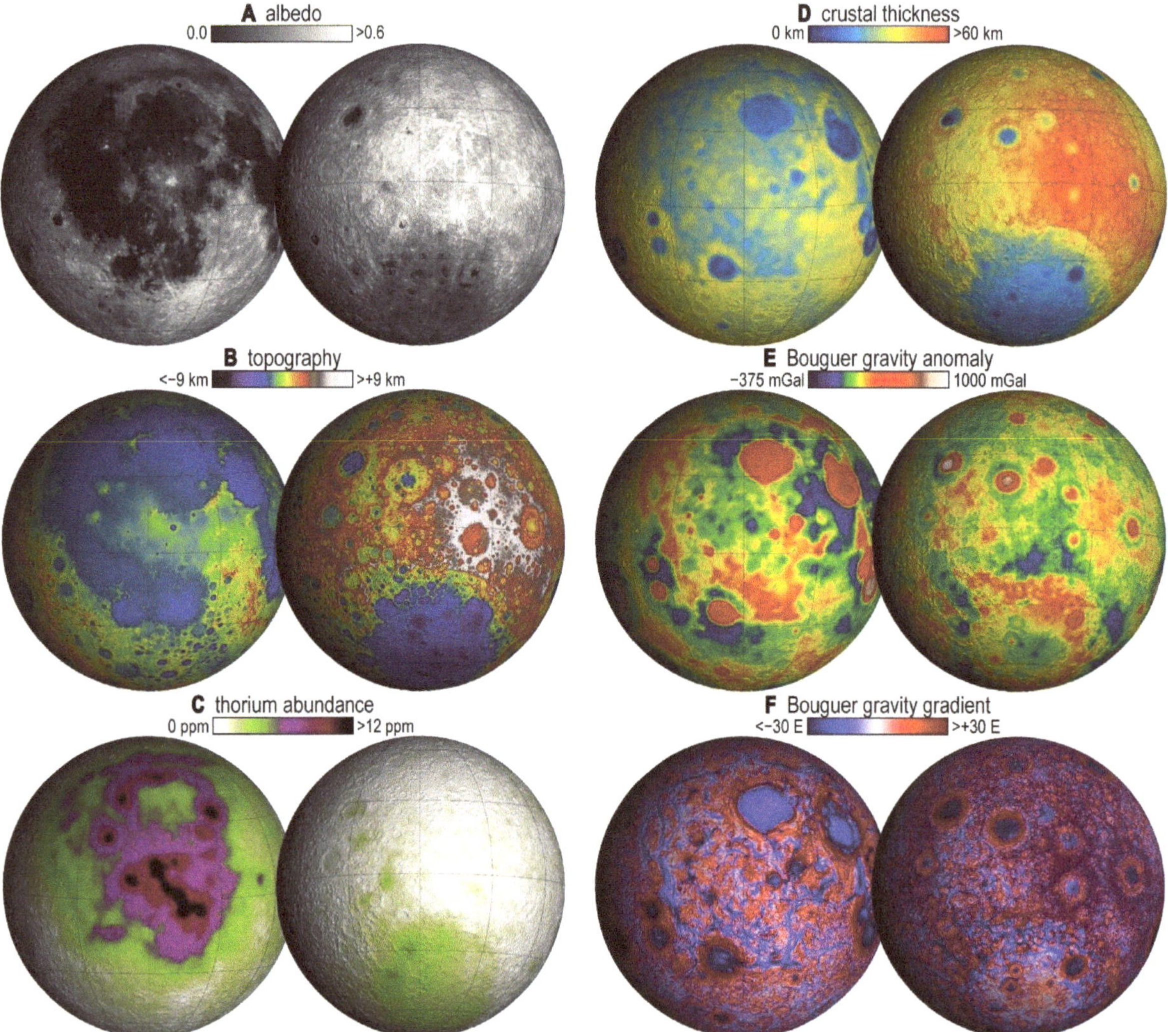

FIGURE 2-4 Large-scale crustal asymmetries on the Moon. (A) Lunar Reconnaissance Orbiter (LRO) Lunar Orbiter Laser Altimeter (LOLA) albedo map of the Moon at 1,064 nm. (B) High-resolution lunar topography, derived from the combination of LRO/LOLA and SELENE Terrain Camera data. (C) Thorium abundance derived from the Lunar Prospector Gamma-Ray Spectrometer. (D) Model for the crustal thickness of the Moon, derived from the Gravity Recovery and Interior Laboratory (GRAIL) mission and LRO/LOLA. (E) Bouguer gravity anomaly (gravity anomaly corrected for surface topography) derived from GRAIL and LRO/LOLA. (F) Bouguer gravity gradient (a measure of the horizontal derivative of the Bouguer gravity anomaly) derived from GRAIL and LRO/LOLA. SOURCES: Courtesy of J.T. Keane with data from (A) Lemelin et al. (2016); (B) Barker et al. (2015); (C) Lawrence et al. (2007); (D) Wieczorek et al. (2013), Goossens et al. (2019); (E) Zuber et al. (2013), Goossens et al. (2019); (F) Andrews-Hanna et al. (2014).

Such large-scale crustal asymmetry is further found in nearside/farside differences in porosity, heat-producing elements, and extent of volcanism; the cause of the Moon's crustal asymmetry remains one of the greatest outstanding mysteries regarding lunar early evolution. The crustal thickness estimates also provide constraints on the bulk composition of the silicate portion of the Moon. In particular, the crustal thickness estimates indicate the Moon has a complement of refractory elements that is a close match to the bulk silicate composition of Earth rather than earlier estimates, based on inferred crustal thickness larger than currently accepted values that indicated the Moon was enriched in refractory elements by up to 50 percent relative to Earth and CI chondrites. GRAIL data, coupled with advances in theoretical models, have also transformed our understanding of how large impact basins (and the associated, mysterious gravity anomalies—"mascons"—that destabilized the orbits of early lunar missions) formed, and how tides shaped the Moon. New insights have been gained into the Moon's deep interior structure as well, from a combination of data from GRAIL and LRO, modern analyses of Apollo seismic data, and continued laser ranging observations, and indicate that the Moon likely has a solid inner core between 130 and 200 km in radius, possibly overlain by a fluid outer core extending out to ~380 km in radius.

Over the past decade, remanent crustal magnetism studies and laboratory analyses of returned lunar samples have revealed that the ancient Moon had an internally generated magnetic field. Between 4.25 and 3.56 Ga, surface field intensities reached values up to 40–110 μT. This high-field period was followed by a lower intensity field of ~5 μT or less that may have persisted beyond 2 Ga. However, no single proposed dynamo mechanism to date satisfactorily explains both the longevity and intensity of the ancient lunar dynamo inferred from the paleomagnetic record. Ongoing research is further exploring hypotheses such as a basal magma ocean dynamo, the possibility that impact effects could have led to transient amplification of a baseline core dynamo field, or a situation wherein multiple dynamo mechanisms with differing predicted field intensities may have operated at different points in time.

The Moon's dominant form of volcanism involved eruption of voluminous basin-filling mare lavas peaking about 3–3.5 Ga, but samples recently returned from Oceanus Procellarum by the Chang'e 5 mission show that the formation of flood basalts continued until at least 2 Ga. Morphological and compositional diversity is expressed through abundant mafic pyroclastic deposits, less common silicic pyroclastics, and complexes of small domes and cones, and evidence for large-scale shields. Over the past decade, new evidence challenges the notion that all lunar volcanism is ancient. The Lunar Reconnaissance Orbiter Camera (LROC) revealed numerous irregular mare patches, depressions containing patchy rough and smooth mafic deposits (Figure 2-5), thought to be remnants of eruptions younger than 100 My, volcanic materials resurfaced by late degassing, or the product of more ancient eruptions that resulted in materials with atypical physical properties. If these features are confirmed to be related to recent eruptions, the occurrence of such young volcanism implies that the lunar mantle was warmer than previously thought, and/or that heterogeneous or localized concentrations of radioactive elements allowed for small-scale eruptions to continue late into lunar evolution. In addition, silicic volcanism appears to be more prevalent than previously appreciated, raising the question of how substantial volumes of evolved magma could be produced on a single-plate planetary body.

GRAIL and LRO data enabled a new global inventory of magmatic intrusions, suggesting that for every volcanic eruption at the surface, there is an order-of-magnitude more magmatism beneath the surface. Analyses of the Moon's Bouguer gravity revealed a global population of previously unseen, 100-km-long igneous bodies distributed across the Moon, likely reflecting a period of global expansion and magmatism early in the Moon's history. On the nearside, even larger dikes and buried rifts were discovered enclosing the Procellarum KREEP Terrane (PKT; a region characterized by low elevations, thin crusts, extensive mare basalts, and abundant heat-producing elements), delineated by a 3.5-ppm-Th contour on the lunar nearside hemisphere. The buried rifts enclosing the PKT may be the result of planetary-scale tectonism and magmatism, and may have been the plumbing for much of the mare basalts that preferentially flooded the lunar nearside terranes. This early period of expansion was succeeded by a period of global contraction as a result of secular cooling. Evidence for this contraction comes from LROC images, which revealed a far greater abundance of contractional tectonic features than previously known, particularly lobate scarps. Similar to those on Mercury, but generally of smaller scale, these features are the surface expression of thrust faults, and their locations suggest that they have been further influenced by stresses from tidal interactions with Earth and from orbital recession away from Earth. Movement along these faults may continue into the present day, and may be the source of the large but rare shallow moonquakes recorded by Apollo seismometers.

The well-preserved state and ancient surface of the Moon has exquisitely preserved the most complete record of the major events that have occurred in our part of the solar system since the formation of the Earth–Moon system.

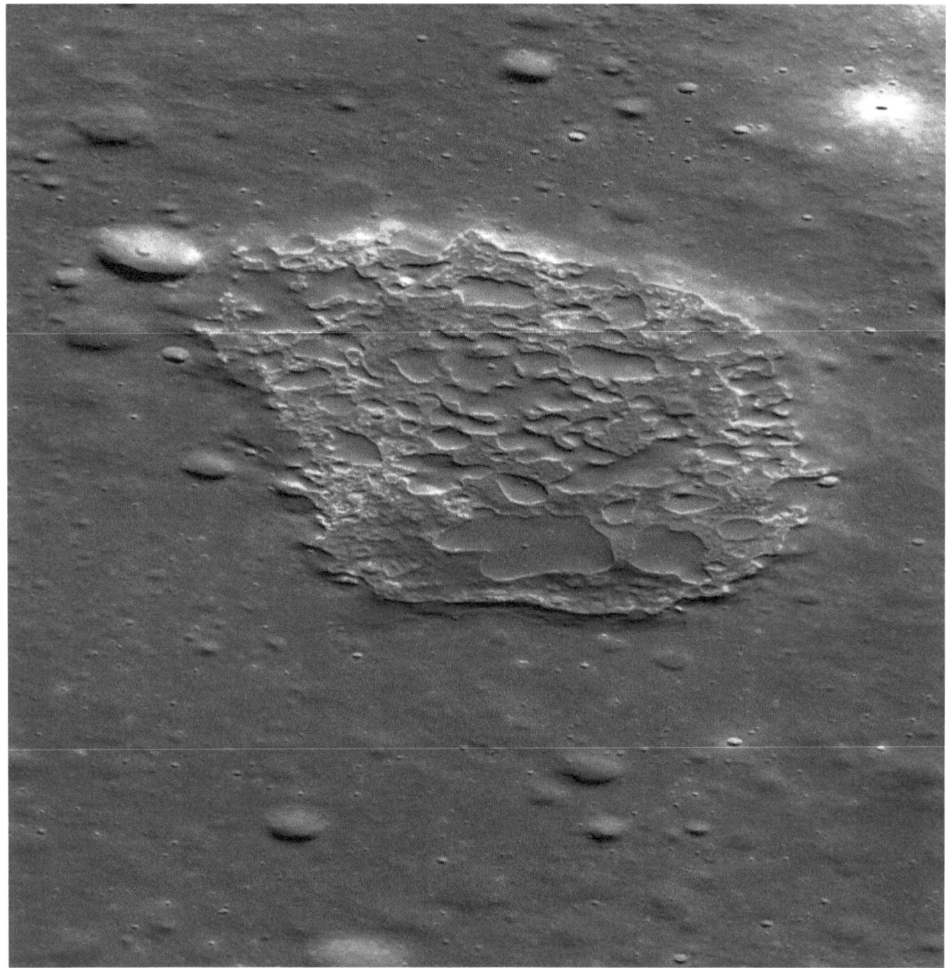

FIGURE 2-5 The Moon's volcanic activity may have extended into geologically recent times to produce features like Ina, thought to be a volcanic caldera some 2 km across. North is to the right in this image from NASA's Lunar Reconnaissance Orbiter. SOURCE: Courtesy of NASA/GSFC/Arizona State University.

Although the early record of the accretion and rate of exogeneous delivery of chemical compounds are poorly preserved on Earth, the flux of chemical inventories important to the emergence and subsequent evolution of Earth's life may be decipherable at the Moon. The Moon's impact history not only provides a means to understand the linked evolution of the Earth-Moon system since antiquity, but also the chronology of events across the solar system. Indeed, significant advances have been made in the past decade by refined laboratory analysis of samples that have been returned and dated from the Moon's surface. An example of such a recent advance is an improved age determination for the important nearside impact basin Imbrium to 3.91–3.94 Ga based on U-Pb dating. The history of impacts on the Moon prior to Imbrium remains a subject of substantial uncertainty. Since the 1970s, it has been hypothesized that there was an intensive peak in the impact bombardment of the Moon in the 3.8–4.1 Ga epoch (known as the late heavy bombardment or lunar cataclysm). Modeling and geochemical studies completed over the previous decade have cast doubt on whether an intense impact cataclysm in this epoch is required, in part because the Imbrium impact event may be radically overrepresented in the existing sample collection. At a minimum, the magnitude of the peak in the impact bombardment during this period might be lower than was once thought. GRAIL and LRO substantially clarified the total number of impact basins whose signatures are preserved on the Moon, but the period over which these basins formed remains to be established. Intense interest in the population and ages of lunar basins persists because they have implications beyond the Moon; an intense period of impact bombardment of the lunar surface would be mirrored across the inner solar system, including on the early Earth. Only the most recent impact events are preserved on Earth, and LRO data combined with the terrestrial evidence suggest that the flux of impactors to the Earth-Moon system may have increased at about 290 Ma, and the breakup of specific asteroid families may be recorded in the Moon's crater population.

The past decade has seen a major advance in direct observations of the ongoing impact process on the Moon. In particular, repeated meter-scale imaging by LROC enabled detection of ~500 impact craters formed over the first 11 years of the LRO mission (Figure 2-6). These observations provide a direct constraint on the flux of small meteoroids to the Moon. Terrestrial ground-based monitoring of the Moon has also captured impact flashes associated with impacts, and direct observations of the dusty ejecta from small impacts seen by LADEE. Integrating these observations has led to a revised picture of the rate at which the uppermost surface of the Moon is gardened, and emphasized the importance of high frequency, small meteoroid impacts and their ejecta for gradually eroding surface topography and transporting materials vertically and laterally.

Intense interest has also focused attention over the past decade on furthering our knowledge of the abundance, distribution, and origin of the Moon's near-surface volatiles and understanding their implications for where habitable environments may exist in the solar system. LRO data continues to unravel the mysteries of polar volatiles (Figure 2-7), as neutron absorption measurements provide increased detail regarding the distribution of hydrogen,

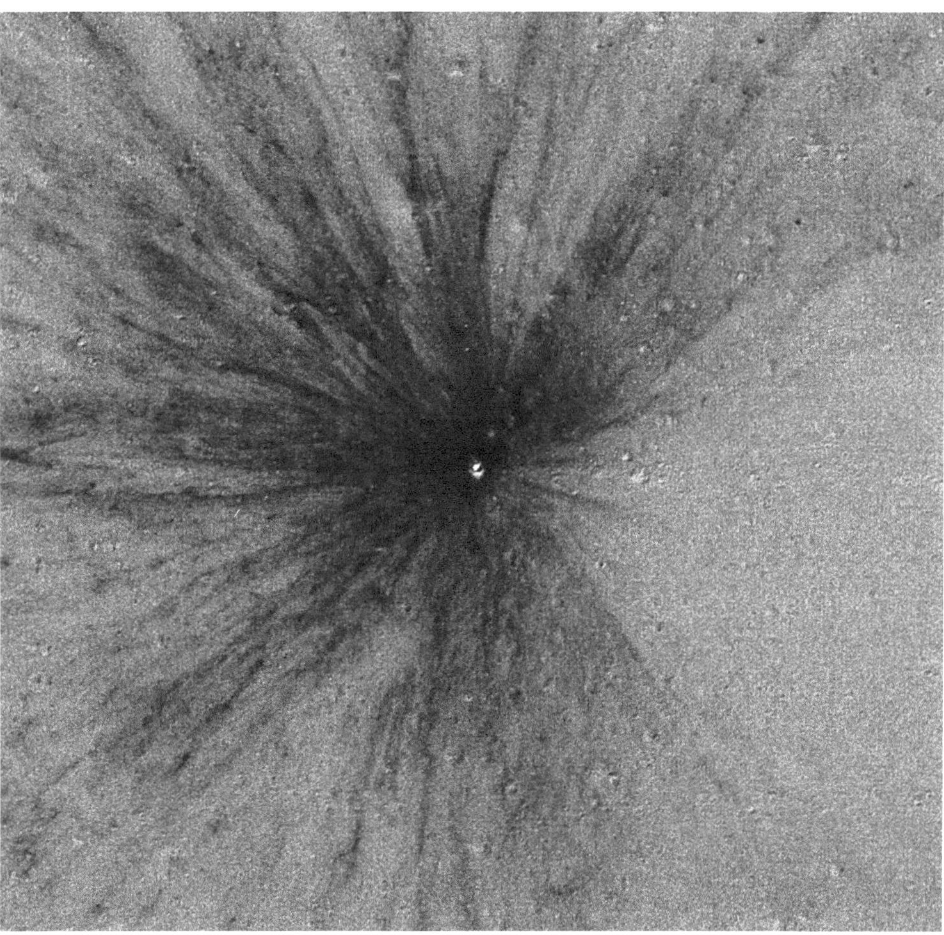

FIGURE 2-6 A new crater on the Moon that formed in 2012 or 2013. Hundreds of new impact craters on the Moon have been detected by NASA's Lunar Reconnaissance Orbiter by comparing images acquired at different times and looking for changes. Here, a ratio of images acquired after and before a 120 m crater formed reveals extensive changes around the crater. The image shows an area approximately 1,200 meters wide. SOURCE: Courtesy of NASA/GSFC/Arizona State University.

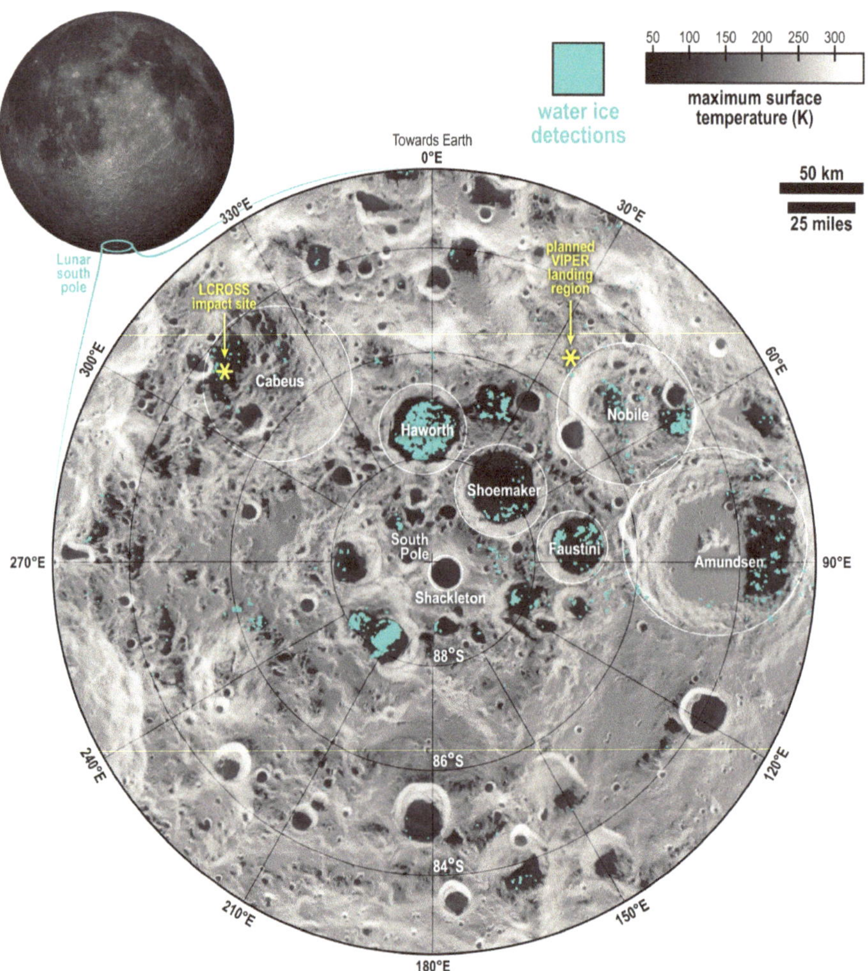

FIGURE 2-7 A map of the lunar south pole (poleward of 87.5°S). Locations consistent with the presence of water ice, based on anomalous ultraviolet reflectance, are indicated in cyan (Hayne et al. 2015a). The grayscale indicates the maximum annual surface temperature; dark grays indicate regions of permanent shadow. Yellow asterisks mark the impact site of the LCROSS mission, and the planned landing region of VIPER. SOURCES: Courtesy of J.T. Keane with data from Hayne et al. (2015a,b).

and far-ultraviolet and laser reflectometry data are consistent with surface frosts in polar shadowed regions. Whereas previous data were ambiguous as to the form in which hydrogen is present, results from Moon Mineralogy Mapper data confirm for the first time that at least some polar H is present as H_2O. Numerical modeling indicates that volatiles are also thermodynamically stable beneath the surface in areas of temporary sunlight near the lunar poles, and in micro-cold traps at higher latitudes where small shadows persist. New measurements from the Stratospheric Observatory for Infrared Astronomy indicate that H_2O is present in near-surface sunlit regions, likely derived from solar wind H^+ and stored within impact glasses known as agglutinates or possibly between regolith grains. Off-sets in the distribution of hydrogen from the poles have been interpreted to suggest that the location of the poles may have changed over time owing to large impacts and/or volcanism that affected the Moon's obliquity, and thus some portion of the Moon's polar volatiles are preserved from those ancient times and recorded volatile delivery from various sources such as comets, asteroids, solar wind implantation, and/or volcanic outgassing. Through such interrelated processes, the distribution and nature of polar volatile deposits may be indicative of the geologic and geophysical evolution of the Moon as well as the bombardment history of the inner solar system.

The study of volatiles at the Moon continues into its exosphere. In the past 10 years, a new appreciation of the exospheric hydrogen budget has been gained, where solar wind implanted protons can be backscattered as neutral hydrogen, energetic protons, or chemically altered in the regolith and released as new H-bearing species, including H_2 and methane. The LADEE mission did not find a substantial and sustained water exosphere. However, transient energetic water plumes were observed to be released from the surface during meteor streams, suggesting that the equatorial surface contains substantial amounts of H-bearing material either as water or OH that can be converted to water during highly energetic impactor events. LADEE also confirmed the presence of argon-40 and helium, found neon in comparable abundances, and discovered regional variations in the sodium and potassium exosphere. Furthermore, a particulate exosphere, produced by micrometeoroid impacts, was found by LADEE to extend many hundreds of kilometers above the surface, although no evidence was found for electrostatic transport of dust at altitudes >1 km.

Over the past decade, numerous lunar space missions, analyses of lunar samples, and modeling approaches have deepened our understanding of the distribution of volatiles, impact history, magnetic field, and geological features not only for the Moon, but for early Earth and planetary objects in general. These observations have also raised new questions that will only be answered by returning to the Moon and making direct in situ measurements and by returning samples to study in laboratories on Earth.

Key Discoveries from the Past Decade

- **GRAIL explored the interior structure of the Moon with unprecedented detail, transforming our understanding of rocky worlds.** GRAIL produced the highest resolution gravity field measurements of any object (including Earth). These data revealed the Moon's crust to be far more fractured and porous than expected, exposed previously undiscovered global-scale magmatic-tectonic features, elucidated the formation and evolution of impact basins, and constrained models for both the bulk composition and formation of the Moon. While GRAIL focused on the Moon, these results have had sweeping implications for the understanding the geophysics of planetary bodies across the solar system.
- **New analyses of samples from the Moon provide key insights into its formation and physicochemical evolution.** Measurements of lunar samples have demonstrated the extent to which volatile elements are depleted in the lunar interior and the isotopic compositions of volatile elements, providing key insights into the process responsible for volatile depletion and at what stage(s) of the Moon's evolution the depletion occurred. Increasing evidence for the essentially identical isotopic compositions between Earth and Moon for nonvolatile elements has strongly challenged prior Earth–Moon origin models. Additionally, paleomagnetic studies of lunar samples have revealed that the Moon once generated a core dynamo that persisted for over 2 billion years and had surface field strengths that, at times, rivaled that of the current Earth. Last, geochronological studies of lunar samples have elucidated the Moon's formation age, which ties into our understanding of the timing of giant impacts during planetary accretion.
- **The Lunar Reconnaissance Orbiter (LRO) revealed the Moon in unprecedented detail, including ways in which its surface has been altered in recent geologic times.** Lunar volcanism was thought to have ended well over 1 billion years ago, but images from LRO revealed irregular patches of basaltic deposits that may have erupted within the past 100 million years. Newly discovered tectonic features indicate their locations may be influenced by tidal interactions with Earth and from orbital recession away from Earth, and movement along these faults likely continues into the present day. Images of craters that have formed after LRO began its mission indicate that impacts affect the surface far from the impact site, and secondary cratering overturns regolith at rates more than 100 times higher than previously thought.
- **Water ice lies at the surface within some regions of permanent shadow at the Moon's poles.** New results confirm that at least some polar H is present as H_2O, although its origin and abundance is still not known. Understanding the nature of the Moon's polar volatiles could provide insight into the origin, timing of delivery, and subsequent evolution of water and volatiles in the inner solar system.

Further Reading

Lunar Exploration Analysis Group. 2017. *Advancing Science of the Moon: Report of the Specific Action Team.* August 7–8. Houston, TX: Lunar and Planetary Institute. https://www.lpi.usra.edu/leag/reports/ASM-SAT-Report-final.pdf.

Mineralogical Society of America. 2023. "New Views of the Moon 2." C.R. Neal, B.L. Jolliff, C.K. Shearer, S. Valencia, L. Gaddis, S. Mackwell, and S.J. Lawrence, eds. *Reviews in Mineralogy and Geochemistry.*

VENUS

A Tale of Two Planets

Viewed from afar—say, a few tens of light-years—the solar system contains two remarkably similar large, rocky worlds. They are close in size: one is only slightly less massive, and slightly smaller, than the other. They are likely the same age, and are presumably made of the same materials in about the same proportions. The smaller of the two orbits the Sun about a third closer than its bigger neighbor.

But actually visit that slightly smaller world and you'll find a dramatic difference. Whereas Earth has blue skies, liquid water oceans that abound with life, and an oxygen-rich atmosphere, Venus has a global layer of yellow sulfuric acid clouds and suffocates under a thick blanket of carbon dioxide so dense that the pressure at the surface is 90 bars—equivalent to almost a kilometer underwater on Earth. The mean surface temperature at Venus is 740 K, about that of a self-cleaning oven (Figure 2-8). And instead of oceans, the second planet has vast lava plains, towering highlands, and an enormous, equator-spanning rift system.

Since NASA's Mariner 2 flyby of Venus in 1962—the first successful encounter with any planet, by any nation—the striking differences in surface conditions between Venus and Earth have motivated a major question in planetary science: Why is Earth's closest sibling not its twin?

Enigmatic Venus

Before the heady days of sustained Mars exploration, Venus attracted substantial attention from the major spacefaring nations of the time, the United States and former Soviet Union. Between the 1960s and 1980s, 35 Venus missions were dispatched (not all successfully reaching their destination). Including those using Venus as a gravity assist during flybys to other destinations, 47 missions visited (or attempted to visit) the second planet since 1961. Since 1991, however, that number has been eight. JAXA's Akatsuki, in orbit since 2015, is the sole spacecraft

FIGURE 2-8 Venus in ultraviolet and infrared from Akatsuki. A composite view of Venus in ultraviolet (*left*) and infrared (*right*) from image data returned by the JAXA Akatsuki orbiter. In ultraviolet, prominent cloud structure is visible; in infrared, the hot, lower atmosphere glows through the cooler clouds above. SOURCE: Composed by P. Byrne with data and Venus image processing from JAXA/ISAS/DARTS/K.M. Gill. CC BY 2.0.

currently operating at Venus. The last dedicated U.S. mission to the planet was Magellan, which launched in 1989 and operated in orbit from 1990 until its decommissioning in 1994.

But during the preparation of this report, the Venus mission landscape changed dramatically. In early June 2021, NASA selected two new Venus missions for its ninth Discovery Program competition—VERITAS (Venus Emissivity, Radio Science, InSAR, Topography, and Spectroscopy) and DAVINCI (Deep Atmosphere Venus Investigation of Noble gases, Chemistry, and Imaging)—and just a few days later the European Space Agency announced the winner of its fifth medium-class mission competition as EnVision, also bound for Venus. In the blink of an eye, the Venus community went from having no missions in play to a *fleet*. VERITAS, DAVINCI, and EnVision will tackle some of the major outstanding questions that are yet to be answered for the second planet—building on the trove of data returned by those early missions that helped paint a picture of Venus as a tortured planet.

The finding by Mariner 2 of a sweltering surface, and later atmospheric measurements by NASA's Mariner 5 and Pioneer Venus missions, motivated the development of a "runaway greenhouse" scenario to explain the Venus climate. Under this view, a perhaps-once-temperate planet lost its oceans to the sustained onslaught of radiation from a star that Venus unfortunately happened to orbit just a little too closely. Later missions helped flesh out that picture: a planet with major geological activity within the past billion years that all but erased any earlier record of surface conditions; a planet that, unlike Earth, lacks an internal magnetic field; a planet with a super-rotating atmosphere that at the cloud tops spins sixty times faster than the surface.

But, as is always the case in planetary exploration, we are left with more questions than answers for Venus. Did the planet indeed have a temperate past? If so, what were conditions there like before the climate catastrophe that befell it? Is Venus volcanically and tectonically active today? What are its rocks made of, and what do those compositions tell us about the planet's formation and chemical evolution? And what lessons might the second planet hold for us regarding the fate that awaits our own, and for our understanding of large rocky worlds orbiting other stars?

A Landscape Both Familiar and Strange

The global cloud layer makes imaging the Venus surface from orbit with conventional cameras extremely difficult. And so Pioneer Venus, Magellan, and the Soviet Venera 15 and 16 missions carried radar instruments, together unveiling Venus as an Earth-sized world quite unlike Earth itself. These missions found no evidence for plate tectonics, the primary way our planet regulates its temperature through geological time via the carbon-silicate cycle. But scientists recognized myriad features from Earth and elsewhere, including huge volcanoes, long lava flows, and an astonishing array of structures attesting to a sustained record of tectonic activity (Figure 2-9). Venus may not have been found to look like Earth, but it clearly had a detailed active geological story all its own.

Indeed, when Magellan acquired almost full global radar coverage of the surface (albeit at a resolution comparable to what the Viking orbiters returned for Mars in the 1970s), one type of feature was conspicuously absent: large impact craters. The planet has fewer than 1,000 craters of any size, and only *one* greater than 200 km across—a far cry from the hundreds of craters of such size on Mercury, the Moon, or Mars. Our best explanation for this unexpected dearth of large craters? Major volcanic resurfacing that served to bury the large impact craters and basins that were once presumably on the surface. Whether this resurfacing occurred in one catastrophic event or in phases remains unknown, but our best estimate is that the average age of the Venus surface is no more than a few hundred million years.

In addition to the volcanoes and lava flows we recognize across Venus, the planet also hosts an unusual type of landform that is both volcanic *and* tectonic in nature: the enigmatic corona. Approximately circular features often accompanied by considerable fracturing of the surrounding plains, coronae are sites where upwelling magma from deep in the planet's interior is thought to impinge on the crust. Some of that material might even return to the interior in a form of local subduction not dissimilar to how tectonic plates are recycled on Earth. Recent computer modeling suggests that at least some of Venus's ~500 coronae are actively forming, and deforming, today.

Venus offers us other hints of ongoing volcanic activity. ESA's Venus Express spacecraft, which orbited the planet from 2006 until 2015, found with its infrared spectrometer that some lavas seem to have barely been weathered—consistent with their having erupted geologically recently. That same mission saw short-lived, highly localized increases in surface temperature in regions of the planet interpreted with Magellan data to be where the crust has pulled apart; some of the most concentrated areas of volcanic activity on Earth are in such rift zones. And Pioneer Venus recorded a dramatic reduction in the abundance at very high altitudes of sulfur dioxide over

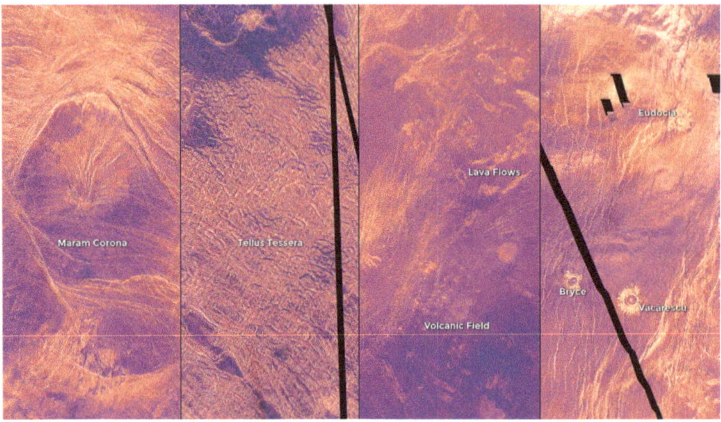

FIGURE 2-9 Examples of volcanic, tectonic, and impact features on Venus. Magellan radar data revealed a remarkably diverse planetary surface at Venus. Some examples are shown here, including (*from left to right*): Maram Corona; a portion of Tellus Tessera; a field of volcanoes and lava flows near Atla Regio, in the planet's southern hemisphere; and the craters Eudocia, Bryce, and Văcărescu at high southern latitudes. The colorized stretch here denotes high (orange) to low (purple) radar backscatter, corresponding primarily to rough to smooth surface textures at the wavelength of the Magellan radar. The black stripes are gaps in radar imagery. Each scene is approximately 400 km across. SOURCES: Figure by P. Byrne based on data taken by the NASA Magellan orbiter, courtesy of NASA/JPL-Caltech.

the 10 years after it made orbit in 1978, as if shortly before the spacecraft arrived a major volcanic eruption had injected a plume rich in that gas high into the atmosphere, which then slowly dissipated.

What of tectonic activity? The surface abounds with fractures, some of which cross the geologically youngest lava plains. Given how widespread apparently well-preserved structures are across Venus (and the planet's sheer size), it would be a surprise were the planet not tectonically active today. But there once may have been even greater tectonic action: a region high in the north, called Lakshmi Planum, is ringed by mountains that bear more than a passing resemblance to those formed when India collided with Asia. Recent mapping with decades-old Magellan imagery has also found indications that some of the oldest terrains on Venus look like they have been pieced together, and that portions of the planet's lowlands have recently jostled and moved like pack ice. Together, these observations strongly point to major horizontal motion of the Venus surface at some point since major volcanic burial of the crust began. Both the VERITAS and EnVision orbiter missions will search for evidence of ongoing volcanic and tectonic activity from space and, with an atmospheric entry probe, the DAVINCI mission will look for chemical signatures of volcanism in the atmosphere itself.

Another mystery of Venus is the nature of the planet's "tesserae," the oldest and most highly deformed rocks on the planet. Tesserae have no obvious counterpart on any other rocky body except, perhaps, heavily tectonically deformed continental rocks on Earth. Indeed, geophysical data from Magellan suggested that those tesserae that constitute the planet's highlands correspond to regions of notably thicker crust, drawing parallels between this terrain and correspondingly thick portions of crust on Earth, including continental crust. The formation of continental crust is thought to require the presence of both oceans and plate tectonics—two properties absent on Venus, at least today. Recently, layering has been documented within parts of several tessera exposures; this layering resembles that formed by massive stacks of lava in vast volcanic regions termed large igneous provinces on Earth but is also consistent with sedimentary rocks—which do not form under present climate conditions on Venus.

Indeed, the details of what exactly the surface of Venus is actually made of still remain largely unknown 40 years after the last visit by the Soviet Venera and Vega landers, the only missions to obtain quantitative chemical measurements and photographs of venusian rocks. These landers were technological marvels of their time, but the chemical data they provided were hardly comprehensive, and they made no measurements of mineralogy. Even so, what measurements they did take were enough to show that much of the surface is weathered basalt—consistent with interpretations made with orbital radar data that the expansive plains occupying almost four-fifths of the

FIGURE 2-10 Venera 13 on the Venus surface. Artist's impression of the Soviet Venera 13 lander sitting in Navka Planitia in March 1982. The landing site was found to be a relatively flat area characterized by blocky slabs and dark, fine-grained soil. SOURCE: Figure by P.K. Byrne. Surface texture courtesy of M. Malmer. CC BY-SA 3.0.

planet surface are basaltic lavas. The images returned of those landing sites, at a resolution very much finer than the relatively coarse Magellan radar imagery, also evoke an alien landscape we barely understand (Figure 2-10).

Establishing what materials make up Venus is key to understanding the planet's formation and evolution: the starting composition of the planet has enormous bearing on the types of rocks we might expect to find on the surface. Modern advances in instrumentation technology hold the promise of incredible breakthroughs for Venus *and* planetary science by carrying out in situ sampling and analysis of Venus surface materials, both in the planet's vast volcanic plains and in the tesserae. Just as in situ measurements of rocks by the Spirit, Opportunity, Curiosity, and Perseverance rovers have dramatically expanded our understanding of Mars's geological character and history, performing such analyses at the Venus surface would rewrite the textbooks about the second planet.

Until we deploy new-generation instruments to the Venus surface, we can take advantage of several "windows" at infrared wavelengths through which an appropriately configured instrument can see all the way to the surface from orbit through the otherwise opaque atmosphere. The Venus Express spacecraft was able to take infrared (IR) measurements of emissivity—the amount of IR radiation emitted by surface materials—of some of the southern hemisphere through one such window. Laboratory work has shown that it is possible to distinguish iron-rich basaltic rocks from iron-poor, silica-rich rocks on the surface, based on their infrared emissivity. Constructing a global map of the distribution of these rock types would represent a major advance in understanding the geological and chemical properties of the planet—which is something both the VERITAS and EnVision missions plan to do. Venus Express found that the largest tessera exposure in the south, Alpha Regio, has a markedly lower emissivity than the surrounding basaltic plains, indicative of it having less iron and more silica than those lavas. There are numerous explanations for this finding, but one intriguing possibility is that Alpha Regio represents material akin to continental crust on Earth—which is marked by relatively low iron and high silica abundances. Excitingly, Alpha Regio is the ultimate destination of DAVINCI's atmospheric entry probe, which will take detailed, close-up images of this enigmatic landscape as it nears the end of its hour-long plunge through the atmosphere.

Rocks at the Venus surface are chemically weathered by the punishing atmospheric conditions there. Those high-temperature, high-pressure surface conditions are unique among the solar system's rocky planets and, unsurprisingly, are challenging to replicate on Earth. But some experimental facilities able to simulate these conditions have come into operation in the past decade, allowing for the study of these weathering processes. Limited by the few constraints we actually have for rocks on the Venus surface, and by even less information about the types of surface-atmospheric interactions that weather rocks there, such studies are nonetheless valuable. For example, laboratory results in the past few years have raised the possibility that mineral grains might weather much faster than had been thought—suggesting that those lightly weathered lava flows seen by Venus Express could be *very* young indeed. Expanding our capabilities to simulate Venus's surface conditions in the lab would substantially increase understanding of the geological and atmospheric processes in this relatively unexplored part of the solar system.

Just as for the surface, major questions regarding the Venus interior remain. Presumably Venus, like Earth, has a metallic core, silicate mantle, and a rocky crust (although we do not know even the *size* of that core: we have to extrapolate our estimates from Earth). Unlike Earth, however, Venus has no strong magnetosphere; Pioneer Venus found that any internally generated magnetic field is at least one hundred thousand times weaker than that of Earth today. And no magnetometer-equipped mission has flown sufficiently close to or landed on the planet to determine if the crust preserves any record of an ancient Venus dynamo. Establishing whether the planet once generated its own magnetic field would help us understand if such a field really is a necessity for a large rocky planet to hold on to its atmosphere, with major implications for our broader understanding of planetary habitability.

A key tool for understanding planetary interiors is seismology, enabled on Earth in no small part because of the large-magnitude quakes here. Whether there are quakes on Venus large enough to probe the interior is an open question. But could we use those putative venusquakes to assess the interior, search for a hypothesized deep molten layer within the planet, or establish whether there is an inner core? Our present capabilities face severe limits to the length of time a seismic station could operate on the infernal surface, but encouraging technological developments promise at least the possibility of long-lived landers operating on Venus, performing not just geophysical investigations but geochemical and atmospheric studies, too. And by taking advantage of just how strongly coupled the thick, lower atmosphere is to the ground—such that seismicity in the crust is transmitted through the Venus air 60 times more effectively than on Earth—it may well be possible to search for seismic signals from balloon-borne instruments, complemented by observations of how the nightside upper atmosphere ripples as it conducts seismic waves.

An Alien Sky … and a Warning?

Among the most striking aspects of Venus is its planet-encompassing layer of sulfuric acid clouds, which reflects 70 percent of incoming sunlight and makes the planet appear so bright in our morning and evening skies. Those clouds are situated in the middle Venus atmosphere, where conditions are relatively hospitable: a balloon-based platform could operate there for weeks or even months. Indeed, the *only* natural shirt-sleeve environment in the solar system beyond Earth is at about 55 km up in the Venus atmosphere—a fact the Soviet Vega 1 and 2 missions took advantage of in 1985 when deploying two balloons into the middle atmosphere. Those aerial platforms beamed information on temperature, pressure, winds, and aerosols directly to Earth as they traveled a third of the planet's circumference from night to day (Figure 2-11).

Even so, we still know remarkably little about the Venus atmosphere. We have the basics: the main cloud layer on Venus consists mainly of droplets of sulfuric acid, together with small particles, possibly of sulfur, and an unknown absorber of ultraviolet light that shares some properties with a photosynthetic pigment. Water vapor is now only a minor component in the Venus atmosphere, and horizontal temperature variations are small. The general circulation of the atmosphere is dominated by super-rotation whereby the upper atmosphere rotates much faster than the solid planet itself, wind speed generally decreases with latitude but increases with altitude, and there are major vortices at the poles. At high altitudes, winds blow from the dayside to the nightside.

Yet several of the main features of the Venus atmosphere remain unexplained. How circulation patterns change with altitude and latitude is not clear, as is why the atmosphere super-rotates in the first place. Solar and thermal radiation play a dominant role in many of the processes that define the Venus climate, including the large greenhouse effect resulting from the abundance of carbon dioxide and other atmospheric absorbers. But this effect is despite the lower atmosphere receiving less solar energy than Earth: half of the solar flux received by Venus is absorbed in the cloud top region, with only a few percent reaching the surface (leading to a hazy surface environment a Soviet optics engineer once compared to "a cloudy day in Moscow"). In short, we do not fully understand how and where solar energy is absorbed and redistributed within the planet's atmosphere.

FIGURE 2-11 The Vega-1 balloon in flight. Artist's impression of the Soviet Vega-1 balloon drifting in the middle Venus atmosphere in June 1985. The aerial platform operated at an altitude of about 54 km; carried westward by zonal winds, the Vega-1 balloon ultimately traversed about 11,600 km over around 46 hours before losing contact with the flyby probe (which continued on for an encounter with Halley's Comet). SOURCES: Figure by P.K. Byrne based on photograph by David Peterson, Naval Research Laboratory. Vega balloon is courtesy of G.A. Landis. CC BY-SA 4.0.

Surprisingly few details of the chemistry of the Venus atmosphere are known. Key chemical interactions include those between sulfur, nitrogen, hydrogen, and oxygen, driven by solar radiation (and possibly by lightning). Reactions between atmospheric gases and sunlight in the upper atmosphere give way to high-temperature chemical processes in the lower atmosphere, and to poorly understood rock-atmosphere interactions at the surface. But the size distribution, shape, and composition of the majority of the clouds are still undetermined. Other such mysteries abound, including whether organic chemistry in Venus's atmosphere can produce enough nutrients to support an aerial biosphere—a decades-old focus of speculation based on the relatively clement conditions in the middle atmosphere. This question in particular is all the more pertinent given the tentative, if contested, detection of phosphine in the Venus atmosphere reported in 2020.

The chemical makeup of the atmosphere offers us another critical piece of the Venus puzzle: the geological history of the solid planet itself. The abundances of the heavy noble gases xenon, krypton, and argon (and their isotopes)—measurements of which form a key objective of the DAVINCI atmospheric probe—directly trace to the history of volcanic degassing of the Venus interior, and even the very components from which the planet was built. And, of course, measurements of sulfur dioxide in the atmosphere, and particularly how the abundance of that gas changes with time, would place a firm constraint on estimates of ongoing volcanic activity.

How the upper atmosphere interacts with Venus's space environment needs to be better described. Early missions, including Mariner 5, the Soviet Venera 9 and 10 orbiters, and Pioneer Venus characterized some of this interaction, establishing the composition of the upper atmosphere and ionosphere, and documenting interactions between the ionosphere and the solar wind (the flowing, magnetized, and ionized tenuous interplanetary medium emanating from the Sun). Another discovery was the magnetization of the Venus upper atmosphere by electrical currents induced by the solar wind—a finding that *some* magnetic fields are present at the planet, just not any intrinsic to the planet's interior.

The Venus Express mission found that the planet may be losing more oxygen to space than had been thought. On the other hand, the loss of yet heavier elements was lower than expected, which has implications for the evolution of the overall atmosphere. But we have still to answer such key questions as, say, the way "space weather" associated with coronal mass ejections from the Sun affect how much atmosphere Venus loses to space. Equally, we do not know how deep those magnetic fields induced by the solar wind penetrate into the atmosphere, nor whether the fields can move around the planet. And what of the history of atmospheric loss through time? Because the young Sun might have been much more efficient at stripping away atmosphere than at present, this question is especially important to our efforts to piece together the ancient history of the Venus climate.

Excitingly, it is this climate history that we have recently begun to rethink. Sophisticated climate simulations have raised the possibility that the Sun was *not* the main driver of Venus's present surface and atmospheric conditions. It seems, instead, that either Venus was *always* wholly inhospitable—with enormous consequences for our understanding of how large rocky planets develop in general—or it hosted a clement climate, oceans, and perhaps even plate tectonics for more than three billion years … until it ran out of luck.

That is because these climate simulations suggest that gigantic volcanic eruptions—of the kind that injected thousands of gigatons of carbon dioxide into the atmosphere in the late Permian on Earth to trigger the largest mass extinction in our planet's history—may have been responsible for dramatically and abruptly changing the Venus climate. But, whereas one such event on its own might not have dealt the climate a fatal blow, several enormous *simultaneous* eruptions could have been enough to overwhelm whatever mechanism(s) Venus used to regulate its climate to that point. With a sudden, rapid increase in the volume of carbon dioxide in the atmosphere, perhaps as geologically recent as ~1 Ga, that potent greenhouse gas raised the surface temperature and, ultimately, may have driven off the planet's putative oceans.

The DAVINCI probe will start to tackle this question, which carries with it a cautionary tale. Huge volcanic eruptions—of the type discussed here, the type that forms what we call on Earth "large igneous provinces"—have taken place throughout the history of our own planet, albeit infrequently. But the factors controlling such eruptions are not well understood for Earth, much less for another planet. It is not clear, therefore, whether Venus was unlucky to have experienced several calamitous events at the same time … or if Earth is lucky that it *has not*.

The Exoplanet in Our Backyard

Over the past few decades, planetary science has been revolutionized with the discovery of more than 4,000 confirmed extrasolar planets of a variety of sizes and masses extending beyond that of the solar system's inventory—enabled in large part by the Kepler mission and the Transiting Exoplanet Survey Satellite. Yet many of these exoplanets may be more Venus-like than Earth-like because our current observation techniques are biased toward detecting and observing planets relatively close to their stars. The James Webb Space Telescope (JWST) will soon observe high-priority exoplanet targets, but the vast distances involved mean that even a telescope as powerful as JWST will still acquire sparse and noisy exoplanet data. Here, we can leverage our decades of solar system insights to anchor and validate the models we will need to make sense of our exoplanet observations.

Venus has a particularly critical role to play in guiding our interpretations of exoplanet data, because although Venus-analog worlds will likely be one of the most common classes of planets observed by Webb, they may also be one of the most difficult types of world to interpret correctly if they, too, possess a global cloud layer that blocks the bulk of the atmosphere from view. To understand the chemical signatures of gases and aerosols detected in the atmospheres of those planets, then, we will need to understand the atmospheric composition and chemical cycles on Venus itself—our knowledge of which is, at present, woefully incomplete.

Venus may even have an important role to play in the search for life beyond the solar system. If we establish that Venus once had oceans, then we will need to redefine the traditional boundaries separating potentially habitable worlds from lifeless solely ones on the basis of distance from the host star. Understanding the processes that enabled—or did not enable—early habitability on Venus, and how those conditions were lost, will surely expand—or contract—the regions we regard as suitable locales for habitable worlds as we observe exoplanet systems with diverse architectures and at various ages and evolutionary stages.

And Venus offers us the opportunity to learn how to distinguish certain signs of life—biosignatures—from abiotic false positives on exoplanets. Considerable recent attention has been devoted to understanding false positives for oxygen, a critical biosignature for modern-Earth-like exoplanets, and we could use Venus to better understand mechanisms that form and destroy oxygen abiotically. The possible loss of oceans in an earlier epoch on Venus has been invoked as a mechanism to generate large quantities of oxygen on exoplanets that endure the same fate. In fact, small quantities of abiotic oxygen are generated on Venus today through atmospheric chemical processes, but the details of this process are poorly understood and could lead to dramatic consequences in other chemical contexts. Establishing the processes that remove the oxygen generated by these mechanisms would place valuable bounds on how much we might expect to find abiotically in exoplanet atmospheres. Desiccated, scorched Venus may seem the last place to look for guidance in the search for life beyond Earth, but we would be foolish to ignore the lessons it can teach us.

The Case for Continued Exploration of Venus

Venus is once again back in the spotlight. In the coming years and decades, we will continue to develop the recently emerging view of Venus as a complex, active world, augmented by new spacecraft data, ever more sophisticated climate modeling, and the finding of an increasing number of Venus-size rocky worlds in close proximity to their host stars. And with more and more attention focused on understanding and mitigating human-driven climate change on Earth, Venus's runaway greenhouse provides a dramatic example of a planet where self-regulating climate feedbacks have failed.

What we have also learned over the past decade in particular is that if we are to fully understand Venus, then we need to study it as a *system*. No one set of measurements will solve the mystery that is Venus, just as no one mission has answered all our questions of Mars, nor of the Moon, nor any planetary body. But as important as new measurements of the close coupling between Venus's surface, atmosphere, and space environment are, it is just as important that we have sufficient laboratory data to predict, calibrate, and make sense of such measurements as well.

The data we will acquire from the VERITAS, DAVINCI, and EnVision missions at the end of this decade will fundamentally alter our understanding of the planet as surely as those early Venus missions did, 6 decades ago. Yet even with the discoveries that await those missions, compelling science questions remain—such as those regarding the chemical and physical cycles of the atmosphere, the interactions between that atmosphere and the surface, and the make-up and structure of the planet itself. Studying Venus past and present is critical for comprehending the second planet in its own right; for comparing the divergent evolutionary paths of Venus, Earth, and Mars; for gaining context for similarly sized exoplanets; and simply for understanding the rules that govern Earth-like worlds in general. Learning why Venus took the path it did—why our sibling is not our twin—will tell us whether we are lucky that the sky over our heads is blue, and not yellow.

Key Discoveries from the Past Decade

- **Venus is likely a geologically active planet today.** Several lines of evidence together suggest that volcanic activity takes place today on Venus, and the planet's record of tectonic deformation speaks to recent and perhaps even ongoing deformation.
- **Models show that the young surface is consistent with Venus being habitable for billions of years.** New climate models indicate that Venus could have had modern-Earth-like conditions until as geologically recently as about a billion years ago, before entering a runaway greenhouse driven by several simultaneous, major volcanic eruptions.
- **Exoplanet discoveries motivate renewed Venus exploration.** The ongoing detection of large, rocky exoplanets close to their host stars, especially those that are amenable to having their atmospheres characterized, increasingly requires that we better understand the atmospheric properties and climate history of the second planet.

Further Reading

Glaze, L.S., C.F. Wilson, L.V. Zasova, M. Nakamura, and S. Limaye. 2018. "Future of Venus Research and Exploration." *Space Science Reviews* 214:89. https://doi.org/10.1007/s11214-018-0528-z.

Kane, S.R., G. Arney, D. Crisp, S. Domagal-Goldman, L.S. Glaze, C. Goldblatt, D. Grinspoon, et al. 2019. "Venus as a Laboratory for Exoplanetary Science." *Journal of Geophysical Research: Planets* 124:2015–2028. https://doi.org/10.1029/2019JE005939.

Marcq, E., F.P. Mills, C.D. Parkinson, and A.C. Vandaele. 2018. "Composition and Chemistry of the Neutral Atmosphere of Venus." *Space Science Reviews* 214:10. https://doi.org/10.1007/s11214-017-0438-5.

Smrekar, S.E., A. Davaille, and C. Sotin. 2018. "Venus Interior Structure and Dynamics." *Space Science Reviews* 214:88. https://doi.org/10.1007/s11214-018-0518-1.

Way, M.J., and A.D. Del Genio. 2020. "Venusian Habitable Climate Scenarios: Modeling Venus Through Time and Applications to Slowly Rotating Venus-Like Exoplanets. *Journal of Geophysical Research: Planets* 125:e2019JE006276. https://doi.org/10.1029/2019JE006276.

MARS

Significant advances in understanding Mars as a system have been made during the previous decade since *Vision and Voyages* (NRC 2011). These advances resulted from a dedicated and coordinated program of exploration involving in situ laboratories on rovers at carefully selected sites, and measurements of a dynamic surface and atmosphere that link the past and the present, and inform investigation of Mars's subsurface. Each new measurement—from spacecraft, Earth-based telescopic observations, and laboratory analysis of rocks from Mars that have made their way to Earth as meteorites, combined with detailed analysis and interpretation—has contributed to understanding different aspects of Mars. *Vision and Voyages* identified three areas in which Mars exploration would provide fundamental new insights:

1. Determine whether life ever arose or existed on Mars.
2. Understand the processes that control weather and climate and the long-term evolution of climate and habitability.
3. Decipher the evolution of the surface and interior and the processes that control them.

These goals were expected to guide exploration for longer than a decade. None of these goals can be addressed fully in isolation from the others; Mars has the complexity of Earth where interior, surface, and atmospheric processes along with solar and impact processes combine to create climate and environments that have changed profoundly over time. Mars's record of solar-system history is unique and special. Moreover, Mars also provides potential records of prebiotic chemistry and of biosignatures. It is the only rocky planet with an atmosphere where a complete 4-billion-year-plus record of its history sits intact, ready for exploration; even Earth does not have such a detailed record for its entire history. Thus, an independent origin of life on Mars is more testable than life's origin on Earth because the martian geological record is preserved. Understanding the trajectory of planetary climate and habitability over long timescales—the effects of a brightening Sun, large impacts, early volcanism, and large-scale, long-term climate change—can be traced on Mars in a way that allows comparison with the history of Earth and extrapolation to planets around other stars.

Collectively over the past decade, new observations, measurements, and models have dramatically changed our views of the evolution of the martian surface, interior, atmosphere, and climate, and of the potential for past surface life, or potentially still extant subsurface life. They reveal the history of the planet from 4.5 Ga to the present, a long and complex story of changing climate and the availability of liquid water, and previously unrecognized dramatic changes at the poles even in recent times. These scientific findings have set us up for the coming decade, in which we expect to return to Earth samples from a location chosen for its relevance to the history of water and its potential to have harbored life in the past.

Through Mars's preservation of a record of its entire history, we can examine the interplay between processes from the deep interior to the upper atmosphere to understand what controls the fate of habitable worlds. Our understanding of Mars feeds into an understanding of terrestrial planets in the solar system and of the possible evolutionary paths of terrestrial planets around other stars (many of which are thought to have undergone extensive loss of their atmospheres and water to space over time analogous to Mars). Here, the committee summarizes some of the major results of the past decade under the same three headings as called out in *Vision and Voyages*. This summary can only begin to scratch the surface on specific results and identify only some of the interconnections between different components of the Mars environmental system.

Did Life Ever Arise on Mars?

New discoveries in planetary sciences and astrobiology have advanced our understanding of key factors that influence whether Mars was habitable in the distant past or might offer habitable refuges at the present day: the availability of liquid water, organic (prebiotic) compounds, and energy sources. These new findings provide multiple recommended lines of investigation for the coming decade as outlined in *An Astrobiology Strategy for the Search for Life in the Universe* (NASEM 2019) and represent considerable progress over a decade ago, when spacecraft observations had revealed evidence of past water on Mars but the best-characterized environments appeared too chemically harsh to support life as we know it.

Preserved Organic Compounds

A groundbreaking discovery of the decade relating to the possibility of past life on Mars was the detection of organic matter in the lake sediments of Gale Crater by the Curiosity rover's SAM instrument (Figure 2-12). The characterization of ringed and straight-chain hydrocarbons and their chlorinated and sulfurized forms, together with the indirect detection of organic matter through chemical analyses of carbon, represents a key milestone in Mars exploration. The refractory macromolecular compounds point to relatively large molecules that are hard to destroy, similar to kerogens in Earth's geologic record. This is not a discovery of past life; rather, the diversity and composition of these compounds are consistent with large, complex organic molecules. Although the origin of the organic molecules is undetermined, their discovery sets the stage for more detailed characterization and astrobiological investigations to come. The discovery also suggests that organics are an important part of an active martian carbon cycle. Even at the planet's surface, organic compounds have not been entirely destroyed by radiation or oxidation chemistry, so greater organic abundances might be found beneath the surface.

In parallel with mission exploration, organic compounds in martian meteorites have been analyzed and their structures and isotopic compositions have been characterized in new ways. Laboratory analytical results from examined meteorites (all igneous rocks) conclude that organics trapped in martian meteorites represent a combination

FIGURE 2-12 Views in Gale Crater from the Curiosity rover. *Upper panel:* A thick sequence of lakebed and lake-margin sedimentary rocks. *Lower panel, left to right:* Drill hole where organics were detected, active dunes, and sulfate veins formed during diagenesis. SOURCE: Courtesy of NASA/JPL-Caltech/MSSS.

of infall of exogenic carbonaceous materials and abiotic synthesis on Mars during water–rock reactions and magmatic processes. Although meteorite results so far have not found any organic biosignatures, they support the widespread occurrence and preservation of prebiotic compounds on Mars.

Liquid Water Stability and Accessibility

New discoveries offer a way that liquid water can exist on present-day Mars. In general, pure liquid water is not stable at the surface today, owing to average temperatures below its freezing point and low atmospheric pressure; water would either freeze or evaporate quickly. Liquid water in the martian subsurface was reported, albeit debated, for the base of the polar caps using data from the Mars Advanced Radar for Subsurface and Ionosphere Sounding (MARSIS) instrument on the Mars Express spacecraft. On the planet's surface, liquid water can exist as a transient phase. At the surface, perchlorate minerals have been discovered in soils. Perchlorates are deliquescent, meaning that they can absorb water vapor from the atmosphere and dissolve in it as a liquid; thus, under conditions that occur within the top meter of the surface, films of liquid water can be stable for a large fraction of a sol (a martian day), for certain times of the year over parts of the planet. Temperature data from the Curiosity rover and its measurements of salt abundances showed the possibility of such a mechanism creating habitats in the near surface.

Recurring slope lineae, recognized in 2010, give the appearance of having been formed by water flowing downhill from beneath or on the surface. As small-scale features (meters to tens of meters across), they form and disappear over the course of martian seasons, changing their appearance most rapidly at seasonally warmer temperatures. New analyses and models make it unclear whether these features are related to liquid water or not; they may form from transient water near the surface, from dust avalanching caused by deliquescing salts, or from dry avalanches. If they are related to liquid water, they are telling us something important about water's availability at or near the surface.

Long-Term Habitable Environments at Gale Crater and Elsewhere on Mars

Exploration of Gale Crater by the Curiosity rover has provided definitive evidence that water flowed and ponded on the surface of Mars billions of years ago. Sedimentary processes that took place over thousands to millions of years indicate that liquid water, although perhaps not necessarily present continuously, was stable and available for considerable periods of time on the planet's surface. A series of neutral-to-alkaline lakes spanning many square kilometers was followed by even longer-lived episodic groundwaters, occurring hundreds of millions of years later (Figure 2-12). The extensive deposits identified in Gale Crater provide evidence for water, organic carbon, and a chemical source of energy for potential microbial metabolism. The inferred aqueous environment featured fluctuations in the saltiness of water and in redox states. In addition, later overprinting by diagenetic events suggests that habitability could have been sustained in the subsurface for millions of years. This remarkable set of discoveries demonstrates there were long-lived habitats on Mars and that Earth-like life could have inhabited this site.

At the time of the last decadal survey (NRC 2011), data from the Mars Reconnaissance Orbiter's Compact Reconnaissance Imaging Spectrometer for Mars (CRISM) instrument had documented a diversity of ancient environments with liquid water, including lakes, rivers, and hydrothermal systems. Key discoveries on this front continued into this decade, revealing specific craters with hydrothermal systems associated with lakes, excavations of deeply buried carbonate deposits, widespread hydrous minerals excavated by craters even below the volcanic northern plains, and thick weathering sequences of clay minerals. Collectively, the data indicate that habitable environments were likely widespread across ancient Mars. Only a subset of such habitats has so far been explored. Dozens of studies pinpointed the Jezero crater-Nili Planum system that was subsequently selected for samples to be cached by the Perseverance rover and eventually returned to Earth. This site encompasses lakebed clays and carbonates, a watershed with strata preserving effects of a large basin-forming impact, early volcanic rocks, and mineralized veins of a hydrothermal groundwater system.

Processes and History of Martian Climate

The martian atmosphere has changed on timescales ranging from seconds to seasons to years to billions of years. Determining the evolution of the atmosphere and climate is necessary to understand the time-varying availability of liquid water, its influence on the geology and geochemistry of the surface, and the potential to support life.

Seasonal Behavior of the Atmosphere

The seasonal and annual behavior of weather in today's lower atmosphere is dominated by the seasonal cycles of dust, water vapor, and CO_2. Although we have known about dust storms for over a century, improved observations and continuous monitoring over the past decade have resulted in greater understanding of how they begin and evolve, and of annual patterns in when and where they occur. Global and large regional dust storms in 2018 became the most comprehensively studied to date, thanks to concurrent observations made by multiple orbiters and two landed missions. At the same time, advances in atmospheric modeling have led to increased realism in simulated storms. Absent a detailed understanding of today's dust storms and the complex coupling to water and CO_2 behavior, extrapolation to other epochs and determination of their long-term impact remains difficult.

Dust behavior in the lower atmosphere affects the dynamics and composition of the upper atmosphere via lower-atmosphere heating, expansion, and the modification of vertically propagating waves. For example, the hydrogen abundance in the extended exosphere previously was thought to be relatively constant, owing to the long timescales of its production from water and its thermal escape to space. New observations, however, show an order-of-magnitude variation in hydrogen abundance, occurring in the seasons in which the atmosphere is dustiest. The mechanism for connecting them, involving heating of the lower atmosphere resulting in changes in atmospheric structure and circulation that enhance hydrogen escape, shows the strong coupling between different components of the atmosphere and provides a mechanism by which short-term weather can affect the long-term evolution of water.

The presence and behavior of methane in today's atmosphere is important as a potential indicator of either ongoing geological or biological activity. In the past decade, Curiosity's Sample Analysis at Mars (SAM) instrument has identified methane at very low abundances that shows a possible seasonal cycle, suggesting thermally controlled release from the regolith, in addition to occasional much higher methane abundances that are very short-lived. Observations from the ESA-Russian Trace Gas Orbiter (TGO) have not detected methane at all, with upper limits well below the abundances measured by SAM. Although TGO instruments cannot see methane that might be in the lowest scale height of the atmosphere, rapid vertical mixing should prevent methane from staying trapped there for long. For these spacecraft measurements to be consistent with each other thus requires either a very localized source in Gale Crater or a trapping mechanism (via chemical and/or atmospheric dynamical process) that has not yet been identified. Getting to the bottom of the Mars methane mystery and the surface-subsurface processes controlling the methane cycle remains an important goal for future investigations.

Polar and Non-Polar Ice Forcing by Obliquity Oscillations

Mars's climate is driven partly by the 10,000- to 1,000,000-year variations in the tilt of the planet's spin axis (axial obliquity), which changes the amount of sunlight received by the poles and the resulting rates of ice loss and accumulation. Estimates based on study of small impact craters identified on the ice show that the north-polar ice deposits, which are many kilometers thick and contain the equivalent of a global layer of water some 20 meters thick, might only have formed around 5 Ma and might be only a recent (and possibly ephemeral) phenomenon. Radar results from the Mars Shallow Radar (SHARAD) instrument on Mars Reconnaissance Orbiter have supported this interpretation based on large-scale growth and retreat cycles for the ice.

Massive deposits of CO_2 ice have been identified within the south-polar cap using radar data. These layers presumably represent atmospheric CO_2 that has condensed to form ice at lower obliquity values and that can be released into the atmosphere at higher values. Estimates of the volume of ice show that its release would more than double the martian atmospheric pressure. If correct, then the observed layers in the polar ice deposits would reflect variations in deposition on much shorter timescales than previously thought; modeling based on the obliquity variations over the past half-million years can successfully replicate some of the observed layer thicknesses. Most importantly, these results demonstrate that the present-day atmospheric pressure may be significantly lower than the average value over the past half-million years or even longer.

A key discovery in the climate history puzzle was the presence of massive mid-latitude deposits of water ice (Figure 2-13). These deposits occur as relatively clean water ice, standing up to hundreds of meters thick and buried by a thin overburden of dust or debris. These deposits recently have been determined to be extensive, based on both

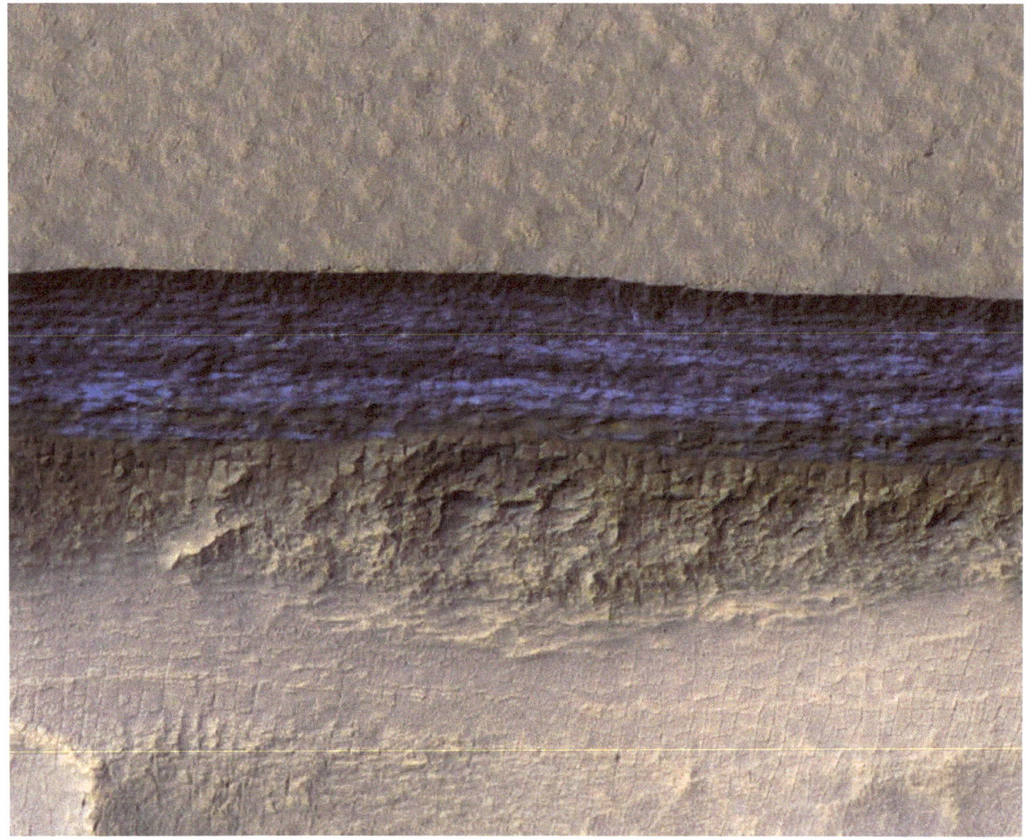

FIGURE 2-13 A layer of underground ice exposed along a steep slope; the ice appears bright blue in this enhanced-color view from the HiRISE camera on NASA's Mars Reconnaissance Orbiter. The scene is about 500 m wide. SOURCE: Courtesy of NASA/JPL/UA/USGS.

high-resolution imaging (from the High-Resolution Imaging Science Experiment (HiRISE)) and radar measurements (from SHARAD). When these ice deposits were emplaced, whether they represent a one-time deposition or a cyclical deposition and removal associated perhaps with obliquity cycles, and what is the total integrated amount of water they contain are uncertain. However, they certainly represent a source that could produce at least transient liquid water that might be able to support microbial life under some conditions, as well as a potential resource for future human missions to Mars.

Drivers of Climate Change over Billions of Years

Significant progress has been made in the past decade on understanding the mechanism for producing a change from an early, warmer and wetter environment to the cold, dry Mars that we see today. Evidence has long pointed to a very different early martian climate—geological and geochemical data both require the occurrence of liquid water in much greater abundance or for longer periods of time prior to about 3.5 Ga, as compared to the present climate in which liquid water can exist only transiently. Although it has been widely accepted that an early, thicker greenhouse atmosphere contributed to the warming, early models had been unable to produce adequate heating. Recent advances in three-dimensional climate modeling have confirmed that CO_2 and H_2O alone are insufficient to warm early Mars, even with a much thicker early atmosphere. However, two recent approaches have demonstrated the viability of greenhouse models. First, it was recognized that the early water cycle would have been dramatically different when the atmosphere was thicker, owing to the cooling of elevated terrain as occurs on present-day Earth;

ice could deposit at higher altitudes and provide a source for liquid water. Second, recent modeling and laboratory experiments have shown that greenhouse warming from a combination of CO_2 and reducing gases such as hydrogen or methane could have warmed early Mars. Reducing gases could have been produced from volcanism, meteoroid impacts or crustal-alteration, water–rock reaction processes. The occurrence of these gases appears to be consistent with the presence of mineral phases identified by the Curiosity rover that require a reducing environment (although the occurrence of both oxidized and reduced minerals in close proximity complicates the potential interplay between oxidizing and reducing processes). Our growing understanding of Mars's early atmosphere will have important implications for astrobiology.

What caused the climate to change from one that was at least intermittently warmer to today's cold and dry environment? Impacts that could cause transient warming, as well as volcanism that could intermittently release reducing gases, were declining from 4.0 to 3.5 Ga. At the same time, the Sun and solar wind were stripping the early, thicker atmosphere to space. Measurements by the Mars Atmosphere and Volatile Evolution (MAVEN) spacecraft allow us to understand how stripping of the atmosphere by the Sun operates today, involving breaking apart of H_2O and CO_2 molecules by ultraviolet photons from the Sun and the escape of the individual atoms to space through multiple processes (Figure 2-14). These processes would have been more effective early in Mars's history when the solar ultraviolet radiation and solar wind were more intense, and we can extrapolate the loss to those earlier times using the derived history of the Sun. The rate of loss of gas to space today is relatively low planetwide; it may have been as much as 10,000 times greater 4 billion years ago.

In addition to loss to space, formation of carbonates within the crust from atmospheric CO_2 and hydration of subsurface minerals by H_2O removed substantial quantities of each. Combined, loss to space, loss of H_2O and CO_2 to the crust, and sequestration in the polar caps can explain the removal of an early thicker atmosphere and accessible water, and thereby explain the transition inferred for climate. The results suggest that loss of both H_2O and CO_2 to space were major factors in the evolution of the atmosphere and climate on early Mars.

Understanding the history of climate and the availability of liquid water requires understanding the complex interplay between processes occurring on all timescales. New discoveries in the past decade have dramatically changed our view of how these processes work, and new and anticipated observations will help us to see how they actually played out through time.

FIGURE 2-14 False-color ultraviolet images showing the extended coronae of carbon, oxygen, and hydrogen atoms that surround Mars and that are contributing to loss of atmospheric CO_2 and H_2O to space. Red circles in each image show location of the surface of the planet, showing that all three coronae extend well above the surface and lower atmosphere. SOURCES: Courtesy of NASA/University of Colorado, adapted from Jakosky et al. (2015).

Evolution of the Martian Surface and Interior

Geologic findings in the past decade have significantly altered our understanding of the first billion years of martian history, as well as revealed active surface and subsurface processes on modern-day Mars.

New Discoveries from Very Old Rocks

The ancient igneous crust of Mars is now recognized to be much more compositionally complex than previously realized. The Curiosity rover found volcanic rocks of unusual compositions shed from the rim of Gale Crater, and laboratory analyses of the Northwest Africa (NWA) 7034 meteorite, first described in 2013, extended the occurrence of similar volcanic rocks back into the earliest period of Mars history. The magmas that formed these rocks indicate geochemical evolution in the first billion years of Mars history was distinct from that of more recent times.

Some ancient rock units previously recognized as volcanic have now been reinterpreted as cemented sedimentary rocks. Curiosity found unaltered olivine in a mudstone, indicating minimal chemical weathering in some sedimentary rocks. Mineral sorting and fractionation during transport by water or wind has resulted in significant geochemical variations across Mars that were previously attributed to chemical weathering. Curiosity also identified sedimentary rocks that contain pieces of earlier sedimentary rocks, requiring rock-forming processes to have occurred over long timescales. The occurrence of sedimentary rocks across Mars indicates that the planet possesses a relatively complete temporal record of surface geologic processes, although the incomplete chemical weathering in many locations may be owing to intermittency of waters or cold temperatures when waters were present.

Near the end of its operational life, the Opportunity rover explored Endeavour Crater, a 22-km-diameter impact structure that excavated and redeposited Noachian strata. Prior to this time, Mars rovers (including Opportunity) had only explored younger Hesperian terrains, so this foray provided the first chance to analyze the oldest rocks yet encountered on the martian surface. The lower units of the crater rim stratigraphy represent ejecta from impacts that predate the Endeavour impact, and superposed layers represent breccias (cemented rock fragments) formed during the impact itself. The compositions of the breccias are basaltic, and chemical alteration of the rocks by aqueous fluids produced clay minerals and crosscutting vein minerals that vary with the age of the units. This investigation confirms interpretations from orbital data that the Noachian was a time characterized by basaltic volcanism, large impacts, and intense hydrothermal activity.

Although some Pre-Noachian rocks have been recognized from orbital observations, our understanding of this earliest period of Mars history has been very limited. However, the NWA 7034 meteorite (and paired stones from the same fall) have radiometric ages of 4.4 billion years, placing them deep within this time period (Figure 2-15). These samples are regolith breccias—cemented soils consisting mostly of fragments of basaltic igneous rocks that comprised the early crust. Research on these meteorites in the years since their recovery provides unique insights into the earliest evolution of the planet.

Global differentiation into core, mantle, and crust represents the most profound geologic process that affected the terrestrial planets. Analyses of radiogenic isotopes in NWA 7034 provide time constraints on the rapid differentiation of Mars, as well as rule out a mantle formed by simple extraction of crustal magmas and instead appear to require an early magma ocean. The origin of magnetized rocks in the ancient crust of Mars has been a puzzle since their discovery. NWA 7034 is magnetic and contains an unusual assemblage of iron minerals that formed by hydrothermal alteration, so a similar subsurface alteration process could account for at least some of the magnetized crustal rocks mapped from orbit. Trace elements and oxygen isotopes in mineral grains of the meteorite further confirm that aqueous fluids circulated in the earliest martian crust. Understanding the origin and evolution of the martian atmosphere is critical to interpreting paleoclimates, and stable isotope measurements of atmospheric gases, as done by MAVEN, suggest that a significant portion was lost to space. Xenon isotopes in NWA 7034 indicate hydrodynamic escape of the earliest atmosphere within only a few 100 million years of the planet's formation.

FIGURE 2-15 The NWA 7034 martian meteorite, a breccia containing fragments of 4.4-billion-year-old volcanic rocks, has provided unique insights into the earliest period of Mars history. SOURCE: Institute of Meteoritics UNM © The University of New Mexico. All rights reserved.

Insights from InSight

The seismometer on the Interior Exploration using Seismic Investigations, Geodesy and Heat Transport (InSight) mission has revealed that Mars is seismically active (Figure 2-16), although the most significant seismic event so far detected is modest by terrestrial standards, at magnitude 4.2. The seismic data have, for the first time, allowed measurement of the thickness of the martian crust, mantle, and core and determined the liquid nature of the core. The low seismic velocity of the crust requires that it is highly fractured and that pore spaces are mostly not filled with ice. InSight's seismic data allows a determination of the size of the martian core and, combined with Mars's moment of inertia, the density and composition of the core were constrained, finding a significant proportion of light elements like sulfur or oxygen in addition to iron. InSight's magnetometer has provided ground truth for orbital measurements by measuring the magnetic field on the planet's surface for the first time.

A related discovery, newly recognized from decades of orbital data from several spacecraft, is that Mars has a Chandler wobble, an oscillation of the rotation axis that can provide information on temperature and composition of the mantle.

Mars Is an Active World

The more than 13-year record of repeated visible observations by HiRISE on the Mars Reconnaissance Orbiter, supplemented by other orbital cameras, and the more than 20-year thermal infrared observations by THEMIS on Mars Odyssey, reveal much more surface activity than previously recognized, leading to a

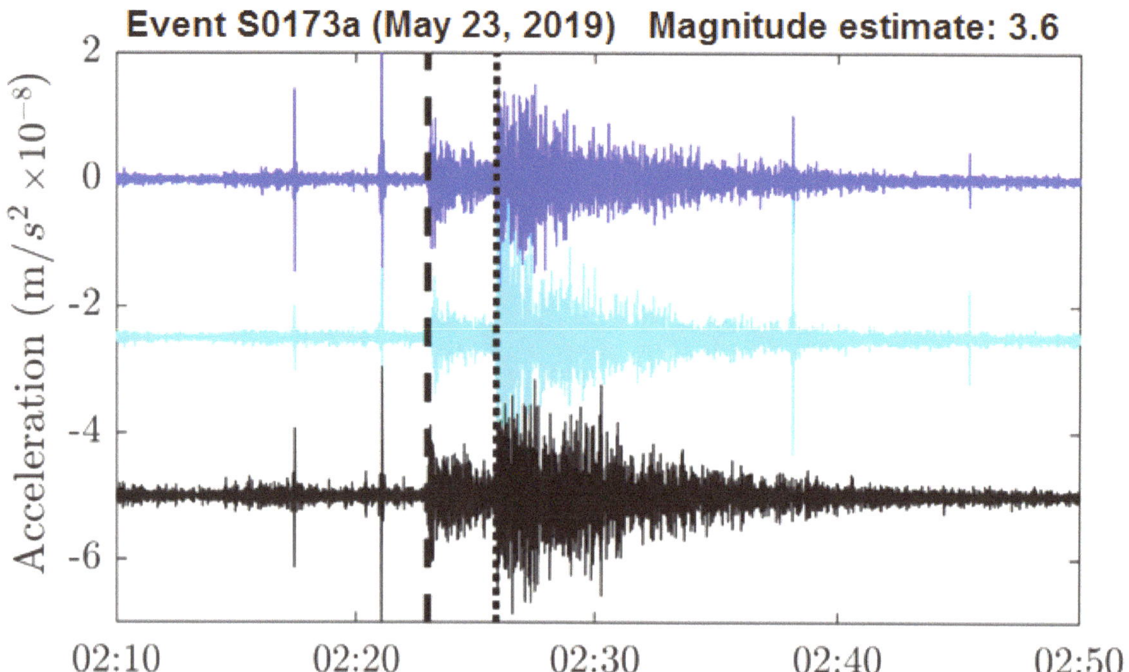

FIGURE 2-16 Ground acceleration as a function of time over a period of 40 minutes recorded by the very broad band seismometer on InSight for a marsquake located near Cerberus Fossae, the site of geologically recent eruptions of magma and discharge of water. The three traces show accelerations in the east, north, and vertical directions, and dashed lines show the arrival of P and S waves and began at about 2:23 (local mean solar time). SOURCE: P. Lognonné, W.B. Banerdt, W.T. Pike, et al., 2020, "Constraints on the Shallow Elastic and Anelastic Structure of Mars from InSight Seismic Data," *Nature Geosciences* 13:213–220. Springer Nature, 2020, reproduced with permission from SNCSC.

new view of surface-atmosphere interactions, especially when combined with the results from Mars Global Surveyor (1997–2006). These activities and interactions include evidence of aeolian transport (including migrating dunes and ripples), frost, gullies, recurring slope lineae, new impact craters and degradation of older craters, sublimation of ice, new insights into dust storms and dust devils, seasonal ice, and growth of unusual geomorphic features within polar ice. These presently active processes have been occurring for millennia, and the long-term observations help constrain their rates.

Measurements from the surface in the past decade have provided more insights into the nature and extent of Mars aeolian activity. Curiosity became the first rover to explore an active dune field on another planet and observed that large ripple bedforms may require a low-density atmosphere, a novel constraint on ancient atmospheric density. Curiosity was also able to characterize daily, as well as seasonal, timescales of aeolian changes. InSight further correlated surface changes with environmental conditions, a key requirement for understanding what drives sand motion and dust lifting.

Mars trembles often, although less vigorously than expected. An average of about one seismic event per day has been detected so far by InSight's seismic instrument. Most of these events are probably actual marsquakes. Detection of some of the strongest tremors—emanating from the Cerberus Fossae region, where large volumes of water and lava erupted through fissures within the past tens of millions of years, and where boulders appear to have been shaken from hillsides—confirms that Mars is tectonically (and perhaps volcanically) active today. Ongoing seismic activity, by regenerating exposed fracture mineral surfaces and pore space, has implications for potential subsurface habitability and astrobiology.

Collectively, these discoveries highlight Mars as a dynamic habitable world and a key current and future destination in the search for life and understanding the evolution of terrestrial planets.

Key Discoveries from the Past Decade

- **Detection of organic matter in the lake sediments of Gale Crater.** Curiosity rover data acquired from sedimentary rocks show similar evidence as data from igneous Mars meteorites (exogenic infall of carbonaceous organic matter and chemical reactions on Mars that synthesize or alter organic matter), indicating past indigenous organic chemistry in a martian habitable environment.
- **Presence of massive mid-latitude deposits of water ice.** Although how and when they formed is not yet known, direct imaging, radar measurements, and crater morphology analyses show large areas in Mars's northern lowlands where hundreds-meter thick ice slabs occur just meters below the surface, preserving a sizeable reservoir of water and a significant record of martian climate change.
- **Possibility of current or recent near-surface liquid water.** Enrichments of salt in near-surface rocks and sediments and features that change seasonally with warm temperatures suggest a potential role for small amounts of liquid water in shaping martian geology even in the modern cold, dry climate regime.
- **Multiple types of habitable environments were widespread across ancient Mars.** Data from orbiters and rovers have revealed evidence for lakes, rivers, playas, groundwater systems, and hydrothermal systems of varying temperatures and water chemistries, preserved in the rock record at thousands of locations, an environmental diversity similar to Earth.
- **Kilometer-thick layers of H_2O ice and CO_2 ice in the martian polar caps formed less than 10 million years ago.** New radar analyses of polar deposits, coupled with climate models, indicate that much of the ice thickness of Mars's poles is not billions of years old but rather a product of recent climate change.
- **Loss of H_2O and CO_2 to space and sequestration in crustal minerals were major factors in the evolution of the atmosphere and climate.** Spacecraft measurements coupled with modeling show that Mars's climate became drier and colder because of both escape of volatiles to space and their sequestration in minerals in its crust. When compared with volatile evolution on Earth and Venus, this points to the important role of volcanic and tectonic processes in replenishing volatiles and regulating planetary climate and long-term habitability.
- **Detection of marsquakes and their use to probe the interior structure of Mars.** Data from the InSight lander recorded frequent marsquakes up to magnitude 4 that allowed probing the structure of the martian interior, revealing a thick, fractured crust, mantle structure, and a liquid core that includes a sizeable fraction of light elements.
- **Ancient alkali-silica-rich igneous rocks.** Data from orbiters and Gale crater have discovered more than just basaltic rocks on Mars. High alkali and silica rocks indicate more differentiated magmas.
- **Mars is active today.** More dynamic activity occurs on Mars today than was previously known, including migrating sand dunes, recurring slope lineae formation, changing ice landforms, methane release, and many marsquakes centered near Cerberus Fossae. Causes of some of these phenomena, including methane release and marsquakes, remain to be determined.

Further Reading

Ehlmann, B.L., F.S. Anderson, J. Andrews-Hanna, D.C. Catling, P.R. Christensen, B.A. Cohen, C.D. Dressing, et al. 2016. "The Sustainability of Habitability on Terrestrial Planets: Insights, Questions, and Needed Measurements from Mars for Understanding the Evolution of Earth-Like Worlds." *Journal of Geophysical Research: Planets* 121(10):1927–1961. https://doi.org/10.1002/2016JE005134.

Jakosky, B.M. 2021. "Atmospheric Loss to Space and the History of Water on Mars." *Annual Reviews of Earth and Planetary Science* 49(1):71–93. https://doi.org/10.1146/annurev-earth-062420-052845.

McLennan, S.M., J.P., Grotzinger, J.A. Hurowitz, and N.J. Tosca. 2019. "The Sedimentary Cycle on Early Mars." *Annual Reviews of Earth and Planetary Science* 47(1):91–118. https://doi.org/10.1146/annurev-earth-053018-060332.

Udry, A., G.H. Howarth, C.D.K. Herd, J.M.D. Day, T.J. Lapen, and J. Filiberto. 2020. "What Martian Meteorites Reveal About the Interior and Surface of Mars." *Journal of Geophysical Research: Planets* 125(12). https://doi.org/10.1029/2020JE006523.

Wordsworth, R. 2016. "The Climate of Early Mars." *Annual Reviews of Earth and Planetary Science* 44(1):381–408. https://doi.org/10.1146/annurev-earth-060115-012355.

Wray, J.L. 2021. "Contemporary Liquid Water on Mars?" *Annual Reviews of Earth and Planetary Science* 49(1):141–171. https://doi.org/10.1146/annurev-earth-072420-071823.

SMALL SOLAR SYSTEM BODIES

Small bodies are rocky and icy worlds that span the solar system. The most commonly known ones are asteroids and comets, and their major reservoirs include the asteroid belt, Trojan populations, trans-neptunian region, and Oort cloud. Small body populations that can potentially strike the planets, such as near-Earth asteroids and comets, are steadily replenished over time by dynamical processes in the major reservoirs. Small bodies also include most meteorite precursors and interplanetary dust particles. For this section, we do not consider objects that have formed equilibrium spherical shapes, roughly >800 km in diameter objects, which are covered in the Ocean Worlds and Dwarf Planets section, or small moons in the outer solar system, which are discussed in the Giant Planet Systems section.

The diversity of small bodies in their sizes, orbits, compositions, and physical natures provides unique scientific opportunities unavailable for larger bodies. First, because many small bodies have undergone minimal processing since formation and had their orbits set in place by early dynamical processes, they are relics of the origin and evolution of the solar system. Second, small bodies are fascinating worlds in their own right, with geologic, geophysical, and geochemical histories that are distinct from those that occur on larger bodies. Third, many have struck the planets over the age of the solar system, yielding beneficial effects, such as the delivery of water and organics, and deleterious ones, such as the destruction of established environments.

The past decade has brought significant scientific discoveries about small bodies that have reshaped our understanding of fundamental planetary processes. Here, the committee highlights several key achievements from the past decade, which have been advanced by complementary discoveries from spacecraft measurements, astronomical observations, theoretical and experimental studies, and laboratory measurements of samples.

In this decade, two NASA Discovery missions will launch to investigate previously unexplored types of solar system small bodies, with Psyche visiting a metal-rich planetesimal in the main belt in 2029 and Lucy encountering the Jupiter Trojan asteroids in 2027. In addition, new space-based telescopic projects, like the James Webb Space Telescope, and Earth-based telescopic facilities, like the Vera C. Rubin Observatory, will begin operations and produce a wealth of new discoveries and data about small solar system bodies.

Early Solar System Formation and Small Body Reservoirs

Computational studies over the past decade have provided new insights into the initial accretion of bodies in the solar system as well as the subsequent structure of the solar system and its small body populations. These dynamical models have used constraints from both meteorite studies and observations of small bodies to develop our latest understanding. A critical step in planet formation is the accretion of planetesimals, defined as asteroid and comet-like bodies. Theoretical studies now suggest that planetesimals are formed by aerodynamical concentration of ~1 cm to 10 cm pebbles in the protoplanetary gas nebula. When the spatial density of pebbles becomes high enough, they gravitationally collapse into 100-km-class bodies, a process referred to as pebble accretion, with some planetesimals turning into binary objects.

The first epoch of giant planet migration may have occurred a few million years after the birth of the Sun, when the protoplanetary gas nebula was still in existence. As the giant planets grew, they traded angular momentum with the gas disk, allowing them to move to new locations. As a consequence, small bodies from the terrestrial planet and giant planet zones were injected into the asteroid belt. The outward migration of Uranus and Neptune then destabilized a massive outer comet disk, leading to new orbits for all of the giant planets and enhanced bombardment rates on all solar system worlds. This event also placed comet-like planetesimals into major reservoirs such as the main belt, Trojan, trans-neptunian region, and Oort cloud populations (Figure 2-17). The giant planet instability may have also ejected one Neptune-size and multiple super-Earth-size worlds from the giant planet zone. The possible capture of an ejected super-Earth in the most distant regions of the solar system at this time may explain the unusual orbits of distant comets and the most distant trans-neptunian objects, but such a body has not been directly observed to date. Last, giant planet migration after the gas disk dissipated may have gravitationally influenced the growth of the terrestrial planets by exciting inner solar system planetesimals. Some of the leftover planetesimals may have played a critical role in the early bombardment of Earth and the terrestrial planets.

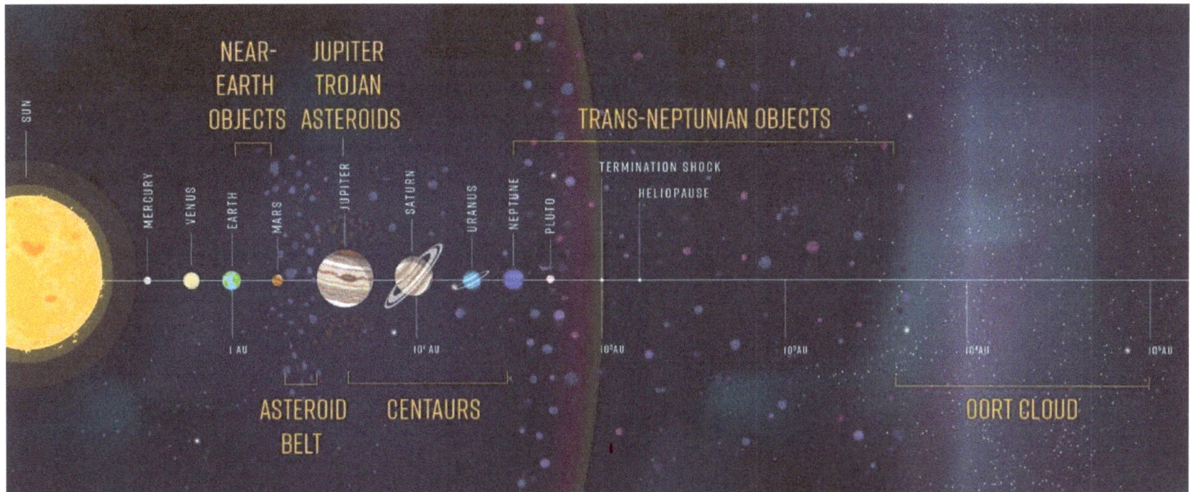

FIGURE 2-17 Major populations of solar system small bodies. Historically, the reservoir of small bodies between ~30 and ~50 AU has been referred to as the Kuiper belt. However, with the discovery of many objects beyond those limits in the past decade, the term of trans-neptunian objects is used to refer to all objects beyond Neptune's orbit up to the Oort cloud. The majority of known Trojan asteroids share the orbit of Jupiter, although a few Trojan asteroids have been detected associated with Mars, Neptune, Uranus, and Earth. SOURCE: Courtesy of Johns Hopkins University Applied Physics Laboratory, CC BY 4.0.

A key factor in our ability to constrain planetesimal and planet formation models is through direct laboratory analysis of extraterrestrial materials. Dramatic advances in spatial resolution and sensitivity for structural, elemental, and isotopic measurements have occurred over the past decade (Figure 2-18). Additionally, we have many new asteroid and comet fragments to explore in the form of more than 25,000 new meteorites discovered since 2011, captured interplanetary dust particles, and ultra-carbonaceous micrometeorites collected from Antarctic ice. When combined with the continued analysis of returned samples from the Stardust and Hayabusa missions, they collectively provide us with a set of diverse planetesimal samples that formed from different compositional reservoirs in a range of solar system regions and eras, and insight into what might have been the prebiotic chemistry of Earth—a key question for astrobiology.

Characterization of newly discovered ungrouped meteorites and some carbonaceous chondrite groups have provided one of our first windows into planetesimals that formed in the giant planet and trans-neptunian regions. Analysis of these samples in state-of-the-art laboratory facilities have revealed the complexity of the history of water and organic chemistry over the age of the solar system, including mixing of materials accreted to comets and asteroids. Measurements of hydrogen and nitrogen isotopic compositions of planetary materials have been crucial for understanding the extent of radial mixing in the solar nebula. Isotopic measurements of meteorites have also enabled new insights into compositional reservoirs in the early solar system. Advances in mass spectrometry and sample preparation have enabled precise measurements of many new isotopic systems (e.g., Mo, Ru, W, and Ti). This has suggested that meteorites (and planetesimals) are potentially derived from two distinct nebular reservoirs: one in the inner and another in the outer solar system, with their separation hypothesized to have been caused by the formation of Jupiter.

Outer Solar System Small Bodies

Major advances in our knowledge of outer solar system planetesimals have been driven primarily by observational surveys. Nearly two-thirds of the known trans-neptunian objects were discovered in the past decade, a striking statistic that illustrates just how new our knowledge is of this population. Some trans-neptunian objects

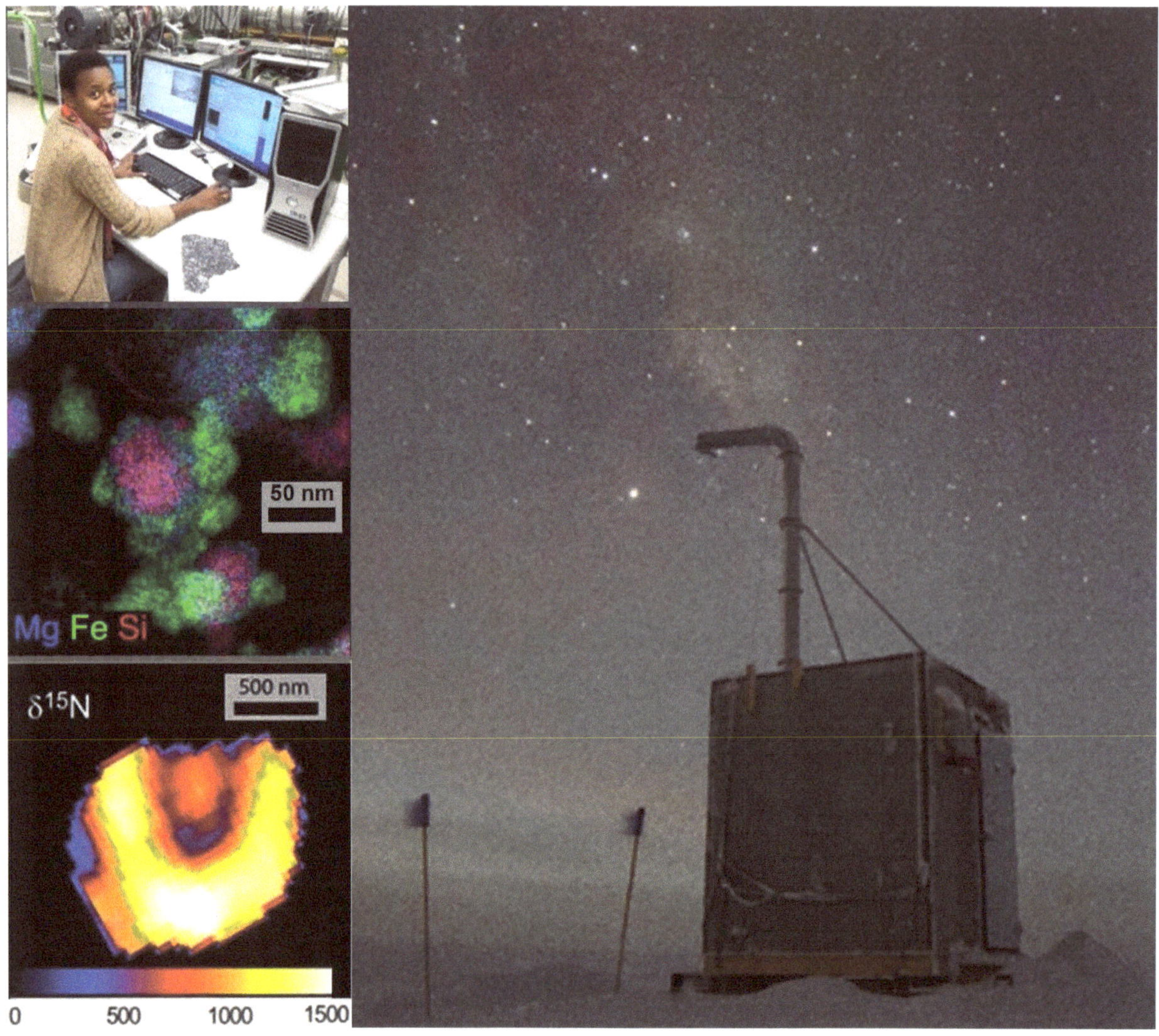

FIGURE 2-18 Analyses of extraterrestrial samples occur in laboratories across the country. *Upper left:* Meteorite analysis being conducted at Carnegie Institution. Such analyses provide countless opportunities for investigations and shown are two examples: (*middle left*) a cometary building block incorporated into a meteorite and (*bottom left*) an organic nanoglobule with an N isotope anomaly returned from comet Wild 2 by the Stardust mission. Antarctica continues to be a major collection area for discovering new extraterrestrial samples, with (*right*) the experimental dust collection station collecting interplanetary dust particles from the air, while tens of thousands of meteorites have been collected from the Antarctic ice fields in the past decade alone. SOURCES: *Top left*: Courtesy of M. Telus; *middle left*: R. Stroud; *bottom left*: L. Nittler and R. Stroud; and *right*: K. West.

discovered have orbits that travel beyond 1,000 astronomical units, more than 30 times farther from the Sun than Neptune. Gravitational interactions with Neptune cause some objects to evolve onto crossing orbits with the giant planets, producing the Centaur population, transitional objects between Jupiter and Neptune on their path to becoming Jupiter-family comets.

The orbits of the most distant trans-neptunian objects (TNOs) appear to have an orbital alignment. This alignment has been suggested to potentially be owing to the presence of an as-yet undiscovered giant planet, probably several times more massive than Earth (Figure 2-19). It is hypothesized that this planet would be on an elliptical

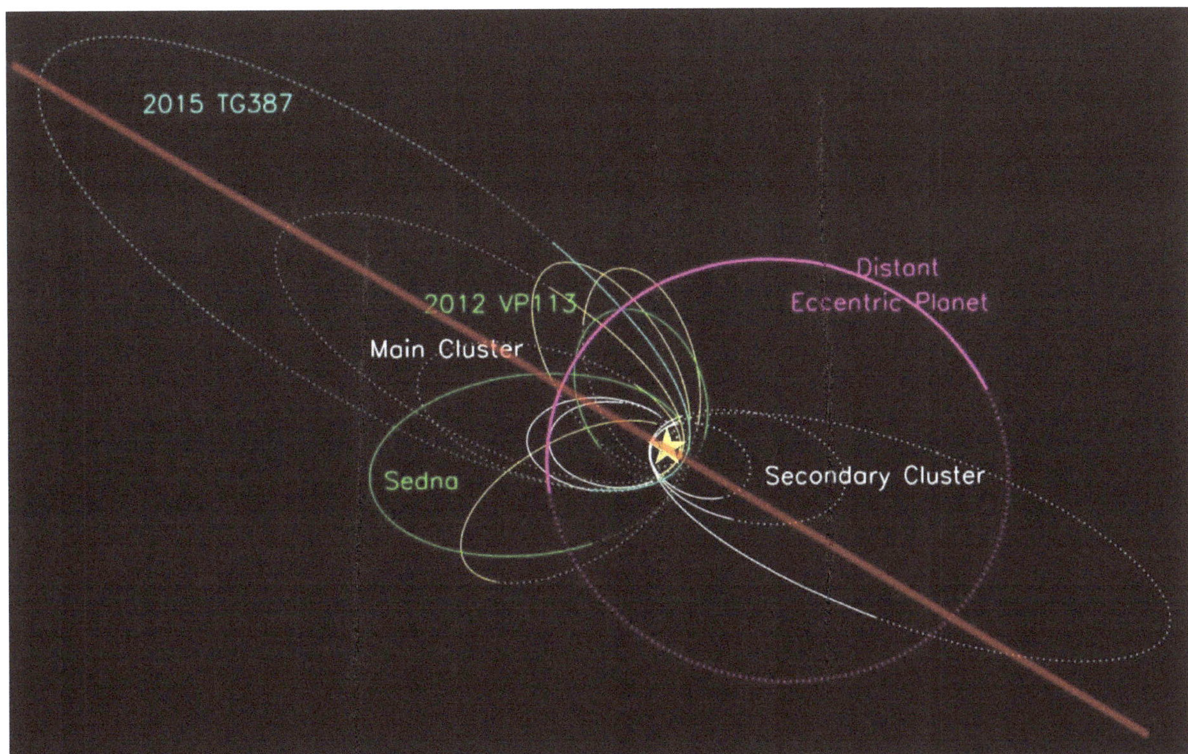

FIGURE 2-19 The orbits of the most distant trans-neptunian objects in the solar system are shown. The magenta orbit is one possible orbit for a distant, as-yet undiscovered, giant planet that dynamical simulations suggest could be responsible for the clustering of trans-neptunian objects. SOURCES: Adapted by C. Trujillo from S. Sheppard, C.A. Trujillo, D.J. Tholen, and N. Kaib, 2019, "A New High Perihelion Trans-Plutonian Inner Oort Cloud Object: 2015 TG387," *The Astronomical Journal* 157(4):139, https://iopscience.iop.org/article/10.3847/1538-3881/ab0895#ajab0895f4, ©AAS, reproduced with permission.

orbit more than 10 times farther away than Neptune. although observational searches have begun, surveys in the next decade may bring additional evidence to support or refute this current hypothesis.

One highlight of small body exploration this decade was the exploration of the TNO 486958 Arrokoth. In 2014, the Hubble Space Telescope discovered Arrokoth, an optimal target for the second flyby of the New Horizons mission in the dynamically cold classical TNO population. Cold classical TNOs are thought to be undisturbed by giant planet migration and formed in situ, and so they should represent pristine planetesimals whose nature can be used to probe the earliest epochs in the primordial trans-neptunian region. From ground-based stellar occultation campaigns, however, all that could be determined was that Arrokoth had a complex shape about 36 km long.

On January 1, 2019, the New Horizons spacecraft flew by Arrokoth, giving the world the first images of a distant TNO (Figure 2-20). The images showed that Arrokoth is a contact binary with two flattened lobes attached by a bright narrow neck. This shape is consistent with theoretical work on planetesimal formation, where the gravitational collapse of a cloud of pebbles can lead to such a two-lobed structure. Arrokoth's surface is red with methanol ice and organic material, both of which tell us the kinds of constituents that formed in the far reaches of the solar nebula. Arrokoth is also lightly cratered, which may indicate that few <1–2 km objects exist in the trans-neptunian region. Arrokoth appears to be a representative cold classical object in both its color and its slow spin, and thus much can be gleaned about planetesimal and planet formation from this intriguing body.

Ground-based observations have also led to the notable discovery of rings around some small bodies. In 2014, Chariklo, a 250-km body, was the first Centaur shown to have rings, detected through stellar occultations.

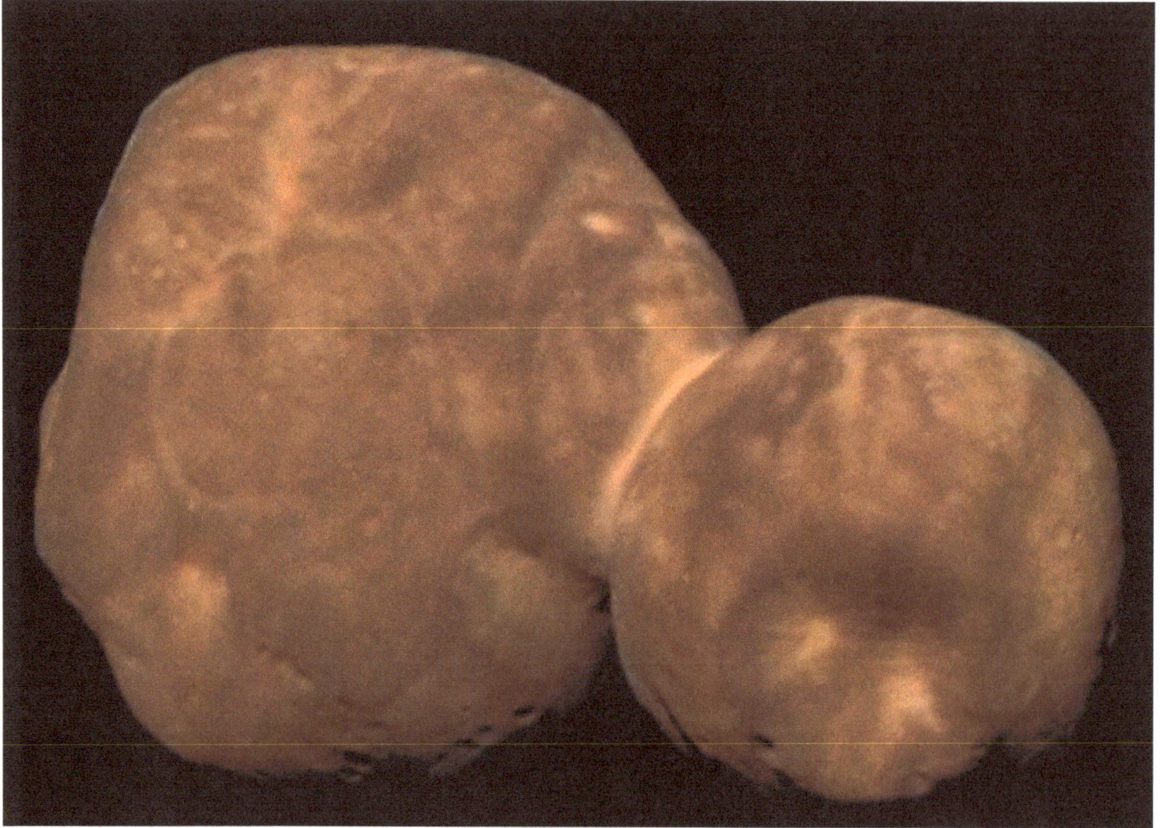

FIGURE 2-20 New Horizons composite view of Arrokoth from images obtained on January 1, 2019. Arrokoth measures approximately 36 km along its longest axis. SOURCE: Courtesy of NASA/Johns Hopkins University Applied Physics Laboratory/Southwest Research Institute/R. Tkachenko.

Since then, rings have been discovered around the dwarf planet Haumea, and the Centaur Chiron is suspected to have rings as well.

This decade also brought the unprecedented exploration of a Jupiter-family comet with the European Space Agency's Rosetta mission. In 2014, Rosetta became the first spacecraft to orbit a comet, conducting 2 years of detailed observations of the comet 67P/Churyumov-Gerasimenko and deploying the Philae lander to the comet's surface. In particular, Rosetta monitored the comet's evolution during its closest approach to the Sun and beyond, providing new insights into the geologic features, surface properties, active outburst processes, and interior structure of the comet. Measurements by Rosetta nearly doubled the inventory of coma organics detected, including the confirmation of the amino acid glycine. This mission provided fundamental insight into the dynamic complexity of cometary comae and nuclei, ground-truth for past missions, and context for both future missions and telescopic observations.

Observational studies of comets revealed new insights into the D/H ratio, which has now been measured in over a dozen objects with remote sensing and in situ techniques, yielding a range of D/H values from equivalent to Earth's ocean to up to a factor of three higher. No consistency has been observed in periodic versus long-period comets; this variability leaves the question of cometary delivery of Earth's ocean water and/or organics still unclear. Remote sensing has advanced to routinely measure the volatile distribution in fainter cometary comae with facilities such as the Atacama Large Millimeter Array, where asymmetric outgassing, distributed and extended source species, and variability have been found. Observational studies have expanded the molecular inventory of complex organics detected in comets, providing new understanding into the materials available during the formation of the terrestrial planets.

Main Asteroid Belt Planetesimals and the Moons of Mars

A highlight of the past decade was the first spacecraft exploration of the asteroid Vesta, the second-largest body in the main belt (Figure 2-21). The Dawn spacecraft's 14-month orbital investigation confirmed the affinity of Vesta's mineralogy and elemental composition to the Howardite-Eucrite-Diogenite (HED) meteorite clan. It also provided constraints on Vesta's internal structure that confirmed its igneous-differentiated nature.

The presence of two large overlapping impact basins in the southern hemisphere of Vesta (Veneneia and Rheasilvia), and the global trough systems (fossae) tied to stresses of those impacts indicate that while Vesta experienced significant impact-induced stress from these titanic blasts, it remained intact. Despite the deep excavation within those basins, no olivine was detected, defying expectations that this mineral, common in

FIGURE 2-21 Dawn at Vesta. *Top left*: Enhanced-color image mosaic on high-resolution shape model (525 km diameter). Light green color signifies diogenite-rich terrain, while deep blue/purple is eucritic. *Top right*: Mosaic of Marcia (*top*; 58 km diameter), Calpurnia, and Minucia craters showing dark and bright material exposures and pitted terrain in the bottom of Marcia. *Bottom left*: Perspective view of the giant Rheasilvia impact basin (500 km diameter). The central peak in Rheasilvia is the second highest mountain in the solar system after Olympus Mons on Mars. *Bottom right*: Cornelia crater (15 km diameter) with dark and bright material and pitted floor. SOURCES: *Top and bottom left*: Courtesy of NASA/JPL-Caltech/UCLA/MPS/DLR/IDA/PSI. *Top right*: NASA/JPL-Caltech/UCLA/MPS/DLR/IDA. *Bottom right*: NASA/JPL-Caltech/UCLA/MPS/DLR/IDA/JHUAPL.

planetary mantles, would be exposed. The missing olivine is likely sequestered in the deep mantle, whereas the upper mantle is dominated by orthopyroxene-rich diogenite. The ~1 billion-year-old crater-based age for the Rheasilvia impact basin is consistent with it being the source of Vesta's dynamical asteroid family and the HED meteorites.

Geochemical modeling based on Dawn constraints concluded that Vesta formed within 1.5 Ma of the first condensates in the solar system. It was likely made of volatile-depleted material with a bulk composition that was ~3/4 H-type ordinary chondrite and ~1/4 carbonaceous chondrite. Unexpectedly, dark, hydrated material covers a large portion of Vesta's surface and is thought to be remnants of carbon-rich low-velocity impactors, possibly including the impactor that created the Veneneia basin. Pitted terrains within young impact craters, as well as curvilinear gully systems, point to hydrated minerals or possibly even ice buried in the subsurface that has been excavated and mobilized by impacts. In addition to hydrated material delivered by the impact of carbonaceous asteroids, ice mixed into near-surface materials might be delivered by ice-rich comet-like bodies that strike Vesta at low impact angles. These significant discoveries at Vesta support models of volatile delivery to many different asteroids by carbon- and possibly ice-rich impactors.

In addition to Dawn's spacecraft encounter, telescopic observations and meteorite studies have continued to expand our understanding of the main belt and associated near-Earth asteroids and comets. Recently, the Spectro-Polarimetric High-Contrast Exoplanet Research (SPHERE) instrument on European Southern Observatory's Very Large Telescope (VLT) has been used to observe large (>100 km diameter) main belt planetesimals. Detailed shape models and density estimates of dozens of objects have been derived, leading to the recognition that 434-km-diameter asteroid 10 Hygiea appears to have retained significant volatiles. Telescopic observations have characterized main-belt comets and active asteroids—small bodies that have asteroid-like orbits but show comet-like visual characteristics—which blur the distinction between comets and asteroids in the small body population and attest to the volatile content of outer main belt objects.

A major advance has come from observations by the Wide-Field Infrared Survey Explorer (WISE), which characterized the sizes and albedos of hundreds of thousands of inner solar system bodies. This has enabled new theoretical studies of asteroid families, clusters of asteroid fragments produced by catastrophic collisions. Using theoretical models, these data can be used to probe the nature of asteroid disruption events, when they took place, and whether they produced surges of impactors capable of striking the terrestrial planets. WISE data has also been used to produce sophisticated models of the near-Earth object population, an important resource because the vast majority of hazardous objects have yet to be discovered.

From meteorite studies, the timescales of water-rock alteration in ice-rich planetesimals, as shown by Mn-Cr age dating of secondary minerals such as carbonates and fayalite, have been crucial for revealing the overall framework of early solar system chronology. Now, instead of dates for primary aqueous alteration within parent objects spanning 10 million years, they are constrained to an interval of 1 million years to 4 million years following solar system formation. Meanwhile, advances in the sensitivity and spatial resolution of magnetometers have enabled the first magnetic measurements of many meteorite groups and identification of an asteroid dynamo and the nebular magnetic field. Such investigations continue to expand our understanding of the formation and evolution of inner solar system planetesimals.

The two moons of Mars, Phobos and Deimos, are irregularly shaped small bodies that resemble asteroids in their shapes, low densities, and spectral characteristics. However, the explanation that the moons are asteroids captured into orbit about Mars has always had dynamical challenges to explain the origin of the moons, given both moons have near-circular, near-equatorial orbits, in contrast to highly elliptical and inclined orbits expected from capture models. In the past decade, there have been multiple new studies investigating an alternate origin for the martian moons—formation from a giant impact into Mars. Giant impacts have been recognized as important events across the solar system, including with the formation of Earth's Moon, and the impact modeled to form the martian moons is envisioned to have been smaller than the event that formed the Moon. Dynamical and geochemical models developed in the past decade have made testable predictions to distinguish between the competing hypotheses for the martian moons. Other studies have focused on explaining the different spectral units and large systems of grooves on Phobos to gain insights into the moons' origins. JAXA's Martian Moons Exploration (MMX) mission, planned for launch in 2024 with NASA as a contributing partner agency, is positioned to test these theories for the origin of the martian moons, through spacecraft measurements of the two moons as well as bringing samples of Phobos to Earth in 2029.

Exploration of Near-Earth Asteroids

Near-Earth asteroids are fragments of main belt asteroids that can approach or possibly strike Earth. They offer unique investigation opportunities owing to their proximity and accessibility. Because the near-Earth population originated elsewhere in the solar system before these objects entered Earth approaching orbits, studying near-Earth objects (NEOs) provides insights into the diversity of asteroids and comets, the properties and evolution of small planetesimals, and the timing and nature of terrestrial planet impacts over time.

Over the past decade, ground- and space-based surveys have discovered more than 17,000 NEOs, while ground-based radar observatories have further characterized more than 700 of these. Direct imaging of NEOs through radar and adaptive optics facilities has highlighted the population's spectrum of sizes and shapes. Radar facilities, such as that at Arecibo Observatory in Puerto Rico and the Goldstone Solar System Radar in California, provided data to generate three-dimensional models of small bodies without requiring a spacecraft encounter. In fact, the radar-derived shape model of asteroid Bennu was utilized by the OSIRIS-REx mission for planning prior to arrival, and comparison with the subsequent high-resolution spacecraft-derived shape model showed excellent agreement.

The simultaneous and complementary spacecraft exploration of two NEOs in 2017–2020 by JAXA's Hayabusa2 and NASA's OSIRIS-REx missions generated surprising results. Each of these missions rendezvoused with sub-kilometer diameter carbonaceous near-Earth asteroids, Ryugu and Bennu, respectively. Initial spacecraft observations of both asteroids revealed them to have similar spinning top-like shapes with low-albedo surfaces covered with large boulders and a surprising paucity of large smooth areas, in contrast to prearrival predictions made based on ground-based observational data.

Detailed reconnaissance confirmed the presence of hydrated surface materials on both bodies. Ryugu's material appears to resemble thermally metamorphosed or shocked carbonaceous meteorites, whereas Bennu's is similar to aqueously altered carbonaceous meteorites. Each spacecraft studied and interacted with its respective target to characterize the surface and collect samples. Hayabusa2 deployed two rovers and a lander to the surface prior to its first touchdown, then deployed a small crater-forming impactor which excavated subsurface material for collection prior to the second touchdown. The impact experiment produced a crater dominated by gravity in a surface of cohesionless materials. During proximity operations, OSIRIS-REx discovered that small particles were being ejected from multiple locations on Bennu; fortunately, this activity posed no danger to the spacecraft but it indicated that Bennu is shedding material, but not from cometary-like outgassing events.

Both missions shared the goal to deliver samples to Earth for coordinated, integrated analysis. Analysis of the returned samples will provide ground truth comparisons with Earth-based and spacecraft observations of the nature of asteroids and with our meteorite collections. It will also provide key constraints into the entire history of both asteroids, from their preserved presolar grains components through to pre- and post-accretion environments and geologic activity, to surface processes and the overall dynamical evolution of each asteroid—advances to be made in the coming decade. Hayabusa2 successfully returned to Earth with approximately 5.4 grams of sample on December 6, 2020. OSIRIS-REx touched down successfully on Bennu to obtain a sample (Figure 2-22) and will return with at least several hundred grams to Earth in September 2023. The upcoming analysis of the asteroid samples delivered to Earth by Hayabusa2 and OSIRIS-REx will bring new discoveries of primitive building block materials.

Our understanding of the lifecycles of small, rubble pile asteroids and comets has seen great advances in the past decade. The decade started with the discovery and analysis of several active asteroids, small bodies that shed material in a variety of contexts (e.g., ice sublimation, collisions, landslides driven by rotational processes). These findings motivated more detailed and precise models of the Yarkovsky-O'Keefe-Radzievskii-Paddack (YORP) effect, a thermal torque produced by the absorption and reemission of sunlight, which can cause some small

FIGURE 2-22 OSIRIS-REx sampling the asteroid Bennu on October 20, 2020. The white circular sampling head is about 30 cm across. SOURCE: Courtesy of NASA/Goddard Space Flight Center/University of Arizona.

bodies to reach disruptive rotation rates. Rotational spin-up from YORP is on par with collisions in influencing the geologic evolution of small gravitational aggregates like Bennu or Ryugu.

Additionally, the active asteroids motivated theoretical studies focused on the mechanical properties and behavior of small rubble pile asteroids. This led to the identification of weak molecular forces between components as an important additional mechanical force that shapes the evolution of these bodies. Constraints on the strength of such forces have been estimated by astronomical observations of active asteroids, buttressed by in situ measurements from the Hayabusa2 and OSIRIS-REx missions. These two missions have further expanded our insight into rubble pile bodies by providing our first direct estimate of the internal mass distribution within the asteroid Bennu, and have placed strong constraints on past spin rates and global failure mechanisms for Ryugu. To further understanding of the processes shaping small rubble pile asteroids, NASA's SIMPLEx-class Janus mission is, at the time of writing, scheduled to launch with Psyche and send twin smallsat to explore two binary asteroid systems in 2026.

Knowledge gained about near-Earth asteroids also has implications beyond the scientific exploration of the solar system. For example, the orbits of both Bennu and Ryugu are Earth-crossing, classifying them as potentially hazardous asteroids to Earth. By characterizing the shapes, sizes, and physical properties of NEOs, we glean insights into the planetary defense mitigation efforts that might be the most effective against such an object in the future. Another critical aspect of planetary defense is to identify near-Earth objects and accurately predict their future orbital pathways. A more in-depth discussion of planetary defense activities is provided in the Planetary Defense chapter.

NEOs in accessible orbits are also targets for human exploration and resource utilization. In the past decade, the NASA Near-Earth Object Human Space Flight Accessible Targets Study (NHATS) identified at least ten objects offering round-trip voyages of less than a year and requiring a ΔV (i.e., velocity change) of less than 6 km/s. For reference, this is less energy than a one-way trip to the lunar surface. At Ryugu and Bennu, Hayabusa2 and OSIRIS-REx characterized the low-gravity asteroid landscapes similar to those human explorers may encounter. Space-suited astronauts in neutral buoyancy simulations prepared for exploration and sampling operations at near-Earth asteroids and the martian moons to realistically simulate operations and surface sampling techniques. Such skills will be needed when astronauts or robots visit a resource-rich NEOs in the future, possibly to extract water from hydrated clay minerals, such as those detected on Bennu by OSIRIS-REx, for use as propellant. The upcoming analysis of returned samples will further characterize this material, telling us whether water can be readily extracted. From a cost-benefit perspective, it may be possible that NEO water resources are more plentiful and easier to access and utilize than those in the permanently shadowed regions on the Moon.

Connections Between Impacts and Astrobiology

Asteroid impacts have had a profound influence on both the habitability of Earth and the evolution of life. For example, 66 Ma, a 10 km body hit in what is now Mexico's Yucatan peninsula and triggered a mass extinction event that ended the reign of the dinosaurs. The heavily cratered Moon is testament that other impact-related catastrophes took place in Earth's deep past, but well-preserved impact effects in the terrestrial geologic record are rare. This raises the question of how else impacts have influenced our biosphere.

One intriguing example comes from the disruption of the L-chondrite parent body in the asteroid belt ~470 Ma. This event sent an enormous shower of small particles and sub-km impactors to Earth, with the flux of fine-grained extraterrestrial material increasing by three to four orders of magnitude. This dust cooled Earth prior to the start of the Ordovician ice age, while the numerous impactors stressed many abodes of life. Together, this bombardment potentially explains the timing and nature of the Great Ordovician Biodiversification Event, a ~30 million year period that produced nearly modern levels of marine invertebrate biodiversity.

Impact craters can also be used to tell us about tremendous changes in the history of Earth. For example, most craters on Earth are found on stable regions called cratons and are younger than 650 million years old. Their near total absence from older terrains takes place at the same time as the so-called Great Unconformity, a large gap in Earth's stratigraphic record. Both these missing records appear linked to heavy erosion that took place in an era known as Snowball Earth, at a time when Earth was cloaked in global ice. Erosion linked to these ice sheets may have removed 3–5 kilometers of worldwide crust and sent it to the ocean floor. The history of life and our biosphere may be determined by such events.

The onset of the Snowball Earth era ~700–800 Ma is poorly understood, but it is fascinating to find that it is close in time to another major impact shower occurring ~800 Ma. This event was identified using lunar impact spherules returned by the Apollo astronauts and the calculated ages of numerous large lunar craters. Like the 470-million-year shower, the ~800-million-year event is marked by a number of changes in the history of life: the return of anoxic conditions to the deep ocean for the first time since ~1.8 Ga, an abrupt decrease in carbon isotopes ($\delta^{13}C$) in Australia's Bitter Springs formation, and major changes in the abundance, diversity, and environmental distribution of marine eukaryotes. It is plausible that the onset of glaciation at this time was brought about by an event similar in character to the 470-million-year event.

Critical connections between life and impacts can also be indirect. For example, large impacts appear to enhance the output of existing volcanic plumes via seismic shaking. This was first identified for the Chicxulub impact 66 Ma, with the erupted volumes of both mid-oceanic ridges and large igneous provinces suddenly increasing at that time. Going back to the Archean era more than 2.5 Ga, impact spherule beds tell us that tremendous impacts took place, with some projectiles being several tens of km in diameter or more. Models show that these events can instigate volcanic plumes from the core-mantle boundary, with eventual consequences for Earth's surface and atmosphere. The largest events may even help to initiate plate tectonics.

Last, impacts likely played a crucial role in the development of life on Earth and possibly Mars as well. For example, water and organics were probably delivered to Earth and Mars during the planet formation era by volatile-rich projectiles that originated beyond the orbit of Jupiter. In addition, early bombardment of these worlds may have also helped produce periods of surface habitability by both exposing and distributing subsurface materials. The question is whether life had enough time to emerge in a sustained manner within these intervals. Early Earth studies allow us to explore a new planet in the solar system that is very different from the current Earth. The search for connections between impacts and astrobiology has only just begun.

Small Bodies from Beyond the Solar System

One of the major discoveries in the past decade was the identification of the first interstellar object detected passing through the solar system. Interstellar object 1I/'Oumuamua, estimated to be between 100–1,000 meters long, was discovered in 2017 by the Pan-STARRS survey. Aptly named 'Oumuamua, meaning "a messenger from afar arriving first" in the Hawaiian language, this first discovery was followed only 2 years later by the discovery of the second interstellar object, 2I/Borisov, roughly half a kilometer in size. The interstellar origins of these objects were determined by their hyperbolic trajectories through the solar system, thus indicating that they are not bound by the Sun's gravity.

While the discovery of two interstellar objects is a historic achievement of this decade, it is also noteworthy how different the two objects are from each other. 1I/'Oumuamua had no measurable activity, and appeared more asteroidal, while 2I/Borisov exhibited comet-like behavior, with its nucleus surrounded by a coma and an extensive tail. However, while cometary in appearance, 2I/Borisov was observed to have an extreme CO abundance with respect to H_2O, unlike most comets from the solar system. These two apparitions, although fleeting, provided the ability to observe the chemistry and physical conditions in small bodies from other planetary systems. Upcoming sky surveys in the next decade, such as planned for the Vera C. Rubin Observatory, are anticipated to find many more interstellar objects that will transit the solar system and continue this new field of discovery.

Although the rapid traverse of interstellar objects through the solar system has given us our first tantalizing glimpses into small bodies from extrasolar worlds, the microscopic traces of many such worlds in the form of interstellar and presolar grains is, and has been, our most accessible source of extrasolar materials for prolonged investigation. This decade, the collector tray from the Stardust mission that was devoted to gathering interstellar grains was investigated in state-of-the-art analytical facilities. The preliminary analysis identified seven pristine interstellar dust grains that exhibited a diversity of crystal structures and elemental composition. The analysis of isotopically anomalous SiC grains isolated from a meteorite provided the first successful dating of interstellar grains, which showed a range of ages that inform models for the origin of these particles beyond the solar system. As analytical techniques continue to advance, new measurements of interstellar materials contained in meteorites and other extraterrestrial samples will continue to provide constraints into the evolution of the rocky components of other solar systems.

Key Discoveries from the Past Decade

- **Distinct chemical and physical reservoirs were produced during the evolution of the early solar system and the formation of the giant planet.** These momentous events distributed small bodies across the solar system; chemical signatures measured in meteorites and remote observations of volatiles point to the extent of mixing between the reservoirs and the compositions of the building blocks available to the terrestrial planets.
- **Outer solar system small bodies display a wide diversity of properties that are distinct from their inner solar system counterparts.** The unprecedented number of outer solar system small bodies discovered

in the past decade and the first spacecraft exploration of a primitive trans-neptunian object suggest distinctive accretion and evolutionary processes, with still much to discover about this population we have only begun to explore.

- **Inner solar system planetesimals provide new insights into planetary evolutionary processes.** The first orbital expedition to the main asteroid belt documented Vesta's very early formation, battering by two giant impacts, and preservation of hydrated materials delivered to its surface by impactors. Meteoritic and telescopic studies investigated processes ranging from early solar system magnetic dynamos to current-day outburst events from active asteroids, and new models for the origin of the martian moons by a giant impact were developed, emphasizing the wide array of processes that affect the evolution of small bodies.
- **The physical properties of near-Earth asteroids indicate that they are complex assemblages that are actively evolving.** Exploration of two near-Earth asteroids revealed rugged surfaces dominated by boulders and provided key ground-truth for models of the shapes, structures, strength, and lifecycle of the near-Earth asteroid population.
- **The first interstellar objects passing through the solar system were identified,** The two interstellar objects discovered were strikingly different from each other. Along with analytical measurements of interstellar grains, these objects provide insights into origin environments unlike those of the solar system.

Further Reading

Bockelée-Morvan, D., U. Calmonte, S. Charnley, J. Duprat, C. Engrand, A. Gicquel, M. Hässig, et al. 2015. "Cometary Isotopic Measurements." *Space Science Reviews* 197:47–83. https://doi.org/10.1007/s11214-015-0156-9.

Fitzsimmons, A., C. Snodgrass, B. Rozitis, B. Yang, M. Hyland, T. Seccull, M.T. Bannister, W.C. Fraser, R. Jedicke, and P. Lacerda. 2018. "Spectroscopy and Thermal Modelling of the First Interstellar Object 1I/2017 U1 'Oumuamua." *Nature Astronomy* 2:133–137. https://doi.org/10.1038/s41550-017-0361-4.

Grundy, W.M., M.K. Bird, D.T. Britt, J.C. Cook, D.P. Cruikshank, C.J.A. Howett, S. Krijt, et al. 2020. "Color, Composition, and Thermal Environment of Kuiper Belt Object (486958) Arrokoth." *Science* 367(6481). https://doi.org/10.1126/science.aay3705.

Jewitt, D. 2012. "The Active Asteroids." *Astronomical Journal* 143(3):66. https://doi.org/10.1088/0004-6256/143/3/66.

Lauretta, D.S., D.N. DellaGiustina, C.A. Bennett, D.R. Golish, K.J. Becker, S.S. Balram-Knutson, O.S. Barnouin, et al. 2019. "The Unexpected Surface of Asteroid (101955) Bennu." *Nature* 568:55–60. https://doi.org/10.1038/s41586-019-1033-6.

Rosenblatt, P., S. Charnoz, K.M. Dunseath, M. Terao-Dunseath, A. Trinh, R. Hyodo, H. Genda, and S. Toupin. 2016. "Accretion of Phobos and Deimos in an Extended Debris Disk Stirred by Transient Moons." *Nature Geoscience* 9:581–583. https://doi.org/10.1038/ngeo2742.

Russell, C.T., and C.A. Raymond. 2015. "The Dawn Mission to Vesta and Ceres." Pp. 419–432 in *Asteroids IV*, P. Michel, F. DeMeo, and W.F. Bottke, eds. Tucson: University of Arizona Press. https://uapress.arizona.edu/book/asteroids-iv.

Spencer, J.R., S.A. Stern, J.M. Moore, H.A. Weaver, K.N. Singer, C.B. Olkin, A.J. Verbiscer, et al. 2020. "The Geology and Geophysics of Kuiper Belt Object (486958) Arrokoth." *Science* 367(6481):eaay3999. https://doi.org/10.1126/science.aay3999.

Taylor, M.G.G.T., C. Alexander, N. Altobelli, M. Fulle, M. Fulchignoni, E. Grün, and P. Weissman. 2015. "Rosetta Begins Its Comet Tale." *Science* 347:387. https://doi.org/10.1126/science.aaa4542.

Thomas, N., B.J.R. Davidsson, L. Jorda, E. Kührt, R. Marschall, C. Snodgrass, and R. Rodrigo. 2020. "Editorial to the Topical Collection: Comets: Post 67P/Churyumov-Garasimenko Perspectives." *Space Science Reviews* 216:107. https://doi.org/10.1007/s11214-020-00727-1.

Vokrouhlický D., W.F. Bottke, S.R. Chesley, D.J. Scheeres, and T.S. Statler. 2015. "The Yarkovsky and YORP Effects." Pp. 509–531 in *Asteroids IV*, P. Michel, F. DeMeo, and W.F. Bottke, eds. Tucson: University of Arizona Press. https://uapress.arizona.edu/book/asteroids-iv.

Wadhwa, M., T.J. McCoy, and D.L. Schrader. 2020. "Advances in Cosmochemistry Enabled by Antarctic Meteorites." *Annual Review of Earth and Planetary Sciences* 48:233–258. https://doi.org/10.1146/annurev-earth-082719-055815.

Watanabe, S., M. Hirabayashi, N. Hirata, N.A. Hirata, R. Noguchi, Y. Shimaki, H. Ikeda, et al. 2019. "Hayabusa2 Arrives at the Carbonaceous Asteroid 162173 Ryugu—A Spinning Top-Shaped Rubble Pile." *Science* 364(6437):268–272. https://doi.org/10.1126/science.aav8032.

GIANT PLANET SYSTEMS

Overall Architecture

The giant planets—Jupiter, Saturn, Uranus, and Neptune—dominate the solar system, containing more than 99 percent of the mass and angular momentum outside of the Sun. Their formation, and likely orbital migration, dictated the structure of the rest of the solar system, controlling terrestrial planet growth rates, locations, and much of their volatile inventories, as well as the ultimate distribution of all small body populations in the solar system. Their ever-changing atmospheres provide windows to the deeper hidden interiors, which lock away the secrets of the origin of the solar system. Each giant planet harbors a rich and diverse system of satellites and rings, embedded within enormous and complex magnetospheres (Figure 2-23). The tidal interactions between many icy satellites

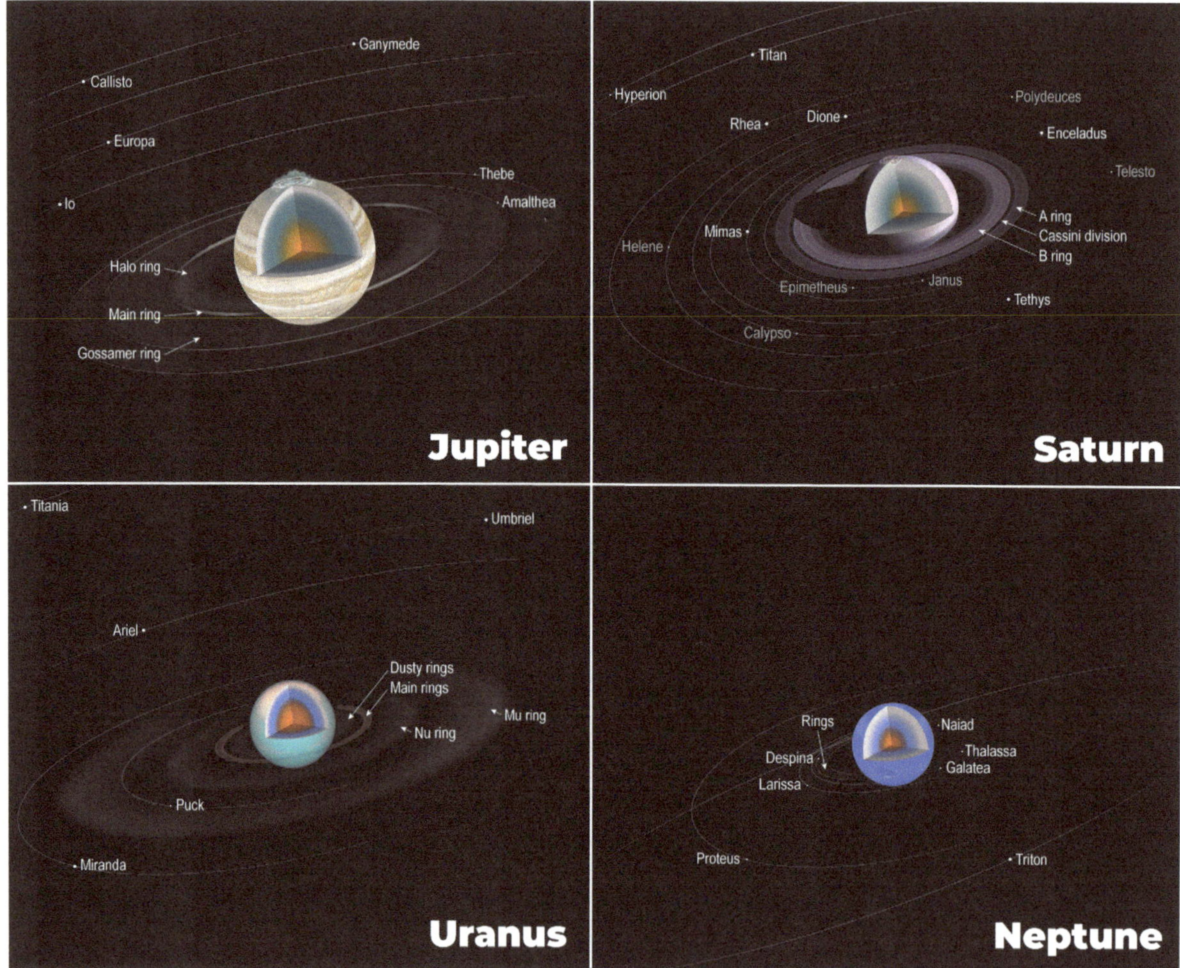

FIGURE 2-23 Schematic of the outer giant planet systems, showing a rich set of moons, rings, and varying interior structure indicated by color, and shown in more detail later in Figure 10-1. Color code for Jupiter and Saturn: red, heavy element core; blue-green, metallic hydrogen envelope; and gray, molecular hydrogen envelope. Color code for Uranus and Neptune: red, rocky core (possibly mixed with ice); blue, water-rich (mixed with hydrogen and helium); and white: predominantly hydrogen and helium. Note that the planets and moons are not drawn to scale and that details of the interiors for Uranus and Neptune are highly speculative at this point. SOURCE: J.T. Keane, Jet Propulsion Laboratory, modified from J. Friedlander/TRAX/GSFC.

and their host planets provide provides a third energy source in addition to energy of accretion and radiogenic heating, helping to maintain subsurface oceans which may harbor life. To understand whether the solar system architecture is typical or unique requires comparing a deep understanding of the formation and evolution of the giant planet systems, while surveying the size, number, and composition of the more than 2,000 giant exoplanets discovered to date.

The outer planets are often divided into two groups: (1) the gas giants, Jupiter and Saturn, which are H-He-dominated; and (2) the "ice giants," Uranus and Neptune, which possess H-He atmospheres containing ~10–20 percent of their total mass and a larger proportion of ices. There are many similarities among the giant planets: all appear to have interiors dominated by a large, dense core; deep, dynamic hydrogen/helium atmospheres; multiple satellites, some of which suffer additional heating by tidal dissipation to have subsurface oceans; multiple rings that interact with small moons; and substantial magnetospheres. The Galileo, Cassini, and Juno missions have shown that the interior structures of gas giants, Jupiter and Saturn, are far more interesting and complex than expected, and have advanced our understanding of their atmospheres, magnetospheres, satellite, and ring systems.

However, even within each sub-group there are clear differences among the planets. Saturn is not simply a smaller version of Jupiter and there are key differences between Uranus and Neptune. Each planet is unique and has different key questions associated with it. These differences give clues to their origin and unique histories, provide key insight into how giant planets are built, and can untangle the dynamic history of our early solar system.

Atmospheres

Giant planet atmospheres are natural planetary-scale laboratories for studying the interplay between dynamics, meteorology, chemistry, and cloud formation, and represent the transition region between the external magnetosphere and the hidden deep interiors. These atmospheres are in a constant state of motion and change, transporting energy and material from place to place in response to long seasonal cycles and meteorological phenomena. Their deeper atmospheres are characterized by horizontal bands of clouds, composition, and temperature, organized by east-west winds and punctuated by spectacular vortices and storms. Their cloud-free stratospheres are warmed by sunlight absorbed by methane gas. The stratospheres interact with the external planetary environments. Material from "ring rain," micrometeoroids, and even large impacts deposits water and other external material into the stratosphere. At the upper edge of the stratosphere, which blends into the thermosphere and ionosphere, energy is also deposited in the form of charged particles streaming in from the magnetospheres, some of which generates delicate auroral patterns. Circulation of the stratospheres and thermospheres then redistributes this energy with latitude.

Clouds condense in the atmospheres of the giant planets. The troposphere is where most of the solar heating occurs, but in all the giant planets except Uranus, more heat comes from the deeper interior than from the Sun. Heat is transported within this "weather layer" by mixing—on small scales by storms and on large scales by atmospheric circulation, both of which vary substantially with time. The composition of clouds in the giant planets is diverse. In addition to clouds formed from water, the giant planets host clouds of ammonia ice and ammonium hydrosulfide, joined by hydrogen sulfide and methane clouds in the even colder atmospheres of Uranus and Neptune. Molecules such as carbon monoxide and phosphine are mixed upward from deeper levels where high-temperature thermochemistry dominates, exposing them to sunlight to help provide the breadth of colors in the visible clouds.

Hubble and Keck observations over many years have shown the appearance of more clouds on Neptune than seen by Voyager, with changes occurring on short timescales. Additional Great Dark Spots akin to that seen by Voyager have been seen by Hubble recently to appear and then go away on a multiyear timescale. Uranus's variability is even more striking, challenging the idea that the atmosphere is driven entirely by seasonal effects. When Uranus approached equinox in 2007, after a long period in which no changes were seen by Hubble, an outbreak of clouds all over its surface occurred. After that outbreak, no additional outbursts were expected, but in fact many large bright cloud outbreaks have since been observed.

Giant planet atmospheres are mostly hydrogen and helium, with other common elements like carbon included as hydrogenated molecules in small amounts, and a general trend of more enriched elemental abundances in the ice giants than the gas giants. The noble gases (helium, neon, argon, krypton, and xenon) are largely unaffected by atmospheric chemistry. Measurement of their elemental and isotopic abundances provide key pieces of the puzzle of how the solar system formed as different models predict a different pattern of abundances.

In the stratosphere, methane is broken apart by solar ultraviolet and energetic particle radiation to recombine as more complex hydrocarbons. Water coming in from external sources can also participate in stratospheric chemistry, and chemical signatures of external inputs such as cometary impacts can persist for centuries. The complex molecules are then redistributed by stratospheric circulation, sediment downward to contaminate the deeper clouds, and can sometimes condense to form thin haze layers.

Studies of the weather, climate, and atmospheric circulation of the giant planets provide many parallels to processes operating in the atmospheres from Earth to exoplanets. Owing to the fast rotation of these planets and the absence of a solid surface, winds blow in mainly the east-west direction. The fastest winds are near the equator, where Jupiter and Saturn have super-rotating (eastward) jets and Uranus and Neptune have retrograde (westward) jets. Gigantic convective storms on Jupiter and Saturn are known for lightning activity and strong updrafts that allow mixing of cloud and precipitation particles of different types (e.g., water and ammonia), and show episodic and nonseasonal behavior which might someday allow us to predict giant planet weather. Earth-size vortices spinning in both cyclonic directions (like hurricanes) and anticyclonic directions have been observed on Jupiter and Saturn. On Uranus and Neptune, only anticyclonic vortices have been detected, with similar properties but shorter lifetimes than Jupiter's Great Red Spot.

The planetary banding on all four giants may demarcate circulations similar to the Hadley and Ferrel cells in Earth's atmosphere, but although this has been studied in depth at Jupiter and Saturn, the applicability to the ice giants remains a mystery. The polar regions differ from world to world, with long-lived clusters of cyclonic vortices on Jupiter, a hexagon on Saturn, a seasonal polar hood on Uranus, and a hot and chemically depleted vortex at Neptune's summertime pole. Understanding how the dynamics, clouds, and composition vary from world to world, particularly via exploration of ice giants, will provide powerful new insights into the formation processes and environmental processes shaping giant planets in all their guises.

Impacts have been observed many times in Jupiter's atmosphere, and streaks from small impacts appear on Saturn's rings in Cassini high resolution images. Whether Uranus and Neptune are impacted by objects from the Kuiper belt is an open question. The observed overabundance of CO and HCN in Uranus and Neptune relative to what chemists predict for these hydrogen-rich atmospheres suggest that cometary and asteroid impacts might be responsible.

Interiors, Including Deep Structure, Circulation, and Heat Balance

The bulk composition and internal structure of the outer planets are still poorly constrained, but knowledge of them is critically important to understanding the formation and evolution histories of the planets. Additionally, various physical and chemical processes and their interplay govern the interiors; dynamics and magnetic fields, dynamo generation and composition, dynamics and rotational contribution to the density distribution, as well as processes like convection, core erosion, and immiscibility. In addition, the deep interiors of the giant planets serve as natural laboratories for materials at high pressures and temperatures, as a result, constraining the interiors of the outer planets is also of interest to the high-pressure physics community. Last, the planets in the solar system serve as prototypes for exoplanetary science. A better understanding of Jupiter, Saturn, Uranus, and Neptune will improve our understanding of distant giant planets elsewhere in the Milky Way galaxy. Therefore, understanding the deep structure and global composition of giant planets is a key objective in planetary science.

In the past decade, the Cassini-Huygens (Figure 2-24) and Juno (Figure 2-25) missions around Jupiter and Saturn, respectively, have led to many exciting and important discoveries. With extremely accurate measurements of the planets' gravity fields, the depth of Jupiter and Saturn's zonal flows has been determined and found to be

FIGURE 2-24 Cassini highlight montage. *Top row*, *left to right*: Red streaks on Tethys; south pole of Enceladus; Saturn's rings seen from high elevation. *Middle row, left to right*: Saturn's rings at high resolution; Saturn's small moon Methone. *Lower row, left to right*: A montage of three ring moons, Atlas (*upper left*), Daphnes (*upper right*), and Pan; Titan's surface beneath its hazy atmosphere; Mimas; and the hexagon at Saturn's north pole. SOURCES: *Top left*: Courtesy of NASA/JPL-Caltech/Space Science Institute/Lunar & Planetary Institute. *Other images*: NASA/JPL-Caltech/Space Science Institute.

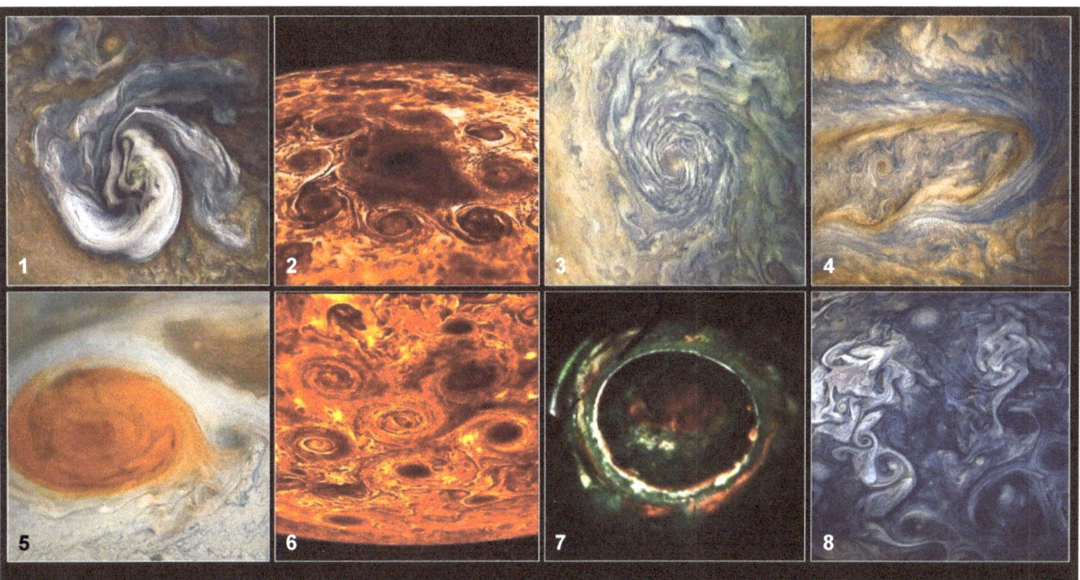

FIGURE 2-25 Juno highlight montage. Juno images of Jupiter's atmosphere. Panels 2 and 6 are JIRAM images of cyclonic storms at the north and south poles; Panel 7 is an ultraviolet image of the southern hemisphere aurora. The remaining panels are JunoCam images processed by citizen scientists highlighting atmospheric features at various scales. SOURCES: (1, 3–5, 8) Courtesy of NASA/JPL-Caltech/SwRI/MSSS/Gerald Eichstädt/Seán Doran, CC BY-NC-SA 3.0. (2) NASA/JPL-Caltech/SwRI/ASI/INAF/ JIRAM. (6) NASA/SWRI/JPL/ASI/INAF/IAPS. (7) NASA/JPL-Caltech/SwRI.

3,000 km and 9,000 km for Jupiter and Saturn, respectively. The flows decay where the electrical conductivity is sufficiently high to drag the flow into nearly uniform rotation at greater depth. With the knowledge of the dynamical contribution to the gravity field, the interior structure can be further constrained. For Jupiter, models matching the Juno measurements indicate that the envelope is not well-mixed and that the core is diluted (i.e., has an extended region of enriched elements heavier than hydrogen and helium in the deep interior). For Saturn, the detection of oscillations in its rings by the Cassini spacecraft has demonstrated that part of its interior is stably stratified, with indications that the core is diluted.

The bulk compositions of Uranus and Neptune are poorly constrained, and accurate measurements of their gravity fields are essential for their characterization. Such measurements can also be used to constrain the depth of the winds on the ice giants. To understand the nature of Uranus and Neptune better, improved measurements of their magnetic fields and thermal fluxes are required. There is a clear need to link the magnetic fields and the deep interiors. A better determination of the internal structure and the variability of the planetary magnetic fields can further constrain the composition and internal structure via the required conditions to sustain a dynamo and the role of ohmic dissipation in decaying the winds. Knowledge of the fluxes and heat transport mechanisms within the outer planets can reveal information not only on their evolution and structure, but also on the link between the atmosphere and deep interior. The existence and nature of the magnetic fields provide important observational constraints on the present-day interior structure and dynamics of the outer planets. Dynamo generation is thought to require large-scale motions in a medium that is electrically conducting, and rapid (at least moderate) rotation. In Jupiter and Saturn, the conducting material is metallic hydrogen in a region where heat is transported by convection. Since metallic hydrogen is associated with a relatively deep region in the planets, the resulting magnetic field is nearly dipolar. Recent Juno observations explored the morphology of Jupiter's magnetic field. When viewed at the dynamo surface, Jupiter's magnetic field is characterized by an intense isolated magnetic spot near the equator with negative flux, an intense and relatively narrow band of positive flux at ~45 degrees latitude in the northern hemisphere, and relatively smooth magnetic field in the southern hemisphere. The north-south dichotomy in Jupiter's magnetic field morphology could be a result of Jupiter's dilute core, which either limits the dynamo action to the upper layer of Jupiter or creates spatially separated active dynamos within the planet. Saturn's intrinsic magnetic field is unusually weak, with surface field strength ranging from 20 to 50 μT. Surprisingly, Saturn's magnetic field seems to be symmetric, to within the accuracy of the data, with respect to the spin-axis. Both the weak strength and the extreme spin axisymmetry of Saturn's magnetic field might be linked to helium rain that could create a stably stratified layer atop the deep dynamo. However, this is only a speculation, and this topic is still being investigated.

The magnetic fields of Uranus and Neptune are poorly understood, and the available measurements are very limited. One possibility is that the magnetic field in these planets is generated by an exotic form of water called "super ionic." This is motivated by models that predict that Uranus and Neptune consist of mostly water in their deep interiors. However, other elements could also lead to high electrical conductivity to generate a dynamo, and therefore the link between the composition and the magnetic field in these planets remains unknown. The location at which the magnetic field is generated in Uranus and Neptune (where the material is electrically conducting and the region is convective) is expected to be in an outer shell (i.e., closer to the "surface"). This could explain the multi-polar nature of the magnetic field of Uranus and Neptune. Understanding the magnetic fields of the giant planets is not only crucial for putting constraints on the composition, but also on the heat transport mechanisms within the planets and the interplay between rotation, interior, and dynamics.

Magnetospheres

Jupiter's magnetosphere—the sphere of influence of its magnetic field—is the largest planetary structure in the solar system, 10 times the volume of the Sun and stretching out past the orbit of Saturn. When we compare the magnetospheres of planets, however, it is usual to compare with the size of the planet (e.g., radius of planet,

R_p) and use the distance from the planet's center to the subsolar boundary (basically the smallest scale) for comparison. At Jupiter, Saturn, Uranus, and Neptune, the magnetospheres are 63–92 R_J, 22–27 R_S, 18 R_U, and 24 R_N, respectively. The range in sizes at Jupiter and Saturn shows the variability owing to changes in the pressure of the solar wind. At Uranus and Neptune, the single Voyager 2 flyby of each did not provide enough data. As the solar wind spreads out with distance, the pressure at farther planets correspondingly decreases, but the major cause of these ranges in size between these magnetospheres comes from the different strengths of the magnetic fields generated by their internal magnetic dynamos. In the case of Jupiter, a major additional factor is the internal pressure of hot ionized gases—plasma—trapped in the magnetic field that further inflate the magnetosphere by about a factor of two.

But size is not everything. The characteristics of the magnetospheres of Uranus and Neptune are radically different from those of Jupiter and Saturn (Figure 2-26) owing to a combination of two factors: the geometry of the internally generated magnetic fields and the sources of plasma. At Jupiter and Saturn, beyond a couple radii away from the planet, the strong internal fields (generated in large volumes of metallic hydrogen) are approximated by a simple dipole—like a bar magnet—with a small tilt (10 degrees for Jupiter, 0 degree for Saturn) from the planet's spin axis. At Uranus and Neptune, the field is much more complicated (probably owing to dynamos generated in a shell of water), with not just a large (50–60 degrees) tilt from the spin axis, but also a highly irregular and nondipolar form. This means that over the Uranus and Neptune spin periods (17 and 16 hours, respectively) the geometry of the planetary field relative to the solar wind, and the magnetic field embedded therein, change dramatically. Consequently, the magnetospheres of Uranus and Neptune are thought to be highly dynamic, and any plasma sources are quickly flushed out of the system.

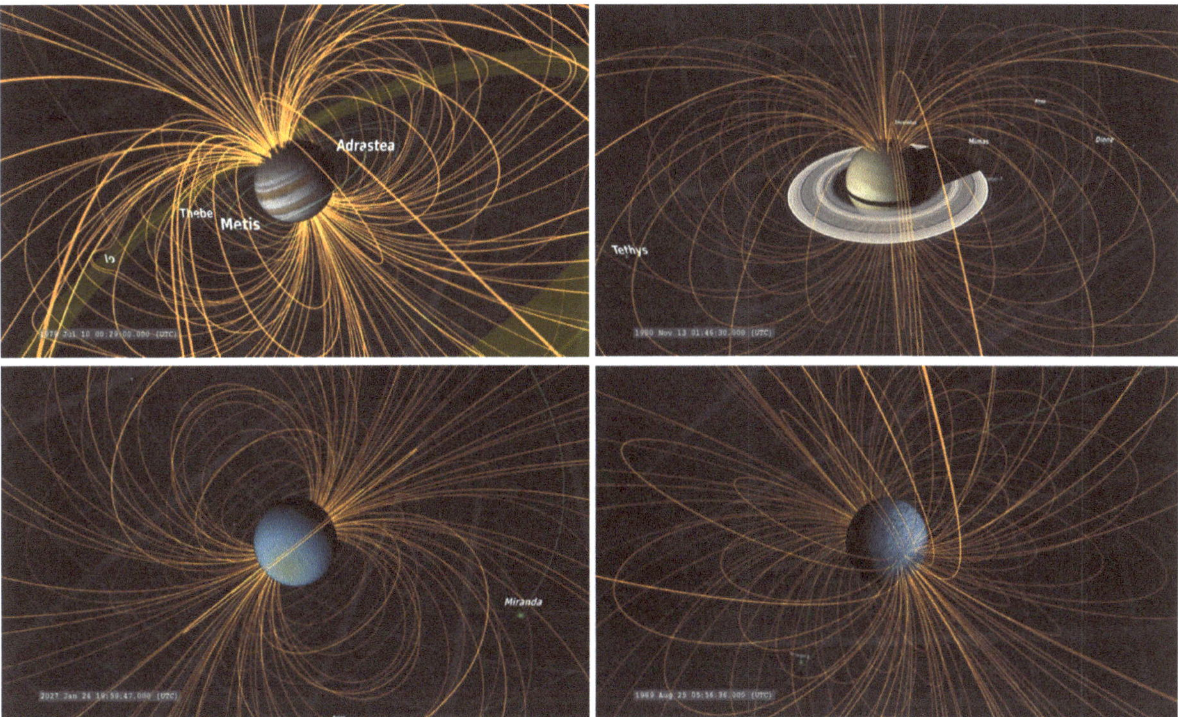

FIGURE 2-26 Outer planet magnetospheres. The gold lines represent the magnetic field structure; the subtle semi-transparent mesh in the distance is the boundary of the magnetosphere. Orbits of selected satellites are included for scale. SOURCE: Courtesy of NASA's Scientific Visualization Studio.

The magnetospheres of Jupiter and Saturn are dominated by plasma sources from their geologically active moons—Io and Enceladus, respectively. The interaction of the surrounding plasma with Io's atmosphere causes tons of atmospheric gases (mostly SO_2 and dissociation products O, S) to escape from the moon every second. This cloud of neutral material along Io's orbit is quickly ionized to produce 260–1,400 kg/s of plasma (dominated by ions of O and S, plus electrons) that forms a torus of plasma that is coupled to Jupiter via the magnetic field and co-rotates around the planet with Jupiter's 10-hour spin period. As this Io-genic plasma moves out into the vast magnetosphere, it becomes heated by processes that are not well determined (to tens to hundreds keV). Ultimately, the plasma is lost mostly via ejection of blobs down the magnetotail, but also through acceleration into Jupiter's atmosphere where it excites intense auroral emissions. Similar processes occur at Saturn with the plasma source (~250 kg/s) being water group ions from ionization of material spewed out by Enceladus's plumes. But lower energy electrons at Saturn mean that the neutrals survive longer and spread out in the system. The resulting ratio of neutral- to ionized-particles is 1:100 at Jupiter and 100:1 at Saturn.

The Voyager 2 flybys of Uranus and Neptune showed weak (<1 kg/s) plasma sources of mostly protons that could originate in the planets' ionospheres, the solar wind, or icy moons. Nevertheless, Uranus showed a remarkably intense radiation belt. There are glimpses of auroral emissions but the physical processes driving these intriguingly complex magnetospheres are undetermined.

Moons

The four giant planet systems each have distinctly different systems of orbiting moons, sometimes thought of as miniature solar systems. Like the incredible variety of exoplanetary systems, the diversity of our neighboring moon systems allows us to explore the diversity and complexity of planet formation and subsequent dynamical processes that modify or disrupt them, right in our "celestial backyard." Satellite systems can also preserve the nongaseous composition of the planets and solar nebula in the region they formed. The satellite systems of the giant planets reveal four very different outcomes of planet formation processes and the subsequent dynamical evolution that can modify or disrupt them, as recorded in their surface features and crater populations. As a result, the thermal and geologic processes that have shaped and sculpted the satellites of these four systems are also distinct.

Jupiter has a primordial system of large moons and an extensive population of smaller captured bodies. Jupiter's four large Galilean moons reveal the fundamental importance of gravitational tides in generating internal heat, producing the giant volcanoes on Io and melting ice on Europa and Ganymede. At Saturn, we see a suite of smaller, ice-rich moons of uncertain age that likely experienced tremendous dynamical upheaval (some of which are still active today), one large ice-rock world, and a number of smaller outer captured moons. Saturn's lone large moon, Titan, has a dense nitrogen-methane atmosphere, with wind-blown dunes and organic precipitation producing Earth-like river channel systems, lakes and seas, suggesting a world akin to an early Earth in hibernation in the colder regions of the outer solar system.

Uranus and Neptune present even more system diversity, evident even in brief flybys by Voyager 2 in 1986 and 1989. At Uranus, a giant impact into the planet itself may have disrupted the initial satellites, leading to the solar system's only example of a second-generation satellite system. The uranian moons are similar in size to the smaller ice-rich moons of Saturn, which range from ancient, battered relics of satellite formation to currently active ocean worlds such as Enceladus. Several moons, such as Saturn's Dione and Tethys, and Uranus's Ariel and Miranda, have been extensively fractured and resurfaced by water- and ammonia-rich lavas, betraying internal sources of heat from gravitational tides and radioactive decay. At Neptune, the capture of Triton (a likely Kuiper belt object) led to the total disruption and/or ejection of the original satellite system. Triton itself is likely an ocean world, with an active surface, complex geology of volcanic resurfacing and crustal convection, and a thin nitrogen atmosphere producing surface frosts and ices that migrate during Triton's long deep seasons. Triton appears to be a near twin of Pluto but has experienced a very different dynamic and geologic history.

Astrobiological Potentials of Giant Planet Moons

Three key ingredients required to support life on Earth are liquid water, source(s) of energy (oxidants and reductants), and core biological elements (C, H, N, O, P, and S or CHNOPS). Strong geophysical evidence exists for subsurface water oceans in the jovian satellites Europa, Ganymede and Callisto and the saturnian satellites Enceladus and Titan. Evidence from surface geology that can be interpreted as owing to a subsurface ocean also exists for Saturn's satellite Dione and Neptune's satellite Triton. Two mid-size moons of Uranus, Titania and Oberon, are large enough to be capable of harboring subsurface oceans, especially if their H_2O layers contain sufficient amounts of ammonia or other antifreeze. A third, Ariel, might have been tidally heated to the point of creating a water ocean in its interior.

Whether an icy satellite would develop and maintain an ocean depends on the amount of initial primordial heat, the rate of internal heating from radiogenic and tidal sources, the rate at which the heat escapes the satellite and the freezing point of the liquid (determined by antifreeze concentrations). The largest of the giant planet satellites such as Ganymede, Callisto, and Titan likely have sufficient remnant primordial heat and current radiogenic heating to maintain internal oceans whereas the primary source of heating in Enceladus is very likely tidal heating. Enceladus also show strong evidence of communication between the ocean and the rocky seafloor, while for Europa little is as yet known about communication between its ocean and the surface. Heating of the ocean and communication between it and the moon's surface are important respectively, for habitability and the search for life. The same source of energy that keeps oceans viable over geological time would also support chemical disequilibrium in the ocean through leaching of new materials from the mantle that can be exploited by life forms. In terms of key biological elements, both oceans of Europa and Enceladus contain salts, while the latter also contains carbon- and nitrogen-bearing molecules. The compositions of potential water oceans within the uranian moons and Triton are unknown.

Water plumes linked to the subsurface oceans have been observed on Enceladus, and tentatively on Europa, and provide the most attractive targets for in situ detection of biosignatures through plume flybys. The detection and interpretation of the biosignatures in ocean-derived materials is aided by the fact that, as highlighted in *An Astrobiology Strategy for the Search for Life in the Universe* (NASEM 2019), "slow" life that is barely able to survive in an austere environment is easier to detect because its environmental noise level is low. The direct detection of biosignatures by a lander on the surface of Europa is challenging because of the harsh surface radiation environment but the surface of Enceladus would preserve biogenic evidence over a long time and is conducive to landed astrobiological science. Expanded understanding of habitability of chemosynthetic subsurface environments, brine stability, and adaptations of life to saline fluids have widespread implications for the search for life on Enceladus, Europa, and other ocean worlds, as discussed in the next section.

Rings

Each of the giant planets is surrounded by a distinctive ring system composed of many small particles orbiting the planet. Jupiter has the most tenuous ring system, which is composed primarily of fine dust grains that were probably knocked off the planet's small inner moons. By contrast, Saturn has the most elaborate ring system, with many different components ranging from extremely tenuous rings of dust-size particles (including one generated by Enceladus's cryovolcanic activity) to the much denser and more massive main rings composed primarily of particles millimeters to meters across. Uranus also has a ring system that includes both dense and dusty components, but its rings also contain a surprisingly large number of exceptionally narrow structures, as well as an unusually blue dusty ring associated with a small moon. Last, Neptune's ring system contains multiple dusty features, along with a set of dense but incomplete ring arcs that have been slowly changing over time.

These rings provide important information about the dynamics and history of their host systems. For example, the extensive data returned by the Cassini mission enabled Saturn's rings to be used as a seismometer that records the oscillations and asymmetries in the planet's gravitational field, thereby providing new insights into Saturn's internal structure and rotation. More dramatically, a variety of measurements of both the mass of Saturn's main rings and the mass fluxes between the rings and planet made around the end of the Cassini mission suggest that the rings may be only about 100 million years old. While the age of Saturn's ring system is still being debated, the possibility that the rings are young, together with the surprisingly rapid tidal evolution of Saturn's moons has led to a reevaluation of the Saturn system's history. At the same time, theoretical investigations of the material around Uranus and Neptune have revealed that some of the solid material around these planets may have cycled back and forth between rings and moons multiple times. The differences between Uranus's and Neptune's ring-moon systems could therefore indicate that they are in different phases of this cycle.

The rings around the giant planets also provide insights into the physical processes that operate astrophysical disks, including the protoplanetary disks that gave rise to planetary systems like our own. For example, the distribution of particle sizes and structures within dense rings depend on fundamental disk processes like particle aggregation and fragmentation. Furthermore, the discovery of embedded objects within Saturn's rings has enabled direct observations of orbital migration arising from interactions with surrounding disk material. At the same time, the discovery of multiple new ring systems around small bodies orbiting among and beyond the giant planets suggests that rings can be found in a broader range of contexts than previously appreciated.

Connection to Exoplanets

The first planet detected around a normal star like the Sun was a giant planet like Jupiter, but in a very close orbit around its parent star. Such were the easiest exoplanets to detect, but with the advent of the Kepler and TESS satellites using the transit technique, we now know that these are not the most abundant planets. Uranus and Neptune mass objects (20 times the mass of Earth) are more plentiful, and bodies of 10 Earth masses and below (the sub-Neptunes and super Earths) perhaps even more so. However, planets in orbits like those of Jupiter and Saturn around the Sun remain notoriously difficult to detect. Those in closer orbits are being studied by Hubble, and soon James Webb Space Telescope, to understand what their atmospheres are made of. This makes comparison with the detailed studies of the solar system's giant planets especially valuable. Comparing the abundances of elements like carbon and oxygen in the atmospheres of giant exoplanets with those in their parent stars gives us clues to how they formed, so that we have a window into whether the way our own giant planets formed is typical of that of their sister planets elsewhere in the Milky Way galaxy. Atmospheric processes (cloud formation and spatial variability) in giant planets allow us to understand clouds in exoplanets and the sources of variable light curves.

Summary

Our knowledge of each of the giant planet systems was enabled by a suite of missions to these bodies, and both Earth orbiting and ground-based telescopes. In situ mission elements, such as the Galileo probe, provided key ground truth for the missions that came both before and after, especially by obtaining measurements not possible via remote sensing. In the past decade, Juno has provided deep insights into our remaining Jupiter knowledge gaps, such as its interior structure and composition, and will continue to provide science throughout the Jupiter system in its extended operations phase. Cassini, as a comprehensive, system-encompassing mission enabled a deeper understanding of both the individual bodies and of the key interactions within the Saturn system. The continued analysis of rich multi-instrument Cassini data informs our understanding of the interplay between the rings, satellites, planetary atmosphere, interior, and magnetosphere and their coupled evolution. While advances in understanding Uranus and Neptune following the Voyager flybys came from Earth-based observing and modeling, they are ready for orbital and in situ exploration and spectacular new discoveries in the upcoming decade.

Key Discoveries from the Past Decade

- **Jupiter and Saturn have dilute cores, not the small well-defined cores assumed by models.** Gravity data from Cassini's final orbits at Saturn, and Juno's high inclination orbits at Jupiter have revealed that the deep interior structure of both planets is not sharply defined as most models had assumed. Rather, they likely have extended envelopes enriched with heavy elements. New models of giant planet formation and evolution are needed to explain this, along with similar data for Uranus and Neptune to understand if this is common.
- **Belt-zone structure of Jupiter and Saturn goes deep, yet Jupiter has polar cyclones very different from the Saturn polar hexagon.** Findings from Juno show that Jupiter likely has deep winds, as does Saturn based on Cassini results. Lab and numerical studies show that circumpolar jets can form vortex streets or polygons depending on the vertical structure and depth of the wind field. However, as they both have deep structure, Juno's observations that Jupiter's poles have cyclones, while Saturn's does not, is surprising and may be owing to subtle differences in the local environment.
- **Saturn's ring-moon system has changed dramatically over time and the change is ongoing, unlike Jupiter's.** Cassini's long exploration of Saturn's rings revealed new ringlets, changes in dust content, vertical features, waves, and dynamical interactions with its many moons. Some of the ring moons show evidence of accretion in the form of equatorial ridges, such as on Daphnis, Pan, and Atlas.
- **Magnetic spots on Jupiter (like sunspots); Saturn's magnetic field is surprisingly symmetric.** Juno's mapping of Jupiter's magnetic field found a patch of intense strength near the equator and strong secular variation. In contrast, Cassini's mapping of Saturn's field showed strong axisymmetry, but with the magnetic equator shifted northward from the planetary equator. However, the many latitudinal variations in Saturn's magnetic field require a complex internal dynamo.
- **Changing ammonia abundance with depth and latitude on Jupiter suggests violent storms and large ammonia "mushballs" bringing ammonia deep into the atmosphere below the cloud base.** Jupiter was thought to have a well-mixed troposphere below the clouds, but Juno data revealed that ammonia varies with latitude and with depth; it is only well mixed in a narrow low latitude band. Combined with observations of lightning, it is theorized that ammonia-rich hailstones in thunderstorms carry the ammonia deeper into that atmosphere than was expected.
- **Changeable Uranus and Neptune: big outburst on Uranus in 2014; brightening of seasonal polar hood on Uranus; Neptune dark spots are frequent and short-lived (years).** Increased seasonal storms were expected around Uranus equinox in 2007, however, much larger bright cloud outbreaks were observed since then. This challenged ideas about solar insolation and convection on Uranus. Additionally, new dark spots were discovered on Neptune in 2015 and 2018, studies of the full lifecycle of large anticyclone formation. New storms occur every few years and last for 3–5 years.
- **Titan's polar seas are deep and mostly methane.** Owing to the extreme surface temperature, ~95 K, and the abundance of atmospheric hydrocarbons, Titan's seas are predominantly composed of methane and ethane. The largest of the seas was mapped by Cassini's RADAR system and found to have a depth of ~160 m.
- **Evidence of ongoing sizable impacts on Uranus and Neptune, and Saturn's rings.** The observed overabundance of CO and HCN in Uranus and Neptune, above thermochemical equilibrium levels, suggests an ongoing, and external, source; cometary and asteroid impacts are likely responsible. Additionally, Cassini's high spatial resolution images of Saturn's ring show streaks from small impacts.

Further Reading

Guillot, T., D.J. Stevenson, S.K. Atreya, S.J. Bolton, and H.N. Becker. 2020. "Storms and the Depletion of Ammonia in Jupiter: I. Microphysics of 'Mushballs.'" *Journal of Geophysical Research: Planets* 125:e2020JE006403. https://doi.org/10.1029/2020JE006403.

Morales-Juberías, R., K.M. Sayanagi, T.E. Dowling, and A.P. Ingersoll. 2011. "Emergence of Polar-Jet Polygons from Jet Instabilities in a Saturn Model." *Icarus* 211:1284–1293. https://doi.org/10.1016/j.icarus.2010.11.006.

Müller, S., R., Helled, and A. Cumming. 2020. "The Challenge of Forming a Fuzzy Core in Jupiter." *Astronomy & Astrophysics* 638:A121. https://doi.org/10.1051/0004-6361/201937376.

Spilker, L. 2019. "Cassini-Huygens' Exploration of the Saturn System: 13 Years of Discovery." *Science* 364(6445):1046–1051. https://doi.org/10.1126/science.aat3760.

OCEAN WORLDS AND DWARF PLANETS

In addition to Earth, we have identified more than 20 worlds throughout the solar system that may once have had or currently support large liquid water oceans. These so-called ocean worlds include several icy moons of Jupiter and Saturn, which harbor confirmed modern oceans, as well as icy moons of Saturn, Uranus, and Neptune and several dwarf planets, including Pluto and Ceres, in which candidate oceans may exist (Figure 2-27). As identified in the high-level recommendations of *An Astrobiology Strategy for the Search for Life in the Universe* (NASEM 2019), exploration of these ocean worlds presents an opportunity to find extant life beyond Earth and may provide natural laboratories from which we can study the prebiotic processes that led to the emergence of life on Earth, as well as chronicle the development and sustainability of habitable environments across the solar system. Over the past decade, the study of ocean worlds has also spurred the development of a new cross-cutting field: comparative oceanography. Oceans, much like planetary atmospheres, now extend to worlds beyond Earth. Studying them will lead to a better understanding of how Earth's oceans and cryosphere work and, hence, how best to protect them.

Starting with the initial discovery of hydrothermal vent ecosystems near the Galapagos Islands coinciding with Voyager 2's first images of Jupiter's moon Europa, exploration of ocean worlds has complemented advances in terrestrial ocean science that are redefining our understanding of how terrestrial life may have emerged. For example, definitive evidence for de novo abiotic synthesis of organic compounds at deep-sea hydrothermal

FIGURE 2-27 Ocean worlds and dwarf planets of the solar system, shown to scale. For the ocean moons (outlined in white), label colors indicate that they orbit Jupiter (red), Saturn (orange), Uranus (teal), or Neptune (blue). For dwarf planets, only objects with an estimated radius greater than 300 km (180 mi) are shown at approximate color and brightness; this arbitrary cutoff size is larger than ocean moon Enceladus. For "confirmed" ocean words (innermost circle), diagnostic ocean signatures have been measured. For "candidate" ocean words, indicative evidence can best be interpreted as owing to an ocean, but other interpretations have not been excluded. "Credible possibilities" may harbor an ocean because of their ice-rich composition and their likely heat inputs, but no strong evidence has been found to date. SOURCE: Data from NASA/JPL-Caltech and Southworth et al. (2019), CC BY 4.0.

vents and continental subsurface settings on Earth informs investigations of hydrothermal activity with similar characteristics within Saturn's moon Enceladus (see below for details). These parallels, along with the broad scientific questions associated with ocean world exploration, have attracted a diverse cohort of scientists that includes those not traditionally involved in solar system exploration. Terrestrial oceanographers, cryosphere scientists, and microbiologists are working alongside planetary scientists and engineers to define the next generation of ocean world exploration missions. The ocean worlds themselves are also extremely diverse, with significant differences in ice shell thickness, geologic history, and surface characteristics that provide a plethora of environments to explore.

In the past decade, the response to discoveries made in ocean world exploration have not only excited the scientific community but have also garnered significant public interest. The New Horizons encounter with Pluto, for example, was mentioned on the front pages of 450 newspapers around the world; on the day of the flyby itself, the mission earned 1.7 billion mentions across 21 social media platforms and reached 144 million people through Facebook alone. Cassini's Grand Finale similarly reached 1.7 billion mentions in social media and was the topic of 791 major media articles. The Jet Propulsion Laboratory won an Emmy award for Outstanding Original Interactive Program for its coverage of the Grand Finale.

The next decade of solar system exploration will be even more exciting and could be the decade in which life beyond Earth is detected (see the Future of Ocean World and Dwarf Planet Exploration section below). The discoveries of the past decade have identified ocean worlds that could be habitable today, and advances in technology have provided the tools to search for evidence of life within these environments. In the following subsections, the committee describes major discoveries made in the past decade and list outstanding questions in planetary sciences and astrobiology associated with ocean moons of the giant planets and dwarf planets found in both the inner and outer solar system. This section ends with a discussion of the future of ocean world and dwarf planet exploration, including major discoveries that could occur over the course of the next decade.

Ocean Moons

Among the confirmed ocean worlds are the jovian satellites Europa and Ganymede, as well as the saturnian satellites Enceladus and Titan. Candidate ocean worlds, where ocean presence is not confirmed but available evidence can best be interpreted as owing to an ocean, include Jupiter's satellite Callisto, Saturn's satellite Dione, and Neptune's satellite Triton. Smaller natural satellites of Saturn—including Mimas, Tethys, Rhea, and Iapetus, as well as the natural satellites of Uranus (Miranda, Ariel, Umbriel, Titania, and Oberon)—are ocean worlds candidates whose composition and proposed heat budget are consistent with the possible presence of an ocean, but no strong evidence has been found to date in their limited observations. These ocean moons represent a diverse set of targets that express a wide variability in ocean depth and seafloor pressures, geologic history, accessibility (e.g., plumes of liquid water erupting from the surface), overlying ice layer thickness, presence or absence of high-pressure ice layers, presence or absence of atmospheres, and surface environments (e.g., Europa's irradiated icy surface versus Titan's hydrocarbon-based sedimentary surface). Of these bodies, the oceans of Europa, Enceladus, and Titan are the best documented and have emerged as key targets in the search for an independent emergence of life and the study of prebiotic and possibly biotic processes across the solar system.

While Voyagers 1 and 2 collectively investigated all the giant planet systems between 1979 and 1989, the acquired data from these flybys were insufficient to determine the presence or absence of subsurface oceans. The subsurface oceans of Europa, Ganymede, and possibly Callisto were discovered via magnetic induction by the Galileo spacecraft, which orbited Jupiter from 1995 to 2003. The oceans of Enceladus and Titan were identified and characterized by the Cassini spacecraft, which orbited Saturn from 2004 to 2017. Following on from the discoveries from Galileo, the Europa Clipper spacecraft will perform multiple flybys of Europa in the early 2030s (launching in the 2024). Dragonfly, selected as NASA's fourth New Frontiers mission, will explore the equatorial region of Saturn's moon Titan with a rotorcraft drone in the mid- to late 2030s (scheduled to launch in 2027). ESA's L-class Jupiter Icy Moons Explorer (JUICE) mission, currently slated to launch in the early to 2023, will fly by Europa and Callisto before entering orbit around Ganymede in the 2030s. Collectively, these upcoming missions are designed to characterize the habitability of these confirmed ocean moons. Follow-on missions, which are actively being designed and proposed today, will address the question of whether they are in fact inhabited (see below).

Europa

Prior to the previous decadal survey, *Visions and Voyages* (NRC 2011), the Voyager and Galileo missions revealed that Europa has a global liquid water ocean that is sandwiched between a dynamic ice shell and a silicate core (Figure 2-28a). The outer surface of water ice was shown to be crisscrossed with long linear features and marked by "chaos" terrain where the surface ice has been disrupted, broken, and in some cases, rotated and frozen into new positions. The presence of non-ice components was determined spectrally and suggests that salt-rich material originating in the ocean may have been emplaced on the surface, with these substances subsequently processed by Jupiter's strong radiation. Much less is known about the interior, although the ocean is expected to be saline and has the potential for active material exchanges at the ice-ocean and ocean-silicate mantle boundaries.

Our understanding of such exchange processes represents a major development over the past 10 years and is significant because the energy needed to power life is potentially sustained through dynamic water-rock interactions on the seafloor coupled with radiolytically produced oxidants created on the surface and cycled into the ocean below (Figure 2-28b). Apparent detection of subduction on Europa has been used to argue that this satellite is the second known body in the solar system to exhibit plate tectonics-like behavior. If confirmed by future exploration, subduction would provide a mechanism by which surface materials may be transported downward through the ice shell. Conversely, evidence for plumes of water that sporadically erupt from the surface have been detected, tentatively, by multiple, independent observations (Figure 2-28d,e). Perhaps relatedly, pockets of liquid water, from which the plume material may be sourced, may be perched within the ice shell above the subsurface ocean. Ground-based observations show that chaos terrains are associated with high abundances of NaCl, or table salt, suggesting the expression of saline liquids (Figure 2-28c). If NaCl is the dominant salt in Europa's ocean (as on Earth) then it, too, may have been subject to extensive high-temperature seafloor water–rock interactions.

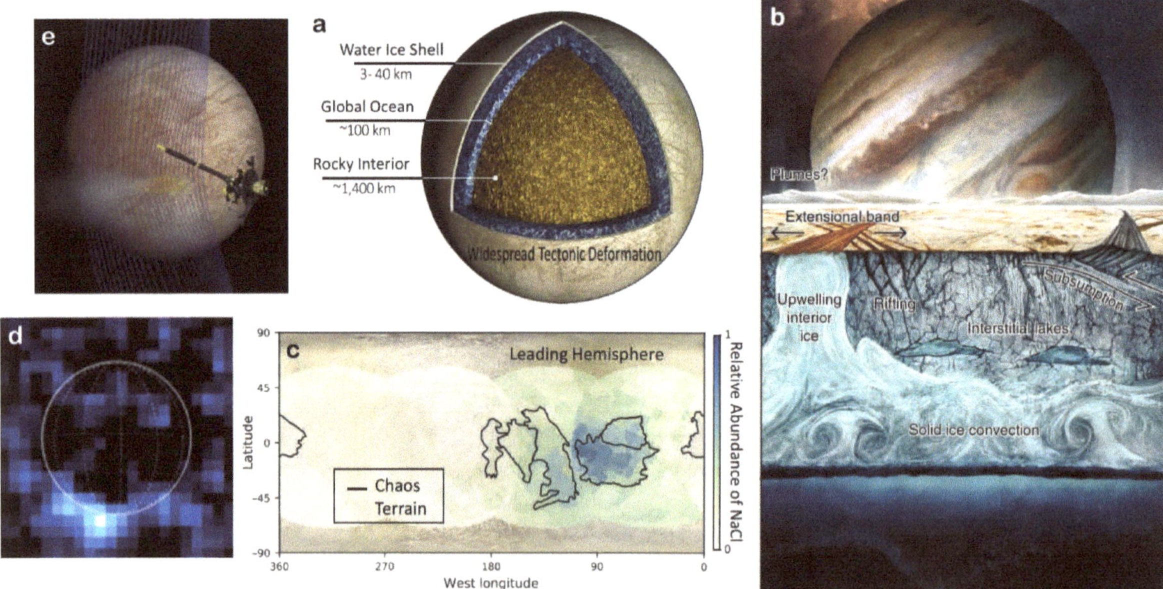

FIGURE 2-28 Europa from the interior to the near-space environment: (a) Europa's interior structure is shown, including the scales of the ice shell, global ocean, and rocky interior, and (b) the likely tectonic and dynamic ice shell processes and associated surface expressions. (c) High relative abundances of NaCl found on the leading hemisphere surface appear correlated with the dominant chaos terrains outlined in black. (d) Plume activity was observed from the excess ultraviolet emission of Lyman-α by the Hubble Space Telescope, and (e) confirmed by modeling the local perturbation of the jovian magnetic field observed by the Galileo spacecraft. SOURCES: (a and b) Howell and Pappalardo (2020), CC BY 4.0; (c) Trumbo et al. (2019), CC BY-NC 4.0, with slight modifications (labels and axes); (d) From L. Roth, J. Saur, K.D. Retherford, D.F. Strobel, et al., 2014, "Transient Water Vapor at Europa's South Pole," *Science* 343(6167):171–174, https://doi.org/10.1126/science.1247051; reprinted with permission from AAAS; (e) NASA/JPL-Caltech/University of Michigan.

In the coming decade, the Europa Clipper mission will begin its exploration of the Jupiter system, conducting dozens of close flybys of Europa (closest approach of each encounter is 35–100 km), with the goal of assessing the moon's habitability and addressing questions such as: How does physical and chemical oceanography affect Europa's past and present state (e.g., ocean thickness and geochemical exchange between the ocean, ice shell, and seafloor)? How does Europa evolve, both internally and on its surface? Tectonism, subduction, and convection are just a few of the wide-scale processes for which Europa offers a second type example, helping to inform not just how Europa works, but also how these processes work on Earth. Do the plumes originate in Europa's ocean, or from water pockets in the icy shell? Beyond plume eruptions, what other modes of surface–interior exchange are most common on Europa? Perhaps the grandest question one can ask, however, is: Does Europa harbor life? This question will likely be the objective of the next mission, beyond Europa Clipper, which may involve landing on the surface and/or directly sampling Europa materials to search for evidence of life. If Europa does harbor life, what regulates its habitability and what biochemistry does it utilize? If Europa does not harbor life, but a habitable ocean exists, what are the limits on the emergence of life itself?

Enceladus

Thanks largely to the Cassini mission, the past decade has revealed Saturn's geologically active moon Enceladus to be a habitable world that contains significant liquid water, energy to sustain metabolism, and conditions favorable for the assembly of complex organic molecules. Much of this evidence has been supplied via analysis of material from a "cryovolcanic" plume of icy particles and gas that erupts from fractures in Enceladus's South Polar Terrain (Figure 2-29).

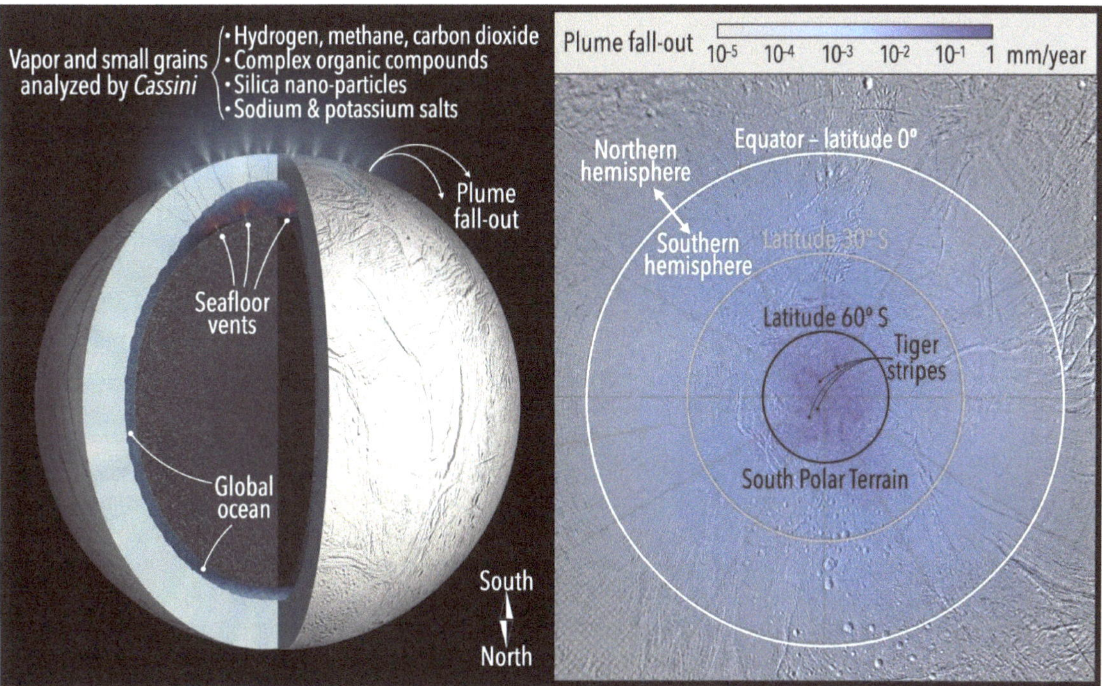

FIGURE 2-29 *Left*: New discoveries in the past decade have established that the ice-shell at Enceladus overlies a global-scale ocean that contains complex organic molecules and hosts submarine venting. Analyses of materials ejected through fractures in its South Polar Terrain (SPT) suggest that vents on Enceladus share key characteristics with newly discovered vents on Earth: vents that can both synthesize simple organic molecules independent of life and sustain the metabolism of some of Earth's most primitive organisms. More recent data suggest a thicker ocean and thinner shell than depicted here. *Right*: Plume fall-out from the SPT can be detected across much of Enceladus's southern hemisphere. SOURCES: *Left*: Courtesy of NASA/JPL-Caltech. *Right*: Data from Southworth et al. (2019), CC BY 4.0.

The plume's source within Enceladus is a global subsurface ocean, as established by Cassini in 2014 from analyses of Enceladus's gravity, topography, and wobble as it rotates. The ocean is deduced to be about 40 km (25 mi) deep, or 10 times deeper than Earth's oceans, although the reduced gravity leads to seafloor pressures that are equivalent to depths of ~1,000 m on Earth. The chemical compositions of icy particles and gas in the plume provide a "free" sample for analysis that can reveal the nature of the underlying ocean, and its potential to sustain life. Tiny silica grains emanating from the plume point to ongoing high-temperature water–rock reactions made possible by tidal heating below the ocean floor. Sodium and potassium salts, an alkaline pH, and a low-density rocky core provide further evidence for water–rock interactions at depth. Importantly, hydrogen gas, methane, and carbon dioxide were also found in the plume—similar to newly discovered hydrothermal systems and subsurface fluids on Earth. Terran systems have recently been demonstrated to sustain the spontaneous synthesis of simple organic compounds, even in the absence of life, and to support some of Earth's most primitive forms of microbial metabolism. If Enceladus's ocean and hydrothermal activity are sufficiently long-lived (current estimates for the age of Enceladus range from 100 million to 4.5 billion years), life could potentially have gained a foothold there.

Cassini showed that the Enceladus plume also contains a diversity of organic molecules spanning a range of sizes and containing, in addition to carbon, hydrogen, and oxygen, the bio-essential element nitrogen. Owing to high-speed collisions between plume particles and the Cassini spacecraft that break down organic matter during collection, however, what were detected are likely to be fragments of even larger molecules. These molecules may have been produced by living organisms, or by abiotic processes; ascertaining their nature and source is a major unanswered question that awaits future exploration.

In general terms, we now know that Enceladus is habitable. But is it inhabited? Answering this civilization-scale question requires a new mission. Modeling from Cassini data indicates that the plume feeds Saturn's E-ring and, thus, is likely to be long-lived. This is important because it would ensure that future missions to Enceladus can investigate subsurface processes by accessing expelled ocean material without the need to dig or drill into or beneath the ice-shell in the decade(s) ahead.

Titan

Titan is unique among the ocean moons: it has two ocean realms (Figure 2-30). The first, like its ocean world siblings, is a liquid water ocean that lies beneath a water-ice shell. The second consists of liquid hydrocarbon (natural gas) lakes and seas that sit on a surface, shrouded by a dense, hazy atmosphere of nitrogen and methane. A decade ago, we knew that this atmosphere drove Titan to be a dynamic world shaped by Earth-like processes and populated by familiar features including dunes, rivers, lakes, and other earthly landscapes. Within this atmosphere, chemical reactions produce a plethora of organic compounds, some as large as terrestrial proteins. Methane rains onto and carves channels into Titan's organic- and water-ice-rich landscape, eventually pooling in lakes and seas. The presence of large equatorial dunes demonstrated that, in addition to rain, wind further modifies and transports this organic material. Modeling showed that impacts may support ephemeral liquid-water environments while geologic evidence hinted at cryovolcanism. Laboratory work had shown that if Titan's organics were to mix with this melt water, amino acids (some of the building blocks of life) would be produced within days.

In the past decade, research has confirmed Titan's global subsurface water ocean. While high-pressure ice may line the ocean floor (see Figure 2-30), such a layer does not prevent exchange (although it does slow) between the ocean and rocky core (a key to its habitability). Titan's icy crust is thought to be convecting, and the temperature and pressure at the top of the convection zone are similar to those within terrestrial deep glacial ice, which hosts diverse microbial life between ice grains. The chemical complexity of Titan's atmospheric constituents is now known to increase in polar winter and with decreasing altitude. Our current list of ~20 atmospheric compounds represent only the tip of the organic factory iceberg. The depth $(100-300+$ m) and major components (methane and ethane) of the hydrocarbon seas are now known and theoretical investigations are exploring if they might support spontaneous assembly of cell membrane-like structures and other elements that could enable a non-water-based alien biology.

These discoveries make Titan a key target for exploration, and many mysteries remain. The processes that create complex species in Titan's atmosphere are not well understood and operate without, presumably, biological

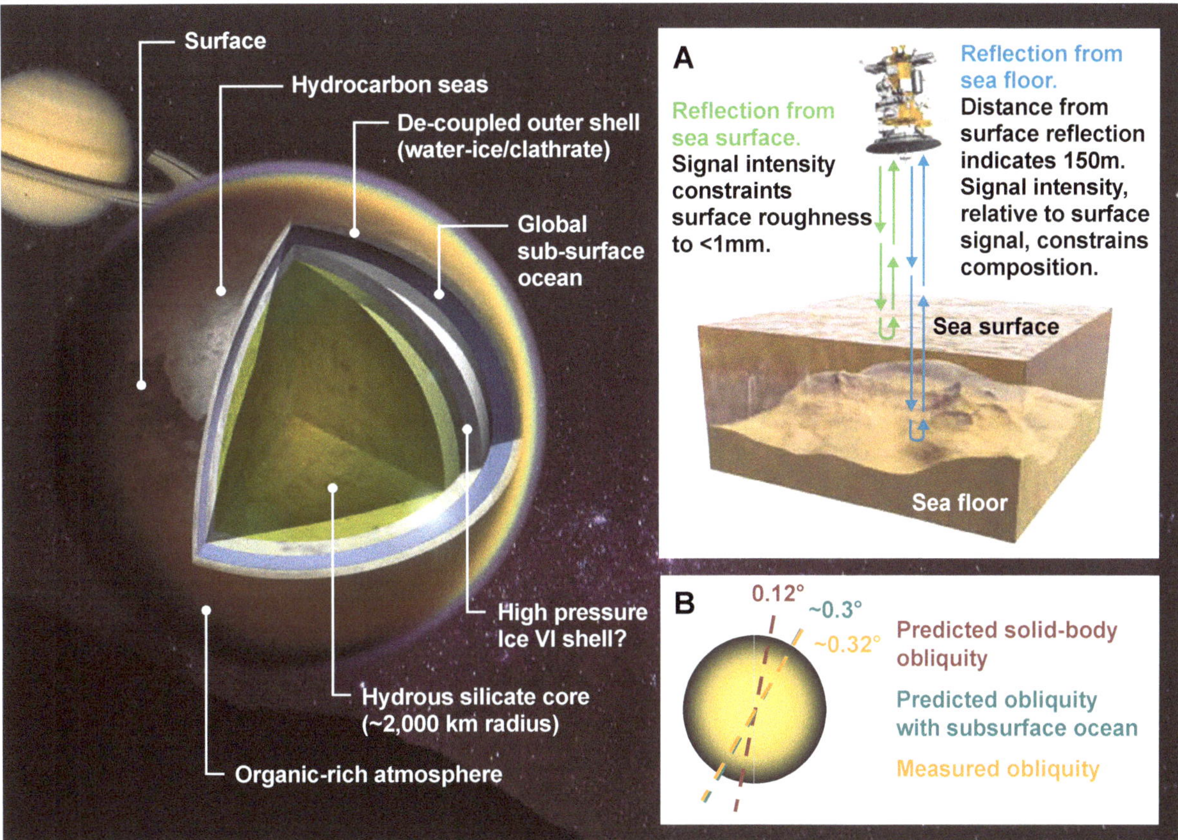

FIGURE 2-30 Titan is a world of two oceans; a subsurface water ocean that lies below a water ice crust and surface lakes and seas of liquid hydrocarbon that lie below a dense nitrogen-based atmosphere. (A) Cassini measured the depth (~200 m) and bulk composition (liquid methane and ethane) of Titan's seas using its RADAR as a depth sounder. (B) Titan's 0.32-degree obliquity (tilt between its equator and orbital plane) is consistent with the expected obliquity in the presence of a global subsurface ocean (~0.3 degree) and inconsistent with the obliquity expected in the absence of a subsurface ocean (0.12 degree). SOURCES: *Left, Saturn*: NASA; *Titan's internal structure*: NASA/JPL-Caltech/A.D. FORTES/UCL/STFC. (A) R. Kelly/*Astronomy* magazine. (B) F. Nimmo and A. Hayes.

catalysts like those responsible for large molecules on Earth. Titan's exploration thus offers fundamental insights into the chemistry that may have preceded and facilitated the rise of biochemistry on early Earth. The Dragonfly mission, which will explore Titan's equatorial regions in the mid-2030s, will be essential to this effort by providing our first detailed understanding of the surface composition. New isotopic measurements of noble gases and methane would resolve key questions concerning the ocean composition, the evolution of the interior and atmosphere, and the formation of Titan, including the age of Titan's atmosphere and how it mysteriously remains methane-rich. Determining if Titan's ocean is interacting with its rocky core would provide a key constraint on the habitability of large ocean worlds both within and beyond the solar system. Global high-resolution imaging and topography would allow us to use Titan's surface and climate system as a natural laboratory; for instance, to study how planetary-scale hydrologic cycles control the physical and chemical evolution of a landscape in an environment akin to, but less complex than, Earth's. Similarly, detailed observations of the composition, physical conditions, and seasonal evolution of Titan's polar lakes and seas would allow us to constrain their role in the hydrologic cycle and climate, and perhaps even their potential habitability.

Other Bodies

The identification of ocean worlds as a new class of planetary bodies brings a new perspective on the diversity of worlds across the solar system. Reexamination of pre-Cassini mission era data from the Galileo and Voyager 2 spacecraft has led to new discoveries at Jupiter's large moons Ganymede and Callisto, the uranian moons, and Neptune's largest moon Triton. Results from Galileo demonstrated the likelihood of oceans beneath the surfaces of Ganymede and Callisto, but it was not until the discovery of a habitable ocean at Enceladus during the Cassini mission that moons of the uranian system and at Neptune became widely recognized as potential ocean worlds.

Ganymede and Callisto

Initial studies of Ganymede's ocean suggested that a layer of high-pressure ice, likely sandwiched between the rocky core and subsurface ocean, would significantly limit water–rock interaction, and hence limit the ocean's habitability potential. Recent modeling work, however, suggests that not only is exchange through a high-pressure ice layer possible, but that Ganymede's interior may instead consist of alternating layers of high-pressure ices and salty oceans that would permit direct water-rock interactions. Since Galileo, observations with the Hubble Space Telescope have further supported the presence of an ocean by documenting the dynamics of Ganymede's aurora. Magnetic induction measurements made by Galileo also indicated that Callisto may have a subsurface ocean, but recent analysis has suggested that some or even all of the induction signal could have instead come from the ionosphere, as opposed to a salty subsurface ocean. At Ganymede, these induction measurements showed that it is the only known satellite with an active magnetic field generated in its metallic core. The Juno mission will study Ganymede's magnetosphere and its interaction with the jovian magnetosphere. In the early to mid-2030s, ESA's JUICE mission will reassess the presence or absence of an ocean within Callisto during multiple close flybys before entering orbit around Ganymede to determine the latter's interior structure and dynamics, map its surface, investigate its tenuous atmosphere, and characterize its intrinsic magnetic field.

Other Saturnian Moons

In addition to Titan and Enceladus, Saturn hosts five moons of intermediate or comparable size: Mimas, Tethys, Dione, Rhea, and Iapetus. In the past decade, the Cassini mission did not detect ongoing geological activity at these moons, although it found indications on Tethys and Dione of past activity (e.g., fractures) and past warm interiors. It also revealed that the interiors of Mimas and Dione comprise a rocky core and ice shell, with possible evidence for an ocean in between arising from their wobble (for Mimas) or gravity and shape (for Dione); however, ocean-free interpretations remain at least as likely. The interiors of Tethys and Iapetus are unconstrained, but likely contain little rock because their densities are similar to that of water ice. In the past decade, the moons' ages have become debated. Their surfaces, covered with impact craters, seem to be billions of years old. While the moons' rapid orbital expansion could indicate recent formation (e.g., from a collision disrupting an older moon), this expansion may instead be a consequence of how tidal energy is dissipated inside Saturn, permitting the moons to be as old as the solar system. The moons' ages thus remain uncertain, which bears on their potential to harbor oceans and on their habitability.

Uranian Moons

Despite a dearth of data, most dating back to Voyager 2, the uranian moons Miranda, Ariel, Umbriel, Titania, and Oberon are considered credible ocean world possibilities for four reasons. First, some moons show possible evidence for cryovolcanism, perhaps sourced from an ocean. Second, some moons show evidence for ammonia, a powerful antifreeze able to sustain liquid water down to 173 K. Third, both Miranda and Ariel are tectonically deformed, with indications of high heat flows at the time of deformation. Last, all five moons are either larger than the confirmed ocean world Enceladus or comparable in size. While only Titania and Oberon would be large enough for oceans to persist solely through radioactive heating, it is likely that all these satellites were heated by tides in the past.

Triton

Triton is a Kuiper belt object that was captured into Neptune's orbit. Voyager images showed a young surface with relatively few craters and discovered active plumes that stand out among the icy satellites of the outer solar system and put Triton in a class with Io, Europa, Enceladus, and Titan—moons with geological processes active today. While error bars on the absolute crater model ages for Triton are large and the origin of the plumes remains unknown, they do tell the story of a young, dynamic surface heavily modified by internal geologic processes. Theoretical modeling of Triton's orbital evolution after capture and tides associated with its tilt (obliquity) suggest that tidal energy could maintain a subsurface ocean to the present-day, adding Triton to the family of candidate ocean words. Triton's unique surface features suggest cryovolcanic processes, which offer the prospect of studying surface–ocean exchange. Earth-based observations show the presence of H_2O and CO_2, which are presumed to form the surface bedrock. Ethane has also been tentatively detected, and the volatile ices N_2, CO, and CH_4 are also present. Recent comparisons to the Enceladus plume suggest that Triton's plumes could be driven by internal processes rather than solar heating. Near-surface mixtures of ammonia and water can freeze and leave ammonia-rich ice near Triton's surface whose antifreeze properties may help facilitate cryovolcanism. These observations show that Triton is both exotic and complex, with an array of unique surface features that, when explored in detail, could indicate ongoing exchange between the surface and a subsurface ocean.

Dwarf Planets

Because the current definition of a dwarf planet includes a requirement on its shape, the number of objects known to be dwarf planets is quite small. The large asteroid Ceres is the only dwarf planet in the inner solar system. The Kuiper belt contains Pluto, the archetypal example of a dwarf planet, and Haumea and Makemake, comparable in size to Pluto's moon Charon. Further out is Eris, the same size as Pluto. Various other candidate dwarf planets exist in the trans-neptunian region.

So far, only Ceres and Pluto have been investigated by spacecraft. Ceres was orbited by the Dawn spacecraft from 2015 to 2018, and New Horizons flew by Pluto in 2015. As a result, these two dwarf planets have been studied intensively, and both are now also regarded as candidate ocean worlds (see below). Other dwarf planets of similar size are credible ocean world possibilities, but there is currently no observational evidence to support this.

Ceres

A decade ago, Ceres—a dwarf planet and the largest object in the main asteroid belt—was but a blurry disk in images from Hubble and other large telescopes. Over the past decade, its exploration by NASA's Dawn mission has revealed it as a potentially once-habitable candidate ocean world and the most water-rich body in the inner solar system after Earth. Ceres has had sufficient water and radioactive heat to potentially host a deep ocean throughout most of its history, leading to a layered interior with the opportunity for extensive water–rock interaction.

The Dawn mission also revealed recent and even ongoing geological activity on Ceres (Figure 2-31), the presence of liquid water below its ice-rich crust, organic matter (locally) and carbon (globally), and the presence of an exosphere and volatile transport. Material, perhaps erupted from a gradually freezing subsurface layer of salty water and mud, can be found as "bright spots" or "faculae" at multiple locations These observations demonstrate that Ceres has harbored liquid water that has driven, and may still be driving, geologic activity, contains organic compounds, and once experienced water-rock interactions akin to those occurring in hydrothermal environments on Earth.

Ceres shares many similarities with the confirmed ocean worlds Europa and Enceladus, including composition and geologically recent extrusions of salt-rich fluids onto its surface. Unlike these moons, however, Ceres lacks tidal heating so water will have frozen more rapidly. Furthermore, Ceres may represent a surviving member of the protoplanets that transported water and organic material into the inner solar system, including Earth. This makes Ceres a prime target for studying the emergence, evolution, and longevity of ocean world habitability (including Earth).

Despite the success of Dawn, many unanswered questions remain. Is Ceres currently habitable? If not, was it habitable in the past? If so, how long ago was it habitable and how long did habitable conditions last?

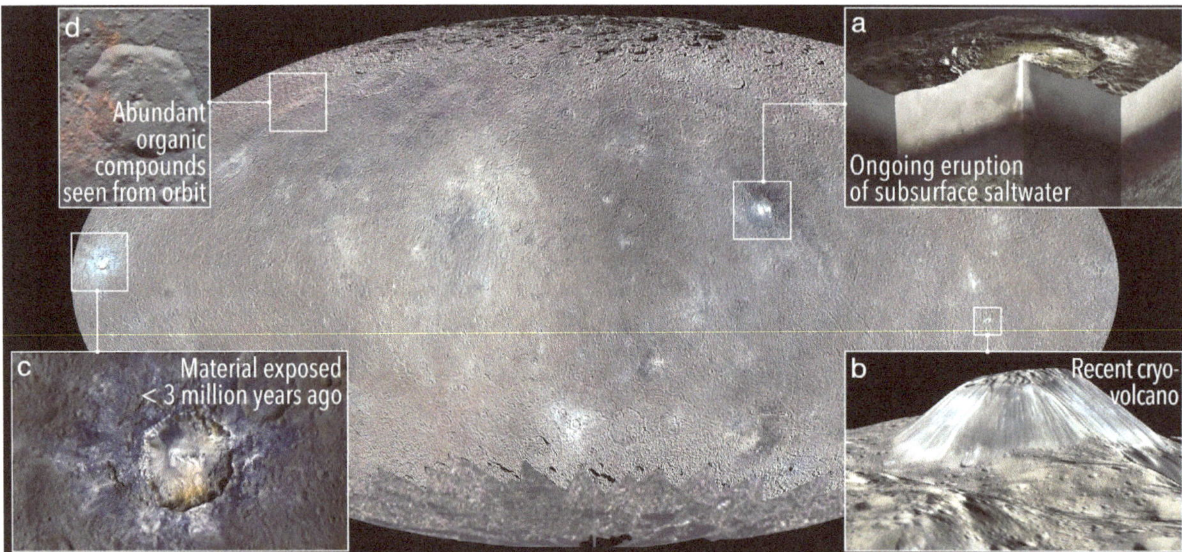

FIGURE 2-31 NASA's Dawn mission revealed Ceres as a carbon-rich world shaped by processes involving liquid water. Against Ceres's relatively uniform surface, which is blanketed by minerals formed through the action of liquid water early in its history, a few landmarks stand out. *Clockwise from top right*: (a) The 92-km-wide Occator crater is the site of multiple styles of eruption of saltwater sourced from the subsurface (artist rendition). (b) The 20-km-wide, 4-km-tall Ahuna Mons, Ceres's best preserved cryovolcanic edifice, has streaks of salt on its sides (perspective reconstruction; elevations exaggerated by a factor of two). (c) The 34-km-wide Haulani crater, another site of newly exposed material. (d) Organic material (red in this false-color image) seen from orbit at the 52-km-wide Ernutet crater. SOURCES: (a) Courtesy of JPL-Caltech; remaining images courtesy of NASA/JPL-Caltech/UCLA/MPS/DLR/IDA.

How extensive are the liquid reservoirs within Ceres today? What does Ceres reveal about the habitability, over time, of ice-rich bodies and of their ability to generate habitable environments? What processes have driven recent geological activity? Did Ceres form at its current location, further out among the giant planets, or did it migrate inward from the outer reaches of the solar system? What does Ceres's origin reveal about the potential for planets to migrate on a large-scale? Answering these questions will require follow-on missions to Ceres, including eventual sample return.

Pluto

At the time of the past decadal survey (NRC 2011), almost nothing was known about Pluto. Its mass and approximate radius were known, as was the existence of one large moon (Charon) and at least two small ones. Its density indicated it was made up of ice and rock. But it was still a point of light in a telescope, more the province of astronomers than geologists. In 2015, NASA's New Horizons spacecraft flew through the Pluto system, revealing an unexpectedly complex and dynamic world (Figure 2-32).

Pluto has active glaciers of nitrogen ice, which carve their way down to fill a plain called Sputnik Planitia. Sputnik Planitia's solid nitrogen appears to be convecting like a pot of oatmeal on the stove, resulting in a very young surface age. In the slow course of Pluto's seasons (248-year orbital period), this nitrogen is thought to vaporize and freeze back on the mountaintops, in a manner analogous to Earth's hydrological cycle. Other materials, like methane, also appear to freeze out only on the peaks of Pluto's highlands.

Pluto's atmosphere is made mostly of nitrogen and methane. New Horizons discovered prominent atmospheric haze layers (see Figure 2-32), made of small particles created by irradiation of the gas molecules. These particles eventually fall out to create the dark red material seen on the surface. The hazes themselves may make Pluto's atmosphere much colder than was expected.

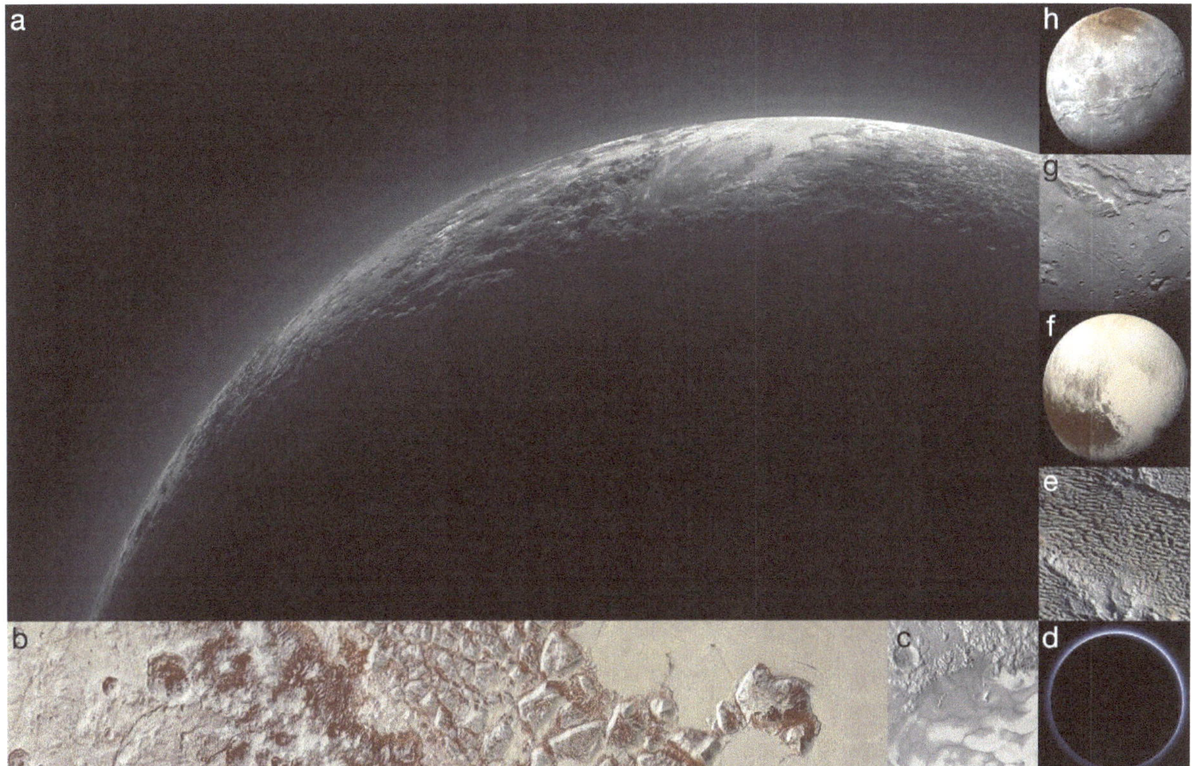

FIGURE 2-32 Pluto and Charon through the eyes of New Horizons. *From top left, counterclockwise*: (a) Pluto's 1,500 km-wide horizon of mountains (up to 3,500 m high), possible cryovolcanoes, frozen plains, and many layers of foggy hazes backlit by the setting Sun. (b) Blocks of water ice (*center*) detach from ancient crust (*left*) into the smooth expanse of Pluto's Sputnik Planitia (*right*); enhanced-color image is 500 km wide. (c) Glaciers of nitrogen ice flow into Sputnik Planitia (image is 100 km wide). (d) Pluto's atmosphere is blue, like Earth's. (e) Bladed terrain on Pluto (image 100 km wide). (f) Pluto in natural colors, with the bright frozen heart of Sputnik Planitia contrasting with a dark equatorial band of fallen haze particles. (g) Smooth plain on Pluto's moon Charon, likely the result of a massive eruption of water onto its surface as its ocean froze billions of years ago (image 200 km wide); near the top is one of the canyons forming a belt that appears to circle Charon. (h) Charon's north pole appears red in this enhanced-color image, likely owing to methane that escaped from Pluto's atmosphere, condensed, and became irradiated. SOURCES: (g) Courtesy of NASA/JHUAPLSwRI/A. Parker; remaining images courtesy of NASA/JHUAPL/SwRI.

In addition to impact craters, Pluto's surface is scarred by large fractures, some of which appear to be quite young. These fractures suggest that the crust is expanding, as would occur if a subsurface ocean has been slowly freezing. They thus also suggest an interior in which the ice and rock have completely separated. Two large, mysterious mountains might be cryovolcanoes built from piled-up, now refrozen water magma, and there are hints of cryovolcanism at some of the young fractures. Enigmatic branching channels might be evidence for ancient glacial activity. Other terrain displays thin blades of hardened snow or ice. Such penitentes are found on Earth, but Pluto's penitentes are 500 times taller.

Theoretical arguments showed that Pluto could retain a subsurface ocean beneath its ice shell. More subtly, Sputnik Planitia's location opposite to Charon could be explained if a subsurface ocean was present. Although not definitive, these arguments are sufficient to classify Pluto as a candidate ocean world. Despite its achievements, New Horizons imaged only half of Pluto and Charon at useful resolutions; the other halves remain terra incognita.

Other Bodies

Many Pluto-scale worlds orbit the Sun beyond Neptune, including all known dwarf planets except Ceres (see Figure 2-32). What little is known about these distant worlds, and the unexpected complexity discovered up close at Pluto, makes them intriguing targets for exploration. Telescopic observations indicate a striking diversity in their basic properties including the existence and relative size of moon(s) or rings, density, brightness, color, and surface composition. Many have densities similar to Pluto, implying a roughly half-rock, half-ice makeup. This significant rock content implies that the decay of radioactive isotopes contained in the rock can heat these bodies, drive geological activity, and, perhaps, generate and maintain subsurface liquid water oceans.

The largest dwarf planets contain ices composed of methane or nitrogen that, at these worlds' frigid surfaces, can condense, vaporize, and flow, enabling processes analogous to those that have sculpted the wide variety of landscapes observed on Pluto. Orcus and possibly Quaoar's surfaces contain ammonia; this potent antifreeze may enable subsurface liquid water, both in the past and perhaps even today. Sedna and Gonggong's red surfaces suggest the presence of complex organic molecules, while the bright surfaces of Eris, Haumea, and Makemake appear refreshed, potentially by ongoing glacial processes as on Pluto. Haumea is unique, with an elongated football shape owing to its rapid rotation (4 hours), a nearby family of small, similarly icy objects on related orbits, and rings. All of these features indicate a past collision that ejected fragments of Haumea and spun it up. Haumea's rings, in particular, are key to understanding how rings can form and evolve around solid bodies. As was the case at Pluto, the diversity of compositions and shapes found among the dwarf planets hints at a trove of unexpected discoveries that await exploration. Such exploration, as pioneered by New Horizons, is essential to understand how these worlds formed and evolved, and whether they (as well as other yet-to-be-discovered dwarf planets) represent habitable oases in the outermost reaches of the solar system.

Future of Ocean World and Dwarf Planet Exploration

The initial ocean world discoveries made by the Voyager, Galileo, and Cassini missions motivated *Vision and Voyages* to recommend a Europa mission to confirm the presence of a subsurface ocean and take the first steps in understanding the potential of the outer solar system as an abode for life. This recommendation is being implemented as the Europa Clipper mission, which will enter orbit around Jupiter in the early 2030s. Europa Clipper will explore the jovian system alongside ESA's JUICE, which will perform flybys of Europa and Callisto before entering orbit around Ganymede in the mid-2030s. *Vision and Voyages* also recommended a Uranus orbiter and probe, which would provide the first detailed investigation of an ice giant system, and, in the event of an optimistic budget environment, an Enceladus orbiter mission. While these last two missions have yet to be implemented, the discoveries made over the past decade have done nothing but strengthen the recommendations put forth by *Vision and Voyages*. Around midway through the past decade, discoveries made during Cassini's second extended mission led to the inclusion of ocean worlds (Enceladus and/or Titan) as a mission theme to the New Frontiers 4 target list. This addition resulted in the selection of Dragonfly, which will explore Titan's equatorial terrains using a drone quadcopter in the mid-2030s.

Over the next decade, we will move toward the next stage in the astrobiological investigation of the outer solar system and continue to investigate the physical and chemical processes that shape the ocean worlds. This will include developing technology and designing missions that will directly search for evidence of life at Europa and/or Enceladus. Exploration of the ice giant systems, Uranus and/or Neptune, remains a priority and would provide an opportunity to confirm the presence or absence of subsurface oceans on the larger uranian satellites and Triton. A Titan orbiter would provide critical context to extend Dragonfly's regional exploration to a global scale, while a lake probe would provide in situ exploration of an environment inaccessible to Dragonfly and directly assess the potential habitability of the liquid hydrocarbon seas. Returning to Ceres, perhaps even to bring a sample back to Earth, would permit investigation of both its habitability, as well as the generation and evolution of habitable environments within ice-rich bodies in general. As highlighted in *An Astrobiology Strategy for the Search for Life in the Universe* (NASEM 2019), the ocean worlds represent a diverse and rich set of targets for exploration over the next several decades. Enough, in fact, to sustain a multi-decade program of ocean world exploration, although much can still be accomplished through directed flagships and competed missions within the Discovery and New Frontiers programs.

Key Discoveries from the Past Decade

- **Intermittent plumes on Europa.** Although Europa has a young surface, space telescope observations suggesting intermittent plumes of water vapor jetting into space were a complete surprise. Any such plumes would allow us to directly sample the subsurface ocean and will be a major focus for Europa Clipper.
- **A habitable ocean at Enceladus.** At the time of the last decadal survey, Enceladus was known to harbor subsurface water, but its distribution was unknown. Geodetic observations confirmed that the liquid layer was global in extent (i.e., an ocean) and in direct contact with the rock beneath, making it a potentially habitable environment.
- **Titan is a world of two oceans.** The existence on Titan's surface of methane seas and lakes was known at the last decadal survey, but subsequent geodetic observations also pointed strongly to a subsurface, liquid water ocean, sandwiched between two layers of ice. The combination of surface seas and a subsurface ocean makes Titan unique.
- **Brine deposits on Ceres.** An unexpected result of the Dawn mission was the detection of surface brine deposits, in some cases associated with recent cryovolcanism. This discovery suggests that water–rock interactions persisted throughout much of Ceres's history, increasing its astrobiological potential.
- **Pluto's unexpected surface diversity.** Pluto hosts solid nitrogen glaciers, channels apparently carved by liquid, an actively convecting ice sheet that is unique in the solar system and several possible cryovolcanoes. Almost none of these features were expected; Pluto demonstrates that even very distant objects can be geologically active and potentially host subsurface oceans.

Further Reading

Hand, K.P., C. Sotin, A.G. Hayes, A. Coustenis. 2020. "On the Habitability and Future Exploration of Ocean Worlds." *Space Science Reviews* 216:95. https://doi.org/10.1007/s11214-020-00713-7.

Hayes, A.G., R.D. Lorenz, and J.I. Lunine. 2018. "A Post-Cassini View of Titan's Methane-Based Hydrologic Cycle." *Nature Geoscience* 11:306–313. https://doi.org/10.1038/s41561-018-0103-y.

Lunine, J.I. 2017. "Ocean Worlds Exploration." *Acta Astronautica* 131:123–130. https://doi.org/10.1016/j.actaastro.2016.11.017.

Nimmo, F., and R.T. Pappalardo. 2016. "Ocean Worlds in the Outer Solar System." *Journal of Geophysical Research: Planets* 121:1378–1399. https://doi.org/10.1002/2016JE005081.

Stern, S.A., W.M. Grundy, W.B. McKinnon, H.A. Weaver, and L.A. Young. 2018. "The Pluto System after New Horizons." *Annual Review of Astronomy and Astrophysics* 56(1):357–392. https://doi.org/10.1146/annurev-astro-081817-051935.

REFERENCES

Andrews-Hanna, J.C., J. Besserer, J.W. Head, C.J.A. Howett, W.S. Kiefer, P.J. Lucey, P.J. McGovern, et al. 2014. "Structure and Evolution of the Lunar Procellarum Region as Revealed by GRAIL Gravity Data." *Nature* 514:68–71. https/doi.org/10.1038/nature13697.

Armytage, R.M., B. Anzures, W.B. Banerdt, J. Benkhoffet, S. Besse, D.T. Blewett, N. Bott, al. 2018. "Landed Mercury Exploration and the Timely Need for a Mission Concept Study," white paper presented to European Space Agency in October 2018.

Barker, M.K., E. Mazarico, G.A. Neumann, M.T. Zuber, J. Haruyama, and D.E. Smith. 2016. "A New Lunar Digital Elevation Model from the Lunar Orbiter Laser Altimeter and SELENE Terrain Camera." *Icarus* 273:346–355. https://doi.org/10.1016/j.icarus.2015.07.039.

Delano, J.W., ed. 2009. "Scientific Exploration of the Moon." *Elements* 5(1):11–46. Mineralogical Society of America. https://doi.org/10.2113/gselements.5.1.11.

Goossens, S., T.J. Sabaka, M.A. Wieczorek, G.A. Neumann, E. Mazarico, F.G. Lemoine, J.B. Nicholas, et al. 2020. "High-Resolution Gravity Field Models from GRAIL Data and Implications for Models of the Density Structure of the Moon's Crust." *Journal of Geophysical Research: Planets* 125:e06086. https://doi.org/10.1029/2019JE006086.

Hayne, P.O., A. Hendrix, E. Stefon-Nash, M.A. Siegler, P.G. Lucey, K.D. Retherford, J.-P. Williams, et al. 2015a. "Evidence for Exposed Water Ice in the Moon's South Polar Regions from Lunar Reconnaissance Orbiter Ultraviolet Albedo and Temperature Measurements." *Icarus* 255:58–69.

Hayne, P.O., P.G. Lucey, T.R. Swindle, J.L. Bandfield, M.A. Siegler, and D.A. Paige. 2015b. "Thermal Infrared Observations of the Moon During Lunar Eclipse Using the Air Force Maui Space Surveillance System." Abstract 1997 in 46th Lunar and Planetary Science Conference. https://www.hou.usra.edu/meetings/lpsc2015/pdf/1997.pdf.

Howell, S.M., and R.T. Pappalardo. 2020. "NASA's Europa Clipper—A Mission to a Potentially Habitable Ocean World." *Nature Communications* 11:1311. https://www.nature.com/articles/s41467-020-15160-9.

Jakosky, B.M., J.M. Grebowsky, J.G. Luhmann, and D.A. Brain. 2015. "Initial Results from the MAVEN Mission to Mars." *Geophysical Research Letters* 42(21):8791–8802.

Lawrence, D.J., R.C. Puetter, R.C. Elphic, W.C. Feldman, J.J. Hagerty, T.H. Prettyman, and P.D. Spudis. 2007. "Global Spatial Deconvolution of Lunar Prospector Th Abundances." *Geophysical Research Letters* 34:L03201. https://doi.org/10.1029/2006GL028530.

Lemelin, M., P.G. Lucey, G.A. Neumann, E.M. Mazarico, M.K. Barker, A. Kakazu, D. Trang, et al. 2016. "Improved Calibration of Reflectance Data from the LRO Lunar Orbiter Laser Altimeter (LOLA) and Implications for Space Weathering." *Icarus* 273:315–328. https://doi.org/10.1016/j.icarus.2016.02.006.

Lognonné, P., W.B. Banerdt, W.T. Pike, et al. 2020. "Constraints on the Shallow Elastic and Anelastic Structure of Mars from InSight Seismic Data." *Nature Geosciences* 13:213–220.

NASEM (National Academies of Sciences, Engineering, and Medicine). 2019. *An Astrobiology Strategy for the Search for Life in the Universe*. Washington, DC: The National Academies Press.

NRC (National Research Council). 2011. *Vision and Voyages for Planetary Science in the Decade 2013–2022*. Washington, DC: The National Academies Press.

Roth, L., J. Saur, K.D. Retherford, D.F. Strobel, et al. 2014. "Transient Water Vapor at Europa's South Pole." *Science* 343(6167):171–174. https://doi.org/10.1126/science.1247051.

Sheppard, S., C.A. Trujillo, D.J. Tholen, and N. Kaib. 2019. "A New High Perihelion Trans-Plutonian Inner Oort Cloud Object: 2015 TG387." *The Astronomical Journal* 157(4):139. https://iopscience.iop.org/article/10.3847/1538-3881/ab0895#ajab0895f4.

Southworth, B.S., S. Kempf, and J. Spitale. 2019. "Surface Deposition of the Enceladus Plume and the Zenith Angle of Emissions." *Icarus* 319:33–42.

Wieczorek, M.A., G.A. Neumann, F. Nimmo, W.S. Kiefer, G.J. Taylor, H.J. Melosh, R.J. Phillips, et al. 2013. "The Crust of the Moon as Seen by GRAIL." *Science* 339(6120):671–675.

Zuber, M.T., D.E. Smith, M.M. Watkins, S.W. Asmar, A.S. Konopliv, F.G. Lemoine, H.J. Melosh, et al. 2013. "Gravity Field of the Moon from the Gravity Recovery and Interior Laboratory (GRAIL) Mission." *Science* 339:668–671. https://doi.org/10.1126/science.1231507.

3

Priority Science Questions

The research strategy described in the prior decadal survey in planetary science, *Vision and Voyages*, was structured largely by destination or destination class. Panels constituted to address the inner planets, Mars, giant planets, icy satellites, and primitive bodies (a.k.a. small bodies) each contributed a chapter that identified science goals, future directions for investigations and measurements, and related missions, technology, and other activities to advance knowledge within their topical scope. In its guidelines to this committee, NASA requested that this decadal survey report instead "be organized according to the significant, overarching questions in planetary science, astrobiology, and planetary defense" (see Appendix A).

Implementing a report structure that would, for the first time, be based on overarching science questions required reassessment of committee organizational structure, as well as new processes to define these science questions and to construct the corresponding report. Creating panels based on science questions was determined to be impractical because it required the priority questions to be defined before the panels could be formed, and defining the questions themselves was to be an important initial deliberative task for the committee. Destination-based panels would, as they had previously, provide target-specific recommendations for missions and other activities throughout the decadal process. Destination-specific expertise also allowed the panels to efficiently identify additional mission studies, beyond those already available, early in the process as required for study completion prior to committee prioritizations.

The committee structure comprised a steering group and six destination-oriented panels that addressed Mercury and the Moon, Venus, Mars, small bodies, giant planet systems, and ocean worlds and dwarf planets. Initial deliberations, led by the steering group, defined the priority science questions. Writing groups—one for each question topic—were created to each contribute a corresponding chapter in the report. The writing groups were informal, fluid structures comprising committee (panel and steering group) members with appropriate expertise and interests, with committee members typically participating in multiple writing groups.[1] This organization effectively drove interaction and discussion across target-focused subcommunities, providing an integrated view of how a question related to, for example, planetary formation or habitability might be addressed at destinations across the solar system.

A primary initial task for the steering group was to define the priority science questions around which the writing groups and the report would be structured. These questions would ultimately provide the intellectual

---

[1] An analogous writing group structure was also adopted to address key topical areas, including state of the profession, research and analysis, planetary defense, human exploration, infrastructure, and technology.

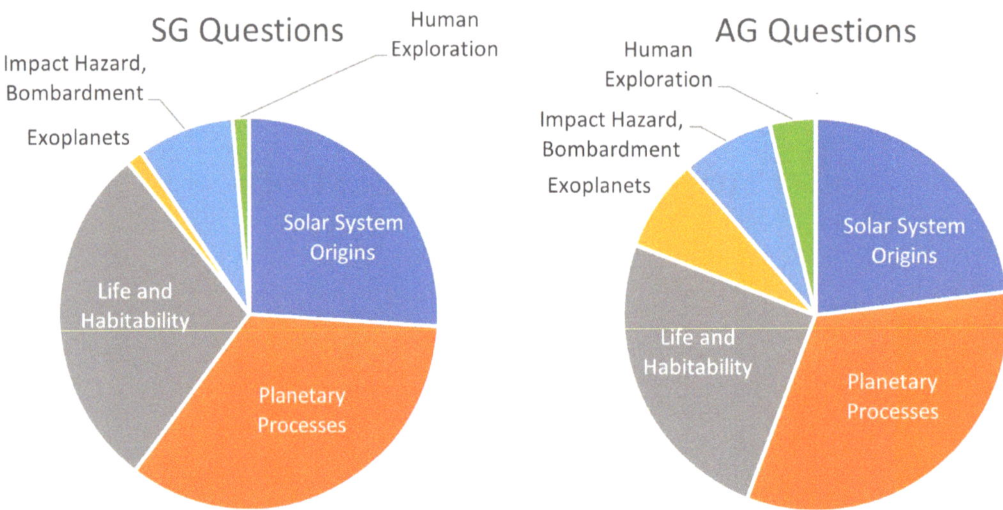

FIGURE 3-1 Comparison of topical distribution of crosscutting questions initially identified by the steering group (*left*) with those previously identified by the analysis and assessment groups, including LEAG, MAPSIT, MEPAG, OPAG, SBAG, MEXAG, and VEXAG (*right*).[2]

framework for the evaluation of potential research activities by the committee. The steering group sought priority question topics that were sufficiently broad and high-level to (1) clearly convey, even to a nonspecialist, a topic of fundamental importance, and (2) emphasize crosscutting connections rather than specific singular objects, in keeping with NASA's request. However, it was clear that the question topics needed to be defined narrowly enough to allow distinguishing between different activities as being more or less broadly impactful, and to keep question-oriented chapters to a manageable scope.

Steering group members began by each submitting a few draft questions that they considered to be significant and overarching for planetary science, astrobiology, and planetary defense. Steering group members considered as inputs to their deliberations white papers submitted to the committee, prior crosscutting themes and priority questions identified in Table 3-1 of *Vision and Voyages*, and some 50 "big questions" that had been previously identified by the various analysis and assessment groups (AGs) in response to a request from PSD Director Dr. Lori Glaze. The ~70 initial steering group questions were compiled and organized by general topic. Figure 3-1 shows the distribution of questions by topic for this initial set of steering group questions and for those previously identified by the AGs. The two distributions were remarkably similar, reflecting an average of 33 percent of the questions relating to planetary processes, 27 percent to habitability and life, 25 percent to solar system origins, 8 percent to bombardment and impact hazard, 5 percent to exoplanets, and 3 percent to human exploration. The main differences between the two distributions were a somewhat greater emphasis on astrobiology-related topics by the steering group, whose membership included multiple astrobiologists in keeping with the decadal scope, and a somewhat greater emphasis on exoplanets by the AGs.

Priority questions were progressively developed through discussion (including feedback from the panels) and polls that assessed, for example, whether draft questions were "appropriate in scope," "too broad," or "too narrow," as well as the subtopics to be included in each. The result was the 12 priority science questions shown in Table 3-1. Each of what the committee henceforth defines as the 12 priority science questions consists of a single overarching topic, as well as a one- to two-sentence description of the question's scope. The 12 priority questions provide the organizational structure around which the scientific portions of the report are organized. The scope of scientific inquiry encompassed by each of the 12 priority questions is of high and comparable priority for optimizing advances

---

[2]A glossary of acronyms and technical terms can be found in Appendix F.

TABLE 3-1 The Twelve Priority Science Questions

Scientific Themes	Priority Science Question Topics and Descriptions
(A) Origins	Q1. *Evolution of the protoplanetary disk.* What were the initial conditions in the solar system? What processes led to the production of planetary building blocks, and what was the nature and evolution of these materials?
	Q2. *Accretion in the outer solar system.* How and when did the giant planets and their satellite systems originate, and did their orbits migrate early in their history? How and when did dwarf planets and cometary bodies orbiting beyond the giant planets form, and how were they affected by the early evolution of the solar system?
	Q3. *Origin of Earth and inner solar system bodies.* How and when did the terrestrial planets, their moons, and the asteroids accrete, and what processes determined their initial properties? To what extent were outer solar system materials incorporated?
(B) Worlds and Processes	Q4. *Impacts and dynamics.* How has the population of solar system bodies changed through time, and how has bombardment varied across the solar system? How have collisions affected the evolution of planetary bodies?
	Q5. *Solid body interiors and surfaces.* How do the interiors of solid bodies evolve, and how is this evolution recorded in a body's physical and chemical properties? How are solid surfaces shaped by subsurface, surface, and external processes?
	Q6. *Solid body atmospheres, exospheres, magnetospheres, and climate evolution.* What establishes the properties and dynamics of solid body atmospheres and exospheres, and what governs material loss to space and exchange between the atmosphere and the surface and interior? Why did planetary climates evolve to their current varied states?
	Q7. *Giant planet structure and evolution.* What processes influence the structure, evolution, and dynamics of giant planet interiors, atmospheres, and magnetospheres?
	Q8. *Circumplanetary systems.* What processes and interactions establish the diverse properties of satellite and ring systems, and how do these systems interact with the host planet and the external environment?
(C) Life and Habitability	Q9. *Insights from terrestrial life.* What conditions and processes led to the emergence and evolution of life on Earth; what is the range of possible metabolisms in the surface, subsurface, and/or atmosphere; and how can this inform our understanding of the likelihood of life elsewhere?
	Q10. *Dynamic habitability.* Where in the solar system do potentially habitable environments exist, what processes led to their formation, and how do planetary environments and habitable conditions co-evolve over time?
	Q11. *Search for life elsewhere.* Is there evidence of past or present life in the solar system beyond Earth, and how do we detect it?
Crosscutting A–C linkage	Q12. *Exoplanets.* What does our planetary system and its circumplanetary systems of satellites and rings reveal about exoplanetary systems, and what can circumstellar disks and exoplanetary systems teach us about the solar system?

in planetary science and astrobiology over the next decade. The first eleven priority science questions are grouped into three scientific themes: Origins (Theme A), Worlds and Processes (Theme B), and Life and Habitability (Theme C). The number of priority questions in each theme approximately reflects the topical distribution shown in the Figure 3-1 plots, with bombardment included in Theme B. The twelfth priority question, which addresses exoplanets, is a crosscutting question that relates to all three themes.

Writing groups focused on each priority question subsequently identified key open issues and subquestions. A central element of this process was identification of strategic research areas that would provide substantial progress in addressing the questions over the next decade. These strategic research areas provided important inputs for assessing potential missions and for identifying key technology development needs for the coming decade. Additionally, they are intended to provide useful guidance and context in support of activities beyond those specifically prioritized in this report—for example, involving Discovery-class or smaller missions, basic research, and instrumentation and technology development. Initial drafts of each science chapter were made available to the full committee for their review and comment.

The resulting 12 science chapters identify the most compelling scientific questions, goals, and challenges that should motivate future strategy in planetary science and astrobiology and provide a comprehensive research strategy to advance the frontiers of planetary science, thus addressing items (3) and (5) of the statement of task (see Appendix A). When considering activities such as future missions, the committee evaluated their potential for addressing the priority science questions and subquestions identified in these chapters, providing direct traceability between recommended activities and science goals (see Table 22-4).

Structuring the report around priority science questions led to greatly increased interactions among committee members across the panels and steering group. It is hoped that this material will provide broadly useful background for the sponsors and the community, with content complementary to that produced by destination-focused AG science goal documents.

Q1 PLATE: An Atacama Large Millimeter/Submillimeter Array (ALMA) image of the protoplanetary disk around the star HL Tau in 2014. The disk is approximately 200 AU in diameter. SOURCE: Courtesy of ALMA (ESO/NOAJ/NRAO). CC BY 4.0.

4

Question 1: Evolution of the Protoplanetary Disk

What were the initial conditions in the solar system? What processes led to the production of planetary building blocks, and what was the nature and evolution of these materials?

This chapter addresses the history of the solar nebula, the protoplanetary disk that evolved into the solar system.[1] Our disk was formed as a by-product of star formation via the collapse of a molecular cloud composed of gas and dust. The evolution of the protoplanetary disk had four sequential, but partially contemporaneous phases: (1) the initial molecular cloud collapse and disk formation; (2) the physical and chemical evolution of the disk; (3) planetesimal formation; and (4) dispersal of the nebular gas (Figure 4-1). The processes that occurred in these phases are foundational to establishing the conditions that led to the physical components of the solar system as we know it today, from primordial presolar grains preserved in comets, to gases accreted to giant planets, to the volatile contents of Earth and the other inner solar system rocky bodies. Moreover, the history of the solar nebula provides a point of comparison for models and astronomical observations of protoplanetary disks in general, with cross-cutting relevance to exoplanet studies (see Question 12, Chapter 15).

That said, understanding this portion of the solar system's history is challenging because the processes are complex and the forces governing disk evolution are not fully identified. Gas and dust, evolving in the protoplanetary disk, are not only coupled to one another but also to the young Sun and the surrounding star-forming region. Solids combine with one another to form small particles while being affected by a host of processes, including gas drag, turbulence within the disk, and the gravitational effects of new worlds plowing new orbital paths. Tiny spherules in meteorites called chondrules tell us that some small particles experienced sudden melting events within the disk, but how and where they did so are critical unsolved problems. Macroscopic small bodies apparently became concentrated enough within certain regions of the disk to form large planetesimals, the building blocks of the planets, but the details of this process remain elusive. The best chance we have to solve these issues is to study them from many different directions, with theory, observations, laboratory studies, and spacecraft missions all playing critical roles.

---

[1] A glossary of acronyms and technical terms can be found in Appendix F.

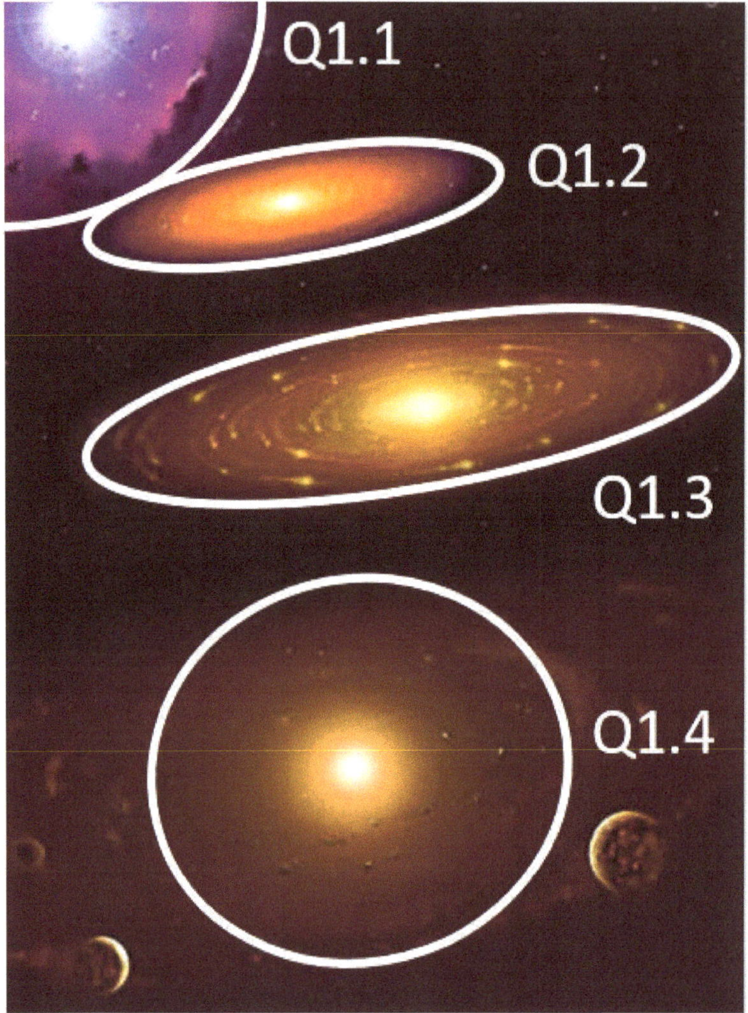

FIGURE 4-1 Schematic of four basic phases of the protoplanetary disk evolution and their relationship to Question 1. From top to bottom: collapse of the molecular cloud core to form an embedded protostar (Q1.1); formation of a disk of gas and dust (Q1.2); accretion and the formation of planetesimals (Q1.3); dispersal of the nebular gas (Q1.4). These phases are highly simplified; in reality, there is overlap between the phases and additional complex evolution occurs, particularly between Q1.3 and Q1.4. SOURCE: Courtesy of M. Garlick/Science Photo Library, adapted by B. Weiss.

Q1.1 WHAT WERE THE INITIAL CONDITIONS IN THE SOLAR SYSTEM?

The story of the protoplanetary disk is one of rapid evolution occurring as new material infalls from the stellar birth cluster and is transported and processed within the disk. Meteorites contain byproducts of prior stellar evolution and other energetic processes that took place millions to billions of years before the protosun formed. The most pristine samples from this epoch may be found in comets, interplanetary dust, and asteroids composed of carbonaceous chondrites. Missions that can return samples of these bodies to Earth for analysis or those that can perform detailed in situ studies provide the best chance of producing breakthroughs in our understanding of the initial conditions of the protoplanetary disk over the near future.

Q1.1a What Initiated the Collapse of the Molecular Cloud?

The solar system likely formed from the collapse of a molecular cloud core within a cluster of perhaps $\sim 10^3$–10^4 stars (Adams 2010). Dense cloud cores are supported against self-gravity by a combination of turbulent motion, magnetic fields, thermal (gas) pressure, and centrifugal forces. As turbulence within a cloud dissipates over time, magnetic fields delay the initial collapse until they are depleted by ambipolar diffusion and rapid collapse begins (Boss and Goswami 2006). Another possible trigger of collapse is a shock front from a nearby supernova (Cameron and Truran 1977), suggested by isotopic anomalies preserved in meteorites. The best chance of better constraining the events that triggered collapse is a combination of astronomical observations and modeling of disks, and measurements of the isotopic record inherited from the molecular cloud and injected nucleosynthetic materials preserved in meteorites. Regardless of the collapse mechanism, the Sun started as a protosolar core supported by thermal pressure while still undergoing infall from the surrounding cloud material.

Subsequent evolution progressed through so-called class I (protostar and gas-dust disk surrounded by infalling spherical cloud), class II (protostar obscured gas-dust disk), and class III (protostar and gas-depleted debris disk) phases. These evolutionary phases can now be observed with facilities such as the Atacama Large Millimeter/submillimeter Array (ALMA) in various star forming regions to conduct detailed studies on conditions at the initial onset of star formation (e.g., Friesen et al. 2018). During the first stages of the Sun's formation, a >30 AU-wide solar nebula of gas (99 wt.% and mostly in the form of hydrogen and helium) and solids (1 wt.% dust) formed because of the conservation of angular momentum during molecular cloud infall. Astronomical observations of hundreds of protoplanetary disks indicate they have a large range in mass with a median of about 1 percent of the Sun's mass and have rich substructures, including holes, gaps, and rings (Andrews 2020).

The best representative of the molecular cloud composition is probably the Sun, which contains >99 percent of the mass of the present solar system. Its composition is inferred from spectroscopic measurements, analyses of refractory elements in primitive meteorites, and direct sampling of the solar wind (Palme et al. 2014). The analysis of chondrites, however, also reveals the presence of nanometer- to submicron-sized grains with extremely variable isotopic compositions that can be related to specific types of parent stars. These data demonstrate that the molecular cloud core was not fully homogenized or had not digested its inherited interstellar components (Nittler and Ciesla 2016), and that laboratory measurements of minute isotopically anomalous materials can be used to investigate the molecular cloud collapse, inherited protosolar components, and disk evolution.

Furthermore, planetary bodies—including primitive meteorites and the inner planets—host daughter products of radioactive isotopes with half-lives shorter than the age of the solar system, the so-called extinct radionuclides (Dauphas and Chaussidon 2011). Their half-lives, denoted as $t_{1/2}$, vary over three orders of magnitude, such that they could not originate from a single stellar event but instead required contributions from several generations of stars having different properties. In particular, the presence of the aluminum isotope ^{26}Al [$t_{1/2}$ = 0.73 million years] suggests that at least one supernova exploded near the parent molecular cloud core. The supernova could have injected ^{26}Al and other radionuclides into the solar system after it had already collapsed (Adams 2010) or, if the injection event occurred earlier, the supernova could have also triggered the initial collapse of the molecular cloud core (Cameron and Truran 1977; Vanhala and Boss 2002). On the other hand, the ^{26}Al may have been sourced from one or more dying giant stars, not necessarily involving a supernova (Tang and Dauphas 2012).

Furthermore, the abundances of isotopes produced by different nucleosynthetic pathways (e.g., those of molybdenum (Mo), ruthenium (Ru), titanium (Ti), and chromium (Cr)) systematically vary among meteoritic groups. The so-called noncarbonaceous chondrites (NC), which include the ordinary and enstatite chondrite meteorites, form a group having variations in slow neutron-capture process (s-process) isotopes, whereas the carbonaceous chondrites (CC)—which include the CI, CM, CV, and CO meteorite groups—form another distinct group richer in rapid neutron-capture process (r-process) isotopes (see Figure 6-1).

The NC versus CC dichotomy in meteorites has been interpreted as resulting from the heterogeneous infall of molecular cloud core material at different radial locations in the disk and/or a different timing for this infall. This scenario is consistent with the idea that the source bodies for the NC meteorites formed in the inner solar system, while those for the CC meteorites formed in the outer solar system. The growth of Jupiter might have then prevented subsequent mixing between NC and CC groups, although a pressure bump in the gas disk near the location of Jupiter's formation may have provided an even earlier barricade (Kruijer et al. 2020).

One way to elucidate this problem is to analyze extinct radioactivity products in various mineral phases whose origin and formation are constrained by independent approaches. The interpretation of the available data in terms of spatial and/or temporal homogeneity of ^{26}Al in the nascent solar system, however, remains controversial (Gregory et al. 2020). At present, the samples of primitive material used to address the ^{26}Al controversy are severely constrained by the availability of natural objects that are known to have originated in select regions of the solar nebula. Knowing that two bodies formed in the same location, even if that location is unknown, would help distinguish spatial versus temporal effects. Interpretation is further complicated by the fact that some meteorite samples have been altered by either planetesimal geologic processes, entry into Earth's atmosphere, and/or their residence time on Earth.

If the collapse of the molecular cloud that made the Sun was triggered by a stellar explosion, the nature and distribution of presolar materials and extinct radionuclides could have been heterogeneous between the inner and outer parts of the solar system. The analysis of extinct radioactivity products and of presolar phases requires, however, analytical precision and spatial resolution that cannot be obtained in situ on planetary objects through missions. Thus, a clear resolution to the question of a supernova trigger versus later injection of isotopes like ^{26}Al requires laboratory-based measurements of additional primitive materials that sample different reservoirs in the nebula from sample return missions and terrestrial collection.

Q1.1b What Were the Original Elemental, Isotopic, and Molecular Compositions of Gas, Dust, and Ice Components Delivered from the Molecular Cloud to the Solar Nebula?

Measurements of the Sun's composition (Palme et al. 2014) and chemical thermodynamics indicate that the most abundant species in the starting material of the solar system and, presumably, of the molecular cloud core were hydrogen (H_2), carbon monoxide (CO), water (H_2O), and nitrogen (N_2) (these species, along with related ones, are collectively referred to as HCON) plus the noble gas helium (He). Despite this, HCON and noble gases only make up <0.1 percent of Earth's mass. The initial depletion on Earth and other terrestrial planets may in fact have been even more severe, with one or more post-accretional processes probably enriching them to these levels.

The relative paucity of the terrestrial planets' initial volatile abundances arises because gaseous HCON host species could not be efficiently trapped in rocky solids during solar system formation. It can be shown that condensation in rocky solids or dissolution in molten silicate/metal are very inefficient processes for trapping volatile elements in planetary bodies, with the possible exception of carbon, and noting that it is necessary to account for the incorporation of oxygen into silicates and oxides (Lodders 2004). In fact, these processes are so inefficient at incorporating volatiles that they also cannot account for the observed low hydrogen, oxygen, nitrogen and noble gas abundances in planetary bodies or primitive meteorites. This implies that other more efficient processes took place that could concentrate and preserve volatile elements during planetary accretion.

The isotopic signatures of HCON in solar system objects and reservoirs are markedly different from those of the protosolar nebula (based on solar composition), and, by inference, from those of the parent molecular core (Figure 4-2). The origin of these large-scale isotopic heterogeneities is debated, with two main scenarios being considered: (1) solid materials grew from a heterogeneous molecular cloud hosting interstellar dust that escaped isotopic homogenization, or (2) the processes that were instrumental in the formation of HCON-bearing solid phases (e.g., photochemical and ion-molecule exchange reactions) were themselves responsible for the large isotopic variations.

The isotopic variations of HCON of solid material originating at different heliocentric distances (which may differ from their current distances) can provide insights into the processes of early solid formation and transport from the molecular cloud to the early solar system. Furthermore, these isotopic fingerprints are tracers for investigating the origin(s) of planetary material. For instance, the isotopic composition of the jovian atmosphere points to a nebular gas origin but those of the other giant planets are unknown. Contributions of cometary-like matter may also imprint significant isotopic shifts, depending on the overall mass of material added, given the large differences between comets and protosolar nebula. Likewise, the composition of the venusian atmosphere, in particular its

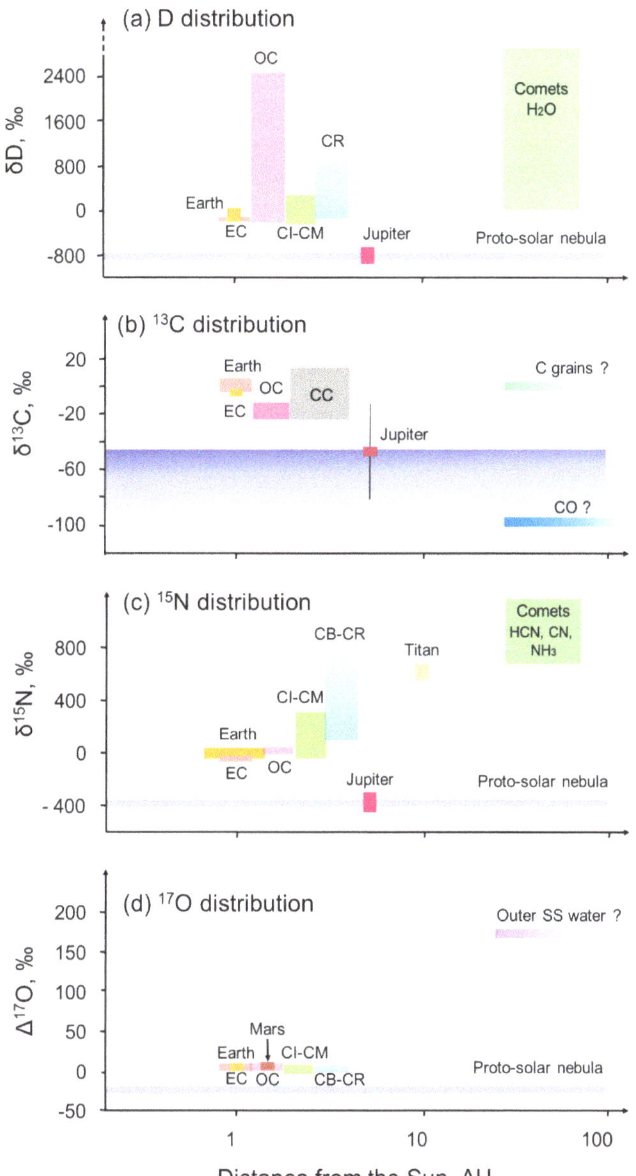

FIGURE 4-2 The isotopic composition of planetary materials in the solar system—hydrogen (H_2), carbon monoxide (CO), water (H_2O), and nitrogen (N_2), collectively referred to as HCON—compared to average nebular values for (a) deuterium (i.e., heavy hydrogen, hydrogen-2), (b) carbon, (c) nitrogen, and (d) oxygen. Chondritic meteorites (EC, enstatite chondrite; OC, ordinary chondrites; CC, carbonaceous chondrites, with subgroups CI, CM, CB, and CR) are shown in approximate relative distance from the Sun of each parent body. Materials from the outer solar system (i.e., beyond the orbit of Jupiter), such as comets, tend to show more drastic enrichments in some heavy isotopes relative to the inner solar system (Earth, Mars, chondrites) and the protosolar nebula. All compositions are plotted as deviations from a terrestrial standard in parts per thousand, and for oxygen, the mass independent value, $\Delta^{17}O$, is shown. NOTE: The vertical scale in (d) masks important differences between Earth, Mars, and different meteorite group values (see text for details). SOURCES: Created by R. Stroud and B. Marty. *Cometary data* from Altwegg et al. (2015) and Bockelée-Morvan et al. (2015). *Solar nebula data—hydrogen,* Geiss and Gloeckler (1998); *carbon,* Lyons et al. (2018); *nitrogen,* Marty et al. (2011); *oxygen,* McKeegan et al. (2011). *Jupiter data—hydrogen*—Lellouch et al. (2001); *nitrogen*—Owen et al. (2001).

$\Delta^{17}O$ composition (where $\Delta^{17}O \equiv \delta^{17}O - 0.56\ \delta^{18}O$; see Figure 4-2), would permit a test for a genetic relationship with documented solar system reservoirs, as already achieved for Earth and Mars.

The distribution of presolar grains in primitive undifferentiated meteorites (chondrites) appears heterogeneous, with CCs being richer in stellar dust than NCs. Because CCs are argued to originate from greater radial distances than NCs (e.g., Desch et al. 2018), this observation suggests a process that separated the inner and outer solar system from mixing at a critical point during early accretion, or an outward radial gradient that could be related to heterogeneous infall of molecular cloud material, as may be suggested by preserved isotopic anomalies. Testing the heterogeneity and dynamics of molecular cloud core contributions, including refractory dust and more easily altered organic matter, would require return of material originating from the outer solar system to terrestrial laboratories or definitive proof that a primitive sample originated in that region.

Heterogeneous infall of stellar material alone cannot produce all of the observed isotopic variability. The variability in HCON of several hundreds to thousands of parts per million throughout the solar system is not observed in the nonvolatile elements. For instance, isotopic variations observed in refractory elements such as Mo, Ti, Cr, and Ru are of the order 1 part per 10,000 or less. Ion-molecule exchange reactions, which occur at low temperature in dense molecular clouds (Terzieva and Herbst 2000; Aikawa et al. 2018) could potentially account for some of the deuterium (D) enrichments found in molecular cloud ices (Sandford et al. 2001; Cleeves et al. 2014) and in organic molecules such as nitrogen (^{15}N)-rich organics (Rodgers and Charnley 2008). Isotopic variations of volatile elements across the solar system could be pristine fingerprints of molecular cloud chemistry, with the outer solar system having better preserved unprocessed (D- and ^{15}N-rich) material than the inner regions of the solar nebula (Pignatale et al. 2018). Laboratory measurements of returned and collected samples of inner and outer solar system materials are needed to address this question.

The observation of mass-independent isotope fractionation of oxygen in planetary bodies (as opposed to mass-dependent fractionation that occurs during ordinary processes such as sublimation and condensation) indicates another possible fractionation process. In particular, self-shielding occurs when stellar photons penetrate into a cloud of gas and become progressively absorbed by photoreactions, as observed in giant molecular clouds (Bally and Langer 1982). Self-shielding of gaseous CO could represent a significant source of atomic carbon (C) and oxygen (O) that could then react with H_2 to form hydrocarbons (C_xH_y molecules) and H_2O molecules. Self-shielding could have taken place in the parent molecular cloud illuminated by nearby stars (Yurimoto and Kuramoto 2004) or at the disk surface of the solar nebula with photons from the nascent Sun (Lyons and Young 2005; Figure 4-3). This process may also account for ^{15}N enrichments in organic molecules (Heays et al. 2014; Garani and Lyons 2020). Noble gases provide a further piece of evidence for irradiation of nebular gas. Chondritic argon (Ar), krypton (Kr) and xenon (Xe) are trapped in specific phases associated with refractory organics and present isotopic and elemental signatures different from those inferred for the solar nebula (Busemann et al. 2000). These signatures have only been reproduced in the laboratory when noble gases are incorporated into organics as ions (Frick et al. 1979; Marrocchi et al. 2011). Spatially resolved astronomical observations of isotopologues (i.e., molecules that differ only in their isotopic composition) of C- and O-bearing species in protoplanetary disks and the interstellar medium can address the role of self-shielding.

Q1.1c How Did the Compositions of the Gas, Dust, Ice, and Organic Components and the Physical Conditions Vary Across the Protoplanetary Disk?

A protoplanetary disk is intrinsically heterogeneous, with large radial gradients in temperature, pressure, and chemical compositions owing to the presence of a growing central protostar, cold interstellar medium at its edges, and nearby massive, luminous stars. Many aspects of the disk are uncertain, including the temperature, pressure, and density of the nebula, its effects on the elemental and isotopic composition of nebular gas and solids, and the petrology and crystal structure of solids. Identifying and quantifying certain species—for example, the major carriers of carbon (CO, CO_2, organics, graphite, etc.) and how these change over time and space—can provide constraints on disk chemical evolution.

In a broad sense, hot and reducing conditions are thought to have prevailed within the inner regions of the disk, whereas oxidizing, cold and volatile-rich environments dominated in the outer disk (Bergin et al. 2007). This heterogeneous distribution of volatile elements across the solar nebula has major implications for the origin of volatile elements in the terrestrial planet-forming region. For example, a key question is whether the majority of

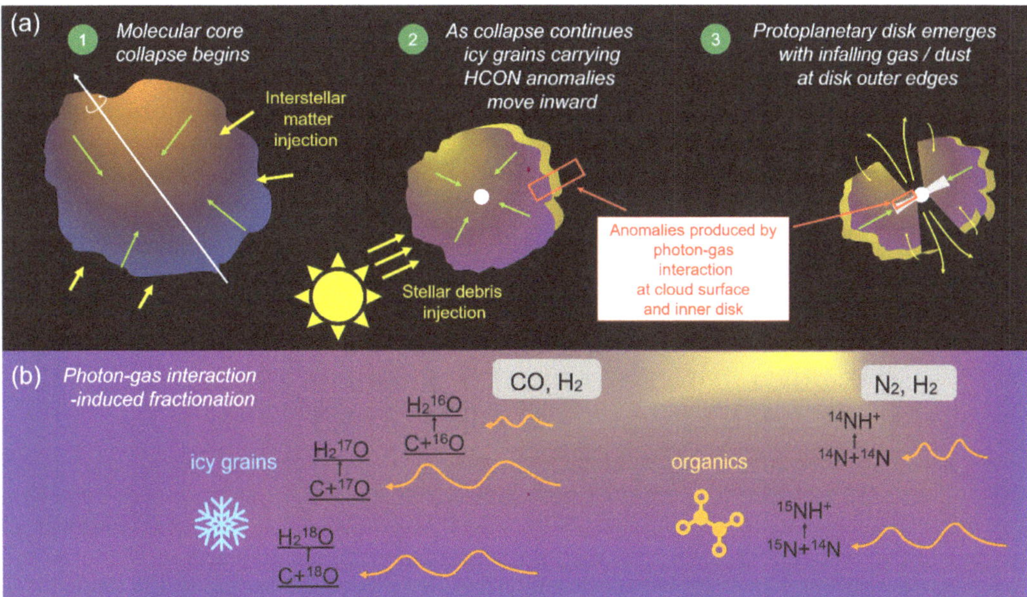

FIGURE 4-3 Sources of isotopic anomalies in the protosolar nebula from hydrogen (H_2), carbon monoxide (CO), water (H_2O), and nitrogen (N_2), collectively referred to as HCON. (a) In the initial stages of the molecular cloud core collapse, (1) anomalous interstellar materials are injected into the cloud and carried inward; (2) as collapse continues, dust and gas from nearby stars are injected, and irradiated organic and icy grains are transported inward toward the core; (3) collapse proceeds until the protoplanetary disk emerges, and material is ejected outward and falls back on the outer disk. Gases are subject to photodissociation by ultraviolet irradiation (2) of the cloud core by nearby stars or (3) irradiation of the disk surface by the protosun. (b) Icy and organic grains form from the products of photodissociation of gaseous H_2, CO, and N_2, and carry the signatures of the isotopic fractionation of the photodissociation. SOURCES: Created by R. Stroud and B. Marty based on data from Yurimoto and Kuramoto (2004) and Lyons and Young (2005). Graphics in the top panel inspired by Lee et al. (2008).

the volatiles in Earth and Mars were indigenous to the formation regions of these worlds or were delivered from outer solar system sources.

Over timescales of several million years, organic-bearing materials and silicate solids agglomerated with variable amounts of ice to form increasingly larger planetary bodies such as asteroids, planetary embryos, and planets. During this period, HCON volatile condensation fronts (both radial and vertical) swept through the disk, changing the density and composition of the residual gas (Owen 2020). This in turn differentially influenced the elemental and isotopic compositions of volatile and refractory components depending on their accretion locations.

Disk processes are recorded in the volatile, organic, and silicate compositions of small bodies and planets, which in turn can be used to help constrain the provenance of the planets' accretionary material over time. Ground- and space-based observations of primitive bodies across the solar system provide constraints on preserved volatile components from the disk. Comparisons to resolved maps of disks at various evolutionary stages will inform compositional tracers of evolution that enable detailed modeling of the presolar disk.

Strategic Research for Q1.1

- **Constrain the variation in physical conditions and distribution of gas and dust components across the nebula** through in situ and orbital measurements of the elemental and isotopic composition of the surface, and, where relevant, atmospheres of bodies formed from different nebular reservoirs (especially Uranus, Neptune, and Mercury, but also Venus, asteroids, Centaurs, and Saturn), laboratory analyses of returned samples of comets and terrestrially collected interplanetary dust and meteorite samples, and ground- and space-based telescopic observations of the composition (gas, ice, dust) of comets, Kuiper belt objects, and protoplanetary disks.

- **Address the timing and role of injection of supernova material on the formation of the solar nebula, and distinguish materials that retain a presolar heritage from the products of nebular and parent body processes** through return of comet samples and laboratory isotopic analyses of returned and terrestrially collected samples.
- **Constrain the role of self-shielding on the formation of O and N isotopic reservoirs in the solar nebula** through spatially resolved astronomical observations of isotopologues C- and O-bearing species in protoplanetary disks and the interstellar medium using ground- and space-based telescopes (e.g., ALMA and the James Webb Space Telescope).

Q.1.2 HOW DID DISTINCT RESERVOIRS OF GAS AND SOLIDS FORM AND EVOLVE IN THE PROTOPLANETARY DISK?

Conventional wisdom in previous decades held that most asteroids formed within the main asteroid belt, and that evolution in the protoplanetary disk between 2 and 3 AU was ultimately responsible for major isotopic differences found between the different meteorite classes. More recently, a new model has been advanced that suggests that the isotopic differences between noncarbonaceous and carbonaceous chondrite meteorites are so pronounced that they likely represent materials formed in vastly different parts of the solar system (e.g., Kruijer et al. 2020).

This idea is provocative, in that it suggests samples from planetesimals that formed in the terrestrial planet, giant planet, and trans-neptunian zones may already be in our possession in the form of meteorites or could be obtained from highly accessible near-Earth objects. It also raises a number of challenging issues, such as ascertaining the degree of mixing that took place in the solar nebula prior to the isolation of the noncarbonaceous and carbonaceous zones, determining the mechanism(s) that brought about the isolation of the inner and outer solar system, and identifying the formation locations of meteoritic materials in our collections. Testing these ideas will require a combination of theory, observations, laboratory studies, and spacecraft missions.

Q1.2a When and How Did the First Macroscopic Solids Form?

The primary rock record for constraining the nature of the first macroscopic solids in the protoplanetary disk are chondrites. Chondrites are largely composed of chondrules, refractory inclusions including calcium aluminum-rich inclusions (CAIs) and finer-grained matrix materials.

CAIs are the oldest known macroscopic solar system solids and contain minerals that are thermodynamically predicted to be the first to condense from a gas of approximately solar composition (Krot 2019). They define the cosmochemical time zero for the solar system at 4,567.3 ±1 Ma (Amelin et al. 2010) or perhaps 4,568.2 ±0.2 Ma (Bouvier and Wadhwa 2010). CAIs are variably solidified melts, condensates, and/or agglomerates (MacPherson et al. 2012; Krot 2019). Some authors have argued based on petrography, abundances in rare-Earth elements (REE) and relative ages derived from aluminum-magnesium (Al-Mg) isotopic systematics that at least some of the nonigneous CAIs may be older than those that are igneous, with the nonigneous potentially being the precursors to the igneous ones (e.g., MacPherson et al. 2012). Some refractory inclusions can be as simple as melted spheres of the minerals spinel ($MgAl_2O_4$) and hibonite ($CaMg_xTi_xAl_{12-2x}O_{19}$), whose precursors were likely direct condensates from a gas phase.

Individual minerals such as platy-hibonite crystals found in the matrices of chondrites are generally accepted as condensates, having model ages equal to that of other CAIs. It can be argued that small refractory inclusions (30–100 µm) in carbonaceous CO_3 chondrites likely represent the earliest stages of coagulation of refractory dust to form refractory inclusions, with larger refractory inclusions forming on the order of 40,000 years after the smaller ones (Liu et al. 2019). Thus, CAIs may record a progression from almost pure condensation to the agglomeration of larger, cm-sized objects. This may explain why some refractory minerals have ages younger than 4,567.3 ±1 Ma, which is either attributed to longer durations of agglomeration and formation or to disturbances in the initial isotope values owing to processing post-formation—either pre- or post-accretion.

Chondrules are tiny, 0.1–1 mm igneous rocks and are the main structural component of chondrites. Various models have been proposed to explain the transient heating necessary for formation of chondrules from their precursor dust particles. For example, bipolar outflows and X-ray flares, observed around young stellar objects, have been postulated as possible mechanisms for CAI and chondrule formation because they are transient and can redistribute dust in the disk (Shu et al. 2001). Nebular shock waves, such as bow shocks produced by planetesimals plowing through nebular gas, is the most developed of the ideas (Boley et al. 2013). Transient heating during impacts between planetesimals has also been recently studied (Johnson et al. 2015). Other putative chondrule formation mechanisms include lightning within the disk and magnetic reconnection events (Joung et al. 2004). These models could be distinguished by measuring chondrule cooling rates, the strength of the ambient magnetic field during chondrule formation, and determining the ages of chondrules relative to that of planetesimals and the nebular lifetime.

Chondrules are generally considered younger than CAIs by at least 1 million years based on Al-Mg isotopic systematics (Kita et al. 2010) and have ages distributed over at least 3 million years. With that said, absolute model ages obtained by lead-lead (Pb-Pb) isotopic systematics of individual chondrules from one class of chondrite yield no gap between the two types of objects (Connelly et al. 2012; Bollard et al. 2017). Resolving this issue is important for two reasons. First, it will elucidate the different formation mechanisms for CAIs. Second, it is unclear how to maintain CAIs and chondrules in the disk until they can be incorporated into chondrites, an interval of several million years, given how fast these small objects can reach the Sun by gas drag, although pressure bumps may inhibit this inward transport (Desch et al. 2018) (see Q1.2b). A related issue is that if chondrules and CAIs are incorporated into large planetesimals very early, the planetesimals could heat up and alter or even destroy these grains. As such, further research is needed to detail the absolute ages of chondrules and CAIs.

As described above, some solar system minerals originally condensed from the gas phase within the nebula (Grossman 1972; Krot 2019) and were likely the precursors to some CAIs. The complex relationships between the condensation of refractory minerals, rocks, and igneous CAIs remain unclear (Yoneda and Grossman 1995; Ebel and Grossman 2000; Connolly et al. 2001; Bollard et al. 2015). Chondrules and some CAIs are igneous, recording high temperatures (>2,000°C) for seconds to minutes. This evidence indicates these objects experienced transient melting events within the disk. The majority of them also experienced cooling rates on the order of 5–100°C per hour, relatively slow when compared to expectations of how fast a molten droplet would cool if it was sitting in empty space. That suggests chondrules, after formation, were surrounded by gas that could allow slow cooling. Another major constraint on the melting of chondrules, and to a lesser extent CAIs, is that they were processed multiple times and preserve petrologic and geochemical evidence of remelting (Connolly and Jones 2016). Satisfying all of these constraints is a challenge for any formation model.

Continued investigation of chondrules and igneous CAIs is important for placing their formation in context with that of other rocky solar system bodies (Connolly et al. 2018). An important implication of such studies is that any igneous rocks in chondrites that formed from some kind of collision mechanism would be byproducts of the growth of planetesimals. By comparison, igneous rocks formed while free-floating in the disk before accretion were on the direct pathway to forming planetesimals. Such mechanisms could be potentially distinguished with measurements of the ambient magnetic field strength (Weiss et al. 2021) and of extinct radioactivity products within chondrules.

The matrices of chondrites are composed of a variety of materials that include mineral grains and organic phases and are enriched in volatile elements compared to chondrules and CAIs (Ehrenfreund and Charnley 2000; Weisberg et al. 2006; Glavin et al. 2021). The origin of organic phases is an issue that requires considerable further research. It is unclear how much of the diversity in the organic matter observed across chondrites and cometary dust samples is owing to differences in hydrothermal parent body processing and how much reflects differences in the materials accreted to different parent bodies. Moreover, the range of formation locations, temperatures, and mechanisms spans accretion as icy mantles on anhydrous minerals in the outer nebula, accretion to metal grains and chondrules in the inner nebula, and as direct nebular condensates (Alexander et al. 2017). Laboratory studies of the organic matter in newly returned comet and asteroid samples from known bodies, in addition to organics in terrestrially collected samples, are needed to deconvolve nebular and parent body histories and terrestrial alteration artifacts.

Q1.2b How, Where, and When Did Radial Mixing and Segregation of Solids and Gas Occur in the Nebula?

During the protoplanetary disk phase as the Sun continued to accrete, the first solids formed and planetesimals assembled and agglomerated into protoplanets. Along the way, some planetesimals experienced large-scale melting, collisional evolution and disruption, and gas-driven orbital migration. At the same time that all of these processes were occurring, the gas and solids of the nebula experienced radial mixing and were likely transported both inward and outward. Inward transport of volatile solids may have led to volatile evaporation at their respective snow lines (e.g., the distances from the young sun at which temperatures are at the condensation temperatures of ices), which in turn could have produced changes in the local nebular gas composition. Outward transport of refractory solids, on the other hand, could have enriched the dust content of planetesimals.

Such mixing played a fundamental role in determining the size, bulk composition, internal structures, and orbital architecture of the solar system today. For example, it may have contributed to Mercury's iron-rich and reduced, moderately volatile-rich composition (Kruss and Wurm 2018; Nittler and Weider 2019). It may have also led to the enhancement of heavy elements and noble gases in the giant planets relative to the bulk composition of the nebula (Mandt et al. 2020) and the presence of chondrule-like and CAI-like grains in comets and interplanetary dust particles (Wooden et al. 2017). Mixing has presumably also influenced the formation of more distal objects like Centaurs and Kuiper belt objects for which we currently have few compositional constraints from remote sensing and no in situ measurements or known samples.

It has long been recognized that gas drag leads to the inward drift of solids, particularly for meter-sized bodies (Weidenschilling 1977a). It has also long been predicted that the disk experienced viscous spreading owing to magnetohydrodynamic and/or purely hydrodynamic turbulence as well as owing to laminar, nonturbulent processes (Weiss et al. 2021). Turbulence from the magnetorotational instability (MRI) and/or torques in a laminar disk owing to a large-scale toroidal field are thought to have transported material radially within the disk plane (Armitage 2015). Meanwhile a magnetized, laminar disk wind may have thrown material upward and outward. Whether these magnetic mechanisms dominantly transported inward or outward was influenced by the relative orientation of the rotation direction of the solar system relative to the direction of the mean vertical magnetic field. A diversity of hydrodynamic instabilities that generate turbulence have also been theorized to transport angular momentum (Fromang and Lesur 2019).

All of these processes are thought to have occurred at the time disk substructures may have formed. Such substructures may have formed as the result of these disk transport mechanisms or alternatively by the accretion of giant planets, and whether planets or substructures came first is unclear. In either case, the formation of substructures themselves would have then influenced subsequent disk evolution, possibly serving as a barrier to further transport in their locations or instead filtering the grain sizes of transported materials. For example, after the giant planets grew to a sufficiently large mass, they should have opened up a gap in the disk that would have inhibited radial transport (Johansen and Lambrechts 2017). Also, snow lines serve as barriers to inward transport of volatile ices (Pontoppidan et al. 2014). Last, magnetized disk winds can also open gaps in the disk and transport disk materials outward.

Evidence for transport of dust outward in the solar system is provided by the presence of crystalline silicates including chondrule-like and refractory inclusions in comets and interplanetary dust particles (Bockelée-Morvan et al. 2002; Joswiak et al. 2017). The presence of clasts, chondrules, and refractory inclusions formed with isotopic compositions distinct from their bulk meteorites provides evidence for dust transport in both directions (Brennecka et al. 2020; Schrader et al. 2020; Williams et al. 2020), although it is possible that some of these isotopic variations could be explained by temporal changes in the nebula's composition. Evidence for inward transport of ices and evaporative concentration of volatiles in the gas phase at snow lines may be provided by enrichments of heavy elements and noble gases in giant planet atmospheres (Mandt et al. 2020).

There are several key unsolved questions relating to disk evolution. First, and foremost, the radial extent of mixing and transport is uncertain. For example, there is a lack of clarity about which materials formed close to the Sun (<0.1 AU), or in the terrestrial planet and asteroid belt region (1–3 AU), and were transported outward to 15 AU and beyond. Similarly, it is unclear which materials formed outside the terrestrial planet region and were transported inward. As a result of this bidirectional transport, many regions of the solar nebula may have served as the source materials for the terrestrial planets, the giant planets, and small bodies like asteroids, comets, Centaurs, and Kuiper belt objects. The issue is to quantify the magnitude of the contribution and the degree of mixing from each zone.

Second, the nature of the mechanisms driving such transport are uncertain. Several of the above processes (or others) were responsible, but the role of hydrodynamic processes and magnetic torques are debated. If magnetic fields were important, certain magnetic mechanisms (MRI, laminar toroidal field, and/or magnetized disk wind) controlled transport at different distances from the Sun, but their strength is unknown. The spin pole and vertical magnetic field of the nebula are also mysterious, and it is unclear whether they were aligned or anti-aligned. If hydrodynamic turbulence was important, certain conditions enabled this, and the mechanisms that generated such behavior need to be explored.

Third, despite astronomical evidence for the presence of holes, rings, and gaps in other protoplanetary disks, it is unknown whether the solar nebula formed such substructures. Accepting that it did, their geometry, location, and timing are critical to understanding planet formation processes, as are the mechanisms that influenced their formation. We also need to know how they may have inhibited or filtered the transport of gas and/or dust in the nebula. Progress on these questions can be achieved with chemical, isotopic, petrographic, and paleomagnetic studies of nebular materials in meteorites and returned samples, in situ isotopic and compositional measurements of planetary and small bodies at a range of distances from the Sun, and astronomical observations of the density, velocity structure and magnetism of disks.

Q1.2c How, Where, and When Were Gas and Solids Processed in the Nebula and During Accretion (by Heat, Aqueous Alteration, and Electromagnetic and Particle Irradiation)?

In addition to constraints from chondrules and CAIs (see Q1.2a), astronomical observations and analyses of meteorites tell us that irradiation also played an important role in the evolution of dust and gas in the nebula. For example, stellar and galactic cosmic rays can cause spallation—fragmentation of larger nuclides (e.g., carbon, oxygen, magnesium, and iron, or C, O, Mg, and Fe, respectively) into smaller ones (e.g., lithium, beryllium, and boron, or Li, Be, B, respectively)—which can result in detectable changes in the composition of dust. Evidence for energetic particle irradiation of nebular dust comes from Be- and B-isotope signatures of CAIs, which are consistent with live ^{10}Be ($t_{1/2}$ = 1.5 million years), a short-lived radionuclide associated with exposure to cosmic ray irradiation.

Ultraviolet and X-ray irradiation contributes to ionization of nebular gas and formation of complex molecules on the surfaces of dust and ice grains. Such irradiation may have influenced the isotopic composition of light elements (HCON) in nebular gas and dust. For example, ultraviolet dissociation of CO has been proposed to explain mass-independent O-isotope variations observed in meteorites and ionization from X-rays may facilitate ion-molecule reactions that could explain enrichment in deuterium in organic material in chondrites (see Q1.1b). Studies of cosmogenic systems using a wide range of planetary material along with observational analyses and models of irradiation from young stellar objects are necessary to further develop our understanding of the chemical evolution of the solar nebula.

Nebular dust is also affected by thermal metamorphism and fluid alteration associated with the accretion of planetesimals. Evidence for thermal processing of dust in planetesimals includes observations of a range in metamorphic grades and an inverse correlation between the ages and cooling rates for chondrites with the same bulk composition. Thermal processing of nebular dust within chondrite parent bodies includes dehydration and recrystallization of matrix minerals, modification of organic material, destruction of presolar grains, chemical equilibration of minerals, and recrystallization of glassy material (e.g., Quirico et al. 2014; Floss and Haenecour 2016).

Early thermal metamorphism of planetesimals is largely controlled by the abundance of ^{26}Al, the time of planetesimal formation, the size of the body formed, and its ratio of ice to rock. Planetesimals that form earlier end up with more active ^{26}Al, larger bodies are better able to insulate materials warming up in the interior, and the temperatures of ice-rich bodies are buffered by the ice's latent heat of melting. Fluid alteration of nebular solids is largely associated with water–rock reactions in chondrite parent bodies. Evidence for these processes include the occurrence of secondary minerals that require precipitation from fluid and varying degrees of fluid alteration for chondrites with the same bulk composition. Melting of accreted ice to facilitate these water-rock reactions likely stemmed from heating from the decay of ^{26}Al. Fluid alteration occurred 2–5 million years after CAIs, based on age dating of secondary minerals such as fayalite and carbonates (Jilly-Rehak et al. 2017). Additionally, thermal metamorphism and fluid alteration of nebular solids within chondritic planetesimals could have also been affected by impacts, which can both heat bodies (through deposition of the impactor's energy) and cool them by unroofing or excavating deep materials in the body.

Deconvolution of parent body and nebular processing enables the identification of common parent bodies for groups of meteorites, and a more comprehensive understanding of the conditions that gave rise to the chemical signatures in planetary materials, including the nature of materials before incorporation into parent bodies. The composition of the accreted ice and that of the fluids associated with water–rock alteration in chondrite parent bodies is still unclear, but necessary for a comprehensive picture about secondary processing of nebular components.

Fundamental questions regarding all forms of nebular processing remain to be answered. For example, the full range of heating mechanisms for processing solids in the nebula is unknown, and there is no consensus on the heating mechanisms for melting refractory inclusions (see Q1.2a). We know that gas, ice, and organics were affected by heating and irradiation in the disk, but the degree is uncertain and could vary with location and other factors. There is also the issue of distinguishing nebular, parent body, and impact processes, all which have shaped the composition of planetary materials. Ideally, we would like to isolate specific effects both to trace conditions in the nebula and to better understand the original composition of dust, gas, ice, and organic matter in the disk.

Progress on these topics will require an interdisciplinary approach that combines astronomical observations of young stellar objects, numerical modeling, and geochemical analyses of chondrites and samples returned from primitive bodies. Observational surveys of disks for detailed studies of density structure that span an appropriate range of environmental and evolutionary states could be used to work out the mechanics of key evolutionary processes. Such studies are now possible with the sensitivity, angular and spectral resolution now offered through millimeter interferometers such as ALMA (Andrews 2020).

Strategic Research for Q1.2

- **Constrain the predominant driving forces for, and the radial extent of, nebular mixing and transport** through spacecraft elemental, isotopic, and magnetic measurements of—most importantly comets, Centaurs, and Mercury—and also of Saturn, Venus, or Kuiper belt objects; return of samples from comet surfaces and, with a lower priority, from asteroids; disk transport modeling; and laboratory petrologic, isotopic, and paleomagnetic analyses of returned and terrestrially collected samples, and telescopic observations of protoplanetary disks.
- **Determine if, how, when, and where gaps, rings, or holes developed in the nebular disk** through spacecraft isotopic and elemental measurements of gas, dust, ice, and organic components in outer and inner solar bodies; return of asteroid and comet surface samples; disk transport modeling; ground- and space-based astronomical measurements of protoplanetary disks; and laboratory petrologic, isotopic, and paleomagnetic analyses of returned and terrestrially collected samples.
- **Constrain the original compositions and processing histories of dust, gas, ice, and organic matter in the solar nebula** through return of asteroid and comet surface samples; astronomical observations of young stellar objects and outer solar system volatiles; modeling of heating and radiation processing; and laboratory petrographic, elemental, and isotopic analyses of returned and terrestrially collected asteroid and comet samples.
- **Determine the timing and range of formation mechanisms of the earliest solids in the solar system,** including gas phase condensation, irradiation, and transient heating by collisional, X-ray, magnetohydrodynamic, or other means, through return of asteroid and comet surface samples; astronomical observations (e.g., ALMA) of dust in young stellar objects; and laboratory petrographic, elemental, isotopic, and paleomagnetic analyses of returned and terrestrially collected asteroid and comet samples.

Q1.3 WHAT PROCESSES LED TO THE PRODUCTION OF PLANETARY BUILDING BLOCKS—THAT IS, PLANETESIMALS?

A common assertion in classical planet formation models is that the initial size of planetesimals—that is, planetary building blocks, was approximately a kilometer, and that they grew from pairwise collisions between smaller objects. As awareness grew that meter-size bodies could not last long in the protoplanetary disk, however, new physics was incorporated, and new accretion models were considered that made the aerodynamic gas-particle processes in the protoplanetary disk a critical component of planetesimal formation. Current models describe how small particles, often called pebbles, can become highly concentrated in zones within the solar nebula. Their mutual gravity can then lead to collapse and the agglomeration of objects that are commonly ~100 km in diameter. The formation and physical properties of planetesimals made this way can be constrained by the small body populations, although some signatures of formation may have been erased by subsequent collisional and thermal evolution.

In addition, we also do not yet understand the role of chondrules—tiny molten droplets found within chondritic meteorites—in creating planetesimals across the solar system. Indeed, chondrule formation may postdate the formation of many planetesimals. While chondrules have been studied in detail for decades, they do not yet have a broadly accepted mode of formation, and their numerous constraints have long confounded understanding. This limitation hinders our ability to interpret what meteorites are actually telling us about planetesimal and planet formation.

Q1.3a How and When Did the First Grains Aggregate and Form Centimeter-Scale Objects?

Clues to how the earliest accretion of solids occurred are recorded by chondrites and other primitive planetary materials. They suggest that small grains came together in the protoplanetary disk under low velocity collisions, where they stuck together. From there, the generally accepted scenario suggests that dust accreted to form millimeter- to centimeter-sized objects that melted and cooled. These bodies may have then accreted with dust and other materials to form centimeter- to meter-sized bodies (see discussion in Q1.3b), which then accrete into ~100-km-size bodies by some concentration mechanism in the protoplanetary disk (e.g., Morbidelli et al. 2009).

With that said, there are several unknowns about the processes that aggregated grains into centimeter-sized objects. First, we do not know what processes are responsible for enabling the first submicron-sized grains to stick together, with van der Waal's forces, electrostatic attraction, and ferromagnetic attraction forces all being possibilities (Dominik et al. 2006). We also do not understand how chondrites and planetesimals were lithified (i.e., turned into coherent, rock-like bodies) and if the lithification process was linked with the accretion process.

Additionally, we do not know the full extent to which the millimeter- to centimeter-scale objects found in these meteorites formed before or after meter-sized to kilometer-sized to ~100-km-scale accretion occurred. Although it is generally argued that CAIs formed, melted, and cooled before the accretion of planetesimals, the same cannot be said for chondrules. Indeed, the measured formation ages of some iron meteorites indicate that some large, differentiated bodies formed contemporaneously with some chondrules (Connelly et al. 2012; Kruijer et al. 2020). If the formation of all igneous objects within chondrites occurred before planetary accretion began in earnest, then their formation is on the pathway to creating larger solid bodies.

A key issue that could help us quantify this scenario is to determine when CAIs, chondrules, and mineral grains formed. Additional constraints on planetesimal formation and evolution would come from the timing of post-accretional aqueous and thermal metamorphic events on their parent bodies. These events can potentially be constrained using radiometric dating with short-lived (e.g., aluminum-magnesium, magnesium-chromium, hafnium-tungsten, and iodine-xenon) and long-lived (e.g., uranium-lead) radionuclides (see Q1.3b). Furthermore, observations of protoplanetary disks by ALMA and the next-generation Very Large Array can provide constraints on the growth of up to centimeter-scale particles.

Q1.3b How and When Did Grains Grow from Centimeter-Size Objects to ~100 Kilometer-Size Planetesimals?

The formation of planetesimals is a necessary step to making the terrestrial planets and giant planet cores, but precisely how this happened is still debated. We start by considering constraints from small bodies, namely the size-frequency distributions of the asteroid and Kuiper belt objects. Both show local maxima for objects ~100 km in diameter. This feature is likely a fossil from their primordial size distributions rather than a signature of collisional evolution over the past several billions of years (e.g., Bottke et al. 2015). This supports the idea that planetesimals were born big, with sub-meter bodies jumping to objects tens to hundreds of kilometers without passing through intermediate stages (Morbidelli et al. 2009; Johansen et al. 2015).

The mechanisms leading to the formation of such bodies does not appear to be pairwise accretion, as suggested by many studies of the past. The first problem is that particles do not reliably stick together during collisions, but instead may bounce off one another (Zsom et al. 2010). Mechanisms have been proposed to enable particles to jump this so-called "bouncing barrier" (Musiolik et al. 2016; Steinpilz et al. 2020), and future progress may be obtained by determining the grain size, composition, ratios of ice to rock, electrostatic forces, and remanent magnetism of primitive accretional grains.

The second problem, however, is more fundamental. Even if bodies find a way to achieve decimeter sizes to objects larger than a meter, the gas drag they experience as they orbit within the nebula leads to two negative effects: (1) it increases their mutual collision velocities with other solids, leading to fragmentation (Birnstiel et al. 2011), and (2) they begin to rapidly spiral in toward the Sun, with timescales of <1,000 years from 1 AU in a nonturbulent disk (Weidenschilling 1977b).

One way to overcome the meter-sized barrier is to concentrate sufficient particles in a small region of the disk so that they undergo gravitational collapse and form an aggregate body. Perhaps the best-studied concentration mechanism in the solar nebula is the streaming instability (SI), which describes how aerodynamic forces cause small particles to collect in regions where the solid-to-gas ratio is enhanced over solar abundances (Youdin and Goodman 2005; Johansen et al. 2007). The pressure-supported gas orbits at a somewhat slower azimuthal velocity than would a particle at the same orbital radius on a purely Keplerian orbit. As such, particles orbiting in the gas experience a "head wind" as they encounter the slower orbiting gas, which causes the particle orbits to lose energy and drift radially inward. If as particles drift inward a local concentration forms, this concentration will accelerate the local gas somewhat, lessening the rate of the concentration's inward drift. The concentration can then continue to grow by accreting outer particles that are drifting inward more rapidly, and as the concentration grows, its effects on the gas strengthen, allowing its drift to slow further and its growth to continue. If the local spatial density of solids becoming sufficiently high, the concentration can rapidly gravitationally collapse to form ~100-km-class planetesimals directly from pebbles. Other proposed concentration mechanisms for particles include turbulent eddies (Cuzzi et al. 2008) and pressure bumps (e.g., associated with volatile snow lines, forming giant planets, and a variety of hydrodynamic or magnetohydrodynamic instabilities [Johansen et al. 2014]). Alternatively, some studies that consider a distribution of velocities and higher sticking efficiencies for ices find pairwise accretion might still form planetesimals. Further progress on this idea will require better models, and expanded studies of sticking in low-temperature, low-pressure environments.

To test different planetesimal formation mechanisms, it is useful to consider constraints from bodies relatively undisturbed by collisional or dynamical evolution. For example, a common outcome of a gravitationally bound clump of particles undergoing collapse is the formation of an equal-sized well separated binary. Many binaries of this nature are found in the Kuiper belt beyond Neptune, with most residing in the cold classical Kuiper belt region. Most of these binaries (~80 percent) have prograde rotation relative to their heliocentric orbits, a prediction fully consistent with the streaming instability, where prograde clumps are more likely to become gravitationally bound than retrograde clumps (Nesvorný et al. 2019, 2021). Individual binary components also have similar colors, again consistent with being formed simultaneously during planetesimal formation (Noll et al. 2020).

Other SI predictions are that primitive planetesimals are often porous, low-strength composites of ~mm–cm size grains with possibly limited evidence of processing by hypervelocity and catastrophic impacts. This is consistent with high-resolution Rosetta images of comet Churyumov-Gerasimenko's surface and measurements of the high porosity and weak strength of comets and Kuiper belt objects, particularly Arrokoth (see Figure 2-20), a contact binary in the cold classical Kuiper belt with modestly flattened lobes ($22 \times 20 \times 7$ km and $14 \times 14 \times 10$ km) that likely formed by the gravitational collapse of a "pebble cloud" formed by the SI or a related aerodynamic concentration mechanism (Blum et al. 2017; McKinnon et al. 2020). Last, SI predictions appear to broadly match the inferred primordial size distributions of the asteroids and Kuiper belts (Johansen et al. 2015).

Q1.3c Which Reservoirs and Materials (Gas, Rock, Ice, Organics) Formed into Planetesimals?

Isotopic and elemental measurements of planetary objects can constrain the provenance of their source materials and the temperatures at which these bodies and their constituents formed. The abundances and isotope compositions of HCNO (see Figure 4-2) and noble gases are among the best available tracers for investigating the origins of planetary material.

The building blocks of the inner planets, as exemplified by the compositions of Earth, Mars, to a lesser extent the venusian atmosphere, and of primitive and differentiated meteorites did not acquire HCNO directly and unaltered from the protosolar nebula, but instead from material that was isotopically distinct. Such heterogeneity might have existed in the interstellar medium or in the parent molecular cloud during the formation of icy, organic, and silicate-bearing dust (see Figure 4-3). Trapping of isotopically processed HCNO into solids might have maintained strong isotopic heterogeneities relative to the nebular gas composition. Processed grains were transported and distributed in the disk where they accreted into forming planetesimals.

The isotopic compositions of HCNO make it possible to trace genetic relationships between the different families of meteorites and the inner planets like Earth and Mars. Earth was mainly sourced by material akin of the so-called NC chondrites, especially the enstatite chondrite group (Kleine et al. 2020), whereas material originating from more volatile-rich, and presumably more distant material akin to CC meteorites might have supplied volatile elements to an initially dry proto-Earth (Alexander et al. 2012; Marty 2012; see Chapter 6). The analysis of cometary matter, especially during the Rosetta mission, strongly suggests that such bodies also contributed volatiles to the terrestrial atmosphere and oceans, although in more limited amounts (Marty et al. 2017).

The case of Mars is more difficult to investigate, partly because the composition of volatile elements in the martian mantle is poorly known owing to lack of samples, but also because the martian atmosphere and hydrosphere evolved drastically as they lost volatiles to space. The combination makes it difficult to identify the original building blocks of Mars (Carr and Head 2003). The composition of venusian atmosphere, crust, and mantle is unknown except for a few low precision measurements of atmospheric stable isotopes and noble gases (Donahue et al. 1982).

Knowledge of the composition of volatile-bearing bodies, such as comets or primitive asteroids not yet sampled in our collections, will permit a major advance in our understanding of the origin of inner planet material as well as of dynamic processes that led to the distribution of dust during solar system formation. New observations of primitive bodies and sample return mission that clarify genetic relationships between meteorite and IDPs already in our collections and specific bodies would be particularly valuable.

In the case of Jupiter, the Galileo probe found that the most volatile elements (e.g., noble gases and nitrogen) appear to be just as enriched in Jupiter's atmosphere as are less-volatile elements (e.g., carbon, sulfur, and phosphorus). These enrichments are relative to the abundance ratio of the element of interest to that of hydrogen in the Sun (Q2.1). The Galileo probe results support the core accretion model and led to the idea that the building blocks of Jupiter formed at very low temperatures (and thus perhaps farther out in the nebula than 5 AU). Otherwise, the noble gases and nitrogen (primarily as N_2) would not have been trapped in solid form nearly as efficiently as carbon, and especially sulfur and phosphorus. Accretion of Jupiter's building blocks at larger orbital distances may introduce greater constraints on accretion efficiency needed to form Jupiter prior to nebular dispersal.

The Juno mission has subsequently confirmed Jupiter's nitrogen enrichment, with measurements of the deep-water abundance in Jupiter's equatorial zone suggesting that oxygen also shows a similar enrichment (by a factor of ~3). Interpreting the apparently uniform element enrichments in terms of temperature-controlled volatile trapping in solids could indicate that the building blocks of Jupiter formed in a more distant region of the protoplanetary disk, where water ice would have been in an amorphous form and cold enough that the trapping efficiency of volatiles was uniform.

Alternatively, Jupiter's elemental enrichments could be telling us more about the evolution of nebular gas rather than the thermal history of small particles and planetesimals. This would apply if Jupiter captured gas at a time when the gas in the protoplanetary disk was already enriched. The enrichment process may have involved a depletion of hydrogen (H_2) in the disk owing to the great difficulty of retaining it in amorphous ice during gas loss. The preferential removal of hydrogen from the disk would make other elements, whose volatiles can be trapped in amorphous ice, appear enriched with respect to hydrogen. The continual removal of hydrogen over time also may produce larger enrichments until the nebular gas is fully dissipated. This scenario would imply that elemental enrichments may provide chronological information that can be linked to disk evolution. Determining the elemental enrichments in the atmospheres of the other giant planets is key to generalizing and discriminating between different scenarios of disk conditions and evolution that led to giant planet formation. The noble gases are particularly useful in this respect (see Q2.2).

Nonradiogenic argon-36 (^{36}Ar) was measured in Titan's atmosphere by the Huygens probe (Niemann et al. 2010). The expectation was that ^{36}Ar would have a high abundance (relative to atmospheric N_2) if Titan formed at low temperatures. Instead, a very low ^{36}Ar abundance was measured, suggesting Titan's building blocks had a relatively warm origin. This putative warm environment was likely an offshoot of the protoplanetary disk, one that formed a protosatellite disk around Saturn (see Q2.3 and Question 8; Chapters 5 and 11, respectively). Other bodies that are likely to have formed directly in the protoplanetary nebula and have N_2-rich atmospheres that are amenable to this test are Triton and Pluto. Voyager 2 and New Horizons were unable to measure their argon abundances or their isotopic compositions. But if the $^{36}Ar/N_2$ ratio were determined by a future mission, Triton or Pluto could provide another valuable snapshot of temperatures in a distant region of the solar system (i.e., the primordial Kuiper belt).

Isotopic measurements can provide powerful constraints on the specific chemical forms of elements that existed in the protoplanetary disk and were incorporated into large bodies. In the outer solar system, the two isotopic systems that have proven to be most useful are hydrogen and nitrogen, owing to relatively large differences in isotopic compositions between various hydrogen and nitrogen- bearing materials. The practical consequence is that a robust conclusion can often be reached, even if in situ or remote-sensing data lack high precision.

Returning to the previous examples, the high $^{15}N/^{14}N$ ratio in Titan's nitrogen supports an alternative source as the most important N-bearing constituent in Titan's building blocks. This isotopic measurement is consistent with the aforementioned warm origin of Titan's building blocks that would have prevented the trapping of both ^{36}Ar and N_2. Plausible alternative sources for Titan's nitrogen are ammonia-bearing ices or salts and N-bearing organic matter (Miller et al. 2019).

The deuterium/hydrogen or D/H ratios in water, the dominant hydrogen reservoir on icy worlds, have been measured for certain bodies (Saturn's midsize icy satellites; Waite et al. 2009; Clark et al. 2019). Values appear to be similar to or slightly above that of Earth's ocean water, and they could help disentangling different sources of water and organic compounds that contributed to the bulk inventory of hydrogen on these bodies and on Titan as well.

For Jupiter, the ratios of both D/H and $^{15}N/^{14}N$ have been measured. They are found to be indistinguishable from the protosolar values, as expected for the dominant inheritance of H_2 and N_2 from the protoplanetary disk. At the other giant planets, higher values of D/H are expected to reflect increasing proportions of water relative to H_2, and this is generally observed with the ice giants having noticeably higher D/H ratios. However, the $^{15}N/^{14}N$ ratio remains to be measured in Saturn, Uranus, and Neptune, which could help constrain the deviation from protosolar values and potential contributions of N-bearing organics or ices.

Jupiter-family and nearly isotropic (i.e., Oort cloud) comets have consistently demonstrated heavy nitrogen isotope ratios. These large enrichments can be explained by interstellar chemistry theories involving ion-molecule ^{15}N fractionation at 10 K (Charnley and Rodgers 2002; Rodgers and Charnley 2008) that has been observed in dark cold cores and modeled in disks. The similarity of the $^{14}N/^{15}N$ ratios found in comets and interplanetary dust particles strengthens a possible link to interstellar chemistry as the origin of isotopically anomalous organic particles in comets.

Strategic Research for Q1.3

- **Infer the compositions and locations of nebular source reservoirs** by return of samples from, especially, comet surfaces as well as from asteroids; measuring the elemental and stable isotopic compositions of refractory and volatile elements with lander and orbiter missions (especially for the ice giants, Centaurs, and Mercury and also Venus, comets, Saturn, and Kuiper belt objects); and ground- and space-based telescopic observations of atmospheric and/or sublimated volatiles toward small bodies in outer solar system, planets across the solar system and their moons; and laboratory petrological, elemental, and isotopic analyses of returned and terrestrially collected samples.
- **Clarify the mechanisms that enabled accretion of objects beyond the "fragmentation barrier" size (~1 m)** through determination of the structure, porosity, magnetization, and size and shapes of grains on small bodies by return of comet surface samples; in situ imaging, strength, and gravity measurements of comets, Centaurs, or Kuiper belt objects; ground- and space-based telescopic observations; accretion modeling; and laboratory petrological and paleomagnetic analyses of returned and terrestrially collected samples.
- **Test models of accretion of micrometer-to-centimeter-scale objects by determining the relative ages of crystallization of chondrules, CAIs, and mineral grains, and thermal and aqueous alteration events** using return of comet surface samples; and laboratory radioisotopic analyses of returned surface samples from comets and asteroids and terrestrially collected samples.
- **Constrain accretion processes in protoplanetary disks by resolved studies of the volatile composition as well as of the composition, sizes and shapes of grains** using ground- and space-based telescopic observations of protoplanetary disks.
- **Understand the processes of accretion, fragmentation, and deformation associated with grain and particle collisions** through laboratory grain accretion experiments, observations of collisions in dense giant planet rings, and modeling.

Q1.4 HOW AND WHEN DID THE NEBULA DISPERSE?

The mechanisms and timescales of solar nebular dispersal remain fundamental open questions. One interpretation of meteorite constraints is that planetesimal formation came to an end when the solar nebula dispersed. With that said, the solar nebula is big enough that this could mean modestly different timescales for the endgame of planetesimal formation in the inner and outer solar systems, and perhaps within those regions as well. For example, photoevaporation of the solar nebula from the outside-in may limit the formation of distant planetesimals in the primordial Kuiper belt, potentially explaining why Neptune's outward migration ground to a halt prior to passing through the most distant indigenous population of icy planetesimals in the so-called cold classical Kuiper belt. The end of the solar nebula also means the termination of gas processes that can damp planetary eccentricities and inclinations. This could set the stage for a period of violent upheaval for the orbits of the giant planets (see Question 2, Chapter 5). The dispersal time of the nebula also affects the final composition of planetary objects by truncating the condensation and accretion sequence at a particular location-dependent temperature and pressure.

Q1.4a What Was the Lifetime of the Solar Nebula?

A key parameter influencing the final architecture of the solar system and the composition and structure of the planets is the lifetime of the nebula. The dispersal time of the nebula sets the timescale for stellar accretion, the formation of the gas giants, and the epoch of gas-driven planetary migration and has major implications for dust dynamics and disk structure (Takeuchi and Artymowicz 2001), the final sizes and eccentricities of the terrestrial planets (Kominami and Ida 2004), and the viability of hypothesized chondrule and planetesimal formation mechanisms involving nebular gas or nebular magnetic fields. For example, disk gravitational instabilities could in principle have formed the giant planets in <0.1 million years, while so-called core accretion is favored by longer (several to perhaps >10 million years) timescales (Helled et al. 2014; see Question 2). The evolving composition of the gas resulting from progressive condensation of increasingly more volatile elements can in turn influence the final compositions of the terrestrial planets (Grossman 1972), giant planets (Guillot and Hueso 2006; Monga and Desch 2015) and their moons (Glein 2017).

The lifetime of the solar nebula is poorly constrained (Figure 4-4) through limited direct and indirect evidence. Astronomical observations of the abundance of dust and gas and of active accretion onto protostars indicate that protoplanetary disks have estimated lifetimes from <1 to ~20 million years with a mean value of

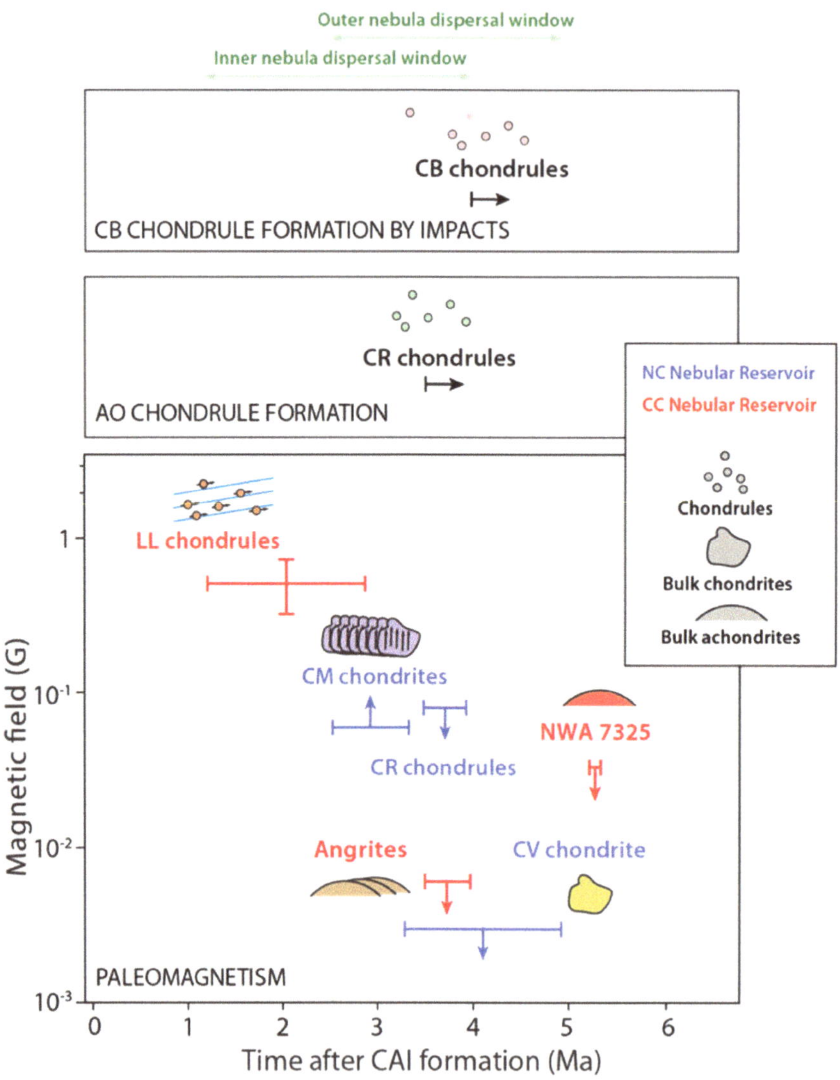

FIGURE 4-4 Meteorite constraints on the lifetime of the nebula and its magnetic field. (*Bottom*) Constraints from meteorite paleomagnetism. All meteorites known to have acquired magnetic records prior to ~3 million years after calcium-aluminum-rich inclusion (CAI) formation (LL chondrites from the noncarbonaceous chondrites [NC] reservoir and CM chondrites from the carbonaceous chondrites [CC] reservoir) show evidence for a substantial nebular magnetic field (10–100 µT). All meteorites formed after 3 million years (angrites and NWA 7325 from the NC chondrites and CV chondrites from the CC reservoir) show no evidence of nebular magnetic fields, suggesting the nebula had largely dissipated by this time. (*Middle*) The presence of agglomeratic olivine (AO) chondrules suggests the persistence of the nebular dust disk until at least ~3 million years after CAI formation. (*Top*) The formation of CB chondrules by impacts on planetesimals suggests the persistence of the nebular gas until at least ~4 million years after CAI formation. SOURCES: Created by B. Weiss based on data from Weiss et al. (2021), Johnson et al. (2016), and Schrader et al. (2018).

2 million years (Mamajek 2009). However, these bounds are themselves uncertain owing to uncertainties in the model ages of pre-main sequence stars that host these disks. Other estimates of the mean lifetime range up to ~6 million years (Bell et al. 2013). Also, there is the additional uncertainty associated with where in this distribution the solar system lies.

Other than inferences inferred from astronomical observations lifetimes of protoplanetary disks, the lifetime of the solar nebula has been constrained mainly using observations of chondrites. The presence of agglomeratic olivine chondrules in CR chondrites has been interpreted as evidence that the nebular dust disk persisted until at least the formation time of CR chondrules (Schrader et al. 2018) (i.e., at >3.46 million years). It has also been proposed that the formation of chondrules in CH and CB chondrites requires the presence of nebular gas under the hypothesis that they are impact melt sprays from planetesimal collisions. Such gas could enable planetesimals to reach the high relative velocities that can produce such impacts (Johnson et al. 2016). This may indicate that the solar nebula persisted until the ~4–6 million years formation age of CB chondrules.

Because the sustenance of magnetic fields requires the existence of a conducting medium, the dispersal of the nebula would have led to dissipation of the nebular field. Therefore, under the assumption that the field is present whenever there is gas (and this might not necessarily be the case), the dispersal time of the nebula could be estimated by establishing when the solar nebula magnetic disappeared as inferred from the absence of paleomagnetism in meteorites younger than a certain age. In fact, paleomagnetic measurements of several meteorite groups indicate that the solar nebula dispersed sometime between 1.2 and 3.9 million years after CAI-formation in the region where ordinary chondrites formed and between 2.5 and 4.9 million years after CAI-formation in the region where carbonaceous chondrites formed (Weiss et al. 2021). The presence of certain chondrules in CR chondrites has been interpreted as evidence that at least the nebular dust disk persisted until at least 3.5 million years after CAI-formation (Schrader et al. 2018). It has also been proposed that the formation of certain metal-rich chondrules may indicate the presence of nebular gas until ~4–6 million years after CAI-formation (Morris et al. 2015; Johnson et al. 2016; Garvie et al. 2017).

Several uncertainties limit our ability to accurately date the nebular lifetime. Current constraints are derived from measurements on only a half dozen meteorite groups with a narrow range of ages. The locations in the nebula for which these constraints apply are highly uncertain (up to tens of AU). Also, the uncertainties from these measurements on the dispersal times have large uncertainties relative to the mean age of dispersal. There are very few direct constraints on the gas density itself, with most indicating simply whether a solar nebula analogous to that observed for actively accreting protoplanetary disks is present or absent. Last, there are virtually no constraints on the evolution of the composition of the residual gas.

Q1.4b What Mechanisms Dispersed the Nebula?

The processes by which the protoplanetary disk dispersed are uncertain. The steady accretion of the Sun fed by viscous disk spreading would have progressively depleted the disk. However, astronomical observations indicate that after slowly evolving for several million years, most disks abruptly disperse at a rate an order of magnitude faster than their earlier accretion rates (Ercolano and Pascucci 2017).

This two-timescale evolution has motivated two theorized disk dispersal mechanisms. A leading candidate is photoevaporation, in which the central star and/or neighboring stars heat the disk atmosphere increasing the velocity of the gas to above the escape velocity (Owen et al. 2010; Weiss et al. 2021). Depending on the source of the radiation and its spectrum, this can lead the disk to disperse from the inside out, outside in, and/or to form gaps (Gorti et al. 2009). Alternatively, the solar wind could be launched by magnetic mechanisms (Shadmehri and Ghoreyshi 2019). If it has a sufficiently high flux, this wind could potentially dominate over photoevaporation in depleting the disk. The role of magnetized disk winds versus photoevaporation could be constrained by paleomagnetic measurements of meteorites, as discussed here. Also, constraints on the gas density as a function of time and distance from the Sun and young stellar objects using meteorite studies and astronomical observations could determine the direction of dispersal (inward, outward, or with gaps), which in turn could distinguish between the stellar sources for photoevaporation and the role of giant planet formation.

Several key unsolved questions thus remain. We do not yet know with confidence how rapidly the solar nebula dispersed, or whether it evolved over different timescales in the inner and outer solar systems. It is also uncertain whether the dispersal of the nebula was associated with long-lived disk substructures. Last, we do not yet understand the role of hydrodynamic winds and/or magnetic fields in a dispersing gas in different regions of the nebula, nor do we know how the composition of the gas and the ratio of gas to dust evolved with time in the same regions. Answers to these questions would enable us to constrain which mechanisms (e.g., photoevaporation, magnetized disk winds, or other mechanisms) led to dispersal of the solar nebula and how the accretion of planets was influenced by the dispersal process.

Strategic Research for Q1.4

- **Constrain the temporal and spatial evolution of the composition of nebular gas** with in situ measurement of the volatile elemental compositions (noble gases and hydrogen) of Saturn, Uranus, and Neptune.
- **Constrain the temporal and spatial evolution of the composition of nebular solids** with return of samples from comet surfaces and laboratory petrologic, elemental, and isotopic studies and analyses of returned and terrestrially collected samples.
- **Measure the intensity of the solar nebula magnetic field as a function of space and time** with return of asteroid and comet surface samples; in situ magnetic measurements at asteroids, comets, Centaurs, and Kuiper belt objects; and laboratory paleomagnetic measurements of returned and terrestrially collected samples.
- **Measure the temporal and spatial evolution of the density, composition, and magnetism of protoplanetary disks** using optical, infrared, millimeter, and radio measurements of nearby young stellar objects.

SUPPORTIVE ACTIVITIES FOR QUESTION 1

- Expanded terrestrial-based extraterrestrial sample collection (especially ANSMET, the Antarctic Search for Meteorites).
- Expanded laboratory instrumentation development and acquisition beyond the support associated with active sample return missions.
- Laboratory observations of returned samples from Ryugu, Bennu, and Phobos.
- Telescopic observations that support cross-disciplinary studies relevant to early solar system processes, particularly protoplanetary disks.

REFERENCES

Adams, F.C. 2010. "The Birth Environment of the Solar System." *Annual Review of Astronomy and Astrophysics* 48:47–85.

Aikawa, Y., K. Furuya, U. Hincelin, and E. Herbst. 2018. "Multiple Paths of Deuterium Fractionation in Protoplanetary Disks." *Astrophysical Journal* 855:119.

Alexander, C.M.O.'D., R. Bowden, M.L. Fogel, K.T. Howard, C.D.K. Herd, and L.R. Nittler. 2012. "The Provenances of Asteroids, and Their Contributions to the Volatile Inventories of the Terrestrial Planets." *Science* 337:721.

Alexander, C.M.O.'D., G.D. Cody, B.T. De Gregorio, L.R. Nittler, and R.M. Stroud. 2017. "The Nature, Origin and Modification of Insoluble Organic Matter in Chondrites, the Major Source of Earth's C and N." *Chemie der Erde/Geochemistry* 77:227–256.

Altwegg, K., H. Balsiger, A. Bar-Nun, J.J. Berthelier, A. Bieler, P. Bochsler, C. Briois, et al. 2015. "67P/Churyumov-Gerasimenko, a Jupiter Family Comet with a High D/H Ratio." *Science* 347(6220):1261952.

Amelin, Y., A. Kaltenbach, T. Iizuka, C.H. Stirling, T.R. Ireland, M. Petaev, and S.B. Jacobsen. 2010. "U-Pb Chronology of the Solar System's Oldest Solids with Variable U-238/U-235." *Earth and Planetary Science Letters* 300:343–350.

Andrews, S.M. 2020. "Observations of Protoplanetary Disk Structures." *Annual Review of Astronomy and Astrophysics* 58:483–528.

Armitage, P. 2015. "Physical Processes in Protoplanetary Disks." Pp. 1–150 in *From Protoplanetary Disks to Planet Formation*, M. Audard, M.R. Meyer, Y. Alibert, eds. Berlin, Germany: Springer.

Bally, J., and W.D. Langer. 1982. "Isotope-Selective Photodestruction of Carbon Monoxide." *Astrophysical Journal* 255:143–148.

Bell, C.P.M., T. Naylor, N.J. Mayne, R.D. Jeffries, and S.P. Littlefair. 2013. "Pre-Main-Sequence Isochrones. II. Revising Star and Planet Formation Time-Scales." *Monthly Notices of the Royal Astronomical Society* 434:806–831.

Bergin, E., Y. Aikawa, G.A. Blake, and E.F. van Dischoeck. 2007. "The Chemical Evolution of Protoplanetary Disks." Pp. 751–766 in *Protostars and Planets V.* B. Reipurth, D. Jewitt, K. Keil, eds. Tucson: University of Arizona Press.

Birnstiel, T., C.W. Ormel, and C.P. Dullemond. 2011. "Dust Size Distributions in Coagulation/Fragmentation Equilibrium: Numerical Solutions and Analytical Fits." *Astronomy & Astrophysics* 525:A11.

Blum, J., B. Gundlach, M. Krause, M. Fulle, A. Johansen, J. Agarwal, I. von Borstel, et al. 2017. "Evidence for the Formation of Comet 67P/Churyumov-Gerasimenko Through Gravitational Collapse of a Bound Clump of Pebbles." *Monthly Notices of the Royal Astronomical Society* 469:S755–S773.

Bockelée-Morvan, D., D. Gautier, F. Hersant, J.-M. Huré, and F. Robert. 2002. "Turbulent Radial Mixing in the Solar Nebula as the Source of Crystalline Silicates in Comets." *Astronomy & Astrophysics* 384:1107–1118.

Bockelée-Morvan, D., V. Debout, S. Erard, et al. 2015. "First Observations of H_2O and CO_2 Vapor in Comet 67P/Churyumov-Gerasimenko Made by VIRTIS Onboard Rosetta," *Astronomy & Astrophysics* 583:A6.

Boley, A.C., M.A. Morris, and S.J. Desch. 2013. "High-Temperature Processing of Solids Through Solar Nebular Bow Shocks: 3D Radiation Hydrodynamics Simulations with Particles." *Astrophysical Journal* 776:101.

Bollard, J., J.N. Connelly, and M. Bizzarro. 2015. "Pb-Pb Dating of Individual Chondrules from the CBa Chondrite Gujba: Assessment of the Impact Plume Formation Model." *Meteoritics and Planetary Science* 50:1197–1216.

Bollard, J., J.N. Connelly, M.J. Whitehouse, E.A. Pringle, L. Bonal, J.K. Jørgensen, Å. Nordlund, et al. 2017. "Early Formation of Planetary Building Blocks Inferred from Pb Isotopic Ages of Chondrules." *Science Advances* 3(8):e1700407.

Boss, A.P., and J.N. Goswami. 2006. "Presolar Cloud Collapse and the Formation and Early Evolution of the Solar System." Pp. 171–186 in *Meteorites and the Early Solar System II*, D.S. Lauretta and H.Y. McSween, Jr., eds. Tucson: University of Arizona Press.

Bottke, W.F., M. Broz, D.P. O'Brien, A. Campo Bagatin, A. Morbidelli, and S. Marchi. 2015. "The Collisional Evolution of the Asteroid Belt." Pp. 701–724 in *Asteroids IV*, P. Michel, F. DeMeo, and W.F. Bottke, eds. Tucson: University of Arizona Press. https://uapress.arizona.edu/book/asteroids-iv.

Bouvier, A., and M. Wadhwa. 2010. "The Age of the Solar System Redefined by the Oldest Pb–Pb Age of a Meteoritic Inclusion." *Nature* 3:637–641.

Brennecka, G.A., C. Burkhardt, G. Budde, T.S. Kruijer, F. Nimmo, and T. Kleine. 2020. "Astronomical Context of Solar System Formation from Molybdenum Isotopes in Meteorite Inclusions. *Science* 370:837–840.

Busemann, H., H. Baur, and R. Wieler. 2000. "Primordial Noble Gases in 'Phase Q' in Carbonaceous and Ordinary Chondrites Studied by Closed System Stepped Etching." *Meteoritics and Planetary Science* 35:949–973.

Cameron, A.G.W., and J.W. Truran. 1977. "The Supernova Trigger for Formation of the Solar System." *Icarus* 30:447–461.

Carr, M.H., and J.W. Head. 2003. "Oceans on Mars: An Assessment of the Observational Evidence and Possible Fate. *Journal of Geophysical Research: Planets* 108:5042.

Charnley, S.B., and S.D. Rodgers. 2002. "The End of Interstellar Chemistry as the Origin of Nitrogen in Comets and Meteorites." *Astrophysical Journal* 569:L133–L137.

Clark, R.N., R.H. Brown, D.P. Cruikshank, and G.A. Swayze. 2019. "Isotopic Ratios of Saturn's Rings and Satellites: Implications for the Origin of Water and Phoebe." *Icarus* 321:791–802.

Cleeves, L.I., E.A. Bergin, C.M.O.'D. Alexander, F. Du, D. Graninger, K.I. Öberg, and T.J. Harries. 2014. "The Ancient Heritage of Water Ice in the Solar System." *Science* 345:1590–1593.

Connelly, J.N., M. Bizzarro, A.N. Krot, Å. Nordlund, D. Wielandt, and M.A. Ivanova. 2012. "The Absolute Chronology and Thermal Processing of Solids in the Solar Protoplanetary Disk." *Science* 338:651.

Connolly, H.C., and R.H. Jones. 2016. "Chondrules: The Canonical and Noncanonical Views." *Journal of Geophysical Research: Planets* 121:1885–1899.

Connolly, H.C., G.R. Huss, and G.J. Wasserburg. 2001. "On the Formation of Fe-Ni Metal in Renazzo-like Carbonaceous Chondrites." *Geochimica et Cosmochimica Acta* 65:4567–4588.

Connolly, H.C., A.N. Krot, and S.S. Russell. 2018. "Summary of Key Outcomes." Pp. 428–436 in *Chondrules: Records of Protoplanetary Disk Processes*. Cambridge Planetary Sciences, S. Russell, H. Connolly, Jr., and A. Krot, eds. Cambridge, UK: Cambridge University Press.

Cuzzi, J.N., R.C. Hogan, and K. Shariff. 2008. "Toward Planetesimals: Dense Chondrule Clumps in the Protoplanetary Nebula." *Astrophysical Journal* 687:1432–1447.

Dauphas, N., and M. Chaussidon. 2011. "A Perspective from Extinct Radionuclides on a Young Stellar Object: The Sun and Its Accretion Disk." *Annual Review of Earth and Planetary Sciences* 39:351–386.

Desch, S.J., A. Kalyaan, and C.M.O'D. Alexander. 2018. "The Effect of Jupiter's Formation on the Distribution of Refractory Elements and Inclusions in Meteorites." *Astrophysical Journal Supplement Series* 238:11.

Dominik, C., J. Blum, J.N. Cuzzi, and G. Wurm. 2006. "Growth of Dust as the Initial Step Toward Planet Formation." Pp. 783–800 in *Protostars and Planets V*, B. Reipurth, D. Jewitt, and K. Keil, eds. Tucson: University of Arizona Press.

Donahue, T.M., J.H. Hoffman, R.R. Hodges, and A.J. Watson. 1982. "Venus Was Wet: A Measurement of the Ratio of Deuterium to Hydrogen." *Science* 216:630–633.

Ebel, D.S., and L. Grossman. 2000. "Condensation in Dust-Enriched Systems." *Geochimica et Cosmochimica Acta* 64:339–366.

Ehrenfreund, P., and S.B. Charnley. 2000. "Organic Molecules in the Interstellar Medium, Comets, and Meteorites: A Voyage from Dark Clouds to the Early Earth." *Annual Review of Astronomy and Astrophysics* 38:427–483.

Ercolano, B., and I. Pascucci. 2017. "The Dispersal of Planet-Forming Discs: Theory Confronts Observations." *Royal Society Open Science* 4:170114.

Floss, C., and P. Haenecour. 2016. "Presolar Silicate Grains: Abundances, Isotopic and Elemental Compositions, and the Effects of Secondary Processing." *Geochemical Journal* 50:3–25.

Frick, U., R. Mack, and S. Chang. 1979. "Noble Gas Trapping and Fractionation During Synthesis of Carbonaceous Matter." *Lunar and Planetary Science Conference Proceedings* 2:1961–1972.

Friesen, R.K., A. Pon, T.L. Bourke, P. Caselli, J. Di Francesco, J.K. Jørgensen, and J.E. Pineda. 2018. "ALMA Detections of the Youngest Protostars in Ophiuchus." *Astrophysical Journal* 869:158.

Fromang, S., and G. Lesur. 2019. "Angular Momentum Transport in Accretion Disks: A Hydrodynamical Perspective." Pp. 391–413 in *Astro Fluid*, A.S. Brun, S. Mathis, C. Charbonnel, and B. Dubrulle, eds. EAS Publications Series 82.

Garani, J., and J.R. Lyons. 2020. "Modeling Nitrogen Isotope Chemistry in the Solar Nebula." *51st Annual Lunar and Planetary Science Conference*, 2540.

Garvie, L.A.J., L.P. Knauth, and M.A. Morris. 2017. "Sedimentary Laminations in the Isheyevo (CH/CBb) Carbonaceous Chondrite Formed by Gentle Impact-Plume Sweep-Up." *Icarus* 292:36–47.

Geiss, J., and G. Gloeckler. 1998. "Abundances of Deuterium and Helium-3 in the Protosolar Cloud." *Space Science Reviews* 84:239–250.

Glavin, D.P., J.E. Elsila, H.L. McLain, J.C. Aponte, E.T. Parker, J.P. Dworkin, D.H. Hill, et al. 2021. "Extraterrestrial Amino Acids and L Enantiomeric Excesses in the CM2 Carbonaceous Chondrites Aguas Zarcas and Murchison." *Meteoritics and Planetary Science* 56:148–173.

Glein, C.R. 2017. "A Whiff of Nebular Gas in Titan's Atmosphere—Potential Implications for the Conditions and Timing of Titan's Formation." *Icarus* 293:231–242.

Gorti, U., C.P. Dullemond, and D. Hollenbach. 2009. "Time Evolution of Viscous Circumstellar Disks Due to Photoevaporation by Far-Ultraviolet, Extreme-Ultraviolet, and X-ray Radiation from the Central Star." *Astrophysical Journal* 705:1237–1251.

Gregory, T., T.-H. Luu, C.D. Coath, S.S. Russell, and T. Elliott. 2020. "Primordial Formation of Major Silicates in a Protoplanetary Disk with Homogeneous $^{26}Al/^{27}Al$." *Science Advances* 6(11):eaay9626.

Grossman, L. 1972. "Condensation in the Primitive Solar Nebula." *Geochimica et Cosmochimica Acta* 36:597–619.

Guillot, T., and R. Hueso. 2006. "The Composition of Jupiter: Sign of a (Relatively) Late Formation in a Chemically Evolved Protosolar Disc." *Monthly Notices of the Royal Astronomical Society* 367:L47–L51.

Heays, A.N., R. Visser, R. Gredel, W. Ubachs, B.R. Lewis, S.T. Gibson, and E.F. van Dishoeck. 2014. Isotope Selective Photodissociation of N_2 by the Interstellar Radiation Field and Cosmic Rays. *Astronomy & Astrophysics* 562:A61.

Helled, R., P. Bodenheimer, M. Podolak, A. Boley, F. Meru, S. Nayakshin, J. Fortney, et al. 2014. "Giant Planet Formation, Evolution, and Internal Structure." Pp. 643–666 in *Protostars and Planets VI*, H. Beuther, C. P. Dullemond, R. S. Klessen, T. K. Henning, eds. Tucson: University of Arizona Press.

Jilly-Rehak, C.E., G.R. Huss, and K. Nagashima. 2017. "$^{53}Mn/^{53}Cr$ Radiometric Dating of Secondary Carbonates in CR Chondrites: Timescales for Parent Body Aqueous Alteration." *Geochimica et Cosmochimica Acta* 201:224–244.

Johansen, A., and M. Lambrechts. 2017. "Forming Planets via Pebble Accretion." *Annual Review of Earth and Planetary Sciences* 45:359–387.

Johansen, A., J.S. Oishi, M.-M. Mac Low, H. Klahr, T. Henning, and A. Youdin. 2007. "Rapid Planetesimal Formation in Turbulent Circumstellar Disks." *Nature* 448:1022–1025.

Johansen, A., J. Blum, H. Tanaka, C. Ormel, M. Bizzaro, and H. Rickman. 2014. "The Multifaceted Planetesimal Formation Process." Pp. 547–570 in *Protostars and Planets VI*, H. Beuther, C.P. Dullemond, R.S. Klessen, and T.K. Henning, eds. Tucson: University of Arizona Press.

Johansen, A., M.-M. Mac Low, P. Lacerda, and M. Bizzarro. 2015. "Growth of Asteroids, Planetary Embryos, and Kuiper Belt Objects by Chondrule Accretion." *Science Advances* 1(3):1500109.

Johnson, B.C., D.A. Minton, H.J. Melosh, and M.T. Zuber. 2015. "Impact Jetting as the Origin of Chondrules." *Nature* 517:339–341.

Johnson, B.C., K.J. Walsh, D.A. Minton, A.N. Krot, and H.F. Levison. 2016. "Timing of the Formation and Migration of Giant Planets as Constrained by CB Chondrites." *Science Advances* 2(12):e1601658.

Joswiak, D.J., D.E. Brownlee, A.N. Nguyen, and S. Messenger. 2017. "Refractory Materials in Comet Samples." *Meteoritics and Planetary Science* 52:1612–1648.

Joung, M.K.R., M.-M. Mac Low, and D.S. Ebel. 2004. "Chondrule Formation and Protoplanetary Disk Heating by Current Sheets in Nonideal Magnetohydrodynamic Turbulence." *Astrophysical Journal* 606:532–541.

Kita, N.T., H. Nagahara, S. Tachibana, S. Tomomura, M.J. Spicuzza, J.H. Fournelle, and J.W. Valley. 2010. "High Precision SIMS Oxygen Three Isotope Study of Chondrules in LL3 Chondrites: Role of Ambient Gas During Chondrule Formation." *Geochimica et Cosmochimica Acta* 74:6610–6635.

Kleine, T., G. Budde, C. Burkhardt, T.S. Kruijer, E.A. Worsham, A. Morbidelli, and F. Nimmo. 2020. "The Non-Carbonaceous-Carbonaceous Meteorite Dichotomy." *Space Science Reviews* 216:55.

Kominami, J., and S. Ida. 2004. "Formation of Terrestrial Planets in a Dissipating Gas Disk with Jupiter and Saturn." *Icarus* 167:231–243.

Krot, A.N. 2019. "Refractory Inclusions in Carbonaceous Chondrites: Records of Early Solar System Processes." *Meteoritics and Planetary Science* 54:1647–1691.

Kruijer, T.S., T. Kleine, and L.E. Borg. 2020. "The Great Isotopic Dichotomy of the Early Solar System." *Nature Astronomy* 4:32–40.

Kruss, M., and G. Wurm. 2018. "Seeding the Formation of Mercurys: An Iron-Sensitive Bouncing Barrier in Disk Magnetic Fields." *Astrophysical Journal* 869:45.

Lee, J.-E., E.A. Bergin, and J.R. Lyons. 2008. "Oxygen Isotope Anomalies of the Sun and the Original Environment of the Solar System." *Meteoritics and Planetary Science* 43:1351–1362.

Lellouch, E., B. Bézard, T. Fouchet, H. Feuchtgruber, T. Encrenaz, and T. de Graauw. 2001. "The Deuterium Abundance in Jupiter and Saturn from ISO-SWS Observations." *Astronomy & Astrophysics* 370:610–622.

Liu, M.-C., J. Han, A.J. Brearley, and A.T. Hertwig. 2019. "Aluminum-26 Chronology of Dust Coagulation and Early Solar System Evolution." *Science Advances* 5(9):eaaw3350.

Lodders, K. 2004. "Jupiter Formed with More Tar Than Ice." *Astrophysical Journal* 611:587–597.

Lyons, J.R., and E.D. Young. 2005. "CO Self-Shielding as the Origin of Oxygen Isotope Anomalies in the Early Solar Nebula." *Nature* 435:317–320.

Lyons, J.R., E. Gharib-Nezhad, and T.R. Ayres. 2018. "A Light Carbon Isotope Composition for the Sun." *Nature Communications* 9:908.

MacPherson, G.J., N.T. Kita, T. Ushikubo, E.S. Bullock, and A.M. Davis. 2012. "Well-Resolved Variations in the Formation Ages for Ca-Al-Rich Inclusions in the Early Solar System." *Earth and Planetary Science Letters* 331:43–54.

Mamajek, E.E. 2009. "Initial Conditions of Planet Formation: Lifetimes of Primordial Disks." *AIP Conference Proceedings* 1158:3–10.

Mandt, K.E., O. Mousis, J. Lunine, B. Marty, T. Smith, A. Luspay-Kuti, and A. Aguichine. 2020. "Tracing the Origins of the Ice Giants Through Noble Gas Isotopic Composition." *Space Science Reviews* 216:99.

Marrocchi, Y., B. Marty, P. Reinhardt, and F. Robert. 2011. "Adsorption of Xenon Ions onto Defects in Organic Surfaces: Implications for the Origin and the Nature of Organics in Primitive Meteorites." *Geochimica et Cosmochimica Acta* 75:6255–6266.

Marty, B. 2012. "The Origins and Concentrations of Water, Carbon, Nitrogen and Noble Gases on Earth." *Earth and Planetary Science Letters* 313:56–66.

Marty, B., M. Chaussidon, R.C. Wiens, A.J.G. Jurewicz, and D.S. Burnett. 2011. "A ^{15}N-Poor Isotopic Composition for the Solar System as Shown by Genesis Solar Wind Samples. *Science* 332(6037):1533.

Marty, B., K. Altwegg, H. Balsiger, A. Bar-Nun, D.V. Bekaert, J.-J. Berthelier, A. Bieler, et al. 2017. "Xenon Isotopes in 67P/Churyumov-Gerasimenko Show That Comets Contributed to Earth's Atmosphere." *Science* 356:1069–1072.

McKeegan, K.D., A.P.A. Kallio, V.S. Heber, G. Jarzebinski, P.H. Mao, C.D. Coath, T. Kunihiro, et al. 2011. "The Oxygen Isotopic Composition of the Sun Inferred from Captured Solar Wind." *Science* 332(6037):1528–1532.

McKinnon, W.B., D.C. Richardson, J.C. Marohnic, J.T. Keane, W.M. Grundy, D.P. Hamilton, D. Nesvorný, et al. 2020. "The Solar Nebula Origin of (486958) Arrokoth, a Primordial Contact Binary in the Kuiper Belt." *Science* 367(6481):aay6620.

Miller, K.E., C.R. Glein, and J.H. Waite. 2019. "Contributions from Accreted Organics to Titan's Atmosphere: New Insights from Cometary and Chondritic Data." *Astrophysical Journal* 871:59.

Monga, N., and S. Desch. 2015. "External Photoevaporation of the Solar Nebula: Jupiter's Noble Gas Enrichments." *Astrophysical Journal* 798:9.

Morbidelli, A., W.F. Bottke, D. Nesvorný, and H.F. Levison. 2009. "Asteroids Were Born Big." *Icarus* 204:558–573.

Morris, M.A., L.A.J. Garvie, and L.P. Knauth. 2015. "New Insight into the Solar System's Transition Disk Phase Provided by the Metal-Rich Carbonaceous Chondrite Isheyevo." *Astrophysical Journal Lettters* 801(2):L22.

Musiolik, G., J. Teiser, T. Jankowski, and G. Wurm. 2016. "Collisions of CO_2 Ice Grains in Planet Formation." *Astrophysical Journal* 818:16.

Nesvorný, D., R. Li, A.N. Youdin, J.B. Simon, and W.M. Grundy. 2019. "Trans-Neptunian Binaries as Evidence for Planetesimal Formation by the Streaming Instability." *Nature Astronomy* 3:808–812.

Nesvorný, D., R. Li, J.B. Simon, A.N. Youdin, D.C. Richardson, R. Marschall, and W.M. Grundy. 2021. "Binary Planetesimal Formation from Gravitationally Collapsing Pebble Clouds." *Planetary Science Journal* 2:27.

Niemann, H.B., S.K. Atreya, J.E. Demick, D. Gautier, J.A. Haberman, D.N. Harpold, W.T. Kasprzak, et al. 2010. "Composition of Titan's Lower Atmosphere and Simple Surface Volatiles as Measured by the Cassini-Huygens Probe Gas Chromatograph Mass Spectrometer Experiment." *Journal of Geophysical Research: Planets* 115:E12006.

Nittler, L.R., and F. Ciesla. 2016. "Astrophysics with Extraterrestrial Materials." *Annual Review of Astronomy and Astrophysics* 54:53–93.

Nittler, L.R., and S.Z. Weider. 2019. "The Surface Composition of Mercury." *Elements* 15:33–38.

Noll, K., W.M. Grundy, D. Nesvorný, and A. Thirouin. 2020. "Trans-Neptunian Binaries (2018)." Pp. 205–224 in *The Trans-Neptunian Solar System*, D. Prialnik, A. Barucci, and L.A. Young, eds. Amsterdam, Netherlands: Elsevier.

Owen, J.E. 2020. "Snow Lines Can Be Thermally Unstable." *Monthly Notices of the Royal Astronomical Society* 495:3160–3174.

Owen, T., P.R. Mahaffy, H.B. Niemann, S. Atreya, and M. Wong. 2001. "Protosolar Nitrogen." *Astrophysical Journal* 553:L77–L79.

Owen, J.E., B. Ercolano, C.J. Clarke, and R.D. Alexander. 2010. "Radiation-Hydrodynamic Models of X-Ray and EUV Photoevaporating Protoplanetary Discs." *Monthly Notices of the Royal Astronomical Society* 401:1415–1428.

Palme, H., K. Lodders, and A. Jones. 2014. "2.2—Solar System Abundances of the Elements." Pp. 15–36 in Vol. 2: Planets, Asteriods, Comets and the Solar System of *Treatise on Geochemistry*, 2nd ed., H.D. Holland and K.K. Turekian, eds. Amsterdam, Netherlands: Elsevier.

Pignatale, F.C., S. Charnoz, M. Chaussidon, and E. Jacquet. 2018. "Making the Planetary Material Diversity During the Early Assembling of the Solar System." *Astrophysical Journal* 867:L23.

Pontoppidan, K.M., C. Salyk, E.A. Bergin, S. Brittain, B. Marty, O. Mousis, and K.I. Öberg. 2014. "Volatiles in Protoplanetary Disks." Pp. 363–386 in *Protostars and Planets VI*, H. Beuther, R.S. Klessen, C.P. Dullemond, and T. Henning, eds. Tucson: University of Arizona.

Quirico, E., F.-R. Orthous-Daunay, P. Beck, L. Bonal, R. Brunetto, E. Dartois, T. Pino, et al. 2014. "Origin of Insoluble Organic Matter in Type 1 and 2 Chondrites: New Clues, New Questions." *Geochimica et Cosmochimica Acta* 136:80–99.

Rodgers, S.D., and S.B. Charnley. 2008. "Nitrogen Isotopic Fractionation of Interstellar Nitriles." *Astrophysical Journal* 689:1448–1455.

Sandford, S.A., M.P. Bernstein, and J.P. Dworkin. 2001. "Assessment of the Interstellar Processes Leading to Deuterium Enrichment in Meteoritic Organics." *Meteoritics and Planetary Science* 36:1117–1133.

Schrader, D.L., K. Nagashima, S.R. Waitukaitis, J. Davidson, T.J. McCoy, H.C. Connolly, and D.S. Lauretta. 2018. "The Retention of Dust in Protoplanetary Disks: Evidence from Agglomeratic Olivine Chondrules from the Outer Solar System." *Geochimica et Cosmochimica Acta* 223:405–421.

Schrader, D.L., K. Nagashima, J. Davidson, T.J. McCoy, R.C. Ogliore, and R.R. Fu. 2020. "Outward Migration of Chondrule Fragments in the Early Solar System: O-isotopic Evidence for Rocky Material Crossing the Jupiter Gap?" *Geochimica et Cosmochimica Acta* 282:133–155.

Shadmehri, M., and S.M. Ghoreyshi. 2019. "Time-Dependent Evolution of the Protoplanetary Discs with Magnetic Winds." *Monthly Notices of the Royal Astronomical Society* 488:4623–4637.

Shu, F. H., H. Shang, M. Gounelle, A.E. Glassgold, and T. Lee. 2001. "The Origin of Chondrules and Refractory Inclusions in Chondritic Meteorites." *Astrophysical Journal* 548:1029–1050.

Steinpilz, T., K. Joeris, F. Jungmann, D. Wolf, L. Brendel, J. Teiser, T. Shinbrot, et al. 2020. "Electrical Charging Overcomes the Bouncing Barrier in Planet Formation." *Nature Physics* 16:225–229.

Takeuchi, T., and P. Artymowicz. 2001. "Dust Migration and Morphology in Optically Thin Circumstellar Gas Disks." *Astrophysical Journal* 557:990–1006.

Tang, H., and N. Dauphas. 2012. "Abundance, Distribution, and Origin of ^{60}Fe in the Solar Protoplanetary Disk." *Earth and Planetary Science Letters* 359:248–263.

Terzieva, R., and E. Herbst. 2000. "The Possibility of Nitrogen Isotopic Fractionation in Interstellar Clouds." *Monthly Notices of the Royal Astronomical Society* 317:563–568.

Vanhala, H.A.T., and A.P. Boss. 2002. "Injection of Radioactivities into the Forming Solar System." *Astrophysical Journal* 575:1144–1150.

Waite, J.H., W.S. Lewis, B.A. Magee, J.I. Lunine, W.B. McKinnon, C.R. Glein, O. Mousis, et al. 2009. "Liquid Water on Enceladus from Observations of Ammonia and ^{40}Ar in the Plume." *Nature* 460:1164.

Weidenschilling, S.J. 1977a. "Aerodynamics of Solid Bodies in the Solar Nebula." *Monthly Notices of the Royal Astronomical Society* 180:57–70.

Weidenschilling, S.J. 1977b. "The Distribution of Mass in the Planetary System and Solar Nebula." *Astrophysics and Space Science* 51:153–158.

Weisberg, M.K., T.J. McCoy, and A.N. Krot. 2006. "Systematics and Evaluation of Meteorite Classification." Pp. 19–52 in *Meteorites and the Early Solar System II*, D.S. Lauretta and H.Y. McSween, Jr., eds. Tucson: University of Arizona Press.

Weiss, B.P., X.-N. Bai, and R.R. Fu. 2021. "History of the Solar Nebula from Meteorite Paleomagnetism." *Science Advances* 7(1):eaba5967.

Williams, C.D., M.E. Sanborn, C. Defouilloy, Q.-Z. Yin, N.T. Kita, D.S. Ebel, A. Yamakawa, et al. 2020. "Chondrules Reveal Large-Scale Outward Transport of Inner Solar System Materials in the Protoplanetary Disk." *Proceedings of the National Academy of Sciences* 117:23426–23435.

Wooden, D.H., H.A. Ishii, and M.E. Zolensky. 2017. "Cometary Dust: The Diversity of Primitive Refractory Grains." *Philosophical Transactions of the Royal Society A* 375:20160260.

Yoneda, S., and L. Grossman. 1995. "Condensation of $CaO-MgO-Al_2O_3-SiO_2$ Liquids from Cosmic Gases." *Geochimica et Cosmochimica Acta* 59:3413–3444.

Youdin, A.N., and J. Goodman. 2005. "Streaming Instabilities in Protoplanetary Disks." *Astrophysical Journal* 620:459–469.

Yurimoto, H., and K. Kuramoto. 2004. "Molecular Cloud Origin for the Oxygen Isotope Heterogeneity in the Solar System." *Science* 305:1763–1766.

Zsom, A., C.W. Ormel, C. Güttler, J. Blum, and C.P. Dullemond. 2010. "The Outcome of Protoplanetary Dust Growth: Pebbles, Boulders, or Planetesimals? II. Introducing the Bouncing Barrier." *Astronomy & Astrophysics* 513:A57.

Q2 PLATE: An enhanced-color image mosaic of Pluto and Charon taken by the New Horizons spacecraft in 2015. SOURCE: Courtesy of NASA/JHUAPL/SwRI.

5

Question 2: Accretion in the Outer Solar System

How and when did the giant planets and their satellite systems originate, and did their orbits migrate early in their history? How and when did dwarf planets and cometary bodies orbiting beyond the giant planets form, and how were they affected by the early evolution of the solar system?

The outer solar system, stretching from Jupiter to the Oort cloud, contains the keys to understanding the formation and early evolution of the solar system.[1] Gas giant planet formation is a primary open question in theoretical astrophysics, with Jupiter and Saturn used to calibrate and test new models. As a result, a detailed characterization of these planets is required. Uranus and Neptune represent a unique class of planets that clearly differ from terrestrial and gas giant planets, and their formation challenges planet formation models because it is unclear whether they are simply failed gas giants or if they formed in a different way. That planets with similar sizes/masses appear abundant in the galaxy suggests that the formation of such intermediate-mass gas planets is common, emphasizing the need to understand the formation of Uranus and Neptune (see Questions 7 and 12; Chapters 10 and 15, respectively). The giant planets possess full planetary systems with rings and regular satellites, as well as irregular satellites thought to have been captured from heliocentric orbit early in solar system history (see Question 8, Chapter 11). The unique characteristics of these systems provide further constraints on giant planet system origin and early evolution.

The outer solar system also has diverse small body populations: comets, irregular satellites, Trojans, Centaurs, and the swarm of dwarf planets and smaller bodies in the trans-neptunian belt. The latter population presents a complex orbital structure whose full extent and character is still being explored to the limits of telescopic capabilities. However, even as they are understood today, the orbital and physical properties of the smaller objects in the outer solar system provide profound and detailed clues to its formation. The emerging view of exoplanetary systems (see Question 12), and how they differ from the solar system, provide additional strong motivations to study the formation of the outer solar system, as it is only the solar system that can be truly explored, and in time understood, in the necessary depth and detail.

---

[1] A glossary of acronyms and technical terms can be found in Appendix F.

127

Q2.1 HOW DID THE GIANT PLANETS FORM?

The main research themes of outer planet formation have been known for decades, but developments in observations and theory continually revise our understanding of how these themes relate and how the planets formed. The leading mechanisms for giant planet formation are core accretion and disk instability (Helled et al. 2014).

In the core accretion scenario, giant planet formation begins with buildup of a heavy-element core (heavy elements referring to all elements heavier than helium), followed by accretion of hydrogen-helium, or H-He, gas. Volatile species that are heavier than helium, such as CO and N_2, can also be accreted in the form of gas (see Question 1 for details, Chapter 4). Heavy elements in the solid form are delivered by objects ranging in size from millimeter-to-decimeter-size pebbles, to planetesimals up to hundreds of kilometers in size, to finally planetary embryos/cores. H-He from the protostellar disk can be accreted during the early stages of core formation; however, as the planet grows in mass, at some point the gas accretion rate increases considerably, nearly at free fall, and a gas giant planet is formed. During the final stages of gas accretion when the planet has sufficiently contracted in size, a circumplanetary disk may form and control the transfer rate of gas to the accreting planet. Gas accretion likely ends as the circumsolar nebula dissipates (Russell et al. 2006; see Question 1), or perhaps by a gap opening in said disk. In contrast, in the disk instability scenario, giant planets form by a local gravitational collapse in the circumstellar disk (Boss 1997). This model is typically applied to giant exoplanets orbiting at tens of AU from their host star and/or of giant planets around M dwarf stars. While we cannot exclude that the gas giants in the solar system formed via disk instability, their complex internal structures are more consistent with formation by core accretion (e.g., Helled and Stevenson 2017).

Q2.1a What Is the Formation Mechanism of Gas Giant Planets? What Were the Accretion Rates of Solids (Planetesimals/Pebbles) and Gas During the Formation Process? How Long Did It Take?

A key uncertainty with the core accretion model is how cores massive enough to prompt runaway gas accretion by Jupiter and Saturn were able to accrete prior to solar nebula dispersal. The early core growth is dominated by heavy-element accretion, with many expected large objects with sizes of 100 to 1,000 km growing in separate feeding zones. At this point, the conventional wisdom has been that the protoplanets gravitationally perturb each other and merge into a few large cores. The problem is that dynamical simulations find that mergers are less common than expected, and gravitational interactions cause the protoplanets to be spread out (Levison et al. 2015), increasing the core accretion timescale to longer than the expected lifetime of the solar nebula (see Question 1 and text below), thereby preventing substantial gas accretion. In addition, it is possible that planetesimals may become isolated in rings between protoplanets, which may also lead to less efficient accretion.

A plausible solution comes from so-called pebble accretion. Small, millimeter-to-decimeter-sized protoplanetary solids ("pebbles"; see Q1.3a–b) lose orbital energy owing to aerodynamic drag by the solar nebula gas. If a pebble's aerodynamic stopping time is less than or comparable to the time for it to encounter a growing protoplanet, then it is decelerated with respect to the protoplanet during the encounter and can become gravitationally bound, eventually spiraling into the protoplanet. This effect allows protoplanets above a threshold size to grow rapidly if they are embedded in a sea of pebbles. Numerical models suggest that pebble accretion in combination with protoplanet mergers can form the giant planet cores before the solar nebula dissipates (Levison et al. 2015; Johansen and Lambrechts 2017). Once cores reach about 10 Earth masses, rapid gas accretion can occur until the local supply of gas is depleted. A sketch of these various growth phases is shown in Figure 5-1.

Because Jupiter and Saturn are H-He dominated, and Uranus and Neptune have H-He atmospheres of a few Earth masses, the formation of the outer planets occurred within the lifetime of the gas-rich protosolar nebula. However, their exact formation timescale remains an open question. The formation timescale of the gas giant planets is thought to be on the order of 10^6 to 10^7 years. This timescale corresponds to the observed lifetimes of protoplanetary disks around other stars (see also Q1.4), which ensures the existence of H-He disk gas that can be accreted by the growing protoplanet. Shorter formation timescales with efficient pebble accretion are possible, but it may then be difficult to explain, for example, why Uranus and Neptune did not accrete more gas. Substantial uncertainty in formation timescale is very significant for formation models, in particular for the implied disk

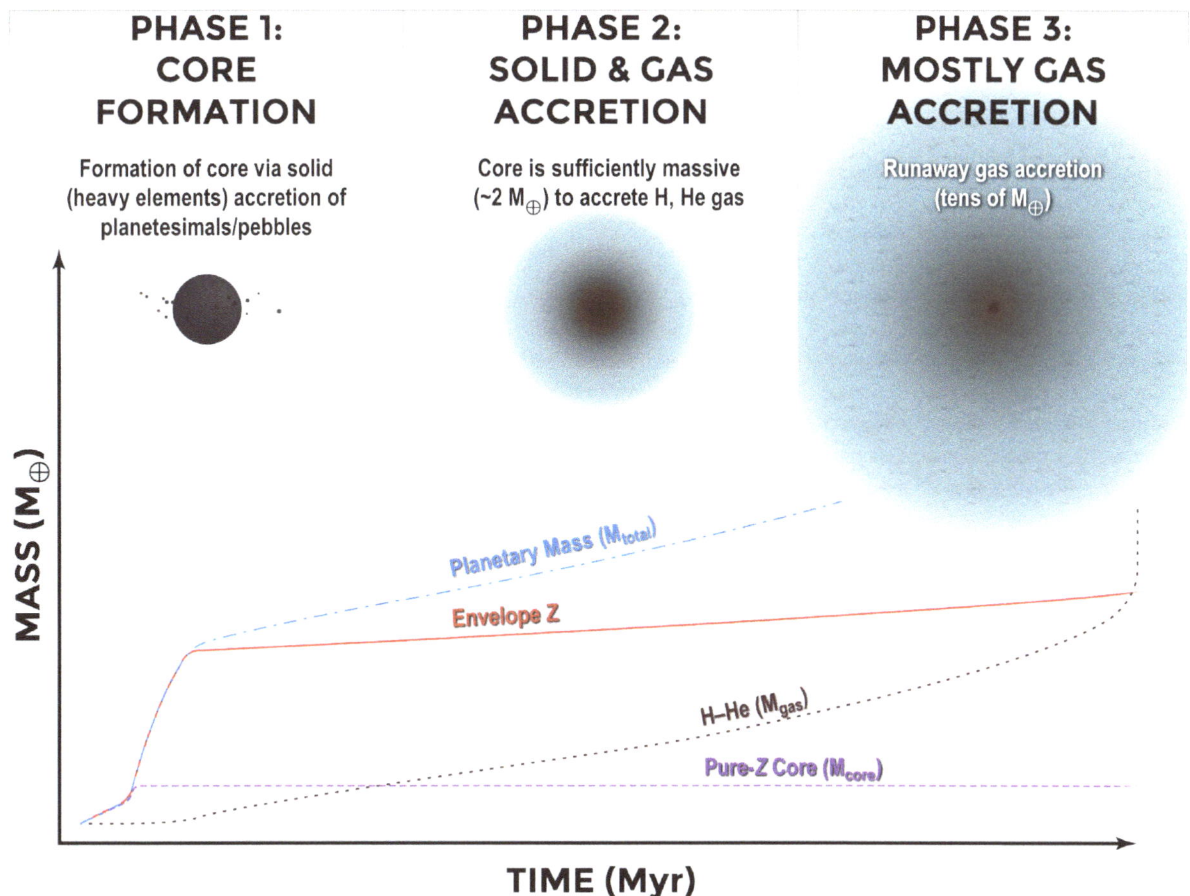

FIGURE 5-1 A sketch of the planetary growth in the core accretion model. Shown is the planet's mass (in Earth masses) versus time up to the onset of Phase 3 (blue line), when runaway gas accretion begins. Purple line: core mass of pure heavy elements (meaning here all elements other than H and He, termed "Z"). Brown line: mass of H and He. Red line: heavy-element mass ("Z") in the envelope. SOURCE: P.K. Byrne and R. Helled.

conditions (which determine the accretion rates), formation efficiency, early evolution, as well as the predicted giant planet masses and compositions. Recent isotopic measurements of meteorites have been interpreted to imply a million-year formation time for Jupiter's core and concomitant gap opening in the solar nebula (Kruijer et al. 2017), but alternative interpretations have been offered (Lichtenberg et al. 2021).

The alternative disk instability model is presently thought to be less relevant for the solar system, especially considering Juno and Cassini results for Jupiter and Saturn, respectively (Q2.1c). However, this model currently cannot be excluded, and it is therefore important to understand predicted differences that could be used to discriminate between the two models, which is an active topic of investigation. In general, core accretion requires a heavy-element core, but the presence of a core in the disk instability model cannot be ruled out. Overall, better understanding of giant planet origin requires improved information on their bulk compositions and internal structures, which can be inferred from structure models that use accurate measurements of their gravitational and magnetic fields, and atmospheric compositions (see Question 7). In combination with planet evolution models, such data can then be used to constrain formation locations, accretion rates, and formation timescales, possibly allowing us to discriminate between these two formation scenarios (see e.g., Helled et al. 2014; Helled and Morbidelli 2021).

Q2.1b How Did Uranus and Neptune Form and What Prevented Them from Becoming Gas Giants?

There are several challenges to explaining the origin of Uranus and Neptune. Planetesimal accretion is predicted to be too inefficient to form these planets at their current locations while the nebula was still present unless extreme local conditions are assumed. Uranus and Neptune could have formed faster in situ via pebble accretion, although this scenario could easily lead to their accreting too much H-He (e.g., Lambrechts and Johansen 2012). Another possibility is that Uranus and Neptune formed by collision and merging of a few low-mass planets accreted from a population of planetary embryos (e.g., Izidoro et al. 2015). As discussed below, dynamical properties of the trans-neptunian belt suggest that Neptune formed considerably closer to the Sun and later migrated outward; forming both Uranus and Neptune and smaller orbital radii would lead to faster formation potentially consistent with limited H-He accretion (see Q2.4 and Q2.6, Chapter 5).

None of the existing models, however, explain all observed properties of Uranus and Neptune, including the heavy-element to H-He ratios inferred by structure models, or even the identity of these heavy elements (i.e., are they mainly rocky, icy, or carbonaceous, or some mixture?). In addition, it is still unknown what prevented these planets from becoming H-He-dominated like Jupiter and Saturn (Helled et al. 2020). It seems unlikely that their gas accretion was truncated by the opening of gaps in the nebula, as this is generally associated with much more massive planets ($\geq$ Jupiter mass). We infer that the growth of Uranus and Neptune was slow (i.e., accretion rates were low) and/or that it occurred late as the gaseous nebular disk had begun to dissipate, preventing substantial accretion of H-He gas. The latter idea could suggest that substantial photoevaporation in the Uranus–Neptune zone limited the size of these bodies; a signature of this could be an enrichment in noble gases, as hydrogen is expected to escape first because it is the lightest gas component (e.g., Guillot and Hueso 2006; Question 1). That Uranus and Neptune could have formed after Jupiter and Saturn is generally consistent with a heavy-element accretion rate that would be lower for larger radial distances, owing to decreasing disk solid surface densities and orbital frequencies. Further constraints on the origin of Uranus and Neptune would be greatly improved by better measurements of their gravitational and magnetic fields and atmospheric compositions.

Q2.1c What Were the Primordial Internal Structures of Giant Planets?

For decades, the giant planets' interiors were generally assumed to be homogeneously mixed, with a compact central core. However, recent formation and internal structure models find that the heavy elements are not uniformly mixed (e.g., Helled and Stevenson 2017), and it is now understood that giant planets probably did not have distinct cores but rather an innermost region that was highly enriched with heavy elements. Once the core mass becomes sufficiently high for rapid H-He accretion, subsequent incoming solids (heavy elements) dissolve in the atmosphere, leading to deep interior compositional gradients in gaseous protoplanets. Indeed, fuzzy cores seem to exist today in both Jupiter and Saturn based on Juno and Cassini data, respectively (Wahl et al. 2017; Mankovich and Fuller 2021, see Question 7). Uranus and Neptune may have also had primordial structures with fuzzy cores and inhomogeneous interiors; however, their internal structures remain poorly constrained, hampering understanding of how they formed and evolved. More information is required to better constrain Uranus and Neptune's formation, evolution and structure, including measurements of their gravitational and magnetic fields, determination of their atmospheric composition (preferably by an entry probe), and improved information on the behavior of planetary elements at relevant pressures and temperatures. Details on the evolution of the internal structures of giant planets are given in Chapter 10 and references therein.

Q2.1d What Were the Roles of Early Giant Impacts and Magnetic Fields in Shaping the Properties of the Outer Planets?

Collisions during the late stages of planet accretion tend to reduce the number of massive objects and increase the masses of the survivors, until a stable configuration is obtained (Izidoro et al. 2015). Simulations of giant impacts (e.g., Reinhardt et al. 2020; Rufu and Canup 2022) show that an oblique impact on Uranus can explain its 98-degree axial tilt and rotation rate (and perhaps the concomitant formation of its regular satellite system; see Q2.3b). Giant impacts are also invoked to explain Neptune's smaller but still substantial 30-degree obliquity, key differences between Uranus and Neptune (e.g., their satellite systems, heat fluxes, and predicted moments of inertia), potential differences between Jupiter and Saturn, and the origin of fuzzy cores.

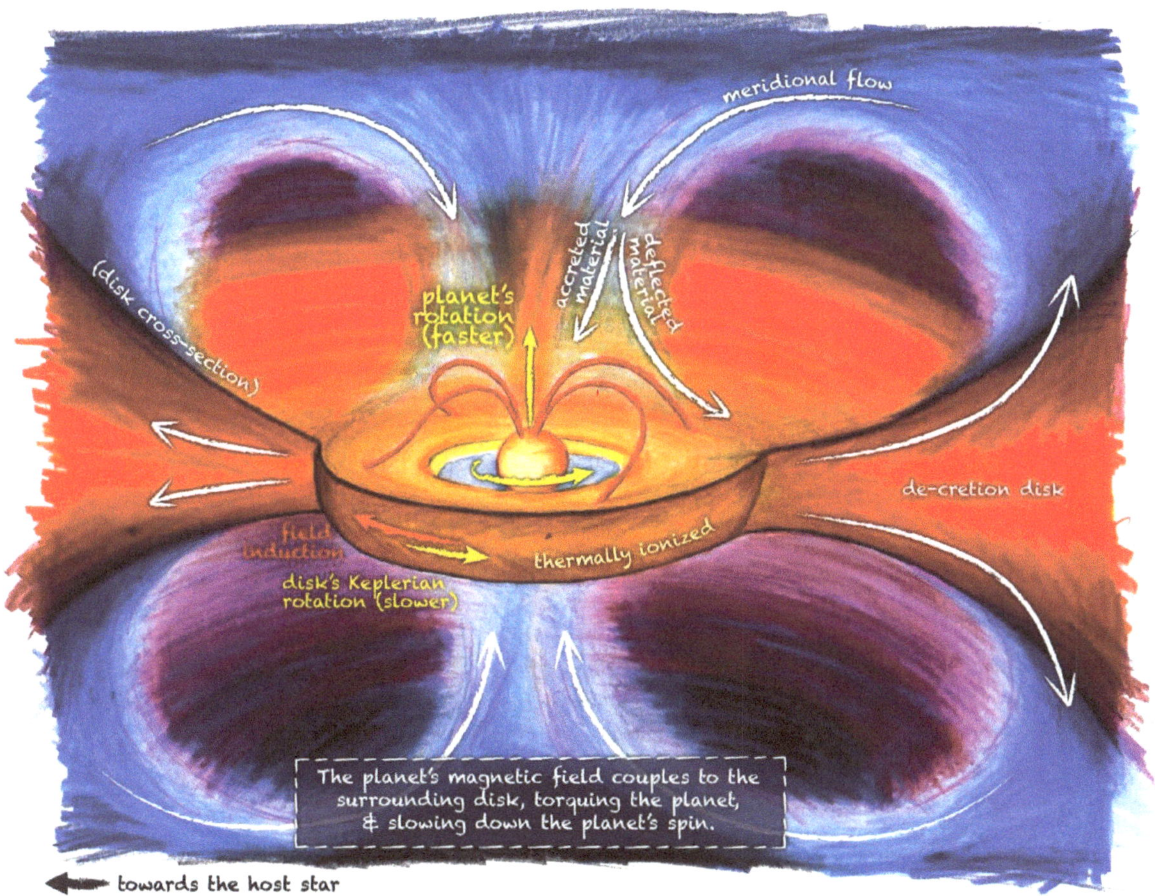

FIGURE 5-2 A cartoon of the formation of a gas giant in a circumplanetary disk in Keplerian rotation. In the inner regions of the disk, thermal ionization imparts conductivity to the disk, coupling the planetary magnetic field to the quasi-Keplerian motion of the gas. A meridional circulation system of gas within the gravitational sphere of influence feeds a "decreting" circumplanetary disk that connects back to the parent protosolar nebula. Net angular momentum is lost by the planet to the circumplanetary disk, resulting in the spin-down of the planet. SOURCE: Courtesy of K. Batygin, 2018, "On the Terminal Rotation Rates of Giant Planets," *The Astronomical Journal* 155(4):178, https://iopscience.iop.org/article/10.3847/1538-3881/aab54e, © AAS, reproduced with permission.

As discussed above, gas giant planets are thought to follow a core-nucleated accretion process whereby cores capture massive gaseous envelopes from the gas nebula. The concurrent accretion of angular momentum is expected to spin the protoplanet to near-breakup speeds. However, Jupiter and Saturn (and most long-period extrasolar planets; Bryan et al. 2020) rotate well below their breakup rotation rate, suggesting that some mechanism expelled angular momentum from the planet-forming region. A leading model redistributes angular momentum from the protoplanet to a circumplanetary disk (CPD) via magnetic coupling/braking (Takata and Stevenson 1996). In one of the versions of such models (Figure 5-2), vigorous convection in the protoplanet's interior generates a strong magnetic field, which in turn couples to the ionized portion of the CPD. The terminal spin of the planet is determined by the terminal radius of the planet and the location of the magnetospheric truncation-radius of the CPD at the time of nebular gas dispersal. Testing such models requires a much better understanding of magnetic field generation within giant planets today (which constrains the internal structure implied by dynamo sustenance), as well as improved knowledge of CPDs and how their evolution controlled the formation of satellites and rings (see Q2.5). Although Uranus and Neptune accreted vastly smaller gas components than Jupiter and Saturn, and per above appear to have had their final spin states affected by late giant impacts, the current rotation periods of all four giant planets are broadly similar (about 10–17 hours), a commonality that may be coincidental.

Strategic Research for Q2.1

- **Determine the atmospheric composition of Saturn, Uranus, and Neptune** via in situ sampling of noble gas, elemental, and isotopic abundances, and remote sensing by spacecraft and ground/space-based telescopes.
- **Determine the bulk composition and internal structure of Uranus and Neptune** via gravity, magnetic field, and atmospheric profile measurements by spacecraft, as well as Doppler seismology.[2]
- **Constrain physical properties and boundary conditions (i.e., tropospheric temperatures, shapes, and rotation rates) for structure models of Uranus and Neptune** via gravity, magnetic field, and atmospheric profile measurements by spacecraft, remote sensing by spacecraft and ground/space-based telescopes.
- **Better determine the formation and early evolution of the outer planets** through improved numerical simulations and theoretical models.
- **Improve our understanding of the behavior of planetary material at high pressures and temperatures** using laboratory experiments and numerical simulations.

Q2.2 WHAT CONTROLLED THE COMPOSITIONS OF THE MATERIAL THAT FORMED THE GIANT PLANETS?

The giant planets accreted both solids and gas, regardless of the specific formation processes involved. Although these primordial components have been at least partially mixed within the planets, the current planetary composition can still be linked back to the composition of accreted primordial solids and gas (see Question 1). For the accreted gas component, current isotope ratios and elemental abundances reflect the balance between planetary growth rates and the evolution of nebular gas (including effects from photoevaporation as well as influxes of material from sources in the Sun's stellar birth cluster). The combination of isotopic/elemental ratios from giant planets—as well as from comets and meteorites—is crucial for determining protosolar values for helium isotopes (^{3}He/^{4}He) and hydrogen isotopes (deuterium/hydrogen, or D/H), because the composition of the Sun itself has evolved after 4.5 Ga of nuclear fusion.

For accreted solids, relative abundances of the elements distinguish between classes of material such as ice, rock, and organics; between levels of physical and thermal processing such as ice crystallization; and between source region temperatures with respect to the ice lines (condensation fronts) for water and other volatiles within the protosolar nebula (described further below).

Q2.2a How Was the Overall Bulk Fraction of Heavy Elements in the Giant Planets Established?

Understanding the origin of giant planet heavy element fractions hinges on our incomplete knowledge of the bulk heavy element fractions themselves. For Jupiter, abundances of most heavy elements detected by the Galileo probe and remote sensing (Figure 5-3) are higher than their protosolar abundances. Assuming that heavy element abundances in the other giant planets scale with carbon (as measured by atmospheric methane), all the giant planets most likely accreted an excess of solid material compared to gaseous material.

In the core accretion scenario (Q2.1a), the observed supersolar enrichment of heavy elements results from both solid material that formed the giant core that was subsequently mixed into the envelope, as well as from solid material that was accreted directly into the envelope, possibly abetted by photoevaporative loss of H-He gas from the protosolar nebula itself. In situ measurements of atmosphere composition, as well as accurate gravity and magnetic field measurements, provide relevant constraints. Noble gas compositions can constrain the composition of the planetary building blocks and reveal information on the chemical and physical properties of the solar nebula, and possibly the planetary formation timescale (e.g., Guillot and Hueso 2006; see Q1.4).

The balance between these different sources is an open question, which can be explored in the coming decade with new data and continuing developments in theory. Adding data points to Figure 5-3 would require atmospheric

---

[2]Trapped normal modes in the interiors of giant planets can create a velocity field that can be sensed remotely. Measurements of the spatial and temporal frequencies of these modes provide information on the deep interior, like seismology in the case of Earth.

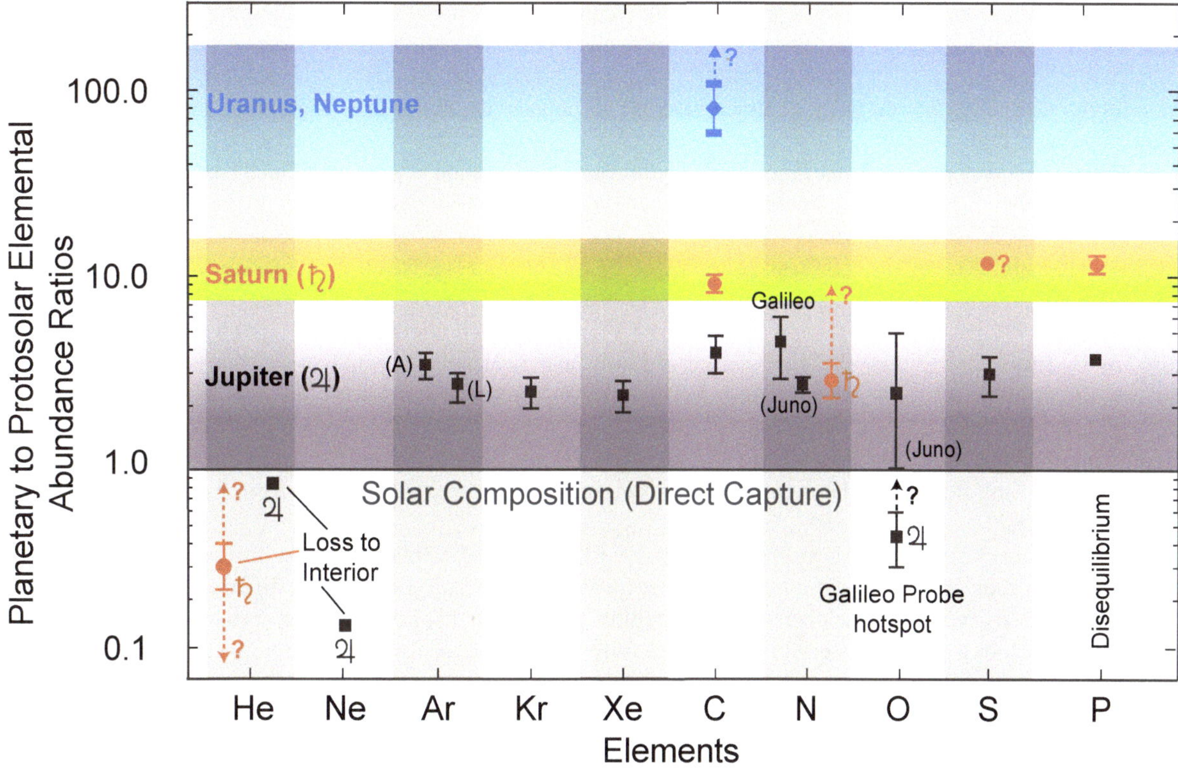

FIGURE 5-3 Elemental abundances in the atmosphere of the four giant planets divided by their protosolar values, with the x-axis showing, from left to right helium, neon, argon, krypton, xenon, carbon, oxygen, sulfur, and phosphorous. Values marked with "?" are either uncertain (N, S, He) or may not be representative of the actual ratio in the deep atmosphere. Because of measurements from the Galileo probe and Juno, Jupiter's atmospheric composition is far better understood than that of any of the other giant planets. SOURCE: Atreya et al. (2022).

probes, especially for the noble gases. Measurements of carbon-to-hydrogen (C/H) and sulfur-to-hydrogen (S/H) ratios at Uranus and Neptune have recently been advanced by submillimeter and microwave spectroscopy, which can probe atmospheric levels beneath the cloud condensation levels of at least methane (CH_4) and hydrogen sulfide (H_2S) (Tollefson et al. 2021). But key ratios such as nitrogen to hydrogen (N/H), oxygen to hydrogen (O/H), and S/H at Saturn, Uranus, and Neptune are difficult to relate to abundances in the deeper atmospheres owing to cloud condensation and chemistry (see Q7.3).

Relating atmospheric composition to bulk composition, even with improved measurements in the coming decade, will require improvements in theory. Models of radial profiles of interior density and heavy element concentrations—driven largely by new gravity field data from Juno and Cassini—achieve better fits using a "fuzzy core" in Jupiter and Saturn (see Q2.1 and Q7.2). These results suggest that heavy element enrichment in the observable envelopes is at least partially owing to mixing of core material, yet models struggle to match gravity field data with envelope enrichments higher than the protosolar abundance (e.g., Wahl et al. 2017).

Q2.2b What Were the Contributions to the Giant Planets from Different Types of Solids (Rock, Ices, and Organics)?

There is no class of solid material, whether icy or rocky, in the current solar system that reflects the generally three-times supersolar enrichment of heavy elements in Jupiter (see Figure 5-3), a fundamental challenge to understanding the types of solids accreted during planetary formation. Equilibrium condensation of a protosolar-abundance mixture would produce more ice than rock, if the condensation temperature is low enough, but verifying a high (or any) ice-to-rock ratio in the giant planets is challenging because rock-forming species condense in deep cloud layers inaccessible to observations (although evidence for such clouds abounds in hot exo-Jupiters). Reality is even more complex because carbonaceous matter and/or graphite could also enrich the planets with carbon.

Thermochemical models constrained by disequilibrium species such as germane (GeH_4) and arsine (AsH_3) may lead to estimates of some rocky element abundances (Wang et al. 2016), and the volatile gas H_2S contains sulfur that may have been accreted in rocky material. Disequilibrium carbon monoxide (CO) abundances in giant planet tropospheres similarly provide constraints on deep O/H as a tracer of accreted icy material (Wang et al. 2016).

The D/H ratio in the giant planets is much lower than observed in comets, which may be analogs of accreted icy materials. For gas-rich Jupiter and Saturn, low D/H relative to comets indicates a high overall gas fraction. But for Uranus and Neptune, the gas fraction is lower, and D/H values, while somewhat elevated, have been interpreted as indicators of a low ice/rock fraction, or of incomplete mixing with low D/H in outer layers owing to accreted protosolar gas (Teanby et al. 2020). Higher-order gravitational moments at Uranus and Neptune, and/or internal structure constraints from Doppler seismology, are needed to constrain more advanced models of density profiles capable of distinguishing between heavy element contributions from rocky, icy, and carbonaceous material.

Advances in our understanding of the composition of Uranus and Neptune in the coming decade may lead to a reevaluation of their bulk compositions and whether the term "ice giant" should be modified to a more appropriate name that properly represents the planetary composition. The idea that solid carbonaceous material may have been as abundant as rocky and icy material has not been fully explored. Constraints on the bulk carbonaceous fraction in comets, trans-neptunian objects, and outer planet moons may test the viability of carbonaceous material as a major component of protoplanetary solid material. Studies of protoplanetary disks may provide important context in this regard (Q1.1).

Q2.2c How Were Compositional Differences Between the Gas Giants and Ice Giants Influenced by the Chemical and Physical Processing of Accreted Solids and Gas?

Composition varied spatially and temporally within the protosolar disk, with effects preserved in the present-day composition of the giant planets. The planets formed at different heliocentric distances and thus sampled spatial variation within the disk, and time variation of disk composition affected the planets differently owing to variation in individual gas accretion timescales. Disk temperature decreased as a function of radius, setting up a sequence of condensation fronts or "ice lines" for different volatile species.

Solid material spiraled inward at the disk midplane owing to gas drag, so these ice lines represent boundaries where cold ices began turning to gas. Cold amorphous water ice went through an additional phase transition to crystalline ice, releasing trapped gases. Thus, compositional differences between the gas giants and ice giants may reflect their formation at different heliocentric distances with respect to condensation temperatures of ices containing oxygen, nitrogen, carbon, noble gases, and possibly other elements such as sulfur.

The composition of gases and solids in the solar nebula was affected by many chemical processes in addition to condensation fronts (e.g., Mousis et al. 2018; Öberg and Bergin 2021). Over time, gas composition in the disk evolved as hydrogen and helium (and perhaps neon) were evaporated under the influence of protosolar ionizing ultraviolet radiation, as well as ultraviolet radiation from massive stars in the Sun's birth cluster. These massive stars may have also affected disk composition via injection of material from stellar winds and supernovae (see Q1).

Individual simulations—including radial and/or temporal variation of gas and solid nebular composition—have matched aspects of planetary composition, but additional effects from cleared gaps, vortices, and other two- and three-dimensional structures have yet to be included in a comprehensive way. New compositional measurements, particularly in Uranus and Neptune as a contrast to Jupiter, are needed in the coming decade to constrain increasingly complex models of giant planet formation. The potential effects of volatile loss from planetesimals before they are accreted by a growing giant planet (e.g., Lichtenberg et al. 2021) should continue to be investigated.

Strategic Research for Q2.2

- **Determine the atmospheric composition of Saturn, Uranus, and Neptune** via in situ sampling of noble gas, elemental and isotopic abundances, and remote sensing by spacecraft and ground- or space-based telescopes.
- **Understand how compositional gradients in the atmosphere and interior of Jupiter, Saturn, Uranus, and Neptune affect the determination of bulk planetary composition based on observed atmospheric composition,** using gravity, magnetic field, and atmospheric profile measurements by spacecraft, Doppler seismology, and laboratory/theoretical studies of physical processes (e.g., turbulent diffusion, moist convection, precipitation, and helium rain).
- **Constrain the composition of early accreted materials by determining abundance and isotopic ratios in diverse present-day objects from dust and meteorites to comets and TNOs,** using sample return, in situ measurements, and remote sensing by spacecraft and ground-/space-based telescopes.
- **Contextualize ice lines and elemental partitioning between gases and solids in the evolving protosolar disk** using comparative observations of protoplanetary disks obtained with ground-/space-based telescopes via coronagraphy, interferometry, and spectroscopy.
- **Continue to improve knowledge of protosolar elemental and isotopic abundances** by study of primitive meteorites and the solar atmosphere.

Q2.3 HOW DID SATELLITES AND RINGS FORM AROUND THE GIANT PLANETS DURING THE ACCRETION ERA?

Each of the giant planets possesses satellites and rings (see Question 8). The main processes thought to have produced circumplanetary disks—gas accretion and giant impacts—would have occurred during or near the end stages of giant planet accretion. Thus, it is probable that all the giant planets had primordial satellites and rings (and perhaps multiple generations of them). A first overall question is the extent to which observed satellites and rings result from a common set of processes, or whether their individual characters derive largely from stochastic or contingent historical events. A second, related question, which has received increasing attention of late, is whether some of the current satellites and rings may have formed much later in solar system history, or even whether formation is ongoing (see Question 8).

Rings and regular satellites lie in or nearly in a given giant planet's equator, and orbit prograde, implying an origin from a dissipative orbiting disk of gas and/or solid particles. Irregular satellites are found in dynamically stable zones that lie at great distances from their primaries, which exist for both prograde and retrograde orbits. They are judged to have been captured from heliocentric orbit (see Q2.6 and Question 8). While not giant planets, trans-neptunian dwarf planets, when telescopic observations are sufficient, are also seen to possess one or more satellites. Pluto has five moons, while Haumea has two known moons and a ring system. Indeed, so common are satellites (and to an extent, rings) about the larger bodies of the outer solar system, that the question might be not so much why they have rings and moons, but why the terrestrial planets are deficient in satellites and lack rings entirely.

Each satellite and ring system is unique in detail, and so untangling the roles of systematic and stochastic processes is challenging. This issue is highlighted by the common occurrence of satellite pairs in a given satellite system, which by rights ought to be similar but are not. The key example is the dichotomy between Ganymede and Callisto, described below. The physical processes that allowed satellites to accrete is unclear, and it has yet to be determined how similar these processes were to those, at a larger scale, that led to the formation of the planets.

Nevertheless, our understanding of giant planet satellite accretion continues to be revolutionized by new spacecraft discoveries, analysis of meteorites, theoretical modeling, and studies of exoplanetary systems. Similarly, dwarf planet satellite formation is tied to and informs us of the dynamical processes that formed the trans-neptunian belt (see Q2.5). Satellite formation in general informs us of the conditions that formed the Moon and may have led, more broadly, to the formation of habitable icy worlds throughout the solar system.

Q2.3a How Did Protosatellite Disks Form, and How Did Disk Structure and Composition Evolve During the Accretion of Primordial Satellite Systems?

Formation of the gas giants Jupiter and Saturn is thought to have been accompanied by that of a CPD surrounding each (e.g., Peale and Canup 2015; see Q2.2). As proto-Jupiter and proto-Saturn grew toward their full masses, they opened gaps in the protosolar nebula, across which gas and coupled dust continued to flow, forming the CPDs (Figure 5-4). The evolving pressure, temperature, dynamical structures, and lifetimes of these disks are actively debated. They were originally thought to be gas-dominated but may instead be gas-starved and solid-particle-rich. They may be accretion disks spreading viscously and contributing to the mass growth of the giant protoplanet or they may be decretion disks spreading outward and returning matter to the protosolar nebula, or both. Here a key uncertainty is the specific angular momentum of the inflowing CPD gaseous material, and how that relates to the supply of CPD solids. The CPDs may or may not have been truncated by inner magnetospheric cavities generated by the primordial magnetospheres of Jupiter and Saturn (Figure 5-5).

The solids (rock, ices, and organics) that accreted to form the regular satellites of Jupiter and Saturn were sourced from heliocentric solids, with a possible contribution to the more volatile components from direct condensation in the CPD, depending on the latter's temperature and pressure history. Pebbles in the protosolar nebula may have been largely blocked from flowing into the CPD by gas pressure gradients, but the gravity of proto-Jupiter and proto-Saturn would have excited the orbits of heliocentric planetesimals, causing them to encounter the CPDs at high speeds (Raymond and Izidoro 2017; Ronnet and Johansen 2020). Ablation, disintegration, and wholesale capture of these planetesimals and recondensation of their component volatiles could have been a major source of a new generation of circum-jovian and circum-saturnian pebbles. Streaming or other instabilities within CPDs could then have formed first-generation "satellitesimals," subsequently growing by hierarchical coagulation and/or accretion of remaining pebbles into protosatellites (Shibaike et al. 2019; Bagytin and Morbidelli 2020; Ronnet and Johansen 2020; see Figure 5-5). Different formation scenarios make different predictions for satellite structure and composition, both chemical and isotopic, and so the more we can learn about the satellites, the better we can test these scenarios. Improved numerical modeling is also key, as it has long been difficult to simultaneously model the vastly different time and spatial scales of planet and satellite formation.

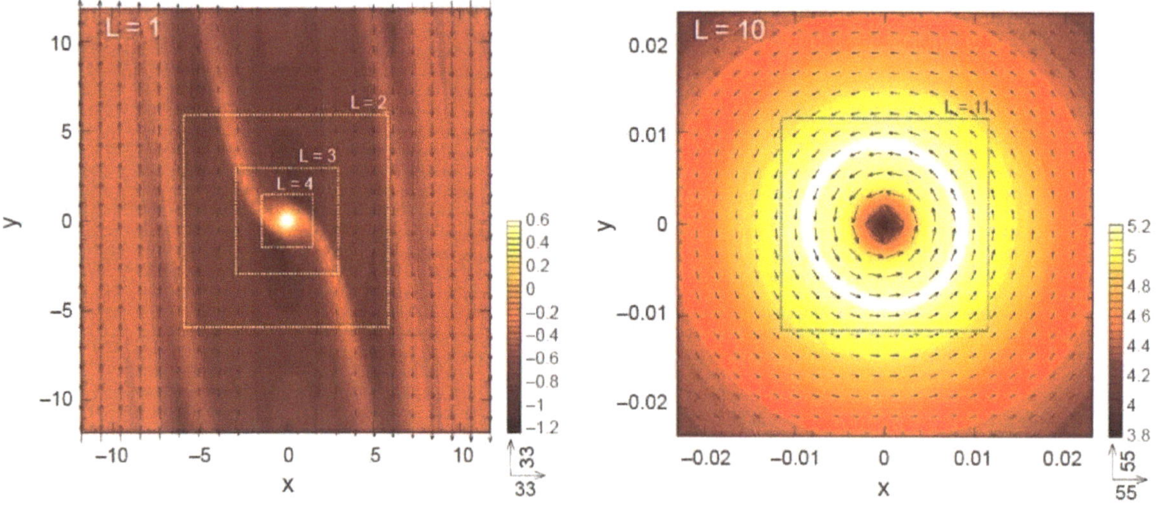

FIGURE 5-4 Velocity (arrows) and log density (color) (both nondimensional) of the flow around an accreting "Jupiter" in the protosolar nebula midplane, and at nested spatial scales (each factor of L is a power of 2); *x* and *y* are distances from the planet in units of the planet's Hill radius (sphere of gravitational influence). SOURCE: T. Tanigawa, A. Maruta, and M.N. Machida, 2014, "Accretion of Solid Materials onto Circumplanetary Disks from Protoplanetary Disks," *The Astrophysical Journal* 784(2):109, https://doi.org/10.1088/0004-637X/784/2/109, © AAS, reproduced with permission.

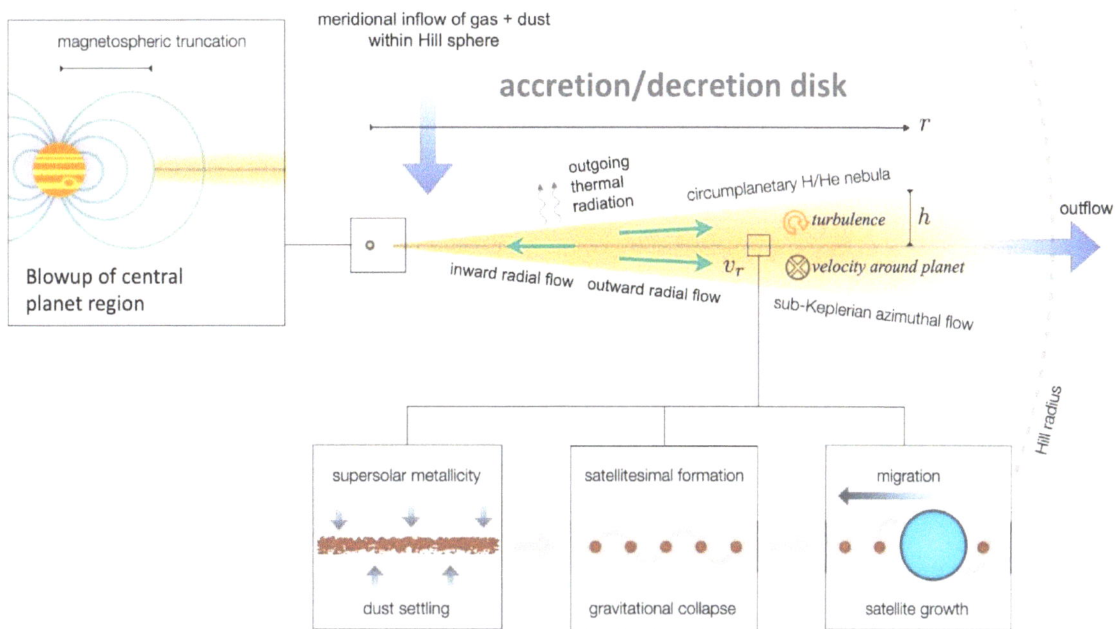

FIGURE 5-5 Schematic cross section of some hypothetical processes within circumplanetary, satellite-forming nebulae. Material flows in from the protosolar nebula and forms a flared disk around a gas giant such as Jupiter. Owing to turbulent motions in the gas, materials flow both inward toward the central planet (accretion) and outward, back toward the solar nebula (decretion). Satellites form from "satellitesimals" and by the same physics that affects the planets in the protosolar nebula, may migrate toward the central planet or form resonant chains. SOURCE: Modified from K. Batygin and A. Morbidelli, 2020, "Formation of Giant Planet Satellites," *The Astrophysical Journal* 894(2):143, https://doi.org/10.3847/1538-4357/ab8937, © AAS, reproduced with permission.

Q2.3b What Were the Roles of Giant Impact and Capture in the Outer Solar System for the Origin of Primordial Satellites and Planetary Rings?

It is unclear whether Uranus and Neptune formed circumplanetary disks in a manner like that posited for Jupiter and Saturn (Peale and Canup 2015). It has been argued to be less likely because they never reached the runaway gas accretion stage (see Q2.1), although conversely their slower rate of gas accretion might have been more favorable to a contracted planet size and CPD formation (Ward and Canup 2010). A related issue is that Uranus and Neptune have substantial spin axis tilts, and it has long been thought that these were owing to late impacts. Gas accretion would generally lead to planets with a very small obliquity, like that of Jupiter. Uranus and Neptune's current spin states (obliquity and rotation period) can be explained by an oblique impact by approximately an Earth-size body. Saturn's tilt is also substantial, Earth-like, and may have been similarly caused by a giant impact, or alternately, by a dynamical resonance in the early solar system (Ward and Hamilton 2004; Vokrouhlický and Nesvorný 2015). Uranus's spin is retrograde with respect to its orbital motion about the Sun, and its regular satellites orbit in the same sense as the planet's rotation, which is in the opposite sense of a CPD formed by gas accretion. The uranian satellites may have formed from a disk produced by a Uranus-tipping giant impact (e.g., Ida et al. 2020; Woo et al. 2022), or from a combination of a CPD produced via gas accretion and a Uranus-tipping impact (Morbidelli et al. 2016; Rufu and Canup 2022; Salmon and Canup 2022). This process could have been repeated multiple times, with each major planetary reorienting impact and new impact-produced satellite-forming disk interacting with and reprocessing the previous generation of satellites. However, whereas much has been learned of the jovian and saturnian satellites from Galileo and Cassini, detailed knowledge of the principal uranian regular satellites is lacking. Improved understanding of these satellites (including, e.g., their moments of inertia and differentiation states; compositions; geological character and signs of past/current activity; constraints on orbital evolution) is the surest way to test giant impact satellite formation models.

The role of impacts in satellite accretion is not limited to the giant impact scale or the protosolar nebula epoch. The early planetesimal (and dwarf planet) reservoir that survived beyond the orbit of Neptune, which is thought to have been the ultimate source of the trans-neptunian belt and Oort cloud (see Q2.6), was also a source for captured satellites and material for planetary rings. During the giant planet instability epoch (see Q2.4), dynamical transport of heliocentric planetesimals and dwarf planets throughout the giant planet region would have occurred but would have been especially important for Neptune as it migrated outward through the primordial Kuiper belt population. Capture into the satellite region of a giant planet could have occurred by direct collision with a preexisting regular satellite or by tidal stripping of a binary (e.g., Agnor and Hamilton 2006; Nogueira et al. 2011); such a process could have led to the capture of Neptune's retrograde Triton, with further catastrophic consequences for any original regular midsize satellite system (McKinnon et al. 1995). Further constraints on Triton's origin depend on a better understanding of the formation of Neptune and the Kuiper belt.

Close passes of a heliocentric body to within a giant planet's tidal Roche limit could have led to tidal disintegration and capture of debris that could have ultimately formed rings close to the planet (Hyodo et al. 2017). Furthermore, the smallest innermost moons are susceptible to catastrophic disruption by early heavy bombardment, as both the flux (impact rate) and speed of heliocentric impactors are concentrated by a giant planet's gravity. Such disrupted moons, however, may have subsequently reaccreted, giving these bodies complicated histories that are not easy to decipher.

Unsteady and evolving conditions at the edge of a gas giant's gravitational sphere of influence (Hill sphere) during the giant planet instability epoch can also lead to the capture of the distant irregular satellites of each (see Q2.6). Here it is suspected that wandering objects located at the right place and time were captured during giant planet encounters. Current models predict that all the irregular satellites derive from the same trans-neptunian source population, a prediction that can be tested. Testing capture scenarios for irregular satellites and ring materials requires improved knowledge of bodies like Triton, the outer irregulars of all the giant planets, and ultimately, isotopic measurements of their components (e.g., D/H and $^{14}N/^{15}N$).

Q2.3c What Are the Expected Properties of Satellites and Rings Formed During the Accretion Era, and Are These Consistent with the Satellite and Ring Systems Today?

It is plausible that the observed diversity of satellite and ring systems arises in large part from their accretion conditions. Primordial rings could be formed by a variety of processes, including satellite collisional or tidal disruption/stripping, or disruption during a close pass within the Roche limit per above. Accordingly, a wide range of initial ring masses and compositions could be envisioned. A ring collisionally spreads with time, causing its mass to decrease owing to accretion onto the planet or the spawning of moons as ring material spreads beyond the Roche limit. As a ring's mass decreases, its rate of spreading slows. A massive primordial ring would after more than 4.5 billion years of evolution asymptotically evolve to a mass that is independent of its initial mass. Remarkably, this theoretically predicted asymptotic mass is essentially equal to the mass of Saturn's rings (Salmon et al. 2010).

That said, rings may dynamically evolve sufficiently fast that none, even Saturn's rings, can be confidently assumed to be primordial. As discussed in Question 8, that Saturn's rings are so bright, despite their being continually bombarded by dark meteoritic material, argues against their being primordial and instead suggests they formed more recently. Whatever the origin of Saturn's rings, satellite formation tied to outward spreading of ring material has also led to the hypothesis that the inner midsize saturnian satellites may have accreted in sequence and tidally evolved away from Saturn, one after another, over perhaps a billion years or more (e.g., Crida and Charnoz 2012; Salmon and Canup 2017). Testing such varied scenarios has proven difficult, but better understanding of giant planet interiors, and the tidal dissipation mechanisms therein, offers some promise for understanding whether satellites can tidally migrate outward by the large distances implied by the late formation hypothesis.

There appear to be two absolute size scales of regular satellites: midsize (diameter 500 to 1,500 km) and large (Moon to Mercury in size). Jupiter's regular satellites, the Galileans, are all large, whereas those for Uranus are all midsize. Saturn's regular satellites are a mix, with midsize moons both inside and outside Titan's orbit. Midsize moons also, generally, increase in mass with distance from the planet. The saturnian satellites, with their mixed character, clearly challenge formation scenarios. Detailed characterization of the midsize moons of Uranus—their internal structures, chemical and isotopic compositions, and geological histories—would greatly advance our understanding of midsize moons through comparisons with those of Saturn.

Despite the many differences seen across the giant planet regular satellite systems, they all have approximately the same total mass when compared to their host planet mass, with this ratio being about 10^{-4} at Jupiter, Saturn, and Uranus. This is remarkable given the presumably varied formation histories across these systems. Satellites are expected to migrate inward within their CPDs owing to unbalanced disk torques. It has been argued that the observed ~10^{-4} mass ratio reflects the balance between satellite growth versus loss owing to inward satellite migration (Canup and Ward 2006), wherein multiple generations of satellites form and are lost, with each having a common total mass compared to the planet's mass. If so, earlier formed satellites may be lost to the parent planet, and even torn apart by tides to form a primordial ring (Canup 2010), which itself can lead to a new generation of inner satellite accretion (Crida and Charnoz 2012; Salmon and Canup 2017). Or, if there is a magnetospheric cavity close to the planet, the innermost migrating satellite may halt at the inner edge of the CPD (as suggested in analogy with exoplanet systems), with the orbits of subsequently forming and migrating satellites forming a resonant chain with the orbit of the first (e.g., Batygin and Morbidelli 2020). It is possible that such cavities could be detectable via polarimetric measurements of accreting extra-solar giant planets (sensitive to magnetic fields) by the James Webb Space Telescope. Alternatively, it has been suggested that the mass of a typical Galilean satellite is set by gap opening in the proto-jovian CPD, which slows further satellite accretion.

The compositional gradient within the Galilean satellite system, from rocky Io through icy Callisto, is an important clue to accretional conditions. This gradient could reflect local temperature conditions as well as pebble and satellitesimal composition. In contrast, the four largest and outermost uranian moons have similar compositions, while inner Miranda appears more ice rich. Significant inward satellite migration within a given circumplanetary nebula may also imply time evolution of pressure and temperature conditions at satellite birth, which can be addressed through detailed modeling of satellite formation. Note that such primordial inward migration (see Figure 5-5) is owing to gravitational interactions with nebular gas, and is distinct from the slower, outward migration of satellites owing to planetary tides over geologic time and as observed today for the Galilean satellites of Jupiter and Saturn's inner midsize satellites (and for the Moon) (see Question 8).

The differences between Ganymede and Callisto, the two largest of the Galilean satellites, are of particular interest and significance. Both are major ice-rock worlds in adjacent orbits. Ganymede has had a complex geological history, major resurfacing, is differentiated, and exhibits an internally generated dynamo magnetic field that implies an inner metallic core. Callisto in contrast appears geologically inert, with no signs of large-scale resurfacing, and gravity data from Galileo flybys implies it may only be partially differentiated. Yet both bodies may possess internal liquid water oceans, as inferred from Galileo magnetic induction measurements.

If Callisto is only partially differentiated it places strong constraints on the timing and character of its accretion. It does not partake in the Laplace resonance between Io, Europa, and Ganymede, and thus has not been substantially tidally heated. Future measurements of its gravitational and induced magnetic field will be critical to resolve this issue and its evolutionary history. Because Callisto rotates, assessing its state of differentiation from gravity data is sensitive to nonhydrostatic effects like those known to exist in other similarly sized bodies (Gao and Stevenson 2013), so that additional approaches (e.g., utilizing shape data and/or pole position analyses) may be needed.

Strategic Research for Q2.3

- **Determine fundamental properties of the midsize uranian moons** through gravity, magnetic field, and geodetic measurements (by spacecraft), surface composition measurements (by remote sensing from spacecraft and ground-/space-based telescopes), and geological characterization (based on remote sensing by spacecraft, including imaging of the hemispheres unseen by Voyager 2).
- **Determine fundamental properties of Neptune's moon Triton** through gravity, magnetic field, and geodetic measurements (by spacecraft), surface composition measurements (by remote sensing from spacecraft and ground-/space-based telescopes), and geological characterization (based on remote sensing by spacecraft, including imaging of the hemisphere poorly seen by Voyager 2).

- **Determine Callisto's state of differentiation to constrain the accretional conditions of large icy moons** via spacecraft geodesy (shape), gravity, pole position, and magnetic field and associated charged particle measurements (the latter necessary for proper interpretation).
- **Study accretion dynamics in ring systems** through time-series imaging by spacecraft of rings and ring-embedded satellites (see Q7.2).
- **Improve the understanding of satellite and planetary ring formation** with improved numerical simulations and theoretical models.

Q2.4 HOW DID THE GIANT PLANETS GRAVITATIONALLY INTERACT WITH EACH OTHER, THE PROTOSOLAR DISK, AND SMALLER BODIES IN THE OUTER SOLAR SYSTEM?

It is now thought that the giant planets did not form where they currently reside, but instead migrated inward and/or outward because of disk torques during the protosolar nebular phase or by later gravitational interactions with remnant planetesimals. Evidence for these interactions is inferred from the orbits of the giant planets in the solar system and in extrasolar planetary systems (see Question 12), as well as how giant planet migration dynamically affected primordial small body populations (see Question 3, Chapter 6) and the bombardment history of the solar system (see Question 4, Chapter 7). The picture that has emerged so far of the solar system's earliest history involves an instability in the orbits of all the giant planets, a hypothesis with far reaching implications, but one that requires much further elaboration, refinement, and testing.

Q2.4a Did the Giant Planets Create Gaps in the Protosolar Nebula, and What Consequences Did This Have for Their Accretion and Potential Orbital Migration?

As described in Q2.1–Q2.3, it is thought that the gas giants were massive enough to open gaps in the protostellar nebula, in analogy with protoplanetary disks around other stars (e.g., as observed by ALMA, see Question 12). Unless torques are exquisitely balanced, giant planets should migrate to some degree, that is, have their orbits move inward or outward with rates that depend on disk properties (e.g., surface density and viscosity) and the presence or absence of gaps.

The timing and extent of giant planet migration during the solar nebula phase is debated. Giant planets may have formed in or evolved into resonant chains, in analogy with some exoplanet systems. Theoretical studies suggest migration of a single body should generally be inward, but it has been shown that the proximity of gaps around proto-Jupiter and proto-Saturn may allow for the reverse, that is, outward migration. This provides a possible explanation for why the solar system does not possess close-in, "hot Jupiters" (Morbidelli and Crida 2007).

Early migration implies that Jupiter and Saturn may have accreted material over a range of heliocentric distances (>1 AU); implications of this for the asteroid belt are discussed in Question 3.

Q2.4b Did the Giant Planets of the Early Solar System Migrate, and If So, How Far, and What Was the Effect of This Migration on Other Outer Solar System Bodies?

The clearest evidence for giant planet migration in the solar system is the dynamical structure of the trans-neptunian belt (Nesvorný 2018). It is most easily explained by the outward planetesimal-driven migration of Neptune out to 30 AU through a primordial Kuiper belt population that started near ~20 AU. Some discussion of the effects of Neptune's migration are included here, while other aspects are discussed in Q2.6.

As Neptune entered the primordial Kuiper belt, it excited the objects residing there, giving many higher eccentricities and inclinations. A small fraction was captured into resonances with Neptune—for example, Pluto's capture into Neptune's 2:3 mean motion resonance. Models suggest that the primordial belt population was reduced by a factor of ~1,000, with roughly this number of Pluto-size bodies and enumerable smaller bodies scattered into the giant planet zone and/or out into a scattered disk of comets associated with Neptune, or even the Oort cloud. The interactions between Neptune and ~1,000 Pluto-size bodies made Neptune's migration grainy rather than smooth, and this in turn prevented the capture of an excess number of objects within Neptune's resonances (Nesvorný and

Vokrouhlický 2016; Morbidelli and Nesvorný 2020). Moreover, a small fraction of the primordial Kuiper belt population ejected by Neptune was potentially retained on distant scattered orbits, where they now await discovery.

It is also possible that an earlier giant planet migration took place in the presence of nebular gas, as suggested in Q2.4a. The most viable model for this is referred to as the Grand Tack (Walsh et al. 2011). It describes how Jupiter and Saturn migrated across the asteroid belt, followed by a reversal of direction back toward their present orbits. This hypothesis derives its justification as an explanation for the low mass (inefficient accretion) of Mars and the broad mixing of compositional types in the asteroid belt, although other explanations exist for both constraints (see Q3.3).

Scenarios of giant planet migration can be tested by comparison with observations of their effects on, for example, small body populations, the formation and characteristics of the terrestrial planets, and the stability of giant planet satellites, which would strongly constrain the character (e.g., timing and physics) of giant planet migration, as well as, potentially, the formation of the Moon and the asteroids. Presently there is no fully accepted, self-consistent model for the simultaneous formation of the giant planets during the nebular era. The most important data to obtain, in the next decade, if possible, would be chronological constraints on the early bombardment of the Moon and the asteroids (Question 4), which would strongly constrain the character (timing, physics) of giant planet migration.

Q2.4c Was There a Global Instability Among the Giant Planets, and If So, When Did It Occur? Were Any Major Planets Ejected (Lost) During This Instability?

Given that migration models for Neptune can reasonably reproduce the orbital structure of the Kuiper belt and scattered disk, it is necessary to consider the broader implications of Neptune's migration for the rest of the solar system. For example, if Neptune entered the primordial Kuiper belt at ~20 AU, it probably had to form near that location. In turn, that means the giant planets once had a different configuration and a migration mechanism had to move them to where we see them today (Nesvorný 2018) (Figures 5-6 and 5-7).

Models indicate that gas accretion of the giant planets may have left them on nearly circular coplanar orbits between ~5 and ~20 AU, with most or perhaps all locked in mutual mean motion resonances with one another. This system, however, eventually went unstable in a violent exchange of orbital energy and angular momentum that is referred to here as a giant planet instability (e.g., Tsiganis et al. 2005; Nesvorný and Morbidelli 2012). Here giant planets encountered one another while also interacting with a sea of objects liberated from the primordial Kuiper belt.

Early versions of such models postulated that the instability and migration could have occurred hundreds of millions of years after the formation of the solar system, but more recent dynamical work favors an earlier instability; as the nebula clears, the giant planets emerge in a spacing too close to remain stable without the eccentricity and inclination damping effects of nebular gas. Regardless, the dispersal of the primordial Kuiper belt led to the heavy bombardment of the solar system worlds that existed at that time by ice-rich planetesimals (see Question 4, Chapter 7). It is possible many basins and large craters on the icy satellites trace back to this epoch.

As Neptune scattered trans-neptunian objects (TNOs) into the giant planet zone, dynamical friction from gravitational interactions between these bodies and the giant planets decreased the orbital eccentricities and inclinations of the latter, but not all the way to zero (Tsiganis et al. 2005). This explains the small but nonzero eccentricities and inclinations of Jupiter, Saturn, Uranus, and Neptune, otherwise expected to be effectively zero from gas interactions or large owing to mutual perturbations.

The number of giant planets gravitationally interacting with one another at the time of the instability is unknown. Dynamical studies suggest systems with 5 or 6 giant planets (e.g., three or four Uranus/Neptune-size bodies) have greater success reproducing dynamical constraints across the solar system than those that start with 4 giant planets (Batygin et al. 2012; Nesvorný and Morbidelli 2012). This extra Uranus/Neptune-size body is useful because it can encounter Jupiter multiple times before being ejected from the solar system. The tiny "jumps" produced in Jupiter's orbit are not only needed to explain its current orbit but also the dynamical structure of the asteroid belt and the putative capture of trans-neptunian objects in several regions (e.g., central and outer asteroid belt, Hilda asteroids, Trojan populations; see Question 3) (Nesvorný and Morbidelli 2012; Vokrouhlický et al. 2016).

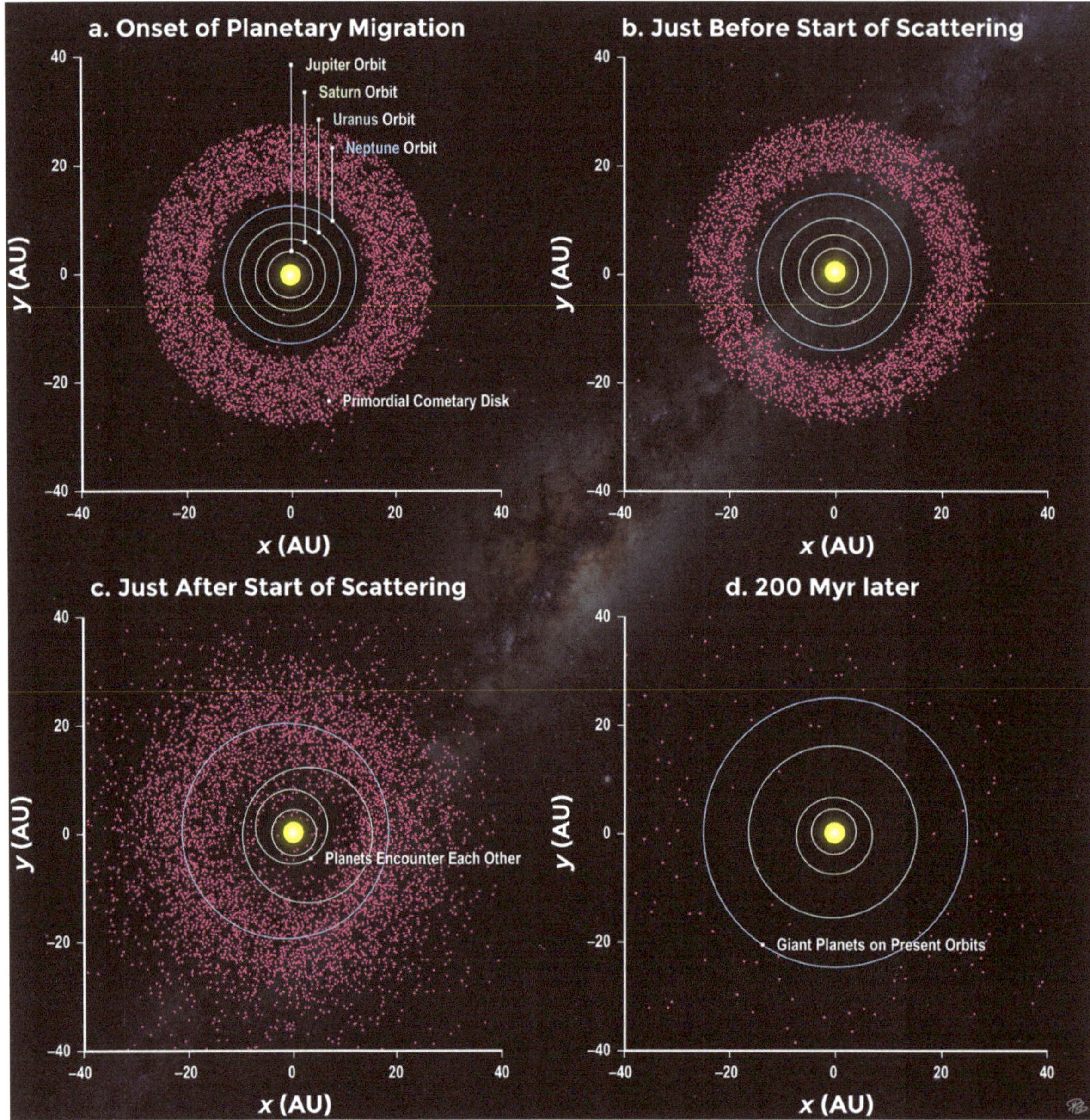

FIGURE 5-6 Gas-free giant planet migration and dynamical instability. The giant planets scatter each other, while Neptune migrates into a massive outer disk of cometesimals (i.e., primordial comets). This migration dynamically affects the terrestrial planets, asteroid belt, and primordial Kuiper belt (e.g., Nesvorný 2018). SOURCE: Figure adapted by P.K. Byrne. Used with permission of Annual Reviews, Inc., from D. Nesvorný, 2018, "Dynamical Evolution of the Early Solar System," *Annual Review of Astronomy and Astrophysics* 56:137–174; permission conveyed through Copyright Clearance Center, Inc.

Further observations to locate and characterize the most distant TNOs are the most important research activity to constrain the behavior of Neptune, along with observations and characterization of other TNOs, Centaurs, and comets to understand the installation of such extreme objects, and by implication, the instability that likely emplaced them.

The timing of the instability can be constrained by determining the chronology of impact bombardment, both in the inner and outer solar system (see Question 4). Bombardment by outer solar system objects should carry

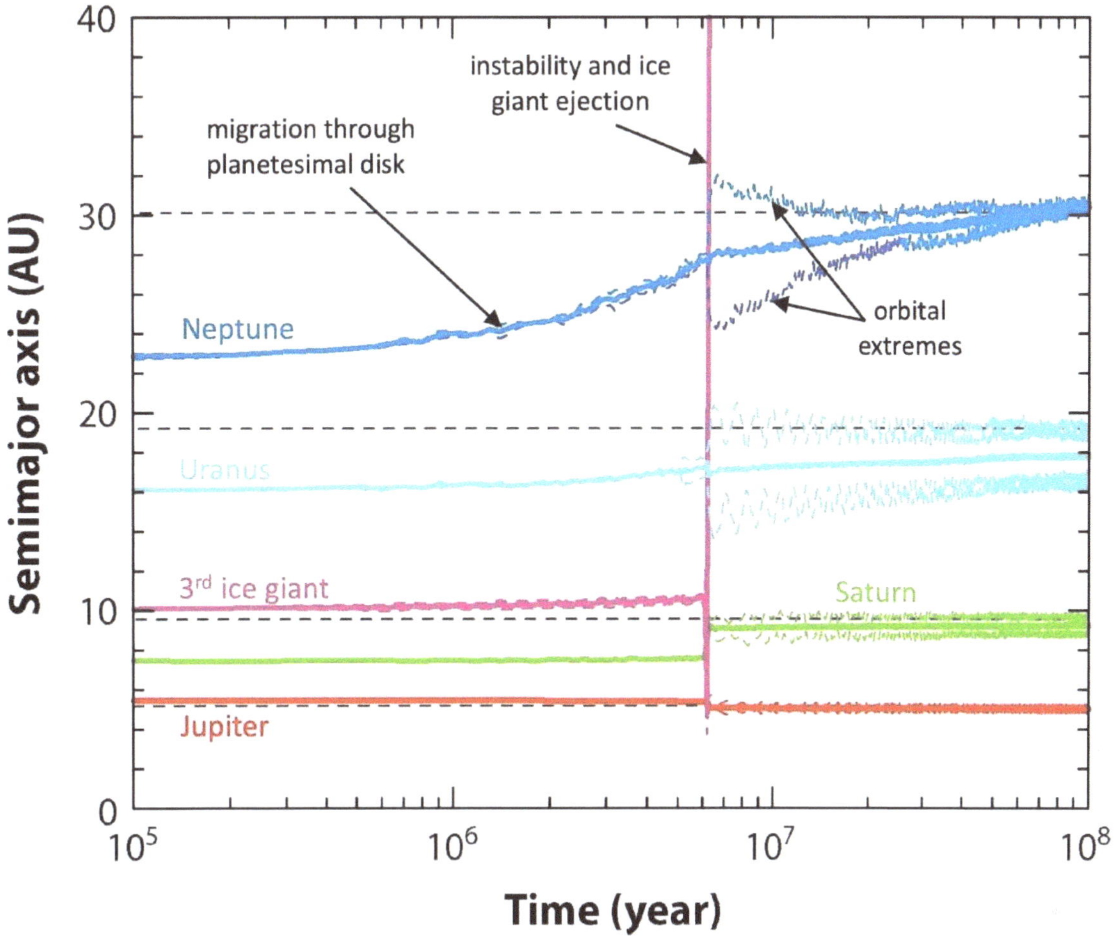

FIGURE 5-7 A possible orbital history of the giant planets. Five planets were started in a 3:2, 4:3, 2:1, and 3:2 mean-motion resonant chain along with a 20 Earth-mass planetesimal disk between 23 AU and 30 AU. The semimajor axes (solid lines) and perihelion and aphelion distances (dashed lines) of each planet's orbit are indicated. The horizontal dashed lines show the semimajor axes of planets in the present solar system. Note that the middle planet in this simulation, a third ice giant, is ejected from the solar system after an encounter with Jupiter. The final orbits obtained in the model are a good match to those in the present solar system. SOURCE: Adapted from D. Nesvorný and A. Morbidelli, 2012, "Statistical Study of the Early Solar System's Instability with Four, Five, and Six Giant Planets," *The Astrophysical Journal* 144(4):117, https://doi.org/10.1088/0004-6256/144/4/117, © AAS, reproduced with permission.

compositional (mainly volatile) and isotopic signals as well that may be discernible on the terrestrial planets and asteroids. Additional clues to the bombardment history of the outer solar system may be found on the cratered surfaces of outer solar system moons and TNOs.

Strategic Research for Q2.4

- **Determine the timing, extent, and effects of giant planet migration** by measurement of impact basin ages on the terrestrial planets; compositional and isotopic constraints on early terrestrial planet evolution, including the origin of the Moon; and studies of impact crater populations on diverse outer solar system bodies.
- **Further constrain the dynamical structure of the distant trans-neptunian population, including classical and resonant objects, so-called detached objects (Q2.6), and any undiscovered planet(s),** through remote sensing by ground-/space-based telescopes (including surveys) and theoretical modeling.

- **Characterize the basic properties of TNOs of diverse size, binarity, and dynamical subpopulations** with flyby(s)/orbital/landed missions to the outer solar system and through remote sensing by ground-/space-based telescopes (including surveys).
- **Improve our understanding of giant planet formation and migration in the early solar system** using improved numerical simulations and theoretical models.
- **Contextualize the early configuration and evolution of the solar system using comparative observations of protoplanetary disks and exoplanets** obtained with ground-/space-based telescopes.

Q2.5 HOW DID PROCESSES IN THE EARLY OUTER SOLAR SYSTEM PRODUCE THE STRUCTURE AND COMPOSITION (SURFACE AND INTERIOR) OF PLUTO AND THE TRANS-NEPTUNIAN OBJECTS?

Trans-neptunian objects, or TNOs (also Kuiper belt objects, KBOs, with the KBOs representing the stable population beyond Neptune) far outnumber any other type of bodies, but little is known about them, with the first TNO other than Pluto-Charon only discovered in 1992. TNOs retain unique information of planet formation in the outer solar system. In the past decade, thanks to an explosion in the number of TNOs being characterized with telescopic observations, combined with the New Horizons spacecraft's flybys of the Pluto system and the small TNO Arrokoth (a bilobate, contact binary object—i.e., two bodies in physical contact but that remain distinguishable), new constraints have enabled a better understanding of how the trans-neptunian belt and, by extension, other solar system objects may have formed and evolved early on.

Q2.5a When and How Did Trans-Neptunian Objects and Cometary Bodies Form?

Crater counts suggest that Pluto, its moons, and Arrokoth developed surfaces able to retain an impact record early in solar system history, perhaps more than 4 Ga ago (Stern et al. 2018). Their bombardment history is consistent with planetesimal formation models where bodies up to ~100 km in size are formed from gravitational instability (e.g., the streaming instability) during the first few million years of solar system history when nebular gas still existed (see Q1.3b).

Until the past decade, it was thought that TNOs were assembled through successive collisions between smaller bodies. Such events can potentially form a few Pluto-sized objects in the region between ~20 and 30 AU in about 100 million years and a size distribution of bodies comparable to that seen today. Features of the trans-neptunian belt size distribution predicted from such models, however, do not produce the more massive and numerous populations of bodies implied by planet migration models and conditions needed to account for the properties of resonant TNOs (Morbidelli and Nesvorný 2020) (see Q2.5b). In contrast, models of planetesimal formation via streaming instabilities reproduce these aspects (see Q1.3b).

A major question concerns the origin of comets. These bodies, as observed, are generally small (~1 to 10 km or so in diameter) and come from the scattered disk associated with Neptune or the Oort cloud. But their ultimate origin is thought to be the primordial Kuiper belt, the same region that birthed Pluto and the other larger members of the TNO population. Debate centers on whether these "proto-comets" formed as primordial small bodies (e.g., Davidsson et al. 2016) or whether they are the fragmented and reaccreted remnants of collisions among the larger TNO planetesimals just described (e.g., Morbidelli and Nesvorný 2020). Further physical study of comets and Centaurs, defined as bodies transitioning from the scattered disk to Jupiter-family comet orbits, should be illuminating regarding the primordial versus collisional origin question. But as Rosetta observations of comet 67P/Churyumov-Gerasimenko made clear, the insolation-driven activity of comets is a complicating or obscuring factor in understanding their formation. Visiting additional primitive, cold classical Kuiper belt objects unaffected by such activity is clearly warranted.

In contrast, of all the subcategories of TNOs, the cold classical TNOs near 45 AU stand apart (see Q1.3b and Q2.6a). Their characteristics (orbital and size distributions, colors, and binarity) as well as the benign impact environment in the cold classical region imply that the cold classicals are the small bodies most closely related to primordial planetesimals. The only cold classical TNO visited to date is Arrokoth, and its singular nature shows that these bodies can teach us a great deal about planetesimal formation (Stern et al. 2019; McKinnon et al. 2020). It is important to confirm and expand on these findings by visiting additional cold classicals as well as TNOs of diverse size, binarity, and dynamical subpopulations for comparison, and to contrast with cometary observations. Such encounters should be a part of future flyby missions to the outer solar system, supported by continuing ground and space-based observations.

Q2.5b How Many Trans-Neptunian Objects Formed, and What Were Their Initial Size Distributions?

The initial mass distribution of the trans-neptunian belt, set during the accretion era, probably exceeded the mass estimated today from observations by a factor of ~1,000 (see Q2.4b). This value is favored for several reasons: (1) it is otherwise difficult or impossible to form Pluto-sized bodies with the current mass (in any formation scenario), (2) ~1,000 Plutos are needed to keep Neptune's migration grainy enough to capture the right number of TNOs in Neptune's mean motion resonances, and (3) that quantity allows the depletion of the primordial Kuiper belt by Neptune's migration to explain various small body populations captured during the giant planet instability (Nesvorný and Vokrouhlický 2016; Vokrouhlický et al. 2016, 2019; Morbidelli and Nesvorný 2020).

The largest body within this mass distribution is not known but may exceed that of Triton, likely a captured TNO and more massive than Pluto. Moving to smaller sizes, if it is inferred there are ~10^5 bodies with diameter $D > 100$ km in the current Kuiper belt, a factor of ~1,000 depletion would suggest there were initially ~10^8 such bodies in the primordial Kuiper belt (Nesvorný and Vokrouhlický 2016). Moreover, if the Kuiper belt and Jupiter's Trojan populations were captured from the same source population, it can be deduced that the cumulative power law slope (q, the exponent in a distribution versus diameter of the number of objects larger than that diameter) of the primordial Kuiper belt for $D < 100$ km objects once followed $q = -2.1$ for $10 < D < 100$ km bodies (Grav et al. 2012; Nesvorný and Vokrouhlický 2016).

The smallest objects ($\lesssim 10$ km, depending on albedo) cannot be observed directly because of their faintness, but as with most solar system populations, there are more small objects than large ones. Cratering records of Pluto/Charon and Arrokoth from New Horizons provide our best estimate of smaller objects, with craters (or pits) identified on Arrokoth as small as 200 m implying even smaller impactors. Taken together, the crater data suggests a shallow cumulative power-law slope of $q \sim -1$ for projectiles between a few tens of meters and ~1 km (Singer et al. 2019; Spencer et al. 2020; Morbidelli et al. 2021; Robbins and Singer 2021). Such impactors are smaller than those ostensibly created by the streaming instability, and that could suggest this shallow slope was produced by collisional fragmentation. A possible analogy would be the shallow $q \sim -1$ slope seen among $0.1 < D < 1$ km bodies in the main asteroid belt (Bottke et al. 2015; Morbidelli et al. 2021).

Comprehensive astronomical surveys would greatly improve our understanding of the population of TNOs greater than 10 km in size, while doing so for smaller TNOs requires relying on crater counts from future spacecraft encounters and/or on stellar occultation surveys, data that will place crucial constraints on planetesimal formation and evolution models.

Q2.5c How Prevalent Were Giant Impacts in the Early Trans-Neptunian Belt?

Nearly all trans-neptunian dwarf planets have satellites, and formation as a binary or multiple system is thought to be typical of TNOs in general (Noll et al. 2020). Many satellite origin scenarios have been proposed, but the dwarf planets are massive enough that the formation of moons and possibly rings around them is thought to result from relatively giant impacts between dwarf planets themselves, the Pluto system being the prime example (Canup et al. 2021).

Questions remain, however, on what led to the variety of satellite systems seen in the trans-neptunian belt. Of the ten largest known TNOs, nine have at least one satellite orbiting relatively close (in terms of gravitational binding). This contrasts with the smallest binary TNOs observed from Earth, most of which have comparably sized components and large relative separations. At the smallest end, Arrokoth is a contact binary with two similarly sized components. This suggests, subject to observational selection effects, different formation mechanisms between large and small TNO systems.

The orbit, masses, and compositions of Pluto-Charon all point to a giant impact origin (Canup et al. 2021) involving a low-velocity collision between similarly sized progenitors. Pluto-Charon could have retained volatiles through such an event, although except for CO, none of their observed volatiles are primordial (see Question 6, Chapter 9). If thousands of Pluto-class bodies formed in the ~20 to ~30 AU region prior to Neptune's migration, giant impacts may have been common. On the other hand, if instead only a few Pluto-size objects formed (e.g., by standard pairwise accretion), giant impacts would have been rare, or possibly some key aspect of the formation problem is missing.

Differences in impact angle and/or velocity between two like-sized dwarf planets can lead to larger ice/rock fractionations, possibly explaining the properties of other dwarf planet systems, or outcomes other than a binary, such as rapidly rotating, triaxially shaped Haumea, which only has small moons, a ring, and a heliocentric family of fragments (the only one known so far among the TNOs). Alternative origins for binary systems have been proposed (Noll et al. 2020), such as three-body capture, but for (at least) dynamically cold, 100-km-class trans-neptunian binaries, these proposed capture mechanisms are inconsistent with the observed distribution of the binaries' mutual orbit inclinations, which instead favors co-formation by streaming and gravitational instabilities (Nesvorný et al. 2019) (see Q1.3b and Q2.5a). Possible origins may also depend on formation location. Further studies of binary properties generally will help constrain these and other binary formation mechanisms.

Q2.5d What Were the Relative Proportions of Ices, Rock, and Organic Materials Accreted by Small Objects (Comets, TNOs, Moons) in the Outer Solar System?

The sizes and masses of TNOs have been refined in the past decade thanks to telescopic observations and—for the Pluto system and TNO Arrokoth—flybys with the New Horizons spacecraft. Dividing an object's mass by its volume provides its bulk density, the first clue as to its composite materials. Mass determination generally depends on a body having a satellite whose orbit can be determined; size is especially difficult to measure for such distant bodies and is often done by thermal modeling. However, stellar occultations, if available, can be quite precise sizes. Thus, only a small minority of TNOs have well constrained bulk densities and compositions; for most others, especially objects smaller than ~500 km in diameter, densities are poorly known.

Density estimates for dwarf planets are intermediate between those of rock and ice, which is consistent with ice being the stable form of water this far from the Sun, and close to the densities of large molecular weight organic compounds; the latter are abundant in dust particles recovered on Earth and that may have originated in the outer solar system, and in comets 1P/Halley and 67P/Churyumov-Gerasimenko (Bardyn et al. 2017). A bulk density can be interpreted with specific proportions of rock and ice, but uncertainties in rock mineralogy, other ices, presence of impurities in the ice (e.g., salts), proportion of organic material (McKinnon et al. 2008, 2017), and porosity (that may exceed 70 percent in small, comet-sized bodies but is debated in larger bodies) translate into uncertain bulk proportions of these materials. Future astronomical measurements should provide a broader set of reliable TNO densities, and radio tracking of future spacecraft encounters should prove decisive in density determinations.

Q2.5e During Accretion, (How) Did the Interiors of Outer Solar System Moons and Dwarf Planets Transition from Homogeneous to Layered?

Constraining a planetary body's internal structure requires precise measurements of shape and either its rotation or Doppler tracking of a spacecraft trajectory near the body. Such measurements have been made at the Galilean satellites, Titan, and several midsize satellites of Saturn, but not yet as satellites of ice giants or on any TNOs.

The degree of layering (differentiation) inside icy moons varies, and the timing of differentiation is unclear. Ice and rock/organics have decidedly separated inside Europa, Ganymede, and tiny Enceladus; but apparently less so inside Callisto, Titan, Rhea, Dione, and Mimas. Differentiation does not seem to depend on size or density as much as distance from the giant planet. This suggests that among accretional, radiogenic, and tidal sources of heat, it is the latter that may have driven ice to melt, letting rock and organics settle to form a core. (Further heating can differentiate metal from rock.) However, one cannot discount the other two heat sources, which may have dominated during giant planet system formation.

Early differentiation of a Titan-like satellite is a requirement of one hypothesis for the formation of Saturn's rings, via tidal stripping of an icy mantle (Canup 2010). Loss of early surface oceans has also been suggested to explain the ice/rock compositional gradient of the Galilean satellites (Bierson and Nimmo 2020), although other explanations exist. Determining outermost Callisto's state of differentiation is a key test for the accretional conditions of large icy moons, as Galileo flyby gravity and magnetic field measurements were not definitive on this point (see Q2.3c). Additional measurements from polar flybys and/or an orbital survey are needed.

We do not currently know whether most trans-neptunian dwarf planets are layered. The above heat sources could have allowed ice-rock/organic differentiation, especially if accretional energy includes that of giant impacts. This is consistent with if not implied by the surface geology of Pluto and Charon (i.e., undifferentiated interiors are inconsistent with observations) (Moore et al. 2016). That said, rock/organic settling may have been impeded by convection in a solid mantle, and rock and ice may not have separated in the outer regions of Charon-sized or smaller TNOs (Figure 5-8). The formation of Pluto's rock-rich moon Charon and retinue of small but apparently very icy satellites is thought to require the giant impact of partially differentiated precursor bodies (Canup et al. 2021), that is, neither undifferentiated ice-rock bodies nor fully differentiated worlds.

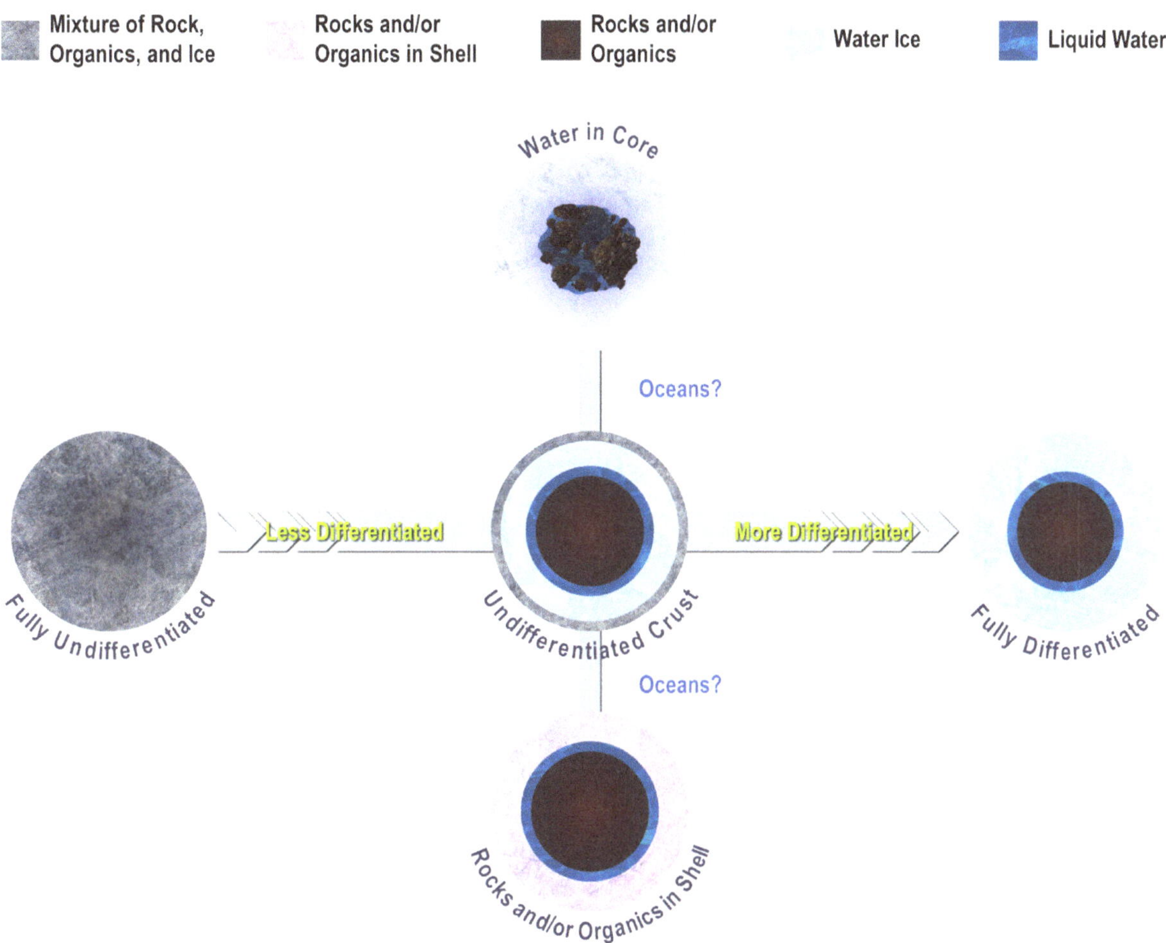

FIGURE 5-8 Possible early internal structures of outer solar system moons and dwarf planets. The degree of separation between rock and organic material is unknown, as is the organic fraction. If hot enough, metal can separate from silicate rock to form an inner core. Deep inside icy objects larger than Pluto, water is in the form of high-pressure ice that may have underlain liquid water oceans. SOURCE: Figure by P.K. Byrne, inspired by Canup et al. (2021).

Strategic Research for Q2.5

- **Characterize the basic properties (size, mass, shape, cratering, rings, and binarity) of diverse TNOs and related bodies (Centaurs and comets)** via remote sensing by spacecraft and ground-/space-based telescopic observations.
- **Improve understanding of giant impacts between ice-rock protoplanets** through state-of-the-art numerical simulations of giant impacts.
- **Improve the understanding of the accretion and internal differentiation, of icy moons and TNOs, including ocean formation,** through interpretation of spacecraft data in terms of consistent physicochemical models and laboratory experiments on ices and carbonaceous materials.
- **Improve the understanding of binary system, TNO family, and ring formation** through theory, observations, and numerical modeling.

Q2.6 HOW DID THE ORBITAL STRUCTURE OF THE TRANS-NEPTUNIAN BELT, THE OORT CLOUD, AND THE SCATTERED DISK ORIGINATE, AND HOW DID GRAVITATIONAL INTERACTIONS IN THE EARLY OUTER SOLAR SYSTEM LEAD TO SCATTERING AND EJECTION?

Small bodies across the solar system can constrain the migration of the giant planets and planet formation. The trans-neptunian belt out to the distant Oort cloud is a complex region with several interacting sub-populations, and they are the ultimate source of Jupiter family and long-period comets (also known as nearly isotropic comets). Irregular satellites of the giant planets, Trojan asteroids of Jupiter and Neptune, and even certain main belt asteroids are part of this cosmogonically linked small-body complex. Recently, unexpected small body properties, which can help us to constrain the formation and evolution of the solar system, have been discovered. They range from ring systems around some small bodies, to comets showing activity at large heliocentric distances, and to the existence of interstellar objects, with two such now known.

Q2.6a How Did the Dynamical Structure of the Trans-Neptunian Belt Originate?

During the giant planets' migration and scattering (see Q2.4 and Q3.3), Neptune's outward movement transformed the early trans-neptunian belt. The discovery of more than 3,000 TNOs in the past 30 years has revealed that this migration dynamically sculpted the trans-neptunian belt into four main groups of objects: (1) classical, (2) resonant, (3) scattered disk, and (4) detached (Gladman et al. 2008). Possibly the Oort cloud is a result of this scattering as well (e.g., Vokrouhlický et al. 2019; Morbidelli and Nesvorný 2020). Numerical simulations provide estimated sizes for these populations, relative to the size of the primordial Kuiper belt, and are given below (Vokrouhlický et al. 2019).

The classical TNOs are located between the 3:2 and 2:1 mean motion resonances with Neptune and reside in the semimajor-axis range of 42–47 AU. The classical group is divided into two sub-populations; the dynamically cold classicals with low eccentricities and inclinations and the hot classicals with higher eccentricities and inclinations.

The cold classicals are some of the most pristine bodies in the solar system. These bodies formed far enough away from Neptune to avoid substantial disturbances as the planet migrated outward (Nesvorný 2018). They are likely to be indigenous to this region. Their initial mass is debated, but it seems probable that the ensemble was relatively small compared to the primordial Kuiper belt. The cold classicals tend to be smaller, redder, and have a higher fraction of wide binaries compared to the hot classicals, suggesting that none of these bodies were affected by Neptune close encounters (Noll et al. 2020). The latter is consistent with dynamically brief events that may have led to ice giant ejection (e.g., Figure 5.7). Arrokoth is a resident of this population (McKinnon et al. 2020).

The hot classicals are thought to have originated in the primordial Kuiper belt and represent $\sim 5 \times 10^{-4}$ of that population (e.g., Nesvorný and Vokrouhlický 2016). They were pushed outward by Neptune scattering events and by interactions with its mean motion resonances. Further interactions with Neptune's resonances, secular resonances, and Pluto-sized objects allowed some to drop out of resonances while also achieving low enough eccentricities to be captured between 42 AU and 47 AU.

The resonant TNOs were trapped within Neptune's mean motion resonances as Neptune migrated outward (e.g., Nesvorný and Vokrouhlický 2016). The size of the captured population in each resonance was regulated by the speed of Neptune's outward migration and interactions with numerous Pluto-size bodies along its path. The resonant TNOs are physically similar to the hot classicals and contain a few $\times\ 10^{-4}$ of the primordial Kuiper belt. The most prominent resonant population is within Neptune's 3:2 resonance, which contains Pluto-Charon, while smaller populations are in additional resonances (e.g., 2:1, 5:3, 7:4, and 5:2).

The major questions for the formation of the various trans-neptunian populations relate directly to Q2.4: precisely when and how did migration of the giant planets occur, and what were the properties (mass, size-frequency distribution, orbital distribution, composition, etc.) of the ancestral planetesimal or cometesimal disk that was the ultimate source of these populations? How do the properties of the various trans-neptunian populations (e.g., number, size-frequency distributions, compositions and colors, and binarity) constrain models of their formation? How have these subpopulations evolved through time? What accounts for the apparent "edge" of the classical Kuiper belt at semimajor axes of ~47 AU? Below the committee addresses the scattered disk and detached populations specifically, in Q2.6b and Q2.6c, respectively.

Q2.6b How Did the Oort Cloud and Scattered Disk Form?

It has been long inferred from the existence of long-period comets (LPCs) that the Sun is surrounded by a spherical cloud of icy small bodies, called the Oort cloud (e.g., Peixinho et al. 2020). It contains billions of objects and is located between an inner boundary of ~2,000–5,000 AU and outer boundary of ~50,000–100,000 AU. All are far beyond the reach of current telescopes.

The Oort cloud has long been thought to have been populated by planetesimals ejected from the giant planet region. Recent models suggest that much of the Oort cloud was constructed during the giant planet instability (see, e.g., Nesvorný 2018). Here small bodies scattered into the giant planet zone from the primordial Kuiper belt experienced giant planet encounters, with many placed onto orbits with very large semimajor axes. These bodies then had their perihelia increased by a combination of gravitational perturbations from galactic tides and close passing stars. Approximately ~5 percent of the primordial Kuiper belt still exists within the Oort cloud (see Vokrouhlický et al. 2019). Oort cloud objects were also potentially captured from the protoplanetary disk(s) of other star(s) when the Sun was still in its stellar birth cluster (Levison et al. 2010). A combination of the scattering and capture scenarios is plausible.

Scattered disk objects (SDOs) are bodies with semimajor axes between ~30 and ~1,000 AU on unstable crossing orbits with Neptune (Nesvorný 2018). SDOs were scattered outward from the primordial Kuiper belt during Neptune's outward migration, leaving them on highly eccentric and inclined orbits. All will eventually be sent deeper into the giant planet zone, making them the primary source of the Centaurs (objects in chaotic orbits between Jupiter and Neptune) and Jupiter Family Comets. Their true population is thought to be larger than the classical and resonant TNOs, but because they are faint, only a few hundred SDOs have been discovered to date near their perihelion. This number should be substantially increased with telescopes coming online (e.g., Vera C. Rubin Observatory). Numerical models suggest that the SDO population includes a factor of ~300 times more objects than have been identified to date (Vokrouhlický et al. 2019). Interactions with Neptune led to the loss of ~two orders of magnitude of SDOs to the giant planet zone over billions of years. Accordingly, SDOs are the primary source of impactors on the giant planet satellites.

Note that it is possible that some objects from the giant planet zone or the inner solar system could have been passed outward through the giant planet region, had a weak interaction with Neptune and achieved SDO orbits. These objects may be rare, but they could potentially be identified by their unusual colors and spectra.

The major questions for the formation of the scattered disk relate to those in Q2.6a. For the Oort cloud, the major questions are its dynamical structure, how it relates to the more distant Detached population (Q2.6c), and whether it is truly, or mainly, sourced from the same primordial planetesimal disk as the SDOs.

Q2.6c How Did the Most Distant (e.g., Detached) Trans-Neptunian Objects Form?

Owing to the extreme distances of the detached TNOs, this population—which is transitional between the SDOs and the Oort cloud—is the least constrained. The detached population, defined as being beyond Neptune's gravitational influence and not on a Neptune-crossing orbit, has several sub-groups, with the extreme TNOs having perihelion distances between ~40 and ~55 AU and the Inner Oort cloud objects having perihelia beyond ~65 AU (Nesvorný 2018). So far, no TNO has been found with a perihelion between ~55 and ~65 AU (Sheppard et al. 2019).

Here the detached TNOs are divided into inner and outer parts. In the inner portion, many have semimajor axis values that are slightly on the sunward side of mean motion resonances with Neptune, and the objects are too close to the Sun to have been influenced by passing stars within the solar system's stellar birth cluster. These constraints indicate that their origin may have been analogous to the hot classicals (see Q2.6a), in that the inner bodies were scattered outward by Neptune as it migrated through the primordial Kuiper belt; eventually, however, some interacted with distant Neptune mean-motion resonances and/or Kozai resonances, either of which lowered their eccentricities enough to remove them from Neptune-crossing orbits (Gomes 2011). These bodies then fell out of resonance while Neptune was still migrating outward, possibly when Neptune experienced tiny kicks from encounters with Pluto-size bodies (Lawler et al. 2019).

The outer portion, or the extreme KBOs, have large perihelia and semimajor axes that are too far from Neptune for Neptune's mean motion resonances to influence their orbits. Several hypotheses have been proposed to explain how these bodies escaped a Neptune-crossing orbit: (1) a passing star from the solar system's stellar birth cluster could scatter TNOs onto extreme TNO orbits, (2) the collective gravity of a distant massive, small body population influenced the objects and affected their orbits, or (3) a very distant planet (or planets) gravitationally shaped the outer edge of the trans-neptunian belt.

For the latter mechanism, only a handful of distant TNOs have been discovered so far, but some argue they display evidence for orbital clustering, which would be surprising if true. This was the original evidence used to suggest the hypothesis of a super-Earth planet (usually called Planet X or Planet 9) in the outer solar system (Figure 2-19 in the Small Body section of Chapter 2, Trujillo and Sheppard 2014; Batygin and Brown 2016). Alternatively, the clustering could be a by-product of observational selection effects by TNO surveys. To resolve this question, we either need to find enough extreme TNOs that the issue of observational bias can be ruled out or we need to discover the putative planet itself (Trujillo 2020).

Q2.6d How Did Scattering, Capture, and Ejection Affect the Small Body Populations?

Two critical components of the giant planet instability are that (1) giant planets are capable of having encounters with one another, which can cause them to migrate via gravitational kicks or "jumps," and (2) these encounters are happening concurrently with Neptune's migration through the primordial Kuiper belt, which sends approximately 20 Earth masses of TNOs into the giant planet zone. Some TNOs could have been located at the right place and time to be captured within stable reservoirs across the solar system by three-body reactions during this time (see also Question 3). Study of current populations in these reservoirs may, for example, provide constrains on the mass of the primordial trans-neptunian population (Nesvorný 2018).

The most well-known captured populations are the Trojans of Jupiter and Neptune, orbiting around the L4 and L5 Lagrange points located 60 degrees in front of and behind each planet, respectively. (Such orbits for Saturn and Uranus are unstable.) The most remarkable dynamical property of the Trojans is that their inclinations range from 0 degrees to 35 degrees. This distribution challenges most origin models that require the Trojans to be captured with low inclination orbits (Emery et al. 2015). However, objects ejected from the primordial Kuiper belt have high inclinations, and giant planet encounters can capture a small fraction of them within Jupiter and Neptune's L4 (i.e., 60 degrees ahead) and L5 (i.e., 60 degrees behind) locations (Vokrouhlický et al. 2019), resolving this issue. Other captured populations are discussed in Question 3.

Giant planet encounters can also lead to irregular satellite capture (see Question 8, Chapter 11). Models show that several $\times\ 10^{-8}$ of the primordial Kuiper belt population can be captured around Jupiter, Saturn, Uranus, and Neptune. The physical properties of the Trojans and irregular satellites should be the same as the resonant TNOs and the hot classicals, but the irregulars are thought to be heavily collisionally evolved (Bottke et al. 2010). Surveys have demonstrated that the Jupiter/Neptune Trojans and irregular satellites have similar surface colors and thus could share the same origin, but they lack the very red bodies found in the scattered disk, resonant TNOs, and hot/cold classicals (Jewitt 2018). One possible explanation is that the transition to very red objects occurs beyond 30 AU in the primordial Kuiper belt, either from sublimation-driven surface depletion in some organic molecules or from collisional evolution (Nesvorný et al. 2020). The alternative is that the existing giant planet instability model is missing something.

The origin of, and dynamical processes affecting, these small body populations (e.g., Trojans, irregular satellites and TNOs) at different heliocentric locations have been, and should continue to be, studied by comparing the results of spacecraft missions and ground- and space-based telescopic observations. Combining these measurements and observations with state-of-the-art theoretical and numerical modeling offers the surest path to more complete understanding.

Q2.6e What Do Active or Unusual Phenomena Among Distant Small Body Populations Tell Us About Accretion in the Outer Solar System?

The Centaur population is mostly composed of objects from the scattered disk and a minority of TNOs from the 2:1 and 3:2 Neptune's resonances. Through stellar occultations, ring systems (unpredicted by theory) were discovered around two Centaurs (Ortiz et al. 2020). A second interesting feature is that Centaurs can be active from a low level up to major outburst-type activity. Active Centaurs have been observed beyond 10 AU, but the physical process leading to activity is still unknown and could include (1) water-ice crystallization from an initially amorphous state providing energy to drive activity; (2) sublimation of super volatiles such as methane, carbon monoxide, or nitrogen ice; and/or (3) rotational fission. Both the distant outgassing and ring systems are unobserved among the comets—a Centaur's final state—nor do inner solar system objects possess rings. Owing to their unusual physical characteristics and because they are in transition from the outer to the inner solar system, Centaurs are of great interest for future missions (Harris et al. 2020).

In contrast, Manx comets are on long-period orbits where other known objects are active, yet the Manx comets show no activity. They could be extinct comets that have had many volatile-depleting passages near the Sun, they could be interloper objects, perhaps formed near the main belt and then placed on extremely distant orbits through an as-yet-unknown process, or they could be planetesimals from the giant planet zone with limited initial volatile content (Meech and Castillo-Rogez 2021).

Q2.6f What Can Interstellar Objects Tell Us About the Formation and Early Evolution of Our Solar System and Others?

Although the possibility of interstellar objects (ISOs) has long been hypothesized, only recently have two been discovered passing through the solar system: 1I/'Oumuamua and 2I/Borisov (see Question 3). These objects likely originated in an exoplanetary system and were flung out (possibly shortly after their formation, if the solar system is any guide) by an encounter with a giant planet or their host star(s).

Obvious ISOs have distinctly hyperbolic orbits. Some objects in the Oort cloud, however, may have been captured from other stars in the solar system's stellar birth cluster. These ISOs, taking the form of near-isotropic comets, would not have hyperbolic orbits. Thus, identification of ISOs may also involve other diagnostics, such as a body's chemical properties (e.g., deuterium to hydrogen (D/H) and other isotopic ratios, presence or outgassing of extremely volatile compounds) and/or physical properties (e.g., surface features or shapes) unlike those of the small bodies in the solar system (Meech et al. 2017). A caveat is that ISOs in the Oort cloud from the solar system's stellar birth cluster, derived ultimately from the same natal molecular cloud, may turn out to be indistinguishable from comets native to the solar system.

The primary difficulty with observation of ISOs is that they only spend a short time in the solar system, requiring rapid-response efforts from ground- and space-based telescopes, and/or a mission designed and launched even before the ISO is discovered (such as the Comet Interceptor mission being developed by ESA and JAXA). It is expected that in the coming decade, the discovery rate of ISOs will increase dramatically. By discovering and characterizing additional ISOs, we will be able to compare them to the small body populations in the solar system, and potentially constrain their birthplace.

Strategic Research for Q2.6

- **Determine the rotational, physical, chemical, geological, and interior properties of a diversity of primitive small bodies (TNOs) in the outer solar system** with spacecraft and/or ground-/space-based observations.
- **Determine the rotational, physical, chemical, geological, and interior properties of a diversity of irregular satellites, Trojans, Centaurs, and comets as well as investigate their ring system(s) and/or activity** with spacecraft and/or ground-/space-based observations.
- **Characterize the rotational, physical, chemical, geological, and interior properties of interstellar objects and comparison with small bodies in the solar system** with spacecraft and/or ground-/space-based observations.
- **Improve our understanding of the formation and evolution of the early solar system (and by extension, exoplanetary systems), planetary migration, including undiscovered planet(s)** with improved numerical simulations, theoretical models, and/or ground-/space-based observations.

SUPPORTIVE ACTIVITIES FOR QUESTION 2

Improvement of computational capabilities for advanced numerical work (e.g., greater resolution in space and time, increased number of particles, faster computation and visualization, and machine learning).

Telescopic observations (spectroscopy, color, light curve, stellar occultation, mutual event, binarity study, etc.) of small bodies across the solar system to infer their properties, ideally utilizing both ground- and space-based assets across multiple wavelengths.

REFERENCES

Agnor, C.G., and D.P. Hamilton. 2006. "Neptune's Capture of Its Moon Triton in a Binary-Planet Gravitational Encounter." *Nature* 441:192–194.

Atreya, S.K., A. Crida, T. Guillot, C. Li, J.I. Lunine, N. Madhusudhan, O. Mousis, and M.H. Wong. 2022. "The Origin and Evolution of Saturn: A Post-Cassini Perspective," in *Saturn: The Grand Finale*, K.H. Baines, et al., eds., Cambridge University Press. Accepted for Publication. Preprint: https://doi.org/10.48550/arXiv.2205.06914.

Bardyn, A., D. Baklouti, H. Cottin, N. Fray, C. Briois, J. Paquette, O. Stenzel, et al. 2017. "Carbon-Rich Dust in Comet 67P/Churyumov-Gerasimenko Measured by COSIMA/Rosetta." *Monthly Notices of the Royal Astronomical Society* 469:S712–S722.

Batygin, K. 2018. "On the Terminal Rotation Rates of Giant Planets." *Astronomical Journal* 155:178.

Batygin, K., and M.E. Brown. 2016. "Evidence for a Distant Giant Planet in the Solar System." *Astronomical Journal* 151:22.

Batygin, K., and A. Morbidelli. 2020. "Formation of Giant Planet Satellites." *Astrophysical Journal* 894:143.

Batygin, K., M.E. Brown, and H. Betts. 2012. "Instability-Driven Dynamical Evolution Model of a Primordially Five-Planet Outer Solar System." *Astrophysical Journal* 744:L3.

Bierson, C.J., and F. Nimmo. 2020. "Explaining the Galilean Satellites' Density Gradient by Hydrodynamic Escape." *Astrophysical Journal Letters* 897:L43.

Boss, A.P. 1997. "Giant Planet Formation by Gravitational Instability." *Science* 276:1836–1839.

Bottke, W.F., D. Nesvorný, D. Vokrouhlický, and A. Morbidelli. 2010. "The Irregular Satellites. The Most Collisionally Evolved Population in the Solar System." *Astrophysical Journal* 139:994–1014.

Bottke, W.F., M. Broz, D.P. O'Brien, A. Campo Bagatin, A. Morbidelli, and S. Marchi. 2015. "The Collisional Evolution of the Asteroid Belt." Pp. 701–724 in *Asteroids IV*, P. Michel, F. DeMeo, and W.F. Bottke, eds. Tucson: University of Arizona Press.

Bryan, M.L., S. Ginzburg, E. Chiang, C. Morley, B.P. Bowler, J.W. Xuan, and H.A. Knutson. 2020. "As the Worlds Turn: Constraining Spin Evolution in the Planetary-Mass Regime." *Astrophysical Journal* 905:37.

Canup, R.M. 2010. "Origin of Saturn's Rings and Inner Moons by Mass Removal from a Lost Titan-Sized Satellite." *Nature* 468:943–946.

Canup, R.M., and W.R. Ward. 2006. "A Common Mass Scaling for Satellite Systems of Gaseous Planets." *Nature* 441:834–839.

Canup, R.M., K.M. Kratter, and M. Neveu. 2021. "On the Origin of the Pluto System." In *The Pluto System after New Horizons*, S.A. Stern, R.P. Binzel, W.M. Grundy, J.M. Moore, and L.A. Young, eds. Tucson: University of Arizona Press.

Crida, A., and S. Charnoz. 2012. "Formation of Regular Satellites from Ancient Massive Rings in the Solar System." *Science* 338:1196–1199.

Davidsson, B.J.R., H. Sierks, C. Güttler, F. Marzari, M. Pajola, H. Rickman, M.F. A'Hearn, et al. 2016. "The Primordial Nucleus of Comet 67P/Churyumov-Gerasimenko." *Astronomy & Astrophysics* 592(A63).

Emery, J.P., F. Marzari, A. Morbidelli, L.M. French, and T. Grav. 2015. "The Complex History of Trojan Asteroids." Pp. 203–220 in *Asteroids IV*, P. Michel, F. DeMeo, and W.F. Bottke, eds. Tucson: University of Arizona Press.

Gao, P., and D.J. Stevenson. 2013. "Nonhydrostatic Effects and the Determination of Icy Satellites' Moment of Inertia." *Icarus* 226:1185–1191.

Gladman, B., B.G. Marsden, and C. VanLaerhoven. 2008. "Nomenclature in the Outer Solar System." Pp. 43–47 in *The Solar System Beyond Neptune*, M.A. Barucci, H. Boehnhardt, D. Cruikshank, and A. Morbidelli, eds. Tucson: University of Arizona Press.

Gomes, R.S. 2011. "The Origin of TNO 2004 XR_{190} as a Primordial Scattered Object." *Icarus* 215:661–668.

Grav, T., A.K. Mainzer, J.M. Bauer, J.R. Masiero, and C.R. Nugent. 2012. "WISE/NEOWISE Observations of the Jovian Trojan Population: Taxonomy." *Astrophysical Journal* 759(1):49.

Guillot, T., and R. Hueso. 2006. "The Composition of Jupiter: Sign of a (Relatively) Late Formation in a Chemically Evolved Protosolar Disc." *Monthly Notices of the Royal Astronomical Society: Letters* 367(1):L47–L51.

Harris, W., Y.R. Fernandez, G. Sarid, J.K. Steckloff, K. Volk, M. Womack, and L.M. Woodney. 2020. "Active Primordial Bodies: Exploration of the Primordial Composition of Ice-Rich Planetesimals and Early-Stage Evolution in the Outer Solar System." White paper #296 submitted to the Planetary Science and Astrobiology Decadal Survey 2023–2032. *Bulletin of the AAS* 53(4).

Helled, R., and A. Morbidelli. 2021. "Planet Formation." Pp. 12-1–12-5 in *ExoFrontiers: Big Questions in Exoplanetary Science*, N. Madhusudhan, ed. Bristol, UK: IOP Publishing.

Helled, R., and D. Stevenson. 2017. "The Fuzziness of Giant Planets' Cores." *Astrophysical Journal Letters* 840(1):L4.

Helled, R., P. Bodenheimer, M. Podolak, A. Boley, F. Meru, S. Nayakshin, J.J. Fortney, et al. 2014. "Giant Planet Formation, Evolution, and Internal Structure." Pp. 643–665 in *Protostars and Planets VI*, H. Beuther, R.S. Klessen, C.P. Dullemond, and T. Henning, eds. Tucson: University of Arizona Press.

Helled, R., N. Nettelmann, and T. Guillot. 2020. "Uranus and Neptune: Origin, Evolution and Internal Structure." *Space Science Reviews* 216:38.

Hyodo, R., S. Charnoz, K. Ohtsuki, and H. Genda. 2017. "Ring Formation Around Giant Planets by Tidal Disruption of a Single Passing Large Kuiper Belt Object." *Icarus* 282:195–213.

Ida, S., S. Ueta, T. Sasaki, and Y. Ishizawa. 2020. "Uranian Satellite Formation by Evolution of a Water Vapor Disk Generated by a Giant Impact." *Nature Astronomy* 4:880–885.

Izidoro, A., A. Morbidelli, S.N. Raymond, F. Hersant, and A. Pierens. 2015. "Accretion of Uranus and Neptune from Inward-Migrating Planetary Embryos Blocked by Jupiter and Saturn." *Astronomy & Astrophysics* 582:A99.

Jewitt, D. 2018. "The Trojan Color Conundrum." *Astronomical Journal* 155:56.

Johansen, A., and M. Lambrechts. 2017. "Forming Planets via Pebble Accretion." *Annual Review of Earth and Planetary Sciences* 45:359–387.

Kruijer, T.S., C. Burkhardt, G. Budde, and T. Kleine. 2017. "Age of Jupiter Inferred from the Distinct Genetics and Formation Times of Meteorites." *Proceedings of the National Academy of Sciences* 114:6712–6716.

Lambrechts, M., and A. Johansen. 2012. "Rapid Growth of Gas-Giant Cores by Pebble Accretion." *Astronomy & Astrophysics* 544:A32.

Lawler, S.M., R.E. Pike, N. Kaib, M. Alexandersen, M.T. Bannister, Y.-T. Chen, B. Gladman, et al. 2019. "OSSOS. XIII. Fossilized Resonant Dropouts Tentatively Confirm Neptune's Migration Was Grainy and Slow." *Astronomical Journal* 157(6):253.

Levison, H.F., M.J. Duncan, R. Brasser, and D.E. Kaufmann. 2010. "Capture of the Sun's Oort Cloud from Stars in Its Birth Cluster." *Science* 329:187–190.

Levison, H.F., K.A. Kretke, and M.J. Duncan. 2015. "Growing the Gas-Giant Planets by the Gradual Accumulation of Pebbles." *Nature* 524:322–324.

Lichtenberg, T., J. Drążkowska, M. Schönbächler, G.J. Golabek, and T.O. Hands. 2021. "Bifurcation of Planetary Building Blocks During Solar System Formation." *Science* 371:365–370.

Mankovich, C., and J. Fuller. 2021. *A Diffuse Core in Saturn Revealed by Ring Seismology.* arXiv:2104.13385.

McKinnon, W.B., J.I. Lunine, and D. Banfield. 1995. "Origin and Evolution of Triton." Pp. 807–877 in *Neptune and Triton*, D.P. Cruikshank, ed. Tucson: University of Arizona Press.

McKinnon, W.B., D. Prialnik, S.A. Stern, and A. Coradini. 2008. "Structure and Evolution of Kuiper Belt Objects and Dwarf Planets." Pp. 213–241 in *The Solar System Beyond Neptune*, M.A. Barucci, H. Boehnhardt, D.P. Cruikshank, and A. Morbidelli, eds. Tucson: University of Arizona Press.

McKinnon, W.B., S.A. Stern, H.A. Weaver, et al. 2017. "Origin of the Pluto-Charon System: Constraints from the New Horizons Flyby." *Icarus*: 287:2–11.

McKinnon, W.B., D.C. Richardson, J.C. Marohnic, J.T. Keane, W.M. Grundy, D.P. Hamilton, D. Nesvorný, et al. 2020. "The Solar Nebula Origin of (486958) Arrokoth, a Primordial Contact Binary in the Kuiper Belt." *Science* 367(6481):aay6620.

Meech, K.J., and J. Castillo-Rogez. 2021. "In-Situ Exploration of Objects on Oort Cloud Comet Orbits: OCCs, Manxes, and ISOs." White paper #282 submitted to the Planetary Science and Astrobiology Decadal Survey 2023–2032. *Bulletin of the AAS* 53(4).

Meech, K.J., R. Weryk, M. Micheli, J.T. Kleyna, O.R. Hainaut, R. Jedicke, R. Wainscoat, et al. 2017. "A Brief Visit from a Red and Extremely Elongated Interstellar Asteroid." *Nature* 552:378–381.

Moore, J.M., W.B. McKinnon, J.R. Spencer, A.D. Howard, P.M. Schenk, R.A. Beyer, F. Nimmo, et al. 2016. "The Geology of Pluto and Charon Through the Eyes of New Horizons." *Science* 351:1284–1293.

Morbidelli, A., and A. Crida. 2007. "The Dynamics of Jupiter and Saturn in the Gaseous Protoplanetary Disk." *Icarus* 191:158–171.

Morbidelli, A., and D. Nesvorný. 2020. "Kuiper Belt: Formation and Evolution." Pp. 25–59 in *The Trans-Neptunian Solar System*, D. Prialnik, M.A. Barucci, and L. Young, eds. Amsterdam, Netherlands: Elsevier.

Morbidelli, A., K. Tsiganis, K. Batygin, A. Crida, and R. Gomes. 2016. "Explaining Why the Uranian Satellites Have Equatorial Prograde Orbits Despite the Large Planetary Obliquity." *Icarus* 219:737–740.

Morbidelli, A., D. Nesvorný, W.F. Bottke, and S. Marchi. 2021. "A Re-Assessment of the Kuiper Belt Size Distribution for Sub-Kilometer Objects, Revealing Collisional Equilibrium at Small Sizes." *Icarus* 356:114256.

Mousis, O., T. Ronnet, J.I. Lunine, A. Luspay-Kuti, K.E. Mandt, G. Danger, F. Pauzat, et al. 2018. "Noble Gas Abundance Ratios Indicate the Agglomeration of 67P/Churyumov–Gerasimenko from Warmed-Up Ice." *Astrophysical Journal Letters* 865:L11.

Nesvorný, D. 2018. "Dynamical Evolution of the Early Solar System." *Annual Review of Astronomy and Astrophysics* 56:137–174.

Nesvorný, D., and A. Morbidelli. 2012. "Statistical Study of the Early Solar System's Instability with Four, Five, and Six Giant Planets." *Astronomical Journal* 14:117.

Nesvorný, D., and D. Vokrouhlický. 2016. "Neptune's Orbital Migration Was Grainy, Not Smooth." *Astrophysical Journal* 825:94.

Nesvorný, D., R. Li, A.N. Youdin, J.B. Simon, and W.M. Grundy. 2019. "Trans-Neptunian Binaries as Evidence for Planetesimal Formation by the Streaming Instability." *Nature Astronomy* 3:808–812.

Nesvorný, D., D. Vokrouhlický, M. Alexandersen, M.T. Bannister, L.E. Buchanan, Y.-T. Chen, B.J. Gladman, et al. 2020. "OSSOS XX: The Meaning of Kuiper Belt Colors." *Astronomical Journal* 160:46.

Nogueira, E., R. Brasser, and R. Gomes. 2011. "Reassessing the Origin of Triton." *Icarus* 214:113–130.

Noll, K.S., W.M. Grundy, D. Nesvorný, and A. Thirouin. 2020. "Trans-Neptunian Binaries. Pp. 205–224 in *The Trans-Neptunian Solar System*, D. Prialnik, M.A. Barucci, and L. Young, eds. Amsterdam, Netherlands: Elsevier.

Öberg, K.I., and E.A. Bergin. 2021. "Astrochemistry and Compositions of Planetary Systems." *Physics Reports* 893:1–48.

Ortiz, J.L., B. Sicardy, J.I.B. Camargo, P. Santos-Sanz, and F. Braga-Ribas. 2020. "Stellar Occultation by TNOs: From Predictions to Observations." Pp. 413–437 in *The Trans-Neptunian Solar System*, D. Prialnik, M.A. Barucci, and L. Young, eds. Amsterdam, Netherlands: Elsevier.

Peale, S.J., and R.M. Canup. 2015. "The Origin of the Natural Satellites." Pp. 559–604 in Vol. 10: *Physics of Terrestrial Planets and Moons* of *Treatise on Geophysics*, 2nd ed., G. Schubert, ed. Amsterdam, Netherlands: Elsevier.

Peixinho, N., A. Thirouin, S.C. Tegler, R. Di Sisto, A. Delsanti, A. Guilbert-Lepoutre, and J.G. Bauer. 2020. "From Centaurs to Comets—40 Years." Pp. 307–329 in *The Trans-Neptunian Solar System*, D. Prialnik, M.A. Barucci, and L. Young, eds. Amsterdam, Netherlands: Elsevier.

Raymond, S.N., and A. Izidoro. 2017. "Origin of Water in the Inner Solar System: Planetesimals Scattered Inward During Jupiter and Saturn's Rapid Gas Accretion." *Icarus* 297:134–148.

Reinhardt, C., A. Chau, J. Stadel, and R. Helled. 2020. "Bifurcation in the History of Uranus and Neptune: The Role of Giant Impacts." *Monthly Notices of the Royal Astronomical Society* 492:5336–5353.

Robbins, S.J., and K.N. Singer. 2021. "Pluto and Charon Impact Crater Populations: Reconciling Different Results." *Planetary Science Journal* 2:192.

Ronnet, T., and A. Johansen. 2020. "Formation of Moon Systems Around Giant Planets. Capture and Ablation of Planetesimals as Foundation for a Pebble Accretion Scenario." *Astronomy & Astrophysics* 633:93.

Rufu, R., and R.M. Canup. 2022. "Coaccretion + Giant-Impact Origin of the Uranus System: Tilting Impact." *Astrophysical Journal* 928(2):123.

Russell, S.S., L. Hartmann, J. Cuzzi, A.N. Krot, M. Gounelle, and S. Weidenschilling. 2006. "Timescales of the Solar Protoplanetary Disk." Pp. 233–251 in *Meteorites and the Early Solar System II*, D.S. Lauretta and H.W. McSween, eds. Tucson: University of Arizona Press.

Salmon, J., and R.M. Canup. 2017. "Accretion of Saturn's Inner Mid-Sized Moons from a Massive Primordial Ice Ring." *Astrophysical Journal* 836(1):109.

Salmon, J., and R.M. Canup. 2022. "Co-accretion + Giant Impact Origin of the Uranus System: Post-Impact Evolution." *Astrophysical Journal* 924(1):6.

Salmon, J., S. Charnoz, A. Crida, and A. Brahic. 2010. "Long-Term and Large-Scale Viscous Evolution of Dense Planetary Rings." *Icarus* 209:771–785.

Sheppard, S.S., C.A. Trujillo, D.J. Tholen, and N. Kaib. 2019. "A New High Perihelion Trans-Plutonian Inner Oort Cloud Object: 2015 TG387." *Astronomical Journal* 157:139.

Shibaike, Y., C.W. Ormel, S. Ida, S. Okuzumi, and T. Sasaki. 2019. "The Galilean Satellites Formed Slowly from Pebbles." *Astrophysical Journal* 885:79.

Singer, K.N., W.B. McKinnon, S. Greenstreet, B. Gladman, E.B. Bierhaus, S.A. Stern, A.H. Parker, et al. 2019. "Impact Craters on Pluto and Charon Indicate a Deficit of Small Kuiper Belt Objects." *Science* 363:955–959.

Spencer, J.R., S.A. Stern, J.M. Moore, H.A. Weaver, K.N. Singer, C.B. Olkin, A.J. Verbiscer, et al. 2020. "The Geology and Geophysics of Kuiper Belt Object (486958) Arrokoth." *Science* 367(6481):eaay3999. https://doi.org/10.1126/science.aay3999.

Stern, S.A., W.M. Grundy, W.B. McKinnon, H.A. Weaver, and L.A. Young. 2018. "The Pluto System After New Horizons." *Annual Review Astronomy and Astrophysics* 56:357–392.

Stern, S.A., H.A. Weaver, J.R. Spencer, C.B. Olkin, G.R. Gladstone, W.M. Grundy, J.M. Moore, et al. 2019. "Initial Results from the New Horizons Exploration of 2014 MU$_{69}$, a Small Kuiper Belt Object." *Science* 364(6481):eaaw9771.

Takata, T., and D.J. Stevenson. 1996. "Despin Mechanism for Protogiant Planets and Ionization State of Protogiant Planetary Disks." *Icarus* 123:404–421.

Tanigawa T., A. Maruta, and M.N. Machida. 2014. "Accretion of Solid Materials onto Circumplanetary Disks from Protoplanetary Disks." *Astrophysical Journal* 784:109. https://doi.org/10.1088/0004-637X/784/2/109.

Teanby, N.A., P.G.J. Irwin, J.I. Moses, and R. Helled. 2020. "Neptune and Uranus: Ice or Rock Giants?" *Philosophical Transactions of the Royal Society of London Series A* 378:20190489.

Tollefson, J., I. de Pater, E.M. Molter, R.J. Sault, B.J. Butler, S. Luszcz-Cook, and D. DeBoer. 2021. "Neptune's Spatial Brightness Temperature Variations from the VLA and ALMA." arXiv:2104.06554.

Trujillo, C. 2020. "Observational Constraints on an Undiscovered Giant Planet in the Solar System." Pp. 79–105 in *The Trans-Neptunian Solar System*, D. Prialnik, M.A. Barucci, and L. Young, eds. Amsterdam, Netherlands: Elsevier.

Trujillo, C.A., and S.S. Sheppard. 2014. "A Sedna-Like Body with a Perihelion of 80 Astronomical Units." *Nature* 507:471–474.

Tsiganis, K., R. Gomes, A. Morbidelli, and H.F. Levison. 2005. "Origin of the Orbital Architecture of the Giant Planets of the Solar System." *Nature* 435:459–461.

Vokrouhlický, D., and D. Nesvorný. 2015. "Tilting Jupiter (a Bit) and Saturn (a Lot) During Planetary Migration." *Astrophysical Journal* 806:143.

Vokrouhlický, D., W.F. Bottke, and D. Nesvorný. 2016. "Capture of Trans-Neptunian Planetesimals in the Main Asteroid Belt." *Astrophysical Journal* 152:39.

Vokrouhlický, D., D. Nesvorný, and L. Dones. 2019. "Origin and Evolution of Long-Period Comets." *Astrophysical Journal* 157:181.

Wahl, S.M., W.B. Hubbard, B. Militzer, T. Guillot, Y. Miguel, N. Movshovitz, Y. Kaspi, et al. 2017. *Geophysical Research Letters* 44:4649–4659.

Walsh, K.J., A. Morbidelli, S.N. Raymond, D.P. O'Brien, and A.M. Mandell. 2011. "A Low Mass for Mars from Jupiter's Early Gas-Driven Migration." *Nature* 475:206–209.

Wang, D., J.I. Lunine, and O. Mousis. 2016. "Modeling the Disequilibrium Species for Jupiter and Saturn: Implications for Juno and Saturn Entry Probe." *Icarus* 276:21–38.

Ward, W.R., and R.M. Canup. 2010. "Circumplanetary Disk Formation." *Astrophysical Journal* 140:1168–1193.

Ward, W.R., and D.P. Hamilton. 2004. "Tilting Saturn. I. Analytic Model." *Astrophysical Journal* 128:2501–2509.

Woo, J.M., C. Reinhardt, M. Cilibrasi, A. Chau, R. Helled, and J. Stadel. 2022. "Did Uranus' Regular Moons Form via a Rocky Giant Impactor?" *Icarus* 375:11842.

Q3 PLATE: An enhanced-color image mosaic of Mercury acquired by the MESSENGER spacecraft in 2013. The colors reveal different surface compositions. SOURCE: Courtesy of NASA/JHUAPL/Carnegie Institution of Washington, adapted by P.K. Byrne.

6

Question 3: Origin of Earth and Inner Solar System Bodies

How and when did the terrestrial planets, their moons, and the asteroids accrete, and what processes determined their initial properties? To what extent were outer solar system materials incorporated?

The inner solar system has the four terrestrial planets, Mercury, Venus, Earth, and Mars, along with Earth's large Moon, two small moons of Mars, dwarf planet Ceres, numerous 100-km-class asteroids, and a multitude of small bodies that populate the asteroid belt and the inner planet region.[1] To what extent does this structure reflect a deterministic outcome of general solar system formation processes, and how much of it is instead a consequence of stochastic events that would change in unpredictable ways if the system's formation could be rerun? This fundamental question motivates scientific investigations of the inner solar system, with implications not only for understanding our own planet's origin but also for the origin of planetary systems and Earth-like planets beyond our own.

For example, general aspects of solid planet formation—including collisional accumulation from initially small planetesimals, temperatures too high for ice condensation in inner disk regions, and minimum orbital separations between large planets needed for stability—suggest that rocky planets close to their parent stars might be a common outcome of planet accretion. In contrast, processes such as giant planet migration and sporadic giant impacts may depend sensitively on conditions that vary greatly from system to system. The interplay between predictable and randomly determined events is an important theme in solar system origin. Study of the inner solar system worlds provides a uniquely valuable means to address this cross-cutting issue, thanks to their accessibility and the powerful constraints provided by the combination of remote observations and analyses of physical samples. Is the inner solar system—and Earth—a typical outcome of planetary system formation, or is it an outlier compared to most systems in the universe?

Q3.1 HOW AND WHEN DID ASTEROIDS AND INNER SOLAR SYSTEM PROTOPLANETS FORM?

The classic model for the formation of asteroids and inner solar system protoplanets involves a transition from a protoplanetary gas nebula characterized by dust and gas to one populated by hundreds of 10^3-km-size bodies. Meteorites analyzed on Earth and asteroids studied in space preserve some of the compositional, mineralogical, and isotopic characteristics of these early formed bodies. The study of these remnants of solar system formation

---

[1] A glossary of acronyms and technical terms can be found in Appendix F.

157

is a critical component for constraining both the physical and chemical characteristics of the terrestrial planets, as well as for understanding the processes that occurred in the protoplanetary gas nebula. However, many fundamental questions remain unanswered. We do not understand the origin of the materials these planetesimals derive from, or even know if the current population of meteorites and asteroids is representative of the population that comprises inner solar system terrestrial bodies. Similarly, although we know that these small bodies undergo varying degrees of heating and associated aqueous alteration and differentiation, we are unsure how this translates into compositional variations observed in larger bodies. In fact, there are even gaps in our knowledge of how accretion works in the context of an evolving protoplanetary gas nebula. Addressing these questions is fundamental to a better understanding of how the solar system evolved into its present state.

Q3.1a What Were the Feedstocks to the Early Inner Solar System and Did Their Compositions Change with Time?

The compositions of planets are set by an array of processes, perhaps the most fundamental being compositional variations in the initial materials. The solar nebula was a dynamic environment, with distinct compositional reservoirs developed early in solar system history. Many properties of these reservoirs may reflect differences in distance from the Sun (Question 1, Chapter 4), while others would have been affected by early disk, planetesimal, and/or protoplanet growth conditions. Isotopic, chemical, and mineralogical heterogeneities in the building blocks of the inner solar system helped produce the diversity of inner solar system bodies seen today, which show strong variations in composition (e.g., Figure 6-1).

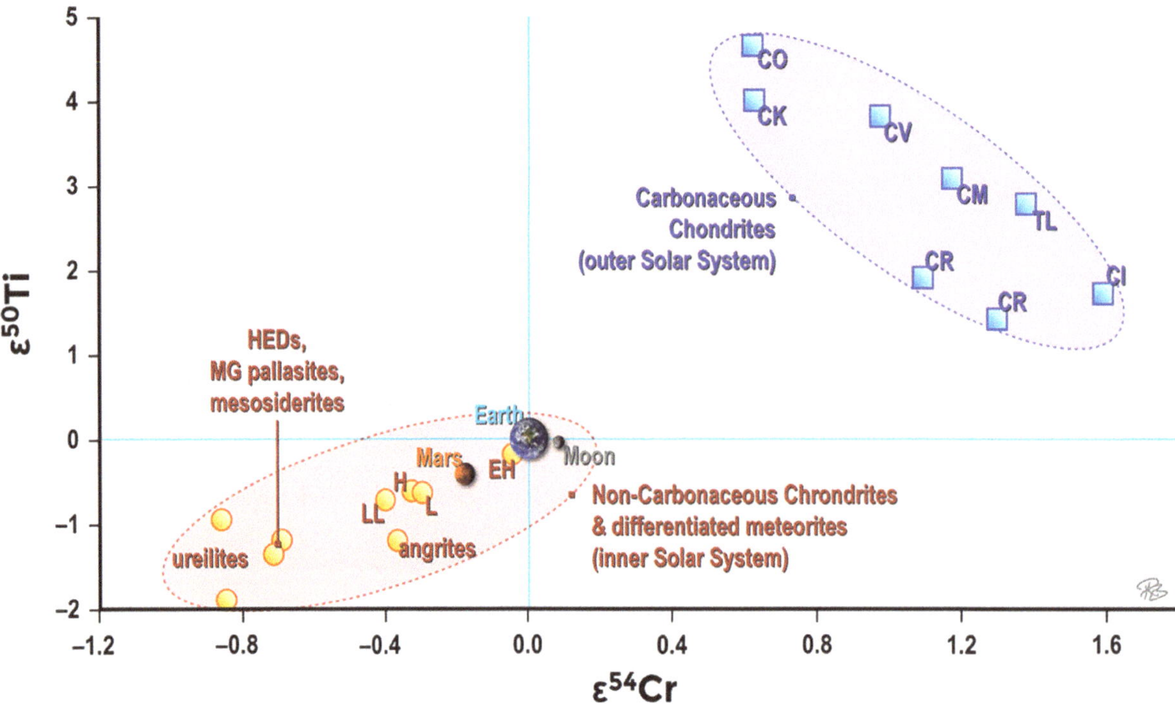

FIGURE 6-1 Plot of the relative abundances of ^{50}Ti versus ^{54}Cr illustrating the differences in the stable isotopic compositions of materials in the inner solar system. The carbonaceous chondrites represent the CC reservoir, whereas the noncarbonaceous chondrites, differentiated meteorites, and lunar and terrestrial samples represent the NC reservoir. Note the isotopic similarity of the Moon and Earth on this plot. Current data and samples may well not reflect the full range of primordial compositions. SOURCE: Figure by Paul K. Byrne, adapted from P.H. Warren, 2011, "Stable-Isotopic Anomalies and the Accretionary Assemblage of the Earth and Mars: A Subordinate Role for Carbonaceous Chondrites," *Earth and Planetary Science Letters* 311:93–100, https://doi.org/10.1016/j.epsl.2011.08.047. Copyright 2011, with permission from Elsevier.

The innermost planet, Mercury, has a highly reduced surface, with low iron contents and unexpectedly high sulfur contents, consistent with a planet formed from highly reduced materials. Information about the composition of Venus is extremely limited, a fundamental gap in our understanding of the compositional variations among the terrestrial planets. Studies indicate that Earth may have formed from more reduced materials in its early accretion stages, followed by the agglomeration of more oxidized materials. Mars, the outermost terrestrial planet, has a more oxidized surface and (by some arguments) interior. If not a coincidence given the small number of inner planets, this could reflect a radial gradient in oxidation fugacity among planetesimals in the early solar system, which in turn would imply limited dynamical mixing among early planetesimal reservoirs, such that the terrestrial planets mainly formed from local materials in the protoplanetary disk.

In addition, two reservoirs of planetary building blocks have been identified through the analysis of both primitive and differentiated meteorites. The first is associated with carbonaceous chondrites (CC) and the second with noncarbonaceous chondrites (NC) (see Q1.1a). The CC reservoir is characterized by more volatile-rich compositions and a higher proportion of isotopes produced by explosive nucleosynthesis—that is, supernovae—prior to solar system formation, whereas the NC reservoir is less volatile element-rich and has a greater proportion of isotopes produced by fusion in the cores of stars prior to solar system formation. The differences in the nucleo-synthetic isotope signatures of these reservoirs, combined with the inference that volatile element abundances reflect condensation location within the protoplanetary disk, indicate that the disk was radially stratified at the time of accretion. Earth appears to represent a compositional endmember, suggesting that a reservoir, representing the solar system inside Earth's orbit, may not yet have been identified or sampled (Dauphas 2017; Mezger et al. 2020). Samples from Mercury and/or Venus would be invaluable in helping to reconstruct the conditions of inner solar system accretion.

Early alteration processes—such as chemical and physical modification of gas and dust by the young Sun, acting on nebular materials—may have been important. Depending on the timing of the initiation of hydrogen fusion in the Sun, this process could have bathed the surrounding gas and dust in light, including ultraviolet radiation, causing chemical changes. As material aggregated into pebbles and planetesimals (see Question 1, Chapter 4), this energy, in combination with energy released by radiogenic decay of short-lived isotopes and gravitational energy, would have melted some solids, potentially mobilizing some volatile elements that could enable aqueous altera-tion of materials. If volatile species were lost during this phase, it could fundamentally modify the composition of these building blocks of larger terrestrial bodies. Likewise, melting and internal differentiation of planetesimals could have led to significant differences in the chemical and physical characteristics of subsequently accreted planetary bodies. Constraining the effects of the alteration and differentiation process on the composition of plan-etary bodies relies on understanding leftover materials from within the solar system, such as primitive chondrites and small planetesimal bodies. The range of evolutionary processes incorporating these ideas is discussed further in section Q3.1c.

Some of the best constraints on the building blocks of the terrestrial planets come from meteorite samples and asteroid studies. Asteroids display considerable diversity in spectral properties (e.g., color, albedo), size, and geophysical characteristics, and may include materials that originated across both the inner and outer solar system (see Question 2, Chapter 5, and Q3.1c). While there are a multitude of asteroids, their total mass is small, 0.04 per-cent of Earth's mass, and the extent to which their compositions are fully representative of materials that formed the terrestrial planets remains unclear. More reduced inner solar system materials that formed near the Sun may be underrepresented in meteorite collection because they were accreted by the Sun, although some were probably scattered into the asteroid belt during planet formation or were captured as Mars Trojans. (Additional examples of the latter may potentially be detectable by the Gaia mission.) Among reduced meteorite samples inferred to derived from the innermost solar system, there is significant isotopic diversity (see Figure 6-1), suggesting that the inner solar system planetary zones were not efficiently mixed. Furthermore, the observation that Earth and the Moon lie on the NC array on the opposite side from Mars, as shown in Figure 6-1, hints that not all NC-like planetary components have been identified.

Earth provides a foundation for all compositional studies, and we are fortunate to have samples from the Moon, specific asteroids—including returned samples from (25143) Itokawa, (162173) Ryugu, and soon (101955) Bennu—meteorites from Mars, and numerous meteorites from other meteorite-parent-body asteroids that enable continual advances in compositional measurements. Samples from Mercury and Venus have hitherto not been obtained or recognized if they exist within our meteorite collection. Refined in situ geochemical characterization of surface materials at Mercury and Venus would transform our knowledge of these bodies, especially in the absence of samples. Although we have a multitude of asteroidal meteorites, extraterrestrial materials delivered naturally to Earth are strongly biased; for example, icy materials are strongly underrepresented in meteorite collections. There potentially remain unsampled asteroidal materials that could be investigated by compositional measurements or sample return from rare or unexplored asteroid spectral types.

Q3.1b What Were the Mechanisms of Accretion from Planetesimals to Larger Bodies?

The standard story of planet formation, told for decades, is that planets formed from the solar nebula cloud of dust and gas in a multi-part process termed accretion (Question 1, Chapter 4). The first solids to condense from nebular gases were composed predominantly of refractory elements and include calcium-aluminum-rich inclusions (CAIs) and more silicate-rich chondrules. These early-formed solids subsequently accreted into small, perhaps 100-km-scale planetesimals (see Q1.3). In the inner solar system, planetesimals collisionally accreted into Moon-to-Mars-size planetary embryos within a few million years. Planetary embryos initially on nearly circular, co-planar orbits began to gravitationally perturb one another once the local mass in planetesimals and embryos became comparable and/or the gaseous nebula dispersed. Orbital eccentricities increased, leading to a phase of orbit crossing and giant collisions over tens to a hundred million years, reducing the number of protoplanets through mergers until only a few planets remained on stable orbits (e.g., Figure 6-2).

The standard model is popular because in numerical simulations it yields a modestly successful set of terrestrial planet analogs. When examined in detail, however, the model planets frequently have the wrong masses, orbits, and compositions. A particular problem involves the masses of Mars analogs, which often rival those of Earth and Venus, while real Mars has only about 10 percent the mass of Earth. The "small Mars" problem is a driver for considerable innovation in current planet formation models. The inability of the numerical simulations to account for many fundamental characteristics of the solar system suggests that we do not fully understand the processes by which material accretes to form the current set of planets and small bodies, including the extent to which pebble accretion alters the process of terrestrial planet formation. Some of this reflects difficulty in implementing a multi-disciplinary approach involving expertise in diverse fields ranging from numerical modeling to chemical behavior and characteristics of potential planetary building blocks. Another issue is the challenge in

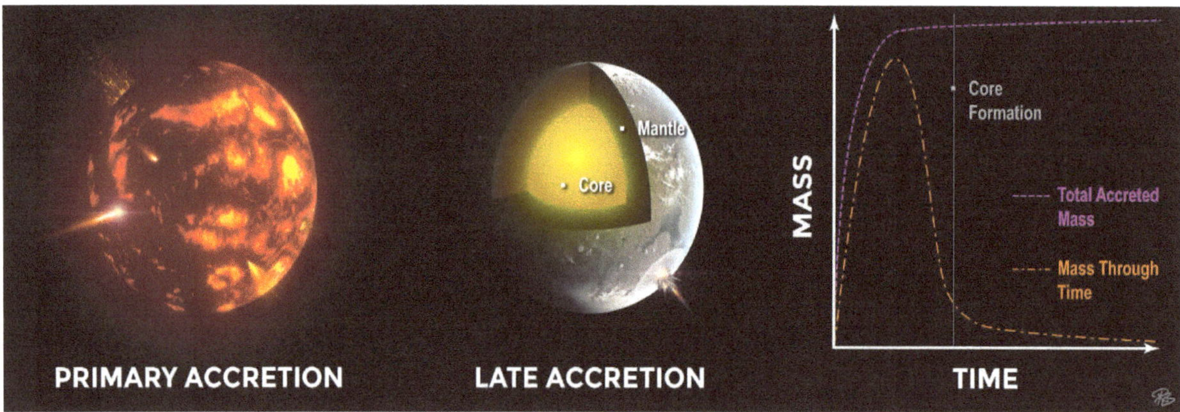

FIGURE 6-2 Growing the terrestrial planets. The illustration and associated timeline show the distinction between the main phase of accretion, when most of a terrestrial planet's mass is added, versus late accretion, which succeeds core formation. Here "Mass Through Time" corresponds to the mass growth rate, and the gray line represents the point at which core formation is complete. SOURCE: Figure by P.K. Byrne, adapted from J.M.D. Day, A.D. Brandon, and R.J. Walker, 2016, "Highly Siderophile Elements in Earth, Mars, the Moon, and Asteroids," *Reviews in Mineralogy and Geochemistry* 81(1):161–238, https://doi.org/10.2138/rmg.2016.81.04. CC BY 3.0.

treating many stages of evolution and planetary size scales simultaneously in numerical models, which may be needed to ultimately link large-scale planet accretion to the meteoritic record and small body populations. We also may not fully understand the dynamic interplay between nebular gas, dust, nascent planetesimals, and the Sun.

Q3.1c What Evolutionary Processes Led to the Initial Diversity of Asteroids and on What Timescales?

According to planet formation and giant planet migration models, the main asteroid belt is a collection zone for planetesimals formed from across the solar system (e.g., Walsh et al. 2011; Morbidelli et al. 2015; Vokrouhlický et al. 2016; see Q3.2). While some asteroids may have originated in the main belt region, the majority likely came from the terrestrial planet zone (predominately the Mars region), the giant planet zone, and the primordial Kuiper belt. This wide variety of source regions means that the asteroids and meteorites potentially reflect a broad range of planetesimal compositions.

Meteorite studies have shown that the decay of short-lived radioisotopes resulted in heating, melting, and in some cases differentiation of early-formed planetesimals within the first few million years of solar system history (see Q1.2c). Chronologic investigations further suggest that differentiation on parent bodies was contemporaneous with, and in some cases earlier than, condensation of some chondrules thought to be one of the earliest solids to form in the solar system (Q1.2 and Q1.3). Thus, at least some of the building blocks of the terrestrial planets and giant planet cores are likely to have experienced melting and differentiation prior to their incorporation into these bodies. However, meteorites display the gamut of heating histories, manifested by varying degrees of alteration and differentiation. It is not clear if there is a continuum of heating and melting related to planetesimal formation ages, sizes, and volatile fractions, or if other events such as early collisions set different planetesimals on alternative evolutionary paths.

Collisions have shattered and scrambled asteroids over billions of years, creating mixed-up bodies with originally deep interior materials now on their surfaces or within ejected fragments (e.g., Q4.1). An example might be the iron-rich asteroid (16) Psyche, whose origin will be explored by the Psyche mission. Asteroid impacts can also mix projectile and target so that some worlds appear to be derived from materials sourced from different locations in the solar system.

Spectral observations of asteroids have identified a wide range of classifications, a few of which have been linked to meteorite types by spacecraft exploration. For example, Hayabusa's sample return from (25143) Itokawa has demonstrated a link to ordinary chondrite meteorites, and Dawn's exploration of (4) Vesta has confirmed that (4) Vesta is the parent-body of the howardite, eucrite, and diogenite (HED) meteorites. Other associations between meteorite types and asteroid classes remain to be tested. For example, the spectral signatures of some partially differentiated asteroids may have exteriors that resemble more primitive bodies. Meteorite samples suggest that many asteroid bodies experienced complete differentiation, yet materials that represent mantles of differentiated asteroids are extremely rare, in both our meteorite collections and in asteroid observations. This may be owing to processes preferentially breaking up these materials (although more fragile chondrites have survived) or changing their spectral signatures through space weathering (a collection of processes that alter the surfaces of material when exposed to the space radiation environment, as described in Q5.5). Linking the diversity observed in meteorite samples to the large-scale diversity observed in the asteroid population is fundamental to advancing our knowledge of both types of objects.

Even primitive chondritic meteorites—thought to best reflect the initial, unaltered composition of the protoplanetary disk—show evidence of parent-body processes that have affected their properties and mineralogy. Hydrothermal alteration, thermal metamorphism, shock heating, and compaction occurred early in the solar system, prior to complete accretion of the terrestrial planets. All these processes played important roles in determining the physical characteristics of chondritic materials and contributed to the variability of the building blocks of terrestrial bodies. As one example, although there is evidence that aqueous alteration can affect the abundances and chirality of amino acids in primitive materials that were potentially delivered to Earth, there are open questions about the nature and extent of aqueous alteration on these bodies (Question 5, Chapter 8), such as the peak temperatures, the lifetimes of liquid water, and the role of gas phases. Understanding the range and extent of these evolutionary processes is important to interpreting the nature of the asteroid population and for constraining the building blocks that formed the inner planets.

Strategic Research for Q3.1

- **Determine the compositional diversity of the terrestrial planets and inner solar system feedstocks** by obtaining mineralogical, geochemical, and isotopic data from the surfaces and atmospheres of Mercury, Venus, the Moon, and the less explored regions of the Moon and Mars, as well as the currently unsampled small body population.
- **Determine the diversity of compositions and nature of remnant planetesimals residing in the inner solar system and establish links between the small body taxonomy and meteorite types** through Earth-based and spacecraft-based remote sensing, in situ measurements, and laboratory analyses of meteorites and returned samples.
- **Evaluate the nature of early projectiles that struck the terrestrial planets and the Moon** by analyzing regolith samples likely to contain remnant clasts from early bombardment impactors and by obtaining isotopic traces of projectile materials from lunar craters and basins.
- **Determine temporal changes to inner solar system feedstock compositions** by coupling geochronological measurements with other geochemical and isotopic measurements for refractory- and volatile-rich materials from a wide range of parent bodies.
- **Determine what secondary processes have led to the diversity of asteroids and planetary feedstocks** by conducting geochemical, petrological, and geophysical investigations of meteorites, asteroids, and samples returned from asteroids.
- **Determine the mechanisms of planetary accretion** by developing and evaluating physical models that link dynamics and chemistry coupled with observational constraints on compositions and distributions of material and planetesimal structures comprising the solar system.

Q3.2 DID GIANT PLANET FORMATION AND MIGRATION SHAPE THE FORMATION OF THE INNER SOLAR SYSTEM?

Historically, it was thought that the giant planets and small bodies formed near their current locations, but it has become increasingly clear that the structure of the solar system changed as the giant planets grew and their orbits migrated. This migration could have been driven by interactions between the giant planets and the gas nebula, or by later interactions with leftover planetesimals after the gas nebula dissipated, or both (see Q2.4 and Q2.6). Depending on when and how it occurred, giant planet migration could have reshaped the terrestrial planet region as the inner solar system was forming.

Q3.2a How Would Early, Nebula-Driven Migration Have Affected the Inner Solar System?

The earliest phase of giant planet migration may have occurred within the protoplanetary gas nebula (see Q2.4 and Q12.2). In the best developed model of this behavior, the so-called Grand Tack model (Raymond and Morbidelli 2014), gas disk interactions led early Jupiter to migrate *inward* to ~1.5 AU from the Sun, at which point Jupiter and Saturn became trapped in a mutual 2:3 resonance and began to move *outward* (Walsh et al. 2011). Outward migration continued until the gas disk dispersed, presumably when Jupiter reached its current distance of ~5.2 AU. The net effect of the migration was to deplete the Mars zone of material, providing an explanation for the small mass of Mars compared to Earth and Venus (Hansen 2009).

A key feature of the Grant Tack is that numerous planetesimals are scattered into the inner solar system from the giant planet zone. Some are captured into the central and outer main asteroid belt, the observed location of the carbonaceous chondrite asteroids (CC bodies). This would link CC asteroids to giant planet zone planetesimals. Other outer solar system planetesimals may have struck the growing terrestrial planets, providing a source of water-rich material (Morbidelli et al. 2015).

There are other, non-Grand Tack scenarios to explain Mars's small mass (e.g., Q3.2b), and how CC bodies may have been delivered to the inner solar system. A key question is whether any CC bodies were formed originally in the main belt. NC and CC meteorites have major differences in titanium, chromium, oxygen, molybdenum, and tungsten isotopic compositions, and this has been interpreted to mean that NC and CC planetesimals had to form in distinct chemical reservoirs (Q1.3). One way to produce and maintain such reservoirs is if Jupiter formed early and acted as a barrier against inward migration of material (Kruijer et al. 2017). In this case, NC and CC bodies would represent inner and outer solar system planetesimals, respectively, implying that only the NC asteroids were originally formed in the inner solar system. Future work can test these scenarios by seeking evidence that Jupiter and Saturn migrated across the primordial main asteroid belt, by exploring a range of possible masses of the asteroid belt prior to giant planet migration, and the fate of bodies potentially lost from the asteroid belt during this migration. Tests of these scenarios will require meteorite studies, ground-based and space-based observations, data returned by small body missions, and modeling.

Q3.2b How Was the Inner Solar System Affected by Giant Planet Migration After the Gas Disk Dispersed?

The orbital properties of Pluto and other Kuiper belt objects provide compelling evidence that early Neptune migrated substantially outward (Q2.6). Decades of theoretical models have shown that after the protoplanetary gas nebula dispersed, the orbits of Saturn, Uranus, and Neptune would have indeed migrated outward owing to gravitational interactions with a remnant outer planetesimal disk, while such interactions would have caused Jupiter's orbit to contract. During such migration, the giant planets often undergo an orbital instability, such as in the widely explored "Nice" model (Gomes et al. 2005; see Q2.4).

A giant planet instability would have affected the terrestrial planet region, depending on when it occurred. The timing of the instability is uncertain because it is controlled by the unknown size of the gap between Neptune's initial orbit and the primordial Kuiper belt's inner edge. The smaller the gap, the earlier the instability. If the gap was <1.5 AU, Neptune's migration would have taken place shortly after dissipation of the protoplanetary gas nebula, within the first few to 10 million years of solar system history. This migration would then have had a major influence on terrestrial planet accretion and could provide an alternative to the Grand Tack for explaining Mars's small mass, with Mars zone planetesimals scattered away before Mars could fully form (Raymond and Izidoro 2017; Clement et al. 2018; Deienno et al. 2018; Nesvorný et al. 2021). Improved isotopic constraints on the timing of Mars's assembly might aid in distinguishing between such scenarios (see Q3.4). Alternatively, a late instability after terrestrial accretion was complete—and initially favored explanation for a clustering of lunar basin ages at about 4 Ga (the so-called late heavy bombardment or terminal lunar cataclysm; see Q4.2b)—would have likely destabilized the orbits of the fully grown terrestrial planets. To date, only a small fraction of existing simulations find that the terrestrial planets could avoid this fate, thus favoring an earlier instability before the terrestrial planets reached their final configuration (Bottke and Norman 2017; Nesvorný 2018).

The number of giant planets gravitationally interacting with one another at the time of the instability is also unknown. It is possible the solar system had additional gas giants early in its history that were subsequently ejected. Dynamical studies suggest systems with five or six giant planets (e.g., three or four Neptune-size bodies) have greater success in reproducing dynamical constraints across the solar system than those that start with four giant planets (Q2.4). In the most successful models, a Neptune-size body interacting with Jupiter causes it to migrate slightly inward via numerous tiny jumps before it is eventually ejected, and these Jupiter "jumping events" strongly affect the dynamical structure of the asteroid belt (Morbidelli et al. 2015; Vokrouhlický et al. 2016).

Giant planet encounters with one another may also allow a small fraction of comets ejected from the primordial Kuiper belt to be captured in stable orbits across the solar system (Q2.6d). This might explain why primitive, comet-like asteroids (D- and P-types) are found in the central and outer main belt as well as in the Hilda and Trojan populations associated with Jupiter, and would imply that the asteroid belt contains planetesimals from the furthest reaches of the solar system (Vokrouhlický et al. 2016). As the giant planets moved to their present-day orbits, dynamical resonances associated with them also moved to their current locations, perhaps depleting or eliminating portions of the primordial inner main belt while perhaps creating the innermost asteroid population, the Hungaria asteroids (Nesvorný et al. 2017). The instability would have also driven impacts within the belt; this could be assessed by studying the impact history of asteroids large enough to have resisted disruption, as well as by identifying evidence for any common shattering/disruption times from meteorite shock degassing ages.

An overall goal is to find evidence for or against these dynamical set pieces through missions to small bodies and/or meteorite analysis. We need to determine precisely how the signatures of post-nebula giant planet migration are recorded in small body populations and whether the nature of the asteroid belt can tell us how many giant planets existed prior to the giant planet instability. We want to determine whether dormant comet-like asteroids were implanted in the main belt, Hildas, and/or Trojans by the giant planet instability and whether the primordial asteroid belt interacted with an "extra Neptune" (Vokrouhlický et al. 2016).

Strategic Research for Q3.2

- **Determine whether Kuiper belt objects, Centaurs, comets, and P- and D-type asteroids within the main belt, Hilda, Trojan asteroid, and irregular satellite populations originated in the primordial Kuiper belt** by measuring their properties via remote observations, in situ studies, and/or sample return.
- **Determine whether C- and B-type asteroids within the main asteroid belt originated within the giant planet region** by assessing their volatile content, using in situ methods or sample return, and identifying whether any provide evidence of parent body origin at low temperatures beyond Jupiter.
- **Determine the timing of the giant planet instability** through evidence of early comet bombardment of the asteroid belt (e.g., impact history of large asteroids that resisted disruption, identification of common shattering/disruption times for asteroids from meteorite shock degassing ages) and constraining the ages of the oldest lunar impact basins.
- **Investigate how planetesimals from the giant planet zone may have reached the inner solar system in the presence of the gaseous nebula** through dynamical models coupled to constraints provided by the orbits and sizes of C- and B-type asteroids within the main belt and the abundance and distribution of inner solar system volatiles.
- **Investigate how giant planet migration after gas nebula dissipation affected the asteroid belt and terrestrial planet accretion** by observing comet-like asteroids (i.e., D- and P-type bodies) within the main asteroid belt and using their properties to constrain if and how migration led to their capture into the main belt, and through dynamical and collisional models coupled to compositional constraints.
- **Constrain the orbit of Jupiter prior to its post-nebula giant planet migration phase, and thereby the nature of the giant planet instability** by identifying putative asteroids that were initially captured in resonances associated with Jupiter's orbit prior to migration.

Q3.3 HOW DID THE EARTH–MOON SYSTEM FORM?

Among our terrestrial planets, only Earth has a large satellite, and the Moon has affected our planet throughout its history. Lunar and terrestrial samples, together with spacecraft data, provide an extensive and powerful set of constraints on the origin of Earth–Moon, making this system unique in its ability to reveal conditions during the final stages of terrestrial planet accretion. The favored giant impact hypothesis proposes that the Moon formed following a collision between the proto-Earth and another protoplanetary embryo, "Theia," at the end of Earth's main accretion (Hartmann and Davis 1975). The impact left Earth with its rapid early rotation rate, and produced a disk of iron-poor, highly heated debris, which later accreted into the Moon, forming a primordial lunar magma ocean (LMO). Although this general model accounts for many characteristics of the Earth–Moon system, some characteristics have proved extremely challenging to explain, spurring development of a wide range of new impact models that are actively debated. Questions also remain regarding the timing of the giant impact.

Q3.3a What Are the Physical Constraints on Earth–Moon System Origin?

Key constraints on how Earth and the Moon formed include their compositional relationships, the Moon's age, and its initial thermal state. The extent of understanding of each of these constraints has evolved substantially in the past decade, yet many open questions remain.

Laboratory analyses continue to provide breakthroughs in understanding of how the compositions of lunar samples relate to those of terrestrial rocks. Much recent work focuses on isotopic compositions thought to best reflect the original feedstock of both bodies. Remarkably, the silicate Earth and Moon have nearly identical isotopic compositions for many elements (e.g., oxygen, titanium, chromium, silicon, and tungsten), implying that they formed from a common source that was isotopically distinct from nearly all meteorite classes. The tungsten isotopic similarity is the most challenging to explain, requiring either a compositional match between the proto-Earth and Theia that is very improbable given their separate core formation histories, or complete mixing of protolunar material with Earth's mantle, potentially within vaporized phases produced during or after the impact. Better understanding of how late accretion (Q3.5d) affected Earth–Moon tungsten isotopic compositions, as well as how lunar and terrestrial rocks compare in highly refractory elements (e.g., Ca) less likely to vaporize, is important to further progress (Zhang et al. 2012). Disagreements remain as to whether Earth and the Moon have identical oxygen isotopic compositions, or whether small differences between the two indicate that the Moon still partially records the composition of Theia (e.g., Young et al. 2016; Cano et al. 2020).

The Moon's volatile content may provide important constraints on the energetics of the giant impact and the conditions of the Moon's formation. Lunar samples are generally dry and depleted in volatile elements compared with terrestrial rocks, which are strongly volatile-element-depleted relative to most other planetary materials that have been analyzed. However, analyses of a small number of lunar samples, notably volcanic glass beads that trapped mantle melt, now show that at least portions of the early Moon were comparably water-rich to Earth's mantle (e.g., Hauri et al. 2015). The stable isotope composition of moderately volatile elements is slightly heavier and significantly more variable in lunar rocks than in terrestrial rocks (e.g., Wang and Jacobsen 2016). Accounting for these observations is challenging and has led to a variety of models of evaporative loss or partial condensation in the prelunar disk or by geologic processing on the early Moon (e.g., Canup et al. 2015; Day et al. 2017; Kato and Moynier 2017; Lock et al. 2018; Nie and Dauphas 2019; Wang et al. 2019).

An overall question is whether the Moon's composition is heterogeneous with depth, which would imply that lunar samples are not representative of the Moon's bulk composition. Improved constraints on the Moon's bulk composition, through additional samples originating in its deep interior (volcanic glasses or mantle exposed from the largest basin-forming impacts, particularly South Pole–Aitken basin) and/or a more thorough understanding of the Moon's geophysical structure, would be of great value.

Current estimates for the age of the Moon-forming giant impact range from 4.52 to about 4.42 Ga, corresponding to about 50 to 150 million years after formation of the earliest solar system solids (e.g., Kruijer and Kleine 2017; Thiemens et al. 2019; Maurice et al. 2020). The large uncertainty stems from the fact that these estimates are based on assumption-laden models of isotopic evolution, which need to be assessed in the context of multiple geologically related events such as accretion, differentiation, and lunar magma ocean solidification (see Q3.6 and Question 5; Chapters 6 and 8, respectively).

A fundamental issue for origin models is whether the Moon was completely or only partially molten when it formed—a so-called magma ocean. Multiple geophysical constraints seem to imply an initial solid lunar interior beneath the lunar magma ocean, including the history of tectonics and strain most recently revealed by GRAIL data (Andrews-Hanna et al. 2013), and a seismic transition that may represent the magma ocean base (Khan et al. 2006). However, alternative explanations exist that could permit a fully molten Moon given current data limitations and uncertainties. Resolving this issue is important because a partially molten Moon is difficult to reconcile with a giant impact origin, and thus would provide strong constraints on lunar origin models. Long temporal baseline geophysical and seismic data are needed to reveal the Moon's interior structure and to evaluate whether, for example, it is consistent with layering expected for solidification of a magma ocean in a high thermal regime. Geochemical analyses of additional and more diverse crustal samples, as well as improved constraints on the Moon's bulk composition, would also contribute significantly to this fundamental issue.

Q3.3b What Was the Nature of a Moon-Forming Giant Impact(s) and Its Implications for the Initial States of Earth and the Moon?

A canonical Moon-forming impact by a Mars-size Theia can account for the masses of Earth and the Moon, and the angular momentum of the system (e.g., Canup 2004). However, it produces a proto-lunar disk that originates primarily from Theia rather than from the proto-Earth. Meteorites that originate from Mars, and nearly all those from the asteroid belt, have substantially different isotopic compositions than Earth. If Theia were similarly non-Earth-like, one would expect measurable differences between Earth and the Moon. Instead, Earth and the Moon are nearly isotopically indistinguishable for all nonvolatile elements (e.g., Figure 6-1). A multitude of new concepts has been proposed to try to resolve this so-called isotopic crisis for the giant impact model. Notable examples include mixing and chemical equilibration of the silicate Earth with protolunar material before the Moon's assembly (Pahlevan and Stevenson 2007; Lock et al. 2018); high-angular-momentum/high-energy impacts that produce "synestias" (Canup 2012; Ćuk and Stewart 2012; Lock et al. 2018); a "hit-and-run" impact (Reufer et al. 2012); formation of the Moon by multiple impacts (Rufu et al. 2017); and a Theia and proto-Earth that had similar isotopic compositions (Dauphas 2017).

The relative merit of these different models is a matter of active debate (e.g., Canup et al. 2021), and progress will require improved understanding on multiple fronts. First, current thinking that Theia would have been isotopically distinct from Earth is heavily influenced by the isotopic composition of martian meteorites, the only other inner planet from which we have samples. Knowledge of the isotopic compositions of Venus and/or Mercury—currently unknown—is needed to address this fundamental issue and reveal the primordial feedstock of the innermost planets. Whether lunar materials could have thoroughly mixed with Earth's mantle prior to the Moon's assembly remains unclear and needs to be assessed through modeling and observational tests. Explaining the Earth–Moon tungsten isotopic similarity (see Q4.3a) remains challenging for all concepts save equilibration.

Another central issue is whether the Earth–Moon system angular momentum was greatly modified after the Moon formed. Tidal interaction between Earth and the Moon conserves their total angular momentum, but certain gravitational interactions involving the Sun could have slowed Earth's spin and reduced the system angular momentum by a factor of 2 or more soon after the Moon formed (e.g., Ćuk and Stewart 2012). If this occurred, it implies that the Moon-forming giant impact was vastly more energetic than originally envisioned. Uncertainty persists because the spin-slowing mechanisms depend sensitively on the nature of early tidal dissipation in both bodies, in turn a function of their early thermal evolution. Additional constraints may arise from better understanding of the Moon's early orbital evolution and the related origin of its fossil figure (the remnant shape of the Moon, frozen in from this early epoch; Gerrick-Bethell et al. 2014; Keane and Matsuyama 2014). More integrated physical-chemical-isotopic models for the outcomes of giant impacts and the evolution of the protolunar disk are also needed to better understand the conditions of the Moon's accretion in varied impact scenarios.

Overall, unraveling how the Earth–Moon system formed will require development of improved observational tests to allow model predictions—for example, for the Moon's composition and initial thermal state—to be compared with past and future data. Primary questions involve the physical processes and mixing associated with the giant impact and their effects on the proto-Earth, conditions in the prelunar disk produced by the impact (including whether it was ionized and affected by magnetic fields), and the nature of subsequent lunar accretion and early orbital evolution.

Strategic Research for Q3.3

- **Determine the internal structure of the Moon with sufficient resolution to constrain its bulk composition and initial thermal state** using geophysical measurements obtained from spacecraft and/or a seismic network and other in situ analyses.
- **Determine the Moon's interior composition** by sample return and/or in situ analysis of materials that reflect the Moon's endogenic composition at depth—for example, glass beads, primitive basalts, and/or exposed lunar mantle.
- **Constrain the physical and chemical characteristics of Theia and the proto-Earth** through sample analysis or in situ isotopic measurements of inner solar system bodies (particularly Venus or Mercury), as well as improved models to explain the isotopic and geochemical constraints of Earth and the Moon.
- **Determine the origin of the Moon's volatile element abundances** by completing additional stable isotopic analyses of volatile elements in lunar samples originating from the lunar interior as well as gamma-ray and neutron spectrometer measurements (e.g., K/Th ratios) by spacecraft.
- **Determine the timing of the Moon-forming giant impact and solidification of the LMO** by isotopic analysis of rocks from the nearside and farside of the Moon and by refining theoretical models to estimate the timescale for the duration of LMO crystallization.

- **Seek evidence for post-giant impact equilibrium between Earth and the Moon** by analyzing terrestrial and lunar samples for stable refractory element isotopic compositions.
- **Differentiate between giant impact concepts** by developing model predictions for observable properties of the Moon and Earth and comparing them with lunar compositional and geophysical data.

Q3.4 WHAT PROCESSES YIELDED MARS, VENUS, AND MERCURY AND THEIR VARIED INITIAL STATES?

The four terrestrial planets—Mercury, Venus, Earth, and Mars—share many similarities. All are solid bodies with a central core composed dominantly of metallic iron surrounded by layers of silicate rocks. However, there are substantial differences among these bodies that, at least in part, appear to reflect substantial differences in their formation histories.

Q3.4a Why Are Earth and Venus So Different?

The contrast between the current Earth and Venus is striking. Earth is an oasis for life, with plentiful liquid water on its surface, a temperate atmosphere, and a protective magnetic field. In contrast, Venus's surface temperatures are the hottest in the solar system (~460°C), its atmosphere is the thickest of any rocky body and is rich in the greenhouse gas carbon dioxide, and it has no global magnetic field. Yet Earth and Venus are similar in size, with Venus only 30 percent closer to the Sun than Earth.

What set Earth and Venus on such different paths? Is this a manifestation of their locations in the solar system or did chance events during their formation play a critical role? Determination of Venus's atmospheric and surface compositions, as well as its geophysical properties, are needed to provide insight into whether Earth and Venus were more similar in the past and evolved differently through time (e.g., see Question 6, Chapter 9, for discussion of their divergent climate evolution), or whether these neighboring planets were made differently from the start. For example, knowing the bulk and isotopic composition of Venus, which along with Earth, makes up >90 percent of the inner solar system's mass, would enable the compositional heterogeneity of the inner solar system to be much better understood, and would reveal Earth–Venus similarities or differences. This in turn, would provide crucial insights into Earth–Moon origin (Q3.3).

Unlike Earth, Venus has a slow retrograde rotation and lacks a moon. These differences, as well as Venus's current lack of a magnetic field, could reflect stochastic differences in the types of impacts experienced by Venus during the end of its accretion. During planetary accretion, relatively large planets, such as Earth and Venus, may establish a compositionally stratified structure, meaning that the outer part of the core is more enriched in light elements and therefore less dense. Such a structure is stable and would prohibit core convection and therefore generation of a dynamo. If the stable structure is reset by a large impact, dynamo generation is possible; this may have happened to Earth but not Venus. Although this could potentially explain why Venus does not have a moon or a magnetic field, it is not clear if Venus could have avoided giant impacts, given that accretion simulations find these are common for planets as massive as Earth or Venus.

Alternatively, some argue that Venus experienced two giant impacts (Alemi and Stevenson 2006). The first generated a moon and provided Venus with a rapid prograde spin, while the second was oppositely oriented and caused Venus to rotate in the reverse direction. The latter caused the prior moon to tidally evolve inward and eventually be lost. If the angular momenta delivered by the first and second impacts nearly cancelled out, the resulting spin angular velocity would have been small, potentially consistent with the long spin period of Venus (243 days); however, such a combination of events may be improbable. Alternatively, Venus's current spin state may reflect a combination of internal and atmospheric tides, core–mantle friction, and planetary perturbations. For an initial Venus day much shorter than its year, dissipation in the planet's interior owing to solar tides is the dominant mechanism that causes the planet's spin rate to slow, and Venus's current state can be achieved if Venus's initial day was >48 hours long and its average tidal dissipation comparable to that of Earth (e.g., Correia and Laskar 2001). Thus, whether Venus experienced a large impact or had a moon remains controversial. Observations to constrain the core structure and other interior layers of Venus would be valuable to gain insight into the planet's evolution and the role of giant impacts in its history.

Q3.4b What Was the Nature of Mars's Formation and How Did Its Small Moons Originate?

Mars's small mass compared to those of Earth and Venus may be evidence that some process depleted material from its orbital region before it formed (e.g., giant planet migration; Q3.2). Martian meteorites display large isotopic variations in now extinct systems (e.g., hafnium-tungsten and samarium-neodymium) compared to the much more limited isotopic compositions observed in lunar and terrestrial samples. Mars's tungsten, neodymium, as well as strontium, isotopic compositions nominally suggest that it accreted earlier than the assembly of Earth and the Moon, although these results are sensitive to the nature of core formation, late accretion, and details of the isotopic evolution models. Its small mass and inferred early formation time have led to the suggestion that Mars is a leftover planetary embryo that may not have experienced an extended phase of giant impacts like the other terrestrial planets. However, tungsten, and to a lesser extent neodymium and strontium, isotopic compositions do not appear to vary systematically with many geochemical characteristics of the samples, indicating that our understanding of Mars's isotopic composition is incomplete. Additional samples from Mars are needed to more accurately decipher the processes responsible, as well as the age of formation of Mars.

One possible proxy for giant impacts are moons. While Mercury and Venus have no moons, Mars has two small irregularly shaped satellites that resemble primitive asteroids. Views on the origin of Phobos and Deimos have changed dramatically over the past decade. Originally, they were thought to be asteroids gravitationally captured intact by Mars. However, Phobos and Deimos have very small eccentricities and inclinations that are not easily explained by the capture model; such moons tend to have large eccentricities and inclinations, and inclinations cannot be sufficiently reduced by Mars's tidal forces.

The most attractive alternative scenario is that the moons accreted from a disk formed by a large impact, as is thought to have been the case for Earth's Moon (e.g., Rosenblatt 2011; Citron et al. 2015; Canup and Salmon 2018). It is plausible that this collision was the same one that formed hypothesized Borealis basin, an approximately 10,600 by 8,500 km impact feature that may be responsible for Mars's crustal dichotomy between the northern lowlands and southern highlands (Andrews-Hanna et al. 2008), or by the impactor that produced the Utopia or Hellas basins. The impactor mass needed to yield a stable Deimos is between 0.0005 to 0.003 times the mass of Mars (Canup and Salmon 2018), proportionally much smaller relative to Mars than Theia compared to Earth. Although the full effects of a Vesta-to-Ceres-size body smashing into Mars have yet to be explored, it is difficult to imagine that martian core and mantle were unaffected. Models suggest that Phobos and Deimos may be the last survivors of multiple satellites originally formed by this impact, and/or that Phobos may have formed more recently (Hesselbrock and Minton 2017; see Question 8, Chapter 11). Other possibilities, such as disintegration from a single progenitor (Bagheri et al. 2021), are speculative but remain in play. A critical step in constraining the origin of Phobos and Deimos will be to determine their bulk compositions and interior structures; determination of their isotopically derived ages would also be very valuable.

Q3.4c What Conditions Led to Mercury's Anomalously High Density and Large Core?

Little is known about the nebular conditions at the innermost edge of the protoplanetary disk, which contributes to uncertainty associated with the origin of Mercury. It has long been known from its high density that Mercury has a large core. Primarily as a result of MESSENGER and Earth-based radar observations, it is now known that Mercury's core comprises between 69 to 77 percent of its mass (Hauck et al. 2013). The origin of its anomalously large core is thought to be key to understanding the formation of Mercury and its subsequent thermochemical evolution. Numerous hypotheses for Mercury's origin have been proposed over the past several decades, but data returned by the MESSENGER mission have narrowed the likely formation mechanisms of Mercury into two competing groups: (1) mantle-stripping giant impact processes and (2) formation from highly reduced metal-rich precursor materials from the inner portion of the protoplanetary disk.

There has long been a proposal that the anomalously high metal content of Mercury was produced by a high-velocity giant impact that blasted away most of Mercury's primordial mantle (e.g., Benz et al. 1988). However, MESSENGER data also show that Mercury's surface is rich in volatiles, such as sulfur, potassium, and chlorine (e.g., Nittler et al. 2011). These data have challenged the idea of forming the high-density planet by a disruptive impact, which is plausibly expected to have resulted in extensive volatile loss. In contrast, the idea of volatile depletion through giant impact processes for large planet-size bodies has not been adequately demonstrated, and Mercury's volatile-rich nature may not disqualify the giant impact model (Ebel and Stewart 2018). Alternatively, it has been suggested that Mercury might have been the smaller of two planets involved in a hit-and-run collision that stripped its mantle while avoiding disruption and extensive heating (Asphaug and Reufer 2014). Another possible explanation for Mercury's origin is that its distinctive composition is an outcome of its accretion from highly reduced, metal-rich materials that differed from the building blocks of the other terrestrial planets simply by virtue of having been sourced from a different location in the disk. However, it is not clear to what extent Mercury's reduced nature can be attributed to impacts onto Mercury's precursor planetesimals or even Mercury itself.

Both formation mechanisms have different implications for the subsequent thermochemical evolution of Mercury, and both possibilities need to be considered and weighed when interpreting Mercury's nature. Geochemical, mineralogical, and isotopic measurements of Mercury materials are needed to constrain the origin of Mercury, as are better constraints on the planet's interior structure.

Mercury's fractionally large core has interesting implications for exoplanets. Two innermost planets of Kepler-107 have large density variations, indicating that one of them is Earth-like and the other one is Mercury-like. This diversity could have been produced by an impact (Bonomo et al. 2019). Understanding the origin of Mercury's core would give us key insights to planetary diversities in extrasolar systems as well.

Strategic Research for Q3.4

- **Determine the interior structure and bulk composition of Venus** by seismic observations, constraints on the moment of inertia and gravity field from spacecraft and Earth-based radar measurements, and through atmospheric and surface observations.
- **Determine the isotopic compositions of Mercury and Venus** through remote sensing, in situ measurement, sample return, and/or identification of meteorites originating from these planets.
- **Determine the origin of the martian moons** by comparing chemical and isotopic ratios of Mars, Phobos, Deimos, and asteroids and constraining their interior structures.
- **Determine the formation time of Mars** through isotopic analyses of diverse martian samples.
- **Determine the origin of the large core of Mercury** by in situ geochemical, mineralogical, and isotopic measurements.

Q3.5 HOW AND WHEN DID THE TERRESTRIAL PLANETS AND MOON DIFFERENTIATE?

Present-day characteristics of all differentiated bodies are closely linked to the mechanisms and timing of their initial differentiation, which involved the formation of metallic cores, rock mantles, crusts, and in some cases volatile-rich atmospheres/hydrospheres. Significant research has been devoted to this topic over the past few decades, and open questions remain.

Q3.5a What Were the Mechanisms of Primordial Differentiation?

Differentiation is inextricably linked to heating during planetary formation, produced by the decay of radiogenic elements and the conversion of gravitational energy to heat during accretion. For planetesimals and small bodies, radiogenic heating dominates (Dodds et al. 2021). Two classes of radiogenic heat sources are short-lived nuclides with half-lives of order a million years (isotopes of aluminum and iron; ^{26}Al and ^{60}Fe), and long-lived radiogenic nuclides with half-lives greater than ~1 billion years (isotopes of uranium, thorium, and potassium; ^{235}U, ^{238}U, ^{232}Th, ^{40}K). Short-lived heating led to differentiation on smaller bodies formed within a few million years of the oldest dated materials in the solar system (calcium-aluminum inclusions, or CAIs), producing the parent bodies of iron meteorites, Vesta's iron core (Zuber et al. 2011), and partial differentiation on Ceres, based on thermal models (Castillo-Rogez et al. 2019). Ancient ages, such as the 4,565.4 ± 0.2 million year age determined for a meteorite sample from Vesta (Wadhwa et al. 2009), suggest that 500- to 1,000-km-class planetesimals were differentiated as they formed. Decay of long-lived radionuclides and accretional heating are thought to be the dominant heat sources for larger bodies.

Differentiation processes involving magma oceans are thought to be responsible for the structure of most large bodies in the inner solar system, including the Moon, Earth, Mars, and asteroids such as Vesta (see also Question 5, Chapter 8). This involves progressive crystallization of minerals: light minerals buoyantly rise to the surface, forming a primary crust, while denser minerals sink. Geologic activity on all of the terrestrial planets has all but destroyed the surface evidence of these primary crusts. However, this record is preserved at the Moon, where plagioclase (the first buoyant minerals that crystallize out of a magma ocean) rose to the surface to form the lunar highlands, providing the most compelling evidence for magma ocean differentiation. A giant impact can produce a global magma ocean, and large impacts may remelt substantial fractions of a planet's surface, so that it may be more accurate to think of a series of magma oceans as opposed to only one. Magma oceans likely survived for millions to tens of millions of years, interspersed by relatively temperate conditions. One notable exception is Jupiter's moon Io, which may have maintained a long-lived magma ocean until today via tidal heating (Question 8, Chapter 11).

Although Venus is similar in size to Earth, its primordial differentiation processes remain unclear. Venus may have experienced an ongoing differentiation process analogous to plate tectonics on Earth or some yet undefined differentiation process. One hypothesized mechanism involves a major crustal instability that rapidly resurfaced the venusian crust (e.g., Strom et al. 1994). The other hypothesized mechanism involves progressive volcanism (e.g., Smrekar et al. 2007). Currently, the surface resolution and crater statistics are unable to distinguish between the two hypotheses; radar mapping and other measurements in upcoming missions to Venus should help to better address this long-standing issue.

Improved determination of how the inner planets differentiated is critical to understanding their formation conditions and early evolution. Measuring noble gas concentrations in the venusian atmosphere would provide an important constraint on degassing from the interior. Determining the ages and composition of the enigmatic tesserae regions on Venus would test the hypothesis that these large plateaus are analogous to continents, suggesting the presence of water when they formed (Gilmore et al. 2017). Another important question is what criteria, such as primary heat source, size and composition of the body, or efficiency of thermal blanketing by an atmosphere, dictate the style of planetary differentiation. Last, better understanding is needed of the relationship between mechanisms of silicate differentiation and the resulting characteristics of planetary cores, mantles, and crusts on differentiated bodies. For example, would some mechanisms lead to more sulfur within planetary cores, as has been suggested to explain the larger and lighter core of Mars relative to Earth? Could the retention or loss of volatile elements and compounds, including water, be related to planetary differentiation mechanisms? While it is thought that magma oceans are effective means for losing volatiles or generating transient or permanent atmospheres via outgassing, it is not clear if other differentiation mechanisms would lead to similar outcomes.

Q3.5b What Was the Timing and Duration of Primordial Differentiation?

All of the inner planets differentiated, but whether they underwent this process at the same time remains unknown. Determining ages for planetary differentiation is challenging because isotope-based chronometers only date the time that parent and daughter isotopes were fractionated from one another. The fractionation event is therefore linked geologically to the differentiation age of the body. Three such events are commonly used to provide constraints on the timing of planet formation: (1) core formation, (2) crystallization of a magma ocean, and (3) the age of the earliest rocks. Dating of these events involves the application of various radiometric chronometers that have underlying assumptions that may be valid only under certain circumstances, date different aspects of differentiation, and/or be disturbed to varying extents by later geologic processes. Consequently, dating events occurring on planetary bodies requires significant interpretation of the data.

Dating core formation is accomplished primarily using the Hf-W (hafnium-tungsten) isotopic system in which ^{182}Hf decays to ^{182}W in 9 million years. The chronometer is based on the principle that W is concentrated in metal phases during core formation whereas Hf remains in silicate phases. A core that formed early will not have much ^{182}W, whereas the remnant mantle will be enriched in ^{182}W, compared to chondritic abundances. Thus, the Hf-W isotopic system has been used to obtain core formation ages from samples derived from both metal cores as well as silicate mantles and crusts. This system has revealed that iron meteorites represent pieces of planetesimal cores formed in the first 1–2 million years of the solar system history (Kruijer et al. 2014). It has also been possible to demonstrate that the cores of Mars and Earth formed within the first 1–10 million years and 30–200 million years of solar system history, respectively (Nimmo and Kleine 2007, 2015). These ages are more uncertain, however, because late accretion can deliver tungsten to planetary mantles after core formation (see section Q3.5d) and core formation on planetesimals, as well as varied degrees of chemical equilibration of such cores with mantle material during large collisions, also potentially contribute to ^{182}W observed isotopic variations.

Primordial silicate differentiation has been dated on large bodies such as the Moon, Earth, and Mars, using a variety of long- and short-lived chronometers. This requires independent knowledge of how a sample, or groups of samples, formed. Most investigations oversimplify the evolution of these bodies by assuming that the silicate portion of large planetary bodies differentiated from an isotopically homogeneous magma ocean at the end of planetary accretion. In addition, these evolutionary models assume solidification of large magma ocean occurs relatively quickly, when in fact thermal modeling is quite ambiguous on the subject. The Moon, for example, is estimated to cool (i.e., experience freezing of its magma ocean) in anywhere from 2 to 200 million years (Elkins-Tanton et al. 2011; Maurice et al. 2020).

Ages for solidification of the lunar magma ocean have been addressed via isotopic ages on three types of crystallization products: (1) mafic cumulate mare basalt sources, (2) crustal rocks of the ferroan anorthosite suite, and (3) late-stage crystallization products of the lunar magma ocean—for example, urKREEP, material rich in potassium (K), rare-earth elements (REE), and phosphorus (P). Detailed reviews of these ages, their merits, and their inconsistencies are provided in Borg et al. (2015), Nyquist et al. (2001), and Papike et al. (2018). The majority of ages fall in the range of 4.30 to 4.38 Ga, and these provide a lower age limit for the age of the Moon. However, all of these age determinations are based on Apollo samples collected from the lunar nearside, and therefore could represent either lunar magma ocean solidification or a widespread, but regional, magmatic event focused on the nearside (e.g., Borg et al. 2011; Tartèse et al. 2019). The age range of lunar magma ocean cumulates is concordant with rocks of the Mg-suite, which are not currently thought to be products of lunar magma ocean solidification, suggesting that the ages may not record lunar magma ocean solidification. Alternatively, the resolution of current chronometers may not be sufficient to distinguish two nearly contemporaneous, but geologically unrelated, events. Last, the committee notes that there are a few seemingly accurate ages that are older than 4.38 Ga (Tartèse et al. 2019; Borg et al. 2020). Additional high-precision isotopic measurements are needed on crustal and mantle rocks collected outside the region sampled by the Apollo astronauts to unravel the chronology of early geologic events in the history of the Earth–Moon system. The age of silicate differentiation of Earth and Mars is constrained by the oldest measured sample ages and the observation that ^{142}Nd, a decay product of the short-lived isotope ^{146}Sm (half-life of 103 million years), is variable. Ages for the Earth range from about 4.38 to about 4.42 Ga (Caro 2011; Valley et al. 2014), whereas ages for Mars range from about 4.50 to 4.53 Ga (Debaille et al. 2007; Borg et al. 2016; Kruijer et al. 2017; Bouvier et al. 2018). Many of the ages for silicate differentiation on Mars are based on the isotopic compositions of shergottite meteorites. However, not all samples (e.g., Nakhlite meteorites) demonstrate similar isotopic systematics, suggesting that the assumption that the sources of the shergottites were produced in a global magma ocean may be incorrect. Furthermore, martian meteorites have large variations in ^{182}W, that perhaps implies even earlier differentiation, or more extreme fractionation of the parent element (Hf) from the daughter element (W) during differentiation. On Earth, variations in ^{182}W isotopic compositions have been found in ancient crustal rocks and modern oceanic lavas, suggesting that differentiation of Earth may have isolated some silicate reservoirs very early on that have been preserved for billions of years until the present day, perhaps even through the Moon-forming impact (e.g., Canup et al. 2021).

While differentiation ages of planetesimals and asteroids have not been as well determined, the observation that basaltic samples from Vesta crystallized within a few million years of the beginning of the solar system indicates that differentiation was very early. This stems from the understanding that basalts are generated by partial melting of more primitive (Mg- and Fe-rich) materials.

Overall, small bodies appear to have differentiated earlier than large bodies, perhaps reflecting a combination of the size of these bodies and their proclivity to lose heat, and the preponderance of ^{26}Al in the early solar system that provided an important heat source during early times and stages of accretion. Large bodies such as Mars seem to have completed differentiation tens of millions of years later. Later differentiation in the Earth–Moon system was probably the result of a combination of stochastic events, including the Moon-forming giant impact, and may have been influenced by early tidal interactions between Earth and the Moon. Owing to the lack of samples, no differentiation ages can be estimated for other bodies like Mercury and Venus. Fundamental questions remain as to how the timing and style of primordial differentiation was linked to a body's size and the physical environment and chemical conditions in which it formed. Additional samples, more detailed thermal modeling, and more thorough understanding of the surface geology and interior structure of differentiated bodies are required to make further progress.

Q3.5c What Were the Causes of Variation in Oxygen Fugacity in Differentiated Planetary Bodies?

A notable feature of Earth is the difference in oxidation between its surface and interior. On a large scale, the core is metallic, requiring progressive oxidation over time. This intrinsic thermodynamic variable is known as oxygen fugacity, sometimes abbreviated to fO_2. Like Earth, Mars has the same general variation in fO_2 with a reduced core and an oxidized crust, giving the latter its moniker as the "Red Planet." In contrast, studies of meteorites from asteroids and Apollo lunar samples suggest that the range of oxidation state within these bodies is more restricted. Data from MESSENGER indicate that Mercury is the most reduced of the terrestrial planets, with an fO_2 well below most asteroids and the Moon, and those data suggest that perhaps the crust is more reduced than its interior, opposite to what is exhibited by Earth and Mars.

Oxygen fugacity changes the geochemical behavior of elements in common geologic processes and therefore has a profound effect on their distribution with a planetary body. On Earth, subduction leads to oxidization of the mantle, changing the valence of elements like uranium (U), which is insoluble in the reduced state (U^{+4}) and soluble in the oxidized state (U^{+6}). This process has had a remarkable effect on Earth's surface and atmosphere. On Mercury, the fO_2 is sufficiently low that many elements that typically are partitioned into the crust during differentiation are instead partitioned into the iron-rich core.

Despite its profound importance for constraining geochemical processes, we do not fully understand the primary drivers of oxidation and reduction. Oxidation and reduction could result from biologic activity, as has been suggested for Earth, or abiotic activity, as appears to be the case for Mars. We also do not understand the relationship between oxidation conditions on the surface of a body and those in its interior. Further studies of the oxidation state of Mercury and Venus will be important to addressing such issues. For example, an oxidized crust on Venus might reveal fundamental processes for how planetary differentiation can—and cannot—lead to life-sustaining environments.

Q3.5d How Did Late Accretion Affect Planet Composition and Chemistry?

Formation of metallic cores had a profound role in shaping the chemistry of planets. In addition to removing iron, core formation would have removed precious metals like platinum or gold from the silicate portions of planets almost entirely. That these siderophile (iron favoring) elements exist in Earth's crust today has been attributed by some to what is termed "late accretion," the addition of material rich in precious metals and volatile compounds to Earth's upper layers by impacts after core formation had essentially ended (see Figure 6-2). For Earth, the precious metal inventory in the crust and mantle can be accounted for by the addition of only about 0.5 to a few percent of Earth's mass (e.g., Kleine and Walker 2017; Marchi et al. 2018). Analysis of martian and lunar samples demonstrate that Mars experienced a similar amount of late accretion, whereas Earth's Moon experienced a far smaller proportion.

Understanding why different bodies in the inner solar system appear to have different proportions of late accreted materials is critically important because it provides insights into the planet formation and differentiation processes. Various arguments now suggest that late accretion was dominated in mass by large, up to lunar-size projectiles left over from planet accretion (e.g., Bottke et al. 2010; Pahlevan and Morbidelli 2015; Brasser et al. 2016; Marchi et al. 2018). Therefore, understanding the size distribution of late accretion projectiles and how the addition of large, differentiated projectiles affect planetary surface composition and chemistry is of great importance. Addition of cometary material with its copious amounts of ices would be substantially different from accreting a volatile element depleted body like the Moon, for example. Likewise, we do not know if the late accreting materials have, on average, fundamentally different compositions from the materials responsible for the earlier main phases of accretion. Such compositionally heterogeneous accretion has been used to explain the volatile content of Earth and the formation of its oceans and seems generally consistent with dynamical models that suggest that late accreting material would increasingly have been derived from the outer solar system. It is also important to account for the effect of additions to Earth by Theia, the Moon-forming impactor. In particular, Theia's core material would have been incorporated into Earth's core and the Moon's core, but it could have also supplied the highly siderophile elements in Earth's mantle and accounted for the high lunar mantle FeO content (Sleep 2016). Last, determining how water and other key volatile elements over a range of volatilities (e.g., carbon, hydrogen, oxygen, phosphorus, and sulfur) were supplied is critical for our understanding of what makes a habitable planet. Means for addressing these issues lie in obtaining a refined understanding of late accretion from study of available materials linked with improved modeling efforts. Determining late accretion components to Venus and Mercury, currently unconstrained, are particularly important needed data.

Strategic Research for Q3.5

- **Determine the age relations between the oldest lunar crustal and mantle rocks** by dating lunar crustal and mantle rocks, which may be potentially found at the South Pole–Aitken basin.
- **Reveal the mechanisms of planetary differentiation on Venus** by measuring ages and composition of the tesserae regions with spacecraft observations or sample return.
- **Determine the age and duration of primordial differentiation on Earth, the Moon, and Mars** through isotopic analysis of samples collected from new locations on the Moon and Mars, and through thermal modeling.
- **Determine and compare the mechanisms of differentiation, size of body, and location of bodies in the solar system** through sample analysis, spacecraft observation, and geochemical/geophysical modeling.
- **Determine the oxidation state of planetary surfaces to understand the primary drivers of redox conditions** with spacecraft observations and sample analysis.
- **Determine the contribution of outer solar system materials to the inventory of the inner solar system planets** through measurements of the volatiles and refractory components of water-rich asteroids and comets by telescopic observations, in situ measurements, and/or analysis of returned samples.
- **Assess the contribution and effects of late accretion on the post-differentiation inner planets** by analysis of ancient terrestrial materials, samples from regions of Mars and the Moon likely to be derived from each world's mantle, and/or samples derived from Venus or Mercury, and through improved dynamical and geochemical modeling.

Q3.6 WHAT ESTABLISHED THE PRIMORDIAL INVENTORIES OF VOLATILE ELEMENTS AND COMPOUNDS IN THE INNER SOLAR SYSTEM?

Volatile elements include a wide range of geologically important elements, including hydrogen, fluorine, chlorine, nitrogen, carbon, and sulfur, as well as noble gases and alkali elements. These elements encompass nearly all of organic chemistry and water, including the life-essential elements. Although both differentiated and undifferentiated rocky planetary bodies in the inner solar system have nearly chondritic abundance ratios of refractory elements typically partitioned into silicate minerals (lithophile elements), differentiated rocky bodies are depleted in volatile elements compared to undifferentiated rocky parent bodies or CC meteorites. The magnitude of volatile depletion seems to correlate very roughly with the 50 percent nebular condensation temperature for a given element, with those with lower condensation temperatures depleted more strongly (Lodders 2003).

There is a wide range of explanations for the origin and timing of observed volatile depletions and isotopic compositions. For example, some models suggest that depletion occurred prior to the assembly of the rocky parent bodies (e.g., owing to partial condensation in the nebula) from which samples derive (Lodders 2003). In contrast, other models call for parent body processes to account for the loss of volatiles (Mezger et al. 2021). The timing of water delivery is one of the most researched topics, and there is still an unsettled debate as to whether water was inherited from the nebula, delivered during the primary stage of accretion, or delivered during the waning stages of late accretion. In addition, the relatively high abundances of water in Earth, relative to the Moon, suggests that processes other than location in the solar system are responsible for producing the volatile element inventories of the terrestrial planets (e.g., Hauri et al. 2015). The timing of planetary volatile depletions and how volatile elements were successfully retained or added to planets, like Earth, remain key issues associated with understanding how, when, and why life began (Question 9, Chapter 12).

Q3.6a What Were the Primordial Sources of Volatiles?

Through determination of the chemical and isotopic makeup of elements like hydrogen, carbon, nitrogen, and oxygen in a variety of materials, including the Sun, atmospheres, planets, moons, and primitive bodies, several key endmember compositions have been identified in the solar system.

The first endmember is the solar composition, which represents that of the early protosolar nebula, the cloud of ~99 percent gas (mainly hydrogen, carbon monoxide, and nitrogen, or H_2, CO, N_2, and noble gases like helium or argon) and ~1 percent dust from which the solar system emerged. Even some rocky planets have compositions approaching this endmember, including Earth. How this composition was obtained is not fully resolved. Volatile elements (H, C, N, and noble gases) trapped in the mantle are derived from planetary sources. One exception is neon, for which the isotopic composition in mantle-derived samples points to the presence of a solar-like component in the solid Earth. This occurrence implies the incorporation of solar gas during terrestrial accretion. In one model, if the planets formed early enough, solar nebula gas could have been gravitationally trapped into primary atmospheres and eventually dissolved into planet interiors (Porcelli et al. 2001). Another model suggests that solar ions could have also been implanted from solar wind irradiation onto dust and/or planetary surfaces later incorporated into the planets (Vogt et al. 2019).

The second endmember is a chondritic component. This endmember is defined from analysis of primitive meteorites (chondrites), which are diverse in composition, but which generally contain volatile elements, either as hydrated minerals or as organic material hosting hydrogen, carbon, and nitrogen, or both. Primitive meteorites that originate from leftover asteroids are our best representatives of early formed planetesimals that contributed to the formation of inner planets. The noncarbonaceous chondrites (NC) are generally (but not always) volatile-poor and might have originated from a region in the disk located between the tar line (a radial distance from the Sun where some refractory organics could have survived) and the snow line (a radial distance beyond which water could exist as ice). Carbonaceous chondrites (CC) can contain up to ~10 percent equivalent water in hydrated minerals and a few percent carbon and nitrogen as organics. Given their generally more abundant water, CC are thought to have formed beyond the snow line and either more distant from the Sun than NC and/or at a different time in the evolution of the disk.

The third endmember composition is cometary. Comets are our best representatives of planetary bodies formed beyond the original orbits of the giant planets and are 10–50 percent water and other compound ices by mass. Given that their trajectories can cross the orbits of the terrestrial planets, they are often regarded as potential sources of inner planet volatiles.

The interplay of these different contributors shaped the volatile inventories of the inner planets. Despite advances in knowledge of what the sources of volatiles were, how these three major reservoirs were established in detail, and what the ultimate interstellar source of these volatiles was, is largely unknown. Analysis of volatile elements from samples returned from asteroids, Mars and the Moon, volatile-rich deposits on the Moon potentially representing cometary compositions, and ultimately comets will provide the principal means of making further progress on these questions through comparison with compositions determined for planetary atmospheres measured in situ and on returned samples.

Q3.6b How Was the Inner Solar System Populated with Volatiles and How Did Volatile Delivery Evolve with Time?

Hydrogen, carbon, nitrogen, and oxygen were likely incorporated into growing planetesimals in the form of ice and organic dust after the solar system had cooled following its initial formation. Volatile-rich dust was inherited from the interstellar medium (ISM) or synthesized by photon–gas interactions in the parent molecular cloud or at the protoplanetary disk surface. Importantly, hydrogen, carbon, nitrogen, and oxygen trapped in chondrites, comets, and the atmospheres of the terrestrial planets have elemental and isotopic compositions markedly different from those of the protoplanetary disk as determined from the ISM or local gaseous reservoirs such as the giant planets, precluding a direct genetic relationship between the building blocks that make up the terrestrial planets and the gaseous solar nebula (e.g., Dauphas and Morbidelli 2014).

Gas and dust were processed and distributed throughout the protosolar nebula by turbulent mixing, with (1) preferential preservation of molecular cloud signatures in the colder regions of the outer solar system, and (2) organo-synthesis mainly occurring within the irradiated regions of the protosolar nebula. The spatial distribution of hydrogen, carbon, nitrogen, and oxygen was then mainly controlled by the location of the tar and snow lines. As discussed in Questions 2 (Chapter 5), 3.1, and 3.2, planetesimal/planet formation and giant planet migration led to the scattering of outer solar system planetesimals, some of which entered the terrestrial planet region and were captured inside the asteroid belt.

Earth is thought to have accreted mainly from dry material of the NC type, although significant contributions from enstatite chondrites, for example, is possible. Volatile elements contributing wetter CC materials, namely planetesimals, comets, and dust from the giant planet zone, were accreted later as these materials became dynamically accessible. This compositional shift is consistent with the isotopic compositions of terrestrial planet noble gases as well as hydrogen and nitrogen stable isotopes observed in the mantle and atmosphere, respectively (e.g., Piani et al. 2020). Although this model is appealing, one inconsistency is that deuterium-to-hydrogen (D/H) ratios determined for a significant subset of comets are about twice as high as D/H values measured on Earth. To what extent this scenario is applicable to the other terrestrial planets is unknown because representative samples of planetary interiors are limited so far to Earth. Likewise, the impact of the early Sun's activity on terrestrial planet atmospheres cannot be assessed without more detailed knowledge of the compositions of the atmospheres of Mercury and Venus. These constraints could be incorporated into numerical simulations of planet formation to test which groups of bodies and processes are the most important for volatile delivery.

The timing of volatile delivery to planets remains poorly known, despite having a framework chronology of solar system events derived from meteorites. There is evidence, however, that both Earth and Mars (or their precursor embryos) incorporated nebula gas during their differentiation from the noble gas compositions trapped within terrestrial rocks and martian meteorites, respectively. Overall, it seems likely that the initial delivery of volatiles into the inner solar system may have occurred when accretion was taking place. Limited volatile delivery from impacting carbonaceous chondrite asteroids and comets continues to the present day (discussed in Questions 4, 5, and 6; Chapters 7, 8, and 9, respectively) indicating volatile elements have been added to the terrestrial planets by this process as well. Further research is needed to constrain the characteristics and abundances of nebular gas added to planetary interiors during initial accretion. In addition, broad and geochemically consistent models to understand potential late addition of CC materials to planetary bodies would be valuable. Polar volatiles in the permanently shadowed regions of the Moon and Mercury may also retain a record of volatile delivery, although the exact age and origin of these volatiles is debated (Question 6, Chapter 9).

Q3.6c How Were Primary Volatile-Rich Reservoirs (Atmospheres and Oceans) Produced?

Of the terrestrial planets, only Mercury does not have a substantial, visible atmosphere. The present-day atmospheres of Venus and Mars have comparable compositions, namely >90 percent carbon dioxide (CO_2) and a few percent N_2, but they are at substantially different pressures and have different water contents. Venus might have had early oceans that subsequently vanished. Geological evidence on Mars attests to the presence of oceans within the first few hundreds of million years.

Noble gas abundance patterns of the three planets are comparable, but this is difficult to reconcile with different atmospheric escape processing otherwise suggested by isotope variations. All three planetary atmospheres show a depletion of xenon relative to lighter noble gases. On Earth, this xenon depletion has been accounted for by prolonged hydrogen escape from photodissociation of water (e.g., Avice et al. 2018). This could also be the case for Mars but at different periods of time. Although D/H ratios have been measured in the Venus atmosphere, their significance remains somewhat ambiguous (e.g., Grinspoon 1993; Krasnopolsky et al. 2013). Earth's atmosphere has been strongly modified by biological processes, and the atmospheres of Venus and Mars have been modified by runaway greenhouse effects or extensive mass loss, respectively. Despite eradication of the primary volatile-rich reservoirs on these bodies, evidence can still be found for these early stages from volcanic rocks arising from deep mantle sources in Earth and from Mars (as meteorites). A major unanswered question is why Earth has solar compositions of light noble gases like helium and nitrogen, but chondritic abundances of heavier noble gases like krypton and xenon, whereas for Mars, all appear solar. This effect may relate to late accretion of NC or CC materials, but exactly how is unclear.

The isotopic composition of cometary noble gases measured on comet 67P/Churyumov-Gerasimenko by the Rosetta mission indicates that Earth's atmosphere, but not its mantle, has a substantial contribution of cometary materials. It is thought that these additions occurred after the Moon-forming impact. The timing and extent of cometary contribution to the terrestrial planets and the Moon is a fundamental problem that will require precise documentation of the compositions of inner planet reservoirs (mantle and atmosphere of Mars and Venus), as well as those of cometary materials returned to Earth.

The elemental and isotopic compositions of terrestrial noble gases indicate that the young Earth experienced several episodes of degassing and volatile loss during the first tens to hundreds of million years. In particular, extant and extinct gas isotopes with variable half-lives permit reconstruction of the degassing/tectonic history of Earth: early isolation (within 100 Ma) of the deep and the convective mantle regions, intense degassing within a few tens to hundreds million years, secular evolution of convection and crustal growth around 3 Ga, and recycling of surface volatiles including water becoming effective during the past 2 Ga.

Mars might have experienced a different volatile history. Early oceans seem to have vanished within the first few hundreds of million years of its history, while its tectonic activity might have been too limited to permit replenishment of the atmosphere. Reconstructing the history of martian volatiles will await the return of martian sediments and igneous rocks. The volatile history of Venus is for the most part unknown and awaits a new generation of in situ missions to that challenging environment.

Strategic Research for Q3.6

- **Determine in situ the origin, degassing history, and mantle–atmosphere exchange rate of Venus's volatile elements** by measuring the noble gas elemental and isotopic compositions of Venus's atmosphere, as well as the abundances and stable isotope compositions of H, C, N, O, S species.
- **Precisely determine the elemental and isotopic compositions of martian mantle and atmospheric volatiles at present and in the past** by analyzing carefully selected atmospheric and solid samples with different ages and provenances returned from Mars.
- **Measure the abundances and isotopic compositions of volatile elements in asteroids from different radial locations and in comets** by analysis of returned samples and/or in situ investigations.
- **Determine the origin and abundances of volatiles in inner solar system bodies** by conducting geochemical, petrologic, and spectral measurements of these bodies and their associated samples and by coupling results from planetary accretion models, laboratory experiments on volatile behavior, observations of volatile distribution in the asteroid and comet populations, and geochemical measurements from a wide range of parent bodies.
- **Study the behavior and the elemental/isotopic fractionation of volatiles in temperature, pressure, chemistry, and ionization conditions relevant to the formation of planetesimals, embryos, Earth, and Venus** using laboratory experiments and modeling.

SUPPORTIVE ACTIVITIES FOR QUESTION 3

- Improve knowledge of chemical and isotopic abundances through analysis of existing samples of meteorites, including martian and lunar meteorites, and continued collection of meteorites, which affords the possibility of finding meteorite samples of the other inner planets.
- Continued observations of the Venus atmosphere from ALMA and other Earth- and space-based observatories to detect species potentially indicative of volcanic processes.
- Studies of exoplanet system architectures to determine size and semi-major axis distribution of terrestrial-sized planets for comparison with the solar system's architecture.

REFERENCES

Alemi, A., and D. Stevenson. 2006. "Why Venus Has No Moon." *AAS/Division for Planetary Sciences Meeting Abstracts* #38, id.07.03. https://ui.adsabs.harvard.edu/abs/2006DPS. . .38.0703A.

Andrews-Hanna, J.C., S.W. Asmar, J.W. Head, W.S. Kiefer, A.S. Konopliv, F.G. Lemoine, I. Matsuyama, et al. 2013. "Ancient Igneous Intrusions and Early Expansion of the Moon Revealed by GRAIL Gravity Gradiometry." *Science* 339:675–678. https://doi.org/10.1126/science.1231753.

Andrews-Hanna, J.C., M.T. Zuber, and W.B. Banerdt. 2008. "The Borealis Basin and the Origin of the Martian Crustal Dichotomy." *Nature* 453:1212–1215. https://doi.org/10.1038/nature07011.

Asphaug, E., and A. Reufer. 2014. "Mercury and Other Iron-Rich Planetary Bodies as Relics of Inefficient Accretion." *Nature Geoscience* 7:564–568. https://doi.org/10.1038/ngeo2189.

Avice, G., B. Marty, R. Burgess, A. Hofmann, P. Philpott, K. Zahnle, and D. Zakharov. 2018. "Evolution of Atmospheric Xenon and Other Noble Gases Inferred from Archean to Paeoproterozoic Rocks." *Geochimica et Cosmochimica Acta* 232:82–100. https://doi.org/10.1016/j.gca.2018.04.018.

Bagheri, A., A. Khan, M. Efroimsky, M. Kruglyakov, and D. Giardini. 2021. "Dynamical Evidence for Phobos and Deimos as Remnants of a Disrupted Common Progenitor." *Nature Astronomy* 5:539–543. https://doi.org/10.1038/s41550-021-01306-2.

Benz, W., W.L. Slattery, and A.G.W. Cameron. 1988. "Collisional Stripping of Mercury's Mantle." *Icarus* 74:516–528. https://doi.org/10.1016/0019-1035(88)90118-2.

Bonomo, A.S., L. Zeng, M. Damasso, Z.M. Leinhardt, A.B. Justesen, E. Lopez, M.N. Lund, et al. 2019. "A Giant Impact as the Likely Origin of Different Twins in the Kepler-107 Exoplanet System." *Nature Astronomy* 3:416–423. https://doi.org/10.1038/s41550-018-0684-9.

Borg, L.E., G.A. Brennecka, and S.J. Symes. 2016. "Accretion Timescale and Impact History of Mars Deduced from the Isotopic Systematics of Martian Meteorites." *Geochimica et Cosmochimica Acta* 175:150–167. https://doi.org/10.1016/j.gca.2015.12.002.

Borg, L.E., W.S. Cassata, J. Wimpenny, A.M. Gaffney, and C.K. Shearer. 2020. "The Formation and Evolution of the Moon's Crust Inferred from the Sm-Nd Isotopic Systematics of Highlands Rocks." *Geochimica et Cosmochimica Acta* 290: 312–332. https://doi.org/10.1016/j.gca.2020.09.013.

Borg, L.E., J.N. Connelly, M. Boyet, and R.W. Carlson. 2011. "Chronological Evidence That the Moon Is Either Young or Did Not Have a Global Magma Ocean." *Nature* 477:70–72. https://doi.org/10.1038/nature10328.

Borg, L.E., A.M. Gaffney, and C.K. Shearer. 2015. "A Review of Lunar Chronology Revealing a Preponderance of 4.34–4.37 Ga Ages." *Meteoritics and Planetary Science* 50:715–732. https://doi.org/10.1111/maps.12373.

Bottke, W.F., and M.D. Norman. 2017. "The Late Heavy Bombardment." *Annual Review of Earth and Planetary Sciences* 45:619–647. https://doi.org/10.1146/annurev-earth-063016-020131.

Bottke, W.F., R.J. Walker, J.M.D. Day, D. Nesvorný, and L. Elkins-Tanton. 2010. "Stochastic Late Accretion to Earth, the Moon, and Mars." *Science* 330:1527. https://doi.org/10.1126/science.1196874.

Bouvier, L.C., M.M. Costa, J.N. Connelly, N.K. Jensen, D.Wielandt, M. Storey, A.A. Nemchin, et al. 2018. "Evidence for Extremely Rapid Magma Ocean Crystallization and Crust Formation on Mars." *Nature* 558:586–589. https://doi.org/10.1038/s41586-018-0222-z.

Brasser, R., S.J. Mojzsis, S.C. Werner, S. Matsumura, and S. Ida. 2016. "Late Veneer and Late Accretion to the Terrestrial Planets." *Earth and Planetary Science Letters* 455:85–93. https://doi.org/10.1016/j.epsl.2016.09.013.

Cano, E.J., Z.D. Sharp, and C.K. Shearer. 2020. "Distinct Oxygen Isotope Compositions of the Earth and Moon." *Nature Geoscience* 13:270–274. https://doi.org/10.1038/s41561-020-0550-0.

Canup, R.M. 2004. "Simulations of a Late Lunar-Forming Impact." *Icarus* 168:433–456. https://doi.org/10.1016/j.icarus.2003.09.028.

Canup, R.M. 2012. "Forming a Moon with an Earth-Like Composition via a Giant Impact." *Science* 338:1052. https://doi.org/10.1126/science.1226073.

Canup, R.M., K., Righter, N. Dauphas, K. Pahlevan, M. Ćuk, S.J. Lock, S.T. Stewart, et al. 2021. "Origin of the Moon." In *New Views of the Moon II*, C.R. Neal, B.L. Jolliff, C.K. Shearer, S. Valencia, L. Gaddis, S. Mackwell, and S.J. Lawrence, eds. Preprint: https://arxiv.org/abs/2103.02045.

Canup, R.M., and J. Salmon. 2018. "Origin of Phobos and Deimos by the Impact of a Vesta-to-Ceres Sized Body with Mars." *Science Advances* 4:eaar6887. https://doi.org/10.1126/sciadv.aar6887.

Canup, R.M., C. Visscher, J. Salmon, and B. Fegley. 2015. "Lunar Volatile Depletion Due to Incomplete Accretion Within an Impact-Generated Disk." *Nature Geoscience* 8:918–921. https://doi.org/10.1038/ngeo2574.

Caro, G. 2011. "Early Silicate Earth Differentiation." *Annual Review of Earth and Planetary Sciences* 39:31–58. https://doi.org/10.1146/annurev-earth-040610-133400.

Castillo-Rogez, J.C., M.A. Hesse, M. Formisano, H. Sizemore, M. Bland, A.I. Ermakov, and R.R. Fu. 2019. "Conditions for the Long-Term Preservation of a Deep Brine Reservoir in Ceres." *Geophysical Research Letters* 46:1963–1972. https://doi.org/10.1029/2018GL081473.

Citron, R.I., H. Genda, and S. Ida. 2015. "Formation of Phobos and Deimos via a Giant Impact." *Icarus* 252:334–338. https://doi.org/10.1016/j.icarus.2015.02.011.

Clement, M.S., N.A. Kaib, S.N. Raymond, and K.J. Walsh. 2018. "Mars' Growth Stunted by an Early Giant Planet Instability." *Icarus* 311:340–356. https://doi.org/10.1016/j.icarus.2018.04.008.

Correia, A.C.M., and J. Laskar. 2001. "The Four Final Rotation States of Venus." *Nature* 411:767–770. https://doi.org/10.1038/35081000.

Ćuk, M., and S.T. Stewart. 2012. "Making the Moon from a Fast-Spinning Earth: A Giant Impact Followed by Resonant Despinning." *Science* 338:1047. https://doi.org/10.1126/science.1225542.

Dauphas, N. 2017. "The Isotopic Nature of the Earth's Accreting Material Through Time." *Nature* 541:521–524. https://doi.org/10.1038/nature20830.

Dauphas, N., and A. Morbidelli. 2014. "6.1: Geochemical and Planetary Dynamical Views on the Origin of Earth's Atmosphere and Oceans." Pp. 1–35 in Vol. 6: *The Atmosphere—History* in *Treatise on Geochemistry*, 2nd ed., H.D. Holland and K.K. Turekian, eds. Amsterdam, Netherlands: Elsevier. https://doi.org/10.1016/B978-0-08-095975-7.01301-2.

Day, J.M.D., F. Moynier, and C.K. Shearer. 2017. "Late-Stage Magmatic Outgassing from a Volatile-Depleted Moon." *Proceedings of the National Academy of Sciences* 114:9547–9551. https://doi.org/10.1073/pnas.1708236114.

Debaille, V., A.D. Brandon, Q.Z. Yin, and B. Jacobsen. 2007. "Coupled ^{142}Nd-^{143}Nd Evidence for a Protracted Magma Ocean in Mars." *Nature* 450:525–528. https://doi.org/10.1038/nature06317.

Deienno, R., A. Izidoro, A. Morbidelli, R.S. Gomes, D. Nesvorný, and S.N. Raymond. 2018. "Excitation of a Primordial Cold Asteroid Belt as an Outcome of Planetary Instability." *Astrophysical Journal* 864:50. https://doi.org/10.3847/1538-4357/aad55d.

Dodds, K.H., J.F.J. Bryson, J.A. Neufeld, and R.J. Harrison. 2021. "The Thermal Evolution of Planetesimals During Accretion and Differentiation: Consequences for Dynamo Generation by Thermally Driven Convection." *Journal of Geophysical Research: Planets* 126:e06704. https://doi.org/10.1029/2020JE006704.

Ebel, D.S., and S.T. Stewart. 2018. "The Elusive Origin of Mercury". In *Mercury: The View After MESSENGER*, S.C. Solomon, B.J. Anderson, L.R. Nittler, eds. Cambridge, UK: Cambridge University Press.

Elkins-Tanton, L.T., S. Burgess, and Q.-Z. Yin. 2011. "The Lunar Magma Ocean: Reconciling the Solidification Process with Lunar Petrology and Geochronology." *Earth and Planetary Science Letters* 304:326–336. https://doi.org/10.1016/j.epsl.2011.02.004.

Garrick-Bethell, I., V. Perera, F. Nimmo, and M.T. Zuber. 2014. "The Tidal-Rotational Shape of the Moon and Evidence for Polar Wander." *Nature* 512:181–184. https://doi.org/10.1038/nature13639.

Gilmore, M., A. Treiman, J. Helbert, and S. Smrekar. 2017. "Venus Surface Composition Constrained by Observation and Experiment." *Space Science Reviews* 212:1511–1540. https://doi.org/10.1007/s11214-017-0370-8.

Gomes, R., H.F. Levison, K. Tsiganis, and A. Morbidelli. 2005. "Origin of the Cataclysmic Late Heavy Bombardment Period of the Terrestrial Planets." *Nature* 435:466–469. https://doi.org/10.1038/nature03676.

Grinspoon, D.H. 1993. "Implications of the High D/H Ratio for the Sources of Water in Venus's Atmosphere." *Nature* 363:428–431. https://doi.org/10.1038/363428a0.

Hansen, B.M.S. 2009. "Formation of the Terrestrial Planets from a Narrow Annulus." *Astrophysical Journal* 703:1131–1140. https://doi.org/10.1088/0004-637X/703/1/1131.

Hartmann, W.K., and D.R. Davis. 1975. "Satellite-Sized Planetesimals and Lunar Origin." *Icarus* 24:504–515. https://doi.org/10.1016/0019-1035(75)90070-6.

Hauck, S.A., J.-L. Margot, S.C. Solomon, R.J. Phillips, C.L. Johnson, F.G. Lemoine, E. Mazarico, et al. 2013. "The Curious Case of Mercury's Internal Structure." *Journal of Geophysical Research: Planets* 118:1204–1220. https://doi.org/10.1002/jgre.20091.

Hauri, E.H., A.E. Saal, M.J. Rutherford, and J.A. Van Orman. 2015. "Water in the Moon's Interior: Truth and Consequences." *Earth and Planetary Science Letters* 409:252–264. https://doi.org/10.1016/j.epsl.2014.10.053.

Hesselbrock, A.J., and D.A. Minton. 2017. "An Ongoing Satellite-Ring Cycle of Mars and the Origins of Phobos and Deimos." *Nature Geoscience* 10:266–269. https://doi.org/10.1038/ngeo2916.

Kato, C., and F. Moynier. 2017. "Gallium Isotopic Evidence for Extensive Volatile Loss from the Moon During Its Formation." *Science Advances* 3:e1700571. https://doi.org/10.1126/sciadv.1700571.

Keane, J.T., and I. Matsuyama. 2014. "Evidence for Lunar True Polar Wander and a Past Low-Eccentricity, Synchronous Lunar Orbit." *Geophysical Research Letters* 41:6610–6619. https://doi.org/10.1002/2014GL061195.

Khan, A., J. Maclennan, S.R. Taylor, and J.A.D. Connolly. 2006. "Are the Earth and the Moon Compositionally Alike? Inferences on Lunar Composition and Implications for Lunar Origin and Evolution from Geophysical Modeling." *Journal of Geophysical Research: Planets* 111:E05005. https://doi.org/10.1029/2005JE002608.

Kleine, T., and R.J. Walker. 2017. "Tungsten Isotopes in Planets." *Annual Review of Earth and Planetary Sciences* 45:389–417. https://doi.org/10.1146/annurev-earth-063016-020037.

Krasnopolsky, V.A., D.A. Belyaev, I.E. Gordon, G. Li, and L.S. Rothman. 2013. "Observations of D/H Ratios in H_2O, HCl, and HF on Venus and New DCl and DF Line Strengths." *Icarus* 224:57–65. https://doi.org/10.1016/j.icarus.2013.02.010.

Kruijer, T.S., C. Burkhardt, G. Budde, and T. Kleine. 2017. "Age of Jupiter Inferred from the Distinct Genetics and Formation Times of Meteorites." *Proceedings of the National Academy of Sciences* 114:6712–6716. https://doi.org/10.1073/pnas.1704461114.

Kruijer, T.S., and T. Kleine. 2017. "Tungsten Isotopes and the Origin of the Moon." *Earth and Planetary Science Letters* 475:15–24. https://doi.org/10.1016/j.epsl.2017.07.021.

Kruijer, T.S., T. Kleine, L.E. Borg, G.A. Brennecka, A.J. Irving, A. Bischoff, and C.B. Agee. 2017. "The Early Differentiation of Mars Inferred from Hf–W Chronometry." *Earth and Planetary Science Letters* 474:345–354. https://doi.org/10.1016/j.epsl.2017.06.047.

Kruijer, T.S., M. Touboul, M. Fischer-Gödde, K.R. Bermingham, R.J. Walker, and T. Kleine. 2014. "Protracted Core Formation and Rapid Accretion of Protoplanets." *Science* 344:1150–1154. https://doi.org/10.1126/science.1251766.

Lock, S.J., S.T. Stewart, M.I. Petaev, Z. Leinhardt, M.T. Mace, S.B. Jacobsen, and M. Ćuk. 2018. "The Origin of the Moon Within a Terrestrial Synestia." *Journal of Geophysical Research: Planets* 123:910–951. https://doi.org/10.1002/2017JE005333.

Lodders, K. 2003. "Solar System Abundances and Condensation Temperatures of the Elements." *Astrophysical Journal* 591:1220–1247. https://doi.org/10.1086/375492.

Marchi, S., R.M. Canup, and R.J. Walker. 2018. "Heterogeneous Delivery of Silicate and Metal to the Earth by Large Planetesimals." *Nature Geoscience* 11:77–81. https://doi.org/10.1038/s41561-017-0022-3.

Maurice, M., N. Tosi, S. Schwinger, D. Breuer, and T. Kleine. 2020. "A Long-Lived Magma Ocean on a Young Moon." *Science Advances* 6:eaba8949. https://doi.org/10.1126/sciadv.aba8949.

Mezger, K., A. Maltese, and H. Vollstaedt. 2021. "Accretion and Differentiation of Early Planetary Bodies as Recorded in the Composition of the Silicate Earth." *Icarus* 365:114497. https://doi.org/10.1016/j.icarus.2021.114497

Mezger, K., M. Schönbächler, and A. Bouvier. 2020. "Accretion of the Earth—Missing Components?" *Space Science Reviews* 216:27. https://doi.org/10.1007/s11214-020-00649-y.

Morbidelli, A., K.J. Walsh, D.P. O'Brien, D.A. Minton, and W.F. Bottke. 2015. "The Dynamical Evolution of the Asteroid Belt." Pp. 493–508 in *Asteroids IV*, P. Michel, F. DeMeo, and W.F. Bottke, eds. Tucson: University of Arizona Press. https://uapress.arizona.edu/book/asteroids-iv.

Nesvorný, D. 2018. "Dynamical Evolution of the Early Solar System." *Annual Review of Astronomy and Astrophysics* 56:137–174. https://doi.org/10.1146/annurev-astro-081817-052028.

Nesvorný, D., F.V. Roig, and W.F. Bottke. 2017. "Modeling the Historical Flux of Planetary Impactors." *Astronomical Journal* 153(3):103. https://doi.org/10.3847/1538-3881/153/3/103.

Nesvorný, D., F.V. Roig, and R. Deienno. 2021. "The Role of Early Giant-Planet Instability in Terrestrial Planet Formation." *Astronomical Journal* 161(2):50. https://doi.org/10.3847/1538-3881/abc8ef.

Nie, N.X., and N. Dauphas. 2019. "Vapor Drainage in the Protolunar Disk as the Cause for the Depletion in Volatile Elements of the Moon." *Astrophysical Journal* 884:L48. https://doi.org/10.3847/2041-8213/ab4a16.

Nimmo, F., and T. Kleine. 2007. "How Rapidly Did Mars Accrete? Uncertainties in the Hf–W Timing of Core Formation." *Icarus* 191:497–504. https://doi.org/10.1016/j.icarus.2007.05.002.

Nimmo, F., and T. Kleine. 2015. "Early Differentiation and Core Formation: Processes and Timescales." Pp. 83–102 in *The Early Earth: Accretion and Differentiation, Geophysical Monograph 212*, 1st ed., J. Badro and M. Walter, eds. Washington, DC: American Geophysical Union Publishing. https://www.wiley.com/en-us/TheEarlyEarth3AAccretionand Differentiation-p-9781118860571.

Nittler, L.R., R.D. Starr, S.Z. Weider, T.J. McCoy, W.V. Boynton, D.S. Ebel, C.M. Ernst, et al. 2011. "The Major-Element Composition of Mercury's Surface from MESSENGER X-Ray Spectrometry." *Science* 333:1847. https://doi.org/10.1126/science.1211567.

Nyquist, L., D. Bogard, and C. Shih. 2001. "Radiometric Chronology of the Moon and Mars." Pp. 1325–1376 in *The Century of Space Science,* Chapter 55, J.A. Bleeker, J. Geiss, and M. Huber, eds. Dordrecht, Netherlands: Kluwer Academic Publishers. https://doi.org/10.1007/978-94-010-0320-9_55.

Pahlevan, K., and A. Morbidelli. 2015. "Collisionless Encounters and the Origin of the Lunar Inclination." *Nature* 527:492–494. https://doi.org/10.1038/nature16137.

Pahlevan, K., and D.J. Stevenson. 2007. "Equilibration in the Aftermath of the Lunar-Forming Giant Impact." *Earth and Planetary Science Letters* 262:438–449. https://doi.org/10.1016/j.epsl.2007.07.055.

Papike, J.J., G. Ryder, and C. Shearer. 2018. "Chapter 5. Lunar Samples." Pp. 719–952 in *Planetary Materials*, J.J. Papike, ed. Berlin, Germany, and Boston, MA: De Gruyter. https://doi.org/10.1515/9781501508806-020.

Piani, L., Y. Marrocchi, T. Rigaudier, L.G. Vacher, D. Thomassin, and B. Marty. 2020. "Earth's Water May Have Been Inherited from Material Similar to Enstatite Chondrite Meteorites." *Science* 369(6507):1110–1113. https://doi.org/10.1126/science.aba1948.

Porcelli, D., D. Woolum, and P. Cassen. 2001. "Deep Earth Rare Gases: Initial Inventories, Capture from the Solar Nebula, and Losses During Moon Formation." *Earth and Planetary Science Letters* 193:237–251. https://doi.org/10.1016/S0012-821X(01)00493-9.

Raymond, S.N., and A. Izidoro. 2017. "Origin of Water in the Inner Solar System: Planetesimals Scattered Inward During Jupiter and Saturn's Rapid Gas Accretion." *Icarus* 297:134–148. https://doi.org/10.1016/j.icarus.2017.06.030.

Raymond, S.N., and A. Morbidelli. 2014. "The Grand Tack Model: A Critical Review." *Proceedings of the International Astronomical Union* 9(S310):194–203. https://doi.org/10.1017/S1743921314008254.

Reufer, A., M.M.M. Meier, W. Benz, and R. Wieler. 2012. "A Hit-and-Run Giant Impact Scenario." *Icarus* 221:296–299. https://doi.org/10.1016/j.icarus.2012.07.021.

Rosenblatt, P. 2011. "The Origin of the Martian Moons Revisited." *Astronomy and Astrophysics Review* 19:44. https://doi.org/10.1007/s00159-011-0044-6.

Rufu, R., O. Aharonson, and H.B. Perets. 2017. "A Multiple-Impact Origin for the Moon." *Nature Geoscience* 10:89–94. https://doi.org/10.1038/ngeo2866.

Sleep, N.H. 2016. "Asteroid Bombardment and the Core of Theia as Possible Sources for the Earth's Late Veneer Component." *Geochemistry, Geophysics, Geosystems* 17:2623–2642. https://doi.org/10.1002/2016GC006305.

Smrekar, S.E., L. Elkins-Tanton, J.J. Leitner, A. Lenardic, S. Mackwell, L. Moresi, C. Sotin, and E.R. Stofan. 2007. "Tectonic and Thermal Evolution of Venus and the Role of Volatiles: Implications for Understanding the Terrestrial Planets." *American Geophysical Union Geophysical Monograph Series* 176:45–71. https://doi.org/10.1029/176GM05.

Strom, R.G., G.G. Schaber, and D.D. Dawson. 1994. "The Global Resurfacing of Venus." *Journal of Geophysical Research* 99:10899–10926. https://doi.org/10.1029/94JE00388.

Tartèse, R., M. Anand, J. Gattacceca, K.H. Joy, J.I. Mortimer, J.F. Pernet-Fisher, S. Russell, et al. 2019. "Constraining the Evolutionary History of the Moon and the Inner Solar System: A Case for New Returned Lunar Samples." *Space Science Reviews* 215:54. https://doi.org/10.1007/s11214-019-0622-x.

Thiemens, M.M., P. Sprung, R.O.C. Fonseca, F.P. Leitzke, and C. Münker. 2019. "Early Moon Formation Inferred from Hafnium-Tungsten Systematics." *Nature Geoscience* 12:696–700. https://doi.org/10.1038/s41561-019-0398-3.

Valley, J.W., A.J. Cavosie, T. Ushikubo, D.A. Reinhard, D.F. Lawrence, D.J. Larson, P.H. Clifton, et al. 2014. "Hadean Age for a Post-Magma-Ocean Zircon Confirmed by Atom-Probe Tomography." *Nature Geoscience* 7:219–223. https://doi.org/10.1038/ngeo2075.

Vogt, M., J. Hopp, H.-P. Gail, U. Ott, and M. Trieloff. 2019. "Acquisition of Terrestrial Neon During Accretion—A Mixture of Solar Wind and Planetary Components." *Geochimica et Cosmochimica Acta* 264:141–164. https://doi.org/10.1016/j.gca.2019.08.016.

Vokrouhlický, D., W.F. Bottke, and D. Nesvorný. 2016. "Capture of Trans-Neptunian Planetesimals in the Main Asteroid Belt." *Astronomical Journal* 152:39. https://doi.org/10.3847/0004-6256/152/2/39.

Wadhwa, M., Y. Amelin, O. Bogdanovski, A. Shukolyukov, G.W. Lugmair, and P. Janney. 2009. "Ancient Relative and Absolute Ages for a Basaltic Meteorite: Implications for Timescales of Planetesimal Accretion and Differentiation." *Geochimica et Cosmochimica Acta* 73:5189–5201. https://doi.org/10.1016/j.gca.2009.04.043.

Walsh, K.J., A. Morbidelli, S.N. Raymond, D.P. O'Brien, and A.M. Mandell. 2011. "A Low Mass for Mars from Jupiter's Early Gas-Driven Migration." *Nature* 475:206–209. https://doi.org/10.1038/nature10201.

Wang, K., and S.B. Jacobsen. 2016. "Potassium Isotopic Evidence for a High-Energy Giant Impact Origin of the Moon." *Nature* 538:487–490. https://doi.org/10.1038/nature19341.

Wang, X., C. Fitoussi, B. Bourdon, B. Fegley, and S. Charnoz. 2019. "Tin Isotopes Indicative of Liquid-Vapour Equilibration and Separation in the Moon-Forming Disk." *Nature Geoscience* 12:707–711. https://doi.org/10.1038/s41561-019-0433-4.

Young, E.D., I.E. Kohl, P.H. Warren, D.C. Rubie, S.A. Jacobson, and A. Morbidelli. 2016. "Oxygen Isotopic Evidence for Vigorous Mixing During the Moon-Forming Giant Impact." *Science* 351:493–496. https://doi.org/10.1126/science.aad0525.

Zhang, J., N. Dauphas, A.M. Davis, I. Leya, and A. Fedkin. 2012. "The Proto-Earth as a Significant Source of Lunar Material." *Nature Geoscience* 5:251–255. https://doi.org/10.1038/ngeo1429.

Zuber, M.T., H.Y. McSween, R.P. Binzel, L.T. Elkins-Tanton, A.S. Konopliv, C.M. Pieters, and D.E. Smith. 2011. "Origin, Internal Structure and Evolution of 4 Vesta." *Space Science Reviews* 163:77–93. https://doi.org/10.1007/s11214-011-9806-8.

Q4 PLATE: An enhanced-color image of Haulani crater on Ceres, taken by the Dawn mission in 2017. The crater has a diameter of 34 kilometers. SOURCE: Courtesy of NASA/JPL-Caltech/UCLA/MPS/DLR/IDA.

7

Question 4: Impacts and Dynamics

How has the population of solar system bodies changed through time owing to collisions and dynamical interactions, and how to match questions in Table 3.1 has bombardment varied across the solar system? How have collisions affected the evolution and properties of planetary bodies?

Much of the story of how the solar system formed and evolved can be told through the history of impacts and dynamics. Questions 1–3, Chapters 4–6, have discussed the growth of planetesimals, protoplanets, and how they merge to make the terrestrial planets and giant planet cores.[1] A poorly understood part of planetary formation, however, is its endgame. Modelers now need to consider how giant planet migration, in conjunction with planet formation processes, affected bodies throughout the solar system (e.g., Nesvorný et al. 2021).

Previous work has also led to a series of tiered tests for solar system planet formation and evolution models. The top tier is to reproduce the orbits, sizes, and angular momentum budgets of the terrestrial and giant planets, including the small size of Mars. Successful models then move on to a second tier, where they need to reproduce small body reservoirs including the asteroid and trans-neptunian (or Kuiper) belts.

Here the committee focuses on a critical third tier of constraints, namely how small bodies—including trans-neptunian objects (TNOs), comets, leftover planetesimals, and asteroids—both dynamically evolved and bombarded the solar system worlds over the past 4.5 billion years. While the impacts discussed here often involve a limited amount of mass compared to planetary-scale collisions (e.g., the Earth–Moon forming impact, Question 3), they are crucial to deciphering critical questions, such as the nature of different episodes of giant planet migration, the primordial histories of solar system worlds, the size and orbital distributions of small body populations, the collisional histories of said populations, and the astrobiological history of Earth and other abodes of life affected by impacts.

Q4.1 HOW HAVE PLANETARY BODIES COLLISIONALLY AND DYNAMICALLY EVOLVED THROUGHOUT SOLAR SYSTEM HISTORY?

The major set pieces of early solar system evolution—namely, planet formation and giant planet migration—depleted early small body reservoirs, created new ones, and reconfigured most through mutual collisions between planetesimals (e.g., see Questions 2 and 3; Nesvorný 2018). The final stages of these processes left us with small

---

[1] A glossary of acronyms and technical terms can be found in Appendix F.

183

body populations both in relatively stable zones and on unstable orbits. Some of the latter found their way onto orbits where they could strike the planets and satellites (e.g., Zahnle et al. 2003; Dones et al. 2009; Bottke and Norman 2017). In addition, all these populations experienced mutual collisions before, during, and after these events. By using models constrained by the bombardment histories of all worlds, we can glean insights into how the planets, dwarf planets, and small body populations formed and evolved.

Many questions and gaps in our knowledge, however, still obscure our understanding of these fundamental planet-building processes. For example, the initial size distributions of planetesimals across the solar system are poorly understood, as are the physical processes that govern small body disruption. Many small body reservoirs might also be created/depleted by different phases of giant planet migration, but the details and precise timing are uncertain. Particular small body populations are recorded in the bombardment history of worlds with ancient surfaces (e.g., Mercury, Moon, Mars, and Iapetus), yet the nature of those populations at different times is unsettled and is difficult to disentangle from subsequent geological or cratering events. Impacts in the asteroid belt produce a collisional cascade of fragments, with many smaller bodies having their physical evolution affected still further by nongravitational forces, yet the relative importance of these processes in breaking down small bodies is unclear (e.g., Bottke et al. 2015). Many small bodies also escape their reservoirs and approach the Sun closely enough to shed mass or disrupt, yet the mechanisms describing this behavior are still enigmatic (Jewitt et al. 2015).

Q4.1a How Were the Last Surviving Planetesimal Populations in Regions Now Populated by Planets Affected by Collisional and Dynamical Evolution?

Planet formation models suggest that there were three major sources of early solar system bombardment (Bottke and Norman 2017). The first was the primordial Kuiper belt, a massive disk of ice-rich planetesimals that formed beyond the original orbits of the giant planets. At an unknown time, Neptune migrated outward through this population and scattered most of the planetesimals onto planet-crossing orbits, thereby triggering the dynamical instability of the giant planets (see Question 2; Tsiganis et al. 2005; Nesvorný and Morbidelli 2012; Nesvorný 2018). These former Kuiper belt planetesimals produced most early impacts on outer solar system worlds, and they were a strong component in the initial bombardment of terrestrial planets and asteroid belt as well. The impact signatures left behind also show intriguing signs that the primordial Kuiper belt population was affected by collisional evolution before, during, and after the giant planet instability (Morbidelli et al. 2021). Key unknowns in deducing the nature and early evolution of the primordial Kuiper belt are its initial size and orbital distributions, the timing of the giant planet instability, and how ice-rich planetesimals are disrupted by various mechanisms.

The bombardment produced by the dispersal of the primordial Kuiper belt is also ongoing. Many ice-rich bodies were left on long-lived but unstable orbits or were placed into the Oort cloud. Over billions of years, some of these bodies have escaped onto giant planet-crossing orbits, where they were able to hit numerous worlds. As shorthand, the committee refers to these escaped bodies as comets, although only some may ultimately develop tails. Accordingly, knowledge of what happened at early times is constrained by what is known of the present-day comet populations, what is known of craters and impact events on outer solar system worlds, and modeling work (Zahnle et al. 2003; Dones et al. 2009; Nesvorný 2018).

The second source of early solar system bombardment are the small bodies from the terrestrial and giant planets zones that were left on highly eccentric and inclined orbits by interactions with the growing planets, which are classified here as leftover planetesimals (Bottke et al. 2007; Morbidelli et al. 2018). These bodies experienced extensive collisional and dynamical evolution, and only a small number lasted for more than a few tens of millions of years. Some impactors even moved to new regions, with an unknown fraction of giant planet zone planetesimals moving into the inner solar system. The survivors, however, had high collision probabilities with the worlds in/near their formation zones, and this may allow them to dominate other sources of early bombardment. Their impacts may also have left behind critical traces of the unknown planetesimals that made the terrestrial planets and giant planet cores.

The third source, referred to here as asteroids, are bodies that formed or were captured near/within the main asteroid belt between Mars and Jupiter (Morbidelli et al. 2015). The effects of giant planet migration moved some asteroids onto planet-crossing orbits over a wide range of timescales, with some objects in mildly unstable zones hitting the Moon and terrestrial planets over a billion years later. Depending on the impact signatures left behind on various worlds, the size and nature of the initial asteroid populations can be discerned by modeling the dynamical process by which these bodies are transported out of the main asteroid belt. While these escaping bodies experienced a lesser degree of collisional evolution than leftover planetesimals, mainly because their collision probabilities are lower than objects that reside closer to the Sun, their net probabilities of striking inner solar system worlds are smaller as well (Bottke et al. 2007).

All three populations struck worlds in the inner solar system, although the degree to which different populations dominated at different times is uncertain. The signatures they produced may be found in a variety of early bombardment traces (e.g., the size distributions of ancient craters and remnant small body populations, meteorite shock degassing ages, impactor fragments, and trace compositions found within terrestrial and extraterrestrial samples).

Q4.1b How Has Collisional and Dynamical Evolution Affected Small Body Populations Now Found in Stable Reservoirs Within the Inner and Outer Solar Systems?

The events involved with planet formation and giant planet migration (see Questions 2 and 3; Chapters 5 and 6, respectively) left the solar system with two primary small body reservoirs within our observational reach: the asteroid and trans-neptunian belts, and several smaller ones, such as those populations captured in mean motion resonance with giant planets (e.g., the Trojan asteroids of Jupiter and Neptune and the Hilda asteroids). It also created a reservoir of comets beyond our observational reach—namely, the Oort cloud of comets (see Q2.6b)—and populations of bodies on planet-crossing orbits that are steadily replenished by the stable reservoirs, such as the near-Earth objects, the ecliptic comets (which include the Jupiter-family comets and the Centaurs), and the nearly isotropic comets (i.e., long-period comets). The committee focuses here on the asteroid and trans-neptunian belts.

For the main asteroid belt, numerous bodies were likely captured from nearly every solar system region (Bottke et al. 2006; Walsh et al. 2011; Morbidelli et al. 2015; Vokrouhlický et al. 2016; Nesvorný et al. 2017; Raymond and Izidoro 2017; see Question 3). This implies that some bodies experienced collisional and physical evolution prior to entering the main belt region. Once in the main belt, asteroids were potentially struck at early times by each other and external impactors, although the precise number of main belt bodies shattered or disrupted at early times is uncertain (e.g., Bottke et al. 2015).

Once the planets settled into their quasi-final configurations ~4 to 4.5 Ga, the asteroid belt was left with a population comparable to the current one. It can be argued that only a few tens of the largest asteroids (diameter, D >100 km) have been disrupted over the past several billions of years (Bottke et al. 2015). These events provided enough smaller fragments to keep the asteroid belt size distribution in a quasi-steady state, with its overall structure controlled by the nature of the disruption laws affecting different asteroid types. Evidence for this behavior may be found in the crater size distributions identified on ancient asteroid surfaces (Marchi et al. 2015).

Asteroids gain mobility in the main belt via collisions, encounters with larger asteroids, and, most importantly, by the Yarkovsky effect, defined as a nongravitational thermal force produced by the anisotropic reemission of energy from sunlight (see also Q4.1c). Over time, the Yarkovsky effect allows many asteroids smaller than a few tens of kilometers to "drift" into dynamical resonances with the planets, where they can be pushed onto planet-crossing orbits (Vokrouhlický et al. 2015). A small fraction then go on to strike the terrestrial planets.

Small asteroids can also have their spin rates and obliquities modified by the Yarkovsky-O'Keefe-Radzievskii-Paddack (YORP) effect, a net torque produced by solar photons that are absorbed and reemitted from a body (Vokrouhlický et al. 2015). Those with an asymmetric shape can have their rotation rates accelerated or decelerated, potentially causing spin rates on the orders of minutes for bodies tens of meters in size, or such slow spin rates that some asteroids enter tumbling rotational states. YORP torques also tilt asteroid spin vectors toward 0 degrees or 180 degrees, which maximizes how fast they migrate by the Yarkovsky effect. The combined behavior of the Yarkovsky/YORP effects make it possible to estimate the timing of many asteroid breakups from the dynamical evolution of their fragments. Accordingly, it is now possible to connect the evidence for putative impact surges on Earth, the Moon, and other terrestrial planets with specific asteroid breakup events from the past (e.g., Terada et al. 2020). This opens new areas of investigation for how changes in the history of various worlds, including life itself, might be linked to impact surges.

For the trans-neptunian belt, the extent of collisional evolution is dependent on the nature of the primordial population, when it was dynamically dispersed by a migrating Neptune (via the giant planet instability; see Questions 2 and 3), how it was collisionally bombarded by external populations, such as the numerous planetesimals residing in the giant planet zone, and the disruption law controlling these bodies (e.g., Morbidelli et al. 2021). Collision probabilities decrease as bodies move away from the Sun, though, so it seems likely the present-day size distribution for bodies larger than a few tens of kilometers was mainly set by planetesimal formation and dynamical depletion processes (see Q1.3b). It is possible, though, that collisional evolution occurring prior to the dynamical excitation of the primordial Kuiper belt influenced the shape of the present-day size distribution.

Curiously, cratering evidence on Pluto's moon, Charon, and the Kuiper belt object (486958) Arrokoth (Figure 2.20) suggests a paucity of impactors between a few tens of meters and 1 km (Singer et al. 2019). The reasons for this deficit are debated; perhaps larger disruptions produce few bodies in this size range, or the disruption law for ice-rich planetesimals produces a collisional equilibrium consistent with such trends. If the latter scenario holds, the size distribution for comets and TNOs would likely become substantially steeper for sizes smaller than a few tens of meters (Morbidelli et al. 2021). Testing this may require visits to additional large TNOs to see primary craters not influenced by superposed secondary impacts (i.e., the impact of fragments of rock or ice ejected during the formatter of larger craters), determination of how many small comets that strike Jupiter and other giant planets, or detection of small TNOs with serendipitous stellar occultations.

Q4.1c What Are the Life Cycles (Physical States and Rotational Properties) of Small Bodies in the Solar System and How Are They Affected by Collisions, Thermal Changes, and Nongravitational Forces?

Collisions are a primary geologic process for small body populations. They break down worlds and create new fragments that can also be disrupted by subsequent impacts. Collisional by-products include craters, asteroid satellites, fragments with rubble-pile structures, and so on. Major collisions also create swarms of fragments on similar orbits, called families, that can tell us about the nature of large-scale impact events occurring in the past. Using models, these kinds of constraints can be used to glean insights into how individual bodies and populations have evolved from their primordial states (e.g., Bottke et al. 2015).

For asteroids smaller than a few tens of kilometers, the evolutionary effects of impacts and gravitational perturbations are expanded to include the Yarkovsky and YORP effects (see Q4.1b) (Vokrouhlický et al. 2015). Concerning YORP, as a body's spin state evolves, its coupled rotational and translational motions become increasingly complex to model. The rubble pile nature of small bodies further complicates their evolution, allowing them to be spun fast enough that they reshape themselves or even fail (e.g., rotational mass shedding). If caught in the act of mass shedding, these bodies are referred to as active asteroids (Jewitt et al. 2015). These effects can lead to the formation of satellites that can evolve into long term stable systems. In turn, these systems are subject to complex internal dynamics driven by their mutual gravity and exogenous nongravitational effects akin to Yarkovsky and YORP (Margot et al. 2015; Walsh and Jacobson 2015).

A full understanding of how rubble pile asteroids evolve in response to Yarkovsky and YORP effects is lacking. While the specific YORP and Yarkovsky effects have been clearly documented—as best exemplified by OSIRIS-REx observations of (101955) Bennu (e.g., Farnocchia et al. 2021)—there are several aspects that are not understood yet that can dominate the eventual evolution of the mechanical state of these bodies.

There are hypothetical stable end states that such bodies may evolve to, preserving their mechanical structure and dynamical state over long-time spans. Conversely, there are other hypothesized effects which ultimately cause rubble piles to disaggregate into their constituent boulders and grains, which would then further evolve as monolithic bodies. The existence and efficiency of these different processes are unknown, yet are crucial to understanding the nature and age of the small bodies in the solar system, and the current rate at which they are created through catastrophic impacts.

Another set of active asteroids, sometimes called main belt comets, show sporadic cometary activity in the main belt (Jewitt et al. 2015). The best-known ones, such as 133P/Elst-Pizarro, are associated with asteroid families, and they show repeated activity near their perihelion. It is possible these bodies were formed by the disruption of a carbonaceous chondrite-like body with an ice-rich mantle or subsurface ocean. This might make some of the resultant family members agglomerations of ice and rock, thereby allowing near-surface ice to sublimate when exposed by impacts, YORP spin-up, or some other process.

It is often challenging to identify the specific process that led to mass loss on a given asteroid. Not only are there several mechanisms to choose from (e.g., rapid spins, released volatiles, impacts, and subdued surface activity for inactive comets), but in some cases, the active asteroid is too small to be resolved using existing telescopes. At present, dust-signature analysis and long-term observations of repeated activity are the primary tools available to distinguish between different mass loss mechanisms (Jewitt et al. 2015).

In parallel to asteroids, the spins and orbits of cometary bodies, are also driven in complex ways by non-gravitational effects. For cometary bodies, these effects are dominated by outgassing and mass ejection, which in general is more effective than solar photons (i.e., Yarkovsky effect) at changing an object's spin state or orbit. Owing to differences in their mechanical properties and the strength of the effects, the response of cometary bodies to rapid rotation and rotational fission are often markedly different than seen in rubble pile asteroids. For example, wide-binary comets are unknown and thought unlikely, even though observed comets include many contact-binary shapes (Keller et al. 2015). On the other hand, rings have been detected around sizable Centaurs (see Question 8, Chapter 11). The exploration of stable orbiting material around ice-rich small bodies is a fascinating problem.

How cometary bodies preserve these morphologies over their lifetimes and yet are susceptible to rapid dissolution requires improved understanding of the mechanics of these primitive bodies. Although these evolutionary effects may be unique to the current epoch of the solar system, similar physics and processes undoubtedly were present, and perhaps persistent, at earlier epochs of the solar system.

Our knowledge of the life cycles of other small body populations—including Trojans, Centaurs, and small trans-neptunian objects—are far more speculative than for asteroids and comets because they have not been as thoroughly explored by robotic missions, and telescopic observations can be more challenging owing to their distance. Small trans-neptunian objects may have comparatively tame life cycles, as the YORP and Yarkovsky effects are not strongly effective far from the Sun, and outgassing likely ceased long ago. This implies that Arrokoth, a ~20 km world with a flattened bilobate shape and enigmatic geologic features, may be largely a byproduct of planetesimal formation rather than post-formation processes (see Question 1, Chapter 4). It remains to be seen if Trojans and Centaurs, also potentially derived from the primordial Kuiper belt, are similar or whether their shapes are dominated by processes connected to Kuiper belt collisions and/or solar-driven outgassing (e.g., volatile outbursts, comet splitting events, and spin up effects).

Q4.1d Which Bodies and Processes Produce the Smallest Particles and How Do the Particles Evolve?

The smallest bodies in the solar system are often grouped as "dust," although the term is not well defined. For dynamical purposes, dust may be used as a stand-in for monolithic grains, allowing the definition to range from micron-sized shards to cm-sized pebbles, depending on the application. Across this size range, different nongravitational effects can affect their dynamics: relativistic Poynting-Robertson drag moves the smallest grains toward the Sun or, when created within a giant planet system, toward the giant planet in the system; larger grains are affected by YORP torques and Yarkovsky thermal drift forces. It is important to note that monolithic grains can spin very fast (much faster than an equivalently sized rubble pile), owing to their internal strength.

The existence and cataloged distribution of dust in the solar system has expanded greatly over time. They run the gamut from the identification of the Zodiacal dust cloud to the detection of fine dust on the surface of many asteroids and comets, to the discovery of ejected particles from (101955) Bennu's surface by the OSIRIS-REx mission (e.g., Jenniskens 2015; Lauretta et al. 2019).

Dust originating within the solar system can be produced by many mechanisms, including cratering and catastrophic impacts, thermal disruption events produced by a body's proximity to the Sun, electrostatic levitation, asteroidal rotational instability, and cometary outgassing. The subsequent orbital migration and accretion of these smallest grains, however, is largely unknown except for the simplest of circumstances. Further, to produce ejected dust grains at (101955) Bennu, fundamental questions remain on whether they are primarily owing to thermal processing, meteoroid impacts, or a synergetic combination of both processes (Lauretta et al. 2019).

The primary source of interplanetary dust particles (IDPs) in the inner solar system is thought to be from Jupiter-family comets that disrupt at relatively small perihelion distances (Nesvorný et al. 2008). (Jupiter-family comets are often defined as comets with short orbital periods, less than ~20 years, whose dynamics are dominated by Jupiter.) Smaller yet important IDP contributions come from nearly isotropic comets and asteroids (Nesvorný et al. 2008). Recent disruption or mass shedding events among small bodies, such as those associated with comet 2P/Encke or asteroid (3200) Phaethon, lead to meteor streams that can strike Earth, the Moon, and other bodies at specific times (Jenniskens 2015).

The meteoroid population produced by these sources is a major source of bombardment for spacecraft and celestial bodies. Dust particles can strike objects at high orbital speeds and are capable of breaking down rocks and producing regolith on airless bodies. On Mercury and the Moon, meteoroid impacts also create exospheres detectable by remote and/or in situ observations (e.g., Colaprete et al. 2015; Killen et al. 2015; Benna et al. 2019; Jauches et al. 2021). Micrometeoroid bombardment can also contribute to space weathering (see Q5.5). Understanding the full role of dust impacts in the evolution of airless planets and small body surfaces will require additional in situ observations and sample return missions.

In the outer solar system, many IDPs originate from collisions among TNOs, with active Centaurs, nearly isotropic comets, and the interstellar medium also making contributions (e.g., Poppe et al. 2019). For giant planet systems, one needs to also consider irregular satellites as a source of dust. For example, dust from Saturn's distant irregular satellite Phoebe likely explains the extreme albedo differences seen on Iapetus. In fact, if the irregular satellite populations were much larger in the past, as suggested by modeling work (e.g., Nesvorný 2018), their debris may explain why dark carbonaceous chondrite-like material coats the surfaces of ancient terrains on worlds ranging from Callisto to the uranian moons (Bottke et al. 2013).

Taken together, dust can not only tell us much about the nature of the original planetesimals in the solar system but also about exoplanetary disks and giant molecular clouds evolving within our galaxy.

Strategic Research for Q4.1

- **Determine how meteoroid bombardment can alter the surfaces and potentially produce exospheres on airless worlds** by characterizing dust populations across the solar system and determining their impact effects through laboratory studies, observations, and numerical experiments.
- **Determine the nature of impactors striking the most ancient regions of the Moon to constrain early bombardment populations** by returning samples of soils/breccias from lunar farside regions where ancient materials have been recently excavated by impacts.
- **Find evidence for the earliest terrestrial impactors** by searching for and analyzing impact spherule beds in early Earth terrains that tell us about massive impacts from the Hadean, Archean, and Proterozoic eons.
- **Constrain the early impact populations striking Mars** by identifying the oldest basins and impact structures (including basins that have been potentially erased by other geologic processes) and determining their age.
- **Determine the Kuiper belt size distribution for objects smaller than 30 meters, and by proxy constrain the production population striking the giant planet satellites**, by obtaining high resolution images of large (>50 km) KBOs, where secondaries and sesquinaries (i.e., those craters formed on planetary satellites by the impact of debris ejected into planetary orbit from larger impacts) are not factors, and by expanded TNO surveys from ground and space-based telescopes.
- **Determine the size-frequency distribution for comets smaller than a few tens of meters in diameter, thereby constraining the nature of the comet disruption law**, by monitoring impact flashes/events on the giant planets from ground-based observations or orbiting spacecraft.
- **Constrain the early dynamical history of the asteroid belt, the nature of impactor populations whose members were captured within existing small body reservoirs, and the nature of existing asteroid families** by characterizing the sizes, orbits, and compositions of small body populations within stable reservoirs and on planet-crossing orbits with a combination of ground- and space-based observations.

- **Benchmark the ages of asteroid families and the nature of family-forming events** by observing asteroid family members in situ, counting craters on their surfaces, and comparing their model ages to dynamical evolution models of how the family members evolve.
- **Determine how carbonaceous chondrite asteroids and comets disrupt as they approach the Sun** by observing primitive asteroids or comets at low perihelion and tracking how they evolve.
- **Determine the processes that create active asteroids and compare their relationship to comets** by in situ observations of active or recently active asteroids.

Q4.2 HOW DID IMPACT BOMBARDMENT VARY WITH TIME AND LOCATION IN THE SOLAR SYSTEM?

The history of impacts on different bodies varies across the solar system and is dependent on impactor populations that evolved over time. The fingerprints of these different impactor populations are recorded in the cratering record of planetary surfaces, so interpreting these crater populations can tell us about small body populations that might no longer exist. Additionally, untangling the temporal evolution of impact bombardment on different bodies has scientific value beyond impact studies, because the accumulated crater populations on landforms is the only way of estimating their age without returning samples and analyzing them in laboratories on Earth.

Despite the usefulness of crater records, there are many potential sources of error—including observational errors and errors caused by geologic disruptions—that need to be better understood if impact crater populations are to be used for reliable age determinations. Most critically, the contribution of secondary craters and sesquinary craters—together with the processes that degrade, relax, or remove craters—needs to be understood to fully analyze planetary crater populations (e.g., Zahnle et al. 2003). Additional analysis is needed to understand these factors across the solar system.

For the Moon, a provisional sample-calibrated impactor rate for objects smaller than a few hundred meters has been established for 3.9 to 3.0 Ga, but the rate of impact bombardment outside this epoch and projectile size range remains uncertain (e.g., Wilhelms 1987; see Bottke and Norman 2017) (Figure 7-1). It is common practice to use models to extrapolate lunar surface ages to other planets and moons, although this practice has yet to be tested by direct sample analysis. It is evident from observations of many planetary bodies that the first 0.5–1 billion years of solar system history had an impact bombardment rate that was much higher than it is at present. However, the sources and forms of this higher early impact flux remain in dispute. More robust geochronological measurements on the Moon and other worlds (e.g., Mercury, Mars, and large main belt asteroids) are needed to fill this gap.

Q4.2a What Small Body Populations Dominated the Early Bombardment of Worlds in the Inner and Outer Solar Systems?

As discussed in Q4.1a, there are three major populations that dominated early bombardment of solar system worlds: comets, leftover planetesimals, and asteroids. Their relative importance depends on how the impact flux from those populations changed with time as well as the context of the worlds in question, namely when they formed and what happened to them when they were struck.

Starting with comets, their impacts dominate the bombardment of outer solar system worlds, with the magnitude of the early flux dependent on the timing of giant planet migration and the size of the primordial Kuiper belt (see Question 2, Chapter 5). With that said, many large impacts may have taken place before certain worlds could achieve a stable crust that could record an impact over geologic time. It is also possible that large impact basins can be removed from icy worlds with high heat flow or near-surface oceans via viscous relaxation of topography (e.g., Marchi et al. 2016). Taken together, it is possible that the largest observed basins on Mercury, the Moon, Mars, large asteroids like Vesta and Ceres, and giant planet satellites did not form until many tens to hundreds of millions of years after accretion.

Early impacts, produced by comets liberated by the depletion of the primordial Kuiper belt (see Question 2) and leftover planetesimals in the terrestrial planet region (see Q4.1), had the potential to scramble and disrupt

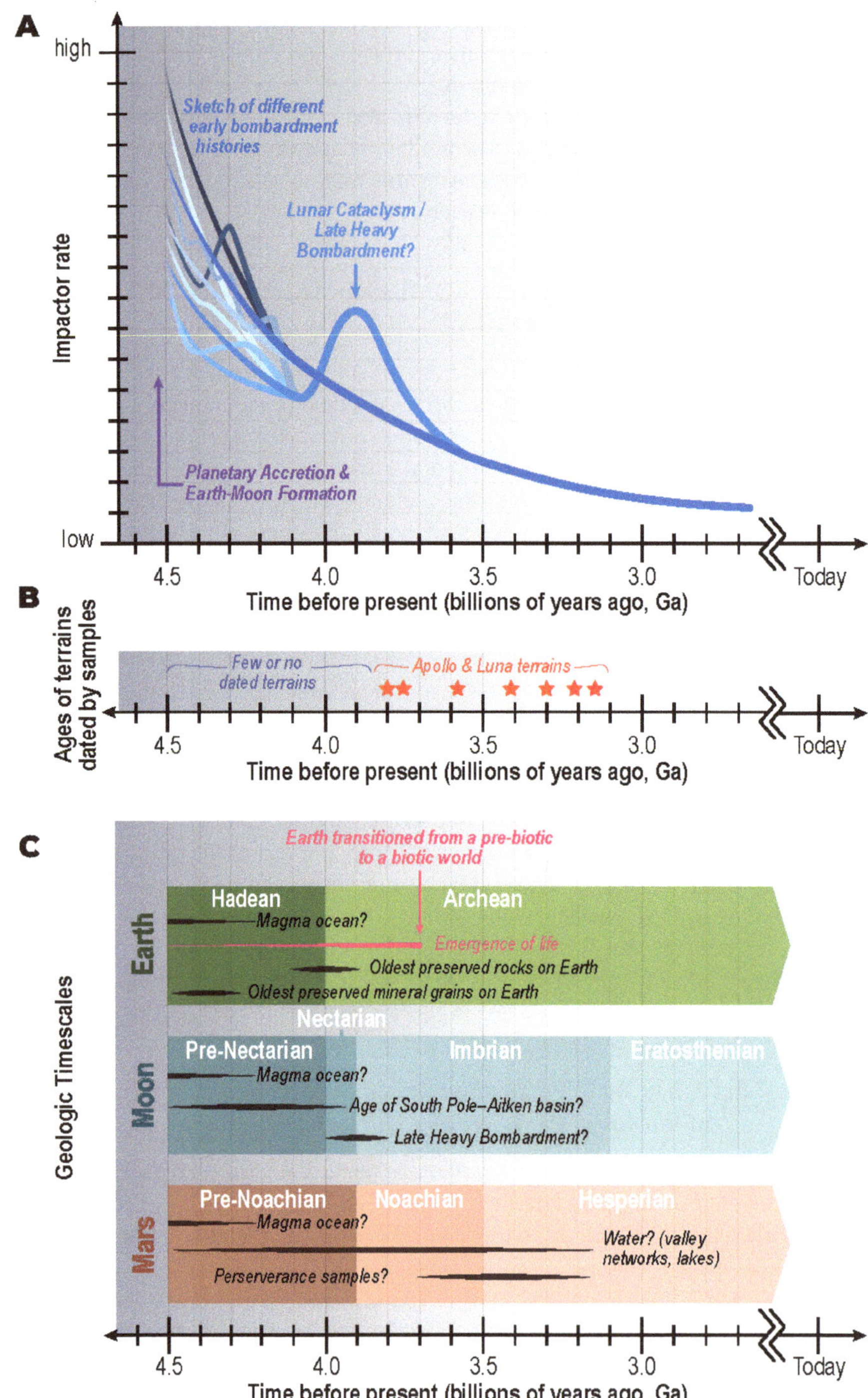

FIGURE 7-1 (*A*) Qualitative illustration of the early impact record of the Moon. As the curves go further back in time (to the left), possible bombardment histories (blue lines) diverge, revealing our lack of knowledge of the earliest epochs of solar system formation and evolution. (*B*) The ages of lunar mare terrains sampled by the Apollo and Luna missions (Stoffler and Ryder 2001). None are older than 3.9 billion years old, making it difficult to interpret what happened at early times. (*C*) A broad schematic of the timing of major events on Earth, the Moon, and Mars. Early bombardment was taking place when life was emerging on Earth (and possible elsewhere in the solar system). SOURCE: Courtesy of J.T. Keane.

main belt asteroids still warm from being heated by the decay of the aluminum isotope [26]Al (Q1.2c) (Bottke et al. 2015). Discriminating these primordial impact events from billions of years of subsequent impacts produced by main belt asteroids is a challenge that has yet to be met.

The signatures of cometary impacts on the terrestrial planets is unclear, except perhaps from constraints provided by noble gas and volatile isotope delivery, which would tell us about the very largest impactors (see Question 3, Chapter 6). If cometary impact basins exist in the inner solar system, the most accessible place to study them may be the ancient farside of the Moon—which has the oldest, most heavily cratered terrains. For leftover planetesimals, it is expected that their flux undergoes a steep decline as the most unstable leftovers are dynamically removed from the system and as the leftovers themselves undergo collisional evolution (Bottke et al. 2007; Nesvorný et al. 2017; Bottke and Norman 2017; Morbidelli et al. 2018). Despite this, some leftovers may be able to strike worlds after hundreds of millions of years of evolution. This suggests some ancient craters identified on Mercury, the Moon, and Mars may come from such impacts. It is also possible that the nature of these projectiles may be discerned from trace materials found within meteorites, lunar/martian samples, and ancient terrestrial rocks.

A poorly characterized component of the inner solar system impact flux is the leftover planetesimal population scattered in from the giant planet zone (Walsh et al. 2011; Morbidelli et al. 2015; Raymond and Izidoro 2017). The size and orbital distributions of these carbonaceous chondrite-like bodies will depend on models of giant planet formation, migration, and evolution (see Questions 2 and 3; Chapters 5 and 6, respectively). While this leftover component may explain volatile delivery to the terrestrial planets, the best existing constraint on this population may come from the quantity of carbonaceous chondrite-like bodies ostensibly captured within the asteroid belt (see Question 3).

Eventually, as the other bombardment sources fade, main belt asteroids will begin to dominate the impact flux of the terrestrial planets (e.g., Nesvorný et al. 2017). This transition for the production of km-sized and smaller craters is thought to take place between ~3 and 3.5 Ga. Even then, however, there are indications that meaningful impact surges took place in this early era, with compelling evidence provided by studies of impact spherules beds on Earth (e.g., Bottke and Norman 2017).

Q4.2b Was There a Late Heavy Bombardment Approximately 4 Billion Years Ago, and If So What Caused It and What Was Its Magnitude?

Shock age estimates on Apollo samples based on the ^{40}Ar-^{39}Ar isotopic systems show clustering around 4.1 to 3.8 Ga (see Figure 7-1). This led to the so-called late heavy bombardment (LHB) hypothesis, which suggests that the lunar surface was heavily bombarded during this short time window (see Wilhelms 1987; Bottke and Norman 2017). To constrain whether the LHB occurred, it is crucial to understand the full impact record of the Moon, given that the Moon is the only planetary object that provides a strong connection between absolute and relative ages based on age dating of lunar rock samples and crater counting. The intensity over time of the Moon's early impact record determines our understanding of the impact history of the solar system.

The LHB hypothesis has been supported by other isotopic measurements, including Rb-Sr, Sm-Nd, and U-Pb. If the LHB was a real uptick in the impact flux hundreds of millions of years after the end of planet formation, the late timing needs an explanation, because this is not expected from the declining remnants of planetesimals (Nesvorný et al. 2017; Morbidelli et al. 2018). This led, in part, to the formulation of the so-called Nice model. The Nice model originally proposed that the giant planets experienced an instability that led to the ejection of numerous comets and asteroids from their stable reservoirs near the proposed time interval (e.g., Gomes et al. 2005; Tsiganis et al. 2005; Nesvorný 2018; see Questions 2 and 3). While the timing of the instability is a free parameter in the model, and recent work favors an instability taking place shortly after the dissipation of the solar nebula (i.e., within a few million years to perhaps a few tens of millions of years; Nesvorný et al. 2018, 2021), a late instability could potentially explain the LHB, if it exists. Even so, there are substantial uncertainties in the magnitude, timing, and nature of the LHB, and a major challenge has been establishing reliable absolute age estimates for the large, early impact basins on the Moon and other bodies.

One of the arguments against the LHB is that the Imbrium basin could have distributed ejecta across much of the lunar nearside, and thus be dominating the observed impact record from the Apollo samples (which are all from the lunar nearside, potentially contaminated by Imbrium). Consequently, the apparent shock age clustering may represent mostly the Imbrium-forming impact, instead of multiple basin-forming events. Additionally, Boehnke and Harrison (2016) suggest that partial resetting of the ^{40}Ar–^{39}Ar system may have led to the observed clustering of ages in that system, which may not be traceable to a specific, narrow basin-forming period.

These challenges and uncertainties in the record of early lunar impact basins have implications across the inner solar system because the lunar crater chronology is used as a basis for understanding the absolute timing of events elsewhere. There are LHB-era constraints from other terrestrial bodies besides the Moon, but their interpretation is challenging (see Bottke and Norman 2017). For example, interpreting Earth's surface condition during the LHB is extremely limited because there are no preserved crustal rocks older than 4.03 billion years to date. The only accessible materials are detrital zircons, which are extremely resilient minerals that formed as long ago as 4.4 Ga. Whether these zircons tell us anything about Earth's ancient impact record, however, has been actively debated. The crater size distributions on Mars and the ^{40}Ar–^{39}Ar shock degassing ages of meteorites from asteroids provide further constraints, but whether they record the LHB remains unclear. New measurements of the ages of ancient basins on the Moon (particularly from the lunar farside, far from the contamination by Imbrium) and other worlds would be valuable to test the LHB hypothesis. Such measurements could be achieved either with new sample return missions or in situ geochronology.

Q4.2c Can Absolute Chronologies Be Derived for the Timing of Events Across the Solar System?

Determining the absolute timing of when events occurred on planets is important for inferring planets' geologic and geophysical histories, as well as the rate of planetary processes. Presently, we have a range of certainty about the timing of events in different settings, because we assess the age of materials and landforms using techniques with widely varying fidelity. Our understanding of planetary histories is often limited by uncertainty in the ages of major events or landforms. Examples include the unknown ages of large ancient impact basins like the South Pole–Aitken (SPA) basin on the Moon and Hellas basin on Mars, giant impact scars on large icy bodies, as well as the unknown average age of Venus's crust.

The cornerstone of geochronology is radiogenic dating of planetary materials in terrestrial laboratories. For lunar samples, ages derived using radiogenic techniques have been tied directly to observations of the crater populations on surfaces to create a model for the absolute cratering rate. With such a crater chronology model in hand, the ages of events and rates of processes can be assessed with crater statistics (i.e., how many craters are found that post-date a landform or geologic unit). For the Moon, the chronology model is arguably well-constrained from ~3.9 to 3 Ga, but is uncertain for impact basins older than Imbrium, a critical knowledge gap for understanding its early history (e.g., Q4.2b) (e.g., Wilhelms 1987; Stöffler and Ryder 2001). For the period younger than 3 Ga, the absolute chronology of lunar surfaces is also less well-established because few geologic units from that period have been dated with samples (e.g., Wilhelms 1987; Stöffler and Ryder 2001).

On planetary bodies besides the Moon, a direct connection between the radiogenic ages of samples and the observed crater population has yet to be established (Zahnle et al. 2003). There have been two primary strategies employed to assess absolute chronology in the absence of this absolute age calibration.

The first is to extrapolate the lunar absolute calibration to other planetary surfaces using models. At least in the inner solar system, the impactor population that affects planets is thought to be a common one, so the observed lunar impact flux can be corrected to determine an absolute chronology by accounting for differences in impact velocity, gravity, and target properties like porosity. The reliability of this extrapolation of the lunar record remains uncertain, however, and needs to be tested.

The second strategy requires determining the cratering rates from observed populations based on impactor dynamics and the physics of impact cratering. Where the lunar record is not expected to be representative, particularly in the outer solar system, only this second strategy can be employed to establish an absolute cratering chronology.

Substantial progress on these issues would be aided by directly measuring the absolute ages of additional samples from the Moon and Mars, either using returned materials or in situ dating methods. A requirement for this type of investigation is to obtain material that can be placed in geologic context; ages derived from lunar or martian meteorites are useful but cannot be fully exploited because they have an unknown provenance. Additional pathways to enhance current knowledge would be to obtain robust techniques for measuring planetary crater populations accurately, improving our understanding of how small body populations evolve and the impact cratering process itself, and establishing independent constraints on outer solar system chronologies.

Q4.2d When Does Recorded History Start and When Did Major Geological and Geophysical Events Occur on Different Worlds?

The dating of major events on planetary bodies is fundamentally important, whether constraining the initial stability of an ancient surface or finding the timing and duration of an internal activity such as volcanism. Ages of these events are usually estimated by the comparison of the spatial density of craters and estimated projectile fluxes, calibrated where possible by lunar sample ages and various theoretical constraints of how the impactor rates vary with time and location in the solar system.

For the Moon, determination of the oldest preserved terrains or events is possible if the right samples can be obtained. For example, SPA basin is likely the oldest distinct impact basin from a stratigraphic standpoint (Bottke and Norman 2017). Samples that date SPA would thus answer whether SPA formed shortly after the Moon formed, implying much of the Moon's early history (and early solar system bombardment history) can be constrained, or that SPA formed hundreds of millions of years later, suggesting that much of the Moon's early history has been erased. Reliable age determinations of lunar impact features in poorly sampled eras would substantially improve our calibration of the impact flux and thus ages of major events across the Moon and solar system.

Model ages of events on Mercury, Venus, and Mars are presently dependent on extrapolation of the lunar (and terrestrial) cratering record, which limits our ability to probe their early and more recent histories. This approach depends on theoretical assumptions and is not yet calibrated by independent constraints (e.g., lunar and Mars meteorite samples have yet to be conclusively linked to a given terrain). Accordingly, like SPA, the oldest basins on Mercury and Mars have indeterminate ages; they could be nearly as old as the planets themselves or hundreds of millions of years younger. For Venus, there is considerable debate about how to interpret the distribution of sizable craters across its surface, which are indistinguishable from random. They either tell a story of catastrophic resurfacing within the past billion years or regional volcanism that is continually erasing swaths of craters (Smrekar et al. 2018). Near-term progress on the chronology of these worlds will likely come from strategically acquired samples from the Moon and Mars, in situ dating of terrains on the Moon and Mars, and/or confirmation that certain dated meteorites come from particular source terrains.

For a given surface, impacts increase the crater population but eventually start erasing existing ones. When the rate of crater production eventually balances with that of crater erasure, the crater population apparently remains in a steady state, referred to as crater saturation equilibrium (e.g., Melosh 1989). This process can make it difficult or even impossible to determine accurate surface ages. For the Moon, researchers debate whether the global crater population up to 100 km in diameter on the most ancient surfaces is in crater saturation equilibrium, while it is in general agreed that craters less than ~200 m in diameter may be at crater saturation equilibrium at local scales on the lunar maria. To quantify when, where, and how crater saturation equilibrium occurs, cross-disciplinary studies may be needed (i.e., a combination of modeling, experimental, and geomorphological approaches).

Determining the ages of the oldest terrains and major events in the outer solar system is even more problematic (e.g., Zahnle et al. 2003; Dones et al. 2009). Impact rates on icy satellites and within the TNO region remain uncalibrated by sample-derived ages. Independent constraints on the impact flux over time mainly come from the theoretical modeling results linked to craters on outer solar system worlds and present-day observations of various small body populations. Even there, crater counting and telescopic observations are limited at smaller sizes by data resolution and completeness.

Interpretations of the impact records for outer solar system worlds are full of challenges. There is a growing recognition that many icy satellites may not retain a full record of early impact basins. As demonstrated by Ceres, worlds with subsurface oceans or zones of weakness may allow viscous relaxation to erase large basins (e.g., Marchi et al. 2016). Impact disruption and/or heating events may also contribute to basin erasure. A second problem for planetary satellites stems from how to interpret the contributions from impactors that orbit the planet (planetocentric), rather than the Sun. Planetocentric impactors can arise from disruption events, basin ejecta, or other processes (e.g., Zahnle et al. 2003; Dones et al. 2009). These populations may dominate small crater production on many icy satellites, yet they are both poorly understood and may be different in each planetary system.

Ultimately, to interpret the earliest history of outer solar system worlds and reduce uncertainties, we need improved models of small body evolution, crater production, crater retention, and new observations of surfaces that have only been poorly explored to date.

Q4.2e What Is the Current Impact Flux on Planetary Worlds, and Has the Flux Changed Substantially over the Past Several Billion Years?

The current impact flux of large bodies across the solar system is largely determined by dynamical models constrained by the direct observations of small bodies. Ground- and space-based surveys have now discovered most km-size asteroids on planet-crossing orbits within the inner solar system and have made considerable progress on understanding comet populations in the outer solar system as well.

For inner solar system worlds, modeling results predict that most impacts come from asteroids; comets are only a minor contributor (Granvik et al. 2018). When asteroids smaller than a few tens of kilometers are formed, they begin to drift inward toward and outward away from the Sun by Yarkovsky/YORP thermal forces, with direction controlled by the body's obliquity (Q4.1b; Vokrouhlický et al. 2015). The process is slow enough that collisional evolution and mass shedding via YORP spin-up processes often disrupt these bodies, although the surviving fragments will continue to migrate via Yarkovsky/YORP forces. Eventually, these collisional cascades will deliver bodies to a main belt escape route, such as a powerful resonance with the giant planets, where they can be driven onto orbits from where they can strike the terrestrial planets. This process keeps the planet-crossing asteroid population fairly steady over time, and it helps explain why the main belt, planet-crossing asteroids, and terrestrial crater populations have similar size frequency distributions (e.g., Bottke et al. 2015).

Large asteroid breakup events in the asteroid belt near powerful resonances can change the terrestrial planet impact flux, but only if the fragments can get out of the asteroid belt before they are decimated by collisional evolution. One example of this is the destruction of the L-chondrite parent asteroid that sent numerous small particles to Earth ~470 Ma (e.g., Terfelt and Schmitz 2021). Other breakups produce so many fragments that they can substantially increase the delivery rate of asteroids to the terrestrial planets for many hundreds of million years. Estimates of lunar crater ages and terrestrial impact spherule ages suggest that several impact surges may have taken place over the past few billions of years (e.g., Terada et al. 2020).

Direct detection of fresh impacts into Earth's atmosphere or onto other nearby worlds is also now possible. For example, repeat imaging has enabled discovery of new impact craters on the Moon and Mars, allowing estimates of the present-day impact flux. Telescopic monitoring for impact flashes on the lunar near-side has also enabled detection of craters in the process of formation. Ongoing telescopic and/or space-based detection of impact flashes during the next decade would be useful to refine the impact flux, better understand the impact process, and would be potentially synergistic with other geophysical observations (e.g., seismology).

The heliocentric impact flux on the giant planets and their satellites is largely set by the escape rate of comets from the TNO region, with a small fraction reaching giant planet-crossing orbits over time. While they keep the ecliptic comet populations in a short-term steady state, this population has been steadily decreasing over billions of years (Nesvorný 2018).

A critical issue today is how to use these model results to estimate absolute surface ages on icy worlds when constraints are limited (Zahnle et al. 2003). Existing model benchmarks include the observed comet populations and the observed present-day impact flux on the giant planets. Unfortunately, at present, these constraints yield results that differ by a factor of ~10 (e.g., Zahnle et al. 2003; Dones et al. 2009; Nesvorný et al. 2019). The reasons are unclear, but it could be because cometary activity makes it difficult to estimate comet nuclei sizes. Two solutions for the next decade would be to find and characterize additional icy bodies far from the Sun, where activity is limited, and to monitor bolide impacts into giant planet atmospheres or any other large witness plates (e.g., Saturn's rings).

With that said, though, a good impact flux model will still leave us with challenges in estimating absolute surface ages. For example, as discussed below, comets striking giant planet satellites often produce enormous ejecta showers, which in turn produce numerous secondary and sesquinary craters. The contribution of these types of craters to small crater populations on a wide range of terrains is uncertain, partly because impact events produce variable ejecta size distributions but also because our understanding of small comet populations is limited. New observational data and laboratory/modeling work are needed to determine more reliable surface ages in the outer system from the spatial densities of small craters.

Strategic Research for Q4.2

- **Determine the age of the SPA basin to determine the beginning of recorded bombardment on the ancient lunar farside** by dating samples formed from or excavated by the SPA basin forming event.
- **Determine a precise absolute chronology for lunar impactors that can be applied to other worlds** by measuring radiometric ages for terrains likely to be much older than 3.9 Ga and younger than 3 Ga, counting superposed small craters on D >10 km craters, and calculating model ages for those craters.
- **Determine the present-day lunar impact rate and better understand the nature of impact mechanics** by coupling seismic monitoring to lunar observations of impact flashes and fresh impact craters.
- **Determine impactor sizes, impactor compositions, and impact ages for all terrestrial impact structures** by identifying and characterizing past impact structures, analyzing samples that contain telltale traces of the projectile, and dating material affected by the impact event.
- **Determine the absolute age of a martian basin or well-defined surface and use it to calibrate the timing of early martian bombardment** by dating a surface whose age can be determined by in situ methods or returned samples.
- **Use crater counts on the ancient surfaces of Kuiper belt objects to constrain the populations of the primordial Kuiper belt and determine the start of the giant planet instability (and post-nebula giant planet migration)** by mapping craters on larger (>100 km) KBO and using them to constrain early bombardment models.
- **Pursue crater counting on icy bodies of the outer solar system to characterize and compare how projectile populations vary radially from the Sun (especially focused on gaps in the observational record on Europa, Ganymede, in the uranian system, and on additional >100 km trans-neptunian objects)** by observing icy satellite surfaces that have yet to be imaged and at higher resolutions than existing images.
- **Determine the nature of early bombardment and the primordial asteroid belt** by observing large intact asteroids that may still have some record of impacts/craters from early bombardment phases, counting craters, and modeling their crater size distributions.
- **Constrain the integrated bombardment history of the asteroid belt over the past few billion years** by performing in situ dating of the largest basins, such as those for which we have samples (e.g., Vesta's Rheasilvia impact basin) whose model age can also be derived from superposed crater records.

Q4.3 HOW DID COLLISIONS AFFECT THE GEOLOGICAL, GEOPHYSICAL, AND GEOCHEMICAL EVOLUTION AND PROPERTIES OF PLANETARY BODIES?

Collisional and accretional events were an integral aspect of the formation of planetary bodies and exercised a controlling influence on their physical and chemical states throughout their evolution (Melosh 1989). In the context of modern-day Earth, collisions and impacts are often viewed as destructive events. Yet collisions have had a profoundly constructive role in the formation and evolution of planetary bodies, beginning with their bulk composition and extending through to the timing and duration of differentiation and exogenous delivery of chemical ingredients essential to life (e.g., Osinski et al. 2020). The scars of collisional events—impact craters—litter the surfaces of planetary bodies across the solar system and chronicle the role collisional events had in shaping large-scale geology, volcanism and magmatism, atmospheres, and core dynamos and magnetic fields.

As we improve our understanding of the formation and evolution of planetary bodies, we continue to discover that collisions were often—although not always—responsible for major planetary-scale events. One of the most poignant examples is the formation of the Earth–Moon system (see Question 3, Chapter 6), canonically considered to be the result of a large-scale collision event early in the solar system history. Additionally, several planetary bodies—such as the Moon, Mars, and Pluto—possess ancient hemispheric asymmetries often hypothesized to have originated with large-scale impact events (e.g., Andrews-Hanna et al. 2008). These hemispheric asymmetries profoundly shape the evolution of these planetary bodies and often plausibly give rise to subsequent hemispheric asymmetries in tectonic and magmatic activity, as is evident for the Moon and Mars. Additionally, collisions can alter the orbital parameters and rotation state of planetary bodies, leave lasting consequences on the surfaces and near-surfaces of planetary bodies (Question 5, Chapter 8), are capable of generating circumplanetary systems (Question 8, Chapter 11), and may even potentially affect core dynamo activity within a planetary body (Question 5).

Q4.3a How Did the Earliest and/or Largest Impact Events Influence the Physical Evolution of Solar System Worlds?

The initial conditions of planets and small bodies are determined by accretional processes, while their subsequent evolution is often shaped by impacts. Among the most famous examples are Earth's Moon (see Question 3) and Pluto's moon Charon (see Question 2, Chapter 5), thought to have formed by large impacts. In the case of the Moon, this event would have provided enough energy to melt a large portion and form a deep magma ocean (see Question 3). Subsequently, the Moon crystallized, allowing less dense minerals (anorthosites) to separate from the residual melt and float to the surface and forming the lunar highlands.

The surface topography of many worlds has also been shaped by numerous impacts at the largest scales, but uncertainties remain. For example, the farside of the Moon has a thicker crust (50–60 km) than the nearside (30–40 km; Wieczorek et al. 2013); on Mars, the crust on the northern hemisphere, where the 10,000 km Borealis basin is located, is much thinner than that on the southern hemisphere (32 versus 58 km, respectively; Andrews-Hanna et al. 2008; Goossens et al. 2017) (see Question 3); and Enceladus possesses a thinner crust in the south pole, associated with the tectonic fractures that source its enormous plume (Hemingway et al. 2018) (see Question 8, Chapter 11). The origin of these dichotomies is still debated and might be explained by impact-induced mantle convection, a large impact transporting the crustal material from one hemisphere to another, or alternative nonimpact hypotheses. One way to constrain this issue may be to explore the martian moons Phobos and Deimos that were plausibly formed by this impact event (see Question 3).

Mercury may represent another world that was substantially affected by a giant impact. Mercury's high density may be explained by a large impact that removed its original crust and much of its mantle (see Question 3). Others suggest that Mercury's building blocks were simply iron-rich. Given that high density planets have been found in extrasolar systems, it is now even more important to determine the origin of Mercury's distinctively high density.

In the asteroid belt and TNO populations, many bodies have been shattered, disrupted, and scrambled by large impacts (Bottke et al. 2015; Nesvorný et al. 2018). In the process, interior materials normally hidden away at depth are now potentially accessible, some on small bodies that may eventually approach Earth. Accordingly, by interpreting the jigsaw puzzles created by impacts, and placing their samples into geologic context, we can probe the origin and evolution of planetesimals to a much greater extent than would be possible with intact bodies.

Large impact events may also have disrupted mid-size icy satellites in the saturnian and uranian systems (e.g., Movshovitz et al. 2015). Given that basins on Iapetus can exceed 500 km diameter, and comparable events would be expected on the inner satellites, it seems plausible that some events completely shattered protosatellites like Mimas or Enceladus, which could reassemble into the current satellites. Whether such events have occurred, or did so multiple times, is not known but could be revealed by detailed mapping of internal mass distributions or by other means. The reader is referred to Question 8 for additional discussion about circumplanetary systems.

Q4.3b How Do Impacts Affect Surface and Near-Surface Properties of Solar System Worlds?

Impacts modify surface morphologies on planetary bodies (see Q5.5b) (Melosh 1989). These modification processes mainly result from crater excavation, ejecta blanketing, and topographic diffusion. They occur on many scales ranging from a microscale, including small boulder erosion, up to surface modification on heavily bombarded planets and moons. While all of these surface modification processes directly connect the history of each planetary body, their stories are incomplete.

Small bodies have low surface gravitational acceleration, so their surface conditions are affected by impacts, dynamic processes, or both. As one example, outgassing and mass ejection events on comet 67P/Churyumov-Gerasimenko have led to the collapse of cliffs and surface changes from the reaccumulation of ejected dust (Pajola et al. 2017). A second example is impact-induced seismic shaking, likely responsible for substantial depletions in small crater populations on asteroids tens of km or smaller across (e.g., Marchi et al. 2015). A third example comes from the asteroids (101955) Bennu and (162173) Ryugu, both of which exhibit surface material flows that follow trends expected from YORP-driven spin-up (e.g., Jawin et al. 2020). A fourth example is Vesta's Rheasilvia basin, whose unique central peak and escarpment were created by a combination of the impact itself and the asteroid's spin. More work is required to understand how impact- and dynamical-related processes trigger mass wasting and crater erasure events on small bodies.

Impact-induced melt sheets play essential roles in altering the target structure, age determination, and inducing hydrothermal activities, and can affect craters formed on top of them. Melt sheets are commonly found in impact basins (Melosh 1989). If a deep melt sheet cools slowly, it can differentiate and form a compositionally stratified structure, altering the original target structure. This behavior is observed in the Sudbury impact basin, and it likely also occurred within the SPA basin on the Moon. Such melt sheets may also affect the later formation of superposed craters, making it more challenging to use crater spatial densities to estimate the ages of their source craters. With that said, melt sheets are ideal locations for sample return missions because the rocks provide accurate and unique impact ages. Melt sheets can also offer a long-term heat source, which contributes to forming hydrothermal systems if subsurface volatiles are present (Q3.4d).

Last, there are cases where planetary atmospheres are sufficiently dense to prevent asteroids and comets from reaching the surface prior to disruption (e.g., Venus, Earth, and Titan). For the case of Venus, radar-dark splotches observed by Magellan could be byproducts of bolide airbursts, with the shock wave from the blast producing fine-grained material on the surface. Modestly larger projectiles that can penetrate more deeply in the atmosphere may produce a shotgun blast of small craters. Insights into how small bolides break apart in atmospheric disruption events can also be gleaned from observations of comet impacts into giant planet atmospheres (e.g., fragments of comet Shoemaker-Levy 9 hitting Jupiter in 1994; small comet impacts into Jupiter whose effects can be observed by both amateur and professional astronomers; Huseo et al. 2018). Ultimately, the effects of bolide disruptions in all atmospheres require further research, in that they can tell us about the strength/nature of asteroids/comets and how atmospheres/surfaces react to energetic blast events.

Q4.3c How Do Impacts Affect the Deep Interior of Solar System Worlds, and Can They Expose Interior Compositions?

Large basin-forming impacts have taken place on many planetary bodies in the solar system (Figure 7-2) (see also Q4.3a). Our direct knowledge of their effects on the interiors of worlds is limited, but models provide several intriguing results. For example, models suggest that major impacts can generate substantial thermal anomalies in the interiors of certain planetary bodies. This can potentially reorganize convection patterns and may even induce the formation of plumes that stretch from the core-mantle boundary to the surface (O'Neill et al. 2020). Large thermal anomalies may even reduce the efficacy of core dynamos (e.g., Roberts and Arkani-Hamed 2012, 2017). For example, it has been argued that the large basin forming impact on Mars heated the outer part of the core and made it thermally stratified, temporarily shutting off martian core convection and its dynamo (see also Question 5).

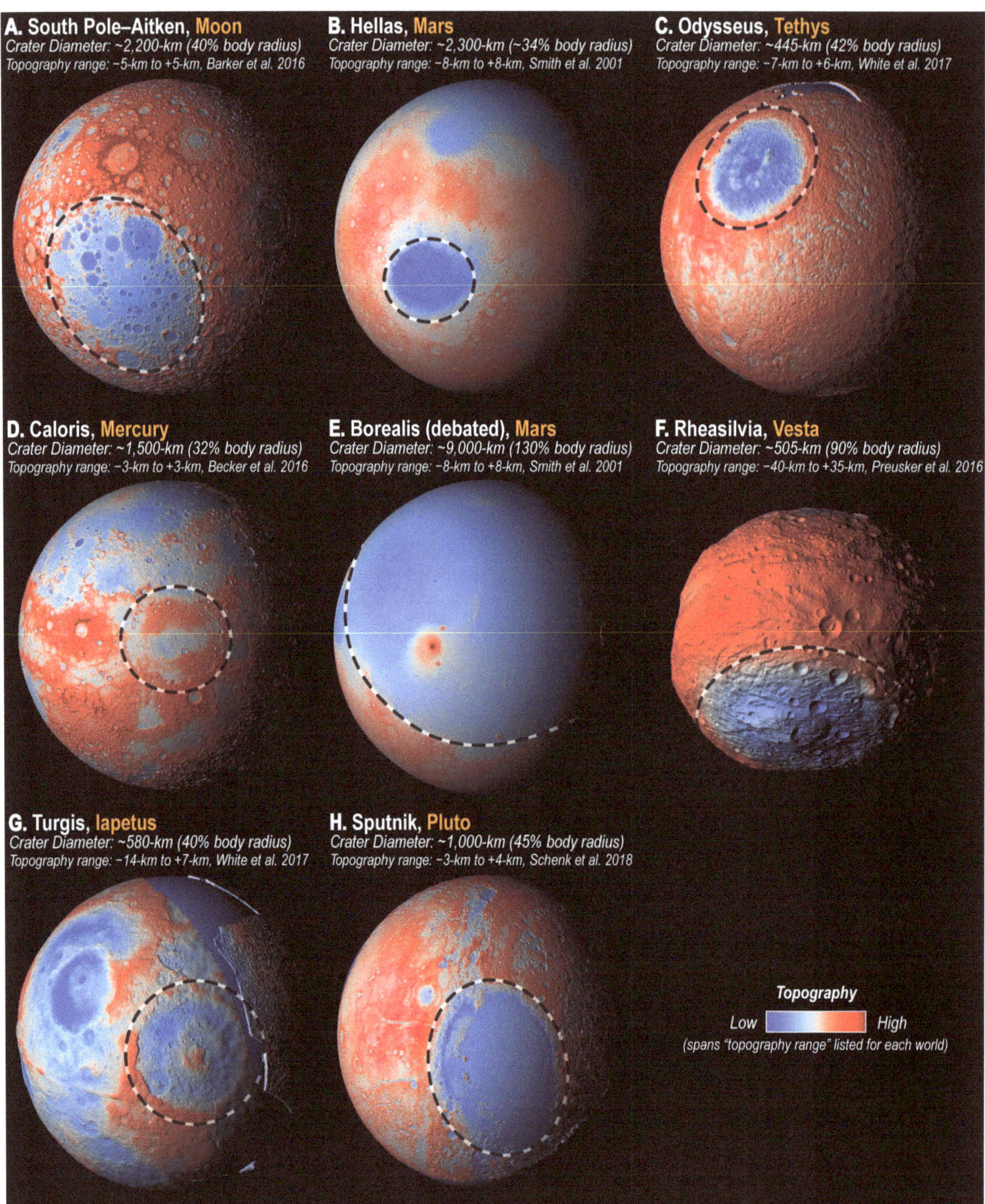

FIGURE 7-2 Topographic perspective views of planetary bodies highlighting many of the large impact events (dashed circles) that have scarred their surfaces and possibly altered their interiors. Maps are color coded to lows as blue and reds as high. SOURCE: Courtesy of J.T. Keane.

Heating produced by large impact events has also been proposed to explain the putative differences in the interior structures of Ganymede and Callisto (see Questions 5 and 8; Chapters 8 and 11, respectively). Ganymede, being closer to Jupiter, should have been hit by more impactors (and with higher impact speeds) than Callisto owing to the gravitational focusing of Jupiter. This may have triggered planetary differentiation in Ganymede while leaving Callisto in a partially differentiated state (Barr and Canup 2010).

Additionally, impact events are capable of excavating material from substantial depths (e.g., Melosh et al. 2017). For example, SPA may have excavated the early lunar mantle, which would provide an unprecedented view into the interior structure and thermochemical evolution of the Moon. More extreme cases of excavation have been suggested for worlds like Mercury and asteroid (16) Psyche, proposed to be remnants of much larger planetary bodies that were stripped in hit-and-run collisions (Asphaug and Reufer 2014). Understanding the viability of such scenarios would provide insights into the magnitude and type of collisional events that occurred early in solar system history.

Q4.3d How Frequently Are Impact-Driven Hydrothermal Systems Formed and Which Processes and Target Properties Control Their Evolution?

Impacts have many ways of changing a planetary surface. Not only do they create craters and ejecta, but they also fragment the surface, changing its porosity, and produce heat that can mobilize volatiles (like water and CO_2). This residual impact heat can lead to the formation of hydrothermal systems beneath impact craters. Through permeable channels, such systems are thought to be possible abodes for life for Earth, Mars, and possibly other worlds such as Ceres. Thus, impact-driven hydrothermal systems may strongly influence geological, geochemical, and biological evolution, especially on those worlds whose internal heat flows and volcanic activity may be modest (see also Q5.3d and Question 10, Chapter 13).

Hydrothermal systems have been observed in ~80 craters out of 180 craters on Earth (see Osinski et al. 2020), where the crater size ranges from 1.8 km to 250 km. They can exist anywhere in the crater, but most commonly are found in the central uplift, crater-fill, ejecta, and rim, and have been identified in drilled core samples from the Chicxulub impact crater (Kring et al. 2021).

Impact heating can be limited when projectile velocities are less than 10 km/s (e.g., Marchi et al. 2013), potentially explaining why little impact melt is seen on Vesta or within asteroidal meteorites, because asteroids in the main belt tend to have impact speeds near 5 km/s. For higher impact velocities, common on Earth because of its 11.2 km/s escape velocity, heating correlates with kinetic energy, and this means large craters on Earth can have long-lived hydrothermal systems. For example, it could take ~1 million years for a Sudbury-sized impact crater to cool.

In addition to endogenic hydrothermal activity, Sudbury-like hydrothermal systems may have been ubiquitous on many planetary surfaces hosting water. For example, given the likely existence of ice in the martian subsurface (e.g., Bramson et al. 2017; Dundas et al. 2018), it is expected that large martian craters hosted numerous hydrothermal systems. Indeed, associations between hydrated rock phases and martian craters have already been identified using spectral data. Another example may be the bright spots on Ceres, which are likely salt and carbonate deposits extruded through channels formed by an impact (Castillo-Rogez et al. 2019). Future sample return missions may provide key constraints on this issue.

The frequency of impacts large enough to sustain hydrothermal processes and environments has potential importance for the origin of life on several water-rich worlds. Large impact events were common in the first billion years, decreasing afterward. During these early times, it is likely that links can be found between hydrothermal processes and the alteration of planetary interiors, heating of planetary crusts, delivery of volatiles, formation of atmospheres, and other processes. But as seen in Q4.2 and Q4.3, there remain significant uncertainties in the flux of impactors during the early epochs of solar system evolution. To understand the role of larger impacts in the development of viable biospheres, we need to understand how early bombardment affected the history of the solar system. Determining how impact parameters and surface geology controlled such environments necessitates a deeper understanding of the mineralogy and chemistry (Figure 7-3) of such deposits on different planets.

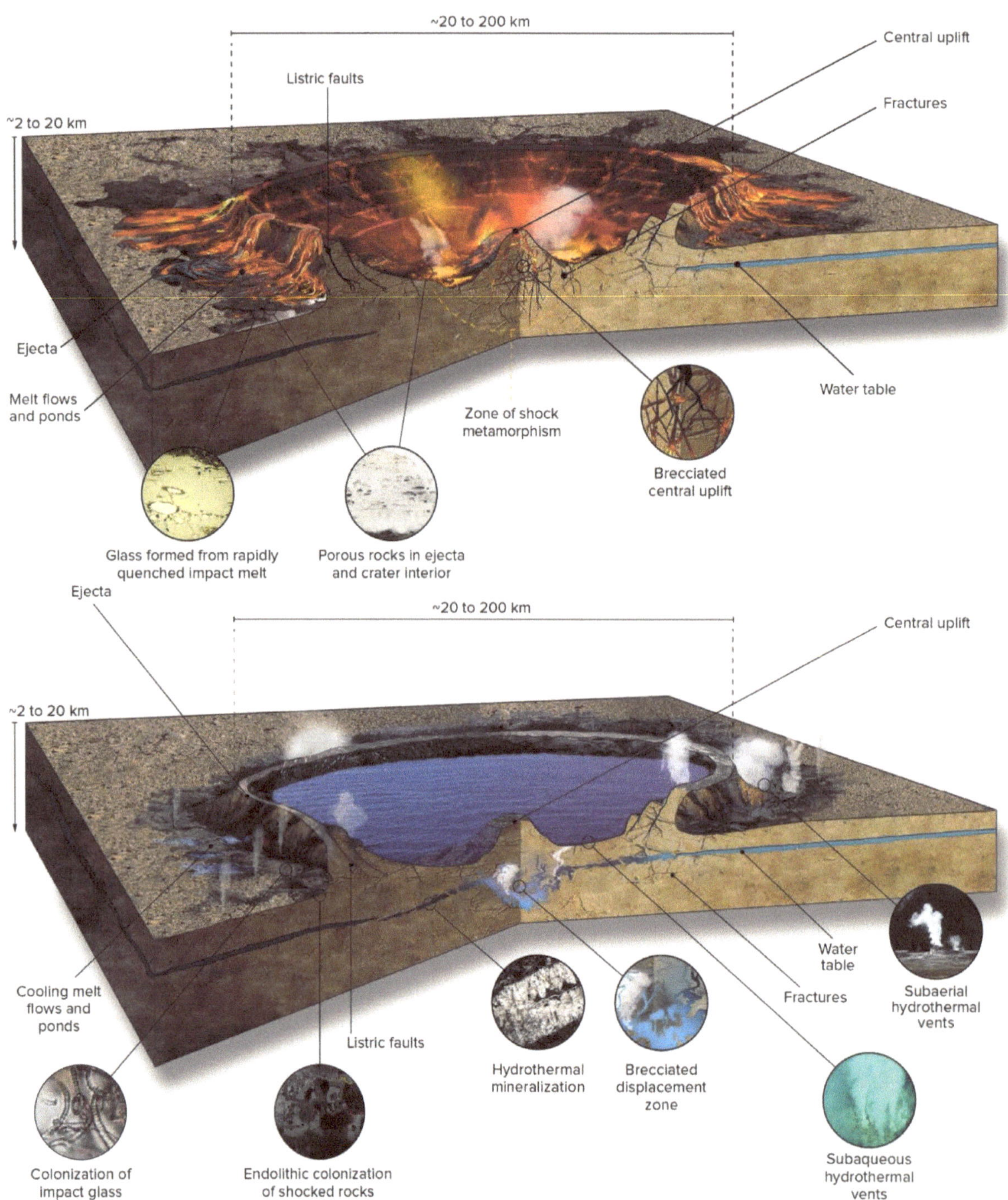

FIGURE 7-3 *Top panel:* An artist's impression of a complex crater shortly after an impact. The crater is fractured by the impact and filled with impact melt. The ballistic components of the ejecta external to the rim (including the ejecta blanket and secondary craters) are omitted for clarity. *Bottom panel:* Schematic view of a hydrothermal system developed after cooling. This can provide a suitable environment for life. SOURCE: Adapted from G.R. Osinski, C.S. Cockell, A. Pontefract, and H.M. Sapers, 2020, "The Role of Meteorite Impacts in the Origin of Life," *Astrobiology* 20(9):1121–1149, https://doi.org/10.1089/ast.2019.2203. © G.R. Osinski et al., 2020. Published by Mary Ann Liebert, Inc. CC BY 4.0.

Q4.3e What Exogenic Volatile and Nonvolatile Materials Are Delivered to Planetary Bodies?

Impactors deliver materials from one world to another, although the extent to which exogenic materials survive an impact event is not well known. Hypervelocity impacts are thought to result in the vaporization of most of the impactor, but this may vary depending on the impactor's velocity, composition, structure, the nature of the target material, and whether the target body has an atmosphere (see Q4.4). For example, bolides with modest mechanical strengths often manage to land on Earth and Mars in the form of meteorites. For larger asteroid or comet strikes on Earth, however, the recovered fragments tend to be tiny.

Fragments of exogenic materials have been observed in lunar samples. The most common type is carbonaceous chondrite-like materials delivered by meteoroids. This steady rain of tiny impactors explains why the lunar regolith is composed of ~2 percent exogenic materials (e.g., Heiken et al. 1991). An investigation of early Earth samples retained as meteorites on the Moon, such as the putative terrestrial meteorite fragment in an Apollo 14 breccia (Bellucci et al. 2019), may provide unique insights into the evolution of our own planet during a time for which scant evidence remains. Evidence for larger-scale impactor debris is also found on the Moon; examples would include the traces of highly siderophile elements found within returned samples (see Question 3, Chapter 6).

Impact velocities are lower among main belt asteroids than on the terrestrial planets, such that projectiles are more likely to survive as small fragments during cratering and catastrophic disruption events. This means the bombardment of a large main belt asteroid by other asteroids will lead to increasing concentrations of exogenic materials near the surface of that body. The mixing of these materials in the regolith, and the subsequent formation of breccias, could explain why exogenic clasts are found in many kinds of meteorites. More broadly, much of the northern hemisphere of Vesta is littered with exogenic carbonaceous material (e.g., Marchi et al. 2015). Apparently impact heating was low enough to preserve much of the volatile content of this debris.

The disruption of a main belt parent body with exogenic materials will produce smaller asteroids with varying concentrations of foreign fragments. The exogenous materials found on the surfaces of Bennu and Ryugu (DellaGiustina et al. 2020; Tatsumi et al. 2021), as well as the plethora of meteorite types associated with the progenitor of the Almahata Sitta bolide (e.g., Goodrich et al. 2015), are telltale signs of impact mixing among main belt asteroids. Similar mixing is likely in the outer solar system, modulated by the lower impact rates.

Comets, primitive asteroids, or meteoroids have also delivered volatiles and organics to the inner solar system over many billions of years (see Question 5, Chapter 8). It is possible that at least some ices preserved at cryogenic temperatures within the permanently shadowed regions near the poles of the Moon and Mercury are from these sources, although volcanic outgassing and/or solar wind implantation are potential competitors. The composition of these ices could provide us with important clues to deduce their origin. The polar volatile inventories of the Moon are also much smaller than Mercury, and it is not known whether this is owing to differences in the volatile sources and stability, the timing of volatile delivery, or local processing.

Strategic Research for Q4.3

- **Constrain the origin of Mercury's high density** by obtaining in situ compositional information from the surface.
- **Determine the composition and depth of the materials, possibly lunar mantle, excavated by the SPA basin formation** by returning samples from near or within the basin with the characteristics of the deep interior.
- **Determine the origin of polar volatiles** by obtaining and analyzing the properties of ices found within the permanently shadowed craters located near the lunar and hermean poles.
- **Determine the nature and global distribution of hydrothermal deposits in large martian or cerean craters that may have been habitable zones** by mapping at high resolution the mineralogy, composition, and distribution of these deposits or returning samples.
- **Identify physical interactions between impacts and Venus's atmosphere** by conducting remote sensing observations of fireball events and by characterizing unique chemical anomalies in the atmosphere owing to such events.

- **Characterize uplifted deeper icy crustal materials and projectile contaminants on icy bodies** by obtaining high-resolution spectroscopic identification of mineralogy, crystallinity, and chemistry of impact crater floors, peaks, and ejecta.
- **Determine the distribution of exogenic materials in comets to identify impactor material conditions** by performing high-resolution spectroscopic and imaging observations, and by identifying exogenic materials in sampled materials.

Q4.4 HOW DO THE PHYSICS AND MECHANICS OF IMPACTS PRODUCE DISRUPTION OF AND CRATERING ON PLANETARY BODIES?

Impact events have been ubiquitous across the solar system. Resulting craters are strongly controlled by target and projectile properties as well as the physical nature of the impact. Target body disruption events, and possible target reaccumulation, are also controlled by these factors but introduce dynamical processes as well. While investigations have extensively explored the physics and mechanics of impacts, there are still numerous questions to be resolved.

Because projectile and target properties can be strongly altered by their formation and subsequent evolution (e.g., Wiggins et al. 2019), resulting impact signatures can be unique and can differ from planet to planet (Zahnle et al. 2003; Robbins et al. 2018) and even from place to place on one planet (e.g., van der Bogert et al. 2017). Impact processes in the inner solar system, where most projectile and target types are rocks, may differ significantly from those in the outer solar system, where most objects are ice-rich bodies. Material and impact variations cause different thermal, chemical, and mechanical processes, including the generation of ejecta. How ejected debris contributes to the reaccumulation of new bodies and ejecta deposition on planetary surfaces is an outstanding question in impact cratering. While high-speed ejecta escapes immediately, low-speed ejecta returns to the target surface after crater formation or is reaccumulated after a catastrophic disruption, depending on the target size. Collectively, impacts are a rich topic for scientific investigations.

Q4.4a How Does the Impact Process Vary with Projectile/ Target Body Properties and Impact Parameters?

Understanding how impact craters form or disruption events occur is fundamental to interpreting the observed crater and small body record, whether it is modeling regolith generation by small impacts, converting crater sizes to projectile sizes, modeling excavation of a world's mantle, or quantifying how an asteroid family formed. Projectile and target body properties, including material strength, porosity, and projectile sizes, speeds, rotations, and orientations, are fundamental parameters controlling the resulting impact craters or disruption events (Melosh 1989). Very low kinetic energy impacts cause particles to bounce against each other or stick together without disruption (Brisset et al. 2018). Higher-energy impacts induce plastic flows, generating craters and ejecta. Impacts catastrophically disrupting both projectiles and targets generate a multitude of new objects (for planet-forming impacts, see Q4.3a) or debris disks around target worlds. The existence of an ocean and an atmosphere can also change impact processes. These widely ranging properties affect the outcomes of impact events (also see Q5.5b), but the physics of this process remain incompletely understood.

Uniquely determining projectile properties and velocities from impact site constraints is challenging, but in some cases, it is possible to identify remnant traces of the projectile. It is easier to find such traces when the impact kinetic energy is relatively low. This allows the projectile materials to be more easily mixed with target materials. An example would be the carbonaceous chondrite-like material observed on Vesta; its context suggests that it was delivered by a large carbonaceous chondrite-like projectile.

Characterizing how target properties control impacts is another key issue. Thermal, chemical, and mechanical conditions can change the amount of melt created and how the material flows (Grieve and Cintala 1992). Porosity can control the compaction of target materials. Mechanical strength and structure can control fragmentation by modifying how shock and rarefaction waves propagate through the target. Overall, the wide range of properties in planets and small bodies helps explain the variations in impact processes observed from world to world across the solar system.

Last, there is still limited knowledge about how the entire impact process plays out to its end. Impacts consist of multiple physical processes (i.e., contact and compression, excavation, and modification, as well as ejecta formation and deposition, and subsequent relaxation). Because each component has a different timescale (ranging from milliseconds to millions of years, depending on the process in question and the size of the impact), investigations of short-term processes do not characterize all cratering mechanics, and state-of-the-art numerical models still have limited capabilities to integrate all physical effects. Alternatively, experimental approaches can yield useful insights, but by necessity they focus on small-scale impacts, and uncertain scaling relationships are needed to compare them to larger impacts. Technological innovations for all these approaches are urgently needed. Furthermore, collaborations in geological studies and numerical/experimental approaches are also recommended to further constrain full-timescale impact processes.

Q4.4b What Materials Ejected from Impact Craters Are Deposited on Planetary Surfaces?

Impact events excavate planetary surfaces and eject materials from depth away from the impact point. They also redistribute target body materials across planetary bodies in the form of particulate and impact melt ejecta deposits, secondary craters, rays, and sesquinaries. Materials pulverized in this process are ejected ballistically as a "curtain" that moves away from the impact point and often results in the emplacement of an ejecta deposit around the crater. The complex interactions among ejected fragments while in motion are not fully understood. The radial extent and thickness of ejecta deposits vary for different planetary bodies and are influenced by the impact parameters (e.g., impact angle, target composition, and target gravity), with implications for the composition and age determinations of lunar samples.

Impact melt deposits are often formed during an impact event and occur both exterior and interior to the final crater. Different impact conditions, such as impact velocity, however, can affect the extent, distribution, and volume of impact melt generated. Given the desire to date impact craters from recrystallized impact melt by in situ methods or sample return missions, this topic needs further study.

Secondary craters form when large ejecta fragments strike the target surface and produce craters. Some ejecta can also escape into deep space before returning or hitting a different body, forming craters known as sesquinaries. Both secondary and sesquinary craters sometimes look like primary craters. This allows them to inflate the crater populations formed by primary impactors, and therefore can complicate the determination of surface ages via the spatial density of craters. Significant issues remain on this difficult problem, but progress can be made by better understanding the morphologies, populations, and spatial distributions of craters formed by ejecta (e.g., Bierhaus et al. 2018), and by improvements in modeling and experimental techniques.

The redeposition and orbital distribution of fragments and ejecta becomes more complex as a body's size decreases, with an increasing portion of an ejecta field escaping or entering into a long-lived orbit. For some impacts, ejecta can become distributed globally across the body and can cause measurable geophysical effects owing to angular momentum transfer. The interaction between ejecta and a small body's gravitational field can become strongly coupled, and through careful observation can provide a natural opportunity for remote determination of small body gravity fields (Chesley et al. 2020). The efficiency with which impact ejecta is lost is also a key question related to the efficiency of kinetic impactors proposed for planetary defense.

Strategic Research for Q4.4

- **Determine how impact physics and mechanics changes at different impact sites** by mapping those sites with high-resolution remote sensing images, identifying target material compositions using in situ and/or remote methods, and applying the information as constraints for numerical impact simulations.
- **Determine the formation of the SPA basin's asymmetric structure** by characterizing the chemical compositions and geologic structures (e.g., shock fragmentation, etc.) of the surface and internal materials returned near or within SPA and by constraining the internal structure beneath SPA.
- **Determine how impacts affect oceans and ices on Mars** by characterizing martian surface conditions (i.e., geologic and chemical compositions, and the existence of water or ice) through time and then by simulating impacts into such target materials.

- **Map structural deformation and characterize impact mechanics in icy target bodies** by performing high-resolution imaging of impact crater morphology, with example high-value targets, including the uranian satellites, Europa, Ganymede, and trans-neptunian objects.
- **Improve crater counts on icy bodies to characterize secondary crater mechanics and constrain smaller projectile populations (especially focused on gaps in the observational record on Europa, Ganymede, in the uranian system, and on additional >100 km trans-neptunian objects)** by performing high resolution imaging of impact crater morphology on various outer solar system bodies
- **Determine how impacts crush porous structures and materials on comets and asteroids in microgravity** by characterizing the density variations beneath impact craters based on impact experiments, high-resolution gravity measurements, and remote sensing observations

SUPPORTIVE ACTIVITIES FOR QUESTION 4

- Establish a well-constrained chronology for events in the solar system through improved cataloging of impactor reservoirs using ground- and space-based assets, improved dynamical simulations of the formation and evolution of small bodies, improved mapping of new craters on all planetary surfaces, more complete observations of present-day small body impacts in different contexts, remote dating of planetary surfaces, dating of samples via in situ methods and/or returned samples from diverse bodies, and improved modeling of crater formation.
- Understand variations in impact mechanics owing to the target and projectile properties and impact conditions at all scales, with geophysical and geochemical constraints coming from geologic mapping efforts and through modeling of impact processes at higher spatial and temporal resolutions.
- Establish better equations of state for potential projectile/target materials as well as improved impact scaling relationships for cratering and disruption events through experimental and numerical work.

REFERENCES

Andrews-Hanna, J.C., M.T. Zuber, and W.B. Banerdt. 2008. "The Borealis Basin and the Origin of the Martian Crustal Dichotomy." *Nature* 453:1212–1215. https://doi.org/10.1038/nature07011.

Asphaug, E., and A. Reufer. 2014. "Mercury and Other Iron-Rich Planetary Bodies as Relics of Inefficient Accretion." *Nature Geoscience* 7:564–568. https://doi.org/10.1038/ngeo2189.

Barker, M.K., E. Mazarico, G.A. Neumann, M.T. Zuber, J. Haruyama, and D.E. Smith. 2016. "A New Lunar Digital Elevation Model from the Lunar Orbiter Laser Altimeter and SELENE Terrain Camera." *Icarus* 273:346–355. https://doi.org/10.1016/j.icarus.2015.07.039.

Barr, A.C., and R.M. Canup. 2010. "Origin of the Ganymede–Callisto Dichotomy by Impacts During the Late Heavy Bombardment." *Nature Geoscience* 3:164–167. https://doi.org/10.1038/ngeo746.

Becker, K.J., M.S. Robinson, T.L. Becker, L.A. Weller, K.L. Edmundson, G.A. Neumann, M.E. Perry, and S.C. Solomon. 2016. "First Global Digital Elevation Model of Mercury." 47th Annual Lunar and Planetary Science Conference: 2959. https://www.hou.usra.edu/meetings/lpsc2016/pdf/2959.pdf.

Bellucci, J.J., A.A. Nemchin, M. Grange, K.L. Robinson, G. Collins, M.J. Whitehouse, J.F. Snape, et al. 2019. "Terrestrial-Like Zircon in a Clast from an Apollo 14 Breccia." *Earth and Planetary Science Letters* 510:173–185. https://doi.org/10.1016/j.epsl.2019.01.010.

Benna, M., D.M. Hurley, T.J. Stubbs, P.R. Mahaffy, and R.C. Elphic. 2019. "Lunar Soil Hydration Constrained by Exospheric Water Liberated by Meteoroid Impacts." *Nature Geoscience* 12:333–338. https://doi.org/10.1038/s41561-019-0345-3.

Bierhaus, E.B., A.S. McEwen, S.J. Robbins, K.N. Singer, L. Dones, M.R. Kirchoff, and J.-P. Williams. 2018. "Secondary Craters and Ejecta Across the Solar System: Populations and Effects on Impact-Crater-Based Chronologies." *Meteoritics and Planetary Science* 53:638–671. https://doi.org/10.1111/maps.13057.

Boehnke, P., and T.M. Harrison. 2016. "Illusory Late Heavy Bombardments." *Proceedings of the National Academy of Sciences* 113:10802–10806. https://doi.org/10.1073/pnas.1611535113.

Bottke, W.F., M. Broz, D.P. O'Brien, A. Campo Bagatin, A. Morbidelli, and S. Marchi. 2015. "The Collisional Evolution of the Asteroid Belt." Pp. 701–724 in *Asteroids IV*, P. Michel, F. DeMeo, and W.F. Bottke, eds. Tucson: University of Arizona Press. https://muse.jhu.edu/book/43354.

Bottke, W.F., H.F. Levison, D. Nesvorný, and L. Dones. 2007. "Can Planetesimals Leftover from Terrestrial Planet Formation Produce the Lunar Late Heavy Bombardment?" *Icarus* 190:203–223. https://doi.org/10.1016/j.icarus.2007.02.010.

Bottke, W.F., D. Nesvorný, R.E. Grimm, A. Morbidelli, and D.P. O'Brien. 2006. "Iron Meteorites as Remnants of Planetesimals Formed in the Terrestrial Planet Region." *Nature* 439:821–824. https://doi.org/10.1038/nature04536.

Bottke, W.F., and M. Norman. 2017. "The Late Heavy Bombardment." *Annual Review of Earth and Planetary* Sciences 45:619–647. https://doi.org/10.1146/annurev-earth-063016-020131.

Bottke, W.F., D. Vokrouhlický, D. Nesvorný, and J.M. Moore. 2013. "Black Rain: The Burial of the Galilean Satellites in Irregular Satellite Debris." *Icarus* 223:775–795. https://doi.org/10.1016/j.icarus.2013.01.008.

Bramson, A.M., S. Byrne, and J. Bapst. 2017. "Preservation of Midlatitude Ice Sheets on Mars." *Journal of Geophysical Research: Planets* 122(11):2250–2266. https://doi.org/10.1002/2017JE005357.

Brisset, J., J. Colwell, A. Dove, S. Abukhalil, C. Cox, and N. Mohammed. 2018. "Regolith Behavior Under Asteroid-Level Gravity Conditions: Low-Velocity Impact Experiments." *Progress in Earth and Planetary Science* 5:73. https://doi.org/10.1186/s40645-018-0222-5.

Castillo-Rogez, J.C., M.A. Hesse, M. Formisano, H. Sizemore, M. Bland, A.I. Ermakov, and R.R. Fu. 2019. "Conditions for the Long-Term Preservation of a Deep Brine Reservoir in Ceres." *Geophysical Research Letters* 46:1963–1972. https://doi.org/10.1029/2018GL081473.

Chesley, S.R., A.S. French, A.B. Davis, R.A. Jacobson, M. Brozović, D. Farnocchia, S. Selznick, et al. 2020. "Trajectory Estimation for Particles Observed in the Vicinity of (101955) Bennu." *Journal of Geophysical Research: Planets* 125:e06363. https://doi.org/10.1029/2019JE006363.

Colaprete, A., M. Sarantos, D.H. Wooden, T.J. Stubbs, A.M. Cook, and M. Shirley. 2015. "How Surface Composition and Meteoroid Impacts Mediate Sodium and Potassium in the Lunar Exosphere." *Science* 351(6270):249–252. https://doi.org/10.1126/science.aad2380.

DellaGiustina, D.N., H.H. Kaplan, A.A. Simon, W.F. Bottke, C. Avdellidou, M. Delbo, R.-L. Ballouz, et al. 2020. "Exogenic Basalt on Asteroid (101955) Bennu." *Nature Astronomy* 5:31–38. https://doi.org/10.1038/s41550-020-1195-z.

Dones, L., C.R. Chapman, W.B. McKinnon, H.J. Melosh, M.R. Kirchoff, G. Neukum, and K.J. Zahnle. 2009. "Icy Satellites of Saturn: Impact Cratering and Age Determination." Pp. 613–635 in *Saturn from Cassini-Huygens*, M.K. Dougherty, L.W. Esposito, S.M. Krimigis, eds. Dordrecht, Netherlands: Springer. https://doi.org/10.1007/978-1-4020-9217-6_19.

Dundas, C.M., A.M. Bramson, L. Ojha, J.J. Wray, M.T. Mellon, S. Byrne, A.S. McEwen, et al. 2018. "Exposed Subsurface Ice Sheets in the Martian Mid-Latitudes." *Science* 359:199–201. https://doi.org/10.1126/science.aao1619.

Farnocchia, D., S.R. Chesley, Y. Takahashi, B. Rozitis, D. Vokrouhlický, B.P. Rush, N. Mastrodemos, et al. 2021. "Ephemeris and Hazard Assessment for Near-Earth Asteroid (101955) Bennu Based on Osiris-REx Data." *Icarus* 369:114594. https://doi.org/10.1016/j.icarus.2021.114594.

Gomes, R., H.F. Levison, K. Tsiganis, and A. Morbidelli. 2005. "Origin of the Cataclysmic Late Heavy Bombardment Period of the Terrestrial Planets." *Nature* 435:466–469. https://doi.org/10.1038/nature03676.

Goodrich, C.A., W.K. Hartmann, D.P. O'Brien, S.J. Weidenschilling, L. Wilson, P. Michel, and M. Jutzi. 2015. "Origin and History of Ureilitic Material in the Solar System: The View from Asteroid 2008 TC3 and the Almahata Sitta Meteorite." *Meteoritics and Planetary Science* 50:782–809. https://doi.org/10.1111/maps.12401.

Goossens, S., T.J. Sabaka, A. Genova, E. Mazarico, J.B. Nicholas, and G.A. Neumann. 2017. "Evidence for a Low Bulk Crustal Density for Mars from Gravity and Topography." *Geophysical Research Letters* 44:7686–7694. https://doi.org/10.1002/2017GL074172.

Granvik, M., A. Morbidelli, R. Jedicke, B. Bolin, W.F. Bottke, E. Beshore, D. Vokrouhlický, et al. 2018. "Debiased Orbit and Absolute-Magnitude Distributions for Near-Earth Objects." *Icarus* 312:181–207. https://doi.org/10.1016/j.icarus.2018.04.018.

Grieve, R.A.F., and M.J. Cintala. 1992. "An Analysis of Differential Impact Melt-Crater Scaling and Implications for the Terrestrial Impact Record." *Meteoritics* 27:526. https://doi.org/10.1111/j.1945-5100.1992.tb01074.x.

Heiken, G.H., D.T. Vaniman, and B.M. French. 1991. *Lunar Sourcebook: A User's Guide to the Moon*. Cambridge, UK: Cambridge University Press. https://www.lpi.usra.edu/lunar_sourcebook.

Hemingway, D., L. Iess, R. Tajeddine, and G. Tobie. 2018. "The Interior of Enceladus." Pp. 57–77 in *Enceladus and the Icy Moons of Saturn*, P.M. Schenk, R.N. Clark, C.J.A. Howett, A.J. Verbiscer, J.H. Waite, eds. Tucson: University of Arizona. https://doi.org/10.2458/azu_uapress_9780816537075-ch004.

Hueso, R., M. Delcroix, A. Sánchez-Lavega, S. Pedranghelu, G. Kernbauer, J. McKeon, A. Fleckstein, et al. 2018. "Small Impacts on the Giant Planet Jupiter." *Astronomy & Astrophysics* 617:A68. https://doi.org/10.1051/0004-6361/201832689.

Janches, D., A.A. Berezhnoy, A.A. Christou, G. Cremonese, T. Hirai, M. Horányi, J.M. Jasinski, et al. 2021. "Meteoroids as One of the Sources for Exosphere Formation on Airless Bodies in the Inner Solar System." *Space Science Reviews* 217:50. https://doi.org/10.1007/s11214-021-00827-6.

Jawin, E.R., K.J. Walsh, O.S. Barnouin, T.J. McCoy, R.-L. Ballouz, D.N. DellaGiustina, H.C. Connolly Jr., et al. 2020. "Global Patterns of Recent Mass Movement on Asteroid (101955) Bennu." *Journal of Geophysical Research: Planets* 125:e06475. https://doi.org/10.1029/2020JE006475.

Jenniskens, P. 2015. "Meteoroid Streams and the Zodiacal Cloud." Pp. 281–295 in *Asteroids IV*, P. Michel, F. DeMeo, and W.F. Bottke, eds. Tucson: University of Arizona Press. https://uapress.arizona.edu/book/asteroids-iv.

Jewitt, D., H. Hsieh, J. Agarwal. 2015. "The Active Asteroids." Pp. 221–241 in *Asteroids IV*, P. Michel, F. DeMeo, and W.F. Bottke, eds. Tucson: University of Arizona Press. https://uapress.arizona.edu/book/asteroids-iv.

Keller, H.U., S. Mottola, B. Davidsson, S.E. Schröder, Y. Skorov, E. Kührt, O. Groussin, et al. 2015. "Insolation, Erosion, and Morphology of Comet 67P/Churyumov-Gerasimenko." *Astronomy & Astrophysics* 583(A34). https://doi.org/10.1051/0004-6361/201525964.

Killen, R.M., and J.M. Hahn. 2015. "Impact Vaporization as a Possible Source of Mercury's Calcium Exosphere." *Icarus* 250:230–237. https://doi.org/10.1016/j.icarus.2014.11.035.

Kring, D.A., M.J. Whitehouse, and M. Schmieder. 2021. "Microbial Sulfur Isotope Fractionation in the Chicxulub Hydrothermal System." *Astrobiology* 21:103–114. https://doi.org/10.1089/ast.2020.2286.

Lauretta, D.S., C.W. Hergenrother, S.R. Chesley, J.M. Leonard, J.Y. Pelgrift, C.D. Adam, M. Al Asad, et al. 2019. "Episodes of Particle Ejection from the Surface of the Active Asteroid (101955) Bennu." *Science* 366(6470):eaay3544. https://doi.org/10.1126/science.aay3544.

Marchi, S., W.F. Bottke, B.A. Cohen, K. Wünnemann, D.A. Kring, H.Y. McSween, M.C. De Sanctis, et al. 2013. "High-Velocity Collisions from the Lunar Cataclysm Recorded in Asteroidal Meteorites." *Nature Geosciences* 6:303–307. https://doi.org/10.1038/ngeo1769.

Marchi, S., C.R. Chapman, O.S. Barnouin, J.E. Richardson, and J.-B. Vincent. 2015. "Cratering on Asteroids." Pp. 725–744 in *Asteroids IV*, P. Michel, F. DeMeo, and W.F. Bottke, eds. Tucson: University of Arizona Press. https://uapress.arizona.edu/book/asteroids-iv.

Marchi, S., A.I. Ermakov, C.A. Raymond, R.R. Fu, D.P. O'Brien, M.T. Bland, E. Ammannito, et al. 2016. "The Missing Large Impact Craters on Ceres." *Nature Communications* 7(12257). https://doi.org/10.1038/ncomms12257.

Margot, J.L., P. Pravec, P. Taylor, B. Carry, S. Jacobson. 2015. "Asteroid Systems: Binaries, Triples, and Pairs." Pp. 355–374 in *Asteroids IV*, P. Michel, F. DeMeo, and W.F. Bottke, eds. Tucson: University of Arizona Press. https://uapress.arizona.edu/book/asteroids-iv.

Melosh, H.J. 1989. *Impact Cratering: A Geologic Process. Oxford Monographs on Geology and Geophysics 11*. New York: Oxford University Press.

Melosh, H.J., J. Kendall, B. Horgan, B.C. Johnson, T. Bowling, P.G. Lucey, and G.J. Taylor. 2017. "South Pole–Aitken Basin Ejecta Reveal the Moon's Upper Mantle." *Geology* 45:1063–1066. https://doi.org/10.1130/G39375.1.

Morbidelli, A., D. Nesvorný, W.F. Bottke, and S. Marchi. 2021. "A Re-assessment of the Kuiper Belt Size Distribution for Sub-Kilometer Objects, Revealing Collisional Equilibrium at Small Sizes." *Icarus* 356:114256. https://doi.org/10.1016/j.icarus.2020.114256.

Morbidelli, A., D. Nesvorný, V. Laurenz, S. Marchi, D.C. Rubie, L. Elkins-Tanton, M. Wieczorek, and S. Jacobson. 2018. "The Timeline of the Lunar Bombardment: Revisited." *Icarus* 305:262–276. https://doi.org/10.1016/j.icarus.2017.12.046.

Morbidelli, A., K.J. Walsh, D.P. O'Brien, D.A. Minton, and W.F. Bottke. 2015. "The Dynamical Evolution of the Asteroid Belt." Pp. 493–508 in *Asteroids IV*, P. Michel, F. DeMeo, and W.F. Bottke, eds. Tucson: University of Arizona Press. https://uapress.arizona.edu/book/asteroids-iv.

Movshovitz, N., F. Nimmo, D.G. Korycansky, E. Asphaug, and J.M. Owen. 2015. "Disruption and Reaccretion of Midsized Moons During an Outer Solar System Late Heavy Bombardment." *Geophysical Research Letters* 42(2):256–263. https://doi.org/10.1002/2014GL062133.

Nesvorný, D. 2018. "Dynamical Evolution of the Early Solar System." *Annual Review of Astronomy and Astrophysics* 56:137–174. https://doi.org/10.1146/annurev-astro-081817-052028.

Nesvorný, D., W.F. Bottke, D. Vokrouhlický, M. Sykes, D.J. Lien, and J. Stansberry. 2008. "Origin of the Near-Ecliptic Circumsolar Dust Band. *Astrophysical Journal* 679(2):L143–L146. https://doi.org/10.1086/588841.

Nesvorný, D., and A. Morbidelli. 2012. "Statistical Study of the Early Solar System's Instability with Four, Five, and Six Giant Planets." *Astronomical Journal* 144(4):117. https://doi.org/10.1088/0004-6256/144/4/117.

Nesvorný, D., F. Roig, and W.F. Bottke. 2017. "Modeling the Historical Flux of Planetary Impactors." *Astronomical Journal* 153(3):103. https://doi.org/10.3847/1538-3881/153/3/103.

Nesvorný, D., F.V. Roig, and R. Deienno. 2021. "The Role of Early Giant-Planet Instability in Terrestrial Planet Formation." *Astronomical Journal* 161:50. https://doi.org/10.3847/1538-3881/abc8ef.

Nesvorný, D., D. Vokrouhlický, W.F. Bottke, and H.F. Levison. 2018. "Evidence for Very Early Migration of the Solar System Planets from the Patroclus-Menoetius Binary Jupiter Trojan." *Nature Astronomy* 2:878–882. https://doi.org/10.1038/s41550-018-0564-3.

Nesvorný, D., D. Vokrouhlický, A.S. Stern, B. Davidsson, M.T. Bannister, K. Volk, Y.-T. Chen, et al. 2019. "OSSOS. XIX. Testing Early Solar System Dynamical Models Using OSSOS Centaur Detections." *Astronomical Journal* 158(3):132. https://doi.org/10.3847/1538-3881/ab3651.

O'Neill, C., S. Marchi, W. Bottke, and R. Fu. 2020. "The Role of Impacts on Archaean Tectonics." *Geology* 48:174–178. https://doi.org/10.1130/G46533.1.

Osinski, G.R., C.S. Cockell, A. Pontefract, and H.M. Sapers. 2020. "The Role of Meteorite Impacts in the Origin of Life." *Astrobiology* 20:1121–1149. https://doi.org/10.1089/ast.2019.2203.

Pajola, M., S. Höfner, J.B. Vincent, N. Oklay, F. Scholten, F. Preusker, and S. Mottola. 2017. "The Pristine Interior of Comet 67P Revealed by the Combined Aswan Outburst and Cliff Collapse." *Nature Astronomy* 1:0092. https://doi.org/10.1038/s41550-017-0092.

Poppe, A.R., C.M. Lisse, M. Piquette, M. Zemcov, M. Horányi, D. James, J.R. Szalay, et al. 2019. "Constraining the Solar System's Debris Disk with In Situ New Horizons Measurements from the Edgeworth-Kuiper Belt." *Astrophysical Journal* 881(1):L12. https://doi.org/10.3847/2041-8213/ab322a.

Preusker, F., F. Scholten, K.-D. Matz, T. Roatsch, R. Jaumann, C.A. Raymond, and C.T. Russell. 2016. "Dawn FC2 Derived Vesta DTM SPG V1.0." NASA Planetary Data System: DAWN-A-FC2-5-VESTADTMSPG-V1.0.

Raymond, S.N., and A. Izidoro. 2017. "The Empty Primordial Asteroid Belt." *Science Advances* 3(9):e1701138. https://doi.org/10.1126/sciadv.1701138.

Robbins, S.J., W.A. Watters, J.E. Chappelow, V.J. Bray, I.J. Daubar, R.A. Craddock, R.A. Beyer, et al. 2018. "Measuring Impact Crater Depth Throughout the Solar System." *Meteoritics and Planetary Science* 53(4):583–637. https://doi.org/10.1111/maps.12956.

Roberts, J.H., and J. Arkani-Hamed. 2012. "Impact-Induced Mantle Dynamics on Mars." *Icarus* 218:278–289. https://doi.org/10.1016/j.icarus.2011.11.038.

Roberts, J.H., and J. Arkani-Hamed. 2017. "Effects of Basin-Forming Impacts on the Thermal Evolution and Magnetic Field of Mars." *Earth and Planetary Science Letters* 478:192–202. https://doi.org/10.1016/j.epsl.2017.08.031.

Schenk, P.M., R.A. Beyer, W.B. McKinnon, J.M. Moore, J.R. Spencer, O.L. White, K. Singer, et al. 2018. "Basins, Fractures and Volcanoes: Global Cartography and Topography of Pluto from New Horizons." *Icarus* 314:400–433. https://doi.org/10.1016/j.icarus.2018.06.008.

Singer, K.N., W.B. McKinnon, B. Gladman, S. Greenstreet, E.B. Bierhaus, S.A. Stern, A.H. Parker, et al. 2019. "Impact Craters on Pluto and Charon Indicate a Deficit of Small Kuiper Belt Objects." *Science* 363:955–959. https://doi.org/10.1126/science.aap8628.

Smith, D.E., M. Zuber, H.V. Frey, J.B. Garvin, J.W. Head, D.O. Muhleman, G.H. Pettengill, et al. 2001. "Mars Orbiter Laser Altimeter: Experiment Summary After the First Year of Global Mapping of Mars." *Journal of Geophysical Research: Planets* 106:23689–23722. https://doi.org/10.1029/2000JE001364.

Smrekar, S.E., A. Davaille, and C. Sotin. 2018. "Venus Interior Structure and Dynamics." *Space Science Reviews* 214:88. https://doi.org/10.1007/s11214-018-0518-1.

Stöffler, D., and G. Ryder. 2001. "Stratigraphy and Isotope Ages of Lunar Geologic Units: Chronological Standard for the Inner Solar System." *Space Science Reviews* 96:9–54. https://doi.org/10.1023/A:1011937020193.

Tatsumi, E., C. Sugimoto, L. Riu, S. Sugita, T. Nakamura, T. Hiroi, and T. Morota. 2021. "Collisional History of Ryugu's Parent Body from Bright Surface Boulders." *Nature Astronomy* 5:39–45. https://doi.org/10.1038/s41550-020-1179-z.

Terada, K., T. Morota, and M. Kato. 2020. "Asteroid Shower on the Earth-Moon System Immediately Before the Cryogenian Period Revealed by KAGUYA." *Nature Communications* 11:3453.

Terfelt, F., and B. Schmitz. 2021. "Asteroid Break-Ups and Meteorite Delivery to Earth the Past 500 Million Years." *Proceedings of the National Academy of Sciences* 118:2020977118.

Tsiganis, K., R. Gomes, A. Morbidelli, and H.F. Levison. 2005. "Origin of the Orbital Architecture of the Giant Planets of the Solar System." *Nature* 435:459–461.

van der Bogert, C.H., H. Hiesinger, C.M. Dundas, T. Krüger, A.S. McEwen, M. Zanetti, and M.S. Robinson. 2017. "Origin of Discrepancies Between Crater Size-Frequency Distributions of Coeval Lunar Geologic Units via Target Property Contrasts." *Icarus* 298:49–63.

Vokrouhlický, D., W.F. Bottke, S.R. Chesley, D.J. Scheeres, and T.S. Statler. 2015. "The Yarkovsky and YORP Effects." Pp. 509–532 in *Asteroids IV*, P. Michel, F. DeMeo, and W.F. Bottke, eds. Tucson: University of Arizona Press.

Vokrouhlický, D., W.F. Bottke, and D. Nesvorný. 2016. "Capture of Trans-Neptunian Planetesimals in the Main Asteroid Belt." *Astronomical Journal* 152:39.

Walsh, K.J., and S.A. Jacobson. 2015. "Formation and Evolution of Binary Asteroids." Pp. 375–393 in *Asteroids IV*, P. Michel, F. DeMeo, and W.F. Bottke, eds. Tucson: University of Arizona Press.

Walsh, K.J., A. Morbidelli, S.N. Raymond, D.P. O'Brien, and A.M. Mandell. 2011. "A Low Mass for Mars from Jupiter's Early Gas-Driven Migration." *Nature* 475:206–209.

White, O.L., P.M. Schenk, A.W. Bellagamba, A.M. Grimm, A.J. Dombard, and V.J. Bray. 2017. "Impact Crater Relaxation on Dione and Tethys and Relation to Past Heat Flow." *Icarus* 288:37–52. https://doi.org/10.1016/j.icarus.2017.01.025.

Wieczorek, M.A., G.A. Neumann, F. Nimmo, W.S. Kiefer, G.J. Taylor, H.J. Melosh, R.J. Phillips, et al. 2013. "The Crust of the Moon as Seen by GRAIL." *Science* 339:671–675.

Wiggins, S.E., B.C. Johnson, T.J. Bowling, H.J. Melosh, and E.A. Silber. 2019. "Impact Fragmentation and the Development of the Deep Lunar Megaregolith." *Journal of Geophysical Research: Planets* 124:941–957.

Wilhelms, D.E. 1987. "The Geologic History of the Moon." Professional Paper 1348. Washington, DC: U.S. Geological Survey.

Zahnle, K., P. Schenk, H. Levison, and L. Dones. 2003. "Cratering Rates in the Outer Solar System." *Icarus* 163:263–289.

Q5 PLATE: A portion of an approximately true-color mosaic of the martian surface near the "Mont Mercou" outcrop, taken by the Curiosity rover in 2021. SOURCE: Courtesy of NASA/JPL-Caltech/MSSS/© T. Appéré, all rights reserved.

8

Question 5: Solid Body Interiors and Surfaces

How do the interiors of solid bodies evolve, and how is this evolution recorded in a body's physical and chemical properties? How are solid surfaces shaped by subsurface, surface, and external processes?

Planetary surfaces and interiors are tapestries on which the history of each body is woven.[1] This chapter focuses on "solid" bodies, meaning those that are primarily solid and monolithic, thus excluding rubble pile asteroids and comets (at the small end) and gas/ice giant planets (at the large end). Terrestrial planets, icy moons, large asteroids, trans-neptunian objects, and comets all have surfaces and interiors that record their evolution, as discussed below.

The first two questions in this chapter address the fundamentals of planetary interiors: what kinds of internal structures exist (Q5.1) and how they have evolved with time (Q5.2). The next three questions address the evolution of planetary surfaces, and how they have been modified by interior (Q5.3), surficial (Q5.4) and external (Q5.5) processes. Last, the special case of present-day activity is discussed (Q5.6). This is deliberately separated from the other questions because the measurement *techniques* used to probe active processes (e.g., seismology, radar interferometry) are fundamentally different from investigations of geologic processes that have happened in the past.

Q5.1 HOW DIVERSE ARE THE COMPOSITIONS AND INTERNAL STRUCTURES WITHIN AND AMONG SOLID BODIES?

The fundamental control on the compositions of solid bodies is the range of their constituent building blocks, including the accreted solid matter that condensed from the solar nebular gas, or incorporated presolar grains (solid material that predates the Sun and the solar system). The composition of these solids varied throughout the solar nebula in response to the local condensation temperature and gas composition (Questions 2 and 3, Chapters 5 and 6).

Depending on the timing, energy, and nature of the accretion process, the body can be partly to completely melted, and this can strongly influence the development of internal structure. The amount of short-lived heat-producing radioactive elements incorporated during accretion, which is a function of accretion time, can also strongly influence compositions, chemical characteristics, and thermal evolution of solid body interiors. As such, structure and evolution are largely controlled by initial composition, formation time, and body size.

---

[1]A glossary of acronyms and technical terms can be found in Appendix F.

209

Q5.1a How Much Variability in Composition and Internal Structure Is There Within and Between Solid Bodies, and How Did Such Variability Arise and Evolve?

Many factors influence the initial structure of silicate bodies, including the interior oxidation state. Under reduced conditions, iron metal can become stable, and under very reducing conditions silicon in SiO_2 can be partitioned into the dense metal phase that can segregate to form a core. The presence and abundance of sulfur, oxygen, and hydrogen can also influence core formation processes. Under moderately reduced conditions, sulfur can combine with iron to form a molten FeS core. However, under more reducing conditions, sulfur will strongly partition into the silicate melt and modify subsequent melting and differentiation.

Oxidation state determines how much iron is metal versus in silicate form. Martian meteorites suggest that Mars is highly oxidized, and that the mantle has varying oxidation states. The recent measurement of Mars's core size by the InSight spacecraft is consistent with an Fe-O-S rich core (Stähler et al. 2021). Spectroscopy from MESSENGER data of Mercury shows little or no oxidized Fe in surface lavas and taken together with its massive core indicates its interior is highly reduced (McCubbin et al. 2012). Venus probably has an Earth-sized core, but we do not know its composition.

Rocky planet mantles can exhibit heterogeneity. Phase transitions or compositional variations (like those arising from magma ocean overturn) can inhibit homogenization by convection. Isotopic and trace element data in martian meteorites demonstrate multiple ancient reservoirs in that planet's mantle, and the same is probably true on the Moon.

Giant impacts and the addition of material in the waning stages of accretion can also influence the compositions of both rocky and icy bodies (Questions 2 and 3, Chapters 5 and 6). A giant impact on Earth is responsible for shaping the unique Fe-metal-poor composition of the Moon. An early giant impact on Mars is one hypothesis for explaining the global hemispheric dichotomy and likely modified the composition of the planet. A giant impact on Mercury has long remained as one of the theories to explain its large core mass (Q5.1). Giant impacts in the outer solar system may also be responsible for the wide range of bulk compositions of outer solar system icy bodies (e.g., Saturn's moon Tethys is almost pure ice, while its similarly sized neighbors are a mix of rock and ice).

Icy bodies can have complicated internal structures, consisting of silicate mantles, sometimes with underlying iron cores, which in turn are overlain by an outer layer that contains one or more water ice phases and, in some cases, include a liquid water ocean. Very distant bodies (e.g., Pluto) might also contain substantial fractions of ices normally regarded as volatile (e.g., N_2 and CH_4). Ganymede is the only icy body where a liquid iron core is certain (based on the observation of a core dynamo); conversely, in the case of Callisto it is unclear whether full separation of rock and ice has taken place. The rock–ice ratio can vary significantly, from Europa (rocky) to Tethys (icy) (Figure 8-1). Oxidation state is not as important a factor for icy bodies, but their bulk compositions could potentially involve substantial amounts of carbon. The thermal evolution of icy bodies is controlled by melting of the ice phases and alteration of silicate minerals to clays, resulting in segregation of an ice-rich surface layer and sometimes a subsurface ocean. Melting is driven either by decay of short-lived radionuclides (which are most abundant in icy satellites that formed earliest), or by tidal dissipation (Q5.2a, Question 8, Chapter 11) and is facilitated by the presence of impurities such as salts. Furthermore, the efficient advection of heat by circulating fluids moderates internal temperatures, potentially allowing volatiles to be retained.

In the solid bodies in the solar system, we have directly measured or inferred the nature of internal layering and compositional variations from remotely sensed physical properties, such as static and time-variable gravity, shape, and seismic velocity. Constraints on core mass/size for planets, and for rock/metal or rock/ice relative abundances in bulk bodies, depend mostly on mean density and moment of inertia factors. Refinement of these, as exemplified most recently by measurement of the mean density of Vesta by the Dawn mission, lead to better estimates (Ermakov et al. 2014). The only extraterrestrial bodies for which limited seismic data are available to constrain interior structures are the Moon (from analysis of Apollo seismometer data) and Mars (from preliminary analysis of InSight data). On Mars the lessons learned on the Moon influenced the design of the InSight seismometer, and after one Mars year the mission has measured crustal thickness and upper mantle velocity structure, as well as determine the core radius (Stähler et al. 2021). Seismology on Venus requires either high-temperature electronics or aerial pressure sensors (Brissaud et al. 2021);

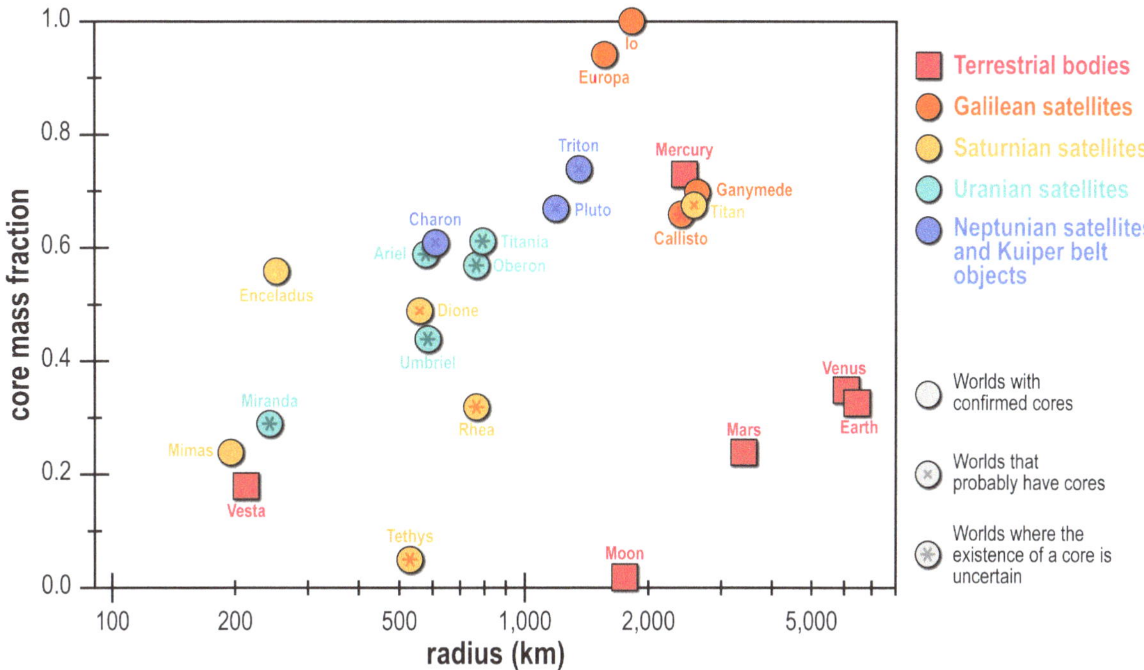

FIGURE 8-1 "Core" mass fraction for rocky (terrestrial) and icy bodies. For terrestrial bodies, the core is the metallic central region, with the exterior mantle being silicate; for icy bodies, it is the rock plus metal central region, with the exterior mantle being ice (for consistency, Io is categorized as an icy body but with an ice fraction of zero). Filled symbols indicate worlds where there is strong evidence for the presence of a core. Symbols with an "x" indicate worlds where there is either ambiguous observational evidence or a strong theoretical argument for the presence of a core. Symbols with asterisks indicate worlds where the existence of a core is uncertain. SOURCE: Figure created by F. Nimmo and J.T. Keane.

technological development for either approach would enable comparably fundamental discoveries to be made in future decades.

Tidally induced heating can also lead to melting and chemical differentiation. This is thought to be important on Io and Europa and also important on Triton during its capture into Neptune's orbit (Question 8, Chapter 11). Measuring gravitational tides can be used to constrain internal structures, especially the presence of liquid layers (e.g., the outer core of Mars or the ocean of Europa).

The smallest bodies of the solar system—asteroids, comets and trans-neptunian objects (TNOs)—are dominantly porous aggregates of variably sized blocks, dust and in many cases, ice. A population of larger (>100 km diameter) planetesimals in the main asteroid belt, exemplified by the differentiated protoplanet Vesta, comprises a record of the original planetary building blocks. The composition of the constituent material across the small body population ranges from relatively primitive in volatile-rich comets and TNOs to highly processed in small asteroids within the main and near-Earth asteroid belts. Rosetta's exploration of Jupiter-family comet 67P/Churyumov-Gerasimenko revealed its affinity to interstellar material, including inherited organic material (Grady et al. 2018). Dawn's mapping of Vesta provided context for laboratory analyses of the howardite-eucrite-diogenite meteorites, revealing the complexities of magmatic evolution on this smallest terrestrial planet (McSween et al. 2013) (Q5.3b). Catastrophic disruption and reaggregation have scrambled the interiors of small bodies in some cases, leading to small-scale compositional diversity as seen in some meteorite breccias, and well as on the surfaces of the bodies. An understanding of the interior composition and structure of larger planetesimals by measuring surface composition, dielectric properties, shape, density, and spin state informs accretionary and dynamical models (Question 2, Chapter 5).

Q5.1b What Kinds of Internal Liquid Layers (e.g., Oceans) or Discrete Regions Occur in Solid Bodies, What Are Their Characteristics, Where Are They, and How Long Do They Persist?

The existence, locations, and compositional variations of liquid layers and discontinuous liquid regions (e.g., molten cores, partly melted magma source regions, brine lakes, subsurface oceans) in solid body interiors and their persistence depend fundamentally on their compositions, the abundance of heat-producing elements, and the influence of internal and external physical processes that can lead to melting.

As for the terrestrial planets, the composition of the core and the inventory of heat-producing radioactive elements combine to determine the longevity of molten metallic cores. Mercury, the Moon, Earth, and Mars all have long-lived molten cores, in some cases consisting of a fluid outer core, and a solid inner core). The state of Venus's core is largely unknown.

Solid state convection of the planetary mantles can lead to decompression melting and the presence of continuous small extents of melting in planet interiors; in Mercury this may be reflected in the continual change in erupted basalt composition over the 700-million-year history inferred for volcanism on this planet. On Mars, the vast Tharsis rise provides a 4-billion-year record of volcanism, most likely arising from melting owing to a long-lived mantle upwelling. The surface of Venus is covered with voluminous lava flows and displays a uniform crater density; whether this uniformity is indicative of a catastrophic resurfacing event or continuous volcano-tectonic resurfacing is currently unresolved. The variability in volcanic styles and longevity among the terrestrial planets is remarkable and reflects the differences in internal compositional layering and their thermal histories, but in ways that are not well understood. The most volcanically active body with the youngest surface in the solar system is Io, Jupiter's innermost large satellite, whose interior is tidally heated by the oscillating pull of Jupiter's gravity. Io may possess a subsurface, silicate magma ocean at the present day.

In the outer solar system, except for Io, internal liquid layers are dominated by water and other volatiles rather than silicate melts. The roster of such "ocean worlds" has grown with time (Nimmo and Pappalardo 2016). Europa, Titan, Enceladus, Ganymede, and Callisto all possess subsurface oceans, and other bodies, including Triton, Dione and Pluto, are also candidate ocean worlds. Ceres, the innermost dwarf planet in the solar system, may have had a subsurface ocean in the past and perhaps contains remnants of that ocean today as subsurface brine pockets. Little is known of the compositions, pH, and ages of these oceans, yet those characteristics are key to assessing their habitability (Question 10, Chapter 13). In the case of Enceladus, the composition of its geyser-like plumes can be used to infer the composition of its ocean. However, the age of its ocean remains largely unconstrained. Ocean worlds are often found by combining theoretical modeling with magnetic induction measurements (for Europa, Ganymede, and Callisto), gravity and/or rotation state measurements (for Enceladus and Titan), and other geodetic methods and geological inferences. However, ocean detection and characterization is challenging. In general, the compositions, thicknesses, and dynamics of subsurface oceans on these worlds are unknown. While we know that tidal heating helps to maintain these oceans (Question 8, Chapter 11), the likely presence of an ocean on Pluto and a putative past ocean on Ceres suggest that such heating is not required. Additional insulating layers such as gas hydrates (clathrates) within these worlds may have helped maintain these oceans. Nevertheless, the ages and long-term evolution of the oceans are largely unconstrained (Q5.2c).

Q5.1c How Does the Presence of Porosity, Ices, Liquids, or Gases Affect the Physical (e.g., Mechanical, Thermal, Electromagnetic) Properties of the Crust?

The outer layers of solid bodies—the crust and regolith—influence their thermal and mechanical properties. Regolith is the loose, porous, unconsolidated rock and dust on the surface of a planetary body. This layer can stabilize volatile ices within centimeters of the surface by protecting them from diurnal heating and sublimation. The composition of the crust determines its thermal and mechanical properties. For instance, clathrates (ices with structural cages containing gas molecules) within the crust of an icy body lower its thermal conductivity and can keep the interior warm, as is suggested to explain the longevity of brine effusion on Ceres (Castillo-Rogez et al. 2019), and the possible persistence of an ocean on Pluto (Kamata et al. 2019). Clathrates are mechanically strong but unstable at low-pressure and can explosively destruct, which may result in lateral heterogeneity in crustal strength and morphology. The viscosity of ice is sensitive both to the local temperature and to the fraction of impurities present; thus, the temperature and ice content of the crust of icy bodies will control its mechanical properties. Particularly on small icy or rocky worlds, the low conductivity of the porous near-surface material will keep the interior warm, although high temperatures will close pores by viscous flow. Porous material will also be mechanically weaker than its intact equivalent, and can lead to enhanced scattering, for both seismic and electromagnetic waves. Even small amounts of melt or other liquid can greatly change the mechanical (e.g., viscosity), seismological (e.g., shear-wave speed) and electromagnetic (e.g., conductivity) properties of the crust.

Strategic Research for Q5.1

- **Probe the internal structures of the Moon, Mars, and Mercury** by establishing a geophysical (seismic/magnetometer) network on the former two bodies and making the first surface seismic/magnetometer measurements on the latter.
- **Investigate the properties of subsurface water or magma oceans and melt reservoirs within Europa, Io, Titan, Enceladus, Triton and the uranian moons** via electromagnetic sounding (active/passive) or induction, or geodetic measurements from orbiting or landed spacecraft.
- **Investigate magmatism, and the effects of interior processes on surface compositions of planetesimals (specifically large asteroids and dwarf planets)** via high-resolution imaging, spectroscopy, and topography.
- **Determine stable mineral assemblages and the pressure-temperature conditions of melt generation in the interiors of Moon, Mars, Venus, Mercury, Io, and other rocky worlds** by carrying out laboratory experiments on returned samples, meteorites, and analog compositions under relevant conditions.

Q5.2 HOW HAVE THE INTERIORS OF SOLID BODIES EVOLVED?

Following their initial differentiation into primordial cores, mantles, ice or rock crusts, and oceans, ongoing mass and energy transport and cycling between surfaces and interiors have led to evolving temperatures and compositions. At the surface, tectonic activity and eruptions are a consequence of internal evolution or external tidal forces. Below the surface, liquid metal cores generate magnetic fields whose histories reflect changes in the temperature and composition of the core and overlying mantle. A great diversity of planetary evolution arises because of different body sizes, different bulk compositions, and the degree of crustal recycling (Figure 8-2). Our ability to reconstruct interior evolution is based on linking sample-based studies, surface observations, and geophysical measurements.

Q5.2a What Mechanisms Have Contributed to Post-Accretion or Post-Differentiation Planetary Cooling and Heating?

The early processes of accretion and differentiation impart a planetary body with an initial heat budget, which can be augmented over time owing to endogenous processes such as radioactive decay and exothermic crystallization, or exogenous processes such as giant impacts and tidal forces. Heat is lost over time via a combination of internal convection, conduction, and either volcanism or cryovolcanism (so-called heat pipe volcanism). The mechanisms through which solid bodies recycle their surfaces have a dominant effect on the composition and thermal evolution of their interior. Plate tectonics, as exemplified by Earth, provides a mechanism for efficient heat loss and direct transport of surface material into the interior. Plate tectonics is largely coupled to mantle convection, wherein cold material buoyantly sinks into the interior, displacing hotter material. Upwelling hot material may also melt by decompression, producing magmas and related volcanism. The reasons why plate tectonics starts and ends remain uncertain. Small rocky worlds such as Mercury, Mars, and the Moon have stagnant lid tectonic surfaces. Venus and ocean worlds such as Europa may also have mechanisms to recycle their surfaces, in the past or possibly at present. Partially mobile surfaces may be possible. On Io, and other planetary bodies earlier in their histories, burial of the surface by volcanism can recycle crusts back into the mantle (Moore et al. 2017). Magmatism and eruptions can also transport heat-producing radiogenic materials to different portions of a solid body, influencing subsequent cooling and volcanism. The various modes of heat transfer in planetary interiors (Figure 8-3)

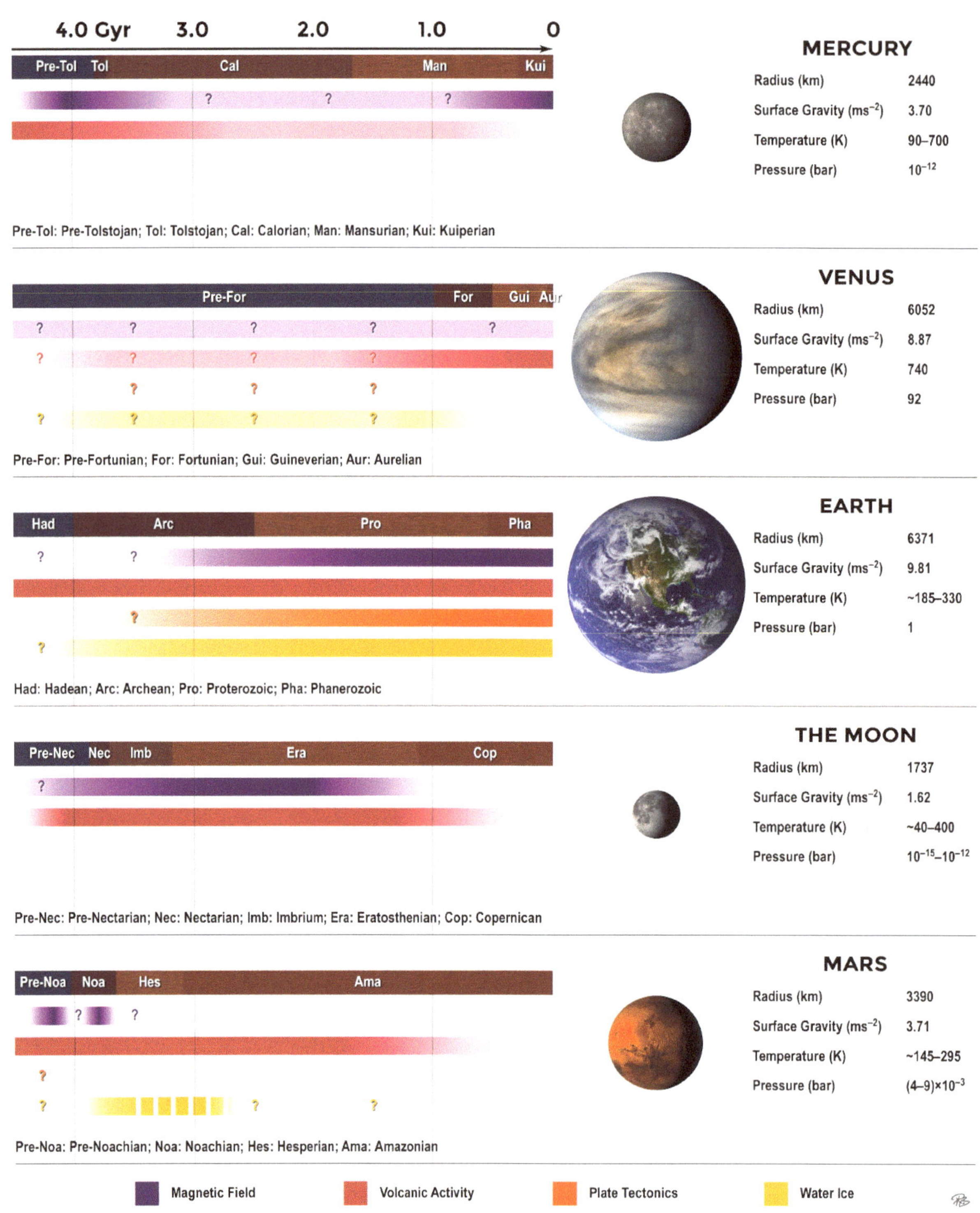

FIGURE 8-2 Internal evolution of selected solid surface bodies in the inner solar system. The conversion from geologic period boundaries to absolute ages uses commonly accepted values based on the estimated frequency of meteoroid impacts but is somewhat uncertain. SOURCE: Adapted by P.K. Byrne from M.G.A. Lapôtre, J.G. O'Rourke, L.K. Schaefer, et al., 2020, "Probing Space to Understand Earth," *Nature Reviews Earth and Environment* 1:170–181, https://doi.org/10.1038/s43017-020-0029-y. Springer Nature, 2020, reproduced with permission from SNCSC.

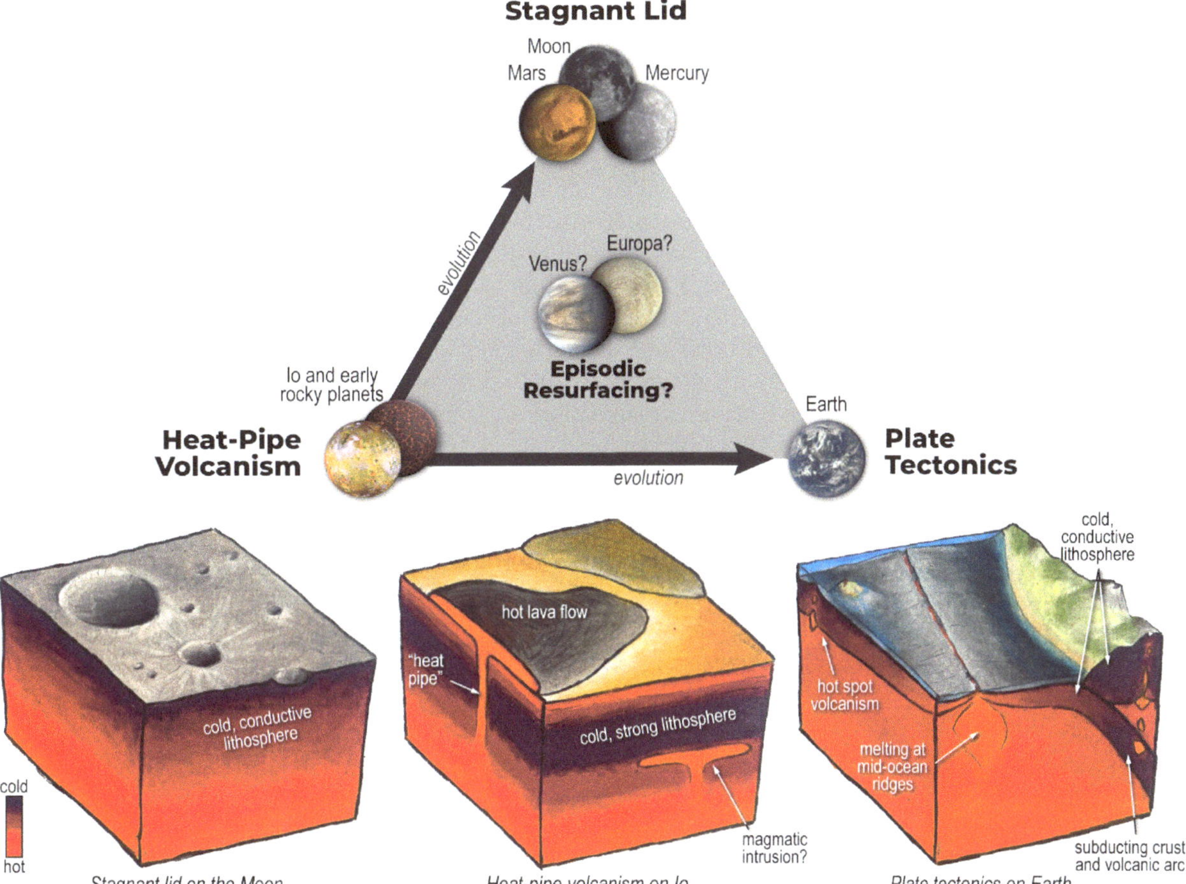

FIGURE 8-3 Diagram showing three different modes of planetary heat transfer, with various bodies as examples. As planets cool, melt production decreases and a heat-pipe planet may transition to plate tectonics or stagnant-lid heat transfer. Note that heat-pipe volcanism is compatible with (but does not require) a subsurface magma ocean. SOURCE: Figure by J.T. Keane from J. Keane, A.A. Ahern, F. Bagenal, et al., 2020, "The Science Case for Io Exploration." White paper #178 submitted to the Planetary Science and Astrobiology Decadal Survey 2023–2032. https://assets.pubpub.org/x6wu6hz2/11617915381503.pdf.

and their effects on surface mobility give rise to a great diversity of ages of planetary surfaces. The stresses and resurfacing from a convecting interior also create large-scale landforms such as rifts and scarps and contribute to crustal thinning and thickening.

Key to understanding the evolution of planetary interiors is determining how planetary materials deform in response to internal (e.g., radioactivity, primordial heat, and ongoing crystallization) and external (e.g., tides and large impacts) forcing and the feedbacks between that forcing and properties of the interior. Rheology, the relationship between applied stresses and the resulting deformation, affects the ability of interiors to convect and thus to transport heat and mass. Rheology also affects the heating produced by tidal forces, hence the coupling between internal and orbital evolution of satellites (see Question 8, Chapter 11).

The rocky planets and most of the medium-to-large satellites are differentiated and have internal liquid layers. These liquid layers can be responsible for generating magnetic fields and producing fluids that can erupt onto planetary surfaces. Liquid metal cores, partially to totally melted layers in mantles, and subsurface oceans require sufficient heat to melt and remain molten, while crystallization of these layers will produce latent heat. It is unclear whether these internal liquid layers are primordial—formed and maintained by heat from accretion and other heat sources—or if they are formed cyclically or incidentally via episodes of tidal heating or large impacts.

Q5.2b What Processes Control the Production and Evolution of Magnetic Fields?

Within solid planetary bodies, dynamo-generated magnetic fields are typically produced by motion of conductive fluid within a metallic core. Therefore, establishing that a planetary body produced a magnetic field during its history implies that the body underwent differentiation to create a fluid core. Spacecraft-based observations and paleomagnetic studies collectively indicate that Mercury, Earth, the Moon, Mars, Ganymede, and some planetesimals (e.g., Vesta and the pallasite parent body) have generated dynamos at some point in their histories. Dynamo fields are generally sustained by thermal or thermochemical convection driven by heat extraction into the overlying mantle. Thus, dynamo activity is usually controlled by the ability of the mantle to extract heat, so that mantle thermal evolution ultimately dictates whether a dynamo will operate. Dynamo fields may also be produced or affected by mechanical perturbations of cores from large impacts or precession, chemical exchange between cores and mantles, and even convection of sufficiently conductive silicate fluids within basal magma oceans (Lapôtre et al. 2020). In contrast, dynamo activity may be inhibited if, during cooling, core fluids achieve certain combinations of composition, temperature, and pressure that result in top-down or iron snow crystallization regimes that can inhibit convection. Therefore, establishing the field intensity histories and lifetimes of dynamo magnetic fields via paleomagnetism (Tikoo et al. 2017) or crustal magnetization studies (Johnson et al. 2020b), in tandem with gaining information regarding internal structure and composition, can reveal the thermochemical evolution of planetary interiors. Our knowledge of the history of planetary magnetic fields is currently too limited to understand the various ways in which planetary magnetic fields start and end.

Mapping of crustal magnetism not only sheds light on the history of the dynamo (if the nature of magnetizations and their ages can be constrained) but can also be used to understand surface and crustal processes. For example, magnetic anomalies associated with crater interiors and ejecta deposits on the Moon and Mercury may provide information regarding impact cratering dynamics, and the fate of iron sourced from the impactor. On Mars, and potentially Mercury, elongate zones of contrasting crustal magnetization may record large-scale tectonics and crust formation (Johnson et al. 2015). On airless bodies, intense remanent crustal magnetization may affect space weathering and/or regolith sorting (e.g., at lunar swirls; Q5.5a). Crustal magnetization may also record hydrothermal alteration and fluid flow within planetary crusts.

Q5.2c What Are the Chemical and Physical Consequences of Cooling and Solidification on Solid Body Crusts?

The composition and physical characteristics of a planetary body's initial, primordial crusts set the stage for the remainder of its evolution. Primordial crusts on rocky bodies generally formed from solidification and overturn of magma oceans. Crusts on differentiated icy satellites may have formed from the solidification of water and other volatiles that were initially liquid owing to accretional heating, or from later melting and mobilization of ice within their interiors. As a result, a great diversity of primordial crusts may have formed across the solar system, with compositions ranging from graphite on Mercury (e.g., Peplowski et al. 2016); anorthosite on the Moon; basalt and/or ultramafic magmas on Earth, Mars, Io, and probably Venus; and to ices on outer solar system satellites. These crusts are physically and chemically modified by secondary processes such as (cryo)volcanism,[2] impacts, weathering, and tectonics.

Magmatism and tectonism are the major internal processes that act to modify primordial crusts through the addition of new material (extrusive or intrusive volcanism), modification of crustal thickness (deformation), or crustal processing (remelting, recycling of material into the interior). The composition of material added to primordial crusts through magmatic and volcanic activity is coupled to the thermal and compositional evolution of the deeper interior (Q5.3b). As planetary objects cool and evolve, their interior compositions will change owing to the progressive extraction of magma at planetary surfaces and freezing of subsurface oceans and cores.

---

[2]While cryovolcanism can be regarded as a subset of volcanism, this report draws a distinction between the two processes for clarity. Thus, (cryo)volcanism should be read as volcanism and/or cryovolcanism.

Studying the composition of materials exposed on planetary surfaces, such as changing compositions of volcanic products erupted on the surfaces of rocky bodies and ocean worlds, thus provides an opportunity to reconstruct internal evolution and ongoing differentiation. For example, remote sensing data, in situ measurements of rocks by Mars rovers and analysis of martian meteorites record changes in composition that have been attributed to progressive cooling of the martian mantle (e.g., Baratoux et al. 2011; Filiberto 2017). Other observable changes in volcanism over time may include the type and quantity of volatiles in magmas, and the degree to which processes such as fractional crystallization occur during magma ascent. In extreme cases, initially basaltic melts can evolve to granitic compositions, as seen in samples and remote sensing data of the Moon (Jolliff et al. 2011), possibly from remelting the crust.

Changes in composition and temperature create stresses and change densities and hence can influence surface tectonics, whether eruptions can bring subsurface materials to the surface, and how surface features will be preserved. Cooling and contraction of the interior of Mercury created some of the largest faults in the solar system and surface compression led to a great reduction in the rate of volcanism (Byrne et al. 2016). Solidification of metallic cores may also enable the eruption of metals onto the surface (Johnson et al. 2020a). Within the ocean worlds, ongoing solidification of subsurface oceans can generate stresses that enable the eruption of water (e.g., Hemingway et al. 2020). The icy satellites preserve a great diversity of tectonic structures that are produced by variable combinations of external tidal forces and internal dynamics within and below their ice shells. Cooling also makes rocks and ice more viscous and hence increases the ability to preserve features on planetary surfaces. Unravelling the processes that create and preserve tectonic structures and their chronology can thus constrain the internal evolution of bodies with solid surfaces.

Combining the various insights offered by the composition of erupted materials, with gravity, topography, radar and electromagnetic mapping, stratigraphy, dated surfaces, and measurements of present-day heat flow, provides opportunities to determine temperature in the past and at present.

Strategic Research for Q5.2

- **Determine the timing and flux of volcanism on Venus, Mars, and Mercury** using orbital and in situ measurements of crustal composition and mineralogy with accompanying in situ radiometric dating.
- **Probe the magmatic history of the Moon** by conducting coordinated high-fidelity geochronology, geochemistry, and petrologic analyses either by in situ exploration or by analyzing samples returned by robotic or crewed missions.
- **Determine crustal composition, heat production, and origin of crustal dichotomies (if any) on the Moon, Mars, and Venus** by in situ geochemical, mineralogical and heat flow measurements by rover(s) or lander(s), by laboratory analyses of returned samples, and by collecting orbital and seismic data.
- **Determine the temperature, depth and timing of chemical differentiation, and the compositional and petrologic characteristics of magmas on different bodies** by a combination of theoretical and experimental studies on samples or analog compositions.
- **Assess the diverse mechanisms that create magnetic fields** by measuring the topology and evolution of active geodynamos in Mercury and Ganymede, determining whether Venus had an active geodynamo, and by studying remanent crust magnetization produced by extinct dynamos on the Moon and Mars, via spacecraft measurements or paleomagnetic studies of returned samples.

Q5.3 HOW HAVE SURFACE/NEAR-SURFACE CHARACTERISTICS AND COMPOSITIONS OF SOLID BODIES BEEN MODIFIED BY, AND RECORDED, INTERIOR PROCESSES?

Planetary surfaces, the most external portion of the solid body, are influenced by complex physical and chemical processes within the body's interior. The evolution of the two regions is therefore linked, and surface features can preserve the history of the interior. The questions that follow highlight multiple ways in which the interior modifies the surface and illustrate the importance of teasing out how different interior processes influence the surface. Constraining a planetary body's unobservable interior using surface materials is inherently challenging, and many questions about how interior processes influence surfaces remain.

Q5.3a What Internal Processes Control Surface Topography and Produce Tectonic Features?

Surface topography is supported through a combination of dynamic, active processes (such as a plume of hot rising mantle material pushing up on the lithosphere, creating a topographic rise) and the strength of the lithosphere. Lithospheric strength depends on (at least) thickness and composition—for example, ice is generally weaker than silicate rock. Tectonic features are produced by lithospheric or crustal deformation, and their presence requires heterogeneous stresses and strains. Deep understanding of topography, tectonics, and geologic structures requires detailed information of the surface structures and their relative ages, combined with knowledge of a body's internal properties (e.g., Black and Manga 2016). In particular, it is necessary to know the thickness and the composition of the lithosphere, and whether those parameters have remained constant with time. The state of the underlying mantle can also have important consequences for surface topography and tectonics, and so it is necessary to understand if mantles were warm and deformable (either from tidal heating or radiogenic heat) or essentially frozen—and if so, when that transition happened. Topography and tectonics features can also record tidal and rotational forces (in both orientation and magnitude) and provide key insights to the evolution of a planetary body's spin and orbit through time. A common question for nearly every solid body is: What is (or was) the duration and magnitude of tectonic activity? The special case of present-day activity is discussed in Q5.6.

The varied activity of solar system bodies gives rise to a diverse array of tectonic features that have yet to be explained. Mariner 10 and MESSENGER revealed that Mercury possesses near-global scarps and folds. The Moon and Mars both have global-scale crustal dichotomies or asymmetries that are still unexplained. Venus's surface is replete with tectonics, yet it is unclear if they are the result of movement of lithospheric plates (analogous to Earth's plate tectonics), or some other process. The age and crustal thickness of Venus's major geologic terranes, the tesserae and the plains, are uncertain. This tectonic menagerie only becomes more varied in the outer solar system. In most cases, the thicknesses of ocean world ice shells are unknown. Plate-tectonic-like motion has been hypothesized for Europa, but still debated. On Enceladus, the formation and evolution of the south polar tiger stripes (the tectonic fractures that source Enceladus's plume) are unknown. The tectonic history of more distant worlds, like the uranian and neptunian satellites and trans-neptunian objects (including Pluto and Charon) remain elusive.

Q5.3b What Controls (Cryo)Volcanic Eruptions on Bodies with Solid Surfaces, and How Does the Composition of Erupted Materials Vary?

Volcanic eruptions shed light on the internal compositions, temperatures, and degree of processing of subsurface materials on solid bodies without having to directly sample their interiors. Eruptions on planetary bodies are controlled by numerous, interrelated factors, including (but not limited to) amount of internal heating (controls melt generation and degree of melting), how easily melts can erupt (influenced by volatile content, composition, and physical properties of melts), melt composition (influenced by conditions at (cryo)magma source), and crustal thickness, stress state, and composition (influencing melt ascent and eruption style) (e.g., Head and Wilson 2017). The ratio of silicate magmas erupted onto the surface versus those intruded into the crust appears to be highly variable and is known only for Earth with any fidelity. However, typical ratios suggest that only a small fraction of magmas are actually able to erupt on the surface of any planet or moon. The terrestrial ratio of extrusive to intrusive volcanism varies with tectonic setting, is estimated to vary inversely with size of the body (Black and Manga 2016) and is related to the structure and stress state of the crust and the efficiency of mantle degassing. Although much less is known about the eruptibility of cryomagmas on ocean worlds and dwarf planets, they may follow similar trends.

Compositional variation in erupted materials is controlled by several factors. On icy bodies, the degree of connectedness between the subsurface ocean and the surface (e.g., through fractures in the ice shell) can influence the degree of processing of cryomagmas during ascent, possibly resulting in higher concentrations of sulfate, carbonate, or chloride salts in eruptions on larger bodies like Europa and Titan (e.g., Quick and Marsh 2016). Materials that erupted onto the surface of dwarf planet Ceres may be representative of the remnants of a muddy ocean, the gradual freezing of which produced salts whose composition may have changed over time (e.g., Raymond et al. 2020). On rocky bodies, silicate magmas produced by partial melting of the mantle (primary magmas) rarely erupt directly onto the surface, although they have been suggested to be more common on some planets and planetesimals (e.g., eucrites on Vesta). Instead, on rocky and ocean worlds, (cryo)magma compositions typically evolve during ascent, while in residence in intermediate staging chambers within the crust, and upon eruption. They commonly experience separation of crystals from melt (fractional crystallization) and sometimes contamination by the enclosing rocks (assimilation). A quantified understanding of compositional changes that (cryo)lavas have experienced is necessary if they are to be used to constrain the internal compositions of planets and satellites. As the interior of bodies change, so does their volcanism (see Q5.2c). For example, if no crustal recycling exists, volcanism can provide volatiles with a one-way ticket from the mantle to the surface. Over time, this would cause mantle material to become volatile poor (i.e., to "dry out"), affecting the nature of volcanic eruptions throughout the planet's history.

Q5.3c How Have Diverse (Cryo)Volcanic Processes Shaped Solid Body Surfaces, Physically and Chemically?

Volcanism is a process that transfers both energy and material from a planetary body's interior to its surface, and is driven by conditions in a body's interior, providing a link between volcanic products and internal conditions. Volcanism results in addition of mass to a surface and a reshaping of the landscape; especially voluminous effusions (e.g., flood basalts) may form the bulk of the crust in some locations. Along with volcanic edifices, a myriad of other volcanic landforms is observed throughout the solar system including flows, channels, lava tubes, domes, rilles, calderas, and pyroclastic deposits. In some cases, such as the Tharsis volcanoes on Mars, the edifices are so large they can deform the crust beneath and surrounding them. As discussed in Q5.6b, Io, the most volcanically active world in the solar system, possesses active, 100-kilometer-wide overturning lava lakes, fire fountains, gigantic plumes, and is globally resurfaced by volcanic processes at a rate of roughly 1 centimeter per year. Venus is similarly covered with volcanic landforms, although it is uncertain if Venus is volcanically active today. Solidified lava flows have been observed to cover older topography on bodies including Mars, the Moon, Mercury, Venus, and Io. This process also occurs with smooth cryolava units on ocean worlds such as Europa, Triton, the uranian satellite Ariel, and Pluto's moon Charon (e.g., Beyer et al. 2019). Pyroclastic deposits have been identified on Mercury, the Moon, and Io, and possibly on Mars and Venus as well, and would serve as a source of sediment and also prevent erosion of the surfaces upon which they are deposited. While we can identify many volcanic features across the solar system, we have nowhere near a complete catalogue of these features. In many cases, we lack high enough resolution imagery of these features to definitively identify them as volcanic in origin. Similarly, understanding emplacement processes is hampered by the difficulty in determining internal flow structures and subtle compositional variations. These issues are particularly problematic on Venus and icy bodies, where an array of unique, likely volcanic landforms have been identified but cannot be fully described, classified, or explained (e.g., Crumpler et al. 1997). The physical expression of volcanism is often controlled by the composition of the lava involved, thus obtaining compositional data of these landforms is also critical.

The addition of fresh (cryo)volcanic material to a planetary surface also alters the surface chemical environment. This has the effect of providing locally distinct chemical regions, or replenishing elements that may have been transported away or taken up by other surface processes. Dustings of fine particles from geyser-like plumes on Triton, Enceladus, and possibly Europa, and brine eruptions on Ceres have produced varying compositions on their surfaces, including local enhancements of nitrogen, to extensive icy mantlings and elevated concentrations of sulfate and/or chloride and carbonate salts, respectively. Much is still to be learned concerning the compositions of flow-like features and airfall deposits on planetary bodies. The majority of volcanic products thus far identified on rocky bodies appear to be basaltic, and on bodies for which we have sufficient compositional data, these basalts are different in composition from the average crust. However, detailed investigations of meteorite samples from Mars and the Moon, as well as exploration by Mars rovers, have revealed local generation of more evolved melt compositions (Q5.2c). This suggests that broad-scale observation of volcanic products may not be sufficient to capture the full story of the chemical modification of planetary surfaces by volcanism.

Q5.3d Where and How Are Surfaces Modified by Hydrothermal/Geothermal Processes?

Materials that have been mineralogically or geochemically altered by hydrothermal processes in the interior can be exposed on the surface, thus modifying its composition. One example is the metamorphic minerals thought to have been formed in hydrothermal systems established below large martian impact craters—systems that can persist for centuries or longer (see also Q4.3d). Such hydrothermally altered rocks in impact ejecta have been analyzed by the Opportunity rover in Endeavour crater on Mars. The occurrence of water ice in permanently shadowed regions of the Moon and Mercury may allow the possibility that hydrothermal systems may have been established below some craters on those bodies as well. Another example is serpentine and clays in carbonaceous chondrite meteorites, formed by early aqueous alteration of ice-bearing asteroid parent bodies heated by rapid decay of short-lived radioisotopes. The emplacement of magmas within planetary crusts containing groundwaters or ices may have also driven hydrothermal reactions. Available observations of Venus are not extensive enough to know whether or not hydrothermal activity may have occurred there, but the ancient tesserae may hold evidence of past fluid-rock interactions. Some geothermal reactions may also involve fluids composed of volatiles that are not water. For example, Mercury and Io may have had fluid-rock interactions involving sulfur, and the Moon shows evidence of rock interaction with H_2 and CO fluids. Although hydrothermal processes may be widespread among the terrestrial planets, our knowledge of them is hampered by limited surface exposures.

Ocean worlds have ongoing geothermal activity. The Cassini spacecraft found silica grains thought to be sourced from erupting plumes on Enceladus, pointing to water–rock interaction at temperatures likely occurring at the seafloor of its subsurface ocean. Melting of carbon- and nitrogen-containing ices could have produced non-aqueous or hybrid fluids in icy worlds. The Dawn mission discovered ammonia-bearing clays and salts formed by extensive aqueous alteration on dwarf planet Ceres. Similar fluid-rock interactions may occur on the ocean floors of Europa, Titan, and other giant planet satellites that are subject to tidal and/or radiogenic heating. As water-rock interactions are only sustainable if fresh rock is available, constraining the timing and duration of water-rock interactions on ocean worlds also provides information on the exposure or production rates of fresh rock surfaces, giving insight into other geological processes. Hydrothermal processes may also be responsible for synthesis of some of the organic compounds on ocean worlds, dwarf planets, and asteroids. Much work remains to be done in deciphering the compositions of subsurface fluids in extraterrestrial bodies and in understanding their origin, chemical evolution, and persistence through time.

Strategic Research for Q5.3

- **Identify and classify tectonic and volcanic landforms (both modern and ancient) on rocky bodies (Venus, Mars, Mercury, the Moon, and Io), and provide fundamental constraints on lithospheric properties such as the thickness of the deforming layer** via high-resolution topography, gravity and images of the surfaces, and chemical and mineralogical measurements.
- **Investigate and classify tectonic and cryovolcanic activity and landforms on icy bodies (Europa, Enceladus, Triton, and Titan) to determine properties such as location of deforming regions and the thickness of deforming layers** via global high-resolution imaging and topography, and constraining the ages of cryovolcanic units/structures.
- **Probe compositional evolution, surface weathering, and fluid–rock interactions on icy moons and dwarf planets such as Europa, Enceladus, Titan, Triton, Ceres, and the uranian moons** using orbiting and landed spacecraft measurements of crustal composition and mineralogy.
- **Determine the range of volatile contents and species in planetary melts in igneous samples from Mars, the Moon, and asteroids, to constrain the range and variety in planetary volatile contents, and factors influencing melt generation, composition, and eruptibility** using Earth-based laboratory measurements of returned samples and/or meteorites.
- **Assess the nature and timing of hydrothermal processes on small bodies and planets** via modeling of aqueous alteration, and a combination of high-resolution spectral data and laboratory analyses of samples of asteroids Ryugu and Bennu returned by Huyabusa2 and OSIRIS-REx, from Mars's moon Phobos from MMX, and from Mars by the Mars Sample Return Program.

Q5.4 HOW HAVE SURFACE CHARACTERISTICS AND COMPOSITIONS OF SOLID BODIES BEEN MODIFIED BY, AND RECORDED, SURFACE PROCESSES AND ATMOSPHERIC INTERACTIONS?

The surfaces of solid bodies record evidence of atmospheric interaction, and thereby provide information about how the atmosphere may have evolved through time (see also Question 6, Chapter 9). Constraining the nature of the interaction between atmospheres and solid surfaces requires detailed morphologic, stratigraphic, and compositional information of the solid surface, and how these characteristics may be expressed in different locations and preserve different times in the past. Generation and transport of sediment, regolith formation, composition (including the products of chemical weathering) and distribution, and how the regolith composition changes with time and place all provide vital details about how the atmosphere shapes a planetary surface. Given that "the present is the key to the past," precise information about the current atmosphere (including composition, temperature, and pressure), and how it varies laterally, vertically, and temporally (on diurnal, seasonal, and longer timescales), is essential for improving our understanding of atmosphere-surface interactions.

Q5.4a Where and How Have Fluvial Processes Sculpted Landscapes?

Landscapes shaped by liquids inherit physical, chemical, and mineralogical properties reflecting the climatic, fluid, and geologic conditions at their formation. Investigation of these fluvial landscapes and subsequently formed sedimentary materials can reveal the history of atmospheric conditions, timing of habitable conditions, and the impact of distinct planetary conditions on fluvial processes. Such landscapes and preserved sedimentary rock records are clearly observed on Mars, where multiple orbital and landed missions have confirmed the presence of liquid water in the past, which formed rivers, deltas, lakes, and possibly oceans (McLennan et al. 2019). However, it is unclear if the current landscapes as observed from orbit form a complete record of Mars history, or if surface-based observations would reveal a much longer history of fluvial activity. Possible records of past surface fluid flow may also exist on Venus and Titan. Venus was not previously considered a candidate for habitable surface conditions owing to its extreme surface conditions, but work in the past decade has suggested that Venus's deviation from Earth-like conditions may have been geologically recent, and some rocks on the venusian surface might record fluvial activity from more clement past surface conditions (Way et al. 2016). Titan has modern fluvial activity, but with organic fluids and a water-ice crust, it is unclear whether a sedimentary record of past fluvial activity is preserved. Some features on the dwarf planets Ceres and Pluto have been ascribed to fluvial activity, although this is more uncertain. For all of these worlds, it remains unclear how planetary hydrologic cycles redistributed, modified, and concentrated materials in the past, at what time, and for how long.

Q5.4b Where and How Have Glacial Processes Sculpted Landscapes?

Glaciers sculpt landscapes through strong erosive processes at the base of massive ice piles and through evaporation, melting, and sublimation of ice deposits. Glacial processes have affected landscapes on terrestrial and icy bodies from Earth to Pluto and possibly beyond. On terrestrial bodies, wet-based glacial activity causes substantial erosion and smoothing of surface topography, as well as smoothing of impact features, and is recognizable long after the activity stops. On Mars, as on Earth, the strong topographic effects of glacial activity are frequently used as a constraint for paleoclimate models. However, the distribution of identified glacial features on the surface of Mars does not match the expected distribution based on current paleoclimate models, so improved understanding of the past extent of glacial surface modification may improve understanding of the martian paleoclimate. Further insight into Mars's present and paleoclimate would be obtained by constraining the current amounts and spatial distribution of near-surface and subsurface ice. A record of past glaciation is also contained in the polar caps on Mars as observed by radar, but we have yet to understand the chronostratigraphy of Mars's polar layered deposits and the nature of the atmospheric-surface interactions preserved in these layers.

Glacial activity on dwarf planets such as Pluto, and on outer-planet moons, generally involves exotic ices rather than water ice. Nitrogen ice glaciers are flowing today on Pluto, filling a vast solid nitrogen ice sheet within an impact basin, and geomorphic evidence exists for more extensive glacial activity in the geologic past. Similar or related glacial activity may be occurring on Triton and on other dwarf planets in the Kuiper belt. These different ices and mixtures of ices create glaciers with distinct physical properties compared to Earth, so further study of such glacial processes is important for understanding and further investigating surface processes and their records on bodies farther out in the solar system. The record of glaciation on icy bodies and dwarf planets remains uncertain, as does the exact composition of the glaciers, and the processes controlling their accumulation, movement, and destruction.

Q5.4c How Has Regolith Generation and Subsequent Gravitational or Aeolian Transport of Material Driven Landscape Evolution?

Regolith, or unconsolidated surface material, is found on nearly all solid solar system surfaces regardless of size, density, solar distance, or age. The generation of such regolith during impact cratering, physical or chemical weathering, precipitation, and/or chemical sintering frequently involves interactions between surfaces, planetary volatiles, atmospheres, and/or the surrounding space environment, and is discussed in more detail in Q5.5b. Further transport of such materials downslope by mass wasting or through aeolian transport has shaped landscapes with bedforms recording ancient processes.

Bodies with even thin, transient atmospheres also contain a record of aeolian transport processes that may record past climates, atmospheres, and atmospheric densities. For example, large aeolian ripple bedforms, which form under atmospheres less dense than Earth's current atmosphere, are recorded in both modern sediments and sedimentary rocks on Mars (Lapôtre et al. 2016) and have been putatively identified on comet 67P/Churyumov-Gerasimenko (Thomas et al. 2015). Other preserved landscape-forming aeolian bedforms, such as transverse aeolian ridges on Mars, Venus's dune fields, extensive fields of longitudinal dunes on Titan, and putative aeolian features on Pluto and Triton are not yet well-understood. Investigation of such aeolian landscapes and those preserved in sedimentary records on Mars and potentially Titan or Venus would help expand our understanding of aeolian systems on Earth to other planetary conditions and improve understanding of aeolian physics more broadly.

Mass wasting—the movement of unconsolidated materials downslope under the force of gravity—depends on composition, preexisting weaknesses, and other factors. All planetary bodies have gravity and at least some topographic relief, if only from impact craters, so mass wasting may be one of the most common surficial processes. Small impacts themselves can give rise to diffusive down-slope motion of material, or trigger landslides. Liquid has prompted mass wasting on Mars and perhaps on Titan. The flanks of volcanoes are also subject to slope failure, and mass wasting is a natural part of volcano evolution. The origin of young martian gullies, mass wasting at a much smaller scale, remains controversial (McEwen et al. 2011). Mass wasting also occurs on scarps on small bodies, such as asteroid Vesta (Krohn et al. 2014).

Q5.4d What Are the Signatures of Chemical Weathering/Alteration, and How/Why Have Surface Mineralogies Varied over Time?

Chemical weathering and alteration of surface materials represent the main way atmospheres and hydrospheres interact chemically with the solid surfaces of planetary bodies and produce secondary minerals from reactions between minerals and species in atmospheric gases or fluids. The specific secondary minerals formed are determined by the initial mineralogy of the surface, chemical species in the atmosphere or fluid, concentration of gas/fluid species, and the pressure and temperature conditions of the near-surface environment. Therefore, chemical weathering is highly dependent on the climate of planetary bodies and its chemical signatures can vary between planets, as well as over time on a single planet. This fact makes preserved weathering products useful tracers of past climate conditions.

Signatures of weathering are contained in both mineralogy and chemistry of weathering products. For example, orbital mineralogical data has been used to identify changing weathering products on Mars over geologic time, by identifying clay minerals in ancient layers, sulfates in progressively younger material, then anhydrous iron oxides in modern regions (Bibring et al. 2006). Alteration at the surface and within the crust may have largely drawn down the planet's surface water inventory (Scheller et al. 2021). However, it remains unclear how extensive this weathering is on Mars, and what it can reveal about the presence (or absence) of surface water or groundwater, and the compounds dissolved into that water through Mars's history. Venus may once have had a more clement surface environment, potentially hosting liquid water for substantial amounts of time (Way et al. 2016). In such a scenario, weathering would likely have occurred there and would be starkly different than the current weathering style occurring under hot, high-pressure, dry conditions. Similarly, weathering processes involving dissolution of soluble organic material by methane rain have been proposed for Titan, but there is currently not enough data to document this process or use it to unravel the history of surface–atmosphere interactions on Titan, including climate change (Neish et al. 2015). Aqueous alteration, which could be considered a form of chemical weathering, has occurred on asteroids or comets, and likely on (or within) icy bodies as well.

Strategic Research for Q5.4

- **Search for evidence of weathering or the presence of ancient water on Venus** by characterizing the chemical compositions and mineralogy of surface rocks paired with high-resolution imaging and topography using global and local scale measurements.
- **Map and measure the geologic, chemical, and mineralogical characteristics of Mars's Noachian stratigraphic record to correlate local and regional sedimentary depositional episodes and provide insight into the range and diversity of environments and their relative timing** via in situ measurements from a long-distance rover or airborne vehicle.
- **Characterize the paleoenvironment, weathering, habitability, geochemistry, petrology, geochronology, and geologic history of returned samples from an ancient martian sedimentary sequence, as well as regolith and any igneous rocks**, via laboratory analyses of samples returned from Mars.
- **Investigate the fundamental processes that govern hydrologic cycles** by investigating Titan's hydrologic cycle via global high-resolution imaging and topographic and mineralogical data.
- **Constrain the history of glaciation and erosion on icy bodies** via experimental studies of the material properties (e.g., strength, density, volatilities, and rheology) of relevant planetary ices, planetary ice-regolith mixtures, and ice clathrates at relevant temperatures and pressures.
- **Constrain weathering rates and regolith formation, including physical, chemical, and mineralogical changes to surface materials during weathering under conditions relevant for Venus, Mars, and Titan** using experimental studies of abrasion, weathering reactions, and kinetics.

Q5.5 HOW HAVE SURFACE CHARACTERISTICS AND COMPOSITIONS OF SOLID BODIES BEEN MODIFIED BY, AND RECORDED, EXTERNAL PROCESSES?

Planetary surfaces often retain records of the external environments in which planetary bodies formed and evolved. Deciphering these records, however, requires detailed knowledge of the processes by which the environment and surface interact. Space weathering influences the physical and chemical properties of planetary bodies, and how they reflect and absorb light. Impacts and volatile sublimation/redistribution significantly modify the surfaces of bodies throughout the solar system. Impacts disrupt and redistribute materials on planetary surfaces, excavate materials from depth and introduce exogenic materials. Volatile activity shapes the surfaces of smaller bodies, drives many tenuous atmospheres, and volatiles trapped on airless bodies provide crucial historic records.

Q5.5a How Do Space Weathering Processes Modify Surface Characteristics and Compositions?

The surfaces of airless bodies are bombarded by micrometeoroids, the solar wind, magnetospheric plasma, solar energetic particles, and galactic cosmic rays (e.g., Pieters and Noble 2016). These processes impart profound changes to the physical properties, chemistry, and spectral properties of airless bodies, and are collectively known as space weathering. Because of the complex interactions of space weathering agents operating in concert, and the difficulty of reproducing the environment, processes, and relevant timescales in laboratory simulations, space weathering is only partly understood.

Charged particles (from the solar wind, and/or the relevant planetary magnetosphere for planetary satellites) collide with individual regolith grains and lose energy as they are implanted into grains, breaking bonds. With increasing radiation exposure, regolith grain surfaces accumulate damage and lose their periodic crystalline structure (become increasingly amorphous). Some fraction of these charged particles also causes sputtering, where atoms are ejected from regolith grains and are either lost to the exosphere or deposited onto nearby grains. Micrometeoroid impacts also result in the production of impact glass (for rocky surfaces) and vapor, which can be deposited onto the surfaces of regolith grains.

For silicate targets, these processes form opaque nanoscale particles (e.g., metallic iron, troilite) within the glass coatings on grains, which strongly affect light absorption and reflection. Micrometeoroid impacts further produce glass-welded agglomerations of opaque-rich soil grains called agglutinates. Because charged particle and micrometeoroid bombardment are both thought to result in similar physical and chemical changes (apart from agglutinates), it is not known whether one process dominates, their relative importance in altering the regolith, or if there are substantial differences when both processes operate together. Further, the relative roles of more-energetic but infrequent particles—solar energetic particles (SEPs) and galactic cosmic rays (GCRs)—are not known. SEPs and GCRs may cause dielectric breakdown ("sparking") that would also result in the production of glass and possibly nanoscale opaques (Jordan et al. 2015); the contribution of dielectric breakdown to space weathering is unknown.

For icy surfaces such as Europa, radiation from magnetospheric corotating plasma converts crystalline ice into amorphous ice, forms radiolytic volatiles (e.g., H_2O_2, O_2) and non-ice compounds (e.g., acid hydrates), and alters other materials (e.g., changing the colors of salts); these processes are thought to produce a gradual brightening of features with increasing age. These processes, however, compete with impact gardening and thermal processing, and can be confused with micrometeoroid-delivered materials. For most planetary satellites, the plasma effects are usually strongest on the trailing hemispheres of synchronous moons (as charged particles caught in the planet's rapidly rotating magnetosphere "run-into" the trailing hemisphere), whereas impact gardening is dominant on the leading hemispheres (as the moon "sweeps up" orbiting debris in the planet's orbit). In some circumstances, such as the inner midsize moons of Saturn, such effects can be reversed, with enhanced MeV electron bombardment on the leading hemispheres of Mimas and Tethys, and enhanced E-ring particle bombardment on Mimas's trailing hemisphere (Schenk et al. 2011). Deciphering these data allows constraints to be placed on the radiation-exposure ages and compositions of surface-feature but requires improved modeling and laboratory data.

These chemical and physical effects of space weathering are of great interest in part because they profoundly affect surface appearance. Newly exposed and comminuted regolith, such as is observed in young impact crater rays, is often substantially brighter than the space-weathered surroundings at optical wavelengths on airless bodies. Significant space weathering effects have been observed across wavelengths, from the ultraviolet to the thermal infrared. These spectral changes are thought to be largely owing to the accumulation of nanoscale opaque particles in grain coatings and agglutinates that are highly efficient light absorbers. Thus, as space weathering proceeds, the composition of a surface can be increasingly masked. The degree to which spectra have been changed can be a powerful tool to understand the duration of surface exposure and date surfaces, although the weathering rates are poorly known and depend on surface composition and space/magnetospheric environment.

Mature regolith samples from the Moon and asteroid Itokawa demonstrate that space weathering products depend on initial composition. However, comparison of space weathering processes across the solar system, or even across a single complex planetary body, needs to also consider environmental differences. For example, the degree to which micrometeoroid impacts produce melt and vapor depends on impact velocity, which varies with heliocentric distance. Solar wind fluence is highest closest to the Sun, but GCRs play a larger role in the outer solar system. The effective solar wind flux scales with latitude, and thus for bodies with minimal axial tilt, space weathering may vary with latitude. Magnetic fields can shield the surface from charged particles to some degree but may also help to transfer species between nearby bodies (e.g., from Earth to the Moon or between the inner Galilean moons). However, the fluxes and energies of species that reach and thus weather the surface and their dependence on magnetospheric conditions are not well known.

Developing a general understanding of how different space weathering processes modify the surfaces of planetary bodies and how those vary with local conditions will help to determine rates of geologic processes, understand duration of surface exposure of materials, understand surface composition, and provide a valuable framework that can be used to better interpret remote sensing observations across the solar system.

Q5.5b How Have Impacts Affected Surface and Near-Surface Properties?

Impacts are one of the most ubiquitous geologic processes, affecting nearly every solar system body and in many cases dominating their landscapes. Impacts modify the compositions of planetary surfaces by redistributing target material and delivering exogenic material. Impacts modify crustal structure through both melting and fracturing of the target material (Melosh 1989), and some models suggest impacts may also induce volcanism. Impact processes are considered in detail in Question 4 (Chapter 7).

Nearly all remote sensing data and returned samples are of regolith (with the notable exception being martian igneous meteorites); thus, it is critical to understanding the formation, structure, and evolution of regolith in order to interpret these data and samples. Impacts play a critical role in the formation of regolith, particularly on airless bodies, via ejecta/catastrophic disruption and micrometeoroid impacts (Q5.6c). The rate at which impacts create regolith relative to other processes such as thermal fatigue, however, remains unknown, particularly on smaller and/or icy bodies. While we know impacts mix or "garden" regolith, the rates and depths of this mixing are not well known but have major implications for radiation processing of near-surface material (e.g., on Europa). Our knowledge of how the regolith varies vertically is poor, even for the Moon, and yet understanding this third dimension is essential for constraining lunar surface evolution, the origin of samples, and for any successful lunar human engineering endeavors.

Regolith can also preserve ancient layers of ejecta that both record the local history of impact events and the ancient space environment (variations in solar/GCR activity; Q5.5a). Impacts also excavate material from depth, providing a window into the composition of the lower crust and the mantle or ocean below; the correlation between crater size and source depth of surface material, however, is poorly constrained, as is the degree to which crater ejecta is deposited versus local material is exhumed with increasing distance from the primary crater. Impacts deliver exogenic material to planetary surfaces. For neighboring bodies, like Phobos and Deimos, or satellite systems in the outer solar system, material can be exchanged through so-called sesquinary impacts, where ejecta from an impact on one satellite can orbit the primary and either reimpact the original satellite or another satellite. Improved modeling of these processes allows observed surface compositions to provide insight into both the materials comprising the original planetary crust, from the surface to the mantle/ocean beneath, and those materials delivered to the surface.

In addition to redistributing material, the energy imparted by impacts and the resulting shockwave modify the existing target material through melting and fracturing. Major outstanding questions exist, however, about the details of these processes. The distribution of melted material is debated (e.g., the melt deposit antipodal to, and possibly originating from, the lunar crater Tycho), and the fate of melted water-ice "bedrock" on icy worlds (e.g., Titan) is not well constrained. Fundamental questions remain regarding the composition (and homogeneity thereof) impact melt composition and if cooling occurs slowly enough to allow for differentiation.

Q5.5c Where and How Do Volatile Deposition, Sublimation, Transport, Redeposition, and Loss Take Place, Now and in the Past?

Volatiles exist and interact with regolith under a wide range of chemical, geological and gravitational conditions, from Mercury to TNOs. Many airless bodies cold trap substantial volatile deposits (including water ice) in permanently shadowed regions (PSRs). PSRs, like those on the Moon and Mercury, may retain a record of volatile transport across the solar system and planetary dynamics, although this is still uncertain and depends on their unknown age. Landscapes and climates of entire worlds (e.g., Mars, Callisto, Pluto, Triton, comets, possibly the dwarf planet Eris), are controlled, in some cases almost exclusively, by volatile interactions (Mangold 2011). On Ganymede, water frost redistribution is modulated by the background magnetic field, resulting in bright polar caps

(Khurana et al. 2007). However, the interaction timescales are generally unknown. For example, it is unclear how TNOs are modified into the Centaur and Jupiter family comet populations—which have different characteristics (e.g., colors)—as they migrate closer to the Sun.

Volatiles trapped on terrestrial bodies (e.g., the Moon, Mercury, asteroids) may retain rich records. For example, lunar PSR deposits may preserve the volatile history of the Earth–Moon system, including the delivery of organics to Earth. The abundances of (moderately volatile) alkali elements on Mercury may reflect magmatic abundances or imply thermal redistribution of material. Fundamental questions remain, however, regarding the transport, retention, physical and chemical alteration, and loss processes operating on such deposits over seasonal, diurnal, and precessional timescales (see also Question 6, Chapter 9). The relative contributions of impacts, volcanism, and solar wind to volatile inventories are poorly constrained. There is a lack of strong constraints on relevant rates, including diffusion, low-temperature chemical reactions, clathrate formation/retention (in thicker deposits), and loss mechanisms. There is little information regarding the chemical, physical, or mechanical/structural evolution of these ices, the distribution of volatiles beyond the poles, and the variability of these characteristics with solar distance (e.g., between Mercury and the Moon).

In the outer solar system, volatiles are not trace species, but rather, substantial components of planetary bodies. The surface materials and activity (e.g., glaciers) observed on these bodies, in particular, provide critical clues to understand the process(es) that affect how these bodies incorporate and retain volatiles and refractory materials (Q5.4b). Investigating the most distant objects in the solar system (e.g., TNOs, Centaurs, interstellar objects, and dynamically new comets) provides insight into the most volatile elements. Studying Jupiter family comets, the most accessible comets, provides insight into less-volatile elements (such as water, semi-volatile organics, and refractory materials) that can inform the dust-to-ice ratio of cometary materials, and test if comets delivered Earth's volatiles and organics. Large advances are possible by exploring these yet unknown bodies (e.g., Centaurs and TNOs) and by investigating the chemical/physical/mechanical alterations of volatile and nonvolatile materials in situ or within a returned sample from a primitive small body.

Surface-volatile sublimation is known to evolve landscapes and climates, from low-volatility ices like water in the inner solar system to high-volatility ices like CO_2 at Mars and Callisto, and CH_4, CO, and N_2 at Triton, Pluto and beyond. The relative importance of sublimation, however, is often poorly constrained due both to limits in observational data and difficulties replicating relevant temperatures and pressures in laboratories, thereby limiting knowledge of relative rates and material properties. It is unknown, for example, if Triton's plumes are driven by cryovolcanism (Q5.6b) or surface/near-surface volatile sublimation, if such plumes exist elsewhere, and where not (e.g., Pluto), why not; how seasonal sublimation of Mars's polar caps influence its climate; how volatiles are transported on icy satellites/dwarf planets, and how this influence their climates; how dominant a geologic process sublimation is on icy satellites/dwarf planets; and how far into the Kuiper belt sublimation-driven activity exists.

Strategic Research for Q5.5

- **Determine the origin, time of delivery, vertical and lateral distribution, and current cycling of cold-trapped lunar volatiles** via in situ analyses of isotopes (e.g., deuterium/hydrogen), sulfur, organics, abundance and distribution of volatiles, and local exospheric measurements.
- **Investigate the effects of sublimation, space weathering and interior processes on the surfaces of ice-dominated worlds (including icy satellites, active asteroids, and comets)** via high-resolution imaging, spectroscopy, and topography.
- **Investigate the role of space weathering processes on airless rocky bodies** using high-resolution imaging and spectroscopy of planetary surfaces coupled with laboratory studies of representative/analog materials and laboratory analyses of returned samples.
- **Determine how small bodies (asteroids and comets) incorporate and retain volatiles** via analyses of returned samples or in situ analyses of volatile and nonvolatile materials.
- **Assess processes producing lunar regolith heterogeneity** by measuring the thickness variations, and vertical and lateral compositional variability of the lunar regolith, using geophysical profiling, high-resolution multi-spectral imaging, and petrologic analyses of in situ or returned samples.

Q5.6 WHAT DRIVES ACTIVE PROCESSES OCCURRING IN THE INTERIORS AND ON THE SURFACES OF SOLID BODIES?

Activity is present on the surfaces and in the interiors of both rocky and icy bodies throughout the solar system. Active processes (volcanism, impact cratering, tectonic, fluvial, and aeolian activity) manifest differently, but all reveal critical information about the evolution of planetary bodies. The driving mechanisms behind geologic activity might be endogenic, driven by the body's internal heat (Q5.2a), or exogenic, controlled by energy derived from the Sun (Q5.5c), or by the influx of impactors (Q5.4c, Q5.5b). Through a more thorough understanding of the process controlling the activity, the better the rate of change, frequency or duration of the activity can be constrained, such as the location of volcanism on Venus, the occurrence of mass wasting events, or the duration of eruptive activity on Enceladus.

Q5.6a Where and How Are Convection and/or Crustal Recycling Taking Place?

Earth is the archetype of one style of planetary crustal recycling via plate tectonics, although plate tectonics has been hypothesized for some other worlds (Q5.2a). Crustal recycling has been proposed for at least two other planetary bodies, Venus and Europa. In both cases, though, the form of crustal recycling differs. On Venus, plume-induced subduction has been proposed to occur at coronae. Regardless of whether subduction is occurring, coronae are identified as the manifestation of plume-lithosphere interactions that is potentially ongoing across Venus. Crustal recycling on Europa appears more like Earth's plate tectonics, with obvious spreading zones and putative subduction zones. The extent of these recycling zones on Venus and Europa is not yet known. A style of crustal recycling that involves remelting or removal of the lower crust has been proposed for both Venus and Io, although the details and tectonic implications are uncertain There is debate about whether Mars experienced "sea-floor spreading" style plate tectonics early in its history, perhaps recorded by Mars's remanent magnetic field.

Other proposed regions of convection include the Sputnik Planitia basin on Pluto, where a cellular floor structure is proposed to form from localized convection in a kilometers-thick surface nitrogen ice sheet, and a similar process may occur in Triton's so-called cantaloupe terrain in the geologic past. Convection and crustal recycling on the icy bodies is poorly understood. Tidal forces acting on Io and the ocean worlds (Europa, Enceladus, Titan, Triton, Pluto, and Charon, in particular) may affect the circulation patterns within the oceans and mantles of these bodies (Question 8, Chapter 11), but the physical characteristics of this forced convection remain mysterious.

In the absence of crustal recycling, any (cryo)volcanic activity will tend to "dry out" planetary interiors such that they become depleted in volatile compounds (such as H_2O and CO_2). Volatiles within the mantle and crust contribute to volcanic and tectonic activity; a loss of volatiles over time, for example, would reduce the likelihood of explosive volcanic eruptions (Q5.3b).

Q5.6b Where and How Are Active Melt Generation, Outgassing and Plume/(Cryo)Volcanic Activity Taking Place, and What Melt and Gas Compositions Are Produced?

Beyond Earth, there are very few places where active outgassing or (cryo)volcanic activity has been directly observed. Jupiter's volcanic moon Io is a testament to the power of tidal forces in driving interior heating and the resulting eruptive activity. At Saturn, diminutive Enceladus (~500 km diameter) is another prime example of the work of tidal heating: the plume of water vapor, ice particles and other compounds discovered by the Cassini mission attests to a body of liquid water beneath the ice shell. Possible plumes of water vapor have also been reported at Europa, another ocean world. Dark, nitrogen-driven plumes were observed by Voyager 2 at Neptune's moon Triton, although it is unclear whether these are internally driven or whether solar heating of sub-surface nitrogen ice is responsible. Quiescent outgassing and outbursts have been observed at active primordial bodies beyond 5 AU (Centaurs and dynamically new comets), while so-called "main belt comets" and even small bodies such as Bennu exhibit activity, but the driving processes, beyond simple solar heating, are poorly understood. In the inner solar system, the combination of primordial heat and radiogenic activity (in different proportions) would have driven endogenic and eruptive activity for at least some portion of each planet's history. Despite abundant evidence for the work of volcanism in shaping the surfaces of the terrestrial planets and our own Moon, only Io is demonstrably volcanically active, although Venus has provided us with tantalizing hints of present-day volcanic eruptions. However, just because we may not have "caught them in the act," it does not necessarily mean that other bodies (such as Mars) do not harbor interior melt and are not capable of eruptions today or in the future. On Earth, we define a volcano as active based on whether it has erupted within the past 10,000 years. Whether we can observe eruptive activity on other bodies, should it be occurring, is a matter of when, how often, and how we look for it.

In the inner solar system, melts are mostly silicate in nature and dominated by mafic compositions. Magmatic volatiles include primarily H_2O, CO_2, SO_2, based on terrestrial experience, but could also include more reduced species like CO, H_2S or other compounds on other bodies (Q5.1a). Io also exhibits dominantly silicate volcanism, inferred to be of mafic to ultramafic composition, with sulfur or SO_2 the primary volatile species. Excluding Io, solid bodies at distances from Ceres to Pluto and beyond generally contain ice as a major constituent. Some of these bodies (Europa, Ganymede, Callisto, Enceladus, and Titan) are thought to possess a subsurface ocean, based on multiple lines of evidence (Q5.1b). Subsurface oceans on other bodies, including Ceres, Pluto, and Triton, are supported by theoretical modeling or a single observation type. Cryomagmas might derive directly from the oceans or result from localized melting within the ice shell. In either case, cryomagma compositions are likely to be dominantly water, possibly containing salts, ammonia, or minor amounts of other constituents. Candidate volatiles driving explosive volcanism include water vapor, CO_2, N_2, SO_2, ammonia, and methane, based on cosmogenic abundance or detections in Enceladus's south polar plumes. Water sublimation drives comet and active asteroid outgassing at distances <5 AU, while activity at primordial Centaurs and dynamically new long-period comets may be driven by annealing and crystallization of amorphous ice and/or sublimation of volatiles such as CO.

The processes of volcanism and cryovolcanism are direct manifestations of the heat sources (e.g., radiogenic and tidal) within a planetary body. Understanding the nature of volcanism through time, and where it is active today, provides information about interior evolution and compositional variability across the solar system. Active cryovolcanism, and the ability to sample plume constituents, has the potential to yield profound discoveries about the habitability (and the potential inhabitants) of icy satellites (Question 10). Sampling the gases given off by primordial active bodies reveals the nature of unprocessed primordial material, the driving mechanisms of the activity, and provides context for understanding the record retained in short-period Jupiter family comets and active asteroids.

Q5.6c Where and With What Intensity Are Tectonic Processes and Deformation Currently Occurring?

Tectonic processes are likely to be active on many planetary bodies in the solar system, from small icy moons to the terrestrial planets, and can shed light on the strength of planetary crusts and the ease with which material may be exchanged between their surfaces and interiors. For icy bodies in particular, tectonic processes can illuminate the evolution of their oceans. Fractures may transport materials from subsurface oceans, or other extensive fluid pockets within these worlds, to their surfaces. For example, the observed brightness variations of Enceladus's plumes appear to be related to motion of south polar faults over a tidal cycle (although the exact link between plume output and tectonism is unclear). Diurnal tides may also be responsible for the active formation of cycloidal ridges on Europa and double ridges on Triton. Europa's cycloidal ridges may form over a matter of days and require an extensive, near-surface liquid layer. As such, they may be indicative of both current tectonic activity and a global subsurface ocean on Europa. Likewise, double ridges are amongst the most youthful features on Triton's surface. Their formation in regions of the surface that have experienced enhanced heating and possibly convective overturn suggests that their presence also requires the existence of a near-surface ductile layer, possibly an internal ocean. Io is currently experiencing both volcanic and tectonic deformation on a massive scale. If a magma ocean is present on Io, tectonic deformation could be caused by diurnal motion as its lithosphere deforms atop this liquid layer. Steadily increasing tidal stresses are plausibly responsible for the grooves on Phobos.

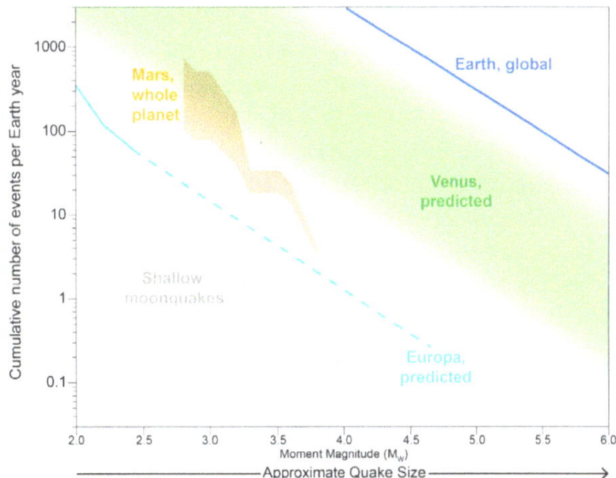

FIGURE 8-4 Frequency of quakes as a function of size for different bodies. No seismometer has been placed on Europa. SOURCES: Europa prediction data extrapolated from M.P. Panning, S.C. Stähler, H.-H. Huang, et al., 2018, "Expected Seismicity and the Seismic Noise Environment of Europa," *Journal of Geophysical Research: Planets* 123:163–179, https://doi.org/10.1002/2017JE005332; Venus prediction data extrapolated from P. Lognonné and C. Johnson, 2007, "Planetary Seismology," pp. 67–122 in Vol. 10: Physics of Terrestrial Planets and Moons of *Treatise in Geophysics*, T. Spohn and G. Schubert, eds., Amsterdam, Netherlands: Elsevier; figure modified from W.B. Banerdt, S.E. Smrekar, D. Banfield, et al., 2020, "Initial Results from the InSight Mission on Mars," *Nature Geoscience* 13:183–189, https://doi.org/10.1038/s41561-020-0544-y. Springer Nature, 2020, reproduced with permission from SNCSC.

All large planetary bodies in the inner solar system are currently understood to be experiencing tectonic deformation. The scale of deformation detected by remote instruments varies from small-scale fractures to large-scale gravity perturbations. For example, on the Moon and Mercury small-scale thrust faults indicate that these bodies are tectonically active today with the fault movement caused by continued cooling and shrinking of these bodies; on the Moon, the Apollo seismic network provided direct evidence of present-day tectonic activity. A combination of analog studies, models, and data observations of larger-scale tectonic landforms on Venus, known as coronae (hundreds of km across), also suggest more recent activity. Active crustal recycling is inferred from the observed tectonic landforms on Venus (Q5.6a).

Understanding of planetary interior processes has been greatly enhanced by in situ data collection, particularly seismic data. The InSight lander is detecting hundreds of marsquakes, some caused by atmospheric phenomenon and others are of tectonic origin. A few large enough to locate are associated with young faults and fissures in the Cerberus Fossae region of Mars (Giardini et al. 2020). These larger marsquakes are of similar magnitude to intraplate tectonic activity on Earth. Having data from several bodies allows for comparative studies that facilitate greater understanding of the formation and state of planetary interiors. Currently available data indicate that Earth is the most seismically active body measured, followed by Mars and then the Moon (Banerdt et al. 2020) (Figure 8-4). The dearth of in situ seismological investigations on Venus, ocean worlds, and dwarf planets, and in many cases, the lack of global, high-resolution imagery, limits our ability to tie tectonic activity and deformation to processes occurring in their interiors.

Q5.6d Where and How Are Active Sedimentary and Regolith Processes Occurring?

The physical and chemical processes controlling the generation and movement of planetary regolith materials vary across the solar system. On airless bodies, thermal stresses and impact-related processes dominate regolith formation (see Q5.5b, and Question 4, Chapter 7). For instance, micrometeoroids pulverize solid materials and chemically weather regolith particles through formation of nanophase iron, also known as space weathering (Q5.5a). At a larger scale, impacts continue to actively mix the regolith. Images of new craters on the Moon show that even decameter-scale impacts affect the surface up to thousands of meters away from the impact event, and secondary cratering is gardening the uppermost regolith at rates >100 times faster than previously thought (Speyerer et al. 2016).

Regolith formation may be intimately linked to dynamic processes occurring on dwarf planets and ocean worlds and can shed light on their internal evolution. Regolith formation occurs on icy bodies by various processes; for instance, when unconsolidated, icy material is emplaced on their surfaces as a result of impact cratering and, in the case of Enceladus, plume eruptions. Because active regolith formation may erase craters and mute surface topography on ice-rich worlds, constraining the composition and thickness of these unconsolidated layers, as well as the rate at which these layers form, allow us to place improved constraints on the evolution of dwarf planet and satellite surfaces. Furthermore, because thin regolith layers (if rocky) act to shield ice-rich layers from sublimation on bodies like Ceres and Callisto and may serve as insulating layers, investigating the properties of planetary regolith can inform thermal models for icy bodies. The presence and composition of regolith on planetary surfaces may also shed light on planetary evolution. For example, compositional analysis of regolith on Ceres revealed that it underwent extensive aqueous alteration, a significant amount of ice-rock fractionation, and that its non-ice regolith has a composition similar to primitive meteorites (CI/CM chondrites) (Prettyman et al. 2017).

Investigations of active sedimentary processes can tell us much about how the climate and surface conditions of planets and satellites have changed over time. Mass wasting of particulate matter down high-slope regions occurs on planetary bodies with and without atmospheres (e.g., Mercury, Moon, Mars, icy satellites). Recent mass movement of material is observed to occur owing to the loss of volatiles from the subsurface. For example, imagery of mass-wasting events along the edges of the martian polar ice caps have been captured. Recurring slope lineae have been observed for many years on Mars, but their exact formation mechanism is still debated. The same is true for movement of materials on comet 67P/Churyumov-Gerasimenko. These mass wasting processes reveal important information about the rate of volatile loss on planetary surfaces.

The presence of an atmosphere facilitates the lofting of fine-grained particles into near-surface planetary atmospheres, creating a suite of ever-changing depositional and erosional landforms. High-resolution imagery shows the movement of dune fields on Mars, indicating large-scale movement of sand-size particles. Venera lander data, both panoramas and light flux data, suggest atmospheric transport of particulates by near-surface winds, although whether these winds were typical or modified by the landing event are unclear. Titan's extensive, mid-latitude dune fields are a testament to the dry climatic conditions in the mid-latitudes and suggest that, like fluvial deposition, aeolian deposition is still ongoing. These landforms and their temporal changes alert us to the most recent weather patterns of these planetary bodies. Comparison of active landforms with inactive dune fields, for example, reveals how recent weather or climate patterns have changed and can reveal the processes that led to the observed changes.

Strategic Research for Q5.6

- **Investigate the potential for active volcanism and deformation, and where and how crustal recycling is happening, on Venus** with synthetic aperture radar, infrared, ultraviolet, or repeat-pass interferometry measurements of the venusian surface and atmosphere.
- **Constrain the rate of active surface changes on Mars related to dune migration, mass movements, sedimentation, or ice sublimation** using either long-term, repeat-pass, high-resolution altimetry or imaging.
- **Characterize present-day plate mobility and recycling on Europa, Titan, and Enceladus** by visible imaging at regional scale, global, high-resolution gravity and topography, and/or repeat-pass interferometry.
- **Characterize individual eruptions and determine the rate of volcanic activity on Io** using repeat high-resolution imagery and/or spectroscopy.
- **Characterize the style and intensity of active tectonism occurring on rocky or icy worlds**, through seismic and other geophysical measurements.

SUPPORTIVE ACTIVITIES FOR QUESTION 5

- Measure optical constants of the range of materials expected in the solar system under relevant pressure and temperature conditions (from Venus to airless icy satellites) to serve as the basis by which we constrain the compositions of planetary surfaces from remote sensing data.
- Continued meteorite collection activities in the Antarctic and associated curation.

REFERENCES

Banerdt, W.B., S.E. Smrekar, D. Banfield, D. Giardini, M. Golombek, C.L. Johnson, P. Lognonné, et al. 2020. "Initial Results from the Insight Mission on Mars." *Nature Geoscience* 13:183–189. https://doi.org/10.1038/s41561-020-0544-y.

Baratoux, D., M.J. Toplis, M. Monnereau, and O. Gasnault. 2011. "Thermal History of Mars Inferred from Orbital Geochemistry of Volcanic Provinces." *Nature* 472:338–341. https://doi.org/10.1038/nature09903.

Beyer, R.A., J.R. Spencer, W.B. McKinnon, F. Nimmo, C. Beddingfield, W.M. Grundy, K. Ennico, et al. 2019. "The Nature and Origin of Charon's Smooth Plains." *Icarus* 323:16–32. https://doi.org/10.1016/j.icarus.2018.12.036.

Bibring, J.-P., Y. Langevin, J.F. Mustard, F. Poulet, R. Arvidson, A. Gendrin, B. Gondet, et al. 2006. "Global Mineralogical and Aqueous Mars History Derived from OMEGA/Mars Express Data." *Science* 312:400–404. https://doi.org/10.1126/science.1122659.

Black, B.A., and M. Manga. 2016. "The Eruptibility of Magmas at Tharsis and Syrtis Major on Mars." *Journal of Geophysical Research: Planets* 121:944–964. https://doi.org/10.1002/2016JE004998.

Brissaud, Q., S. Krishnamoorthy, J.M. Jackson, D.C. Bowman, A. Komjathy, J.A. Cutts, Z. Zhan, et al. 2021. "The First Detection of an Earthquake from a Balloon Using Its Acoustic Signature." *Geophysical Research Letters* 48:e93013. https://doi.org/10.1029/2021GL093013.

Byrne, P.K., L.R. Ostrach, C.I. Fassett, C.R. Chapman, B.W. Denevi, A.J. Evans, C. Klimczak, et al. 2016. "Widespread Effusive Volcanism on Mercury Likely Ended by about 3.5 Ga." *Geophysical Research Letters* 43:7408–7416. https://doi.org/10.1002/2016GL069412.

Castillo-Rogez, J.C., M.A. Hesse, M. Formisano, H. Sizemore, M. Bland, A.I. Ermakov, and R.R. Fu. 2019. "Conditions for the Long-Term Preservation of a Deep Brine Reservoir in Ceres." *Geophysical Research Letters* 46:1963–1972. https://doi.org/10.1029/2018GL081473.

Crumpler, L.S., J.C. Aubele, D.A. Senske, S.W. Keddie, K. Magee, and J.W. Head. 1997. "Volcanoes and Centers of Volcanism." Pp. 697–756 in *Venus II*, S. W. Bouger, D. M. Hunten, and R. J. Phillips, eds. Tucson, AZ: University of Arizona Press.

Ermakov, A.I., M.T. Zuber, D.E. Smith, C.A. Raymond, G. Balmino, R.R. Fu, and B.A. Ivanov. 2014. "Constraints on Vesta's Interior Structure Using Gravity and Shape Models from the Dawn Mission." *Icarus* 240:146–160. https://doi.org/10.1016/j.icarus.2014.05.015.

Filiberto, J. 2017. "Geochemistry of Martian Basalts with Constraints on Magma Genesis." *Chemical Geology* 466:1–14. https://doi.org/10.1016/j.chemgeo.2017.06.009.

Giardini, D., P. Lognonné, W.B. Banerdt, W.T. Pike, U. Christensen, S. Ceylan, J.F. Clinton, et al. 2020. "The Seismicity of Mars." *Nature Geoscience* 13:205–212. https://doi.org/10.1038/s41561-020-0539-8.

Grady, M.M., I.P. Wright, C. Engrand, and S. Silijeström. 2018. "The Rosetta Mission and the Chemistry of Organic Species in Comet 67P/Churyumov-Gerasimenko." *Elements* 14:95–100. https://doi.org/10.2138/gselements.14.2.95.

Head, J.W., and L. Wilson. 2017. "Generation, Ascent and Eruption of Magma on the Moon: New Insights into Source Depths, Magma Supply, Intrusions and Effusive/Explosive Eruptions (Part 2: Predicted Emplacement Processes and Observations)." *Icarus* 283–223. https://doi.org/10.1016/j.icarus.2016.05.031.

Hemingway, D.J., M.L. Rudolph, and M. Manga. 2020. "Cascading Parallel Fractures on Enceladus." *Nature Astronomy* 4:234–239. https://doi.org/10.1038/s41550-019-0958-x.

Johnson, B.C., M.M. Sori, and A.J. Evans. 2020a. "Ferrovolcanism on Metal Worlds and the Origin of Pallasites." *Nature Astronomy* 4:41–44. https://doi.org/10.1038/s41550-019-0885-x.

Johnson, C.L., A. Mittelholz, B. Langlais, C.T. Russell, V. Ansan, D. Banfield, P.J. Chi, et al. 2020b. "Crustal and Time-Varying Magnetic Fields at the InSight Landing Site on Mars." *Nature Geoscience* 13:199–204. https://doi.org/10.1038/s41561-020-0537-x.

Johnson, C.L., R.J. Phillips, M.E. Purucker, B.J. Anderson, P.K. Byrne, B.W. Denevi, J.M. Feinberg, et al. 2015. "Low-Altitude Magnetic Field Measurements by MESSENGER Reveal Mercury's Ancient Crustal Field." *Science* 348:892–895. https://doi.org/10.1126/science.aaa8720.

Jolliff, B.L., S.A. Wiseman, S.J. Lawrence, T.N. Tran, M.S. Robinson, H. Sato, B.R. Hawke, et al. 2011. "Non-Mare Silicic Volcanism on the Lunar Farside at Compton-Belkovich." *Nature Geoscience* 4:566–571. https://doi.org/10.1038/ngeo1212.

Jordan, A.P., T.J. Stubbs, J.K. Wilson, N.A. Schwadron, and H.E. Spence. 2015. "Dielectric Breakdown Weathering of the Moon's Polar Regolith." *Journal of Geophysical Research: Planets* 120:210–225. https://doi.org/10.1002/2014JE004710.

Kamata, S., F. Nimmo, Y. Sekine, K. Kuramoto, N. Noguchi, J. Kimura, and A. Tani. 2019. "Pluto's Ocean Is Capped and Insulated by Gas Hydrates." *Nature Geoscience* 12:407–410. https://doi.org/10.1038/s41561-019-0369-8.

Khurana, K.K., R.T. Pappalardo, N. Murphy, and T. Denk. 2007. "The Origin of Ganymede's Polar Caps." *Icarus* 191:193–202.

Krohn, K., R. Jaumann, K. Otto, T. Hoogenboom, R. Wagner, D.L. Buczkowski, B. Garry, et al. 2014. "Mass Movement on Vesta at Steep Scarps and Crater Rims." *Icarus* 244:120–132. https://doi.org/10.1016/j.icarus.2014.03.013.

Lapôtre, M.G.A., R.C. Ewing, M.P. Lamb, W.W. Fischer, J.P. Grotzinger, D.M. Rubin, K.W. Lewis, et al. 2016. "Large Wind Ripples on Mars: A Record of Atmospheric Evolution." *Science* 353:55–58. https://doi.org/10.1126/science.aaf3206.

Lapôtre, M.G.A., J.G. O'Rourke, L.K. Schaefer, K.L. Siebach, C. Spalding, S.M. Tikoo, and R.D. Wordsworth. 2020. "Probing Space to Understand Earth." *Nature Reviews Earth and Environment* 1:170–181. https://doi.org/10.1038/s43017-020-0029-y.

Mangold, N. 2011. "Ice Sublimation as a Geomorphic Process: A Planetary Perspective." *Geomorphology* 126:1–17. https://doi.org/10.1016/j.geomorph.2010.11.009.

McCubbin, F.M., M.A. Riner, K.E. Vander Kaaden, and L.K. Burkemper. 2012. "Is Mercury a Volatile-Rich Planet?" *Geophysical Research Letters* 39:L09202. https://doi.org/10.1029/2012GL051711.

McEwen, A.S., L. Ojha, C.M. Dundas, S.S. Mattson, S. Byrne, J.J. Wray, S.C. Cull, et al. 2011. "Seasonal Flows on Warm Martian Slopes." *Science* 333:740. https://doi.org/10.1126/science.1204816.

McLennan, S.M., J.P. Grotzinger, J.A. Hurowitz, N.J. Tosca. 2019. "The Sedimentary Cycle on Early Mars." *Annual Review of Earth and Planetary Sciences* 47:91–118. https://doi.org/10.1146/annurev-earth-053018-060332.

McSween, H.Y., R.P. Binzel, M.C. de Sanctis, E. Ammannito, T.H. Prettyman, A.W. Beck, V. Reddy, et al. 2013. "Dawn; the Vesta-HED Connection; and the Geologic Context for Eucrites, Diogenites, and Howardites." *Meteoritics and Planetary Science* 48:2090–2104. https://doi.org/10.1111/maps.12108.

Melosh, H.J. 1989. *Impact Cratering: A Geologic Process.* New York: Oxford University Press; Oxford, UK: Clarendon Press.

Moore, W.B., J.I. Simon, and A.A.G. Webb. 2017. "Heat-Pipe Planets." *Earth and Planetary Science Letters* 474:13–19. https://doi.org/10.1016/j.epsl.2017.06.015.

Neish, C.D., J.W. Barnes, C. Sotin, S. MacKenzie, J.M. Soderblom, S. Le Mouélic, R.L. Kirk, et al. 2015. "Spectral Properties of Titan's Impact Craters Imply Chemical Weathering of Its Surface." *Geophysical Research Letters* 42:3746–3754. https://doi.org/10.1002/2015GL063824.

Nimmo, F., and R.T. Pappalardo. 2016. "Ocean Worlds in the Outer Solar System." *Journal of Geophysical Research: Planets* 121:1378–1399. https://doi.org/10.1002/2016JE005081.

Peplowski, P.N., R.L. Klima, D.J. Lawrence, C.M. Ernst, B.W. Denevi, E.A. Frank, J.O. Goldsten, et al. 2016. "Remote Sensing Evidence for an Ancient Carbon-Bearing Crust on Mercury." *Nature Geoscience* 9:273–276. https://doi.org/10.1038/ngeo2669.

Pieters, C.M., and S.K. Noble. 2016. "Space Weathering on Airless Bodies." *Journal of Geophysical Research: Planets* 121:1865–1884. https://doi.org/10.1002/2016JE005128.

Prettyman, T.H., N. Yamashita, M.J. Toplis, H.Y. McSween, N. Schörghofer, S. Marchi, W.C. Feldman, et al. 2017. "Extensive Water Ice Within Ceres' Aqueously Altered Regolith: Evidence from Nuclear Spectroscopy." *Science* 355:55–59. https://doi.org/10.1126/science.aah6765.

Quick, L.C., and B.D. Marsh. 2016. "Heat Transfer of Ascending Cryomagma on Europa." *Journal of Volcanology and Geothermal Research* 319:66–77. https://doi.org/10.1016/j.jvolgeores.2016.03.018.

Raymond, C.A., A.I. Ermakov, J.C. Castillo-Rogez, S. Marchi, B.C. Johnson, M.A. Hesse, J.E.C. Scully, et al. 2020. "Impact-Driven Mobilization of Deep Crustal Brines on Dwarf Planet Ceres." *Nature Astronomy* 4:741–747. https://doi.org/10.1038/s41550-020-1168-2.

Scheller, E.L., B.L. Ehlmann, R. Hu, D.J. Adams, and Y.L. Yung. 2021. "Long-Term Drying of Mars by Sequestration of Ocean-Scale Volumes of Water in the Crust." *Science* 372:56–62. https://doi.org/10.1126/science.abc7717.

Schenk, P., D.P. Hamilton, R.E. Johnson, W.B. McKinnon, C. Paranicas, J. Schmidt, and M.R. Showalter. 2011. "Plasma, Plumes and Rings: Saturn System Dynamics as Recorded in Global Color Patterns on Its Midsize Icy Satellites." *Icarus* 211:740–757. https://doi.org/10.1016/j.icarus.2010.08.016.

Speyerer, E.J., R.Z. Povilaitis, M.S. Robinson, P.C. Thomas, and R.V. Wagner. 2016. "Quantifying Crater Production and Regolith Overturn on the Moon with Temporal Imaging." *Nature* 538:215–218. https://doi.org/10.1038/nature19829.

Stähler, S.C., A. Khan, W.B. Banerdt, P. Lognonné, D. Giardini, S. Ceylan, M. Drilleau, et al. 2021. "Seismic Detection of the Martian Core." *Science* 373:443–448. https://doi.org/10.1126/science.abi7730.

Thomas, N., H. Sierks, C. Barbieri, P.L. Lamy, R. Rodrigo, H. Rickman, D. Koschny, et al. 2015. "The Morphological Diversity of Comet 67P/Churyumov-Gerasimenko." *Science* 347:aaa0440. https://doi.org/10.1126/science.aaa0440.

Tikoo, S.M., B.P. Weiss, D.L. Shuster, C. Suavet, H. Wang, and T.L. Grove. 2017. "A Two-Billion-Year History for the Lunar Dynamo." *Science Advances* 3:e1700207. https://doi.org/10.1126/sciadv.1700207.

Way, M.J., A.D. Del Genio, N.Y. Kiang, L.E. Sohl, D.H. Grinspoon, I. Aleinov, M. Kelley, et al. 2016. "Was Venus the First Habitable World of the Solar System?" *Geophysical Research Letters* 43:8376–8383. https://doi.org/10.1002/2016GL069790.

Q6 PLATE: A view of Venus's atmosphere in the ultraviolet from the Akatsuki mission in 2016. Venus's north pole is to the top in this image. SOURCES: ©PLANET-C Project Team. Courtesy of ISAS/JAXA, adapted by DARTS/K.M. Gill. CC BY 2.0.

9

Question 6: Solid Body Atmospheres, Exospheres, Magnetospheres, and Climate Evolution

What establishes the properties and dynamics of solid body atmospheres and exospheres, and what governs material loss to and gain from space and exchange between the atmosphere and the surface and interior? Why did planetary climates evolve to their current varied states?

The nebular accretion processes outlined in Questions 1–3 (Chapters 4–6) gave rise to a diverse array of planets and moons that eventually evolved into the ones that we observe today.[1] In this chapter, the committee focuses on the atmospheres of these bodies, specifically those with solid surfaces, including the four terrestrial planets, the various moons, and the dwarf planets (e.g., Pluto). Gas and ice giant planets are discussed in Question 7 (Chapter 10). The atmospheres discussed here include dense atmospheres (e.g., Venus, Earth, Titan, and early Mars), atmospheres dominantly controlled by vapor-pressure equilibrium (current Mars, Triton, Pluto, and some other Kuiper belt objects, or KBOs), and collisionless exospheres (e.g., Mercury and the Moon), shown in Figure 9-1. Collisional atmospheres also have exospheres that separate them from the near vacuum of space. In the remainder of this chapter, the term "atmosphere" includes these exospheres. The study of planetary atmospheres is critical to understanding past and current habitability throughout the solar system, including the prebiotic processes that led to the emergence of life on early Earth. It also provides natural laboratories that we can use to better understand the processes governing Earth's past and current climate.

How did the atmospheres of solid bodies form, and why did some of them end up with dense atmospheres while others did not? What processes contributed to atmospheric accumulation, and what processes led to atmospheric loss? How do climates evolve on solid bodies with atmospheres, and were any of them besides Earth capable of supporting life? This last question is revisited in more detail in Questions 9–11 (Chapters 12–14) of this report, but the foundations for that discussion are laid here.

Atmospheres also change on a variety of shorter timescales ranging from daily to seasonal to those encompassing orbital (Milankovitch) cycles. Some changes are closely linked to variations in solar forcing, while others (such as dust storms on Mars or changes in SO_2 abundance at cloud deck levels on Venus) are episodic and hard to predict given current understanding of these atmospheres. Long-term observations are needed to make progress toward understanding the processes responsible for such changes. Further changes (such as polar layered deposits on Mars or the migration of methane lakes between hemispheres on Titan) occur on much longer timescales, yet insight can still be gained by measurements of current conditions or through changes observed over several decades.

---

[1]A glossary of acronyms and technical terms can be found in Appendix F.

FIGURE 9-1 Overview of atmospheric characteristics for the solid planets and moons discussed in this chapter. *Top panel*: Bodies with collisionless exospheres only. *Bottom panel*: Bodies with dense, collisional atmospheres and atmospheres in vapor pressure equilibrium. SOURCE: Courtesy of K. Kostadinova.

Many mysteries also have yet to be solved regarding the present atmospheres of solid planets and moons. What accounts for the ultraviolet absorption in Venus's clouds and the reported occurrence of methane in Mars's atmosphere? What is the detailed composition of the thick organic haze on Saturn's moon Titan, and how does liquid methane cycle between Titan's lower atmosphere and surface/near-subsurface? How do volatiles migrate on bodies with atmospheres dominated by condensation-sublimation flows, such as Triton and Pluto? How and to what extent do volatiles migrate within exospheres, such as on Mercury and the Moon? And what drives atmospheric superrotation on bodies like Venus and Titan, or dust storms on Mars and Titan? How do atmospheres interact with the space environment, and how is this interaction mediated by the presence of ionospheres and magnetospheres?

These are but a few of a fascinating array of questions that remain to be answered about the atmospheres of the planets and moons in today's solar system. The committee highlights the major outstanding issues below.

Q6.1 HOW DO SOLID BODY ATMOSPHERES FORM AND WHAT WAS THEIR STATE DURING AND SHORTLY AFTER ACCRETION?

The formation of solid body atmospheres depends on an array of processes, including supply and removal of volatiles by impacts, atmospheric escape, and exchange with the interior. Many of these processes remain poorly understood. All solar system atmospheres have evolved significantly since their formation; thus, to investigate their earliest states we need to gather clues from a wide variety of sources.

In the inner solar system, the major differences in the atmospheres of Mercury, Venus, Earth, and Mars arise in large part from differences in the planets' masses, distance from the Sun, and the presence or absence of a significant magnetosphere. Mars may also have formed as volatile-rich, but because it is small, it has lost much of its atmosphere to space over time. Mars's small size may have also affected the chemical composition of its mantle, and hence its early atmosphere (Wade and Wood 2005; Deng et al. 2020), with important implications for climate (Q6.2).

Farther out in the solar system, the rich atmospheric diversity of the icy moons and dwarf planets poses fascinating challenges in comparative planetology. Titan stands out because of its thick nitrogen and methane atmosphere, in stark contrast to similarly sized objects such as Ganymede, which has only a tenuous oxygen exosphere. Understanding why Titan has retained a thick atmosphere while other outer solar system bodies did not remains a key challenge. Pluto and Triton, similarly sized bodies, have N_2 atmospheres in vapor pressure equilibrium with surface ices but the structure of their atmospheres is quite different. The oxygen exospheres of Europa and Ganymede are likely produced by sputtering of surface ice, but our understanding of this process has yet to be validated.

Progress in understanding atmospheric formation across the solar system requires a range of approaches, but from an observational standpoint, obtaining more precise isotopic data on volatiles and noble gases in solid body atmospheres, chondritic meteorites, and comets is particularly important. Noble gases—particularly He, Ar, and Xe—have both radiogenic and nonradiogenic isotopes that can be used to discriminate between early and late outgassing of a planet's atmosphere (Avice and Marty 2020).

Q6.1a What Was the Role of Accretion and Meteoroid Impacts in Sculpting Early Atmospheres?

The earliest atmospheres of the inner solar system planets formed while the planets were still accreting, and the accompanying impacts both delivered and ejected volatiles. The volume and flux of impactors, their volatile content, and their degree of differentiation are all critical to determining what types of atmospheres first formed on the terrestrial planets. At the Moon, asteroids are believed to have delivered water during the lunar magma ocean period, suggesting that they also delivered water to early Earth (Barnes et al. 2016). Recent work suggests that transient, hydrogen-rich atmospheres caused by the thermochemistry of large impacts may have been common on planets like Earth and Mars (Haberle et al. 2019a; Zahnle et al. 2019). This process has important implications for early habitability (Q6.2) and prebiotic chemistry (Question 9; Question 11), but it remains poorly understood.

Q6.1b What Was the Role of Hydrodynamic Escape in Early Atmospheric Evolution?

Early atmospheres of planets and moons were influenced by many of the same thermal and nonthermal escape processes that occur today. (See Q6.5 for a discussion.) But early atmospheres were also likely influenced by an additional escape process, hydrodynamic escape, that is not observed on present solar system bodies, with the possible exception of Titan (Strobel 2009; Schaufelberger et al. 2012). Hydrodynamic atmospheric escape occurs when the upper regions of a planet's atmosphere are dense enough that the pressure force remains important at all altitudes. The frequency of collisions also ensures that this is a form of thermal escape. In the extreme case, the density is high enough that the atmosphere can expand outward to space as a collisional fluid, so the process can be simulated using the standard equations of hydrodynamics, albeit formulated in such a way as to be able to handle transonic flow (Johnstone et al. 2018). The high levels of extreme ultraviolet radiation (XUV) produced by the young Sun would have made this process particularly important to atmospheric evolution in the first few hundred million years after the planets formed (Lammer et al. 2018). This process likely drove loss of the primordial atmospheres of Earth, Venus, Mars, and Titan, setting the stage for later atmospheric and climate evolution (Q6.2).

When one studies this process in more detail, one finds that hydrodynamic escape grades continuously into a more familiar form of thermal escape, Jeans escape, that operates on multiple bodies today (e.g., Earth and Mars). In Jeans escape, the atmosphere is modeled as being collisionless above some level, termed the exobase. Ballistic particle trajectories are assumed above the exobase, and the atmosphere is assumed to be hydrostatic below it. But there is an intermediate regime in which an exobase occurs, yet the bulk atmosphere is *not* hydrostatic, and the pressure force is *not* negligible above it. In this regime, more complicated kinetic treatments of the Boltzmann equation are needed to accurately calculate escape rates (Johnson et al. 2013).

Hydrodynamic escape causes fractionation and hence can be investigated empirically by making precise measurement of isotopic variations in atmospheric noble gases, especially Xe (Zahnle et al. 2019). Interpreting such data requires a better understanding of solar extreme ultraviolet variation through time and of upper-atmospheric chemical and radiative processes. Additional modeling work considering multiple species and isotopes and comparing different numerical approximations are needed to understand thermal escape of early atmospheres.

Q6.1c What Was the Role of Magma Oceans in Early Atmospheric Evolution?

During formation, the surfaces of the terrestrial planets were likely molten owing to heat from accretion and blanketing by thick atmospheres, leading to formation of a local or global magma ocean (Elkins-Tanton 2012). Even when the magma ocean lifetime is short in geologic terms (tens to hundreds of million years), its influence on subsequent atmospheric evolution can be profound. For example, Earth's mantle is thought to be more oxidized than that of Mars because of processes that occurred within its deeper, higher-pressure magma ocean (Wade and Wood 2005; see also Figure 9-2). Consequently, Mars's early atmosphere may have been more H_2-rich than Earth's

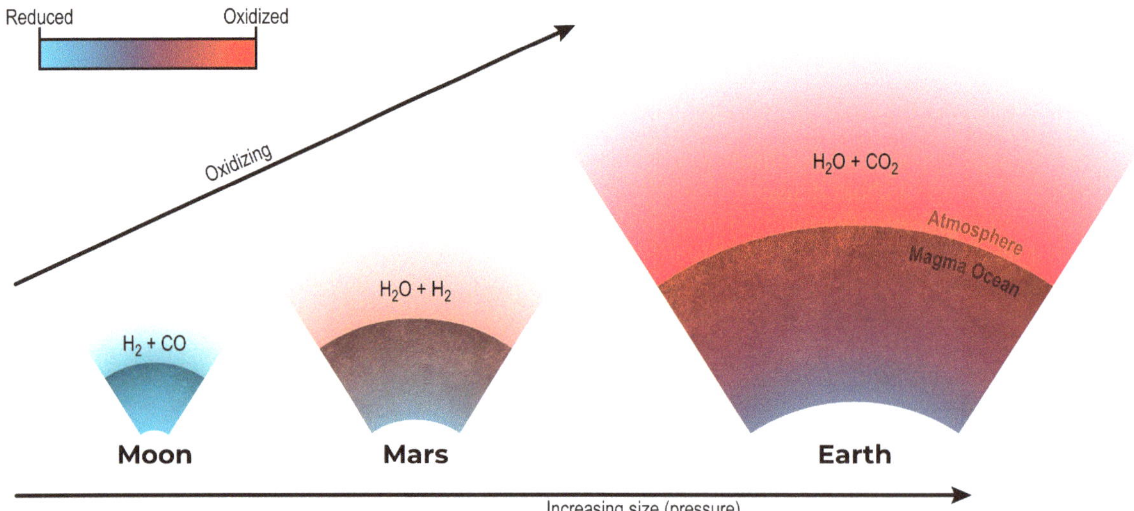

FIGURE 9-2 Magma ocean depth can play a major role in the composition of early atmospheres, as shown here by the calculated abundances of major constituents overlying magma oceans on the early Moon, Mars, and Earth. For simplicity, a C–O–H atmosphere with an H/C mass ratio of 0.5 at 1 bar and 1,800 K has been assumed in each case. Actual magma oceans should have been thicker on Earth and Mars and thinner on the Moon. Magma oceans are predicted to become more oxidized at their surfaces as their depth increases. Note that the thicknesses and oxidation states of atmospheres are not scaled.
SOURCE: Courtesy of J.T. Keane, adapted from J. Deng, Z. Du, B.B. Karki, et al., 2020, "A Magma Ocean Origin to Divergent Redox Evolutions of Rocky Planetary Bodies and Early Atmospheres," *Nature Communications* 11:2007, https://doi.org/10.1038/s41467-020-15757-0. CC BY 4.0.

early atmosphere, and this may help explain why Mars's early climate was relatively warm despite the faintness of the young Sun (Ramirez et al. 2014). Measurements of noble gases in planetary atmospheres and crustal samples are needed to constrain the earliest outgassing rates (particularly for Mars, Venus, and Titan), while laboratory experiments are needed to determine the chemistry and physics of interior exchange processes at temperatures and pressures relevant to magma ocean conditions.

Q.6.1d What Role Does the Space Environment Play in Forming and Liberating the Volatiles Contained Within Surface Bounded Exospheres Like That at the Moon and Mercury? What Role Do the Magnetospheres of Jupiter and Saturn Have on the Development of Exospheres at Ganymede, Europa, Dione, and Rhea?

On bodies with surface boundary exospheres, the solar wind and micrometeoroids are directly incident on the surface and are capable of altering surface material to create exospheric volatiles and to influence the cycling of the released volatile species. For example, the solar wind is directly incident at the Moon for three quarters of its orbit and implants protons in the top ~30 nm of grains. This proton implantation has been suggested to be the origin of a thermally modulated hydroxyl signature observed in the infrared (Li and Milliken 2017). Water has been observed to be released from the lunar surface during meteor streams and is believed to be manufactured from the solar wind implanted hydroxyls when the local surface near an impact is undergoing "flash" heating (Benna et al. 2019). It remains unclear if this transient water exosphere can migrate to the poles. Molecular hydrogen is also emitted from the lunar surface as part of the solar wind proton/hydrogen diffusion process (Hurley et al. 2016; Tucker et al. 2019). The solar wind-implanted hydrogen and carbon atoms may find each other via surface diffusion to form the observed methane emitted during the surface warming at dawn (Hodges 2016). We anticipate similar complex surface-exosphere interactions at other solid bodies that are directly exposed to the space environment, such as Mercury, or Phobos. We are still advancing our understanding of the complex surface-exosphere interactions for moons directly exposed to the energetic magnetoplasma environment of their parent planets such as Ganymede, Europa, Dione, Rhea. Both planetary and moon magnetic fields will also affect the processes (see also Q6.5, Q5.5a,c).

Q6.1e Was There an Early Collisional Atmosphere on Exposed Solid Bodies Like the Moon, Mercury, and Europa?

It has been hypothesized that the Moon had a collisional atmosphere shortly after the Moon-forming event during the lunar magma ocean period (Stern 1999). Needham and Kring (2017) further suggested that transient collisional atmospheres formed episodically in response to lunar mare volcanism. Some of the released mantle material may have been cold-trapped and sequestered within lunar polar crater deposits, especially the Cabeus region (Siegler et al. 2016). It remains unclear why the collisional atmospheres at the Moon dissipated. Similarly, it is unknown whether bodies that lack collisional atmospheres today, like Mercury or Europa, had significant collisional atmospheres early in their evolution and, if so, how those initial atmospheres dissipated over time.

Strategic Research for Q6.1

- **Constrain the earliest stages of atmospheric evolution on Venus, Mars, and Titan** by measuring noble gas abundances and isotopic fractionation to sufficient precision to quantify their minor isotopes.
- **Derive the sources of exospheric volatiles** by measuring the distribution, composition, and abundance of surface volatiles (including in permanently shadowed regions) on solid bodies including the Moon, Mercury, Ceres, and outer planet satellites such as Europa.
- **Improve understanding of the initial states of planetary atmospheres** by developing models of magma oceans coupled to processes such as delivery, loss of volatiles from impacts, chemistry, dynamics, and atmospheric escape owing to both thermal and nonthermal escape mechanisms.

- **Develop physical and chemical constraints on early atmosphere–interior exchange** by performing laboratory experiments on volatile partitioning in silicate melts, meteorite shock chemistry, and related phenomena.

Q6.2 WHAT PROCESSES GOVERN THE EVOLUTION OF PLANETARY ATMOSPHERES AND CLIMATES OVER GEOLOGIC TIMESCALES?

Planetary (and satellite) atmospheres probably formed during accretion, but they have continued to evolve over time as volatiles are added by volcanism or impacts or lost by escape to space. Volatiles can also be exchanged with a planet's surface, especially on bodies like Earth or Titan on which liquids are present to facilitate erosion and weathering. Earth is also inhabited, so its atmospheric composition is affected by biogeochemical cycles and, more recently, by anthropogenic inputs.

Planetary climates are linked tightly to atmospheric density and composition. For the three terrestrial planets with substantial atmospheres, this is sometimes called the "Goldilocks problem": Why is Venus too hot, Mars too cold, and Earth just right? Part of the answer clearly lies in their relative distances from the Sun. But a planet's surface temperature also depends on the greenhouse effect of its atmosphere: clouds and hazes cool the surface by reflecting incoming sunlight, while various gases (e.g., CO_2 and CH_4) warm the surface by retarding the emission of thermal-infrared radiation. The density and composition of terrestrial planet atmospheres depend on a host of factors, including planetary size, volcanic activity, mantle redox state, and surface weathering processes. Exploring those factors can yield useful insights into why Earth remained habitable while its nearby neighbors did not and can help us understand whether Earth-like planets may exist elsewhere in the galaxy.

Outer solar system moons (Titan) and dwarf planets (Pluto) have their own volatile cycles that are controlled by a different set of processes. Titan, with its dense N_2-CH_4 atmosphere, has a climate analogous to Earth's, but with the condensable volatile being methane instead of water. The climates of Titan, Triton, and Pluto, like those of Mars and Earth, are expected to vary in response to orbital forcings (e.g., obliquity and eccentricity), as volatiles are locked up in surface deposits and released in different epochs. Pluto's orbit is so highly eccentric that its climate changes dramatically as it passes between perihelion and aphelion. In this section, the committee explores questions related to how planetary climates evolve.

Q6.2a What Processes Have Kept Earth's Climate (Mostly) Clement over Geological Time, and What Has Been the Relative Role of Biological Versus Abiotic Feedbacks?

The Sun was ~30 percent less bright early in solar system history, yet the early Earth was not frozen. Evidence for surface liquid water dates back to almost 4.4 Ga (Valley et al. 2002). What greenhouse gases helped keep the surface warm? CO_2 and H_2O are thought to have been the two main candidates, but CH_4 could have played a role as well, especially prior to the rise of atmospheric O_2 (Haqq-Misra et al. 2008). CH_4 is largely biogenic, so this implies a degree of biological control of the evolution of Earth's climate. A few percent of today's CH_4 is produced by abiotic processes such as serpentinization. CO_2 is produced by volcanoes and is removed by weathering of silicate minerals, followed by deposition of carbonate sediments or veined carbonate in the seafloor. Its removal rate slows as the climate cools; thus, high CO_2 concentrations are an expected consequence of, and solution to, the faint young Sun problem (Walker et al. 1981). But what were early CO_2 levels and how rapidly did they decline with time? What geochemical proxies can we use to test this climate control hypothesis? Similar questions apply to atmospheric CO_2 and climate evolution on Venus and Mars, as discussed in Q6.2b,c.

Equally important to eukaryotic organisms, including higher plants and animals, was the evolution of atmospheric O_2. Prebiotic O_2 levels are believed to have been extremely low, of the order of 10^{-13} times the present atmospheric level, or PAL (Catling and Kasting 2017). Atmospheric O_2 remained low throughout the Archean, then rose abruptly during the Great Oxidation Event at ~2.4 Ga (Catling and Kasting 2017). Although the reasons why O_2 rose at this time continue to be debated, researchers agree that it was the evolution of oxygenic photosynthesis by cyanobacteria that ultimately led to this event. Cyanobacteria (formerly known as blue-green algae) are the only prokaryotic organisms that can perform oxygenic photosynthesis. The subsequent history of atmospheric O_2

during the ensuing Proterozoic and Phanerozoic eons is a research topic of great interest to astrobiologists because it is poorly understood and because it is relevant to the question of whether O_2 is a good biomarker on extrasolar planets (see Question 12, Chapter 15).

Q6.2b How Have Climate Conditions on Venus Evolved over Time, and Did the Planet Ever Have Surface Liquid Water in the Past?

Venus has a dense, 92-bar atmosphere consisting of 97 percent CO_2 but only 30 parts per million by volume of H_2O. The surface temperature on Venus is ~730 K, well above the critical temperature for water (647 K), so liquid water could not exist even if H_2O was abundant. Although Venus and Earth probably received an abundance of volatiles from planetesimals originating from farther out in the solar system (see Question 3, Chapter 6), Venus was closer to the Sun and experienced a higher solar flux, which could have triggered a runaway greenhouse (Ingersoll 1969). The steam atmosphere was photodissociated, hydrogen was lost to space, and the leftover oxygen either escaped along with the hydrogen or reacted with the crust. The buildup of CO_2 after that time was inevitable, as the weathering of silicate rocks to form carbonates was not possible once liquid water was no longer present (Urey 1952; Q6.1a). That said, the details of this process remain controversial. One hypothesis is that Venus developed a steam atmosphere during accretion and never had surface liquid water (Hamano et al. 2013). This model offers a straightforward explanation for how oxygen was lost: the greenhouse effect was so large that it kept Venus's surface molten, allowing oxygen to react directly with a convecting magma ocean. This hypothesis is strengthened by recent 3D climate model calculations that support the idea that an initial steam atmosphere on Venus should never have condensed (Turbet et al. 2021). Both calculations conflict with a second hypothesis, also bolstered by 3D climate simulations, which suggest that if Venus started out as a slowly rotating planet with an ocean already present on its surface, dense cloud cover on the dayside could have kept surface temperatures well below the critical point of water (Way et al. 2016; Way and Del Genio 2020). In this latter scenario, liquid water could have remained on Venus's surface for as long as 4 billion years until either increasing solar luminosity or some other planetary event such as enhanced volcanic outgassing destabilized it. Apparently, the assumed initial conditions can make a big difference in such calculations.

One way of distinguishing between these two competing scenarios is by determining the mineralogic composition of surface tesserae on Venus. Some investigators (e.g., Gilmore et al. 2015) have argued that the tesserae have a felsic composition that could only have been generated if liquid water was present. Evaluating this claim is a high priority for future Venus missions.

Q6.2c What Was the Nature of the Early Martian Climate, and How Were Conditions Allowing Rivers, Lakes, and Similar Features on the Surface Maintained?

While Mars today is cold and dry, overwhelming evidence indicates that large amounts of liquid water flowed over its surface for extended periods in the past (3–4 Ga). Mars's more distant orbit makes this evidence even harder to explain than in the corresponding "faint young Sun problem" for Earth (Q6.2a). Various mechanisms have been proposed, including transient steam atmospheres from meteoroid impacts, sulfur-bearing gases from volcanism, or reducing gases such as hydrogen and methane from various sources, but all have challenges (Haberle et al. 2017). Understanding this period of martian history is essential for developing robust theories of exoplanet habitability (see Question 12, Chapter 15). The early evolution of Mars's climate is also of key importance to planetary astrobiology because in situ analysis has shown that the planet was habitable to microbial life at this time (Grotzinger et al. 2014).

Current research on this problem is focused on understanding the timing and duration of these early warm episodes, and the nature of the water cycle at that time. Martian geochemistry provides a rich record to constrain early atmospheric composition (Ehlmann and Edwards 2014), but access to a range of samples from different environments is required. Isotopic analysis of rock and atmospheric samples returned to Earth, enabling far higher precision measurements than are possible in situ, will be particularly vital for constraining the chemical state of the early atmosphere. Some noble gas isotopes in the present-day atmosphere may be interpreted to infer the efficacy of early loss processes, while comparing elemental and isotopic composition of noble gases and light elements between samples of ancient Mars (in returned rock samples and meteorites) and the present-day atmosphere provides information on atmospheric source and sink processes over time. Geochronological analysis of a range of surface samples is also needed to ascertain the timing of flowing water on Mars, and research into solar evolution is required to better constrain how the Sun's bolometric and XUV luminosity has changed over geologic time. Last, further advances in atmospheric modeling are needed to understand the warming potential of various gases and aerosols and their impact on the observable rock record.

Q6.2d How Does Orbital Forcing, Including Obliquity and Eccentricity Changes, Govern Climate Change and Surface Volatile Redistribution on Extraterrestrial Bodies with Volatile Cycles Like Modern Mars, Triton, and Pluto?

Mars, Triton, and Pluto, and possibly large KBOs, have atmospheres controlled, or strongly influenced, by vapor pressure equilibrium—that is, the main atmospheric constituent (CO_2 on Mars, N_2 on Pluto and Triton) can partially or entirely condense on the surface; hence, the pressure of the atmosphere is regulated by the stability of surface ice temperature. In the case of Mars, CO_2 may also be exchanged with the regolith, greatly impacting pressure variations on orbital timescales (Buhler and Piqueux 2021). The temperature of surface ice varies primarily with insolation, which in turn varies seasonally but also with orbital parameters such as eccentricity and obliquity (Laskar et al. 2002). Changes in orbital parameters owing to Milankovitch cycles occur over tens of thousands of years and can impact the stability of ice formed from minor atmospheric constituents (e.g., H_2O on Mars, CH_4 on Triton and Pluto). Mars's polar layered deposits are thought to record this orbital cycling of water ice and dust deposition, as dust abundance and transport also vary on orbital timescales. However, establishing a direct correlation is challenging because of uncertainties in how the water, CO_2, and dust cycles varied over time, the depositional/removal processes involved, and the detailed layer structure and composition (Byrne 2009). The presence of the vast N_2 ice deposit in Sputnik Planitia on Pluto stabilizes the pressure of Pluto's atmosphere, although it is still subject to diurnal, seasonal and Milankovitch cycles (Bertrand et al. 2018). The distribution of ices on Triton remains largely unknown but could potentially provide information about volatile cycling in and out of the atmosphere. Large KBOs may have similar nitrogen atmospheres to Pluto and Triton and experience similar cycles over diurnal, seasonal and astronomical timescales (see also Q6.3b, Q6.4c,d).

Q6.2e What Is the Role of Transient Climate Forcing Owing to Meteoroid Impacts, Large Volcanic Eruptions, and Other Episodic Atmospheric Processes in Climate?

Transient climate forcing is important to many solar system objects. On Earth, large igneous province volcanic eruptions and meteoroid impacts are implicated in many past mass extinction events, while on Mars meteoroid impacts may have driven warming via direct thermal effects or alteration of the atmospheric composition (Q6.1). For Venus and Titan, it remains unclear if geologically recent transient outgassing events are responsible for their current climates, or if steady-state processes maintain the abundance of destructible species in their atmospheres. Changes in the rate of H_2O outgassing on Venus can have strong effects on albedo, and hence on climate, through coupling with the sulfur cycle (Bullock and Grinspoon 2001). The SO_2 atmosphere of Jupiter's moon Io is supplied by volcanic eruptions discussed further in 6.4a. In all cases, it has become increasingly clear that transient processes play a major role in long-term climate evolution. Future progress in this area requires targeted geologic analyses to constrain the timing, duration, and volatile release of major volcanic and impact events, as well as laboratory experiments and numerical modeling to understand the physical and chemical behavior of atmospheres under the extreme conditions expected following a major volcanic eruption, meteoroid impact, or similar transient process.

Strategic Research for Q6.2

- **Assess whether surface liquid water existed for an extended period on Venus** by determining if the tesserae are felsic in composition.

- **Constrain the timing of martian climate transitions** by performing geochronological dating of samples from multiple locations on the planet's surface.
- **Constrain atmospheric evolution processes on Mars** by returning samples of the atmosphere to Earth of sufficient concentration and fidelity to allow noble gas abundance and isotopic fractionation to be measured.
- **Determine how and why Mars's climate has changed over orbital timescales** by performing radar and spectroscopic mapping of the polar layered terrain and by making in situ measurements of their structure and composition (thickness of layers, dust content, and isotope ratios) and their local meteorology (including volatile and dust fluxes).
- **Study the annual climate cycles and long-term evolution of atmospheres controlled by vapor pressure equilibrium** by measuring the isotopic composition and fractionation of atmospheric gases and, if possible, of surface ices and atmospheric haze, on Triton and Pluto.
- **Study surface-exchange processes across a diverse range of atmospheric compositions** by developing one-, two-, and three-dimensional models of past and present planetary climates.
- **Determine accurate models of the extreme ultraviolet solar luminosity evolution over geologic time** by observing a variety of solar-type stars using space-based observatories.
- **Investigate the radiative forcing potential, chemistry and microphysics of greenhouse gas, haze, and cloud combinations relevant to climate evolution processes** by performing laboratory studies.

Q6.3 WHAT PROCESSES DRIVE THE DYNAMICS AND ENERGETICS OF ATMOSPHERES ON SOLID BODIES?

The dynamics and energetics of solid-surface atmospheres are largely driven by the distribution of solar radiation, which is governed by the solid body's orbital parameters. Atmospheric composition, surface properties, and magnetic environment also impact how solar radiation forces atmospheric circulations. Many outstanding questions relating to atmospheric dynamics and energetics remain unanswered (see Figure 9-3).

Large-scale, meridional (latitude-height) overturning circulations, such as Hadley cells, transport heat, momentum, and tracers between different atmospheric regions. Surface topography, boundary-layer convection, and daily patterns of surface heating also produce various atmospheric waves, which can travel large distances before breaking and depositing their energy and momentum, modifying the angular momentum structure of an atmosphere. Waves are thought to be responsible for strong atmospheric superrotation and for large perturbations to the dynamics of upper atmospheres, but the proposed mechanisms remain mostly unexplored.

The surface-atmosphere transfer of heat, momentum, gases, volatiles, and dust, including their mixing within the lowest portion of the atmosphere (the planetary boundary layer), has a major impact on much of the observed weather and climate, yet has rarely been measured other than on Earth. This is especially problematic when such processes dominate the atmospheric circulation and its variability, as is the case for lofted dust on Mars or condensation-sublimation flows on Pluto, Triton, and Io. Aerosols (dust, hazes, and clouds) are key features of many planetary atmospheres and affect the atmospheric absorption and scattering of solar radiation. Aerosol transport, microphysics, and radiative processes are often coupled via complex feedbacks, yet few measurements related to such processes exist for atmospheres other than Earth's.

At higher altitudes, an ionosphere is generated by the interaction between high-energy particles from the Sun and the thermosphere. The ionosphere and thermosphere are thus coupled chemically and dynamically via energy and momentum transfer (ion-neutral drag). At still-higher altitudes, the ionosphere is coupled to the magnetosphere and through it to the surrounding space plasma. Magnetospheres are driven from above by the highly variable solar wind and from below by particles escaping from the planetary atmosphere and ionosphere, with these interactions varying if a planetary magnetic field exists. These upper-atmosphere couplings and interactions have never been fully characterized for any planetary body, thus their full impact remains poorly known.

Surface-bounded exospheres have their own unique dynamics based on the specific atomic and molecular species present, surface temperatures, surface solar-wind and meteoroid exposure, and the loss processes for the species. The dynamic formation and migration of exospheric volatiles may have been critical in forming the polar volatile deposits at the Moon, Mercury, and Ceres (see also Q5.5c). However, it remains unclear whether some

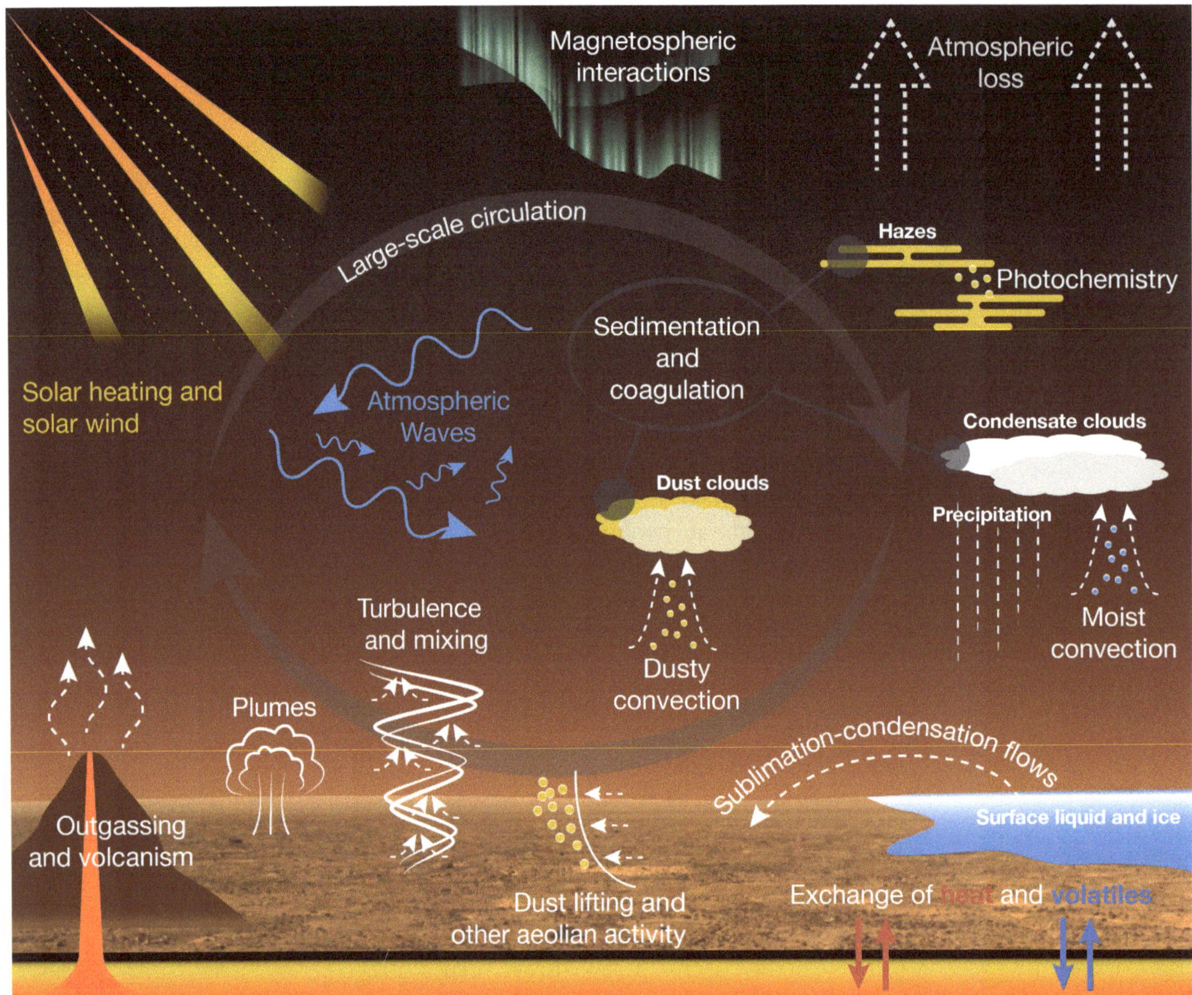

FIGURE 9-3 Processes influencing solid body collisional atmospheres, including interior-surface-atmosphere and solar environment interactions. Note that the relative heights of where processes are shown to occur are not drawn to scale. SOURCE: Courtesy of K. Kostadinova, C. Newman, and A. Brecht.

volatiles, such as water, can efficiently migrate to polar cold traps and/or have a surface-exosphere cycle that operates on short timescales.

Studying present-day atmospheric circulations, surface fluxes, and solar inputs on bodies with differing orbital settings, atmospheric compositions, and surface properties thus provides valuable insight into a huge range of fundamental dynamical processes. This helps us to not only understand the atmospheres of present-day Earth and other solid bodies in the solar system, but also to extrapolate this knowledge to the past climate states of those bodies (see Q6.1, 6.2) and to the atmospheres of rocky exoplanets (see Q12.6).

Q6.3a How Do Horizontally and Vertically Propagating Waves Drive Planetary Atmosphere Dynamics?

The atmospheres of the solar system's slowest rotators, Venus and Titan, exhibit strong equatorial superrotation, with weaker equatorial jets also found on the more rapidly rotating Earth and Mars (Read and Lebonnois 2018). Based on dynamical arguments, equatorial superrotation requires the wave-driven transport of energy up the angular-momentum gradient. Yet while mechanisms for driving this have been proposed, most of the atmospheric waves involved have not been observed. Furthermore, it is unclear whether Venus's upper atmosphere superrotation is driven by the same mechanism as the observed superrotation at the cloud tops and below.

Recurring orographic gravity waves above venusian mountain ranges substantially torque the solid body (Navarro et al. 2018), while possibly reaching Venus's upper atmosphere and may in turn impact atmospheric dynamics at high altitudes. Even less information is available for vertical wave propagation on Titan. On Mars, increased thermal tides and convectively driven gravity waves during dust storms likely help to drive water to higher altitudes and result in greater atmospheric loss (Yiğit et al. 2021). The influence of lower atmosphere waves on thermospheric and ionospheric dynamics is beginning to be examined in models, but more data are needed to test these predictions.

Crucially, the episodic and often unpredictable nature of wave-driven dynamical changes can only be understood via long-term atmospheric measurements of planetary atmospheres, combined with theoretical/modeling studies.

Q6.3b What Controls the Onset, Evolution, and Year-to-Year Variability of Dust Storms on Mars and Titan?

Dust clouds have long been observed in the martian atmosphere, and more recently in Titan's. While the impact on Titan's thick atmosphere is unknown, regional and global dust storms have a huge impact on radiative heating in the thin martian atmosphere, and hence on temperature and winds (e.g., Kahre et al. 2017), as well as on atmospheric loss rates (see Q6.5b). The trigger for global storms, which shroud Mars in dust and completely change the atmospheric circulation for several months, is not currently understood. Modeling suggests that surface-dust availability, water-ice nucleation on dust particles, and complex feedbacks between dust lifting, transport, and circulation patterns (including waves) may all be important (e.g., Newman et al. 2016). Fundamental unknowns include how and where dust is lifted from the surface (e.g., whether direct lifting or saltation of sand particles is involved), how that dust is then raised through the boundary layer and free atmosphere, and how the dust and water cycles are coupled. The infrequency (typically three per 20 Earth years) and significant variability of global storms further complicate understanding their origins and motivates long-term measurements of Mars. Additionally, the presence of high-altitude dust layers outside of dust storms remains largely a mystery (e.g., Heavens et al. 2014).

Q6.3c How Do Sublimation–Condensation Flows Drive Circulation in Thin, Transient Atmospheres? What Is the Nature of This Circulation and How Does It Relate to Surface-Ice Distributions?

Far from the Sun, the surface pressures of Pluto and Triton vary by orders of magnitude over their long orbits as their N_2 atmospheres largely condense out onto the surface around aphelion. Modeling suggests their atmospheric circulations are dominated by surface topography and the distribution of surface ice, which control both thermally driven and sublimation-condensation driven flows (e.g., Bertrand et al. 2018).

On Io, the contribution of volcanic emission versus sublimation of SO_2 surface ices to the varying surface pressure, which varies spatiotemporally over more than five orders of magnitude, is vigorously debated (e.g., Tsang et al. 2016). Day-night, volcanic, and eclipse-driven temperature differences cause huge pressure differentials and fast, complex circulations, complicated further by plumes (McDoniel et al. 2017). On Triton, such phenomena may also be important. Constraints on these models, however, are very limited as the distribution of surface ices is basically unknown for Triton and Io, and only partially known for Pluto, and the predicted volatile fluxes and complex circulations are unmeasured (see also Q6.4d).

Q6.3d How Do Haze and Cloud Processes Impact Atmospheric Dynamics, and What Are the Key Couplings and Feedbacks?

Haze layers impact radiative heating via absorption and scattering, and are present in at least Venus, Titan, Pluto, and Triton's atmospheres. Models suggest that haze-dynamical feedback may be an important driver of Titan's stratospheric circulation, but this is unclear from observations. On Venus, unknown ultraviolet absorbers in the thick clouds produce significant albedo variations (hence amount of solar energy absorbed) on decennial timescales, potentially modifying superrotation (Pérez-Hoyos et al. 2018). However, despite decades of study, the nature of these absorbers remains unknown, as does their impact on dynamics. On Titan, latent heat release and strong convection are associated with methane clouds. The processes by which hydrocarbons form clouds (Q6.6) and evaporate from Titan's surface seas and lakes (Q6.4)—analogous to those involved in Earth's water cycle—have never been observed in situ and are unlikely to be encountered by the Dragonfly mission, which will target the dry low latitudes during a period expected to have minimal precipitation.

Q6.3e What Determines the Effectiveness of Ion-Neutral Drag on Augmenting Upper Atmospheric Circulation?

For Venus, Mars, and Titan, the transfer of energy and momentum between the neutrals and the ionospheric plasma is not well quantified. The measurement of upper-atmosphere neutral and ionospheric winds by current and past missions has been incomplete and inconsistent. At Venus, polar asymmetric ionospheric ion flow has been observed, but it is unknown if this asymmetry drives upper-atmosphere superrotation through ion-neutral drag (Lundin et al. 2011). Moreover, the interaction of the neutral gas and ionosphere in the southern hemisphere of Mars is mediated by strong remanent crustal magnetic fields, but the dynamical effects of these fields for both the neutrals and ions is poorly understood. Understanding the ionosphere structure, variation, and drivers feeds back into knowing how it connects to the neutral atmosphere.

Q6.3f How Do the Structure and Dynamics of Planetary Magnetospheres Vary with Season and Solar Inputs?

Planetary magnetospheres represent a transitional region between the upper atmosphere and the interplanetary-space environment (Russell 2001; Kivelson and Bagenal 2014). The nature of the interface depends critically on the properties of the solar wind and the interplanetary magnetic field (IMF) that it carries. For intrinsic magnetospheres, such as those of Earth and Mercury, the IMF orientation controls the flow of matter and energy from the solar wind to the magnetosphere by magnetic reconnection and boundary layer processes such as the Kelvin-Helmholtz instability. For induced magnetospheres, such as those of Mars and Venus, the IMF determines the orientation of the entire magnetosphere. In either case, the magnetosphere is driven both from above by the highly variable solar wind energy, momentum, and electromagnetic fields, and from below by escaping particles from the planetary atmosphere (see Q6.5c). Magnetospheric structure and dynamics thus vary both with season and solar cycle. The relative importance of these competing influences depends on a wide range of parameters and leads to a complex and variable interaction that has yet to be well characterized for most solar system objects.

Q6.3g What Controls the Transport and Sequestration of Volatiles in Solid-Surface Exospheres?

Molecular hydrogen and ^{40}Ar in the Moon's atmosphere may undergo numerous surface adsorption-desorption sequences to effectively migrate across the body and sequester in cold regions, such as the lunar nightside and polar cold traps. Sequestered nightside volatiles can be rereleased via thermal desorption when the surface rotates into daylight, forming a volatile cycle. However, it is unclear if released water molecules, an important lunar trace species, undergo this same migration and cycling. Early modeling suggested that solar-wind protons convert to regolith water, which is eventually released and migrates to polar cold traps, implying that the cold traps are currently active and dynamic. However, the lack of a detectable stable water exosphere on the Moon suggests that released water may be dissociated/destroyed at the surface after a single adsorption-desorption hop, preventing water from migrating to the poles (Benna et al. 2019). This raises the question of whether water and other volatiles (such as CO, CO_2, and methane) can migrate on exposed rocky bodies such as the Moon, Mercury, Ceres, and Phobos, and whether they form dynamic cycles or are lost to the surface immediately upon release. Recent research suggests water may be released from Mercury's regolith and may migrate to the poles (Jones et al. 2020), but it remains unclear if surface formation/release, transport, and sequestration/trapping of volatiles are universal processes across all exosphere-only rocky bodies.

Strategic Research for Q6.3

- **Determine how atmospheric waves drive atmospheric dynamics and energetics, especially phenomena such as supperrotation and lower-upper atmosphere coupling** by observing wave amplitudes, periods, phases, and spatiotemporal distributions in thermal and direct wind measurements over multiple annual cycles on Venus, Mars, and Titan.
- **Determine how the surface is coupled to the main atmosphere** by measuring the transport of heat, momentum, volatiles, and dust through the planetary boundary layer, via in situ and remote sensing observations of fluxes covering key time periods or environmental conditions, on bodies with collisional atmospheres such as Venus, Mars, and Titan.
- **Investigate the cause of variability in martian dust storms and hence climate** by making in situ measurements of surface dust and sand fluxes simultaneous with environmental conditions, and in situ and orbital measurements of surface dust and sand availability.
- **Determine how the atmospheric circulation is driven on bodies with thin, transient atmospheres** by measuring the thermal state and winds, and distribution of surface topography, ices, and (where relevant) plumes, via remote sensing of Pluto, Triton, and Io.
- **Determine how aerosols influence atmospheric dynamics and energetics** by measuring their properties and spatiotemporal distributions, simultaneous with the thermal and circulation response of the atmosphere, on diurnal, seasonal, and multi-annual timescales on Venus (clouds and hazes), Mars (dust and clouds), and/or Titan (hazes, clouds, and dust).
- **Determine the effectiveness of ion-neutral drag on augmenting upper atmospheric circulation** by performing in situ measurements of ion and neutral winds, as well as ion electron densities, plasma distribution functions, and magnetic fields, on Venus, Mars, and Titan.
- **Determine how the magnetospheres of solid bodies are driven by solar inputs from plasma timescales to solar cycle timescales** by simultaneously measuring the upstream solar wind input and the magnetospheric response at Mercury, Venus, and Ganymede.
- **Investigate the mechanisms by which waves are generated and interact with the mean flow in planetary atmospheres to better understand phenomena including equatorial superrotation (e.g., Venus and Triton), lower-upper atmosphere coupling, atmospheric loss, and martian dust storms,** using numerical models.
- **Explore the connections between the solar wind, magnetic fields, and the neutral atmosphere/exosphere** through numerical modeling.

Q6.4 HOW DO PLANETARY SURFACES AND INTERIORS INFLUENCE AND INTERACT WITH THEIR HOST ATMOSPHERES?

Volcanism and other forms of interior outgassing have played a pivotal role in the formation of atmospheres on various bodies, including Earth, Mars, Venus, Io, and Titan (see Q6.1), and have continued to influence those atmospheres throughout their history. Io, for example, is the most volcanically active world in the solar system, and near-continuous volcanic outgassing is the source of both its SO_2 atmosphere and a global sulfuric acid cloud layer. On other bodies, including Mars, Venus, and Titan, the importance of modern outgassing remains a mystery. While some observations point to modern and active volcanism on these worlds, definitive evidence for these processes is lacking (Q6.4a). In addition to volcanic processes, plumes can deliver particulates that affect the atmospheric chemistry, composition, and dynamics of bodies such as Europa, Enceladus, and Triton (Q6.4b). On Mars, CO_2 jets loft dust into the atmosphere each spring, but the importance and detailed mechanics of this source remain unknown (Q6.3b). The extent to which Triton's plumes influence atmospheric composition and the moon's haze layers similarly remains a mystery. Images of several of these bodies are shown in Figure 9-4.

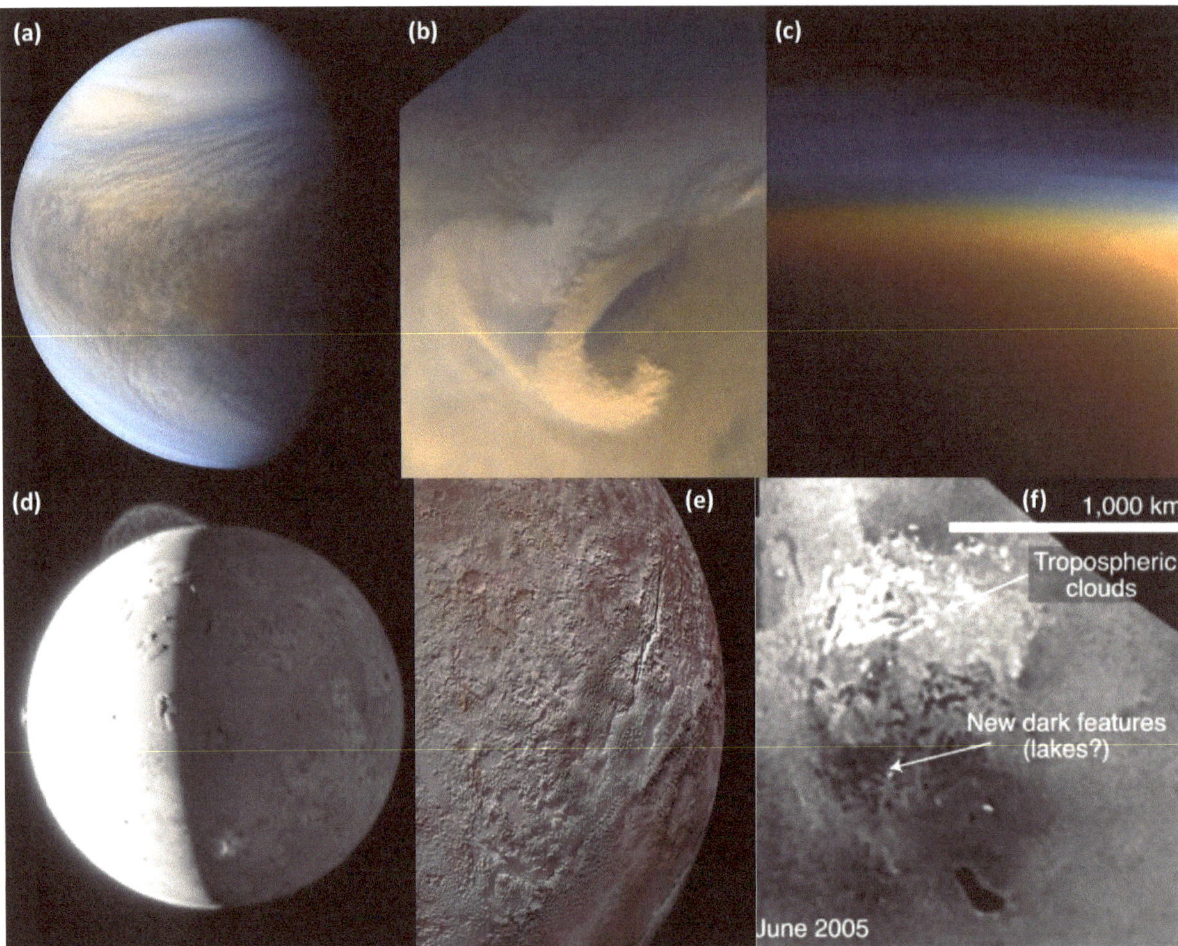

FIGURE 9-4 (a) Dynamic clouds on Venus as seen in composite Akatsuki UVI camera images. (b) A frontal dust storm forming just off the north polar cap of Mars seen in a Mars Orbiter Camera image. (c) Detached haze layers on Titan as seen in natural color using Cassini wide-angle camera images with red, green, and blue spectral filters. (d) Io imaged by the Long-Range Reconnaissance Imager on New Horizons at 11:04 Universal Time on February 28, 2007. This processed image provides the best view yet of the enormous 290-kilometer high plume from the volcano Tvashtar, in the 11 o'clock direction near Io's north pole. (e) Pluto's bladed terrain as seen from New Horizons during its July 2015 flyby. These jagged geological ridges formed of methane ice are found at the highest altitudes on Pluto's surface, near its equator, and can soar hundreds of meters into the sky. Modeling shows that these dunes could be formed by sand-size grains of solid methane ice transported in typical Pluto winds. (f) Cassini Imaging Science Subsystem mosaic of Titan's south pole acquired on 6 June 2005, showing Ontario Lacus, tropospheric clouds, and small dark features interpreted as ponded liquid hydrocarbons following a precipitation event. SOURCES: Courtesy of (a) JAXA/ISAS/DARTS/Damia Bouic; (b) NASA/JPL/Malin Space Science Systems; (c) NASA/JPL/Space Science Institute; (d) NASA/Johns Hopkins University Applied Physics Laboratory/Southwest Research Institute; (e) NASA/Johns Hopkins University Applied Physics Laboratory/Southwest Research Institute (f) NASA/JPL-Caltech/Space Science Institute.

Surface-atmosphere interactions encompass a multitude of processes that link the surfaces of planetary bodies to their atmospheric boundary layers or surface-bounded exospheres. Interactions are bi-directional: they include energy and material exchanges (e.g., condensation and sublimation) and can also include feedback mechanisms that amplify or attenuate coupled processes (e.g., positive feedbacks between dust lifting and circulation strength).

Observations of surface changes caused by aeolian processes often constitute our only information on near-surface wind patterns and environmental conditions. The composition of volatile species in planetary atmospheres drives the formation, evolution, and stability of ices across planetary surfaces. The mechanisms by which condensable volatile reservoirs drive material transport, govern the evolution of large deposits including polar caps, and generate clouds and weather within their host atmospheres remain poorly understood.

In addition to Earth, Titan and early Mars are known to support, or have supported, active hydrologic systems that incorporate stable bodies of surface liquid. Together, they provide natural laboratories for investigating how hydrologic cycles operate and exchange material between atmospheric, surface, and subsurface reservoirs over diurnal, seasonal, and orbital timescales. On airless bodies such as the Moon, Mercury, and Ceres, polar cold traps are of compelling interest to investigations of how volatiles inform the processes of prebiotic chemistry, as well as the history of volatiles throughout the age of the solar system. Our understanding of the volatile/organic composition, evolution, and astrobiological potential of the deposits in these regions, however, remains incomplete. For example, despite observations that surface ice is sparse in lunar polar cold traps (Hayne et al. 2015), it is believed that the ice in these cold traps should be in sublimation–condensation vapor-pressure equilibrium (Zhang and Paige 2009). Understanding this will require comparing the structure, content, and concentration of the trapped volatiles on the Moon to those at Mercury and Ceres (see also Q5.5c).

Q6.4a How Does Modern Volcanic Outgassing from a Planet's Interior Influence the Composition of Its Current Atmosphere?

Following initial volcanic outgassing, SO_2 on Io subsequently experiences diurnal condensation and sublimation. The dynamics of SO_2 exchange processes, including supersonic flow from the dayside to the nightside, is not well understood. On Mars, Venus, and Titan, it is unknown how important modern outgassing might be. The youngest lava flows and meteoritic evidence on Mars suggest it was active as recently as ~50,000 years ago (Horvath et al. 2021). The InSight lander has measured hundreds of "marsquakes" and continues to monitor seismic activity that may be related to active volcanism. There is circumstantial evidence for ongoing volcanic activity on Venus, including transient thermal emissions from areas of stratigraphically young volcanic deposits and residual heat emitting from young lava flows that lack evidence of weathering. Decadal changes in atmospheric composition suggest that SO_2 may be reinjected into Venus's upper atmosphere by major volcanic eruptions, and the presence of a global sulfuric acid cloud is, as on Io, thought to be maintained by continuous volcanic outgassing. While Doom Mons and Sotra Patera on Titan have been interpreted as a cryovolcano and cryovolcanic caldera, respectively (Lopes et al. 2013), individual flows have not been identified. Regardless, modern outgassing is one hypothesis for the replenishment of methane in Titan's atmosphere.

Q6.4b How Do the Particulates Delivered by Plumes Affect the Atmospheric Chemistry, Composition, and Dynamics of Their Host Bodies?

Every spring, cold jets of CO_2 sublimating from seasonal polar caps loft dust into the martian atmosphere (Kieffer et al. 2006). Understanding how these plumes affect the martian climate will require long-duration monitoring observations that encompass multiple dust storm events. At least 2 and possibly as many as 14 plumes erupting on Triton were observed by Voyager 2, along with more than 120 dark streaks and fan features thought to be the remains of plumes (Smith et al. 1989). While the mass flux of these plumes was calculated, other basic properties such as composition remain unknown. The origin of the plumes is similarly unknown, with hypotheses ranging from surficial N_2 sublimation to eruption of subsurface water. Their influence on Titan's atmosphere is also unknown. Understanding plume composition can reveal fundamental properties that will reveal their source and constrain their influence on Triton's atmosphere.

While evidence supports the possible eruption of water vapor plumes from Europa, we know very little about them if they do exist. Are particulates present? If so, what is their composition? Once deposited, do they leave an identifiable mark on the surface, such as color or photometric properties?

Q6.4c What Can Aeolian Features and Their Temporal Variation Tell Us About Near-Surface Winds and Other Atmospheric Variables? What Are the Mechanisms That Connect Wind-Driven Surface Modifications to Environmental Conditions?

Aeolian processes are important for lofting dust into the atmosphere of Mars and for modifying its circulation (see Q6.3b). Dust clouds have also been observed on Titan and possibly Venus. Aeolian processes are also vital for shaping and reshaping planetary surfaces by modifying albedo, forming, and migrating dunes, sculpting rocks, and more (see Question 5, Chapter 8), with such features observed on bodies as diverse as Venus, Pluto, and even comet 67P Churyumov-Gerasimenko. While aeolian-driven surface changes typically have only minimal impact on the state of the atmosphere, observing such features is often the only way by which the near-surface circulation of a planetary atmosphere can be inferred. Therefore, understanding the mechanisms that connect atmospheric variables (such as wind stress and direction) to aeolian features on bodies other than Earth is vital for extracting information on the modern near-surface atmosphere, while ancient aeolian features provide insight into the past. Major unknowns remain, such as how particulate lofting and transport varies with material composition and environmental factors (e.g., gravity and atmospheric density), along with the processes that lead to dune formation in varying environmental conditions.

Q6.4d How Do the Major Constituents in Planetary Atmospheres Drive the Formation, Evolution, and Stability of Polar Caps and Other Reservoirs over Seasonal Timescales? How Do These Reservoirs Drive Volatile Transport and Generate Winds That Can Result in Aeolian Activity?

The sublimation and condensation of volatiles in atmospheres controlled, or strongly influenced, by vapor pressure equilibrium (e.g., Mars, Triton, and Pluto) (e.g., Leighton and Murray 1966) are driven predominantly by energy balance. There are nuances (such as albedo of frost versus ice), however, that are not well understood. These factors can affect the rates of sublimation and condensation that, in turn, drive winds as the volatiles move between cold traps (e.g., polar caps, Sputnik Planitia; see also Q6.3b). The nature of the surface topography also affects local energy balance. These subtleties can have profound effects on polar cap boundaries, weather, and climate. For example, regional dust storms on Mars affect the amount of spring sublimation activity, while spring sublimation drives the initiation of local dust storms along the edge of the retreating seasonal cap and winds associated with cap thermal contrasts may be key to generating regional and even global dust storms (see also Q6.3b). Does the level of dust in the atmosphere affect the amount of snowfall, which has been observed to vary from year to year? Over orbital timescales encompassing obliquity and eccentricity cycles, how does the energy balance change, and are polar cap boundaries affected?

Q6.4e How Do the Minor Constituents in Planetary Atmospheres Drive the Distribution of Surface and Near-Surface Volatiles, Such as the Distribution of H_2O Ice on Mars or CH_4 and CO Ices on Triton/Pluto?

For atmospheres dominantly controlled by sublimation/condensation equilibrium, volatile ices on the surface are also present as gases in the atmosphere. Depending on the nature of surface-atmosphere interactions, however, the abundances of these species can vary by orders of magnitude. The abundance of minor constituents influences atmospheric processes, including cloud generation and weather, and determines the redistribution of surface volatiles over seasonal and orbital timescales. Unknowns include how stable water ice deposits form and survive on Mars, as well as the distribution and migration of volatile methane and carbon monoxide ices on bodies with tenuous atmospheres like Triton and Pluto. Over orbital timescales, the role of minor constituents also changes. Understanding the influence of water ice and dust cycles in the polar layered deposits in response to variations in, for example, obliquity and eccentricity, analogous to Milankovitch cycles on Earth, is important for interpreting how climate differed over orbital timescales (Q6.2d). The layering seen in the martian polar caps is assumed to reflect such changes; however, it is challenging to identify a direct correspondence (Becerra et al. 2019). To extrapolate back in time, the microphysics of the current climate needs to be better understood. For example, how much ice is condensed every year, how much is sublimed, and what sort of "dust lag" (if any) is left behind?

Q6.4f What Do the Present-Day Methane Cycle on Titan and Past Water Cycle(s) on Mars Tell Us About How Hydrologic Cycles Operate and Exchange Material Between Surface and Atmospheric Reservoirs?

Titan is the only solar system body other than Earth to have stable surface liquid and major latent heat exchange within its hydrologic cycle, with methane and ethane taking the place of water. As a result, Titan's methane-based hydrologic cycle is an extreme analogue to Earth's water cycle and represents a natural laboratory to study how planetary climate and hydrologic cycles operate and are maintained (Hayes et al. 2018). Many aspects of Titan's hydrologic system remain unknown, including the composition and full seasonal variability of clouds, the importance of surface-atmosphere exchange and surface/subsurface transport, and the importance of ethane. For example, clouds were expected to form on Titan at northern mid-latitudes in the period following spring equinox, but their appearance was delayed for unknown reasons (Turtle et al. 2018). Better identification of cloud formation mechanisms, their links to surface sources, and whether clouds are composed of ice, liquid, or a binary methane-nitrogen solution, are required to understand the present climate and interpret surface evidence for paleo-seas and fluvial activity. Titan's present hydrologic cycle is also a natural laboratory for investigating exchange processes between atmospheric, surface, and subsurface reservoirs over multiple timescales. Evidence suggests that Mars also had a water-based hydrological cycle in the past (Q6.2c), but it is unclear how the location and size of a possible ocean would have affected the nature of this hydrological cycle, especially during periods when other surface liquid was marginal or short-lived, or how water ice clouds would have affected the radiation balance in a warmer and wetter climate.

Strategic Research for Q6.4

- **Determine if modern volcanic or tectonic activity is influencing Venus's atmosphere** using thermal emission, infrasound, and/or ultraviolet emission from breaking atmospheric waves.
- **Investigate the impact of volcanic outgassing on Io's atmospheric composition** by documenting volcanic activity on Io over multiple seasonal cycles using ground- or space-based observatories.
- **Investigate the redistribution and transport of liquid and solid hydrocarbons on Titan over time** by mapping the distribution of geologic features, volatile ices, and tropospheric clouds over multiple Titan seasons, utilizing ground- and space-based observatories to provide a more extended temporal dataset.
- **Investigate the source of Triton's plumes and their contribution to its atmosphere** by measuring the composition of plume material.
- **Study the properties of Triton's vapor-pressure atmosphere, including why/how it is different from Pluto,** by measuring the distribution of surface ices, as well as atmospheric pressure and temperature as a function of altitude.
- **Investigate how and where stable water ice deposits form on Mars** by measuring their distribution through radar and spectroscopic mapping from orbit, and by measuring the ice vertical distribution, volatile fluxes, and environmental drivers at the surface.
- **Determine how dust lifting and sand motion are linked to the state of the near-surface atmosphere** by making simultaneous, in situ measurements of dust and sand fluxes, surface properties, and environmental conditions (e.g., winds and electric fields) at the surface of Mars, Venus, and Titan.
- **Test and improve theories relating dust lifting, sand motion, and aeolian feature characteristics to environmental conditions (e.g., winds, electric fields, grain sizes) on other planetary surfaces** by performing laboratory, numerical, and terrestrial analog field studies.
- **Infer near-surface wind patterns and environmental conditions, and how they change over time,** by documenting aeolian features and how they vary over seasonal and annual cycles via high-resolution imaging on Mars and Titan.

Q6.5 WHAT PROCESSES GOVERN ATMOSPHERIC LOSS TO SPACE?

Atmospheres and exospheres of every variety escape to space through a richly varied set of processes (Figure 9-5). Atmospheric escape is inextricably linked to evolution (Q6.2), with loss processes coupled to interiors and surfaces (Q6.4), as well as to the Sun and the space environment and to impacts in the early solar system (Q6.1).

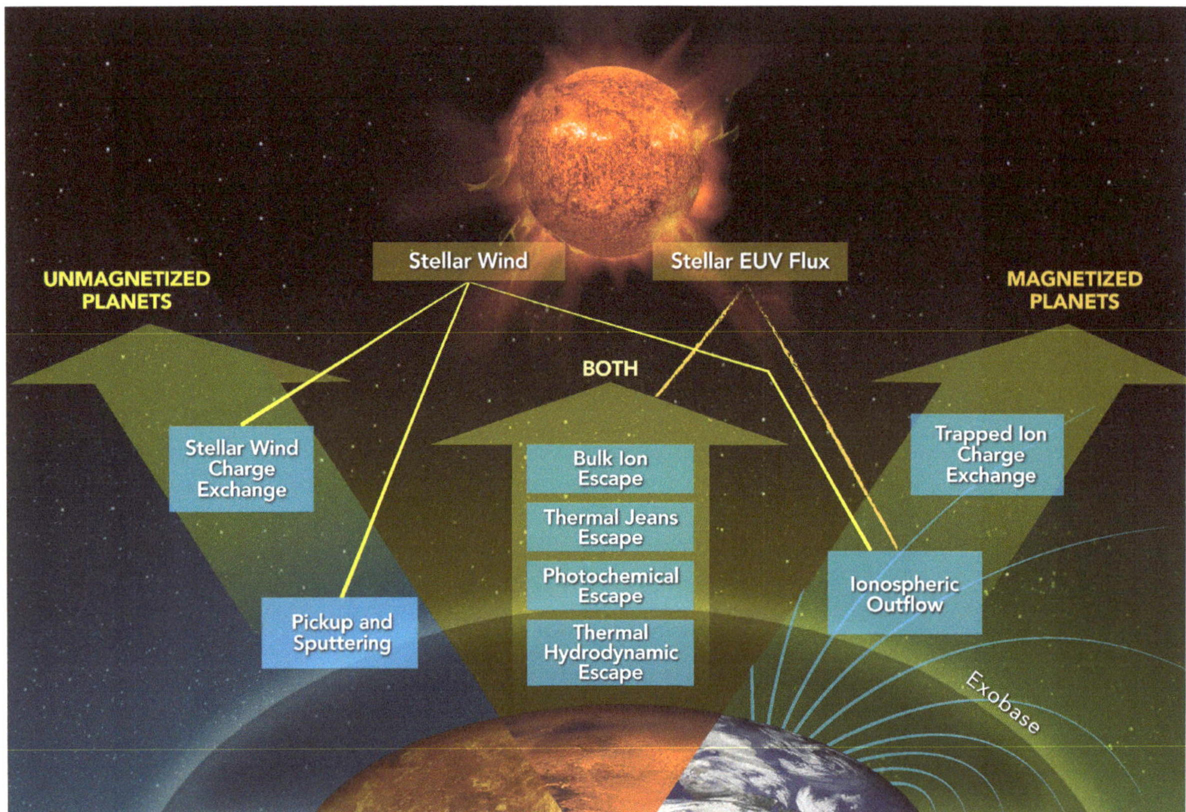

FIGURE 9-5 Atmospheric escape processes. Those on the left act on unmagnetized planets; those on the right act on magnetized planets. Processes in the center operate on both types of planets. SOURCE: Figure adapted from G. Gronoff, P. Arras, S. Baraka, et al., 2020, "Atmospheric Escape Processes and Planetary Atmospheric Evolution," *JGR Space Physics* 125(8):e2019JA027639, https://doi.org/10.1029/2019JA027639.

Escape can alter the chemistry of an atmosphere by preferentially removing some species or isotopes (typically the less massive constituents), or it can transform an early thick atmosphere to a thin atmosphere, as is thought to have occurred on Mars (Jakosky et al. 2018). The net escape rate depends on both the reservoir of particles and the external energy inputs, and thus may be either supply-limited or energy-limited. Atmospheric escape pathways include both thermal and nonthermal processes (Lammer et al. 2008; Gronoff et al. 2020). The relative importance of thermal mechanisms such as hydrodynamic and Jeans escape depends on object size (through gravity) and upper atmospheric temperature, and thus on the evolutionary stage of the body and the Sun (see Q6.1b). Many nonthermal processes involve both neutral and charged particles, thereby coupling atmospheric escape to the ambient plasma environment.

The presence of a magnetic field alters the interaction with the solar wind (see Figure 9-5), reducing the efficiency of some escape processes, but increasing the efficiency of others; however, the net effect of a magnetic field on atmospheric escape remains unquantified. Although all escaping particles ultimately pass through the collisionless exosphere, the structure, composition, chemistry, and dynamics of the collisional atmosphere constrain what species can travel from the surface to the exosphere and escape. Escaping particles from planets and their satellites interact with the solar wind and other plasmas to drive the structure of planetary magnetospheres, in turn affecting the escape pathways.

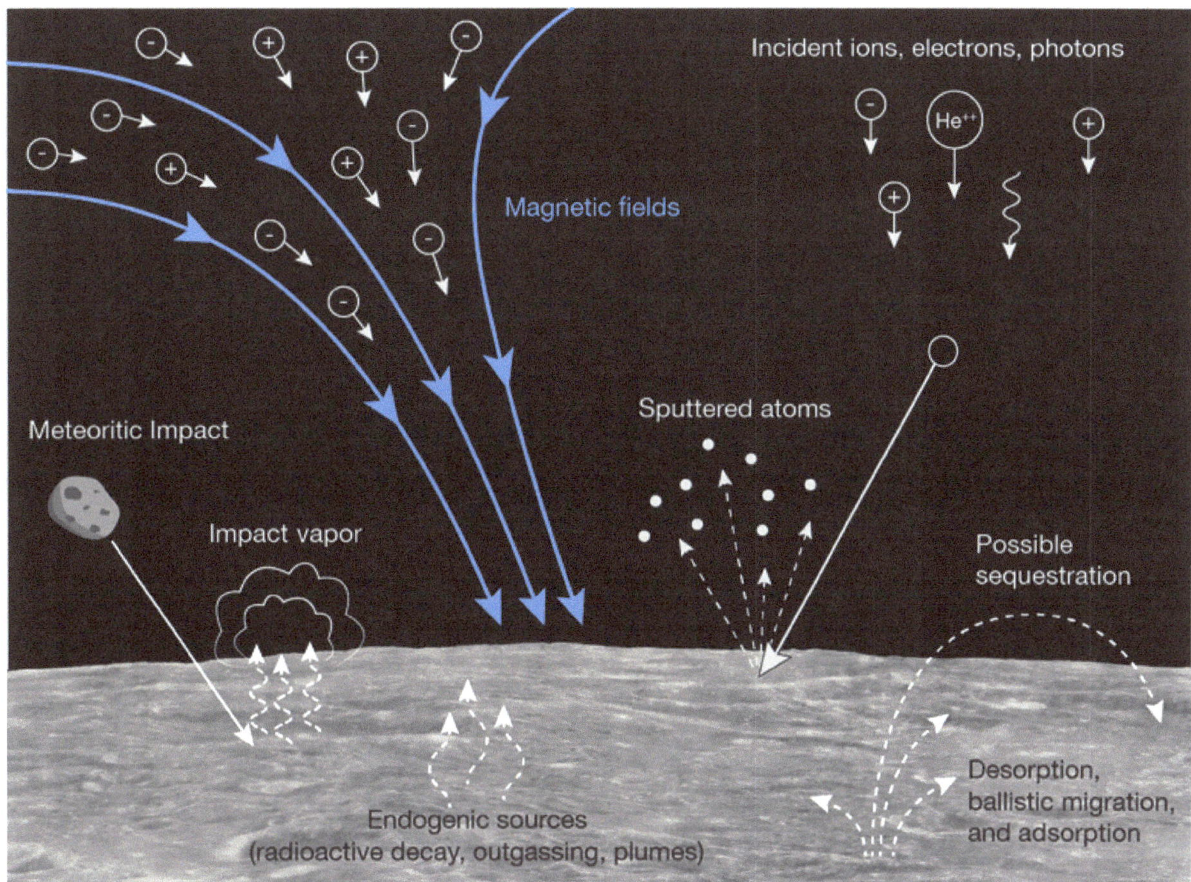

FIGURE 9-6 Multiple processes contribute to the formation and evolution of surface-boundary exospheres. Source terms include both endogenic processes such as natural radioactivity in the crust and outgassing/plumes from localized regions, and exogenic processes such as photon, micrometeoroid, and charged particle bombardment. Magnetic fields affect charged particle motion and can therefore shield and/or divert charged particle flux from more strongly magnetized regions of the surface. Once liberated from the surface, some exospheric constituents can escape directly to space. Others can be ionized, picked up in the ambient plasma, and accelerated to escape velocity and beyond. Still others follow ballistic trajectories and may interact repeatedly with the surface, potentially migrating to colder regions and becoming at least temporarily sequestered.
SOURCE: Courtesy of K. Kostadinova.

Surface boundary exospheres involve their own unique set of processes (Dukes and Hurley 2016). In contrast to the collisional case, the space environment directly interfaces with the surface of these bodies, influencing both the delivery and loss of volatiles (Figure 9-6). Magnetic fields, both global and local, impact plasma-driven delivery to and loss of volatiles from the surface. To fully understand surface boundary exospheres, their equilibrium states, and their long-term evolution, we need a better understanding of the surface physics and chemistry and quantify the relative contributions from all relevant energy sources, including their seasonal and solar cycle variations. Atmospheric loss discussed herein also applies to those bodies having quasi-collisional or transient atmospheres, temporarily evolving from a collisionless exosphere to a collisional atmosphere and back. The early Moon and large KBOs are possible examples of such systems. Additional remarks on escape from quasi-collisional atmospheres can be found in Q6.1b.

Q6.5a How Does the Presence or Absence of Intrinsic or Parent Body Magnetic Fields Influence the Escape of Gases from Solid Planets and Satellites?

The hypothesis that a global magnetic field shields the atmosphere from solar wind-driven loss processes seems intuitively obvious, as the solar wind plasma and its electromagnetic fields cannot then interact directly with the upper atmosphere. However, a global field also increases the interaction cross section of a planet, causing it to intercept more solar wind energy that can drive outflow through the polar cap and cusps, competing with shielding effects and in some scenarios even enhancing loss (Gunell et al. 2018; Gronoff et al. 2020). Similarly, parent body magnetic fields can channel magnetospheric plasma to satellites (see Q8.4), potentially enhancing atmospheric loss. Among our limited sample of terrestrial planets, Venus has a thick atmosphere but no global magnetic field, Earth has a global field and a substantial atmosphere, and Mars has no present-day magnetic field and a much thinner atmosphere. Yet all three have comparable charged particle escape rates. In the outer solar system, Io, Titan, and Triton are located within giant planet magnetospheres and thus are subject to very different incident plasma than Pluto in the solar wind. Although some constraints on thermal escape exist for Pluto and Titan, the nonthermal escape rates and drivers of escape remain poorly constrained for all four bodies. Ganymede presents a unique scenario, with a significant intrinsic magnetic field embedded in the intense jovian field. While the moon's intrinsic field could potentially shield its atmosphere, reconnection with the parent body field may enable escape. Ultimately, while the planets and satellites in the solar system provide a variety of endmembers, they have so many confounding differences (notably size and atmospheric composition) that the importance of magnetic fields remains uncertain. Thus, we do not yet know whether an Earth-like atmosphere can accumulate and remain stable without a global magnetic field, or whether the cessation of the martian dynamo early in its history played a significant role in the loss of its early atmosphere (Lillis et al. 2008).

Q6.5b How Do Atmospheric Dynamics, Such as Martian Dust Storms, Affect the Escape of Gases from Solid Planets and Satellites?

While atmospheric gases, by definition, escape through the exosphere, processes that occur below the exobase can still have significant effects on loss rates. At Mars, although diffusion-limited transport of H_2 would lead to only slow variations in hydrogen loss, we now know that hydrogen escape rates can vary by up to an order of magnitude during dust storms (see also Q6.3b). The observed correlation between dust storms, increased high-altitude water content in the upper atmosphere, and enhanced hydrogen escape suggests that coupling between the lower and upper atmosphere may enable the transport of water to high altitudes, thereby driving more rapid variations in hydrogen escape (Heavens et al. 2018). Atmospheric dynamics may facilitate escape directly and/or by coupling to ionospheric and magnetospheric processes (see Q6.3e,f), with the latter particularly relevant for unmagnetized bodies, such as Venus, Mars, and Pluto (see Q6.5c). Upper atmospheric processes also influence escape through photochemistry (see Q6.6d), with dissociative recombination capable of providing particles with escape energy at less massive bodies, such as Mars, Titan, Triton, and Pluto. The complex coupling from the lower atmosphere to the upper atmosphere to the exosphere implies that physics in multiple regions may affect the total escape, but in many cases the limiting processes remain unknown. To advance our understanding of atmospheric escape, we therefore need to make more complete measurements of the energetic inputs and develop coupled models capable of capturing the physics in the different regions of the atmosphere.

Q6.5c How Do Escaping Gases Influence the Structure of Planetary Magnetospheres?

Escaping neutral and charged particles influence the structure of planetary magnetospheres. Escape from planetary satellites provides an important (even primary, for some giant planets) source of mass and momentum in their parent body magnetospheres (Q8.4). At bodies with induced magnetospheres, such as Venus, Mars, Titan (when in the solar wind), Pluto, and even comets, the upper atmosphere interfaces directly with the space environment, providing the primary obstacle to the solar wind (Luhmann et al. 2004). However, the relative importance of induced currents that produce a magnetic barrier and thus an obstacle to the solar wind flow, as opposed to mass loading in the extended exosphere that transfers momentum from the upstream solar wind, remains poorly quantified and may vary significantly with, for example, object size and solar inputs. For bodies with intrinsic magnetic fields, such as Earth (Toledo-Redondo et al. 2021) and Mercury, plasma derived from the atmosphere and/or exosphere can also affect the magnetosphere by influencing large-scale current systems and altering local dynamics (e.g., magnetic reconnection at the magnetopause). However, the processes by which escaping gases influence magnetospheric structure and dynamics, and the details of the feedback between solar wind energy input, atmospheric escape, and magnetospheric structure remain incompletely understood for both intrinsic and induced magnetospheres.

Q6.5d How Do Magnetic Fields Influence the Loss of Volatiles from Objects with Surface Boundary Exospheres?

The presence of local and/or global magnetic fields can also affect the escape of volatiles from objects with surface boundary exospheres. By shielding portions of the surface from charged particle bombardment, magnetic fields can reduce the release of volatiles, thereby decreasing the net loss rate by direct escape to space and/or subsequent loss processes (Poppe et al. 2014). Magnetic fields can also regulate the loss of charged particles by shielding them from the ambient plasma and preventing them from being picked up. At Mercury, some portions of the surface routinely experience plasma bombardment, with others only exposed during more extreme solar wind conditions. At the Moon, localized portions of the surface have significant magnetic fields capable of deflecting or reflecting the solar wind. At Ganymede, the polar and equatorial portions of the surface interact with very different incident charged particle populations (as demonstrated by surface color variations). What are the local and global effects on escape from these bodies?

Q6.5e How Is the Escape of Volatiles from the Moon, Mercury, and Other Bodies with Surface Boundary Exospheres Driven by Photon, Charged Particle, and Micrometeorite Influx?

The escape of volatiles from bodies with surface boundary exospheres is driven by a variety of energetic inputs (Dukes and Hurley 2016). Photons stimulated desorption and charged particle sputtering remove particles from the surface in both neutral and charged form, and micrometeoroid impacts also liberate material. Solar photon irradiation stimulates thermal desorption of atoms and molecules from the surface. Photons can also facilitate the escape of some exospheric species by exerting radiation pressure, and by ionizing and/or dissociating neutral particles. Once ionized, exospheric constituents can be picked up and accelerated to escape energy and beyond by the ambient plasma, via electromagnetic fields. These processes occur to some degree at all surface boundary exospheres, but our understanding of their relative importance, their similarities and differences, and their dependence on parameters such as surface binding energy, desorption activation energy, solar activity, body mass (gravity), and distance from the Sun, is just beginning to take shape (Killen et al. 2018). We have not yet quantified the relative importance of these escape processes in comparison to sequestration in the regolith or in cold traps, nor understood the variations in the efficiency of these processes with season and solar cycle.

Strategic Research for Q6.5

- **Trace the flow of energy and escaping gases through collisional atmospheres (e.g., Venus, Mars, Triton, Titan, Pluto) to diagnose lower–upper atmosphere coupling,** utilizing simultaneous measurements of the lower and upper atmosphere and exosphere.
- **Discover how escaping ions influence the magnetospheric current systems** by performing multi-point measurements in induced (e.g., Venus and Mars) and intrinsic (e.g., Mercury) magnetospheres.
- **Relate the loss of volatiles from surface boundary exospheres to solar wind dynamics and quantify the effects of magnetic fields** by measuring escaping, migrating, and bound species in regions of different magnetic topology (e.g., Mercury and Ganymede polar and equatorial regions and lunar magnetic anomalies).

- **Reveal the factors that control the structure, composition, and dynamics of surface boundary exospheres** (e.g., Mercury, Moon, Ceres, and Europa) by simultaneously measuring energetic inputs and escaping species for at least one orbit and preferably for a substantial portion of the solar cycle.
- **Establish the influence of magnetic fields in constraining atmospheric loss** by simulating escape from bodies with and without intrinsic and/or parent body magnetic fields with sufficient fidelity to resolve pertinent loss processes.
- **Investigate how photons, charged particles, and micrometeoroids drive escape from different types of surfaces** by performing desorption, sputtering, and impact vaporization (i.e., space weathering) laboratory experiments.

Q6.6 WHAT CHEMICAL AND MICROPHYSICAL PROCESSES GOVERN THE CLOUDS, HAZES, CHEMISTRY, AND TRACE GAS COMPOSITION OF SOLID BODY ATMOSPHERES?

The major gas composition of solid body atmospheres is controlled by processes such as volcanic outgassing, surface weathering, and escape to space, whereas much of their trace gas composition is controlled by in situ atmospheric photochemistry. Photochemistry can create hazes that absorb visible and ultraviolet radiation, and it can produce trace gases that might be misconstrued as biosignatures. Methane on Mars and phosphine on Venus, if indeed present, are possible examples. Trace gases can also catalyze reactions that affect major gas composition. For example, the relatively undissociated nature of the upper atmospheres of Mars and Venus is explained by catalytic chemistry that allows CO to recombine with O to reform CO_2 (McElroy and Donahue 1972).

Clouds themselves are of crucial importance in planetary atmospheres. They modify the radiative balance, most significantly in atmospheres with thick, planetwide cloud layers (e.g., sulfuric acid clouds on Venus), although even tenuous clouds can affect the radiative balance of thin atmospheres (e.g., water ice clouds on Mars). Hazes—aerosols formed from photochemical processes in the atmosphere—can also strongly modify radiative balance and may control much of the atmosphere's vertical thermal structure (Titan and Pluto). In addition, heterogeneous chemistry in clouds and aerosols may be important in some atmospheres (Venus). More data are needed, however, on cloud and haze particle properties and chemical/microphysical formation processes. Figure 9-7 shows clouds and hazes, along with atmospheric temperature profiles, on the four rocky bodies with the densest atmospheres.

Saturn's largest moon, Titan, is the site of other intriguing atmospheric chemistry. Titan is completely shrouded in organic haze produced from photolysis of methane in its atmosphere. The complicated chemistry leading to its formation is incompletely understood but is thought to include formation of polyacetylenes—long polymers made up of carbon atoms linked by alternating single and triple bonds. Similar chemistry may account for the hazes observed in Pluto's atmosphere. This chemistry is also of interest to astrobiologists studying the early evolution of life on Earth, as organic haze may have also been present in Earth's atmosphere during the Archean eon, prior to the rise of atmospheric oxygen.

Q6.6a What Processes Are Important in Controlling the Trace Gas Composition of the Martian Atmosphere?

CH_4 has been reported in Mars's atmosphere from ground-based spectroscopic measurements (Mumma et al. 2009) and mass spectrometer measurements made by NASA's Curiosity rover (Webster et al. 2018). These measurements suggest that CH_4 is present at levels of a few parts per billion by volume (ppbv) and that its concentration exhibits both spatial and temporal variations, with maximum values up to ~7 ppbv. But ESA's Trace Gas Orbiter, which should be capable of detecting CH_4 levels of ~10 parts per trillion, found no evidence for it (Korablev et al. 2019). What is the resolution to this conundrum? Is the methane that Curiosity detects produced locally and destroyed rapidly without having sufficient time to spread throughout the atmosphere? And, if so, is its source biotic or abiotic? A recent report (Civis and Knizek 2021) suggests that CH_4 is produced by ultraviolet-driven surface chemistry, but its rapid removal remains unexplained. An alternative explanation is that methane seeps into Gale Crater from underground sources at night when the planetary boundary layer (i.e., the bottom part of the atmosphere) is stable; then, when daylight arrives, atmospheric mixing dilutes the methane signal below the

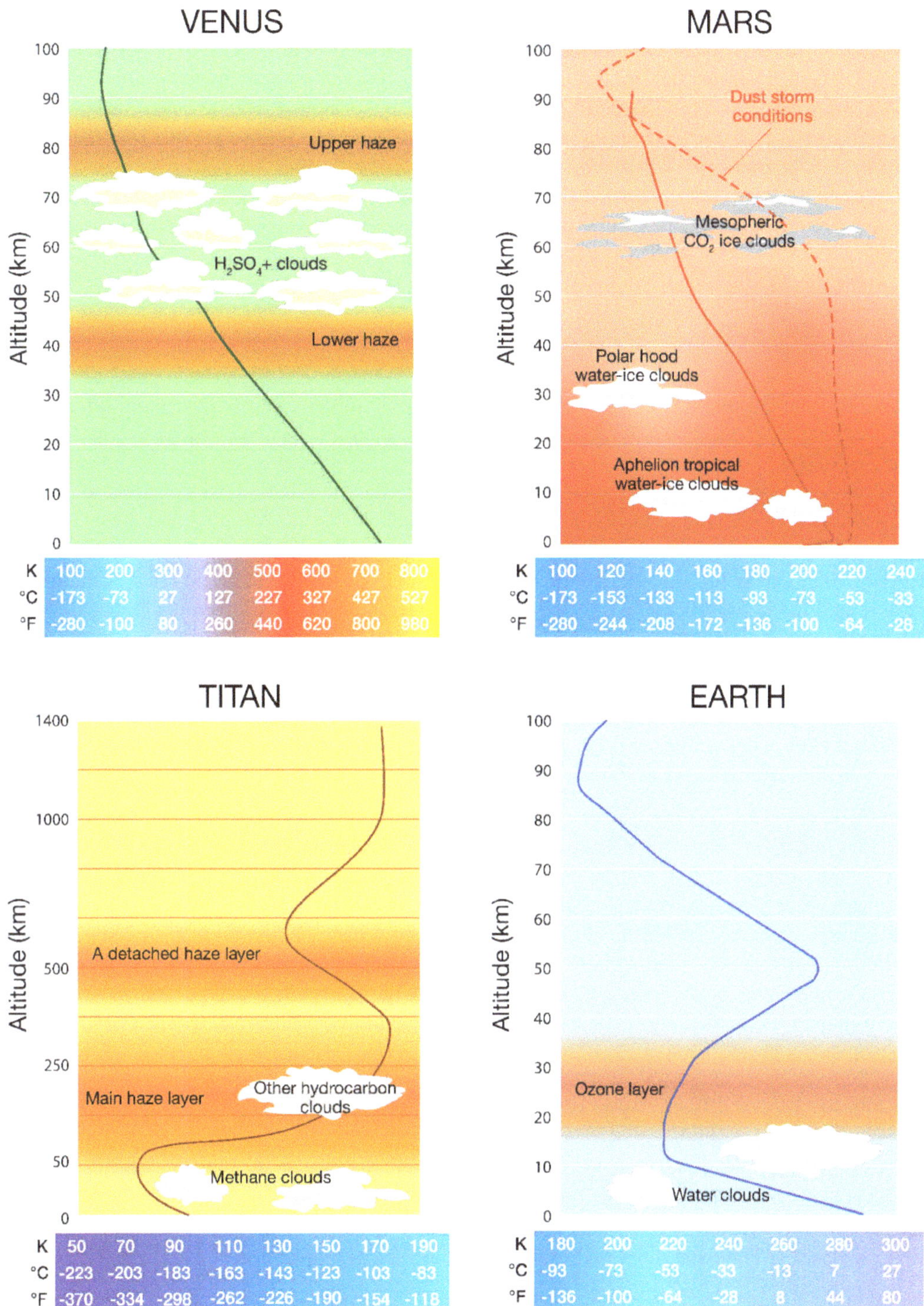

FIGURE 9-7 Temperature profiles on three terrestrial planets and one moon, all of which have relatively dense atmospheres. Condensation clouds are present on Earth and Mars, photochemical haze is present on Venus, and Titan has both. The top of each figure represents the homopause, where the light gases begin to separate out from the heavier ones. Below that altitude, the gases are well mixed. SOURCE: Courtesy of K. Kostadinova.

detection limits of either the rover or the orbiter (Webster et al. 2021). The mixing ratio of O_2 also shows unexpected seasonal and interannual variability, again suggesting an unknown atmospheric or surface process at work (Trainer et al. 2019). The atmosphere of Mars is expected to contain other trace gases, notably SO_2 and halogens, especially HCl. These gases can provide additional insight into the formation, source, and outgassing mechanisms of methane, as well as any current fumarolic activity on Mars. These gases have not yet been detected on Mars; they could be present at abundances lower than the current detection limit of 0.3 ppbv. Detailed measurements of trace species, especially to determine their isotopic fractions with sufficient accuracy to discriminate between a geological and biological origin, will require the return of a large (in volume or concentration), dedicated sample of atmospheric gas (Beaty et al. 2019).

Q6.6b What Processes Control the Trace Gas Composition of Venus's Atmosphere at and Below Cloud Level?

Phosphine has been reported in Venus's atmosphere at cloud-deck level, based on submillimeter-wave observations from Earth (Greaves et al. 2020); however, this detection has been challenged (Snellen et al. 2020). Phosphine is a potential biosignature, so it would be interesting if it were present. Other longstanding questions concerning Venus's atmospheric composition remain to be answered. What is the mysterious ultraviolet (UV) absorber that blocks roughly half of the solar UV radiation incident on the Venus cloud deck (Esposito 1980)? Sulfur compounds—for example, elemental sulfur chains—are likely candidates, but this hypothesis remains to be tested. Also, what drives the fluctuations of SO_2 in the venusian atmosphere? SO_2 is the main trace gas in the atmosphere and the main progenitor of the sulfuric acid clouds. Observations have detected two-orders-of-magnitude fluctuations of SO_2 in the cloud deck over hours to decades. Do these variations result from volcanic outgassing or from dynamical processes within the atmosphere?

Q6.6c What Photochemical Pathways Take Methane to More Complex Hydrocarbons, Including Hazes, in the Atmospheres of Titan, Triton, Pluto, and Possibly Other Kuiper Belt Objects?

The photolytic destruction of atmospheric methane initiates a chain of photochemical reactions (Hörst 2017) resulting in a plethora of organic species that comprise atmospheric hazes and surface deposits like those observed on Titan by Cassini-Huygens and by ground-based facilities. Similar photochemistry is thought to occur on Triton, Pluto, and perhaps other Kuiper belt objects. This chemistry likely exhibits notable similarities to the chemistry that preceded the emergence of life on Earth. While a great deal about this chemistry has been learned from Cassini-Huygens data, fundamental questions remain.

What is the composition of Titan's negative ions with mass-to-charge ratios comparable to terrestrial proteins? What are the implications of changes in concentrations of molecules such as HC_3N, C_6H_6, C_4H_2, C_3H_4, and HCN owing to seasonal changes in Titan's atmosphere (i.e., downwelling in the polar regions of Titan's stratosphere, leading to enrichment of these species in the stratosphere) and how might this affect the type and extent of different chemical reactions at various levels in the atmosphere? Can abundances of organics detected in Titan's atmosphere with ground-based telescopes (e.g., C_2H_4, C_4H_2, and C_3H_2) be validated with in situ measurements? What co-crystalline materials or polymorphs might be forming in the haze layers of Titan's atmosphere as species condense, and how might this inform surface composition and physical properties? What is the composition of the haze in Triton's atmosphere and how is it expected to evolve with time? And what commonalities exist between the atmospheric hazes at Titan and other bodies such as Pluto (e.g., composition and stratification.)?

Q6.6d How Does Atmospheric Composition, via Ionospheric Chemistry, Affect the Structure and Composition of Exospheres?

Planetary ionospheres and neutral exospheres and thermospheres are closely linked (Schunk and Nagy 2009). Ionization and dissociation, both by solar radiation and by auroral precipitation of energetic charged particles from the external environment, alter the exospheric composition and raise its temperature, on which Jeans escape of atomic hydrogen and other light species is critically dependent. Ion-neutral chemistry is another key process in the interaction of the neutral atmosphere and exosphere with the ionosphere. At Venus and Mars, ionization of the major neutral species CO_2 produces CO_2^+ ions, which chemically react with atomic oxygen to produce the main ionospheric species O_2^+. The O_2^+ ions dissociatively recombine, generating suprathermal O atoms which populate extensive exospheres at these two planets. At Mars, a large fraction of the atomic oxygen produced then escapes. But the ionospheric chemistry at Venus and Mars has other pathways that are not yet understood and that can produce hot neutral atoms (Bougher et al. 2014). And the nonthermal exospheres of other solar system bodies (e.g., Titan, Triton, Pluto), and their ionospheric source, are poorly understood. The question remains: What superthermal neutral species are produced by ionospheric chemical reactions, thus populating each planetary exosphere?

Q6.6e Do Exospheric Volatile Gases Like Hydrogen, Methane, and Water at the Moon, Mercury, and Other Exposed Solid Bodies Result from Regolith-Grain Chemistry That Converts Atomic Species, Like Solar Wind-Implanted Hydrogen and Carbon, to New Molecules?

In the past decade, there has been a new appreciation that oxygen-rich regolith at exposed rocky bodies can act as chemical conversion surfaces that take in material delivered from the solar wind and micrometeoroids and release new surface-manufactured products into the exosphere. As an example, the Moon has a substantial molecular hydrogen exosphere that results from solar wind proton implantation, hydrogen diffusion, and H_2 formation and release via recombinative desorption (Tucker et al. 2019). The solar wind is collisionless, but the H atoms congregating at the surface of regolith grains interact to form H_2. Methane observed at the Moon by LADEE has also been hypothesized to result from solar wind implantation of ionized hydrogen and carbon (Hodges 2016). LADEE also detected water from the lunar surface during meteor streams, and this has been suggested to be manufactured from impact-related heating of the regolith that is rich in oxygen and implanted ("doped") hydrogen (Benna et al. 2019). Similar surface chemical processing is expected at Mercury, Phobos, and other solid bodies.

Q6.6f What Processes Control the Formation and Composition of Clouds in the Atmospheres of Venus, Mars, Titan, Triton, and Pluto?

Clouds are seen in the atmospheres of most planetary bodies, from Venus and Mars to Titan, Triton and Pluto (Montmessin et al. 2018), and beyond (see Question 12, Chapter 15). They can affect the atmospheric state via latent heat release, radiative heating, impact on chemistry, and by sequestering atmospheric dust or haze particles as condensation nuclei. However, many uncertainties exist regarding their composition and formation mechanisms. For Venus, better knowledge of cloud processes and composition are needed to explain the mysterious ultraviolet absorbers found there (Q6.6b). On Titan, a major unknown is the distribution, nature, and abundance of the tropospheric clouds over a full annual cycle and whether they are composed of ice, liquid, or a binary methane-nitrogen solution (see also Q6.4e). Studies of the present day coupled dust-volatile cycles of these bodies show that the clouds produced are sensitive to microphysical parameters, such as the heterogeneous nucleation rate of water ice on dust, while the impact on atmospheric state is also sensitive to radiative parameters, controlled at least partly by the size distribution of particles (e.g., Haberle et al. 2019b).

Strategic Research for Q6.6

- **Determine the source location and origins of Mars methane** by making rapid, accurate measurements of methane fluxes at the surface of Mars on hourly to annual timescales, and by returning an atmospheric sample of sufficient concentration to measure methane isotopic fractions indicative of a biotic or abiotic origin.
- **Determine the chemical nature of the unknown ultraviolet absorber in the Venus's clouds** by measuring the composition of the venusian atmosphere and aerosols, particularly at cloud-deck level.

- **Constrain the photochemical paths that take methane to more complex hydrocarbons in methane-rich atmospheres (primarily Titan, but also including Triton, Pluto, and other KBOs)** by measuring the composition of hazes and aerosols in the atmosphere and on the surface.
- **Investigate the microphysical parameters that influence the formation of clouds in planetary atmospheres (primarily Venus, Mars, and Titan, but also including Triton, and Pluto)** by determining the distribution, nature and abundance of clouds and the composition and particle size of the droplets comprising them and cloud condensation nuclei around which they form.
- **Determine possible pathways for haze formation** from experimental study of the chemistry of methane polymerization under both Titan (cold, no oxygen) and early Earth (warm, high CO_2 abundance) conditions.
- **Investigate the conversion of regolith-implanted H and C ions into more complex molecules like water, molecular hydrogen, and methane** via laboratory studies and molecular dynamic simulations.

REFERENCES

Avice, G., and B. Marty. 2020. "Perspectives on Atmospheric Evolution from Noble Gas and Nitrogen Isotopes on Earth, Mars & Venus." *Space Science Reviews* 216. https://doi.org/10.1007/s11214-020-00655-0.

Barnes, J.J., D.A. Kring, R. Tartèse, I.A. Franchi, M. Anand, and S.S. Russell. 2016. "An Asteroidal Origin of Water in the Moon." *Nature Communications* 7(11684). https://doi.org/10.1038/ncomms11684.

Beaty, D.W., M.M. Grady, H.Y. McSween, E. Sefton-Nash, B.L. Carrier, F. Altieri, Y. Amelin, et al. 2019. "The Potential Science and Engineering Value of Samples Delivered to Earth by Mars Sample Return." *Meteoritics & Planetary Science* 54(3):667–671. https://doi.org/10.1111/maps.13242.

Becerra, P., M.M. Sori, N. Thomas, A. Pommerol, E. Simioni, S.S. Sutton, S. Tulyakov, and G. Cremonese. 2019. "Timescales of the Climate Record in the South Polar Ice Cap of Mars." *Geophysical Research Letters* 46:7268–7277, https://doi.org/10.1029/2019GL083588.

Benna, M., D.M. Hurley, T.J. Stubbs, P.R. Mahaffy, and R.C. Elphic. 2019. "Lunar Soil Hydration Constrained by Exospheric Water Liberated by Meteoroid Impacts." *Nature Geoscience* 12:333–338. https://doi.org/10.1038/s41561-019-0345-3.

Bertrand, T., F. Forget, O.M. Umurhan, W.M. Grundy, B. Schmitt, S. Protopapa, A.M. Zangari, et al. 2018. "The Nitrogen Cycles on Pluto over Seasonal and Astronomical Timescales." *Icarus* 309:277–296. https://doi.org/10.1016/j.icarus.2018.03.012.

Bougher, S.W., T.E. Cravens, J. Grebowsky, and J. Luhmann. 2014. "The Aeronomy of Mars: Characterization by MAVEN of the Upper Atmosphere Reservoir That Regulates Volatile Escape." *Space Science Reviews* 195:423–456. https://doi.org/10.1007/s11214-014-0053-7.

Buhler, P.B., and S. Piqueux. 2021. "Obliquity-Driven CO_2 Exchange between Mars' Atmosphere, Regolith, and Polar Cap." *Journal of Geophysical Research: Planets* 126:e2020JE006759. https://doi.org/10.1029/2020JE006759.

Bullock, M.A., and D.H. Grinspoon. 2001. "The Recent Evolution of Climate on Venus." *Icarus* 150:19–37. https://doi.org/10.1006/icar.2000.6570.

Byrne, S. 2009. "The Polar Deposits of Mars." *Annual Review of Earth and Planetary Sciences* 37(1):535–560. https://doi.org/10.1146/annurev.earth.031208.100101.

Catling, D., and J.F. Kasting. 2017. *Atmospheric Evolution on Inhabited and Lifeless Worlds*. Cambridge, UK: Cambridge University Press.

Civis, S., and A. Knizek. 2021. "Abiotic Formation of Methane and Prebiotic Molecules on Mars and Other Planets." *ACS Earth and Space Chemistry* 5:1172–1179. https://doi.org/10.1021/acsearthspacechem.1c00041.

Deng, J., Z. Du, B.B. Karki, D.B. Ghosh, and K.K.M. Lee. 2020. "A Magma Ocean Origin to Divergent Redox Evolutions of Rocky Planetary Bodies and Early Atmospheres." *Nature Communications* 11(2007). https://doi.org/10.1038/s41467-020-15757-0.

Dukes, C., and D. Hurley. 2016. "Sampling the Moon's Atmosphere." *Science* 351:230–231. https://doi.org/10.1126/science.aad8245.

Ehlmann, B.L., and C.S. Edwards. 2014. "Mineralogy of the Martian Surface." *Annual Review of Earth and Planetary Sciences* 42:291–315. https://doi.org/10.1146/annurev-earth-060313-055024.

Elkins-Tanton, L.T. 2012. "Magma Oceans in the Inner Solar System." *Annual Review of Earth and Planetary Sciences* 40:113–139. https://doi.org/10.1146/annurev-earth-042711-105503.

Esposito, L.W. 1980. "Ultraviolet Contrasts and the Absorbers Near the Venus Cloud Tops." *Journal of Geophysical Research: Space Physics* 85:8151–8157. https://doi.org/10.1029/JA085iA13p08151.

Gilmore, M.S., N. Mueller, J. Helbert. 2015. "VIRTIS Emissivity of Alpha Regio, Venus, with Implications for Tessera Composition." *Icarus* 254:350–361. https://doi.org/10.1016/j.icarus.2015.04.008.

Greaves, J.S., A.M.S. Richards, W. Bains, P.B. Rimmer, H. Sagawa, D.L. Clements, S. Seager, et al. 2020. "Phosphine Gas in the Cloud Decks of Venus." *Nature Astronomy* https://doi.org/10.1038/s41550-020-1174-4.

Gronoff, G., P. Arras, S. Baraka, J.M. Bell, G. Cessateur, O. Cohen, S.M. Curry, et al. 2020. "Atmospheric Escape Processes and Planetary Atmospheric Evolution." *Journal of Geophysical Research: Space Physics* 125:e2019JA027639. https://doi.org/10.1029/2019JA027639.

Grotzinger, J.P., D.Y. Sumner, L.C. Kah, K. Stack, S. Gupta, L. Edgar, D. Rubin, et al. 2014. "A Habitable Fluvio-Lacustrine Environment at Yellowknife Bay, Gale Crater, Mars." *Science* 343(6169). https://doi.org/10.1126/science.1242777.

Gunell, H., R. Maggiolo, H. Nilsson, Stenberg, G. Wieser, R. Slapak, J. Lindkvist, M. Hamrin, and J. De Keyser. 2018. "Why an Intrinsic Magnetic Field Does Not Protect a Planet Against Atmospheric Escape." *Astronomy & Astrophysics* 614(L3). https://doi.org/10.1051/0004-6361/201832934.

Haberle, R.M., R.T. Clancy, F. Forget, M.D. Smith, and R.W. Surek, eds. 2017. *The Atmosphere and Climate of Mars.* Cambridge, UK: Cambridge University Press. https://doi.org/10.1017/9781139060172.

Haberle, R.M., K. Zahnle, N.G. Barlow, and K.E. Steakley. 2019a. "Impact Degassing of H_2 on Early Mars and Its Effect on the Climate System." *Geophysical Research Letters* 46:13355–13362.

Haberle, R.M., M.A. Kahre, J.L. Hollingsworth, F. Montmessin, R.J. Wilson, R.A. Urata, A.S. Brecht, et al. 2019b. "Documentation of the NASA/Ames Legacy Mars Global Climate Model: Simulations of the Present Seasonal Water Cycle." *Icarus* 333:130–164.

Hamano, K., Y. Abe, and H. Genda. 2013. "Emergence of Two Types of Terrestrial Planet on Solidification of Magma Ocean." *Nature* 497:607–611.

Haqq-Misra, J.D., S.D. Domagal-Goldman, P.J. Kasting, and J.F. Kasting. 2008. "A Revised, Hazy Methane Greenhouse for the Early Earth." *Astrobiology* 8:1127–1137.

Hayes, A.G., R.D. Lorenz, and J.I. Lunine. 2018. "A Post-Cassini View of Titan's Methane-Based Hydrologic Cycle." *Nature Geoscience* 11:306–313.

Hayne, P.O., A. Hendrix, E. Sefton-Nash, M.A. Siegler, P.G. Lucey, K.D. Retherford, J.-P. Williams, B.T. Greenhagen, and D.A. Paige. 2015. "Evidence for Exposed Water Ice in the Moon's South Polar Regions from Lunar Reconnaissance Orbiter Ultraviolet Albedo and Temperature Measurements." *Icarus* 255:58–69.

Heavens, N.G., M.S. Johnson, W.A. Abdou, D.M. Kass, A. Kleinböhl, D.J. McCleese, J.H. Shirley, and R.J. Wilson. 2014. "Seasonal and Diurnal Variability of Detached Dust Layers in the Tropical Martian Atmosphere." *Journal of Geophysical Research: Planets* 119(8):1748–1774. https://doi.org/10.1002/2014je004619.

Heavens, N.G., A. Kleinböhl, M.S. Chaffin, J.S. Halekas, D.M. Kass, P.O. Hayne, D.J. McCleese, S. Piqueux, J.H. Shirley, and J.T. Schofield. 2018. "Hydrogen Escape from Mars Enhanced by Deep Convection in Dust Storms." *Nature Astronomy* 2:126–132. https://doi.org/10.1038/s41550-017-0353-4.

Hodges, R.R. 2016. "Methane in the Lunar Exosphere: Implications for Solar Wind Carbon Escape." *Geophysical Research Letters* 43:6742–6748.

Hörst, S.M. 2017. "Titan's Atmosphere and Climate." *Journal of Geophysical Research: Planets* 122(3):432–482. https://doi.org/10.1002/2016JE005240.

Horvath, D.G., P. Moitra, C.W. Hamilton, R.A. Craddock, and J.C. Andrews-Hanna. 2021. "Evidence for Geologically Recent Explosive Volcanism in Elysium Planitia, Mars." *Icarus* 365:114499.

Hurley, D.M., J.C. Cook, K.D. Retherford, T.K. Greathouse, G.R. Gladstone, K. Mandt, et al. 2016. "Contributions of the Solar Wind and Micro-Meteoroids to Molecular Hydrogen in the Lunar Exosphere." *Icarus* 283:31–37.

Ingersoll, A.P. 1969. "The Runaway Greenhouse: A History of Water on Venus." *Journal of Atmospheric Sciences* 26:1191–1198.

Jakosky, B.M., D. Brain, M. Chaffin, S. Curry, J. Deighan, J. Grebowsky, J. Halekas, et al. 2018. "Loss of the Martian Atmosphere to Space: Present-Day Loss Rates Determined from MAVEN Observations and Integrated Loss Through Time." *Icarus* 315:146–157. https://doi.org/10.1016/j.icarus.2018.05.030.

Johnson, R.E., A.N. Volkov, and J.T. Erwin. 2013. "Molecular-Kinetic Simulations of Escape from the Ex-Planet and Exoplanets: Criterion for Transonic Flow." *Astrophysical Journal Letters* 768(1).

Johnstone, C.P., M. Gudel, H. Lammer, and K.G. Kislyakova. 2018. "Upper Atmospheres of Terrestrial Planets: Carbon Dioxide Cooling and the Earth's Thermospheric Evolution." *Astronomy & Astrophysics* 617:A107.

Jones, B.M., M. Sarantos, and T.M. Orlando. 2020. "A New in Situ Quasi-Continuous Solar-Wind Source of Molecular Water on Mercury." *Astrophysical Journal Letters* 891:L43.

Kahre, M.A., J.R. Murphy, C.E. Newman, R.J. Wilson, B.A. Cantor, M.T. Lemmon and M.J. Wolff. 2017. "The Mars Dust Cycle." Pp. 295–337 in *The Atmosphere and Climate of Mars.* Cambridge Planetary Science, R.M. Haberle, R.T. Clancy, F. Forget, M.D. Smith, and R.W. Zurek, eds. Cambridge, UK: Cambridge University Press.

Kieffer, H.H., P.R. Christensen, and T.N. Titus. 2006. "CO_2 Jets Formed by Sublimation Beneath Translucent Slab Ice in Mars' Seasonal South Polar Ice Cap." *Nature* 442:793.

Killen, R.M., M.H. Burger, and W.M. Farrell. 2018. "Exospheric Escape: A Parametrical Study." *Advances in Space Research* 62:2364–2371.

Kivelson, M.G., and F. Bagenal. 2014. "Planetary Magnetospheres." Pp. 137–157 in *Encyclopedia of the Solar System*, 3rd ed., T. Spohn, D. Breuer, and T.V. Johnson, eds. Amsterdam, Netherlands: Elsevier.

Korablev, O., A.C. Vandaele, F. Montmessin, A.A. Fedorova, A. Trokhimovskiy, F. Forget, F. Lefèvre, et al. 2019. "No Detection of Methane on Mars from Early ExoMars Trace Gas Orbiter Observations." *Nature* 568:517–520.

Lammer, H., J.F. Kasting, E. Chassefière, R.E. Johnson, Y.N. Kulikov, and F. Tian. 2008. "Atmospheric Escape and Evolution of Terrestrial Planets and Satellites." *Space Science Reviews* 139:399–436. https://doi.org/10.1007/s11214-008-9413-5.

Lammer, H., A.L. Zerkle, S. Gebauer, N. Tosi, L. Noack, M. Scherf, E. Pilat-Lohinger, et al. 2018. "Origin and Evolution of the Atmospheres of Early Venus, Earth and Mars." *Astronomy and Astrophysics Review* 26:2.

Laskar, J., B. Levrard, and J.F. Mustard. 2002. "Orbital Forcing of the Martian Polar Layered Deposit." *Nature* 419:375–377.

Leighton, R.B., and B.C. Murray. 1966. "Behavior of Carbon Dioxide and Other Volatiles on Mars." *Science* 153:136–144. https://doi.org/10.1126/science.153.3732.136.

Li, S., and R.E. Milliken. 2017. "Water on the Surface of the Moon as Seen by the Moon Minerology Mapper: Distribution, Abundance, and Origins." *Science Advances* 3:e1701471.

Lillis, R.J., H.V. Frey, and M. Manga. 2008. "Rapid Decrease in Martian Crustal Magnetization in the Noachian Era: Implications for the Dynamo and Climate of Early Mars." *Geophysical Research Letters* 35:L14203. https://doi.org/10.1029/2008GL034338.

Lopes, R.M.C., R.L. Kirk, K.L. Mitchell, A. LeGall, J.W. Barnes, A. Hayes, J. Kargel, et al., 2013. "Cryovolcanism on Titan: New Results from Cassini RADAR and VIMS." *Journal of Geophysical Research: Planets* 118:1–20. https://doi.org/10.1002/jgre.20062.

Luhmann, J.G., S.A. Ledvina, and C.T. Russell. 2004. "Induced Magnetospheres." *Advances in Space Research* 33:1905–1912. https://doi.org/10.1016/j.asr.2003.03.031.

Lundin, R., S. Barabash, Y. Futaana, J.-A. Sauvaud, A. Fedorov, and H. Perez-de-Tejada. 2011. "Ion Flow and Momentum Transfer in the Venus Plasma Environment." *Icarus* 215:751–758. https://doi.org/10.1016/j.icarus.2011.06.034.

McDoniel, W.J., D.B. Goldstein, P.L. Varghese, and L.M. Trafton. 2017. "The Interaction of Io's Plumes and Sublimation Atmosphere." *Icarus* 294:81–97. https://doi.org/10.1016/j.icarus.2017.04.021.

McElroy, M.B., and T.M. Donahue. 1972. "Stability of the Martian Atmosphere." *Science* 177:986–988.

Montmessin, F., and A. Määttänen. 2018. "Temperature, Clouds, and Aerosols in the Terrestrial Bodies of the Solar System." Pp. 235–263 in *Handbook of Exoplanets*, H. Deeg and J. Belmonte, eds. Cham, Switzerland: Springer. https://doi.org/10.1007/978-3-319-55333-7_48.

Mumma, M.J., G.L. Villanueva, R.E. Novak, T. Hewagama, B.P. Bonev, M.A. Disanti, A.M. Mandell, and M.D. Smith. 2009. "Strong Release of Methane on Mars in Northern Summer 2003." *Science* 323(5917):1041–1045. https://doi.org/10.1126/science.1165243.

Navarro, T., G. Schubert, and S. Lebonnois. 2018. "Atmospheric Mountain Wave Generation on Venus and Its Influence on the Solid Planet's Rotation Rate." *Nature Geoscience* 11:487–491. https://doi.org/10.1038/s41561-018-0157-x.

Needham, D.H., and D.A. Kring. 2017. "Lunar Volcanism Produced a Transient Atmosphere Around the Ancient Moon." *Earth and Planetary Science Letters* 478:175–178.

Newman, C.E., M.I. Richardson, Y. Lian, and C. Lee. 2016. "Simulating Titan's Methane Cycle with the TitanWRF General Circulation Model." *Icarus* 267:106–134. https://doi.org/10.1016/j.icarus.2015.11.028.

Pérez-Hoyos, S., A. Sánchez-Lavega, A. García-Muños, P.G.J. Irwin, G. Holsclaw, W.M. McClintock, and J.F. Sanz-Requena. 2018. "Venus Upper Clouds and the UV Absorber from MESSENGER/MASCS Observations." *Journal of Geophysical Research: Planets* 123:145–162. https://doi.org/10.1002/2017JE005406.

Poppe, A.R., M. Sarantos, J.S. Halekas, G.T. Delory, Y. Saito, and M. Nishino. 2014. "Anisotropic Solar Wind Sputtering of the Lunar Surface Induced by Crustal Magnetic Anomalies." *Geophysical Research Letters* 41:4865–4872. https://doi.org/10.1002/2014GL060523.

Ramirez, R.M., R. Kopparapu, M.E. Zugger, T.D. Robinson, R. Freedman, and J.F. Kasting. 2014. "Warming Early Mars with CO_2 and H_2." *Nature Geoscience* 7:59–63.

Read, P., and S. Lebonnois. 2018. "Superrotation on Venus, on Titan, and Elsewhere." *Annual Review of Earth and Planetary Sciences* 46:175–202. https://doi.org/10.1146/annurev-earth-082517-010137.

Russell, C.T. 2001. "The Dynamics of Planetary Magnetospheres." *Planetary and Space Science* 49:1005–1030.

Schaufelberger A, P. Wurz, H. Lammer, and Y.N. Kulikov. 2012. "Is Hydrodynamic Escape from Titan Possible?" *Planetary and Space Science* 61:79–84.

Schunk, R., and A. Nagy. 2009. *Ionospheres: Physics, Plasma Physics, and Chemistry*, 2nd ed. Cambridge, UK: Cambridge University Press.

Siegler, M.A., R.S. Miller, J.T. Keane, M. Laneuville, D.A. Paige, I. Matsuyama, D.J. Lawrence, A. Crotts, and M.J. Poston. 2016. "Lunar True Polar Wander Inferred from Polar Hydrogen." *Nature* 531:480–484.

Smith, B.A., L.A. Soderblom, D. Banfield, C. Barnet, A.T. Basilevsky, R.F. Beebe, K. Bollinger, et al. 1989. "Voyager 2 at Neptune: Imaging Science Results." *Science* 246:1422.

Snellen, I.A.G., L. Guzman-Ramirez, M.R. Hogerheijde, A.P.S. Hygate, and F.F.S. van der Tak. 2020. "Re-analysis of the 267 GHz ALMA Observations of Venus: No Statistically Significant Detection of Phosphine." *Astronomy & Astrophysics* 644.

Stern, S.A. 1999. "The Lunar Atmosphere: History, Status, Current Problems and Context." *Reviews of Geophysics* 37:453–491.

Strobel, D.F. 2009. "Titan's Hydrodynamically Escaping Atmosphere: Escape Rates and the Structure of the Exobase Region." *Icarus* 202:632–641.

Toledo-Redondo, S., M. Andrè, N. Aunai, C.R. Chappell, J. Dargent, S.A. Fuselier, A. Glocer, et al. 2021. "Impacts of Ionospheric Plasma on Magnetic Reconnection and Earth's Magnetosphere Dynamics." *Reviews of Geophysics* 59(3):e2020RG000707.

Trainer, M.G., M.H. Wong, T.H. McConnochie, H.B. Franz, S.K. Atreya, P.G. Conrad, F. Lefèvre, et al. 2019. "Seasonal Variations in Atmospheric Composition as Measured in Gale Crater, Mars." *Journal of Geophysical Research: Planets* 124:3000–3024. https://doi.org/10.1029/2019JE006175.

Tsang, C.C., J.R. Spencer, E. Lellouch, M.A. Lopez-Valverde, and M.J. Richter. 2016. "The Collapse of Io's Primary Atmosphere in Jupiter Eclipse." *Journal of Geophysical Research: Planets* 121:1400–1410. https://doi.org/10.1002/2016JE005025.

Tucker, O.J., W.M. Farrell, R.M. Killen, and D.M. Hurley. 2019. "Solar Wind Implantation into the Lunar Regolith: Monte Carlo Simulations of H Retention in a Surface with Defects and the H_2 Exosphere." *Journal of Geophysical Research: Planets* 124:278–293. https://doi.org/10.1029/2018JE005805.

Turbet, M., E. Bolmont, G. Chaverot, D. Ehrenreich, J. Leconte, and E. Marcq. 2021. "Day-Night Cloud Asymmetry Prevents Early Oceans on Venus But Not on Earth." *Nature* 598:276–280.

Turtle, E.P., J.E. Perry, J.M. Barbara, A.D. Del Genio, S. Rodriguez, S. Le Mouélic, C.J. Sotin, et al. 2018. "Titan's Meteorology Over the Cassini Mission: Evidence for Extensive Subsurface Methane Reservoirs." *Reviews of Geophysics* 45:5320–5328. https://doi.org/10.1029/2018GL078170.

Urey, H.C. 1952. "On the Early Chemical History of the Earth and the Origin of Life." *Proceedings of the National Academy of Sciences* 38:351–363.

Valley, J.W., W.H. Peck, E.M. King, and S.A. Wilde. 2002. "A Cool Early Earth." *Geology* 30:351–354.

Wade, J., and B.J. Wood. 2005. "Core Formation and the Oxidation State of the Earth." *Earth and Planetary Science Letters* 236:78–95.

Walker, J.C.G., P.B. Hays, and J.F. Kasting. 1981. "A Negative Feedback Mechanism for the Long-Term Stabilization of Earth's Surface Temperature." *Journal of Geophysical Research: Oceans* 86(C10):9776–9782.

Way, M.J., and A.D. Del Genio. 2020. "Venusian Habitable Climate Scenarios: Modeling Venus Through Time and Applications to Slowly Rotating Venus-Like Exoplanets." *Journal of Geophysical Research: Planets* 125:e2019JE006276.

Way, M.J., A.D. Del Genio, N.Y. Kiang, L.E. Sohl, D.H. Grinspoon, I. Aleinov, M. Kelley, and T. Clune. 2016. "Was Venus the First Habitable World of Our Solar System?" *Geophysical Research Letters* 10.1002/2016gl069790. https://doi.org/ 2016GL069790.

Webster, C.R., P.R. Mahaffy, S.K. Atreya, J.E. Moores, G.J. Flesch, C. Malespin, C.P. McKay, et al. 2018. "Background Levels of Methane in Mars' Atmosphere Show Strong Seasonal Variations." *Science* 360:1093–1096.

Webster, C.R, P.R. Mahaffy, J. Pla-Garcia, S.C.R. Rafkin, J.E. Moores, S.K. Atreya, G.J. Flesch, et al. 2021. "Day-Night Differences in Mars Methane Suggest Nighttime Containment at Gale Crater." *Astronomy & Astrophysics* 650:A166. https://doi.org/10.1051/0004-6361/202040030.

Yiğit, E., A.S. Medvedev, M. Benna, and B.M. Jakosky. 2021. "Dust Storm-Enhanced Gravity Wave Activity in the Martian Thermosphere Observed by MAVEN and Implication for Atmospheric Escape." *Geophysical Research Letters* 48:e2020GL092095. https://doi.org/10.1029/2020GL092095.

Zahnle, K.J., M. Gacesa, and D.C. Catling. 2019. "Strange Messenger: A New History of Hydrogen on Earth, as Told by Xenon." *Geochimica et Cosmochimica Acta* 244:56–85.

Zhang, J.A., and D.A. Paige. 2009. "Cold-Trapped Organic Compounds at the Poles of the Moon and Mercury: Implications for Origins." *Geophysical Research Letters* 36:L16203.

Q7 PLATE: One of the poles of Jupiter, as imaged by the JunoCam instrument onboard the Juno spacecraft in 2019. SOURCE: Courtesy of NASA/JPL-Caltech/SwRI/MSSS/G. Eichstädt/P.K. Byrne (https://www.missionjuno.swri.edu/junocam/processing?id=3080).

10

Question 7: Giant Planet Structure and Evolution

What processes influence the structure, evolution, and dynamics of giant planet interiors, atmospheres, and magnetospheres?

The giant planets comprise 99.5 percent of the mass of the solar system, apart from the Sun, and 96 percent of the solar system's total angular momentum.[1] They are the most massive remnants of accretion from the Sun's nebular disk and to have played a substantial role in shaping the overall architecture of the solar system. Each giant planet hosts what is akin to a solar system in miniature, including a substantial, and sometimes complicated, planetary magnetic field and a diverse collection of satellites and rings. Each planet system is unique, a complex byproduct of both common and differing formation and evolutionary processes. Comparative planetology between the hydrogen-rich "gas giants" (Jupiter and Saturn), and the intermediate-size and heavier element-enriched "ice giants" (Uranus and Neptune), is crucial for understanding the processes that govern their present-day interiors, atmospheres, and magnetospheres. Time-resolved, multi-wavelength remote sensing coupled with in situ atmospheric, gravitational, and magnetospheric measurements provide essential tools to reveal the properties of these four worlds. Studies of our giant planets provide the scientific template for understanding a broad class of astrophysical objects. A large fraction of known exoplanets falls in the giant planet size class, and among these, the ice giants appear to be a particularly abundant sub-class in the galaxy whose composition and structure remain poorly understood.

Q7.1 WHAT ARE GIANT PLANETS MADE OF AND HOW CAN THIS BE INFERRED FROM THEIR OBSERVABLE PROPERTIES?

Determining the composition of the giant planets is fundamental for understanding their diversity and distinguishing between the gas giant and ice giant classes. Giant planet atmospheres are comprised primarily of hydrogen and helium, like the Sun itself and the protosolar disk (Question 1, Chapter 4), but their detailed elemental abundances vary owing to different formation conditions, planet masses, and evolutionary paths. Apart from helium and neon (thought to sink deeper into the interiors of Jupiter and Saturn in the form of heavy droplets), most elements observed in giant planets to date are enriched compared with their protosolar values (Question 2, Chapter 5;

---

[1] A glossary of acronyms and technical terms can be found in Appendix F.

265

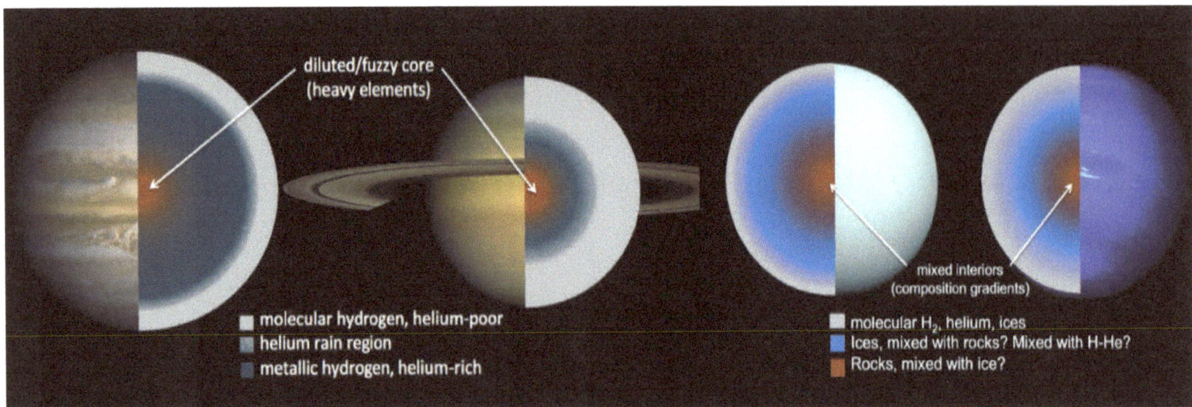

FIGURE 10-1 Interiors of Jupiter, Saturn, Uranus, and Neptune (not to scale). Jupiter and Saturn's interiors are characterized by a deep metallic hydrogen region and a phase separation of helium in hydrogen leading to helium rain. Recent investigations indicate that their central dense cores are diluted into the metallic hydrogen envelope. The hydrogen envelopes of Uranus and Neptune are much less extended and overlay a mixture of ices and rocks; it is unknown whether these ices and rocks form distinct layers. SOURCE: Courtesy of R. Helled.

see Figure 5-2), implying that heavy elements were preferentially delivered during or after the planets' formation, or that they have been dredged-up from a heavy-element core (Moll et al. 2017). While for Jupiter's atmosphere the abundance of most major species is known relatively accurately, only sparse remote sensing data are available for Saturn, and even less is available for Uranus and Neptune. Relative to the gas giants, Uranus and Neptune contain more heavy elements with only a modest envelope of hydrogen and helium (Figure 10-1). However, there remain major uncertainties in the ice giant global compositions, and the total mass of heavy elements within them is poorly constrained (Helled et al. 2020).

Q7.1a Are the Helium and Noble Gas Abundances Across the Giant Planets Consistent with Interior and Solar Evolution Models?

Helium and noble gases bear essential information for understanding giant planet origin and evolution. Helium is the second most abundant species after hydrogen in the gaseous envelopes of Jupiter, Saturn, Uranus and Neptune. Although helium contributes substantially to the mean molecular weight of their atmospheres, the determination of its abundance from stellar and radio occultations and thermal emission spectroscopy has proven difficult. The only accurate and reliable determination is from the Galileo probe into Jupiter (Young 2003). In Jupiter and Saturn, helium droplets form in the deep metallic hydrogen envelopes and are transported downward. Measuring He abundance precisely in Saturn is key to constraining Saturn's evolution and cooling for comparison with Jupiter (Mankovich and Fortney 2020). The shallow envelopes of Uranus and Neptune imply that helium rain does not occur, so that measuring the He abundance in these planets would provide a direct determination of the protosolar He abundance, which is presently only inferred from models of the Sun's evolution.

Neon, thought to dissolve into helium droplets (Wilson and Militzer 2010), is depleted in Jupiter's atmosphere compared to its protosolar abundance and a depletion is similarly expected in Saturn's atmosphere. In Uranus and Neptune, measurement of a low neon abundance would directly constrain models of the early formation of the solar system as well as interior evolution. Measurement of the abundances of heavier inert gases (argon, krypton, and xenon) is key to understanding the mechanisms that led to the formation of giant planets (see Questions 1 and 2) as well as planetary envelope mixing. The Galileo probe found that these gases are enriched by a factor of 2 to 3 in Jupiter compared to their protosolar values (Mahaffy et al. 2000), while abundances in the other giant planets remain unknown. Because noble gases are trapped into solids only at very low temperatures and they have few chemical interactions, determination of their enrichment compared to that of the major gases is an essential clue to understanding giant planet evolution (Guillot and Gautier 2015). For example, a uniform enrichment in noble gases and major species would indicate an early enrichment with little mixing from the deeper interior. Although it would be desirable to determine helium and noble gases for all the giant planets, having even a second point of comparison from Saturn, Uranus, or Neptune to supplement the Galileo probe measurements in Jupiter would be a tremendous advance.

Q7.1b How Do Bulk Abundances of Major Species and Ice-to-Rock Ratios Compare with Nebular Models?

The gas fractions and ice-to-rock ratios of the giant planets provide fundamental constraints on how these planets formed and evolved, and more broadly, on how different classes of giant planets—for example, gas giant, ice giant, rock giant, super-Earth, and sub-Neptune (Q2.2; Question 12)—are established. The main compositional difference among the giant planets is a higher gas fraction in Jupiter and Saturn compared with Uranus and Neptune. The ice-to-rock ratios in the giant planets are more difficult to constrain, and current data and models are unable to conclusively determine whether Uranus and Neptune are rock-dominated or ice-dominated (Helled et al. 2020). Resolving this question may provide important context for understanding why super-Earths and sub-Neptunes—frequently detected in exoplanet systems—are missing in the solar system. Gravity field data from orbiters and composition data from atmospheric probes and remote sensing, particularly at Uranus and Neptune, would provide the greatest advances in our understanding of giant planet bulk compositions.

Among the giant planets, C/H is the best-constrained compositional ratio thanks to plentiful spectral signatures of atmospheric methane. This ratio is commonly used as a proxy for the overall heavy-element enrichments of Uranus and Neptune, based on the roughly equal enrichments of C, N, S, P, and heavy noble gases seen at Jupiter (Atreya et al. 2022). However, uncertainties arise because carbon could have been incorporated as part of the nebular gas (as CO), as ices (e.g., in water ice clathrates, or CH_4 and CO_2), or in an additional organic solid phase. It is likely that relative abundances may vary significantly in the giant planets, given their large differences in gas fraction. Compositional measurements are needed for all four giant planets to unravel the numerous nebular processes and sources of primordial materials during giant planet formation (Question 2).

Q7.1c How Are Condensable Species and Disequilibrium Species Distributed and Thus Transported in the Planetary Atmospheres and Interiors?

Apart from noble gases, all species in giant planets either condense or undergo major chemical reactions in the atmosphere or interior. Composition in the observable part of giant planet atmospheres varies with height (Figure 10-2), owing to the interplay between chemical processes (i.e., thermochemistry and cloud chemistry in the troposphere, and photochemistry in the stratosphere) and dynamical transport (e.g., global circulation, diffusive mixing, storms, and vortices). Because giant planets are fluid and mostly convective, it has long been thought that abundance variations are primarily owing to radial changes in pressure and temperature. However, measurements by the Galileo probe and Juno's microwave radiometer, as well as remote spectroscopy data, provide evidence for latitudinal and longitudinal variations in water and ammonia abundance even in Jupiter's deep atmosphere at tens of bars (Bolton et al. 2017). The ice giant atmospheres also show compositional contrasts with latitude down to tens of bars, as shown by Earth-based infrared-radio remote sensing (Molter et al. 2021; Tollefson et al. 2021). The implication is that abundance variations are likely to be common, but the main transport mechanisms are unknown. While large-scale circulation shapes the appearances of all giant planets, storms and small-scale processes are known to be significant and may, for example, account for the transport of Jupiter's intrinsic heat flux. Mapping of abundances of key condensable species, including methane and H_2S, in Uranus and Neptune is needed to understand transport mechanisms in their atmospheres. Disequilibrium species such as CO and PH_3 are also particularly important because their observed abundances provide insights into the composition of the deep atmosphere, at pressures of hundreds of bars (Fouchet et al. 2009; Moses et al. 2020).

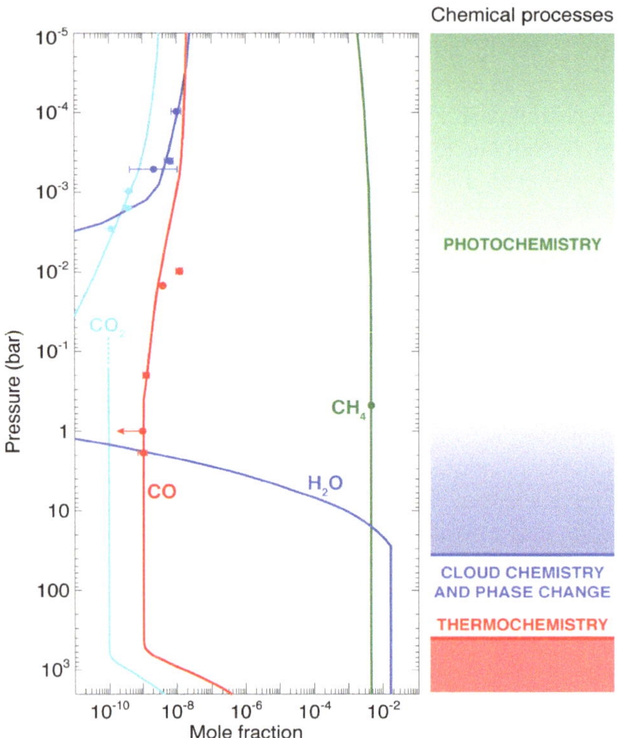

FIGURE 10-2 Atmospheric composition varies in the vertical direction owing to thermochemistry in the deep troposphere, cloud chemistry in the upper troposphere, and photochemistry in the stratosphere. Stratospheric water comes from external (ring rain and cometary) sources. SOURCES: Reproduced from S.K. Atreya, A. Crida, T. Guillot, C. Li, J.I. Lunine, N. Madhusudhan, O. Mousis, and M.H. Wong, "The Origin and Evolution of Saturn—A Post-Cassini Perspective," to be published by Cambridge University Press as part of a multivolume work edited by K. Baines, M. Flasar, N. Krupp, and T. Stallard, entitled *Saturn: The Grand Finale*, Cambridge University Press." Based on Saturn data and models from Wang et al. (2016) and Moses and Poppe (2017).

Improved measurements of chemically active species help advance the understanding of physical and chemical processes in the stratosphere, troposphere, and deep atmosphere, illuminating similarities and differences between the giant planets and their exoplanet cousins. Chemically active species also touch on a wide range of broader science questions. For example, measurements of the CO mixing ratio (interpreted by realistic models) provide constraints on the O/H ratio in the deeper atmosphere. This provides an important check on abundances calculated from the H_2O mixing ratio directly, because CO is unaffected by water cloud condensation. Condensable species such as ammonia, H_2S, and water trace three-dimensional dynamical flows in the weather layer, and photochemical species are sensitive to external influences (Q7.3b).

Q7.1d How Are Atmospheric and Interior Abundances Related?

Our knowledge of the interior, evolution, and formation of giant planets depends on our ability to link atmospheric and interior abundances. The importance of this link is illustrated by helium in Jupiter and Saturn, in which an atmospheric abundance lower than the protosolar value demonstrates the presence of a phase separation of helium in hydrogen. This separation, and the sinking droplets, profoundly modifies the cooling of these planets (Mankovich and Fortney 2020). The unmixing of other elements is also possible. In Uranus and Neptune, water may separate from hydrogen (Bailey and Stevenson 2021), which would affect the structure, chemistry, and evolution of these planets. In addition, in giant planets all condensing species are heavier than the atmospheric gas and there is no planetary surface, making it presently unclear how far the condensates sink (Guillot et al. 2020). The implication is that the high-pressure behavior of matter and related transport processes both need to be characterized (Q7.1c, Q7.2b–c, Q7.3a–c). Such issues will become increasingly important as abundances are measured in exoplanetary atmospheres. Given the limited spatial information available for exoplanets, interpretation of such abundances will need to rely on understanding obtained from study of our giant planets for which much more detailed spatial and temporal information can be obtained (Question 12, Chapter 15).

Q7.1e What Are the Isotopic Compositions of the Giant Planets and How Do They Compare with Formation and Interior Evolution Models?

Because the giant planets formed in the presence of the solar nebula, measurements of isotope ratios in their atmospheres are extremely important as points of comparison with other objects in the solar system and with each other. D/H has been measured in many objects in the solar system—including Jupiter, Saturn, Uranus, and Neptune—but most other isotopes have not. The jovian D/H value is used as a proxy for the protosolar value, because deuterium is destroyed in the Sun. Because D/H in Uranus and Neptune is much larger than on Jupiter, but still relatively low compared to cometary values, it can also be used to understand mixing in planetary interiors as well as planetary ice-to-rock ratios (Feuchtgruber et al. 2013). As with D/H, the protosolar ^{3}He/^{4}He ratio is based on probe measurements at Jupiter, but there remain unexplained differences between these values and those in meteorites (Mandt et al. 2020). Additionally, heavier element ratios such as ^{12}C/^{13}C, ^{14}N/^{15}N, ^{18}O/^{17}O/^{16}O, ^{20}Ne/^{22}Ne, or the isotopes and relative abundances of heavier noble gases, can be used to test formation models, infer characteristics of the carriers that delivered these elements to the giant planets, and constrain evaporation processes in the early solar system (Question 1, Chapter 4).

Strategic Research for Q7.1

- **Constrain the interior evolution of Saturn, Uranus, and Neptune** via in situ sampling of noble gases, elemental, and isotopic abundances, in combination with remote sensing composition observations.
- **Determine the bulk compositions of Saturn, Uranus, and Neptune, and the ice-to-rock ratios in Uranus and Neptune,** from gravity field and elemental abundance measurements.
- **Constrain chemical processes, vertical mixing, and dynamical transport in all four giant planets** by simultaneously measuring multiple tracers (e.g., temperature and condensable and disequilibrium species) over varied temporal, vertical, and horizontal scales, from both remote sensing (all giant planets) and in situ measurements at Saturn, Uranus, and Neptune.

Q7.2 WHAT DETERMINES THE STRUCTURE AND DYNAMICS DEEP INSIDE GIANT PLANETS AND HOW DOES IT AFFECT THEIR EVOLUTION?

The deep interiors of the giant planets consist of mixtures of common materials (hydrogen, helium, water, methane, ammonia, and silicates) under a wide range of pressures and temperatures. These materials behave in complex and dynamic ways as they mix, separate, settle, and flow, giving rise to the planets' intrinsic magnetic fields, as well as asymmetries and variations in the planets' gravitational fields (Figure 10-3). Jupiter and Saturn have very different gravitational and magnetic fields, despite both being composed primarily of hydrogen and helium, suggestive of distinct histories. The much higher abundances of heavier elements in Uranus and Neptune give rise to distinctive internal structures and dynamics, most dramatically their unusual magnetic fields. While currently available data on the internal structures of the ice giants is very limited, the fact that Uranus emits much less internal heat than Neptune may imply distinct internal structures and histories for these planets as well.

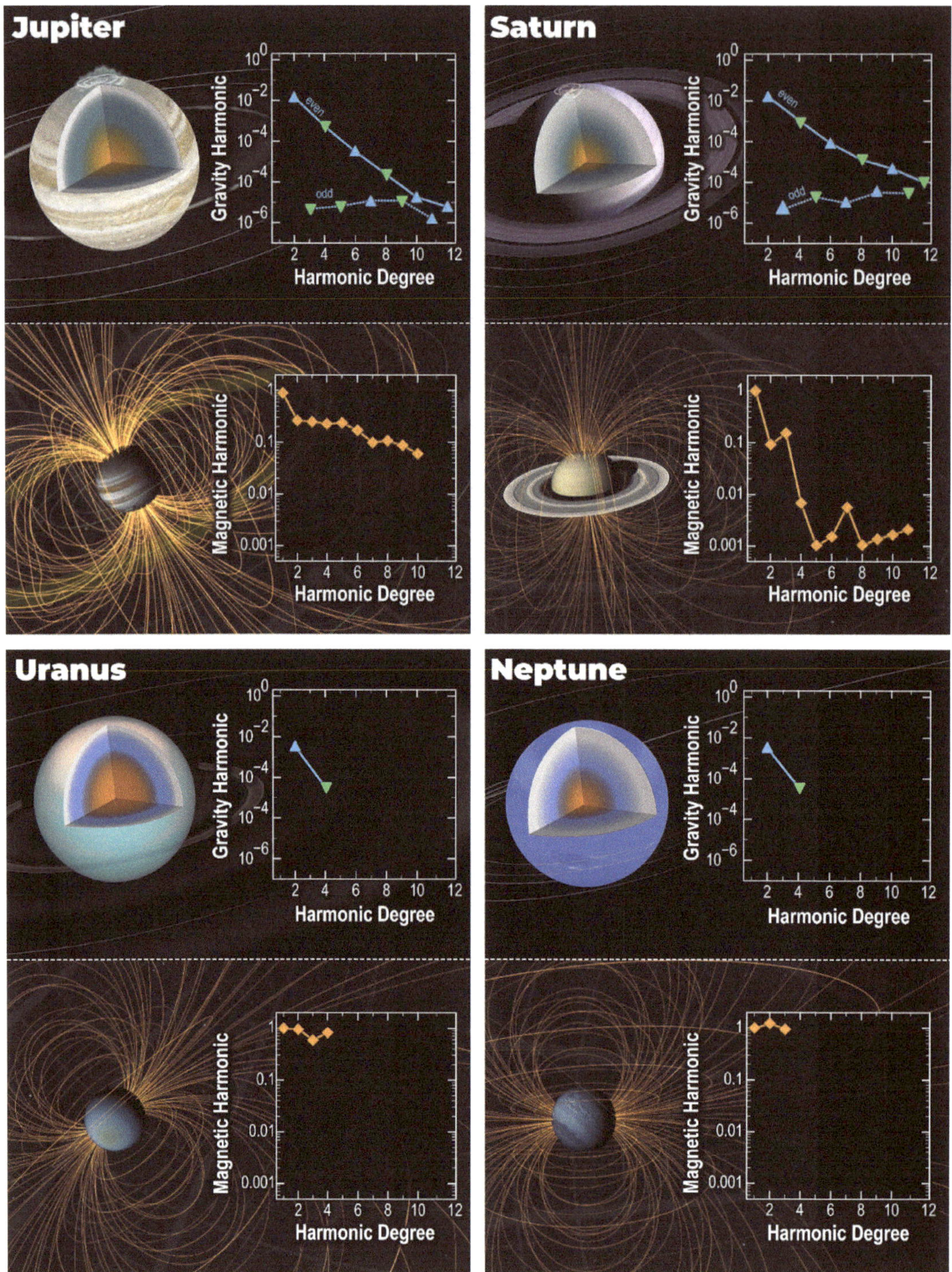

FIGURE 10-3 Visualizations of the interior structure (top panels) and magnetic fields (bottom panels) of the outer planets. Inset plots show power spectra of the observed gravity and magnetic fields. Power spectra of magnetic fields are normalized to the dipole term. In gravity power spectra, upward pointing triangles correspond to positive values, while downward pointing triangles correspond to negative values. SOURCES: Created by J.T. Keane based on data from M. Hedman. Magnetic field diagrams courtesy of NASA's Scientific Visualization Studio.

Q7.2a How Does Composition Change with Depth in Giant Planet Interiors?

Recent measurements of Jupiter's and Saturn's gravitational fields, as well as observations of waves in Saturn's rings, have revealed that both planets have dilute cores. Instead of a discrete core (composed primarily of rock and ice) surrounded by a distinct envelope (composed primarily of hydrogen and helium), both planets exhibit a smooth change in composition and density with depth (Wahl et al. 2017; Mankovich and Fuller 2021). This reflects both on how these planets originally formed and to their long-term evolution. For example, the formation and stability of these gradients depend on how well rock and ice mix with metallic hydrogen under high pressures, and their existence can influence how quickly the planet's interior cools over time (Helled and Stevenson 2017). Further theoretical study, together with improved characterization of material properties under appropriate conditions, are needed to determine which evolutionary histories are consistent with the dilute core structures.

The internal structures of Uranus and Neptune are far less well constrained, and it is not yet certain whether these planets have distinct or diffuse layers in their interiors. At least Uranus's interior may have been stirred by a primordial giant impact(s) that tilted its rotational axis. High-pressure experiments involving mixtures of materials and pressures relevant to Uranus and Neptune would clarify the conditions under which rock, ice, hydrogen, and helium mix, providing needed information to reconstruct the histories of these planets (Helled and Fortney 2020). Measurements of Uranus's and Neptune's gravity and magnetic fields are also needed to make progress on this question, with further improvements also possible from seismology if normal modes are detectable on these planets (Markham and Stevenson 2018).

Q7.2b How Are Elements and Heat Transported from the Deep Interior to the Atmosphere?

With the possible, still unexplained, exception of Uranus, all giant planets emit more heat than they receive from the Sun. This is thought to drive convection and mixing, leading to relatively homogeneous interiors in longitude and latitude. However, recent analyses suggest a more complex picture: the abundances of major chemical species (water in Jupiter, ammonia in Jupiter and Saturn, methane and H_2S in Uranus and Neptune) exhibit significant spatial and temporal variability (Fletcher et al. 2020), indicating that mixing is far from complete and/or is competing with other processes such as precipitation during storms (Guillot et al. 2020). The presence of vortices and waves also indicates that at least part of the deep atmosphere is, on average, stable against convection. This issue is not limited to the so-called "weather layer"—that is, the region characterized by condensation and latent heat release: it also extends to deeper regions, where strong compositional gradients may reflect phase separation (e.g., with helium and hydrogen at Mbar pressures in Jupiter and Saturn), phase transition (e.g., the transition to superionic water in Uranus and Neptune), or a compositional gradient leftover from the formation era.

The question of the transport of heat and elements in giant planets is a major one for the next decade. Solving it will require combining new observations, experiments, and models. The recent revision of Jupiter's heat flux (Li et al. 2018) shows the need for revisiting thermal balance analyses for all four giant planets, particularly for Uranus and Neptune for which only partial observations are available. Large-scale observations of global circulation, as well as small-scale, high-resolution observations of storm activity, will need to be coupled to the constraints on key abundances and interior structure (Q7.1) to clarify the nature of giant planet internal transport.

Q7.2c What Is the Deep Rotational and Dynamical State of Giant Planets?

Analyses of giant planet gravitational fields have revealed interiors that are more complex and dynamic than previously expected. Recent measurements have constrained how deep the visible winds extend within each planet. While currently available data only place limits on the depths of the winds on Uranus and Neptune (Kaspi et al. 2013), Juno and Cassini data reveal that Jupiter's winds reach a depth of about 3,000 km (Guillot et al. 2018; Kaspi et al. 2018), while Saturn's extend to depths of about 8,000 km (Iess et al. 2019; Galanti and Kaspi 2021). Furthermore, the saturnian gravitational field is surprisingly time-variable (Iess et al. 2019; Markham et al. 2020) and contains asymmetries that rotate at about the same rate as its surface winds. Based on ring structures, Saturn is also known to exhibit normal-mode oscillations with a complex excitation spectrum. At Jupiter, zonal flows at depth can advect the magnetic field, leading to detectable secular variations on timescales of tens of years.

Given the spatial complexity of the ice giant dynamos, measurements of Uranus's and Neptune's magnetic fields would likely be a powerful tool to constrain their internal dynamics (Soderlund and Stanley 2020).

The properties of the interior oscillations and asymmetries inside Saturn have been used to constrain aspects of the planet's internal structure and rotation state (Mankovich et al. 2019; Mankovich and Fuller 2021); more observations and modeling work are needed to ascertain how these deep oscillations and flows are generated and maintained (see also Q7.2d). Deep fluid flows provide opportunities to examine how the fundamental processes that underlie atmospheric dynamics operate under extreme conditions that are otherwise inaccessible. Available data on Uranus and Neptune are very limited (see Figure 10-3), and additional measurements (gravity and magnetic field, planetary shape) are needed to ascertain whether the distinctive compositions of these bodies shape their internal dynamics. Finer details (e.g., direct measurements of differential rotation and detection of interfaces) will require planetary seismology.

Q7.2d How Are the Complex Magnetic Fields of the Giant Planets Generated?

Planetary magnetic fields show remarkable variations that reveal clues to their deep interiors and internal histories. Saturn and Mercury have strikingly axisymmetric fields and slow secular variation. Jupiter and Earth have dipole-dominated fields with ~10-degree tilts, prominent regions of enhanced intensities, and measured secular variation. Uranus and Neptune have multipolar (i.e., nondipole dominated) surface fields with comparable intensities, no clear symmetries along any axis (Soderlund and Stanley 2020), and a true rotation rate that is uncertain (Helled et al. 2020). Secular variations of the magnetic field compatible with some advection by the deep zonal flow have been measured at Jupiter (Moore et al. 2019a). (Unfortunately, these measurements are not possible on Saturn because of the axisymmetry of the field and on Uranus and Neptune because of the singular set of field measurements provided by the Voyager 2 flybys.) These observations lead to fundamental questions about planetary magnetic field generation, including the processes that control field morphology, strength, and temporal evolution, and the aspects of planetary interiors responsible for observed variations across the terrestrial, gas giant, and ice giant planets, as well as within each of these classes. These questions can be answered by determining the detailed configurations of the ice giants' magnetic fields and their temporal variation, the internal density and composition distribution, whether layers of stable stratification and/or double-diffusive convection exist, the thermodynamic and transport properties of the planets as a function of radius and over time as the planets evolve, the characteristics of zonal winds, meridional circulations, and turbulent convective flows in the deep interior, and the dynamo characteristics of exoplanets to further test hypotheses developed to explain the planetary magnetic fields within the solar system (see also Question 12, Chapter 15).

Q7.2e How Are the Interiors of the Giant Planets Evolving Today?

After formation, the giant planets cool and lose the internal heat acquired during gravitational collapse. However, uncertainties in the current radiated heat flux at each planet, and in equations of state, lead to uncertainties in interior and evolution models (Fortney et al. 2011). For Jupiter, the thermal heat flux value has been revised by 30 percent recently (Li et al. 2018). For Saturn, uncertainty on the atmospheric helium abundance, and therefore on the amplitude of the helium rain process, is too large to constrain models. Last, the extremely limited data from Uranus and Neptune suggest that these two planets are losing heat at very different rates from each other for reasons that are still unclear (e.g., Kurosaki and Ikoma 2017). In addition, discoveries from the past decade have revealed that both Jupiter and Saturn have interiors that may not be fully convective (Leconte and Chabrier 2012; Wahl et al. 2017; Mankovich and Fuller 2021), requiring reassessments of their thermal histories. Mixing in the planetary interior (Vazan et al. 2018) and evolution in the aftermath of a possible giant impact (Liu et al. 2019) are also processes that need to be considered in evolution calculations. The recently discovered rapid orbital migration of some of Saturn's moons could be owing to ongoing changes in the planet's internal structure (Q7.5a). Little is currently known about Uranus and Neptune's interior structure, and ongoing evolution, for comparison with Jupiter and Saturn. Progress on this question requires accurate measurements of heat fluxes, a much better assessment of transport processes in the atmospheres and interiors (Q7.2b), a determination of the deep interior structure (Q7.2a), and, for the case of Saturn, a determination of the helium abundance in its atmosphere.

Strategic Research for Q7.2

- **Determine the interior structure and composition of Uranus and Neptune** using gravity and magnetic field mapping.
- **Search for the locations and extent of discrete layers in the deep interior in all four giant planets,** using planet/ring seismology (i.e., the ability to detect planetary seismic waves from perturbations in the motion of ring particles).
- **Constrain the rate of heat transport in Jupiter, Saturn, Uranus, and Neptune** by measuring thermal balance and vertical temperature profiles.
- **Measure vertical mixing in Jupiter, Saturn, Uranus, and Neptune** from measurements of deep vortices, storm and wave activity, and disequilibrium species distribution.
- **Understand the deep rotation rate and dynamics in Uranus and Neptune** from time-resolved gravity and magnetic field mapping, radio occultations, planet/ring seismology measurements, and deep circulation modeling.
- **Characterize the intrinsic magnetic fields of Uranus and Neptune** through magnetic field mapping and dynamo modeling.
- **Constrain the ongoing interior evolution of Saturn, Uranus, and Neptune** from helium abundance, thermal balance, satellite tidal evolution, occultations, and gravity and magnetic field measurements.

Q7.3 WHAT GOVERNS THE DIVERSITY OF GIANT PLANET CLIMATES, CIRCULATION, AND METEOROLOGY?

Comparative planetology among the four giant planets affords study of diverse regimes of planetary rotation, size, chemical enrichment, condensation processes, atmospheric mixing, seasonal influences, exterior influences (e.g., auroral and chemical) and interior connections (from metallic hydrogen to watery oceans). Exploration of the giants can reveal how these regimes vary from world to world, setting our terrestrial atmosphere into a broader context (Question 6, Chapter 9), and providing ground truth data for understanding atmospheres on giant and sub-giant exoplanets (Question 12, Chapter 15).

Q7.3a What Processes Maintain Banded Patterns and Unique Polar Regions on Each Giant Planet, How Do They Connect with the Deep Interior, and What Controls Their Variability?

As shown in Figure 10-4, all four giants exhibit banded atmospheres, with east-west zonal winds separating domains of different temperatures, aerosols, and chemical composition (Fletcher et al. 2020). Differing conditions within adjacent latitudinal bands influence the prevalence of convection, lightning, and vertical mixing, and on Jupiter these circulations appear to penetrate deeply (at least to the ~hundred bars levels; Ingersoll et al. 2017). The bands are punctuated by large-scale vortices, some of which drift with latitude, particularly on Uranus and Neptune (Hueso and Sánchez-Lavega 2019), while others remain fixed in their respective bands, for example, on Jupiter. Juno and Cassini data have provided important clues on the depths of these bands, but significant questions remain, particularly for Uranus and Neptune for which few data are available (see Q7.2c,d and Figure 10-4). Still uncertain are what determines the wind strength, the energy sources that maintain the zonal flows and thermal gradients, and how these vary with depth (Kaspi et al. 2020), why ice giant banding differs from that seen on the gas giants, and how and why do banded patterns change over quasi-predictable timescales, such as the multi-year cycles of "upheavals" in Jupiter's belts and zones at cloud level.

Each giant planet appears to display different circulation regimes as a function of latitude, from chemically enriched equatorial plumes and neighboring chemically depleted belts, to finer-scale circulations at the scale of the zonal jets at mid-latitudes. This well-organized banding gives way, at higher latitudes, to more chaotic and turbulent regimes nearer the poles, exhibiting filamentary cloud complexes and organized patterns of cyclones on Jupiter and wave phenomena on Saturn (Adriani et al. 2018). What controls this transition, how it differs between the gas and ice giants, and whether these polar meteorological phenomena are long-lived are key open issues. Unraveling how energy, momentum, and material are transported vertically and horizontally in giant planet atmospheres requires observations to constrain giant planet circulation, from the equator to the poles, the depth and asymmetries in the flows, and the vertical structure of the powerful winds.

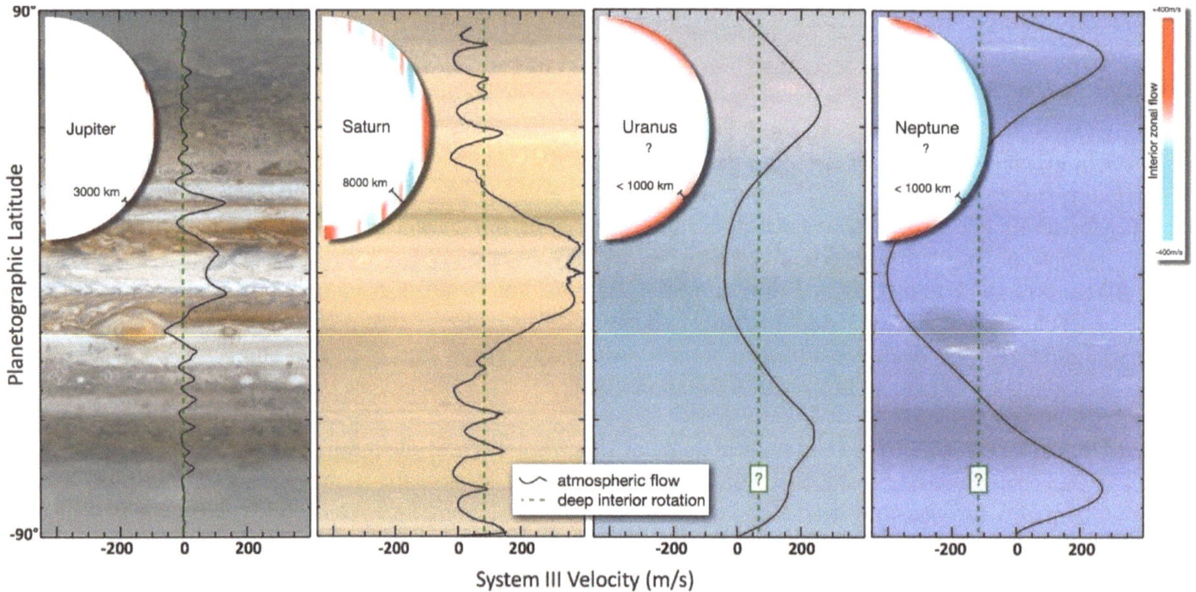

FIGURE 10-4 Cloud bands, zonal winds, and deep rotation on the outer planets. The wind profiles are shown at the same scale for each planet and show zonal jets reaching 200–400 m/s; Jupiter and Saturn exhibit prograde equatorial jets, while Uranus and Neptune's are retrograde. Deep rotation speeds (dashed lines) are determined from magnetic field rotation in Jupiter, seismology in Saturn (Mankovich et al. 2019), and estimated from planet shape arguments in Uranus and Neptune (Helled et al. 2020). Interior wind profiles (insets) are inferred from gravity data (Guillot et al. 2018; Kaspi et al. 2018; Iess et al. 2019). In Uranus and Neptune, only an upper limit to zonal flow depth can be determined (Kaspi et al. 2013). SOURCES: Wind data illustrated by T. Guillot, M.M. Hedman, L.H. Fletcher, and A.A. Simon, based on data from Simon et al. (2015), García-Melendo et al. (2011), Sromovsky et al. (2015), and Sánchez-Lavega et al. (2019). Map images: *Jupiter*: https://photojournal.jpl.nasa.gov/catalog/PIA07782; credit: NASA/JPL/Space Science Institute. *Saturn and Neptune*: https://bjj.mmedia.is; created by Björn Jónsson. *Uranus*: https://supernova.eso.org/exhibition/images/uranusmap-10x5k-CC; credit: NASA.

Q7.3b How Do Stratospheric Properties Trace Interactions with Internal and External Phenomena?

Planetary stratospheres are transitional domains between the meteorology of the deeper troposphere, and interactions with the external environment. They are sensitive to influences from below, such as rising storm plumes from the troposphere, mixing of gases upwelling from deeper levels, and potentially equator-to-pole gradients of the source material for photochemistry on Uranus and Neptune (Moses et al. 2020). They are also sensitive to influences from above, such as auroral energy deposition and resistive Joule heating in connection with the planetary magnetosphere, and via the influx of exogenic materials from rings, satellites, interplanetary dust, and impacts (cometary and asteroidal). This mix of chemical compounds is then redistributed vertically and horizontally by large-scale and seasonally variable interhemispheric circulation patterns, causing variations in chemical abundances and radiative energy balance from place to place (Hue et al. 2018). Equatorial oscillations of the stratospheric temperatures and winds have been discovered on Jupiter and Saturn and are prone to disruption by meteorological processes (Antuñano et al. 2021)—it is unknown whether Uranus and Neptune exhibit the same phenomenon. Understanding the redistribution of energy is a key challenge for the giant planets, where the middle and upper atmospheres are much warmer than expected from solar heating alone, an imbalance known as the energy crisis. Last, atmospheric chemistry is sensitive to the strength of vertical mixing, such that comparison of Uranus and Neptune, where the strength of mixing differs substantially (Moses et al. 2020), would provide a unique test of how photochemistry operates under very different conditions.

Q7.3c How and Why Do Discrete Meteorological Features (e.g., Storms and Vortices) Evolve?

Large-scale planetary bands are interrupted by, but intricately connected to, meteorological phenomena on smaller scales, including discrete convective clouds and storm plumes, waves in all their guises, and anticyclones and cyclones of various scales (Ingersoll et al. 2004). Some anticyclonic features on Jupiter are long-lived, but vary in size, color, and energetics via interactions with storms and vortices at their edges. Cyclonic features appear to show increased prevalence for moist convection and lightning. On the other hand, some vortices on Uranus and Neptune are short-lived and migrate with latitude, dissipating as they approach the equator (Hueso and Sánchez-Lavega 2019). The mechanisms that control the life cycles of convective storms are poorly understood, including why some (like Saturn's storms and Jupiter's bright plumes) appear to show predictable cycles; some of these may be influenced by motions and episodic heat transport deep below the clouds (Li and Ingersoll 2015). The depths of these vortices are unknown, along with what causes the long-term stability or short-term susceptibility of storms. Small-scale eddies and plumes may also be responsible for the maintenance of the zonal jets at larger scales. Lastly, some vortices, like the Great Red Spot, exhibit colorful hazes that appear to be lacking in others, like Neptune's Great Dark Spot. Understanding how and why meteorological features differ from world to world requires long-term, multi-wavelength datasets to develop a better understanding of how planetary climates influence the types of "weather" that we observe.

Q7.3d What Chemical and Physical Processes Influence the Gas and Aerosol Absorbers That Produce the Diverse Colors and Spectral Properties of the Giant Planets?

The composition of planetary tropospheres and stratospheres is determined by thermochemistry, condensation processes, photochemistry, and the redistribution of material by circulation and meteorological phenomena. The different clouds, hazes, and chemical compositions on each planet reflect differences in these underlying processes, but the basic question of cloud and aerosol composition remains unresolved (e.g., West et al. 2009). We do not yet know how disequilibrium species (see Figure 10-2), which should be sequestered in the deeper troposphere, are transported to the observable weather layer, and how they contribute to observed color. Given the long seasons on these planets, the mechanisms that govern the formation of clouds and hazes as a function of depth, beyond simple thermochemical equilibrium (Guillot et al. 2020), and how the hazes (and their associated colors) change with seasons and auroral influences are not well studied. The nonsolar nitrogen-to-sulfur ratio observed in Uranus and Neptune may represent the bulk abundances of these species, or chemistry in deep water clouds may preferentially sequester ammonia. Additionally, the impact of discrete aerosol layers on the thermal structures of the upper troposphere and stratosphere, and how they differ in each planet, remain unknown. Other dynamical processes, such as precipitation and subsidence surrounding storm plumes, may also lead to dramatic changes in cloud colors. It is particularly important to understand the chemical compositions of major cloud features and dark vortices on the ice giants (Hueso and Sánchez-Lavega 2019), and why they differ substantially from similar features on the gas giants. The complex interplay of dynamics, thermal structure, and composition need to be more thoroughly explored to understand the differences observed at each planet.

Q7.3e How Does Moist Convection Shape Atmospheric Structure in Hydrogen-Dominated Atmospheres?

Convective processes are fundamental to understanding the motions of atmospheres and oceans, and the giant planets provide ideal testbeds for understanding the influences of buoyancy, stratification, rotation, and the importance of different condensable species under conditions not found on Earth—namely, a hydrogen-dominated atmosphere in which condensables are heavier than the surrounding environment (Hueso and Sánchez-Lavega 2019). While water-dominated convection can be studied on Jupiter and Saturn, this is largely hidden from remote sensing by overlying ammonia and NH_4SH clouds. Methane-driven convection (and, to a lesser extent, hydrogen sulfide) is more accessible on the ice giants but might be restricted to thin layers. Uranus and Neptune's high methane abundance implies that they lie in a regime in which moist convection is locally inhibited by the stabilizing effects of molecular weight gradients (Guillot 1995). Convective inhibition in the ice giants may lead to centuries-long intervals between convective outbursts, a potential explanation for apparent differences in the intrinsic luminosities of Uranus and Neptune (Smith and Gierasch 1995; Li and Ingersoll 2015). Comparing and contrasting convective processes—their vertical velocities, advected chemicals, thermal and chemical gradients, potential for lightning generation, and their relationship to latitudinal bands—between the differing environments will reveal how moist convection works in diverse planetary environments.

Strategic Research for Q7.3

- **Determine what processes maintain tropospheric circulation and meteorology of Jupiter, Saturn, Uranus, and Neptune** from in situ measurements of the vertical wind field and from multi-wavelength remote sensing measurements of temperatures, composition, and lightning frequency as a function of latitude, longitude, and over many timescales.
- **Understand stratospheric coupling and seasonal influences on Uranus and Neptune** from multi-wavelength remote sensing of stratospheric temperatures, trace gas composition, and aerosols over long timescales, in connection with ionospheric and magnetospheric variability.
- **Constrain storm evolution on Jupiter, Saturn, Uranus, and Neptune** with frequent, multi-wavelength remote sensing over multiple timescales from hours to days to years.
- **Determine what governs cloud top color and how it ties to transport and chemistry in the atmospheres of Saturn, Uranus, and Neptune** from in situ sampling of composition and particle properties, coupled with global imaging in reflected sunlight (at multiple phase angles) and thermal emission.
- **Elucidate how convection works on Uranus and Neptune** from multi-wavelength remote sensing over many timescales and depths, combined with analytical, radiative transfer, and general circulation modeling.

Q7.4 WHAT PROCESSES LEAD TO THE DRAMATICALLY DIFFERENT
OUTCOMES IN THE STRUCTURE, CONTENT, AND DYNAMICS OF THE
OUTER PLANETS' MAGNETOSPHERES AND IONOSPHERES?

The interaction between the strong magnetic fields of outer planets and the solar wind creates fast rotating magnetospheres that serve as natural laboratories for studies of astrophysical plasma processes accessible to in situ measurements from space missions. Our understanding of plasma physics processes such as collisionless shocks, magnetic reconnection, plasma pick-up, field-aligned plasma acceleration, plasma interchange, auroral emissions (visible, ultraviolet, X-ray, and radio), generation of plasma waves, and the formation of radiation belts owes a lot to in situ observations within these giant magnetospheres. A magnetosphere involves a planetary magnetic field, the magnetized flowing solar wind that interacts with it, and a source of magnetospheric plasma (often a moon, but the planet's atmosphere is also a significant source). Why the jovian magnetosphere is the largest, extremely dense (with $>10^9$ kg of plasma), and hot, while the magnetospheres of Uranus and Neptune are near vacuum, is not well understood (Figure 10-5). Most of our current knowledge of the jovian and saturnian magnetosphere comes from orbiting spacecraft (Galileo, Juno, and Cassini) whereas Uranus and Neptune have been visited only once by Voyager 2 flybys and are relatively unexplored. One of the main obstacles to making progress in studies of the magnetospheres is their vast scale and a lack of simultaneous multiple-point measurements to distinguish between temporal and spatial changes.

Q7.4a What Processes Govern the Content and Dynamics of the Giant Planet Magnetospheres?

Sources of plasmas in planetary magnetospheres are outflow from the planetary ionosphere (ions are mostly protons), leakage in from the solar wind (protons and alpha particles), and outgassing of moons with ion species that are source dependent. For example, Enceladus's water vapor plumes provide O^+, OH^+, H_2O^+. and H_3O^+ ions; SO_2 from Io's volcanoes provide S^+, S^{2+}, S^{3+}, O^+, and O^{2+} ions; and Triton's and Titan's atmospheres source nitrogen ions (N^+ and N_2^+). The relative importance of moon sources and their associated mass loading to the

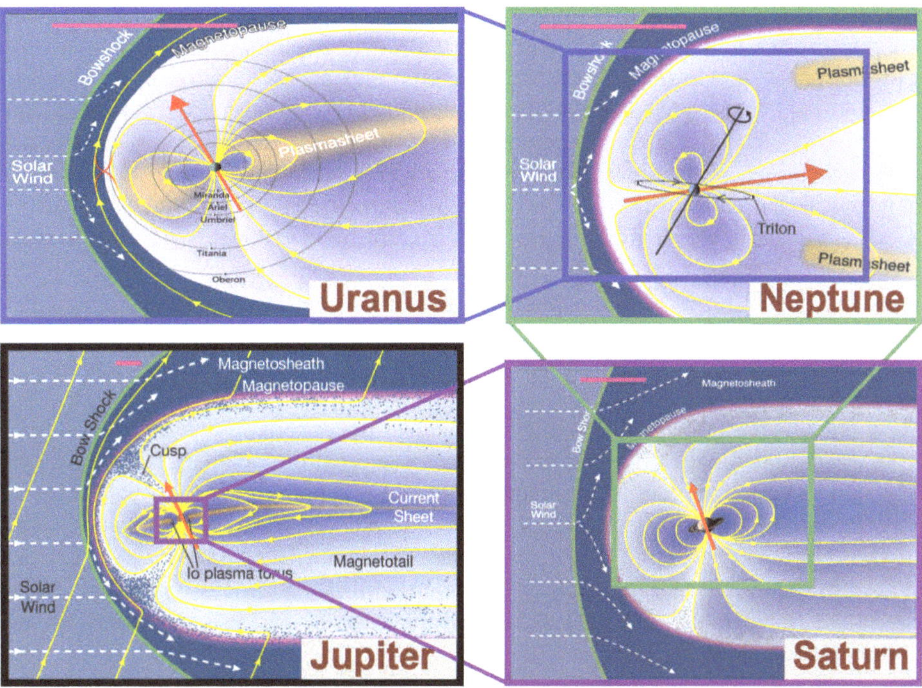

FIGURE 10-5 Comparison of size and geometry of the giant planet magnetospheres. For scale, the purple box at Jupiter spans ~1 million km, and the Uranus box spans ~650,000 km. SOURCES: *Uranus:* Modified from S. Plainaki, J. Lilensten, A. Radioti, et al., 2016, "Planetary Space Weather: Scientific Aspects and Future Perspectives," *Journal of Space Weather and Space Climate* 6:A31. CC BY 4.0. Image credit: Chris Arridge, Fran Bagenal, and Steve Bartlett. *Neptune:* From P. Kollmann, I. Cohen, R.C. Allen, et al., 2020, "Magnetospheric Studies: A Requirement for Addressing Interdisciplinary Mysteries in the Ice Giant Systems," *Space Science Reviews* 216:78. CC BY 4.0. Image credit: Fran Bagenal and Steve Bartlett. *Jupiter and Saturn:* F. Bagenal, 2013, Planetary Magnetospheres, Pp. 251–307 in *Planets, Stars and Stellar Systems*, T.D. Oswalt, L.M. French, and P. Kalas, eds., Volume 3, Dordrecht, Netherlands: Springer. Reproduced with permission from SNCSC.

magnetosphere depends on the volcanic and cryovolcanic activity of moons orbiting within the magnetosphere and the local plasma environment responsible for ionizing outgassed neutrals. The size of the magnetosphere (see Figure 10-5) depends strongly on the strength and spatial configuration of the internal magnetic field generated by the planetary dynamo (Q7.2d). But strength and configuration is also influenced by internal dynamics that heat the plasma (increasing the internal plasma pressure and "inflating" the magnetosphere), as well as by the strength of the solar wind (that decreases with distance from the Sun).

Magnetospheric dynamics depend on the coupling between the solar wind, the planetary ionosphere, and the magnetosphere which comprises the planetary magnetic field and plasma sources described above. Earth-like magnetospheric convection is expressed by coupling between the solar wind and the planet's ionosphere and driven by the process of magnetic reconnection at the dayside magnetopause. The characteristic solar wind Alfvén speed decreases with distance from the Sun, and, for a constant solar wind speed, the Alfvén Mach number increases. This results in the boundary between the solar wind and giant planet magnetospheres (particularly if inflated with hot plasma) being frequently dominated by a viscous coupling (e.g., Kelvin-Helmholtz instability) across the magnetopause boundary layer rather than by global magnetic reconnection (Masters 2018).

For the rapidly rotating gas giants, the coupling between the ionosphere (that is collisionally coupled to the spinning gas planet) and the magnetosphere produces rotation-dominated plasma flows. The energy of the plasma in the jovian and saturnian magnetospheres is derived mostly from the rotation of the planet with some contributions coming from the solar wind imposed electric field (Khurana et al. 2004; Kivelson and Bagenal 2007; Gombosi et al. 2009). Both Jupiter's and Saturn's magnetospheres are also marked by large plasma acceleration events connected to magnetotail reconnection and plasma interchange (Gombosi et al. 2009; Bagenal 2013; Vogt et al. 2014). The excited aurorae in the waveband spanning radio to X-ray spectrum are both a window into the workings of these magnetospheres and one of the main loss processes of the energy of the confined plasma. A combination of in situ plasma and magnetic field observations with simultaneous or near simultaneous remote observations of the aurora would help to determine the nature of magnetospheric energy transport and quantify the competing processes governing magnetospheric dynamics and structure in the outer solar system.

Q7.4b What Is Responsible for the Differences Between the Magnetospheres of the Gas Giants and Ice Giants?

The active moons Io and Enceladus dominate the magnetospheres of Jupiter and Saturn, respectively, filling their large magnetospheres with plasma disks that co-rotate with the planet. The strong magnetic field of Jupiter harnesses the spin momentum of the planet and heats the plasma as it moves out (to >20 keV in the middle magnetosphere), inflating the magnetosphere to 100 times the radius of the planet on the sunward side and stretching past the orbit of Saturn downstream in the solar wind. The weaker Enceladus source coupled to the magnetic field of Saturn produces a similarly rotation-dominated magnetosphere that is smaller in extent. The plasma is colder (temperatures below 1 keV), and the magnetosphere is dominated by neutral material sourced from Enceladus's plumes.

The highly tilted and off-centered dipole moments of Uranus and Neptune generate asymmetric magnetospheres and twisted magnetotails which were only fleetingly explored by Voyager 2 flybys. Because of this geometry, their magnetospheric interaction with the solar wind varies greatly with rotation and with season (Paty et al. 2020). Their magnetospheric plasma composition is dominated by protons (along with N^+ ions in the case of Neptune sourced from Triton), suggesting that solar wind coupling to their ionospheres produces Earth-like global convection. However, the process whereby a tenuous solar wind couples to these asymmetric magnetospheres, and the dynamic roles of their sparsely observed satellites in terms of sculpting the radiation belts and sourcing plasma to the magnetosphere, remain unclear. One of the major unresolved questions in planetary magnetospheres is how the extremely low plasma density in the uranian magnetosphere maintains electron radiation belts similar in intensity to Earth's. These differences illuminate relative roles of the strengths of planetary fields, plasma sources (including their composition), corotation breakdown from finite effective conductivities of the ionospheres, and plasma processes such as interchange and reconnection operating in these magnetospheres.

Q7.4c How Is Energy Redistributed with Latitude and Altitude Within Giant Planet Ionospheres/Thermospheres, and What Is Responsible for Their High (and Variable) Temperatures?

Giant planet thermospheres are influenced by both external processes (e.g., solar heating and auroral heating) and by internal processes (e.g., waves and circulation patterns in the middle atmosphere), such that tracing the redistribution of energy can reveal insights into thermospheric and ionospheric circulation. All four giants exhibit thermospheric temperatures far warmer than can be explained by solar heating alone (García Muñoz et al. 2017), hinting that dynamical energy redistribution may be a significant contributor, and evidence of rapid thermospheric flows associated with auroral ovals has been observed on Jupiter and Saturn. How energy is redistributed latitudinally through the upper atmosphere, from the auroral domains to the lower latitudes, and whether tropospheric meteorology influences thermospheric temperatures as a function of latitude, is not well constrained. Nor is the mechanism responsible for the slow cooling trend of Uranus's ionosphere over multiple decades: does its weak atmospheric mixing and low homopause, coupled with the extreme axial tilt, produce a uniquely variable ionosphere? Further exploration of the ice giant systems is needed to compare the redistribution of energy in their ionospheres/thermospheres with those of Jupiter and Saturn.

Q7.4d How Do External Inputs and Local Ion Chemistry
Produce the Complex Variability Observed in Ionospheres?

Although overall composition should be similar in the ionospheres of the giant planets, the insolation, seasonal forcing, magnetic field configuration, and ring influxes are very different. At Saturn, variation in H_3^+ emission and electron density may relate to variable influxes of ring material (including neutral nanograins near the equator and charged grains at mid-latitudes), to gravity waves breaking in the lower thermosphere, and/or to other extreme ionization or transport processes. At Uranus, nonseasonal thermospheric temperatures vary on timescales greater than a solar cycle. But these temperatures are derived from H_3^+, implying that seasonal change and spatial and temporal variability in this ion could influence derivations of temperature (Moore et al. 2019b). For Neptune, H_3^+ has not yet been detected, and models of the ionosphere predict H_3^+ concentrations greater than the current upper limit. This may suggest that the thermospheric temperature and/or methane homopause level differ from what was observed by Voyager, or that an influx of external material, perhaps from Triton/Neptune's rings, has depleted ionospheric H_3^+ densities. While many other ionic species are predicted to be present at the ice giants, none have yet been detected. Radio occultations, ultraviolet/infrared spectral limb scans, and in situ mass spectral measurements from deep-dive orbiter passes would provide crucial new data for understanding ionospheric variability.

Strategic Research for Q7.4

- **Determine the dominant processes governing the magnetospheres of Uranus and Neptune** via plasma, particle, and magnetic field observations with simultaneous or near simultaneous remote observations of the aurora.
- **Constrain the structure, dynamics, and temporal evolution of the magnetospheres of Uranus and Neptune** with long-term magnetic field measurements.
- **Investigate the evolving composition of the ionospheres of Uranus and Neptune** from remote sensing, magnetospheric plasma, and in situ ion/neutral composition measurements.
- **Determine the variability and thermal structure of the thermospheres and ionospheres of Uranus and Neptune** from radio occultations and infrared and ultraviolet spectral limb scans, distributed in latitude and time.

Q7.5 HOW ARE GIANT PLANETS INFLUENCED BY, AND HOW DO
THEY INTERACT WITH, THEIR ENVIRONMENT?

Giant planet systems have many components: the planet and its rings, satellites, and magnetosphere. Each element interacts with the others (Question 8, Chapter 11), but also with external influences from the Sun and cosmic rays. Inputs into the giant planets (both energetic and mass/compositional) affect upper atmospheric composition, which then precipitates to lower levels over time; this input needs to be understood to relate current compositional measurements of the giant planets to current and primordial bulk abundances. The planets themselves also lose energy to the satellites and rings through tidal interactions, which can further elucidate the history of their internal structure.

Q7.5a How Is Angular Momentum Lost, and Tides Dissipated, from the Giant Planets?

Recent observations reveal that all of Saturn's moons are migrating outward at surprisingly fast rates (Lainey et al. 2017, 2020), which means that angular momentum is being transferred from Saturn to its moons much more efficiently than previously expected. The observed orbital evolution rates suggest that they are driven by ongoing changes in the planet's internal structure (Fuller et al. 2016; Lainey et al. 2020). It is not yet clear what is happening inside Saturn to produce these changes, nor whether similar processes are operating within the other giant planets. Planned missions to Ganymede and Europa can be used to refine their orbital evolution and constrain Jupiter's tidal dissipation. Clarifying the orbital history of the giant planet moons via precise astrometry and long-term monitoring promises to provide more insights into the recent history and evolution of the giant planets.

Q7.5b How Is Atmospheric Composition Influenced by Ring Rain, Large Impacts, and Micrometeoroids?

Measurements made by Earth-based telescopes and the Cassini spacecraft reveal a large flux of material into Saturn's atmosphere from the rings, which may provide a total water flux to Saturn of 10,000 kg/s (Moore et al. 2015). While no other planet is surrounded by as massive a ring system as Saturn, some fraction of circumplanetary debris surrounding all the giant planets can be transported along magnetic field lines into the planet. Additionally, exogenic material is continuously delivered to the planets via impacts. Larger impacts (e.g., Shoemaker-Levy 9 at Jupiter) are governed by physics similar to terrestrial airbursts, but with very different chemistry. New observations taken soon after large impacts in giant planet atmospheres are needed to test these models across a range of impact geometries and impactor properties. Observations and models of stratospheric chemical signatures in Uranus and Neptune have begun to elucidate the balance between influx from micrometeoroids versus major cometary impacts within the past several centuries (Cavalié et al. 2014; Moreno et al. 2017). Monitoring of smaller impacts on Jupiter (but not yet on the other giant planets) currently provides the only constraints on the small end of the impactor size distribution, a parameter important for establishing surface ages near the planets.

**Q7.5c How Does Seasonally Variable Solar Insolation Influence
Middle Atmosphere Chemistry and Haze Production?**

Despite their large heliocentric distances of 10 AU to 30 AU, the giant planets beyond Jupiter show significant seasonal influence from the Sun. Saturn and Neptune have obliquities like that of Earth, and Uranus's axis is tilted by an extreme 98° from the ecliptic, leading to hemispheres bathed in sunlight for long portions of an orbital period. Indeed, changes in Uranus's polar hazes have been observed over the uranian year, and there is some evidence for seasonal change in Neptune's brightness and banding, as well. Thus, very low solar irradiation can still cause photochemical change over time, modulated by the atmospheric circulation, and these processes, and the chemical/dynamics pathways involved, are not yet well understood.

Strategic Research for Q7.5

- **Determine the role of tidal dissipation in angular momentum transfer at Jupiter, Uranus, and Neptune** from measurements of satellite orbital migration.
- **Quantify the diverse external influences on the atmospheric chemistry and dynamics of Jupiter, Saturn, Uranus, and Neptune** via time-series imaging and spectral measurements of the effects of impacting objects from micrometeoroids to comets.
- **Constrain the influence of seasonal solar insolation on Uranus and Neptune's atmospheric chemistry and hazes** from measurements of temperature, haze optical depth, and gas abundances over long time periods.

SUPPORTIVE ACTIVITIES FOR QUESTION 7

- Laboratory measurements and numerical simulations of opacities under outer planet conditions (both at high pressures and for long columns at low pressures/temperatures), high-pressure equations of state, chemical reaction rates, and transport properties (i.e., viscosity, thermal and electrical conductivity, and diffusion coefficients).
- Numerical and analytical models of dynamic processes in the atmosphere, interior, and magnetosphere of giant planets, coupled to an investment in high-performance computing.
- Continued data analysis from ongoing and past missions, along with acquisition and archiving of planetary datasets from ground-based facilities to enable maximum science return, as well as generate high level science products to inform the development of future missions.
- Long-term monitoring of atmospheric dynamics, waves and oscillations, auroras, and impacts, ideally by a space-based telescope.

REFERENCES

Adriani, A., A. Mura, G. Orton, C. Hansen, F. Altieri, M.L. Moriconi, J. Rogers, et al. 2018. "Clusters of Cyclones Encircling Jupiter's Poles." *Nature* 555:216–219.

Antuñano, A., R.G. Cosentino, L.N. Fletcher, A.A. Simon, T.K. Greathouse, and G.S. Orton. 2021. "Fluctuations in Jupiter's Equatorial Stratospheric Oscillation." *Nature Astronomy* 5:71–77.

Atreya, S.K., A. Crida, T. Guillot, C. Li, J.I. Lunine, N. Madhusudhan, O. Mousis, and M.H. Wong. 2022. "The Origin and Evolution of Saturn: A Post-Cassini Perspective," in *Saturn: The Grand Finale*, K.H. Baines, F.M. Flasar, N. Krupp, T.S. Stallard, eds. Cambridge, UK: Cambridge University Press.

Bagenal, F. 2013. "Planetary Magnetospheres. Planets, Stars and Stellar Systems." Pp. 251–307 in Volume 3: *Solar and Stellar Planetary Systems* of *Planets, Stars and Stellar Systems*, T.D. Oswalt, L.M. French, and P. Kalas, eds. Dordrecht, Netherlands: Springer.

Bailey, E., and D.J. Stevenson. 2021. "Thermodynamically Governed Interior Models of Uranus and Neptune." *Planetary Science Journal* 2(2):64.

Bolton, S.J., A. Adriani, V. Adumitroaie, M. Allison, J. Anderson, S. Atreya, J. Bloxham, et al. 2017. "Jupiter's Interior and Deep Atmosphere: The Initial Pole-to-Pole Passes with the Juno Spacecraft." *Science* 356:821–825.

Cavalié, T., R. Moreno, E. Lellouch, P. Hartogh, O. Venot, G.S. Orton, C. Jarchow, et al. 2014. "The First Submillimeter Observation of CO in the Stratosphere of Uranus." *Astronomy & Astrophysics* 562:A33.

Feuchtgruber, H., E. Lellouch, G. Orton, T. de Graauw, B. Vandenbussche, B. Swinyard, R. Moreno, et al. 2013. "The D/H Ratio in the Atmospheres of Uranus and Neptune from Herschel-PACS Observations." *Astronomy & Astrophysics* 551:A126.

Fletcher, L.N., Y. Kaspi, T. Guillot, and A.P. Showman. 2020. "How Well Do We Understand the Belt/Zone Circulation of Giant Planet Atmospheres?" *Space Science Reviews* 216:30.

Fortney, J.J., M. Ikoma, N. Nettelmann, T. Guillot, and M.S. Marley. 2011. "Self-Consistent Model Atmospheres and the Cooling of the Solar System's Giant Planets." *Astrophysical Journal* 729(1):32.

Fouchet, T., J.I. Moses, and B.J. Conrath. 2009. "Saturn: Composition and Chemistry." Pp. 83–112 in *Saturn from Cassini-Huygens*, M.K. Dougherty, L.W. Esposito, and S.M. Krimigis, eds. Dordrecht, Netherlands: Springer.

Fuller, J., J. Luan, and E. Quataert. 2016. "Resonance Locking as the Source of Rapid Tidal Migration in the Jupiter and Saturn Moon Systems." *Monthly Notices of the Royal Astronomical Society* 458:3867–3879.

Galanti, E., and Y. Kaspi. 2021. "Combined Magnetic and Gravity Measurements Probe the Deep Zonal Flows of the Gas Giants." *Monthly Notices of the Royal Astronomical Society* 501:2352–2362. https://doi.org/10.1093/mnras/staa3722.

García-Melendo, E., S. Pérez-Hoyos, A. Sánchez-Lavega, and R. Hueso. 2011. "Saturn's Zonal Wind Profile in 2004–2009 from Cassini ISS Images and Its Long-Term Variability." *Icarus* 215:62–74.

García Muñoz, A., T.T. Koskinen, and P. Lavvas. 2017. "Upper Atmospheres and Ionospheres of Planets and Satellites." Pp. 349–374 in *Handbook of Exoplanets*, H. Deeg and J. Belmonte, eds. Cham, Switzerland: Springer. https://doi.org/10.1007/978-3-319-55333-7_52.

Gombosi, T.I., T.P. Armstrong, C.S. Arridge, K.K. Khurana, S.M. Krimigis, N. Krupp, A.M. Persoon, et al. 2009. "Saturn's Magnetospheric Configuration." Pp. 203–255 in *Saturn from Cassini-Huygens*, M.K. Dougherty, L.W. Esposito, and S.M. Krimigis, eds. Dordrecht, Netherlands: Springer.

Guillot, T. 1995. "Condensation of Methane, Ammonia, and Water and the Inhibition of Convection in Giant Planets." *Science* 269:1697–1699.

Guillot T., and D. Gautier. 2015. Giant Planets. Pp. 529–557 in Volume 10: *Physics of Terrestrial Planets and Moons* of *Treatise in Geophysics*, T. Spohn and G. Schubert, eds. Amsterdam, Netherlands: Elsevier.

Guillot, T., Y. Miguel, B. Militzer, W.B. Hubbard, Y. Kaspi, E. Galanti, H. Cao, et al. 2018. "A Suppression of Differential Rotation in Jupiter's Deep Interior." *Nature* 555:227–230.

Guillot, T., D.J. Stevenson, S.K. Atreya, S.J. Bolton, and H.N. Becker. 2020. "Storms and the Depletion of Ammonia in Jupiter: I. Microphysics of 'Mushballs.'" *Journal of Geophysical Research: Planets* 125(8):e2020JE006403.

Helled, R., and J.J. Fortney. 2020. "The Interiors of Uranus and Neptune: Current Understanding and Open Questions." *Philosophical Transactions of the Royal Society of London Series A* 378(2187).

Helled, R., and D. Stevenson. 2017. "The Fuzziness of Giant Planets' Cores." *Astrophysical Journal Letters* 840(1):L4. https://10.3847/2041-8213/aa6d08.

Helled, R., N. Nettelmann, and T. Guillot. 2020. "Uranus and Neptune: Origin, Evolution and Internal Structure." *Space Science Reviews* 216:38.

Hue, V., F. Hersant, T. Cavalié, M. Dobrijevic, and J.A. Sinclair. 2018. "Photochemistry, Mixing and Transport in Jupiter's Stratosphere Constrained by Cassini." *Icarus* 307:106–123.

Hueso, R., and A. Sánchez-Lavega. 2019. "Atmospheric Dynamics and Vertical Structure of Uranus and Neptune's Weather Layers." *Space Science Reviews* 215:A52.

Iess, L., B. Militzer, Y. Kaspi, P. Nicholson, D. Durante, P. Racioppa, A. Anabtawi, et al. 2019. "Measurement and Implications of Saturn's Gravity Field and Ring Mass." Science 364(6445):eaat2965.

Ingersoll, A.P., T.E. Dowling, P.J. Gierasch, G.S. Orton, P.L. Read, A. Sánchez-Lavega, A.P. Showman, A.A. Simon-Miller, and A.R. Vasavada. 2004. "Dynamics of Jupiter's Atmosphere." Pp. 105–128 in *Jupiter: The Planet, Satellites and Magnetosphere*. Cambridge, UK: Cambridge University Press.

Ingersoll, A.P., V. Adumitroaie, M.D. Allison, S. Atreya, A.A. Bellotti, S.J. Bolton, S.T. Brown, et al. 2017. "Implications of the Ammonia Distribution on Jupiter from 1 to 100 Bars as Measured by the Juno Microwave Radiometer." *Geophysical Research Letters* 44:7676–7685.

Kaspi, Y., A.P. Showman, W.B. Hubbard, O. Aharonson, and R. Helled. 2013. "Atmospheric Confinement of Jet Streams on Uranus and Neptune." *Nature* 497:344–347.

Kaspi, Y., E. Galanti, W.B. Hubbard, D.J. Stevenson, S.J. Bolton, L. Iess, T. Guillot, et al. 2018. "Jupiter's Atmospheric Jet Streams Extend Thousands of Kilometres Deep." *Nature* 555:223–226.

Kaspi, Y., E. Galanti, A.P. Showman, D.J. Stevenson, T. Guillot, L. Iess, and S.J. Bolton. 2020. "Comparison of the Deep Atmospheric Dynamics of Jupiter and Saturn in Light of the Juno and Cassini Gravity Measurements." *Space Science Reviews* 216:84.

Khurana, K.K., M.G. Kivelson, V.M. Vasyliunas, N. Krupp, J. Woch, A. Lagg, B.H. Mauk, et al. 2004. "The Configuration of Jupiter's Magnetosphere." Pp. 593–616 in *Jupiter: The Planet, Satellites and Magnetosphere*. Cambridge, UK: Cambridge University Press.

Kivelson, M.G., and F. Bagenal. 2007. "Planetary Magnetospheres." Pp. 519–540 in *Encyclopedia of the Solar System*, 2nd ed., L.-A. McFadden, P.R. Weissman, and T.V. Johnson, eds. Amsterdam, Netherlands: Elsevier.

Kollmann, P., I. Cohen, R.C. Allen, G. Clark, E. Roussos, S. Vines, W. Dietrich, et al. 2020. "Magnetospheric Studies: A Requirement for Addressing Interdisciplinary Mysteries in the Ice Giant Systems." *Space Science Reviews* 216:78. https://doi.org/10.1007/s11214-020-00696-5.

Kurosaki, K., and M. Ikoma. 2017. "Acceleration of Cooling of Ice Giants by Condensation in Early Atmospheres." *Astronomical Journal* 153:260.

Lainey, V., R.A. Jacobson, R. Tajeddine, N.J. Cooper, C. Murray, V. Robert, G. Tobie, et al. 2017. "New Constraints on Saturn's Interior from Cassini Astrometric Data." *Icarus* 281:286–296.

Lainey, V., L.G. Casajus, J. Fuller, M. Zannoni, P. Tortora, N. Cooper, C. Murray, et al. 2020. "Resonance Locking in Giant Planets Indicated by the Rapid Orbital Expansion of Titan." *Nature Astronomy* 4:1053–1058.

Leconte, J., and G. Chabrier. 2012. "A New Vision of Giant Planet Interiors: Impact of Double Diffusive Convection." *Astronomy & Astrophysics* 540:A20.

Li, C., and A.P. Ingersoll. 2015. "Moist Convection in Hydrogen Atmospheres and the Frequency of Saturn's Giant Storms." *Nature Geoscience* 8:398–403.

Li, L., X. Jiang, R.A. West, P.J. Gierasch, S. Perez-Hoyos, A. Sanchez-Lavega, L.N. Fletcher, et al. 2018. "Less Absorbed Solar Energy and More Internal Heat for Jupiter." *Nature Communications* 9:3709. https://doi.org/10.1038/s41467-018-06107-2.

Liu, S.-F., Y. Hori, S. Müller, X. Zheng, R. Helled, D. Lin, and A. Isella. 2019. "The Formation of Jupiter's Diluted Core by a Giant Impact." *Nature* 572:355–357. https://doi.org/10.1038/s41586-019-1470-2.

Mahaffy, P.R., H.B. Niemann, A. Alpert, S.K. Atreya, J. Demick, T.M. Donahue, D.N. Harpold, et al. 2000. "Noble Gas Abundance and Isotope Ratios in the Atmosphere of Jupiter from the Galileo Probe Mass Spectrometer." *Journal of Geophysical Research: Planets* 105:15061–15072.

Mandt, K.E., O. Mousis, J. Lunine, B. Marty, T. Smith, A. Luspay-Kuti, and A. Aguichine. 2020. "Tracing the Origins of the Ice Giants Through Noble Gas Isotopic Composition." *Space Science Reviews* 216:99.

Mankovich, C.R., and J.J. Fortney. 2020. "Evidence for a Dichotomy in the Interior Structures of Jupiter and Saturn from Helium Phase Separation." *Astrophysical Journal* 889(1):51.

Mankovich, C., and J. Fuller. 2021. "A Diffuse Core in Saturn Revealed by Ring Seismology." arXiv e-prints: arXiv:2104.13385.

Mankovich, C., M.S. Marley, J.J. Fortney, and N. Movshovitz. 2019. "Cassini Ring Seismology as a Probe of Saturn's Interior. I. Rigid Rotation." *Astrophysical Journal* 871(1). https://10.3847/1538-4357/aaf798.

Markham, S., and D. Stevenson. 2018. "Excitation Mechanisms for Jovian Seismic Modes." *Icarus* 306:200–213.

Markham, S., D. Durante, L. Iess, and D. Stevenson. 2020. "Possible Evidence of p-modes in Cassini Measurements of Saturn's Gravity Field." *Planetary Science Journal* 1(2):27.

Masters, A. 2018. "A More Viscous-Like Solar Wind Interaction with All the Giant Planets." *Geophysical Research Letters* 45:7320–7329.

Moll, R., P. Garaud, C. Mankovich, and J.J. Fortney. 2017. "Double-Diffusive Erosion of the Core of Jupiter." *Astrophysical Journal* 849(1):24.

Molter, E.M., I. de Pater, S. Luszcz-Cook, J. Tollefson, R.J. Sault, B. Butler, and D. de Boer. 2021. "Tropospheric Composition and Circulation of Uranus with ALMA and the VLA." *Planetary Science Journal* 2:3.

Moore, K.M., H. Cao, J. Bloxham, D.J. Stevenson, J.E.P. Connerney, and S.J. Bolton. 2019a. "Time Variation of Jupiter's Internal Magnetic Field Consistent with Zonal Wind Advection." *Nature Astronomy* 3:730–735.

Moore, L., J. O'Donoghue, I. Müller-Wodarg, M. Galand, and M. Mendillo. 2015. "Saturn Ring Rain: Model Estimates of Water Influx into Saturn's Atmosphere." *Icarus* 245:355–366.

Moore, L., H. Melin, J. O'Donoghue, T.S. Stallard, J.I. Moses, M. Galand, S. Miller, et al. 2019b. "Modelling H_3^+ in Planetary Atmospheres: Effects of Vertical Gradients on Observed Quantities." *Philosophical Transactions of the Royal Society of London Series A* 377(2154):20190067.

Moreno, R., E. Lellouch, T. Cavalié, and A. Moullet. 2017. "Detection of CS in Neptune's Atmosphere from ALMA Observations." *Astronomy & Astrophysics* 608:L5.

Moses, J.I., and A.R. Poppe. 2017. "Dust Ablation on the Giant Planets: Consequences for Stratospheric Photochemistry." *Icarus* 297:33–58.

Moses, J.I., T. Cavalié, L.N. Fletcher, and M.T. Roman. 2020. "Atmospheric Chemistry on Uranus and Neptune." *Philosophical Transactions of the Royal Society of London Series A* 378(2187): 20190477.

Paty, C., C.S. Arridge, I.J. Cohen, G.A. DiBraccio, R.W. Ebert, and A.M. Rymer. 2020. "Ice Giant Magnetospheres." *Philosophical Transactions of the Royal Society of London Series A* 378(2187): 20190480.

Plainaki, C., J. Lilensten, A. Radioti, M. Andriopoulou, A. Milillo, T.A. Nordheim, I. Dandouras, et al. 2016. "Planetary Space Weather: Scientific Aspects and Future Perspectives." *Journal of Space Weather and Space Climate* 6:A31. https://doi.org/10.1051/swsc/2016024.

Sánchez-Lavega, A., L.A. Sromovsky, A.P. Showman, A.D. DelGenio, R.M.B. Young, R. Hueso, E. Garcia-Melendo, et al. 2019. "Gas Giants." Pp. 72–103 in *Zonal Jets: Phenomenology, Genesis, and Physics*, B. Galperin and P.L. Read, eds. Cambridge, UK: Cambridge University Press.

Simon, A.A., M.H. Wong, and G.S. Orton. 2015. "First Results from the Hubble OPAL Program: Jupiter in 2015." *Astrophysical Journal* 812(1):55.

Smith, M.D., and P.J. Gierasch. 1995. "Convection in the Outer Planet Atmospheres Including Ortho-Para Hydrogen Conversion." *Icarus* 116:159–179.

Soderlund, K.M., and S. Stanley. 2020. "The Underexplored Frontier of Ice Giant Dynamos." *Philosophical Transactions of the Royal Society A* 378(2187):20190479.

Sromovsky, L.A., I. de Pater, P.M. Fry, H.B. Hammel, and P. Marcus. 2015. "High S/N Keck and Gemini AO Imaging of Uranus During 2012–2014: New Cloud Patterns, Increasing Activity, and Improved Wind Measurements." *Icarus* 258:192–223.

Tollefson, J., I. de Pater, E.M. Molter, R.J. Sault, B.J. Butler, S. Luszcz-Cook, and D. DeBoer. 2021. "Neptune's Spatial Brightness Temperature Variations from the VLA and ALMA." *Planetary Science Journal* 2:105.

Vazan, A., R. Helled, and T. Guillot. 2018. "Jupiter's Evolution with Primordial Composition Gradients." *Astronomy & Astrophysics* 610:L14.

Vogt, M.F., M.G. Kivelson, K.K. Khurana, R.J. Walker, M. Ashour-Abdalla, and E.J. Bunce. 2014. "Jupiter's Evolution with Primordial Composition Gradients." *Journal of Geophysical Research: Space Physics* 119:1925–1950.

Wahl, S.M., W.B. Hubbard, B. Militzer, T. Guillot, Y. Miguel, N. Movshovitz, Y. Kaspi, et al. 2017. "Comparing Jupiter Interior Structure Models to Juno Gravity Measurements and the Role of a Dilute Core." *Geophysical Research Letters* 44:4649–4659.

Wang, D., J.I. Lunine, and O. Mousis. 2016. "Modeling the Disequilibrium Species for Jupiter and Saturn: Implications for Juno and Saturn Entry Probe." *Icarus* 276:21–38.

West, R.A., K.H. Baines, E. Karkoschka, and A. Sánchez-Lavega. 2009. "Clouds and Aerosols in Saturn's Atmosphere." Pp. 161–179 in *Saturn from Cassini-Huygens*, M.K. Dougherty, L.W. Esposito, S.M. Krimigis, eds. Dordrecht, Netherlands: Springer.

Wilson, H.F., and B. Militzer. 2010. "Sequestration of Noble Gases in Giant Planet Interiors." *Physical Review Letters* 104:121101.

Young, R.E. 2003. "The Galileo Probe: How It Has Changed Our Understanding of Jupiter." *New Astronomy Reviews* 47:1–51.

Q8 PLATE: Cassini's final image mosaic of Saturn, taken 48 hours before the mission ended with the spacecraft entering Saturn's atmosphere in 2017. SOURCE: Courtesy of NASA/JPL-Caltech/Space Science Institute.

11

Question 8: Circumplanetary Systems

What processes and interactions establish the diverse properties of satellite and ring systems, and how do these systems interact with the host planet and the external environment?

Circumplanetary systems—where a system of moons and/or rings orbit a central body—are seen throughout the solar system, and in some cases are akin to mini-solar systems with numerous and varied orbiting bodies (Figure 11-1).[1] A distinguishing feature of these systems is the importance of coupled interactions, which can occur on shorter timescales and have larger effects than on bodies orbiting the Sun, owing to the compactness of circumplanetary orbits, shorter orbital periods, and relatively close moon–moon and moon–ring orbital separations. For example, while tidal dissipation is a general process operating throughout the solar system, in circumplanetary systems it has heightened strength and importance as it can drive substantial and ongoing heating and activity within moons (leading in some cases to active volcanism, subsurface oceans, plumes, and distinct tectonic features), as well as substantial moon orbital migration and associated passage through orbital resonances. Other notable coupled processes include interactions of the planetary magnetic field with a satellite's intrinsic field and/or its induced field, supply of plasma and neutral content to a planet's magnetosphere from embedded satellites, and complex dynamics within a planetary magnetosphere that affects the material interacting with embedded moons and rings. Collisional ejecta and sputtered particles that escape a moon typically continue to orbit the planet and may be reaccreted or reimpacted onto neighboring moons and rings, affecting surface compositions and properties. Inner moons can be particularly vulnerable to disruption over their histories, owing to tides and the strong gravitational focusing and high impact speeds set by their host planet's gravity. Disrupted moons may form rings, and subsequently reaccrete, resetting the moon's thermal and physical state. Ring systems remain dispersed owing to their proximity to the planet, providing a unique means to study a variety of problems, including the evolution of a self-colliding particulate disk (analogous to early protoplanetary and circumplanetary disks), ring–moon and ring–planet gravitational interactions and resulting wave features, collisionally driven spreading of ring material, and ongoing accumulation into new moons as ring material spreads away from the planet. At the outer edge of some circumplanetary systems are irregular satellites whose complex populations are strongly affected by both the Sun's and the host planet's gravity and may record these interactions.

---

[1] A glossary of acronyms and technical terms can be found in Appendix F.

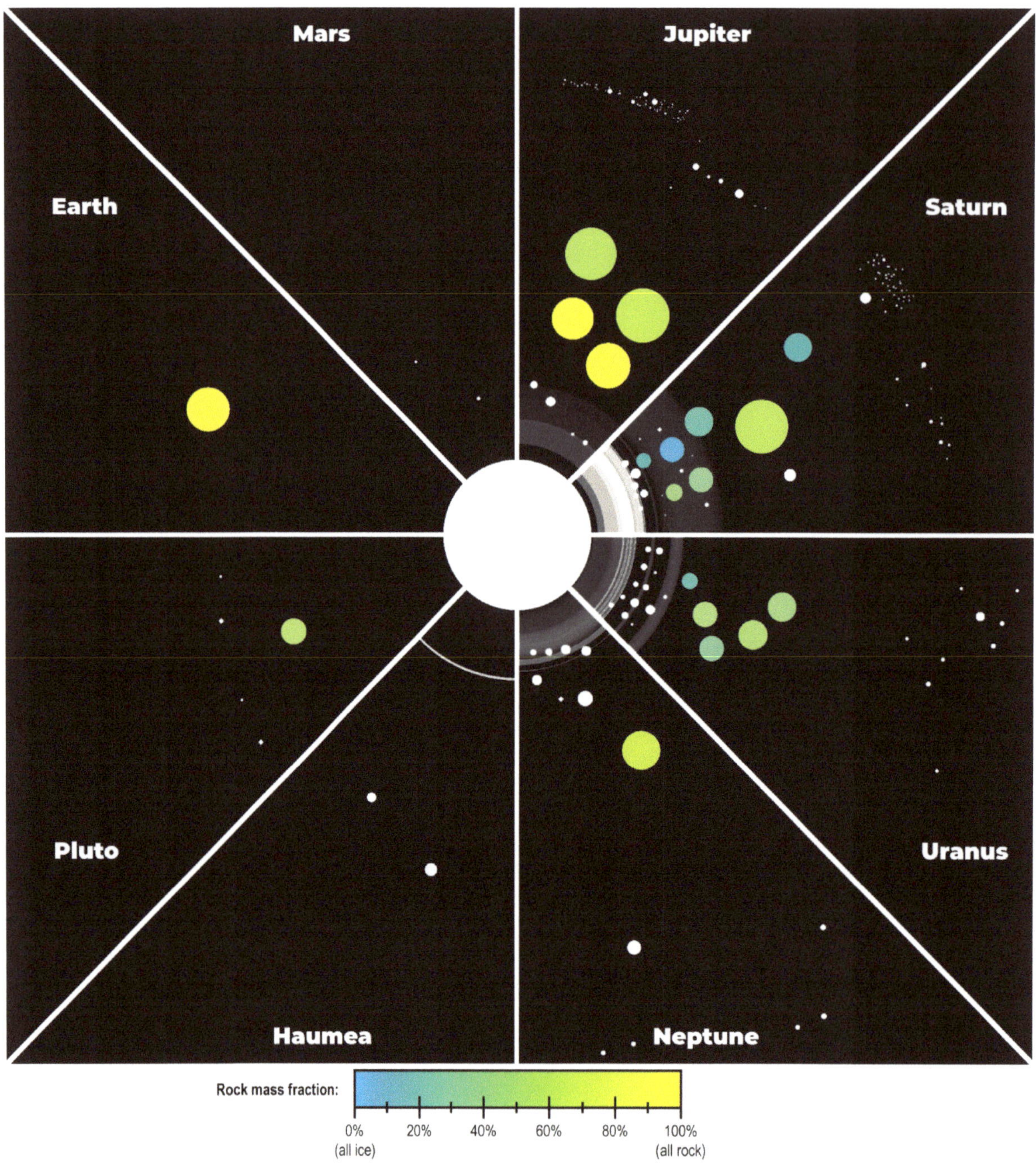

FIGURE 11-1 A selection of circumplanetary systems of the solar system. Distances are normalized to the size of each primary body. Satellite sizes are normalized across the figure. Color indicates the silicate mass fraction for the major satellites and is inferred from the bulk density assuming a mixture of rock and water ice. SOURCE: Courtesy of M. Hedman and J.T. Keane.

Q8.1 HOW DID CIRCUMPLANETARY SYSTEMS FORM AND EVOLVE OVER TIME TO YIELD DIFFERENT PLANETARY SYSTEMS?

The physical, orbital, and compositional characteristics of the present-day circumplanetary material provide clues as to how these systems, and their parent planets, have formed and evolved. For instance, many moons (e.g., Moon, Mars's moons, Pluto's moons, and Haumea's rings and moons) have been attributed to giant impacts, whereas the large moons of the gas giants are believed to have formed directly from the primordial circumplanetary disks, and others may have been captured objects formerly orbiting the Sun (e.g., Triton). The diverse characteristics of the present-day circumplanetary material provide clues as to how these systems have formed and evolved. The regular progression of bulk densities of the Galilean satellites of Jupiter is not duplicated at Saturn or Uranus (see Figure 11-1), indicating some fundamental difference in how these systems formed or evolved. A crucial aspect of moons is that their orbital and thermal evolution are intimately coupled. Furthermore, in some cases the moons have developed subsurface oceans (Q8.1d), which makes them high-priority targets for astrobiology (Questions 9, 10, and 11; Chapters 12, 13, and 14, respectively). Details of the Moon's formation are covered in Question 3 (Chapter 6).

Q8.1a What Processes Have Led to the Diversity of Bulk Properties of Satellite and Ring Systems?

The regular ring–moon systems of the giant planets (i.e., the rings and moons orbiting in or near the planet's equatorial plane) probably originated from a disk of gas and solid material that arose as each planet formed. Thus, the bulk composition of these ring–moon systems can constrain the material that flowed into the giant planets during the last phases of their accretion. Meanwhile, comparisons between the compositions of solid bodies and their circumplanetary systems can reveal the conditions and circumstances under which they formed, including whether they were produced by giant impacts.

Recent investigations of the ring–moon systems surrounding Saturn, Uranus, and Mars indicate that the configuration of material orbiting these bodies may have changed substantially over the history of the solar system, with rings accreting into moons and/or moons breaking down into rings at various times (Q8.1c). For example, there is active debate about the age of the Saturn's rings (e.g., Crida et al. 2019), and what role the hypothesized Uranus-tilting giant impact played in forming or modifying the uranian satellites (e.g., Kegerreis et al. 2018). On one hand, this complicates efforts to reconstruct the initial configuration of the material surrounding these planets. On the other hand, detailed investigations of compositional, geological, and structural variations within and among these systems provide an opportunity to test how fundamental processes such as accretion and fragmentation have operated at different times and under different environmental conditions.

Q8.1b What Determines the Composition and Rock–Ice Ratios of Moons and Rings?

The bulk composition of moons and rings (i.e., the relative abundances of ice, rock, and metal) provides clues to their formation. Bodies built from material excavated in a giant impact are expected to be made of the outer layer of the impacted world—for example, the Moon is presumed to be made in part from Earth's mantle (Question 3, Chapter 6). Conversely, bodies that formed together with their host planet would have bulk compositions initially set by the protoplanetary and/or circumplanetary disk(s). Last, the composition of a captured body would not be expected to resemble that of its host planet. However, reality is more complicated than this simple analysis. The regular density progression of the Galilean satellites (see Figure 11-1) might suggest a compositional gradient in the protoplanetary disk, or loss of volatiles from the innermost moons. There is no such regular progression at Saturn or Uranus. At Saturn, Tethys, whose low density suggests that it may have formed from debris from an icy mantle produced by an impact or tidal stripping (e.g., Canup 2010), has orbital neighbors Enceladus and Dione, both moderately rocky bodies that seem to have formed from the disk. At both Uranus and Saturn, the ring compositions appear quite different from many of the satellites. Similarly, Phobos and Deimos are spectrally distinct from Mars. Pluto's small satellites look like impact fragments, but Charon's bulk composition is similar, but not identical, to Pluto's outer layers. For trans-neptunian objects, there is an enigmatic trend where smaller bodies are less dense than larger bodies (Brown 2012), but our understanding of their formation and evolution is severely limited by a lack of basic information, including their sizes, masses, and densities. At present, the combination of factors leading to the various bulk composition patterns observed across the solar system (see Figure 11-1) is not understood.

Q8.1c How Old Are Moons and Rings, and Do They Undergo Cyclic Formation and Destruction?

The mid-size moons of Saturn have received particular attention in the past decade owing to the detailed datasets acquired by the Cassini mission. The moons vary in composition, geological activity, and internal heating with little apparent regard for their distances from Saturn—in contrast to Galilean satellites where geologic activity decreases (and ice content increases) as a function of distance from Jupiter. In order to address these peculiarities, several models have emerged that suggest not all of Saturn's moons are primordial. In some hypotheses, large impacts led to the destruction of one or more moons, the remains of which accreted to form the current moons (e.g., Asphaug and Reufer 2013), while others suggest the innermost moon or moons grew from material expanding from a massive initial ring, perhaps with rocky "kernels" within Saturn's rings accreting outer layers of ice (e.g., Canup 2010; Charnoz et al. 2011). Although still an area of active investigation, these different ideas all point to the Saturn system undergoing a much more dynamic evolution than once thought. The life cycle of ring–moon systems may involve several iterations of formation and destruction. Similar hypotheses have been put forward for cyclical formation and destruction of Mars's satellites, Phobos and Deimos, leading to the hypothesis that Mars may have once had rings (e.g., Hesselbrock and Minton 2017).

Looking outward to the satellite systems of Uranus and Neptune, where data are comparatively sparse, there is further evidence of systems once in upheaval, sculpted by impacts. At Uranus, models suggest that a giant impact titled the planet, destabilizing and possibly destroying its original, primordial satellites. Uranus's present satellite system thus may be a second-generation of planetary satellites. Uranus's small moons and rings are darker than the larger moons. The ring particles may consist of heavily processed material which was initially similar to that of the inner moons. At Neptune, the capture of Triton, and its subsequent tidal evolution, likely wiped out much of the original satellite system, the remains of which seem not to have coalesced into a set of mid-sized moons, unlike at Saturn and Uranus. Neptune's rings, and its ring arcs, consist primarily of fine dust that is sculpted by a variety of dynamical processes, which give clues to how material is aggregating/fragmenting in circumplanetary systems. A major implication is that the evolution of these systems may have been intimately linked to the architecture of the giant planets whose motions governed the timing and extent of material being flung at their developing satellite systems. For both Uranus and Neptune, there are fundamental knowledge gaps in our understanding of the compositions of the small moons and rings.

Q8.1d What Determines the Internal Structure of Moons, and Which Moons Have Subsurface Oceans?

The moons of the solar system have tremendously varied interior structures (Figure 11-2), ranging from largely solid ice balls, to partially molten rocky worlds, and worlds with plausibly habitable subsurface water oceans—so-called ocean worlds. Much of the basic structure of a moon is determined by its initial bulk composition, and the energy imparted from formation, decay of radioactive material, subsequent impacts, and tides. Most large satellites are differentiated, with dense phases (metal and rock) forming a core, overlain by a mantle or crust of low-density phases (rock and ice). However, while this is the norm, it is not the rule. Callisto may be largely undifferentiated, despite having a comparable size and bulk composition to several other differentiated icy worlds, like Ganymede and Titan. Determining the differentiation state of large planetary satellites provides a key window into the earliest epochs of a circumplanetary system (Question 2, Chapter 5).

A handful of icy satellites are heated enough that their outer layers can melt, producing subsurface oceans of liquid water hidden beneath insulating icy shells that are either in contact with the rocky seafloor or high-pressure ices at their base. Ganymede may have several ice polymorphs form with perched oceans between them (Journaux et al. 2020). These ocean worlds may be abodes for life (Questions 9, 10, and 11; Chapters 12, 13, and 14, respectively). Despite their importance, the solar system's inventory of ocean satellites is uncertain and likely incomplete. There is direct evidence for subsurface oceans within Europa, Ganymede, Callisto, Titan,

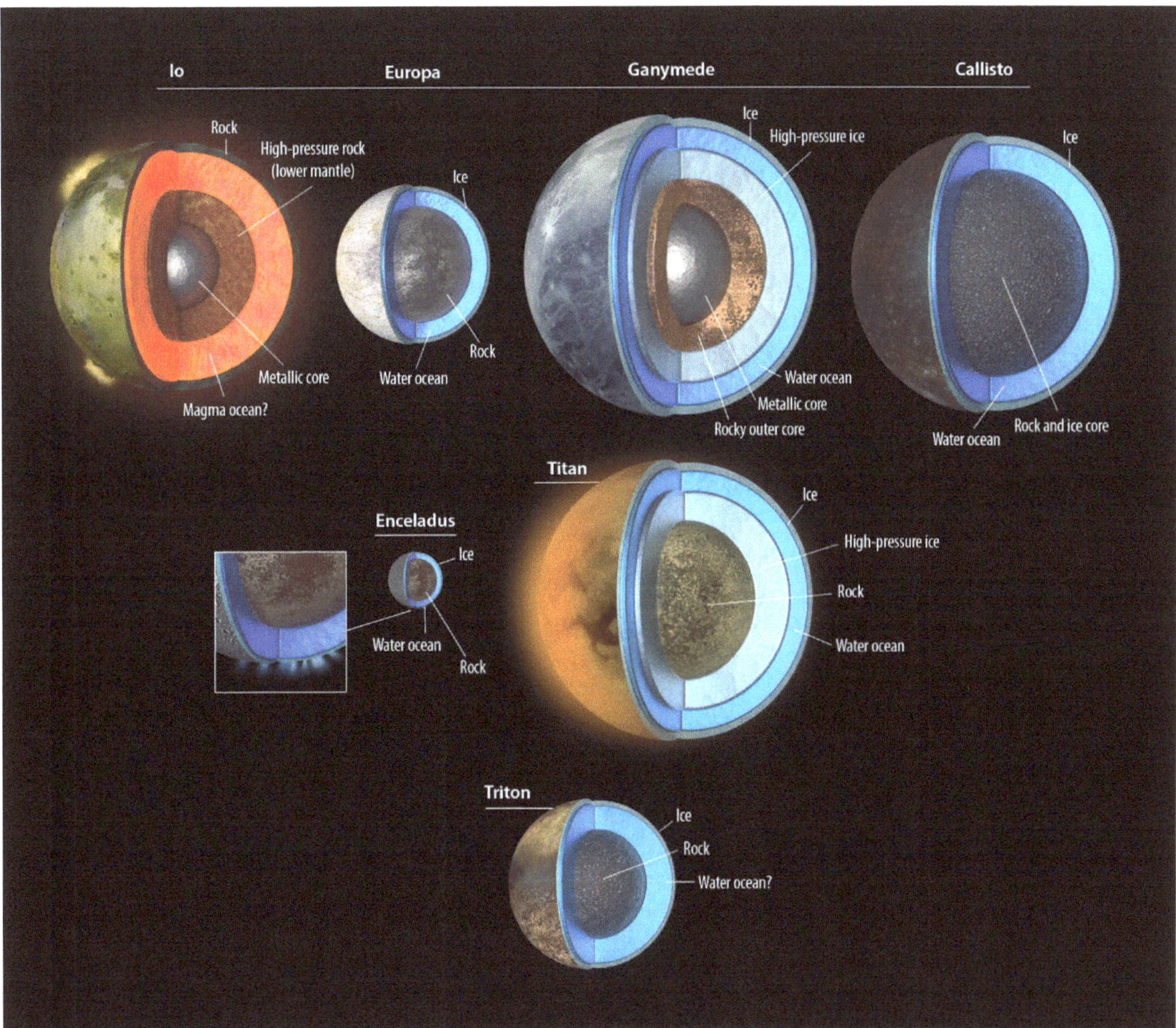

FIGURE 11-2 The inferred interior structures of select moons in the solar system. These structures are subject to uncertainties on material porosity and composition (e.g., metal content and degree of hydration of the silicate rock, impurity content in the ice, abundance of refractory organic material of intermediate density, relative thicknesses of oceans and ice shells). New measurements are required to test these interior structure models. SOURCE: Courtesy of C. Carter/J.T. Keane/Keck Institute for Space Studies.

and Enceladus, and circumstantial evidence for oceans within several other worlds, including Pluto and Triton (Nimmo and Pappalardo 2016). There is also evidence for a subsurface *magma* ocean within Io, plausibly formed and maintained by similar processes (Khurana et al. 2011). Subsurface oceans may be persistent and long-lived, or they may go through episodic or cyclic periods of freezing and melting. This is complicated because the thermal evolution of these worlds is likely coupled to their orbital evolution and the gravitational interactions between neighboring satellites and their host planets (Q8.2). For example, the existence, longevity, and evolution of Europa's ocean is likely tied to the evolution of the orbital resonance between Io, Europa, and Ganymede, their interior structures, and the interior structure of Jupiter (Hussmann and Spohn 2004; de Kleer et al. 2019a). This highlights the need for investigations that span entire circumplanetary systems.

Strategic Research for Q8.1

- **Determine the differentiation state, radial interior structure, tidal response, and presence/absence of water and magma oceans and reservoirs within the moons of Jupiter, Saturn, Uranus, and Neptune** by measuring their gravity fields, shape, induced magnetic field and plasma environment, and other geophysical quantities.
- **Determine the masses, densities, and bulk compositions of Kuiper belt objects and their satellites** with surveys for multiple systems and satellites, and characterization of their bulk properties and orbital motion.
- **Determine the composition of rings and small moons at Uranus and Neptune to elucidate their origin, evolution, and present-day balance between exogenic and endogenic processes** through a combination of geophysical measurements, imaging, and spectroscopic observations, including at spatial resolution sufficient to resolve regional variations and layering.
- **Constrain the origin of Phobos and Deimos, including whether they arose from past martian rings,** by determining their bulk composition and interior structure via in situ geochemical and geophysical measurements.

Q8.2 HOW DO TIDES AND OTHER ENDOGENIC PROCESSES SHAPE PLANETARY SATELLITES?

A host of processes beneath the surfaces of planetary satellites shape their diverse properties and evolution over time (Figure 11-3). Tides are a prominent such process that are largely responsible for the presence of liquid water oceans and volcanism in the outer solar system, may drive flows in planetary oceans and cores, and fracture the ice shells of ocean worlds. Tectonics, (cryo)volcanism, hydrothermal activity, and magmatism can further shape these worlds and are ultimately the consequence of heat and materials transported from the deep interior outward. Beyond their surface expression, these processes can also have magnetic signatures, both in terms of intrinsic magnetic fields generated by dynamo action and induced magnetic fields owing to interactions between subsurface oceans and the host planet's magnetosphere. By understanding the diversity of these processes across planetary satellites throughout the solar system, we will learn how they operate on each body as well as collectively. Understanding these processes has broad impact across all of planetary science, including the ability assess the astrobiological potential of planetary bodies and the interplay of oceanography, glaciology, and hydrology with life (Question 9, Chapter 12), the time-dependent supply of liquid water, energy, and nutrients (Question 10, Chapter 13), and how these processes affect biosignature detectability, survivability, and reliability (Question 11, Chapter 14).

Q8.2a How and Where Is Tidal Heat Dissipated Within Circumplanetary Systems?

As a moon orbits its parent world, it is distorted by tides. If the moon's orbit were perfectly circular and unchanging, its shape would adjust to the tidal forces—producing a tidal bulge. Tidal dissipation in the moon ultimately drives the spin period of the moon to match its orbital period (i.e., becoming "tidally locked"), with the bulge constantly facing its host planet. However, if the moon is on an eccentric orbit, if it spins at a different rate than it orbits, or if the axes of its spin and orbit are not aligned, it may never reach a static shape. Instead, the moon will continuously and repeatedly deform in response to tides, leading to friction and heating within the moon (e.g., Peale et al. 1979). Long-term tidal heating requires maintenance of the orbital eccentricity, which in turn requires tidal dissipation within the planet. The long-term rate of satellite tidal heating depends on how dissipative the planet is, so there is an intimate connection between planetary and satellite evolution. Tidal heating is the main driver of recent geologic activity on outer solar system satellites, including supporting long-lived subsurface water oceans within many satellites and a possible magma ocean within Io, and it may have played a role in shaping other bodies in the early solar system, including the Moon. Despite its importance, there are still many fundamental questions about how tidal heating operates, where the heat is dissipated (be it in the icy shells, water or magma oceans, or deeper rocky layers), whether eccentricity or obliquity tides dominate, how tidal heating affects the surface geology, and how it ultimately escapes to space (Beuthe 2013; de Kleer et al. 2019a).

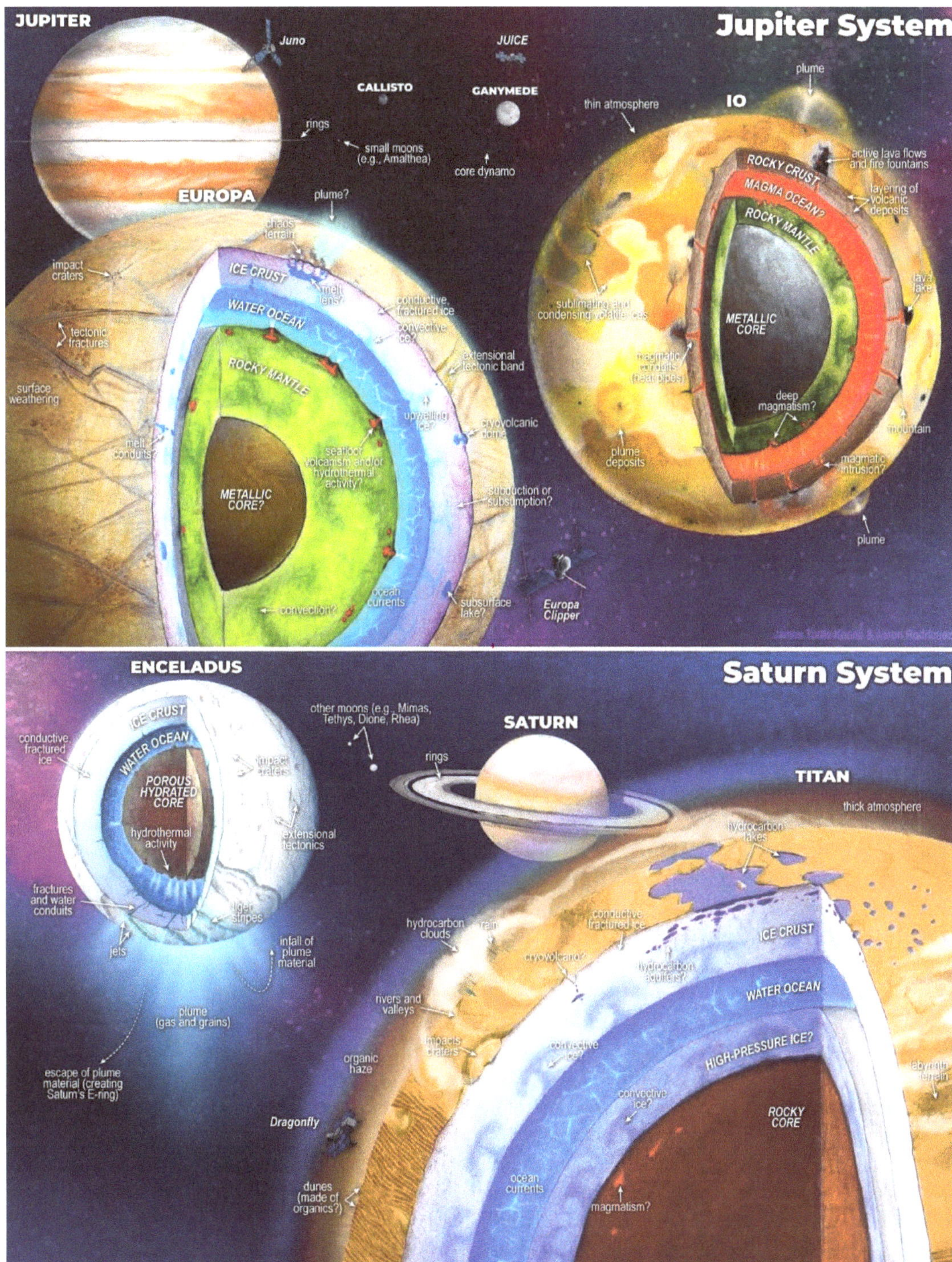

FIGURE 11-3 Schematic illustration of the interior structures and proposed processes responsible for the transport of heat and material through planetary satellites—focusing on Io, Europa, Enceladus, and Titan. While these worlds represent four well-studied endmembers, similar processes may occur on a variety of planetary satellites across the solar system, including the satellites of Jupiter, Saturn, Uranus, and Neptune. SOURCE: Illustration by J.T. Keane and A. Rodriquez.

Matching observations with theoretical predictions for tidal hearing is difficult because of a lack of observational constraints of the interior structures of tidally heated worlds, incomplete theoretical models, and the lack of knowledge of how relevant planetary materials (e.g., both within the rock/ice/melt of the satellite, and within the giant planet cores and fluid envelopes) behave under relevant pressure, temperature, and forcing conditions.

Q8.2b What Is the Thermal and Orbital Evolution of Planetary Satellites, and What Is the Role of the Host Planet and Other Satellites in That Evolution?

Under the action of tides, the orbits of satellites in circumplanetary systems change with time. Exactly how a satellite's orbit changes depend on the internal structure and thermal state of both the satellite and its tide-raising parent body—which determine how and where energy is transferred and dissipated. This results in a complex coupling between satellite orbital evolution and thermal evolution. These coupled processes affect the timing and likelihood of capture into resonances between orbital and/or spin periods, orbital eccentricities and inclinations, and the magnitude and duration of tidal heating—which governs the formation and sustainability of subsurface oceans. Although solid planets and moons cool over time, the coupled evolution of satellites could lead to phases of high heat flows, transient or late oceans, and episodic geologic activity (Ojagankas and Stevenson 1989; Hussmann and Spohn 2004). For example, the Moon's outward migration owing to tides may have been quite complicated, plausibly leading to periods of high eccentricity, obliquity, and tidal heating, which may leave an imprint on the overall shape of the Moon (its so-called fossil figure—e.g., Gerrick-Bethell et al. 2014; Keane and Matsuyama 2014). How moons respond geophysically to temporal changes in heating and deformation, how changes in their host planets might affect their long-term orbital evolution, and the potential for such dynamic environments to host life are all active areas of research.

Q8.2c What Is the Expression of Tectonic Activity on Planetary Satellites, and Why Is It So Varied Between Circumplanetary Systems?

Tectonic activity on planetary satellites is varied in both style and extent, which can provide insight into satellite evolution, interiors, and the geophysical mechanisms at play in the solar system (Collins et al. 2009). Moons with the potential for tidal activity because of their eccentric orbits tend to have younger surfaces with more diverse geology (e.g., Europa and Enceladus), although they display curious differences in feature types and distributions. For example, Europa possesses "cycloids," arcuate fractures that can be explained by diurnal tidal stresses (e.g., Rhoden et al. 2021), although similar fractures are not observed on other worlds, and other tectonic structures on Europa remain unexplained. Enceladus possesses a long-lived cryovolcanic plume which is sourced from jets originating from a set of tectonic fractures on the moon's south pole—the "tiger stripes." While these faults are the conduits by which Enceladus's ocean material is ejected to space, we do not know how they form, evolve, or what modulates their behavior (e.g., Běhounková et al. 2017). However, some moons with eccentric orbits display little evidence for tectonism (e.g., Mimas), and some moons with circular orbits display extremely young surfaces and tectonic features similar to their eccentric cousins (e.g., Triton). Enceladus's long-lived plume system and Europa's plate-tectonic-like behavior and disrupted "chaos" terrain are unique among explored satellites. Other moons, such as Tethys, Miranda, and Charon, for which tidal deformation likely ceased billions of years ago, display large canyon systems that might be linked to volume expansion during the freezing of ancient oceans, but the processes responsible for the distribution and shapes of these features (e.g., location, orientation) remain enigmatic. For outer solar system satellites, our understanding of tectonic processes is limited by a dearth of data on the response of ice to mechanical stress under relevant conditions and limited investigations into the complex interactions between fractures and other processes, such as cratering. In addition, sources of stress beyond eccentricity or obliquity tidal stresses (e.g., nonsynchronous rotation, physical libration, true polar wander) have been suggested but, for the most part, not confirmed for these moons. Expanding our theoretical models and obtaining additional observational constraints on these processes would help clarify their importance in the evolution and activity of planetary satellites.

The surface manifestations of tectonic stresses are incredibly varied across the circumplanetary systems in the solar system. Many icy moons show evidence of extensional tectonics, such as cycloids, lineaments, and extensional bands on Europa, fractures on Enceladus, grooved terrain on Ganymede, wispy terrain on Dione, fractures on Triton, and canyons on Tethys, Miranda, Pluto, and Charon. Tectonic activity from shear stresses is less prevalent but is observed on Io, Europa, Ganymede, and Enceladus in the form of strike-slip faults, deformation within bands, and deformation of existing features (e.g., craters). Compressional tectonics is prevalent on many rocky worlds, including lobate scarps and wrinkle ridges on the Moon, and enormous mountains on Io. Compressional tectonics can be more difficult to identify on icy worlds, although compressional bands have been definitively identified on Europa, and mountain ranges on Titan may form from compression. The relationships between local and global stress regimes are not well-understood. Additionally, the level of present-day tectonic activity is uncertain on most worlds. A notable exception is the Moon, where Apollo seismometers and observations of boulder falls suggest that the Moon's lobate scarps are seismically active today (Senthil Kumar et al. 2019). Theoretical models strongly suggest that many icy worlds are tectonically and seismically active today, including Enceladus—where tides may cause meters of displacement on the tiger stripes every Enceladus orbit (33 hours), and possibly modulate plume output (e.g., Běhounková et al. 2017).

Q8.2d Where Does Liquid Water Exist in the Ice Shells of Ocean Worlds?

We now know that several icy moons in the solar system possess subsurface liquid water oceans (Q8.1d), and there is increasing evidence that liquid exists *within* their ice shells as well (Cable et al. 2020; Journaux et al. 2020; Schmidt 2020). Enceladus's plumes have provided proof that icy satellites can develop fracture systems that directly link subsurface liquid water reservoirs with the surface, but the evolution that led to this system is not well understood. In contrast, liquid water may not completely penetrate Europa's ice shell, but instead collect in melt pockets within the ice shell (i.e., in sills), which may play a key role in the formation of ubiquitous double ridges, disrupted "chaos" terrains, relatively small, distinct pits, uplifts, and smooth dark regions dubbed "spots," and the purported plumes—although convection of warm ice has also been proposed for most cases. More generally, subsurface water kept liquid by antifreezes (e.g., ammonia) may occur in warmer ice (e.g., upwelling convection cells or sites of localized tidal heating), and brines may be retained in a mushy layer at the base of the ice shell. As oceans freeze, they become pressurized and may fracture the overlying ice shell to allow ocean infiltration. In large ocean worlds (e.g., Ganymede and Titan), liquid water may also exist in high-pressure ice layer(s) at great depth. The extent to which liquid water can move into and out of ice shells has important implications for habitability and the search for life (see Questions 10 and 11, Chapters 13 and 14, respectively). In particular, the habitability of in situ melt pockets depends on the source of liquid water; melt pockets that are not sourced from the ocean are less likely to contain biosignatures. Melting in high-pressure ices may enable silicate–ocean exchange of salts and volatiles, which are necessary for life, but their ability to cycle resources in a way that supports life may be limited.

Q8.2e What Causes Plumes, (Cryo)Volcanism, Hydrothermal Activity, and Magmatism in Planetary Satellites?

Active volcanic or cryovolcanic eruptions (where melted or vaporized ices are extruded; Geissler 2015) occur on Io, Enceladus, Triton, and possibly Europa. The Moon is not volcanically active today, but possesses a rich volcanic history as revealed by both orbital datasets and returned samples—most recently with the samples collected by the Chang'e-5 mission which reinforce the concept of long-lived lunar magmatism (Che et al. 2021). Beyond the Moon, past volcanic and cryovolcanic activity is hard to identify with certainty, although many satellites—including Ganymede, Titan, Ariel, and Charon—appear to have experienced (cryo)volcanic activity in their past. But some satellites, like Callisto and Mimas, have apparently never experienced such activity—or such activity happened early enough (billions of years ago) that it was erased by subsequent impacts. This variability in behavior is not simply a function of distance from the planet, orbital eccentricity, or body size, and thus represents a major puzzle. Tides are presumably the dominant heat source for recent volcanic activity, and they are known to modulate the plumes on Enceladus (Hedman et al. 2013) and possibly Io (de Kleer et al. 2019b). However, geologically recent (<1 Ga) volcanic and cryovolcanic constructs on worlds where tidal heating is either insignificant (e.g., Ahuna Mons on Ceres), or ceased long ago (e.g., Wright and Piccard Montes on Pluto, irregular mare patches on the Moon [see Figure 2-5]), suggest that tidal heating is not always required to power volcanic activity on planetary satellites. For large and rock-rich bodies, such as Europa, the contribution from radiogenic heating alone may be sufficient to drive volcanism.

On Earth and the Moon, intrusive magmatism (where the magma never reaches the surface) dominates the volcanic history. On the Moon, the ratio between the volume of intruded magmas to erupted lavas may be as high as ~50:1 (Head and Wilson 1992). However, the extent and role of intrusive magmatism within other planetary satellites, and in the rocky mantles of ocean worlds, is harder to conclusively identify. Io-like volcanic activity on Europa's rocky seafloor cannot be ruled out; it would lead to vigorous hydrothermal activity and could alter ocean chemistry. Silica particles in Enceladus's plume may be evidence for high-temperature water–rock interactions (Sekine et al. 2015), but evidence for water–rock interactions within other worlds is lacking. Understanding intrusive magmatism on tidally heated worlds is a key factor in determining the heating, chemical inventory, and habitability of ocean worlds.

Q8.2f How Are Heat and Material Transported Through—and Ultimately Out of—Satellite Interiors?

Fundamental characteristics, such as the presence or absence of a magnetic dynamo and the survival of a subsurface ocean, depend on how heat is transferred from a body's interior to its surface (Soderlund et al. 2020). In smaller bodies (e.g., asteroids), heat tends to be transferred by conduction, possibly affected by porosity (which disappears at high enough temperatures). In larger bodies (e.g., planets and some large satellites), conduction can be augmented by convection (where warm solid rock or ice rises and transports heat closer to the surface—as in Earth) or "heat pipe" volcanism (where melt rises to the surface, transporting heat with it—as in Io). For most planetary bodies, the balance between conduction, convection, and heat pipe volcanism is not well known.

Large icy bodies like Callisto or Titan may experience convection in their outer ice shells, but beneath a thick, rigid, cold, conductive lid. Conversely, Europa's ice shell may experience "mobile lid" tectonics (i.e., similar to Earth's plate tectonics), which would increase the rate of heat transfer, while at Enceladus the ice shell appears static, but cryovolcanism enhances heat transfer. Ocean circulation further controls how heat and material are exchanged between rocky seafloors and the overlying ice shells, depending on the composition and temperature structure of the ocean. If the rocky core is porous (as at Enceladus), its heat may be transported convectively by hydrothermal circulation. Io is certainly in the heat-pipe regime, as was the early Moon (and plausibly other early terrestrial worlds), but the details of this process are largely unknown (Moore et al. 2017).

Q8.2g Which Planetary Satellites and Dwarf Planets Have or Had an Intrinsic Magnetic Field, and What Processes Govern These Dynamos?

Ganymede is the only known satellite with an intrinsic magnetic field at present (Kivelson et al. 1996). The Moon records the presence of an ancient dynamo through remanent magnetization of crustal rocks (Wieczorek et al. 2017). Paleomagnetic studies of meteorites further demonstrate that small bodies, such as the asteroid Vesta, may also have had core dynamos early in their histories (Scheinberg et al. 2017). It is not yet known whether dynamos—past or present—await discovery in the satellites and dwarf planets that have not yet been explored in detail. An active dynamo within moons orbiting within a planetary magnetosphere is expected to be detectable from magnetometer measurements during a single low-altitude flyby, and distinguishable from crustal or induced fields. The rare exception to this being a dipolar field observed at a moon that contains a similar phase but lower magnitude than that anticipated from the magnetic-wave-driven induction response. Weak dynamo fields on dwarf planets or satellites residing in the solar wind may require several encounters or orbits to definitively characterize as sourced from a dynamo or from induction, while mapping of crustal fields would require much greater spatial coverage. Disentanglement of these fields and any induced magnetic fields driven by an ocean could be achieved by comparing observations with models of the structure, amplitude, and temporal variability of the magnetic field.

Beyond knowing what bodies have dynamos, it is also important to characterize the magnetic fields. Ganymede's intrinsic magnetic field is known to be dipole-dominated, but its detailed configuration and evolution over time have not yet been determined (Journaux et al. 2020). Conversely, few constraints exist on the Moon's ancient dynamo field morphology. Samples returned by the Apollo missions revealed that it was anomalously strong, and a weaker dynamo may have been active until relatively recently (Wieczorek et al. 2017). For both bodies, the mechanisms of magnetic field generation—and their evolution over time—are not well understood. For ocean worlds such as Ganymede, there may further be interactions between the intrinsic core field and the oceanic induced field together with the magnetosphere of the host planet. By understanding these processes in planetary satellites, we will not only learn how intrinsic magnetic fields operate in these bodies specifically, but also facilitate tests of dynamo theory more broadly through comparative planetology.

Strategic Research for Q8.2

- **Characterize the spin states and orbital and rotational evolution of planetary satellites (including at Jupiter, Saturn, Uranus, and Neptune)** with spacecraft and radar observations, and long temporal baseline astrometry.
- **Characterize the current orbital evolution of planetary satellite systems across the solar system (including Earth, Jupiter, Saturn, Uranus, and Neptune), and determine if they are in thermal equilibrium,** by measuring how the satellite orbits are currently evolving, how their host planet responds to satellite tides (including phase lag), and by measuring satellite heat flows.
- **Assess the past and present geologic activity of the uranian satellites, Triton, and large Centaurs and trans-neptunian objects with moon or ring systems—including their cratering record, tectonic, and cryovolcanic activity—and understand how and why they differ from satellites in other systems** by imaging their surfaces with resolution, coverage, and spectral range that is at least comparable to those of the Galilean satellites and mid-sized moons of Saturn.
- **Determine the differentiation state, radial interior structure, tidal response, and presence/absence of water and magma oceans and reservoirs within the Galilean, saturnian, and uranian moons, as well as Triton,** by measuring their gravity fields, shapes, induced magnetic fields and plasma environments, and other geophysical quantities.
- **Determine if/how tides have shaped the crustal structure, tectonics, and (cryo)volcanism of the large/mid-sized saturnian (e.g., Enceladus, Titan) and Galilean (Io, Europa, Ganymede, Callisto) satellites** by characterizing the three-dimensional structure of their crusts through topography, gravity, ice-penetrating radar, and other geophysical methods.
- **Determine if/how tides have shaped the crustal structure of the Moon,** by characterizing the three-dimensional structure of its crust through seismology, electromagnetic sounding, heat flow, and other geophysical methods.
- **Characterize the thermophysical processes of the icy shells of the large/mid-sized satellites of Jupiter and Saturn (e.g., Europa, Enceladus, and Titan), and determine their present-day activity** via in situ geophysical analyses, including seismology, electromagnetic sounding, heat flow measurements, and other methods.
- **Characterize tectonic and eruptive processes on Io, Europa, Enceladus, and other active bodies, and assess their relationship to tides, crustal structure, and interior processes** with high-resolution imaging over a range of illumination conditions, tidal stress conditions, and observational cadences, and long-term monitoring of activity at various temporal/spatial scales and wavelengths.
- **Characterize volcanic and magmatic processes on the Moon and assess their relationship to tides, crustal structure, and interior processes** via in situ geochemical and geophysical analyses (including seismology, electromagnetic sounding, and heat flow measurements), and/or returned samples from key volcanic and magmatic sites across the Moon.
- **Determine the size and state of the Moon's solid inner core** through seismology measurements, electromagnetic sounding, and other geophysical investigations.

- **Determine the rheological behavior of icy crusts, including determining how and when they fracture** with laboratory and analogue studies of ice failure and rheological behavior at the conditions relevant to icy satellites.
- **Determine how, why, and when tectonic and (cryo)volcanic surface features form on rocky and icy satellites** through modeling of fractures, fracture systems, fracture-water interactions, and (cryo)magmatic processes in ice and rock shells.
- **Determine the nature and origin of Ganymede's intrinsic magnetic field** through magnetic field mapping over time, improved knowledge of its interior structure, and dynamo modeling.
- **Understand how subsurface water and magma oceans are affected by convection, orbital forcing, magnetism, and salinity, and how those processes modify heat fluxes and circulation patterns** with fluid dynamical models.

Q8.3 WHAT EXOGENIC PROCESSES MODIFY THE SURFACES OF BODIES IN CIRCUMPLANETARY SYSTEMS?

Satellite surface landforms and composition provide a record of the processes that shaped each of them, key to understanding their formation and evolution. Some of this record is owing to exogenic processes, such as radiation, impacts, and the exchange of material between bodies within circumplanetary systems. It is important to understand how these processes affect what we see today on each satellite, as exogenic processes can be superposed on intrinsic and endogenic processes—obscuring more primitive surface composition and morphologies, leading to misinterpretation.

Q8.3a How Do Weathering and Atmospheric Loss Affect the Environment and Objects in Circumplanetary Systems?

Weathering is the general alteration of surface materials owing to prolonged exposure to both small impacts and energetic radiation. Sputtering is a subset of this, covering the ejection of surface particles at the molecular level upon radiation exposure, which can be the source of tenuous satellite atmospheres, like those on the Moon, Europa, and Enceladus (Question 6, Chapter 9).

Radiation-induced weathering is observed on nearly all planetary bodies without a substantial atmosphere (e.g., so-called airless bodies: Mercury, the Moon, asteroids, and many icy satellites) all show evidence of such alteration. But the details of this process depend on surface composition, and the object's location in the solar system. For the inner solar system, exposed minerals containing iron or sulfur darken with prolonged exposure to radiation, as confirmed by study of returned samples. In the outer solar system, radiation-induced weathering is largely controlled by the desaturation of organics, which reddens surface coloration. More information about space weathering can be found in Question 5 (Chapter 8). Radiation also affects ice grain size and can brighten or darken the surface of icy satellites.

In circumplanetary systems, the loss of material from one body can act as a source to neighboring bodies. This can be the result of atmospheric loss owing to radiative processes, as seen from Earth to Moon, and Pluto to Charon. Sputtering of material off the surface can also contribute to "polluting" the surrounding environment. In addition, endogenic activity such as plumes or volcanism can launch material into space (e.g., Enceladus and Io), creating diffuse rings at Jupiter and Saturn, and contaminating other neighboring surfaces. Saturn's rings also act as a volatile source to Saturn itself; volatile "ring rain" was observed by Cassini (see Q8.4c), a loss mechanism that may suggest the rings are young. Some radiation and mass exchange processes may be in steady state, or if not, can be used as chronometers to constrain event histories.

Q8.3b What Is the Role of Circumplanetary and Heliocentric Impacts in Creating the Observed Structures on Moons and Rings?

Impact craters are the most common landform on planetary surfaces, seen on nearly every major satellite surface (Question 4, Chapter 7). As older surfaces have been exposed to more impacts, the cratering record can be used as a chronometer for dating surfaces (Figure 11-4). This is one of the primary tools used to constrain both absolute and relative histories of planetary and satellite surfaces. However, calibration of the cratering record

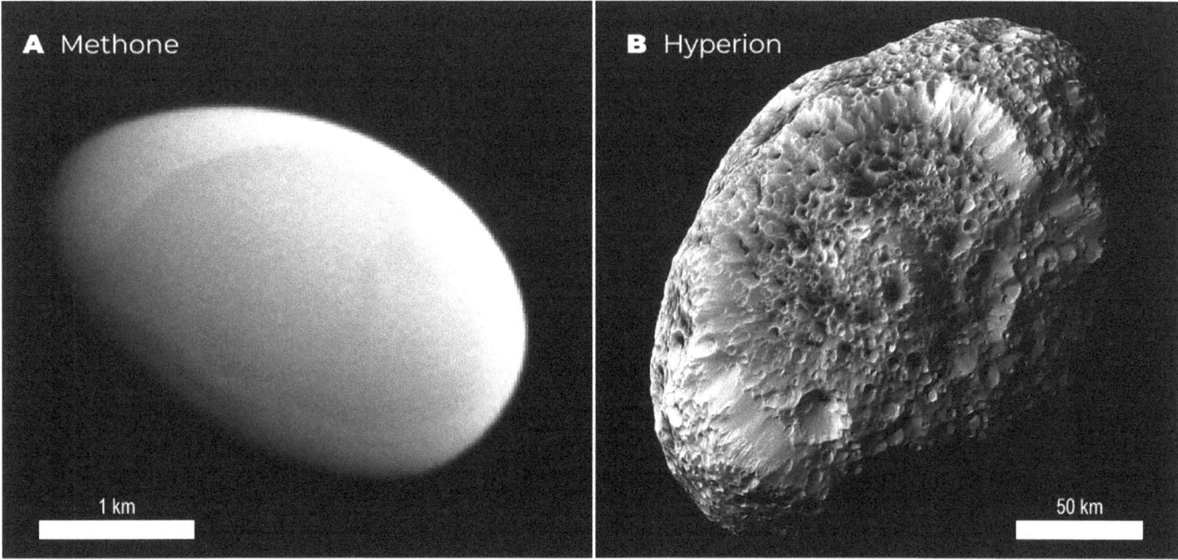

FIGURE 11-4 Two satellites of Saturn imaged by Cassini show the extremes of surface crater density. *Left:* Methone has a smooth surface with no visible impact craters, as it is very small and young. *Right:* Hyperion is saturated with craters owing to its age. SOURCE: Courtesy of NASA/JPL-Caltech/Space Science Institute.

to absolute age is not uniform across the solar system and is complicated by secondary impacts (craters formed from the impact of suborbital ejecta from a single impact) and sesquinary impacts (craters formed from the impact of ejecta that initially escaped the target body, orbited its host body, and then reimpacted the target body, or another body in the circumplanetary system). The inner and outer solar system have different size frequency distributions (SFD) of impactors, as the source population is asteroidal for the inner solar system, and cometary for the outer solar system. There is a larger uncertainty in the SFD for the outer solar system. Detailed observations of uranian satellites would provide useful new cratering data to help constrain outer solar system chronology, as some surfaces appear to be quite old, perhaps primordial. Future missions could determine whether unseen terrains on the satellites of the ice giants are consistent with known processes, features, and statistics.

Planetary rings can also be shaped by interplanetary impacts. At least four such individual events were seen during the Cassini tour. Disruption from these events was very localized, but models show that interplanetary impactors sculpt some aspects of ring structure, for example, the ramps observed near Saturn's ring edges. Interplanetary impacts are likely the major source of small dust particles in the rings, because collisions within the ring are too gentle to generate such dust.

Q8.3c What Is the Fate of Ejecta from Impacts onto Moons and Rings?

The fate of ejecta from interplanetary impacts in circumplanetary systems encompasses a wide range of possible outcomes, depending on ejecta velocity relative to the escape velocity of the impacted moon (Question 4, Chapter 7). If moving much slower than escape velocity, ejecta will remain localized to the impact region of the target body. If close to escape velocity, ejecta will cover the impacted moon more broadly. Above escape velocity, ejecta will orbit around the primary planet, ultimately dispersing into a diffuse, dusty ring owing to the range of initial individual debris velocities. Examples of such dusty rings include those coincident with Phoebe at Saturn, and likely Mab at Uranus. Phobos and Deimos may have exchanged surface material through a similar process. Such dust will slowly spiral inward, likely coating the surfaces of other satellites (Q8.3d).

Ejecta owing to collisional "grinding" of captured, irregular satellites is likely a primary source of dust in giant planet systems, acting as the major source of surface deposition on outer, regular satellites—for example, Phoebe dust on Iapetus. Internal collisions in planetary rings will also create a population of dust that can be readily ionized, enabling it to then be carried along magnetic field lines into the atmosphere, as seen on Saturn. Deeper searches at Uranus and Neptune may show new dusty rings. In the inner solar system, Phobos and Deimos may be the result of reaccretion of material from a major Mars impact, or they may be captured asteroids. How does the likely ongoing exchange of surface dust affect our ability to discriminate between these origin theories?

Q8.3d What Causes the Global-Scale Asymmetries Observed on Moons in Circumplanetary Systems?

On tidally locked satellites, the surface can be geographically divided between the nearside (i.e., the planet-facing hemisphere) and the farside, or alternatively, the leading hemisphere (the hemisphere facing the direction of travel along the orbit) and trailing hemisphere. Leading/trailing hemispherical albedo and color asymmetries are evident on most tidally locked satellites in the outer solar system. Examples include dark trailing hemispheres on Io and Europa, reddened trailing hemispheres on several saturnian satellites, and the archetype of this feature, Iapetus, with a dark leading and bright trailing hemisphere. The processes that lead to such leading-trailing dichotomies are generally understood, arising from exogenic processes. Dust sourced from a larger orbital distance tends to migrate inward, preferentially depositing on the leading hemisphere of any satellite that encounters it (e.g., dust from Phoebe coating Iapetus, E-ring dust coating the inner saturnian satellites). Trailing hemispheres of outer planet satellites tidally locked outside of synchronous orbit are exposed to an enhanced flux of energetic particles, as the planet's magnetosphere is rotating faster than the orbital velocities of the satellites. Despite this understanding, the extent and thickness of dust coverage on most satellite surfaces are poorly constrained. Quantified measurements of this coverage can help to infer the source of this material, and when the deposition occurred.

Not all global-scale asymmetries are well-understood, and many may arise from endogenic processes. The Moon's nearside-farside asymmetry—the nearside has a thinner crust and more extrusive volcanism—has remained unexplained since its discovery at the dawn of the space age. Hypotheses range from asymmetric thermal evolution owing to Earthshine, to asymmetric convection of the mantle, asymmetric crystallization of the magma ocean, tidal processes, and giant impacts including South Pole–Aitken or even larger. Enceladus also exhibits a pronounced global asymmetry: with most present-day activity concentrated near to the south pole (Hemingway et al. 2018). Io also exhibits an unusual leading-trailing asymmetry in volcanic output; there are more, smaller volcanoes on the leading hemisphere, and fewer, larger volcanoes on the trailing hemisphere (although the total volcanic output is comparable; de Kleer and de Pater 2016). The cause of these asymmetries (endogenic or exogenic), their relationship to the circumplanetary environment, and how deep into the body they extend, remain unclear.

Strategic Research for Q8.3

- **Determine the shape, structure, and composition of Uranus's and Neptune's rings and small moons to elucidate their origin, evolution, and present-day balance between exogenic and endogenic processes** through a combination of geophysical measurements, imaging, and spectroscopic observations, including at high spatial resolution sufficient to resolve regional variations and layering.
- **Determine the history of weathering and impact processes shaping planetary rings and satellites at Jupiter, Saturn, Uranus, and Neptune** from remote sensing and direct measurement of ejected dust and gas.
- **Determine the locations and distributions of source bodies within dusty rings around Uranus and Neptune, and how fine particles are generated, lost, and transported throughout these systems** with a combination of high-resolution imaging and measurements of the mass flux and composition flowing into and out of the rings.
- **Test the hypotheses for the origin of planetary asymmetries, including leading-trailing asymmetries on planetary satellites around Jupiter and Saturn, the nearside–farside asymmetry on the Moon, and elsewhere,** by characterizing their magnetospheric and dust environment, and by geological, geochemical, and geophysical investigations of the dichotomies themselves.
- **Determine how radiation affects materials in the atmospheres and on the surfaces of planetary satellites and characterize the associated chemical pathways and yields** with laboratory studies of relevant materials, at relevant conditions.

Q8.4 HOW DO PLANETARY MAGNETOSPHERES INTERACT
WITH SATELLITES AND RINGS, AND VICE VERSA?

The co-rotating charged particles of outer planet magnetospheres interact with the surfaces and atmospheres of the moons and rings embedded in the magnetospheres and erode them. The sputtered volatile molecules and dust particles are ionized by solar extreme ultraviolet light and electron impact processes and get picked up by the magnetospheric plasma. Three of the giant planet magnetospheres (Jupiter, Saturn, and Neptune) derive most of their plasma from their satellites (Io, Enceladus, and Triton, respectively). Much of the kinetic and thermal energy of the magnetospheric plasma in the outer planet magnetospheres is derived from the rotations of their parent bodies. How planetary ionospheres lose their angular momentum to the outflowing plasma is a topic of great interest in both understanding the dynamics and energetics of the current day magnetospheres and the angular momentum budget histories of the giant planets.

Studying the magnetospheric interaction with orbiting moons and rings can elucidate the mechanisms behind the evolutions of their surfaces and provide insight into the gain and loss of plasma by the magnetospheres and the sculpting of their radiation belts. Magnetospheric interaction is the dominant mechanism by which material is exchanged between satellites. The rotating magnetic fields of giant planets and their magnetospheres also provide cyclical sounding signals that can be used to probe the interiors of the satellites for electrically conductive liquid water and magma oceans.

Q8.4a How Do Magnetospheric Magnetic Fields, Neutral Atoms and Molecules,
Plasma, and Dust Populations Interact with Moons and Rings?

The magnetospheres of the giant planets provide a time-variable magnetic environment at the orbiting moons. The periods and amplitudes of these magnetic waves depend on the rotation of the planet, the intensity and tilt of the planet's intrinsic magnetic field, magnetospheric morphology, and the orbital geometry of the moons (e.g., eccentricity and inclination of the orbits). These magnetic waves and their harmonics can generate induced dipole magnetic fields in moons harboring global subsurface water or magma oceans. Careful analysis of these induction fields can characterize the oceans in terms of their depths, thicknesses, conductivities, and potentially, fluid circulations (Vance et al. 2021).

The interaction of the magnetospheric plasma and fields with orbiting moons affect the local structure and strength of the magnetic fields at the moons through several contributions from the planetary magnetic field, the plasma dynamic interaction of the magnetosphere with the moon and its ionosphere, the induced magnetic moment from the magnetic waves driving electrical currents in subsurface oceans, the induced magnetic fields sourced from oceanic fluid motion, and any magnetic fields intrinsic to the moon (as with Ganymede). The magnetospheric plasma density, plasma velocity, and magnetic field strength local to the moons determine the nature of the interaction with the moon. These parameters define the local Alfvén speed, and the local flow of the magnetized plasma determines if the interaction is sub-, trans-, or super-Alfvénic. The stability and organization of the upstream flow at the moons may be predictable (as in the cases of the Galilean moons) or may be highly variable (as is the case with Titan) depending on where in the magnetosphere the moons reside. Depending on the energy spectra of the incident ions and electrons and the local magnetic field geometry, some fraction of the magnetospheric plasma may access the exosphere and the surfaces of the moons. This process of plasma precipitation can lead to local aurorae, ionization of the atmosphere and formation of localized ionospheric populations, and implantation, sputtering, and radiolytic processing of the surface. This same process can structure the planet's radiation belts (Figure 11-5, top panel). Moreover, these processes can vary over geologic timescales if the host planet's intrinsic magnetic field changes in intensity and/or direction. Such changes may be reflected in the surface characteristics of the moons and rings.

The magnetospheric environment surrounding dense rings determines whether vapor and small charged particles released from the rings remain trapped in the vicinity of the rings or are transported away from the rings into the planet. The environment therefore strongly affects how quickly massive rings (like those of Saturn) evolve compositionally and erode over time. Cassini measurements revealed that large amounts of particles and vapor are flowing between the rings and the planet, but only directly sampled a small part of the

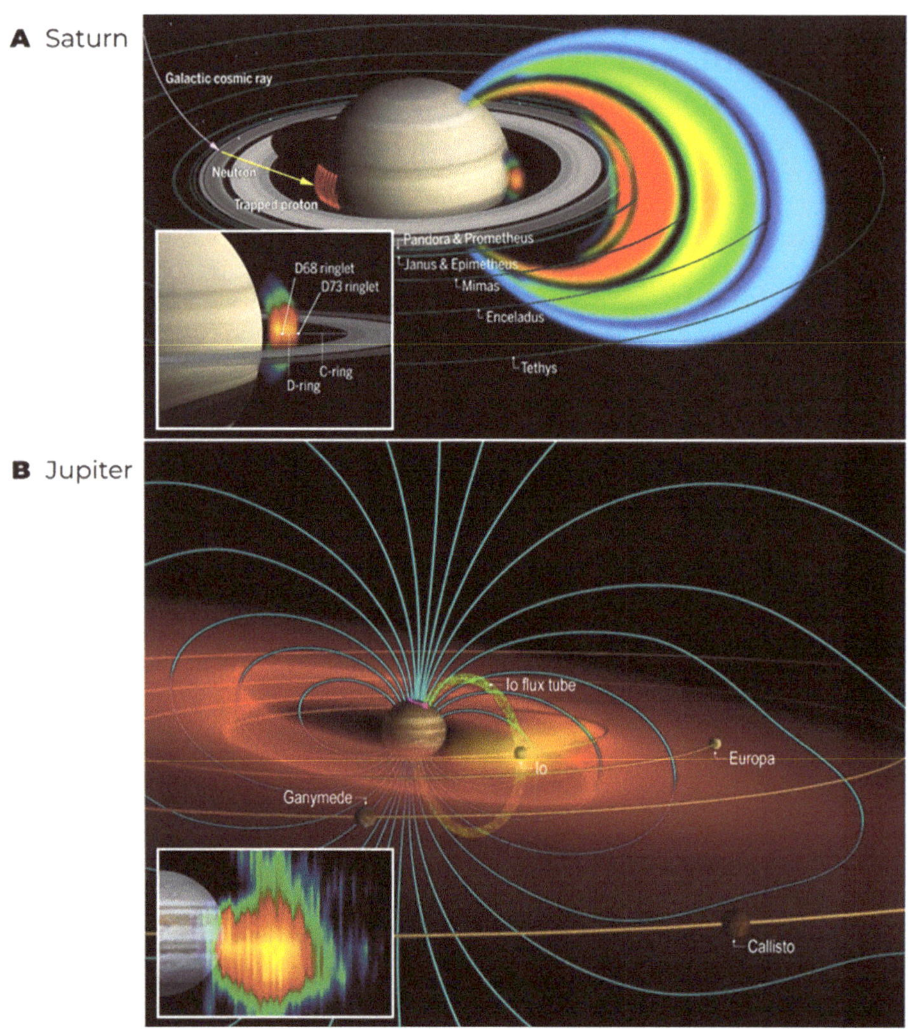

FIGURE 11-5 *Top panel*: Saturn's radiation belts as structured by the rings and inner moons, with gaps owing to the absorption of energetic protons by the moons and rings, and with the highest-energy (giga-electron volt [GeV]) protons confined to radial distances inside the main rings (inset). *Bottom panel*: Jupiter's magnetosphere with Io's volcanic activity as the primary magnetospheric plasma source. Red indicates the Io plasma torus, yellow indicates Io's neutral cloud, green indicates the Io flux tube, cyan lines indicate magnetic field lines, and purple indicates aurorae on Jupiter. The inset shows jovian radiation belt synchrotron emissions from relativistic electrons as observed by Cassini. SOURCES: *Top panel*: From E. Roussos, P. Kollmann, N. Krupp, et al., 2018, "A Radiation Belt of Energetic Protons Located Between Saturn and Its Rings," *Science* 362:aat1962, https://doi.org/10.1126/science.aat1962. Reprinted with permission from AAAS. *Bottom panel*: From J. Spencer, Southwest Research Institute; inset image courtesy of NASA/JPL and S.J. Bolton, M. Janssen, R. Thorne, et al., 2002, "Ultra-Relativistic Electrons in Jupiter's Radiation Belts," *Nature* 415:987-991. https://doi.org/10.1038/415987a, reproduced with permission from SNCSC.

rings' environment, leaving open many questions about the structure and magnitude of this flow, especially given that the plasma environment surrounding the rings almost certainly changes with seasons and local time. More detailed information about the rings' magnetospheric environment is needed to properly estimate the current evolution rates of these dense rings.

The dust-sized particles that form more tenuous rings like Saturn's E-ring or Uranus's μ-ring are strongly perturbed by nongravitational forces, so the plasma environment strongly affects both the particle size distribution and morphology of these rings. Similarly, both the prevalence and evolution of the dusty spokes over Saturn's B-ring are clearly influenced by seasonal changes in the magnetospheric environment. These diverse systems provide opportunities to better understand the sources, sinks and transport rates of small particles in various contexts, ultimately informing how dust coats icy moons (Q8.3d). Furthermore, these insights can shed light on how other dusty systems operate, including the complex debris disks recently seen around distant stars.

Q8.4b How Do Moons and Rings Contribute to the Neutral, Plasma, and Dust Populations Surrounding Their Host Planets and Influence the Dynamics of the Magnetosphere?

Moons can contribute to the dust, neutral, and plasma environment in the magnetosphere in direct and indirect ways. Volcanic moons like Io, which reside in high-radiation environments, can provide more than 1,000 kg/s of gas that is rapidly ionized and creates a plasma torus in the inner magnetosphere (Figure 11-5 bottom panel; Bagenal and Dols 2020; de Pater et al. 2021). In contrast, the plumes of Enceladus provide ~500 kg/s of water vapor and ice grains to Saturn's orbital environment, sourcing the E-ring and an extended neutral cloud that slowly ionized and provides a distributed cold source of plasma to the magnetosphere. Europa contributes sputtered neutral atoms and molecules to a comparably less extensive but stable and observable neutral cloud at Jupiter. At both Jupiter and Saturn, the location and extent of the neutral and plasma sources shape radial mass transport in the magnetosphere, determining the edge of its inner portion that co-rotates with the planet, and, at least in part, the structure and intensity of the planetary aurora.

The inward diffusion of plasma in a magnetosphere leads to the formation of energetic (mega-electron volt, MeV) ion and electron radiation belts. The rings and moons of the jovian and saturnian system sculpt these radiation belts by absorbing these particles. An additional source of energetic protons very close to the planet is the cosmic ray albedo neutron decay process that creates MeV protons from fast neutrons generated from the bombardment of atmospheres of the planets and their rings by cosmic rays. Electric fields related to wave processes, convection of plasma (bursty flows), and temporal variations of the ambient magnetic field are required to energize and facilitate inward convection of charged particles, however the details of these processes are poorly understood. Plasma waves also lead to the losses of these particles into the atmosphere of the planet as they scatter the stably trapped bound belt populations into trajectories that can enter the atmosphere.

Q8.4c How Does "Ring Rain" Affect Planetary Atmospheres and the Circumplanetary Environment?

Observations from both ground-based telescopes and the Cassini spacecraft have revealed that the rings, and material ejected from Enceladus, have profound effects on the environment around Saturn. Most dramatically, as much as 10,000 kilograms of ions, molecules, and small dust grains appear to be flowing from the rings into the planet every second. Some material flows into the mid-latitude ionosphere along magnetic field lines, some sediments downward over the equator. This flow affects not only the long-term evolution of the rings, but also the composition of the upper atmosphere by introducing large amounts of water and organics into regions where these molecules would otherwise be rare. Oxygenated materials may contribute to photochemistry, haze production, and the radiative balance of Saturn's middle and upper atmosphere. This exogenic contamination of Saturn's atmosphere may even complicate attempts to determine Saturn's bulk composition remotely, a key measurement for understanding planetary origins. The discovery of Saturn's substantial planet–ring connection begs the question of whether a similar connection exists on the other giant planets, particularly for the substantial rings of Uranus and Neptune.

Material flow between the rings and planet also affects the configuration, composition, and plasma density of the inner magnetosphere, producing a highly interconnected system. The complexity of this system is demonstrated by the perturbations to the magnetic fields measured when Cassini flew between the rings and the planet, as well as the clear influence of magnetospheric asymmetries on multiple dust populations around the main rings. Detailed examinations of this region promise new insights into material transport and electromagnetic connections within planetary magnetospheres.

Strategic Research for Q8.4

- **Quantify material sources, sinks, and mass transport between Jupiter's magnetosphere and moons** via in situ magnetic field and plasma measurements.
- **Characterize the magnetospheric interactions between Uranus and Neptune's atmosphere, moons, and rings** with observations of planetary aurorae, measurements of satellite and ring exospheres and ionospheres, in situ measurements of the global distribution of the plasma composition, density, velocity and temperature, neutral density and composition, and charged dust in both the orbital plane of the moons and rings and at high inclinations.
- **Characterize how Saturn's rings, neutral clouds, and charged dust are coupled with Saturn's magnetosphere and atmosphere** with high-resolution, high-cadence observations of Saturn's rings, including imaging, composition, magnetic field, and plasma measurements.
- **Develop a more complete framework for integrating and assimilating sparse plasma and magnetic field observations and determining the interior structures of planetary satellites (including recovery of ocean depth, thickness, salinity, and dynamics), and ionospheric/magnetospheric processes (including plasma interactions, ion-neutral, mass-momentum transport, field-aligned current, and induced/ intrinsic magnetic fields)** with advanced analytical and numerical approaches.

Q8.5 HOW DO RINGS EVOLVE AND COALESCE INTO MOONS?

The solar system contains numerous collections of many small objects in orbit around larger bodies in the form of either small moons or planetary rings (Tiscareno and Murray 2018). Recent work has shown that both individual objects and the overall distribution of material within these systems have complex and dynamic histories. At the individual particle level, objects can either accrete into larger bodies or erode away, while on larger scales material can either be transported or confined by a wide variety of internal processes and external forces. The two most important processes involved in these phenomena are collisions and mutual gravitational attraction, both of which have complex and context-dependent outcomes. Similarly, it is not yet clear how fast particle populations spread and disperse under different situations. Hence major uncertainties and controversies remain about the conditions needed for colliding particles to accrete into larger bodies or to break apart into smaller fragments, and about the current evolution rates and even the overall history of these systems of rings and small moons. Improving our understanding would also clarify the formation and early evolution of planetary systems (including the solar system), because it too is/was shaped by particle accretion, transport, and erosion. For example, protoplanets in the protoplanetary disk resemble moonlets and other objects forming now in planetary rings. These embedded objects perturb the disk to create gaps and edges and can trigger accretion. The disk reacts back on the embedded objects to cause them to migrate.

Q8.5a What Determines the Distribution of Particle Properties Within Rings and Other Systems of Small Bodies?

Most parts of dense planetary rings are evolving over timescales that are long compared to their orbital periods, and so in these regions the distributions of particle sizes, densities, relative velocities, and rotation states can be approximated as a nearly steady-state system where smaller particles are being assembled into larger bodies at roughly the same rate as larger bodies are being broken apart (Cuzzi et al. 2009; Colwell et al. 2018). Important aspects of collisional outcomes among these bodies can therefore be determined by quantifying variations and trends in the particle property distributions across the rings. For example, correlations between the typical particle size and the mean relative velocity can constrain the critical impact speeds where aggregation transitions to fragmentation. On the other hand, more dynamic regions of the rings where parameters like density or vertical extent change rapidly owing to external forces can reveal how changes in one aspect of the dynamical environment can affect other particle properties. Furthermore, the magnitude of the rings' response to the driving forces can reveal nonlinear processes or instabilities relevant to the formation of larger structures. Meanwhile, the properties of more tenuous ring systems dominated by fine debris released from larger bodies can provide information about dust production, transport, and loss in different environments (Hedman et al. 2018).

Q8.5b How Are Ring Particles Assembled into Moons and More Transient Aggregates?

Cassini observations of a variety of structures within the rings (Figure 11-6) revealed that Saturn's rings contain a population of exceptionally large particles (up to several hundred meters in diameter, Spahn et al. 2018) that can be regarded as transitional between typical ring particles (which range between millimeters and a few meters in size) and small moons (which are several kilometers across). These objects may be analogous to growing objects in proto-planetary disks. At the same time, other observations show that different parts of these rings contain a wide variety of transient agglomerations of particles that may form quickly and last for only a few orbit periods. Such agglomerations can explain the overall brightness of the ring in different viewing geometries (Schmidt et al. 2009; Salo et al. 2018; Tiscareno et al. 2019). The formation of transient aggregates and more permanent larger bodies are likely related processes, and so we may be able to better understand how particles accrete into larger bodies if we can determine under what conditions different types of transient aggregates form, as well as the sizes, lifetimes, and ultimate fates of the larger objects.

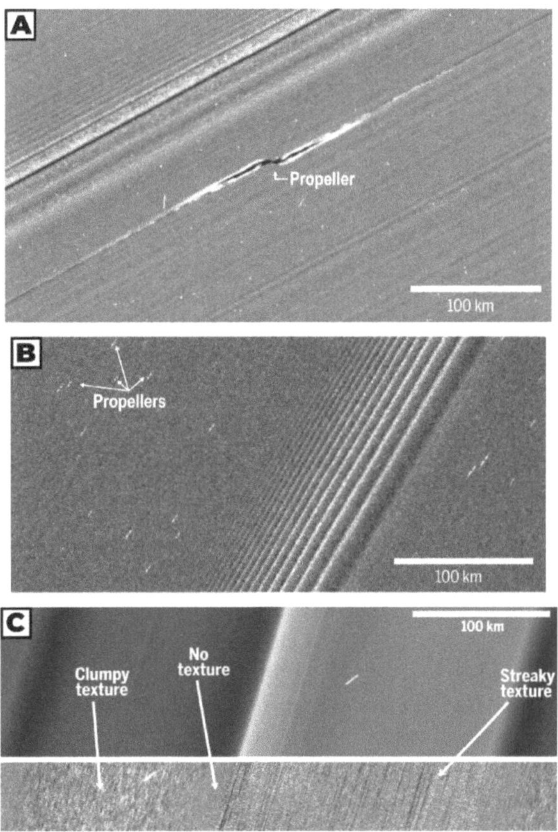

FIGURE 11-6 Images of selected structures and textures in Saturn's rings. *Top panel*: an example of a large "propeller" structure surrounding a body hundreds of meters across. *Middle panel*: a fleet of smaller propellers. *Bottom panel*: a high-resolution image of part of the rings containing various textures that can be seen clearly in the high-pass-filtered portion of the lower portion of the figure. SOURCE: Adapted from M.S. Tiscareno, P.D. Nicholson, J.N. Cuzzi, et al., 2019, "Close-Range Remote Sensing of Saturn's Rings During Cassini's Ring-Grazing Orbits and Grand Finale," *Science* 364(6445):eaau1017, https://www.science.org/doi/10.1126/science.aau1017. Reprinted with permission from AAAS.

Q8.5c What Is the Life Cycle of Planetary Rings?

Recent measurements of the total mass of Saturn's rings and the mass flux between those rings and the planet indicate that the composition and mass of Saturn's rings may change substantially on timescales of hundreds of millions of years, which is much shorter than the age of the solar system (Waite et al. 2018; Crida et al. 2019; Iess et al. 2019; O'Donoghue et al. 2019). The histories of rings and moons seem to be closely intertwined. Theoretical studies have shown that several of Saturn's and Uranus's moons (including the innermost mid-sized moons Mimas and Miranda) may have spawned from the rings well after the solar system formed (Charnoz et al. 2018; Hesselbrock and Minton 2019; Neveu and Rhoden 2019). Even Mars's moon Phobos may be the latest incarnation of material that has cycled between rings and a moon multiple times over the planet's history (Hesselbrock and Minton 2017). The origin of Mars's moons—be it from capture of passing asteroids or accretion from a disk of material in Mars orbit—is still uncertain (e.g., Ramsley and Head 2021). Material in the rings can be perturbed and confined by nearby shepherding moons and mean-motion resonances with more distant bodies to produce arcs, gaps and other features that evolve over timescales of years to decades. Hence, it may be possible to use the structure, composition, and shape of ring-generated moons to ascertain how large objects grow from rings and, based on how quickly the structure and composition of the rings change over time, to constrain the timescales relevant to the evolution of these systems.

Q8.5d Which Worlds in the Solar System and Beyond Have (or Had) Rings?

The past decade revealed that giant planets are not the only objects in the solar system with rings. The recent discovery of rings around small bodies in the outer solar system, along with evidence that several of Saturn's moons might have had rings in the past, has greatly expanded the known environments where rings can occur (Ortiz et al. 2017; Sicardy et al. 2018). However, relatively few small bodies have rings, and it is not yet clear what conditions favor the existence of rings around a given body. Additionally, some worlds may have once possessed rings that were subsequently lost, like Mars (whose moons Phobos and Deimos may have originated from a ring), or Iapetus (whose prominent equatorial ridge is hypothesized to be a collapsed ring). Measuring the properties of these rings would help us better understand the conditions and/or materials that favor ring formation. This information will not only help us to better understand the solar system, but also clarify which planets outside the solar system are likely to have rings.

Strategic Research for Q8.5

- **Determine the prevalence of rings and satellites around small bodies, including Centaurs and transneptunian objects, to understand how circumplanetary systems form and evolve** with high-resolution telescopic observations and occultation campaigns, in situ exploration, and other methods.
- **Determine the composition of Uranus and Neptune's rings and small moons to elucidate their origin, evolution, and present-day balance between exogenic processes and endogenic processes** through a combination of geophysical measurements, imaging, and spectroscopic observations, including at high spatial resolution sufficient to resolve regional variations and layering.
- **Determine the locations and distributions of source bodies within Uranus's and Neptune's dusty rings, and how fine particles are generated, lost, and transported throughout the uranian and neptunian systems** with a combination of high-resolution imaging and measurements of the mass flux and composition flowing into and out of the rings.
- **Elucidate the origin of Jupiter, Saturn, Uranus and Neptune's small regular satellites and ring-moons, and their relationship and interactions with their rings,** by measuring their composition and structure.
- **Characterize the present-day evolution of Jupiter, Saturn, Uranus, and Neptune's rings** by measuring the mass flux and composition flowing into and out of the rings.
- **Observe how particles in dense rings around Saturn, Uranus, and other objects aggregate into larger bodies and fragment into smaller bodies** by observing dense rings with sufficient spatial resolution and imaging cadence to resolve the individual ring particles and aggregates and their evolution over various dynamical timescales (~minutes to observe individual collisions, ~years to observe their orbital evolution).

- **Constrain the origin of Phobos and Deimos, including whether they arose from past martian rings,** by determining their bulk composition and interior structure by in situ geochemical and geophysical measurements.
- **Determine whether equatorial ridges on worlds like Iapetus, Pan, and Atlas are produced by the deposition of ancient rings or by other processes,** with a combination of high-resolution remote sensing observations of equatorial ridges, and theoretical models for ring collapse and other competing hypotheses.
- **Determine the long-term evolution of planetary rings** with high-resolution imaging, stellar occultation techniques, and computer simulations of their orbital evolution and dynamics.
- **Quantify the evolution of ring structures (including arcs, gaps, and edges) over timescales of years to decades** using stellar occultations and high-resolution images.
- **Determine the collisional properties of ice as relevant to understanding impact processes on icy satellites, and the aggregation/disruption of ring particles** with laboratory studies of ice under outer solar system conditions.

SUPPORTIVE ACTIVITIES FOR QUESTION 8

- Determine how planetary materials with relevant compositions and melt fractions behave under the temperatures, pressures, and forcing conditions relevant to understanding processes within circumplanetary bodies—including tidal dissipation, convection, melting and melt transport, and magnetic induction, with laboratory and ab initio studies of planetary materials.

REFERENCES

Asphaug, E., and A. Reufer. 2013. "Late Origin of the Saturn System." *Icarus* 223:544–565. https://doi.org/10.1016/j.icarus.2012.12.009.

Bagenal, F., and V. Dols. 2020. "The Space Environment of Io and Europa." *Journal of Geophysical Research: Space Physics* 125(5):e2019JA027485. https://doi.org/10.1029/2019JA027485.

Běhounková, M., O. Souček, J. Hron, and O. Čadek. 2017. "Plume Activity and Tidal Deformation on Enceladus Influenced by Faults and Variable Ice Shell Thickness." *Astrobiology* 17:941–954. https://doi.org/10.1089/ast.2016.1629.

Beuthe, M. 2013. "Spatial Patterns of Tidal Heating." *Icarus* 223:308–329. https://doi.org/10.1016/j.icarus.2012.11.020.

Bolton, S.J., M. Janssen, R. Thorne, S. Levin, M. Klein, S. Gulkis, T. Bastian, R. Sault, C. Elachi, M. Hofstadter, et al. 2002. "Ultra-Relativistic Electrons in Jupiter's Radiation Belts." *Nature* 415:987–991. https://doi.org/10.1038/415987a.

Brown, M.E. 2012. "The Compositions of Kuiper Belt Objects." *Annual Review of Earth and Planetary Sciences* 40:467–494. https://doi.org/10.1146/annurev-earth-042711-105352.

Cable, M.L., M. Neveu, H.-W. Hsu, T.M. Hoehler, and R. Dotson. 2020. "Enceladus." Pp. 217–246 in *Planetary Astrobiology*, V.S. Meadows, G.N. Arney, B.E. Schmidt, and D.J. Des Marais, eds. Tucson: University of Arizona Press.

Canup, R.M. 2010. "Origin of Saturn's Rings and Inner Moons via Mass Removal from a Lost Titan-Sized Satellite." *Nature* 468:943–946.

Charnoz, S., A. Crida, J.C. Castillo-Rogez, V. Lainey, L. Dones, Ö. Karatekin, G. Tobie, S. Mathis, C. Le Poncin-Lafitte, and J. Salmon. 2011. "Accretion of Saturn's Mid-Sized Moons During the Viscous Spreading of Young Massive Rings: Solving the Paradox of Silicate-Poor Rings Versus Silicate-Rich Moons." *Icarus* 216:535–550. https://doi.org/10.1016/j.icarus.2011.09.017.

Charnoz, S., R.M. Canup, A. Crida, and L. Dones. 2018. "The Origin of Planetary Ring Systems." Pp. 517–538 in *Planetary Ring Systems: Properties, Structure, and Evolution*, M.S. Tiscareno and C.D. Murray, eds. Cambridge, UK: Cambridge University Press. https://doi.org/10.1017/9781316286791.018.

Che, X., A. Nemchin, D. Liu, T. Long, C. Wang, M.D. Norman, K.H. Joy, et al. 2021. "Age and Composition of Young Basalts on the Moon, Measured from Samples Returned by Chang'e-5." *Science* 374:887–890. https://doi.org/10.1126/science.abl7957.

Collins, G.C., W.B. McKinnon, J.M. Moore, F. Nimmo, R.T. Pappalardo, L.M. Prockter, and P.M. Schenk. 2009. "Tectonics of the Outer Planet Satellites." Pp. 264–350 in *Planetary Tectonics*, T.R. Watters and R.A. Schultz, eds. Cambridge, UK: Cambridge University Press. https://doi.org/10.1017/CBO9780511691645.

Colwell, J.E., J. Blum, R.N. Clark, S. Kempf, and R.M. Nelson. 2018. "Laboratory Studies of Planetary Ring Systems." Pp. 494–516 in *Planetary Ring Systems: Properties, Structure, and Evolution*, M.S. Tiscareno and C.D. Murray, eds. Cambridge, UK: Cambridge University Press. https://doi.org/10.1017/9781316286791.017.

Crida, A., S. Charnoz, H.W. Hsu, and L. Dones. 2019. "Are Saturn's Rings Actually Young?" *Nature Astronomy* 3:967–970. https://doi.org/10.1038/s41550-019-0876-y.

Cuzzi, J., R. Clark, G. Filacchione, R. French, R. Johnson, E. Marouf, and L. Spilker. 2009. "Ring Particle Composition and Size Distribution." Pp. 459–509 in *Saturn from Cassini-Huygens*, M. Dougherty, L.W. Esposito, and S. Krimigis, eds. Dordrecht, Netherlands: Springer. https://doi.org/10.1007/978-1-4020-9217-6_15.

de Kleer, K., and I. de Pater. 2016. "Spatial Distribution of Io's Volcanic Activity from Near-IR Adaptive Optics Observations on 100 Nights in 2013–2015." *Icarus* 280:405–414. https://doi.org/10.1016/j.icarus.2016.06.018.

de Kleer, K., A.S. McEwen, R.S. Park, C.J. Bierson, A.G. Davies, D.N. DellaGiustina, A.I. Ermakov, et al. 2019a. *Tidal Heating: Lessons from Io and the Jovian System*. Final Report for the Keck Institute for Space Studies. https://www.kiss.caltech.edu/final_reports/Tidal_Heating_final_report.pdf.

de Kleer, K., F. Nimmo, and E. Kite. 2019b. "Variability in Io's Volcanism on Timescales of Periodic Orbital Changes." *Geophysical Research Letters* 46:6372–6332. https://doi.org/10.1029/2019GL082691.

de Pater, I., J.T. Keane, K. de Kleer, and A.G. Davies 2021. "A 2020 Observational Perspective of Io." *Annual Review of Earth and Planetary Sciences* 49:643–678. https://doi.org/10.1146/annurev-earth-082420-095244.

Geissler, P. 2015. "Chapter 44—Cryovolcanism in the Outer Solar System." Pp. 763–776 in *The Encyclopedia of Volcanoes*, 2nd ed., H. Sigurdsson, ed. Cambridge, MA: Academic Press. https://doi.org/10.1016/B978-0-12-385938-9.00044-4.

Gerrick-Bethell, I., V. Perera, F. Nimmo, and M.T. Zuber. 2014. "The Tidal-Rotational Shape of the Moon and Evidence for Polar Wander." *Nature* 512:181–184. https://doi.org/10.1038/nature13639.

Head III, J.W., and L. Wilson. 1992. "Lunar Mare Volcanism: Stratigraphy, Eruption Conditions, and the Evolution of Secondary Crusts." *Geochimica et Cosmochimica Acta* 56(6):2155–2175. https://doi.org/10.1016/0016-7037(92)90183-J.

Hedman, M.M., C.M. Gosmeyer, P.D. Nicholson, C. Sotin, R.H. Brown, R.N. Clark, K.H. Baines, et al. 2013. "An Observed Correlation Between Plume Activity and Tidal Stresses on Enceladus." *Nature* 500:182–184. https://doi.org/10.1038/nature12371.

Hedman, M.M., F. Postberg, D.P. Hamilton, S. Renner, and H.-W. Hsu. 2018. "Dusty Rings." Pp. 308–337 in *Planetary Ring Systems: Properties, Structure, and Evolution*, M.S. Tiscareno and C.D. Murray, eds. Cambridge, UK: Cambridge University Press. https://doi.org/10.1017/9781316286791.012.

Hemingway, D., L. Iess, R. Tajeddine, and G. Tobie. 2018. "The Interior of Enceladus." Pp. 57–77 in *Enceladus and the Icy Moons of Saturn*, P.M. Schenk, R.N. Clark, C.J.A. Howett, A.J. Verbiscer, and J.H. Waite, eds. Tucson: University of Arizona Press.

Hesselbrock, A.J., and D.A. Minton. 2017. "An Ongoing Satellite–Ring Cycle of Mars and the Origins of Phobos and Deimos." *Nature Geoscience* 10:266–269. https://doi.org/10.1038/ngeo2916.

Hesselbrock, A.J., and D.A. Minton. 2019. "Three Dynamical Evolution Regimes for Coupled Ring-Satellite Systems and Implications for the Formation of the Uranian Satellite Miranda." *Astronomical Journal* 157:30. https://doi.org/10.3847/1538-3881/aaf23a.

Hussmann, H., and T. Spohn. 2004. "Thermal-Orbital Evolution of Io and Europa." *Icarus* 171(2):391–410. https://doi.org/10.1016/j.icarus.2004.05.020.

Iess, L., B. Militzer, Y. Kaspi, P. Nicholson, D. Durante, P. Racioppa, A. Anabtawi, et al. 2019. "Measurement and Implications of Saturn's Gravity Field and Ring Mass." *Science* 364:aat2965. https://doi.org/10.1126/science.aat2965.

Journaux, B., K. Kalousová, C. Sotin, G. Tobie, S. Vance, J. Saur, O. Bollengier, et al. 2020. "Large Ocean Worlds with High-Pressure Ices." *Space Science Reviews* 216(1):1–36. https://doi.org/10.1007/s11214-019-0633-7.

Keane, J.T., and I. Matsuyama. 2014. "Evidence for Lunar True Polar Wander and a Past Low-Eccentricity, Synchronous Lunar Orbit." *Geophysical Research Letters* 41(19):6610–6619. https://doi.org/10.1002/2014GL061195.

Kegerreis, J.A., L.F.A. Teodoro, V.R. Eke, R.J. Massey, D.C. Catling, C.L. Fryer, D.G. Korycansky, M.S. Warren, and K.J. Zahnle. 2018. "Consequences of Giant Impacts on Early Uranus for Rotation, Internal Structure, Debris, and Atmospheric Erosion." *Astrophysical Journal* 861:52. https://doi.org/10.3847/1538-4357/aac725.

Khurana, K.K., X. Jia, M.G. Kivelson, F. Nimmo, G. Schubert, and C.T. Russell. 2011. "Evidence of a Global Magma Ocean in Io's Interior." *Science* 332:1186. https://doi.org/10.1126/science.1201425.

Kivelson, M.G., K.K. Khurana, C.T. Russell, R.J. Walker, J. Warnecke, F.V. Coroniti, C. Polanskey, et al. 1996. "Discovery of Ganymede's Magnetic Field by the Galileo Spacecraft." *Nature* 384(6609):537–541. https://doi.org/10.1038/384537a0.

Moore, W.B., J.I. Simon, and A.A.G. Webb. 2017. "Heat-Pipe Planets." *Earth and Planetary Science Letters* 474:13–19. https://doi.org/10.1016/j.epsl.2017.06.015.

Neveu, M., and A.R. Rhoden. 2019. "Evolution of Saturn's Mid-Sized Moons." *Nature Astronomy* 3:543–552. https://doi.org/10.1038/s41550-019-0726-y.

Nimmo, F., and R.T. Pappalardo. 2016. "Ocean Worlds in the Outer Solar System." *Journal of Geophysical Research: Planets* 121:1378–1399. https://doi.org/10.1002/2016JE005081.

O'Donoghue, J., L. Moore, J. Connerney, H. Melin, T.S. Stallard, S. Miller, and K.H. Baines. 2019. "Observations of the Chemical and Thermal Response of 'Ring Rain' on Saturn's Ionosphere." *Icarus* 322:251–260. https://doi.org/10.1016/j.icarus.2018.10.027.

Ojakangas, G.W., and D.J. Stevenson. 1989. "Thermal State of an Ice Shell on Europa." *Icarus* 81(2):220–241. https://doi.org/10.1016/0019-1035(89)90052-3.

Ortiz, J.L., P. Santos-Sanz, B. Sicardy, G. Benedetti-Rossi, D. Bérard, N. Morales, R. Duffard, et al. 2017. "The Size, Shape, Density and Ring of the Dwarf Planet Haumea from a Stellar Occultation." *Nature* 550:219–223. https://doi.org/10.1038/nature24051.

Peale, S.J., P. Cassen, and R.T. Reynolds. 1979. "Melting of Io by Tidal Dissipation." *Science* 203:892–894. https://doi.org/10.1126/science.203.4383.892.

Ramsley, K.R., and J.W. Head III. 2021. "The Origins and Geological Histories of Deimos and Phobos: Hypotheses and Open Questions." *Space Science Reviews* 217(86):1–35. https://doi.org/10.1007/s11214-021-00864-1.

Rhoden, A.R., K.J. Mohr, T.A. Hurford, W. Henning, S. Sajous, D.A. Patthoff, and D. Dubois. 2021. "Obliquity, Precession, and Fracture Mechanics: Implications of Europa's Global Cycloid Population." *Journal of Geophysical Research: Planets* 126:e06710. https://doi.org/10.1029/2020JE006710.

Roussos, E., P. Kollmann, N. Krupp, A. Kotova, L. Regoli, C. Paranicas, D.G. Mitchell, et al. 2018. "A Radiation Belt of Energetic Protons Located Between Saturn and Its Rings." *Science* 362:aat1962. https://doi.org/10.1126/science.aat1962.

Salo, H., K. Ohtsuki, and M.C. Lewis. 2018. "Computer Simulations of Planetary Rings." Pp. 434–493 in *Planetary Ring Systems: Properties, Structure, and Evolution*, M.S. Tiscareno and C.D. Murray, eds. Cambridge, UK: Cambridge University Press. https://doi.org/10.1017/9781316286791.016.

Scheinberg, A., R.R. Fu, L. Elkins-Tanton, B.P. Weiss, and S. Stanley. 2017. "Magnetic Fields on Asteroids and Planetesimals." Pp. 180–203 in *Planetesimals: Early Differentiation and Consequences for Planets*, L.T. Elkins-Tanton and B.P. Weiss, eds. Cambridge, UK: Cambridge University Press.

Schmidt, B.E. 2020. "The Astrobiology of Europa and the Jovian System." P. 185 in *Planetary Astrobiology*, V.S. Meadows, G.N. Arney, B.E. Schmidt, and D.J. Des Marais, eds. Tucson: University of Arizona Press. https://doi.org/10.2458/azu_uapress_9780816540068.

Schmidt, J., K. Ohtsuki, N. Rappaport, H. Salo, and F. Spahn. 2009. "Dynamics of Saturn's Dense Rings." P. 413 in *Saturn from Cassini-Huygens*, M. Dougherty, L.W. Esposito, and S. Krimigis, eds. Tucson: University of Arizona Press. https://doi.org/10.1007/978-1-4020-9217-6_14.

Sekine, Y., T. Shibuya, F. Postberg, H.-W. Hsu, K. Suzuki, Y. Masaki, T. Kuwatani, et al. 2015. "High-Temperature Water-Rock Interactions and Hydrothermal Environments in the Chondrite-Like Core of Enceladus." *Nature Communications* 6:1–8. https://doi.org/10.1038/ncomms9604.

Senthil Kumar, P., R. Mohanty, K.J.P. Lakshmi, S.T.G. Raghukanth, A.C. Dabhu, R.P. Rajasekhar, and R. Menon. 2019. "The Seismically Active Lobate Scarps and Coseismic Lunar Boulder Avalanches Triggered by 3 January 1975 (MW 4.1) Shallow Moonquake." *Geophysical Research Letters* 46:7972–7981. https://doi.org/10.1029/2019GL083580.

Sicardy, B., M. El Moutamid, A.C. Quillen, P.M. Schenk, M.R. Showalter, and K. Walsh. 2018. "Rings Beyond the Giant Planets." Pp. 135–154 in *Planetary Ring Systems: Properties, Structure, and Evolution*, M.S. Tiscareno and C.D. Murray, eds. Cambridge, UK: Cambridge University Press. https://doi.org/10.1017/9781316286791.007.

Soderlund, K.M., K. Kalousová, J.J. Buffo, C.R. Glein, J.C. Goodman, G. Mitri, G.W. Patterson, et al. 2020. "Ice-Ocean Exchange Processes in the Jovian and Saturnian Satellites." *Space Science Reviews* 216(5):1–57. https://doi.org/10.1007/s11214-020-00706-6.

Spahn, F., H. Hoffmann, H. Rein, M. Seiss, M. Srem evi , and M.S. Tiscareno. 2018. "Moonlets in Dense Planetary Rings." Pp. 157–197 in *Planetary Ring Systems: Properties, Structure, and Evolution*, M.S. Tiscareno and C.D. Murray, eds. Cambridge, UK: Cambridge University Press. https://doi.org/10.1017/9781316286791.008.

Tiscareno, M.S., and C.D. Murray, eds. 2018. *Planetary Ring Systems: Properties, Structure, and Evolution*, M.S. Tiscareno and C.D. Murray, eds. Cambridge, UK: Cambridge University Press. https://doi.org/10.1017/9781316286791.

Tiscareno, M.S., P.D. Nicholson, J.N. Cuzzi, L.J. Spilker, C.D. Murray, M.M. Hedman, J.E. Colwell, et al. 2019. "Close-Range Remote Sensing of Saturn's Rings During Cassini's Ring-Grazing Orbits and Grand Finale." *Science* 364:aau1017. https://doi.org/10.1126/science.aau1017.

Vance, S.D., M.J. Styczinski, B.G. Bills, C.J. Cochrane, K.M. Soderlund, N. Gómez-Pérez, and C. Paty. 2021. "Magnetic Induction Responses of Jupiter's Ocean Moons Including Effects from Adiabatic Convection." *Journal of Geophysical Research: Planets* 126(2):e2020JE006418. https://doi.org/10.1029/2020JE006418.

Waite, J.H., R.S. Perryman, M.E. Perry, K.E. Miller, J. Bell, T.E. Cravens, C.R. Glein, et al. 2018. "Chemical Interactions Between Saturn's Atmosphere and Its Rings." *Science* 362:aat2382. https://doi.org/10.1126/science.aat2382.

Wieczorek, M.A., B.P. Weiss, D. Breuer, M. Fuller, J. Gattacceca, J. Halekas, L. Hood, et al. 2017. "Recent Advances in Lunar Magnetism." In *New Views of the Moon 2—Europe*. LPI Contributions 1998:6036.

Q9 PLATE: Earth, viewed from above the Moon's surface by the Lunar Reconnaissance Orbiter in 2015.
SOURCE: Courtesy of NASA/GSFC/Arizona State University.

12

Question 9: Insights from Terrestrial Life

What conditions and processes led to the emergence and evolution of life on Earth; what is the range of possible metabolisms in the surface, subsurface, and/or atmosphere; and how can this inform our understanding of the likelihood of life elsewhere?

Astrobiology is a holistic field of research into the origin, evolution, and distribution of life in the universe.[1] As such, planetary science and astrobiology encompass a continuous spectrum spanning the multidisciplinarity of life and physical sciences and investigate the codependence and coevolution of life and the environment. Habitability refers to a set of environmental conditions capable of supporting life (see also Question 10, Chapter 13). Dynamic habitability as defined in recommendations from *An Astrobiology Strategy for the Search for Life in the Universe* recognizes that the combined effects of multiple parameters (e.g., T, P, salinity, pH; Table 12-1) define whether or not life can emerge and persist, and that while one or more parameters may vary outside the canonical limits to life, it is their combined effect that causes an environment to be habitable or not (NASEM 2019a).

There is an extensive literature (and debate) about the major milestones in the history of Earth and the history of life, the nature of the earliest life as discerned from fossils and chemical signatures, and the pathways and timing for the evolution of life and for key metabolisms (Betts et al. 2018; Benner et al. 2020); therefore, the committee does not attempt to summarize them here. Continued research into Earth's physiochemical properties, geologic history, and the evolution of habitability and life through deep time—through theoretical, field, and experimental investigations—informs and inspires strategies for planetary and astrobiological exploration of the solar system and universe.

Earth represents our sole reference point, to date, for what constitutes an inhabited world (Figure 12-1). How prebiotic pathways coevolved with the environment to give rise to life on this planet remains an active area of inquiry. But research in the field and the laboratory, as well as the analysis of astromaterials—that is extraterrestrial materials collected on Earth (e.g., meteorites and cosmic dust) and from other planetary bodies via sample return missions—continue to shed light on the conditions and processes that can lead from abiotic chemical reactions to biochemistry. It is important to note that our understanding of life on Earth continues to evolve rapidly thanks to

---

[1] A glossary of acronyms and technical terms can be found in Appendix F.

309

TABLE 12-1 Extremophiles Nomenclature and Ranges

			Low → High[a]		
pH	Hyperacidophile (<pH 3)	Acidophile (<pH 5)	Neutrophile (pH 5–9)	Alkaliphile (>pH 9)	Hyperalkaliphile (>pH 11)
Temperature		Psychrophile (<20°C)	Mesophile (20–45°C)	Thermophile (45–80°C)	Hyperthermophile (>80°C)
Salinity[b]		Non-halophile (<1.2%)	Halotolerant (1.2–2.9%; tolerate ≤14.6%)	Halophile (>8.8%)	Extreme halophile (>14.6%, cannot grow <8.8%)
Pressure			Piezotolerant or barotolerant (0.1–10 MPa)	Piezophile or barophile (10–50 MPa)	Hyperpiezophile or hyperbarophile (>50 MPa)
Water activity			Xerophile (a_w <0.7)		
Polyextremophile			Tolerance or preference for multiple parameters combined		

[a] The distinction between an extremotolerant microbe and an extremophile is based on the location of the optimum along the specific parameter range. See main text for discussion.
[b] Salinity expressed as percent of NaCl (w/v). Specific resistance to more chaotropic salts has been tested for some strains, for instance in the presence of $MgCl_2$.
SOURCE: Merino et al. (2019), https://doi.org/10.3389/fmicb.2019.00780, CC BY 4.0.

paradigm-changing discoveries of life's ability to survive, and even thrive, in multi-extreme environments. Recent discoveries have demonstrated that life can exist in a myriad of subsurface, icy, and water-limited environments, with a previously unrecognized range of metabolic strategies, and adaptations to both rapidly changing conditions and catastrophic events, as well as to long timescales of quiescence and isolation. This evolving knowledge of Earth's biosphere, from its origins to its current state, directly informs and influences the search for life elsewhere.

The diversity of terrestrial life—and its plasticity to respond to different levels of nutrient and energy availability, transport, and flux—informs the detectability of biospheres that might exist in habitable environments beyond Earth. Insights gained from the biochemistry, structure, and physiology of terrestrial organisms—and how they have evolved through deep time—are the basis to develop comprehensive frameworks to search for evidence of life on other worlds. Those frameworks are informed by our knowledge of terrestrial life, but not limited by the assumption that all life will necessarily follow the terran model (see also Question 10 and Question 11, Chapter 14). Major activities in the coming decade will focus on developing frameworks for robust identification of signs of life, but equally to differentiate life from abiosignatures, false negatives and false positives, and to advance detectability through validation of more comprehensive agnostic signatures of life, irrespective of its origin or molecular makeup.

Q9.1 WHAT WERE THE CONDITIONS AND PROCESSES CONDUCIVE TO THE ORIGIN AND EARLY EVOLUTION OF LIFE ON EARTH, AND WHAT DO THEY TEACH US ABOUT THE POSSIBLE EMERGENCE AND EVOLUTION OF LIFE ON OTHER WORLDS?

Understanding the boundaries between abiotic and biotic systems is an essential part of searching for life elsewhere. These boundaries depend on geochemical context and therefore need to be investigated in a range of potentially relevant environments. With few exceptions, modern Earth settings no longer reflect prebiotic conditions, because biology permeates the planet, destroying or influencing organic chemical reactions and the long-term evolution of Earth's surface, subsurface, and atmosphere. But recent discoveries in subsurface environments wherein products of abiotic organic synthesis can be investigated provide exciting new opportunities to investigate potential prebiotic processes (McDermott et al. 2015; Lang et al. 2018; Sauvage et al. 2021; Sherwood Lollar et al. 2021). Controlled laboratory settings represent another productive strategy, where the influence of biology can be eliminated and the geologic conditions and atmospheric, crustal, and oceanic compositions of early Earth environments can be more closely replicated.

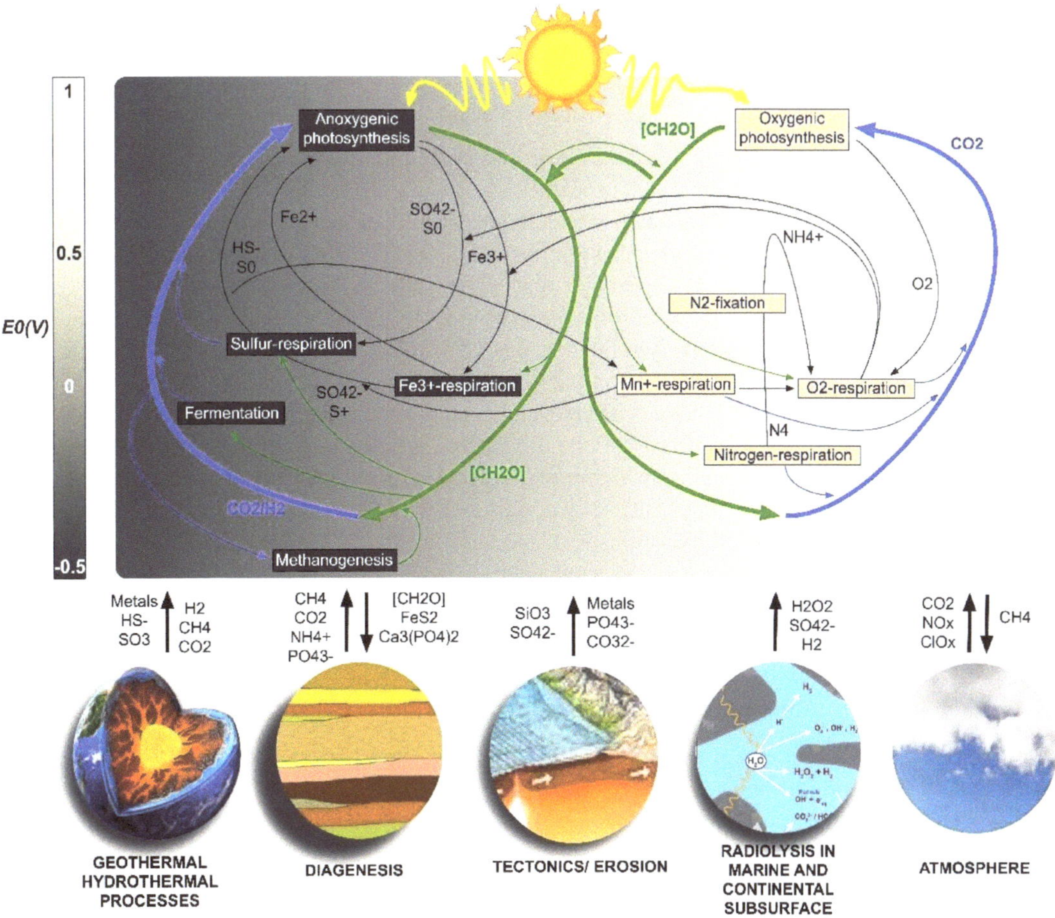

FIGURE 12-1 Generalized model of biogeochemical cycling of energy and essential elements (CHNOPS) on Earth. Abiotic transformations are represented at the bottom, while a subset of biotic transformations, and their connectedness, are represented at the top. Most biotic transformations depend, directly or indirectly, on products of photosynthesis, except for methanogenesis. Abiotic redox reactions from water-rock interactions and radiolysis become progressively more important in the subsurface biosphere. Electron acceptors (oxidants) in the respiratory processes have been arranged from left to right according to increasing capacity to accept electrons. SOURCE: Modified from Falkowski et al. (2008). Images: *Geothermal Hydrothermal* from iStock.com/setixela; *Diagenesis* from Lu et al. (2021), CC BY 4.0; *Tectonics/Erosion* from the U.S. Geological Survey, the Smithsonian Institution, and the U.S. Naval Research Laboratory; *Radiolysis* reprinted from B. Sherwood Lollar, V.B. Heuer, J. McDermott, et al., 2021, "A Window into the Abiotic Carbon Cycle—Acetate and Formate in Fracture Waters in 2.7 Billion Year-Old Host Rocks of the Canadian Shield," *Geochimica et Cosmochimica Acta* 294:295–314, Copyright 2021, with permission from Elsevier; *Atmosphere* from ESA-Thomas Reiter.

The analysis of asteroids, meteorites, comets, and other rocky bodies of the inner solar system can provide key insights into how the chemical inventories (e.g., organics and volatiles) of early Earth may have evolved (NASEM 2019a; see Table 14-1). Astromaterials also offer additional clues regarding the sources, inventories and abundances of key building block elements present during Earth's accretion and its early evolution. The subsequent emergence of life and its evolution from the first self-replicators to the last universal common ancestor of all modern life (LUCA), represents a critical period when geochemistry and biochemistry were likely inseparable. Understanding this interdependence of the earliest forms of life and their environment helps constrain the range of geochemical conditions whence life could emerge.

Q9.1a How Was the Emergence and Evolution of Life on Earth Influenced by Volatiles, Impacts, and Planetary Evolution in Early Solar System Environments?

What were the principal components of the early solar system environment and how did controls on these components (such as available volatile inventory, volatile delivery via impacts and comets, and compositional changes in response to large impacts and planetary evolution) affect the emergence and evolution of life on Earth?

A vital aspect of understanding the physical processes of the early Earth, from its prebiotic state through to the emergence and subsequent evolution of life, is the character, evolution, and abundance of Earth's chemical inventories, especially those deemed essential to terrestrial life (e.g., key building block elements like C, H, N, O, P, S, and redox couples whose relative biochemical importance may have varied in time and space) (also see Question 10). The earliest possible timing for sustainable prebiotic chemistry that led to life's origins followed the moon-forming event and replacement of the magma ocean by solid crust at ~4.5 Ga (Onstott et al. 2019). Over the next half billion years, the onset of plate tectonics, formation of continents, and formation of the atmosphere and oceans, as well as ongoing bombardment, gave rise to the dynamic, diverse, and interconnected environments from which the chemistry that led to life's origins arose (NASEM 2019a). While the Hadean is often thought to be the most dynamic period in the planet's history, it is also the period for which proxies are the sparsest and models are the least reliable. Even constraints on the volatile delivery of organics from asteroids and comets, and whether impacts cause a net gain or loss of volatiles, remain unresolved (see Questions 3, 4, and 6; Chapters 6, 7, and 9, respectively). The next decade requires field work, modeling, and laboratory experiments to expand understanding of processes of abiotic organic synthesis and to decipher the relative role of exogenous versus indigenous building blocks for life.

Q9.1b What Can We Learn from the Moon About the Conditions, Emergence, and Evolution of Life on Early Earth?

Owing to substantial modification by plate tectonics, erosion, and life over the eons, Earth's record of the environments and conditions during and preceding the emergence of life is less preserved than that of some other bodies in the solar system (e.g., the Moon, Mars, and Mercury). Critically, asteroidal and cometary bombardment of the inner solar system during the first billion years served as a mechanism for geologic and climate evolution and for the delivery of water and organics to early Earth and the inner solar system (NASEM 2019a; see also Question 4), likely having a significant influence on the origin and early evolution of life. Most records of ancient bombardment have been erased from Earth's surface, limiting our capability to understand those critical events, including their possible link to the orbital evolution of giant planets in the outer solar system (see Question 7, Chapter 10). Significantly, the well-preserved surfaces, deposits, and rocks of the Moon provide one of the best chronicles of the major events that have occurred in the inner solar system (Bottke and Norman 2017; see Figure 7-1). In addition, the lunar record provides key insights into deciphering the linked dynamical evolution of the Earth–Moon system (see Questions 3 and 8, Chapter 11) that has had a prominent influence on Earth's environment (e.g., orbital parameters, climate, and tides) throughout time. Importantly, the comparatively well-preserved ice deposits and rocks of the Moon (and to some extent on Mercury and Mars) can inform reconstructions of early Earth environments through a more complete record of the asteroidal and cometary bombardment history and delivery of volatiles and organics relevant to the origin of life.

Q9.1c What Is the Boundary Between Abiotic and Biotic Phenomena and How Does That Boundary Change with Earth's Overall Geochemical and Biological States?

Abiotic chemical reactions provided the feedstock for prebiotic chemical evolution that ultimately gave rise to the first terrestrial organisms. Although abiotic and biotic chemistries have commonalities at the level of individual building blocks, research on the origin of life and extant biochemistry can guide system-level approaches to differentiate between biotic and abiotic chemistry. While abiotic chemistry can exhibit multiple pathways toward a particular compound, biochemistry capitalizes on a distinct subset of precursors among the abiotic possibilities. In the past decade, new approaches have emerged to understand complex networks of interacting molecules in chemical and biological systems. These approaches, which form the basis of systems chemistry and systems biology, can enhance our knowledge of abiotic synthetic routes to the building blocks of life and to possible pathways that the earliest forms of life on Earth, or life elsewhere, might have adopted. Nonetheless, differentiating between abiotic and biotic sources and processes remains a major challenge in planetary sciences and astrobiology and an area of focus for the next decade. The recent literature regarding reports of methane in the martian atmosphere provide a case in point (see Q6.6). Even if the isotopic composition of this methane could be measured, kinetic fractionations associated with abiotic organic synthesis have been shown to be of comparable scale to those produced by kinetically controlled biological processes (Etiope and Sherwood Lollar 2013). Similarly, for both terran (Ménez et al. 2012) and meteorite studies (Steele et al. 2012; Ménez et al. 2018) it has been demonstrated that even organic carbon with "light" carbon isotope values requires careful contextual investigation of the microstructure of minerals and fracture infillings between and across those mineral boundaries to determine the abiotic versus biotic nature of macromolecular carbon (Ménez et al. 2012, 2018). Claims for the "oldest life" on Earth or for life detection beyond Earth require integration of the information from both abiotic and biotic phenomenon (see also Question 11; see Figure 12-7 below), and field, laboratory and modeling investigations provide the critical foundation to further develop those frameworks (NASEM 2019a) in preparation for return of data and/or samples from Venus (DAVINCI and VERITAS), Mars (Curiosity, Perseverance, and MSR), Europa (Clipper), Titan (Dragonfly) and other bodies (see Question 11, Chapter 14) and for interpretation of exoplanet atmospheres (see Question 12, Chapter 15). It is not sufficient that life is known to produce a certain compound on Earth; robust life detection requires sufficient investigation to eliminate abiotic sources and processes (see Question 11 for more on false positives). The significance of any potential biosignature comes from not only the probability of life having produced it, but also from the improbability of nonbiological processes producing it, underscoring the need in the coming decade to further investigate the range and breadth of abiosignatures in parallel with biosignatures (NASEM 2019a).

Q9.1d How Did Early Earth Environments and Prebiotic Pathways Coevolve and Give Rise to Life and What Major Milestones in Earth History Were Coincident (or Causative?) with Major Transitions in the Abundance, Quality, and/or Complexity of Life?

Studies of the chemical origins of life start with the hypothesis that mixtures of simple small molecules under the influence of various energy sources and early Earth environments created the building blocks of life and that interactions among these molecules eventually lead to life itself (Figure 12-2). The earliest signs of life on Earth indicate that the prebiotic chemistry that led to life's origins most likely commenced prior to 4 Ga, which overlaps with a period when Earth experienced major transformations in its geological state, such as the transition to stable liquid water (oceans) at the surface and substantial cooling and crustal formation (Olson et al. 2018; Question 5). It is crucial not only to understand the early state of Earth's environments but also the magnitude and timing of transitions that occurred, including catastrophic events such as large impacts, as well as the effect of lengthy quiescent periods (see also Questions 4 and 6). The specific location where life originated on early Earth (whether a surface environment in contact with the atmosphere and incoming ultraviolet radiation (Damer and Deamer 2020), or a subsurface system (Martin et al. 2008) is still vigorously debated. Smith and Morowitz (2016) suggest that the chemical potential of planetary scale disequilibrium on Earth (and hence life's origin) is focused *"to an extreme degree on the rock/water interface and in the mixing chemistry of fluids and volatiles in and near the crust."* If so, these environments are particularly pertinent for astrobiological exploration of both rocky planets and moons such as Mars (Onstott et al. 2019) and the ocean worlds (Hendrix et al. 2019; Hand et al. 2020; Sittler et al. 2020; MacKenzie et al. 2021) (see also Questions 10 and 11). For systems ranging from rocky planets to the ocean worlds and exoplanets there is strong interest in the role of water–rock reactions producing precursor molecules for potential life as well as substrates for habitability (MacKenzie et al. 2021). Both planetary-scale and local-scale disequilibria could be important for prebiotic synthesis and life's coevolution with the planet; therefore, consideration of both planetary scale habitability and localized environments are important in designing successful strategies for searching for life beyond Earth (NASEM 2019a).

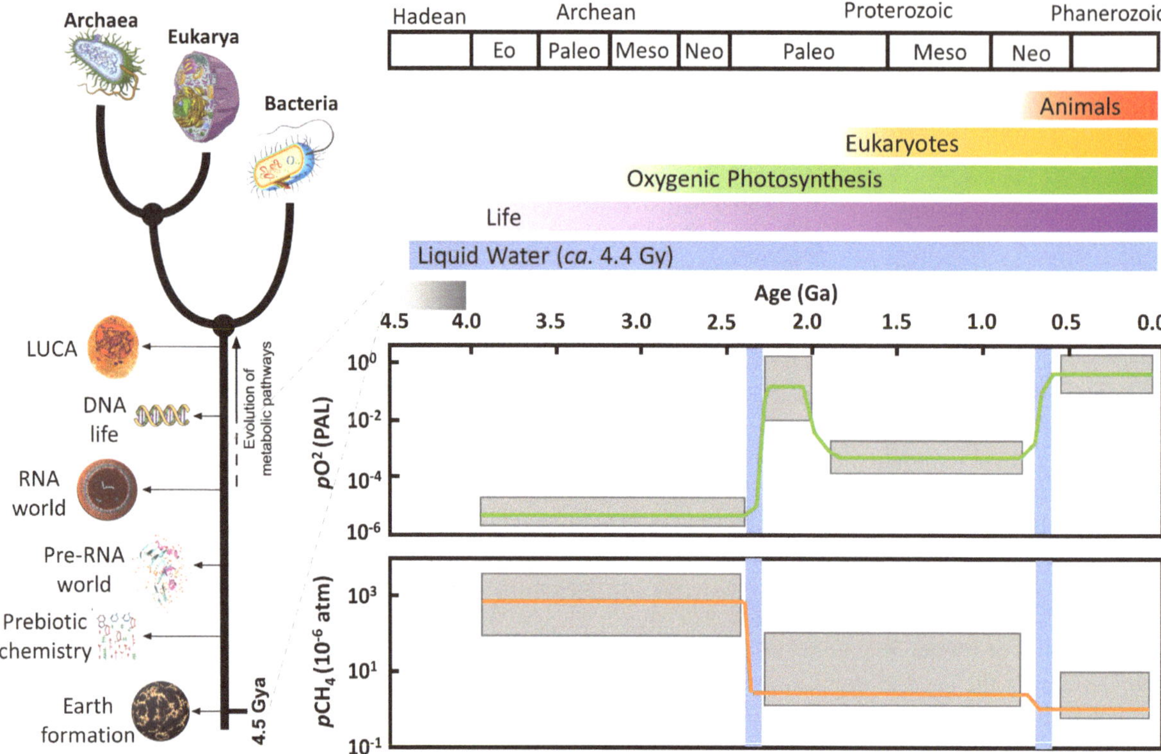

FIGURE 12-2 *Question 9.* Order of prebiotic and origin of life events (*left*), and approximate timing of major transitions—for example, in the abundances of methane (*bottom right*) and oxygen (*middle right*) in Earth's atmosphere—in the history of the biosphere (*top right*). SOURCES: *Left—Evolution of metabolic pathways* and *Pre-RNA world* image modified from Fani (2012), CC BY 2.0; *Archaea* and *Eukarya* from FreeImages.com/oguzaral; *Bacteria* from iStock.com/jack0m; *Luca* from iStock.com/Gunay Abdullayeva; *DNA life* from Stock.com/AlonzoDesign; *RNA world* from the National Science Foundation, credit Janet Iwasa, CC BY 3.0; *Prebiotic chemistry* from G.F. Joyce, 2002, "The Antiquity of RNA-Based Evolution," *Nature* 418:214–221. Springer Nature, 202, reproduced with permission from SNCSC; *Earth formation* from NASA/JPL-Caltech/R. Hurt (SSC). *Right*—From S.L. Olson, E.W. Schwieterman, C.T. Reinhard, and T.W. Lyons, 2018, "Earth: Atmospheric Evolution of a Habitable Planet," pp. 2817–2853 in *Handbook of Exoplanets* (Deeg and Belmonte, eds.), Cham, Switzerland: Springer, reproduced with permission from SNCSC.

Q9.1e How Do Phylogenetic, Structural, and Biophysical Studies Inform Understanding of the Origin and Coevolution of the Genetic and Translational Machinery in Terrestrial Life, and What Was the Timeline for Their Development and Diversification?

The genetic and translational machinery were two of the earliest and most fundamental biochemical inventions of Earth's life (Baross et al. 2020). The ancient and highly conserved nature of these two systems provide fundamental constraints on the evolutionary pathway of terrestrial biochemistry (NASA 2015; Figure 12-3), including the selection of the 20 proteinogenic amino acids and the evolution of proteins with incremental folding ability. Extant proteins utilize a small fraction of the combinatorial sequence space available for 20 proteinogenic amino acids. The ribosome (within its own proteins) contains a structural record of the earliest history of protein folding, and the special family of proteins that recognize and link cognate tRNA with the correct amino acid (i.e., aminoacyl-tRNA synthetases, or aaRS) can provide the basis for comparative evolutionary studies and reconstructing the pre-LUCA history of protein synthesis. Phylogenetic, structural, and biophysical

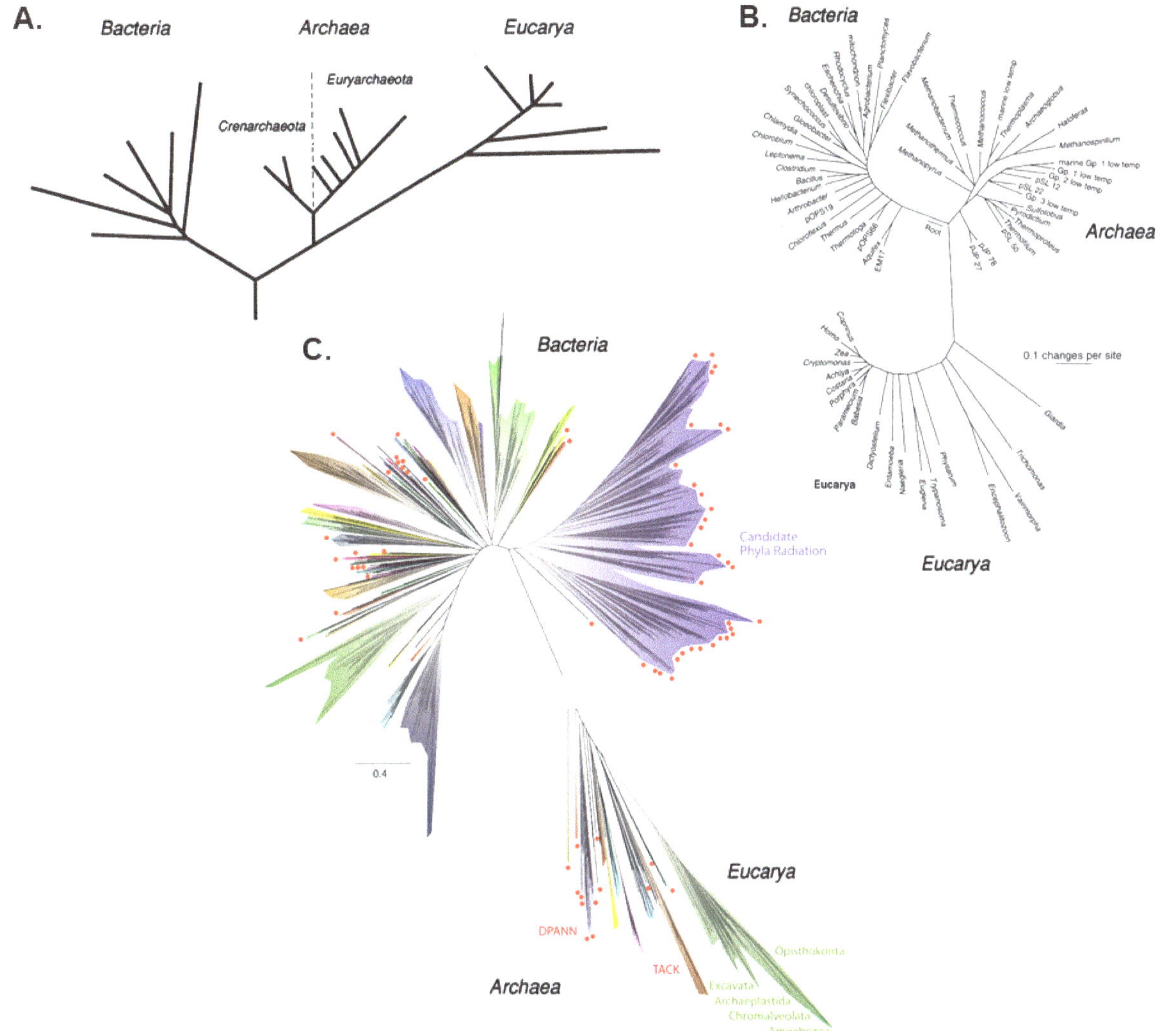

FIGURE 12-3 Universal phylogenetic trees over the past 30 years. (A) 1990 universal phylogenetic tree in rooted form, showing the three domains. Branching order and branch lengths are based on rRNA sequence comparisons; (B) 1997 universal
phylogenetic tree based on SSU rRNA sequences. The scale bar corresponds to 0.1 changes per nucleotide; (C) 2016 view of
the tree of life, encompassing the total diversity represented by sequenced genomes and constructed from ribosomal proteins.
The tree includes 92 named bacterial phyla, 26 archaeal phyla, and all five of the Eukaryotic supergroups.
SOURCES: (A) Adapted from Woese et al. (1990). (B) From N.R. Pace, 1997, "A Molecular View of Microbial Diversity and
the Biosphere," *Science* 276(5313):734–740, https://doi.org/10.1126/science.276.5313.734, reprinted with permission from
AAAS. (C) Hug et al. (2016), CC BY 4.0.

studies of modern organisms can shed light on the nature and functionality of the decoding-competent LUCA
ribosome, and more primitive precoding ribosomes. In the next decade, with further improvements in ancestral
sequence reconstruction algorithms and experimental methods, pre-LUCA aaRS ancestors and other biochemical lineages such as cytochromes, ATP synthases, and biolipids may yield even more secrets, including the
fundamental drivers for the origin and evolution of the genetic code. By analogy, these investigations can offer
insights into the conditions and processes necessary for the emergence and evolution of similar genetic and
translation machinery on other worlds.

Q9.1f What Does the Last Universal Common Ancestor (LUCA) Represent (e.g., a Single Individual, a Species, or a Population of Species), How Does This Inform the Emergence of Core Biological Systems, and What Is the Geological and Evolutionary Context of LUCA's Emergence?

All extant terrestrial life shares a set of genetic commonalities that indicate the existence of a hypothetical common ancestor—LUCA—linked to the core mechanisms of cellular machinery, the structure and function of biomolecules, and interactions and dependencies within cells (NASA 2015). What LUCA represents remains uncertain. Comparative genomics and cell biology suggest that the organism(s) represented by LUCA were likely cellular and contained many genes, proteins, and biological functions present within modern lineages. Further biological investigation of LUCA will require advances in paleogenomics and molecular evolutionary biology as a complement to ongoing theoretical and experimental research in geochemistry, organic chemistry, and planetary science. Reliable reconstructions of LUCA, including the nature of its cell membrane and of important membrane-related protein families, may help explain why the transition toward organismal individuality and vertical inheritance (as opposed to horizontal genetic transfer) became predominant, and whether we should expect a similar transition for other forms of life elsewhere in the universe (Figure 12-3).

Strategic Research for Q9.1

- **Characterize the surface and subsurface processes (e.g., impactor flux, atmospheric conditions, volcanism, and tectonism) and the range of chemical inventories (e.g., volatiles and organics) present during the emergence of Earth's nascent biosphere** through modeling, analyses, and measurements of solar system materials (asteroids, comets, interplanetary dust particles, and meteorites), investigation of Earth's isotopic record, and estimation of fluxes recorded and volatiles deposited on ancient, well preserved inner solar system planetary surfaces (e.g., Moon and Mercury).
- **Characterize how the early, dynamic solar system environment shaped Earth's environments and the subsequent emergence and evolution of life therein** by determining a reliable absolute chronologic record of the early bombardment of the Earth-Moon system, especially prior to 3.7 Ga.
- **Determine what combinations of physical and chemical processes could give rise to conditions that would promote habitability and be able to sustain life beyond Earth** through theoretical modeling, field work, and laboratory experimentation informed by current knowledge of the distribution of life on Earth and Earth processes (e.g., geosphere, hydrosphere, cryosphere, and atmosphere).
- **Identify and characterize chemical processes and pathways that could enable the transition from abiotic reaction networks to biochemical reaction networks** through experimental, modeling and field studies of prebiotic condensation, catalysis, and self-assembly processes and the preservation/diagenesis of those signals with time.[2]
- **Determine the evolutionary history of life's biochemistry through deep time** through genomic/proteomic and phylogenetic/metabolic analyses and reconstructions.
- **Investigate the interplay between availability and biological function of essential elements and how their accessibility and biochemical efficiency changed with environmental abundances** through laboratory experiments, biochemical network models, analyses of Earth's early geologic record, and geochemical studies of other rocky worlds and astromaterials.

Q9.2 WHAT IS THE DIVERSITY, DISTRIBUTION, AND RANGE OF POSSIBLE METABOLIC STRATEGIES OF LIFE IN TERRESTRIAL ENVIRONMENTS (SURFACE, SUBSURFACE, AND ATMOSPHERE), AND HOW DID THEY EVOLVE THROUGH TIME?

Earth, with its rich diversity of habitable spaces, provides a plethora of environments to test models of biological potential—that is, a qualitative measure of the potential for life and for habitability as indicated by factors such as biogeochemical fluxes, organism/metabolic variety, and other features of the physical and chemical environment that can inform the most likely places in the solar system to find evidence of life (Figures 12-3, 12-4, and 12-5).

---

[2]Note: The transition from abiotic to biotic conditions—including key reactions of prebiotic organic chemistry, the formation of abiotic polymers, the origin of replicating heteropolymers, the beginnings of genetics, and the dawn of Darwinian evolution—is discussed in Chapter 14 (see Q11.1).

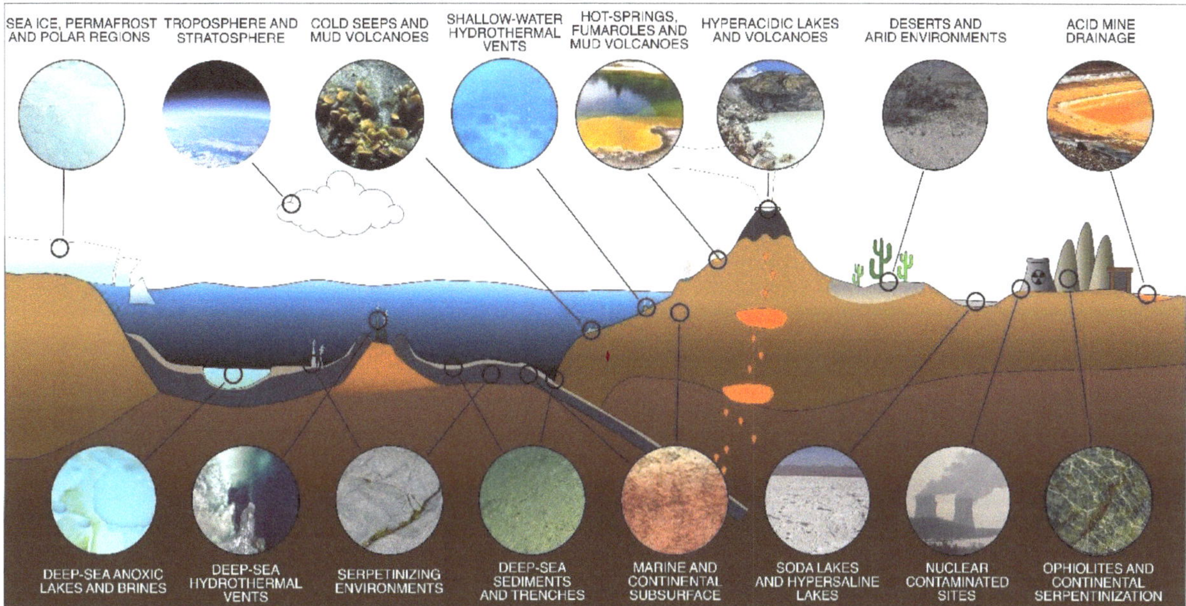

FIGURE 12-4 Representative idealized cross section of Earth's crust showing the diversity of so-called extreme habitable environments and their approximate location. SOURCE: Merino et al. (2019), CC BY 4.0.

Targeted exploration on Earth is a proven, highly effective mechanism by which to validate hypotheses relevant to the search for life, from the deep ocean and subsurface to polar ice caps and deserts (Wierzchos et al. 2018). Terrestrial life has evolved a tremendous diversity of metabolic strategies that tap into an energy spectrum that spans many orders of magnitude, from the low end to the high end (see Figure 12-1).

As a result, biological potential in different environments can also vary significantly. At the high end of the spectrum, light-dependent ecosystems can support large amounts of biomass recycled in short periods of time. At the low end of the spectrum, dark ecosystems subsist in conditions that provide only marginal energy for cell growth and division and seem to barely support the maintenance of basic cellular functions (Hoehler and Jorgenson 2013; Hoehler et al. 2020). Ecosystems elsewhere in the solar system, if they exist, will likely lie closer to the low-energy end of the spectrum (see Figure 12-1), and therefore might be difficult to detect. Understanding the biochemical underpinnings of the different metabolic strategies realized on Earth, their evolutionary history, and their dependence on sources of electron donors and electron acceptors (e.g., H_2, CH_4, SO_4^{2-}), is key to assessing the likelihood that similar strategies might have evolved elsewhere, as well as the sizes of the biospheres those metabolisms might support (see Figure 12-1). Each metabolic strategy is optimized for a specific range of physical and chemical parameters, such as salinity, pH, or temperature. For some parameters such as high temperature, pH, or salinity, there appear to be boundaries beyond which the metabolic activity of all but the most specialized organisms is severely impaired, or entirely disrupted. For other parameters, such as low temperature, the boundaries are more diffuse. An area of focus is understanding the cellular and metabolic impacts of poly-extremes (e.g., an organism experiencing multiple extreme conditions at once is called a polyextremophile), particularly those that might act synergistically, to inform the most likely environments beyond Earth where life might be detectable.

Q9.2a What Are the Different Energy Sources That Life Can Exploit on Earth and Other Planets?

Understanding what limits the wavelengths and intensities of light that can be utilized by life, finding possible alternatives to light energy, and characterizing the spectrum of redox pairs that terrestrial life can exploit, informs the quantity and quality of energy sources that could potentially be utilized by life elsewhere (Hoehler et al. 2020). While thermodynamic disequilibrium can arise in the absence of life, all life that we know on Earth exploits the chemical

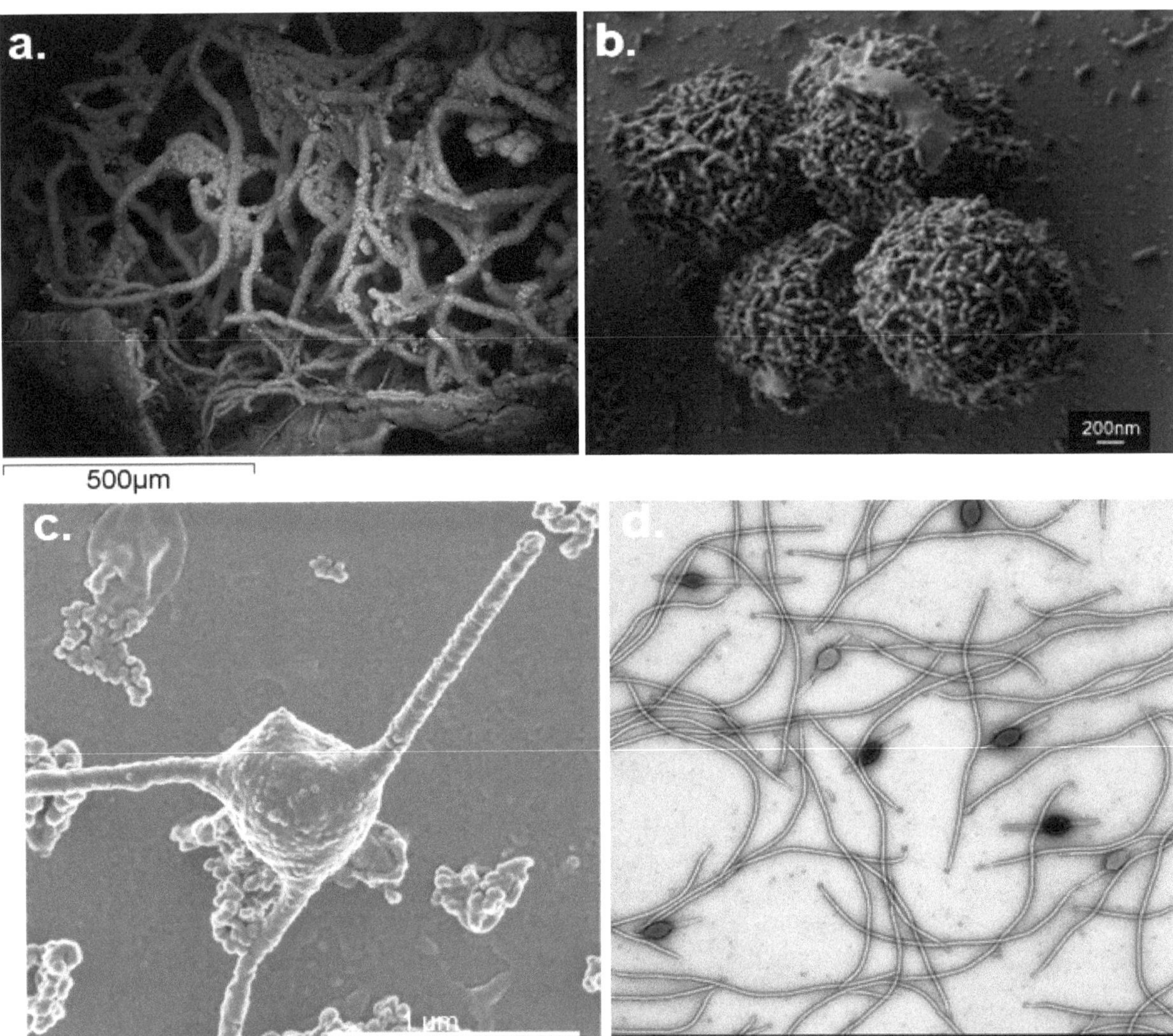

FIGURE 12-5 Microscopic images of life at environmental extremes on Earth, including (a) a fungal-prokaryote colony in vesicular basalt, Koko Seamount, Pacific Ocean, 67.5 m below the seafloor; (b) *Planococcus halocryophilus* Or1, a halopsychrophile isolated from Arctic sea ice, growing at −15°C in 18 percent NaCl and 7 percent glycerol, encrusted in dense nodular material; (c) Asgard archaeon "*Candidatus* Prometheoarchaeum syntrophicum" strain MK-D1, an anaerobic, extremely slow-growing microbe isolated from deep marine sediments; and (d) a mixture of a filamentous virus (family Lipothrixviridae) and a two-tailed virus (family Bicaudaviridae) isolated from the thermophilic acidophile archaeon, *Acidianus*. SOURCES: (a) Magnus and Holmström (2012), BY CC 3.0. (b) N. Mykytczuk, S. Foote, C. Omelon, et al., 2013, "Bacterial Growth at −15°C; Molecular Insights from the Permafrost Bacterium Planococcus Halocryophilus Or1," *ISME Journal* 7:1211–1226, https://doi.org/10.1038/ismej.2013.8, reproduced with permission from SNCSC. (c) H. Imachi, M.K. Nobu, N. Nakahara, et al., 2020, "Isolation of an Archaeon at the Prokaryote–Eukaryote Interface," *Nature* 577:519–525, https://www.nature.com/articles/s41586-019-1916-6/figures/3, reproduced with permission from SNCSC. (d) Courtesy of V. Cvirkaite-Krupovic, Institut Pasteur, Paris, France.

energy released through organisms' metabolic functions, as chemicals react back toward equilibrium (see Figure 12-1; NASEM 2019a). Indeed, in the past decade it has even been hypothesized that life should be envisaged as a "fourth geosphere"—a natural consequence of the processes and fluxes that drive thermodynamic disequilibria (Smith and Morowitz 2016). While some organisms carry out oxygenic photosynthesis—the biological process in which water is split for cellular energy, generating O_2—other organisms extract energy from redox reactions between a broad range of oxidants (e.g., O_2, NO_3^-, SO_4^{2-}, CO_2) and reductants (e.g., Fe^{2+}, H_2, H_2S, CH_4, and more complex organics) (chemotrophy) (see Figure 12-1). In the case of earliest Earth, we have evidence of ancient microbial life associated with geothermal settings, akin to modern day hot springs. There, thermally driven convection gives rise to water–rock reactions that, in turn, can act as a source for, and bring together, reactants that can participate in a variety of redox chemical reactions that release energy in a form that can be exploited by a diversity of microbial metabolisms (see Figure 12-1). In addition to calculating how much energy might be available for life to exploit for new growth, studies are needed that constrain the amount of energy that life expends to repair damaged cellular components and biomolecules, or synthesize new ones (maintenance energy, Hoehler et al. 2020). Coupling the quantity and quality of known energy sources for life with the amount of energy per unit time (power) required by organisms to perform basic biological functions, also informs the biological potential of a habitable environment.

Q9.2b How Does the Biological Potential (i.e., Abundance, Productivity, and Diversity) of Light-Dependent Ecosystems Compare to That of Light-Independent Ones?

The amount of energy delivered by visible and near-infrared light far exceeds the amount of energy delivered by the chemical reactions that life can exploit in dark environments (Hoehler et al. 2020). The detectability of light-dependent and light-independent ecosystems varies accordingly, both in terms of the types and abundances of biosignatures they can generate. Quantitative estimates are needed that compare the biological potential of ecosystems that rely on sunlight as an energy source (directly or indirectly), from those that are truly light independent. The former are especially pertinent to discussions of the potential for life in (near)surface environments and atmospheres of other planetary bodies (e.g., Mars, Venus, and Titan). The latter are relevant to the discussion of the potential for life in dark, subsurface environments in multiple solar system targets (e.g., Mars, Europa, Titan, and Enceladus). Understanding the size of a biosphere that can be sustained by light-dependent and light-independent primary production pathways will provide a baseline to inform science and instrument requirements for life-detection missions that target atmospheric/surface or subsurface environments, respectively. It will be equally important to characterize additional light-independent chemotrophic pathways that also rely on products of radiolysis or water–rock reactions. These pathways can coexist with and complement light-independent primary production, enabling the recycling of products and reactants, and increasing the overall productivity and diversity of a subsurface biosphere whether capable of direct investigation on Earth (Lin et al. 2006; D'Hondt et al. 2009, 2019; Sauvage et al. 2021; Sherwood Lollar et al. 2021) or through habitability models and designing exploration strategies for Mars (Onstott et al. 2019), Europa (Hand et al. 2017), Enceladus (Vance et al. 2007; Waite et al. 2017), Titan (Cable et al. 2018), or the small bodies such as Ceres, for example.

Q9.2c How Do Environmental Factors and Fluxes Control or Limit the Different Metabolic Strategies, Growth Rates, or Productivity in Different Planetary Analogs on Earth, and How Do They Co-Vary?

Together, physical and chemical conditions—including salinity, temperature, pressure, pH, ultraviolet, and ionizing radiation—define a multi-dimensional space that represents the habitability envelope of life on Earth (see also Question 10). Individually, each environmental parameter affects one or more aspects of cellular function, from osmotic regulation and transport across the cell membrane, to the stability and functionality of biomolecules and the overall structure of the cell (see Figure 12-5). The magnitudes of these physical and chemical parameters, and their synergistic interactions, affect overall productivity, cellular abundance, and phylogenetic diversity in a given environment. At the high- or low-end values of those parameters, cellular countermeasures are required to mitigate negative effects, which impose higher energetic demands on the individual cells and the overall ecosystem. Because the extent of that effect is likely to depend on the types and diversity of metabolic strategies realized in the environment, the flux of nutrients and availability of energy also determine the ability of individual organisms and groups of organisms to adapt to poly-extremes. Laboratory and modeling studies, along with research in Earth environments, allow us to probe the fringes of this multi-dimensional space, and the effect on biological potential. Field studies of nutrient and energy resources and fluxes in so-called extreme environments (subterranean, oceanic and/or atmospheric; see Figure 12-4) over different timescales, complemented with theoretical and laboratory studies on the tolerance of terrestrial microorganisms to multiple environments stressors (e.g., temperatures, desiccation, ultraviolet radiation, extreme pH, and salinity) are essential to determine the requirements for the survival of life in similar extraterrestrial settings. Results from these studies can be used to assess whether physicochemical conditions in other solar system environments could support Earth-like life, and to constrain the sizes of the biospheres they might contain.

Q9.2d How and When Did Viruses Originate and What Role Have They Played and Continue to Play in the Evolution of Life on Earth?

Viruses are key contributors to Earth's ecosystems, but there is much yet unknown regarding their influence on cellular life, their role in evolutionary history, their physical interactions with the Earth system, and their persistence and decay under various environmental conditions (see Figure 12-5). Viruses and virus-like replicators are the only known biological entities that contain all types of genomes, including single-stranded and double-stranded RNA and DNA. RNA viruses may serve as models for how RNA and ribonucleoproteins could have propagated via simple self-replicating RNA structures and ribozyme activity; thus, elucidating the history of viruses could provide important clues regarding the emergence and evolution of life on Earth (Forterre 2006; Koonin et al. 2021). In addition, viruses seem to be ubiquitous in terrestrial ecosystems, where they act as agents of microbial evolution and contribute to microbial fitness by moving ecologically important genes from host to host.

Strategic Research for Q9.2

- **Characterize light-dependent and light-independent life and its metabolisms and assess its contribution to biological diversity, productivity, and abundance in Earth environments** using sequence-based molecular studies, field observations, and metabolic models.
- **Investigate how multiple environmental factors (e.g., pH, temperature, pressure, salinity, and redox potential) and fluxes (e.g., energy and nutrients), affect different metabolic strategies in Earth environments** through field and laboratory observations and metabolic models of microbial community responses to environmental stress, including multi-dimensional factors.
- **Elucidate the evolutionary history (origin and divergence) of metabolic pathways as a guide for the interpretation of biosignatures in terms of coevolution of life and its host world** through sequence-based molecular clock studies and modeling of life's biochemistry.
- **Investigate the roles of viruses in Earth environments** through sequence-based molecular studies, long-term environmental monitoring of viruses, and laboratory studies of virus-host interactions.

Q9.3 HOW DO INVESTIGATIONS OF EARTH'S SUBSURFACE ENVIRONMENTS INFORM WHAT HABITABILITY AND/OR LIFE ON OTHER WORLDS MIGHT LOOK LIKE?

On Earth, the discoveries of active microbial communities existing in the subsurface of the ocean floor and continental lithosphere, often far from the influence of the Sun's energy, provide new models for understanding rock-hosted, chemosynthetic life that may exist on other worlds (see Figure 12-4). Many outstanding scientific issues remain unexplored in the subsurface, including quantifying the total biomass and its distribution, the degree to which subsurface ecosystems on Earth exist independently of surface energy sources, the stability of these ecosystems over time and space, and how life adapts when subject to its environmental and energetic limits (see recommendations from NASEM 2019a). Growing sophistication in our understanding of subsurface life and its trajectory on this planet could reveal much about how life could persist on other worlds. We know that subsurface communities on Earth range from energy-rich environments where "fast" life predominates (e.g., hydrothermal vents) to energy-limited environments where "slow" life is barely able to survive and can be difficult to detect (e.g., continental rocks, Magnabosco et al. 2018; deep-sea sediments and rocks, Trembath-Reichert et al. 2017; Suzuki et al. 2020) with the latter providing critical endmembers likely to be applicable to the search for life off-Earth (Onstott et al. 2019). A range of strategies for accessing subsurface samples using both in situ and remote capabilities is needed to investigate subsurface processes and inform our understanding of the controls on interactions between a planet's surface and subsurface. The investigation of subsurface habitats on Earth is one of the most readily actionable strategies to address questions about the processes governing habitability and the nature, diversity, and preservation of both extant and extinct subsurface communities on Earth and other worlds (see also Question 10).

Q9.3a What Are the Physical and Chemical Processes and Spatial and Temporal Controls Sustaining Subsurface Life on Earth and How Does This Expand Concepts of Habitability Elsewhere in the Universe?

Research into the physical and chemical subsurface environment is critical to understanding how subsurface life and its host environment coevolve. Investigations of rock-hosted and other subsurface environments on Earth make up the critical test bed for informing the search for life on multiple targets in the solar system, including sub-ocean silicate crusts of Europa, Enceladus, or other ocean worlds (Hand et al. 2020), and Mars. On Mars, where the loss of its dynamo-driven magnetic field between 4.1 to 3.9 Ga and subsequent loss of atmosphere (Ehlmann et al. 2016) led to surface conditions less hospitable for life, subsurface environments may have provided a widespread stable refuge for life (and signatures of life) to persist (Ehlmann et al. 2011; Onstott et al. 2019; Stamenkovic et al. 2019). Studies of geophysical, geochemical, geological, and hydrogeological processes are vital to understanding subsurface habitability on Earth and beyond at both local and global scales. This involves determining the spatial and temporal distribution of subsurface water—the inventories, sources, and sinks of energy—that could generate habitability in the subsurface, and the processes that sustain these inventories over time and space (Schrenk et al. 2013; Li et al. 2016; D'Hondt et al. 2019; LaRowe and Amend 2019; Onstott et al. 2019; Sauvage et al. 2021; Sherwood Lollar et al. 2021). For instance, shallow subsurface life ranges from cryptoendolithic communities existing only millimeters below rock surfaces and typically consisting of cyanobacteria and algae (Wierzchos et al. 2018) to complex aerobic and anaerobic communities in near-surface groundwaters and extending to kilometers depth in sedimentary basin systems (Head et al. 2014). Deeper yet, in the oceans, studies have focused both on marine sediments (Inagaki et al. 2015; Heuer et al. 2020), and on microbial ecosystems permeating the ocean lithosphere and hydrothermal vents and seeps, reflecting a broad range of microbial metabolisms ranging from heterotrophs degrading residual photosynthate to chemolithoautotrophic communities deriving their energy from water–rock reactions such as serpentinization and radiolysis. In continental settings, the past decade has seen a shift from investigations of sedimentary systems to studies of crystalline rocks and fracture waters ranging from thin ophiolite wedges (Schrenk et al. 2013) to kilometers deep in more than billion-year-old rocks of the crystalline cratons of, for example, the Canadian Shield, South Africa, and Fennoscandian Shield (Sherwood Lollar et al. 2014). These studies have expanded our understanding of the range of water–rock reactions and energy sources (magmatic, serpentinization, and radiolysis) capable of sustaining life in isolation from the surface photosphere, and as such, are particularly relevant to developing a more comprehensive understanding of planetary habitability (see Question 10) and the search for life elsewhere (see Questions 11 and 12, Chapter 15).

Modeling of anticipated water–rock reaction conditions and the available energy for life offers great potential for exploring the past and present habitability of other planetary bodies, but with two important caveats. First, theoretical modeling of water–rock reactions is most robust for situations in which equilibrium is anticipated to have been obtained, whereas in a dynamic setting, reaction kinetics might dominate. Second, the maximum chemical energy that might be available from any set of reaction conditions does not necessarily imply that life is able to exploit it. Figure 12-6 uses deep-ocean hydrothermal systems to show how the cumulative catabolic energy available from subsurface water–rock reactions, at any given location, exhibits *multi*-parameter co-variability. In this illustration, a broad array of parameters combine in multiple ways, providing diverse and alternate pathways to high energy yields or, conversely, to low energy yields. Pressure and hydrogeology each play a role in controlling the maximum temperatures that fluids can reach within such a water-rock reaction-path system. Furthermore, the rates of fluid flow through the system determines the amount of time available for chemical reactions to proceed, as well as the rate at which reactants are mined from the system. In parallel, the bulk composition and mineralogy of the host rock influences which key chemical species are removed from or enriched in the circulating fluid relative to the original source fluid. Note that the composition of that source fluid would also be expected to vary among other (ocean) worlds.

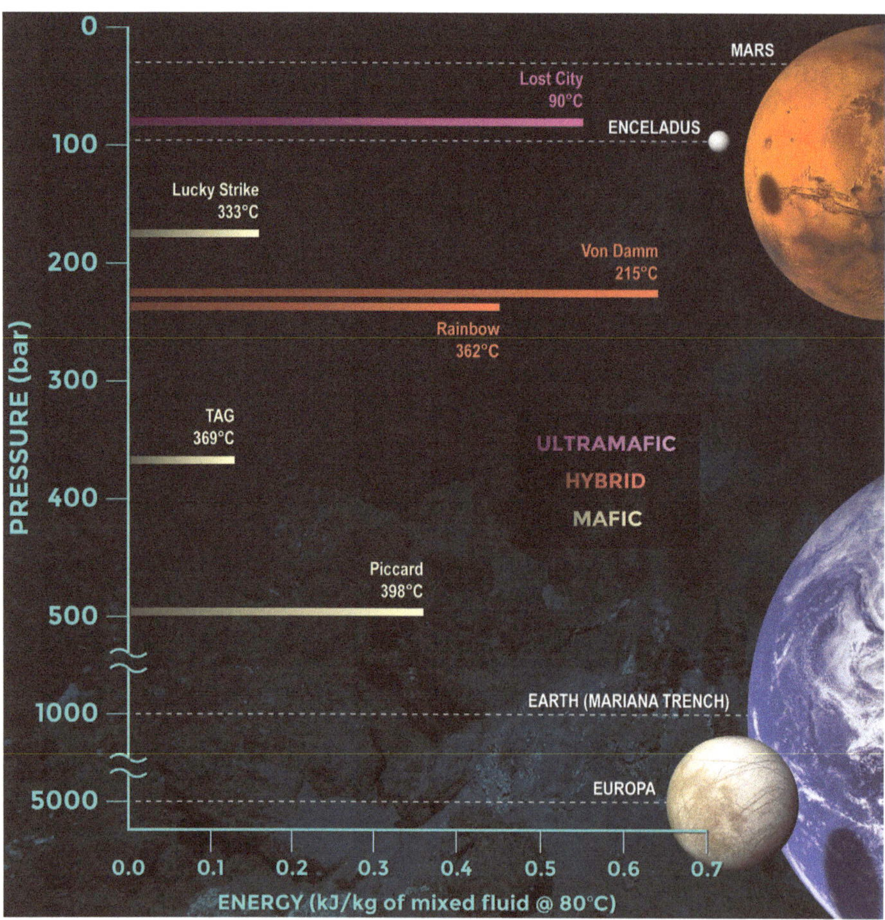

FIGURE 12-6 Plot using submarine vent data from the mid-ocean ridges in the Atlantic Ocean and Caribbean Sea to illustrate how the cumulative energy theoretically available from water–rock interactions for chemolithotrophic microbial metabolism is subject to multilateral co-variability. Key parameters include physical conditions under which water and rocks react in the subsurface (shown = pressure, temperature; not shown = flow rate); variations in host-rock lithology (color key); mixing temperatures at which redox-active species are reacted together when subsurface fluids emerge from host rock (not shown); range and concentration of substrates (chemical redox pairs) present (not shown). Vent-specific energy calculations (horizontal bars) from Reveillaud et al. (2016). Horizontal dashed lines represent maximum seafloor pressures inferred for ancient Mars and present-day Enceladus, Earth, and Europa. SOURCE: Composed by P.K. Byrne; seafloor from MARUM—Center for Marine Environmental Sciences, University of Bremen (CC BY 4.0); *Mars*, *Earth*, and *Europa* from NASA; *Enceladus* from NASA/JPL-Caltech/Space Science Institute.

Q9.3b Can Subsurface Life Exist in the Absence of Surface Life and Can We Test This on Earth to Inform the Search for Life Elsewhere?

While many subsurface ecosystems exist without direct influence of the Sun, life in these environments may still use the products of photosynthetic life as oxidants, reductants, or carbon sources. Studies of subsurface environments on Earth where chemolithoautotrophic life predominates will help inform what a subsurface biosphere might look like on another rocky planet or ocean world without energy from the Sun, while also helping to distinguish whether the surface biosphere on Earth is an outgrowth of the subsurface biosphere, or if the colonization of the subsurface is facilitated by surface phototrophy. There are a range of rock-hosted environments in both the continental and marine subsurface on Earth where water-rock reactions such as serpentinization or radiolysis create energy sources (e.g., methane and molecular hydrogen) to support life (Schrenk et al. 2013; Sherwood Lollar et al. 2014). However, observations of pertinent biomarkers and direct evidence for microbial ecosystems are insufficient to establish the indigenicity of life in the subsurface or its relationship to the surface biosphere (Onstott et al. 2019). Signs of extant life and indicators of past life may have penetrated the subsurface long after the host rock formed, and the permeation of the geologic setting by younger groundwaters needs to be addressed to constrain the timing and history of subsurface life (Lin et al. 2006; Becraft et al. 2021). "Follow the rock, and the water" is a necessary strategy that needs to be as deeply embedded in terran studies as it is in the search for life elsewhere (Lollar et al. 2019). Exploration of the deep seafloor and continental lithosphere, together with parallel investigation of the rock record and hydrogeologic cycle through deep time, are a key strategy for the next decade to expand our picture of the coevolution of the subsurface biosphere and Earth and the degree of interconnection and/or isolation between the surface and subsurface habitable zones.

Q9.3c How Does Subsurface Life Adapt to "Extreme" Environmental Parameters and Variation in Available Energy Sources and/or Supply?

Subsurface life on Earth thrives in nominally "extreme" environments—a term that is less in vogue now that we recognize it embeds observational bias in the sense that we reset the "limits" based on what we have observed or chosen to investigate. An environment "extreme" to our perspective is habitable for the organisms adapted to live therein (see Figure 12-4). The tremendous variation in temperature, pressure, pH, salinity, radiation, and energy/nutrient supply that terran life has evolved to successfully exploit, means that the current "observed limits" for each parameter (see Table 12-1) remain an important feature of on-going research.

Importantly, there is an urgent requirement to understand life's response and ability to adapt to the multiple parameters that organisms encounter in Earth's subsurface (e.g., combined effects of extreme temperature and pressure, or T-P and low water activity combined). A key focus for future work involves investigation of life and habitability both where energy available for life is abundant and where it is limited by paucity of source or by transport (NASEM 2019a). This includes organisms living in deeply buried, energy-starved sediments, with estimated metabolic turnover rates as slow as one cell division every thousand years (Trembath-Reichert et al. 2017; Hoehler et al. 2020; Sauvage et al. 2021), as well as life in other subsurface environments such as the continental rocks, marine ocean crust, hypersaline habitats, and subglacial ice. Such observations of life in its natural habitat are critical, as are mechanistic studies to understand the biological responses to different stressors and how organisms adapt to survive and thrive in these habitats.

It is important, also, to remember that our predictions of where life could exist beyond Earth need not be restricted to environments that we have already identified on Earth but also to where theory and experiments predict that life might also be present but awaiting discovery. In many cases, the planetary science and astrobiology community's priority for exploration of such predicted environments might converge with those of scientists studying terrestrial systems, but that need not necessarily be the case. One example of successful convergence that emerged in the past decade was ultramafic-hosted submarine hydrothermal systems and mafic and ultramafic hosted systems in the continental crust. These systems are relevant to plumes detected on Enceladus (Waite et al. 2017) and suggested for Europa (Sparks et al. 2017). Such environments were among those originally postulated to have the capacity to host abiotic organic synthesis on early Earth, Mars, and Europa (Shock and Schulte 1998). Subsequently, targeted field exploration has led to the demonstration of de novo subsurface abiotic organic synthesis (McDermott et al. 2015; Sherwood Lollar et al. 2021), but—in the same locations—the concomitant discovery of novel subsurface chemolithoautotrophic life forms (Reveillaud et al. 2016; Onstott et al. 2019; see Figure 12-6).

Q9.3d How Does the Biological Potential of Life Vary in the Subsurface, How Much of the Total Planetary Biomass Is Represented by Subsurface Communities, and What Are the Mechanisms for Life's Dispersal and Transport Within and Out of the Subsurface?

The total volume of subsurface habitats on Earth is immense, but life is heterogeneously distributed, and the biological potential (abundance, productivity, diversity, and others) can vary by orders of magnitude as a function of environmental factors (Magnabosco et al. 2018; Onstott et al. 2019; Trembath-Reichert et al. 2021).

Processes such as transport, flux, preservation, degradation, concentration, and dilution can modify and impact the distribution and activity of life in the subsurface, and habitable niches may be ephemeral or isolated (Lin et al. 2006; Lollar et al. 2019). Three-dimensional global assessments of the biological potential of the subsurface on Earth are only roughly quantified and severely lacking in many volumetrically large environments, particularly in marine and continental basement rock, where studies have only begun in the past two decades (NASEM 2019a).

Strategic Research for Q9.3

- **Assess controls on the distribution of subsurface water, the inventories, sources, and sinks of energy, and the processes that sustain these inventories through time and space** through field and laboratory studies of subsurface waters (including saline and hypersaline), and targeting both continental subsurface, and marine subsurface habitats on Earth.
- **Determine to what degree subsurface ecosystems exist independently of surface energy sources** through field work, modeling, and laboratory studies of chemolithoautotrophic life in subsurface environments on Earth.
- **Examine how subsurface life adapts when subject to environmental and energetic limits, as well as life's response and ability to adapt to multiple parameters at the same time** with theoretical, field, and laboratory studies of subsurface environments on Earth.
- **Determine the total planetary biomass in continental and marine subsurface environments and how biomass is maintained and preserved in habitability conditions that are both heterogeneous in space and time as well as stable on long spatial and temporal scales** through theoretical and field studies of Earth's subsurface environments.
- **Determine the technical limits to the collection, preservation, and detection of cells and/or biological/ organic compounds and their delivery to the surface** by developing new technologies and strategies to enable remote, automated, and in situ tools for accessing, sampling, and measuring subsurface processes and life through laboratory studies, and field deployments in subsurface environments on Earth, including examining known nonhabitable environments on Earth.

Q9.4 HOW CAN OUR KNOWLEDGE OF LIFE AND WHERE AND HOW IT ARISES AND IS SUSTAINED ON EARTH ILLUMINATE THE SEARCH FOR LIFE BEYOND EARTH?

Life on Earth currently provides the only reference for a habitable *and* inhabited world. Challenges include an incomplete understanding of how the essential traits of life (e.g., metabolism, Darwinian evolution, and bioenergetics) arose from the geochemical environment and uncertainty concerning the factors essential for the most basic living system (Figures 12-7 and 12-8). A canonical biosignature is defined as an object, substance, and/or pattern whose origin specifically requires a biological agent. To qualify as biosignatures, these "features must be sufficiently complex and/or abundant so that they retain a diagnostic expression of some of life's universal attributes" (Des Marais et al. 2003, 2008). Universal attributes are defined under the broad discipline of "Universal Biology," which leverages universal laws of physics and chemistry to understand the processes that might lead to the development of life beyond Earth, without Earth-centric biases (Des Marais et al. 2008). Further, the significance of any potential biosignature comes not only from the probability of life having produced it but also from the improbability of nonbiological processes producing it (NASEM 2019a). Recent and continuing Earth-based research into characterizing universal or agnostic chemical, morphologic, or physiologic/metabolic biosignatures and abiosignatures, and characterizing reliability, survivability and preservation,[3] and detectability have been recommended as essential in the next decade to advance the search for life beyond Earth (NASA 2015; Meadows 2017; Meadows et al. 2018; NASEM 2019a) (see also Questions 10 and 11).

---

[3]Note: The report treats survivability and preservation as synonyms, as is common in the literature.

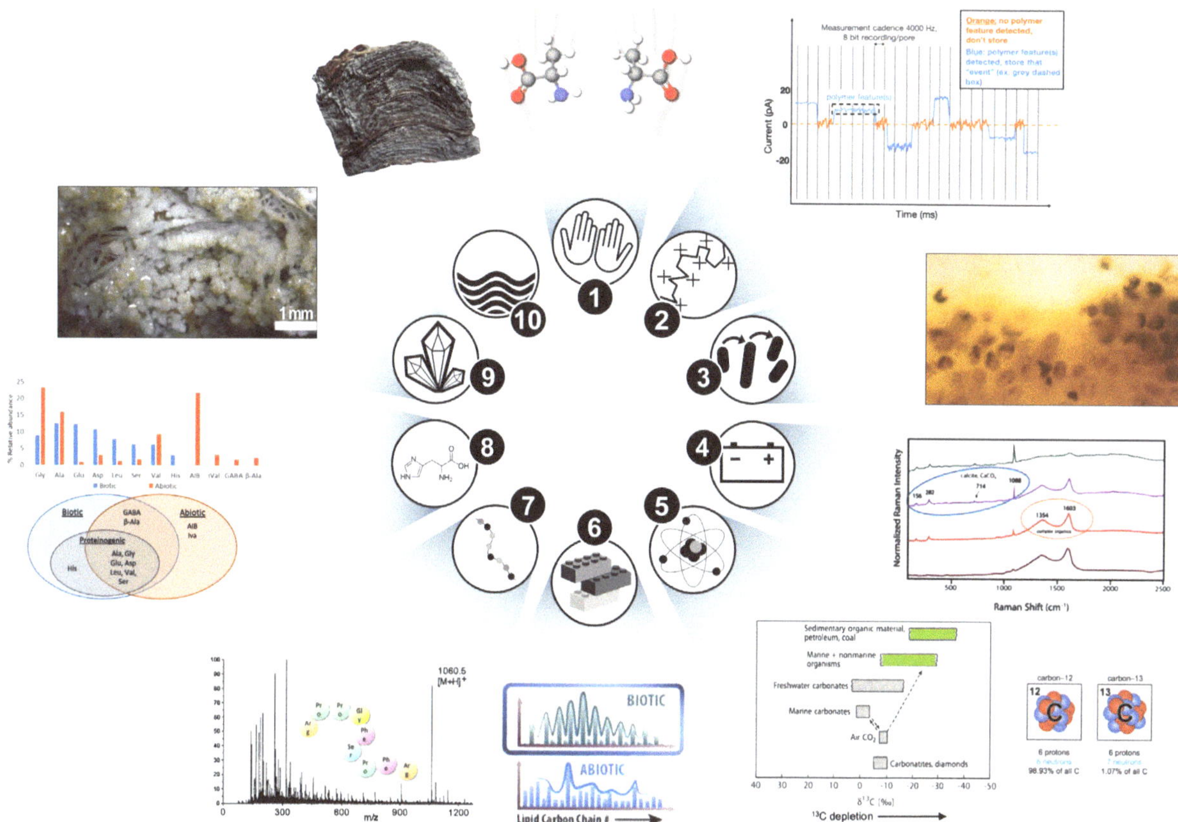

FIGURE 12-7 Biosignature features and measurements ranging in strength of evidence and ease of measurement. *Clockwise from top*: Structural preferences in organic molecules (nonrandom and enhancing function), including (1) enantiomeric excess >20 percent in multiple amino acids and (2) polymers with repeating charge. (3) Preservation of concurrent, variable life stages or localized complex morphology. Metabolisms that may be driven by (4) co-located reductants and oxidants and traced in (5) major element or isotope fractionations indicative of metabolisms, both of which can drive deviation from an abiotic distribution controlled by thermodynamic equilibrium and/or kinetics. (6) Patterns of organic complexity measured as a deviation from equilibrium or abiotic kinetic distribution. The presence of potential biomolecule components such as (7) complex organics (e.g., nucleic acid oligomers, peptides, and polycyclic aromatic hydrocarbons) and (8) monomeric units of biopolymers (nucleobases, amino acids, lipids). (9) Presence of potential metabolic byproducts such as distribution of select metals (e.g., Fe, Ni, Mo/W, Co, S, Se, P). (10) Biofabrics and textures that represent biologically mediated morphologies. SOURCES: Composed by P.K. Byrne. *Inner circle images* from Neveu et al. (2018); for images 1–8 and 10, the publisher for this copyrighted material is Mary Ann Liebert, Inc. Publishers. *Outer circle images:* (1, 4, 5-right, 8-bottom) Hand et al. (2017); (2) NASA (2020); (3) Mustard (2013); (5-left) J. Hoefs, 2004, "Isotope Fractionation Mechanisms of Selected Elements," Ch. 2 in *Stable Isotope Geochemistry*, 5th Edition, Berlin, Heidelberg: Springer, reproduced with permission from SNCSC; (6) NASA GSFC/Ghannadian/Eigenbrode; (7) Aksyonov and Williams (2001), https://doi.org/10.1002/rcm.470, CC-BY-NC 4.0, modified by M. Neveu with addition of amino acids; (8) Adapted from Creamer et al. (2017), https://doi.org/10.1021/acs.analchem.6b04338, CC-BY-NC; (9) Stevens et al. (2015), https://doi.org/10.1111/gbi.12154CC-BY-NC; (10) Mustard et al. (2013).

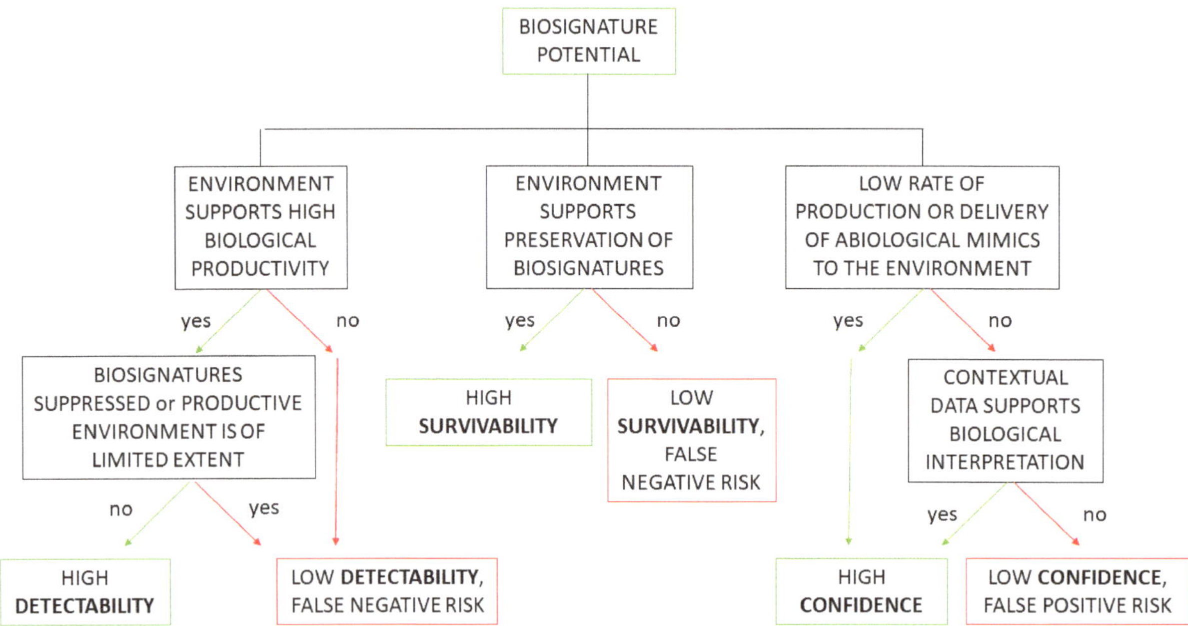

FIGURE 12-8 Decision tree describing the biosignature potential of a given environment (terrestrial or nonterran) in terms of measurable properties that might result in the environment being associated with detectable probable biosignatures that can be confidently ascribed to biology and have a high probability of preservation, along with those environmental properties that might result in a higher probability of false positive or false negative detections. SOURCE: Courtesy of J. Hurowitz.

Q9.4a How Did the Essential Traits of Terrestrial Life (Such as Metabolism and Bioenergetics) Arise from Geochemical Environments, and What Factors Can We Consider Essential for the Most Basic Living System?

All extant life on Earth shares a common biochemistry based on a relatively small set of organic molecules, the same mechanisms of information storage and inheritance (RNA, DNA, and protein), ATP energy currency, a dependence on water, a related cellular organization, and a handful of core metabolic pathways with reactions performed by proteins with a shared ancestry. Fundamental questions remain concerning how these essential traits arose, the factors that are essential to the most basic living system, whether there are alternatives in a planetary context, and whether/ why there is not (apparently) any life on Earth that does not use this common scaffolding.

Q9.4b What Chemical, Morphologic, or Physiologic/Metabolic Biosignatures Are Likely to Be Prevalent in Earth Life, Irrespective of Its Origin, Molecular Makeup, or Metabolism? What Is the Range of Agnostic Biosignatures on Earth?

Examples of terrestrial biosignatures include but are not limited to cellular and extracellular morphologies, biogenic fabrics in rocks, bio-organic molecular structures and biomarkers, chirality, biogenic minerals, biogenic stable isotope patterns in minerals and organic compounds, atmospheric gases, remotely detectable features on planetary surfaces, and temporal changes in global planetary properties (NASA 2015; Meadows 2017; Meadows et al. 2018; NASEM 2019a). Molecular biosignatures alone may include enantiomeric excess, diastereoisomeric and structural isomer preference, repeating constitutional sub-units or atomic ratios, systematic isotopic ordering at molecular and intramolecular levels, and uneven distribution patterns (e.g., carbon number, concentration, and $\delta^{13}C$) of structurally related compounds (Summons et al. 2008; see Figure 12-7).

In the coming decade, biosignature research needs to include a concerted effort to better understand abiosignatures (i.e., a signature of abiotic processes and phenomena), in particular those that may mimic biosignatures (see also Question 11). A continued focus on characterizing agnostic (or universal) chemical, morphologic, or physiologic/metabolic biosignatures, and a decoupling of expectations based on terran life is warranted to improve the capability to search for life beyond Earth (Kempes et al. 2021; Marshall et al. 2021). Agnostic biosignatures are those that manifest themselves as unexpected complexity, and can present themselves at multiple scales, for example, from sustained redox disequilibria in a planetary atmosphere to patterns in the abundance of organic molecules that would not occur by chance (NASEM 2019a). An example of an agnostic atmospheric biosignature on Earth is provided by the high concentrations of CH_4 in Earth's O_2-rich atmosphere (see also Questions 6 and 12). This redox disequilibrium cannot be sustained by known abiological fluxes of methane and is therefore the product of a biosphere that supplies CH_4 at a rate that exceeds its rapid photochemical and oxidative destruction rates (Question 11). Research in the next decade on universal biosignatures will seek to identify attributes that are highly prevalent or even universally present across biological systems and not unique to Earth. There are two aspects of universal biosignatures that need to be considered: Is the proposed universal signature, or combination of signatures, robust (i.e., can it be used to distinguish life from nonlife) and is it truly universal (i.e., can it be generalized to life beyond Earth)? Universal principles of biology, physical laws of biology, complexity, and emergence of life (Kempes et al. 2019; Kim et al. 2019; Walker 2019) can guide theoretical, field, and experimental approaches to search for distinguishable features of terrestrial life and abiotic chemistry by examining data from various types of biotic and abiotic samples to reveal key differences stemming from biology versus abiotic baseline processes (Chan et al. 2019).

Q9.4c What Is the Spectrum of Abiotic Processes for Which Abiosignatures Exist and What Subset of Those Abiosignatures Can Mimic Biosignatures on Earth?

The literature related to claims regarding the discovery of the earliest signs of life on Earth is replete with examples of false positive detections. For example, initial claims regarding the biological origin of isotopically light graphite in ~3.8 Ga rocks of the Akilia and Isua terranes of Greenland (Schidlowsk et al. 1979; Mojzsis et al. 1996), were based in part on a lack of recognition that metamorphic decomposition of iron carbonate to form graphite and magnetite imparted an isotopic fractionation between graphite and carbonate that mimics that of photosynthetic metabolisms. Initial claims regarding the biogenicity of these graphite occurrences have largely been overturned in favor of an abiotic origin (van Zuilen et al. 2003). This interpretation is supported by the development and refinement of a geological context indicating a metamorphic origin for carbonate rocks and associated graphite (Rosing et al. 1996). Such context is critical in cases where biogenicity is in question, and in which plausible abiotic processes could mimic a purported biological feature (e.g., stromatolite morphologies; Grotzinger and Knoll 1999). Research in the somewhat younger Pilbara region of Western Australia provides excellent examples of how confidence in the assessment of biogenicity is improved through a complete understanding of the context in which a biosignature is found (NASEM 2019a).

Q9.4d How Do Nutrient and Energy Flux Affect Metabolic and Biosynthetic Rates as Well as Rates of Abiotic Destruction and Attrition of Biosignatures for Earth Life, and What Is the Impact of Resource Limitation on Biosignature Detectability?

Potential biosignatures are assessed by how well they fulfill three criteria: reliability, survivability, and detectability (Meadows 2017; Meadows et al. 2018; NASEM 2019a). Resource limitation, including nutrient and energy flux, can affect these latter two criteria of survivability (metabolic and biosynthetic rates) and detectability (rates of destruction and attrition of biosignatures) (see also Question 11). Life that may grow/replicate quickly in a nutrient- and energy-rich environment may be detectable because its biosignature signal is elevated (e.g., in soils). Resource-limited life may grow/replicate slowly, and may either be detectable because the noise is diminished or may not be detectable without appropriately sensitive instruments (e.g., in crustal rocks). Assessing the relative signal-to-noise of each type of population in its given environmental context would help to identify corresponding biosignatures that are most relevant and distinctive (Hoehler and Jorgenson 2013; NASEM 2019a).

Strategic Research for Q9.4

- **Assess how the essential traits of Earth life, such as metabolism or bioenergetics, arose from the geochemical environment** through theoretical, field, and laboratory studies of the connectivity and stoichiometry of metabolic networks, and the geological availability of exploitable redox gradients.
- **Develop a comprehensive framework for biosignature categories of Earth life to guide the understanding of what biosignatures will be prevalent in life** through community-level dialog and consensus, supported by laboratory/experimental and modeling/theoretical research as well as field work on environmentally relevant biosignature classes as well as abiosignatures.
- **Elucidate the survivability and detectability of biosignatures and abiosignatures on Earth** with theoretical, field, and laboratory studies of the impact of resource limitation, and relative signal-to-noise ratios of biological versus abiotic processes in environmental context.

Q9.5 HOW DO RECORD BIAS, PRESERVATIONAL BIAS, FALSE NEGATIVES, AND FALSE POSITIVES PLAY A ROLE IN BIOSIGNATURE DETECTABILITY AND RELIABILITY ON EARTH AND WHAT ARE THE IMPLICATIONS FOR TARGETS BEYOND?

The history of Earth's earliest life and environments is written in the regions of the planet where rocks have managed to survive the ravages of time, tectonism, and weathering, leaving behind a fragmental record of life that starts ~3.5–3.8 Ga (see Figure 12-2). This record provides numerous examples of how biases, and false negatives and false positives, can act to confound efforts at biosignature detection, with important lessons for the search for life beyond Earth. Overcoming these biases requires a thorough understanding of the physical and chemical properties operating in the system in question that act to enhance, preserve, mask, or destroy biosignatures produced in them, and the limits of how far lessons from modern systems or more recent, better preserved geological terrains can be extrapolated in time and space (see Figure 12-8; see also Question 11).

Q9.5a How Do Different Physical Environments Modulate the Detectability of Potential Biosignatures on Earth (i.e., Their Nature, Abundance, Diversity, and Survivability)?

There is a broad range of factors to contend with in considering how an environment might modulate the detectability of a biosignature within it, with a major concern being the possibility that the biosignature will get "lost in the noise" of the environment in which it resides, rendering it undetectable and generating a "false negative" for the presence of life in the environment. The history of atmospheric O_2 is part of the coevolution of the planet and life (Question 6) and discussed in the context of its potential as a biosignature for exoplanets (Question 12). Yet the delayed rise in the concentration of O_2 in Earth's atmosphere following the advent of cyanobacterial oxygenic photosynthesis (the biological process in which water is split for cellular energy, generating O_2), spanning up to a billion years of Earth history (Lyons et al. 2014), is perhaps the most well-known example of an environment suppressing the rise of a biosignature. The exact timing of oxygenic photosynthesis and the detection of O_2 in Earth's atmosphere remain controversial (Lyons et al. 2014). In-depth study across a range of inhabited environmental settings, on both the modern and ancient Earth, will yield important insight into the search for life beyond Earth, especially low-energy terrestrial biospheres that can only support a low total biomass that may be difficult to detect against the backdrop of abiological physical and chemical processes. Biomass distribution in these settings is highly heterogeneous; key is the use of preserved environmental guideposts—that is, mineralogic or physical interfaces, to first identify the energetically favorable locales for life and then search for the biosignatures at microscopic scales (Onstott et al. 2019; Table 12-2). Earth's subsurface provides important opportunities to test and develop understanding of biosignature detectability—where the goal is to "find the biologic needle in the abiotic haystack"—the inverse problem to the norm of our "life-saturated" planet where abiotic analogs of prebiotic processes are the proverbial needles in a biological haystack (Sherwood Lollar et al. 2021; Smith et al. 2021).

TABLE 12-2 Biosignature Reliability, Detectability, and Preservation Vary with Time in the Geologic Record

Age (Ma)	Biosignature Type Reported					Forms	Formation Type (location)	Ref.
	Isotope	*Geochemistry/ Mineralogy*	*Morphology*	*Organic Carbon*	*Lipid*			
1			X	X		filaments	Ries Impact Crater (Germany)	a, b
2	X	X	X			concretions	Navajo Sandstone (U.S.)	c
15		X	X			microcolonies	Columbia River Basalt (U.S.)	d
60	X	X				concretions	Moeraki Formation (New Zealand)	e
84–95	X	X	X			concretions	Gammon Shale, Mancos Shale Formation, Fronter Formation (U.S.)	f, g
115–400	X	X	X	X	X	microcolonies	Fennoscandian shield granite (N. Europe)	h–j
120	X	X	X	X	X	microcolonies	Southern Iberia Abyssal Plain (Atlantic Ocean)	k
152	X	X				concretions	Kimmeridge Clay (England)	l
315	X	X				concretions	Lower Westphalian coal (Scotland)	o
355	X			X	X	filaments	Fennoscandian shield granite (Europe)	m
385	X	X	X			filaments	Arnstein pillow basalt (Germany)	p, q
388			X			filaments	Tynet Burn limestone (Scotland)	r
443	X	X				ichnofossils	Caledonian ophiolite (New Caledonia)	s
458			X	X		filaments	Lockne Impact Structure (Sweden)	t
551	X		X			concretions	Doushantuo Formation (China)	u
1,175		X				reduction spheroids	Bay of Stoer Formation (Scotland)	n
1,950	X	X				ichnofossils	Jormua ophiolite complex (Finland)	v
2,900–3,350	X		X	X		ichnofossils	Euro Basalt (Australia)	w, y
3,240			X			filaments	Sulphur Springs Group (U.S.)	y
3,460	X						Dresser Formation (Australia)	z
3,465	X		X	X		microcolonies	Apex Chert (Australia)	za–zc
3,465	X		X			ichnofossils	Hooggenoeg Formation (Africa)	zd, ze
3,770–4,280			X			filaments	Nuvvuagittuq belt (Canada)	zf

[a] Sapers et al. (2014), [b] Sapers et al. (2015), [c] Loope et al. (2010), [d] McKinley et al. (2000), [e] Thyne and Boles (1989), [f] Coleman (1993), [g] McBride et al. (2003), [h] Pedersen et al. (1997), [i] Drake et al. (2015a), [j] Drake et al. (2015b), [k] Klein et al. (2015), [l] Irwin (1980), [m] Drake et al. (2017), [n] Spinks et al. (2010), [o] Curtis et al. (1986), [p] Drake et al. (2018) and Eickmann et al. (2009), [q] Peckmann et al. (2007), [r] Trewin and Knoll (1999), [s] Furnes et al. (2002), [t] Ivarsson et al. (2013), [u] Dong et al. (2008), [v] Furnes et al. (2005), [w] Banerjee et al. (2007), [x] McLoughlin et al. (2012), [y] Rasmussen (2000), [z] Shen and Buick (2004), [za] Schopf et al. (2018), [zb] Pinti et al. (2009), [zc] Ueno et al. (2006), [zd] Banerjee et al. (2006), [ze] Furnes et al. (2004), [zf] Dodd et al. (2017).
SOURCE: Modified from Onstott et al. (2019), CC BY 4.0.

Q9.5b Which Biosignatures Are Most Likely to Survive in the Environment, and at What Timescales of Preservation?

The timescales and mechanisms of survivability make up a core research area that includes the survivability of a modern biosignature against processes that would act to destroy it as it forms; the survivability of a fossil biosignature against geological forces such as weathering, diagenesis, influx of brines or groundwaters; and metamorphism, and/or the survivability of abiosignatures (inherently the study of the relative rates of abiotic and biological, and potentially cryptic, processes) (Hoehler and Jorgenson 2013; NASEM 2019a). At one end of the spectrum, informational biomolecules, such as DNA, are highly reliable biosignatures, but they are susceptible to rapid chemical and enzymatic degradation, making their detection in geological samples older than 1 million years highly problematic (NASEM 2019a). Similarly, fossilized biological materials are biased toward multicellular life, on Earth preserved primarily from the Phanerozoic. Table 12-2 demonstrates the categories of biomarkers and the shift in the balance of detectability and reliability as a function of preservation time (Onstott et al. 2019). In older materials, light stable isotopic biosignatures or organic biomarker compounds as examples, have greater survivability, although perhaps less specificity with respect to the nature of the processes that produced them. However, emerging techniques for determining isotopic fractionation of specific atoms within organic compounds offer an approach to overcoming these shortcomings (Hofmann et al. 2020). Morphological and mineralogical biosignatures (e.g., stromatolite morphology and biogenicity) can survive many of the geological forces that act to remove less robust biosignatures from the rock record but are considered much less reliable (Allwood et al. 2018). Continued field, modeling and laboratory-based research is required to inform our understanding of how to reliably probe Earth's geobiological record and apply those lessons to other habitable worlds (see also Question 11).

Q9.5c What Taphonomic Processes and Environmental Conditions Are Particularly Favorable for Biosignature Preservation on Earth?

Taphonomy is the field of study concerned with processes that result in the formation, preservation, alteration, and destruction of biosignatures. Generally speaking, the taphonomic conditions that favor biosignature preservation are those that promote fossilization through rapid authigenic mineralization and occlusion of porosity during burial, which act to isolate the biosignature from the forces (e.g., oxidation, erosion, and dissolution) that act to destroy them (NASEM 2019a). The taphonomic processes that serve to enhance the preservation of a biosignature, however, also impart biases in Earth's geobiological record. *Record bias* can occur when a particular environment and geologic repository are targeted for investigation to the exclusion of other lithologies (NASEM 2019a). *Preservation bias* is the tendency for environmental and geological forces to favor one "taphonomic window" over another. Together, these two tendencies can create an overarching sense that Earth's fossil record is written almost entirely in the form of sedimentary rock because of its inherently higher preservation and taphonomic potential (i.e., a preservation bias). This results in a strong tendency to look for the remains of organisms that thrive in surface and near-surface sedimentary environments (i.e., a record bias). Attempts to search for ancient forms of life outside of sedimentary systems (e.g., hydrothermal or deep subsurface settings; ancient continental systems versus younger marine crust and sediments) are faced with distinct challenges, although there are (sometimes controversial) examples that provide useful guideposts (e.g., Staudigel et al. 2015; Djokic et al. 2021) (see also Questions 5 and 6).

Q9.5d How Do "Unknown Unknowns" Impact the Search for Life Beyond Earth?

Identifying life in isolated refugia or ephemeral habitats on Earth (e.g., in the Atacama Desert; ice-free polar regions; hydrothermal vents, arctic and sea ice, and subsurface fracture fluids) has demonstrated that habitability, rather than being a binary state, is a continuum defined over varying time and spatial scales (see also Questions 5 and 10). Increased understanding of life's limits in so-called extreme environments (see Figure 12-3) has led to a resurgence of interest in adaptations of life to saline fluids and multi-parameter space (e.g., T-P-pH-Eh). The recent discovery of communities existing in the subsurface of the ocean floor and continental lithosphere, away from the influence of the Sun's energy, has provided new models for rock-hosted, chemosynthetic life that may exist on other worlds (Lin et al. 2006; Onstott et al. 2019; Dunham et al. 2021; Sauvage et al. 2021; Sherwood Lollar et al. 2021). Expanded understanding of habitability of subsurface environments, brine stability of chemosynthetic organisms, and adaptations of life to saline fluids, through continued field, laboratory, and modeling studies have widespread implications for the search for life in the solar system (NASEM 2019a).

Strategic Research for Q9.5

- **Assess the ways in which physical and chemical processes in Earth's habitable environments affect biosignature and abiosignature detectability and whether those processes favor the detection of certain biosignatures over others** using field studies, laboratory, theoretical, and remote sensing approaches designed to investigate the rates of biosignature production and destruction and rates of physical and chemical processes.

- **Investigate the processes and environmental conditions on Earth that are most likely to result in the preservation of biosignatures** with field studies of biosignature preservation in relevant planetary environments and laboratory studies of the alteration of proposed biosignatures under planetary environmental conditions.

- **Characterize on Earth the range of abiotic processes that define the abiotic (nonliving) baseline (abiosignatures), as well as those that are capable of generating false positives from mimics (e.g., inorganically generated morphological biosignatures) on Earth** with field/geological studies, laboratory studies focused on generating and characterizing abiosignatures, and analysis of meteorite falls and finds to assess their inventory of chemical compounds that could be mistaken for false positives by a planetary life-detection mission.

- **Explore the full range of habitable environments known on Earth and their preserved rock record to develop a more complete understanding of what it means to be "habitable," reduce record bias in our understanding of how the record of life is preserved, and mitigate against "unknown unknowns" that might confound the search for life outside of Earth** by conducting exploratory field studies of the biology, biomass, and processes of biosignature generation and preservation in isolated refugia, low-energy systems, and ephemeral environments and investigations of environments where the biotic signal to abiotic noise or baseline may be particularly low (e.g., so-called extreme, oligotrophic, or subsurface environments).

REFERENCES

Aksyonov, S.A., and P. Williams. 2001. "Impact Desolvation of Electrosprayed Microdroplets—A New Ionization Method for Mass Spectrometry of Large Biomolecules." *Rapid Communications in Mass Spectrometry* 15(21):2001–2006.

Allwood, A.C., M.T. Rosing, D.T. Flannery, J.A. Hurowitz, C.M. Heirwegh. 2018. "Reassessing Evidence of Life in 3,700-Million-Year-Old Rocks of Greenland." *Nature* 563:241–244.

Banerjee, N., H. Furnes, K. Muehlenbachs, H. Staudigel, and M. de Wit. 2006. "Preservation of ca. 3.4-3.5 Ga Microbial Biomarkers in Pillow Lavas and Hyaloclastites from the Barberton Greenstone Belt, South Africa." *Earth and Planetary Science Letters* 241:707–722.

Banerjee, N., A. Simonetti, H. Furnes, K. Muehlenbachs, H. Staudigel, L. Heaman, and M.J. Van Kranendonk. 2007. "Direct Dating of Archean Microbial Ichnofossils." *Geology* 35:487–490.

Baross, J.A., R.E. Anderson, and E.E. Stüeken. 2020. "The Environmental Roots of the Origin of Life." Pp. 71–92 in *Planetary Astrobiology*, V.S. Meadows, G.N. Arney, B. Schmidt, and D.J. Des Marais, eds. Tucson: University of Arizona Press.

Becraft, E.D., M.C.Y. Lau Vetter, O.K.I. Bezuidt, J.M. Brown, J.M. Labonté, K. Kauneckaite-Griguole, et al. 2021. "Evolutionary Stasis of a Deep Subsurface Microbial Lineage." *ISME Journal* 15:2830–2842.

Benner, S.A., E.A. Bell, E. Biondi, R. Brasser, T. Carell, H.-J. Kim, S.J. Mojzsis, A. Omran, M.A. Pasek, and D. Trail. 2020. "When Did Life Likely Emerge on Earth in an RNA-First Process?" *ChemSystemsChem* 2:e2000010.

Betts, H.C., M.N. Puttick, J.W. Clark, et al. 2018. "Integrated Genomic and Fossil Evidence Illuminates Life's Early Evolution and Eukaryote Origin." *Nature Ecology and Evolution* 2:1556–1562.

Bottke, W.F., and M.D. Norman. 2017. "The Late Heavy Bombardment." *Annual Review of Earth and Planetary Sciences* 45:619–647. https://doi.org/10.1146/annurev-earth-063016-020131.

Cable, M.L., H.V. Tuan, H.E. Maynard-Casely, M. Choukroun, and R. Hodyss. 2018. "The Acetylene-Ammonia Co-Crystal on Titan." *ACS Earth and Space Chemistry* 2:366–375.

Chan, M.A., N.W. Hinman, S.L. Potter-McIntyre, K.E. Schubert, R.J. Gillams, S.M. Awramik, P.J. Boston, et al. 2019. "Deciphering Biosignatures in Planetary Contexts." *Astrobiology* 19(9):1075–1102. https://doi.org/10.1089/ast.2018.1903.

Coleman, M.L. 1993. "Microbial Processes: Controls on the Shape and Composition of Carbonate Concretions." *Marine Geology* 113:127–140.

Creamer, J.S., M.F. Mora, and P.A. Willis. 2017. "Enhanced Resolution of Chiral Amino Acids with Capillary Electrophoresis for Biosignature Detection in Extraterrestrial Samples." *Analytical Chemistry* 89(2):1329–1337.

Curtis, C.D., M.L. Coleman, and L.G. Love. 1986. "Pore Water Evolution During Sediment Burial from Isotopic and Mineral Chemistry of Calcite, Dolomite and Siderite Concretions." *Geochimica et Cosmochimica Acta* 50:2321–2334.

Damer, B., and D. Deamer. 2020. "The Hot Spring Hypothesis for an Origin of Life." *Astrobiology* 20:429–452.

Des Marais, D.J., L.J. Allamandola, S.A. Benner, A.P. Boss, D. Deamer, P.G. Falkowski, J.D. Farmer, et al. 2003. "The NASA Astrobiology Roadmap." *Astrobiology* 3(2):219–235.

Des Marais, D.J., J.A. Nuth, L.J. Allamandola, A.P. Boss, J.D. Farmer, T.M. Hoehler, B.M. Jakosky, et al. 2008. "The NASA Astrobiology Roadmap." *Astrobiology* 8:715–730. https://doi.org/10.1089/ast.2008.0819.

D'Hondt, S., A.J. Spivack, R. Pockalny, T.G. Ferdelman, J.P. Fischer, J. Kallmeyer, L.J. Abrams, et al. 2009. "Subseafloor Sedimentary Life in the South Pacific Gyre." *Proceedings of the National Academy of Sciences* 106:11651–11656.

D'Hondt, S., R. Pockalny, V.M. Fulfer, and A.J. Spivack. 2019. "Subseafloor Life and Its Biogeochemical Impacts." *Nature Communications* 10:3519.

Djokic, T., M.J. Van Kranendonk, K.A. Campbell, J.R. Havig, M.R. Walter, and D.M. Guido. 2021. "A Reconstructed Subaerial Hot Spring Field in the ~3.5 Billion-Year-Old Dresser Formation, North Pole Dome, Pilbara Craton, Western Australia." *Astrobiology* 21(1):1–38.

Dodd, M.S., D. Papineau, T. Grenne, J.F. Slack, M. Rittner, F. Pirajno, J. O'Neil, and C.T.S. Little. 2017. "Evidence for Early Life in Earth's Oldest Hydrothermal Vent Precipitates." *Nature* 543:60–64.

Dong, J., S. Zhang, G. Jiang, Q. Zhao, H. Li, X. Shi, and J. Liu. 2008. "Early Diagenetic Growth of Carbonate Concretions in the Upper Doushantuo Formation in South China and Their Significance for the Assessment of Hydrocarbon Source Rock." *Science in China Series D: Earth Science* 51:1330–1339.

Drake, H., M.E. Åström, C. Heim, C. Broman, J. Åström, M. Whitehouse, M. Ivarsson, S. Siljeström, and P. Sjövall. 2015a. "Extreme ¹³C Depletion of Carbonates Formed During Oxidation of Biogenic Methane in Fractured Granite." *Nature Communications* 6:7020. https://doi.org/10.1038/ncomms8020.

Drake, H., E.L. Tullborg, M. Whitehouse, B. Sandberg, T. Blomfeldt, and M.E. Åström. 2015b. "Extreme Fractionation and the Microscale Variation of Sulphur Isotopes During Bacterial Sulphate Reduction in Deep Groundwater Systems." *Geochimica et Cosmochimica Acta* 161:1–18.

Drake, H., C. Heim, N.M.W. Roberts, T. Zack, M. Tillberg, C. Bromane, M. Ivarsson, M.J. Whitehouse, and M.E. Åström. 2017. "Isotopic Evidence for Microbial Production and Consumption of Methane in the Upper Continental Crust Throughout the Phanerozoic Eon." *Earth and Planetary Science Letters* 470:108–118.

Drake, H., M.J. Whitehouse, C. Heim, P.W. Reiners, M. Tillberg, K.J. Hogmalm, M. Dopson, C. Broman, and M.E. Åström. 2018. "Unprecedented 34S-Enrichment of Pyrite Formed Following Microbial Sulfate Reduction in Fractured Crystalline Rocks." *Geobiology* 16:556–574.

Dunham, E.C., J.E. Dore, M.L. Skidmore, E.E. Roden, and E.S. Boyd. 2021. "Lithogenic Hydrogen Supports Microbial Primary Production in Subglacial and Proglacial Environments." *Proceedings of the National Academy of Sciences* 118(2):e2007051117.

Ehlmann, B.L., J.F. Mustard, S.L. Murchie, J.P. Bibring, A. Meunier, A.A. Fraeman, and Y. Langevin. 2011. "Subsurface Water and Clay Mineral Formation During the Early History of Mars." *Nature* 479:53–60.

Ehlmann, B.L., F.S. Anderson, J. Andrews-Hanna, D.C. Catling, P.R. Christensen, B.A. Cohen, C.D. Dressing, et al. 2016. "The Sustainability of Habitability on Terrestrial Planets: Insights, Questions, and Needed Measurements from Mars for Understanding the Evolution of Earth-Like Worlds." *Journal of Geophysical Research: Planets* 121. https://doi.org/10.1002/ 2016JE005134.

Eickmann, B., W. Bach, S. Kiel, J. Reitner, and J. Peckmann. 2009. "Evidence for Cryptoendolithlc Life in Devonian Pillow Basalts of Variscan Orogens, Germany." *Palaeogeography Palaeoclimatology Palaeoecology* 283:120–125.

Etiope, G., and B. Sherwood Lollar. 2013. "Abiotic Methane on Earth." *Review of Geophysics* 51:276–299.

Falkowski, P.G., T. Fenchel, and E.F. Delong. 2008. "The Microbial Engines That Drive Earth's Biogeochemical Cycles." *Science* 320:1034–1039. https://doi.org/10.1126/science.1153213.

Fani, R. 2012. "The Origin and Evolution of Metabolic Pathways: Why and How Did Primordial Cells Construct Metabolic Routes?" *Evolution: Education and Outreach* 5:367–381.

Forterre, P. 2006. "Three RNA Cells for Ribosomal Lineages and Three DNA Viruses to Replicate Their Genomes: A Hypothesis for the Origin of Cellular Domain." *Proceedings of the National Academy of Sciences* 103:3669–3674.

Furnes, H., K. Muehlenbachs, T. Torsvik, O. Tumyr, and L. Shi. 2002. "Biosignatures in Metabasaltic Glass of a Caledonian Ophiolite, West Norway." *Geological Magazine* 139:601–608.

Furnes, H., N.R. Banerjee, K. Muehlenbachs, H. Staudigel, and M. de Wit. 2004. "Early Life Recorded in Archean Pillow Lavas." *Science* 304:578–581.

Furnes, H., N.R. Banerjee, K. Muehlenbachs, and A. Kontinen. 2005. "Preservation of Biosignatures in Metaglassy Volcanic Rocks from the Jormua Ophiolite Complex, Finland." *Precambrian Research* 136:125–137.

Grotzinger, J.P., and A.H. Knoll. 1999. "Stromatolites in Precambrian Carbonates: Evolutionary Mileposts or Environmental Dipsticks?" *Annual Review of Earth and Planetary Sciences* 27:313–358.

Hand, K.P., A.E. Murray, J.B. Garvin, W.B. Brinckerhoff, B.C. Christner, K.S. Edgett, B.L. Ehlmann, et al. 2017. *Report of the Europa Lander Science Definition Team.* https://solarsystem.nasa.gov/docs/Europa_Lander_SDT_Report_2016.pdf.

Hand, K.P., C. Sotin, A. Hayes, and A. Coustenis. 2020. "On the Habitability and Future Exploration of Ocean Worlds." *Space Science Reviews* 216(5):95. https://doi.org/10.1007/s11214-020-00713-7.

Head, I.M., N.D. Gray, and S.R. Larter. 2014. "Life in the Slow Lane: Biogeochemistry of Biodegraded Petroleum Containing Reservoirs and Implications for Energy Recovery and Carbon Management." *Frontiers in Microbiology* 5. https://doi.org/10.3389/fmicb.2014.00566.

Hendrix, A.R., T.A. Hurford, L.M. Barge, M.T. Bland, J.S. Bowman, W. Brinckerhoff, B.K. Buratti, et al. 2019. "The NASA Roadmap to Ocean Worlds." *Astrobiology* 19. https://doi.org/10.1089/ast.2018.1955.

Heuer, V.B., F. Inagaki, Y. Morono, Y. Kubo, A.J. Spivack, B. Viehweger, T. Treude, et al. 2020. "Temperature Limits to Deep Subseafloor Life in the Nankai Trough Subduction Zone." *Science* 370:1230–1234.

Hoefs, J. 2004. *Stable Isotope Geochemistry.* Berlin: Springer-Verlag.

Hoehler, T.M., and B.B. Jørgensen. 2013. "Microbial Life Under Extreme Energy Limitation." *Nature Reviews in Microbiology* 11:83–94.

Hoehler, T.M., W. Bains, A. Davila, M.N. Parenteau, and A. Pohorille. 2020. "Life's Requirements, Habitability, and Biological Potential." Pp. 37–70 in *Planetary Astrobiology*, V.S. Meadows, G.N. Arney, B. Schmidt, and D.J. Des Marais, eds. Tucson: University of Arizona Press.

Hofmann, A.E., L. Chimiak, B. Dallas, J. Griep-Raming, D. Juchelka, A. Makarov, J. Schwieters, et al. 2020. "Using Orbitrap Mass Spectrometry to Assess the Isotopic Compositions of Individual Compounds in Mixtures." *International Journal of Mass Spectrometry* 457:116410. https://doi.org/10.1016/j.ijms.2020.116410.

Hug, L., B. Baker, K. Anantharaman, C.T. Brown, A.J. Probst, C.J. Castelle, C.N. Butterfield, et al. 2016. "A New View of the Tree of Life." *Nature Microbiology* 1:16048. https://doi.org/10.1038/nmicrobiol.2016.48.

Imachi, H., M.K. Nobu, N. Nakahara, Y. Morono, M. Ogawara, Y. Takaki, Y. Takano, et al. 2020. "Isolation of an Archaeon at the Prokaryote–Eukaryote Interface." *Nature* 577:519–525.

Inagaki, F., K.-U. Hinrichs, Y. Kubo, M.W. Bowles, V.B. Heuer, W.-L. Hong, T. Hoshino, et al. 2015. "Exploring Deep Microbial Life in Coal-Bearing Sediment Down to 2.5 km Below the Ocean Floor." *Science* 349:420–424.

Irwin, H. 1980. "Early Carbonate Precipitation and Pore Fluid Migration in the Kimmeridge Clay of Dorset, England." *Sedimentology* 27:577–591.

Ivarsson, M., C. Broman, E. Sturkell, J. Ormö, S. Siljeström, M. van Zuilen, and S. Bengtson. 2013. "Fungal Colonization of an Ordovician Impact-Induced Hydrothermal System." *Science Reports* 3. https://doi.org/10.1038/srep03487.

Kempes, C.P., M.A.R. Koehl, and G.B. West. 2019. "The Scales That Limit: The Physical Boundaries of Evolution." *Frontiers in Ecology and Evolution* 7. https://doi.org/10.3389/fevo.2019.00242.

Kempes, C.P., S.I. Walker, H.V. Graham, C.H. House, and H.L. Smith. 2021. "Generalized Stoichiometry and Biogeochemistry for Astrobiological Applications." *Bulletin of Mathematical Biology* 83:73. https://doi.org/10.1007/s11538-021-00877-5.

Kim, H., H.B. Smith, C. Mathis, J. Raymond, and S.I. Walker. 2019. "Universal Scaling Across Biochemical Networks on Earth." *Science Advances* 5(1):eaau0149.

Klein, F., S.E. Humphris, W. Guo, F. Schubotz, E.M. Schwarzenbach, and W.D. Orsi. 2015. "Fluid Mixing and the Deep Biosphere of a Fossil Lost City-Type Hydrothermal System at the Iberia Margin." *Proceedings of the National Academy of Sciences* 112:12036–12041.

Koonin, E.V., V.V. Doija, M. Krupovic, and J.H. Kuhn. 2021. "Viruses Defined by the Position of the Virosphere Within the Replicator Space." *Microbiology and Molecular Biology Reviews* 85(4):e00193-20.

Lang, S.Q., G.L. Früh-Green, S.M. Bernasconi, W.J. Brazelton, M.O. Schrenk, and J.M. McGonigle. 2018. "Deeply-Sourced Formate Fuels Sulfate Reducers But Not Methanogens at Lost City Hydrothermal Field." *Scientific Reports* 8:755. https://doi.org/10.1038/s41598-017-19002-5.

LaRowe, D., and J. Amend. 2019. "Energy Limits for Life in the Subsurface." Pp. 585–619 in *Deep Carbon: Past to Present*, B. Orcutt, I. Daniel, and R. Dasgupta, eds. Cambridge, UK: Cambridge University Press.

Li, L., B.A. Wing, T.H. Bui, J.M. McDermott, G.F. Slater, S. Wei, G. Lacrampe-Couloume, et al. 2016. "Sulfur Mass-Independent Fractionation in Subsurface Fracture Waters Indicates a Long-Standing Sulfur Cycle in Precambrian Rocks." *Nature Communications* 7:13252.

Lin, L.-H., P. Wang, D. Rubmle, J. Lippmann-Pipke, E. Boice, L.M. Pratt, B. Sherwood Lollar, et al. 2006. "Long-Term Sustainability of a High-Energy, Low-Diversity Crustal Biome." *Science* 314:479–482.

Lollar, G.S., O. Warr, J. Telling, M.R. Osburn, and B. Sherwood Lollar. 2019. "'Follow the Water': Hydrogeochemical Constraints on Microbial Investigations 2.4 km Below Surface at the Kidd Creek Deep Fluid and Deep Life Observatory." *Geomicrobiology Journal* 36(10):859–872.

Loope, D.B., R.M. Kettler, and K.A. Weber. 2010. "Follow the Water: Connecting a CO_2 Reservoir and Bleached Sandstone to Iron-Rich Concretions in the Navajo Sandstone of South-Central Utah, USA." *Geology* 38:999–1002.

Lu, R., C. Cheng, T. Nagel, H. Milsch, H. Yasuhara, O. Kolditz, and H. Shao. 2021. "What Process Causes the Slowdown of Pressure Solution Creep." *Geomechanics and Geophysics for Geo-Energy and Geo-Resources* 7:57. https://doi.org/10.1007/s40948-021-00247-4.

Lyons, T., C. Reinhard, and N. Planavsky. 2014. "The Rise of Oxygen in Earth's Early Ocean and Atmosphere." *Nature* 506:307–315. https://doi.org/10.1038/nature13068.

MacKenzie, S.M., S.P.D. Birch, S. Horst, C. Sotin, E. Barth, J.M. Lora, M.G. Trainer, et al. 2021. "Titan: Earth-Like on the Outside, Ocean World on the Inside." *arXiv* 2102.08472v1.

Magnabosco, C., L.-H. Lin, H. Dong, M. Bomberg, W. Ghiorse, H. Stan-Lotter, K. Pedersen, et al. 2018. "The Biomass and Biodiversity of the Continental Subsurface." *Nature Geosciences* 11:707–717.

Magnus, I., and S. Holmström. 2012. "The Use of ESEM in Geobiology." *In Scanning Electron Microscopy*, K. Viacheslav, ed. https://www.intechopen.com/chapters/30964.

Marshall, S.M., C. Mathis, E. Carrick, P.S. Gromski, G.J.T. Cooper, G. Keenan, H.V. Graham, et al. 2021. "Identifying Molecules as Biosignatures with Assembly Theory and Mass Spectrometry." *Nature Communications* 12:3033.

Martin, W., J. Baross, D. Kelley, and M.J. Russell. 2008. "Hydrothermal Vents and the Origin of Life." *Nature Reviews Microbiology* 6:805–814.

McBride, E.F., M.D. Picard, and K.L. Milliken. 2003. "Calcite Cemented Concretions in Cretaceous Sandstone, Wyoming and Utah, U.S.A." *Journal of Sedimentary Research* 73:462–483.

McDermott, J.M., J.S. Seewald, C.R. German, and S.P. Sylva. 2015. "Pathways for Abiotic Organic Synthesis at Submarine Hydrothermal Fields." *Proceedings of the National Academy of Sciences* 112:7668–7672.

McKinley, J.P., T.O. Stevens, and F. Westall. 2000. "Microfossils and Paleoenvironments in Deep Subsurface Basalt Samples." *Geomicrobiology Journal* 17:43–54.

McLoughlin, N., E.G. Grosch, M.R. Kilburn, and D. Wacey. 2012. "Sulfur Isotope Evidence for a Paleoarchean Subseafloor Biosphere, Barberton, South Africa." *Geology* 40:1031–1034.

Meadows, V.S. 2017. "Reflections on O_2 as a Biosignature in Exoplanetary Atmospheres." *Astrobiology* 17:1022–1052.

Meadows, V.S., C.T. Reinhard, G.N. Arney, M.N. Parenteau, E.W. Schwieterman, S.D. Domagal-Goldman, A. Lincowski, et al. 2018. "Exoplanet Biosignatures: Understanding Oxygen as a Biosignature in the Context of Its Environment. *Astrobiology* 18:630–662.

Ménez, B., V. Pasini, and D. Brunelli. 2012. "Life in the Hydrated Suboceanic Mantle." *Nature Geoscience* 5:133–137.

Ménez, B., C. Pisapia, M. Andreani, F. Jamme, Q.P. Vanbellingen, A. Brunelle, L. Richard, et al. 2018. "Abiotic Synthesis of Amino Acids in the Recesses of the Oceanic Lithosphere." *Nature* 564:59–63.

Merino, N., H.S. Aronson, D.P. Bojanova, J. Feyhl-Buska, M.L. Wong, S. Zhang, and D. Giovannelli. 2019. "Living at the Extremes: Extremophiles and the Limits of Life in a Planetary Context." *Frontiers in Microbiology* 10:780. https://doi.org/10.3389/fmicb.2019.00780.

Mojzsis, S.J., G. Arrhenius, K.D. McKeegan, T.M. Harrison, A.P. Nutman, and C.R.L. Friend. 1996. "Evidence for Life on Earth Before 3,800 Million Years Ago." *Nature* 384:55–59.

Mustard, J.F., M. Adler, A. Allwood, D.S. Bass, D.W. Beaty, J.F. Bell III, W.B. Brinckerhoff, et al. 2013. *Report of the Mars 2020 Science Definition Team*. Mars Exploration Program Analysis Group (MEPAG). http://mepag.jpl.nasa.gov/reports/MEP/Mars_2020_SDT_Report_Final.pdf.

Mykytczuk, N.C.S., S.J. Foote, C.R. Omelon, G. Southam, C.W. Greer, and L.G. Whyte. 2013. "Bacterial Growth at −15°C; Molecular Insights from the Permafrost Bacterium *Planococcus halocryophilus* Or1." *The ISME Journal* 7:1211–1226.

NASA (National Aeronautics and Space Administration). 2015. *NASA Astrobiology Strategy 2015*. Washington, DC: NASA Astrobiology Program. https://astrobiology.nasa.gov/nai/media/medialibrary/2015/10/NASA_Astrobiology_Strategy_2015_151008.pdf.

NASA. 2020. *Enceladus Orbilander: A Flagship Mission Concept for Astrobiology*. Planetary Mission Concept Study for the 2023–2032 Decadal Survey. Johns Hopkins University Applied Physics Laboratory. https://science.nasa.gov/files/atoms/files/Enceladus%20Orbilander.pdf.

NASEM (National Academies of Sciences, Engineering, and Medicine). 2019a. *An Astrobiology Strategy for the Search for Life in the Universe*. Washington, DC: The National Academies Press.

Neveu, M., L.E. Hays, M.A. Voytek, M.H. New, and M.D. Schulte. 2018. "The Ladder of Life Detection." *Astrobiology* 18:1375–1402.

Olson, S.L., E.W. Schwieterman, C.T. Reinhard, and T.W. Lyons. 2018. "Earth: Atmospheric Evolution of a Habitable Planet." Pp. 2817–2853 in *Handbook of Exoplanets*, H. Deeg and J. Belmonte, eds. Cham, Switzerland: Springer. https://doi.org/10.1007/978-3-319-55333-7_189.

Onstott, T.C., B.L. Ehlmann, H. Sapers, M. Coleman, M. Ivarsson, J.J. Marlow, A. Neubeck, and P. Niles. 2019. "Paleo-Rock-Hosted Life on Earth and the Search on Mars: A Review and Strategy for Exploration." *Astrobiology* 19(10):1230–1262. https://doi.org/10.1089/ast.2018.1960.

Pace, N.R. 1997. "A Molecular View of Microbial Diversity and the Biosphere." *Science* 276:734–740.

Peckmann, J., W. Bach, K. Behrens, and J. Reitner. 2007. "Putative Cryptoendolithic Life in Devonian Pillow Basalt, Rheinisches Schiefergebirge, Germany." *Geobiology* 6:125–135.

Pedersen, K., S. Ekendahl, E.-L. Tullborg, H. Furnes, I. Thorseth, and O. Tumyr. 1997. "Evidence of Ancient Life at 207 m Depth in a Granitic Aquifer." *Geology* 25:827–830.

Pinti, D.L., R. Mineau, and V. Clement. 2009. "Hydrothermal Alteration and Microfossil Artefacts of the 3,465-Million-Year-Old Apex Chert." *Nature Geosciences* 2:640–643.

Rasmussen, B. 2000. "Filamentous Microfossils in a 3,235-Million-Year-Old Volcanogenic Massive Sulphide Deposit." *Nature* 405:676–679.

Reveillaud, J., E. Reddington, J. McDermott, C. Algar, J.L. Meyer, S. Sylva, J. Seewald, et al. 2016. "Subseafloor Microbial Communities in Hydrogen-Rich Vent Fluids from Hydrothermal Systems Along the Mid-Cayman Rise." *Environmental Microbiology* 8:1970–1987.

Rosing, M.T., N.M. Rose, D. Brigwater, and H.S. Thomsen. 1996. "Earliest Part of Earth's Stratigraphic Record: A Reappraisal of the 3.7 Ga Isua (Greenland) Supracrustal Sequence." *Geology* 24(1):43–46.

Sapers, H.M., G.R. Osinski, N.R. Banerjee, and L.J. Preston. 2014. "Enigmatic Tubular Features in Impact Glass." *Geology* 42:471–474.

Sapers, H.M., N.R. Banerjee, and G.R. Osinski. 2015. "Potential for Impact Glass to Preserve Microbial Metabolism." *Earth and Planet Science Letters* 430:95–104.

Sauvage, J.F., A. Flinders, A.J. Spivack, R. Pockalny, A.G. Dunlea, C.H. Anderson, D.C. Smith, et al. 2021. "The Contribution of Water Radiolysis to Marine Sedimentary Life." *Nature Communications* 12(1):1297. https://doi.org/10.1038/s41467-021-21218-z.

Schidlowski, M., P.W.U. Appel, R. Eichman, and C.E. Junge. 1979. "Carbon Isotope Geochemistry of the 3.7×10^9-Yr-Old Isua Sediments, West Greenland: Implications for the Archaean Carbon and Oxygen Cycles." *Geochimica et Cosmochimica Acta* 43:189–199.

Schopf, J.W., K. Kitajima, M.J. Spicuzza, A.B. Kudryavtsev, and J.W. Valley. 2018. "SIMS Analyses of the Oldest Known Assemblage of Microfossils Document Their Taxon-Correlated Carbon Isotope Compositions." *Proceedings of the National Academy of Sciences* 115:53–58.

Schrenk, M.O., W.J. Brazelton, and S.Q. Lang. 2013. "Serpentinization, Carbon, and Deep Life." *Reviews in Mineralogy and Geochemistry* 75:575–606.

Shen, Y., and R. Buick. 2004. "The Antiquity of Microbial Sulfate Reduction." *Earth-Science Reviews* 64:243–272.

Sherwood Lollar, B., T.C. Onstott, G. Lacrampe-Couloume, and C.J. Ballentine. 2014. "The Contribution of the Precambrian Continental Lithosphere to Global H_2 Production." *Nature* 516:379–382.

Sherwood Lollar, B., V.B. Heuer, J. McDermott, S. Tille, O. Warr, J.J. Moran, J. Telling, and K.-U. Hinrichs. 2021. "A Window into the Abiotic Carbon Cycle—Acetate and Formate in Fracture Waters in 2.7 Billion Year Old Host-Rocks of the Canadian Shield." *Geochimica et Cosmochimica Acta* 294:295–314.

Shock, E.L., and M.D. Schulte. 1998. "Organic Synthesis During Fluid Mixing in Hydrothermal Systems." *Journal of Geophysical Research* 103(E12):28513–28527.

Sittler, E.C., J.F. Cooper, S.J. Sturner, and A. Asgraf. 2020. "Titan's Ionospheric Chemistry, Fullerenes, Oxygen, Galactic Cosmic Rays and the Formation of Exobiological Molecules on and Within Its Surfaces and Lakes." *Icarus* 344:113246.

Smith, E., and H.J. Morowitz. 2016. "The Origin and Nature of Life on Earth: The Emergence of the Fourth Geosphere." *Contemporary Physics* 58:115–116.

Smith, H.L., A.K. Hyde, D.N. Simkus, H.V. Graham, and C.H. House. 2021. "The Grayness of the Origin of Life." *Life* 11:498. https://doi.org/10.3390/life11060498.

Sparks, W.B., B.E. Schmidt, M.A. McGrath, K.P. Hand, J.E. Spencer, M. Cracraft, and S.E. Deustua. 2017. "Active Cryovolcanism on Europa?" *Astrophysical Journal Letters* 839(2):L18. https://doi.org/10.3847/2041-8213/aa67f8.

Spinks, S.C., J. Parnell, and S.A. Bowden. 2010. "Reduction Spots in the Mesoproterozoic Age: Implications for Life in the Early Terrestrial Record." *Astrobiology* 9:209–216.

Stamenkovic, V., L.W. Beegle, K. Zacny, D.D. Arumugam, P. Baglioni, N. Barba, J. Baross, et al. 2019. "The Next Frontier for Planetary and Human Exploration." *Nature Astronomy* 3:116–120. https://doi.org/10.1038/s41550-018-0676-9.

Staudigel, H., H. Furnes, and M. DeWit. 2015. "Paleoarchean Trace Fossils in Altered Volcanic Glass." *Proceedings of the National Academy of Sciences* 112:6892–6897.

Steele, A., F.M. McCubbin, M. Fries, L. Kater, N.Z. Boctor, M.L. Fogel, P.G. Conrad, et al. 2012. "A Reduced Organic Carbon Component in Martian Basalts." *Science* 337:212–215.

Stevens, E.W.N., J.V. Bailey, B.E. Flood, et al. 2015. "Barite Encrustation of Benthic Sulfur-Oxidizing Bacteria at a Marine Cold Seep." *Geobiology* 13:588–603.

Summons, R.E., P. Albrecht, G. Mcdonald, and J.M. Moldowan. 2008. "Molecular Biosignatures." *Space Science Reviews* 135:133–159. https://doi.org/10.1007/s11214-007-9256-5.

Suzuki, Y., S. Yamashita, M. Kouduka, Y. Ao, H. Mukai, S. Mitsunobu, H. Kagi, et al. 2020. "Deep Microbial Proliferation at the Basalt Interface in 33.5–104 Million-Year-Old Oceanic Crust." *Communications Biology* 3:136. https://doi.org/10.1038/s42003-020-0860-1.

Thyne, G.D., and J.R. Boles. 1989. "Isotopic Evidence for Origin of Moeraki Septarian Concretions, New Zealand." *Journal of Sedimentary Petrology* 59:272–279.

Trembath-Reichert, E., Y. Morono, A. Ijiri, T. Hoshino, K.S. Dawson, F. Inagaki, and V.J. Orphan. 2017. "Methyl-Compound Use and Slow Growth Characterize Microbial Life in 2 km-Deep Subseafloor Coal and Shale Beds." *Proceedings of the National Academy of Sciences* 114:44.

Trewin, N.H., and A.H. Knoll. 1999. "Preservation of Devonian Chemotrophic Filamentous Bacteria in Calcite Veins." *Palaios* 14:288–294.

Ueno, Y., K. Yamada, N. Yoshida, S. Maruyama, and Y. Isozaki, Y. 2006. "Evidence from Fluid Inclusions for Microbial Methanogenesis in the Early Archaean Era." *Nature* 440:526–519.

Van Zuilen, M.A., A. Lepland, J. Teranes, J. Finarelli, M. Wahlen, and G. Arrhenius. 2003. "Graphite and Carbonates in the 3.8 Ga Old Isua Supracrustal Belt, Southern West Greenland." *Precambrian Research* 126:331–348.

Vance, S., J. Harnmeijer, J. Kimura, H. Hussmann, B. Demartin, and J.M. Brown. 2007. "Hydrothermal Systems in Small Ocean Planets." *Astrobiology* 7:987–1005.

Waite, J.H., C.R. Glein, R.S. Perryman, B.D. Teolis, B.A. Magee, G. Miller, J. Grimes, et al. 2017. "Cassini Finds Molecular Hydrogen in the Enceladus Plume: Evidence for Hydrothermal Processes." *Science* 356(6334):155–159.

Walker, S.I. 2019. "The New Physics Needed to Probe the Origins of Life." *Nature* 569:36–38.

Wierzchos, J., M.C. Casero, O. Artieda, and C. Ascaso. 2018. "Endolithic Microbial Habitats as Refuges for Life in Poly-Extreme Environment of the Atacama Desert." *Current Opinion in Microbiology* 43:124–131.

Woese, C.R., O. Kandler, and M.L. Wheelis. 1990. "Towards a Natural System of Organisms: Proposal for the Domains Archaea, Bacteria, and Eucarya." *Proceedings of the National Academy of Sciences* 87(12):4576–4579. https://doi.org/10.1073/pnas.87.12.4576.

Q10 PLATE: A composite image of Saturn's moons, Enceladus *(foreground)* and Titan *(background)*, as seen by the Cassini spacecraft. The moons are not shown to scale.
SOURCE: Composed by P.K. Byrne. Images: Courtesy of NASA/JPL-Caltech/Space Science Institute.

13

Question 10: Dynamic Habitability

Where in the solar system do potentially habitable environments exist, what processes led to their formation, and how do planetary environments and habitable conditions co-evolve over time?

The past several decades of exploration of Earth's biosphere have expanded our knowledge of the range of environments with liquid water, nutrients, and energy sources that sustain life.[1] Simultaneously, the past several decades of planetary exploration have revealed multiple ancient and modern potential habitable environments across the solar system. The study of planetary habitability requires understanding the factors controlling habitability, the evolutionary pathways by which an environment became habitable, and the processes that sustain, enhance, diminish, or even extinguish habitability. Increasingly, we understand that habitability is not a yes/no proposition but a continuum—that is, more habitable, less habitable, not habitable—and that planetary environments transition from habitable to not habitable, and vice versa, over space and time. Feedbacks and interrelationships between stellar, dynamical, and planetary evolution drive environmental change and changes in habitability. Life itself can also modify environments and their habitability (Question 9, Chapter 12). Understanding the dynamic habitability of bodies in the solar system through study of their past, current, and future evolutionary trajectories lays the foundation for the study of potentially habitable worlds beyond the solar system.

Here, the committee reviews the state of knowledge of habitability, including habitability of solar system worlds. The committee then examines key aspects controlling habitability: the availability of water, organics, nutrients (e.g., the main chemical elements, CHNOPS), and energy as well as the role of time and the stability or continuity of habitability—that is, the dynamical component of habitability. The chapter concludes with recommended activities and measurements to advance knowledge of solar system habitability.

Q10.1 WHAT IS HABITABILITY?

Planetary habitability is the measure of a body's potential to develop and sustain life. Because there is no current example of life beyond Earth, planetary habitability is largely an extrapolation of conditions under which we find life on Earth (Question 9), as well as knowledge of the characteristics of the Sun and solar system that appear favorable for life to develop and flourish.

---

[1] A glossary of acronyms and technical terms can be found in Appendix F.

339

At present, with a single example of an inhabited world (Earth), we lack the information to fully understand the conditions that lead to—or prohibit—the origin and sustenance of life on worlds in the solar system. Through our study of life on Earth, however, we can measure the conditions under which life as we know it survives, thus providing a benchmark for studying habitability. The search for current or past habitable conditions on other solar system bodies is thus essential to understand both the fundamental processes governing habitability as well as whether life exists on worlds beyond our own.

Q10.1a What Are the Environmental Characteristics Required for Habitability?

The principal habitability criteria at the planetary scale are "the presence of liquid water, conditions favorable for the assembly of complex organic molecules at some time during the planet's history, and energy sources to sustain metabolism" (NASEM 2019a). Favorable conditions include both the materials necessary for life and the maintenance of environmental conditions conducive to life over time (Figure 13-1). The concept of dynamic habitability extends this framework, incorporating the exogenic and endogenic processes that change over time to generate and sustain—or not—the principal habitability criteria. At the planetary scale, dynamic habitability includes "stellar evolution and its impact on the presence of liquid water over time; the evolving structure of the planet interior influencing the magnetic field and plate tectonics; and the state of the atmosphere, which might, for example, redistribute metabolic energy" (NASEM 2019a) (Figure 13-2).

A traditional definition of the habitable zone of a planetary system centers on surface liquid water: the radial distance from a star where incident irradiance, potentially aided by an atmospheric greenhouse, results in temperatures that (given sufficient atmospheric pressure) permit liquid water to remain stable on a planetary surface. The stability of habitable surface environments depends on feedbacks between stellar, orbital, atmospheric, and geological evolution (see Q10.7). Although this definition is useful to identify potentially habitable worlds by remote sensing (a necessity for exoplanets), habitable worlds need not have stable surface liquid water at the surface. Subsurface habitable environments exist on Earth and subsurface liquid water exists on other solar system bodies, mostly independently of solar heating, with water kept liquid by geothermal heating provided from accretion, impacts, radioactivity, volcanism,

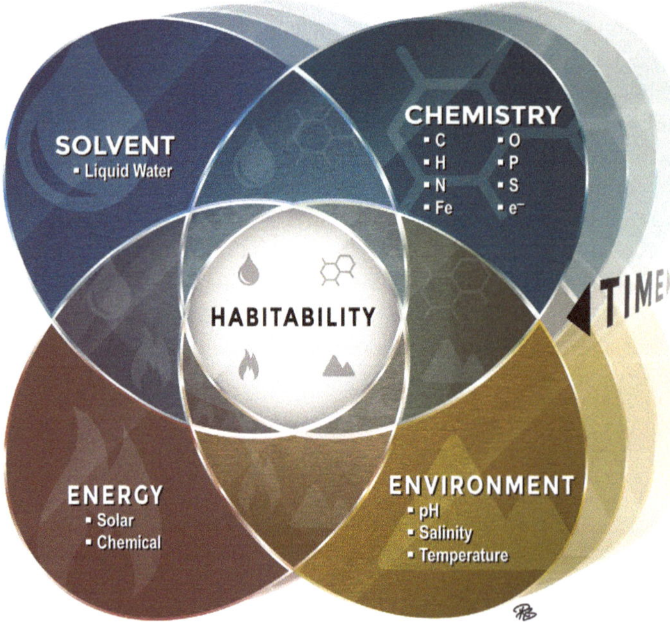

FIGURE 13-1 *Question 10.* The essential components of habitability. Time, and the persistence of these four conditions for a habitable environment, could be considered a fifth keystone. SOURCE: Figure by P.K. Byrne.

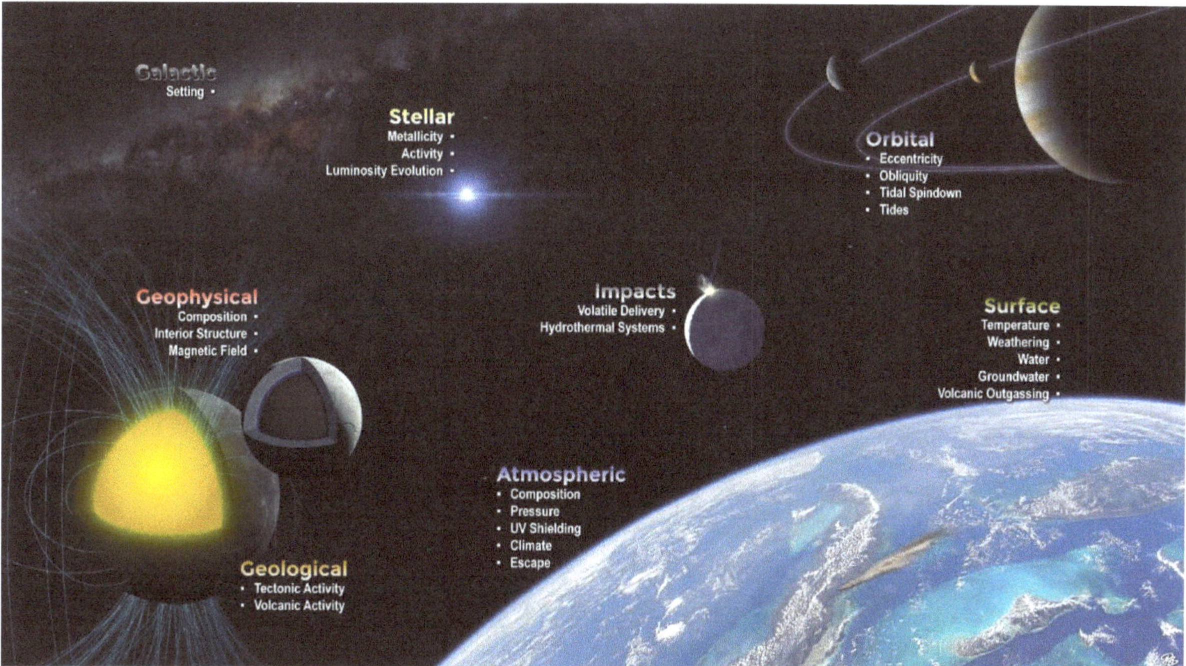

FIGURE 13-2 *Question 10.* Some of the major components of what constitutes a habitable environment. Habitability represents the confluence of a complex, changing array of stellar, galactic, and planetary factors, including properties of a planet's atmosphere, surface, and interior. SOURCES: Figure composed by P.K. Byrne. Images: *Earth* courtesy of the Earth Science and Remote Sensing Unit, NASA Johnson Space Center, NASA Photo ID ISS062-E-117852, https://eol.jsc.nasa.gov. *Planetary bodies Enceladus, Io, Europa, Jupiter, and Mercury* courtesy of NASA. *Starfield background* courtesy of ESO/S. Brunier. CC BY 4.0. *Asteroid* courtesy of NASA/ESA/K. Meech/J. Kleyna (University of Hawaii)/O.Hainaut (European Southern Observatory). *Spring waning crescent Moon with Earthshine* by R. Pettengill, http://astronomy.robpettengill.org. CC BY-NC-ND 2.0. *Dust plume on the Moon* is "Brown Dust Explosion Png Clipart" by transparentpng.com, CC BY 4.0. *Lens flare* by L. Simmons, https://freepngimg.com, CC BY-NC 4.0.

or tidal dissipation, and with geochemical sources of energy for metabolic activity. On rocky bodies, liquid water can be hosted in regolith and in the deeper subsurface. On icy bodies, liquid water can exist in subsurface oceans that are directly in contact with a rocky core or sandwiched between two ice layers, or in reservoirs within the outer ice shell. Liquid water also exists as droplets within atmospheres and, perhaps, within the subsurface of large asteroids (Q10.3). As of 2020, several dozen exoplanets have been identified within the habitable zone of their parent star, where surface temperatures may allow liquid water to be present (Question 12, Chapter 15). Many more may harbor liquid water in their subsurface, but this property would be difficult to detect remotely (Quick et al. 2020).

The habitability of these aqueous environments depends on the supply (abundance and fluxes) of organic matter, chemical nutrients, and energy (Q10.4–10.6), as well as on their longevity (reviewed in Q10.7). Crucially, the exploration of the past, current, and future habitability of a world is distinct from the question of whether that world has actually harbored life. Understanding the factors that generate habitability and how often a *habitable* environment is or can become *inhabited* is central to understanding the prospect for how prevalent and distributed life is in the universe.

Q10.1b How Is Habitability Sustained, Changed, or Lost Over Time?

The solar system is rife with examples of changing habitability over time. The sustainability of habitability on terrestrial planets is determined by planetary size, starting composition, interior evolution (including, but not limited to, dynamo generation), impact bombardment, changes in stellar flux, and orbital dynamics (see Figure 13-2).

For example, any early ocean and atmosphere on Earth would have been sterilized by the Moon-forming impact. Subsequently, as Earth cooled and water oceans became reestablished, early Earth may have had a CO_2- or a CH_4-rich atmosphere akin to Titan today, depending on the balance of input from volcanism and impacts. One of the biggest atmospheric changes on Earth, the rise of oxygen ~2.4–2.0 Ga, was biologically driven. (Additionally, Earth's geologic record indicates periods where much larger portions of Earth's land and ocean were ice-covered during profound glaciations, coined "Snowball Earth" episodes.) On Mars and Venus, substantial changes in habitability are expressed in atmospheric isotopes and geological records. Mars's current 6-mbar CO_2 atmosphere does not sustain stable surface liquid water, but water-deposited sediments, hydrous minerals, and fluvial and glacial landforms visible from orbit attest to the past presence of lakes, rivers, aquifers, and hydrothermal systems, as well as past glaciations. Venus's current dense CO_2 atmosphere sustains a high-temperature greenhouse effect at the surface where liquid water cannot exist, but atmospheric isotopes and some evolutionary models hint at a more clement past (Q6.2b). Oceans have been confirmed on jovian and saturnian satellites, although it is not understood whether, for example, the ocean on tiny Enceladus has the same prospect for long-term heating as larger Europa. On icy worlds, the extent to which accretional, differentiation, radiogenic, and tidal heating can sustain subsurface oceans supplied with chemical energy and nutrients varies over time and can reflect the coevolution of satellites' orbits and interiors in a circumplanetary system (Question 8, Chapter 11).

Strategic Research for Q10.1

- **Characterize conditions on known and candidate modern habitable environments in the solar system—for example, on Mars, Enceladus, Europa, and other bodies**—by measuring water chemistry, mineralogy, ice composition, gases, and organic molecules to assess whether conditions exist that could support life.
- **Determine whether there are modern habitable environments in atmospheres** by characterizing chemistry, including organic molecules, in the atmospheres of Venus and Titan.
- **Determine the character, timing, and duration of past habitable environments on Mars** using chemical, mineralogical, textural, isotopic, and organic measurements from orbit, in situ, and on returned samples.
- **Determine whether Venus ever hosted liquid water on its surface** by geomorphic mapping to search for water-formed landforms as well as mineralogy, chemistry, and isotopic measurements in situ or with samples that may record crust interaction with water and volatile evolution over time.
- **Understand interior structures, tidal dissipation dynamics, and surface-interior exchange for icy shells of ocean worlds** via measurement by spacecraft, theory, and modeling to determine the magnitudes and timescales of heating and persistence of liquid water.

Q10.2 WHERE ARE OR WERE THE SOLAR SYSTEM'S PAST OR PRESENT HABITABLE ENVIRONMENTS?

The solar system offers a broad range of locales to investigate the dynamic nature of habitable environments (Table 13-1): planetary bodies recognized as habitable today, those that were perhaps once habitable, and those that provide information on the framework of habitability. Below, the committee identifies priorities for developing a deeper understanding of pathways toward habitability.

Q10.2a What Can Terrestrial Planetary Bodies Reveal About Habitable Environments?

Rocky worlds in the inner solar system provide key context for understanding interior–surface–atmosphere interactions, including the role played by water in contributing to planetary habitability.

Mercury and the Moon

Although Mercury and the Moon do not possess habitable environments, they provide insights on the processes controlling habitability. Both host ice deposits in the permanently shadowed regions of their poles. MESSENGER revealed that Mercury's poles also host organic-rich deposits that can inform sources of organics on inner planets.

TABLE 13-1 *Question 10.* The Factors That Govern Planetary Habitability, and Whether Those Factors Are Present for Select Planetary Bodies Across the Solar System

		TERRESTRIAL BODIES					OCEAN WORLDS									DWARF PLANETS		SMALL BODIES	
		Mercury	Venus	Earth	Moon	Mars	Europa	Ganymede	Callisto	Enceladus	Titan	Mid-Size Saturnian Moons	Uranian Moons	Triton	Charon	Ceres	Pluto	TNOs	Small Bodies
WATER	Surface Liquid	✗	✗	✓	✗	?	✗	✗	✗	✗	✗	✗	✗	✗	✗	✗	✗	✗	✗
	Subsurface Liquid	✗	✗	✓	✗	?	✓	✓	?	✓	✓	?	?	?	✓?	✓?	✓	?	✗
	Ground Ice	✓	✗	✓	✓	✓	✓	✓	✓	✓	✓	✓	✓	✓	✓	✓	✓	?	✓
	Water Vapor	✗	✓	✓	✗	✓	▨	▨	▨	✓	▨	▨	?	?	▨	✗	?	?	✓
CHEMISTRY	CHNOPS[1]	▨	▨	✓	▨	✓	?	▨	▨	✓	✓	?	✓?	✓	▨	✓	✓	?	✓
	Complex Organics	✓	▨	✓	▨	✓	✓	▨	▨	✓	✓	▨	▨	▨	▨	✓	✓	?	✓
ENERGY	Solar Heating	✓	✓	✓	✓	✓	✗	✗	✗	✗	✗	✗	✗	✗	✗	✓	✗	✗	✓?
	Interior Heating[2]	✓	✓	✓	▨	✓	✓	✓	✓	✓	✓	✓?	✓?	▨	✓	✓	✓	▨	✗
	Redox[3]	▨	?	✓	▨	✓	?	▨	▨	✓	✓	▨	▨	▨	▨	✓	✓	▨	✗
BODY	Atmosphere[4]	✗	✓	✓	✗	✓	✗	✗	✗	✗	✓	✗	✗	✗	✗	✗	✓	?	✗
	Magnetic Field[5]	✓	✗	✓	✗	✗	✗	✓	✗	✗	?	✗	?	✗	✗	✗	?	✗	✗
	Present Habitability	✗	?	✓	✗	?	?	?	?	✓	?	?	?	?	?	?	?	?	✗
	Past Habitability	✗	?	✓	✗	✓	?	?	?	?	?	?	?	?	?	?	?	?	✗

Legend: ✓ Yes/Present; ? Unknown/Uncertain; ✗ No/Absent; ▨ Insufficient Information.

[1] The life-supporting elements carbon, hydrogen, nitrogen, oxygen, phosphorus, or sulfur (not all need be present)
[2] Interior heating is that energy derived from accretion, differentiation, radiogenic decay, and/or tidal dissipation
[3] The prospect for any element or molecule to be reduced or oxidized as a source of chemical energy for life
[4] Subsantial atmospheres only; exospheres (formed by, e.g., impact sputtering) are not included
[5] Intrinsically generated magnetic fields only

NOTE: Cells with a check and question mark signify likely/probable.
SOURCE: Courtesy of P.K. Byrne.

Both airless worlds can help us understand the role of volcanism in the accumulation and nature of an atmosphere, and how such an atmosphere can be lost. Their cratering records establish a flux-calibrated chronology (Question 4, Chapter 7) informing how changing impact bombardment has influenced habitable conditions, especially in the early inner solar system when larger impacts were more frequent. Mercury's current intrinsic magnetic field enables the study of how interior evolution sustains such fields over time, as well as how a magnetic field moderates physical and chemical interactions between the surface and the space environment.

Venus

Venus holds critical clues for understanding the role of climate change for a world becoming uninhabitable. Venus's 740 K surface is clearly uninhabitable today. Measurements of its atmosphere by previous missions, including most recently Venus Express, have suggested that Venus lost perhaps as much as an ocean on Earth's worth of water over its lifetime. When, how, and how much water was lost remain key open questions. One possibility is that shortly after formation, accretional and impact-generated heat, aided by close proximity to the Sun, led Venus to have a magma ocean that outgassed a dense steam atmosphere (e.g., Kasting 1988; Hamano et al. 2013). Under this scenario, Venus's surface was never habitable. However, if Venus escaped the magma ocean phase without experiencing such a runaway greenhouse effect, a cloudy subsolar hemisphere and slow rotation could have helped keep the planet cool and clement for potentially billions of years (Way and Del Genio 2020). In this model, Venus only underwent runaway greenhouse warming within perhaps the past billion years, triggered by a massive release of CO_2 into the atmosphere by volcanic eruptions. If so, there may have been two Earth-sized habitable worlds for much of solar system history, as seen in some exoplanetary systems (Question 12, Chapter 15). Whether Venus's surface was once habitable, and when it ceased to be, could be determined by studying atmospheric isotopes and the mineralogy, chemistry, and isotopes of the rock record.

Even today, Venus's atmosphere has been suggested as a prospective abode for life; at altitudes of 55–60 km, the atmospheric pressure and temperature are comparable to Earth at sea level. The presence of a time-variable ultraviolet-absorbing material and unexplained aspects of measured gas chemistry have led to speculation about a role for biology (Q9.1c). Chemical reactions in a very acidic, low-water, high ultraviolet atmosphere and potential for its habitability needs to be quantified by in situ compositional measurements.

Mars

Mars offers a basis for understanding rock-water-atmosphere interactions and long-term climate change on a body that hosted habitable environments for the first 1–2 billion years of its history and may still host habitable environments. Morphological and mineralogical orbital data reveal a rich record of liquid water with a variety of habitable environments: past hydrothermal springs, groundwater aquifers, deep open and closed basin lakes, sulfate and chloride playa lakes, weathered soils, and possibly oceans, varying geographically and temporally (e.g., Ehlmann and Edwards 2014). In situ results from landed missions have investigated a few of these past environments in detail. The Opportunity rover found evidence for past sulfate playa lakes with acidic waters, which on Earth can be inhabited by organisms capable of tolerating low pH. The Spirit rover identified silica deposits interpreted to have formed in an ancient volcanic hydrothermal spring; such environments are populated with microbial life on Earth. The Curiosity rover found evidence for a ~3.5 Ga habitable lake system within Gale crater with low salinity, neutral pH, and all chemical elements essential to life, as well as episodic pulses of groundwater, until at least ~2.7 Ga. Curiosity also identified long-chain organic material within the sedimentary rocks (e.g., Eigenbrode et al. 2018; Rampe et al. 2020). What processes sustained these habitats with what continuity, and why did conditions change? Surface water in the past was supported by a thicker and warmer atmosphere, since lost. Results from the MAVEN orbiter indicate loss of the martian atmosphere to space (Jakosky et al. 2018), and chemical weathering was a major sink for water and other volatiles (Scheller et al. 2021). The habitability of early Mars motivates the ongoing search for biosignatures within its multiple, diverse ancient aqueous environments (Question 11, Chapter 14). The Mars Sample Return campaign will allow for martian samples acquired by the Perseverance rover to be returned to Earth laboratories for detailed analyses that address the open questions about what processes supported the formation of habitats in Mars's past.

Today, Mars is cold and arid, and whether it hosts modern habitable environments is a key question. Unlike Earth, Mars has large, obliquity (spin axis tilt) variations that drive substantial climate excursions on hundred-thousand-year timescales. These excursions periodically release volatiles into the atmosphere, raising atmospheric temperature and pressure above what we observe today and potentially making ephemeral occurrence of liquid water possible. Orbiters have identified ice sheets and glaciers distributed from the poles to the mid-latitudes and ground ice within 1 m of the surface (Morgan et al. 2021). Whether liquid water exists beneath such ice is not yet known. Features such as recurring slope lineae and other downslope debris flows that preferentially occur during warm months have characteristics suggesting the involvement of liquid water, even if a purely dry process is considered the most likely explanation (McEwen et al. 2021). The martian regolith contains salts, some geologically young (Wray 2020), which may melt ice (Stillman and Grimm 2011) or liquefy in the presence of atmospheric water vapor (i.e., deliquescence) (Rivera-Valentín et al. 2020). At Gale crater, continued in situ detection and cyclical seasonal variation of parts-per-billion traces of methane (Webster et al. 2018)—a nonequilibrium gas that can provide a source of energy for life and, on Earth, can be produced by life—contrasts with its nondetection at parts-per-trillion level from orbit (Korablev et al. 2019). A localized source and differences in time of day of measurement may reconcile these measurements (Moores et al. 2019; Webster et al. 2021), but the process generating methane is not yet known. Collectively, these observations sustain the debate on the habitability of present-day Mars and whether it could be inhabited.

Q10.2b Which Icy Ocean Worlds and Dwarf Planets Are Habitable?

Missions to the outer solar system have demonstrated the presence of organic chemistry and liquid water currently or in the past under the surfaces of many objects, including in subsurface oceans within icy moons around the giant planets. The habitable zone, as defined by the presence of liquid water (Q10.1), thus extends considerably farther out in the solar system than suggested by insolation limits on surface water (Coustenis and Encrenaz 2013).

Although key questions concerning their longevity and habitability remain outstanding, identifying the distribution of subsurface oceans is an important first step. Confirmed or strongly suspected ocean worlds, determined primarily from Galileo and Cassini-Huygens measurements, include (Lunine 2017; Hendrix et al. 2019; Hand et al. 2020):

- **Enceladus:** Plumes of water from fractures at this moon's south pole escape from a global liquid water ocean, bearing silica, salts, and organics that indicate chemical reactions between that ocean and a silicate interior (Postberg et al. 2018a). Key ingredients for habitability appear to be met for this body (Figure 13-1). A porous, rocky interior may sustain these environments through time. Key remaining questions concern the longevity of the ocean and its variability with time.
- **Europa:** Observations of a young and smooth surface with crisscrossing fractures and magnetic induction measurements together provide strong evidence for a global ocean beneath a geologically active ice shell (e.g., Kivelson et al. 2009). Key questions concern the composition of the ocean, the nature of its interactions with the silicate interior, the presence and inventory of organic compounds and bioessential elements, and how the ocean undergoes physical and geochemical exchange with the overlying ice-shell.
- **Ganymede and Callisto:** Magnetic induction measurements on Ganymede revealed a global ocean, likely sandwiched between two or more ice layers with deep, high-pressure ice phases. Ganymede also has an intrinsic magnetic field. Although its precise internal structure is not known today, an induction response in Callisto has been suggested, but may be due either to an ocean or to ionospheric interference (e.g., Hartkorn and Saur 2017).
- **Titan:** A global deep ocean is thought to be present within Titan between ice layers (e.g., Iess et al. 2012). The surface of Titan's ice shell is covered with deposits of liquid hydrocarbons. The nitrogen–methane atmosphere has a rich methanological cycle and supports complex organic chemistry that starts in the ionosphere and diffuses down to the surface, forming prebiotic molecules and aerosols/hazes (e.g., Coustenis 2021). Cryovolcanism has been hypothesized and could deliver methane from the interior to the atmosphere (Lopes et al. 2013).

More putative ocean worlds await evidence to conclusively confirm suggestions from modeling and observations of having had subsurface oceans. Beyond liquid water oceans, findings over the past decade have revealed a continuum that may include "mudballs"—that is, porous, rocky interiors filled with silicate-brines, which increase water–rock interfaces at which energy and nutrients may be supplied. Candidate ocean worlds include:

- **Triton:** The geologically young surface of Triton has been linked to convective and/or cryovolcanic processes, consistent with the observation by Voyager 2 of active nitrogen plumes and, potentially, a subsurface ocean (e.g., Hansen et al. 2021).
- **Ceres:** With a pervasively hydrous and icy crust, and evidence for modern-day brines or diapirs generating young salty deposits on its surface, Ceres once had, and may still have, subsurface liquid water (e.g., Castillo-Rogez et al. 2020). Ceres appears to have an abundant supply of organic compounds and nitrogen at the surface, but its energy supply may be scarce as it is not subject to tidal heating.
- **Uranian and other saturnian moons (especially Ariel and Miranda, or Dione):** These moons have dynamic geological histories preserved on their surfaces, and modeling efforts suggest the possibility that liquid water or mud oceans may exist in the subsurface (e.g., Bierson and Nimmo 2022). Carbon and nitrogen have been detected at Ariel's surface.
- **Pluto:** There is evidence that Pluto currently possesses a liquid water ocean beneath its thick frozen ice shell, which developed when ice melted owing to heat from radioactive elements in Pluto's core, long after the body itself formed (e.g., Nimmo and McKinnon 2021). Pluto's surface and atmosphere are also rich in carbon and nitrogen.

Q10.2c What Can Small Bodies Reveal About Habitability, and Are There Habitable Dwarf Planets Today?

Results over the past decade by NASA's Dawn and New Horizons missions revolutionized our understanding of habitability related to small solar system bodies, such as asteroids, comets, and trans-neptunian objects (TNOs). In addition to potential subsurface water at Ceres and Pluto, ground-based observations of dwarf planets hint at the potential for more ocean worlds in the trans-neptunian population. Determining the occurrence and characteristics of liquid water environments in this population of small bodies remains an open area of investigation.

Additionally, the meteorite record suggests that some asteroids may have experienced aqueous alteration, which could have supported prebiotic chemistry. The presence of water on some asteroids was confirmed this past decade (Campins et al. 2010). Given their observed comet-like activity (e.g., dust tails and other mass loss events), some active asteroids may also presently possess water ice. Although small bodies themselves are unlikely to sustain life, the chemical reactions facilitated on them may have led to the delivery of volatile and organic species to render other worlds habitable.

Strategic Research for Q10.2

- **Determine the extent of present and former habitable environments across the solar system** by making measurements that determine the past and present existence of liquid water, the organic content, and the availability of nutrients and metabolic energy sources on terrestrial planets and ocean worlds.
- **Determine the distribution of past and present subsurface oceans—fully liquid and muddy—and their historical evolution** through detailed geological/geophysical investigations and modeling efforts coupled with a search for oceans by remote sensing.
- **Determine the evolution of the climate of Mars and Venus and the timing of changes** by measurement of atmospheric gases, chemistry, and isotopes in the atmosphere and rocks, and climate modeling.

Q10.3 WATER AVAILABILITY: WHAT CONTROLS THE AMOUNT OF AVAILABLE WATER ON A BODY OVER TIME?

Water can be mineral-bound in rocky materials or occur as ice, liquid, or atmospheric water vapor. The initial inventory of water for a planetary body is provided during and shortly after planetary formation, and its later availability is then a balance between subsequent supply and loss processes.

The presence of liquid water depends on distance from the Sun, the mode(s) of heat and volatile transport on a planet, and composition; for example, salts and volatile compounds (e.g., ammonia) can act as "anti-freeze" to lower melting temperatures. For the terrestrial planets, water availability in the crust or at the surface depends on the history of delivery during accretion, outgassing from the deep interior, physical properties of the subsurface such as porosity and permeability, loss to space, and transport and phase changes within and on the planet driven by temperature and pressure. For outer solar system objects with large volatile volumes, liquid water can be available beneath and within the ice shell, and in some cases in direct contact with rock, depending on the subsurface temperature and pressure conditions.

Q10.3a What Sets Initial Limits on Planetary Liquid Water and Water Inventories?

The initial water inventory depends on volatiles accreted during planetary formation by accretion and impact bombardment and is influenced by the position of the body in the solar system. Water is supplied from accreting planetesimals, asteroids, and comets, incorporated into coalescing objects or delivered later as a surface veneer (Question 3, Chapter 6). In the outer solar system, the strong gradient of water abundance in the Galilean satellites with distance from Jupiter suggests that water also can be accreted directly from a protosatellite disk surrounding giant planets.

Attempts to quantify the source(s) of Earth's water and the history of accretion from the isotopic composition of its water and that in asteroids and comets have so far been inconclusive, given the large variations observed within measured planetary bodies and the uncertainty of how water is accreted into their source materials (see Question 3.6). In addition to accretion from the planetary nebula and subsequent aqueous alteration, a contribution has been suggested for highly hydrated dust grains resulting from the interaction of the solar wind with planetesimals (Daly et al. 2021). Dynamical models of primordial water supply demonstrate the importance of planet formation location, planet size, and random variations in impacts by volatile-rich objects.

In the terrestrial planets and large moons, water can be released to the surface catastrophically during accretion and core formation. Liberated water can oxidize iron, releasing hydrogen; the hydrogen can then escape to space in hydrodynamic loss that can drag other gases with it (Zahnle et al. 2019). The difficulty in establishing robustly how much water has reacted with planetary materials, and thus how much water or hydrogen has been lost to space, challenges efforts to determine a priori how much water a planetary body was born with (Q6.1a).

Q10.3b What Are the Long-Term Endogenic and Exogenic Controls on the Presence of Liquid Water on Terrestrial Planets?

Aside from Earth, we have the most information to disentangle processes controlling availability of water on Mars (Figure 13-3). Major advances were made this decade thanks to MAVEN observations, integrated with compositional data from meteorite studies as well as orbiting and surface missions over two decades of observations that "follow the water," but some questions remain at Mars, as well as for Venus.

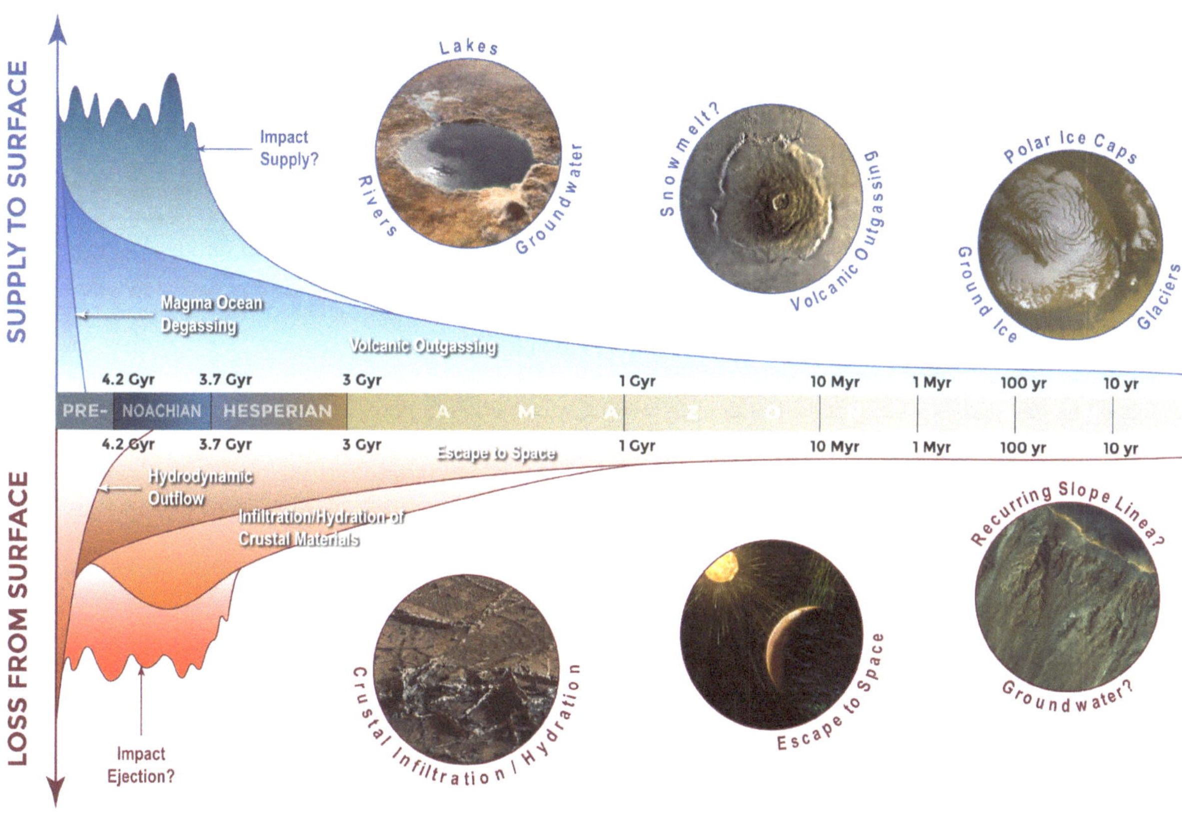

FIGURE 13-3 *Question 10.* The evolution of sources (*top*) and sinks (*bottom*) of water on Mars. The curves are qualitative representations based on the geomorphological record of martian history, quantification of crustal water content, current measured loss rates and water inventories, and models of planetary evolution. Question marks identify processes and water bodies that remain debated. SOURCES: Figure composed by P.K. Byrne. Images: *Top left and middle:* Courtesy of NASA/JPL-Caltech. *Top right and bottom left:* Courtesy of NASA/JPL/Malin Space Science Systems. *Bottom middle:* Courtesy of NASA/GSFC. *Bottom right:* Courtesy of NASA/JPL-Caltech/University of Arizona.

The availability of liquid water at the surface of the terrestrial planets is determined in part by surface temperature, which itself is a function of distance from the Sun and atmospheric greenhouse warming. Although only small amounts of water are present in the clouds today, early Venus may have had a much more clement climate before a greenhouse effect drove liquid water into the atmosphere then away to space (Q6.2b). We know relatively little about early Venus because the mineralogy and chemistry of the rocks that record it are not known. On Mars, whereas the presence of abundant liquid water at the surface early in its history is undisputed (Q10.2), the mechanism for providing greenhouse warming to allow liquid water, either globally or locally, remains poorly understood (e.g., Wordsworth et al. 2017). Uncertainties include the amount of water present as ice or liquid within the martian crust both through time and today (Q.10.2), whether there was an early ocean, and the timing of water loss and climate change.

Volcanism supplies water and other gases to terrestrial planet surfaces and atmospheres through much of their history. The release of volatiles is directly correlated with the amount of volcanism, but the amount of water supplied depends critically on the amount of water incorporated during accretion, the redox state of the interior, and styles of volcanism, particularly temperature and degree of partial melting. Early voluminous volcanism on Mars and most other solar system bodies waned over time, resulting in progressively less water released, although net quantities are not well understood (Figure 13-3). Venus, too, experienced volcanism that drove outgassing, but activity prior to about 1 Ga is no longer preserved, making the history of volatile release there a key unknown.

In addition to endogenous supply from volcanism, impacts might have contributed to transiently thicker atmospheres that supported liquid water. Impacts producing craters hundreds of kilometers across and larger occurred for up to a billion years after planetary formation. Their role in water loss or gain and their environmental effects on atmospheric composition and climate remain key questions. These impactors delivered water and other volatiles to planetary surfaces, and impact heating and ejection may have liberated water ice or CO_2 volatiles. Yet, impact shocks may also have removed substantial atmospheric water and other volatiles. On the whole, the net role of large impacts in water loss or gain and their environmental effects on atmospheric composition and climate remains an open question.

Liquid water can be removed from a planetary surface by freezing, infiltrating the crust to become groundwater, or hydrating rocks (i.e., chemical weathering). On Mars, the volume of water ice beneath the surface and at the poles is equivalent to a ~40 m global-equivalent layer. Such ice could melt via geothermal-, volcanic-, or climate-driven heating. A more permanent means of removal from a hydrologic system is chemical weathering, especially if the planet lacks a means of crustal recycling such as plate tectonics. On Mars, hydrated minerals comprise a global equivalent layer of water 150 to 1,500 m thick (e.g., Scheller et al. 2021). On Venus, we do not yet know whether the crust includes hydrous phases (or their alteration products) that sequestered water or other volatiles.

Atmospheric escape processes deplete water from planetary atmospheres by a variety of thermal, photochemical, and charged particle interactions. Extreme ultraviolet solar radiation plays a role by the photodissociation of H_2O into component ions; the resulting hydrogen escapes to space as H or H_2. Ions also can be picked up by the solar wind and carried to space directly by sputtering, or the solar wind can accelerate molecules in the upper atmosphere and carry other atoms or molecules to space. Many of these processes have been observed directly at Mars and Venus and are inferred to have operated there over extended geological time, albeit at largely unconstrained rates. Calculated loss rates at Mars suggest a ~20 m global equivalent layer of water lost over time and possibly as much as ten times greater than that (Jakosky et al. 2018). The extent to which these processes affected the volatile inventory at Venus remains unclear. For Venus, the measured atmospheric deuterium–hydrogen ratio—more than 100 times that of Earth—is indicative of the loss to space of potentially substantial volumes of water (Donahue et al. 1982). Whether this elevated ratio represents a dynamic balance between supply of water by impacts with escape to space, the loss of steam from a primordial atmosphere, or a long-sustained vanished ocean driven away by a late-onset runaway greenhouse effect, remains a key open question.

Q10.3c What Are the Long-Term Endogenic and Exogenic Controls on the Presence of Liquid Water on Icy Bodies?

There is a strong coupling between the thermal history of icy bodies and the occurrence and persistence of subsurface liquid water. Oceans occur beneath icy shells because of heat supplied from planetary body cores, ultimately derived from accretional, radiogenic, and tidal heating, and potentially facilitated by the presence of antifreeze in the oceans themselves. Ammonia and salts can facilitate melting by freezing point lowering, but an interior heat source is still required. Tidal dissipation accounts for subsurface liquid water oceans on some of the icy bodies, for example, via the Laplace resonance maintaining forced eccentricity among three of the Galilean satellites. A key question for Enceladus is the duration of its ocean, where interior cooling should outpace energy supply by tidal dissipation, unless the core is porous (Choblet et al. 2017). In contrast, Callisto, for example, experiences essentially no tidal energy dissipation, but it has a sufficiently high interior rock mass to sustain an ocean by radiogenic decay. Triton may have experienced considerable internal heating from tides during the early evolution of its orbit around Neptune. Most of the numerous known and suspected subsurface water oceans in the outer solar system are almost certainly in direct contact with rock, for example, at Enceladus and Europa, and thus could facilitate hydration of that rock. How far this liquid percolates into underlying rock depends on poorly known factors such as the porosity, permeability, and fracture density of the seafloor (Q10.6b,c).

For liquid water to reach the surface requires some means to overcome the higher density of liquid water relative to water ice. Cryovolcanism has also been suggested on Titan (e.g., Lopes et al. 2013) as an interpretation of surface features observed by Cassini, but on Titan—and Ganymede—where the subsurface oceans are trapped between two ice layers, the connection to either the silicate interior or the surface is not yet known. The only confirmed ocean–surface link is at Enceladus, where the composition of plumes emanating from south pole fractures indicates that they are sourced from a salty, subsurface ocean in contact with a rocky interior. By contrast, it is not clear that plumes on Triton are in contact with an ocean as those plumes could be modulated by seasonal solar irradiance and a solid-state greenhouse effect (Hansen et al. 2021). Hubble Space Telescope observations and a reanalysis of Galileo data have shown evidence, still debated, for plume activity on Europa (e.g., Paganini et al. 2020). A major question for Ceres is whether it presently has liquid water or brines beneath its surface, or if recently emplaced (~10 Ma) salts discovered by the Dawn spacecraft are mobilized remnants of an ancient ocean (Castillo-Rogez 2020). Liquid water may have been transiently present on many icy bodies following the heating associated with impact events, lasting for up to 10,000 years.

<h3 style="text-align:center">Strategic Research for Q10.3</h3>

- **Establish whether liquid water is present on Mars today in the subsurface** by geochemical measurements of ices and recent hydrous minerals and geophysical measurements to probe the upper crust.
- **Determine the distribution, history, and processes driving the availability of ice and liquid water on Mars over time,** combining mapping stratigraphy and mineralogy, measurements of chemical, mineralogic, and isotopic measurements in situ and from returned samples, sounding of the subsurface, models for geomorphic features and climate processes, and constraints on chronology from in situ radiometric dating and measurements on returned samples.
- **Determine the availability through time of liquid water on Venus** using measurements of present-day escape rates, isotopes and mineral phases in the crust, as well as atmospheric models integrating loss to space with interior and surface evolution.
- **Identify the amounts and locations of any past or present liquid water beneath or emplaced on the surfaces of Enceladus, Europa, Titan, Ceres, and candidate ocean worlds within the satellite systems of Neptune and Uranus** using radar, gravity, topography, magnetic field (induction), and surface spectral measurements, combined with models of tidal deformation and the formation and evolution of surface features.
- **Determine the amount and origin of water ice on the Moon and Mercury** by sampling ice, determining its spatial distribution, measuring H and O isotopes, and determining the nature and abundance of contaminants within the ice as a means of understanding sources of water in the inner solar system.

Q10.4 ORGANIC SYNTHESIS AND CYCLING: WHERE AND HOW ARE ORGANIC BUILDING BLOCKS OF LIFE SYNTHESIZED IN THE SOLAR SYSTEM?

Organic molecules are an essential component of planetary habitability, as they enable a variety of chemical structures and specific reactivity that is foundational to the biochemistry of life as we know it. Abiotic processes can produce diverse, large organic molecules, which are found in many places beyond Earth including Mercury,

Mars, chondrite parent bodies, Ceres, Enceladus, Titan, Pluto, comets, the atmospheres of the giant planets, and possibly Venus (Table 13-1). Of particular relevance to the study of habitable solar system environments are the recent discoveries of organic matter in water-deposited sediments on Mars, organic chemistry and prebiotic molecules on Titan, and organic matter in plume particles from the interior ocean of Enceladus.

To understand how abiotic organic synthesis occurs, it is important to have an inventory of organic products, reaction rates, and isotopic tracers, and their dependence on conditions in the present or past solar system. Such knowledge provides insights into prebiotic organic inventories that might factor into the origin of life, as well as which organic molecules can contribute to habitability by serving as carbon and energy sources for life. Moreover, a comprehensive assessment of abiotic organic processes will lead to an understanding of how signatures of biotic and abiotic products can be distinguished in the search for life on other worlds (Question 11).

Q10.4a Where Were Organic Compounds in the Early Solar System, How Did They Form, and Did They Contribute to Prebiotic Chemistry?

Organic molecules have been available since the birth of the solar system, formed abiotically via chemical processes. The high abundances of organic materials in carbonaceous chondrites, and especially in comets, show that the outer region of the early solar system was a site of organic molecule formation. These organics formed mainly via ion–molecule reactions at low temperatures in the outer protoplanetary disk and/or in presolar interstellar environments and include insoluble organic matter that is found in chondrites, as well as a diversity of smaller organic building blocks bearing nitrogen and oxygen atoms. These organics were transported in the disk (e.g., by turbulent mixing and gravitational scattering) and accreted by larger objects, including those in the inner solar system. Where exposed to warm liquid water—for example, in carbonaceous chondrite parent body asteroids, planet surfaces, and icy world interior oceans and early atmospheres—accreted organics underwent reactions that produced more complex compounds, including key building blocks of life such as amino acids, nucleobases, and sugars.

Q10.4b What Processes Have Enabled Abiotic In Situ Organic Synthesis and Cycling on Planetary Bodies?

Organic synthesis requires carbon and energy. In planetary surface environments, simple organic molecules like methane (CH_4) or formaldehyde (CH_2O) initially seed organic chemistry driven by ultraviolet light from the Sun, charged particles, and cosmic rays. In subsurface environments, the usual carbon source is CO_2; the key energetic driver is chemical disequilibrium with coexisting molecular hydrogen, and transition metals can catalyze synthesis.

Exploration of the solar system, as well as laboratory experiments and chemical kinetic modeling, have revealed how abiotic organic synthesis occurs in methane-bearing atmospheres. We have learned the most about this process on Titan, complemented by additional data from studies of Pluto, Triton, and the atmospheres of the giant planets. Abiotic organic synthesis in these environments is initiated by interaction of methane molecules with high-energy particles that enter the upper atmosphere, leading to the production of an array of radicals and electrically charged chemical species that combine in myriad ways to produce a diversity of organic molecules from simple gases to sooty aerosols (Figure 13-4). When nitrogen and oxygen sources are available, even more complex compounds can be formed by the incorporation of these elements into organic structures, leading to the formation of prebiotic molecules.

Direct sounding of Titan's upper atmosphere by Cassini allowed for the detection of numerous species (e.g., Coustenis 2021). Higher molecular weight species (up to several 1,000 Da) are present in the ionosphere, including hydrogenated amorphous carbons and polycyclic aromatic hydrocarbons (PAHs), which can become long enough to form fullerenes. Methane exists in different phases on Titan, supporting a cycle similar to that of water on Earth and likely being replenished from the interior over long periods of time. The cycling of volatile chemicals starts with the dissociation of N_2 and CH_4 through electron, photon impacts, and cosmic rays higher in Titan's atmosphere, leading to the formation of acetylene (C_2H_2) and hydrogen cyanide (HCN). Several other hydrocarbon compounds (e.g., C_2H_6, C_4H_2, C_2H_4, C_6H_6) and nitriles (e.g., HCN, HC_3N, C_2N_2,) were detected in Titan's stratosphere from space or from Earth. Once formed, these molecules diffuse downward, forming haze and higher hydrocarbons and

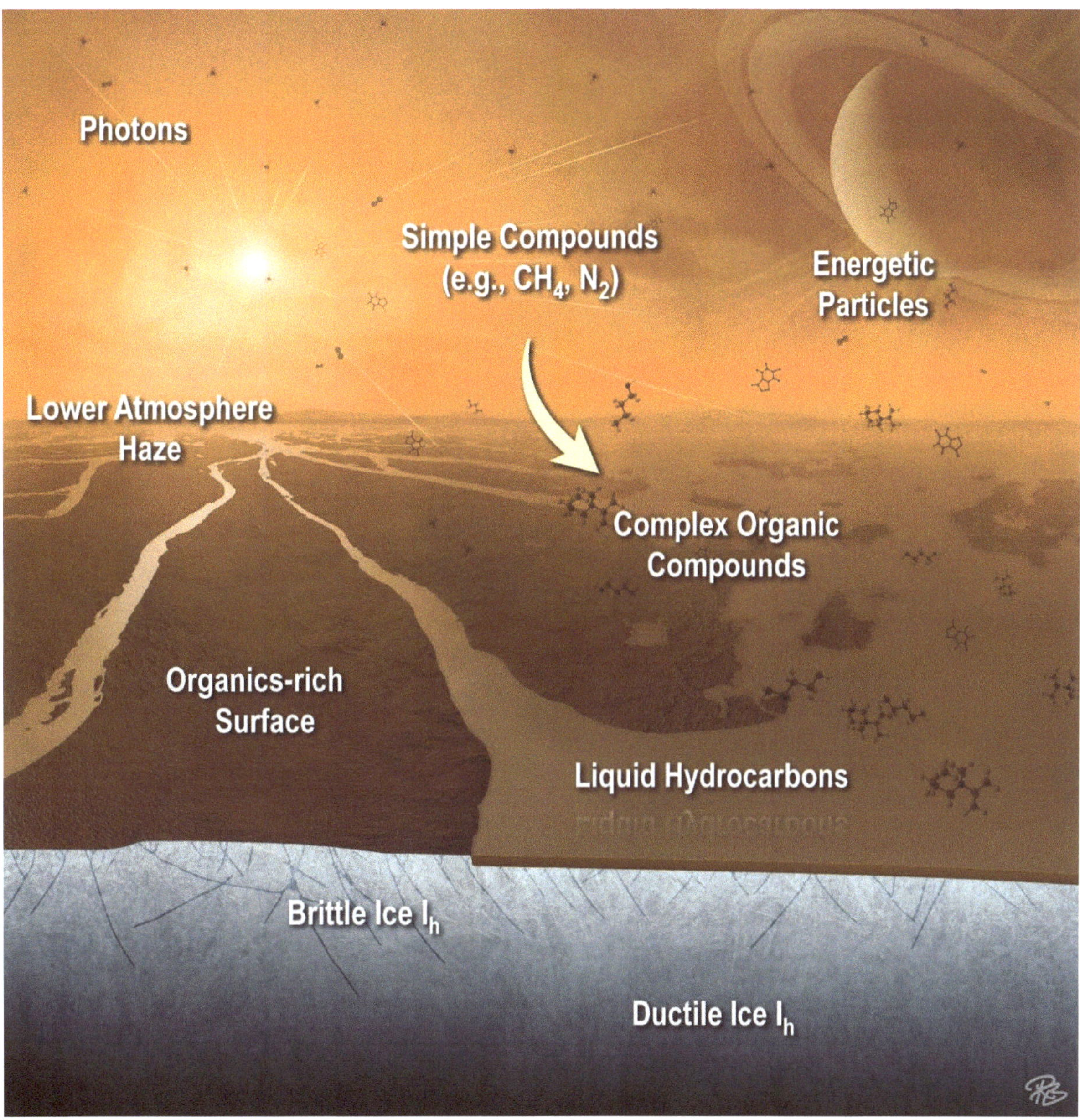

FIGURE 13-4 *Question 10.* Atmospheric chemistry leads to formation of complex organics on Titan. SOURCE: Figure by P.K. Byrne. Surface rendering courtesy of Getty/Mark Garlick/Science Photo Library.

nitriles, eventually depositing on the surface (see Figure 13-4). The extent to which they find their way to Titan's internal ocean remains a matter of debate.

Analogous photochemical processes produce organic compounds on Triton and Pluto. Both worlds have N_2–CH_4 atmospheres, but are colder than Titan, so methane gas is less abundant relative to nitrogen. Pluto's atmosphere contains C_2H_6, C_2H_4, C_2H_2, and HCN, and its surface also hosts organic compounds (e.g., Cruikshank et al. 2019). A key consequence of the different relative abundances of atmospheric N_2 and CH_4 is that the types of photochemically synthesized organic compounds on Triton and Pluto are predicted to be richer in prebiotically relevant nitriles than at Titan (Wong et al. 2015). The relative abundance of nitriles can constrain past levels of atmospheric methane.

FIGURE 13-5 *Question 10.* Oxidation and hydration of silicate rocks in reactions that produce iron-bearing phases, serpentine, and other phyllosilicates also generate hydrogen, which can serve as an energy source for life. The hydrogen can react with oxidized carbon from CO_2 or carbonates to abiotically generate organic matter, as occurs on Earth, occurred on volatile-rich asteroids, and potentially occurred on Mars and in the cores of ocean worlds. SOURCE: Courtesy of P.K. Byrne.

Water–rock reactions are also key sources of solar system organics (Figure 13-5). Abiotic, seafloor H_2 and CO_2 reactions in hydrothermal systems on Earth produce organics, and there is potential for similar processes in the subsurface of icy ocean worlds, particularly Enceladus and Europa (Q10.5). Ice particles from the plumes of Enceladus contain silica as well as Na, Cl, and CO_3, derived from ocean waters reacting with rock in the core (Hsu et al. 2015). Cassini sounded Enceladus's plumes and detected water vapor, molecular nitrogen and hydrogen, ammonia, carbon dioxide, and traces of several organic components: methane, propane, acetylene, benzene, and formaldehyde, similar to what is found in most comets. Importantly, complex macromolecular organics were identified in emitted ice grains (e.g., Postberg et al. 2018a) and amines are also indicated, providing key building blocks for life.

Abiotic organic synthesis at lower water–rock ratios, such as those of rock hydration, was important on Mars and some asteroids. Studies of carbonaceous chondrites suggest that such meteorites contain diverse organic compounds sourced from presolar origins that have undergone evolution in hydrothermal environments (e.g., Vinogradoff et al. 2018). Although difficult to characterize, the insoluble, macromolecular fraction of this material is composed of small PAHs (hydrocarbon rings) with abundant nitrogen, oxygen, and sulfur functions, cross-linked by short, highly branched carbon and carbon–oxygen chains. The soluble fraction comprises amino acids, chained amines ($C–NH_2$) and amides ($C–N=O$), chained and ringed hydrocarbons, alcohols and polyols ($C–OH$), aldehydes and ketones ($C=O$), chained carboxylic and hydroxy acids ($O=C–OH$), nitrogen-bearing rings and nucleobases, alkyl sulfonic ($O=S=O$), and phosphonic ($–PO_3$) acids, which are all of interest to astrobiology (Glavin et al. 2018) and may have contributed to the origin of life on Earth. The analysis of organics in carbonaceous chondrites gives insights on the prebiotic chemistry that occurred in a natural system before life. Telescopic surveys have confirmed infrared absorptions related to C–H in select large asteroids; more recently, the Dawn mission at Ceres found evidence for a high abundance of carbon in the regolith, interpreted as a mix of carbonates and amorphous carbon. Select locations on the surface are enriched in organic chains of atypical composition or abundance, thought to be generated from endogenous hydrothermal processes (De Sanctis et al. 2019).

On Mercury, large swaths of the crust contain graphitic carbon, inferred to be magmatic in origin. Mars also has indigenous organics (found in martian meteorites), and there are hints at a complex organics cycle involving (1) infall of carbonaceous meteorites and interplanetary dust particles; (2) primary igneous reduced carbon—that is, magmatic graphite; (3) primary hydrothermally formed organic carbon/nitrogen-containing species; (4) secondary hydrothermally generated macromolecular carbon and graphite; (5) impact-generated graphite; and atmospheric and aqueous processing of these (Steele et al. 2016). Select sedimentary rocks from a martian paleolake in Gale crater contain organic rings and chains bearing sulfur (Eigenbrode et al. 2018), suggesting transformation and preservation in the presence of H_2S. So far, the martian organics can all be explained by abiotic processes, although our understanding of organic compounds on Mars is nascent, and strategies are being developed to discriminate between abiotic and biotic organics (Question 11).

Q10.4c How Have Organic Fluxes to Planetary Bodies Changed Over Time and Influenced Habitability?

Outer solar system planetesimals and protoplanets delivered both volatiles and evolved primordial organic material to growing worlds in the inner part of the system via impacts (Question 3). This delivery has generally decreased over geological time as leftover materials have been gradually swept up into planets or the Sun or scattered out into interstellar space. However, delivery continues, and there may have been punctuated events of increased supply during episodes of enhanced bombardment. The availability of these compounds and their concentrations in select environments could have played a role in the origin of life by providing building blocks for complex organic chemistry (Question 9).

Strategic Research for Q10.4

- **Determine the complexity attained by organic chemistry in Titan's atmosphere, its sources and sinks, and its role in producing a potentially habitable environment by entering the subsurface ocean,** through in situ and remote spectral imaging and mass spectrometry investigations.
- **Determine the chemical composition and structural characteristics of organic compounds in both Europa's and Enceladus's oceans to understand organic synthesis,** using in situ techniques (chemical derivatization, gas chromatography, capillary electrophoresis, pyrolysis, and mass spectrometry) on plume or extruded materials, and detailed characterization by sample return and analyses in Earth laboratories.
- **Understand the formation and alteration of organics on small bodies (asteroids, comets, meteorites, and dwarf planets such as Ceres) and their potential contributions to the origin of life on Earth or elsewhere** by in situ observations of organics and isotopic composition, laboratory analog experiments, detailed investigation of samples returned to Earth and meteorites.
- **Characterize organic molecules present on Mars for determination of type, distribution, and source of organic materials to understand organic synthesis there** by in situ measurements and sample return for investigation in Earth laboratories.

Q10.5. WHAT IS THE AVAILABILITY OF NUTRIENTS AND OTHER INORGANIC INGREDIENTS TO SUPPORT LIFE?

Six main chemical elements make up life—carbon (C), hydrogen (H), nitrogen (N), oxygen (O), phosphorus (P) and sulfur (S), or "CHNOPS"—along with some "micro-nutrient" metals, for example, Fe and Mo (Question 9). H, C, O, and N are commonly available in terrestrial planet atmospheres, starting out as H_2O, CO_2, and N_2, and can be exchanged with nonatmospheric reservoirs. In contrast, sources of any P, S, and "micronutrient" metals required to sustain habitability are more commonly expected to be sourced from minerals present in rocky silicate materials.

Q10.5a What Are the Inventories, Forms, and Distribution of Life-Supporting Elements on Planetary Bodies?

On Venus, the available inventory of key life-supporting elements remains uncertain, in large part because of limited chemical data for the surface and atmosphere. From those few geochemical measurements made at the surface, Venus's expansive lowland plains appear basaltic, and thus those flows and their physical weathering products are presumably potential sources of Mg, Fe, Ca, K, as well as P and S, which are typically present in basalts as minor constituents. Chemical alteration pathways on Venus are poorly understood, although they are likely dominated by sulfatization and oxidation (Zolotov 2019). Carbon and oxygen abound because the atmosphere is 96.5 percent CO_2; nitrogen and sulfur are also present (the latter as SO_2); H is absent but for trace amounts of

H_2O and H_2S. Whether there is phosphorus in the Venus atmosphere remains unclear: P sourced from volcanic eruptions could exist as P_4O_6 in the lower atmosphere, before being lofted to middle- and upper atmospheric levels and converted to PH_3; this latter phenomenon was recently tentatively observed. There is virtually no information regarding micronutrients in the Venus environment at present.

All of the essential life-supporting elements are present on Mars and have also been detected in forms accessible to life. Oxidized carbon is broadly available; 95 percent of the atmosphere is carbon dioxide, and localized carbonates have been detected by orbital and landed missions. Reduced carbon is also available in a variety of organic compounds detected in drilled samples by Curiosity in Gale crater, in Mars meteorites (Q10.4b), and perhaps locally in methane (Q10.2a). Nitrogen, long known to be present as N_2 in the martian atmosphere, is also present as fixed nitrate in surficial samples analyzed by Curiosity (Stern et al. 2015). H_2O is present in solid and gaseous forms today with abundant evidence for its presence as a liquid in the past. Phosphorus is enriched on Mars's surface relative to Earth's by about an order of magnitude, as measured in the dust and by every landed mission. Sulfates are present in the ubiquitous dust, commonly found in fracture fills, present in sand deposits near the north pole and even form thick surface deposits; sulfides and metals in other minerals are found in Mars's igneous rocks.

Among the key major elements required for life (CHNOPS), all but N and P have already been identified on Europa and all but P in the plume of Enceladus (Hand et al. 2009; Postberg et al. 2018b). Given the abundance of P and metals in chondritic materials (assumed to be representative of the rocky interiors of these ocean worlds), however, neither is likely to be limiting to life on either Enceladus or Europa (e.g., Cable et al. 2020). On Europa, surface S is pervasive as magnesium sulfate salts, at least in part owing to an exogenous flux from Io's volcanoes. On Enceladus, H_2S has tentatively been measured by Cassini in the plume gas (Postberg et al. 2018b).

Q10.5b What Processes Govern the Inventories, Forms, and Distribution of Life-Supporting Elements on Planetary Bodies?

A key factor that affects the availability of life-supporting elements on a world is their initial inventory, determined by the nature of accreted materials and how they were delivered. Because H, C, N, O, and S tend to form species that are solid only at low temperature (e.g., CO_2, NH_3, CO, CH_4, N_2, H_2S), these compounds were depleted in inner solar system solids and enriched in outer solar system materials (as observed in comets today). However, large-scale mixing in the protoplanetary disk helped to blur some of these differences (Questions 3 and 1, Chapter 4). Cometary species such as CO_2, NH_3, and organic matter can provide large inputs of carbon and nitrogen as well as CO, CH_4, N_2, and H_2S to planetary bodies.

The early evolution of a planetary body also factors into the contemporary availability of life-supporting elements. For example, the formation of a metal core sequesters iron, and CHNOPS elements can partition with the iron if conditions are sufficiently reducing. Iron in the silicate portion of a body is predominately in the divalent state, a key source of reducing power. Both early outgassing during a magma ocean phase and later volcanic outgassing can provide volatiles to the surface-atmosphere system. The nature and abundances of these outgassed volatiles depend on temperature, pressure, and compositional characteristics (e.g., oxygen fugacity) of the source region. Once at the surface or in the atmosphere, the volatile composition can be modified by photochemical reactions, escape to space, reactions with surface materials, and the activity of life.

Radiation-induced breakdown fundamentally alters the chemistry, and particularly the redox state, of C, N, S, P, and metals. The formation of highly oxidized species can contribute to the generation of potent sources of chemical energy for life (Q10.6). On the inner planets, the dominant source of radiation is ultraviolet light (Q10.3b). When water molecules are broken apart, H_2 and O_2 or hydrogen peroxide are produced. Mars provides an example where the escape of H_2 to space led to the net oxidation of the near-surface environment and production of other oxidized species, including, for example, sulfates, ferric iron, that are derived from the oxidation of volcanic or crustal reduced species (e.g., sulfides and ferrous iron). In the outer solar system, water radiolysis is mainly driven by charged particles in planetary magnetospheres. At Europa, the population of charged particles consists of electrons with energies of a few electron volts (eV) to MeV, and heavy ions (mostly S^+ and O^+) with energies between a few keV and 1 MeV. Direct radiolysis occurs to a depth of tens of centimeters, but its products are mixed to several meters by impact gardening. Further recycling processes like convection and subsumption of ice may transport oxidants to Europa's ocean (e.g., Hand et al. 2020). At Enceladus, radiolytic processing of surface ice is diminished by approximately an order of magnitude relative to Europa, and thus there may not be a comparably efficient mechanism for generating new oxidants at the surface.

Water–rock reactions, such as dissolution and mineral formation, exert a major influence on the abundances of CHNOPS elements in liquid water. A source for these elements is dissolution of primary rocks and minerals, in which they may only be present as minor constituents. Balancing supplies from water–rock reactions, minerals that sequester these elements include carbonates (for C), ammonium salts and clay minerals (for N), phosphates (for P), and sulfates or sulfides (for S). Once formed, the solubilities driving the bioavailability of elements in these minerals depend on the composition of the coexisting water, and particularly on its pH. H_2 in fluids can be generated by water radiolysis (Q10.6) or from oxidation of Fe and S during alteration (Q10.6) and react with carbon dioxide or carbonate in waters and rocks to produce organic molecules (Sherwood Lollar et al. 2021) (see Figure 13-5).

Critically, for water–rock interactions to occur in the first place, these two environments need to be in direct physical contact. Worlds that likely have high-pressure ices at the base of their oceans, such as Titan or Ganymede, may not host extensive interactions (except, perhaps, where localized heating is sufficient to melt overlying ice). The situation is more promising on Europa, where a silicate seafloor is almost certainly in contact with a subsurface ocean, and even more so within Enceladus—where ice grains containing sodium or silicon, diagnostic of water-rock interactions at depth, are ejected in the south polar jets (Figure 13-6).

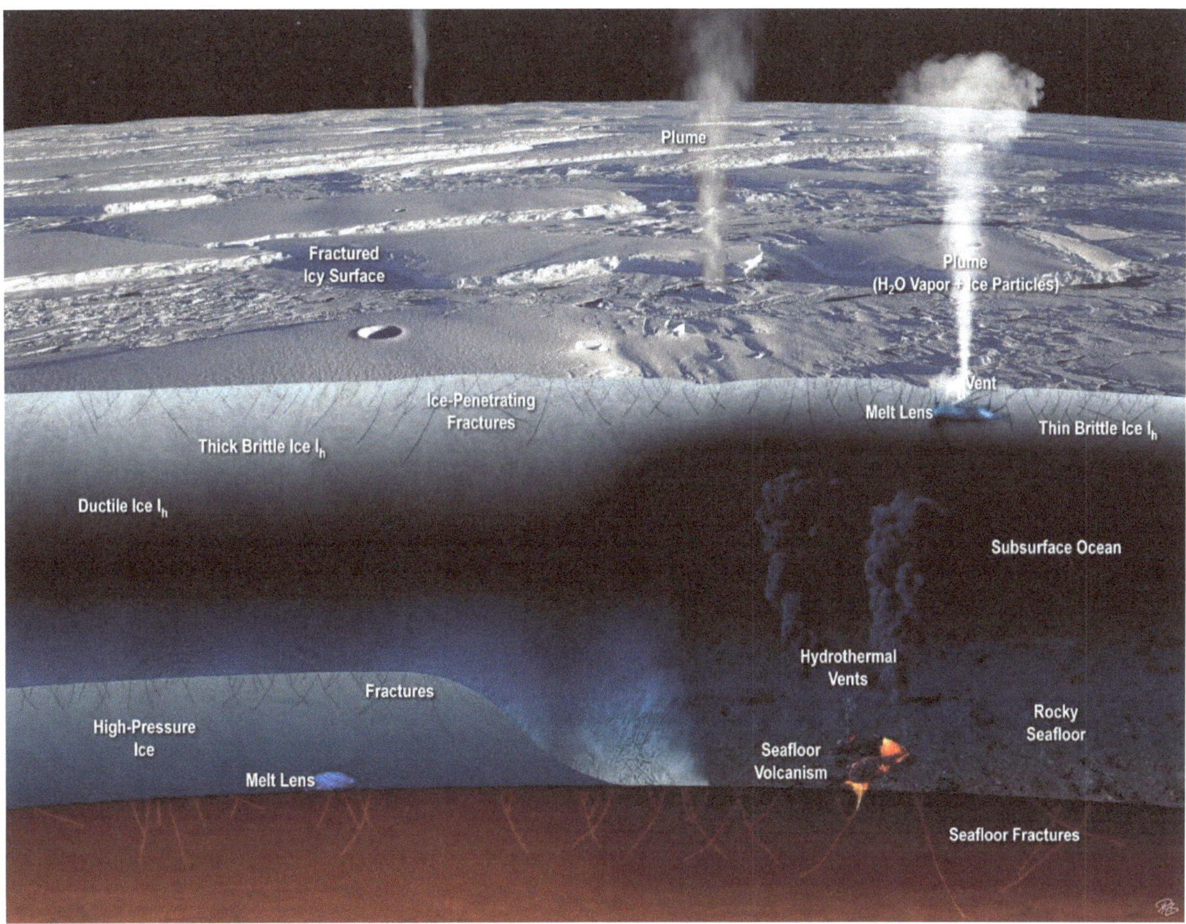

FIGURE 13-6 *Question 10.* Plumes and overturn of a fractured ice shell are means of exchanging matter between surface and ocean; and volcanism, vents, and infiltration of fractured crust are means of exchange between ocean and core within icy satellites. SOURCES: Figure by P.K. Byrne. Glacial landscape courtesy of D.G. Vaughan (BAS)/Photo: Jemma Cox.

Q10.5c How and Why Have the Inventories, Forms, and Distribution of Life-Supporting Elements Changed Through Time?

The availability of CHNOPS on a planet varies with time by mechanisms similar to those controlling the availability of water (Q10.3). Impacts and outgassing by volcanoes supply new gases with CHNOPS to the surface and supply new rocks for weathering. Water–rock reactions during weathering can liberate CHNOPS from rocks, for example, via mineral and glass dissolution, or lock them away in an unusable form, for example, by crustal hydration and formation of carbonate, sulfate, or nitrate minerals. On Earth, the bulk of the planet's CO_2 inventory is stored as carbonate minerals in the crust; CO_2 stored there is unavailable for use by organisms until the carbonates dissolve in water or are subducted into the mantle and released via volcanism. It is not clear whether such crustal recycling mechanisms have existed on Venus or Mars. Escape to space can remove considerable volumes of atmosphere (Q10.3) with a variety of potential consequences: directly decreasing the abundance and availability of gases that might otherwise be bioavailable; changing climate, leading to phase changes that make material more or less bioavailable; and changing redox conditions. For example, the time-integrated impact of loss of H_2 to space from radiolysis of water molecules leads to a buildup of O_2, which can be an important source of energy for life in chemical reactions (Q10.6). Last, seasonal or diurnal cycles can affect the distribution and availability of life-supporting constituents, as on Titan, where the atmospheric composition varies with time—for example, the poles become enhanced during hemispheric winter, as on Earth, leading to more complex molecules (Coustenis 2021)—and on Mars with seasonal cycles in atmospheric H_2O, O_2, CO_2, and methane, as observed by Curiosity. Such cycling can generate disequilibrium, for example, subsurface buildup of reduced gases from water–rock reactions and periodic release into a more oxic atmosphere.

Strategic Research for Q10.5

- **Determine if life-supporting chemical species, including reduced carbon- and phosphorus-bearing molecules, are present in the atmosphere of Venus,** via in situ atmospheric measurements.
- **Determine the diurnal and seasonal variability of Titan's atmospheric and surface chemical composition** by long-term monitoring from space and in situ measurements with support from ground and Earth-bound observatories and modeling.
- **Constrain CHNOPS speciation in liquid water environments on Europa, Enceladus, and other ocean worlds where materials sourced from an ocean can be accessed,** using measurements of tracer species in plume gases and particulates, volatiles in the ambient exosphere, and surface mineralogy/composition associated with materials sourced from the ocean.
- **Determine the inventory and bioavailability of CHNOPS, particularly reduced carbon and fixed nitrogen, for candidate current and ancient habitable environments on Mars** by measurements of mineralogy, chemistry, and isotopes from high-resolution remote sensing, surface missions, and returned samples as well as modeling exchanges.
- **Quantify the impact of the seasonal cycles in the martian atmosphere on the formation and long-term evolution of CHNOPS species,** using telescopic, spacecraft, and in situ observations of the behavior of aerosols, trace gas abundances, and isotopes, as well as observations and models of meteorological behavior and surface fluxes of gases.

Q10.6 WHAT CONTROLS THE ENERGY AVAILABLE FOR LIFE?

Habitability requires that the environment supply energy to the biological system to meet organism demand. Life on Earth utilizes a variety of energetically favorable reactions for maintenance and growth, and there are even examples of community metabolisms where organisms employing families of redox reaction couples exist symbiotically (Question 9). The nature of energy sources available for life, whether there is adequate power (i.e., energy per unit time), and the longevity of energetically favorable conditions (e.g., long-term maintenance of fluxes that sustain disequilibria) are key considerations for dynamic habitability.

Q10.6a What Are the Available Energy Sources for Life?

All life as we know it exploits energy via thermodynamic disequilibria. The presence or absence of disequilibrium conditions was initially considered as a basis for a life-detection strategy. We now recognize that the relationship between life, energy, and thermodynamic disequilibrium is more nuanced. Although thermodynamic disequilibrium can cause life to emerge, promote the growth of living organisms, sustain extant life, and be generated as a byproduct of life, such disequilibria also exist independently of life (NASEM 2019a).

Three primary energy sources power reactions that can support life on a planetary body:

- **Solar radiation:** Light is a dominant source of energy for life on Earth via photosynthesis. In oxygenic photosynthesis, photoreactive molecules are excited by absorbed light at wavelengths characteristic of a pigment (e.g., chlorophyll), triggering biochemical reactions that split water molecules, transform CO_2 to carbohydrate, and release O_2 as a by-product. In anoxygenic photosynthesis, different electron donors (e.g., hydrogen sulfide) are used instead of water for reactions with carbon dioxide. On Earth, oxygenic photosynthesis evolved relatively late, but anoxygenic photosynthesis likely sustained the earliest photosynthetic life forms, including early Archean shallow-water stromatolites (Q9.3). The discovery of diverse pigments exploiting wavelengths from the visible to near-infrared and many electron donors (Kiang et al. 2007) indicates that photosynthesis could be a viable energy source in a wide range of solar-illuminated planetary surface environments, although a challenge elsewhere in the solar system is meeting the other conditions for surface habitability.

- **Chemical energy:** The biomass of chemotrophic organisms on Earth is comparable to the photosynthetic biosphere, sustained by energetically favorable chemical reactions that support a variety of aerobic and anaerobic metabolisms: sulfide and iron oxidation; methanotrophy; methanogenesis; acetogenesis; nitrate, sulfate, and iron reduction; and fermentation (Onstott et al. 2019). These metabolisms are widespread in hydrothermal and groundwater systems, at volcanic vents, and in lakes and oceans at surface-atmosphere and water-bottom interfaces. Chemical energy sources are viable on other worlds; simple, widely available species participate in energy-supplying redox reactions, including H_2O, H_2, O_2, CO_2, CH_2O, CH_4, organic C, SO_4^{2-}, HS^-, NO_2^-, NH_4^+, and $Fe^{2+,3+}$. Environments with these redox reactions existed and may yet exist on Mars, water-altered asteroids like Ceres, and ocean worlds including Europa, Enceladus, Ganymede, and Titan.

- **Radiolysis:** Exposure to radiation breaks chemical bonds, creating new species that supply chemical energy for the redox reactions above. In *endogenic* radiolysis, radiation from radioactive uranium, thorium, and potassium isotopes in rock breaks apart water molecules in rock-permeating fluids to generate reductants (H_2) and oxidants (H_2O_2). Subsequent redox reactions involving these products sustain life kilometers underground, largely disconnected from other water reservoirs. By analogy with Earth, calculations of the energetic fluxes provided by radiolysis in the subsurface on Mars (Tarnas et al. 2021) and ocean worlds (Bouquet et al. 2017; Altair et al. 2018) suggest that those fluxes may be sufficient to sustain organisms. On airless bodies, additional *exogenic* radiolysis arises from solar irradiation or magnetospheric plasmas. For example, particles incident from Jupiter and from solar irradiation break water molecules at the surface of Europa's ice shell. H_2 is generated and typically lost to space, enriching oxidizing species such as O_2 and H_2O_2, and setting up a redox disequilibrium with materials in the interior. An outstanding question is whether these materials cycle back into Europa's subsurface ocean.

Q10.6b What Geophysical and Geochemical Processes Determine Power Available for Life?

Geological processes on a planetary body drive energy availability. Some sources of energy are crucial in sustaining elements of habitability but do not themselves directly act to sustain life. For example, tidal energy sustains liquid water oceans in the subsurface of icy satellites, and heat from geothermal and hydrothermal energy can maintain liquid water in the subsurface of Mars and perhaps even Ceres today. For the sources of energy critical for life directly, the major question is whether the environmental energy flux is sufficient to support life. Voltage (i.e., energy per reaction quantum) and power (i.e., energy per unit time) are the key parameters to not only sustain life in a dormant state against environmental stressors but to support metabolism, mobility, growth, and reproduction (Hoehler 2007). The energy for life (J/mol) available in planetary environments can be calculated, given sufficient compositional data, from fluid chemistry (at the present) or mineral assemblages (in the past). The power for life, or timescale of the energy supply (J/s or W), depends on the fluxes and the energetic reactions available (Q10.6a). There is a minimum maintenance power to keep the cells viable, probably between 1–$1,000 \times 10^{-21}$ W/cell (LaRowe and Amend 2015).

For photosynthetic metabolisms, power is based on the luminosity of the Sun at the orbital radius of the planetary body, accounting for attenuation by an atmosphere and for diurnal, seasonal, and longer-term (e.g., precessional) cycles. In the past, Mars's surface may have had punctuated periods of habitability (Q10.2a) to support photosynthetic life; key questions include how long these surface habitats persisted and whether life took advantage of these settings (Q10.7). To what extent evolutionarily advantageous energy storage or periods of dormancy could take advantage of light energy on bodies spinning slower than Earth's 24-hr cycle (e.g., the 243-day cycle of Venus) remains unclear. More importantly, surface environments where photosynthetic energy for life is readily available face other habitability challenges. Liquid water availability is the challenge on Mars, availability of organic matter and water and very low pH is the challenge in Venus's upper atmosphere, and availability of solvents for non-water-based life is the challenge for Titan.

Investigating whether energy is the limiting factor for life in potentially habitable subsurface environments found throughout the solar system is a priority (NASEM 2019a). At Europa and Enceladus, a key question is whether disequilibrium between water in their oceans and their rocky cores persists despite billions of years of chemical reactions. Alternatively, radiolysis (Q10.6a) may supply necessary redox couples to power life (Ray et al. 2021). At Enceladus, evidence for thermodynamic disequilibrium consistent with such processes has already been reported (Hsu et al. 2015; Waite et al. 2017). There, CO_2, H_2, and CH_4 co-occur in abundances where a strong energetic drive exists for CO_2 and H_2 to combine to form more CH_4 and water. Enceladus's energy supply—which is likely the limiting factor, rather than liquid water or nutrients—could sustain a biomass density comparable to that of subglacial lakes on Earth (Cable et al. 2020). In the inner solar system, fluid circulation in rock was prevalent in the past on Mars and Ceres, and may still occur on both, as well as on other large hydrous asteroids. On Mars, silicate hydration, iron oxidation, sulfate formation, and several other chemical processes operated in the past to supply energy to subsurface life (e.g., Onstott et al. 2019). Radiolytic energy appears to be sufficient to power life (Q10.6a), and there may even be modern hydrogen or methane fluxes locally (Moores et al. 2019).

In situ measurements or sample return are required to gain data on reactants and products for quantitative constraints. Then, modeling of anticipated water-rock reaction conditions and their timescales offers potential for exploring the past and present habitability of other planetary bodies, albeit with caveats. First, theoretical modeling is most robust for situations in which equilibrium is anticipated to have been obtained, whereas, in a dynamic setting, reaction kinetics might dominate; this necessitates complementary study of Earth analogs and laboratory experiments (Question 9). Second, the maximum chemical energy available from any set of reaction conditions does not necessarily imply that life is able to exploit it, as other habitability conditions have also to be met. Third, the supply of energy is highly heterogeneous on a planetary body, and, therefore, so would be the distribution of any biomass. Focused reaction zones are where redox chemistry occurs, and the flux of reactants and products is controlled by permeability, fracturing, and hydrological or geophysical processes that refresh products and reactants. "Follow the interfaces" is a guiding principle in Mars exploration of the subsurface for life (Onstott et al. 2019). This approach is also applicable to ocean worlds where shell-ocean and ocean-seafloor interfaces are expected to be the zones with concentrated energy fluxes that could provide favorable habitats (see Figure 13-6).

Q10.6c What Processes Govern Energy Availability Over the Long Term?

Over geological timescales, the sustained release of chemical energy requires mechanisms for renewing reactants to prevent equilibration. In addition to radiolytic oxidant supply (Q10.6a, Q10.7b), these mechanisms include fracturing and erosion to expose fresh surface materials to chemical reaction; impacts, tectonic activity, and magmatism/volcanism to emplace or redistribute material out of equilibrium with its surroundings; and hydrological parameters that control fluid circulation. In some cases, such processes might become transiently self-sustaining: for example, volumetric expansion associated with mineral hydration, as observed in serpentinization systems, can lead to fresh fracturing and, hence, further introduction of fluid. Presently, the rates of most such processes are unknown on most planetary bodies.

Strategic Research for Q10.6

- **Geochemically characterize past and present environments with liquid water to determine whether there is/was energy to sustain metabolic processes of life** by in situ or sample analysis of waters and preserved water-formed mineral and chemical species to determine concentrations of major ions, electron donors and acceptors, mineral products, and other relevant chemical species.
- **Characterize the compositional and geological heterogeneity of potentially habitable worlds at progressively smaller scales (km- to cm-scale) to identify locales where chemical energy is or was more available** by identification of mineralized fractures, reaction fronts, permeability boundaries, sites of ocean-ice exchange, and other interfaces via orbital remote sensing and in situ landscape- and microscopic-scale characterization.
- **Determine the geophysical parameters that control past and present material fluxes in rocky subsurfaces, such as porosity, permeability, heat flux, volcanic flux, and tectonics** by geophysical measurement, drilling/coring, change-detection experiments, seismic experiments, and modeling.
- **Determine the kinetics of chemical reactions relevant to energy supply and material availability for life under conditions (past or present) on planetary bodies in the solar system** by conducting laboratory experiments.

Q10.7 WHAT CONTROLS THE CONTINUITY OR SUSTAINABILITY OF HABITABILITY?

The continuity of habitable conditions is likely important for both starting and sustaining life on other worlds. If periods of habitability are too short to allow life to take root, a body or environment may never be capable of supporting life, even if it meets the definition of "habitable." On the other hand, environments presently on the edge of habitability (e.g., Venus's atmosphere) could sustain life that started under more clement conditions (e.g., on the Venus surface during an earlier habitable phase) or provide changing conditions conducive to the emergence of life, as has been suggested for episodically wet/dry surface environments on early Earth and Mars.

Q10.7a What Exogenous Factors Control the Continuity of Habitability?

Exogenous factors that affect planetary habitability include (but are not limited to) the properties of the host star(s), the orbital dynamics of the planet, and the impact bombardment flux (see Figure 13-2).

Worlds with slightly elliptical orbits such as Mars are exposed to a wider range of stellar fluxes over the course of a year than those with near-circular orbits such as Venus. Orbital resonances can provide heat and energy for ocean worlds such as Europa and Enceladus through tidal heating. A natural satellite that is large relative to its planet, such as Earth's Moon, can stabilize obliquity (spin axis tilt) and climate; by contrast, Mars, with only relatively small satellites, experiences larger variations leading to ice being redistributed between polar and equatorial regions, and perhaps liquid water becoming transiently stable, over thousands to millions of years because of periodic variations of tens of degrees in axial tilt (Q10.2).

The relative contribution of impact bombardment to degrading or enhancing habitability on any given body remains to be fully understood. Extremely large impacts can strip planetary atmospheres or melt lithospheres, destroying prevailing habitable environments as well as any record of their presence. However, bombardment also delivered volatiles to the inner solar system, including Earth, that were critical to generating our hydrosphere and atmosphere (Q10.3). Impacts can destroy surface habitable environments (e.g., Earth's K-T impact, which caused mass extinctions), but impacts can also simultaneously maintain or even create subsurface habitable environments, such as possible transient hydrothermal systems on Mars, Ceres, and Titan. The relative contribution of impact bombardment to degrading or enhancing habitability on any given body remains to be deciphered.

Q10.7b What Endogenous Factors Control the Continuity of Habitability?

The duration of habitability may be most strongly affected by the processes associated with heat transfer and loss over time, which are themselves a function of planetary size and starting composition (e.g., Ehlmann et al. 2016). For worlds that are sufficiently large and differentiated to have an intrinsically generated dynamo, a magnetic field could provide adequate protection of the surface and atmosphere from stellar activity and atmospheric stripping, or could alternately contribute to atmospheric loss (e.g., Gunell et al. 2018).

The tectonic regime of the terrestrial bodies and icy shells (Question 5, Chapter 8) can play a major role in establishing, sustaining, or destroying habitability. Surface mobility enables the exchange of nutrients, energy, and water between the interior and exterior. Numerous tectonic regimes have been proposed, including active-lid (a subset of which defines Earth's present plate tectonics paradigm), heat pipe, sluggish-lid, episodic-lid, plutonic-squishy lid, and stagnant-lid modes, which differ in how heat is lost and how tectonic and volcanic activity is manifested (e.g., Lourenço et al. 2020; Byrne et al. 2021). This, in turn, determines how volatiles are cycled between the interior, surface, and atmosphere or sequestered, drawing down the atmosphere over time (Question 5; Q10.3). Bodies may transition between tectonic regimes over their lifetime (e.g., Weller and Lenardic 2018). Internal processes that redistribute mass can change a body's orbital and rotational dynamics (e.g., the Tharsis Rise on Mars), as well as elemental availability and oxidation state, hence altering conditions in both surface and subsurface environments.

A tectonic regime conducive to long-term volcanic activity, as is likely the case for Venus, can replenish an atmosphere that might otherwise be lost to weathering or escape. Bursts of volcanic activity can have major effects on climates on short and long timescales (e.g., the Permian-Triassic mass extinction event on Earth). Volcanic activity likely helped Earth transition out of a past snowball state and may have precipitated Venus's runaway greenhouse atmosphere and loss of surface habitability (Way and Del Genio 2020). On ocean worlds, material exchange between the surface, ice shell, and any underlying ocean is still poorly understood, as is the role of such cycling on habitability by promoting exchange of materials.

Strategic Research for Q10.7

- **Assess surface–ocean exchange and the dynamics and long-term evolution of oceans and ice shells on ocean worlds, including Europa and Enceladus,** by measuring the chemistry and speciation of plumes and surfaces with orbital remote sensing, in situ, or sample data, and by models that couple orbital and internal evolution and dynamics.
- **Determine the nature, timing, and processes controlling the existence of past habitable environments on Mars** by measurements in situ and in returned samples of stratigraphy, petrology, organic content, isotopes in rock, and geochronology in multiple environments covering multiple time periods.
- **Identify the effects of large impacts on local and planetary habitability** by investigating impact sites on potentially habitable bodies, analyzing chemical and isotopic signatures in rock/ice records before and after impacts, modeling the thermal and compositional effects of impacts on planetary atmospheres and surfaces, and improving knowledge of impact flux with time.
- **Understand the diversity and controls on rates and styles of recycling of surface materials** by remote sensing of evidence of these processes on relevant worlds, and by developing models for lithosphere dynamics and (cryo)volcanism that account for the thermal and orbital evolution of bodies, and the rheological and compositional evolution of their interiors.

SUPPORTIVE ACTIVITIES FOR QUESTION 10

- Improved radiative transfer modeling and photochemical modeling in planetary atmospheres for models of climate and organics production by fundamental laboratory measurements and computational work to obtain photochemistry reaction coefficients and gas absorption parameters.
- Improve the characterization of diverse ices and the mineral and organic products of water–rock interaction that are fingerprints of habitability by fundamental laboratory measurements to obtain and compile libraries of ice, mineral, and organic spectra and optical constants at ultraviolet to far-infrared wavelengths.

- Develop technologies for in situ measurement of light and radiogenic isotopes that trace the fluxes of key CHNOPS species and geological evolution in absolute time by miniaturization and maturation of instruments for isotope measurement on landed missions to terrestrial and ocean worlds.
- Increasingly high-fidelity analyses of organics and metal isotopes that determine abiogenic versus biogenic organic production as well as reservoirs and fluxes of key elements for biology by development of sample return technology for silicate, ice, and atmospheric samples and advanced facilities for returned sample analysis and curation.
- Determine the chemical and mineralogical outcomes of weathering, other aqueous alteration, and abiotic organic synthesis under non-Earth conditions by reaction path kinetic and thermodynamic modeling and with experiments.
- Characterize the limits of habitable conditions for solar system bodies by field work in terrestrial extreme environments, experimental studies, and modeling.

REFERENCES

Altair, T., M.G.B. de Avellar, F. Rodrigues, and D. Galante. 2018. "Microbial Habitability of Europa Sustained by Radioactive Sources." *Scientific Reports* 8:260. https://doi.org/10.1038/s41598-017-18470-z.

Bierson, C.J., and F. Nimmo. 2022. "A Note on the Possibility of Subsurface Oceans on the Uranian Satellites." *Icarus* 373:114776.

Bouquet, A., C.R. Glein, D. Wyrick, and J.H. Waite. 2017. "Alternative Energy: Production of H_2 by Radiolysis of Water in the Rocky Cores of Icy Bodies." *Astrophysical Journal Letters* 840(1):L8. https://doi.org/10.3847/2041-8213/aa6d56.

Byrne, P.K., R.C. Ghail, A.M.C. Şengör, P.B. James, C. Klimczak, and S.C. Solomon. 2021. "A Globally Fragmented and Mobile Lithosphere on Venus." *Proceedings of the National Academy of Sciences* 118(26):e2025919118. https://doi.org/10.1073/pnas.2025919118.

Cable, M., M. Neveu, H.-W. Hsu, and T. Hoehler. 2020. "Enceladus." Pp. 217–246 in *Planetary Astrobiology*, V.S. Meadows, G.N. Arney, B.E. Schmidt, and D.J. Des Marais, eds. Tucson: University of Arizona Press.

Campins, H., K. Hargrove, N. Pinilla-Alonso, E.S. Howell, M.S. Kelley, J. Licandro, T. Mothé-Diniz, et al. 2010. "Water Ice and Organics on the Surface of the Asteroid 24 Themis." *Nature* 464:1320–1321. https://doi.org/10.1038/nature09029.

Castillo-Rogez, J.C., M. Neveu, J.E. Scully, C.H. House, L.C. Quick, A. Bouquet, K.E. Miller, et al. 2020. "Ceres: Astrobiological Target and Possible Ocean World." *Astrobiology* 20(2):269–291. https://doi.org/10.1089/ast.2018.1999.

Choblet, G., G. Tobie, C. Sotin, M. Běhounková, O. Čadek, F. Postberg, and O. Souček. 2017. "Powering Prolonged Hydrothermal Activity Inside Enceladus." *Nature Astronomy* 1(12):841–847. https://doi.org/10.1038/s41550-017-0289-8.

Coustenis, A. 2021. "The Atmosphere of Titan." In *Oxford Research Encyclopedia of Planetary Science*, P. Read, ed. Oxford, UK: Oxford University Press. https://doi.org/10.1093/acrefore/9780190647926.013.120.

Coustenis, A., and T. Encrenaz. 2013. *Life Beyond Earth: The Search for Habitable Worlds in the Universe*. Cambridge, UK: Cambridge University Press.

Cruikshank, D.P., C.K. Materese, Y.J. Pendleton, P.J. Boston, W.M. Grundy, B. Schmitt, C.M. Lisse, et al. 2019. "Prebiotic Chemistry of Pluto." *Astrobiology* 19(7):831–848. https://doi.org/10.1089/ast.2018.1927.

Daly, L., M.R. Lee, L.J. Hallis, H.A. Ishii, J.P. Bradley, P.A. Bland, D.W. Saxey, et al. 2021. "Solar Contributions to Earth's Oceans." *Nature Astronomy* 5:1275.

De Sanctis, M.C., V. Vinogradoff, A. Raponi, E. Ammannito, M. Ciarniello, F.G. Carrozzo, S. De Angelis, et al. 2019. "Characteristics of Organic Matter on Ceres from VIR/Dawn High Spatial Resolution Spectra." *Monthly Notices of the Royal Astronomical Society* 482(2):2407–2421. https://doi.org/10.1093/mnras/sty2772.

Donahue, T.M., J.H. Hoffman, R.R. Hodges, Jr., and A.J. Watson. 1982. "Venus Was Wet: A Measurement of the Ratio of Deuterium to Hydrogen." *Science* 216:630–633. https://doi.org/10.1126/science.216.4546.630.

Ehlmann, B.L., and C.S. Edwards. 2014. "Mineralogy of the Martian Surface." *Annual Review of Earth and Planetary Sciences* 42:291–315. https://doi.org/10.1146/annurev-earth-060313-055024.

Ehlmann, B.L., F.S. Anderson, J. Andrews-Hanna, D.C. Catling, P.R. Christensen, B.A. Cohen, C.D. Dressing, et al. 2016. "The Sustainability of Habitability on Terrestrial Planets: Insights, Questions, and Needed Measurements from Mars for Understanding the Evolution of Earth-Like Worlds." *Journal of Geophysical Research: Planets* 121(10):1927–1961. https://doi.org/10.1002/2016JE005134.

Eigenbrode, J.L., R.E. Summons, A. Steele, C. Freissinet, M. Millan, R. Navarro-González, B. Sutter, et al. 2018. "Organic Matter Preserved in 3-Billion-Year-Old Mudstones at Gale Crater, Mars." *Science* 360(6393):1096–1101. https://doi.org/10.1126/science.aaq0131.

Glavin, D.P., C.M.O'D. Alexander, J.C. Aponte, J.P. Dworkin, J.E. Elsila, and H. Yabuta. 2018. "The Origin and Evolution of Organic Matter in Carbonaceous Chondrites and Links to Their Parent Bodies." Pp. 205–271 in *Primitive Meteorites and Asteroids*, N. Abreu, ed. Amsterdam, Netherlands: Elsevier.

Gunell, H., R. Maggiolo, H. Nilsson, G.S. Wieser, R. Slapak, J. Lindkvist, M. Hamrin, et al. 2018. "Why an Intrinsic Magnetic Field Does Not Protect a Planet Against Atmospheric Escape." *Astronomy & Astrophysics* 614:L3. https://doi.org/10.1051/0004-6361/201832934.

Hamano, K., Y. Abe, and H. Genda. 2013. "Emergence of Two Types of Terrestrial Planet on Solidification of Magma Ocean." *Nature* 497:607–610. https://doi.org/10.1038/nature12163.

Hand, K.P., C.F. Chyba, J.C. Priscu, R.W. Carlson, and K.H. Nealson. 2009. "Astrobiology and the Potential for Life on Europa." Pp. 589–629 in *Europa*, R.T. Pappalardo, W.B. McKinnon, and K. Khurana, eds. Tucson: University of Arizona Press.

Hand, K.P., C. Sotin, A. Hayes, and A. Coustenis. 2020. "On the Habitability and Future Exploration of Ocean Worlds." *Space Science Reviews* 216(5):95. https://doi.org/10.1007/s11214-020-00713-7.

Hansen, C.J., J.C. Castillo-Rogez, W.M. Grundy, J.D. Hofgartner, E.S. Martin, K. Mitchell, F. Nimmo, et al. 2021. "Triton: Fascinating Moon, Likely Ocean World, Compelling Destination!" *Planetary Science Journal* 2:137. https://iopscience.iop.org/article/10.3847/PSJ/abffd2.

Hartkorn, O., and J. Saur. 2017. "Induction Signals from Callisto's Ionosphere and Their Implications on a Possible Subsurface Ocean." *Journal of Geophysical Research: Planets* 122:11677–11697. https://doi.org/10.1002/2017JA024269.

Hendrix, A.R., T.A. Hurford, L.M. Barge, M.T. Bland, J.S. Bowman, W. Brinckerhoff, B.J. Buratti, et al. 2019. "The NASA Roadmap to Ocean Worlds." *Astrobiology* 19:1–27. https://doi.org/10.1089/ast.2018.1955.

Hoehler, T.M. 2007. "An Energy Balance Concept for Habitability." *Astrobiology* 7(6):824–838. https://doi.org/10.1089/ast.2006.0095.

Hsu, H.W., F. Postberg, Y. Sekine, T. Shibuya, S. Kempf, M. Horányi, A. Juhász, et al. 2015. "Ongoing Hydrothermal Activities Within Enceladus." *Nature* 519(7542):207–210. https://doi.org/10.1038/nature14262.

Iess, L., R.A. Jacobson, M. Ducci, D.J. Stevenson, J.I. Lunine, J.W. Armstrong, S.W. Asmar, et al. 2012. "The Tides of Titan." *Science* 337(6093):457–459. https://doi.org/10.1126/science.1219631.

Jakosky, B.M., D. Brain, M. Chaffin, S. Curry, J. Deighan, J. Grebowsky, J. Halekas, et al. 2018. "Loss of the Martian Atmosphere to Space: Present-Day Loss Rates Determined from MAVEN Observations and Integrated Loss Through Time." *Icarus* 315:146–157. https://doi.org/10.1016/j.icarus.2018.05.030.

Kasting, J.F. 1988. "Runaway and Moist Greenhouse Atmospheres and the Evolution of Earth and Venus." *Icarus* 74:472–494. https://www.sciencedirect.com/science/article/pii/0019103588901169?via%3Dihub.

Kiang, N.Y., J. Siefert, Govindjee, and R.E. Blankenship. 2007. "Spectral Signatures of Photosynthesis. I. Review of Earth Organisms." *Astrobiology* 7(1):222–251. https://doi.org/10.1089/ast.2006.0105.

Kivelson, M.G., K.K. Khurana, and M. Volwerk. 2009. "Europa's Interaction with the Jovian Magnetosphere." Pp. 545–570 in *Europa*, R.T. Pappalardo, W.B. McKinnon, and K.K. Khurana, eds. Tucson: University of Arizona Press. http://www.igpp.ucla.edu/people/mkivelson/Publications/316-Europa%20Chapt23%20.pdf.

Korablev, O., A.C. Vandaele, F. Montmessin, A.A. Fedorova, A. Trokhimovskiy, F. Forget, F. Lefèvre, et al. 2019. "No Detection of Methane on Mars from Early ExoMars Trace Gas Orbiter Observations." *Nature* 568(7753):517–520. https://doi.org/10.1038/s41586-019-1096-4.

LaRowe, D.E., and J.P. Amend. 2015. "Power Limits for Microbial Life." *Frontiers in Microbiology* 6:718. https://doi.org/10.3389/fmicb.2015.00718.

Lopes, R.M.C., R.L. Kirk, K.L. Mitchell, A. LeGall, J.W. Barnes, A. Hayes, J. Kargel, et al. 2013. "Cryovolcanism on Titan: New Results from Cassini RADAR and VIMS." *Journal of Geophysical Research: Planets* 118(3):416–435. https://doi.org/10.1002/jgre.20062.

Lourenço, D.L., A.B. Rozel, M.D. Ballmer, and P.J. Tackley. 2020. "Plutonic-Squishy Lid: A New Global Tectonic Regime Generated by Intrusive Magmatism on Earth-Like Planets." *Geochemistry, Geophysics, Geosystems* 21(4):e2019GC008756. https://doi.org/10.1029/2019GC008756.

Lunine, J.I. 2017. "Ocean Worlds Exploration." *Acta Astronautica* 131:123–130. https://doi.org/10.1016/j.actaastro.2016.11.017.

McEwen, A.S., E.I. Schaefer, C.M. Dundas, S.S. Sutton, L.K. Tamppari, and M. Chojnacki. 2021. "Mars: Abundant Recurring Slope Lineae (RSL) Following the Planet-Encircling Dust Event (PEDE) of 2018." *Journal of Geophysical Research: Planets* 126:e2020JE006575. https://doi.org/10.1029/2020JE006575.

Moores, J.E., P.L. King, C.L. Smith, G.M. Martinez, C.E. Newman, S.D. Guzewich, P.-Y. Meslin, et al. 2019. "The Methane Diurnal Variation and Microseepage Flux at Gale Crater, Mars as Constrained by the ExoMars Trace Gas Orbiter and Curiosity Observations." *Geophysical Research Letters* 46(16):9430–9438. https://doi.org/10.1029/2019GL083800.

Morgan, G.A., N.E. Putzig, M.R. Perry, H.G. Sizemore, A.M. Bramson, E.I. Petersen, Z.M. Bain, et al. 2021. "Availability of Subsurface Water-Ice Resources in the Northern Mid-Latitudes of Mars." *Nature Astronomy* 5(3):230–236. https://doi.org/10.1038/s41550-020-01290-z.

NASEM (National Academies of Sciences, Engineering, and Medicine). 2019a. *An Astrobiology Strategy for the Search for Life in the Universe*. Washington, DC: The National Academies Press. https://doi.org/10.17226/25252.

Nimmo, F., and W.B. McKinnon. 2021. "Geodynamics of Pluto." Pp. 89–103 in *The Pluto System After New Horizons*, S.A. Stern, J.M. Moore, W.M. Grundy, L.A. Young, and R.P. Binzel, eds. Tucson: University of Arizona Press. https://websites. pmc.ucsc.edu/~fnimmo/website/pluto_geodynamics.pdf.

Onstott, T.C., B.L. Ehlmann, H. Sapers, M. Coleman, M. Ivarsson, J.J. Marlow, A. Neubeck, and P. Niles. 2019. "Paleo-Rock-Hosted Life on Earth and the Search on Mars: A Review and Strategy for Exploration." *Astrobiology* 19(10):1230–1262. https://doi.org/10.1089/ast.2018.1960.

Paganini, L., G.L. Villanueva, L. Roth, A.M. Mandell, T.A. Hurford, K.D. Retherford, and M.J. Mumma. 2020. "A Measurement of Water Vapour amid a Largely Quiescent Environment on Europa." *Nature Astronomy* 4:266–272. https://doi. org/10.1038/s41550-019-0933-6.

Postberg, F., N. Khawaja, B. Abel, G. Choblet, C.R. Glein, M.S. Gudipati, B.L. Henderson, et al. 2018a. "Macromolecular Organic Compounds from the Depths of Enceladus." *Nature* 558(7711):564–568. https://doi.org/10.1038/ s41586-018-0246-4.

Postberg, F., R.N. Clark, C.J. Hansen, A.J. Coates, C.M. Dalle Ore, F. Scipioni, M.M. Hedman, and J.H. Waite. 2018b. "Plume and Surface Composition of Enceladus." Pp. 129–162 in *Enceladus and the Icy Moons of Saturn*, P.M. Schenk, R.N. Clark, C.J.A. Howett, A.J. Verbiscer, and J.H. Waite, eds. Tucson: University of Arizona Press. https://www.geo.fu-berlin. de/en/geol/fachrichtungen/planet/projects/habitat_oasis/_layout/postberg_18b.pdf.

Quick, L.C., A. Roberge, A. Barr-Mlinar, and M.M. Hedman. 2020. "Forecasting Rates of Volcanic Activity on Terrestrial Exoplanets and Implications for Cryovolcanic Activity on Extrasolar Ocean Worlds." *Publications of the Astronomical Society of the Pacific* 132(1014):084402. https://doi.org/10.1088/1538-3873/ab9504.

Rampe, E.B., D.F. Blake, T.F. Bristow, D.W. Ming, D.T. Vaniman, R.V. Morris, C.N. Achilles, et al. 2020. "Mineralogy and Geochemistry of Sedimentary Rocks and Eolian Sediments in Gale Crater, Mars: A Review After Six Earth Years of Exploration with Curiosity." *Geochemistry* 80(2):125605. https://doi.org/10.1016/j.chemer.2020.125605.

Ray, C., C.R. Glein, J.H. Waite, B. Teolis, T. Hoehler, J.A. Huber, J.I. Lunine, and F. Postberg. 2021. "Oxidation Processes Diversify the Metabolic Menu on Enceladus." *Icarus* 364:114248. https://doi.org/10.1016/j.icarus.2020.114248.

Rivera-Valentín, E.G., V.F. Chevrier, A. Soto, and G. Martínez. 2020. "Distribution and Habitability of (Meta)Stable Brines on Present-Day Mars." *Nature Astronomy* 4:756–761. https://doi.org/10.1038/s41550-020-1080-9.

Scheller, E.L., B.L. Ehlmann, R. Hu, D.J. Adams, and Y.L. Yung. 2021. "Long-Term Drying of Mars by Sequestration of Ocean-Scale Volumes of Water in the Crust." *Science* 372(6537):56–62. https://doi.org/10.1126/science.abc7717.

Sherwood Lollar, B., V.B. Heuer, J. McDermott, S. Tille, O. Warr, J.J. Moran, J. Telling, and K.U. Hinrichs. 2021. "A Window into the Abiotic Carbon Cycle–Acetate and Formate in Fracture Waters in 2.7-Billion-Year-Old Host Rocks of the Canadian Shield." *Geochimica et Cosmochimica Acta* 294:295–314. https://doi.org/10.1016/j.gca.2020.11.026.

Steele, A., F.M. McCubbin, and M.D. Fries. 2016. "The Provenance, Formation, and Implications of Reduced Carbon Phases in Martian Meteorites." *Meteoritics & Planetary Science* 51:2203–2225. https://doi.org/10.1111/maps.12670.

Stern, J.C., B. Sutter, C. Freissinet, R. Navarro-González, C.P. McKay, P.D. Archer, A. Buch, et al. 2015. "Evidence for Indigenous Nitrogen in Sedimentary and Aeolian Deposits from the Curiosity Rover Investigations at Gale Crater, Mars." *Proceedings of the National Academy of Sciences* 112(14):4245–4250. https://doi.org/10.1073/ pnas.1420932112.

Stillman, D.E., and R.E. Grimm. 2011. "Dielectric Signatures of Adsorbed and Salty Liquid Water at the Phoenix Landing Site, Mars." *Journal of Geophysical Research: Planets* 116:E09005. https://doi.org/10.1029/2011JE003838.

Tarnas, J.D., J.F. Mustard, B. Sherwood Lollar, V. Stamenkovi, K.M. Cannon, J.P. Lorand, T.C. Onstott, et al. 2021. "Earth-like Habitable Environments in the Subsurface of Mars." *Astrobiology* 21(6):741–756. https://doi.org/10.1089/ ast.2020.2386.

Vinogradoff, V., S. Bernard, C. Le Guillou, and L. Remusat. 2018. "Evolution of Interstellar Organic Compounds under Asteroidal Hydrothermal Conditions." *Icarus* 305:358–370. https://doi.org/10.1016/j.icarus.2017.12.019.

Waite, J.H, C.R. Glein, R.S. Perryman, B.D. Teolis, B.A. Magee, G. Miller, J. Grimes, et al. 2017. "Cassini Finds Molecular Hydrogen in the Enceladus Plume: Evidence for Hydrothermal Processes." *Science* 356(6334):155–159. https://doi. org/10.1126/science.aai8703.

Way, M.J., and A.D. Del Genio. 2020. "Venusian Habitable Climate Scenarios: Modeling Venus through Time and Applications to Slowly Rotating Venus-Like Exoplanets." *Journal of Geophysical Research: Planets* 125(5):e2019JE006276. https:// doi.org/10.1029/2019JE006276.

Webster, C.R., P.R. Mahaffy, S.K. Atreya, J.E. Moores, G.J. Flesch, C.A. Malespin, C.P. McKay, et al. 2018. "Background Levels of Methane in Mars' Atmosphere Show Strong Seasonal Variations." *Science* 360(6393):1093–1096. https://doi. org/10.1126/science.aaq0131.

Webster, C.R., P.R. Mahaffy, J. Pla-Garcia, S.C. Rafkin, J.E. Moores, S.K. Atreya, G.J. Flesch, et al. 2021. "Day-Night Differences in Mars Methane Suggest Nighttime Containment at Gale Crater." *Astronomy & Astrophysics* 650:A166. https://doi.org/10.1051/0004-6361/202040030.

Weller, M.B., and A. Lenardic. 2018. "On the Evolution of Terrestrial Planets: Bi-Stability, Stochastic Effects, and the Non-Uniqueness of Tectonic States." *Geoscience Frontiers* 9(1):91–102. https://doi.org/10.1016/j.gsf.2017.03.001.

Wong, M.L., Y.L. Yung, and G.R. Gladstone. 2015. "Pluto's Implications for a Snowball Titan." *Icarus* 246:192–196. https://doi.org/10.1016/j.icarus.2014.05.019.

Wordsworth, R., Y. Kalugina, S. Lokshtanov, A. Vigasin, B.L. Ehlmann, J. Head, C. Sanders, and H. Wang. 2017. "Transient Reducing Greenhouse Warming on Early Mars." *Geophysical Research Letters* 44:665–671. https://doi.org/10.1002/2016GL071766.

Wray, J.J. 2020. "Contemporary Liquid Water on Mars?" *Annual Reviews of Earth and Planetary Sciences* 49:141–171. https://doi.org/10.1146/annurev-earth-072420-071823.

Zahnle, K.J., M. Gacesa, and D.C. Catling. 2019. "Strange Messenger: A New History of Hydrogen on Earth, as Told by Xenon." *Geochimica et Cosmochimica Acta* 244:56–85. https://doi.org/10.1016/j.gca.2018.09.017.

Zolotov, M. 2019. "Chemical Weathering on Venus." In *Oxford Research Encyclopedia of Planetary Science*, P. Read, ed. Oxford, UK: Oxford University Press. https://doi.org/10.1093/acrefore/9780190647926.013.146.

Q11 PLATE: A true-color "selfie" of the Perseverance rover, accompanied by the Ingenuity helicopter, on the surface of Mars in 2021. SOURCE: Courtesy of NASA/JPL-Caltech/MSSS.

14

Question 11: Search for Life Elsewhere

Is there evidence of past or present life in the solar system beyond Earth, and how do we detect it?

Building on insights from terrestrial life and our understanding of the diversity of habitable environments elsewhere, as well as significant advancement in biosignature detection technologies, we are poised to conduct a rigorous, systematic search for life beyond Earth in the solar system.[1] Four hundred years after Galileo revolutionized our understanding of our place in the universe, ours could realistically be the generation that triggers another scientific revolution, this time in biology.

The search for evidence of life is part of the systematic progression in our understanding of planetary environments, which also encompasses detailed environmental characterization (Figure 14-1). Hence, a well-conceived arc of life detection activities serves to expand our knowledge of planetary environments, whether evidence of life is found or not. In the past decade, past and presently habitable environments beyond Earth were identified, providing a rich spectrum of worlds to explore in the context of astrobiology (Figure 14-2). Some of these environments introduce the potential to understand a biochemistry and/or emergence distinct from that of life on Earth, and thus we might begin to develop a universal theory of living systems.

In the coming decade, biosignature searches will require evolution in how planetary systems are studied by new missions, technologies, and approaches to data analysis, shaped by our expanding knowledge of the range of habitats and biochemistries that can produce detectable biosignatures. A comprehensive framework is required to interpret potential biosignatures, abiosignatures, false positives, and false negatives, and promote confidence and consensus in interpretations. This framework needs to leverage past experiences with previous searches for signs of life in Earth and planetary environments and materials (e.g., Viking and ALH84001) to improve our approach of data interpretation and how results are communicated and evaluated, both within the planetary science community and by the general public. Our understanding of Earth life provides a starting point for investigation (Question 9, Chapter 12), contrasting the signatures of life on Earth with the prebiotic/abiotic chemistry of primitive bodies. Key areas of research also include the search for and identification of potential biosignatures that are agnostic to life's molecular makeup or metabolism, and studies of signatures of abiotic processes and phenomena that may define the nonbiological baseline and/or mimic biosignatures, affecting biosignature reliability (NASEM 2019). Expanding the search to include the possibility of nonterran life (i.e., life "not as we know it") (NRC 2007) requires

---

[1] A glossary of acronyms and technical terms can be found in Appendix F.

367

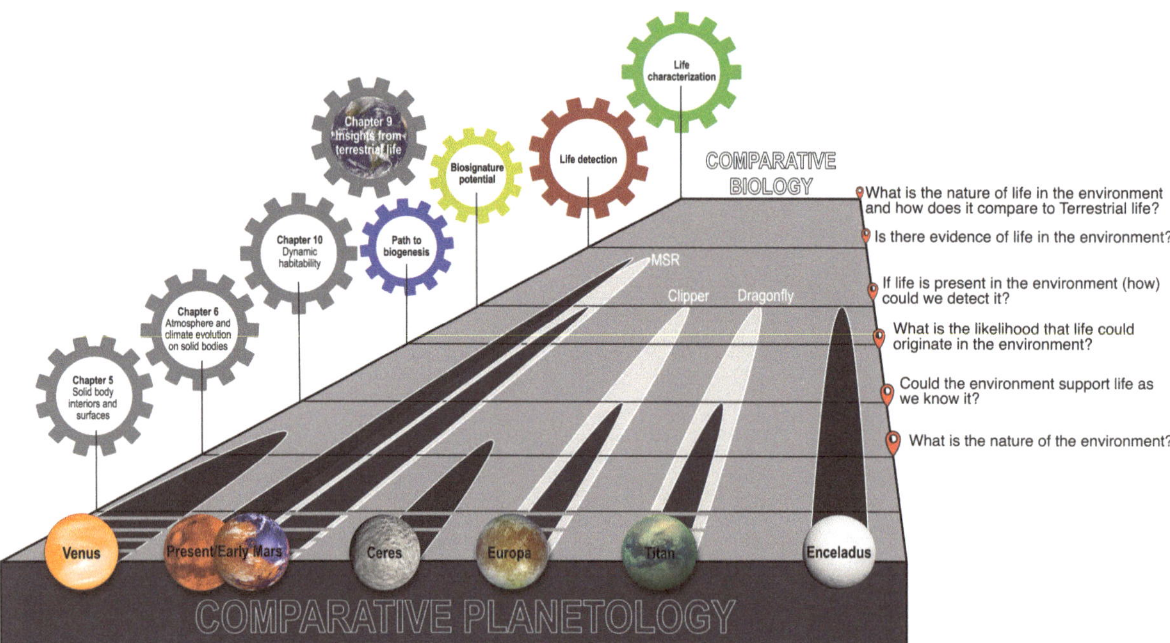

FIGURE 14-1 The search for life builds on previous discoveries that span body interiors and surfaces to atmospheres and dynamic habitability (vertical axis). Investigations at multiple habitable worlds (horizontal axis) allow for comparative planetology, leading to comparative biology if life is discovered elsewhere. For further information on where these and other worlds lie on the habitability spectrum, see Chapter 13. SOURCES: *Venus* from NOAA based on JAXA data; *present Mars* from NASA; *early Mars* from Lunar and Planetary Institute and NASA's MAVEN mission; *Ceres* from NASA/JPL-Caltech/UCLA/MPS/DLR/IDA; *Europa* from NASA/JPL/DLR; *Titan* from NASA/JPL/University of Arizona/University of Idaho; *Enceladus* from NASA/JPL-Caltech/Space Science Institute; *Earth* from NASA/NOAA/GOES Project.

further technical and conceptual maturation, including advances in statistical methods, scaling laws, information theory, and probabilistic approaches. Understanding the relationship between the geochemical environment and the prebiotic pathways that can give rise to life requires cooperation among diverse disciplines that extends beyond the traditional platform to include geochemists, atmospheric chemists, geologists, geophysicists, astronomers, mission scientists and engineers, and astrobiologists, among others (NASEM 2019; Lyons et al. 2020).

Q11.1 PATH TO BIOGENESIS: WHAT IS THE EXTENT AND HISTORY OF ORGANIC CHEMICAL EVOLUTION, POTENTIALLY LEADING TOWARD LIFE, IN HABITABLE ENVIRONMENTS THROUGHOUT THE SOLAR SYSTEM? HOW DOES THIS INFORM THE LIKELIHOOD OF FALSE POSITIVE LIFE DETECTIONS?

Presumably, life originates from a series of abiotic reactions in an organic chemical system, in which prebiotic synthesis of building blocks increases in complexity to ultimately reach biotic synthesis in a living organism. This process of organic chemical evolution develops gradually from simple to complex in terms of both the types of chemical reactions that occur and the diversity of compounds produced. Organic chemical evolution is highly dependent on the geochemical environment: factors such as pH, salinity, temperature, pressure, and the abundance of chemical precursors influence which reactions are favored, the rates at which they proceed, and the lifetimes of reactive intermediates and products available for subsequent chemistry. By understanding the extent and history of organic chemical evolution in the various habitable environments throughout the solar system, we can identify where other worlds might lie along the path to biogenesis (Q10.4), with a spectrum of possible endmembers ranging from habitable (but uninhabited) environments to fully inhabited worlds like Earth (Q9.1d; Sutherland 2017).

FIGURE 14-2 Target environments and possible architectures of missions to search for life on other planetary bodies of the solar system. *Left*: Mission architectures to rocky planets and bodies with an atmosphere include (but are not limited to) atmospheric probes, drones, landers, and rovers. *Right*: Mission architectures to ocean worlds include (but are not limited to) flyby missions, orbiters, landers, and submersibles. SOURCES: *Left*: Courtesy of Keck Institute for Space Studies/C. Carter. *Right*: Modified from Hand et al. (2017).

A thorough understanding of abiotic organic chemical evolution across a diverse set of geochemical conditions informs the likelihood of an independent emergence of life, and helps reduce the likelihood of a false positive result in the search for life beyond Earth.

Research in the emergence of life on Earth and knowledge of extant biochemistry can guide system-level analyses to differentiate between extraterrestrial biotic and abiotic chemistry (Q9.1c). While abiotic chemistry can utilize multiple pathways to generate a particular compound, biotic chemistry is characterized by evolved chemical systems that capitalize on a distinct subset of precursors among the abiotic possibilities. In the past decade, new approaches to understanding complex networks of interacting molecules in chemical and biological systems have emerged. These approaches, which form the basis of systems chemistry and systems biology, need to be incorporated into astrobiology-focused missions by including analyses of potential precursors associated with any molecular target. It is not enough to simply search for a biosignature; an understanding of environmental processes that might result in false positives, such as high concentrations of abiotic organic compounds, is needed to interpret potential biosignatures as a robust biological signal. To assist in these analyses, significant support for research in prebiotic and general organic chemistry will be needed to enhance our knowledge of abiotic synthetic routes to the building blocks of life and to uncover possible pathways that other forms of life might have adopted. In concert with such efforts, contextual environmental and/or geological information needs to be gathered in order to understand what other features in the environment might provide added confidence to the assessment of the biogenicity of a potential biosignature.

A complete picture of the organic chemical evolution in habitable environments requires a thorough understanding of the organics present in and on planetary bodies as they formed (Questions 2 and 3, Chapters 5 and 6), and processes that might have modified them over time (Questions 4–6 and 10, Chapters 7–9 and 13). The organic inventories of meteorites, interplanetary dust particles (IDPs), asteroids, and comets are also important, in that these exogenous sources of organic matter can act as "chemical seeds" for the organic evolution of habitable environments, and also represent important abiotic chemical backgrounds against which signs of life need to be identified. Laboratory experiments designed to simulate the conditions and processes that occur in extraterrestrial material environments, such as cosmic ice irradiation and low-temperature photochemistry and geochemistry, are an important complement to the characterization of astromaterials (Hoehler et al. 2020). These experiments are advantageous in that they can help create a system-level understanding of the dynamic processes that give rise to organic inventories in astromaterials (Ditzler et al. 2020).

Q11.1a What Is the Organic Molecule Inventory of Habitable Environments Throughout the Solar System, Including Complex Organic Molecules That Can Serve as Prebiotic Building Blocks of Life?

The first step to assessing the extent of organic chemical evolution in a habitable environment is understanding what organic molecules exist there now, and in what geochemical context. Such work is under way (Table 14-1)

TABLE 14-1 Organic Inventory of Worlds Throughout the Solar System to Date

Body	Organics[a] Detected or Predicted to Date	Complexity	Reference(s)
Mercury	Possibly cold-trapped volatiles (C, H, O, and N-bearing species), aldehydes, amines, alcohols, cyanates, ketones and organic acids, refractory (tholin-like) organic materials	Low	Zhang and Paige, 2009; Delitsky et al. 2017; Hamill et al. 2020
Venus	Possibly HCN, methane, ethane, ethene, and benzene	Low	Johnson and de Oliveira 2019, Mogul et al. 2021
The Moon	Possibly methane and other cold-trapped volatiles (C, H, O, and N-bearing species)	Low	Zhang and Paige, 2009; Colaprete et al. 2010
Mars	Methane, chlorobenzene, dichloroalkanes; thiophenic, aromatic, and aliphatic compounds	Low/ Moderate	Freissinet et al. 2015; Eigenbrode et al. 2018
Ceres	Aliphatic organic, possibly amines	Low	De Sanctis et al. 2018; Raponi et al. 2021
Europa	$C\equiv N$, C-H functional groups, refractory (tholin-like) organic materials	Low	McCord et al. 1998; Chyba and Phillips 2002
Ganymede and Callisto	$C\equiv N$, C-H functional groups, possibly refractory (tholin-like) organic materials	Low	McCord et al. 1997, 1998
Enceladus	Methane, other hydrocarbons, aromatics, amino acids, small and large (macromolecular) O-, N-bearing organics with ethoxy, hydroxyl, and carbonyl functional groups	Moderate	Postberg et al. 2018; Waite et al. 2006; Steel et al. 2017
Titan	Hydrocarbons, nitriles, aromatics, heterocyclic species, acetylene, ethylene, cyanoacetylene, other N- and O-bearing species, nucleobases, amino acids, heteropolymeric species up to 10,000 Da, refractory (tholin-like) organic materials	High	Brown et al. 2010; Lunine et al. 2020
Uranian satellites	Possibly refractory (tholin-like) organic materials (could also be amorphous pyroxene)	Low	Cartwright et al. 2018
Triton	Ethane, hydrocarbons, acetylene, nitriles, heteropolymers, refractory (tholin-like) organic materials	Low	Thompson et al. 1989; Quirico et al. 1999
Pluto	Methane, acetylene, ethylene, HCN, cyanoacetylene, amino acids, nucleobases, refractory (tholin-like) organic materials	Low/ Moderate	Cruikshank et al. 2019
Comets	Acetylene, methane, methanol, formate, methylamine, ethylamine, PAHs, aromatic nitriles, amino acids, refractory (tholin-like) organic materials	Low/ Moderate	Wickramasinghe and Allen, 1986; Clemett et al. 2010; Altwegg et al. 2016
Asteroids	Hydrocarbons, polyaromatic carbon, refractory (tholin-like) organic materials	Low	Cruikshank et al. 1987; Chan et al. 2021; Fink et al. 1992

[a] An organic molecule is one that contains at least one carbon atom bonded to hydrogen.

NOTE: Black text represents organics detected via remote sensing or in situ measurements; blue text represents compounds predicted by laboratory experiments/modeling. Complexity is compared to Earth as baseline (see Chapter 12).

but incomplete (Q10.4). Of great value would be a comprehensive inventory of organic molecules that includes both volatile and condensed phases, as well as soluble and insoluble (e.g., kerogen) fractions and abundant and trace species. Knowledge of the chemical structures (i.e., functional groups and binding motifs) of these organic compounds is important to ascertain synthesis mechanisms and reactivity. The distribution of organics within the habitable environment, particularly with respect to key habitability parameters such as energy sources, pressure, or temperature, can provide important clues regarding their origin.

The co-location of certain organic molecules with others, or depletion of one molecule with respect to another, may be diagnostic of the chemical pathways that dominate in that region of the environment, although such interpretations are necessarily convolved with careful assessment of the physicochemical properties of that region (see Q11.1c), which can affect the sorption and volatility/fugacity/solubility (and hence detectability) of organic molecules. This also applies to inorganic species, which may catalyze or otherwise facilitate certain organic chemical reactions.

Q11.1b What Is the Extent of Molecular Complexity (e.g., Size, Heteroatom Diversity, Structure, Pathway Assembly Index) and Degree of Organization (e.g., Isomeric Preference, Polymerization) That Can Be Generated Abiotically Under Habitable Conditions? How Does This Compare to Prebiotic Experiments to Date?

The degree of molecular complexity reflects the extent of organic chemical evolution in an environment and can possibly serve to discriminate biotic from abiotic systems. Key indicators of molecular complexity include, but are not limited to, the size and composition (arrangement, type, and ratio of atoms, including diversity of heteroatoms) of molecules, as well as their chemical structure (e.g., molecular geometry, presence of functional groups, intramolecular interactions such as folding conformations, and binding motifs). The pathway assembly index—a probabilistic measure of the number of steps needed to construct a molecule (Marshall et al. 2017)—can also be useful in distinguishing biotic from abiotic molecules, although more work is needed to determine if all complex abiotic molecules will pass this test. Abiotic generation of organic molecules in unhabitable conditions also needs to be considered, as such materials could be inherited by habitable environments at a later stage.

Biology is the dominant source of organic matter on Earth, where it has typically masked the signal from prebiotic/abiotic chemistry in all surface and near-surface environments (Q9.1). Therefore, the extent of molecular complexity that can be achieved through purely abiotic means, under widespread habitable conditions, is a challenge to assess through the study of natural terrestrial samples (Barge et al. 2020), although in the subsurface of both marine and continental systems, environments exist where low rates of biotic processes are such that abiotic signatures (particularly for abiotic H_2, CH_4, ethane, acetate, formate, and other simple organic compounds) have been identified (Sherwood Lollar et al. 2014; McDermott et al. 2015; Lang et al. 2018). Work in the laboratory—where the influence of biology can be controlled, reduced, or eliminated—can serve to inform this question, notwithstanding the challenge of mimicking a broad diversity of geochemical conditions in a laboratory setting. Other habitable environments in the solar system, particularly those with no detectable evidence of biology, can serve as prebiotic laboratories on a planetary scale to understand the extent of molecular complexity and degree of organization that can be achieved without the influence of life (Q10.4). Analyses by the Perseverance Mars rover in situ, and of samples to be returned by the Mars Sample Return campaign, may prove key in this regard. Methods to evaluate the degree of organization of abiotic systems include, but are not limited to, isomeric preference—whether molecules will react with or bind to another molecule of only a certain stereoisomeric conformation—as well as the type and number of monomeric units in any macromolecules (i.e., polymerization) and any intermolecular interactions (i.e., quaternary structure).

If endmember cases of abiotic organic complexity and degree of organization can be established, they would serve as a metric against which habitable environments could be compared in the chemical evolution spectrum spanning fully abiotic to fully biotic. Any unambiguous life detection discovery will require that all plausible abiotic and prebiotic formation mechanisms for the potential biosignature(s) be ruled out (Hoehler et al. 2020); the endmember case of greatest abiotic molecular complexity would serve as an important test for such a claim.

Q11.1c What Are the Relevant Chemical Pathways That Can Lead from Prebiotic Chemistry to Biochemistry, and How Does That Transition Depend on the Geochemical State of the Environment?

Chemical pathways (i.e., links of chemical reactions) in chemical networks are condition-specific; that is, certain pathways will proceed only under certain conditions, such as in the presence of a certain reactant or catalyst/substrate, within, for example, a specific pH, temperature, or pressure range. The physicochemical conditions of a habitable environment are intrinsically linked to the geochemical state of that environment, and as such the progression from prebiotic chemistry to biotic chemistry through evolution of specific chemical pathways is likely to be mediated by geochemical factors (Q9.2). Mapping which chemical pathways are most likely to lead to biochemistry requires a deep understanding of chemical reaction networks in a variety of geochemical regimes spanning past and present conditions of habitable environments in the solar system.

Strategic Research for Q11.1

- **Determine the organic molecule inventory, including extent of molecular complexity and degree of organization, of currently or previously habitable environments throughout the solar system** (e.g., Enceladus, Europa, Titan, Ceres, Mars—see Question 10 for full list) with spacecraft in situ and/or remote sensing observations, telescopic observations, or sample return.
- **Identify relevant chemical pathways that can lead from prebiotic chemistry to biochemistry in currently or previously habitable environments throughout the solar system as a function of geochemical state** by characterizing each environments' geochemistry, physicochemical conditions, and associated prebiotic (or potentially biotic) reactants, intermediates, and products using spacecraft in situ and/or remote sensing observations, telescopic observations, and via laboratory/experimental and modeling/theoretical research.
- **Characterize the extent of molecular complexity and degree of organization that can be achieved for an organic molecule inventory in the absence of life in currently or previously habitable environments throughout the solar system** using laboratory/experimental and modeling/theoretical research as well as field work and astromaterial analysis to explore the parameter space (e.g., temperature, pressure, pH, reactants, and catalysts/substrates).
- **Ascertain how levels of organic inventory, molecular complexity, and/or degree of organization that can be achieved in the absence of life might inform the search for life elsewhere** using laboratory/experimental and modeling/theoretical research as well as field work and astromaterial analysis.
- **Evaluate possible sources of false positives for organic chemical biosignatures** with spacecraft in situ and/or remote sensing observations, telescopic observations, and sample return from worlds unlikely to host life (e.g., the Moon, comets, and asteroids) and uninhabited regions of previously or currently habitable environments (e.g., the surface of Mars) in an effort to assess the production and/or delivery rate of abiotic organic molecules that might mimic biological organic molecules and establish a baseline against which measurements in habitable environments can be compared. This also includes analysis of astromaterials, as well as theoretical and laboratory approaches.

Q11.2 BIOSIGNATURE POTENTIAL: WHAT IS THE BIOSIGNATURE POTENTIAL (I.E., THE RELIABILITY, DETECTABILITY, AND SURVIVABILITY OF BIOSIGNATURES) IN HABITABLE ENVIRONMENTS BEYOND EARTH? WHAT ARE THE POSSIBLE SOURCES OF FALSE POSITIVES AND FALSE NEGATIVES?

Here, the committee defines "biosignature potential" as relating to the properties of a given habitable environment that act to enhance the probability of biosignature preservation and detection, including whether the properties of the habitable environment lend themselves in some way to the discrimination of biosignatures from a background of nonbiological processes that might otherwise render them improbable or nonunique. Several important traits to be considered when assessing the functional merit of a potential biosignature include its reliability, detectability, and survivability (Meadows 2017; Meadows et al. 2018; NASEM 2019). The reliability of a biosignature weighs the probability of life having produced it (i.e., its biological prevalence) against the improbability of nonbiological processes producing it (i.e., its abiotic prevalence). Currently, the basis set for evaluating the biological prevalence of a potential biosignature is limited to our experience with the terrestrial biosphere, which is potentially limiting when considering potential biosignatures of life on non-Earthlike worlds (see Q11.3c). On the other hand, the abiotic prevalence of a potential biosignature (or abiosignature) relies on our understanding of abiotic processes operating over a broad range of geochemical conditions (see Q11.1). The detectability of a biosignature depends on the quality and magnitude of its signal, which are functions of the biological potential of the environment (see Question 10), and the level to which that biosignature can be quantified with current technology. For example, the reported detection of phosphine (PH_3), a potential biosignature, in the atmosphere of Venus (Greaves et al. 2020) has come under scrutiny as the absorption feature in the reprocessed data does not meet standard criteria for being statistically significant (Snellen et al. 2020). The survivability of a biosignature depends on whether it can be preserved long enough for analysis, or whether its decomposition products can be traced back to the identity of the original biosignature. Biosignature detectability and survivability are environment-specific; therefore, a thorough understanding of the environmental context is critical for biosignature evaluation.

Q11.2a What Characteristics Indicative of Life Can Serve as Definitive Biosignatures in Environments Beyond Earth?

As described in Question 9, biosignatures can be defined broadly as objects, substances, and/or patterns whose origin specifically requires a biological agent (Des Marais et al. 2003), and can manifest in a broad diversity of observables including, but not limited to, cell-like morphologies, sedimentary fabrics, organic molecular structures, abundance distributions of organic molecules, chirality in classes of molecules, mineral properties, stable isotope patterns in minerals and organic compounds, relative abundance of atmospheric gases, or temporal changes (e.g., daily and/or seasonal) in global planetary properties (Des Marais et al. 2008). Experience from the study of life on Earth suggests that these potential biosignatures can be grouped into three generic types: (1) chemical, (2) morphological, and (3) physiologic/metabolic. Chemical biosignatures result from life's selectivity toward specific organic and inorganic compounds and from life's capability to "invent" complex organic molecules that fill important biological roles (e.g., biopolymers). Morphological (i.e., structural) biosignatures may emerge from life's fundamental cellular architecture (though may not be required; see Q11.4a), and/or from the complex web of competitive and cooperative interactions among organisms (e.g., sedimentary fabrics), and their interactions with the environment. Physiologic/metabolic biosignatures are direct manifestations of the activity of living organisms (e.g., mobility and growth) including when they have the capacity to interact with and alter the environment at rates that are significantly faster or different than equivalent abiotic processes (e.g., chemical catalysis). For a given element, the relative alteration of different isotopic forms for that element—most commonly ^{13}C and ^{12}C—have frequently been used as biosignatures, despite an increasing recognition of the range of abiotic processes that may produce similar depletions in the ^{13}C signatures and hence "mimic" life (NASEM 2019). Studies of biosignatures produced by life on Earth provide a foundation for the search for similar biosignatures in other habitable environments of the solar system (Question 9).

The building blocks of life can have various abiotic origins and are likely to permeate the solar system and other planetary systems in the Milky Way galaxy. This is to be expected, as abiotic chemistry provided the feedstock for prebiotic chemical evolution that ultimately gave rise to life at some point in the early history of Earth. A complex abiotic organic chemical background represents one possible form of a biosignature—that is, chemicals, features, and/or processes that define the nonbiological background against which signs of life needs to be resolved. This is made even more challenging by the fact that, in some cases, such abiosignatures may mimic biosignatures. Thorough characterization of the environment, including the extent of molecular complexity and degree of organization achievable in abiotic conditions for that environment (Q11.1b), can support assessment of the biogenicity of a potential biosignature even in a complex abiotic chemical background.

Q11.2b What Are the Relationships Between the Physical and Chemical Properties and Processes Operating in a Habitable Environment and the Potential Amount of Biomass That Might Be Present? How Might This Drive the Detectability of Any Biosignatures Present?

The chemical inventory and fluxes of metabolically useful redox couples, the environmental conditions (e.g., pH, Eh, temperature, and the presence/absence of sunlight), and availability of substrates that can facilitate catalysis are among the important factors that will determine whether a habitable world supports the production of a large (in terms of total biomass) and productive biosphere, or a biosphere that is limited in terms of its productivity, total biomass, and areal/geographic extent. Predictions regarding the potential size and extent of the biosphere on a habitable world would provide useful guidance for mission planning, sampling, and instrument design, and inform assessments of the susceptibility of a biosignature search to false negative outcomes.

Studies of Earth's biosphere provide a framework for thinking about biological productivity, and ultimately the rate of biomass production that can be supported by an environment. A useful metric of biological productivity is "net primary productivity" (NPP): the quantity of CO_2 fixed into organic matter per unit time, minus the quantity of organic carbon oxidized to CO_2 by autotrophic respiration. Focusing specifically on Earth's oceans, modern marine photoautotrophs can access virtually unlimited sources of water, sunlight, and inorganic carbon to support an NPP estimated at $\sim 10^{13}$ kg of carbon/year (Field et al. 1998; Bender et al. 1994), such that marine NPP is limited by the availability of other nutrients including dissolved phosphorus, fixed nitrogen, and sometimes iron. Models of the early marine biosphere suggest that NPP would have been limited early on by the availability of inorganic geochemical electron donors, such as Fe^{2+} and H_2, limiting NPP to $\sim 10^{10}$–10^{12} kg of carbon/year (Kharecha et al. 2005; Canfield et al. 2006; Ward et al. 2019). The geological record of carbon isotopes in carbonate and organic carbon indicates that the fraction of Earth's surface carbon budget buried as organic matter in sediments has increased through time (Krissansen-Totton et al. 2015), consistent with the notion of an increasingly productive biosphere. Similar assay and bioenergetics-based approaches to modern subsurface environments provide additional useful guidance for estimates of nonphotosynthetic subsurface biological productivity and biomass (e.g., Spear et al. 2005; Lollar et al. 2019), and these concepts have been extended to estimate the potential biomass of, for example, the modern martian subsurface (Weiss et al. 2000; Sholes et al. 2019), and other habitable environments beyond Earth (e.g., Hand et al. 2017; Cable et al. 2020; Higgins and Cockell 2020). While such modeling efforts require assumptions about the metabolic strategies employed by organisms living in the system, and the fluxes of nutrients available to them, they have the potential to yield estimates of biological productivity and perhaps even predictions regarding the composition of relevant stable isotopic reservoirs on other worlds. Such predictions can be useful in developing measurement requirements (e.g., biosignature type and expected concentration) for astrobiology-focused science objectives.

Q11.2c How Do the Physical and Chemical Properties and Processes in a Habitable World Affect the Survivability of Biosignatures?

When considering the likelihood that a biosignature can be detected in a habitable environment, it is also important to consider how the physical and chemical conditions that make an environment habitable also act to modify, enhance, or erase evidence of inhabitation. NASEM (2019) speaks to this issue in the context of the survivability of biosignatures, or "taphonomy," which is the field of study concerned with those processes that result in the formation, preservation, alteration, and destruction of biosignatures. This issue is also discussed in detail in Question 9.5. Our understanding of the processes that affect the biosignature potential of a habitable environment necessarily have "baked in" assumptions about the nature of the biological system one suspects is or was present. The search for the remnants of an extinct, nonphotosynthetic, lithotrophic microbial biosphere on Mars, for example, would require an exploration of the unique chemical and physical conditions favoring biosignature production and preservation in that environment, distinct from those that might favor the successful search for an atmospheric biosignature on Venus or an exoplanet. Accordingly, the trade space for understanding how the chemical and physical processes in a habitable world affect its biosignature potential is vast, and needs to be narrowed on the basis of assumptions about the biosphere that produced it, and the environment(s) in which the signature(s) of that biology have resided in order to avoid false negatives that result from an under-appreciation of the specific environmental factors that might act to modify or erase a potential biosignature.

Q11.2d How Can We Best Devise a Formal Framework for the Interpretation of a Set of Biosignature Measurements in a Given Environment?

To account for uncertainties in biosignature potential and the possibly unique nature of life elsewhere (Q11.3c), a search for evidence of life is more likely to succeed if it is based on multiple, mutually supporting lines of evidence as well as on their environmental context (Figure 14-3). Building on well-established tenets such as multiple working hypotheses (where every rational explanation of the phenomenon is included and impartially evaluated; Chamberlin 1890, 1897), frameworks that allow correct interpretation of the significance of each potential biosignature, according to their own possible sources of false positive and false negative signals,

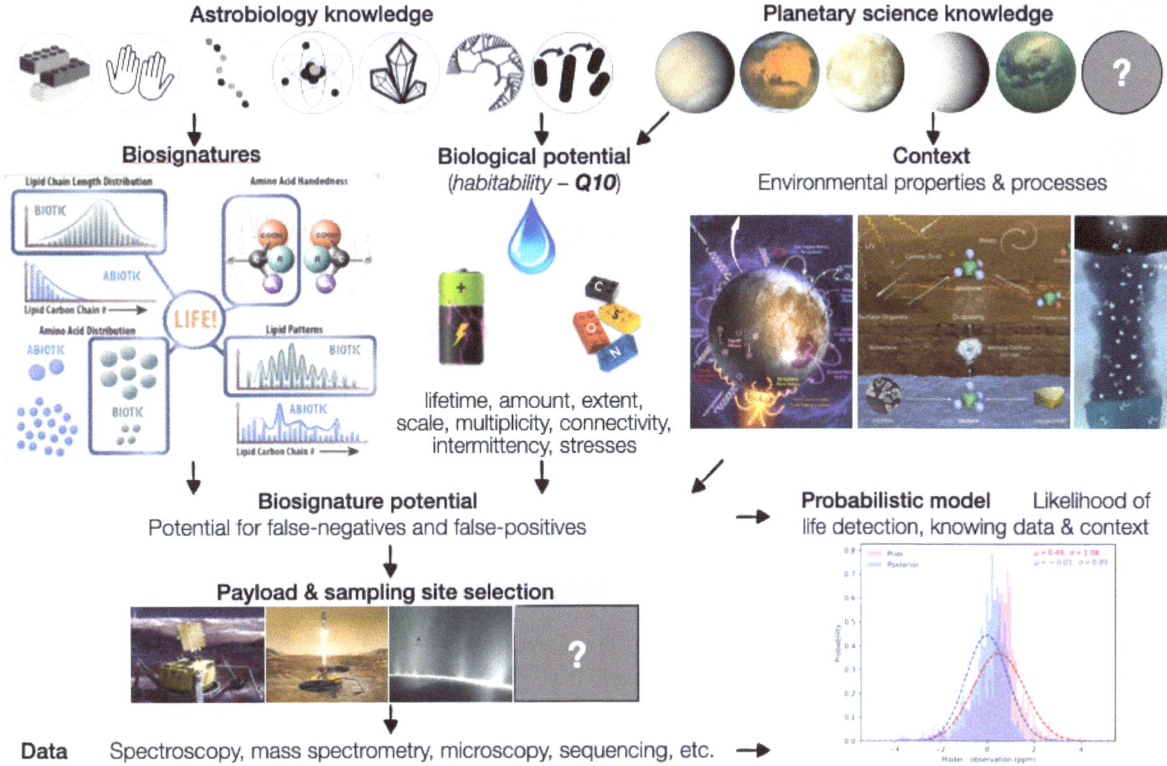

FIGURE 14-3 A formal framework for interpretation of biosignature measurements would build on prior knowledge of biological and abiotic signatures and utilize data, environmental context, and biological potential to construct a probabilistic model robust to false positives and false negatives. SOURCES: *Astrobiology knowledge* images from Neveu et al. (2018); the publisher for this copyrighted material is Mary Ann Liebert, Inc., Publishers. *Planetary* images from NASA/JPL-Caltech; NASA, ESA, the Hubble Heritage Team (STScI/AURA), J. Bell (ASU), and M. Wolff (Space Science Institute); NASA/JPL-Caltech/Space Science Institute; NASA/JPL/University of Arizona/University of Idaho. *Biosignatures* images from NASA and NASA GSFC/Ghannadian/Eigenbrode. *Context* images from Teolis et al. (2017), https://doi.org/10.1016/j.icarus.2016.10.027. CC BY 4.0; NASA's Goddard Space Flight Center/JPL-Caltech/University of Michigan; NASA/JPL-Caltech. *Payload and sampling* images from Hand et al. (2017) and NASA/JPL-Caltech. *Probabilistic model* image from © Villalobos et al. (2020), https://doi.org/10.5194/acp-20-8473-2020, CC BY 4.0.

provide a more rigorous standard of proof. In addition, such frameworks can help identify sets of measurements that together discriminate between a biotic or abiotic origin for potential biosignatures with high statistical significance (NASEM 2019).

Some preliminary concepts for interpretive frameworks have been posited recently for possible life detection mission scenarios, considering uncertainties arising from environmental or instrumental factors and where valuable but limited opportunities exist to collect additional samples or conduct follow-up analyses. At the most basic level, a framework may be represented as a decision tree, wherein questions about acquired data may lead to increased or decreased probabilities or "scores" associated with life detection (e.g., Vago et al. 2017). Thorough, yet flexible, decision trees may prove especially useful during a mission, where selection of samples or measurement parameters can be guided by "real-time" analysis of collected data. Where data increase in volume and complexity and become increasingly cumbersome to manage, such as on remote missions with restricted communication, science autonomy will increasingly be required for life detection (Theiling et al. 2021). Machine learning and other tools may provide an expanded practical framework for interpreting biosignature measurement data in such conditions.

Strategic Research for Q11.2

- **Determine which attributes of terrestrial life can serve as definitive biosignatures in the search for life beyond Earth** by theoretical, field, and laboratory research activities that inform on the range of morphological and chemical (e.g., organic, isotopic, mineralogical, and atmospheric) signatures produced by living systems, with a complementary emphasis on what contextual evidence is required to confidently recognize a potential biosignature.
- **Constrain the biomass and/or bioenergetic potential of habitable environments throughout the solar system** through measurements of environmental factors (e.g., pH, temperature, salinity, and redox potential) and fluxes (e.g., energy and nutrients) with spacecraft in situ and/or remote sensing observations, or returned sample analysis; and theoretical, field, and laboratory research activities to understand how modern and ancient environmental properties and processes on Earth relate to biomass production to inform the development of life detection strategies and technology (see also Question 9).
- **Characterize the range of processes that affect the production and preservation of detectable biosignatures in habitable environments** by theoretical, field, and laboratory research activities that inform us about the pathways and rates of biosignature production, preservation, and destruction, be they morphological or chemical (e.g., organic, isotopic, mineralogical, and atmospheric) in nature. Tie those studies to in situ, remote, and/or telescopic measurements of specific environmental properties of habitable worlds in the solar system (e.g., Enceladus, Europa, Titan, Ceres, and Mars; see Question 10 for full list) that might act to enhance, preserve, or destroy biosignatures.
- **Establish a comprehensive, standardized framework for evaluation of biosignatures, including the potential for abiosignatures, false positives, and false negatives,** through community-level dialog and consensus, supported by laboratory/experimental and modeling/theoretical research as well as field work.

Q11.3 LIFE DETECTION: IS OR WAS THERE LIFE ELSEWHERE IN THE SOLAR SYSTEM?

Recent decades of planetary exploration have witnessed a small revolution in our understanding of habitability conditions elsewhere in the solar system (Figure 14-4). It is now widely accepted that long-lived habitable environments existed on the surface of Mars early in the planet's history, and that habitable conditions may have persisted until the present in the subsurface. There is compelling evidence that a habitable ocean exists beneath the surface of Enceladus, and possibly also beneath the surfaces of Europa and Titan. Motivated by these discoveries, we are poised to address the question of whether there is, or ever was, life elsewhere in the solar system, with scientific rigor.

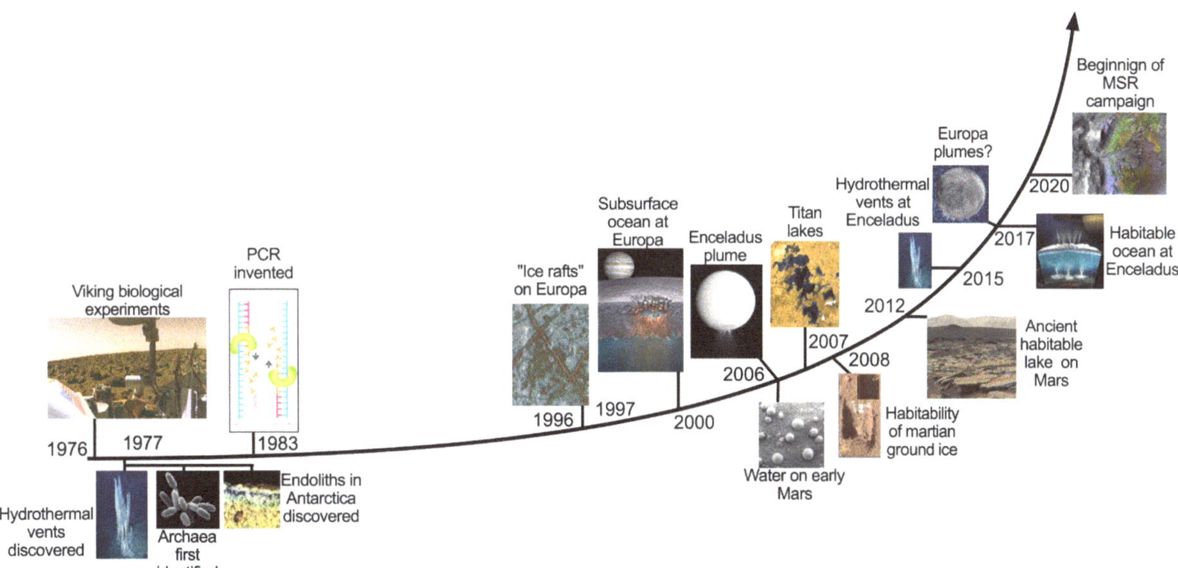

FIGURE 14-4 The search for life has been revolutionized in the past decades by pivotal discoveries such as the identification of what is now recognized as a third branch in the tree of life (Archaea) and confirmation of subsurface liquid water oceans within several moons of Jupiter and Saturn. SOURCES: *Viking* image from NASA/JPL; *PCR* image from National Human Genome Research Institute, genome.gov; *hydrothermal vents* images from NOAA; *Archaea*, *Endoliths*, and *Enceladus* images from NASA; *Europa ice rafts* image from NASA/JPL/University of Arizona; *Europa subsurface ocean* image from Britney Schmidt/Dead Pixel VFX/The University of Texas at Austin; *Titan lakes* image from NASA/JPL-Caltech/ASI/USGS; *early Mars water* image NASA/JPL/Cornell/USGS; *Martian ground ice* image NASA/JPL-Caltech/University of Arizona/Texas A&M University; *Europa plumes* image from NASA/ESA/W. Sparks (STScI)/USGS Astrogeology Science Center; *Ancient Mars lake* image from NASA/JPL-Caltech/MSSS; *MSR campaign* image from NASA/JPL-Caltech/ASU; *Enceladus ocean* image from NASA/JPL-Caltech/Southwest Research Institute.

The search for evidence of life beyond Earth builds on and extends the traditional objectives of planetary sciences, complementing comparative planetology and extending to a new potential field: comparative biology (Q11.4a). The magnitude of the question requires the coming together of scientists from a wide range of disciplines, and the combination of a diversity of perspectives and approaches. The search also needs to satisfy the scientific criteria of experimentation, address testable hypotheses, and allow modification of those hypotheses based on observation and measurement (NASEM 2019). This applies equally to cases where life may have gone extinct, detectable through its imprint preserved over time, or cases where life is currently viable, and may even comprise a population of metabolizing organisms, albeit not necessarily in a way immediately familiar to us (Q11.3c).

To minimize the likelihood of a false negative or a false positive result, optimal life-detection strategies would target multiple, orthogonal, mutually supporting lines of evidence, and carefully select sets of candidate biosignatures based on several key criteria (Q11.2). Assessments of the presence or absence of potential biosignatures in a habitable environment (present or past) need to be congruent with the physicochemical setting and ought to be informed by the biological potential of that environment (Figure 14-5 and Question 10). In addition, life-detection mission designs need to include comprehensive contamination control approaches informed by acceptable levels and types of contamination derived from top-level science and mission requirements. This effort would start as part of mission concept development, and contamination awareness and control would be addressed in detail throughout all mission phases. Last, the novelty of the measurements needed for life detection and the rigorous standards of evidence mandate tailored approaches in the conception, maturation, and deployment of instruments and instrument suites (Hoehler et al. 2020).

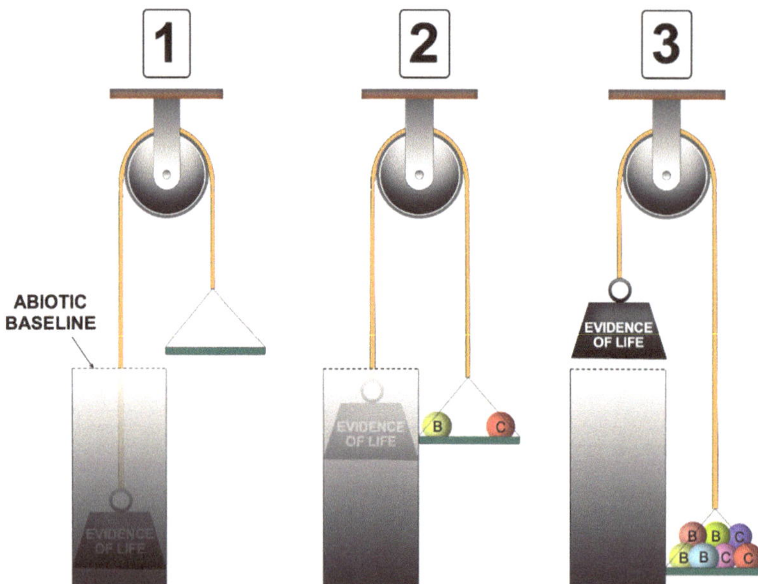

FIGURE 14-5 To provide evidence of life above the threshold established by abiotic processes in a given environment, multiple independent biosignatures (B) are needed, as well as a thorough understanding of the physicochemical setting (C). SOURCE: Modified by P. Byrne from iStock.com/blueringmedia.

Q11.3a Are There Chemical, Morphological, and/or Physiologic/Metabolic or Other Biosignatures in Currently Habitable Environments in the Solar System?

The continued exploration of planetary bodies of the solar system is revealing a broader range of potentially habitable solar system environments than previously anticipated (Question 10). Data gathered by the Cassini spacecraft suggests that the subsurface ocean of Enceladus currently meets the requirements to sustain life (Cable et al. 2020). The Europa Clipper and Dragonfly missions will help constrain the biological potential of Europa's and Titan's subsurface oceans, respectively. The exploration of Venus (e.g., by VERITAS and DAVINCI) and Mars (e.g., by Curiosity and Perseverance) will help establish whether localized habitable regions currently exist within these seemingly uninhabitable worlds. Once habitable environments are identified, the search for evidence of life represents the logical next step and the greatest challenge.

The search needs to be conducted thoughtfully and with an open mind concerning potential outcomes, balancing the *stringency* and *inclusivity* of the observational strategy applied to a given environment. Stringency sets criteria for the quality and robustness of a biosignature detection, amidst potentially confounding conditions or background signals from the planetary environment, and thus seeks to minimize potential false positive results such as a "life-like" abiotic pattern or response. Inclusivity emphasizes consideration of a wide range of possible alien biosignatures (e.g., chemical, morphological, and/or physiologic/metabolic), not relying solely on Earth life as a guide, as well as their prevalence and detectability in the given environment. As such, inclusivity seeks to minimize potential false negative results, where life could be "missed" for lack of the ability to detect or recognize it. These concepts apply equally to cases where life may have gone extinct but remain detectable through its imprint preserved over time (Q11.3b).

The search for evidence of life requires tailored technology solutions (see Chapter 21). The novelty of the measurements needed for life detection and the rigorous standards of evidence we will demand of them mandate new approaches in the conception, maturation, and deployment of instruments and instrument suites. Specifically, because the preferred route to arriving at defensible conclusions is to employ multiple orthogonal lines of life detection, future missions will need to be designed holistically, potentially having instrument providers engage with each other very early in the process (i.e., pre-Phase A) to propose integrated instrument suites. The most sensitive life-detection experiments require sample acquisition, sample handling, and sample analysis (Figure 14-6);

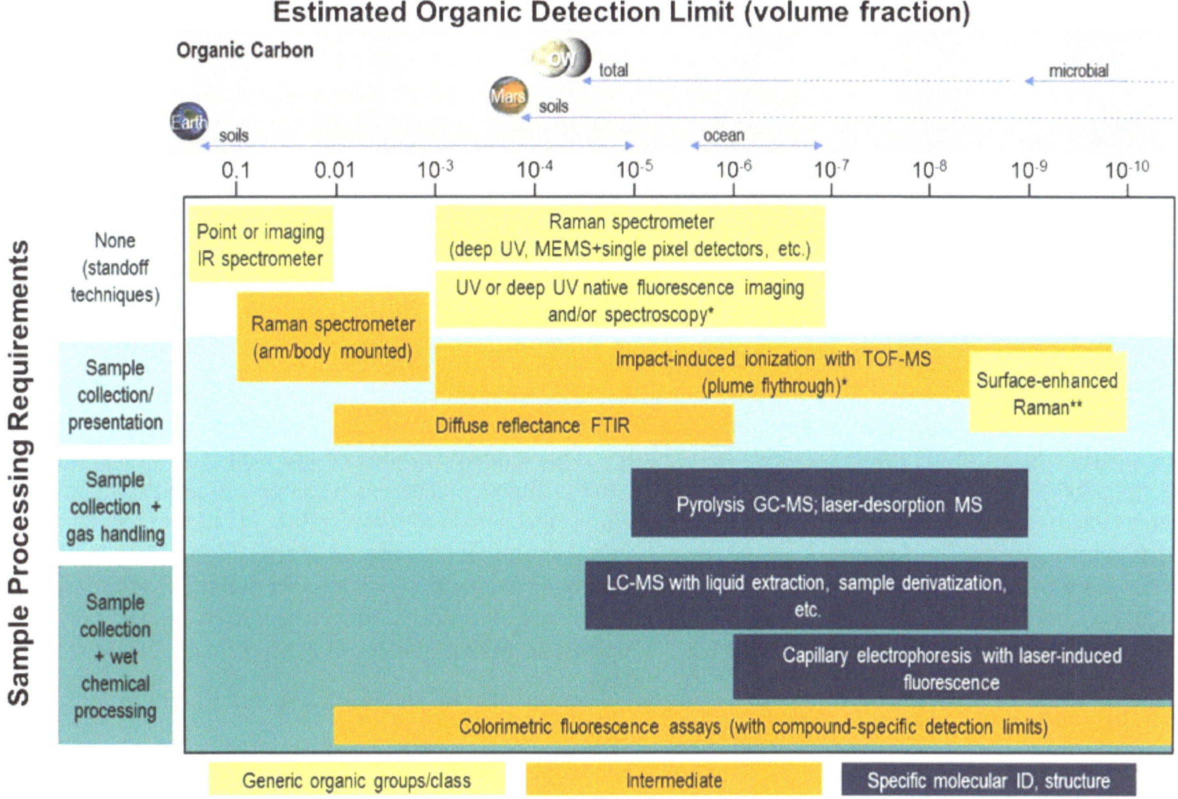

FIGURE 14-6 Examples of instruments that can be used to search for chemical/molecular biosignatures, organized by their range of detection limits and sample processing requirements. Note that sample preconcentration can improve detection limits for some instruments. NOTES: FTIR, Fourier transform infrared; GC, gas chromatography; LC, liquid chromatography; MEMS, microelectromechanical system; MS, mass spectrometry; OW, ocean worlds; TOF, time-of-flight; UV, ultraviolet. *Detection limits depend on sample matrix and instrument architecture. **Some instrument designs require wet chemical processing. SOURCES: Figure updated from Mustard et al. (2013). Images: *Earth* from NASA/NOAA/GOES Project; *Mars* from NASA, ESA, the Hubble Heritage Team (STScI/AURA), J. Bell (ASU), and M. Wolff (Space Science Institute); *Europa* from NASA Visualization Technology Applications and Development; *Enceladus* from NASA/JPL-Caltech/Space Science Institute.

therefore, in situ detection of life is best advanced by integrated suites of instruments or single instruments that permit multiple analytical techniques, including nondestructive approaches, to be applied to the same collected materials (NASEM 2019) (see Chapter 17). Each step forward in this effort will also require a continuing vigilance against forward contamination from Earth (McKay et al. 2020).

Q11.3b Are There Chemical, Morphological, or Other Biosignatures in Previously Habitable Environments in the Solar System?

Mars witnessed a protracted period of habitable conditions early in its history, with relatively mild surface temperatures and widespread bodies of liquid water with physical and chemical conditions compatible with biology. A detailed geologic record of this early habitable period has been preserved on the surface, as shown by data gathered by the Curiosity rover at Gale crater. A search for evidence of past life on Mars is now under way at Jezero crater with the Perseverance rover and is a top priority objective for the analysis of samples cached by the rover, which will eventually be brought to Earth as part of the Mars Sample Return campaign. While physiologic/

metabolic biosignatures (which would indicate organisms that are alive today) may no longer be present, evidence of past life in the returned samples can be ascertained from chemical and morphological fingerprints left behind. To support that objective, data gathered by Perseverance may help constrain biosignature survivability (Q11.2) and offer a deeper understanding of the geologic evolution of that environment, particularly in terms of physical (e.g., temperature, pressure, and transport) and chemical (e.g., oxidation, photodissociation, and radiolysis) processing that biosignatures may have undergone over geologic timescales. Studies of biosignatures produced by ancient life on Earth (Question 9), such as in Archaean sediments, can inform the search for similar biosignatures in martian sedimentary deposits.

Q11.3c How Might We Develop the Scientific Understanding to Recognize Life "Not as We Know It"?

The search for evidence of life beyond Earth is necessarily informed by terran life (i.e., life "as we know it," the particular case of terrestrial life, Question 9), which increases the potential likelihood that we would not detect nonterran life in an otherwise inhabited world. Acknowledging that terrestrial life might be but one example of all possible forms of life in the universe, we need to recognize the possibility of life with biology that differs in a significant way from the terrestrial model in form and/or function. There is a recognized need to improve our understanding of which biosignatures are likely to be prevalent across all forms of life, in order to reduce the risk that our missions and technologies will be too narrowly focused on the features of terran life . The goal of such research is to expand the ability to search for any life through exploration of a broader definition of life based on activity, with less dependence on assumptions about structure and specific biogeochemistry—specifically, more universal or agnostic biosignatures (NASEM 2019). From least to most radical (and therefore most to least likely), potential biochemical differences include (NRC 2007):

- Different lipid molecules for membranes. Major categories of living organisms on Earth are distinguished in part by their use of different lipids, suggesting that alternative patterns compatible with other metabolic pathways are likely.
- Different sets of amino acids or nucleobases other than the ~20 amino acids and 5 nucleobases used by terrestrial life, to encode information (Young and Schultz 2010; Hoshika et al. 2019).
- The opposite enantiomer of amino acid and sugar chirality (i.e., mirror life), and/or a genetic polymer other than a polyanionic backbone.
- Different genetic molecules. Life on Earth has switched genetic molecules at least once (from RNA to DNA); alternatives may be numerous (Cleaves et al. 2019; Zhou et al. 2021).
- Alternative essential elements with similar chemical properties (e.g., elements within the same column of the periodic table or transition metals with similar oxidation states, reactivities, and/or binding affinities).
- A solvent that is liquid at temperatures and pressures other than those of liquid water. In the solar system, two environments with liquids not dominated by water are the hydrocarbon (methane-ethane) lakes found in Titan's polar regions and sulfuric acid clouds in the atmosphere of Venus. Both have been hypothesized to be able to harbor life as we do not know it (Limaye et al. 2018; Lunine et al. 2020), although limitations of organic compounds in these environments (such as low solubility in Titan's lakes and instability in the Venus's atmosphere) may pose insurmountable challenges even to novel/alien biochemistries. Other potential solvents include formamide, ammonia, and carbon dioxide, but no planetary environments are yet known to contain such liquids as the dominant component.

Accordingly, nonterran life can fall along a sliding scale from less to more radical departures from the basic terrestrial model. Strategies to search for evidence of life need to strike the right balance between inclusiveness of alternative biologies (e.g., by considering novel and/or agnostic biosignatures) while adhering to known biological principles (e.g., a chemical system that can undergo Darwinian evolution). Understanding the spectrum of possible departures from the terrestrial model, and how those might impact our selection of potentially habitable environments and the interpretation of potential biosignatures therein, is an important aspect of astrobiology.

Q11.3d If We Do Not Find Evidence of Life in a Habitable Environment, What Would It Take to Convince Ourselves That There Truly Is or Was No Life Present There, Rather Than Possibly Not Having Detected It (a False Negative)?

When searching for evidence of life, the probability of a false negative result is highest in environments where potential biosignatures occur at very low abundance (e.g., owing to low productivity or to degradation/destruction processes), operating at a very low (or even dormant) metabolic state, or where life is not distributed homogeneously (i.e., biological oases amidst an abiotic landscape). False negatives owing to low biological signals can be constrained based on contextual information. But false negatives owing to heterogeneous distributions of biological signals are more difficult to recognize, because spatial heterogeneity can occur at any scale, up to planet wide. Research in various environments on Earth suggests that heterogeneous distributions of life are nevertheless not arbitrary. Biological oases typically occur in areas where resources (e.g., water, nutrients, and energy) are locally more abundant, or where lethal environmental conditions (e.g., radiation and excessive temperatures) are somehow mitigated. Life signatures can be relatively diverse and abundant in those oases, but quickly vanish with distance or time. Often, biological oases are associated with specific substrates or physical environments (e.g., rocks, sediments, subsurface layers, and fracture surfaces) whose chemical or physical properties provide a survival advantage to organisms. As such, correct interpretations of a negative result require adequate understanding of the spatial variability in resources and environmental conditions. Research in terrestrial environments can inform how spatial variations in resources and environmental conditions can shape the distribution of life in the landscape (Question 9). From these studies, models can be developed that predict potential "hotspots" or blooms of life as a function of resources and environmental conditions. Such models can then inform the most likely locations to find evidence of life on a planetary body, and how the likelihood of finding evidence of life changes spatially. Similarly, models considering putative metabolisms in a certain environment could inform protocols for instigating a "bloom" in a collected sample, if nutrient-starved organisms were lying dormant.

Strategic Research for Q11.3

- **Develop and validate effective life detection payloads that support the search for evidence of life beyond Earth** by maturing end-to-end technologies for sample acquisition, sample handling/preparation, and sample analysis, and by prioritizing the early integration and validation of these technologies and instrument suites.
- **Search for evidence of present life in environments beyond Earth that currently have a high biological potential** by looking for multiple, independent biosignatures with spacecraft in situ observations or in samples returned to Earth, informed by laboratory/experimental and modeling/theoretical research as well as field studies of currently habitable environments on Earth.
- **Search for evidence of past life in environments beyond Earth that had a high biological potential in the past** by looking for multiple, independent biosignatures with spacecraft in situ observations or in samples returned to Earth, informed by laboratory/experimental and modeling/theoretical research as well as field studies of ancient life on Earth.
- **Search for evidence of nonterran life in environments beyond Earth** by looking for multiple, independent biosignatures with spacecraft in situ observations or in samples returned to Earth, informed by laboratory/experimental and modeling/theoretical research of putative organisms that might utilize unique biochemistry or alternative essential elements or solvents.
- **Determine the optimal sampling strategy to minimize the likelihood of a false negative life detection measurement (or set of measurements) owing to heterogeneous distributions of biological signals or other factors** by employing laboratory studies and modeling/theoretical research in conjunction with field studies of biological oases and environments with severe nutrient/energy limitations on Earth.
- **Develop and implement analysis techniques to evaluate the likelihood of forward contamination in life-detection missions** through models of contamination transfer from spacecraft surfaces and by improving hardware protection/cleaning methods, in order to meet the stringent contamination requirements imposed by life-detection science.

Q11.4 LIFE CHARACTERIZATION: WHAT IS THE NATURE OF LIFE ELSEWHERE, IF IT EXISTS?

With terrestrial life as the sole example, biology is unlikely to evolve into a universal science in the same way as physics, chemistry, and geology have evolved with the respective advent of telescopic astronomy (17th century), spectroscopy (19th century), and planetary exploration by spacecraft (1960s). For these fields, we now have a predictive understanding. But the discovery of life on another world would allow us, for the first time, to conduct comparative biology beyond the boundaries of Earth, and to address some of the most fundamental questions in biology: Is there a general theory of living systems? Is life a common cosmic phenomenon? What is life, how does it emerge, and what are its universal means of operation?

All life on Earth is genetically related and likely represents only one data point among the possible instances of biology. Characterizing life elsewhere, if it exists, would allow us to begin distinguishing attributes specific to life on Earth from those that are shared across instances of biology (Figure 14-7). An important challenge in this process will be to recognize life beyond Earth as a truly independent biology, with no ancestral links to life on Earth due, for example, to biological exchange by impacts. Our ability to characterize life beyond Earth if/once it is discovered will be key to addressing this challenge. Three scenarios could be possible. First, life could be discovered that is so similar to Earth life that it would be challenging to rule out terrestrial contamination without extensive phylogenetic analysis. Second, identification of extraterrestrial life whose biochemistry overlaps significantly (but not completely) with terrestrial life might point to a shared origin and later dispersion through planetary exchange. Third, we may find extraterrestrial life that has sufficiently different biochemical attributes from any known Earth organism. This case may point to a separate origin of life and imply that living systems can readily emerge under appropriate physical and chemical conditions; alternatively, the two systems could share a common origin but have diverged so greatly over time that any trace of common evolutionary heritage has been obscured. Each scenario would have significantly different implications about the prevalence of life in the universe or the likelihood of it being transported across planetary havens. Coupled with a detailed understanding of its environment, inspection of biological properties and processes beyond Earth would provide insight into how a biosphere arises and co-evolves with its host world. Such insights into universal properties, emergence, prevalence, and workings at scales from microscopic to global could bring within reach a definition of life and a universal theory of biology.

Since the onset of robotic exploration of the solar system, the knowledge gained via such exploration and concomitant advances in the understanding of what life requires, what constitutes evidence of life, and how to build instruments to search for it has poised humanity on the verge of finding and characterizing either life beyond Earth, or habitable worlds with no evidence of life. Either way, these discoveries, for which the groundwork can be undertaken in this coming decade, are bound to bring answers to some of humanity's most profound questions.

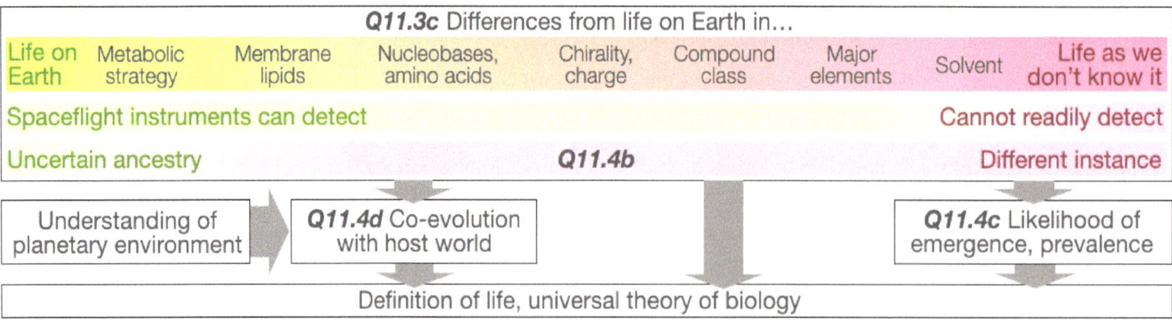

FIGURE 14-7 Development of the scientific understanding to recognize and characterize nonterran life would help us understand the nature of life via comparative biology.

Q11.4a How Might Comparative Biology Be Implemented on a Small Sample Size (N = 2) If Evidence of Life Is Discovered Elsewhere in the Solar System?

Characterizations of life beyond Earth would take place at multiple levels (e.g., biochemically, structurally, and metabolically), each one likely to produce profound insights on the nature of biological systems. For example, the cell is the basic structural, functional, and biological unit of all modern terrestrial organisms. However, life elsewhere could exist in an acellular evolutionary form (e.g., with only a protein coat as in viruses, see Q9.2), or it could be based on a different cellular architecture. It is unknown whether all living systems require a cellular organization altogether. Similarly, for all their metabolic diversity, terrestrial organisms extract energy from a limited subset of physical and chemical sources (see Question 9). However, life elsewhere might be able to exploit other sources of energy (e.g., different wavelengths and intensities of radiation) reflecting adaptation to their local environment. Last, all life on Earth shares the same biochemical framework based on a universal set of molecular building blocks (e.g., amino acids, nucleobases, sugars, and lipids) decorated with essential elements such as N, P, and S that confer a broad range of molecular structures and functionalities. This particular biochemical framework operates optimally within a range of physical and chemical conditions (e.g., temperature, salinity, and pH) (Q9.2). The biochemistry of life elsewhere could be identical to Earth's and be bound by a similar range of environmental conditions, or could be based on a different subset of building blocks and essential elements, and operate optimally under conditions that are prohibitive to terrestrial life. To understand this potential variability, the discovery of evidence of life beyond Earth ought to be followed by dedicated investigations into the nature of those life forms.

Q11.4b How Do We Discriminate Between an Independent Origin of Life Elsewhere in the Solar System and a Shared Ancestry with Earth Life?

The above alternative means by which biological structure and function may be achieved could be used to distinguish life that shares ancestry with Earth life (up to the point of a last universal common ancestor—LUCA; Q9.1f) from life with a more distant relationship or independent emergence. Finding a new metabolic strategy but the same biomolecules as on Earth would suggest adaptation of terrestrial life transported to a new world. The identification of a cell membrane with new lipids would suggest a new category of organism, but one not necessarily more distinct than the three categories of organisms on Earth are from one another. In contrast, alternative biomolecules or elements would point to a more distantly related or unique instance of life.

More generally, among the different types of biosignatures (e.g., chemical, structural, and metabolic/physiological; Q11.2a), only certain chemical (organic) biosignatures, or informational evidence derived from them, might truly reveal an independent emergence. The latter include biopolymers whose sequence changes randomly through time (e.g., nucleic acids, peptides, and membrane lipids) and their specific building blocks (including specific molecular properties such as chirality). Even if alternatives as described in Q11.3c are found, there will be a degree of uncertainty in our assessment of an independent origin of life owing to our limited understanding of the pre-LUCA biochemical nature of life on Earth.

Another clue can be taken from the likelihood of material transfer (and of biological viability in this material during the transfer) between Earth and another inhabited world. The presence of rocks from Mars and other bodies in the meteorite collection show that this likelihood is nonzero; in the case of Mars specifically, the time when Mars was most habitable (late Noachian) would have coincided with a period of large impacts occurring on Earth, meaning that life could have spread from one potentially habitable world to another. Transfer between Earth and any inhabited environments in outer solar system ocean worlds and dwarf planets is less likely, however, owing to low impact probabilities and high impact velocities of ejected material, as well as the barrier presented by ice shells that encase subsurface oceans (Question 5).

Q11.4c What Does the Presence or Absence of Life on Other Worlds Tell Us About the Emergence of Life in General?

Proposed environments for the emergence of life on Earth, likely during the early Archaean, include surface aquatic environments (e.g., lakes and shallow marine settings), ephemeral playas with wet-dry cycling, hydrothermal volcanic fields at the surface, hydrothermal vents on the ocean floor, and subsurface fracture networks (Q9.1). All have attributes that would be favorable or challenging to prebiotic chemical evolution (e.g., stability, longevity, energy supply, and dilution), and the scarcity of early Archaean geologic and fossil records limits our capacity to constrain the actual environmental setting where life could have begun on our planet. The search for evidence of life on other worlds is a means to test some of these origin of life hypotheses. For example, the absence of life in an otherwise habitable ocean world would rekindle ideas for an origin of life in environments with low water–rock ratios (Deamer and Damer 2017). But the discovery of life in an ocean world would lend credibility to theories that life on Earth could have originated in deep hydrothermal vents on the ocean floor. Therefore, characterizing life beyond Earth could shed light on the unknown circumstances of life's emergence on our own planet, at least in terms of likelihood and environmental setting. The likelihood of life emerging and persisting on a habitable world is unknown, but our presence on Earth shows that it is not zero. An instance of biology on another world would help us understand whether life is an unavoidable outcome in the evolution of worlds that harbor habitable environments for sufficient time (1 billion years or less; van Kranendonk et al. 2018), whether it requires an improbable combination of circumstances (Ward and Brownlee 2000), and/or how frequently life transported from another world takes hold.

Q11.4d How Have Life and Its Host World Co-Evolved?

Determining the biochemical and metabolic characteristics of any extraterrestrial life would help us understand how it may have adapted to its environment, and perhaps the impact it has had on that environment from microscopic to global scales. Life on Earth has been shaped by the environment, including multiple mass extinction events that profoundly changed the phylogeny and ecology of the surviving groups. But terrestrial life has also had a profound impact on its environment, causing significant changes in the composition of the atmosphere, ocean sediments, and the upper layers of the crust (Q9.4). Is this co-evolution of life and the environment an intrinsic characteristic of inhabited worlds? If life ever existed on Mars, whose geologic history is in some sense comparable to Earth's, was it able to significantly modify the environment? Did it retreat to cryptic niches as the planet became increasingly colder and dryer? Or did it succumb to one or more mass extinction events? If life exists in an ocean world, has it modified the chemistry of the ocean? Has it been shaped by events outside its sheltered confines? Or does it represent an endmember instance of static biology in a largely unchangeable environment?

The extent of a biosphere and of its effects likely depends on its overall energy and nutrient supply (see Question 10). Evidence to date suggests that any life elsewhere in the solar system has not, and cannot have, had as much of an effect on its host world as—even microbial—life on Earth, which has shaped our oxygen-rich atmosphere (Q6.2). However, the global influence of biology is the prime line of evidence being pursued to search for life on extrasolar planets. The next (ultimate?) level of study would examine the biochemistry and physiology of any discovered organisms. Given the experimental rather than observational or analytical nature of such investigations, current and foreseeable technology likely limits them to approaches involving humans, either with returned samples or, for Mars, in situ human exploration, rather than solely via in situ robotic exploration.

Strategic Research for Q11.4

- **Investigate the chemical environments of Europa and Enceladus that are relevant to potential biochemistry** through measurements of possible metabolic reactants and products (e.g., organic and inorganic compounds), reaction conditions (e.g., temperature, pressure, Eh, and pH), rates and catalysis (e.g., mineral surfaces and trace elements), as well as the structure of biomolecules at the surface and in waters sourced from the subsurface.
- **Determine whether any life present in martian materials might share ancestry with Earth** through measurement of any biomolecules as part of the organic chemical inventory to assess their function (including, for ancient life, their predegradation form) and what they reveal about the co-evolution of Mars's life and climate.
- **Prepare for characterizing life in the subsurface of ocean worlds** by determining the heterogeneity of thicknesses of ice shells via planetary mission data as well as validating and deploying emerging technologies for life characterization, and maturing technology for accessing the subsurface for exploration, by work in the field and in the laboratory.

- **Develop viable solutions to technical challenges for curation of samples to be returned from Mars and ocean worlds that will preserve their scientific integrity,** including biosafe curation coupled (for ocean worlds) with cold conditions.
- **Characterize any form of life discovered beyond Earth, and investigate its relatedness, or lack thereof to life on Earth** through measurements of its biochemical, morphological, and/or physiological/metabolic traits with spacecraft in situ observations or in samples returned to Earth.

REFERENCES

Altwegg, K., H. Balsiger, A. Bar-Nun, J.-J. Berthelier, A. Bieler, P. Bochsler, C. Briois, et al. 2016. "Prebiotic Chemicals— Amino Acid and Phosphorus—in the Coma of Comet 67P/Churyumov-Gerasimenko." *Science Advances* 2(5):e1600285. https://doi.org/10.1126/sciadv.1600285.

Barge, L.M., L. Rodriguez, J.M. Weber, and B. Theiling. 2020. "Beyond 'Biosignatures': Importance of Applying Abiotic/ Prebiotic Chemistry to the Search for Extraterrestrial Life." White paper #173 submitted to the Planetary Science and Astrobiology Decadal Survey 2023–2032. *Bulletin of the AAS* 53(4). https://doi.org/10.3847/25c2cfeb.b66ffca4.

Bender, M., T. Sowers, and L. Labeyrie. 1994. "The Dole Effect and Its Variations During the Last 130,000 Years as Measured in the Vostok Ice Core." *Global Biogeochemical Cycles* 8(3):363–376. https://doi.org/10.1029/94GB00724.

Brown, R.H., J.-P. Lebreton, and J.H. Waite, eds. 2010. *Titan from Cassini-Huygens*. Dordrecht, Netherlands: Springer. https://doi.org/10.1007/978-1-4020-9215-2.

Cable, M.L., M. Neveu, H. Hsu, and T. Hoehler. 2020. "Enceladus." Pp. 217–246 in *Planetary Astrobiology*, V.S. Meadows, G.N. Arney, B.E. Schmidt, and D.J. Des Marais, eds. Tucson: University of Arizona Press. https://doi.org/10.2458/azu_uapress_9780816540068-ch009.

Canfield, D.E., M.T. Rosing, and C. Bjerrum. 2006. "Early Anaerobic Metabolisms." *Philosophical Transactions of the Royal Society of London B: Biological Sciences* 361:1819–1836. https://doi.org/10.1098/rstb.2006.1906.

Cartwright, R.J., J.P. Emery, N. Pinilla-Alonso, M.P. Lucas, A.S. Rivkin, and D.E. Trilling. 2018. "Red Material on the Large Moons of Uranus: Dust from the Irregular Satellites?" *Icarus* 314:210–231. https://doi.org/10.1016/j.icarus.2018.06.004.

Chamberlin, T.C. 1890. "The Method of Multiple Working Hypotheses." *Science* (old series) 15:92–96. Reprint *Science* 148(3671):754–759, May 1965. https://doi.org/10.1126/science.148.3671.754.

Chamberlin, T.C. 1897. "The Method of Multiple Working Hypotheses." *Journal of Geology* 5:837–848.

Chan, Q.H.S., A. Stephant, A. Franchi, X. Zhao, R. Brunetto, Y. Kebukawa, T. Noguchi, et al. 2021. "Organic Matter and Water from Asteroid Itokawa." *Scientific Reports* 11:5125. https://doi.org/10.1038/s41598-021-84517-x.

Chyba, C.F., and C.B. Phillips. 2002. "Europa as an Abode of Life." *Origins of Life and Evolution of the Biosphere* 32:47–67. https://doi.org/10.1023/A:1013958519734.

Cleaves, H.J., C. Butch, P.B. Burger, J. Goodwin, and M. Meringer. 2019. "One Among Millions: The Chemical Space of Nucleic Acid-Like Molecules." *Journal of Chemical Information and Modeling* 59:4266–4277. https://doi.org/10.1021/acs.jcim.9b00632.

Clemett, S.J., S.A. Sandford, K. Nakamura-Messenger, F. Horz, D. and McKay. 2010. "Complex Aromatic Hydrocarbons in Stardust Samples Collected from Comet 81P/Wild 2." *Meteoritics and Planetary Science* 45:701–722. https://doi.org/10.1111/j.1945-5100.2010.01062.x.

Colaprete, A., P. Schultz, J. Heldmann, D. Wooden, M. Shirley, K. Ennico, B. Hermalyn, et al. 2010. "Detection of Water in the LCROSS Ejecta Plume." *Science* 330:463–468. https://doi.org/10.1126/science.1186986.

Cruikshank, D.P., and R.H. Brown. 1987. "Organic Matter on Asteroid 130 Elektra." *Science* 238(4824):183–184. https://doi.org/10.1126/science.238.4824.183.

Cruikshank, D.P., C.K. Materese, Y.J. Pendleton, P.J. Boston, W.M. Grundy, B. Schmitt, C.M. Lisse, et al. 2019. "Prebiotic Chemistry of Pluto." *Astrobiology* 19(7):831–848. https://doi.org/10.1089/ast.2018.1927.

De Sanctis, M.C., V. Vinogradoff, A. Raponi, E. Ammannito, M. Ciarniello, F.G. Carrozzo, S. De Angelis, et al. 2018. "Characteristics of Organic Matter on Ceres from VIR/Dawn High Spatial Resolution Spectra." *Monthly Notices of the Royal Astronomical Society* 482(2):2407–2421. https://doi.org/10.1093/mnras/sty2772.

Deamer, D., and B. Damer. 2017. "Can Life Begin on Enceladus? A Perspective from Hydrothermal Chemistry." *Astrobiology* 17(9):834–839. https://doi.org/10.1089/ast.2016.1610.

Delitsky, M.L., D.A. Paige, M.A. Siegler, E.R. Harju, D. Schriver, R.E. Johnson, and P. Travnicek. 2017. "Ices on Mercury: Chemistry of Volatiles in Permanently Cold Areas of Mercury's North Polar Region." *Icarus* 281:19–31. https://doi.org/10.1016/j.icarus.2016.08.006.

Des Marais, D.J., L.J. Allamandola, S.A. Benner, A.P. Boss, D. Deamer, P.G. Falkowski, J.D. Farmer, et al. 2003. "The NASA Astrobiology Roadmap." *Astrobiology* 3(2):219–235. https://doi.org/10.1089/153110703769016299.

Des Marais, D.J., J.A. Nuth, L.J. Allamandola, A.P. Boss, J.D. Farmer, T.M. Hoehler, B.N. Jakosky, et al. 2008. "The NASA Astrobiology Roadmap." *Astrobiology* 8(4):715–730. https://doi.org/10.1089/ast.2008.0819.

Ditzler, M.A., A.C. Rios, M. Nuevo, M. Popovic, R. Mancinelli, D. Summers, J.T. Broddrick, et al. 2020. "Beyond Targeted Searches: The Need for System-Level Approaches to Understanding the Connection Between Astrochemistry and the Emergence of Life." White paper #269 submitted to the Planetary Science and Astrobiology Decadal Survey 2023–2032. *Bulletin of the AAS* 53(4). https://doi.org/10.3847/25c2cfeb.7a06f84d.

Eigenbrode, J.E., R.E. Summons, A. Steele, C. Freissinet, M. Millan, R. Navarro-Gonzalez, B. Sutter, et al. 2018. "Organic Matter Preserved in 3-Billion-Year-Old Mudstones at Gale Crater, Mars." *Science* 360:1096–1101. https://doi.org/10.1126/science.aas9185.

Field, C.B., M.J. Behrenfeld, J.T. Randerson, and P. Falkowski. 1998. "Primary Production of the Biosphere: Integrating Terrestrial and Oceanic Components." *Science* 281(5374):237–240. https://doi.org/10.1126/science.281.5374.237.

Fink, U., M. Hoffmann, W. Grundry, M. Hicks, and W. Sears. 1992. "The Steep Red Spectrum of 1992 AD: An Asteroid Covered with Organic Material?" *Icarus* 97(1):145–149. https://doi.org/10.1016/0019-1035(92)90064-E.

Freissinet, C., D.P. Glavin, P.R. Mahaffy, K.E. Miller, J.L. Eigenbrode, R.E. Summons, A.E. Brunner, et al. 2015. "Organic Molecules in the Sheepbed Mudstone, Gale Crater, Mars." *Journal of Geophysical Research: Planets* 120(3):495–514. https://doi.org/10.1002/2014JE004737.

Greaves, J.S., A.M.S. Richards, W. Bains, P.B. Rimmer, H. Sagawa, D.L. Clements, S. Seager, et al. 2020. "Phosphine Gas in the Cloud Decks of Venus." *Nature Astronomy* 5:655–664. https://doi.org/10.1038/s41550-020-1174-4.

Hamill, C.D., N.L. Chabot, E. Mazarico, M.A. Siegler, M.K. Barker, and J.M. Martinez Camacho. 2020. "New Illumination and Temperature Constraints of Mercury's Volatile Polar Deposits." *Planetary Science Journal* 1(3):57. https://doi.org/10.3847/PSJ/abb1c2.

Hand, K.P., A.E. Murray, J.B. Garvin, W.B. Brinckerhoff, B.C. Christner, K.S. Edgett, B.L. Ehlmann, et al. 2017. *Report of the Europa Lander Science Definition Team*. Washington, DC: National Aeronautics and Space Administration. https://solarsystem.nasa.gov/docs/Europa_Lander_SDT_Report_2016.pdf.

Higgins, P.M., and C.S. Cockell. 2020. "A Bioenergetic Model to Predict Habitability, Biomass and Biosignatures in Astrobiology and Extreme Conditions." *Journal of the Royal Society Interface* 17(171):1–13. https://doi.org/10.1098/rsif.2020.0588.

Hoehler, T., W. Brinckerhoff, A. Davila, D. Des Marais, S. Getty, D. Glavin, A. Pohorille, et al. 2020. "Groundwork for Life Detection." White paper #202 submitted to the Planetary Science and Astrobiology Decadal Survey 2023–2032. *Bulletin of the AAS* 53(4). https://doi.org/10.3847/25c2cfeb.bd9172f9.

Hoshika, S., N.A. Leal, M.J. Kim, M.S. Kim, N.B. Karalkar, H.J. Kim, A.M. Bates, et al. 2019. "Hachimoji DNA and RNA: A Genetic System with Eight Building Blocks." *Science* 363(6429):884–887. https://doi.org/10.1126/science.aat0971.

Johnson, N.M., and M.R.R. de Oliveira. 2019. "Venus Atmospheric Composition in Situ Data: A Compilation." *Earth and Space Science* 6(7):1299–1318. https://doi.org/10.1029/2018EA000536.

Kharecha, P., J. Kasting, and J. Siefert. 2005. "A Coupled Atmosphere-Ecosystem Model of the Early Archean Earth." *Geobiology* 3(2):53–76. https://doi.org/10.1111/j.1472-4669.2005.00049.x.

Krissansen-Totton, J., R. Buick, and D.C. Catling. 2015. "A Statistical Analysis of the Carbon Isotope Record from the Archean to Phanerozoic and Implications for the Rise of Oxygen." *American Journal of Science* 315(4):275–316. https://doi.org/10.2475/04.2015.01.

Lang, S.Q., G.L. Früh-Green, S.M. Bernasconi, W.J. Brazelton, M.O. Schrenk, and J.M. McGonigle. 2018. "Deeply-Sourced Formate Fuels Sulfate Reducers But Not Methanogens at Lost City Hydrothermal Field." *Scientific Reports* 8:755. https://doi.org/10.1038/s41598-017-19002-5.

Limaye, S.S., R. Mogul, D.J. Smith, A.F. Ansari, G.P. Slowik, and P. Vaishampayan. 2018. "Venus' Spectral Signatures and the Potential for Life in the Clouds." *Astrobiology* 18(9):1181–1198. https://doi.org/10.1089/ast.2017.1783.

Lollar, G.S., O. Warr, J. Telling, M.R. Osburn, and B. Sherwood Lollar. 2019. "'Follow the Water': Hydrogeochemical Constraints on Microbial Investigations 2.4 km Below Surface at the Kidd Creek Deep Fluid and Deep Life Observatory." *Geomicrobiology Journal* 36:859–872. https://doi.org/10.1080/01490451.2019.1641770.

Lunine, J.I., M.L. Cable, S.M. Horst, and M. Rahm. 2020. "The Astrobiology of Titan." Pp. 247–266 in *Planetary Astrobiology*, V.S. Meadows, G.N. Arney, B.E. Schmidt, and D.J. Des Marais, eds. Tucson: University of Arizona Press. https://doi.org/10.2458/azu_uapress_9780816540068-ch010.

Lyons, T.W., K. Rogers, R. Krishnamurthy, L. Williams, S. Marchi, E. Schwieterman, N. Planavsky, and C. Reinhard. 2020. "Constraining Prebiotic Chemistry Through a Better Understanding of Earth's Earliest Environments." White paper #143 submitted to the Planetary Science and Astrobiology Decadal Survey 2023–2032. https://doi.org/10.3847/25c2cfeb.7a898b.

Marshall, S.M., A.R.G. Murray, and L. Cronin. 2017. "A Probabilistic Framework for Identifying Biosignatures Using Pathway Complexity." *Royal Society Philosophical Transactions A* 375(2109). https://doi.org/10.1098/rsta.2016.0342.

McCord, T.B., R.W. Carlson, W.D. Smythe, G.B. Hanson, R.N. Clark, C.A. Hibbitts, F.P. Fanale, et al. 1997. "Organics and Other Molecules in the Surfaces of Callisto and Ganymede." *Science* 278(5336):271–275. https://doi.org/10.1126/science.278.5336.271.

McCord, T.B., G.B. Hansen, R.N. Clark, P.D. Martin, C.A. Hibbitts, F.P. Fanale, J.C. Granahan, et al. 1998. "Non-Water-Ice Constituents in the Surface Material of the Icy Galilean Satellites from the Galileo Near-Infrared Mapping Spectrometer Investigation." *Journal of Geophysical Research: Planets* 103(E4):8603–8626. https://doi.org/10.1029/98JE00788.

McDermott, J.M., J.S. Seewald, C.G. German, and S.P. Sylva. 2015. "Pathways for Abiotic Organic Synthesis at Submarine Hydrothermal Fields." *Proceedings of the National Academy of Sciences* 112(25):7668–7672. https://doi.org/10.1073/pnas.1506295112.

McKay, C., A. Davila, J. Eigenbrode, C. Lorentson, R. Gold, J. Canham, A. Dazzo, et al. 2020. "Contamination Control Technology Study for Achieving the Science Objectives of Life-Detection Missions." NASA/TM-20205008709. Washington, DC: National Aeronautics and Space Administration.

Meadows, V.S. 2017. "Reflections on O_2 as a Biosignature in Exoplanetary Atmospheres." *Astrobiology* 17(10):1022–1052. https://doi.org/10.1089/ast.2016.1578.

Meadows, V.S., C.T. Reinhard, G.N. Arney, M.N. Parenteau, E.W. Schwieterman, S.D. Domagal-Goldman, et al. 2018. "Exoplanet Biosignatures: Understanding Oxygen as a Biosignature in the Context of Its Environment." *Astrobiology* 18(6):630–662. https://doi.org/10.1089/ast.2017.1727.

Mogul, R., S.S. Limaye, M.J. Way, and J.A. Cordova. 2021. "Venus' Mass Spectra Show Signs of Disequilibria in the Middle Clouds." *Geophysical Research Letters* 48(7). https://doi.org/10.1029/2020GL091327.

Mustard, J.F., M. Adler, A. Allwood, D.S. Bass, D.W. Beaty, J.F. Bell III, W.B. Brinckerhoff, et al. 2013. *Report of the Mars 2020 Science Definition Team*. Pasadena, CA: Mars Exploration Program Analysis Group. https://mars.nasa.gov/mars2020/files/mars2020/SDT-Report%20Finalv6.pdf.

NASEM (National Academies of Sciences, Engineering, and Medicine). 2019. *An Astrobiology Strategy for the Search for Life in the Universe*. Washington, DC: The National Academies Press. https://doi.org/10.17226/25252.

NRC (National Research Council). 2007. *The Limits of Organic Life in Planetary Systems*. Washington, DC: The National Academies Press. https://doi.org/10.17226/11919.

Postberg, F., N. Khawaja, B. Abel, G. Cholet, C.R. Glein, M.S. Gudipati, B.L. Henderson, et al. 2018. "Macromolecular Organic Compounds from the Depths of Enceladus." *Nature* 558:564–568. https://doi.org/10.1038/s41586-018-0246-4.

Quirico, E., S. Doute, B. Schmitt, C. de Bergh, D.P. Cruikshank, T.C. Owen, T.R. Geballe, and T.L. Roush. 1999. "Composition, Physical State, and Distribution of Ices at the Surface of Triton." *Icarus* 139(2):159–178. https://doi.org/10.1006/icar.1999.6111.

Raponi, A., M.S. De Sanctis, F.G. Carrozzo, M. Ciarniello, B. Rousseau, M. Ferrari, E. Ammannito, et al. 2021. "Organic Material on Ceres: Insights from Visible and Infrared Space Observations." *Life* 11(1):9. https://doi.org/10.3390/life11010009.

Sherwood Lollar, B., T.C. Onstott, G. Lacrampe-Couloume, and C.J. Ballentine. 2014. "The Contribution of the Precambrian Continental Lithosphere to Global H_2 Production." *Nature* 516:379–382. https://doi.org/10.1038/nature14017.

Sholes, S.F., J. Krissansen-Totton, and D.C. Catling. 2019. "A Maximum Subsurface Biomass on Mars from Untapped Free Energy: CO and H_2 as Potential Antibiosignatures." *Astrobiology* 19(5):655–668. https://doi.org/10.1089/ast.2018.1835.

Snellen, I.A.G., L. Guzman-Ramirez, M.R. Hogerheijde, A.P.S. Hygate, and F.F.S. van der Tak. 2020. "Re-Analysis of the 267 GHz ALMA Observations of Venus." *Astronomy & Astrophysics* 644:1–3. https://doi.org/10.1051/0004-6361/202039717.

Spear, J.R., J.J. Walker, T.M. McCollom, and N.R. Pace. 2005. "Hydrogen and Bioenergetics in the Yellowstone Geothermal Ecosystem." *Proceedings of the National Academy of Sciences* 102(7):2555–2560. https://doi.org/10.1073/pnas.0409574102.

Steel, E.L., A. Davila, and C.P. McKay. 2017. "Abiotic and Biotic Formation of Amino Acids in the Enceladus Ocean." *Astrobiology* 17(9):862–875. https://doi.org/10.1089/ast.2017.1673.

Sutherland, J.D. 2017. "Opinion: Studies on the Origin of Life—the End of the Beginning." *Nature Reviews Chemistry* 1:0012. https://doi.org/10.1038/s41570-016-0012.

Teolis, B.D., D.Y. Wyrick, A. Bouquet, B.A. Magee, and J.H. Waite. 2017. "Plume and Surface Feature Structure and Compositional Effects on Europa's Global Exosphere: Preliminary Europa Mission Predictions," *Icarus* 284:18–29. https://doi.org/10.1016/j.icarus.2016.10.027.

Theiling, B., W. Brinckerhoff, J. Castillo-Rogez, L. Chou, V. Da Poian, H. Graham, S.S. Hosseini, et al. 2021. "Non-Robotic Science Autonomy Development." White paper #048 submitted to the Planetary Science and Astrobiology Decadal Survey 2023–2032. *Bulletin of the AAS* 53(4). https://doi.org/10.3847/25c2cfeb.ee4e6b64.

Thompson, W.R., S.K. Singh, B.N. Khare, and C. Sagan. 1989. "Triton: Stratospheric Molecules and Organic Sediments." *Geophysical Research Letters* 16(8):981–984. https://doi.org/10.1029/GL016i008p00981.

Vago, J.L., F. Westall, Pasteur Instrument Teams, Landing Site Selection Working Group, et al. 2017. "Habitability on Early Mars and the Search for Biosignatures with the ExoMars Rover." *Astrobiology* 17(6–7):471–510. https://doi.org/10.1089/ast.2016.1533.

van Kranendonk, M.J., V. Bennett, and E. Hoffmann, eds. 2018. *Earth's Oldest Rocks*, 2nd ed. Cambridge, MA: Elsevier Science. https://doi.org/10.1016/B978-0-444-63901-1.12001-5.

Villalobos, Y., P. Rayner, S. Thomas, and J. Silver. 2020. "The Potential of Orbiting Carbon Observatory-2 Data to Reduce the Uncertainties in CO_2 Surface Fluxes Over Australia Using a Variational Assimilation Scheme," *Atmospheric Chemistry and Physics* 20(14):8473–8500. https://doi.org/10.5194/acp-20-8473-2020.

Waite, J.H., M.R. Combi, W.-H. Ip, T.E. Cravens, R.L. McNutt, W. Kasprzak, R. Yelle, et al. 2006. "Cassini Ion and Neutral Mass Spectrometer: Enceladus Plume Composition and Structure." *Science* 311(5766):1419–1422. https://doi.org/10.1126/science.1121290.

Ward, L.M., B. Rasmussen, and W.W. Fischer. 2019. "Primary Productivity Was Limited by Electron Donors Prior to the Advent of Oxygenic Photosynthesis." *Journal of Geophysical Research: Biogeosciences* 124(2):211–226. https://doi.org/10.1029/2018JG004679.

Ward, P.D., and D. Brownlee. 2000. *Rare Earth: Why Complex Life Is Uncommon in the Universe*. New York: Copernicus Books. https://doi.org/10.1007/b97646.

Weiss, B.P., Y.L. Yung, and K.H. Nealson. 2000. "Atmospheric Energy for Subsurface Life on Mars?" *Proceedings of the National Academy of Sciences* 97(4):1395–1399. https://doi.org/10.1073/pnas.030538097.

Wickramasinghe, D.T., and D.A. Allen. 1986. "Discovery of Organic Grains in Comet Halley." *Nature* 323:44–46. https://doi.org/10.1038/323044a0.

Young, T.S., and P.G. Schultz. 2010. "Beyond the Canonical 20 Amino Acids: Expanding the Genetic Lexicon." *Journal of Biological Chemistry* 285(15):11039–11044. https://doi.org/10.1074/jbc.R109.091306.

Zhang, J.A., and D.A. Paige. 2009. "Cold-Trapped Organic Compounds at the Poles of the Moon and Mercury: Implications for Origins." *Geophysical Research Letters* 36(16):6. https://doi.org/10.1029/2009GL038614.

Zhou, Y., X. Xu, Y. Wei, Y. Cheng, Y. Guo, I. Khudyakov, F. Liu, et al. 2021. "A Widespread Pathway for Substitution of Adenine by Diaminopurine in Phage Genomes." *Science* 372(6541):512–516. https://doi.org/10.1126/science.abe4882.

Q12 PLATE: Artist's depiction of Kepler-186f, an Earth-size planet orbiting a distant star in the habitable zone, discovered in 2014. Additional exoplanets can be seen closer to the star. SOURCE: Courtesy of NASA Ames/SETI Institute/JPL-Caltech/T. Pyle.

15

Question 12: Exoplanets

What does our planetary system and its circumplanetary systems of satellites and rings reveal about other planetary systems, and what can disks and exoplanets orbiting other stars teach us about the solar system?

The past decade has seen extraordinary growth in our knowledge of planetary systems around other stars, as well as in the conditions of planet formation and the remarkable diversity and abundance of exoplanets.[1] These advances have been supported by continuing exoplanet discoveries, ground-breaking observations afforded in part by new facilities, and a progression from planet detection to detailed study and characterization of individual planets. Observations of protoplanetary disks around young stars have now begun to reveal the entire planet formation process, from the earliest accumulation of grains into large agglomerates to accretion onto growing protoplanets.

The observational approaches for exoplanets are quite disparate compared to those for solar system science, as illustrated in Figure 15-1, which highlights some key examples of this remarkable progress. High-resolution images of protoplanetary disks show evidence for thin disks and multiple rings and gaps, some of them owing to the presence of forming planets. Two accreting planets were discovered around young star PDS 70, with evidence for a circum*planetary* disk around PDS 70 c. The orbital motions of directly imaged planetary systems can be followed accurately, as in the case of Beta Pictoris and HR 8799. We have discovered that there are more planets than stars in our galaxy, as exemplified by the discovery of Proxima Centauri b, a ~3 Earth mass planet orbiting within the "habitable zone" (i.e., a distance from the star where water at the planetary surface, if present, may be liquid) of the nearest star to the Sun. Detailed exoplanetary spectra have been obtained, both for young and massive directly imaged planets and for close-in planets. The presence of clouds and differential rotation is becoming discernable, as shown for the close brown dwarf Luhman 16B. Similarly, high-spectral resolution observations with the Doppler imaging technique can be used to measure wind speeds in hot Jupiters or the spin of planet Beta Pictoris b.

The planetary systems that have been discovered—as exemplified by the seven-planet system around TRAPPIST-1, only about 12 pc away—are extremely diverse, and study of the demographics of large numbers of exoplanets has led to several advances in understanding, including the recognition that many small-mass planets possess hydrogen atmospheres. Progress in observational technology will enable discovery of an even greater number of systems, as shown in Figure 15-2 for the forthcoming Nancy Grace Roman Space Telescope, and much expanded characterization of individual exoplanets by, for example, the James Webb Space Telescope. Overall, the ability to image exoplanets both

---

[1] A glossary of acronyms and technical terms can be found in Appendix F.

391

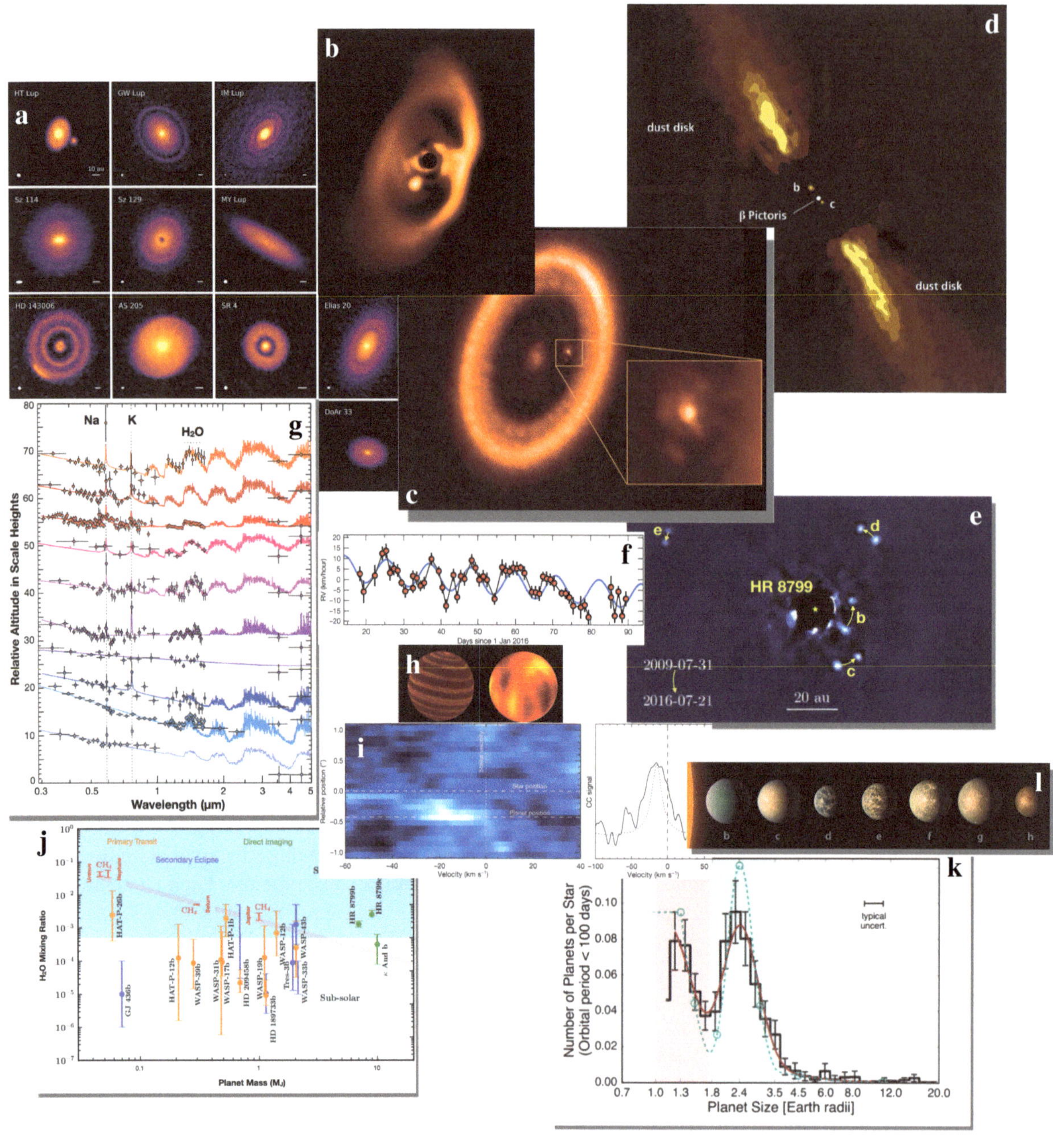

FIGURE 15-1 A selection of discoveries in exoplanet research in the past decade, with a layout intended to capture the explosive recent expansion of scientific understanding. (a) Young protoplanetary disks seen by the Atacama Large Millimeter Array (ALMA) from the Disk Substructures at High Angular Resolution Project survey (Andrews et al. 2018). (b) Accreting protoplanet PDS 70 b discovered by the Very Large Telescope (VLT)/Spectro-Polarimetric High-contrast Exoplanet Research (Müller et al. 2018). (c) First protosatellite disk imaged around protoplanet PDS 70 c using ALMA (Benisty et al. 2021). (d) Planets b and c in the Beta Pictoris system as seen by VLT/GRAVITY (Nowak et al. 2020; see also Lagrange et al. 2019). (e) Orbital motions of the HR 8799 multiplanet system as monitored with the Gemini Planet Imager (Wang et al. 2016). (f) Temperate Earth-mass planet discovered by radial-velocimetry with High Accuracy Radial Velocity Planet Searcher around Proxima Centauri, the closest star to the Sun (Anglada-Escudé et al. 2016), a striking consequence of the fact that 25 percent of M-dwarfs harbor a rocky planet orbiting in their habitable zone (Dressing and Charbonneau 2015). (g) Spectra of hot giant exoplanets showing signatures of water, sodium, potassium, and clouds as seen by the Hubble Space Telescope (HST) and Spitzer (Sing et al. 2016). (h) Clouds and bands in brown dwarf Luhman 16B as inferred by Doppler imaging on VLT/CRyogenic high-resolution InfraRed Echelle Spectrograph (CRIRES; Crossfield et al. 2014) (right) and by long-duration photometry from the Transiting Exoplanet Survey Satellite (Apai et al. 2021) (left). (i) Spin of exoplanet Beta Pictoris b as measured by Doppler imaging on VLT/CRIRES compared to solar system planets (Snellen et al. 2014). (j) H_2O mixing ratios derived for a variety of transiting and directly imaged exoplanets compared to mixing ratios derived from CH_4 in solar system giant planets (Madhusudhan 2019). (k) Distribution of exoplanetary radii from the Kepler survey highlighting the radial gap observed around 1.8 Earth-radii ($R_\oplus$), indicating that half of super-Earths possess hydrogen atmospheres (Fulton et al. 2017). (l) Multiple seven-planet transiting system TRAPPIST-1 discovered by the photometric Transiting Planets and Planetesimals Small Telescope (TRAPPIST; Gillon et al. 2017). Among these, planets d, e, and f have been characterized by HST to have hydrogen-free atmospheres (de Wit et al. 2018). SOURCES: (a) Andrews et al. (2018). © AAS. Reproduced with permission. (b) Courtesy of ESO/A. Müller et al. (eso.org/public/images/eso1821a). CC BY 4.0. (c) ALMA (ESO/NAOJ/NRAO)/Benisty et al. (www.eso.org/public/images/eso2111a). CC BY 4.0. (d) GRAVITY Collaboration/A.M. Quetz, MPIA Graphics Department. (e) Courtesy of J. Wang (Northwestern)/C. Marois (NRC Herzberg). (f) ESO/G. Anglada-Escudé (https://www.eso.org/public/images/eso1629d). CC BY 4.0. (g) D.K. Sing, J.J. Fortney, and N. Nikolov, 2016, "A Continuum from Clear to Cloudy Hot-Jupiter Exoplanets Without Primordial Water Depletion," *Nature* 529(7584):59–62. Springer Nature, reproduced with permission from SNCSC. (h) D. Apai, University of Arizona; Courtesy of ESO/I. Crossfield (eso.org/public/images/eso1404a). CC BY 4.0. (i) I.A. Snellen, B.R. Brabdl, R.J. de Kok, M. Brogi, J. Birkby, and H. Schwarz, 2014, "Fast Spin of the Young Extrasolar Planet β Pictoris b," *Nature* 509:63–65. Springer Nature, reproduced with permission from SNCSC. (j) From N. Madhusudhan, 2019, "Exoplanetary Atmospheres: Key Insights, Challenges, and Prospects," *Annual Review of Astronomy and Astrophysics* 57:617–663. Used with permission of Annual Reviews, Inc.; permission conveyed through Copyright Clearance Center, Inc. (k) Fulton et al. (2017). © AAS. Reproduced with permission. (l) Courtesy of NASA/JPL-Caltech.

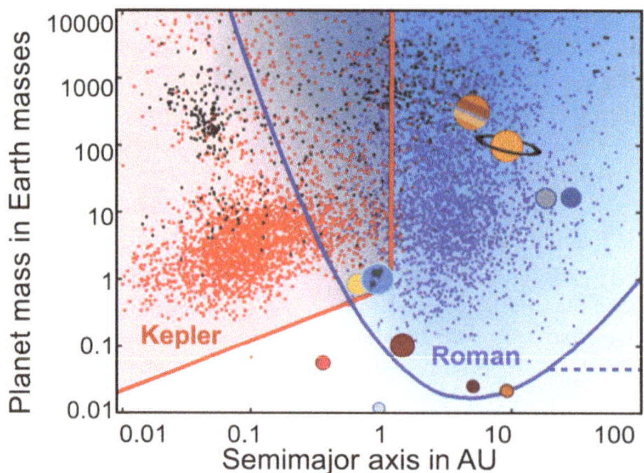

FIGURE 15-2 Exoplanets currently detected by Kepler (red) and other means (black) compared with projected detections by the Nancy Grace Roman Space Telescope (blue). SOURCE: Planets courtesy of NASA. Simulated Roman data from NASA and Penny et al. (2019). © AAS. Reproduced with permission.

when they are forming and in their mature stage, the ability to characterize these exoplanets and their atmospheres, and the immense wealth of data available on more than 4,000 exoplanets (and counting), is providing us with new opportunities to understand planetary systems in the universe and to compare and contrast them with the solar system's planets.

The *Exoplanet Science Strategy* (ESS; NASEM 2018) identified two overarching goals:

- Understand the formation and evolution of planetary systems as products of the process of star formation, and characterize and explain the diversity of planetary system architectures, planetary compositions, and planetary environments produced by these processes.
- Learn enough about the properties of exoplanets to identify potentially habitable environments, their frequency, and connect these environments to the planetary systems in which they reside. Furthermore, researchers need to distinguish between the signatures of life and those of nonbiological processes, and search for signatures of life on worlds orbiting other stars.

Questions 1 through 11 in this report have identified priority science questions for exploring the solar system over the next decade. Exoplanetary data, which increasingly reflect a multiplicity of objects, are an invaluable complement to the detailed observations and theories developed for the solar system's planets. Conversely, understanding derived from solar system studies can be compared to and contrasted with wider-reaching exoplanet observations. The synergy between the ESS goals and the goals for planetary science and astrobiology outlined in this decadal survey provides strong motivation for a new era of collaborative research. This can occur not only through existing cross-disciplinary research and analysis programs, but also through collaborative efforts in mission design and implementation, telescope observations, data analysis, and laboratory and experimental research.

In this chapter, the committee structures its discussion around the 11 priority science questions for (Chapters 4–14) solar system science, providing for each question examples of how it relates to and can benefit from exoplanetary studies, including the identification of strategic research activities.

Q12.1 EVOLUTION OF THE PROTOPLANETARY DISK

The discovery and characterization of disks around other stars have revolutionized our understanding of disk evolution and planet accretion, providing a unique window into the processes that occurred within the Sun's protoplanetary nebula. Coordinating research between studies of these disks and models of solar system formation can greatly advance our understanding of planetary system formation.

Q12.1a How Does a Disk's Bulk Composition Affect the Diversity of Resulting Planetary Materials?

Meteoritic studies tell us that the bulk composition of the solar nebula was similar to that of the Sun's photosphere, providing direct evidence that stellar composition can be used to constrain overall circumstellar disk composition (e.g., Johnson et al. 2012). However, we know from the diversity of meteorites that the composition of early accreted solids and planetesimals in the solar system also depended on localized disk conditions such as temperature and oxygen fugacity. Observations of protoplanetary disks can reveal disk properties across a wide range of stellar types and ages, including variations in dust and gas compositions at different evolutionary stages, providing a powerful complement to meteoritic studies and solar nebula compositional models. Key open issues include the role of stellar metallicity and disk structure in determining planetesimal composition.

The conditions in the solar nebula can only be inferred from what remains today of the gas and dust. We do not know how well-mixed the other heavy elements were in the disk from which the solar system formed (see Q3.2). Observations of bands in cometary material suggest the existence of crystalline silicate material in the cold outer regions where comets formed. The distinct oxygen isotope signatures of terrestrial, asteroidal, and martian materials indicate that the solar system was not well-mixed. Meteorites represent (at least) two distinct reservoirs of material, and perhaps the early formation of Jupiter may have played a role in isolating these reservoirs. Mapping radial compositional gradients in the dusty component of young disks can potentially constrain the origin and mixing of compositional reservoirs.

The surprising discovery of two interstellar objects (ISOs) passing through the inner solar system, 1I/'Oumuammua and 2I/Borisov, raises the possibility of direct analysis of materials formed in the disks of other stars. 1I/'Oumuammua appeared asteroidal, in that it did not show comet-like activity (Meech et al. 2017), while 2I/Borisov exhibited evidence of a cometary color and tail (Guzik et al. 2020). Constraints on the mineralogy and elemental abundances of ISOs could provide insight into the composition of their parent disks. Improvements to Pan-STARRS1 and observations by the Vera C. Rubin Observatory will support detection of additional ISOs in coming years, and development of rapid launch capabilities could enable future spacecraft encounters with one of these extrasolar visitors.

Q12.1b How Do Disks Evolve with Time, and How and When Do Macroscopic Particles and Planetesimals Begin to Form?

Dating of meteoritic materials implies that macroscopic solids and planetesimals formed within the first few million years of solar system formation, with the latter established by the age of the oldest materials, the CAIs (calcium-aluminum-rich inclusions). Multi-wavelength observations of molecular clouds and young stellar objects and their disks allow us to develop a comprehensive picture of the timescales and conditions of disk formation and early evolution. Such observations can help address outstanding issues including the survival of molecular cloud components during disk formation, heating mechanisms responsible for CAIs and chondrules, and planetesimal formation mechanisms. Observations can also track variations in disk structure with time, as well as variations in nebular gas and ice, where the latter are not directly recorded in rocky meteoritic materials but are essential for understanding early planetary accretion mechanisms. Such observations can address fundamental open issues including the range of nebular properties and lifetimes and the mechanisms responsible for gas dissipation.

Meteorites show evidence for irradiation of early dust, but questions remain concerning the source, extent, and influence of irradiation on gas, dust, ice, and organics. Observations of radiation sources (e.g., cosmic and stellar winds) for young stellar objects and tracking the effect of radiation on dust and gas can provide insight into the role irradiation plays in the composition of nebular components. This will require stronger constraints on the ages of young stellar objects, and better constraints on the link between CAIs and the astronomical timescale of stellar evolution.

Disk observations have begun to probe the earliest stages of planet formation. Estimates of the total mass in sub-cm-size particles in 1- to 3-million-year-old disks (Ansdell et al. 2016) appear too low to explain observed exoplanet system masses, suggesting that solids may have already accreted into larger ($\geq$ cm) sizes by this time (Manara et al. 2018). Some observations point to earlier growth of mm-size pebbles in the envelopes and disks of ~10^5-year-old protostars (Miotello et al. 2014), while isotopically derived ages of the oldest chondrules now suggest that accretion of mm-sized bodies in the solar system began within ~10^5 years of CAI formation (Bollard et al. 2017). The interplay of meteoritic data with observations of young disks will provide crucial constraints on the initial and perhaps least understood stages of planet assembly.

Strategic Research for Q12.1

- **Determine how the nascent planets acquired material from the protosolar disk** by measuring abundances and isotopic compositions of noble gases and other key elements (e.g., H, C, O, N, and S) from major planets (e.g., Venus, Mars, and the giant planets), satellites (e.g., Titan), and small bodies via spacecraft data, in situ probes, sample return, and telescopic observations.
- **Determine the composition of primordial material that preserves chemical signatures of the protoplanetary disk** via sample return, spacecraft data, and telescopic observations.
- **Characterize protoplanetary disks around young stars—including their lifetimes, structures, gas versus solid compositions, and properties of any accreted components—**with observations and theoretical and modeling studies of disk processes, to include the survival of components of molecular clouds, irradiation and delivery of disk material, formation and evolution of disk components (e.g., dust, ice, and gas), early stages of planetary accretion, and gas disk dissipation.
- **Assess the range of conditions and mechanisms that may lead to the commencement of accretion in disks around young stars** by model development constrained by diverse disk observations.
- **Identify and characterize interstellar objects (e.g., size, dynamical origin, and composition)** with spacecraft data, telescopic observations, theoretical and modeling studies of their formation and evolution, and laboratory studies of analogue materials.

Q12.2 ACCRETION IN THE OUTER SOLAR SYSTEM

Outer solar system temperatures allowed for the formation of ice-rock planetesimals, the feedstock for cometary bodies and ice-rich dwarf planets, the solid components of the giant planets, and the outer satellite and ring systems. Delivery of this material to the inner solar system was also a source of terrestrial planet volatiles. Gas accretion led to gas-dominated Jupiter and Saturn, but only modest gas components at Uranus and Neptune. Giant planet gravitational interactions with the background gaseous disk and/or other solid bodies led to dynamical effects felt across the whole solar system, including large-scale giant planet orbital migration. These processes are relevant to exoplanet systems, where diverse system properties demonstrate a complexity of possible outcomes that remain incompletely understood.

Q12.2a How Do Giant Planets Form, and How Does Their Origin
Compare to the Formation of Super-Earths and Sub-Neptunes?

Traditional concepts for giant planet growth invoke collisional accretion of a solid core beyond the water ice-line, followed by gas accretion beyond a critical core mass. Meteoritic evidence suggests that Jupiter began forming (up to ~20 Earth-masses) within ~1 million years after CAIs (Kruijer et al. 2017). That Uranus and Neptune accreted only a minority of their mass in gas is attributed to their later formation as the solar nebula was dissipating, and/or to depletion of the local supply of gas owing to its accretion by Jupiter and Saturn. Diverse exoplanet systems display evidence of more complex planetary outcomes. For example, a recent discovery is a relative dearth of planets with radii near 1.8 Earth radii ($R_\oplus$) (see Figure 15-1k), suggesting a difference in either formation or evolution of so-called super-Earths ($<1.8\ R_\oplus$) compared to sub-Neptunes (~1.8 to 3.5 $R_\oplus$). Neither class of exoplanet has a clear solar system analog. This distribution of exoplanet radii may reflect a critical core mass needed for gas accretion, and/or the effects of escape of primary atmospheres (see also Q12.6a).

Q12.2b Is the Solar System Architecture, with Multiple Outer Giant Planets That Formed Beyond the Water Ice Line, a Common or Uncommon Outcome of Planet Formation in the Galaxy? How Common Was Giant Planet Migration?

The majority of known giant exoplanets orbit much closer to their stars than the giant planets in the solar system, many within the orbit of Mercury (see Figure 15-2). Additionally, many exoplanetary systems host several terrestrial planets and no known giants (e.g., the TRAPPIST-1 system; Gillon et al. 2017). A key question is whether the apparent prevalence of close-in giant planets and the existence of systems without giant planets is primarily owing to bias associated with current detection methods, or whether it reflects formation and evolutionary processes distinct from those in the solar system

In the solar system, the dynamical properties of the trans-neptunian objects provide strong evidence for giant planet orbital migration. Multiple types of gravitational interactions are now known to be capable of driving large-scale changes in giant planet orbits, including interactions with the gas disk, interactions with a planetesimal disk, and scattering. This migration in turn may affect the dynamical properties of small body populations. Exoplanetary systems offer a unique dataset to study such processes and improve understanding of the conditions that establish the overall properties of a planetary system, including when giant planets survive versus are lost to collision with their host star or ejection.

Strategic Research for Q12.2

- **Determine how the nascent planets acquired material from the protosolar disk** by measuring abundances and isotopic compositions of noble gases and other key elements (e.g., H, C, O, N, and S) from major planets (e.g., Venus, Mars, and the giant planets), satellites (e.g., Titan), and small bodies via spacecraft data, in situ probes, sample return, and telescopic observations.
- **Improve knowledge of the inventory, composition, and dynamical states of small bodies in the outer solar system** with spacecraft flybys and telescopic observations.
- **Determine noble gas abundances and isotope ratios, and stable isotope ratios (e.g., $^{12}C/^{13}C$ and $^{14}N/^{15}N$) in multiple comets** with spacecraft flybys, sample return, and telescopic observations.
- **Reveal the nature of outer planet accretion in exoplanetary systems** through an observational census of planetary dynamical parameters (e.g., eccentricity and inclination distributions and period ratios), dust produced by planetesimal collisions, young planets recently formed, the demographics of planets in the regions of exoplanetary systems analogous to the solar system, the composition of giant exoplanet atmospheres, and mass-radius relationships for sub-Neptunes.
- **Assess how giant planets may form and migrate** using theoretical and modeling studies constrained by diverse exoplanetary system properties.

Q12.3 ORIGIN OF EARTH AND INNER SOLAR SYSTEM BODIES

Understanding the origin of our inner planets benefits from observations of rocky exoplanets and studies of the mechanisms controlling the final architecture of planetary systems. Remarkably, the proportion of exoplanetary systems observed thus far that have both rocky and gas giant planets appears to be low. Several lines of exploration within the solar system and of exoplanet systems will help to answer fundamental questions about rocky planet formation.

Q12.3a How Do Disk Conditions and Stellar Metallicity Affect Rocky Planet Origin, Composition, and Volatile Content?

Large planets (≥ 4 $R_\oplus$), sub-Neptunes (radii between roughly 1.8 and 4 $R_\oplus$), and rocky planets having radii ≤ 1.8 $R_\oplus$ may occur more frequently around stars that are rich in elements heavier than helium, although the statistics of exoplanet populations remains an area of active research. In the solar system, the spectral composition of the Sun matches the overall composition of elements heavier than helium in primitive meteorites. Whether stellar composition in general is a proxy for planet composition can be begun to be assessed by measuring volatile-poor and volatile-rich exoplanet bulk densities.

Earth's composition and volatile inventories have been essential to its habitability and the development of life (Question 9, Chapter 12), and volatile content varies across the terrestrial planets. Do such differences in abundances of materials with relatively low condensation temperatures, such as those rich in S, O, and K, reflect nebular conditions as these bodies formed, or do they instead reflect stochastic events, for example, mixing induced by giant planet migration? Open issues include whether extrasolar systems without gas giants have volatile-rich inner planets, and the circumstances in which inner planets may be destroyed by giant planet migration and/or dynamical instabilities. For example, how similar are the compositions of the five planets in the TRAPPIST-1 system (see Figure 15-1, panel l), which has no known giant planets, and how do they compare with planets in the Cancri 55 system, which has both rocky and giant planets? Atmospheric emission, and/or thermal emission and reflectance spectra may constrain models of the interior compositions of rocky exoplanets—and enable comparisons between the compositions of these exoplanets and the photospheric abundances of their host stars.

Q12.3b Are Giant Impacts and Magma Oceans Common During Rocky Exoplanet Formation, and If So, What Are Their Observational Signatures?

Models of terrestrial planet accretion predict a final stage of giant impacts during the assembly of Earth-sized planets, including the impact thought to have produced the Earth-Moon system. These singular events establish initial planetary properties including rotation rate, obliquity, and the presence of an impact-generated moon(s). Giant impacts also deliver prodigious heat, and the terrestrial planets were likely melted one or more times during their formation, yielding periodic magma oceans. The complex chemical and dynamical aspects of magma ocean cooling are central elements of planetary formation and differentiation processes (Q3.5).

Observations can help further reveal the role of these processes in exoplanetary systems. Giant impacts can produce highly heated planet-disk systems with vaporized silicate that persist for $\sim 10^2$ years, followed by a protracted, $\sim 10^6$-years phase in which the planet retains a hot magma ocean beneath a volatile-rich, blanketing atmosphere (Zahnle et al. 2015). Giant atmosphere-stripping impacts might produce temporary incandescence and/or atmospheric signatures that could potentially be observed (e.g., observations of Earth shortly after the Moon-forming event would reveal a hot Earth in the habitable zone). Thermal emission from such structures may be observable in surveys of young disks. Detecting exomoons around older, rocky exoplanets may become possible (see Q12.8b).

Young terrestrial planets with magma oceans and early flotation crusts may have detectable surface and/or atmospheric properties. Giant impacts during planet formation may also be inferred from anomalous rocky planet bulk densities (perhaps analogous to metal-rich Mercury or Psyche), or through detection of impact-produced debris (see Q12.4).

Q12.3c What Processes Produce Compact Exoplanet Systems, and How Do They Differ from Conditions in the Inner Solar System That Yielded No Surviving Planets Interior to Mercury?

The existence of compact systems of Earth to sub-Neptune sized planets has been one of the most intriguing discoveries of the Kepler mission. These systems (e.g., Kepler-11 or TRAPPIST-1) typically have multiple planets on low-eccentricity, low-inclination orbits, all located at distances between about 0.01 to 0.3 AU, and with orbital periods of only a few to 100 days. Compact systems are common, occurring around tens of percent of M-dwarf stars, and they differ from the solar system in two key respects. First, they include many planets with estimated masses greater than 1 and less than 10 Earth masses, which have no analog in the solar system. Second, they are remarkably compact, with multiple planets interior to Mercury's orbit, perhaps analogous to the compact orbits of the Galilean satellites around Jupiter. Such characteristics may suggest different accretional conditions than in the solar system. Indeed, forming compact systems may require very massive compact disks and/or inward planet migration. Unraveling the conditions that led to compact systems will help us better understand the origin of the inner solar system, and perhaps why there are no planets interior to Mercury today.

Strategic Research for Q12.3

- **Determine noble gas abundances and isotope ratios, and the stable isotope ratios in the atmospheres of Venus and Mars** with spacecraft observations and in situ probes.
- **Characterize exoplanets smaller than sub-Neptunes, including their mass–radius relations, surface or atmospheric emission, and orbital period ratios** with telescopic observations and modeling studies.
- **Determine how inner exoplanetary systems with and without giant planets form and evolve** using compositional analyses of a wide range of planetary material and theoretical and modeling studies, including detailed predictions for young exoplanets where observational signatures of giant impacts and/or magma oceans could be detected.
- **Assess how compact exoplanetary systems originated and how their origin conditions differed from accretion in the inner solar system** by combining observational constraints (e.g., mass–radius relations and orbital period ratios) with theoretical models.

Q12.4 IMPACTS AND DYNAMICS

The dynamical properties of exoplanets have important implications for their overall stability and potential for habitability. Surveys of thousands of exoplanets show a wide diversity from hot-Jupiters to super-Earths. Understanding the history of these systems compared to the solar system has important implications for planetary formation, evolution, and habitability.

Q12.4a How Do Exoplanet Properties Constrain Their Collisional and Dynamical Histories, and What Can These Properties Tell Us About the Dynamical Histories of Multi-Planet Systems?

There are >700 known multi-planet systems, most more tightly packed than the solar system with multiple planets orbiting close to their parent stars, as perhaps expected given observational biases. They likely experienced different dynamical and collisional evolutionary histories with important implications for the nature of the planet's surface, atmosphere dynamics, and the bulk composition of a planet (e.g., Bonomo et al. 2019; Q12.3). A more comprehensive picture of such processes and their influence on planets requires a census of the physical properties of exoplanets, such as size, mass, composition, orbital eccentricity, and, if possible, spin state (e.g., Gratia and Lissauer 2021). Dynamical history is also constrained by orbital properties, for example, the prevalence of mean-motion resonances among planets. Understanding such features in multi-planet systems is crucial for understanding orbital architecture, giant planet migration, planet–planet interactions, and long-term system evolution (e.g., Tamayo et al. 2020).

Q12.4b How Does Impact History Influence Planetary Bulk and Atmosphere Composition?

The surface conditions of a planet are significantly affected by its susceptibility to impacts, which can potentially strip a planet of its atmosphere or add volatile components (e.g., Hirschmann et al. 2009; Schlichting and Mukhopadhyay 2018). The heterogeneity of material in the asteroid belt points to radial mixing (e.g., DeMeo and Carry 2014), perhaps associated with giant planet orbital migration that triggered delivery of volatile-rich material from the outer to the inner solar system (e.g., Gomes et al. 2005; Walsh et al. 2011). Volatiles are a crucial factor for habitability and life.

Giant planet migration is likely a common process given the prevalence of hot Jupiters, but the extent of radial mixing and volatile delivery to the habitable zone may differ significantly from system to system. Collisions associated with giant planet migration may have produced substantial dust, which could potentially be observed using multi-wavelength analyses (e.g., Youdin and Rieke 2017). Analyses of dust around exoplanetary systems at different evolutionary stages may shed light on the influence of impacts induced by giant planet migration.

Strategic Research for Q12.4

- **Determine the physical and orbital properties of exoplanets in multi-planet systems to understand their dynamical histories** with telescopic observations.
- **Characterize impact conditions that would produce detectable dust signatures in exoplanetary systems** with numerical simulations and models.
- **Determine how impacts contribute volatiles to (or, in some cases, remove volatiles from) planetary bodies** via compositional analyses of a wide range of planetary materials, especially volatile-rich, primordial small bodies in the outer solar system, and by modeling.

Q12.5 SOLID BODY INTERIORS AND SURFACES

Exoplanets offer a huge sample size to provide statistical information about the factors that govern planetary evolution. An overarching goal for the next decade is to characterize first-order properties of solid exoplanet interiors and surfaces. Interdisciplinary connections between experimental, theoretical, and modeling studies will be key to developing "exogeoscience."

Q12.5a What Are Observable Signatures of Bulk Composition and Surface Processes on Solid Exoplanets?

Characterizing exoplanetary surfaces and interiors is challenging, especially if they are shrouded in (perhaps cloudy) atmospheres. Solid bodies that are, for example, extremely close to their parent star might have bare surfaces that are amenable to direct observation. "Polluted" white dwarfs can reveal the bulk compositions of solid bodies that have been accreted by their parent star (e.g., Figure 15-3). Important open issues are how to assess exoplanet bulk and surface compositions (beyond the use of just mean density), the potential for characterization of volcanic activity and/or crustal recycling and assessing observable signs of atmosphere–surface interactions that control the cycling of volatiles through the planetary system.

Q12.5b What Are the Operative Modes of Solid-State Dynamics in the Rocky and/ or Icy Interiors of Exoplanets and Their Observational Signatures?

Rocky bodies are engines that convert internal heat into interesting surface phenomena (e.g., Foley et al. 2020). Their solid components can be stagnant and transport heat mostly by thermal conduction with little, if any, melting and recycling of surface material. Alternatively, solid-state convection enables more rapid cooling and faster transport of volatiles into and out of the interior (see Q5.2). Important issues for exoplanets include the factors that control whether rocky bodies exhibit Earth-like plate tectonics or other modes of mantle convection (e.g., the stagnant-lid regime for Mars today versus the poorly understood regime of mantle dynamics seen on Venus)—and the dynamical regimes of solid bodies with large proportions of ices and/or liquid water. Modeling efforts are needed to predict the observable consequences of different dynamical regimes.

Q12.5c What Factors Control the Prevalence of Dynamos in Exoplanets Smaller Than Sub-Neptunes?

Magnetic fields are unique windows into planetary interiors. There are no unambiguous patterns in the solar system governing which solid bodies host long-lived dynamos and the magnetic histories of some worlds, especially Venus, are poorly known (Laneuville et al. 2020). Are there correlations between the occurrence rates of expected exoplanetary dynamos with, for example, planet–star distance, stellar composition, system age, planetary mass, planetary rotation rate, or planetary atmospheric properties? Metallic cores are considered the usual location of dynamos, but liquid silicates have sufficient electrical conductivity under high pressure/temperature conditions, suggesting that dynamos could also be generated in magma oceans of, for example, super-Earths. Currently, brown dwarfs and gas giants serve as testing grounds for observations of, for example, radio emission and star-planet interactions related to magnetic fields. Eventually, these techniques can be refined and extended to search for magnetic fields around smaller exoplanets.

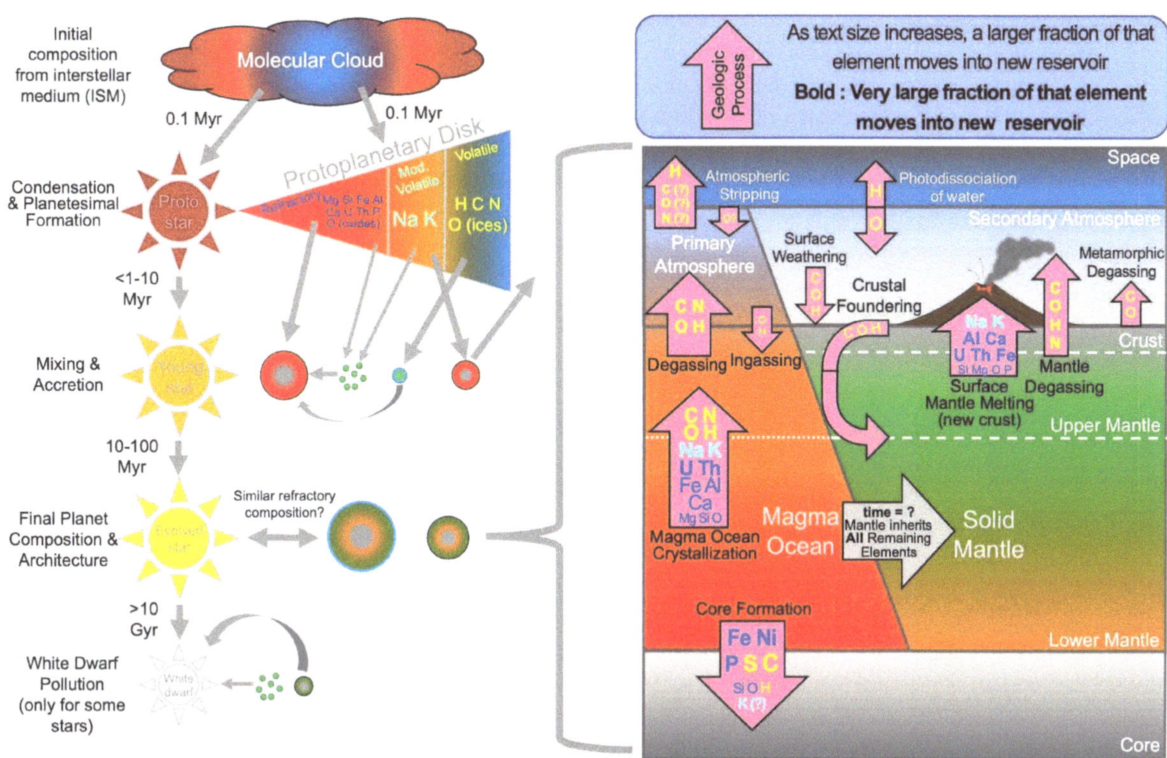

FIGURE 15-3 Origin of solid bodies (Q12.3) and their evolution (Q12.5) from birth to death. Atmospheres and interiors are linked in a planetary system. SOURCE: C. Unterborn, L. Schaefer, and S. Krijt, 2020, "The Composition of Rocky Planets," Pp. 5-1–5-52 in *Planetary Diversity*, E.J. Tasker, C. Unterborn, M. Laneuville, Y. Fujii, S.J. Desch, and H.E. Hartnett, eds., Bristol, England: IOP Publishing. Copyright 2020. Reproduced with permission of the licensor through PLSclear.

Q12.5d How Do Thermophysical Properties of Planetary Materials Affect the Interior and Surface Evolution of Solid Bodies?

Solid bodies in the solar system vary considerably in bulk compositions and internal pressures and temperatures. Exoplanets span an even broader range of conditions (Unterborn et al. 2020). Properties of planetary interiors are typically quite different from those observed at ambient conditions on Earth's surface. Some key quantities (e.g., the viscosity of solid silicates at extreme pressures) are uncertain by many orders of magnitude. To build realistic models of solid-body exoplanet evolution, accurate mineral physics data relevant to exoplanet conditions are needed. Ideally, data-collection efforts would be integrated in close collaboration with modeling and observational studies to build a framework for interpreting observations and generating new, testable predictions.

Strategic Research for Q12.5

- **Characterize the surfaces and interiors of solid body exoplanets** by constraining key properties (e.g., bulk composition, rotation rate, and atmospheric signatures of surface and interior conditions) with telescopic observations.
- **Search for magnetospheric activity at exoplanets** with remote sensing of phenomena associated with magnetic fields, potentially including radio emission, far-ultraviolet auroral emission, infrared H_3^+ auroral emission, variations in transit light curves, and star–planet interactions.

- **Determine how solid bodies over a broad range of conditions relevant to exoplanets evolve through geologic time** with modeling and theoretical research on processes including, for example, solidification of magma oceans; solid-state mantle convection; thermally and chemically driven dynamics in metallic cores; and dynamos generated in liquid metals and silicates.
- **Determine thermophysical properties of ices, silicates, and metal alloys under the ranges of pressure and temperature conditions and compositions relevant to exoplanets** with first-principles simulations and laboratory experiments.

Q12.6 ATMOSPHERE AND CLIMATE EVOLUTION ON SOLID BODIES

The present-day atmospheres of solid bodies in the solar system are snapshots in time of continuously evolving systems. While continued research in the solar system is essential to understand this evolution (see Q.6, Chapter 9), exoplanets will extend our understanding of the diversity of planetary atmospheres across space as well as time. Today, observational constraints on the atmospheres of small exoplanets are limited, but over the next few years major progress is expected. A coordinated program of solar system and exoplanet atmospheric evolution research would maximize advance in understanding.

Q12.6a What Determines the Division Between Gas-Rich and Solid-Body Planets?

In the solar system, there is a clear distinction between the four giant planets with puffy hydrogen-dominated atmospheres and all other solid planetary bodies. However, observations of exoplanet radii for planets intermediate in mass between Neptune and Earth indicate that this boundary is not as clear in many cases (Fulton et al. 2017 and see Figure 12-1k). The gaseous component of planets is thought to be established both by early nebular capture and longer-term loss processes, and continued research is required to understand this interplay. Observations of systems of different ages are particularly important to discriminate between different evolutionary possibilities. These issues are also relevant to understanding the early evolution of rocky planets in the solar system (see Q6.1, Chapter 9).

Q12.6b Which Rocky Exoplanets Have Retained Atmospheres? Can We Use This Information to Constrain Atmospheric Loss History on Solar System Objects?

Once an exoplanet has been identified as solid, based on its mass and radius, perhaps the next most basic question is whether it possesses an atmosphere. Initial constraints on rocky exoplanet atmospheres have been achieved using existing observatories (e.g., Diamond-Lowe et al. 2018; Kreidberg et al. 2019), but tests for the presence of atmospheres on a much wider range of objects in the next decade are expected as new space- and ground-based observatories come online. Such observations will test and constrain theories of solid-body atmospheric loss, in turn helping us better understand the evolution of solar system objects with thin atmospheres today, such as Mars.

Q12.6c For Exoplanets That Have Retained Atmospheres, What Are Trends in Atmospheric Composition with Planet Mass, Orbital Distance, and Host Star Type?

In the next decade, spectroscopic characterization of solid body exoplanets that have atmospheres will allow us to begin to identify trends with key parameters such as orbital distance and planet mass. In the solar system, the divergent evolution of Venus, Earth, and Mars is clearly a result of their differing bulk properties, but with such a small sample size, understanding which properties are most crucial is extremely difficult. While the data on any individual exoplanet will be extremely limited initially, the ability to observe a large number of targets would provide completely new constraints on atmospheric evolution.

Q12.6d How Does the Evolution of the Host Star Affect Planetary Atmospheres, Including Photochemistry and Escape Processes?

The Sun's changing output has had a dramatic impact on atmospheric evolution in the solar system, and the effect of host stars on exoplanets is likely to be similarly important. Indeed, for exoplanets orbiting low-mass M stars (the most observable type in the next decade), atmospheric loss is likely to be much more severe than in the solar system, owing to the high extreme ultraviolet output, frequent coronal mass ejection, and long pre-main sequence stage of these stars (Baraffe et al. 1998; Tarter et al. 2007). High-energy stellar emissions are also important drivers of atmospheric chemistry, both in the steady state and episodically owing to transient events. Characterizing host star properties and exoplanet atmospheric composition simultaneously would provide important new data on the coupling between stellar radiation and atmospheric chemistry in a range of different contexts, strengthening our understanding of processes in the solar system.

Q12.6e What Processes Impact the Evolution of Atmospheric Chemistry, Cloud, and Haze Formation in Diverse Planetary Atmospheres?

The number of planetary atmospheres in the solar system is limited, but even so great diversity in composition and chemistry is seen. By observing exoplanets, we can begin to build and test generalized models of atmospheric chemical pathways important in maintaining any atmospheric composition. Besides the stellar effects mentioned above, key questions include how surface processes and trace gas species affect disequilibrium chemistry and how cloud and photochemical haze formation proceed in diverse situations and impact the atmospheric energy budget.

Q12.6f What Is the Diversity of Atmospheric Circulation Patterns Among Solid Planets, and How Does Circulation Vary with Atmospheric and Planetary Properties?

Atmospheric circulation is known to vary strongly with atmospheric mass and composition, planetary rotation rate, received solar flux, and other factors. While constraining atmospheric circulation on solid-body exoplanets observationally is extremely challenging, future progress may be possible via techniques such as thermal phase curve analysis and high-resolution Doppler spectroscopy (Snellen et al. 2010; Showman et al. 2015; Kreidberg et al. 2019; see Figures 15-1h and 15-1i). Many characterizable solid-body exoplanets are in close orbits around low-mass stars, which means they may have low rotation rates and permanent day and night sides owing to tidal locking. The circulation on such planets is predicted to be very different from that of Earth, Venus, Mars, and Titan, and so studying them in parallel with solar system objects is likely to lead to rich insights into atmospheric circulation generally.

Strategic Research for Q12.6

- **Characterize the atmospheres of solid-body exoplanets** by conducting transit spectroscopy, high-dispersion spectroscopy, and thermal phase curve observations, and compare them with atmospheres of solid bodies in the solar system.
- **Determine past atmospheric mass and composition in the solar system** by measuring and/or collecting noble gas abundance and isotopic fractionation from solid- body atmospheres within the solar system (i.e., for Venus, Mars, and Titan).
- **Determine the properties of the atmospheres of terrestrial planets (i.e., Earth, Venus, and Mars) that would be observable on exoplanets to build a foundation for atmospheric characterization of analogue exoplanets** through coordinating in situ/remote sensing measurements and theoretical studies of wind velocities, radiative balance, cloud dynamics, and atmospheric compositing as function of orbital phase, local time, and solar conditions.

- **Determine the connection between exoplanet observables and atmospheric properties and dynamics by conducting theoretical and modeling studies** to include the following: simulations (1D and 3D) with hazes and clouds; radiative-microphysical feedbacks; volatile transfer between atmospheres and surfaces; and interactions with the solar wind including the influence of magnetic fields on atmospheric escape processes.
- **Determine key radiative properties, gas absorption, and other quantities of interest to understand feedbacks on planetary atmospheres for the solar system and exoplanets** through targeted laboratory studies, including of atmospheres with different primary constituents (e.g., N_2, CO_2, and CH_4) and temperatures.

Q12.7 GIANT PLANET STRUCTURE AND EVOLUTION

The interior structure of a giant planet is the result of both formation and evolution processes. For many giant exoplanets, both mass and radius have been measured, providing estimates of the density of these planets. In some cases, molecules have been observed in the atmosphere, providing hints to the planet's bulk composition. This information enables new modeling of interior structure and evolution, leveraging what is understood about the giant planets in the solar system. Although extrasolar Neptune-mass planets could have very different bulk compositions and interior profiles from Neptune and Uranus, our current limited knowledge of them strongly limits our ability to even begin to understand one of the most commonly detected classes of exoplanets: sub-Neptunes (see Figure 15-1k). Additionally, current understanding is limited by the lack of solar system analogs for some types of exoplanets (e.g., hot Jupiters and super-Earths). Dedicated missions to answer fundamental questions about Uranus and Neptune will provide a needed basis to significantly advance our understanding of this class of exoplanets.

Q12.7a What Does the Discovery of Dilute Cores Inside Saturn and Jupiter Mean for the Interior Structures of Giant Exoplanets? Can We Expect the Cores of Giant Planets Closer in Size to Uranus and Neptune to Be Compact or Dilute?

The internal structures of Jupiter and Saturn show the cores are diffuse—that is, they are partially diluted into the outer layers (Wahl et al. 2017; Mankovich and Fuller 2021). However, modeling of giant exoplanets to date has assumed compact cores. Furthermore, we do not know whether Uranus and Neptune also are partially or fully differentiated (see Q7.1, Q7.2, and Q7.4, Chapter 10). Future studies should address what the interior structures of the solar system's giant planets can tell us about the internal structure, formation, and evolution of giant exoplanets.

Q12.7b How Did Thermal Evolution of the Giant Planets in the Solar System Progress Over Time and What Can This Tell Us About the Current State and Future of Directly Imaged Young Exoplanets?

Recently formed giant exoplanets radiate their heat of formation, making them the easiest planets to directly image thanks to the planet's own emitted infrared radiation. Nevertheless, one needs to separate the radiation from the planet from that of the much brighter star, as well as suppress the diffracted light from the star. The thermal evolution of the giant planets in the solar system is important for interpreting such observations and their connections to future direct imaging campaigns (Berardo et al. 2017). Key open issues include initial planet entropy (e.g., hot versus cold start models), how quickly a young giant planet cools, differences in the thermal evolution of gas versus ice giants, and whether the difference in intrinsic heat between Uranus and Neptune is common among exoplanets of this size, and if so, what this implies for understanding the intrinsic heat flux of hydrogen-dominated planets (see Q7.5).

Q12.7c What Can Uranus and Neptune Teach Us About Magnetic Field Generation and Configuration in Neptune-Size Exoplanets?

The magnetic fields of Uranus and Neptune are radically different from those of Jupiter and Saturn and from each other, suggesting that planets of this size may display a wide diversity of interior structures and magnetic fields (e.g., Soderlund and Stanley 2020; see also Q7.2 and Q7.4). Where the magnetic field is generated within

Uranus and Neptune and what this can tell us about sub-Neptune magnetic field generation remains uncertain, owing to very incomplete knowledge of the Uranus and Neptune field configurations and properties.

Q12.7d What Factors in the Formation and Evolution of Planets Define the Crossover Regime Where a Planet Becomes Either a Super-Earth or a Sub-Neptune?

As noted in Q12.2, the most commonly detected exoplanets are super-Earths and sub-Neptunes (see Figure 15-1k and Fulton et al. 2017). Is the observed "gap" in intermediate planet radii for close-in exoplanets the result of formation processes, evolution, or a combination of both? One possibility to explain this gap is that super-Earths have lost their primary atmospheres (or perhaps never accreted one) while sub-Neptunes retained theirs. Better understanding of the formation and evolution of Uranus and Neptune are critical for answering this question (see Q7.1, Q7.2, and Q7.5). An overall issue is over what range of planetary masses, stellar insolation, and formation timescales are planets able to retain hydrogen-rich atmospheres, accounting for competing processes such as outgassing, accretion, and atmospheric escape.

Q12.7e How Are Heat and Chemicals Transported in the Atmospheres of Uranus and Neptune, and What Does This Mean for Interpreting Future Spectra of Spatially Unresolved Exoplanets with Hydrogen Atmospheres?

Future observations of giant exoplanets will seek to constrain the bulk composition of the atmosphere (see Figures 15-1g and 15-1j), and to understand atmospheric processes at work (e.g., Madhusudhan 2019). Both are critical for determining how these planets formed and evolved and for understanding processes currently at work in their atmospheres. Most of our understanding of hydrogen-dominated atmospheres comes from Jupiter and Saturn. However, Uranus and Neptune are very different, and many questions remain about their atmospheric processes (see Q7.2, Q7.3, and Q7.4). A better understanding of the atmospheres of Uranus and Neptune is required to interpret future spectral observations of sub-Neptunes.

Strategic Research for Q12.7

- **Determine the most likely formation mechanism and interior structure of Saturn, Uranus, and Neptune** by measuring noble gas abundances and isotope ratios, stable isotope ratios (e.g., $^{12}C/^{13}C$, $^{14}N/^{15}N$, and the noble gas isotopes) in their envelopes, and the bulk composition and interior structure of Uranus and Neptune with spacecraft observations and in situ probes.
- **Determine the current thermal state and variability of Jupiter, Saturn, Uranus, and Neptune** by conducting long-time duration thermal infrared observations of these four planets with spacecraft and telescope observations.
- **Determine the contribution of solid materials to the giant planets** by measuring cometary and interstellar object noble gas abundances and isotope ratios, and stable isotope ratios (e.g., $^{12}C/^{13}C$, $^{14}N/^{15}N$, and the noble gas isotopes) in multiple comets by spacecraft flybys and telescopic observations.
- **Create a census of a large population of young planets recently formed, of the composition of giant exoplanet atmospheres, of magnetospheric activity in exoplanets, and of mass-radius relations of sub-Neptunes** by radio and telescopic observations.
- **Observe planets forming within a disk** by telescopic observations.

Q12.8 CIRCUMPLANETARY SYSTEMS

The solar system displays a myriad of circumplanetary systems around both gas and solid planets. The giant planet systems demonstrate that even when planets themselves are not habitable, their moons may be. Moons may not only feed material into their ring systems, but may also "shepherd" rings, clear gaps, and/or gravitationally confine ring material. Detection and characterization of exomoons and circumplanetary disks would enhance our understanding of the formation, evolution, and lifetime of circumplanetary systems, as well as the potential for habitable exomoons.

Q12.8a What Can Observations of Circumplanetary Material Reveal About the Formation and Evolution of Circumplanetary Systems?

The composition of rings and moons hinges on conditions in their precursor circumplanetary disks, as well as their subsequent evolution. In the solar system, protosatellite disks are thought to be produced by two general processes: giant impacts (for solid planets) and gas co-accretion (for giant planets). Uranus may represent a case in which both processes were involved in the origin of the current satellite system, and Neptune likely provides an example of an original satellite system that was largely destroyed by the intact capture of a large and retrograde orbiting moon, Triton. Characterization of circumplanetary disks in exoplanetary systems and detection of exomoons would greatly expand our knowledge of ring and moon origin and evolution, including both an improved understanding of circumplanetary disk properties (e.g., radial scale, density, and lifetime) and of the balance of moon formation versus loss. A recent breakthrough has been the clear detection of a circumplanetary disk orbiting a gas giant in a 10-million-year-old stellar system, with a disk diameter of roughly 1 AU (Benisty et al. 2021).

Q12.8b How Common Are Exomoons and What Are Their Properties?

Exomoon detection is extremely technically challenging, requiring, for example, very accurate photometric light curves and long-baseline observations in combination with numerical data analyses. To date, the primary results have been upper limits on exomoon-to-planet mass ratios, in addition to some inconclusive data (e.g., for Kepler-1625b; Kreidberg et al. 2019; Teachey et al. 2020). Microlensing techniques to be used by the Roman Space Telescope could provide detections of exomoons (Liebig and Wambsganss 2010). The presence of volcanically active exomoons might be inferred from enhancements of elements such as sodium and potassium in the spectra of the giant planets that they orbit (Oza et al. 2019), while the presence of cryovolcanially active exomoons could be detected from enhancements in hydrogen, oxygen, and/or water vapor, as suggested by cryovolcanically active moons in the solar system (e.g., Quick et al. 2020). While similar moon formation mechanisms to those envisioned for the solar system are expected in exoplanetary systems, there may be notable differences too—for example, where moon loss may be more important for exoplanets in compact orbits and/or that have undergone large-scale orbital migration.

Strategic Research for Q12.8

- **Detect and constrain the properties of circumplanetary disks and exomoons** using telescopic observations and theoretical modeling.
- **Constrain the formation and lifetime of exoplanet ring and moon systems** via telescopic observations, and by comparing these observations with spacecraft and long-term telescopic observations of the ring and moon systems of Jupiter, Saturn, Uranus, and Neptune, and by theoretical modeling for ring and satellite formation and evolution.

Q12.9 INSIGHTS FROM TERRESTRIAL LIFE

As the only known habitable and inhabited world, Earth offers invaluable insights into life's requirements, how life coevolves with its environment, and the remotely observable signatures of a global biosphere. But Earth would have looked quite different at different times in its history. Atmospheric O_2, for example, only reached its present high level (21 percent by volume) sometime within the past 500 million years. And prior to ~2.4 Ga, O_2 would have been nearly absent from the lower atmosphere, despite Earth having been inhabited for at least a billion years before then. Early Earth is thus our best example of a planet that supported life but that would have borne little resemblance spectroscopically to the world we inhabit today. Studying the history of life on Earth is fundamentally important for determining whether exoplanets provide habitable environments and potentially support life similar to what once existed or now exists on Earth.

Q12.9a Which Among the Necessary Ingredients and Conditions Required by Life on Earth Can Be Inferred or Detected on Exoplanets?

All known life requires the presence of liquid water as a solvent in which oxidation and reduction reactions can occur. It also requires the presence of carbon-based macromolecules, metals, and trace elements, along with energy to drive metabolisms (e.g., Baross et al. 2020). Even more fundamentally, life requires a stable pressure-temperature environment in which it might originate. Imaginative researchers have proposed that organisms could exist at certain heights within giant planet atmospheres if they had mechanisms, like air bladders in certain fish, to maintain their altitude (Sagan 1995). But such organisms would have to be highly evolved, multicellular organisms. If life originated as single-celled organisms, as is widely believed, it would not have been able to do this; instead, any such organisms would have been wafted up into the cold upper reaches of the planet's atmosphere or down into the hot interior. Hence, the most basic requirement for a habitable planet is the presence of a solid (or liquid) surface. Gas (or ice) giants can be effectively ruled out.

While in situ searches for life in the solar system may be able to seek evidence of all these properties in detail and on local scales, observations of exoplanets will be limited to globally observable planetary properties. In the solar system, liquid water and thus habitable environments are found in the subsurface of icy bodies in the outer solar system (e.g., Europa and Enceladus). For exoplanets, the boundaries of the habitable zone are generally premised on a requirement of *surface* liquid water (Kasting et al. 1993, 2014). Planets within that zone can maintain an active photosynthetically based biosphere that is capable of altering the planet's atmosphere in a way that is remotely detectable. More generally, habitability depends on a complex web of planetary, stellar, and planetary system parameters (e.g., Meadows and Barnes 2018; Q10.1 and Q12.10), and the environmental limits of life on Earth are still under investigation (Question 9). Understanding these properties, and which can be remotely observed or inferred, is imperative for defining the search space for potentially habitable exoplanets. While the molecular building blocks and required chemicals for life may themselves be unobservable on exoplanets, other observable phenomena related to life's requirements might be sought. For example, ultraviolet radiation is thought to play an important role in synthesis of molecules related to the origin of life (e.g., amino acids and ribonucleotides), and insights from synthesis of terrestrial biomolecules and prebiotic molecules under varied ultraviolet irradiation may provide insight for exoplanets under varied stellar spectra (e.g., Ranjan and Sasselov 2016, 2017). As another example, dry-wet cycles on Earth are one method known to aid polymerization of organic molecules, and while such molecules are unlikely to be observed on an exoplanet, land masses on which such cycles could occur may be observable through diurnal light/color curves (Cowan et al. 2009). Detections of such properties would not prove that a planet is habitable (and, conversely, their nondetection would not prove that it is uninhabitable), but they would provide additional context upon which to interpret potential biosignatures.

Q.12.9b What Biosignatures Can Be Sought on Exoplanets Analogous to Earth, Including Past Phases of Earth History?

The metabolisms powering life on Earth may produce detectable byproducts, which could be sought as biosignatures in exoplanet atmospheres with future telescopes. Additionally, life on Earth can produce surface reflectance biosignatures (e.g., the sharp increase in reflectivity produced by plants known as the "red edge"), and analogous features might be observable in exoplanet spectra (e.g., Schwieterman et al. 2018). A thorough understanding of the biosignatures on Earth through time can guide our understanding of the types of features that could be observed on habitable worlds—and also the observational requirements to detect these features. For instance, oxygenic photosynthesis is the dominant metabolism on modern Earth. Oxygen and its photochemical by-product ozone (O_3) produce prominent spectral features that could be sought at ultraviolet, visible, and infrared wavelengths in the spectra of modern Earth analog planets (Meadows et al. 2018). However, over geological history, Earth's observable biosignatures have varied considerably. For instance, Archean Earth (~4–2.5 Ga) had an anoxic atmosphere, but other biosignatures could still be sought, (e.g., methane) (Kaltenegger et al. 2007).

Q12.9c What Can We Learn About False Positive and False Negative Detection of Life on Exoplanets from Earth History?

Environmental contextual information will be critical to distinguish true biosignatures from abiotic "false positive" mimics. For example, while most of Earth's methane is biological, abiotic methane is also produced (e.g., Etiope and Sherwood Lollar 2013), so methods of distinguishing true biological methane from abiotic methane on exoplanets are needed. More broadly, this type of analysis is needed for *all* biosignatures, including for planets with different environmental contexts than Earth. For instance, while oxygen has no significant abiotic sources on Earth, theoretical research has discovered several paths to abiotic oxygen formation that might occur on exoplanets (e.g., Meadows 2017). However, debate continues as to which of these might actually operate (Harman et al. 2018).

Earth history also presents examples of possible "false negatives"—that is, periods when life may be difficult to detect remotely. For example, the Mid-Proterozoic eon, 1.8–0.8 Ga, may have been a time when atmospheric O_2, O_3, and CH_4 may all have been too low to be detected by existing or planned space telescopes (Reinhard et al. 2017). For these periods, it is important to determine what biosignatures—if any—could be detected and the observing requirements to sense them so that similar exoplanets are not incorrectly excluded in the search for life.

Q12.9d What Are "Novel" Biosignatures Not Expressed in Earth's Spectrum Over Geologic Time That Might Be Detected on Exoplanets?

The biosignatures expressed by Earth through time are a function of not just what life produces, but also are dependent on the planetary and stellar environment. Different biosignatures might be more or less prominent for planets with different dominant metabolisms, orbiting different types of stars (which drive different types of photochemistry), with, for example, different levels of background atmospheric gases. To evaluate which of the diverse suite of molecules produced by life (e.g., Seager et al. 2016) could serve as biosignatures for exoplanets, laboratory, field, and theoretical work is needed to evaluate the plausibility of these biosignatures to survive and be detected in varied exoplanet environments. Ideally, one would wish to identify so-called agnostic biosignatures, that is, combinations of gases—Earthlike or not—that could only be produced by a biosphere (NASEM 2019). This remains an outstanding challenge for exoplanet astrobiologists.

Strategic Research for Q12.9

- **Determine the environmental requirements of life on Earth to inform the limits of habitability on exoplanets** through field and laboratory investigations, and studies of "extreme" forms of Earthly organisms (e.g., organisms that survive at very low or high temperatures or in high radiation environments).
- **Study the observable properties of Earth's modern biosphere** through field and laboratory studies that measure the features of Earth's biota accessible to remote sensing (e.g., reflectance spectra of pigments).
- **Observe and characterize modern Earth as an exoplanet analog** through remote sensing observations across a wide wavelength range, and via observing techniques that emulate exoplanet observing methods (e.g., direct imaging and/or transit observations, time resolved to examine temporal variability).
- **Assess the remotely observable properties (e.g., spectral indications of habitability and biosignatures) of an Earth-like planet across varied planet ages, and environmental and stellar conditions** through theoretical modeling and laboratory studies that emulate possible exoplanet conditions, informed by remote sensing and field work data.
- **Study the coevolution of life on Earth and its environment, including biosignatures and biosignature false positives and false negatives through time,** by conducting geological field studies that examine past epochs of Earth history.

Q12.10 DYNAMIC HABITABILITY

The geologic records of Earth, Mars, and Venus show us that planetary habitability is highly variable in space and time. Exoplanet observations soon will offer us the opportunity to address fundamental questions about how a planet's habitability depends on orbital distance, mass, stellar type and other factors. In turn, more detailed study of the dynamic habitability of solar system objects is essential to ensure that insights we gain from exoplanets are well-grounded. Because the presence or absence of an atmosphere is an important constraint on surface habitability, the questions in Q12.6 are highly relevant to this section. Airless exoplanets or exomoons may have subsurface habitable regions (e.g., Europa), but subsurface biospheres are unlikely to be remotely detectable, and so the current focus for exoplanets is on surface habitability.

Q12.10a How Can Solar System Objects Be Used to Determine the Boundaries of Exoplanet Habitability as a Function of Orbital Distance, Planetary Mass, and System Age?

Exoplanet orbital distance, mass, and age can be determined (with varying accuracy), and so assessing the likelihood of habitability as a function of these parameters is important for planning future observations of potentially habitable exoplanets. This will require generalization of insights from solar system planets, particularly Earth, Mars, and Venus. Earth and Venus have similar masses, but Earth's surface is habitable while Venus's is not. Does the inner edge of the habitable zone generally lie between the orbits of these two planets, or were there unique aspects to Venus's evolution (e.g., very early loss of water; Gillmann et al. 2009; Hamano et al. 2013) that were more important? Smaller Mars has a rich geologic record that indicates past surface habitability and atmospheric loss, illustrating the potential for dynamic and intermittent habitability in the early evolution of a planetary environment (e.g., Ehlmann et al. 2016).

Q12.10b What Constraints Can Be Placed on the Presence of Liquid Water on Exoplanets?

The composition of a planet's atmosphere determines surface temperatures via the greenhouse effect, which in turn determines whether surface liquid water can exist. After atmospheric characterization (Q12.6), the next step in assessing exoplanet habitability is to search for surface (or, if possible, subsurface) liquid water. Determining if oceans are present on exoplanets in the next decade will be difficult through transit spectroscopy that probes typically dry stratospheric altitudes, but indirect methods such as testing for the absence of water-soluble species such as sulfur dioxide in the upper atmosphere will allow constraints to be placed (e.g., Loftus et al. 2019). Glint may also be detectable from the reflection of starlight off exoplanetary oceans (e.g., Robinson et al. 2010). Understanding how many exoplanets possess water will allow us to better assess the evolution of oceans, lakes, and rivers on the solar system planets across geologic time, in a direct complement to the aims of Q12.10a. Last, techniques to search for other surface liquids (e.g., hydrocarbon lakes) would also be valuable to explore, in order to keep a broad view of the potential for habitability with solvents other than H_2O.

Q12.10c What External Factors Influence the Loss or Maintenance of Surface Habitability Over Time on Rocky-Type Exoplanets?

Besides direct observations, studying exoplanet habitability will also require observations and modeling of the factors that drive climate evolution on long timescales. A priority in this area is stellar emission: characterizing the high-energy radiation emitted from host stars (particularly extreme ultraviolet radiation and stellar winds) is vital to determining how rapidly exoplanet atmospheres are lost to space (Q12.6b), and what drives their chemistry (Shields et al. 2016). Other poorly understood factors of key importance for habitability include the dynamical and orbital history of a system, the flux of meteoroid impacts on planetary surfaces with time, and exchange of surface volatiles with the interior. Constraining all of these factors for a given exoplanet system based on observations alone will be difficult or impossible, so detailed study in the solar system to provide ground-truth validation for limited cases is important.

Q12.10d How Does Atmospheric Chemical Evolution Affect Habitability?

Atmospheric chemistry is critical to habitability and the emergence of life. The composition of a planet's atmosphere determines surface temperatures via the greenhouse effect, but it also determines the extent to which prebiotic chemistry can develop. Rich organic chemistry is possible in reducing atmospheres that are no longer H_2-dominated, while oxidizing atmospheres are hostile to the emergence and survival of primitive Earth-like life. Oxidation via hydrogen loss to space is extremely important to early atmospheric chemical evolution, as is exchange with the planet's interior. Studying the chemical state of exoplanet atmospheres will allow us to determine how common reducing, or weakly reducing, atmospheres are on young rocky planets. (A weakly reducing atmosphere is one dominated by N_2 and CO_2 but containing smaller amounts of reduced gases such as H_2 and CO.) This will have implications for understanding how life emerged on Earth, as well as whether it could also have emerged on early Mars and/or Venus. Last, the chemical characterization of exoplanets does not need to rely on direct analogies to Earth's history. Such an agnostic approach may ultimately allow insights into the possibility of life emerging under conditions very different from those of the early Earth.

Strategic Research for Q12.10

- **Determine the presence or absence of atmospheres on potentially habitable rocky exoplanets** via telescopic observations, including transit spectroscopy, thermal phase curve analyses, and reflected light spectroscopy (NASEM 2023), and via comparisons to solar system planets.
- **Determine the abundances of water-soluble species (e.g., SO_2 or NH_3) and aerosols in the atmospheres of potentially habitable exoplanets to constrain the presence of surface liquid water** via a combination of space- and ground-based transit spectroscopy.
- **Assess the frequency of planetary conditions conducive to prebiotic chemistry** by determining stellar activity, atmospheric chemistry (i.e., reducing versus oxidizing), and impact/dynamical history on rocky exoplanets via space- and ground-based transit spectroscopy and modeling.
- **Constrain the inner edge of the solar system's habitable zone by studying the surface geomorphology and geochemistry of Venus to assess whether it ever possessed oceans** via orbital radar observations, spectroscopy, in situ analysis, and associated chemical modeling.
- **Improve exoplanet habitability predictions for cold, low-mass planets by determining the key factors that made Mars habitable 3–4 Ga,** via a combination of in situ geological and atmospheric analysis and sample return, orbital observations, and climate modeling.
- **Determine if subsurface exoplanet biospheres could ever be detected remotely** via in situ and/or remote sensing study of potentially habitable "ocean worlds" and accompanying theory and modeling.

Q12.11 SEARCH FOR LIFE ELSEWHERE

Given limitations of exoplanet data, it is crucial to consider how the worlds of the solar system, including Earth (Q12.9) can help guide our search for life on exoplanets. Three major criteria important when searching for biosignatures are reliability (i.e., the likelihood of a potential biosignature being produced by life), survivability (i.e., the likelihood of a potential biosignature surviving long enough to be detected in the context of its environment), and detectability (i.e., the likelihood of actually being able to detect a given biosignature with a given technology) (e.g., NASEM 2019; Questions 9 and 11, Chapter 14). Relevant issues have also been discussed above in Q12.9 and Q12.10.

Q12.11a Can Formal Frameworks Be Devised for Interpreting
Biosignatures on Exoplanets, Given Their Unique Challenges?

Efforts are under way to devise formal frameworks for evaluating and interpreting biosignatures (e.g., Neveu et al. 2018; Question 9). Such frameworks are needed not only in the solar system—where claims of biosignatures have generated considerable discussion in the literature in recent decades (e.g., McKay et al. 1996)—but also for exoplanets, given the inherent limitations and challenges to observing worlds light years away.

Evaluating potential solar system biosignatures is a useful template to develop lessons for evaluating possible biosignatures in exoplanet atmospheres. As a highly irradiated planet, Venus may help us better understand false positive O_2 produced by water loss (that Venus may have experienced in the past) or through CO_2 photolysis (that Venus experiences today; NASEM 2019). Trace quantities of methane on Mars (Mumma et al. 2009; Webster et al. 2018), and phosphine on Venus (Greaves et al. 2020) have been claimed and suggested as possible biosignatures. However, continuing vigorous discourse (e.g., Zahnle et al. 2011; Snellen et al. 2020) on the presence and potential sources of these gases underscores the significant challenge we will face when evaluating possible biosignatures on distant planets. Differentiating true biosignatures from abiotic false positives critically hinges on understanding the context of their environments so that the plausibility of abiotic explanations can be evaluated, and their environments' habitability assessed.

Q12.11b Are Biosignatures Observable on Exoplanets in the Near Future?

Just as comparative planetology is the study of natural phenomena and processes across and between multiple worlds, the discovery of one or more inhabited planets beyond Earth would create the new science of comparative astrobiology. The launch of the James Webb Space Telescope (JWST) will provide a new window into terrestrial exoplanet atmospheres, but JWST's targets will be limited to transiting planets orbiting the lowest mass stars, M dwarfs. These stars often exhibit extreme levels of stellar activity, which may adversely impact the habitability of orbiting planets. However, M dwarfs comprise 75 percent of all stars in the galaxy, so understanding whether life can persist around them is critical to understanding the distribution of possible life in the galaxy, and they provide examples of planets with significantly different star-planet evolutionary histories compared to the worlds of the solar system. Beyond JWST, understanding whether there are biosignatures on planets orbiting around more massive "Sun-like stars" will require new types of facilities capable of suppressing their stars' light so that their orbiting planets can be observed directly, allowing us to study worlds (and possibly biospheres) in the context of systems with planet–star evolutionary histories more akin to our own. There are important synergies with the 6m, space-based ultraviolet/optical/near-infrared telescope capable of directly detecting and characterizing planets in reflected light recommended in NASEM (2021).

Strategic Research for Q12.11

- **Study methods to discriminate past and present false positive biosignatures on solar system bodies (e.g., abiotic O_2 on Venus and Mars) from true biosignatures to inform false positive discrimination methods for exoplanets** through in situ, remote sensing, theoretical/modeling studies, analog field research, and laboratory studies that characterize remotely observable properties of these features.
- **Study whether methods exist to remotely observe and correctly interpret biosignatures from subsurface biospheres (e.g., as on icy moons) or other potential habitable environments that might present false negative detections of life for exoplanets** through in situ, remote sensing, theoretical/modeling studies, analog field research, and laboratory studies that characterize remotely observable properties of these features.
- **Obtain an inventory of properties of solid body exoplanets (i.e., mass, composition, bulk atmospheric chemistry and abundance of clouds and hazes, potential biosignatures, rotation rates, relative distance from host star, and type of host star)** through telescopic observations including radial velocity measurements, transit spectroscopy, high-dispersion spectroscopy, thermal phase curve observations, secondary eclipse analyses, and, eventually, direct image spectroscopy.
- **Devise metrics and frameworks to establish confidence in interpretation of biosignatures in the solar system and exoplanetary systems,** informed by a synthesis of relevant in situ, remote sensing, theoretical modeling, field, and laboratory data.

SUPPORTIVE ACTIVITIES FOR QUESTION 12

- Observations of solar system planets and moons through transit spectroscopy and direct-imaging as analogs to exoplanet observations, including hemispherically averaged fluxes as a function of orbital phase and time; observations of particle and gas opacity in the giant planets and Venus as a function of phase angle to help determine the dependence of reflectivity and scattering on particles and clouds in exoplanet atmospheres; and ultraviolet-/near-infrared-scattered light observations from the poles of the giant planets for comparison with future direct imaging of giant exoplanets.
- A census of protoplanetary disks, young planets, and mature planetary systems across a wide range of planet–star separations to determine how the initial composition and conditions in a protoplanetary disk influence the diversity of resulting planets.
- Improved spatial resolution of telescopic techniques to determine variations in the structure and composition of circumstellar disks, as well as the next-generation telescopes recommended by NASEM (2021) that will allow for observations of circumplanetary disks, detection of exomoons and ring systems, and characterization of exoplanets around sun-like stars.
- Laboratory studies to understand the relationship between the bulk composition of a planet and its atmosphere, and to determine the optical properties of clouds and hazes relevant to exoplanet atmospheres.
- Increased interactions between the astronomy and planetary science and astrobiology communities (supported under, e.g., NASA's Planetary Science and Astrophysics divisions) are needed to maximize advances in exoplanetary science and to address the questions identified in this chapter. This point was emphasized in multiple white papers received by the committee.

REFERENCES

Andrews, S.M., J. Huang, L.M. Pérez, A. Isella, C.P. Dullemond, N.T. Kurtovic, V.V. Guzmán, et al. 2018. "The Disk Substructures at High Angular Resolution Project (DSHARP). I. Motivation, Sample, Calibration, and Overview." *Astrophysical Journal Letters* 869(2):L41.

Anglada-Escudé, G., M. Tuomi, P. Arriagada, M. Zechmeister, J.S. Jenkins, A. Ofir, S. Dreizler, et al. 2016. "No Evidence for Activity Correlations in the Radial Velocities of Kapteyn's Star." *Astrophysical Journal* 830(2):74.

Ansdell, M., J.P. Williams, N. van der Marel, J.M. Carpenter, G. Guidi, M. Hogerheijde, G.S. Mathews, et al. 2016. "ALMA Survey of Lupus Protoplanetary Disks. I. Dust and Gas Masses." *Astrophysical Journal* 828:46.

Apai, D., D. Nardiello, and L.R. Bedin. 2021. "TESS Observations of the Luhman 16 AB Brown Dwarf System: Rotational Periods, Lightcurve Evolution, and Zonal Circulation." *Astrophysical Journal* 906(1):64.

Baraffe, I., G. Chabrier, F. Allard, and P.H. Hauschildt. 1998. "Evolutionary Models for Solar Metallicity Low-Mass Stars: Mass-Magnitude Relationships and Color-Magnitude Diagrams." *Astronomy & Astrophysics* 337:2.

Baross, J.A., R.E. Anderson, and E.E. Stüeken. 2020. "The Environmental Roots of the Origin of Life." Pp. 71–92 in *Planetary Astrobiology*, V.S. Meadows, G.N. Arney, B.E. Schmidt, and D.J. Des Marais, eds. Tucson: University of Arizona Press.

Benisty, M., J. Bae, S. Facchini, M. Keppler, R. Teague, A. Isella, N.T. Kurtovic, et al. 2021. "A Circumplanetary Disk Around PDS70c." *Astrophysical Journal Letters* 916(1):L2.

Berardo, D., A. Cumming, and G.-D. Marleau. 2017. "The Evolution of Gas Giant Entropy During Formation by Runaway Accretion." *Astrophysical Journal* 834:149.

Bollard, J., J.N. Connelly, M.J. Whitehouse, E.A. Pringle, L. Bonal, J.K. Jørgensen, Å. Nordlund, et al. 2017. "Early Formation of Planetary Building Blocks Inferred from Pb Isotopic Ages of Chondrules." *Science Advances* 3(8):e1700407.

Bonomo, A.S., L. Zeng, M. Damasso, A.M. Leinhardt, A.B. Justesen, E. Lopez, M.N. Lund, et al. 2019. "A Giant Impact as the Likely Origin of Different Twins in the Kepler-107 Exoplanet System." *Nature Astronomy* 3:416–423. https://doi.org/10.1038/s41550-018-0684-9.

Cowan, N.B., E.E. Agol, V.S. Meadows, T. Robinson, T.A. Livengood, D. Deming, C.M. Lisse, et al. 2009. "Alien Maps of an Ocean-Bearing World." *Astrophysical Journal* 700:915 923.

Crossfield, I.J.M., B. Biller, J.E. Schlieder, N.R. Deacon, M. Bonnefoy, D. Homeier, F. Allard, et al. 2014. "A Global Cloud Map of the Nearest Known Brown Dwarf." *Nature* 505(7485):654–656.

De Wit, J., H.R. Wakeford, N.K. Lewis, L. Delrez, M. Gillon, F. Selsis, J. Leconte, et al. 2018. "Atmospheric Reconnaissance of the Habitable-Zone Earth-Sized Planets Orbiting TRAPPIST-1." *Nature Astronomy* 2(3):214–219.

DeMeo, F., and B. Carry. 2014. "Solar System Evolution from Compositional Mapping of the Asteroid Belt." *Nature* 505: 629–634. https://doi.org/10.1038/nature12908.

Diamond-Lowe, H., Z. Berta-Thompson, D. Charbonneau, and E.M.-R. Kempton. 2018. "Ground-Based Optical Transmission Spectroscopy of the Small, Rocky Exoplanet GJ 1132b." *Astronomical Journal* 156(2):42.

Dressing, C.D., and D. Charbonneau. 2015. "The Occurrence of Potentially Habitable Planets Orbiting M Dwarfs Estimated from the Full Kepler Dataset and an Empirical Measurement of the Detection Sensitivity." *Astrophysical Journal* 807(1):45.

Ehlmann, B.L., F.S. Anderson, J. Andrews-Hanna, D.C. Catling, P.R. Christensen, B.A. Cohen, C.D. Dressing, et al. 2016. "The Sustainability of Habitability on Terrestrial Planets: Insights, Questions, and Needed Measurements from Mars for Understanding the Evolution of Earth-Like Worlds." *Journal of Geophysical Research: Planets* 121(10):1927–1961.

Etiope, G., and B. Sherwood Lollar. 2013. "Abiotic Methane on Earth." *Reviews of Geophysics* 51(2):276–299. https://doi.org/10.1002/rog.20011.

Foley, B.J., C. Houser, L. Noack, and N. Tosi. 2020. "The Heat Budget of Rocky Planets." Chapter 4 in *Planetary Diversity: Rocky Planet Processes and Their Observational Signatures*, E.J. Tasker, C. Unterborn, M. Laneuville, Y. Fujii, S.J. Desch, and H.E. Hartnett, eds. Bristol, UK: IOP Publishing. https://doi.org/10.1088/2514-3433/abb4d9ch4.

Fulton, B.J., E.A. Petigura, A.W. Howard, H. Isaacson, G.W. Marcy, P.A. Cargile, L. Hebb, et al. 2017. "The California-Kepler Survey. III. A Gap in the Radius Distribution of Small Planets." *Astronomical Journal* 154(3):109.

Gillmann, G., E. Chassefière, and P. Lognonné. 2009. "A Consistent Picture of Early Hydrodynamic Escape of Venus Atmosphere Explaining Present Ne and Ar Isotopic Ratios and Low Oxygen Atmospheric Content." *Earth and Planetary Science Letters* 286(3–4):503–513.

Gomes, R., H.F. Levison, K. Tsiganis, and A. Morbidelli. 2005. "Origin of the Cataclysmic Late Heavy Bombardment Period of the Terrestrial Planets." *Nature* 435:466–469.

Gratia, P., and J.J. Lissauer. 2021. "Eccentricities and the Stability of Closely-Spaced Five-Planet Systems." *Icarus* 358:114038. https://doi.org/10.1016/j.icarus.2020.114038.

Gillon, M., A.H.M.J. Triaud, B.-O. Demory, E. Jehin, E. Agol, K.M. Deck, S.M. Lederer, et al. 2017. "Seven Temperate Terrestrial Planets Around the Nearby Ultracool Dwarf Star TRAPPIST-1." *Nature* 542:456–460. https://doi.org/10.1038/nature21360.

Greaves, J.S., A.M.S. Richards, W. Bains, P.B. Rimmer, H. Sagawa, D.L. Clements, S. Seager, et al. 2020. "Phosphine Gas in the Cloud Decks of Venus." *Nature Astronomy* 5:655–664. https://doi.org/10.1038/s41550-020-1174-4.

Guzik, P., M. Drahus, K. Rusek, W. Waniak, G. Connizzaro, and I. Pastor-Marazuela. 2020. "Initial Characterization of Interstellar Comet 2I/Borisov." *Nature Astronomy* 4:53–57. https://doi.org/10.1038/s41550-019-0931-8.

Hamano, K., Y. Abe, and H. Genda. 2013. "Emergence of Two Types of Terrestrial Planet on Solidification of Magma Ocean." *Nature* 497:607–610.

Harman, C.E., R. Felton, R. Hu, S. Domagal-Goldman, A. Segura, F. Tian, and J.F. Kasting. 2018. "Abiotic O_2 Levels on Planets Around F, G, K, and M Stars: Effects of Lightning-Produced Catalysts in Eliminating Oxygen False Positives." *Astrophysical Journal* 866:56.

Hirschmann, M.M., and R. Dasgupta. 2009. "The H/C Ratios of Earth's Near-Surface and Deep Reservoirs, and Consequences for Deep Earth Volatile Cycles." *Chemical Geology* 262(1–2):4–16.

Johnson, T.V., O. Mousis, J.I. Lunine, and N. Madhusudhan. 2012. "Planetesimal Compositions in Exoplanet Systems." *Astrophysical Journal* 757(2):192. https://doi.org/10.1088/0004-637x/757/2/192.

Kaltenegger, L., W.A. Traub, and K.W. Jucks. 2007. "Spectral Evolution of an Earth-Like Planet." *Astrophysical Journal* 658:598–616.

Kasting, J.F., D.P. Whitmire, and R.T. Reynolds. 1993. "Habitable Zones Around Main Sequence Stars." *Icarus* 101:108–128.

Kasting, J.F., R. Kopparapu, R.M. Ramirez, and C.E. Harman. 2014. "Remote Life-Detection Criteria, Habitable Zone Boundaries, and the Frequency of Earth-like Planets Around M and Late K Stars." *Proceedings of the National Academy of Sciences* 111:12641–12646.

Kreidberg, L., D.D.B. Koll, C. Morley, R. Hu, L. Schaefer, D. Deming, K.B. Stevenson, et al. 2019. "Absence of a Thick Atmosphere on the Terrestrial Exoplanet LHS 3844b." *Nature* 573(7772):87–90.

Kruijer, T.S., C. Burkhardt, G. Budde, and T. Kleine. 2017. "Age of Jupiter Inferred from the Distinct Genetics and Formation Times of Meteorites." *Proceedings of the National Academy of Sciences* 201704461. https://doi.org/10.1073/pnas.1704461114.

Lagrange, A.M., N. Meunier, P. Rubini, M. Keppler, F. Galland, E. Chapellier, E. Michel, et al. 2019. "Evidence for an Additional Planet in the β Pictoris System." *Nature Astronomy* 3(12):1135–1142.

Laneuville, M., C. Dong, J.G. O'Rourke, and A.C. Schneider. 2020. "Magnetic Fields on Rocky Planets." Chapter 3 in *Planetary Diversity: Rocky Planet Processes and Their Observational Signatures*, E.J. Tasker, C. Unterborn, M. Laneuville, Y. Fujii, S.J. Desch, and H.E. Hartnett, eds. Bristol, UK: IOP Publishing. https://doi.org/10.1088/2514-3433/abb4d9ch3.

Liebig, C., and J. Wambsganss. 2010. "Detectability of Extrasolar Moons as Gravitational Microlenses." *Astronomy & Astrophysics* 520:A68.

Loftus, K., R. Wordsworth, and C. Morley. 2019. "Sulfate Aerosol Hazes and SO_2 Gas as Constraints on Rocky Exoplanets' Surface Liquid Water." *Astrophysical Journal* 887(2):231.

Madhusudhan, N. 2019. "Exoplanetary Atmospheres: Key Insights, Challenges, and Prospects." *Annual Review of Astronomy and Astrophysics* 57:617–663.

Manara, C.F., A. Morbidelli, and T. Guillot. 2018. "Why Do Protoplanetary Disks Appear Not Massive Enough to Form the Known Exoplanet Population?" *Astronomy & Astrophysics* 618:L3. https://doi.org/10.1051/0004-6361/201834076.

Mankovich, C., and J. Fuller. 2021. "A Diffuse Core in Saturn Revealed by Ring Seismology." *Nature Astronomy* 5:1103–1109. https://doi.org/10.1038/s41550-021-01448-3.

McKay, D.S., E.K. Gibson, K.L. Thomas-Keprta, C.S. Vali, H., Romanek, S.J. Clemett, X.D.F. Chillier, et al. 1996. "Search for Past Life on Mars: Possible Relic Biogenic Activity in Martian Meteorite ALH84001." *Science* 273(5277):924–930. https://doi.org/10.1126/science.273.5277.924.

Meadows, V.S. 2017. "Reflections on O_2 as a Biosignature in Exoplanetary Atmospheres." *Astrobiology* 17(10):1022–1052.

Meadows, V.S., and R.K. Barnes. 2018. "Factors Affecting Exoplanet Habitability." Pp. 2771–2794 in *Handbook of Exoplanets*, H. Deeg and J. Belmonte, eds. Cham, Switzerland: Springer. https://doi.org/10.1007/978-3-319-55333-7_57.

Meadows, V.S., C.T. Reinhard, G.N. Arney, M.N. Parenteau, E.W. Schwieterman, S.D. Domagal-Goldman, A.P. Lincowski, et al. 2018. "Exoplanet Biosignatures: Understanding Oxygen as a Biosignature in the Context of Its Environment." *Astrobiology* 18(6):630–662.

Meech, K.J., R. Weryk, M. Micheli, J.T. Kleyna, O.R. Hainaut, R. Jedicke, R.J. Wainscoat, et al. 2017. "A Brief Visit from a Red and Extremely Elongated Interstellar Asteroid." *Nature* 552(7685):378–381. https://doi.org/10.1038/nature25020.

Miotello, A., L. Testi, G. Lodato, L. Ricci, G. Rosotti, K. Brooks, A. Maury, and A. Natta. 2014. "Grain Growth in the Envelopes and Disks of Class I Protostars." *Astronomy & Astrophysics* 567:A32.

Müller, A., M. Keppler, T. Henning, M. Samland, G. Chauvin, H. Beust, A.-L. Maire, et al. 2018. "Orbital and Atmospheric Characterization of the Planet within the Gap of the PDS 70 Transition Disk." *Astronomy & Astrophysics* 617:L2.

Mumma, M.J., G.L. Villanueva, R.E. Novak, T. Hewagama, B.P. Bonev, M.A. DiSanti, A.M. Mandell, and M.D. Smith. 2009. "Strong Release of Methane on Mars in Northern Summer 2003." *Science* 323(5917):1041–1045. https://doi.org/10.1126/science.1165243.

NASEM (National Academies of Sciences, Engineering, and Medicine). 2018. *Exoplanet Science Strategy*. Washington, DC: The National Academies Press. https://doi.org/10.17226/25187.

NASEM. 2019. *An Astrobiology Strategy for the Search for Life in the Universe*. Washington, DC: The National Academies Press. https://doi.org/10.17226/25252.

NASEM. 2023. *Pathways to Discovery in Astronomy and Astrophysics for the 2020s*. Washington, DC: The National Academies Press. https://doi.org/10.17226/26141.

Neveu, M., L.E. Hays, M.A. Voytek, M.H. New, and M.D. Schulte. 2018. "The Ladder of Life Detection." *Astrobiology* 18(11):1375–1402. https://doi.org/10.1089/ast.2017.1773.

Nowak, M., S. Lacour, P. Mollière, J. Wang, B. Charnay, E.F. van Dishoeck, R. Abuter, et al. 2020. "Peering into the Formation History of β Pictoris b with VLTI/GRAVITY Long-Baseline Interferometry." *Astronomy & Astrophysics* 633:A110.

Oza, A.V., R.E. Johnson, E. Lellouch, C. Schmidt, N. Schneider, C. Huang, D. Gamborino, et al. 2019. "Sodium and Potassium Signatures of Volcanic Satellites Orbiting Close-In Gas Giant Exoplanets." *Astrophysical Journal* 885:168.

Penny, M.T., B.S. Gaudi, E. Kerins, N.J. Rattenbury, S. Mao, A.C. Robin, and S.C. Novati. 2019. "Predictions of the WFIRST Microlensing Survey. I. Bound Planet Detection Rates," *Astrophysical Journal Supplement Series* 241(1):3. https://doi.org/10.3847/1538-4365/aafb69.

Quick, L.C., A. Roberge, A. Barr Mlinar, and M.M. Hedman. 2020. "Forecasting Rates of Volcanic Activity on Terrestrial Exoplanets and Implications for Cryovolcanic Activity on Extrasolar Ocean Worlds." *Publications of the Astronomical Society of the Pacific* 138:084402. https://doi.org/10.1088/1538-3873/ab9504.

Ranjan, S., and D.D. Sasselov. 2016. "Influence of the UV Environment on the Synthesis of Prebiotic Molecules." *Astrobiology* 16(1):68–88.

Ranjan, S., and D.D. Sasselov. 2017. "Constraints on the Early Terrestrial Surface UV Environment Relevant to Prebiotic Chemistry." *Astrobiology* 17(3):169–204. https://doi.org/10.1089/ast.2016.1519.

Reinhard, C.T., S.L. Olson, E.W. Schwieterman, and T.W. Lyons. 2017. "False Negatives for Remote Life Detection on Ocean-Bearing Planets: Lessons from the Early Earth." *Astrobiology* 17(4):287–297.

Robinson, T.D., V.S. Meadows, and D. Crisp. 2010. "Detecting Oceans on Extrasolar Planets Using the Glint Effect." *Astrophysical Journal Letters* 721:L67.

Sagan, C. 1995. *Cosmos*. Avenel, New Jersey: Wings Books.

Schlichting, H.E., and S. Mukhopadhyay. 2018. "Atmosphere Impact Losses." *Space Science Reviews* 214(1):1–31.

Schwieterman, E.W., N.Y. Kiang, M.N. Parenteau, C.E. Harman, S. DasSarma, T.M. Fisher, G.N. Arney, et al. 2018. "Exoplanet Biosignatures: A Review of Remotely Detectable Signs of Life." *Astrobiology* 18(6):663–708. https://doi.org/10.1089/ast.2017.1729.

Seager, S., W. Bains, and J.J. Petkowski. 2016. "Toward a List of Molecules as Potential Biosignature Gases for the Search for Life on Exoplanets and Applications to Terrestrial Biochemistry." *Astrobiology* 16(6):465–485. https://doi.org/10.1089/ast.2015.1404.

Shields, A.L., S. Ballard, and J.A. Johnson. 2016. "The Habitability of Planets Orbiting M-Dwarf Stars." *Physics Reports* 663:1–38.

Showman, A.P., N.K. Lewis, and J.J. Fortney. 2015. "Three-Dimensional Atmospheric Circulation of Warm and Hot Jupiters: Effects of Orbital Distance, Rotation Period, and Nonsynchronous Rotation." *Astrophysical Journal* 801(2):95.

Sing, D.K., J.J. Fortney, N. Nikolov, H.R. Wakeford, T. Kataria, T.M. Evans, S. Aigrain, et al. 2016. "A Continuum from Clear to Cloudy Hot-Jupiter Exoplanets Without Primordial Water Depletion." *Nature* 529(7584):59–62.

Snellen, I.A.G., R.J. de Kok, E.J.W. de Mooij, and S. Albrecht. 2010. "The Orbital Motion, Absolute Mass and High-Altitude Winds of Exoplanet HD 209458b." *Nature* 465(7301):1049–1051.

Snellen, I.A.G., B.R. Brandl, R.J. de Kok, M. Brogi, J. Birkby, and H. Schwarz. 2014. "Fast Spin of the Young Extrasolar Planet β Pictoris b." *Nature* 509(7498):63–65.

Snellen, I.A.G., L. Guzman-Ramirez, M.R. Hogerheijde, A.P.S. Hygate, and F.F.S. van der Tak. 2020. "Re-Analysis of the 267 GHz ALMA Observations of Venus." *Astronomy & Astrophysics* 644:L2. https://doi.org/10.1051/0004-6361/202039717.

Soderlund, K.M., and S. Stanley. 2020. "The Underexplored Frontier of Ice Giant Dynamos." *Philosophical Transactions of the Royal Society A* 378(2187):20190479.

Tamayo, D., M. Cranmer, S. Hadden, H. Rein, P. Battaglia, A. Obertas, P.J. Armitage, et al. 2020. "Predicting the Long-Term Stability of Compact Multiplanet Systems." *Proceedings of the National Academy of Sciences* 117(31):18194–18205. https://doi.org/10.1073/pnas.2001258117.

Tarter, J.C., P.R. Backus, R.L. Mancinelli, J.M. Aurnou, D.E. Backman, G.S. Basri, A.P. Boss, et al. 2007. "A Reappraisal of the Habitability of Planets Around M Dwarf Stars." *Astrobiology* 7(1):30–65. https://doi.org/10.1089/ast.2006.0124.

Teachey, A., D. Kipping, C.J. Burke, R. Angus, and A.W. Howard. 2020. "Loose Ends for the Exomoon Candidate Host Kepler-1625b." *Astronomical Journal* 159(4):142.

Unterborn, C., L. Schaefer, and S. Krijt. 2020. "The Composition of Rocky Planets." Chapter 5 in *Planetary Diversity: Rocky Planet Processes and Their Observational Signatures*, E.J. Tasker, C. Unterborn, M. Laneuville, Y. Fujii, S.J. Desch, and H.E. Hartnett, eds. Bristol, UK: IOP Publishing. https://doi.org/10.1088/2514-3433/abb4d9ch5.

Wahl, S.M., W.B. Hubbard, B. Militzer, T. Guillot, Y. Miguel, N. Movshovitz, Y. Kaspi, et al. 2017. "Comparing Jupiter Interior Structure Models to Juno Gravity Measurements and the Role of a Dilute Core." *Geophysical Research Letters* 44(10):4649–4659.

Walsh, K.J., A. Morbidelli, S.N. Raymond, D.P. O'Brien, and A.M. Mandell. 2011. "A Low Mass for Mars from Jupiter's Early Gas-Driven Migration." *Nature* 475:206–209. https://doi.org/10.1038/nature10201.

Wang, J.J., J.R. Graham, L. Pueyo, P. Kalas, M.A. Millar-Blanchaer, J.-B. Ruffio, R.J. De Rosa, et al. 2016. "The Orbit and Transit Prospects for β Pictoris b Constrained with One Milliarcsecond Astrometry." *Astronomical Journal* 152(4):97.

Webster, C.R., P.R. Mahaffy, S.K. Atreya, J.E. Moores, G.J. Flesch, C. Malespin, C.P. McKay, et al. 2018. "Background Levels of Methane in Mars' Atmosphere Show Strong Seasonal Variations." *Science* 360(6393):1093–1096. https://doi.org/10.1126/science.aaq0131.

Youdin, A.N., and G.H. Rieke. 2017. "Planetesimals in Debris Disks." Pp. 340–362 in *Planetesimals*, L.T. Elkins-Tanton and B.P. Weiss, eds. Cambridge, UK: Cambridge University Press. https://doi.org/10.1017/9781316339794.016.

Zahnle, K., R.S. Freedman, and D.C. Catling. 2011. "Is There Methane on Mars?" *Icarus* 212(2):493–503. https://doi.org/10.1016/j.icarus.2010.11.027.

Zahnle, K.J., R. Lupo, A. Dobrovolskis, and N.H. Sleep. 2015. "The Tethered Moon." *Earth and Planetary Science Letters* 427:74–82.

16

State of the Profession

INTRODUCTION

The decadal survey's statement of task charged the committee with addressing the state of profession (SoP), including issues of diversity, equity, inclusion, and accessibility (DEIA; see Appendix A). The committee regards the inclusion of these issues, *for the first time in a planetary science decadal survey*, to signal their importance and urgency. The committee's work engaged with available SoP data (such as they are at present), white papers submitted to the decadal, several invited speakers, robust internal discussions, as well as the best social, behavioral, and neuroscience evidence about the causes and outcomes that affect the quality of science, technology, engineering, and mathematics (STEM) professions, of which planetary science and astrobiology (PS&AB) are a part. The findings and recommendations that flow from the committee's analysis can strengthen PS&AB if NASA, as the primary funding agency of PS&AB, acts on these recommendations to build a strong system of equity and accountability that is necessary to locate, recruit, and retain the best talent, and to nurture and sustain the work they do. As requested in the statement of task, these recommendations are actionable and intended to help PSD bring the full force of its leadership and the engagement of the community to advance the profession as part of its scientific mission.

NASA's aspirational nature, built on the idea of limitless exploration, provides a fitting backdrop to develop initiatives that will seek solutions to the issues that concern the state of the profession. This chapter is intended to assist NASA's PSD to boldly address issues that concern its most important resource: the people who propel its planetary science and exploration missions.

Ensuring the broadest level of participation is necessary to produce high-quality science in an environment of fierce competition for limited human resources. The rich and unparalleled diversity of the people in the United States is NASA's strongest advantage, but only if such diversity is tapped by robust procedures for identification and recruitment, onboarding and promotion, and consistent nurturing and fair and equitable reward structures. The committee acknowledges and applauds the hard-earned progress on diversification that has been made in STEM fields including PS&AB, especially with respect to the entry of women into the profession and the growing number of women in positions of leadership and prominence (including at the helm of NASA's PSD). The committee recognizes that the goals and intentions of NASA science leadership with respect to DEIA are exemplary. However, much work remains to be done on gender parity broadly, with a singular focus needed on issues of basic representation by race/ethnicity which demonstrate a shocking lack of change over the past decade.

The chapter is organized to first provide background and present the available data and the inputs the decadal survey received from community white papers. The committee then presents its analysis of these data and its findings and concludes with a set of recommendations. This chapter begins with research on how unintended bias can impact decisions and a growing recognition of systemic bias embedded in individual minds, communities, and institutions. This discussion provides a backdrop against which issues specific to PS&AB can addressed to ensure diverse and robust fields of planetary science and astrobiology going forward.

IMPLICIT AND SYSTEMIC BIAS

Overall progress on matters of equity and accountability in STEM fields has been slow, and in the case of advancing some underrepresented racial/ethnic communities (URC), it has been surprisingly stagnant. New paths to advancement on this issue merit consideration, including the science of implicit bias and the notion of systemic bias that pervades all human activity, individual and institutional. The concept of implicit bias may explain why social change in STEM fields has been slow: the attitudes and stereotypes that contribute to bias can be hidden from conscious awareness. That which is unknown cannot, by definition, be a candidate for change or progress. If NASA PSD embraces evidence from the behavioral and social sciences and works to mitigate bias both in the minds of individuals and the structures of institutions, NASA PSD can move PS&AB forward in unprecedented ways. To be clear, explicit forms of bias although on a downward slope are not nonexistent in the daily lives of scientists. (As a widely discussed example, see *Picture a Scientist*, Cheney and Shattuck 2020.) Rather, there are two reasons for focusing on implicit bias: implicit bias by its nature is hidden and as such is likely to remain unidentified, posing unique resistance to change. Second, a focus on implicit bias naturally includes addressing explicit bias as tackling it requires setting a higher standard to address not only obvious and explicit forms of bias but also those that escape recognition.

Human social groups are a defining part of every society. Variations in gender identity and sexuality, race and ethnicity, age, socioeconomic class, religion, physical ability and features, personality, culture, national origin and many more, deeply determine life's opportunities and outcomes. They do so because of the histories of intergroup contact and conflict, age-old and present-day attitudes (preferences) and beliefs (stereotypes) about social groups and the individual members of those groups. Attitudes and beliefs are the building blocks of all human interactions, and they carry with them the power to imbue individuals with certain essential features, whether such attitudes and beliefs are true or not, such as who is *good* and *bad* and specific traits such as who is naturally *competent* or *incompetent.* Because intergroup attitudes and stereotypes were first empirically investigated at the turn of the 20th century, survey evidence shows that they have consistently moved toward greater and greater neutrality—that is, a belief that all social groups deserve equal opportunity and equal treatment. This provides grounds for optimism in the ability to effect positive change. Yet, objective measures of behavior in the workplace reveal consistent evidence of discrimination; two people can perform the same action (produce the same quality of work) but are judged differently based on beliefs about what their group, not they, is capable of. How is this possible, today, given a fundamental belief in fairness and a commitment to pursuing the best talent?

The concept of implicit bias was introduced in the 1990s to suggest why conscious and explicit statements of egalitarianism may abound, while continued discrimination exists. Substantial and growing evidence shows that implicit forms of bias exist even in individuals who sincerely endorse egalitarian values. Implicit bias has been shown to play a role in the perception of potential, talent, deservingness, rewards, and so on. The disparity between what is explicitly expressed (no bias) and implicitly revealed (existence of bias) makes the problem challenging, because bias is hidden. Moreover, implicit biases influence behavior and become enmeshed in the practices and policies of organizations in ways that may be imperceptible but that can play an active role in shaping a system of discrimination, as signaled by terms such as "systemic racism" (Banaji et al. 2021). Specifically, the microcosm of individual minds reflects the macrocosm of culture and society and studies now demonstrate the shaping of bias as a function of simply geographic location (Charlesworth and Banaji 2021). Newer data suggest that even implicit or hidden bias can be changed (Charlesworth and Banaji 2022), and Cox and Devine (2019) demonstrate strategies that lead to change. From the white papers, the invited speakers, and the committee's deliberations it appears that the community of PS&AB is optimistic and ready to do the work that is necessary at the individual level. By adding the strength of NASA as an organization to such an effort, the future holds increased promise and can be brighter than ever before. A convergence of views from across several sciences has led to the recognition that unless bias operating at many different levels is identified and analyzed as a whole, the possibility for change is unlikely in society at large. The analysis reveals built-in systems that undermine life's opportunities and outcomes by racial category in particular, although extensions to the constraints placed by demographic variables such as socioeconomic status, age, disability, immigrant status and nationality, sexuality and gender identity, and religion are all a part of understanding the nature of systems of bias. The idea of systemic bias takes the long view, starting with American colonial history, and explicit practices and policies that reinforced disadvantage across all domains of life (Banaji et al. 2021). The analysis reveals that racial/ethnic segregation and isolation have led to disproportionate costs to Native Americans' and Black Americans' opportunities among other groups, whether in the form of social networks, education, physical health, or financial resources. Although a miniscule segment of American society, the fields of PS&AB have inherited this history and any attempt to analyze the state of the profession needs to recognize that it is part of the larger culture in which these sciences operate. Recognition of built-in inequities motivates actions to mitigate the effects of this history at least to the extent that it affects progress in PS&AB. NASA PSD is a part of the government of the United States and shares in the responsibility to objectively analyze itself and its procedures, processes, and practices to ensure that its own environment is free of systemic and institutional bias.

Creating the Conditions Necessary for Effecting Change

Once we recognize that bias is at least partly unintentional or implicit, implementing objective measures of self-examination that acknowledge this situation is surest path to progress. Many organizations routinely engage in such analyses (see, for example, Lauer and Roychowdhury, 2021). Groups that collect evidence on all aspects of their activities and use those data to improve procedures and make those data publicly available gain in credibility and engender the trust of the community.

Uncovering hidden bias often involves gathering evidence about disparities well beyond simply measuring the demographic composition of fields. If such evidence is responsibly collected and publicly disseminated, existing disparities will be brought to light and can be corrected. If the evidence shows that hiring, retention, promotion, grants, salaries, mission opportunities, leadership preparation, and recognitions are indeed distributed unevenly and not always based on merit, then legitimate action to correct such inequities in the interest of the best science and applications can be initiated. Among the main messages of this report is that such data are lacking today, and without them NASA PSD is hampered in its mission to effect positive change, no matter how motivated it is to do so. Further, knowing the evidence, while necessary, is not sufficient. There will be, as there always is, resistance to evidence that challenges the status quo. As such, the next step will involve educating members of the community about the corresponding cost borne by the science because of disparities in opportunity and treatment. Websites, seminars, and tests make it possible to educate about bias effectively by showing its effects on the quality of the work that emerges, so long as such education is conducted in an ongoing and rigorous way, in addition to annual reports that can serve to track change over time. Education and awareness of leadership and the workforce (community engagement in these issues is vital to success) are necessary but not sufficient. Concurrent with the education of individuals, organizations need to engage with experts to assist in transforming existing practices and policies to better ensure equity. Based on the best evidence from STEM fields in general, and to the extent available about PS&AB in particular, NASA PSD needs to actively seek to debias its procedures and policies. This includes analyzing every step of decision-making processes involved in advertising, recruitment, selection, onboarding, retention, promotion, compensation, symbolic recognitions, team dynamics, research and analysis, space mission opportunities and outcomes, succession planning, and engaging with the community and the public.

Last, change is unlikely to manifest unless a system of accountability is put into place very much like the accountability scientists regularly exert on their own scientific work. Increasingly, attention to inequities and lack of transparency is being paid, even demanded, by a new generation of scientists. In our scientific work, we use the best measures to find the evidence we are seeking, we report our results publicly and transparently, we allow others to evaluate and challenge it, we improve it and get back to the drawing board. Why are we, members of STEM fields, slow to do the same to measure ourselves on matters concerning the SoP? Our answer is that it is because we regard ourselves to be generally fair and unbiased in matters of interpersonal evaluations. That assumption needs to be set aside because we now know better based on the science of implicit bias (Jost et al. 2009). As such, it will be important for NASA PSD to scrutinize its own procedures and policies in an open and accountable manner. There are indications of progress of this very nature in many STEM fields, and the committee aims to support NASA PSD to usher in practices and policies that ensure the advancement and well-being of a diverse community suited to pursue the deepest knowledge about the diverse worlds that lie beyond our own.

THE EVIDENCE

The impetus to deal with DEIA issues has increased dramatically over the past decade. Scientists understand that the diversity they observe and strive for in their methods and locations are critical to the success of the profession. Analyzing and understanding issues surrounding DEIA are just as important for an adaptive, high-functioning science community. NASA PSD recognizes the importance of diversity in target bodies, methods, and techniques (because no one method, however powerful, can provide a full understanding as can be achieved with a collection of diverse methods), and it is time to similarly recognize the importance of diversity of the members of its profession for the same reason—*to improve the quality of the science and therefore the progress it hopes to achieve.*

The first step in assessing "the state of the profession" is to look at accurate and sufficient evidence about trends in its demographics and culture. *This decadal survey did not collect any original data. It utilized existing data but did so in a comprehensive manner that makes this repository of demographics data the most exhaustive collection anywhere within the PS&AB community.*[1] That said, very sparse data exist on the planetary and astrobiology science communities. Partly, this is because of the complexity of defining the boundaries of such multi- and interdisciplinary fields. The profession is drawn from the geological sciences, physics, astronomy, chemistry, biology, and other disciplines. Moreover, the new field of exoplanets has greatly expanded the population of planetary scientists and further blurred the boundary with astrophysics. These are positive features scientifically speaking, but they create greater complexity when measuring the demographics given the difficulty of categorization.

Until the first planetary science workforce survey was carried out in 2010–2011, there existed no data from which any estimates of the demographics, within an order of magnitude, could be made. Notably, no concerted effort has been made by NASA over the past decades to collect, analyze, and present self-identified data or to support the routine collection of rigorous data of the profession that encompasses PS&AB. The data that do exist are largely owing to (1) the commitment of individuals (including within NASA) who pursued data collection about the state of academic planetary and other scientists and (2) community-driven surveys. Recent efforts to collect data are helpful but there is a need for collection of better data and for continuous monitoring and improvement of procedures and policies beyond what is being done today. As a relatively newer subfield of study, there are essentially no data on astrobiology, so the bulk of the data presented here focus on planetary science (Table 16-1). This situation will need to be rectified in order to conduct a comprehensive review. Open session meetings with the committee and astrobiology representatives suggest a heightened awareness of issues and ongoing work to address this current lack of information.

The data presented in this chapter are a comprehensive collection of the available information on the field of PS&AB. The data have been used to provide insights into the SoP. The committee emphasizes that rigorous data collection and analyses are required to draw robust conclusions, which is not currently possible.

University Departments Surveys, 2011 and 2018

University department surveys were initiated and collected by Fran Bagenal at the University of Colorado's Laboratory for Atmospheric and Space Physics. This is an example of the goodwill of individual members who have collected such data, but this cannot be relied upon as a method for future data gathering. Ensuring proper

---

[1] State of Planetary Science Profession: a presentation of data on demographics on the planetary science research community that were gathered to support this decadal survey are available via University of Colorado's public document depository, CU Scholar, at https://doi org/10.25810/VNTG-FK10.

TABLE 16-1 Overview of the Datasets on the Planetary Science Workforce

Abbreviation	Name of Survey	Year of Survey	Fields Surveyed	Response Rate	Total Number Sent to (N)
Dept-1	2011 Survey of U.S. Planetary Science University Departments[a]	2011 (Academic Years 2008–2009, 2009–2010)	U.S. university departments that include planetary science	54 departments at 40 universities	
Dept-2	2018 Survey of U.S. Planetary Science University Departments[b]	2018 (Academic Years 2016–2017 and 2017–2018)	U.S. university departments that include planetary science	36 departments at 29 universities	
AIP-1	2011 Survey of the Planetary Workforce[c]	2011	AAS Division for Planetary Science, LPSC, AGU Section on Planetary Science	62%	4252
AIP-2	2020 Survey of the Planetary Workforce[d]	2020	AAS Division for Planetary Science, LPSC, GSA Planetary Geology Division	48%	4965
NSPIRES	NASA Science Mission Directorate Demographics Data[e]	2021	Personal profiles on NSPIRES		31,172 PIs and Co-Is of planetary proposals
Missions	NASA Announcement of Opportunity Science Team Demographics[f]	2021	Mission leadership of competed missions		933 proposals

[a] Laboratory for Atmospheric and Space Physics (LASP), University of Colorado (CU) Boulder, 2015, "Summary of Results of Survey of U.S. Academic Departments that Include Planetary Science," https://lasp.colorado.edu/home/mop/files/2015/08/DeptSummary3.pdf.

[b] F. Bagenal, 2021, "Planetary Science Academic Department Surveys 2011 & 2018," LASP, CU Boulder, https://lasp.colorado.edu/home/mop/files/2021/07/DeptComparison2011-2018.pdf.

[c] S. White, R.Y. Chu, and R. Ivie, 2011, "Results from the 2011 Survey of the Planetary Science Workforce," Statistical Research Center, American Institute of Physics (AIP), https://lasp.colorado.edu/home/mop/files/2015/08/Report.pdf.

[d] A.M. Porter, S. White, and J. Dollison, "2020 Survey of the Planetary Science Workforce," Statistical Research Center, AIP, https://dps.aas.org/sites/dps.aas.org/files/reports/2020/Results_from_the_2020_Survey_of_the_Planetary_Science_Workforce.pdf.

[e] L. Barbier and C. Wilson, 2021, "Demographic Data Presentation to: Planetary Science Advisory Committee," NASA, https://science.nasa.gov/science-red/s3fs-public/atoms/files/07-Barbier-Demographics-061421.pdf.

[f] H. Jensen and L. Pappas, NASA, 2021, "NASA Announcement of Opportunity Science Team Demographics," Presentation at the NASEM Increasing Diversity and Inclusion in the Leadership of Competed Spaced Missions Committee Meeting #6, June 16, Washington, DC, https://www.nationalacademies.org/event/06-16-2021/increasing-diversity-and-inclusion-in-the-leadership-of-competed-space-missions-meeting-6.

and regular data collection and consistent standards on composition of the community and taking the pulse on climate are going to be essential.

Small Number and Diverse Programs

There are only ~13 U.S. university departments, with "planetary" in the department's name. Furthermore, being such an interdisciplinary field, it is difficult to track down all the academic institutions in the United States where planetary scientists do research and teach courses. Many of the planetary researchers at universities are not in academic departments but in independent research labs. Here the committee uses survey results of academic departments to seek information about the career pathway from bachelor's to PhD and beyond. The 2011 and 2018 surveys were conducted by an ad hoc committee with questionnaires sent to departments sought largely by word-of-mouth. Some universities had multiple departments that housed planetary science faculty.

In 2011, only 39 universities stated that they had any faculty who identified as planetary scientists, with 105 of the 233 faculty positions located at just six universities. The surveys focused on the previous two academic years: fall 2008 to spring 2010 for the 2011 survey and fall 2016 to spring 2018 for the 2018 survey. Neither survey asked about astrobiology and only the 2018 survey asked if there were people in the department who studied exoplanets (41 faculty).

Demographics of Bachelors, PhDs, Faculty

Few universities have bachelor's degree programs in planetary science and/or astrobiology but many teach upper-division courses in these fields. Accordingly, the survey asked, "How many undergraduates completed a bachelor's degree in your department with a concentration in planetary science (took 2 or more upper division courses in planetary science) during the past two academic years?" The total is about 100 bachelor's degrees per year, of which about one-third are women. Figure 16-1 shows the top 10 departments for total faculty as reported to the committee. The change between decades is highly variable between institutions, and only goes to show that the data available today are insufficient to draw accurate conclusions about the state of the profession.

Regarding PhDs, 50–65 are awarded per year in planetary science, of which 40–45 percent were awarded to women. These data suggest the gender gap is narrowing. The 2018 survey asked departments for data on students' U.S. citizenship or permanent resident status and found that 30 percent of graduate students were non-U.S. citizens. The departments also reported a mere 7 percent of graduate students identifying as members of underserved racial/ethnic communities.

The departmental surveys suggest that the total number of women faculty (tenured and tenure-track) in planetary science has increased marginally over the past decade from 233 to 250—that is, 14 percent to ~20 percent. The percentage of tenure-track women faculty increased from 24 to 35 percent showing increasing gender diversity in the academic career pathway. Figure 16-1 shows the top 10 departments for total faculty. The change between decades is highly variable between institutions.

The AIP Surveys, 2011 and 2020

The Statistical Division of the American Institute of Physics (AIP) has developed a survey protocol that takes email lists from multiple organizations, sorts them into unique addresses and sends out a questionnaire designed (e.g., by a committee of planetary scientists guided by the AIP's experience) to gather demographics and workforce climate data. The AIP follows up with reminders to produce response rates often more than 50 percent. The responses remain anonymous, with reporting shielding personal identification when bin sizes are too small. This is commendable.

In 2011, the survey was sent to 4,252 unique contacts via conference mailing lists (AIP 2011). Responses were received from 62 percent of these people. Because a major goal of the survey was to find out the number of active planetary researchers in the United States, the survey focus was on the responses from 1,518 participants who had earned a doctorate and lived in the United States. Among them, 946 identified as planetary scientists actively working full-time and on research (rather than mission operations, management, or teaching). Assuming the 62 percent response rate applies equally across all populations, one may infer that 1,525 planetary PhD scientists were actively working full-time on research in 2011.

In anticipation of this decadal survey, work started in 2017 on a follow-up to the 2011 survey. The American Astronomical Society's Division for Planetary Sciences (DPS) led this survey and partnered with AIP for data collection and analysis. The 2020 Planetary Science Workforce Survey was sent to 4,965 members of the planetary science community in April 2020 (AAS-AIP 2020). Of the 2,367 responses (48 percent response rate), about 941 have PhDs and are working in the United States in planetary research. Again, assuming the 48 percent response rate applies equally across all surveyed populations, one may infer that 1,960 planetary PhD scientists are actively working on research in 2020. This suggests a ~25 percent increase over the decade or 2.5 percent/year.

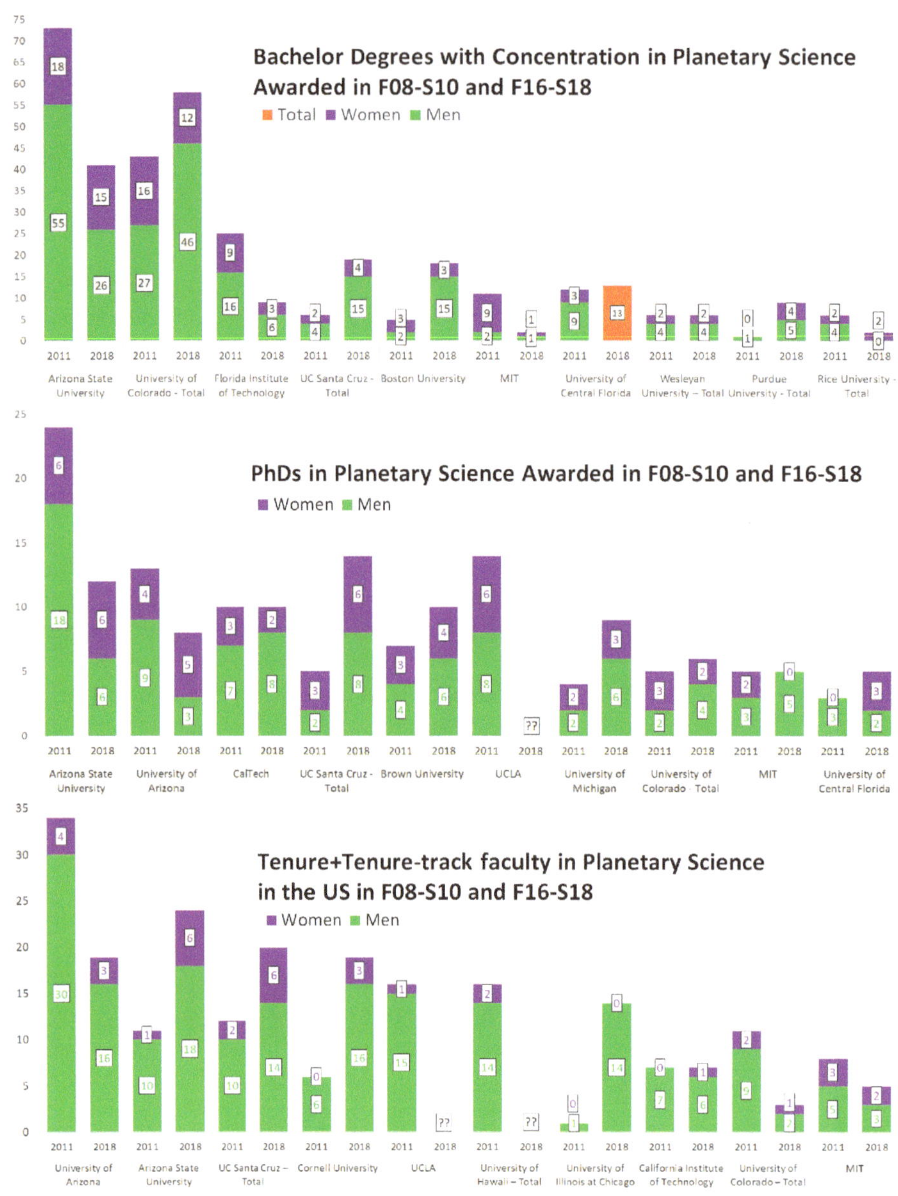

FIGURE 16-1 Gender of planetary scientists from bachelor's degree candidates, PhDs, and tenure-track faculty by major academic institutions.

Area of Degree

The 2011 survey showed that only 2 percent of planetary scientists received their undergraduate degree in planetary sciences, while 37 percent were in physics, 22 percent in geology and geophysics, 7 percent in chemistry, 8 percent in astronomy and astrophysics, 5 percent in Earth science, 4 percent in biology, and 2 percent in math (AIP 2011). When considering PhDs, 40 percent of working planetary scientists received their PhD in planetary science, 15 percent in geology and geophysics, 13 percent in astronomy and astrophysics, 12 percent in physics, and just a few from other areas. Questions about the area of degree were not asked in the 2020 survey.

Employment

Relatively few planetary scientists work at NASA centers: 14 percent in 2011, and only 7 percent in 2020. The largest portion of planetary scientists work at a university or college (48 percent in 2011, 41 percent in 2020). The next largest employers are research institutes/federally funded research and development centers (FFRDCs)/nonprofits which employ about 33 percent. This third of the community depends on research grants to support themselves. This is in contrast with the field of astronomy where 56 percent are employed at a university/college, while only ~14 percent are at research institutes/FFRDCs/nonprofits (14 percent of astronomers are employed at NASA centers). Overall, according to the AAS-AIP 2020 survey, approximately half of the planetary science workforce (post-PhD) are in tenured or hard-money/permanent jobs (varying by about ±10 percent depending on gender and/or race/ethnicity). The other half are in soft-money, tenure-track, post-doctoral, freelance or "other" positions.

Current Demographics

The gender, race, and ethnic diversity of the planetary science workforce is shown in Figure 16-2, which compares the 2011 and 2020 survey numbers as well as the demographics of the U.S. workforce, U.S. STEM workforce, and U.S. physical sciences workforce. Data intentionally focusing on the representation of communities who have been historically underrepresented in PS&AB is presented to draw attention to the needs of these communities. Currently, 37 percent of the planetary science workforce are women and 1 percent are nonbinary or another gender. This is in contrast with a Pew Research Center report (Fry et al. 2021), that found that 50 percent of those employed in STEM jobs are women and 40 percent of those employed in the physical sciences are women. Asian Americans comprise 6 percent of the U.S. workforce, 13 percent of all STEM jobs, 18 percent of physical science jobs, and 13 percent of the planetary workforce.

The recent survey identified a severe underrepresentation of Black and Latinx researchers in planetary science. Although Latinx/Hispanics are 17 percent of the U.S. workforce, 8 percent of all STEM jobs, and 8 percent of physical science jobs, only 5 percent of respondents to the 2020 AAS-AIP survey were Hispanic or Latinx. Moreover, although Black/African Americans are 11 percent of the U.S. workforce, 9 percent of all STEM jobs, and 6 percent of physical science jobs, only 1 percent of respondents to the 2020 AIP survey were Black or African American.

There is little data on Indigenous researchers. Following data reported by NASA on its workforce, American Indian/Alaskan Natives comprise 1.1 percent of the national civilian labor force, 1 percent of all NASA employees, and 0.8 percent of the agency's science and engineering employees (NASA 2019). The 2020 AAS-AIP survey indicates that 10 percent of the field is LGBTQ+, with 1 percent of respondents identifying as nonbinary or another gender. Disability is claimed as an identity by 15 percent respondents to the 2020 AAS-AIP survey. These figures are not comparable to other population surveys because the questions in the AAS-AIP survey are more inclusive of a broad range of identities. In general, indigenous, LGBTQ+, nonbinary, and people with disabilities are examples of demographics that are not consistently reported nor consistently defined in surveys. This hinders the ability to support these communities.

Demographics Over the Past Decade

Over the past decade, the field has made some improvements in the representation of historically excluded people. Notably, the percentage of women in the planetary science workforce has increased from 25 percent in 2011 to 37 percent in 2020, and the representation of Asian Americans has increased from 7 percent in 2011 to 13 percent in 2020 (see Figure 16-2). Asian Americans comprise only 3 percent of planetary scientists having earned their degree on or before 1970, but 17 percent of those who earned their degree between 2011 and 2020 (Table 16-2).

Although remaining below national workforce numbers, the representation of Latinx/Hispanics has increased from 1 percent in 2011 to 5 percent in 2020. In fact, although 0 percent of planetary scientists who earned their PhD on or before 1980 are Hispanic or Latinx, 6 percent of those who earned their degree between 2011 and 2020 are Latinx/Hispanic. Crucially, the representation of Black/African Americans has shown no growth over the past decade, continuing at 1 percent, well below the national workforce value. This is consistent with analysis

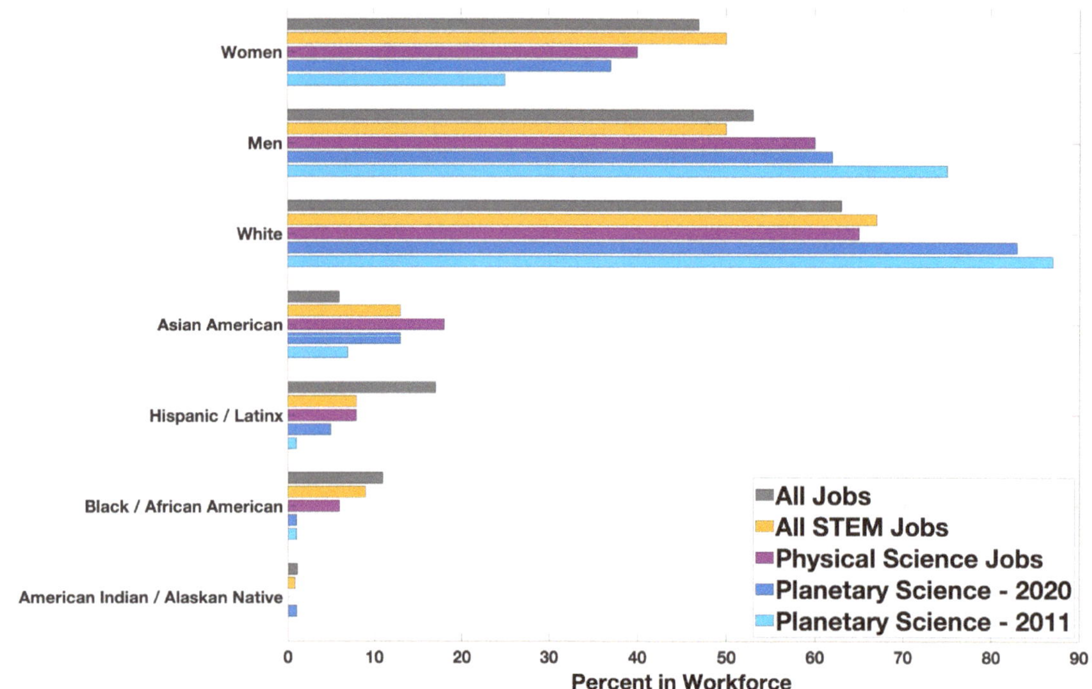

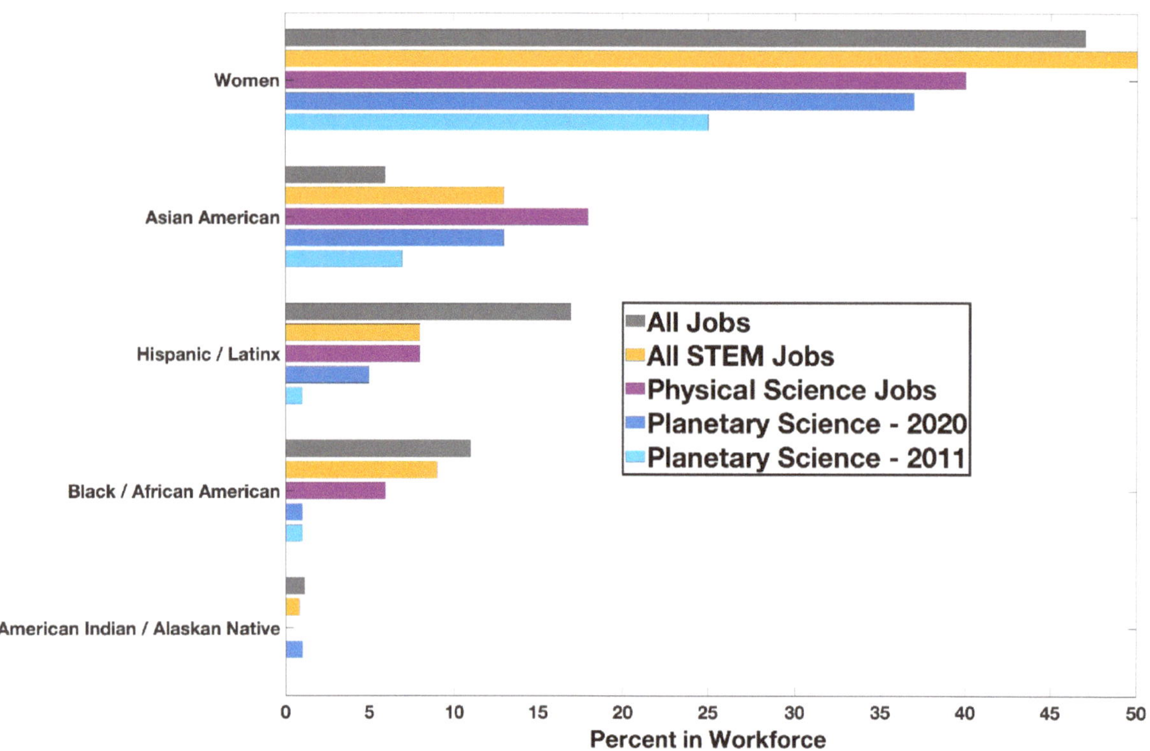

FIGURE 16-2 Representation of men and women, and demographics by race and ethnicity as reported in the 2011 and 2020 planetary science workforce surveys along with the demographics of physical science jobs, all STEM jobs, and all U.S. jobs as reported by a Pew Research Center study. The STEM jobs data for American Indian/Alaskan Native are from a NASA report on demographics of its workforce. Data are lacking for physical sciences and from the 2011 planetary science survey for American Indian/Alaskan Native.

TABLE 16-2 Demographics of the Planetary Science Workforce by Year of PhD Degree Conferral as Presented in the 2020 AAS-AIP Planetary Science Workforce Study

Year of Degree	Black, African American or Another Race/Ethnicity[a]	Hispanic/Latinx	Asian American	White	LGBTQ+[b]
1970 or earlier	5%	0%	3%	92%	4%
1971–1980	3%	0%	5%	92%	4%
1981–1990	3%	3%	4%	90%	2%
1991–2000	4%	2%	11%	83%	6%
2001–2010	4%	5%	10%	81%	6%
2011–2020	5%	6%	17%	72%	12%

[a] Other race/ethnicity included respondents who are Native American/Alaskan Native, Native Hawaiian/Other Pacific Islander, or wrote in another race/ethnicity.

[b] LGBTQ+ included respondents who are lesbian, gay, bisexual, transgender, nonbinary or another noncisgender identity, and other non-heterosexual orientations.

of the geoscience's profession (Bernard and Cooperdock 2018), as well as physics and geoscience PhD recipients (Rivera-Valentín et al. 2020).

The 2020 AAS-AIP survey further showed that the percentage of PhD degree recipients identifying as LGBTQ+ has been increasing over the past 20 years. Although only 4 percent of planetary scientists who earned their PhD on or before 1970 self-identified as LGBTQ+, 12 percent of those who earned their degree between 2011 and 2020 self-identified as LGBTQ+.

Age Structure of the Profession

At present, NASA does not report age data of proposal submitters and awardees. Data on the average age of planetary scientists and astrobiologists as well as the full age distribution are necessary. Such data can then serve as the baseline to compare to further data on (a) the age distribution of proposal submitters and (b) the age distribution of proposal awardees. Given the tendency in science to continue to fund those who have previously been funded (and often with good reason), such data will allow NASA to monitor whether its funding is sufficiently nurturing and growing the field of new proposers. Perhaps NASA is already collecting and analyzing such data; however, in the absence of published data on this demographic variable, it leaves the field wondering about potential age-based bias.

Given the lack of data for proper characterization and assessment of even the age of the profession (as NSF does not track PhDs in planetary science), the only available data the committee could piece together on the age of the profession (not proposal submitters or awardees) were from the two DPS surveys (AIP 2011; AAS-AIP 2020). The 2011 workforce survey provides an age profile (AIP 2011). It shows the age range of the profession to be between 35 and 54 with a median age of about 44 years. The 2020 workforce survey only provided the age profile information for DPS members; it did not provide data for all respondents (AAS-AIP 2020). It reports that 28 percent of DPS members are 60 and older and that about 60 percent of DPS members are between 30 and 60. In comparison, some 20 percent of the U.S. population is 60 and older. The data provided by these two workforce surveys are not directly comparable (one speaks to the age of the field, the other speaks to the age of the DPS members only). The difficulty the committee encountered in responding to questions about the age of the profession and award recipients was a common occurrence as it attempted to provide a description of the state of the profession in the absence of evidence. If the committee's main recommendation is to begin collecting and reporting on the many variables that can illuminate the state of the profession and the life of working PS&AB, it is to avoid a future decadal from making the same discovery of a lack of evidence. The committee's emphasis on this issue of high-quality data-gathering and reporting is not unique. The state of the profession chapter in the most recent astronomy and astrophysics decadal survey (NASEM 2023) raises the same issues.

Comparison across Sub-Fields of Planetary Science: Geology, Astronomy, and Physics

Both the Geological Society of America and the American Geophysical Union gather data from their members and publish gender demographics as a function of career stage (AGU 2018; GSA 2021). Like planetary science, there are clear indications of trends of increasing percentage of women across these fields. The geosciences, astronomy, and planetary science each have ~35 percent women while in physics the change is slower, barely reaching 20 percent. The numbers of Latinx/Hispanic Americans are also increasing across the fields, reaching 5–7 percent. Of greater concern are the numbers for Black/African Americans, which have remained around a percent or two for astronomy and the geosciences for the past two decades (Bernard and Cooperdock 2018; Rivera-Valentín et al. 2020).

Astrobiology

No data have systematically been collected on the astrobiology community. Although encompassing similarities in disciplines to planetary science, astrobiology in the United States has a single flagship conference, the Astrobiology Science Conference or AbSciCon, which began in 2000. AbSciCon has had 600 to 800 attendees roughly every other year before 2010, and 700 to 900 attendees in the 2010s. Additional regular astrobiology conferences are the Gordon Research Conference on the Origins of Life (biennial small-format meetings since 1982) and meetings of the International Society for the Study of the Origins of Life (ISSOL; every 3 years since 1957, gathering about 150 participants in post-2000 meetings). The 2020 AIP survey found that some 26 percent of respondents' primary research interest was astrobiology. Of these respondents, the AIP reported to the committee that 40 percent were women and 2.5 percent were nonbinary or another gender. There are no reliable and public data on racial/ethnic representation, members of LGBTQ communities, or other demographic axes in astrobiology.

NASA PSD Proposal Submissions and Selections

In 2016, NASA began collecting demographic data of Principal (PI) and Co-Investigators (Co-I) on proposals submitted through the NASA Solicitation and Proposal Integrated Review System (NSPIRES). Collected demographic information included binary gender, race, ethnicity, career stage, and questions related to disabilities and serious health conditions; all questions provided a "prefer not to answer" (PNA) option. Demographic information for PIs and Co-Is submitting proposals to NASA PSD was then "back-casted" to 2014 where possible. Here the gender, race, and ethnic demographic data reported for proposal PIs to the Planetary Science Advisory Committee (PAC) on June 14, 2021, is shown (Barbier and Wilson 2021).

Demographic of Proposal Submissions

In Table 16-3, the committee presents PI demographics in terms of binary gender, race, and ethnicity from proposals submitted to NASA PSD between 2014 to 2020. Based on these data, it presents trends in proposal submission rates from different groups. To preserve respondent anonymity, information for PIs who identified as American Indian/Alaskan Native, Black/African American, Latinx/Hispanic, Native Hawaiian or other Pacific Islander, Multi-Racial, and other race/ethnicity is reported as a single category, Under-Represented (Racial/Ethnic) Community (URC).

As can be seen in Table 16-3, between 2014 and 2020, 63 percent of proposal submissions were led by men and 25 percent by women, with 13 percent of respondents choosing PNA. This contrasts with the finding of the 2020 AAS-AIP workforce survey that 37 percent of planetary scientists are women. Additionally, 5 percent of PIs identified as URC between 2014 and 2020, which contrasts with the AIP survey's suggestion that 8 percent of the field is URC. Both points are, however, influenced by the identity of those that chose not to respond.

TABLE 16-3 Demographics of Proposal Submissions by Binary Gender, Race, and Ethnicity

	Identity	2014	2015	2016	2017	2018	2019	2020	Combined
Gender	Woman	22%	25%	23%	26%	24%	27%	28%	25%
	Man	64%	62%	65%	60%	62%	63%	64%	63%
	Gender-PNA	15%	13%	12%	15%	13%	10%	8%	13%
Race-Ethnicity	White	63%	64%	64%	62%	62%	62%	65%	63%
	Asian American	9%	9%	9%	9%	12%	12%	12%	10%
	URC	4%	5%	6%	5%	5%	6%	7%	5%
	Race/Ethnic-PNA	23%	22%	21%	24%	22%	21%	16%	22%

Selection of Proposals

Figure 16-3 shows the percentage of proposals selected for funding (selection rate) in terms of reported binary gender, race, and ethnicity of proposals submitted to NASA PSD from 2014 to 2020. Overall, the selection rates of women and men are reasonably comparable, averaging 23 and 20 percent respectively, although the selection rate for women has in all but one year been higher than that of men. Overall, the selection rate of proposals led by White PIs is 22 percent while for URC PIs it is 15 percent (i.e., the selection rate of White PIs is 1.5 times higher than for URC PIs).

Furthermore, Figure 16-3 shows that every year the selection rate of URC PIs was less than the overall selection rate and the selection rate for proposals with White PIs. To illustrate this point, the committee followed the framework established by Reid (2014). Reid found that women were found to be consistently less likely to be awarded time on the Hubble Space Telescope by comparing the number of selections made (N_a) to the number of expected selections (N_e) given the demographics of proposal submissions. In Reid (2014), this difference was normalized by the square root of N_e, such that a positive quantity shows over-selection, a negative quantity under-selection, and ratios much greater than ±1 indicate selections beyond reasonable expectation. The committee replotted the data presented to the PAC for grant selections made by race and ethnicity following this framework in Figure 16-4. Every year, URC-led proposals have been under-selected, and selections were below reasonable expectation (<−1) in 2014, 2016, and on average over the years. Proposals led by Asian Americans are not consistently under-selected, but selections were below reasonable expectation in 2015 and

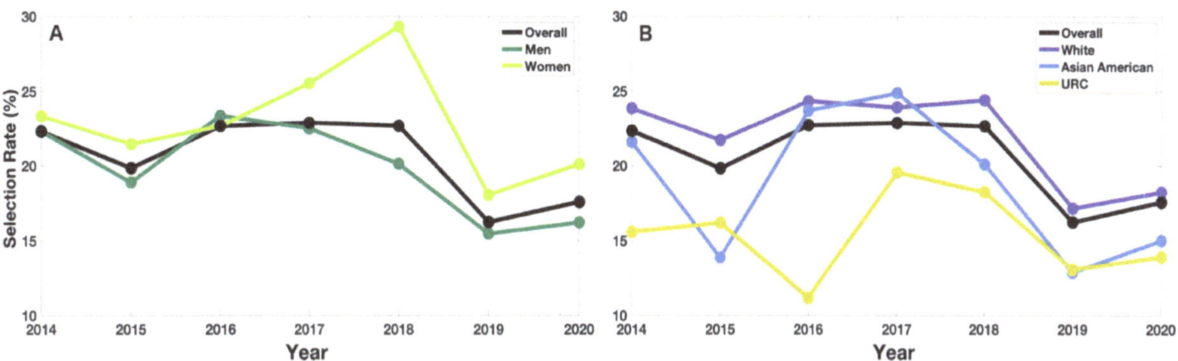

FIGURE 16-3 NASA PSD R&A proposal selection rates per year binned by (A) reported binary gender, and (B) race and ethnicity.

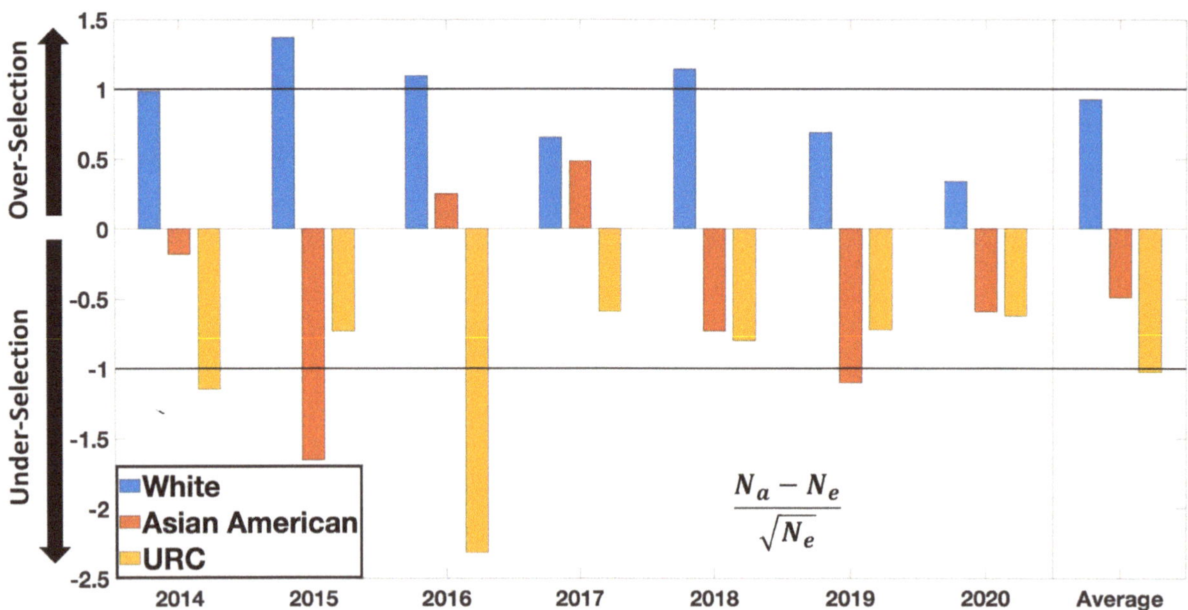

FIGURE 16-4 NASA PSD grant selection rate by race and ethnicity between 2014 and 2020 plotted following the framework of Reid (2014). The average selection rate over all years is shown on the far right. Solid black lines at ±1 show when selection rates were above (>1) or below (<−1) reasonable expectation.

2019. In contrast, proposals led by White PIs are consistently over-selected, and selections were above reasonable expectation (i.e., >1) in 2015, 2016, and 2018.

WHITE PAPERS SUBMITTED TO THE SURVEY

White papers submitted to the committee in the summer to fall of 2020 are the main source of evidence to hear directly and more broadly from the profession. Moreover, the white papers represent a labor of love, by working planetary scientists and astrobiologists seeking to improve the quality of the profession both in terms of the work that is produced and their lives. To derive the best-fidelity evidence from the white papers, the committee evaluated evidence referenced in the white papers by reaching the original source for verification. In addition to shaping the topics addressed in the Summary of Findings, the white papers on the SoP were quantitatively analyzed for additional data on the profession. A total of 36 papers were read and interpreted for this chapter, specifically referencing white papers that included verifiable data and/or data sources.

White paper authors provide insights into the geographical distribution of institutions that make the PS&AB community: 45.8 percent are from institutions in the western United States, 34.7 in the south, 8 percent in the northeast, 4.4 percent from the midwest, and 7.1 percent were international. On the gender axis, 52.5 percent of white paper authors were women, 37.9 percent were men, and 9.6 percent were nonbinary.

The white papers reveal that interest in SoP concerns is robust and high but underscore a sparse evidence base. Almost half the papers offered only an occasional statistic. The white papers discussed concerns about race/ethnicity (50 percent), general issues concerning URCs (41 percent), and gender (31 percent). A few also raised issues of disability, neurodiversity, socioeconomic class, and sexual identity and orientation. The content of the white papers covered a range of topics relevant to the State of the Profession ranged from issues concerning precollege and college education, PhD, tenure track and career pipeline, grants and funding, authorship, collaboration, mentorship, mission and field sites, conferences/workshops, child raising and work family balance, sense of belonging and workplace culture, recognition of service, a need for education and awareness of bias, dual anonymous peer review (DAPR), no proposal due dates, and broadening opportunities

SUMMARY OF FINDINGS

Urgent and Important Need for Evidence About (1) the Size and Identity of PS&AB, (2) Demographic Data of PS&AB, and (3) Workplace Climate

NASA's Science Mission Directorate has expanded its core values to embrace inclusion (NASA 2020): "dedicated to creating a multi-pronged approach that brings systemic and lasting change in this area by fostering inclusion, diversity, equity and accessibility across all elements of our work through dedicated activities and sustained engagement." Actions that bridge opportunity gaps are applauded. However, although NASA fulfills its fiduciary and legal responsibilities with respect to DEIA, it falls short and behind the times for a world leader in planetary science and astrobiology. Data gathering needs to be undertaken with a sense of evolving attempts to get at the most important information. The methodical approach undertaken by the Space Telescope Science Institute in implementing DAPR, and the NIH approach to evolve and address DEIA issues in their programs are excellent examples of the importance of iterative and continuous efforts to reach better clarity of and address the underlying issues. Progress toward equity and accountability can be assessed by evidence that supports the themes and basic tenets shown in Box 16-1 below.

The systems and mechanisms to define, measure, and publicly report on the profession are currently inadequate. The data presented above represent a tremendous investment of time and resources by individuals and organizations who have sought to understand the SoP. Yet, these data do not result from a common framework and community agreement on the scope of PS&AB, including consensus about the items for data gathering and analysis. NASA PSD stands to gain tremendously from such knowledge and the data will have credibility if they are accompanied by NASA's imprimatur.

Finding: Data on field identity and size, demographics, and workplace climate are lacking. Once obtained and regularly updated, they will rationally guide NASA's monitoring of the SoP and assist to maximize equity and accountability goals. The first wave will serve as the baseline for tracking progress and serve as a core vehicle of accountability.

Education of Individuals and Changes in Institutional Procedures, Practices, and Polices

Data collected in response to the above will serve to shape the nature and scope of (a) education of individuals within NASA and associated institutes and (b) initiating and institutionalizing efforts to transform procedures,

---

BOX 16-1
Themes and Basic Tenets for Prompt and
Demonstrable Action to Drive Visible Results

Obtain, Create, Engage, Promote and Report

- Obtain evidence about the nature of the fields of PS&AB (size, identity).
- Obtain evidence about the demographics of the fields of PS&AB.
- Obtain evidence about the quality of workplace climate for PS&AB.
- Engage in scientifically grounded education about factors that detract from bringing and retaining the strongest work force (e.g., implicit bias education).
- Promote practices and policies that lead to fair and equitable access and treatment of all individuals inside NASA and affiliated institutions.
- Create mechanisms for focus on dimensions that are relevant to evaluation of merit regardless of age, socioeconomic class, family background, learning styles, individual personality, gender identity, sexual orientation, race, ethnicity, religion, national origin, and disability status.
- Publicly make available the data and new practices and policies generated to achieve the above goals.

practices, and policies that constitute systemic barriers to scientific progress. This major initiative will require convening meetings consisting of experts in PS&AB as well as behavioral science to identify the specifics behind SoP issues or that hinder scientific progress. Examples of processes to examine and improve are advertising, recruiting, selection, hiring, onboarding, retention, promotion, lab, and field team dynamics. Likewise, processes involved in tenure, publications, special recognitions (such as keynote addresses), funding, missions, administrative loads, and teaching loads, can be unpacked to identify where and how bias can enter and to develop strategies to mitigate its effects. The findings that follow in this section are examples of issues and paths to improve the core processes involved in PS&AB.

Dual Anonymous Peer Review

Analysis of NSPIRES data collected over the past six years shows that while proposals with female PIs have been more likely to be selected than those of male PIs, proposals submitted by URC PIs are consistently less likely to be selected than those submitted by a White PI (see Figure 16-3). Several factors may contribute to this disparity, including (1) biases during the proposal review by the review panel, (2) biases during decision making process for selections and funding, and (3) opportunity gaps for URC PIs (e.g., access to mentorship and service on review panels).

The experience at the Space Telescope Science Institute (STScI) illustrates the impact and value of removing sources of bias. Prior to 2018, a systematically higher success rate in Hubble proposals for male PIs over female PIs was observed (Reid 2014). It might appear that this could reflect a bias based on seniority, not on gender, because proposals written by more senior investigators might be stronger and more likely to be selected on average, and senior scientists are proportionally more male-dominated. However, the Reid analysis found that the under selection of female PIs compared to male PIs was more pronounced for senior women than for junior women, inconsistent with this explanation. To combat implicit bias affecting the results of proposal selection, STScI instituted stepped adjustments to obscure the identity of the proposing team. The Hubble proposal review went to a fully DAPR system in Cycle 26 (Strolger and Natarajan 2019), with positive impact on the selection rate of female PI proposals (Reid 2018). Dual anonymous peer review has also helped close the success gap for proposals led by new PIs. New PIs represented, on average, only 6 percent of proposal awards for Cycles 19–25. In Cycles 26–28, after the institution of DAPR, this fraction jumped to approximately 25 percent of awards. DAPR has even been observed to increase author diversity of papers (Budden et al. 2008; Darling et al. 2015) and so has utility beyond the proposal selection process.

At STScI, DAPR was achieved through an iterative process of deliberately implementing a change in the process, measuring its impact, noting that little to no progress had occurred, and returning with further interventions, one step at a time, until DAPR had achieved its goal of greater fairness in the review process. This approach and experience can serve as a model for all procedures and processes going forward.

Bias exists in other selection programs beyond Hubble. A study of National Radio Astronomy Observatory and Atacama Large Millimeter Array (ALMA) proposals has indicated a significant gender-related bias affecting proposals (Lonsdale et al. 2016). ALMA has instituted DAPR in 2021 (Cycle 8). For European Southern Observatory (ESO) proposals, the success rates are 16.0 ± 0.6 percent for women and 22.0 ± 0.4 percent for men, respectively (Patat 2016; ESO 2021).

NASA has begun experimenting with DAPR in a few of its ROSES programs. Initial data appear promising: for example, in the Astrophysics Data Analysis Program, the percentage of proposals with female PIs in the top two proposals ranked from each panel rose from about 15 percent in 2018 (no DAPR) to about 30 percent in 2020 (with DAPR), bringing the success rate of female-led proposals in line with the fraction of proposals led by women in the total pool (per presentation by Michael New to the committee).

Finding: DAPR mitigates bias in proposal selections. The process by which it was achieved at STScI is a model for improving other procedures and policies.

No Proposal Due Dates

Having immovable proposal due dates has led to numerous issues. Examples of the issues created by rigid due dates for proposal submissions include (1) requests to shift due dates for proposal submissions owing to natural and/or social disasters, which can delay other programs; (2) program officers have high-amplitude, long-duration spikes of effort centered on reviews, which have created workflow challenges; (3) owing to budget cycles misaligning with proposal cycles, programs with due dates late in the year can be disproportionately affected to correct for cost overruns during a fiscal year; (4) overburden of smaller institutions with more limited institutional support; (5) disproportionate pressure on heads of family, caregivers, and women.

Agencies such as NSF and DOE are experimenting with a no due date (NoDD) format for some of its programs. NASA began to implement NoDD in 2021 for several solicitations within ROSES. NASA consulted with NSF to understand the benefits and pitfalls of NoDD implementation. Motivators expressed to NASA for NoDD programs include (NASA 2021) (1) illness, natural phenomena, family circumstances; (2) eliminating conflicts between due dates; (3) flexibility for small institutions (important for diversity, increased time flexibility for thinly staffed AOR departments); (4) separating inspiration from the proposal cycle (less time between having a new idea and proposing it); (5) allowing proposers to participate in reviews more readily; (6) providing additional flexibility for program officers to manage workload; and (7) spreading budget risks naturally across programs. Several challenges have also been identified, including (1) extra efforts to avoid conflicts of interest during reviews and (2) increases in workload, albeit spread out in time, for program officers for forward planning of budgets and timely awards.

> **Finding:** NASA has begun experimenting with NoDD in the ROSES program. Initial data are not yet available. Increased flexibility for PIs and program officers could be of benefit to the community overall (e.g., through improved selection rates). Monitoring whether NoDD has the intended effect over deadlines as a motivator, or whether it creates differential impact on submissions or challenges in administration of proposal reviews will be important (see also Research and Analysis chapter).

Lack of Diversity in Mission Teams

The AAS-AIP 2020 demographic survey found that men were a PI or Co-I on mission proposals significantly more than women (Figure 16-5). Additionally, this survey found that non-LGBTQ+ individuals were a PI or

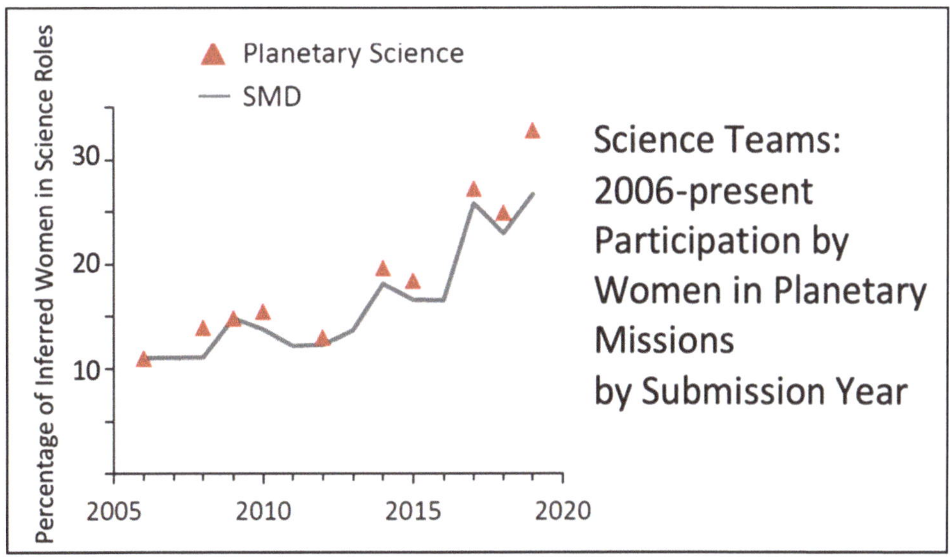

FIGURE 16-5 Percentage of women on competed planetary mission teams from 2006 to 2019. Here, Rathbun (2017) inferred gender from names and pictures of team members (i.e., inferred gender). The inferred percentage of women on mission teams is lower than the representation of women in the field but is increasing, particularly for planetary science compared with the other divisions in the NASA Science Mission Directorate (SMD).

Co-I more frequently than LGBTQ+ respondents. No significant differences were observed in mission proposal participation based on disability. While the AAS-AIP 2020 demographic survey report found no significant race/ethnicity differences in being a mission PI, it does note that Hispanic/Latinx respondents served as Co-Is less frequently compared to other racial groups (AAS-AIP 2020). The combined under-selection of proposals led by URC PIs, and, as shown by the AIP 2020 survey, the higher representation of URC planetary scientists in nonuniversity jobs that depend more on grant funding, may contribute to the lower number of URC researchers in planetary science and, possibly, astrobiology.

Finding: NASA mission team demographics do not reflect the broader planetary science community.

Bridging Mission Opportunity Gaps with Participating Scientist Programs

Rathbun (2017) highlighted that NASA's participating scientist and guest investigator programs have increased the inclusion of women on mission teams (see Figure 16-5). As of 2017, the average percentage of women selected through these programs, 24 percent, is higher than the fraction of women on the original teams, although still lower than the 2017 fraction of women in the field (about 30 percent). For the Mars Science Laboratory (Curiosity) mission, the participating scientist program brought the percentage of women on the science team from 12 percent on the initial 2012 team to 31 percent in 2018, in line with the fraction of women in the community (Zorzano 2020). Participating scientist programs have also been highlighted to provide mission experience for early-career scientists. Prockter et al. (2020) found that almost one-third of participating scientists are within 7 years of their PhD, and half are within 10 years of their PhD.

Finding: Participating scientist programs can be a vehicle to achieve a measurable and positive impact on the demographic diversity of mission teams if the imperative to do so is emphasized. The improvements achieved in the participation of women in PS&AB are encouraging, but do not necessarily imply that all aspects of gender issues have been corrected.

Work-Life Balance

Work-life balance is voiced as the leading cause negatively impacting career satisfaction in for all researchers in the PS&AB community, especially for women and LGBTQ+ individuals (AAS-AIP 2020). Brue (2019) highlighted the challenges women in leadership positions face when managing work and life obligations. Surveys show that both men and women look for jobs that provide flexibility to balance work and family, but women are more likely to cite work/family balance difficulties as a major reason behind lack of gender diversity in STEM (40 percent versus 28 percent), with women frequently citing work-life balance issues as a primary reason (Funk and Parker 2018).

A 2019 colloquium presented by the associate administrator for the NASA Science Mission Directorate, to advise prospective mission PIs on "Writing Successful Proposals: Observations from NASA," indicated an expectation that "[A PI's] life will be taken over, including evenings and holidays," reflecting perhaps the reality of a highly competitive process, and once selected, mission cost and schedule constraints. However, such unbalanced work-life expectations clash with the importance of work-life-balance for a diverse mission PI pool, and thus can be a deterrent to broadening the diversity of the mission PIs, and counter to the colloquium's intended goal.

Finding: Lack of work-life balance in PS&AB can deter community members from staying in the field and leading NASA PSD missions.

A Closer Look at Publications as Key Indicators of Merit

Publications are the vehicle by which scientific progress is primarily documented and the central factor in assessing the merit of researchers. It is the life blood of any science. Ensuring the highest standards by which research is produced and evaluated merits attention in any discussion of the SoP. Studies have shown bias in the peer review process and opportunity gaps that may lead to differences in publishing that reflect the role of social rather than scientific factors. For example, the American Geophysical Union (AGU) has tracked authors' gender, age, race, and ethnicity at its meetings and in publications (Lerback et al. 2020). Data suggest gender bias in who is invited to review. AGU addressed the reviewer issue by encouraging a more diverse reviewer pool on the manuscript submission form, which resulted in an increase in women reviewers. Women corresponding authors have a higher acceptance rate on their submissions, however, they submit fewer papers overall. Author networks are important in professional development and career advancement. Analysis of author networks in AGU publications (Hanson et al. 2020) showed that male scientists tend to have a higher proportion of male co-authors and more international collaboration than women. Because most planetary research is conducted in teams, addressing the source of such differences may assist early-career scientists in improving their scientific output. Tracking authorship, reviewer, and citation statistics by professional societies and organizations that manage publications can help identify and correct potential sources of bias.

Finding: There are important factors in addition to publications that merit consideration when assessing professional merit.

Equity at Conferences

Selection of session conveners, as well as oral versus poster assignments at conferences, have demonstrated biases in the planetary sciences, thereby creating measurable inequities. Attendees from URC at the American Geophysical Union winter meeting, which has a significant attendance by planetary scientists (AIP 2011), are the least likely to have their work selected for oral presentation. URC women are consistently the least likely to be selected to present when compared to URC men and non-URC women and men (Ford et al. 2019). Many scientific societies analyze who speaks at conferences, especially in high-status events like, for example, presidential symposia, keynotes, and awardees of prizes. They proceed over time to rectify bias. In PS&AB, such efforts to document these biases have been undertaken but methods have relied on assumed gender by either name or appearance, which is problematic for many reasons including accuracy and disadvantaging nonbinary and trans members of the community (Strauss et al. 2020). Conference environments also present challenges to members of the community through barriers to accessibility and harassment at meetings. A general awakening around equity is occurring that involves not only gender but race/ethnicity, disability status, and the treatment of sexual and gender minorities, who face higher rates of harassment.

Finding: There is a need for systematic tracking and reporting by institutions that organize conferences of the demographics of conference attendees and of the visibility given to their scientific contribution.

Tenured Positions

The demographic composition of present tenure/high-status academic positions tend to be dominated by white males. Current recruitment and selection processes may continue to limit opportunities unless sources of explicit and implicit bias are addressed. Tenure is based on the university's needs, not solely the achievements of those seeking tenure, and the university sets the rules and controls the outcome. Changing budgets and administrations vary the standards for those receiving tenure over time, making comparisons with earlier cases difficult. In addition, when new faculty members differ from current members, establishing relationships can be more difficult, creating potential challenges for new faculty members, particularly members from underrepresented groups (Phinney 2009). Within the broad umbrella of higher education, URC members held 12.7 percent of faculty positions in 2013, an increase from 8.6 percent in 1993, but only 10.2 percent of those faculty positions were tenured positions. Similarly, in 2013 women held 49 percent of all faculty positions, an increase from 39 percent in 1993, but only 38 percent of those faculty positions were tenured positions in 2013. (Finkelstein et al. 2016). The 2020 AIP survey showed that URC plus the Latinx community compose 9.9 percent of planetary scientists, 12.6 percent of tenure-track faculty, and 8.9 percent of tenured faculty. Similarly, women accounted for 35 percent of planetary scientists, 52 percent of tenure-track faculty, and 24 percent of tenured faculty.

> **Finding:** Tenured/high-status positions are under-populated by URC members and women. Keeping track and understanding the reasons for the imbalance is critical and to the extent that a disparity exists—that is, the pipeline contains greater representation of particular groups than the selection rates reveal, the reasons for this disparity need to be understood and corrected.

Broadening Opportunities to Advance the State of the Profession

Recognizing the Value of Community Service Work

Work through professional organization committees (e.g., Professional Culture and Climate Subcommittee of the DPS, Committee for the Status of Women in Astronomy, and Committee for the Status of Minorities in Astronomy) is a valuable component of a vibrant community that is often undervalued. URC and women are burdened with an uneven share of community work because the inequities they face heighten their drive to serve, and/or because they are solicited to be the face of the community.

> **Finding:** Community service and administrative duties are important contributions, but ones that tend to be distributed inequitably across individuals and appear to fall disproportionately on members of particular groups. Given historical traditions of who performs such work, members of some groups (e.g., women) often willingly take on such work and do it effectively. This places a disproportional burden on URC members and women by virtue of their smaller numbers and other (number, culture/family, etc.) responsibilities.

Service work also includes education and public outreach (EPO). During the past decade, EPO efforts at NASA have become centralized; while this has been beneficial in some respects (Erickson 2021), it has also resulted in loss of opportunity for engagement of planetary scientists and astrobiologists in EPO. Grantees and NASA mission teams no longer receive funding for EPO activities. Instead, grantees are left to volunteer time (i.e., conduct unpaid work) to conduct these important activities as subject-matter experts involved by dedicated funded EPO teams. The former direct link between researchers and EPO audiences is important for advancing the profession (Schmidt et al. 2020), for example, as researchers from underrepresented backgrounds serve as role models and inspiration for future generations. They also can help serve as mentors and guides for first-generation college and graduate school students of all backgrounds.

> **Finding:** Funding for EPO can provide researchers from underrepresented backgrounds the opportunity to serve as role models and provide inspiration for future generations as mentors and guides for first-generation college and graduate school students of all backgrounds.

Inclusion of URC Members in Initiatives to Improve the Diversity of the Community

Interest in physical sciences seems evenly distributed among the U.S. population in childhood and teenage years. However, the representation of women and URC in the physical sciences starts to decrease in college (see Figure 16-1). Despite improvement along the gender axis, the PS&AB communities still have low racial and ethnic diversity, especially regarding Black/African American scientists (Bernard and Cooperdock 2018; Rivera-Valentín et al. 2020), indicating that current and past initiatives have not helped increase membership of some underrepresented racial groups (Rathbun et al. 2020).

Researchers who encounter few community members sharing traits of their identity, especially those at the intersection of two or more underrepresented axes (Cole 2021), may experience a lower sense of belonging in PS&AB. Additionally, faculty and staff at institutions serving URC are insufficiently represented in the community of PIs and Co-Is. Yet, these faculty train the next generation of scientists who could help address existing workforce imbalances. Many URC faculties are passionate about returning knowledge to and empowering their communities, but as they are few, they also tend to receive many more requests from students as well as committees (Schmidt et al. 2020), which can be detrimental to their research and teaching activities. Although it is becoming increasingly clear that traditional social categories are becoming more complex, the committee has retained their usage given the sparsity of available data. For instance, in the domain of race/ethnicity, various types of bi-racial and multiracial identifications are on the rise. In addition, gender which was largely viewed as a binary can no longer be viewed as such. Additionally, when crossing two or more levels of race, gender, socioeconomic status, sexuality, or age, it becomes clear that the effects of bias may fall disproportionately on those who sit within particular intersectionalities (URC sexual minorities for instance).

There is limited participation of primarily undergraduate institutions (PUI) and URC researchers in NASA PSD research, partially owing to lower numbers of faculty at these institutions who do research. Engagement of URC in research and education has been fostered through the Minority Institute Astrobiology Collaborative Program, as well as the previous Astrobiology Faculty Diversity Program and its precursor Minority Institution Research Support Program (MIRS); however, the culture in which they have been implemented has limited the efficacy of such dedicated programs. Additionally, such dedicated programs are prone to cancellation. A previous program that NASA supported and has since eliminated for unknown reasons is the Harriett G. Jenkins Predoctoral Fellowship Project (JPFP), which supported students from underrepresented communities, students with disabilities, and low income/first generation students. In addition to graduate stipends and summer research opportunities, the award provided a community of awardees that would meet and share their research annually. Today, there is still an active JPFP alumni network.

A key stage for career development is the period just after earning the PhD. Many budding planetary scientists and astrobiologists need information about available job opportunities; how to function as part of a larger research group or team; how to succeed in academia; and how to develop skills to write and review papers and proposals. NASA's Planetary Science Division, including its Astrobiology Program, manage successful programs (FINESST, NASA Postdoctoral Program, and Early Career Award) to help students and early-career researchers establish themselves. In NASA PSD mission announcements of opportunity, inclusion of early-career scientists, engineers, and managers is expected and evaluated (New 2021). NASA's Science Mission Directorate has sponsored workshops on proposal writing at conferences and institutions.

NASA's Astrobiology Program, too, has funded such summer and winter programs: the Nordic/NASA summer and winter schools in Iceland and Hawaii, respectively, the NExSS winter school, and the International Summer School in Astrobiology hosted in Santander, Spain. These and the graduate student and postdoc-led Astrobiology Graduate Conference have shaped a collegial and closely networked astrobiology community. However, opportunities to attend such schools have decreased in the past decade.

Finding: NASA's engagement programs have supported the increased representation of women in planetary science over time, but to date have had a lesser impact on URCs. Measures to increase participation of URC students in NASA's student and early-career fellowship funding programs, and in fellowship programs that facilitate engagement of NASA-funded PS&AB researchers with faculty URCs, are crucial to improving racial and ethnic diversity in PS&AB. Previous NASA programs—such as NASA's JPFP and MIRS programs and the National Research Council Resident Research Associateships—were valuable activities to this end. Long-term commitment to these efforts is essential to achieve measurable progress.

Improving NASA Mission Team Diversity

NASA PSD has initiated efforts to diversify the pool of possible mission PIs through the PI Launchpad workshop program (NASA 2022). A first, invitation-only workshop was held in 2018 in Washington, DC, followed by a workshop at the University of Arizona in 2019 with an open application call. In 2021, a virtual workshop was held at no cost to participants. Other programs aimed at broadening the pool of NASA PSD mission participants include the Planetary Science Summer School (PSSS) hosted by JPL and sponsored by NASA PSD, where teams of students develop mission concepts. Data show most alumni from the PSSS are employed or conducting postdoctoral research at NASA centers; FFRDCs; universities; or other research or aerospace centers (Budney et al. 2017). NASA Goddard Space Flight Center (GSFC) offers a similar Planetary Science

Winter School to participants located at GSFC. Such programs help bridge opportunity gaps by increasing access to information about mission concept development, design, and proposal. Last, internship positions, like NSF Research Experience for Undergraduates to work with mission teams, can provide valuable mission experience at even earlier career stages.

> **Finding:** The demographic makeup of the U.S. population is evolving toward increased racial and ethnic diversity. The demographics of the PS&AB community, as is the case with some other sciences, are substantially out of sync with the U.S. demographics. Increasing diversity and representation in NASA PSD missions requires a concerted effort to engage members of URCs at early stages of career/education. The PI Launchpad, Planetary Science Summer School, and other similar programs are excellent vehicles to broaden the pool of participants.

Creating an Inclusive and Inviting Community Free of Hostility and Harassment

Codes of Conduct as Tools to Promote Equity and Inclusion

Members of research and mission teams are expected to follow "Rules of the Road" (ROTR) policies, which outline topics such as team membership, authorship, and data sharing. NASA PSD opportunities also reinforce anti-harassment policies in opportunities. A code of conduct (CoC) augmenting ROTR documents can help enable a safe and equitable environment on mission teams (Diniega et al. 2020). A CoC is also indispensable to promote a culture of safety and inclusion for conferences, workshops, and research teams. NASA PSD is implementing a CoC for peer reviews (New 2021).

An effective CoC documents acceptable behavior, describes how policies will be enforced, provides clear instructions on how to report incidents, and explains the consequences/enforcement mechanisms for rule violations. It is important that CoCs be reviewed regularly and updated, if necessary, particularly for long duration missions and other long-lived teams.

> **Finding:** Codes of conduct enable a culture of safety and inclusion.

Enabling Safety, Equity, and Inclusion in Field Work

Terrestrial field work is a crucial component of planetary and astrobiology research. Current barriers to effective and safe field research include physical safety hazards inherent to field sites, accessibility for persons with disabilities, and harassment in field settings (Richardson et al. 2020). An egregiously high 64 percent (N = 423/658) of participants in the Clancy et al. (2014) study on field research report personal experiences with sexual harassment, with 70 percent of women and 40 percent of men in the study reporting sexual harassment. LGBTQ+ scientists in physics have been found to experience isolation exacerbated by remote environments and related to their sexual orientation or gender identity (Atherton et al. 2016; Vander Kaaden et al. 2020; USAP 2022).

Field work often involves access to indigenous lands. Ongoing and substantial cultural changes aim to honor and equitably engage native peoples, who request prior and informed consent for access to their sacred sites. Both long-term and modern events continue to highlight the importance in recognizing the mores of Indigenous cultures (Alpaslan-Roodenberg et al. 2021; Sahagún, 2021). These concerns can be addressed with thoughtful CoC and training that require field safety plans outlining physical and mental safety, strategies for reducing barriers to field work, and the need for coordination with local communities, including Indigenous community leaders. This can be facilitated by funding such training, funding students and early-career scientists performing field work, and providing avenues for appeals to report issues and concerns over unsafe or disrespectful practices in the field.

> **Finding:** Relevant focused education about changing norms and greater sensitivity to history, and reflecting these in codes of conduct, can make field work safer and more inclusive to all parties involved (participants and communities on whose land field work is carried out).

Finding: Engagement of Native communities require thoughtful engagement and the creation of genuine relationships that are respectful of traditions and gratitude for their contribution to the scientific process (Kaluna et al. 2020). This goes beyond field work but is particularly salient for field work conducted on Native lands.

Improving the Conference and Workshop Experience

Ensuring safe conference environments continues to be a community concern (Bennett et al. 2020; Diniega et al. 2020; Vander Kaaden et al. 2020). NASA deserves credit for describing its policy on discrimination and harassment, including reporting procedures, in its funding solicitation for topical workshops, symposia, and conferences, and in its solicitations and mission announcements of opportunity. Similar policies are merited for any programmatic funding that would result in conference support. It is incumbent upon NASA PSD to continually remind the PS&AB communities and institutions that support these fields that everyone is welcome as we strive to create an environment free of harassment and discrimination of any form.

Finding: Harassment and concerns for safety are most prevalent amongst groups underrepresented in STEM (Clancy et al. 2017).

Finding: Initial LGBTQ+ inclusive programs have made positive change in the planetary science and astrobiology community and their continuation will increase their impact (Vander Kaaden et al. 2020).

RECOMMENDATIONS

An Evidence Gathering Imperative about the State of the Profession

The committee recommends that NASA PSD create a three-part foundation of evidence to examine and understand the community so that it can confidently proceed to advance the SoP. With this, NASA PSD can ensure that it is in command of facts and figures of the PS&AB communities. The expectation is for NASA to augment and deploy the resources necessary to effect progress based on evidence and proactive steps to address this recommendation and the recommendations that follow. The following three types of data are essential:

- *Disciplinary size and identity.* The identity and boundary conditions of PS&AB are amorphous, given their interdisciplinary nature. Interdisciplinarity is a strength, but accurate information is still needed about (1) the size of these disciplines, (2) the feeder disciplines from which PS&AB identities develop over time, and (3) the spread and location of individuals and institutions that identify as PS&AB (e.g., data on colleges, universities, other institutions with programs in PS&AB and individuals who identify as such). Data are needed on the number and rank of employees of these communities at each institution, the evolution of the field from bachelor's through PhD programs and faculty, annual numbers of PhDs in relevant disciplines, and percentage of U.S. and foreign nationals contributing to the profession, and other appropriate dimensions of data-gathering.
- *Demographic composition.* At present the demographic variations that make up PS&AB are poorly documented, their data collection depending on the goodwill of individual scientists or professional organizations. Without concrete data about the demographic variations that make up PS&AB, NASA cannot know if it is utilizing the best possible talent available and, by extension, cannot ensure the competitiveness of American PS&AB. With concrete demographic data on all dimensions, including but not restricted to age, gender identity, sexual orientation, race/ethnicity, citizenship/residency status, and disability status, NASA's work to improve the quality of PS&AB will be targeted and precise.
- *Workplace climate.* NASA PSD has not conducted climate surveys of the disciplines that make up PS&AB. As such it currently lacks knowledge of the working conditions of the scientists involved regarding (1) professional issues (e.g., grants, conferences, mentorship) and (2) social factors that are part of the fabric of all sciences and affect the quality of science (e.g., work-life balance, workplace safety, mental health), particularly during

stressful periods such as the present pandemic. Climate surveys are now routine in many organizations. A focused effort by NASA PSD to produce evidence about the climate of PS&AB would enable them to reveal hidden issues that unintentionally, yet adversely, impact American competitive advantage in PS&AB.

The committee recognizes the complexity of obtaining such data, given the many academic institutions, research institutes, and for-profit corporations that house PS&AB communities. However, NASA PSD as a major supporter and funder of individuals and groups at these institutions has convening power to engage associated institutions in regular discussions of matters concerning equity and accountability. Equity and accountability require accurate and complete data about the facts at the ground level. NASA PSD can create a pathway for the highest quality data on the professions to be regularly collected by engaging with the best behavioral science knowledge about such data collection. An overall goal would be to commence work immediately to have a complete set of data within a reasonable timeframe of 3 years into the period of this decadal. These data, providing an accurate sense of (1) size and identity, (2) demographic composition and (3) workplace climate, will become the foundation on which NASA PSD can successfully identify where resources need to be invested. This in turn will allow NASA PSD to assure itself that it is exploring and finding and retaining the best talent in the world, bringing the same attitude it brings to the science of planetary exploration to matters of SoP.

It is important that such data collection be conducted with proper input to ensure quality through an appropriately constituted advisory body with the authority to recommend actions as required. Given that this effort will take time to set up and implement, it is advisable to continue other activities that can be undertaken alongside the preparation, administration, and analysis of these data. Additionally, some effort is warranted to persuade the community that participation in climate surveys is important to improve the state of the profession.

Recommendation: NASA PSD and NSF, with its wide experience with programs such as the Louis Stokes Alliances for Minority Participation and Organizational Change for Gender Equity in STEM Academic Professions (ADVANCE), should make it a priority to obtain currently lacking evidence about fundamental aspects of the state of planetary science and astrobiology communities. NASA PSD and NSF should engage with experts to undertake data collection on 3- to 5-year cycles with a focus on obtaining accurate data on

- **The size and identity of PS&AB, given their deeply interdisciplinary nature;**
- **The demographic composition of PS&AB along all relevant dimensions; and**
- **The workplace climate at NASA PSD and affiliated institutions, as well as the social issues that facilitate or impede scientific progress in PS&AB.**

Undertaking Education of Individuals About the Costs of Bias and Improvement of Procedures, Practices, and Policies to Create Institutional Change

By engaging in education about bias of all forms (explicit and implicit) that holds back progress in PS&AB, NASA PSD will be able to promote changes in existing policies and procedures to remove or compensate for bias. When people are persuaded about the need for change, they will participate in creating that change.

By engaging in the work of the above recommendation, NASA PSD will have in hand a valuable map of the interdisciplinary fields that constitute PS&AB. It will then possess full and accurate knowledge of the demographic character of the population of PS&AB, and the community's assessment of the climate issues. NASA PSD can assess where bias is evident and where it does not exist. Institutions of every kind, including organizations devoted to STEM fields, are transforming themselves by engaging with newly developing ideas about implicit bias and where such bias unintentionally plays out in the critical work of any scientific organization. It is important that this effort be conducted in parallel with that of the recommendation above.

It is important that NASA, NSF, affiliated institutions, and professional societies work to mitigate bias at all levels. NASA PSD may provide leadership in creating, sharing, and encouraging the propagation of educational experiences about bias, both within NASA and in the organizations, NASA supports and funds. Human beings and their workplace circumstances are such that they lead them to be both creators and receivers of bias. That is, although groups surely differ in the degree to which they have historically experienced discrimination and experience it today, each person can also be the propagator as well as the target of bias. Moreover, bias does not lie just in the minds of the human perceiver. Members of historically disadvantaged groups are known to hold themselves and others of their group back because of biased expectations of their own abilities. To add to this, bias is systemic, by which we understand that in addition to bias being hosted in the minds of individuals, it is also embedded in the social infrastructure of organizations. High quality education, credible to scientists and engineers, is necessary to confront something so hidden that it is often only apparent following the application of the best methodologies available.

It is sometimes wrongly assumed that implicit bias education is sufficient to reduce and remove bias. Implicit bias education is both deeply necessary and wholly insufficient to reduce the impact of bias. While NASA engages in high-quality education on implicit bias to change hearts and minds, a review of its own procedures and policies is also merited. NASA has set an example in its tackling of DAPR following the pioneering effort at the Space Telescope Science Institute for the Hubble Space Telescope program. This type of continuous effort to improve a procedure by obtaining data regularly and returning to the drawing board to improve the process is needed in all aspects of the work, well beyond grant reviewing. The improved situation on gender diversity can serve as an impetus for additional, broader diversities. With proof that we are capable of change, steps can be taken to reach the next stages of equity and accountability. Continued attention to issues of gender is merited, given the data in presented earlier in this chapter. Starkly, involvement of members of underserved communities, especially African Americans, shows a deeply troubling stagnation at all levels. It is for NASA leadership to step in immediately and decisively to understand and improve this state of affairs. Evidence on LGBTQ+ communities and those with disability is sorely lacking but underrepresentation, based on data from STEM fields broadly, is likely.

Recommendation: NASA PSD should adopt the view that bias can be both unintentional and pervasive. To address potential bias issues, NASA should:

- **Seek the expertise of behavioral scientists to develop methods for analyzing its decision-making practices and procedures (e.g., advertising, recruiting, selection, hiring, onboarding, promotion, compensation, managing teams, fieldwork, and mission planning).**
- **Determine where bias does, and does not, play a role and work with the evidence to reduce and eliminate bias from its procedures wherever it is found to exist.**
- **Proactively engage with the PS&AB community in the development of creative initiatives to uncover and mitigate bias in existing processes.**
- **Consider evidence-based bias education for itself and associated institutions. Honest discussions of policies and practices that no longer serve the functioning of modern scientific enterprises should be sought with enthusiasm that mirrors the enthusiasm NASA PSD brings to its scientific innovation.**
- **Follow education at a foundational level with discussions among individuals within NASA PSD with authority to effect change.**
- **Include regular focus on different aspects of the issues, for example, opportunities for tenure of NASA-funded PS&AB members in academia, advancement to senior civil service positions at NASA centers, peer-reviewed research funding opportunities, addressing climate issues, participation in space mission teams, keynote presentation opportunities at scientific conferences, and awards by professional societies.**
- **Publicize the procedures and policies that have been reviewed and transformed each year.**

Broadening Opportunities to Advance the State of the Profession

While there have been benefits to centralizing public engagement in NASA's Science Activation Program initiatives (Erickson 2021), education and public outreach and engagement activities by members of the community have been left unfunded. Engaging URCs at the pinch point of transition from high school to college and providing support systems (including introductory courses) to encourage and retain them along the path of PS&AB is going to be essential to create and grow a diverse community. For example, the opportunity to propose outreach activities as an optional extension to funded R&A grants would allow grantees to make a positive impact on community diversity and inclusion activities.

NASA missions, particularly those to distant parts of the solar system, can span multiple decades from initial planning through launch, cruise, and operations, to end of mission. Multiple generations of scientists and engineers are involved in their planning, development, and operations. Succession plans offer an opportunity to grow the diversity of the community as part of a long term and sustained effort.

Recommendation: NASA PSD should revisit the centralization policy on public engagement and consider mechanisms to support direct engagement of planetary scientists with members of society, particularly students in STEM fields.

Recommendation: PSD should regularly evaluate programs that enhance participation of students and faculty from URCs; fellowship programs that facilitate engagement of NASA-funded planetary scientists and astrobiologists with faculty at URC institutions; and mechanisms for supporting education and outreach as an integral part of research via, for example, the inclusion of outreach activities as optional add-ons to R&A grants, or as a requirement for missions or cooperative agreements.

Recommendation: PSD should strengthen and expand programs aimed at educating the community about the mission proposal process (e.g., PI Launchpad) and actual mission operations (e.g., participating scientist programs), particularly to reach out to URCs. Providing access to personnel or tools that can help guide investigators through the process should be considered, including participation as contributing members of the mission teams.

Recommendation: NASA and PSD should reinstate the Harriett G. Jenkins and similar predoctoral fellowship projects as part of an effort to retain members of URC in the fields of PS&AB prior to them reaching existing pinch points at which substantial decline in URC representation is seen in both fields.

Creating an Inclusive and Inviting Community Free of Hostility and Harassment

Creating an environment that reinforces welcoming and inclusive behaviors, and where members of the communities (i.e., researchers, support staff, local communities) feel safe at work and are treated with respect and appreciation is essential to a healthy scientific community. PS&AB research can involve field work in planetary analogue environments where physical and mental isolation and interpersonal frictions can arise and infringe on CoC policies and agreements. Local community concerns and the cultural mores of indigenous communities can also be a source of friction between researchers and local communities. Basic principles for the success of field campaigns that acknowledge the importance of these stakeholder communities include the following (Appaslan-Roodenberg et al. 2021):

1. Ensuring that all regulations are followed in the places where they work and from which field samples are derived;
2. Preparing a detailed plan prior to beginning any field study;
3. Minimizing damage to study areas and their environs;
4. Ensuring that data are made available following publication to allow critical reexamination of scientific findings; and
5. Engaging with other stakeholders from the beginning of a field study and ensure respect and sensitivity to stakeholder perspectives.

Recommendation: PSD should implement codes of conduct (CoC) for funded field campaigns, conferences, and missions, and should expect acknowledgement of receipt and understanding. The CoC should be codified, reviewed, and updated at regular intervals. An effective CoC should outline expected behavior, explain unacceptable behavior, explain how policies will be enforced, provide clear instructions on how to report incidents, and explain consequences of violations. The process should demonstrate sensitivity to the difficulty of bringing forward accusations and to the rights of the accused.

Recommendation: NASA PSD and affiliated institutions should clearly identify a point of contact or ombudsperson as part of the CoC to provide access to individuals who experience violations to the CoC. The egregious nature of sexual harassment reported in field work requires immediate attention by NASA.

REFERENCES

AAS (American Astronomical Society). 2021. "Partnership Between the DPS and the National Society of Black Physicists." Division for Planetary Sciences. https://dps.aas.org/leadership/nsbp_parnership.

AAS-AIP (American Institute of Physics). 2020. A.M. Porter, S. White, and J. Dollison, "2020 Survey of the Planetary Science Workforce," American Astronomical Society and American Institute of Physics. https://dps.aas.org/sites/dps.aas.org/files/reports/2020/Results_from_the_2020_Survey_of_the_Planetary_Science_Workforce.pdf.

AGU (American Geophysical Union). 2018. "AGU Sections Demographic Breakdown by Gender and Career Level." https://honors.agu.org/files/2018/09/2018-section-membership-by-gender-and-career-stage_Sept12.pdf.

AIP. 2011. S. White, R.V. Chu, and R. Ivie. "2011 Survey of the Planetary Science Workforce." American Institute of Physics. https://lasp.colorado.edu/home/mop/files/2015/08/Report.pdf.

Alpaslan-Roodenberg, S., D. Anthony, H. Babiker, E. Bánffy, T. Booth, P. Capone, A. Deshpande-Mukherjee, et al. 2021. "Ethics of DNA Research on Human Remains: Five Globally Applicable Guidelines." *Nature* 599:41–46. https://doi.org/10.1038/s41586-021-04008-x.

Atherton, T.J., R.S. Barthelemy, W. Deconinck, M.L. Falk, S. Garmon, E. Long, M. Plisch, E.H. Simmons, and K. Reeves. 2016. *LGBT Climate in Physics: Building an Inclusive Community*. College Park, MD: American Physical Society. https://www.aps.org/programs/lgbt/upload/LGBTClimateinPhysicsReport.pdf.

Banaji, M.R., S.T. Fiske, and D.S. Massey. 2021. "Systemic Racism: Individuals and Interactions, Institutions and Society." *Cognitive Research: Principles and Implications* 6:82. https://doi.org/10.1186/s41235-021-00349-3.

Barbier, L. and C. Wilson. 2021. "Demographic Data Presentation to: Planetary Science Advisory Committee." NASA Office of the Chief Scientist. https://science.nasa.gov/science-pink/s3fs-public/atoms/files/07-Barbier-Demographics-061421.pdf.

Bennett, K., M. McAdam, M. Milazzo, P. Garcia, J. Shelton, P. Gardiner, S. Diniega, et al. 2020. "The Preventing Harassment in Science Workshop: Summary and Best Practices for Planetary Science and Astrobiology." White paper #474 submitted to the Planetary Science and Astrobiology Decadal Survey 2023–2032. *Bulletin of the AAS* 53(4).

Bernard, R.E., and E.H.G. Cooperdock. 2018. "No Progress on Diversity in 40 Years." *Nature Geosciences* 11:292–295. https://doi.org/10.1038/s41561-018-0116-6.

Brue, K.L. 2019. "Work-Life Balance for Women in STEM Leadership." *Journal of Leadership Education*. https://doi.org/10.12806/v18/I2/R3.

Budden, A.E., T. Tregenza, L.W. Aarssen, J. Koricheva, R. Leimu, C.J. Lortie. 2008. "Double-Blind Review Favours Increased Representation of Female Authors." *Trends in Ecology & Evolution* 23(1):4–6.

Budney, C.J., L.L. Lowes, K.L. Mitchell, A.S. Wessen, and C.D. Bowman. 2017 (March 20–24). "Updated Career and Workforce Impacts of the NASA Planetary Science Summer Seminar (PSSS): Team X Model 1999–2016." Presentation at the Lunar and Planetary Science XLVIII Conference. The Woodlands, TX. https://www.hou.usra.edu/meetings/lpsc2017/pdf/2828.pdf.

Charlesworth, T.E.S., and M.R. Banaji. 2021. "The Relationship of Implicit Social Cognition and Discriminatory Behavior." Chapter in *Handbook on Economics of Discrimination and Affirmative Action*, A. Deshpande, ed. New York: Springer.

Charlesworth, T.E.S., and M.R. Banaji. 2022. "Patterns of Implicit and Explicit Attitudes IV: Change and Stability from 2007–2020. *Psychological Science* 33(9). https://doi.org/10.1177/09567976221084257.

Cheney, S., and I. Shattuck, co-directors/producers. 2020. "Picture a Scientist." Recorded October 8, University of California Television (UCTV), show ID 36511. YouTube video, 54:30, posted October 22. https://www.youtube.com/watch?v=62qVQPe1sVc.

Clancy, K.B.H., K.M.N. Lee, E.M. Rodgers, and C. Richey. 2017. "Double Jeopardy in Astronomy and Planetary Science: Women of Color Face Greater Risks of Gendered and Racial Harassment." *Journal of Geophysical Research: Planets* 122(7):1610–1623. https://doi.org/10.1002/2017JE005256.

Clancy, K.B.H, R.G. Nelson, J.N. Rutherford, and K. Hinde. 2014. "Survey of Academic Field Experiences (SAFE): Trainees Report Harassment and Assault." *PLOS ONE* July 16. https://doi.org/10.1371/journal.pone.0102172.

Cole, E.R. 2021. "The Intersectionality Framework and Its Importance in DEIA Work." Presentation to the Committee on the Planetary Science and Astrobiology Decadal Survey. Washington, DC: National Academies of Sciences, Engineering, and Medicine. https://www.nationalacademies.org/event/02-08-2021/docs/D3B1990EE01654A1446998C1A6A744682768622D21D0.

Cox, W.T.L., and P.G. Devine. 2019. "The Prejudice Habit-Breaking Intervention: An Empowerment-Based Confrontation Approach." Pp. 249–274 in *Confronting Prejudice and Discrimination: The Science of Changing Minds and Behaviors*, R.K. Mallett and M.J. Monteith, eds. Cambridge, MA: Academic Press.

Darling, E.S. 2015. "Use of Double-Blind Peer Review to Increase Author Diversity." *Conservation Biology* 29:297–299. https://doi.org/10.1111/cobi.12333.

Diniega, S., J. Castillo-Rogez, I. Daubar, J. Filiberto, T. Goudge, K. Lynch, A. Rutledge, et al. 2020. "Ensuring a Safe and Equitable Workspace: The Importance and Feasibility of a Code of Conduct, Along with Clear Policies Regarding Authorship and Team Membership." White paper #448 submitted to the Planetary Science and Astrobiology Decadal Survey 2023–2032. *Bulletin of the AAS* 53(4).

Erickson, K.J. "NASA's Science Activation Program: Achievements and Opportunities." Appendix C in National Academies of Sciences, Engineering, and Medicine. 2020. *NASA's Science Activation Program: Achievements and Opportunities.* Washington, DC: The National Academies Press. https://doi.org/10.17226/25569.

ESO (European Southern Observatory). 2021. "ESO Proposal Anonymisation," https://www.eso.org/sci/publications/announcements/sciann17380.html.

Finkelstein, M.J., V.M. Conley, and J.H. Schuster. 2016. "Taking the Measure of Faculty Diversity." TIAA Institute. https://www.tiaainstitute.org/publication/taking-measure-faculty-diversity.

Ford, H.L., C. Brick, M. Azmitia, K. Blaufuss, and P. Dekens. 2019. "Women from Some Under-Represented Minorities Are Given Too Few Talks at World's Largest Earth-Science Conference." *Nature* 576(7785):32–35. https://doi.org/10.1038/d41586-019-03688-w.

Fry, R., B. Kennedy, and C. Funk. 2021. "STEM Jobs See Uneven Progress in Increasing Gender, Racial and Ethnic Diversity." Pew Research Center. https://www.pewresearch.org/science/wp-content/uploads/sites/16/2021/03/PS_2021.04.01_diversity-in-STEM_REPORT.pdf.

Funk, C., and K. Parker. 2018. "Women and Men in STEM Often at Odds Over Workplace Equity." Pew Research Center, January 9 Report.

GSA (Geological Society of America). 2021. "GSA Membership Demographics." https://www.geosociety.org/documents/gsa/about/MbrDemographics.pdf.

Hanson, B., P. Wooden, and J. Lerback. 2020. "Age, Gender, and International Author Networks in the Earth and Space Sciences: Implications for Addressing Implicit Bias." *Earth and Space Science* 7(5). https://doi.org/10.1029/2019EA000930.

Jost, J.T., L. Rudman, I.V. Blair, D.R. Carney, N. Dasgupta, J. Glaser, and C. Hardin. 2009. "The Existence of Implicit Bias Is Beyond Reasonable Doubt: A Refutation of Ideological and Methodological Objections and Executive Summary of Ten Studies That No Manager Should Ignore." *Research in Organizational Behavior* 29:39–69.

Kaluna, H., C.K. Baybayan, and B. Kamai. 2020. "Creating Spaces for Indigenous Voices Within Planetary Science—Part 1." White paper #502 submitted to the Planetary Science and Astrobiology Decadal Survey 2023–2032. *Bulletin of the AAS* 53(4).

Lauer, M.S., and D. Roychodhury. 2021. "Inequalities in the Distribution of National Institutes of Health Research Project Grant Funding." *eLife* 2021 10:e71712. https://doi.org/10.7554/eLife.71712.

Lerback, J.C., B. Hanson, and P. Wooden. 2020. "Association Between Author Diversity and Acceptance Rates and Citations in Peer-Reviewed Earth-Science Manuscripts." *Earth and Space Science* 7(5). https://doi.org/10.1029/2019EA000946.

Lonsdale, C.J., F.R. Schwab, and G. Hunt. 2016. "Gender-Related Systematics in the NRAO and ALMA Proposal Review Processes." *arXiv* preprint arXiv:1611.04795.

NASA (National Aeronautics and Space Administration). 2019. "NASA Equal Employment Opportunity Strategic Plan: FY 2017–19." FY 2018 Annual Report and Update for FY 2019 (FY 2018 MD-715 Report). Office of Diversity and Equal Opportunity. https://www.nasa.gov/sites/default/files/atoms/files/2018_nasa_md_715_report_5-15-2019_tagged.pdf.

NASA. 2020. *NASA's Explore: Science 2020–2024: A Vision for Science Excellence.* Washington, DC. https://science.nasa.gov/science-pink/s3fs-public/atoms/files/2020-2024_Science.pdf.

NASA. 2021. SMD Townhall Presentation. https://science.nasa.gov/science-pink/s3fs-public/atoms/files/Town_hall-1-21-21_v5.pdf.

NASA. 2022. "For Researchers: PI Launchpad Workshop Content." NASA Science. https://science.nasa.gov/researchers/pi-launchpad.

NASEM (National Academies of Sciences, Engineering, and Medicine). 2023. *Pathways to Discovery in Astronomy and Astrophysics for the 2020s.* Washington, DC: The National Academies Press. https://doi.org/10.17226/26141.

New, M.H. 2021. "NASA/SMD Policies and Practices to Sustain, Grow, and Strengthen the Space Science Community." Decadal Survey on Planetary Science and Astrobiology: State of the Profession Writing Group, Virtual. February 18. https://webapp.nationalacademies.org/napar/projectview.aspx?key=51903.

Patat, F. 2016. "Gender Systematics in Telescope Time Allocation at ESO." *arXiv* preprint arXiv:1610.00920.

Phinney, L.M. 2009, "What I Wish I'd Known About Tenure," *Inside Higher Education* March 27.

Prockter, L.M, M. Aye, K. Baines, M. Bland, D. Blewett, S. Diniega, L. Feaga, et al. 2020. "The Value of Participating Scientist Programs to NASA's Planetary Science Division." White paper #452 submitted to the Planetary Science and Astrobiology Decadal Survey 2023–2032. *Bulletin of the AAS* 53(4).

Rathbun, J. 2017. "Participation of Women in Spacecraft Science Teams." *Nature Astronomy* 1:0148. https://doi.org/10.1038/s41550-017-0148.

Rathbun, J., E.G. Rivera-Valentín, J.T. Keane, K. Lynch, S. Diniega, L.C. Quick, C. Richey, J. Vertesi, O.J. Tucker, and S.M. Brooks. 2020. "Who Is Missing in Planetary Science: Strategic Recommendations to Improve the Diversity of the Field." White paper #435 submitted to the Planetary Science and Astrobiology Decadal Survey 2023–2032. *Bulletin of the AAS* 53(4).

Reid, I.N. 2014. "Gender-Correlated Systematics in HST Proposal Selection." *Publications of the Astronomical Society of the Pacific* 126:923.

Reid, I.N. 2018. "Hubble Cycle 26 TAC and Anonymous Peer Review." *Space Telescope Science Institute Newslettters* 35(4). https://www.stsci.edu/contents/newsletters/2018-volume-35-issue-04/hubble-cycle-26-tac-and-anonymous-peer-review.

Richardson, J., N. Whelley, P. Whelley, M. Milazzo, C. Knudson, R. Romo, and S.K. Nawotniak. 2020. "Building Safer and More Inclusive Field Experiences in Support of Planetary Science." White paper #447 submitted to the Planetary Science and Astrobiology Decadal Survey 2023–2032. *Bulletin of the AAS* 53(4).

Rivera-Valentín, E.G., J. Rathbun, J.T. Keane, K. Lynch, C. Richey, S. Diniega, and J. Vertesi. 2020. "Who Is Missing in Planetary Science: A Demographic Study of the Planetary Science Workforce." White paper #443 submitted to the Planetary Science and Astrobiology Decadal Survey 2023–2032. *Bulletin of the AAS* 53(4).

Sahagún, L. 2021. "Caltech Says It Regrets Drilling Holes in Sacred Native American Petroglyph Site." *Los Angeles Times*, July 19. https://www.latimes.com/environment/story/2021-07-19/caltech-fined-for-damaging-native-american-cultural-site.

Schmidt, B., S. Som, E. Quartini, J. Buffo, C. Chivers, K. Soderlund, C.E. Carr, B. Klempay, and J. Bowman. 2020. "Diversity in Action: Solutions for a More Diverse and Inclusive Decade of Planetary Science and Astrobiology." White paper #426 submitted to the Planetary Science and Astrobiology Decadal Survey 2023–2032. *Bulletin of the AAS* 53(4). https://doi.org/10.3847/25c2cfeb.f220b3a3.

Strauss, B., S.R. Borges, T. Faridani, J.A. Grier, A. Kiihne, E.R. Maier, C. Olsen, et al. 2020. "Nonbinary Systems: Looking Towards the Future of Gender Equity in Planetary Science." White paper #442 submitted to the Planetary Science and Astrobiology Decadal Survey 2023–2032. *Bulletin of the AAS* 53(4).

Strolger, L., and P. Natarayan. 2019. "Doling Out Hubble Time with Dual Anonymous Evaluation." *Physics Today* (March 1):1945–0699. https://doi.org/10.1063/PT.6.3.20190301a.

USAP (United States Antarctic Program). 2022. *Sexual Assault/Harassment Prevention and Response (SAHPR): Final Report.* Alexandria, VA: National Science Foundation. https://www.nsf.gov/geo/opp/documents/USAP%20SAHPR%20Report.pdf.

Vander Kaaden, K., C. Ryan, E.G. Rivera-Valentín, C.B. Phillips, J. Haber, J. Filiberto, and A. Denton. 2020. "Creating Inclusive, Supportive, and Safe Environments in Planetary Science for Members of the LGBTQ+ Community." White paper #411 submitted to the Planetary Science and Astrobiology Decadal Survey 2023–2032. *Bulletin of the AAS* 53(4).

Zorzano, M.P. 2020. "Gender Balance in Mars Exploration: Lessons Learned from the Mars Science Laboratory." *Sustainability* 12(24):10658. https://doi.org/10.3390/su122410658.

17

Research and Analysis

The stated goal of NASA's Planetary Science Division (PSD) is to advance scientific knowledge of the origin and history of the solar system,[1] the potential for life elsewhere, and the hazards and resources present as humans explore space. These goals are achieved primarily using space-based assets—that is, robotic space missions. A critical activity that supports and enables this pursuit is PSD's research and analysis (R&A) portfolio, which yields cutting-edge science from the data these missions return. However, direct analysis of spaceflight data is only one part of the crucial role played by R&A in NASA's PSD. Mission data analyses inevitably lead to new questions that require, for example, theoretical, laboratory, field work, and/or ground-based observations to interpret to ensure that mission data ultimately lead to substantial advances in overall knowledge. Scientific and technical advances arising from R&A programs are used to identify important goals for future exploration, determine the most suitable targets for future exploration, develop and refine needed instrument and analytical techniques, derive the greatest benefit from data returned by past and ongoing missions, and, through the direct involvement of students and young investigators, train future generations of space scientists and engineers (NRC 2011). R&A supports the study of the origin and early evolution of life on Earth, and its implications for the potential for life elsewhere and for detecting its presence. R&A supports national priorities such as the human spaceflight program and enables the R&A-supported community of researchers to communicate to the broader scientific community, policy makers, other stakeholders, and the public at large new discoveries that come from the investment of tens of billions of dollars per year in the world's premier space agency.

> **Finding:** R&A provides the intellectual foundation for NASA's exploration endeavors, ensuring that they are designed and utilized in a manner that maximizes the expansion of knowledge. R&A is thus fundamental to the current and future success of NASA's planetary science and astrobiology program.

Together, PSD's missions and its R&A program are the primary funding mechanisms that allow scientists to enter the field and to enhance their expertise throughout their careers, providing the nation with a highly skilled, stable workforce. R&A supports an unusually large portion of the planetary science and astrobiology community in comparison with other fields, for example, astronomy, which have a larger fraction of their community supported by academic positions (see State of Profession chapter). The broad access provided by the openly competed R&A programs supports the continued diversification of the field along many axes. These programs also incentivize

---

[1] A glossary of acronyms and technical terms can be found in Appendix F.

innovation and allow for flexible response to changing scientific priorities. Maintaining an appropriate balance of support across both missions and R&A is thus needed to ensure the future success of PSD programs.

> **Finding:** R&A is central to sustaining the nation's planetary science and astrobiology community and expertise. Openly competed programs provide broad access to PSD funding and are a key element both for driving innovation and for advancing state of profession issues, including diversity and inclusion.

By any measure, R&A is a *vital* component of NASA's scientific enterprise. Yet, how a "well-balanced and appropriately funded R&A portfolio" is defined may vary depending on one's role in the R&A endeavor. For the government and the public, this definition may be an assessment of whether NASA is achieving the best return on its investment in robotic flight missions. For a scientist, this may instead reflect the relative time spent pursuing funding versus that spent doing scientific research. While acknowledging such varied perspectives, the primary goal in this chapter is to offer findings and recommendations to maximize the utility of R&A within NASA's PSD, to ensure that spacecraft mission data in hand are utilized to their full discovery-making potential, and that breakthrough science continues to drive future mission design and implementation.

This chapter is written against a backdrop of intense concern for the health of PSD's R&A program, as evidenced by numerous white papers and presentations to the committee. The per-year investment by PSD in

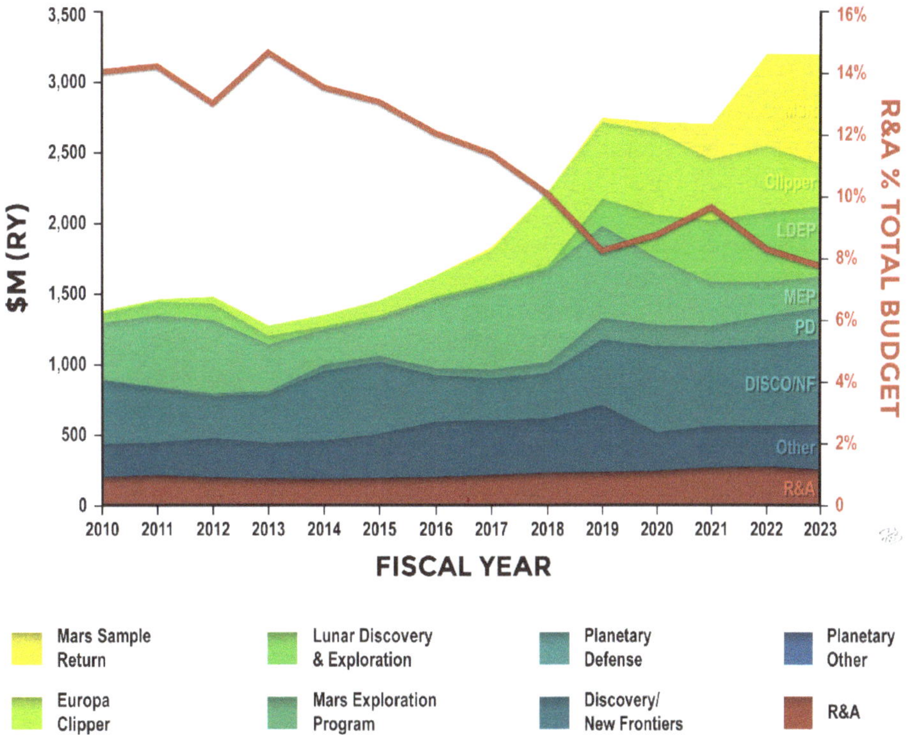

FIGURE 17-1 The NASA PSD budget from 2010 to 2023, in real-year dollars. The R&A wedge here includes research from the Mars, New Frontiers, Discovery, Outer Planets, and Lunar Science funding lines, as well as the Planetary Science Research and Analysis line (that includes openly competed programs, nonopenly competed programs, and various other activities). Although the overall PSD budget over this timeframe has increased considerably, the R&A budget fraction (shown by the red line) has not kept pace—dropping from 14 percent of the budget in 2010 to <8 percent projected for FY 2023. The R&A funding levels shown by the red wedge (in $ million) are 194 (2010), 208 (2011), 193 (2012), 187 (2013), 183 (2014), 190 (2015), 197 (2016), 209 (2017), 225 (2018), 228 (2019), 239 (2020), 262 (2021), 267 (2022), and 249 (2023). SOURCE: Created by P. Byrne based on data presented to the committee by L. Glaze, NASA PSD director, July 28, 2021.

BOX 17-1
Organizing Theme for Issues in Research and Analysis

The committee's ASPIRE concept encapsulates key factors crucial for R&A:

Augment the R&A budget to ensure that NASA PSD gets the science it needs

Survey the community to identify existing expertise, capability and knowledge gaps, and research trends

Promote cross-cutting research that tackles systems-level science questions

Integrate multiple disciplines such as: data analysis, theory, laboratory studies, field work, and ground-based observations to enable the best systems-level science

Revise the review and proposal process to make it more accurate and efficient

Equalize access and opportunity to ensure no pockets of excellence are left untapped

Together, these factors provide a framework for NASA's PSD to enact necessary changes to its R&A investments to support and enhance its overall mission. The ASPIRE concept addresses the key current issues facing the PSD R&A landscape: very low proposal selection rates, a portfolio of programs that may not be fully responsive to NASA's needs, and ensuring an efficient and fair proposal review process that maximizes open access and innovation.

R&A activities has increased from \$194 million in 2010 to \$251 million currently budgeted for 2023,[2] a 30 percent increase in real year (not inflation adjusted) dollars, and a flat funding level in inflation-adjusted dollars (adopting the NASA inflation index of 1.332 from 2010 to 2023). However, during that time the PSD budget has approximately doubled in real year dollars, and with it, the number of missions, volume of new data produced, and needs for new analyses and approaches have greatly increased as well. The net result is that investment in R&A as a percentage of the total PSD program has substantially decreased (from >14 percent in 2010 to <10 percent from 2019 onward; Figure 17-1), leading to a steadily growing portion of highly rated research proposals that cannot be funded—a trend that risks reducing the effectiveness of R&A programs in achieving NASA's overarching goals, and potentially to the loss of outstanding, highly trained individuals from its scientific workforce to other competing fields. Furthermore, inefficiencies and inaccuracies in how proposals are solicited and reviewed, and uncertainties in the composition and capabilities of the R&A-supported planetary and astrobiology community, create additional difficulties and challenges.

> **Finding:** The decreasing fractional investment in PSD R&A activities threatens the continued success and vitality of the nation's planetary science and astrobiology program.

The recommendations herein address these critical, near-term issues and lay the groundwork for a healthy R&A environment through the next decade and beyond, to ensure that the nation achieves the greatest return on its investment in robotic spaceflight missions. These recommendations are contextualized through an organizing theme, ASPIRE, developed by the committee (Box 17-1). The discussion begins with a description of NASA's PSD R&A program, including a discussion of key large programs. A discussion and recommendations for optimizing the scientific return from the R&A program and for improving the proposal submission and evaluation process follows. The committee then discusses recent trends in R&A funding, leading to its most important recommendation: that R&A funding be increased to ≥10 percent of the annual PSD budget in order optimize scientific return and resource allocation for the remainder of the PSD portfolio. The committee closes with a discussion of the role of the National Science Foundation in advancing planetary and astrobiology R&A.

---

[2] All dollar amounts quoted are real-year dollars unless stated otherwise.

WHAT IS R&A?

Because R&A is a vital component of the PSD portfolio, assessing the health and productivity of the R&A programs, individually and collectively, is of central importance. Such evaluation requires transparent reporting of key metrics for R&A programs, and yet no standard definition for which programs comprise the R&A portfolio currently exists. Resulting variability in reporting and in defining what constitutes R&A has led to difficulty in assessing NASA's level of R&A investment and whether it is compliant with prior decadal survey recommendations, as highlighted by the *V&V* midterm review (NASEM 2018). Further, such variability has led to confusion among the science community at a minimum—and, in the worst case, has had the unintended consequence of giving some the impression that PSD is attempting to obfuscate the status of its R&A programs. Because R&A programs are bookkept in different places within the PSD budget, their metrics have been difficult to track, and a standardized definition is needed.

Unfortunately, arriving at a standardized definition of R&A is not as simple as looking at what is labeled as "Planetary Science Research" in the PSD budget. In addition to supporting scientific research, that budget line also includes support for numerous activities that are not pure research or analysis, such as management, the Planetary Data System, and support for proposal review activities, although these elements are clearly necessary for an effective R&A program. Further, there are scientific research programs that are not funded under the "Planetary R&A" budget line, such as Data Analysis Programs (DAPs) and Participating Scientist Programs (PSPs) that are instead supported by a mix of program and mission funding.

The committee's view is that to understand the health of the PSD's R&A portfolio, what matters most are the openly competed programs that fund observations, data and sample analyses, and fundamental research. These programs are open to all proposers, allowing for broad access, and are highly competitive, which drives innovation. Table 17-1 identifies these programs from the ROSES 2021 research announcement and virtual institutes solicited

TABLE 17-1 FY 2021 NASA PSD Openly Competed Research and Analysis Programs

Core Programs	Data Analysis	Astrobiology	Technology	Laboratory	Participating Scientist	Other
EW	NFDAP	EXO	PICASSO	LARS	Juno	SSERVI
SSW	LDAP	PSTAR	MatISSE		VIPER	PMEF**
PDART	MDAP	ICAR	DALI		MSL	CS
SSO	CDAP	HW				
PPR	DDAP*					
ECA						
YORPD						
FINESST						
XRP						

NOTE: CDAP: Cassini Data Analysis Program; CS: Citizen Science; DALI: Development and Advancement of Lunar Instrumentation; DDAP: Discovery Data Analysis Program,* includes data from the ESA Rosetta and ESA–JAXA BepiColombo missions; ECA: Early Career Award; EW: Emerging Worlds; EXO: Exobiology; FINESST: Future Investigators in NASA Earth and Space Science and Technology; HW: Habitable Worlds; ICAR: Interdisciplinary Consortia for Astrobiology Research; Juno: Juno Participating Scientist Program; LARS: Laboratory Analysis of Returned Samples; LDAP: Lunar Data Analysis Program; MatISSE: Maturation of Instruments for Solar System Exploration; MDAP: Mars Data Analysis Program; MSL: Mars Science Laboratory Participating Scientist Program; NFDAP: New Frontiers Data Analysis Program; PDART: Planetary Data Archiving, Restoration, and Tools; PICASSO: Planetary Instrument Concepts for the Advancement of Solar System Observations; PMEF: Planetary Major Equipment and Facilities, ** now Planetary Science Enabling Facilities Program; PPR: Planetary Protection Research; PSTAR: Planetary Science and Technology Through Analog Research; SSERVI: Solar System Exploration Research Virtual Institute; SSO: Solar System Observations; SSW: Solar System Workings; VIPER: VIPER mission Co-Investigator Program, XRP. Exoplanets Research Program; YORPD: Yearly Opportunities for Research in Planetary Defense.

BOX 17-2
Definition of the Research and Analysis Budget Line

The PSD Research and Analysis budgetary wedge in Figure 17-1 is the sum of five budgetary components: (1) Planetary Science Research and Analysis (the main budgetary line within the Planetary Science Research category); (2) Discovery Research (within the Discovery Program budget); (3) New Frontiers Research (within the New Frontiers Program budget); (4) Mars Research and Analysis (within MEP); and (5) Outer Planets Research (within the Outer Planets and Ocean Worlds budget). These five items include a variety of activities that totaled $209 million in FY 2017 and $240 million in FY 2020.

The committee's recommendations highlight the importance of the openly competed R&A programs (see Table 17-1). The portion of the FY 2020 R&A wedge that supported openly competed programs can be (roughly) estimated by deducting the FY 2020 expenditures in the following other budgetary items: ISFM ($21M), the Joint Robotics Program for Exploration ($13.5 million), Science Enabling Research Activities at NASA Centers (SERA; $7.4 million), StratComm ($3.4 million), R&A program support ($3.5 million), Lunar and Planetary Institute ($3.1 million), Decadal Survey Support ($12 million, a once-per-decade expense), program management ($2.1 million), external support ($1.8 million), NRESS contract ($1.5 million), admin support ($1.3 million), and other supportive activities (e.g., discretionary funds, Emerging Topics in Planetary Sciences (ETPS), PSD travel, and detailees; $1.7 million total). These items total about $70 million, implying that in FY 2020 about $170 million supported openly competed programs, or about 71 percent of the R&A wedge. For FY 2017 (prior to the creation of the ISFM and a non-decadal survey year), an analogous estimate implies that approximately 85 percent of the R&A wedge went to openly competed programs that year.

---

via Cooperative Agreements. This proposed definition of what *functionally* constitutes openly competed R&A necessarily crosses PSD budget lines. Among other things, these programs support flight mission data analysis, fundamental scientific research, observing, field work, and technology development programs, as well as participating scientist programs, programs that cover laboratory resources in support of spaceflight data, virtual institutes, and one-off programs of opportunity.

Not included in the committee's definition of openly competed R&A programs is funding received by co-investigators on directed or principal investigator (PI)-led mission science teams—even if membership on those teams was initially competed, such as for the Volatiles Investigating Polar Exploration Rover (VIPER) mission Co-I program. Programs whose funding is intended for the development of hardware components that are mission-enabling for specific destinations (COLDTech and HOTTech) are also excluded because these programs do not fund science or science instrumentation (unlike PICASSO, MatISSE, or DALI that do). Internally allocated civil servant funding (ISFM) is also excluded. Box 17-2 details the items included in the total R&A wedge in Figure 17-1, as well as primary activities included in this wedge that are outside the openly competed R&A programs.

Recommendation: NASA's PSD should adopt a consistent definition of what is included in the Division's R&A portfolio, including, in particular, an easy-to-distinguish category for the openly competed programs as defined in Table 17-1. This definition should be communicated to the science community and utilized in publicly reported metrics, tracking, and so on, which should be made readily available on an annual basis. As programs are added and removed these changes should be advertised clearly.

THE INTERNAL SCIENTIST FUNDING MODEL

The civil service workforce at NASA (hereafter referred to as civil servant(s)) provides substantial value by supporting research, engineering, and programmatic activities that enable the core work of the Agency. NASA defines the duties of its civil servant scientists as including participation in mission management, conducting mission-enabling technology development and research, serving in governmental leadership roles (within and outside of NASA), and leading large, complex collaborations with the broader scientific community. An internal study conducted in 2015 by the Agency Competition Team (created by then-Associate Administrator Robert Lightfoot) found that of ~1,000 NASA-employed civil servants, about 350 scientists were funded at ~150 full-time equivalent (FTE) through competed NASA PSD R&A awards. These proposals were competing with those of external scientists and civil servants not employed by NASA. The case was made that NASA was spending money (unnecessarily) competing for funding that would already be appropriated for civil servant salaries.

This led to the creation of a new structure in 2018, the Internal Scientist Funding Model (ISFM), which now provides PSD support for R&A work by civil servants and related contractors at a level of ~20 million per year. Funding for the initiation of the ISFM was obtained by deducting funds from openly competed R&A programs at a level commensurate with awards previously awarded to civil servants in those programs.

Objectives of the ISFM include the following: reducing the amount of time NASA civil servants spend writing proposals to do strategic and other high-value research that they were, in part, hired to conduct; enabling strategic alignment between NASA Headquarters and the various NASA centers in terms of hiring; providing on-ramps for early-career scientists outside the standard R&A programs; and implementing this funding approach consistently across SMD divisions. The ISFM is not intended to entirely replace the submission of R&A proposals from civil servants, although nominally the balance between the levels of support for NASA and non-NASA researchers should be unperturbed by the ISFM, per the 15 July 2021 presentation by Stephen Rinehart, director of Planetary Research Programs, to the committee.

Overall, it is NASA's prerogative to support its civil servants in whatever manner it deems optimal. Publicly available presentations by NASA SMD/PSD officials state that the Agency has identified eight criteria for the success of the ISFM to ensure that high standards are being maintained, and that civil servants are providing service to the broader scientific community. A recent review of the ISFM included external reviewers who found that, by and large, ISFM projects are generally productive and valuable. However, this is a different evaluation process than is conducted for R&A proposals submitted to openly competed programs. Ensuring that ISFM research maintains top scientific relevance commensurate with its PSD funding level may become increasingly challenging as the time lengthens since it has last been peer-reviewed alongside research proposed in openly competed R&A programs.

Perceived benefits of the ISFM include offering more funding stability, providing an on-ramp for early career scientists, and reducing the burden on resources associated with the submission of proposals to openly competed R&A programs. One of the primary goals of the ISFM was to reduce time spent by NASA civil servants in writing R&A proposals, and in his July 2021 presentation to the committee, Stephen Reinhart stated that this has, on average, occurred. The impact of ISFM on total R&A proposal submissions is unclear; these continued to increase in number from 2018 to 2020 (see Figure 17-2). Areas of concern with the ISFM include maintaining the cutting-edge nature of ISFM science; ISFM support that is being allocated to contractors and other persons who are not NASA civil servants (estimated to be about 10 percent of ISFM funds per the 15 September 2021 public discussion with Stephen Rinehart), which represents a separate system of funding access and allocation from that available to such individuals outside of NASA centers; civil servants potentially having increased difficulty in obtaining supplemental funding from standard R&A programs; confusion about how ISFM funding can be used to support (or not) other proposed work (e.g., ICAR proposals); and a lack of a standardized, well documented approach to the implementation of the ISFM and the evaluation of its funded projects. Clear communication of the standards and mean by which the success of the ISFM will be measured will ensure that NASA civil servants and the community understand the objectives and implementation of the ISFM and that each center adopts a consistent approach to the solicitation, selection, and evaluation of research conducted through this program.

Finding: NASA PSD's investment in the ISFM is substantial, and as such it is in the agency's interest to ensure that this new funding structure plays an appropriate role in its planetary science and astrobiology programs. This would include evaluation of how, for example, scientific productivity, the fraction of early-career civil servant scientists funded by R&A programs, civil servant community service, and the fraction of R&A funding awarded to civil servants versus the external community through open access programs have been affected by the ISFM program.

Recommendation: ISFM funds should only be used to pay NASA civil servant salaries. Funding for other individuals should be pursued through standard R&A proposal processes.

Recommendation: For greater transparency, NASA should document and communicate to its civil servants and the broader community how the ISFM is managed, and the processes by which proposals are solicited and evaluated to ensure the most meritorious civil servant science is supported.

VIRTUAL INSTITUTES AND RESEARCH COORDINATION NETWORKS

The Solar System Exploration Research Virtual Institute (SSERVI), which was founded in 2008 as the NASA Lunar Science Institute before being broadened in 2013 to include other human exploration targets, is intended to enable both science and human exploration through a combination of basic and applied research. The focus of SSERVI is limited to near-term potential targets for human exploration, and currently includes the Moon, near-Earth asteroids, and the moons of Mars. Open calls for proposals for large, 5-year grants are staggered by several years, to allow overlap between established and new teams. As of 2020, there were 13 active teams.

SSERVI is a substantial program within NASA's PSD R&A portfolio, with an average of $15 million/year budget from 2018 to 2021. Over the past decade, PSD has funded ~90 percent of SSERVI's budget, with the Human Exploration and Operations Mission Directorate (HEOMD) contributing the remaining ~10 percent. Of late the HEOMD contribution has increased, to 20 percent in 2020–2021. The recent reorganization of HEOMD into the Space Operations and the Exploration Systems Development mission directorates—SOMD and ESDMD, respectively—will likely mean that ESDMD is the directorate participating in SSERVI going forward.

The virtual institute model allows for longer-term grants that provide stability and continuity to pursue new questions as discoveries are made, promoting flexibility within the proposed research projects. Large awards allow for interdisciplinary and multidisciplinary teams to address broad scientific problems in a manner that is essentially impossible through much smaller individual R&A grants. SSERVI has particularly emphasized the funding of graduate students and postdoctoral researchers, and the community building enabled by SSERVI teams is widely praised by those involved. On the other hand, many in the community—including current and former SSERVI team members—note that the present SSERVI structure is insular and fosters a sense of exclusivity for those inside the virtual institute. Further, there may be substantial overlap between science projects credited to SSERVI versus those funded through separate NASA grants or mission lines, making the scientific return from the program difficult to assess (see, e.g., the 2021 Report of the SSERVI Senior Review Panel[3]).

> **Finding:** SSERVI has played an influential role in community building through its support of early-career researchers and the continuity provided by large, 5-year grants to each SSERVI team. However, there is value in considering whether offering opportunities for guest investigators or similar means to introduce additional scientists would enable SSERVI to leverage relevant expertise from community members who were not part of originally selected teams.

The initial cohort of SSERVI teams, selected in 2013, was perceived by the community to reflect a balance between decadal-level science[4] goals that could be addressed through human exploration, and the typically more applied science supportive of human exploration needs. In response to the initiation of the Artemis program, the most recent call for SSERVI proposals emphasized exploration science, a pivot away from fundamental science confirmed by the SSERVI director in a presentation to the committee. As a result, the current mix of teams skews toward applied research of benefit to HEOMD (now ESDMD), but that is less relevant to conducting decadal-level science. This is at odds with program funding, which is supplied predominantly by PSD rather than by HEOMD/ESDMD. Although providing a framework to work effectively with the new ESDMD is an important objective of SSERVI, SSERVI team members noted that they have struggled to receive adequate input from the ESDMD (and its precursor directorate) to ensure that exploration-focused work serves that directorate's needs. Further, and as is highlighted by multiple findings and recommendations in other chapters in this report, infusing decadal-level

---

[3] The committee obtained a copy of the SSERVI Senior Review Panel report in late January 2022 as this report was being revised in response to reviewer comments. Owing to this timing, the committee was not able to consider the specific findings and recommendations in that report in its deliberations.

[4] Decadal-level science is that which results in significant, unambiguous progress in addressing at least one of the survey's 12 priority science questions.

science goals into both the Lunar Discovery and Exploration Program in general, and into the Artemis program in particular (see Recommended Program and Human Exploration chapters), is viewed by the committee as an essential priority for the next decade.

> **Finding:** Although facilitating HEOMD/ESDMD and PSD collaboration on the science of human exploration targets is a central objective of SSERVI, other Directorates' engagement with SSERVI has to date been limited. The resulting burden on the PSD R&A budget is not commensurate with the shift away from fundamental planetary science and toward exploration questions evinced by recent team selections and statements from senior SSERVI personnel.

> **Recommendation: SSERVI represents a valuable and potentially powerful means to foster important interactions between PSD and ESDMD. As a primarily PSD-funded program, SSERVI should emphasize decadal-level science that can be enabled by human exploration activities, in addition to science needed to support exploration goals. Team selections and program activities (including redirection of existing nodes) should reflect a balance between science and exploration that is consistent with the relative PSD and ESDMD contributions to SSERVI program funding. This balance should be evaluated by an appropriately constituted group mid-decade.**

The Interdisciplinary Consortia for Astrobiology Research (ICAR) was created in 2019 as a successor to the NASA Astrobiology Institute (NAI). An intended benefit of the ICAR model is a reduction in the organizational overhead associated with the prior NAI. Like SSERVI, ICAR grants are 5 years in duration and tend to be considerably larger than typical R&A awards. Proposers to ICAR need to affiliate themselves with one of a predetermined set of Research Coordination Networks (RCNs), which focus on topics such as exoplanets, prebiotic chemistry, ocean worlds, and life detection. For some in the astrobiology community, this focusing of RCNs has had the unintended consequence of limiting rather than promoting interdisciplinary collaboration, especially because of a lack of opportunity to provide input on the initial definition of the RCNs. Additionally, each ICAR call requires proposals to address a specified subset of the RCNs, constraining the topics to which proposals may respond. This may have the unanticipated result of excluding studies that can address the recommendations of the National Academies' *An Astrobiology Strategy for the Search for Life in the Universe* (NASEM 2019).

Further, the rollout and funding details of this new program have been opaque. For example, in addition to proposals selected for funding through a formal review process, the RCN webpage (astrobiology.nasa.gov/about/faq/what-is-rcn) states that as of October 22, 2021, the "NASA Astrobiology Program, along with representatives of relevant research elements and SMD Divisions, will identify co-leads and potential members of the RCN and provide funding to support the logistical requirements of the RCN." The specifics of this approach, including the metrics by which RCN co-leads and members are selected and the extent to which they are funded, are unclear.

> **Finding:** The current implementation of ICAR and the component RCNs is confusing, including the way in which collaborations are constructed around the RCNs—the strict definitions of which may prevent the funding of other interdisciplinary astrobiology efforts that developed organically. As a result, the ICAR model may not be maximally effective at achieving broad astrobiology goals.

> **Recommendation: Given the scale and strategic importance of ICAR to NASA's astrobiology efforts, immediate evaluation by an appropriate external body to ensure that it is optimally designed to maximize desired return to NASA and to PSD is warranted. Particular issues that should be addressed include, but are not limited to, the best mechanisms for generating RCNs, whether and how proposals should be topically constrained, and how the program structure should evolve in response to scientific advances and community input.**

IS THE R&A PORTFOLIO OPTIMIZED FOR NASA'S SCIENTIFIC NEEDS?

Does NASA have the correct balance of research programs to enable the Agency to fulfill its science mission? Answering this question is challenging because it requires both a clear understanding of what NASA needs, as well as a means of determining whether the current portfolio of programs effectively addresses those needs. The latter requires metrics, data, and assessment of the output of R&A programs beyond that currently available or that could reasonably be obtained within the confines of the decadal process. Therefore, this section largely offers observations related to a subset of issues important for NASA to consider as it determines how best to assess and optimize its PSD R&A programs.

Scope of Openly Competed Programs

Many of the key questions we presently face—such as those discussed in the priority science question[5] chapters—require that they be tackled from multiple perspectives with multiple techniques. The field of astrobiology, for instance, exemplifies a discipline that encompasses diverse approaches including geology, biology, chemistry, and engineering (see Q9, Q10, and Q11; Chapters 12, 13, and 14, respectively).

Within the PSD R&A portfolio, the Data Analysis Programs (see Table 17-1) are well positioned to support cross-cutting questions that encompass planetary bodies throughout the solar system—reflecting a key focus of the 2020 NASA Science Mission Directorate's *Science 2020–2024: A Vision for Scientific Excellence* science plan to "exploit interdisciplinary opportunities between traditional science disciplines" (Strategy 1.3). The DAPs play a key role in promoting and supporting analysis of Discovery- and New Frontiers-class mission data, receiving collectively 17 percent of the total number of submitted PSD proposals in 2019. Recently, NASA has solicited proposals to study data from the ESA Rosetta mission to comet 67P/Churyumov-Gerasimenko in parallel with the Discovery Data Analysis Program (DDAP) and, in the ROSES 2021 call, data from the ESA/JAXA BepiColombo mission are within the scope of DDAP. Such inclusions are an effective step in enabling comparative studies that cut across traditional boundaries.

However, not all programs, and notably some DAPs, presently support or encourage analyses of datasets from multiple missions, risking the possibility that NASA may not obtain the cutting-edge, interdisciplinary science it otherwise might. Including standardized language in all DAP solicitations, and those of other programs as appropriate, to clearly state that comparative planetary studies are within scope, and indeed encouraged, as well as giving examples of supported research activities such as laboratory studies, and numerical modeling, would meaningfully enhance researchers' ability to propose comparative data analyses and other science. The planetary science and astrobiology community would thus be able to respond strongly to the cross-cutting themes identified within this decadal survey. By implementing these changes, NASA would address the "Promote" element of the ASPIRE concept (see Box 17-1).

> **Finding:** Explicitly encouraging researchers to propose to use multiple mission datasets within a single data analysis program, and/or to conduct studies with relevance to multiple planetary bodies to a single call, would be a major enabler of cross-cutting, comparative planetary science.

As presently designed, the DAPs facilitate the analysis of mission data acquired by recent or ongoing missions, yet there remains substantial value in revisiting older datasets. For example, recent studies have used data from the Voyager (e.g., Beddingfield et al. 2015), Magellan (Byrne et al. 2021), and Galileo (Mishra et al. 2021) missions returned in the 1980s and 1990s (two of which are no longer operating in any capacity). Yet those and similarly old mission datasets are not presently within scope of an existing DAP, and so are eligible for funded study only in proposals submitted to broader programs such as SSW. Making such data available either in existing DAPs or in a new, general planetary data analysis program would further enable comparative planetary studies and ensure continued value to NASA from missions long since flown.

> **Finding:** The inability to analyze data from missions no longer in operation and/or that falls outside the currently defined scopes of R&A programs is a weakness in NASA's data analysis strategy and program.

The 2014 PSD R&A reorganization effort saw multiple legacy openly competed programs, including Cosmochemistry, Planetary Geology and Geophysics, Planetary Atmospheres, Lunar Advanced Science and

---

[5]Priority science questions are referred to by number, with Q1 referring to Question 1.

Exploration Research, Outer Planets Research, and Mars Fundamental Research largely combined into a new, omnibus program, Solar System Workings (SSW). SSW was deemed to be aligned with one of PSD's five science goals from the 2014 NASA science plan, to "Advance the understanding of how the chemical and physical processes in the solar system operate, interact, and evolve" (NASEM 2017).

Unsurprisingly, the nature of SSW as the amalgamation of multiple antecedent programs means that it has received a plurality of all PSD R&A proposals each year since its inception: 25 percent (2014), 21 percent (2015), 19 percent (2016), 26 percent (2017), 21 percent (2018), 24 percent (2019), and 23 percent (2020). This has, in turn, posed a considerable logistical challenge to PSD program officers as they organize multiple review panels and work to avoid often complex conflicts of interest that can limit reviewer availability. Given these constraints, and that SSW review panels are typically grouped by science theme, the value to NASA of a single, expansive program—instead of multiple, thematic programs that together are just as responsive to the NASA's science plans as SSW—is not self-evident.

Another consequence of the 2014 reorganization was the lack of an explicit focus on fundamental research, one example of which was the Mars Fundamental Research Program (MFR)—which offered "opportunities for Mars research beyond those available from analyses of spacecraft data alone" (per the final MFR solicitation in 2013). By its very definition, fundamental research focuses on questions that might not yet be answerable with mission data—indeed, such research frequently *drives* new mission concepts. Fundamental science is technically within scope of SSW, yet the word "fundamental" appears but twice in the 2021 SSW solicitation. Since the 2014 reorganization, there has also been an increased hardening of defined boundaries between the R&A programs, which further constrains the ability to perform cross-cutting science. Accommodating proposals that address systems-level scientific questions, whether for individual bodies or for phenomena or properties that are common to some or many exploration targets, would allow scientists to explore foundational solar system processes more fully—satisfying the "Integrate" element of ASPIRE (see Box 17-1).

Finding: More than 8 years after its establishment, the community remains unconvinced that SSW provides a greater benefit to NASA than the multiple individual programs it replaced.

Finding: NASA's scientific focus on interdisciplinary, comparative science would be further enhanced by explicitly supporting and encouraging fundamental research within SSW or through a new, dedicated program. Additional flexibility in consideration of proposals that involve elements represented by multiple R&A programs would also be beneficial.

There is also a strategic need for NASA to ensure that scientific expertise is sustained generationally in the R&A-supported community. One way to ensure this capability is through the annual SMD solicitation for the Future Investigators in NASA Earth and Space Science and Technology (FINESST) program (formerly the NASA Earth and Space Science Fellowship), which supports graduate students pursuing PhD-level research. PSD-relevant FINESST selections are funded from (i.e., tied to) thematically relevant mainstay R&A programs, such as SSW. But fluctuations in funding levels of these programs can have commensurate effects on FINESST selections, affecting the types of research in which early-career scientists can be trained. Supporting FINESST as an independent program, free of the influence of varying funding levels for other programs, would enable Future Investigator candidates and their PIs to propose projects that extend beyond the scope and remit of existing, mainstream R&A calls, thus offering to NASA a broader range of research topics than might otherwise be possible.

Participating scientist programs (PSPs) enhance both intellectual and demographic diversity on a mission team and are an effective vehicle for training and networking scientists who might not otherwise have opportunities to be involved with missions and flight investigations (Prockter et al. 2020). PSPs are thus able, at a relatively modest cost, to substantially augment the science return of a spacecraft mission. It is to the benefit of both NASA and the community to solicit PSPs, preferably for every planetary mission. Importantly, diversifying spacecraft teams also addresses another component of SMD's *Science 2020–2024* science plan, Strategy 4.1, to "Increase the diversity of thought and backgrounds represented across the entire SMD portfolio through a more inclusive environment" (see Chapter 16).

Presentations to the committee demonstrated that the nominal 3-year duration of ROSES awards poses difficulties for those projects that require longer lead times for producing results, such as laboratory-focused studies (outside the scope of PSTAR for which several successive field seasons may be beneficial). In some cases where 4-year durations are already permitted (e.g., Emerging Worlds and Solar System Workings), the ROSES elements explicitly stipulate those longer durations "must be [well] justified," even though proposers are already required to justify their schedule, level of effort, and expenses; the practical effect of this language is to discourage such proposals. Yet there can be benefits to giving proposers the flexibility of proposing work over longer durations.

> **Finding:** To maximize the return to NASA on R&A projects where longer durations are particularly beneficial, there is value in allowing for, and not discouraging, grants with performance periods of up to 4, or even 5, years, with the latter especially for some laboratory- and field-based work.

Fourth year funding may also offer the benefit of reducing the number of no cost extension (NCE) requests (which would benefit from tracking by NASA). The committee envisions that most proposals will continue to request 3 years of funding, although the number of proposed 3-year-long projects versus 4- and 5-year projects could be evaluated annually by program managers. Importantly, any increase in NASA's flexibility regarding the duration of funded projects does not alter the requirement for proposers to continue to provide robust justification for any duration of funding, and proposals should continue to be evaluated for cost realism and reasonableness, including whether the proposed duration is commensurate with and appropriate for the work proposed.

Last, the planetary science and astrobiology community can be most responsive to NASA's needs when there is a clear understanding as to what programs will be solicited, under what terms, and when. For example, at least partially in response to a shortfall in funding, NASA elected not to solicit proposals for four PSD programs in ROSES 2021 (MatISSE, PSEFP, ICAR, and MMX PSP). Many scientists plan up to several years in advance the proposals they will write; it can be problematic, therefore, to learn only a few months in advance that a particular program will not be solicited.

> **Finding:** It is to the benefit of all that the cadence by which programs are solicited is predictable. Necessary changes would ideally be communicated to the community at least 1 year in advance and, whenever possible, 2 years in advance, so that researchers (especially early career scientists) have time to strategize their proposal plans.

Assessing Scientific Return from R&A Programs

Without explicit metrics, it is difficult to ascertain which R&A programs are more responsive to NASA's needs than others. For example, is one DAP returning more value for investment than another? Should one core program be prioritized over another based on how it enables NASA to meet its planetary science objectives, or on changing scientific discoveries and/or priorities? Are some programs no longer as critical to NASA's strategic goals as they once were, or are others even more important now? What is the scientific return from virtual institutes or team proposals, per dollar invested, compared to those of the individual-investigator programs?

Remarkably, there is currently no consistent, systematic, and accessible system for tracking the scientific products of NASA's PSD R&A programs. One means of quantifying the scientific return of each PSD R&A program would be to periodically assess the scientific, peer-reviewed publications and other products that result from the awards they support. Proposers generally include a plan—and a request for funds—to publish one or more papers for each project, and collating information on those publications would be a valuable proxy indicator of the return to NASA from the R&A activities it supports. Of course, not all projects necessarily do or ought to result in a peer-reviewed publication, or a project can result in some other type of product, and so other metrics (e.g., published maps or new delivered datasets) may be equally appropriate for some programs.

R&A PIs are required to submit annual and final reports, including information on publications and other products supported by the program. This material is typically e-mailed to a NASA program officer as a separate Word or pdf file, a structure dating back decades that makes tracking the papers produced by each program extremely cumbersome. Enabling funded investigators to instead upload annual and final report information into a NASA-managed portal that populates an associated database of research products would provide a ready means for the

Agency to compile such information and to assess the health and productivity of its component R&A programs.[6] Entry of research products produced after final reports are submitted would also be important to include. PSD would then have better insight into whether its R&A portfolio is optimally constructed, as well as ready information to quantify the impact of R&A to all stakeholders. Further, this would allow PSD to assess the science-return-per-dollar of larger team programs versus individual investigator grants.

Finding: Tracking the science produced by its R&A programs is important to allow PSD to assess the state of these programs and to demonstrate their importance and value to its overall program. This could be efficiently accomplished by development of a NASA portal into which required annual and final report information for standard R&A grants will be entered, allowing for tracking of publications and other research products (including map and data products) that result from each of the PSD R&A programs.

Finally, different scientific topics may require different program structures to optimize scientific return. SSERVI and ICAR are examples of team-oriented R&A programs intended to support broad science beyond that which can be tackled by individuals or small teams of collaborators and that instead require interdisciplinary and/or multi-institutional teams. Progress on other important and cross-cutting scientific topics—such as those that address multiple priority science questions and/or strategic near-term NASA activities—might also be enhanced by larger R&A projects than are feasible within standard R&A grant programs. Assessing this important issue will require both data on the scientific products of NASA's R&A programs, and regular comparison of these products with the scientific priorities identified in this report.

Recommendation: NASA should regularly (i.e., every few years) assess the PSD R&A portfolio to establish if the component programs are optimized for meeting PSD's science objectives. That assessment should consider (1) how the record of research products produced by each program compares with its funding level and strategic importance, (2) whether the existing mix of programs encourages cross-cutting science, and (3) the balance of team versus individual investigator programs. Changes in program structure should be announced with significant lead time to allow ongoing research programs to adjust.

R&A PROPOSAL REVIEW PROCESS

In the openly competed R&A programs, peer review is the central mechanism utilized by NASA to identify the most meritorious proposals that are the highest priority to support. The goal of peer review is to provide an independent expert evaluation and critique, including an articulation of strengths and weaknesses that provide both the rationale for prioritization and essential feedback for improving the work being reviewed. The process is, by design, rigorous, time-consuming, and extremely competitive. It best serves all stakeholders—proposers, reviewers, and NASA itself—by being efficient, accurate, and unbiased. Inefficiencies, inaccuracies, and biases lead to decreased incentive to innovate, loss of productivity, and missed opportunities for NASA to identify and benefit from the highest quality science.

The Review Process

The review of submitted proposals requires a substantial time and financial outlay by NASA to manage and conduct. The time required to write a proposal, even if based on an earlier, unsuccessful attempt, is also considerable. This section discusses several issues relevant to improving the efficiency and fairness of the overall process.

A recurring issue is inconsistency in review panel feedback in successive years. It is NASA PSD's policy to treat proposals as new each time they are submitted, and review panel membership generally changes from year to year. Inconsistencies are especially problematic against the current backdrop of very low selection rates because they may lead to proposals being resubmitted more often, which is inefficient for both proposers and NASA.

---

[6]This reporting measure is distinct from the requirement that all NASA-funded authors and co-authors deposit copies of NASA supported peer-reviewed scientific publications and associated data into the PubSpace repository, which has the objective of increasing public access to scientific research.

Enabling proposers who are resubmitting a previously declined proposal to directly address, either within the proposal itself or as separate information entered into NSPIRES, the weakness(es) identified during the predecessor proposal review would allow them to respond to or refute critiques in a manner like that applied to the review of manuscripts submitted to scientific journals. Other federal agencies, such as the National Science Foundation (NSF), permit this approach. Responding to prior criticism would not guarantee selection nor preclude the identification of new deficiencies, in keeping with the analogy to peer-review of manuscripts.

> **Finding:** Allowing proposers to formally resubmit a proposal at the next opportunity and respond to feedback from the prior year's review would (1) provide review panels with context and allow them to assess whether revised proposals have materially addressed previously identified deficiencies, (2) decrease the occurrence of contradictory reviews, and (3) better incentivize the considered improvement of proposed science. This would likely ultimately reduce the number of proposals resubmitted year after year.

> **Recommendation: To improve the proposal review process, NASA should establish a mechanism to permit PIs to respond to major weaknesses from previous submission rounds.**

Another issue is a challenge in identifying qualified reviewers, both to serve on review panels (internal reviews) and to provide written reviews as input to such panels (external reviews). The pandemic forced the community to institute remote review panels, and, anecdotally, this appears to have increased the portion of scientists willing and able to serve on review panels (Box 17-3). There are, however, concerns that while both all-remote or all in-person panels are efficient and able to fairly incorporate all reviewer inputs, hybrid panels in which some members are in-person while others are remote are more challenging and can alter group dynamics in favor of those in-person. In-person panels also enable early-career participants to network more directly with other scientists and NASA officials and more fully engage in the process.

Review service could be strongly encouraged so that PIs funded by a given program subsequently review other proposals to that program during the period of performance of their own award. There is precedent for this: for instance, the PDART program at present expects PIs of selected mapping proposals to provide, as part of the terms of their award, peer reviews for two other PDART-supported map projects. Some level of organizational memory also would meaningfully help with continuity of reviews from one year to the next. Reviewers might be asked to serve for several successive years, with different reviewer cohorts cycling on and off a given program panel each year (with care being taken to prevent panelist bias in successive reviews).

> **Finding:** Virtual review panels can enable scientists to participate who might otherwise be unable to. However, some programs in the post-COVID-19 world may be better served by in-person reviews, at the discretion of NASA program officers. Mixed-mode panels are best avoided, to ensure as level a playing field as possible for both reviewers and reviewed proposals.

> **Finding:** Encouraging funded PIs to review proposals to that same program for each year of their own award, either in person, virtually, or externally, would help ensure that appropriate expertise is available for the review process, and lessen the workload of program officers in formulating panels. This would not mitigate the need to empanel reviewers who are not currently funded by the program, to ensure a diversity of opinions.

Reducing some of the effort required to produce proposals could also be of benefit to the review process. The ROSES 2021 Discovery Data Analysis Program, for example, requires detailed budgetary information only for those proposals that are considered selectable (with only an approximate total budget amount specified at the time of proposal submission). This procedure is advantageous because assembly of formal institutional budgets can be onerous, notably for projects with numerous Co-Is and for those PIs at smaller institutions. Extending this approach to other programs could be beneficial.

> **Finding:** All actions that NASA can take to ease the burden on proposers, such as only requiring detailed budgetary information for selected/selectable proposals and rapidly returning review panel feedback, could have a positive effect on the quality of science for which PSD R&A money is requested.

BOX 17-3
The COVID-19 Pandemic and NASA's R&A Response

The COVID-19 pandemic caused by SARS-CoV-2 has upended the world. The myriad disruptions to familiar ways of life of course impacted the planetary science and astrobiology community, especially those with children, who are caregivers, who are early in their career, or who are in temporary, soft-money, or untenured positions—with many of them in more than one of these categories.

In response, NASA's Science Mission Directorate solicited its "Call for COVID Augmentations and Funded Extensions" under ROSES Appendix E.10, seeking to relieve some of the pandemic-induced pressure by extending award periods of performance and/or supplementing awarded amounts. This solicitation prioritized those awards supporting junior scientists, noting that

> Due to budgetary limitations, proposals to support the completion of research or technology development without any support for graduate students, postdocs or nontenured or soft-money early-career researchers shall not be considered.

Further, augmentations to existing awards were to be taken from funds for future starts, rather than money appropriated for this particular purpose. This approach, especially to give preference to junior scientists, was motivated by NASA's recognition that

> The future of [NASA's] research and technology development enterprise depends on the contributions of new researchers, with their new ideas, entering and being retained in the national workforce.

Prioritizing vulnerable, less-established members of the community at a time of unprecedented difficulty was laudable. However, reducing funds for new starts may have negatively impacted these members. Further, established members of the community also face many of the challenges affecting their more junior colleagues, such as being in soft money positions or having care-giving responsibilities—and may have fewer opportunities for alternative employment.

It is of little surprise that a sensitive topic such as this augmentation program—where limited financial support should be directed, especially in a time of reduced funding—would elicit multiple viewpoints. But the E.10 solicitation merely highlights an underlying tension in the planetary science and astrobiology community regarding how scarce resources should be deployed generally, and during disruptive events. The COVID-19 pandemic illustrates the need for NASA to have a clearly delineated set of decision rules for how to allocate monies during circumstances that put everyone under pressure—whether it be a government shutdown, a period of substantial budgetary cuts, the next pandemic, or an event we have yet to consider.

A critical component of NASA's R&A program is the peer review performed by members of the community—but reviews take time, both during the time a panel meets (whether in person or virtually), and beforehand as panelists prepare preliminary findings. Rules governing conflicts of interest can also act to impede the review process, substantially decreasing the pool of eligible reviewers, most acutely for major programs again such as SSW. Although honoraria are offered for review panelists, not all panelists can accept them, and external reviewers are not eligible for them at all. Extending honoraria to eligible external reviewers could help increase the uptake of external review requests.

Finding: In addition to policy changes or other incentives NASA might develop and consider adopting, increasing the honorarium for panelists, including enabling those who nominally cannot accept honoraria to include their time on panels in their grants, as well as paying external reviewers, would likely increase the uptake of review invitations and thereby broaden the pool of available reviewers and reduce instances of conflicts of interest. The budgetary impact of such a policy would need evaluation.

Last, there is longstanding awareness that some categories of proposals—namely those that are riskier but that would have high impact if successful—may be disadvantaged by the standard R&A proposal review process (NASEM 2017). In a highly competitive review process, proposals judged the most likely to succeed and/or that pursue well-established approaches are typically favored. However, this tendency may select against innovative— that is, "disruptive" proposals that explore the most novel ideas and/or techniques, which may be of great ultimate importance to science and to NASA. SMD has implemented a "blue ribbon" panel to address this issue, a promising step, although it remains unclear whether high-risk/high-impact proposals are being consistently identified and assessed across PSD programs.

> **Recommendation: NASA should undertake a process to continuously evaluate and improve its R&A proposal review and selection procedures such as, for example, review efficiency; optimizing information collected through NSPIRES; review panel formulation, implementation, and oversight; reviewer incentivization; factors that influence proposal selection; and ensure an appropriate balance between high and low risk proposals.**

Together, the above findings and recommendations constitute the "Revise" element of the ASPIRE concept (see Box 17-1).

Recent Changes Designed to Promote Fairness and Efficiency

At time of writing, the PSD has instituted (or announced plans to effect) changes to the proposal review process including the removal of due dates for select programs (termed "no due date," or NoDD), and the use of dual-anonymous peer review (DAPR). The programs first implementing NoDD include SSW, PICASSO, and LARS. The goal of NoDD is to reduce the burden on proposers (and research institutional staff) by removing hard deadlines and thus allowing scientists to take the time they need to fully develop their proposals. In turn, NASA will fund proposals over the course of a given ROSES year, balancing the selection of compelling science with a fixed budget for each program to enable highly ranked proposals submitted later in that year to be selected. Anticipated advantages to NoDD include additional temporal flexibility to proposers and to small institutions that may have limited proposal support resources, minimizing time between when a new concept is conceived and when it can be proposed, removal of conflicts between overlapping or closely spaced due dates for different grant programs, and overall reduction of proposal pressure. The rationale for DAPR is to remove biases associated with any aspect of proposer identity (see Chapter 16). NASA has stated that all ROSES programs are likely, eventually, to transition to DAPR (and most to NoDD).

These initiatives have the potential to improve review fairness, and NoDD may also reduce overall proposal pressure. Yet the mechanics of both DAPR and NoDD remain to be fully fleshed out—for example, ensuring that feedback is returned to proposers in time for them to revise declined proposals before the one-year resubmission moratorium expires, and ensuring that principles of DAPR to remove bias are applied throughout the review process, including the potential for bias on the part of selecting officials. Further, there are differences between typically short-term, focused Hubble Space Telescope observing awards (to which DAPR has been previously applied) and multi-year R&A grants, as the latter by design confer substantial flexibility to the investigator(s) in determining the nature (and quantity) of work that is ultimately performed. Consideration of a PI's prior performance –requiring knowledge of identity—at some stage of the R&A grant proposal evaluation process will be important for maximizing scientific return on NASA's R&A investment. NoDD may be challenging to programmatically implement and maintain and could have unintended consequences. At a January 2021 town hall, PSD indicated its intent to utilize rolling evaluation panels (REPs) and triage to review NoDD proposals, with review panel members serving for a 6-month term. This process could lead to substantial variations in the time between submission and review across different proposals, which would be problematic. Proposals with longer times between their submission and review would be at a distinct disadvantage because science advances rapidly and a proposal may no longer be judged to be accurate and cutting-edge if its review occurs more than a few months after it was submitted, at no fault of the proposer. Maintaining review consistency across the REPs will be important, as will the continued use of external, in additional to panel, reviews; regarding the latter, the committee reaffirms the importance of external reviews highlighted in the NASEM (2017) report. The committee notes that NSF's planetary R&A program utilized a NoDD structure for a time, before deciding to return to a standard annual due date. An independent assessment of the success of both DAPR and NoDD several years hence would help quantify the extent both to which they meet their stated goals, and to which they should be extended to the rest of the PSD R&A portfolio.

Finding: Successful implementation of NoDD will require mechanisms to ensure a rigorous and fair review process, including constituting review panels with appropriate expertise, continuing the use of multiple external reviews per proposal to provide targeted knowledge, and maintaining a consistent time-until-review for all submitted proposals.

Recommendation: An appropriately constituted independent group should evaluate the impact of DAPR and NoDD on R&A program outcomes, including proposal pressure, proposer and grantee demographics, proposal review ease and fairness, and overall R&A program functionality, before these policy changes are implemented across the full R&A program.

As discussed in detail in the State of Profession chapter, removing bias in proposal reviews is supportive of ensuring diversity, equity, inclusion, and access (DEIA) in planetary science and astrobiology. It is crucial that everyone competing for R&A funding has an equal opportunity to apply for and receive funding (a fundamental rationale for DAPR) and is welcome and encouraged to compete. Various groups, including professional organizations, have performed analyses of demographic information through self-identification within surveys, but not all of these studies may contain sufficiently inclusive datasets, and the conclusions presented from organizationally conducted surveys can be biased if they do not include data from a wide cross-section of the community.

Finding: The PSD and the science community could improve DEIA by working together to develop a mutually agreed upon matrix of variables that records, and which can be used to communicate in a consistent way, data on diversity within the R&A-supported community. Those data could be used to understand the state of DEIA specifically within PSD R&A programs and develop a comprehensive plan for addressing DEIA issues in those programs.

Together, these findings and recommendations constitute the "Equalize" element of the ASPIRE concept (see Box 17-1).

TRENDS IN PSD R&A FUNDING AND PROGRAMS THROUGH TIME

The 2013–2022 planetary science decadal survey *Vision and Voyages* (hereafter *V&V*) stressed the importance of a strong research and analysis program and recommended that R&A be increased by 5 percent in the first year of the decade, followed by annual increases of 1.5 percent above inflation for each successive year (NRC 2011). The *V&V* midterm review (NASEM 2018) collaborated with PSD to develop a rigorous analysis across the R&A portfolio to establish whether this recommendation had been met. That committee determined that, for fiscal year (FY) 2016, R&A spending levels had risen 32 percent relative to FY 2011 spending levels, the year for which *V&V* had budget information, and that NASA had therefore exceeded the *V&V* recommendation. Nevertheless, the midterm review found that analyzing R&A budget levels was difficult because PSD does not track spending on R&A in the way *V&V* had defined it (NASEM 2018). Indeed, the midterm review recommended that:

The next decadal survey committee should work with NASA to better understand the categorization and tracking of the budget for (1) principal investigator-led, competed, basic research and data analysis; (2) ground-based observations; (3) infrastructure and management; and (4) institutional or field center support. Also, the next decadal survey should be unambiguous when stipulating programs and recommended levels of spending.

The committee worked to meet this recommendation but faced similar challenges, as reflected in the first recommendation in this chapter. Without a consistent definition of R&A, PSD will continue to face challenges in assessing the success and impact of its R&A programs and transparently communicating these to all stakeholders. The director of Planetary Research Programs at NASA stated to the committee that R&A (defined to include openly competed programs as well as other programs and activities) funding in FY 2020 exceeded the *V&V* recommendation, and that growth in year-on-year R&A funding exceeded this recommendation in all years except 2019 and 2020 (Stephen Rinehart, January 22, 2021, presentation). Indeed, in the R&A budget as defined by NASA (see Figure 17-1), funding increased from $197 million to $239 million from 2016 to 2020, which met the *V&V* recommended growth of 1.5 percent/year above inflation (assuming 2 percent/year inflation).

However, as emphasized at the beginning of this chapter, the fraction of the PSD budget invested in R&A has substantially decreased, from 13–15 percent in 2010–2015 to <8 percent planned in 2023 (see Figure 17-1). The committee requested and received information for a subset of R&A programs and collected data from publicly available NASA sources. Unfortunately, available information lacks comprehensive data regarding the proposals received and selected, making it impossible to attribute trends to any particular source(s). For example, no information regarding the PI and Co-I institutions (e.g., university versus soft money), anonymized overhead rates, anonymized burdened rates per FTE, and so on are collected. Some data may not be collected by law, and some are not collected because it is not NASA's policy to do so. However, the available data do allow some trends to be examined. These show that since 2010, the number of R&A proposals submitted increased by >30 percent and the number of selected proposals *decreased* by >40 percent (Figure 17-2). The growth in total submissions is attributable at least in part to the combination of the expansion of NASA's spaceflight program as the scale of PSD approximately doubled and increasing data returned by missions.

The quality of proposals (as evinced by the percentage of those submissions rated "selectable"—that is, obtaining an adjectival rating during peer reviewed as "Good" or better) has not varied substantially year on year. Of course, reviewer perception of what constitutes an "Excellent," "Very Good," or "Good" proposal can vary, but the statistical aggregate indicates no long-term trend (e.g., grade inflation) based on the available data. As such, the downward trends in proposal selection rates seen in Figures 17-2 and 17-3 indicate that the PSD is almost certainly losing out on high-quality, high-impact science. By committing future funds for some programs with low selection rates in 2019, NASA was able to select a larger number of high merit/value

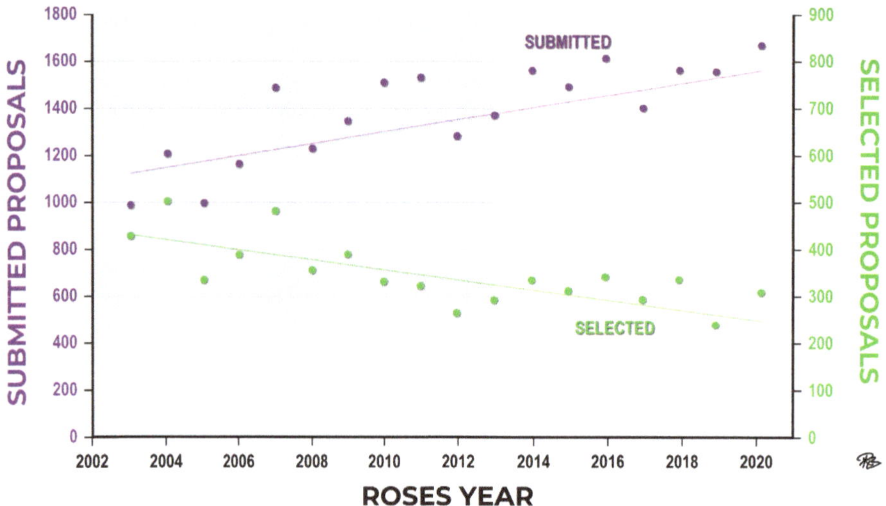

FIGURE 17-2 The total number of proposals submitted to PSD R&A programs (purple) increased by >30 percent from 2003 to 2020, whereas the total number of proposals selected (green) decreased by >40 percent over that same time. The period before the planetary R&A reorganization in 2013 is shaded in light blue. The ROSES calls analyzed here include technology programs, but not SSERVI. Linear best-fit trendlines are shown for both datasets. SOURCE: Created by P. Byrne based on data from the NASA SARA website and presentations to the committee by S. Rinehart, NASA PSD R&A director.

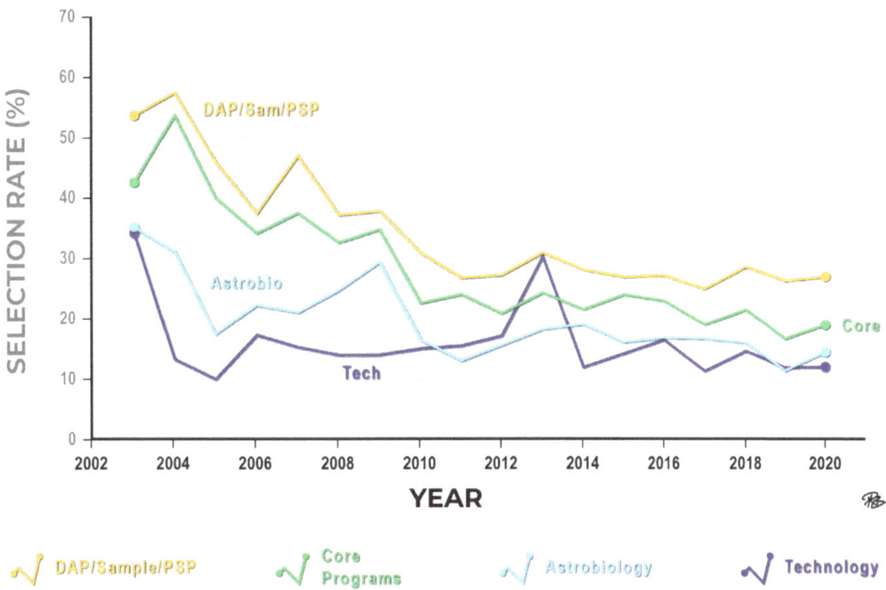

FIGURE 17-3 Selection rates for several program types, including DAPs, sample science, and participating scientist programs (yellow), core programs (green), astrobiology programs (blue), and technology programs (navy) as designated in Table 17-1. Selection rates have decreased since 2003 (except for PSPs), and the reorganization in 2013 (to the right of the blue shaded region) did not notably affect this. SOURCE: Created by P. Byrne based on data from the NASA SARA website.

proposals that year. This action, however, resulted in a worse shortfall of available funding the following year and selection rates dropped even further.

Except for participating scientist programs, average selection rates have decreased steadily since 2003 across the R&A portfolio (Figure 17-3). There has been a perception in the community that this trend was exacerbated by the 2013 reorganization of the PSD R&A programs. However, the committee did not find evidence that the reorganization affected selection rates. Selection rates for all R&A programs (except for NFDAP) have been equal to or less than 30 percent since 2018 and, for most programs, have been less than 25 percent. Although differences in selection rates for what NASA terms planetary R&A (21% ± 3%), the DAPs (26% ± 2%), and PSPs (23% ± 4%) are not statistically significant (to within 1 standard deviation) since 2014, selection rates for astrobiology (16% ± 3%) and technology (13% ± 2%) programs are consistently lower than the others over that same time. Proposals with higher average award amounts (i.e., in dollars per year) have lower selection rates; these awards are typically in technology and astrobiology R&A programs. The Exoplanet Research Program has also since 2014 displayed a lower-than-average selection rate (16.5% ± 2%), despite tremendous developments in this area (see Q12, Chapter 15).

The average number of unique PIs submitting proposals increased over the period 2010–2020 by almost 20 percent, while the number of unique PIs selected for funding decreased, with 30 percent fewer unique PIs selected in 2020 compared to 2010 (Figure 17-4). It is almost certainly the case that, because of natural attrition and growth of the planetary science and astrobiology community over the decade, the population of submitting PIs in 2010 was not the same as that in 2020. Nonetheless, this finding is generally consistent with the decrease in selection rates over that period. The average proposal team size and maximum team size have both increased marginally from 2010 to 2020; meanwhile, the average requested budget for a proposal has increased greatly over inflation (Figure 17-5).

Finding: Planetary science and astrobiology researchers are writing more proposals (with only a marginal increase in average team size) and requesting more funding in recent years than before; meanwhile, the number of selected proposals has decreased, despite the average proposal quality remaining essentially constant. These trends apply across all of PSD R&A but are particularly pronounced for astrobiology, technology, and exoplanets programs. The R&A reorganization in 2013 did not materially change these trends and NASA does not collect sufficient data to enable a full understanding of their origins.

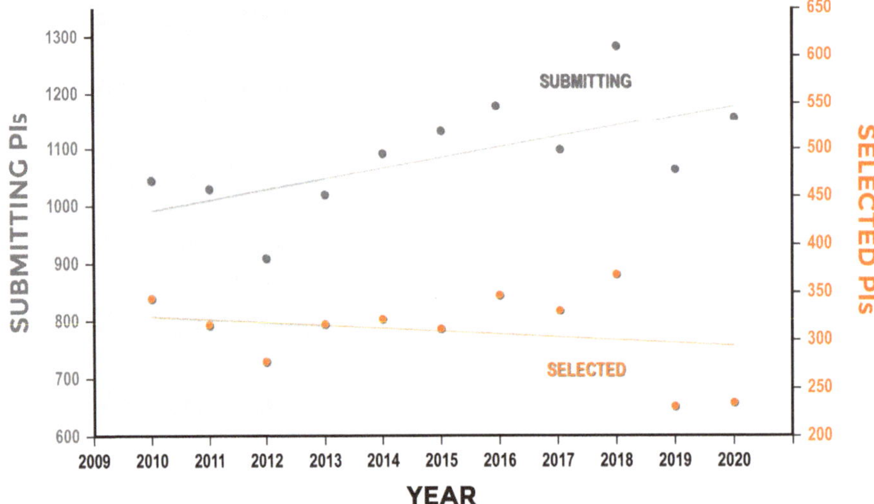

FIGURE 17-4 The number of unique PIs submitting proposals has increased by almost 20 percent (grey) from 2010 to 2020. Over that same time, the number of unique PIs selected for funding (orange) has decreased considerably: 30 percent fewer unique PIs were selected in 2020 than in 2010. The period before the PSD R&A reorganization in 2013 is shaded in light blue. Linear best-fit trendlines are shown for both datasets. SOURCE: Created by P. Byrne based on data from the NASA SARA website.

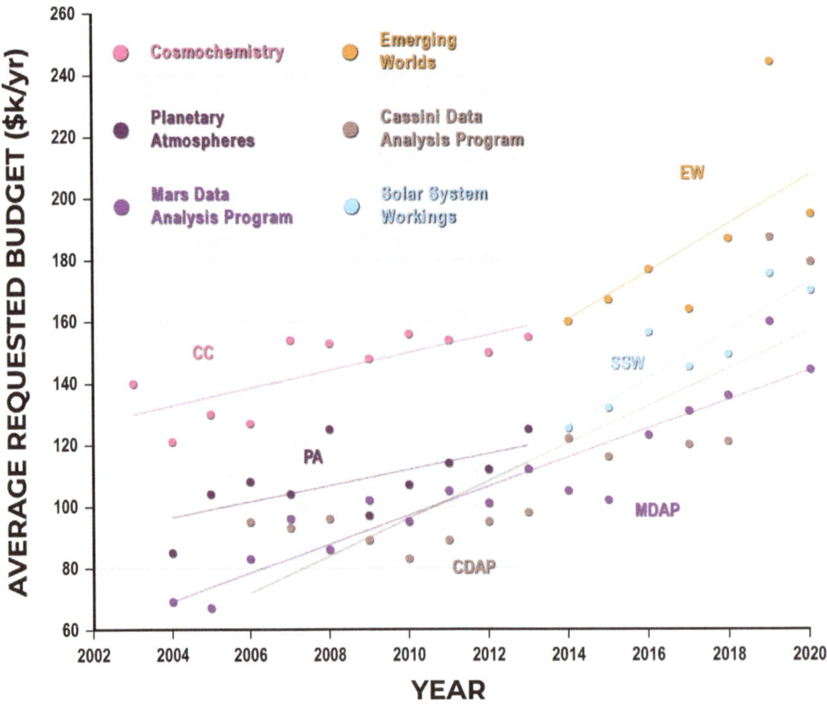

FIGURE 17-5 Average requested budgets for select programs from 2002 to 2020. Average requested budgets in two pre-reorganization programs, Cosmochemistry and Planetary Atmospheres, increased by 11 percent between 2003 and 2013, and by 47 percent between 2004 and 2013, respectively. From 2006 to 2020, CDAP average requested budgets increased 88 percent, and for MDAP essentially doubled from 2005 to 2020 (both CDAP and MDAP remained through the period of reorganization). From 2014 to 2020, the post-reorganization program Emerging Worlds increased by 22 percent; the average SSW proposal budget climbed by about 35 percent from 2014 to 2020. The period before the R&A reorganization in 2013 is shaded in light blue. Linear best-fit trendlines are shown for all datasets. SOURCE: Created by P. Byrne based on data from the NASA SARA website.

Finding: Although the committee could not isolate the reason(s) why proposal budgets have increased so substantially (e.g., increased overhead rates, increased salaries, growth in the proportion of soft money researchers), it is clear that increases to the R&A budget have not kept pace with demand.

Recommendation: NASA should collect comprehensive (as legally permitted) information on proposers and submitted proposals as needed to support internal and external assessments of the health of its R&A programs, addressing issues that include, for example, proposing team demographics and employment trends, and factors affecting proposal pressure and budgets.

RECOMMENDED FUNDING FOR NASA PLANETARY R&A

Figure 17-3 shows a progressive decrease in PSD R&A proposal selection rates, which for core programs have gone from >40 percent at the beginning of the past decade to <20 percent. The very low current selection rates adversely undercut innovation, efficiency of scientific return, and the training and sustaining of the workforce. A Proposal Pressures Study Group commissioned by the Astronomy and Astrophysics Advisory Committee (AAAC), which advises NSF, concluded that a 20 percent overall selection rate was unhealthy for the field of astronomy (Cushman et al. 2015), because it:

> Precludes stable, long-term support for students, postdocs, or researchers on soft money, and it preferentially discourages young researchers from remaining in the field.

Proposal writing takes time and, although having some intrinsic value, writing too many unsuccessful proposals and serving on the review panels that assess them come at the expense of scientific productivity. Viewed from NASA's perspective, this loss of scientific productivity means that fewer discoveries are made from, for example, spaceflight mission data analysis and fundamental research, and the research community is spending less time feeding new knowledge into NASA's planning for future missions than it otherwise could be.

The AAAC study utilized a statistical model (von Hippel and von Hippel 2015) of NSF astronomy grant proposers and success rates, which indicated that a selection rate of 30–35 percent for a given program leads to a manageable level of risk (~30 percent) of no funding after three attempts, representing a healthy competitive environment. For a 20 percent selection rate (very close to the current average selection rate of 21 percent for Planetary R&A), this statistical model demonstrated that presently unfunded and new investigators would compete with one another for an even lower effective funding rate of only 12 percent. Using conditional probabilities for three consecutive attempts, 80 percent of proposers would be unable to secure funding for their research in a 3-year funding cycle for a program with a selection rate of 20 percent. The AAAC study concluded that selection rates below approximately 20 percent likely drive outstanding researchers away from these research programs, again depriving NASA of highly trained and diverse expertise needed to fulfill its mission. If, for example, selection rates are lower than one in three, with substantial numbers of highly desirable and highly ranked ("Excellent" and "Very Good") proposals—based on historical funding data—not being selected by program officers because of budgetary rather than programmatic considerations, then NASA has reason to be concerned. The data obtained by the AAAC indicates that, even if 33 percent selection rates were achieved at present proposal submission numbers, there would still be some proposals rated "Very Good" and above that would not be selected (Figure 17-6).

The 2021 decadal survey in astronomy and astrophysics (NASEM 2023), when reviewing this and other data, advocated for a proposal selection rate of 30 percent, finding that this percentage "strikes an appropriate balance between a healthy competitive environment and a good chance of eventual success with resubmission." The committee reaffirms their finding.

Finding: Current PSD R&A proposal selection rates near or below 20 percent are strongly detrimental to scientific return and to sustaining and growing PSD's overall program. A selection rate near 30 percent would appropriately balance competitiveness with scientific efficiency while maintaining high standards among selected proposals.

Setting the PSD R&A budget to that needed to achieve a specific desired selection rate could, however, be problematic. The number of proposal submissions reflects, to some degree, the science community's interests and

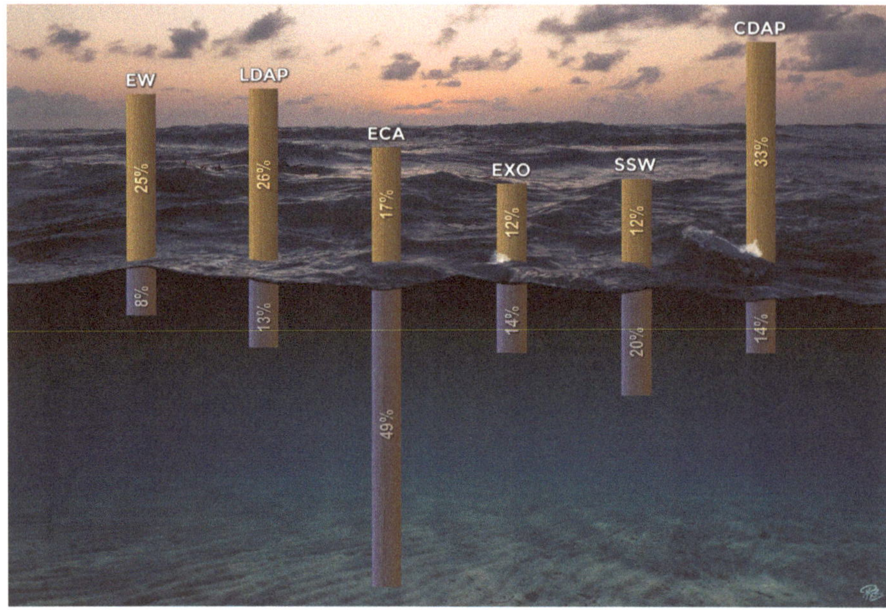

FIGURE 17-6 At present selection rates, NASA is missing out on substantial high-quality science. Shown here are the proportions of selected proposals (above the waterline), and non-selected proposals (below the waterline) among proposals that received an adjective merit rating of "Very Good" or better, for six programs in ROSES 2019. From left: Emerging Worlds (33 percent of all proposals received a "Very Good" or higher grade, with 25 percent selected and 8 percent non-selected); Lunar Data Analysis Program (26 percent and 13 percent, respectively); Early Career Award (17 percent and 49 percent, respectively); Exobiology (12 percent and 14 percent, respectively); Solar System Workings (12 percent and 20 percent, respectively); and Cassini Data Analysis Program (33 percent and 14 percent, respectively). SOURCE: Created by P. Byrne based on data from the NASA SARA website, with waves and sky courtesy of D. Billings and seabed courtesy of Damedias.

perceived likelihood of proposal success. Implementing a fixed selection rate thus risks driving up demand and proposal pressure, potentially requiring ever larger R&A budgets to maintain, to the detriment of the rest of the PSD program.

The committee advocates that PSD's investment in R&A activities should be proportionate to the overall scale of the PSD program, a logic similar to that used in industry to establish investment levels for supportive research and development activities. Clearly the past decade has been enormously successful for PSD, with an approximate doubling in the total PSD annual budget. Recent budgetary expansion reflects substantial growth across most PSD budgetary categories, with flagship (MSR and Europa Clipper) and the Discovery/New Frontiers mission lines approximately doubling from 2018 to 2023, in addition to proportionally even larger growth in planetary defense and LDEP during this period. However, investments in R&A have not proportionately kept up with this expansion of activities, which is a damaging trend that needs to be reversed to ensure that NASA maximizes science from its past and ongoing missions, optimizes design of future missions, and trains its needed workforce. The overall trend of decreasing proportional investments by PSD into R&A—as illustrated in Figure 17-1—needs to be reversed to maintain the appropriate balance between missions and basic scientific research and analysis.

Technology-focused companies typically invest 10 to 15 percent of their annual revenue into research and development (R&D) activities. Industry R&D is not a perfect analog for the full scope of PSD R&A work, because the latter supports analyses based on current and past projects in addition to preparing for the future. Nonetheless, the committee notes that at the beginning of the prior decade, R&A similarly comprised 14 percent of the PSD budget. Recent levels in the 8 to 9 percent range have led to very low selection rates and threaten the continued scientific payoff from NASA's mission investments. Although the percentage difference between 8 to 9 percent and ≥10 percent can seem small, the difference in absolute dollars is substantial with respect to R&A investments.

The current FY 2023 PSD budget is $3.2 billion, and the currently planned FY 2023 R&A level is $249 million (7.9 percent), with approximately $185 million of this for openly competed programs (based on a 2 percent/year increase in non-openly competed activities for nondecadal years; see Box 17-2). Increasing the FY 2023 R&A budget percentage to ≥10 percent would be a ≥$70 million addition to R&A, and if directed to the openly competed programs, would be a ≥37 percent increase in support for those programs. This illustrates that a relatively minor increase in fractional PSD investment into R&A would strongly (and disproportionately) enhance the value delivered by R&A to its flight programs. Yearly progressive augmentations to the R&A budget to reinstate a minimum 10 percent investment percentage by mid-decade would enable PSD program officers to manage their capacity to support higher selection rates sustainably and strategically. During periods of budget contraction, it is critical that the absolute level of support for R&A be maintained, if at all possible, in keeping with decision ules outlined in Chapter 22, so as to maintain NASA PSD's core human capital and expertise. Together, these actions represent a strong response to the ASPIRE's "Augment" component (see Box 17-1).

Finding: NASA PSD's scientific needs and aspirations for the coming decade will require substantially augmenting its investment in supporting R&A.

Recommendation: NASA PSD should increase its investment in what this decadal survey defines as R&A activities to achieve a minimum annual funding level of 10 percent of the PSD total annual budget by mid-decade. This increase should be achieved through a progressive ramp-up in funding allocated to the openly competed R&A programs. Mid-decade, NASA should work with an appropriately constituted independent group to assess progress in achieving this recommended funding level.

This recommendation will lead to a substantive increase in funding of openly competed R&A programs compared with recent funding levels and future plans, as currently understood and detailed next. Adopting the R&A yearly budget for FY 2023 through FY 2026 from the budget provided by the PSD director in July 2021, followed by 2 percent/year inflationary increases through FY 2032, the implied currently planned total PSD R&A investment over the next decade would be $2.6 billion. Analysis of FY 2020 expenditures (see Box 17-2) shows that out of a total of $240 million in the NASA-defined R&A line that year (i.e., the R&A wedge shown in Figure 17-1), approximately $170 million supported openly competed programs, with the remaining $70 million supporting non-openly competed programs and other activities. Assuming a 2 percent per year increase in non-openly competed activities, and treating the decadal survey as a one-time, inflation-adjusted cost in 2030, implies that out of a currently planned $2.6 billion total, ~$1.9 billion would support openly competed programs.

The above recommendation directs R&A funding increases to the openly competed programs. In the Recommended Program chapter, the committee presents two representative programs for *Recommended* and *Level* PSD budgetary profiles (see Table 22-2). The proposed levels of support for R&A in both budgetary profiles adopt a 2 percent/year inflationary increase for the non-openly competed elements of the R&A portfolio, extrapolating from the FY 2020 costs associated with those elements per above (which is ideal but may be unrealistic given recent trends). However, starting in FY 2023, the yearly funding for openly competed elements is increased by 10 percent per year for the first 8 years of the decade (in the *Recommended Program*) or for the first 4 years of the decade, followed by inflationary increases after that time (in the *Level Program*). Both budgetary profiles return the annual R&A investment to 10 percent or more of the annual PSD budget by mid-decade, leading to a decade total R&A investment of $3.9 billion (with $3.2 billion for openly competed programs) in the *Recommended Program* or $3.4 billion (with $2.7 billion for openly competed programs) in the *Level Program*. In real year dollars, these are decade increases in support for openly competed programs of $1.3 billion [$800 million] for the committee's *Recommended* [*Level*] programs relative to current R&A budgetary plans.

THE SIZE OF THE PLANETARY RESEARCH COMMUNITY

There is no single solution to addressing the concerns of all R&A stakeholders. Although additional support for R&A, as recommended above, is one major part of the solution, money alone is not a panacea because of the risk of induced demand—that a larger R&A portfolio without a commensurate plan to manage the appropriate size of the planetary science and astrobiology community could lead to more people submitting proposals, greater pressure on available funds, and a repeating cycle of declining selection rates. In this situation, it is appropriate to consider how we can best shepherd sustainable growth of the community without being unwitting gatekeepers and propagating structural inequities that may contribute to a less diverse workforce.

To maximally benefit from its past, present, and future spaceflight missions, NASA relies on a stable, diverse community of expert scientists to analyze and interpret data, develop new mission concepts, and train the next generation of researchers. Without a stable community, the Agency loses organizational memory and expertise, ultimately risking a decline in U.S. leadership in science. Indeed, the NASA-funded science community arguably fulfills a strategic national interest by helping maintain that science leadership. However, the PSD does not need—nor can it afford to support—a science community of unconstrained size.

Understanding the Demographics of the Community

Establishing the origin(s) of the proposal and funding trends described above requires, in part, collecting and analyzing demographic data, such as how much R&A funding the typical scientist receives, as well as the proportions of those researchers who are teaching or research faculty, either tenured or untenured; those who are research or staff scientists, civil servants, or contractors, either permanent or temporary; those who are postdoctoral research scientists; and the number of graduate or undergraduate students receiving support. Unfortunately, such information is not generally available. As a result, there exists no single, comprehensive snapshot of who comprises the scientific community at a given time, nor what pressures are placed on R&A programs as the demographics of the community evolve (see chapter on the State of the Profession).

Notably, some such data are available. For example, the American Astronomical Society's Division for Planetary Sciences (DPS) conducted a survey of the planetary science workforce in 2020 (Porter et al. 2020; Hendrix and Rathbun 2020). Preliminary results from the DPS study included findings that the planetary science and astrobiology community is growing, that about three quarters of workers are at universities or research institutions, and that more than a third of nonfaculty respondents receive most of their funding from NASA grants. These data have also been used to investigate the demographics of the planetary community (Rathbun et al. 2021; Rivera-Valentín et al. 2020).

However, various organizations collect different data and may not sample the full breadth of the community involved in PSD-supported R&A; those organizations also cannot reasonably collect data as a function of specific funding programs or on anything other than sporadic timescales. Limited data are available from publicly accessible records through NASA's Senior Advisors for Research and Analysis (SARA) website, and indeed underpin some of the findings described in this chapter. Additional information regarding the number of unique PIs, the size of proposal teams, and requested funding levels over time has been made available to the committee during the preparation of this chapter, but with differing levels of completeness due in part to the vagaries of record keeping (especially prior to the availability of digital records) and federal regulations regarding data preservation and privacy.

Collection of information about the skills, expertise, and research foci of the planetary science and astrobiology community would offer insight into current and future trends—with implications, for example, for forecasting how and why proposal pressures may change, and for responding to demographic changes in career types or expertise (e.g., the proportion of soft- versus hard-money scientists in the field, or a loss of research skills in specific areas). The status quo approach leaves NASA in a reactive mode, responding to trends after the fact as best it can. It is this situation that motivates the survey element of ASPIRE (see Box 17-1), and that motivates the earlier recommendation in this chapter on the importance of collecting relevant proposal and proposer demographic data.

Finding: Detailed and inclusive proposal, professional, and demographic information is needed to allow NASA to understand and communicate to stakeholders the factors driving pressures on specific R&A programs, anticipate future trends and problems in a timely manner, and determine whether responses and/or corrective actions are effective.

How Big a Planetary Science and Astrobiology Community Does NASA Need?

A key question that underpins any effort to ensure a healthy R&A program is: what is the appropriate size of the PSD-supported community? In principle, the answer is that the required community size is that which allows the nation's most talented researchers to produce sufficient high-priority and high-value science across the topical scope needed to support and further NASA's science goals.

As detailed above, the committee recommends that the R&A annual budget be increased to ≥10 percent of the PSD annual budget by mid-decade, implying that by the decade's end, annual funding for openly competed programs would be >$400 million [>$300 million] in the *Recommended [Level]* program (see Table 22-2), a factor of 2.4 [1.8] increase relative to the $170 million estimated funding level for openly competed programs in FY 2020 (see Box 17-1). The optimal size of the research community needed to meet NASA's needs will depend on a variety of complex and evolving factors, including, for example, the average portion of time each researcher spends on research, the seniority distribution of researchers optimal for producing the science NASA needs, and breadth of research expertise needed across different disciplines.

However, it is anticipated that there will always be more scientific ideas to explore than can be supported, and there will always be an imbalance between demand (submitted proposals) and supply (funding). In other words, there will always be more people that wish to be supported to do research than can be supported by NASA's PSD R&A portfolio. The relationship between NASA and the research community is symbiotic, but asymmetric: NASA benefits from the work of planetary scientists and astrobiologists in many crucial ways, and its support of scientists allows them to pursue topics of great scientific interest and to participate in remarkable opportunities to explore the universe. Thus, planetary scientists and astrobiologists provide a service for NASA and do not have any a priori right to PSD R&A funding. It is thus necessary to frame this issue in terms of the needs of NASA, and what mix of expertise, capabilities, and career types and stages is necessary to fulfill those needs. Yet, it is vital to recognize that NASA has, both directly and indirectly, invested substantial resources in training the present planetary science and astrobiology community, and care is merited to avoid needlessly losing existing expertise and talent.

Finding: Ongoing evaluation of the necessary and sustainable size of the PSD R&A-supported science community will require NASA, scientists, and other stakeholders to work together in a transparent manner to identify mutual needs and constraints, in order to align expectations, reduce inefficiencies, and support positive relationships between NASA and its scientific workforce.

Expanding Career Options for Planetary Scientists and Astrobiologists

There are several complementary means by which NASA and the scientific community can collaboratively determine the size of a sustainable planetary science and astrobiology workforce. As discussed above, detailed and inclusive demographic and professional information will help paint a much more comprehensive picture of the field as it currently stands, including where there are gaps in expertise, and where the diversity of the community needs to be improved (see Chapter 16). An independent study to examine this issue in detail by acquiring sufficient historical data for R&A programs (e.g., the number and cost of proposals per program per year, and the size of the proposing community through time) would be well placed to draw firm conclusions regarding the sustainable size of an R&A-supported workforce.

At the core of this issue is the fact that while NASA's planetary science activities have grown and the number of people graduating each year with PhDs in planetary science, astrobiology, and related fields has increased, the availability of faculty and research scientist positions has not increased at the same rate.

One consequence of this imbalance is growth of the soft-money community, in which scientists have nominally few or no teaching and service obligations but are required to source up to 100 percent of their salary each year, much of it through the PSD R&A budget. Soft-money scientists play a key role in the planetary science and astrobiology community (Bottke et al. 2022), and there are numerous advantages to the soft-money route, including a flexibility in working location and style that an individual's needs might warrant (e.g., if they are a caregiver/parent or require flexibility to work where no relevant employment possibilities are available). Because of their reliance on R&A programs as a primary or sole source of support, this cohort of the community is particularly vulnerable to decreases or instability in R&A investment, with commensurate negative impacts to the PSD through potential loss of the expertise and diversity this group brings to the planetary workforce (see Chapter 16).

One key action NASA can take to address this issue is to help improve awareness among undergraduate and graduate students, and the faculty who train them, of the breadth of career options after graduation. For example, the analytical, numerical, programming, and presentation skills with which planetary students are equipped when they graduate are readily applicable to a variety of fields distinct from planetary science and astrobiology, as well as those more closely related—including, but not limited to, the quickly growing commercial space sector. Educating planetary science and astrobiology graduate students on career paths outside the traditional academic track (i.e., government, nonprofit, and industry sectors), and providing training for how to prepare for, find, and apply to those jobs, would do much to reduce the pressure on the funds available from NASA for R&A. More broadly, by normalizing the pursuit of careers outside academia, such efforts would help scientists (especially early-career scientists) avoid internalizing the stigma that leaving academia is some sort of failure (Frank et al. 2020).

It is important to recognize that NASA alone cannot address this issue. Although graduate and postdoctoral training programs such as FINESST and NPP support early-career researchers, the commercial sector (both space and technology companies) could also help by, for example, sponsoring student research and conferences. And the science community (including researchers, faculty, and academic institutions) needs to acknowledge the responsibility it has to its junior colleagues when they begin their scientific training—to be honest regarding the career options ahead of them, and not hide from the fact that the prospect of securing a permanent position as a faculty member, mission scientist, or civil servant is far from assured, and that by design securing R&A funds will always been extremely competitive.

Supporting a diversity of career options in the space sciences for young researchers beyond traditional R&A support might not appear to be NASA's responsibility. However, the success of NASA's planetary science program relies substantially on graduate student and young scientist contributions. If such positions become increasingly unlikely to lead to sustainable employment, the number of people undertaking planetary science and astrobiology studies will decrease. This would undercut both the supply of young researchers that NASA needs, as well as the overall quality and diversity of NASA's workforce if the best and the brightest increasingly choose other fields. In addition, individuals that participate in industry and may ultimately return to NASA or research positions, bringing with them new skills, expertise, and connections not readily achieved through standard graduate or post-graduate research training.

> **Finding:** There are other paths beyond those supported by PSD R&A that planetary science and astrobiology students can pursue, to the benefit of the students, NASA, and the community. The fostering of connections between industry and students by NASA and its partners would enhance awareness of such alternatives.

NASA-NSF PARTNERSHIPS

NSF also provides support for planetary science activities. The annual NSF investment in planetary R&A is much smaller than NASA's (e.g., the FY 2017 estimated NSF budget for research grants in both solar and planetary sciences was $10 million; see https://www.nsf.gov/pubs/2016/nsf16602/nsf16602.htm). However, effective partnering between NASA and the National Science Foundation (NSF) in the allocation of R&A resources is important for meeting many decadal survey-recommended research objectives. In the past, NASA and NSF supported the Arecibo Observatory and obtained one-of-a-kind radar observations of a wide array of solar system objects. Heading into the next decade, the NASA-NSF partnership will enable the return of exciting solar system science from existing, as well as new, ground- and space-based telescopes, such as the Vera C. Rubin Observatory (formerly the Large Synoptic Survey Telescope). Many potential new targets for NASA missions reside within the discovery and characterization space for these assets. For example, Rubin Observatory is projected to discover millions of asteroids and tens of thousands of trans-neptunian objects in its first year of operations (currently anticipated to be 2024: cf. Chapter 18), discoveries that would be directly relevant to the PSD.

Similarly, Antarctica is a unique and valuable location for fieldwork relevant to both planetary and Earth sciences. As both NSF and NASA have interests in promoting high-science-return fieldwork from Antarctica, a sustainable pathway for collaboration between the agencies is essential. At present, policies regarding proposals to conduct NASA PSD-funded research in Antarctica vary between programs, which can cause confusion for researchers. Improvements to these policies would help preserve a role for planetary science and astrobiology research in Antarctica, ensuring that such work can be conducted in a sustainable partnership with polar sciences, the United States Antarctic Program, and NSF.

Another example where NSF and NASA could benefit from closer cooperation is in analogue research through the infrastructure, data, and samples offered by scientific ocean drilling programs (Neal et al. 2020). Analog research that expands our understanding of the inner workings of planet Earth, its climate and habitable environments and their evolution through time, and the effects of planetary processes such as impacts, all contribute to our understanding of planets broadly. Developments in the technology associated with scientific ocean drilling also have relevance to planetary exploration at many destinations. The potential exists to further broaden NASA-NSF partnerships on multiple programs relevant to astrobiology, including hydrothermal vent systems and subsurface biospheres, as well as those relevant to ocean worlds, such as ocean–ice and ocean–seafloor interactions (see Chapter 13), physical and chemical oceanography, and marine and ice microbiology.

Despite these many opportunities for collaboration benefiting both NASA and NSF, there is also a recognition that planetary astronomers have had trouble securing funding from NSF in, for example, the NSF Division of Astronomical Sciences program, where work that uses NASA mission data or data from NASA-supported activities/facilities is not permitted or is strongly discouraged. At present, this arrangement means that both NASA and NSF risk losing out on innovative, high-quality scientific research.

In summary, increased NASA and NSF collaborations could enhance both agencies' missions. This could involve activities such as, but not limited to, development of software tools for use with solar system data from existing and future facilities (e.g., the Rubin Observatory); improving coordination between relevant stakeholders when reviewing proposals to conduct fieldwork in Antarctica and other relevant sites; integrating research from scientific ocean drilling projects into planetary science, and more generally between targets of interest to planetary science, astronomy, astrobiology, and/or geoscience; and providing opportunities for joint support of R&A projects relevant to both agencies.

Recommendation: NASA and NSF would realize greater return on their R&A investments by working together to streamline the mechanisms by which researchers can propose and conduct science that is of benefit to both agencies.

REFERENCES

Beddingfield, C.B., D.M. Burr, and J.P. Emery. 2015. "Fault Geometries on Uranus' Satellite Miranda: Implications for Internal Structure and Heat Flow." *Icarus* 247:35–52.

Bottke, W., and H.F. Levison. 2020. "Tenets of an Effective and Efficient Research and Analysis Program for NASA." White paper #463 submitted to the Planetary Science and Astrobiology Decadal Survey 2023–2032. *Bulletin of the AAS* 53(4).

Byrne, P.K., R.C. Ghail, A.M.C. Şengör, P.B. James, C. Klimczak, and S.C. Solomon. 2021. "A Globally Fragmented and Mobile Lithosphere on Venus." *Proceedings of the National Academy of Sciences* 118:e2025919118. https://doi.org/10.1073/pnas.2025919118.

Cushman, P., J.T. Hoeksema, C. Kouveliotou, J. Lowenthal, B. Peterson, K.G. Stassun, T. von Hippel 2015. "Impact of Declining Proposal Success Rates on Scientific Productivity." *arXiv* eprint arXiv:1510.01647.

Frank, E., P.K. Byrne, S.Z. Weider, and L. Elkins-Tanton. 2020. "Normalizing Non-Academic Career Paths in Planetary Science." White paper #405 submitted to the Planetary Science and Astrobiology Decadal Survey 2023–2032. *Bulletin of the AAS* 53(4).

Hendrix, A., and J. Rathbun. 2020. "Results of the 2020 Planetary Science Workforce Survey Conducted by the AAS-DPS." White paper # 473 submitted to the Planetary Science and Astrobiology Decadal Survey 2023–2032. *Bulletin of the AAS* 53(4).

Mishra, I., L. Nikole, J. Lunine, K.P. Hand, P. Helfenstein, R.W. Carlson, and R.J. MacDonald. 2021. "A Comprehensive Revisit of Select Galileo/NIMS Observations of Europa." *Planetary Science Journal* 2:183. https://doi.org/10.3847/PSJ/ac1acb.

NASEM (National Academies of Sciences, Engineering, and Medicine). 2017. *Review of the Restructured Research and Analysis Programs of NASA's Planetary Science Division*. Washington, DC: The National Academies Press. https://doi.org/10.17226/24759.

NASEM. 2018. *Visions into Voyages for Planetary Science in the Decade 2013–2022: A Midterm Review.* Washington, DC: The National Academies Press. https://doi.org/10.17226/25186.

NASEM. 2019. *An Astrobiology Strategy for the Search for Life in the Universe.* Washington, DC: The National Academies Press. https://doi.org/10.17226/25252.

NASEM. 2023. *Pathways to Discovery in Astronomy and Astrophysics in the 2020s*. Washington, DC: The National Academies Press. https://doi.org/10.17226/26141.

Neal, C., S.P.S. Gulick, B. Baker, S. D'Hondt, N. Eguchi, T. Gregg, F. Inagaki, et al. 2020. "Forging Partnerships with Other Federal Programs: NASA and the National Science Foundation (NSF) Through Scientific Ocean Drilling." White paper #510 submitted to the Planetary Science and Astrobiology Decadal Survey 2023–2032. *Bulletin of the AAS* 53(4).

NRC (National Research Council). 2011. *Vision and Voyages for Planetary Science in the Decade 2013–2022*. Washington, DC: The National Academies Press. https://doi.org/10.17226/13117.

Porter, A.M., S. White, and J. Dollison. 2020. "2020 Survey of the Planetary Science Workforce," American Astronomical Society and American Institute of Physics, https://dps.aas.org/sites/dps.aas.org/files/reports/2020/Results_from_the_2020_Survey_of_the_Planetary_Science_Workforce.pdf.

Prockter, L., M.S. Tiscareno, E.J. Grayzeck, C.H. Acton, R.E. Arvidson, J.M. Bauer, R. Beebe, S. Besse, et al. 2020. "The Value of Participating Scientist Programs to NASA's Planetary Science Division." White paper #452 submitted to the Planetary Science and Astrobiology Decadal Survey 2023–2032.

Rathbun, J., E.G. Rivera-Valentín, J.T. Keane, K. Lynch, S. Diniega, L.C. Quick, C. Richey, J. Vertesi, O.J. Tucker, and S.M. Brooks. 2020. "Who Is Missing in Planetary Science: Strategic Recommendations to Improve the Diversity of the Field." White paper #435 submitted to the Planetary Science and Astrobiology Decadal Survey 2023–2032. *Bulletin of the AAS* 53(4).

Rivera-Valentín, E. G., J. Rathbun, J.T. Keane, K. Lynch, C. Richey, S. Diniega, and J. Vertesi. 2020. "Who Is Missing in Planetary Science: A Demographic Study of the Planetary Science Workforce." White paper #443 submitted to the Planetary Science and Astrobiology Decadal Survey 2023–2032. *Bulletin of the AAS* 53(4).

von Hippel, T., and C. von Hippel. 2015. "To Apply or Not to Apply: A Survey Analysis of Grant Writing Costs and Benefits." *PLOS ONE* 10(3):e0118494. https://doi.org/10.1371/journal.pone.0118494.

18

Planetary Defense: Defending Earth Through Applied Planetary Science

Planetary defense is an international cooperative enterprise aimed at providing protection to the nations of the world from devastating asteroid and comet impacts. Through the use of the knowledge and tools gained through planetary science and exploration, it is now possible to develop realistic and cost-effective detection and mitigation strategies against these natural disasters. As awareness of the hazard posed to life and property by Earth-approaching asteroids and comets has grown, the U.S. Congress and presidential administrations have directed NASA, the National Science Foundation (NSF), and other government agencies (e.g., Department of Energy [DOE], Department of Defense [DoD], Department of Homeland Security [DHS], etc.) to pursue activities in support of planetary defense. This chapter discusses the status of planetary defense activities and recommends directions for the next decade.

Our planet orbits the Sun within a swarm of cosmic debris in the form of asteroids and comets, collectively called near-Earth objects (NEOs).[1] The scars of previous NEO collisions in the form of impact craters are evident on Earth (e.g., Meteor Crater in Arizona) and major collisions in the past have substantially altered the course of life on the planet. The most famous of these collisions is the one that created the 150 km diameter Chicxulub crater 66 Ma in the Yucatan Peninsula of Mexico. A consequence of this event was the end to the reign of the dinosaurs.

Fortunately, large NEOs are very much less numerous than small NEOs (Figure 18-1). Smaller, more frequent events can also produce serious local to regional (i.e., an area equal to several U.S. states or a small nation) damage. For example, the disintegration in 1908 of a ~40–80 m NEO in the atmosphere above Tunguska, Siberia delivered between 3 and 20 megatons of TNT-equivalent energy to that remote region, scorching and blasting down trees across 2,000 square kilometers (Boslough and Crawford 2008; Morrison 2018). Although this event led to minimal loss of life, it was only because it occurred over a sparsely populated area. A similar but smaller event took place in 2013 over the Siberian city of Chelyabinsk. In this case a 20-meter-diameter asteroid detonated in an airburst, releasing nearly 440 kt of energy, and injuring more than 1,600 people.

Such impacts occur across Earth with an average interval of several decades for Chelyabinsk-sized events to a few millennia for Tunguska-sized events. However, there are considerable uncertainties regarding how much damage is produced by a bolide's[2] energy, considering that a projectile passing through the atmosphere distributes energy along its path as it disintegrates and that the blast energy itself has downward momentum. This means bolide airbursts are more akin to a descending linear detonation in the sky than the explosion of a nuclear weapon.

---

[1] A glossary of acronyms and technical terms can be found in Appendix F.

[2] A bolide is a term describing an object that burns up in the atmosphere, creating an exceptionally bright "fireball."

471

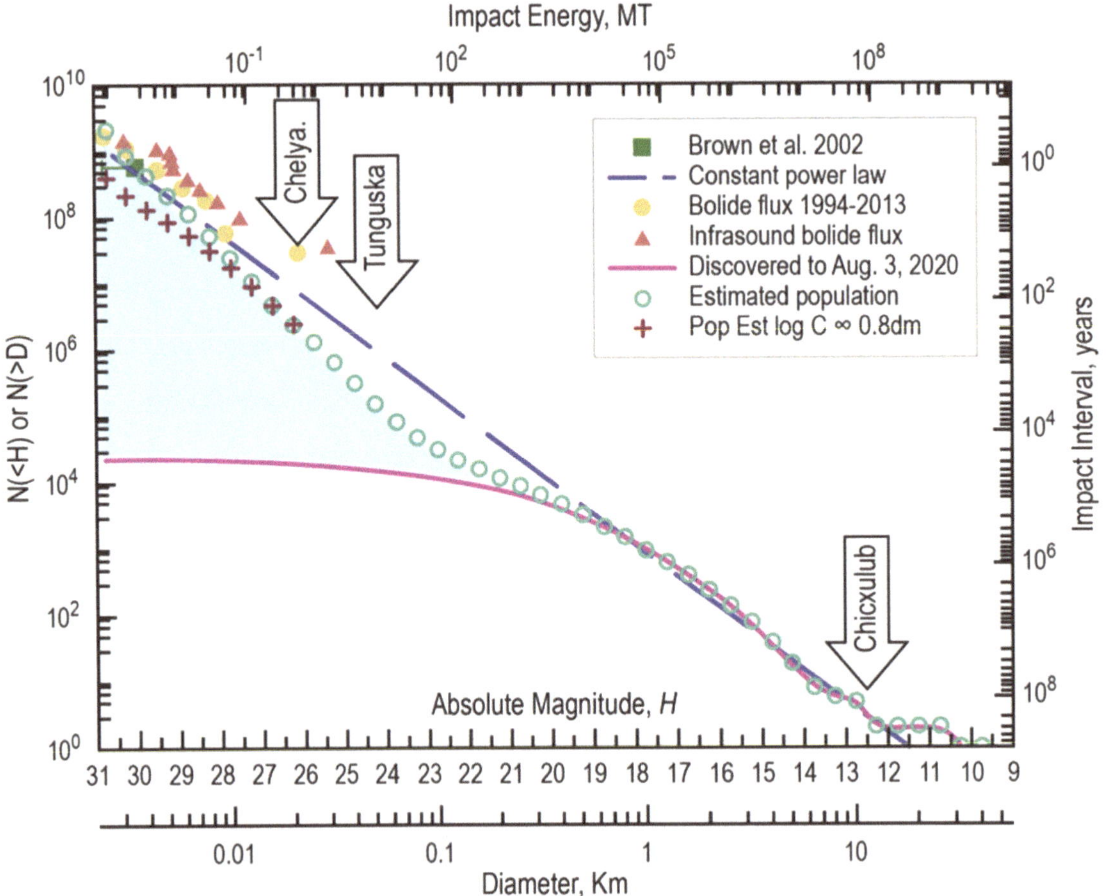

FIGURE 18-1 The known and extrapolated population of near-Earth asteroids (N) in terms of estimated diameter (D) and absolute magnitude (H). The blue shaded area represents the estimated population that remains to be discovered. Collisions with the millions of ~20 m objects, the size of the Chelyabinsk asteroid, will likely occur every 50–100 years, compared to ~10 km Chicxulub-size impactors, which strike Earth every 100 million years on average. The various symbols used are explained in the box at upper right. SOURCE: Reprinted and adapted from A.W. Harris and P.W. Chodas, 2021, "The Population of Near-Earth Asteroids Revisited and Updated," *Icarus* 365:114452, https://doi.org/10.1016/j.icarus.2021.114452. Copyright 2021, with permission from Elsevier.

The amount of damage produced by such events is not easily predicted owing to uncertainties related to the object's physical properties and structure, the breakup process, and atmospheric energy deposition. Some smaller objects may penetrate deeper into the atmosphere and deliver more energy than anticipated.

Orbiting spacecraft routinely detect the high-altitude break-up of smaller NEOs entering Earth's atmosphere, the vast majority of which are harmless (Figure 18-2). Although the annual probability of Earth being struck by a larger asteroid or comet is small, the consequences of such a collision are so serious that prudence dictates that society assess the nature of the threat and be prepared to respond.

To date, numerous NEOs have been found, mainly by U.S.-funded surveys, and cataloged by NASA (see Figures 18-1 and 18-3). Characteristics of the NEO population include:

- There are 1,000 or so NEOs greater than 1 km in diameter that are potentially capable of causing global impact effects. Approximately 95 percent of these bodies have been found, and fortunately none of them are a current threat.

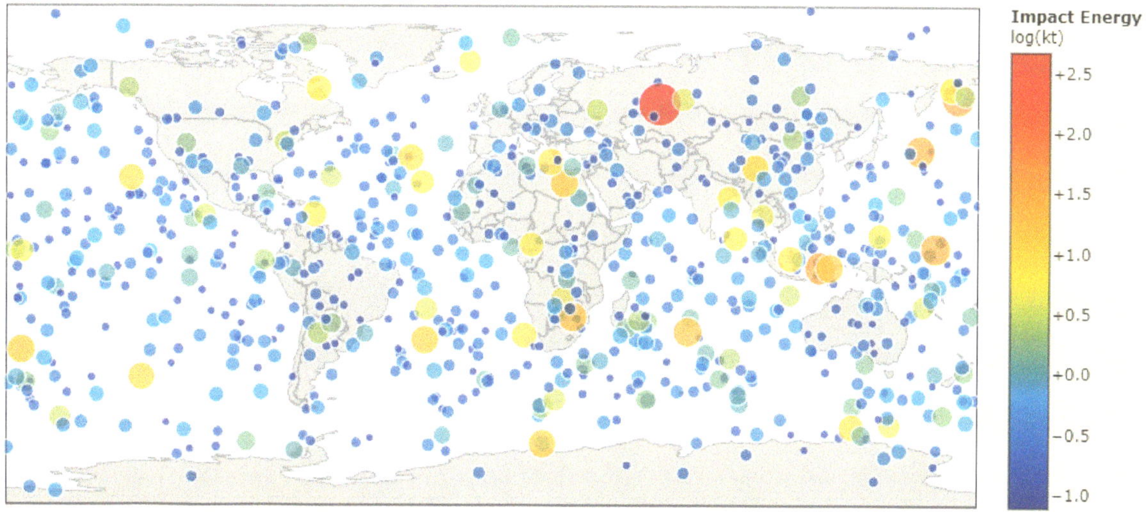

FIGURE 18-2 The chart above shows reported fireball events for which geographic location data are known. Each event's calculated total impact energy is indicated by its relative size and color. The 2013 Chelyabinsk event is the largest recorded over this timeframe (marked by the large red circle on the map), with an energy of approximately 440 kt. For reference, the atomic bomb dropped on Hiroshima released some 15 kt of energy and the above data show that these events deposit roughly 1.5 to 2 times that amount into Earth's atmosphere every year. Data are current through February 2, 2023. SOURCE: Courtesy of A.B. Chamberlin, NASA/JPL-Caltech, https://cneos.jpl.nasa.gov/fireballs.

- There are close to 25,000 objects larger than 140 m in diameter that could cause regional devastation. Only about one-third have been detected and tracked to date.
- An estimated 100,000 or more objects exist that are equal to or larger than 50 m in diameter and could destroy a concentrated urban area. It is estimated that fewer than 2 percent of these have been detected.
- There are millions to tens of millions of NEOs that if they struck Earth would likely break up in Earth's atmosphere, and could potentially cause surface damage, although atmospheric screening does provide partial shielding from blast effects. Relatively few of these small bodies have been discovered.

The public has consistently ranked planetary defense (henceforward, "PD") as one of NASA's top priorities, supported in 2019 by 62 percent of poll respondents (Johnson 2019). Over the past three decades, congressional action has spurred NASA to improve its PD programs.

The Spaceguard Survey, aimed at discovering and tracking NEOs 1 km and larger, was initiated in 1998 and largely completed by 2009. The 2005 George E. Brown, Jr. NEO Survey Act directed NASA to catalog, by 2020, 90 percent of NEOs larger than 140 m; impacts of this size may cause regional devastation. As shown in Figure 18-1, NASA has not yet completed the George E. Brown survey goal. NASEM (2019) stated that NASA was provided inadequate funding to accomplish this task by the 2020 date. Guided by continuing PD technology and policy studies, NASA in 2016 established a Planetary Defense Coordination Office (PDCO), and the National Science and Technology Council in 2018 published the *National Near-Earth Object Preparedness Strategy and Action Plan* (NSTC 2018), followed by a 2021 *Report on NEO Impact Threat Emergency Protocols* (NSTC 2021).

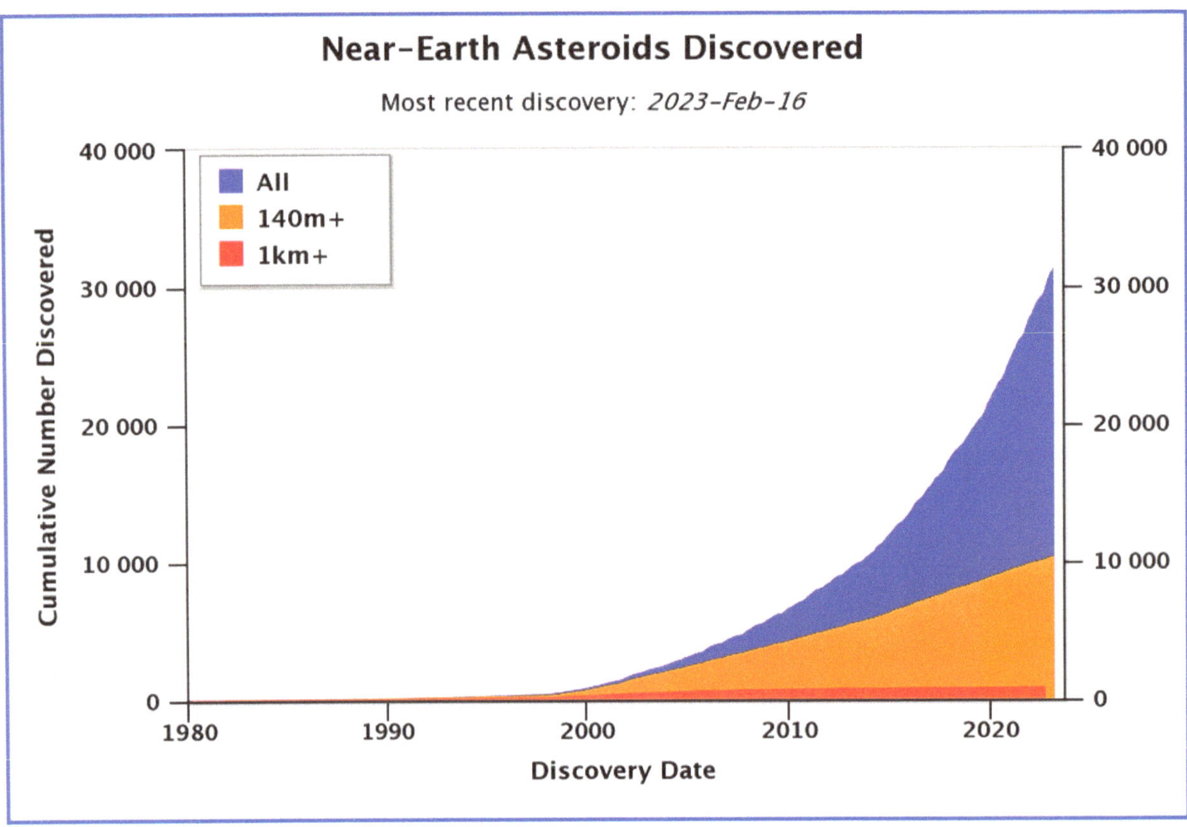

FIGURE 18-3 Near-Earth object discoveries over time, showing the accelerating rate of NEO detection from ground-based search programs. Data current as of February 16, 2023. SOURCE: Courtesy of A.B. Chamberlin, NASA/JPL-Caltech, https:// cneos.jpl.nasa.gov/stats/totals.html.

NASA's annual PD budget within the Science Mission Directorate has risen to more than $160 million.[3] These resources have enabled NASA and its partners to catalog, as of February 2023, more than 30,000 NEOs using ground-based telescopes and the NEOWISE space-based, infrared telescope (NASA 2021b) (Figure 18-3). The discovery rate is increasing, yet millions of NEOs remain undiscovered. NASA is developing the Near-Earth Object Surveyor (NEO Surveyor) infrared space telescope to complete the congressionally mandated 140 m NEO survey; a 2026 launch is planned. Additionally, the agency launched the Double Asteroid Redirection Test (DART) kinetic impact deflection demonstration mission in 2021; the spacecraft will impact the small asteroid Dimorphos, a satellite of the asteroid Didymos, in 2022. Last, outside of NASA's PD budget line, the NSF-funded 8.4-meter Vera C. Rubin Observatory (VCRO), previously referred to as the Large Synoptic Survey Telescope, plans to begin operations in roughly two years and will discover numerous NEOs.

The International Astronomical Union Minor Planet Center (MPC), hosted at the Harvard and Smithsonian Center for Astrophysics, is the key repository for all NEO observations collected worldwide. The MPC conducts an initial analysis of observations to confirm and identify NEOs, computes orbits based on these observations, and highlights objects of interest. MPC transmits this information to the Center for Near-Earth Object Studies (CNEOS) at NASA's Jet Propulsion Laboratory, which calculates high-precision orbits and scans all confirmed

---

[3] All dollar amounts are in real-year dollars unless stated otherwise.

NEOs for any potential impacts via its Sentry impact monitoring system, and monitors potential newly discovered NEOs via its Scout hazard assessment system.

NASA and NSF have broad expertise in both the scientific exploration and characterization of NEOs. NASA's deep space operations experience also enables the development of deflection technologies for future diversion of a threatening NEO. The decadal survey is tasked with making recommendations on future steps and the level of resources needed to meet PD preparedness goals.

The PD findings and recommendations in this report are presented in the framework of the *National Near-Earth Object Preparedness Strategy and Action Plan*, which identifies NASA as the key U.S. government agency to lead such activities and directs NSF to provide support. The plan's five strategic goals underpin the nation's effort to enhance preparedness for dealing with the threat of future NEO impacts (NSTC 2018):

1. Enhance NEO detection, tracking, and characterization capabilities.
2. Improve NEO modeling, prediction, and information integration.
3. Develop technologies for NEO deflection and disruption missions.
4. Increase international cooperation on NEO preparation.
5. Strengthen and routinely exercise NEO impact emergency procedures and action protocols.

Given the statement of task and its purview (see Appendix A), the committee's findings and recommendations will concentrate on the first three items in the list above.

Finding: The threat to Earth from potentially hazardous objects is a low-probability yet high-consequence hazard recognized and documented in several academic and government reports. Efforts to address this threat are most effective when undertaken on a continuous, long-term basis through cooperation among NASA, NSF, U.S. Space Command, and other U.S. government entities.

Finding: The establishment in 2016 of NASA's PDCO has brought leadership and strategic direction to U.S. planetary defense efforts.

Recommendation: NASA's PDCO should be robustly supported and sustained as the critical organization to advance U.S. planetary defense capabilities and initiatives in the next decade and beyond.

NEO DETECTION, TRACKING, AND CHARACTERIZATION

NEO Surveys

NASA's current efforts to detect and track NEOs are guided by the Brown Act, which requires NASA to "detect, track, catalogue, and characterize . . . near-Earth objects ≥140 meters in diameter to assess the threat of such objects to Earth." It further set a goal of detecting 90 percent of this population of objects within 15 years— that is, by 2020. However, by 2021 only an estimated one-third of all NEOs ≥140 m have been discovered (NSTC 2018, 2021; see Figure 18-3).

These NEO survey goals are designed to quantify the risk of regional to global catastrophic impact damage by establishing a degree of completion that is both timely and cost effective. Impacts from 140 m objects would be damaging on a regional level. Such an impact would be equivalent to an explosion of more than 60 megatons of TNT, larger than the most powerful nuclear device ever tested (NASEM 2019). While such events are rare (the average interval between impacts of 140 m objects is ~20,000 years), the consequence of such an event can be large, causing mass casualties and catastrophic loss of infrastructure, particularly if the impact occurs near a populated area (NASEM 2019). A recent study estimated that the number of casualties from such an event—when averaged over the entire surface of Earth—would exceed 10,000 (NASA 2017), although the actual number of casualties will vary given the warning time and civil defense preparations. The primary benefit from a comprehensive NEO survey is to reduce the risk uncertainty associated with such high-consequence events by early identification of potentially hazardous objects—finding them before they find us.

The Brown Act survey goals are based on the findings of a 2003 NASA science definition team study (NASA 2003). A follow-on report (NASA 2017) reconfirmed these goals as a reasonable baseline to understand the current impact risk at an acceptable level within a decade after survey initiation. Although the Brown Act survey's completion deadline was not met, its accomplishment in roughly a decade is still crucially important. Meeting such a deadline dramatically reduces the uncertainties in the risk of impact from medium to large ($\geq$140 m in diameter) NEOs, optimizes the use of facilities and trained personnel, and obtains maximum benefit from the lessons provided by other current space-based detection tools, for example, NEOWISE.

Even though objects $\geq$140 m in size are the priority for detection and cataloging owing to their grave impact consequences, objects smaller than 140 m can also cause significant damage (e.g., the Tunguska and Chelyabinsk events). They represent a tangible threat to public safety, and yet owing to their small size, are some of the most challenging objects to detect prior to impact. It should be noted that small airburst events are by far the likeliest NEO impacts that Earth will encounter in the coming century (see Figures 18-1 and 18-2).

> **Finding:** The Brown Act survey goals still provide a reasonable size threshold ($\geq$140 m), completion level (>90 percent), and timetable (survey completion in <10 years from now). However, the most likely impactors are NEOs less than 140 m in diameter and these can pose a significant threat of local damage. It is therefore important to detect, track, and characterize as much of this smaller population as possible in addition to meeting the act's goals.

Current assets used by NASA for NEO surveys include both ground-based, visible-wavelength telescopes (e.g., Pan-STARRS, Catalina Sky Survey, ATLAS) and the repurposed NEOWISE space-based, infrared telescope (NASA 2021b). The rate of NEO discoveries has increased steadily over the past decade as more capable telescopes have been added to this network. However, current observational capabilities will only achieve ~50 percent completion over the next decade in detection of objects $\geq$140 m (NASEM 2019).

The addition of NSF's 8.4 m, ground-based Vera C. Rubin Observatory, planned for first light in 2023, will augment existing detection capabilities. However, even if this telescope was fully dedicated to NEO search and detection operations, it would still not achieve 90 percent detection completion for decades. Visible-wavelength telescopes are unable to distinguish between small, bright objects and large, dark ones, leading to an uncertainty in the size of detected NEOs that translates into an uncertainty concerning potential impact energy. Follow-up observations would be needed to better quantify the hazard from detected objects (NASEM 2019).

NEO Surveyor

Space-based telescopes have distinct advantages over their ground-based counterparts—e.g., uninterrupted operations, a more optimal viewing geometry yielding higher detection rates, and the ability to operate at wavelengths that are blocked by Earth's atmosphere. Space-based telescopes operating in the mid-infrared detect energy radiating directly from the object, not reflected sunlight. This resolves the uncertainty in size versus reflectivity (albedo) and enables a much more accurate determination of an object's diameter. A recent National Academies study (NASEM 2019) concluded that a space-based mid-infrared survey is the most effective, timely option for meeting the congressional NEO survey completeness and size determination requirements.

> **Finding:** A dedicated space-based mid-infrared survey is the most effective architecture to accomplish congressionally directed NEO survey goals. A space-based mid-infrared survey also provides real-time information on object diameter, critical for rapid impact hazard assessment.

NASA has for several years been developing a space-based, mid-infrared telescope mission, NEO Surveyor. Its precursor concept, NEOCam, was funded to mature the proposed infrared detector technology and refine the mission design. NEO Surveyor is now in Phase B aimed at launch by 2026. However, while NASA has requested funding for full-scale development in the 2022 President's Budget Request, funding is not yet assured, and the mission is not yet confirmed.

Finding: The first priority in planetary defense is early detection, tracking, and characterization of NEOs, whose impact may cause widespread regional damage. NEO Surveyor, in development by NASA for this purpose, is the most timely and effective means to complete the survey goal of detecting 90 percent of NEOs greater than 140 m in diameter.

Recommendation: NASA should fully support the development, timely launch, and subsequent operation of NEO Surveyor to achieve the highest priority planetary defense NEO survey goals.

Vera C. Rubin Observatory

The Vera C. Rubin Observatory (VCRO) is a multi-disciplinary facility funded by NSF and DOE and currently under construction on Cerro Pachón in north-central Chile. Although VCRO alone, or in combination with existing ground-based observatories, cannot meet the completion goals for the NEO survey within the next decade, it does complement NEO Surveyor. The observation cadence of VCRO is also optimal for sampling the rotational light curves of most asteroids, important for constraining asteroid shapes and spin states (Jones et al. 2020).

VCRO's search area is centered away from the Sun, optimal for detection of objects in opposition, while NEOS's search area is oriented toward the near-Sun regions of the sky (Mainzer et al. 2020). NEOS's search strategy is optimized to increase its overall NEO detection rates. In addition, the different search areas provide opportunities for follow-up observations over longer orbital arcs, reducing errors in orbit estimation, which is important for tracking potentially hazardous objects.

Although funding for VCRO is in place, no focused programs exist specifically for analysis of solar system data, such as NEO characterization. Dedicated funding is needed to ensure maximum use of this facility for PD purposes. Because most VCRO discoveries are expected to occur during the first few years of the survey, readiness for handling NEO data prior to expected full operations as early as late 2023 is critical.

Finding: Although the VCRO cannot alone meet the Brown Act's goals, it is complementary to NEO Surveyor. A robust, focused program for VCRO dedicated to NEO detection, follow-up tracking, and characterization, and a well-coordinated data pipeline between VCRO, the Minor Planet Center, and the Center for Near-Earth Object Studies, would provide optimal benefits for planetary defense.

NEO discovery rates have been increasing for more than two decades, taxing the capabilities of existing follow-up facilities to keep pace with NEO detection confirmations, arc extensions, and characterization. NEO detection rates are expected to increase ten-fold when VCRO and NEOS come on-line (Jones et al. 2020) and will put additional stress on follow-up observations. Although VCRO and NEOS can conduct self-follow-up observations, the demand for time on other telescopes to follow-up new NEO discoveries is likely to sharply increase. Follow-up observations of interesting, and often faint, objects discovered by VCRO and NEOS may require access to the new generation of extremely large telescopes now under development, or possibly even large space telescopes such as James Webb Space Telescope or Nancy Grace Roman Space Telescope. New, dedicated, large-aperture follow-up facilities and/or significant access to existing large-aperture telescopes will be required to complement the surveys' self-follow-up plans (Seaman et al. 2020). International collaboration and cooperation will also play a key role in tracking and observing these objects (see below).

Finding: Existing observatories dedicated to NEO discovery and follow-up will still be needed after VCRO and NEO Surveyor come on-line. In addition, time on larger telescopes (both existing and in development) coordinated by NASA and NSF will be needed for tracking and characterization of faint objects that pose a nonnegligible impact probability.

Finding: The data pipeline for detection, tracking, impact assessment, and reporting to the public will be tested by the ten-fold increase in NEO detection rates when VCRO and NEO Surveyor become operational. Both facilities will benefit from coordinated rehearsals and operational readiness reviews well prior to achieving their full operational capabilities.

Mega-Satellite Constellations and NEO Surveys

An increasing threat to the ability of ground-based telescopes to detect and characterize NEOs comes from existing and planned mega-constellations of low-Earth-orbiting (LEO) satellites (e.g., Starlink). These constellations, particularly those below 600 km, are highly visible during twilight when NEO surveys begin and end nightly operations. Accordingly, they are contaminating a fraction of each night's survey observing window with reflected light from their surfaces, making it more difficult to detect potentially hazardous objects.

The consequences of existing LEO satellites are starting to be felt by the leading NEO search facilities such as Pan-STARRS in Hawaii and the Catalina Sky Survey in Arizona. Concern is also mounting among the astronomical community over the proposed networks of 100,000 LEO satellites. If such networks come to fruition, no combination of even "best-case" mitigation measures (e.g., material selection and coatings to reduce reflective light, and orientation of solar arrays) will be capable of fully countering the constellations' negative impact on the search for NEOs (Walker et al. 2021).

> **Finding:** Mega-constellations of commercial satellites will likely impact the efficiency of ground-based searches for NEOs. Efforts by NASA, NSF, and the astronomical community to monitor these constellations and encourage development of mitigation mechanisms to reduce light contamination from these satellites, as well as support software algorithm improvement of NEO search pipelines, would minimize impacts to ground-based NEO surveys.

Inter-Agency Synergies

One of the key areas of overlap for NEO discoveries exists between space situational awareness (i.e., tracking objects in space, identifying them, and predicting where they will be at any given time), monitoring and characterizing orbital debris, and planetary defense. U.S. Space Command's role in space situational awareness is to monitor near-Earth and cislunar (i.e., the space in the immediate vicinity of the Earth-Moon system) space to maintain U.S. national security interests. Some of the assets it uses are optimized for detecting small human-made objects in near-Earth space but can also detect natural objects as a byproduct of their mission objectives. Similarly, NASA's Orbital Debris Program Office uses its own assets to characterize the debris environment in Earth orbit and has capabilities to distinguish natural objects from human-made space debris. Data from these respective organizations have the potential to provide insight into the NEO population in and around near-Earth/cislunar space and contribute to important PD initiatives.

> **Finding:** U.S. Space Command and NASA's Orbital Debris Program Office have capabilities to discover NEOs in cislunar and very near-Earth space environments. Such discoveries would provide useful information concerning the general NEO population. Increased cooperation between NASA and U.S. Space Command to exchange information on NEOs from their respective organizations would aid planetary defense objectives.

NEO Characterization

Policy makers, scientists, and engineers require knowledge of the physical characteristics of a NEO to determine the safest and most efficient measures to deflect a threatening object away from Earth.

The most important factor in NEO characterization is determining the object's orbit. When detected, the trajectories of most objects often have substantial uncertainties, such that Earth impact probability is poorly constrained. More accurate knowledge of the orbit is then required from follow-up observations to determine whether that object poses a true impact threat to Earth. In such circumstances, radar observations are extremely useful to determine precise positions and velocities of these objects.

Beyond orbit determination, characterization encompasses the measurement of all properties of a NEO relevant to planetary defense (NRC 2010). The severity of a NEO impact is determined by an object's mass and velocity at Earth encounter, as these two characteristics define the impact kinetic energy. Other factors influencing impact effects include the NEO's diameter, density, composition, material strength, and its approach angle to the surface. These dynamical and physical characteristics determine how the energy is partitioned upon Earth impact and influence whether the object breaks up harmlessly in the upper atmosphere or whether aerodynamic stress will cause it to detonate near/at Earth's surface. The NEO's physical characteristics also have bearing on the effectiveness of various space deflection/mitigation techniques. For example, a high-velocity projectile launched into a weak rubble pile object may produce unexpected results in deflection efficiency compared to striking a more consolidated asteroid.

Ground-based remote sensing techniques measure and constrain important properties such as diameter, shape, composition, and density. However, in situ spacecraft observations offer precise knowledge of these properties as well as additional ones such as mass, porosity, strength, and the presence of small companions. Such characterization is desired to develop robust mitigation strategies for different NEO object classes and may be needed for specific NEOs if it becomes necessary to deflect or disrupt them (see Next Steps for Planetary Defense Demonstration Missions section below).

Ground-based characterization efforts can deliver insights into the physical nature of individual NEOs and the population in general. However, as the catalog of discovered NEOs grows and becomes more diverse, characterization efforts can lag far behind. This is in part owing to the limited number of ground-based observatories available, the challenges of observing small, distant NEOs, and the personnel required to obtain, analyze, and interpret NEO data. This means that broad-scale characterization efforts of certain properties (e.g., color surveys, such as those performed by the Sloan Digital Sky Survey in New Mexico; diameter and albedo surveys, such as those by NEOWISE) combined with detailed observations of individual NEOs are needed to evaluate which mitigation strategies will be the most effective and to assess the specific risk from the most hazardous NEOs.

When a NEO is discovered by a ground-based survey like Pan-STARRS, the only property immediately known apart from the orbit is its optical brightness. By assuming an albedo (e.g., typical values are 0.04 for carbonaceous chondrites and 0.20 for noncarbonaceous chondrites), it is possible to get a rough estimate of the object's diameter. Uncertainties in albedo, and limited observations of small (dim) NEOs that lead to uncertainties in brightness, can easily result in diameter estimates being off by a factor of two or more. Radar measurements obtain more precise diameter information without any need for brightness/albedo estimates, but such opportunities are limited (see next section).

Spectral similarities between a NEO and specific meteorite compositions can be used to estimate an object's mineralogical makeup. NASA's Infrared Telescope Facility in Hawaii has made key contributions to the spectral characterization of NEOs, and continuing such efforts is important for refining mineralogic classifications. These classifications are then used to infer bulk densities, physical strengths, and assumptions about internal structure, albeit with significant uncertainty, because porosity is not known.

Observed light curves can deliver estimates of a NEO's physical properties such as shape, rotation period, pole orientation, and the inferred presence of satellites. Radar observations can yield those same characteristics as well as NEO size and surface roughness, but for a limited number of targets.

Finding: Because NEOs are best observed closest to Earth, when optically bright (usually soon after discovery), ground-based characterization efforts depend on obtaining scarce telescope observing time, along with availability of favorable observing geometry and sky conditions. Such challenges result frequently in missed opportunities to physically characterize many NEOs, and the lagging pace of characterization will reduce abilities to respond effectively to a future impact threat.

Research on meteors, meteorites, fireballs, and bolides also provides valuable opportunities to characterize NEOs. Data from these studies include estimates of NEOs' compositions, strengths, internal structures, and mechanical properties. These factors are important to assess the potential damage from a NEO impact and to inform any mitigation techniques that could be employed to prevent an impact with Earth. Meteorites provide unique insights into the range of materials that make up NEOs and main belt asteroids, as well as providing scientific knowledge about the early solar system, planetesimal formation, and small body evolution. Currently, more than 65,000 meteorites are cataloged. The most common type of meteorites, ordinary chondrites, are composed of both rocky and metallic components and can have a compressive strength similar to concrete or granite when coherent (Popova et al. 2011). In contrast, other meteorite types are known to have quite different properties, from iron meteorites composed dominantly of an Fe-Ni alloy to highly friable carbonaceous chondrites (Brown et al. 2000). Small asteroids entering the atmosphere typically don't cause direct damage, but they are the most abundant type and are visible by many people. Modeling and accurately predicting the visibility range and brightness of these small impacting NEOs will enhance the credibility of the NEO activities and improve the predictions of the magnitude of expected explosions.

However, it is not possible to directly infer the strength of an NEO from the compressive strength of spectrally similar meteorites. Meteorites are pervasively fractured down to centimeter scales, and bolide observations have repeatedly shown that these objects are considerably weaker than coherent rocks, often breaking up high in the atmosphere. Estimates of the bulk strength of ordinary and carbonaceous chondrite bolides suggest a wide range of values, with some akin to a crumbly clod of dirt and others strong enough to reach deep in the atmosphere (Popova et al. 2011; Brown et al. 2016). The behavior of bolides is attributed to their macroscopic structure; some are highly fractured or simply have a weak structure, while others have some internal integrity. For planetary defense purposes, it is crucial to characterize an object as a whole, including its large-scale structure, especially as that structure may be a controlling factor in the strength of the object.

> **Finding:** Meteor, fireball, and bolide events offer naturally occurring opportunities to characterize atmospheric energy deposition processes, elucidate the mechanical properties of NEO materials, and investigate the break-up process. Such knowledge informs and assists planetary defense activities related to NEO characterization, mitigation, and modeling.

Unique perspectives can also be gained from fireball and bolide events if they can be subsequently linked to a recovered meteorite. One of the most insightful examples of this paired knowledge stemmed from the airburst of 2008 TC_3, which delivered the Almahata Sitta meteorites to Sudan (Jenniskens et al. 2009). The roughly 4-meter 2008 TC_3 asteroid was discovered about 19 hours prior to impact and observed by numerous telescopes worldwide. After its explosion as a bright fireball, a dedicated field search collected numerous meteorites. These meteorites revealed a surprising degree of mineralogical and compositional diversity, illuminating the potential compositional heterogeneity exhibited by a single NEO. This event demonstrated the scientific value of linking and interpreting telescopic observations of small NEOs prior to Earth impact.

Conversely, although the roughly 20-meter Chelyabinsk NEO was not detected prior to atmospheric entry in 2013, the numerous samples that were collected, all ordinary chondrites (Popova et al. 2013), provide key knowledge that can be used to calibrate impact hazard models against a known event and to inform future mitigation strategies.

> **Finding:** Meteorites provide samples of NEOs for investigation in state-of-the-art laboratories, adding to our understanding of compositional and mechanical properties and variations among the NEO population. The rapid recovery of meteorite falls linked to atmospheric entry events is uniquely valuable to combined meteorite and bolide studies.

A challenge with fireball, bolide, and atmospheric entry observations/research is that these events are largely unpredictable. Being in position to observe and follow-up when opportunities arise is essential to such studies. The flexibility to respond rapidly to a given event through international collaboration is important to obtain PD knowledge from these naturally occurring but infrequent events.

> **Recommendation: NASA should support planning, monitoring, and coordination among the global planetary defense, NEO observing, meteor/bolide, and meteoritics communities to take advantage of the opportunistic events provided by atmospheric entry of NEO materials, and to collect any associated meteorites in order to advance planetary defense objectives.**

Importance of Radar Observations for Planetary Defense

Ground-based radar observations of NEOs provide invaluable information for long-term tracking through ultra-precise (1 part in 100 million) measurements of line-of-sight distance and velocities. Radar astrometry of NEOs routinely reduces the orbital uncertainties by several orders of magnitude after only a few minutes of observation, preventing the loss of newly discovered objects and the need for their subsequent rediscovery. This precision also enables radar to accurately predict NEO/Earth encounters on average 400 years into the future. For example, initial orbit determinations of potentially hazardous NEOs (99942) Apophis and 2020 NK$_1$ indicated a significant chance of Earth impact, but radar observations were able to quickly rule out a collision hazard.

Depending on the observing circumstances (e.g., NEO distance, size, and viewing geometry), radar imaging can provide highly detailed characterization of a NEO's physical properties, including direct size, shape, and rotation state measurements, as well as constraints on near-surface density and roughness, surface geology, and identification of satellites (radar has discovered more than 70 percent of all known NEO satellites). Radar-enabled three-dimensional shape modeling of NEOs has proven reliable and precise; for example, the shape of Bennu determined by OSIRIS-REx was within 2 percent of the radar shape model (Nolan et al. 2019).

The level of characterization afforded by radar provides crucial information for impact mitigation strategies. Because NEO impact energy scales with density, diameter, and velocity, and radar can constrain all of these, planetary radar observations are an important post-discovery characterization technique. As such, radar plays a "unique role" in achieving the tracking and characterization goals of the Brown Act (NRC 2010).

> **Finding:** Ultra-precise radar measurements are the most accurate ground-based means for refining NEO orbits and retiring the risk of future impacts. Furthermore, radar-enabled post-discovery characterization permits direct measurements of size and shape, as well as constrains other physical properties important for impact mitigation strategies.

A vital asset for radar observations was the NASA-funded Planetary Radar Project at the NSF's Arecibo Observatory in Puerto Rico. The December 2020 loss of Arecibo Observatory's 305 m radio telescope created an urgent need to recover this critical capability. Currently, the remaining key facility for planetary radar observations is NASA's Goldstone Solar System Radar (GSSR), specifically the Deep Space Station facilities DSS-14 at X-band and DSS-13 at C-band. GSSR is a fully steerable radar telescope that can deliver more precise line-of-sight distances than could Arecibo. Although GSSR can produce longer observing tracks, Arecibo was 15× more sensitive than GSSR; in a typical year, Arecibo could observe twice as many NEOs as GSSR (Naidu et al. 2016). Thus, Arecibo's loss has drastically reduced the capability for follow-up radar characterization of NEOs (see also Infrastructure chapter).

> **Finding:** The loss of the Arecibo Observatory planetary radar greatly inhibits the ability to perform follow-up NEO characterization. Existing radar infrastructure can observe only half the asteroids once observable with Arecibo.

> **Recommendation: NASA and NSF should support studies to develop a plan for ground-based planetary radar capabilities comparable to or exceeding those of the Arecibo Observatory necessary for achieving planetary defense objectives.**

The Green Bank Telescope (GBT) can operate as part of a bistatic radar facility where GSSR transmits and the GBT acts as a receiving station. The GBT is considering the addition of a radar transmitter (Bonsall et al. 2020), adding a complementary capability and some redundancy to GSSR. However, even with two active radar observatories, the national planetary radar infrastructure lacks two-fault-tolerant redundancy: a system able to survive two faults, as experienced in 2020 when both Arecibo and GSSR were inoperative owing to klystron failures (Adamo et al. 2020). Furthermore, these facilities are all in the northern hemisphere, leaving a significant gap in follow-up capability of NEOs at southern declinations.

> **Finding:** In order to conduct the required NEO follow-up characterization observations to meet key planetary defense objectives, it would be valuable to expand and extend planetary radar capabilities to obtain coverage over the northern and southern celestial hemispheres.

NASA-NSF Cooperation

Because ground-based planetary radar observations have been conducted at shared-use facilities, in particular at NSF facilities such as Arecibo and the Green Bank Telescope, improved collaboration between NASA and NSF is needed to ensure the nation meets its PD strategic goals. In particular, the *National Near-Earth Object Preparedness Strategy and Action Plan* requires that NASA and NSF work together to "identify opportunities in existing and planned telescope programs to improve detection and tracking" of NEOs, as well as to identify "opportunities in existing and new telescope programs to enhance characterization of NEO composition and dynamical and physical properties" (NSTC 2018).

> **Finding:** The *National Near-Earth Object Preparedness Strategy and Action Plan* recommends NASA and NSF work together to support and fund opportunities in existing and planned telescope programs to improve the detection, tracking, and characterization of NEOs. Such efforts would improve characterization of objects with a nonnegligible impact probability and the NEO population as a whole.

> **Recommendation: As the steward of ground-based observatories with NEO observing capabilities, NSF should support and prioritize critical planetary defense observations of NEOs at its ground-based facilities.**

Although NASA and NSF have signed a memorandum of understanding to advance space, Earth, biological, and the physical sciences, closer cooperation is needed to ensure that shared-use facilities can be leveraged effectively to meet PD goals. For example, prior to the collapse of Arecibo's platform in December 2020, NSF asked NASA to investigate the "root cause of the Auxiliary M4N cable failure" that occurred in August 2020, which ultimately left the facility vulnerable to collapse (Harrigan et al. 2021). However, the agency's June 2021 report observes that "Additional data and hardware were available to the NSF but were not available to (NASA) for examination" (Harrigan et al. 2021). Thus, a more detailed collaborative agreement that enables direct communication and input between NASA (particularly the PDCO) and NSF for decision-making on shared-use facilities is required.

> **Finding:** NASA and NSF have largely informally cooperated at shared-use facilities, such as Arecibo and GBT, by leveraging their grantees and contractors as intermediaries. A more formalized agreement is required to ensure appropriate collaborations for planetary defense.

While the NSF is tasked to support ground-based observing infrastructure, NASA has been tasked with leading key planetary defense objectives. This leaves ground-based observatories that support PD goals in a nebulous situation in terms of funding. A collaborative agreement between NASA and NSF would enable the effective communication and cooperation required to facilitate critical planetary defense work at existing and future shared-use ground-based facilities. Such a collaboration would also facilitate the development of a ground-based planetary radar facility with planetary defense capabilities comparable to or exceeding those of the Arecibo Observatory planetary radar. Such an endeavor would also require adequate support for the development and maturation of associated radar technologies. If both agencies formally work together, it would also further the nation's ability to expand the planetary radar infrastructure in the northern hemisphere, as well as its expansion, through partnerships or new facilities, into the southern hemisphere in order to achieve key planetary defense objectives.

Apophis, a Unique Characterization Opportunity in the Coming Decade

The close approach of asteroid (99942) Apophis to Earth in 2029, which will pass within Earth's belt of geostationary satellites and about 6 Earth radii from the center of Earth, presents an unprecedented opportunity for both planetary science and global planetary defense awareness. Apophis is 370 m in diameter and its flyby will be the closest approach of an asteroid that size ever recorded, making for a truly exceptional event. At closest approach—occurring on April 13, 2029, at 21:46 UT—the asteroid will be visible with the naked eye or binoculars over Europe, Africa, and Western Asia. The flyby will alter Apophis's heliocentric orbit, transitioning it from an Aten-class to an Apollo-class asteroid (Giorgini et al. 2008). While Apophis remains a potentially hazardous object, any impact with Earth has been ruled out for more than 100 years into the future by recent planetary radar observations (NASA 2021a).

During the flyby, the spin state of Apophis is also expected to undergo a relatively large change owing to Earth's gravitational forces, which will torque the body (Scheeres et al. 2005). Despite these significant effects, the flyby is not expected to cause widespread shifts of surface or interior material, which would require a flyby a few Earth radii closer than will occur in 2029 (Hirabayashi et al. 2021).

The extremely precise Apophis flyby prediction provides an unprecedented opportunity for multiple observatories to coordinate simultaneous pre- and post-encounter observations of the NEO from multiple viewing geometries and at wavelengths from optical to radar. Owing to its inherent interest, Apophis is already a target of opportunity for astronomers, with every apparition leading to significant observations focused on determining its orbit, shape, spin state, and thermal and spectral properties. These have created a precise baseline for the physical state of Apophis prior to its 2029 close approach (Pravec et al. 2014; Brozovic et al. 2018). Precision observations of Apophis leading up to and through its 2029 close approach to Earth offer a unique opportunity to better understand potential changes in surface morphology, track changes in spin state through close approach, and compare pre- and post-flyby spectra to ascertain if the close passage has disturbed space weathered materials on the asteroid's surface.

Finding: Obtaining the best measurements during Apophis's close flyby of Earth in 2029 requires a coordinated, international observational response.

Apophis's close approach has also stimulated discussion of visiting the body through either a flyby or rendezvous spacecraft mission. As one example of a flyby mission, the precise prediction of its 2029 trajectory would enable a spacecraft to be placed on an orbit with apogee beyond the GEO belt with only a small ΔV (i.e., velocity change). The spacecraft could be positioned to take observations of the asteroid as it passes by on either its inbound or outbound leg. This type of flyby space mission would not require the observing spacecraft to leave Earth's gravitational influence. If equipped with an appropriate imaging suite, the craft could observe Apophis under complementary viewing geometries to ground-based observatories and at much higher resolution.

Requiring significantly greater capability would be a spacecraft rendezvous arriving before Apophis's close approach to Earth. A rendezvous well in advance of the Earth flyby could map the asteroid in detail before and after its Earth passage (Binzel et al. 2020). Although significant physical changes in the body (except for its spin state) are not expected, this prediction could be tested by scrutinizing the surface for changes in regolith placement and distribution. If the spacecraft could station-keep with Apophis through the close approach (Scheeres 2019), it would provide novel opportunities for bi-static radar tomography around closest approach, utilizing ground and space-based radio telescopes (Cheng et al. 2020).

A spacecraft arriving at Apophis after its close Earth approach could still carry out significant and scientifically noteworthy observations. Apophis has had numerous close approaches to Earth, which may explain its current complex rotation state (Pravec et al. 2014). A rendezvous mission could map the asteroid in detail to determine if there are signatures of past, closer Earth flybys that may have caused more dramatic changes to the body's morphology. In addition, precise determination and monitoring of Apophis's highly excited spin state over an extended period of time could constrain, or possibly detect, the expected energy dissipation experienced by such a tumbling body.

Finding: The Apophis flyby of Earth creates an opportunity to observe a potentially hazardous asteroid via a coordinated ground-based campaign, potentially supplemented via space-based observations from flyby or rendezvous missions.

A flyby mission to Apophis could be carried out with a SIMPLEx-class mission. A rendezvous mission that could acquire more detailed observations would need a spacecraft with greater ΔV capability, likely requiring a Discovery or medium-class mission. Of particular utility would be a spacecraft capable of performing detailed mapping and characterization, followed by long-term monitoring. It is thus significant to note the effect of solar illumination on the Apophis orbit and spin state to better constrain its future impact potential. These measurements would have overlapping scientific and planetary defense implications. One option would be the OSIRIS-REx spacecraft, which could a rendezvous with Apophis following the delivery of its Bennu sample canister to Earth, although it would only encounter Apophis shortly after its 2029 Earth flyby (Dellaguistina et al. 2021).

Recommendation: NASA should study all relevant observing opportunities surrounding the unique Apophis encounter, using both ground and space-based assets. To maximize the scientific and planetary defense return, NASA should develop plans for making the best use of these identified assets during the Apophis encounter and support international cooperation in carrying out these valuable observations.

NEO MODELING, PREDICTION, AND INFORMATION INTEGRATON

Goal 2 of the *National Near-Earth Object Preparedness Strategy and Action Plan* focuses on advancement of modeling and analysis capabilities, including the assessment of impact probabilities, location, consequences, and mitigation options. The initial conditions and attendant uncertainties for these modeling efforts depend heavily on the detection, tracking, and characterization of NEOs (see NEO Detection, Tracking, and Characterization section). Corresponding modeling expertise is distributed among many communities, including NASA centers, DOE's National Nuclear Security Administration (NNSA) laboratories, the National Oceanic and Atmospheric Administration (NOAA), and academic institutions. Hence, the first objective to meet Goal 2 is to "establish an interagency NEO impact modeling group" (NSTC 2018). The NEO Action Plan Modeling Working Group (MWG) was established in November 2019. The second and third objectives, "Establish an integrated suite of computational tools for modeling NEO impact risks and mitigation techniques" and "Exercise, evaluate, and continually improve modeling and analysis capabilities" are currently in development within the MWG.

NEO Consequence Assessment Modeling

Consequence assessment calculations can help minimize loss of life and property when an incoming NEO cannot be deflected by advising appropriate emergency responses, such as evacuations, safety precautions, and infrastructure protection. For a smaller NEO, these assessment calculations may predict details of atmospheric entry and breakup, and accompanying ground blast effects. For larger NEOs, which deliver most of their energy to the surface, assessment calculations may address ocean wave generation and subsequent coastal flooding, or on land, the expected fireball radiation and blast damage. NEO impact casualties can be caused by blast, ignition, seismic, cratering, or fireball effects. Longer term atmospheric perturbations can be estimated by handing off to a global circulation model, tracking the effects of ejected volatiles and dust.

The increase since 2013 in expertise and resources aimed at impact consequence modeling by NASA and its collaborators has significantly advanced U.S. capabilities to simulate NEO impact effects. That investment has also yielded a more nuanced yet quantitative understanding of impact risk and attendant uncertainties. The Asteroid Threat Assessment Project (ATAP) established in 2015 at NASA Ames Research Center has taken a multidisciplinary approach in developing higher-fidelity airburst modeling and probabilistic risk assessment approaches (Mathias et al. 2017).

An early focus of ATAP has been on smaller NEOs, such as those causing the Tunguska or Chelyabinsk events, which are statistically the most likely damaging objects to strike Earth. These types of events are especially demanding to simulate because these objects deposit most of their energy into the atmosphere and their break-up is sensitive to a bolide's specific material properties. Future work will address impacts from NEOs in the hundreds-of-meters size range, where potential damage transitions from regional to global scales. Even though the probability of these larger events is lower than Chelyabinsk-like airbursts, the maximum probability for an impact with global cataclysmic effects is calculated to be in the 500 to 700 m size range (Reinhardt et al. 2016). Quantification of the damage from these large impacts requires engagement with experts on agriculture, supply chains, and other downstream effects.

Another critical area of impact consequences study are oceanic impacts and their generation of large water waves. This multiphysics problem requires a hand-off between a shock-physics code to an elastic-wave code that can simulate wave propagation over hundreds of kilometers and many hours (Ezzedine and Miller 2014). Coastal inundation can be further simulated by codes designed to handle fluid-structure interactions.

The Second International Workshop on Asteroid Threat Assessment (2016) focused on water impact modeling, with participation from many of the world's experts (Boslough et al. 2016). Significant differences in scientific judgement of the ocean impact risk remain within the water impact modeling community, and will require continued collaborative engagement, including transparent presentation of the methods, assumptions, and limitations of each numerical approach necessary to make progress on ocean impact modeling and risk quantification of water-wave effects.

Finding: Significant differences in technical results and methodology remain within the consequence assessment simulation community. Although a diversity of approaches can provide helpful perspectives on impact problems, more collaborative work is needed to understand the sources of these differences and to refine simulation capabilities.

The interagency NEO MWG, led by the NASA Ames Research Center, has begun working through representative impact studies. The MWG's multidisciplinary activities include consequence assessment, mitigation, and information integration, with the goal of providing rapid assessments to enable decision making and to enhance public awareness in a NEO emergency. While early efforts have focused on code comparisons and probabilistic risk assessments for specific problems of interest, future work will include validation and verification of computational tools against experimental and observational data.

Ongoing support of NEO MWG is critical for meeting the next decade's planetary defense objectives. The MWG currently exists as an "unfunded mandate"—without agency funding—yet is still expected to hold regular meetings to address strategic objectives for NEO modeling, predictions, and information integration. Given the national security value of rapid, trusted, and centralized assessments alongside the ever-present risk of misinformation during a NEO emergency, reliable funding for the MWG is a wise investment.

Finding: NASA's leadership is vital for integrating consequence assessment calculations across the diverse communities capable of this work, particularly when their risk assessments differ. NASA-sponsored workshops and meetings provide an important venue for technical discussions of the assumptions and limitations underpinning various numerical approaches.

NEO Mitigation Modeling

If an asteroid or comet is calculated to be on a likely collision course with Earth, several mitigation techniques are available. Ideally, the NEO will be detected with sufficient warning time to conduct a deflection, in which a modest change in velocity (ΔV), is applied, either slowing down or speeding up the object in its orbit. The integrated change in position over many years will result in the NEO missing Earth entirely. In a successful deflection, the bulk of the asteroid stays intact (i.e., it is not fragmented) and misses Earth. Deflections can be applied using impulsive or "fast push" methods like a kinetic impactor or a standoff nuclear explosion,[4] or via "slow push/pull" methods such as a gravity tractor or ion beam deflection (see NEO Deflection and Disruption Missions section below). For the most challenging scenarios, when warning times are less than a few decades, the kinetic and nuclear approaches are assessed to be the most effective (NRC 2010).

As warning time decreases, the required ΔV to achieve a successful deflection increases. Although nuclear standoff deflections can deliver momentum changes in a mass-efficient payload, once the required deflection ΔV increases to a significant fraction of the NEO's escape velocity, any attempted deflection will risk unintentional disruption of the body, which could exacerbate the threat. Shorter warning times (less than a few years to a decade, depending on the details of the NEO) may require a disruption mission to prevent an Earth impact (Dearborn and Miller 2014). Disruption can be carried out by detonating a nuclear device near the NEO. Although a buried burst would couple a larger fraction of its energy to the body, burial would require increased mission complexity. A close-proximity standoff burst simplifies execution significantly, and the energy coupled into the NEO is sufficient to generate a strong shock wave which shatters the body and robustly disperses the fragments (King et al. 2021).

Both kinetic and nuclear deflection require numerical simulations to assess NEO response to these momentum impulses. In low-gravity environments, understanding late-time crater formation from a kinetic impactor deflection may also require discrete element method modeling approaches. Nuclear deflection or disruption simulation requires radiation transport and hydrodynamic modeling capabilities. Propagating the motion of fragments forward in time to ensure they miss Earth typically requires an N-body gravity code. All of these simulations are collectively represented by the term "multiphysics."

Multiphysics simulations of NEO response to deflection/disruption efforts have matured significantly over the

---

[4]Discussion of the political, policy, and treaty implications of the deployment and use of nuclear explosive devices in space is beyond the scope of this report.

past decade, owing to advances in high performance computing, improved understanding of NEO initial conditions from spacecraft data, and focused research. Improvements include more realistic modeling of rubble pile structures, variable distributions of macro- and microporosity, improved strength and damage models representing asteroidal material, and simulation of a wider range of three-dimensional NEO shapes. Further, new and rapidly developing computational capabilities are now regularly incorporated into both kinetic and nuclear deflection modeling.

Characterization data returned from small body missions like OSIRIS-REx (Lauretta et al. 2019) and Hayabusa2 (Watanabe et al. 2019), have improved the fidelity of simulations. However, mission results always challenge modeling techniques in surprising new ways. For example, the small carry-on impactor experiment on Hayabusa2 produced a crater size on asteroid Ryugu that exceeded predictions, suggesting that gravity, not strength, controlled crater formation (Arakawa et al. 2020). In the coming decade, additional insights from DART and other missions will further refine models, if support for such studies is available.

Mitigation modeling to support end-to-end NEO impact case studies has been a central part of interagency work between NASA and NNSA, such as the ongoing collaboration between NASA Goddard Space Flight Center and Lawrence Livermore National Laboratory, Los Alamos National Laboratory, and Sandia National Laboratories. These case studies are designed to stress the tools and methods used to recommend an optimal mitigation approach. An important element of these studies is the coupling of multiphysics simulations with mission design options for a given NEO. The efficacy and risks associated with a given mitigation depend on the details of an NEO's orbit, the warning time, and derived launch opportunities (Barbee et al. 2017). Because this problem is so scenario-dependent, end-to-end case studies are a critical tool for preparedness. This collaboration has also enabled thorough code comparison efforts for kinetic and nuclear mitigation modeling (Dearborn et al. 2020).

There are scenarios in which a kinetic impactor would be unable to divert a NEO from Earth impact. In these cases, simulations and data from decades of nuclear tests indicate that a standoff nuclear explosion could successfully deflect or disrupt the NEO. Accurate modeling of nuclear deflection or disruption requires an understanding of nuclear device output, radiation transport, hydrodynamics, and material properties at extreme temperatures and pressures. The NNSA laboratories are positioned to provide this expertise, along with the computational resources required for large three-dimensional simulations.

> **Finding:** The expertise to model nuclear mitigation techniques resides at DOE's NNSA laboratories. Preparation for short-warning-time and/or larger-diameter asteroid threats requires effective partnership and open communication between NASA and the NNSA laboratories through joint activities under their interagency agreement.

Deflection and disruption modeling is also a key component of the NEO MWG. Similar to consequence assessment calculations to-date within the MWG, early work has focused on code comparisons for specific scenarios. Addressing the Goal 2 objective "Develop and validate a suite of computer simulation tools for assessing the outcome of deflection or disruption techniques applied to a NEO," will require future work focusing on validation of tools against available experimental data.

The MWG work is addressing the Goal 2 objective, "Assess the sensitivities of these models to uncertainties in NEO dynamical and physical properties." Some of this work is ongoing as part of DART impact preparations and general research on sensitivities and uncertainty quantification for kinetic and nuclear deflection. The MWG will play a key role in pulling state-of-the-art approaches together, to enable risk-informed decisions for asteroid mitigation. In particular, subject matter experts for both modeling and mission design need to work together to resolve difficult NEO scenarios, such as when it is unclear whether a kinetic impact approach may be sufficient, or when a required deflection velocity may inadvertently disrupt an asteroid. Future efforts may also benefit from closer partnership with mission design and technology development groups, who are addressing Goal 3.

Although simulations are an essential tool for predicting NEO response to impulsive deflection or disruption methods, validation of numerical simulations against experimental data provides an important measure of confidence for future mitigation missions. Validation can also illuminate sources of uncertainty (e.g., material model or numerical method shortcomings). For example, important validation of multiphysics X-ray ablation calculations can be achieved at laser facilities like the National Ignition Facility and the University of Rochester's OMEGA laser. Experiments also frequently reveal fruitful new research directions. Without access to key experimental facilities, including NASA's Ames Vertical Gun Range (AVGR) and NASA's Johnson Space Center's Experimental Impact

Laboratory (EIL), many planetary defense questions will remain unaddressed. Additional, complementary insights can be gained from field studies of Earth impact structures.

Over the past decade, R&A support for experimental impact work at NASA facilities has become more difficult to secure. Programs such as Yearly Opportunities for Research in Planetary Defense play an increasingly important role in supporting the continued operation of these laboratories and extraction of new impact results.

Finding: Although multiphysics modeling of impact events has advanced over the past decade, continued experimental validation of material models and numerical approaches is necessary. Natural experiments like the Chelyabinsk and Comet Shoemaker-Levy 9 collisions are relatively rare opportunities; in contrast, timely laboratory-scale experiments can offer fundamental insights and pathways to increase confidence in multiphysics modeling.

Although funding of numerical studies can be seen as logistically and financially simpler than funding of experiments, confidence in multiphysics simulations depends on validation against reality. The accessibility of NASA's AVGR and EIL facilities to qualified researchers is an essential piece of a healthy and effective planetary defense program.

Finding: Establishing a credible and timely national capability for both consequence assessment and mitigation modeling requires enhanced support for NEO Modeling Working Group activities.

Recommendation: NASA should increase levels of support for multiphysics modeling and laboratory experiments necessary to meet the Goal 2 objectives described in the *National Near-Earth Object Preparedness Strategy and Action Plan*.

Information Integration for Planetary Defense Simulations

As emphasized in the NSTC 2018 document, the MWG will need to stand up a capability to provide verified data to decision-makers, using "an integrated suite of computational tools for modeling NEO impact risks and mitigation techniques." Although modeling tools for NEO impact risks, consequences, and mitigation techniques currently exist, they are widely distributed between agencies and institutions, and interconnections between the various codes, models, and results are just beginning to form. Much of this nascent activity takes place within the MWG.

One way to provide verified and timely results to decision makers and the general public in advance of a real emergency is to build a national planetary defense assessment pipeline (Stickle et al. 2020). This pipeline requires a framework for sharing of characterization data, modeling parameters, validation data, calculation results, and analysis tools. Such a pipeline could enable version control, consistent updating of characterization data, sharing of geometry and material property files common to multiple NEO case studies, probabilistic treatment of risk, robust uncertainty quantification, and simpler hand-offs between tools designed to tackle each phase of an impact problem. The pipeline should also connect to international partners to enable common data exchange and formats, communication standards, modeling tools, and so on, while serving as a repository for a broad range of impact effects and mitigation calculation results.

Even with such a pipeline, it is important to obtain "multiple votes" on problems of interest by using different numerical approaches. Strengthened by diverse users and disciplines, the pipeline would streamline the process for validating multiple codes against common sets of benchmarking data, critical to developing a credible and responsive PD capability. The pipeline could also enable faster turnarounds for assessing model sensitivities to uncertainties in NEO properties.

Finding: Integrated modeling assessment across all aspects of planetary defense, including characterization data, modeling parameters, validation data, calculation results, and analysis tools, using uncertainty-aware methods, is needed to establish an operations-ready suite of computational tools for evaluating NEO impact risks and mitigation techniques.

Recommendation: To achieve the modeling, prediction, and information integration objectives listed under Goal 2, NASA should allocate resources for the establishment of a planetary defense modeling pipeline, including support for collaboration between modeling teams and software developers to establish initial requirements.

NEO DEFLECTION AND DISRUPTION MISSIONS

One of the overarching strategic goals of the *National Near-Earth Object Preparedness Strategy and Action Plan* is developing capabilities for NEO impact prevention. Goal 3 of the action plan outlines the activities that NASA will lead to strengthen the U.S. response to NEO impacts, and enable activities to prevent or minimize the damage inflicted by future, NEO-caused natural disasters. Specifically, the emphasis for NASA is to focus on the development and design of rapid reconnaissance mission technologies for NEO characterization, and deflection/disruption mission technologies for NEO mitigation.

The operational threat spectrum also includes potential impacts from long-period comets (e.g., C/Hale-Bopp) and inter-stellar objects (ISOs), which allow little warning time for mitigation efforts. In both cases, the statistical probability of Earth impact by a comet or ISO (e.g., the recently discovered 1I/Oumuamua and 2I/Borisov) is much lower (~1 percent) than from NEOs (NASA 2017). Thus, for the next decade, PD efforts focused on NEO impact prevention are the priority. However, as technologies continue to improve, defense capabilities from ISOs and long-period comets will also need to be developed and matured. Studies of rapid response mission architectures may be useful for helping identify such capabilities for future development (see the Planetary Defense Rapid Mission Response Strategies section).

Planetary Defense Missions Within NASA

NEO characterization and mitigation efforts are essential to understand the range of potential impact scenarios posed by NEOs approximately 50 m in diameter and larger, the size of object determined to potentially invoke spacecraft mission activities in the *Report on Near-Earth Object Impact Threat Emergency Protocols* (NSTC 2021). Most NEOs of this size range have yet to be discovered, and little is known about their dynamical and physical characteristics. Knowledge of orbits, masses, sizes, and other physical attributes of these NEOs will help bound the range of dynamical and physical characteristics to be considered in formulating appropriate NEO mitigation options. Some of this information is provided by ongoing surveys and will be supplemented by more capable NEO search facilities becoming operational in the mid-2020s. However, much of the required information can only be collected in situ via spacecraft missions dedicated to high-priority PD objectives.

NASA has significant scientific expertise and institutional knowledge regarding NEOs that are applicable to PD characterization and mitigation objectives. The agency also has detailed experience developing and operating spacecraft missions to investigate NEOs. Thus, NASA has been identified by the U.S. government's *National Near-Earth Object Preparedness Strategy and Action Plan* (NSTC 2018) as best suited to lead the development, testing, and flight of technology demonstration missions aimed at proving NEO mitigation techniques. Such missions are, by definition, not focused on science return, but rather utilize NASA's experience and expertise to accomplish high-priority PD objectives. Although scientific exploration of the solar system and PD share common measurements and spacecraft implementations, missions focused on accomplishing PD objectives are a valid priority for NASA.

> **Finding:** The recommendation of the 2019 NASEM report *Finding Hazardous Asteroids Using Infrared and Visible Wavelength Telescopes* remains valid and important to follow for the next decade and beyond: "Missions meeting high-priority planetary defense objectives should not be required to compete against missions meeting high-priority science objectives."

Double Asteroid Redirection Test Mission

NASA's DART mission is scheduled to impact the 160 m asteroid Dimorphos in 2022, demonstrating kinetic impact technology as one approach to accomplish asteroid deflection. ESA's Hera mission is scheduled to rendezvous with the Didymos-Dimorphos system in 2026, providing further insight into the results of DART's kinetic impact demonstration. NASA's DART mission is an essential first test of asteroid deflection technology and will significantly increase U.S. and international preparedness for future NEO impacts. However, given the diversity of possible NEO characteristics, this kinetic impact demonstration is simply the initial step in expanding NASA's abilities to divert a threatening object. NASA and the extended PD community need to take advantage of the DART results and apply them to broader mitigation technology efforts to achieve an effective and versatile PD capability. Similarly, the launch of NASA's NEO Surveyor in 2026 will greatly expand our knowledge of the NEO population over the next decade, which will help inform the PD activities that follow.

Finding: Owing to the diversity of possible NEO threats, including variation across individual object characteristics and differences in warning time before Earth impact, significant deflection technology questions will remain even after a full analysis of a successful DART experiment in 2022.

Finding: Sustained investment in planetary defense mission technology and development of additional demonstration missions beyond DART would enable NASA to accomplish critical planetary defense characterization and mitigation objectives for a variety of impact scenarios and build on the lessons learned from the DART mission.

Next Steps for Planetary Defense Demonstration Missions

The committee commissioned a rapid mission architecture (RMA) study to examine a range of spacecraft concepts that would adequately address the needs for PD technology demonstration missions focused on NEO characterization and mitigation objectives (NASA 2021c). The goal of the RMA study was to identify and prioritize PD demonstration missions to be flown in the upcoming decade. More than 30 representative demonstration missions were examined that included a variety of characterization and mitigation objectives designed to advance development of PD mitigation capabilities. This section summarizes the main results of the RMA study, while the RMA study report provides many more details.

The RMA study focused on missions that demonstrated critical techniques for risk reduction, operational readiness, and expanding the knowledge base of NEO characteristics. The study assumed the desired mission cost to be <$500 million (including the launch vehicle and operations but excluding foreign contributions). Other assumptions included the successful launch of the DART and Hera missions to Didymos, the successful launch and operation of NEO Surveyor, completion of VCRO, and the continued operation of ground-based NEO discovery assets (e.g., Catalina Sky Survey and Pan-STARRS).

Two basic mission types were examined: first, *characterization* missions, designed to obtain key information about the dynamical and physical characteristics of NEOs necessary to inform mitigation approaches; and second, mitigation missions, which would demonstrate mitigation technologies and improve operational readiness to prevent a NEO impact.

Both mission types would collect information designed to fulfill planetary defense objectives, particularly to inform the development and implementation of mitigation strategies and techniques. Significantly, the RMA study concluded that increased detection and characterization of the NEO population is critical to reducing risks and ensuring successful mitigation, given the diversity of physical characteristics among these objects. The study also recognized that the most likely object to pose an impact risk was from the undiscovered population of objects greater than approximately 50 m in diameter.

Data from ground-based planetary radars, ground-based telescopes, and the few in situ missions conducted to date (e.g., NEAR, Hayabusa, Hayabusa2, and OSIRIS-REx) suggest that the NEO population is very heterogeneous, varying from unconsolidated "rubble piles" to heavily fractured bodies with some degree of physical integrity. Knowing a NEO's characteristics is crucial because the efficacy of any deflection method depends on the body's mass, cohesiveness, and associated physical properties. For example, an intended deflection may disrupt a loosely bound NEO into multiple objects, and inadvertently increase the probability of impact (albeit with smaller pieces).

In addition, it is estimated that approximately 15 percent of the NEO population are binary systems—that is, they contain two gravitationally bound objects (Pravec et al. 2006)—further complicating any mitigation scenario. Without a sufficiently broad data base of NEO characteristics, it would be difficult to predict the properties of a newly discovered threat without actually observing the object in situ. The best strategy is a two-step process: first, characterization; and second, mitigation efforts appropriate for that specific NEO and its attendant warning time.

Finding: There is much to be learned about the physical characteristics of the NEO population. Only a handful of NEOs have been observed in situ and there are many unknowns concerning the range of physical properties that may be relevant for planetary defense. In addition, smaller NEOs (>50 m in diameter) are challenging to detect and characterize via ground-based methods, and represent the least understood, but statistically the most likely subset of the NEO population to require mitigation actions.

For NEOs with short warning times, rapid characterization may be required in order to implement appropriate mitigation measures. A rendezvous reconnaissance mission would be preferred because it could provide critical knowledge of the NEO's physical properties and monitor it post-mitigation. But if the warning time is short, and there is not enough time to implement a rendezvous, a fast flyby mission would be highly useful to provide at least rough information on the object's key properties and to refine impact probability estimates. Note that in most instances, flyby opportunities are more prevalent than rendezvous mission opportunities to the same NEO. However, such flyby reconnaissance missions are challenging given the high encounter velocities involved, limiting the time available for collecting detailed characterization data.

Finding: Prior characterization of a hazardous NEO via an in situ reconnaissance mission is advisable to determine its physical characteristics and to develop an appropriate mitigation response based on the available warning time. Although rendezvous missions are preferred, fast flyby missions may be required to obtain timely characterization data for short warning time scenarios.

Recommendation: The highest priority planetary defense demonstration mission to follow DART and NEO Surveyor should be a rapid-response, flyby reconnaissance mission targeted to a challenging NEO, representative of the population (~50–100 m in diameter) of objects posing the highest probability of a destructive Earth impact. Such a mission should assess the capabilities and limitations of flyby characterization methods to better prepare for a short-warning-time NEO threat.

For any given impact scenario, selection of appropriate mitigation technologies depends on knowledge of the NEO's physical characteristics, precise trajectory, and available warning time. The choice of mitigation technology requires a balance between generating an adequate amount of deflection/disruption without causing deleterious results (e.g., unwanted disruption or ineffective deflection). Several mitigation techniques have been proposed based on current technologies. The most technically feasible of these are, in no particular order, kinetic impact, nuclear, ion beam deflection, and gravity tractor. All have specific advantages and disadvantages for NEO mitigation and are briefly described below:

Kinetic Impact: This is a relatively straightforward, high-impulse technique that transfers momentum to a NEO, altering its trajectory via a direct hypervelocity impact. However, the effectiveness of the deflection varies with the amount of momentum enhancement generated by the post-impact ejection of material from the NEO. The amount and direction of material ejected is dependent on the physical properties of the NEO and the intercept geometry. Kinetic impact is not effective against larger, more massive NEOs, and for some smaller objects may result in unwanted disruption with subsequent unpredictable outcomes (Figure 18-4). However, kinetic impact may be useful for deflecting NEOs with short warning times, and in cases (very short warning times and/or relatively small NEOs) where deliberate disruption may be desired (see Figure 18-4).

Nuclear: This high-impulse method relies on the radiation from a detonated nuclear explosive device (NED) to deliver an impulse that deflects the NEO and is effective over a wide range of NEO physical characteristics. The NED is triggered near the object's surface; the explosion generates X-rays and neutrons which vaporize the exposed surface layer of the NEO. The burst of vaporized material imparts momentum to the NEO, altering its orbit. Because the detonation of the NED is timed to occur at an optimum distance from the NEO, the NED is best deployed via a rendezvous, which also enables precise directional control of the deflection. However, in cases of short warning times, hypervelocity intercepts could be employed, but attaining the optimum stand-off distance requires precise detonation timing. Nuclear explosives can transfer significant momentum instantaneously, which may be required to deflect large NEOs. When very short warning times make deflection impractical, nuclear disruption of the object may be the only option to prevent Earth impact.

Figure 18-4 demonstrates the effective parameter space of the various mitigation techniques with respect to NEO diameter and warning time. High impulse techniques such as kinetic impact and nuclear are broadly useful in situations with short warning times and/or when disruption of a NEO is desired. However, nuclear methods may be the only suitable option for deflecting large NEOs. Slow, controlled methods such as ion beam deflection

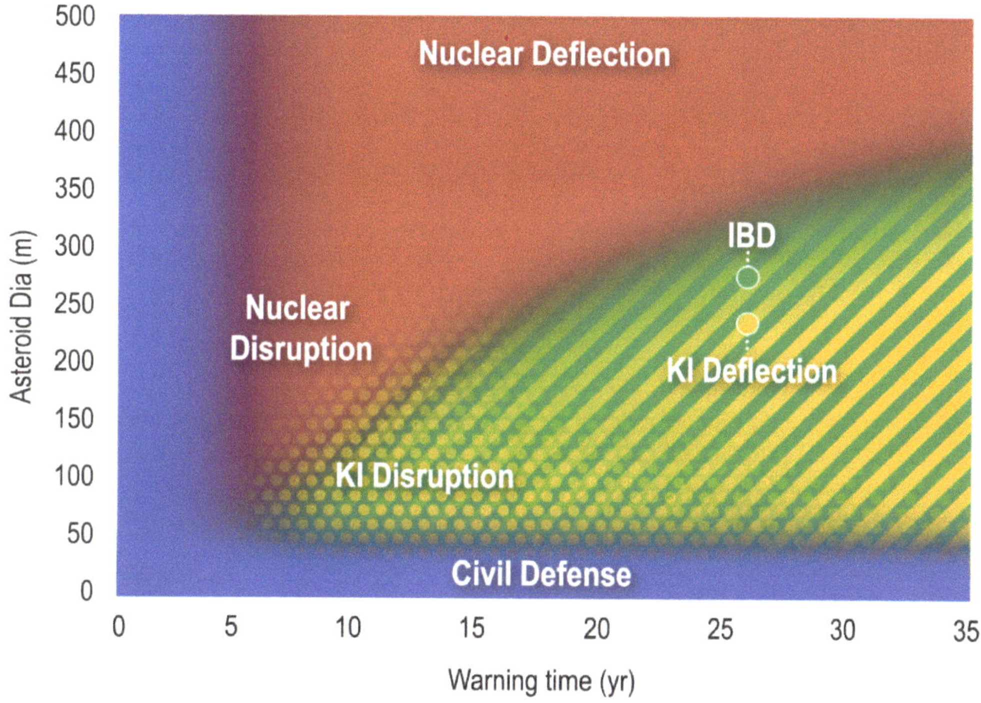

FIGURE 18-4 Numerical results from simulating deflection capabilities of various techniques across a variety of asteroid sizes following fifteen different Earth-impacting orbits. Kinetic impact (KI) techniques (yellow) largely overlap the region where ion beam deflection (IBD) is effective (green). If the warning time is very short and/or the asteroid relatively small, deliberate KI disruption may be the only viable nonnuclear technique (yellow dots). Gravity Tractor not shown. See RMA study for additional details. Note that Civil Defense is a mitigation technique designed to reduce the number of casualties in the projected impact affected area. This can be achieved either through evacuation of the population or by issuing warnings to shelter-in-place and seek cover. SOURCE: Modified from National Research Council, 2010, *Defending Planet Earth*, Washington, DC: The National Academies Press.

and gravity tractor are more applicable given longer warning times and may offer sufficient control to deflect a NEO in an optimal direction.

Ion Beam Deflection: Deflection via use of ions discharged from solar electric propulsion engines is a promising technique. A spacecraft near a NEO delivers momentum by spraying its surface with thruster-generated ions, enabling a slow and controlled deflection. Optimal deflection requires that the spacecraft rendezvous with the NEO and be able to operate autonomously in its vicinity for an extended period. Hence, this technique is useful for objects with longer warning times. The rendezvous offers the potential for detailed NEO characterization and monitoring, while offering more time to reconfigure from critical flight system faults than during a high-speed flyby/intercept. Rendezvous is also more capable of dealing with unexpected NEO physical characteristics. Ion beam deflection is not effective against large NEOs and may not be suitable for deflection of NEOs with natural satellites.

Gravity Tractor: This technique uses the mutual gravitational attraction between the spacecraft and the NEO to slowly alter the latter's trajectory. This technique is similar to ion beam deflection, offering fine deflectional control and extended opportunities for detailed characterization and monitoring. To be effective, however, it is necessary for the spacecraft to perform close, autonomous, and extended proximity operations to station keep at a predetermined distance from the NEO. The required guidance, navigation, and control capabilities are challenging, and the technique may not be suitable for NEOs with irregular shapes, chaotic rotations, or natural satellites. It requires long warning times and is also less tolerant of technical system faults and unexpected NEO physical characteristics.

The gravity tractor technique is not shown on Figure 18-4 since its performance overlaps regions of the ion beam deflection parameter space; its lower efficiency requires longer operation times. Note the overlap in the parameter space between the kinetic impact and ion beam deflection techniques. Please see the RMA study for additional details (NASA 2021c).

Finding: Several proposed planetary defense mitigation techniques are technically feasible, but none has been demonstrated in practice. Each has its own advantages and disadvantages depending on the physical characteristics of the NEO and the available warning time to impact.

Recommendation: Following a rapid-response, flyby reconnaissance mission demonstration, the next highest priority planetary defense mission would be a characterization and/or mitigation mission.

The suggested candidates are, in no particular order:

- A characterization tour mission to exercise characterization capabilities and gain characterization information required for future deflection/disruption missions across a range of NEO targets.
- A kinetic impact mission on a smaller NEO and at a higher closing speed than DART to acquire experience needed for more challenging mitigation missions.
- A slow-push/pull mitigation demonstration, such as ion beam deflection, to develop several different technologies that may be employed against future hazardous objects.

Which technique to be demonstrated will depend on an assessment of the current state of knowledge. The RMA study report contains additional engineering details to demonstrate that all these mission options are technically feasible at a total mission cost of ~$500 million (FY 2025) or less; a characterization tour may be feasible in the $200 million to $430 million (FY 2025) range. Additionally, the RMA study demonstrates that certain mission architectures have the potential to accomplish multiple planetary defense mission objectives at the same target(s) or with the same spacecraft. In particular, mission concepts that share fundamentally similar flyby/intercept or rendezvous implementations could be effectively and efficiently designed with capabilities that test both characterization and mitigation technologies at a relevant NEO.

Finding: Mission concepts that address multiple characterization and mitigation objectives in future planetary defense technology demonstrations would potentially maximize results.

Given the large number of important PD objectives that need to be addressed and are technically and financially feasible, funding and launching a series of modest missions demonstrating capabilities and technologies essential to implementing an effective planetary defense against a range of possible NEO impact scenarios is both important and achievable within the next decade. The RMA study provided cost estimates for the studied missions. Appraisal of those costs shows that it is realistic for the DART mission to be completed in 2023, NEO Surveyor to launch in 2026, and at least one new PD focused mission (a rapid-response, flyby reconnaissance mission) to start prior to the end of 2032. An additional new start from the above list of characterization and mitigation missions may also be possible, depending on the costs of the new missions. A regular cadence of launches will advance key PD technologies while also regularly exercising required capabilities.

Recommendation: NASA's Planetary Defense Coordination Office should be funded at adequate levels to conduct a robust program of necessary planetary defense-related activities, technologies, and demonstration missions launching on a regular cadence.

NASA has a history of soliciting spacecraft mission concepts to increase knowledge of the solar system, explore new destinations, and ensure the safety of humanity. Such solicitations have supported open competition, which has been instrumental in providing a pool of promising cost-effective mission concepts for future development.

Finding: Making planetary defense demonstration mission opportunities, with well-defined objectives, open to industry, academia, U.S. government institutions, and NASA centers would ensure that the most promising and cost-effective concepts are considered and developed.

Promising Planetary Defense Mission Technologies

There are several promising technologies whose pursuit would help attain specific goals of PD characterization and mitigation missions. These technologies can be broadly grouped into several categories and are utilized in mission architectures for flyby/intercept and rendezvous PD concepts. These technology categories include: NEO reconnaissance; impulsive mitigation; slow-push mitigation; and, guidance, navigation, and control (GNC) systems.

- *NEO Reconnaissance Technologies:* Development of these technologies is needed to make necessary measurements during either flyby or rendezvous missions, and are focused on determining key NEO physical characteristics that will inform the required subsequent mitigation efforts given the available warning time. Such technologies include spacecraft systems (e.g., high-speed gimbals) to enable instruments to track NEOs during fast flybys and instruments/systems with the ability to determine the mass of the NEO during such high-speed encounters. Additional development of technologies that involve landed or deployed instruments to determine geophysical and geotechnical properties (e.g., strength, cohesion, and internal structure) would be useful to further enhance NEO characterization objectives and could be used during either flyby or rendezvous missions.
- *Impulsive Mitigation Technologies:* Technologies for impulsive mitigation involve those needed for kinetic impact and nuclear methods. Such missions would benefit from improved visible and thermal-infrared imaging systems to help with targeting NEOs during hypervelocity intercepts. In addition, nuclear mitigation methods would benefit from improvement in sensor technologies (e.g., radar ranging systems) that at very high approach velocities enable accurate and reliable triggering of a NED at a precise time and distance from a NEO's surface. Such sensors would ensure that detonation occurs at the optimum location for deflection or disruption. A radar ranging system, for example, can be flown as a ride-along payload on a future kinetic impact demonstration. Note that testing of a nuclear device in space is neither needed nor advised.
- *Slow-Push/Pull Mitigation Technologies:* Slow-push/pull ion beam deflection and gravity tractor techniques would both benefit from improved solar electric propulsion (SEP) technologies. Both methods rely on low-thrust, high efficiency propulsion systems to deflect a NEO during precise station keeping maneuvers, ideally suited to SEP. Hence development of high-power, long-life SEP systems is an enabling technology for these concepts. Ion beam deflection would also benefit from new thruster technologies which would better focus the ion beam, minimizing the divergence angle from the thruster. This would reduce ion plume losses and increase the impulse imparted to the NEO.
- *Guidance, Navigation, and Control Technologies:* Advancement of these technologies has the broadest impact for all future characterization and mitigation techniques and are applicable to both flyby/intercept and rendezvous missions. Improvement in precision terminal GNC algorithms and associated spacecraft systems for hypervelocity flybys/intercepts would enable accurate and reliable targeting of small NEOs (~50 m and larger) at closure speeds of up to ~15–20 km/s. These improvements would enable both rapid reconnaissance NEO flyby characterization missions, and kinetic impact/nuclear mitigation missions to reach their desired targets and achieve their PD objectives. Further improvement of autonomous GNC systems and associated algorithms for extended, long-duration proximity operations would enable both ion beam deflection and gravity tractor mitigation methods. These techniques require autonomous, real-time sensing of the spacecraft's relative position and attitude with respect to the NEO for appropriate throttling of the SEP thrusters.

In addition to the technology categories discussed above, overall advancement in instrument designs would be beneficial to increase capability and reduce mission costs. Similarly, continued investment in smaller spacecraft and maturation of flight systems will incrementally help reduce costs for deep space missions with PD objectives.

Finding: Impact scenarios may vary widely given the diverse range of NEO physical characteristics and potential warning times. As such, it is important to have several mature technologies available and optimized for possible planetary defense characterization and mitigation situations before they arise.

Finding: Promising new technologies for both characterization and mitigation demonstration missions could be tested on relevant NEOs. Technologies such as those for guidance, navigation, and control (GNC) and sensor instrumentation appropriate for hypervelocity flybys/intercepts, as well as autonomous GNC for long-duration proximity rendezvous operations at NEOs, are fundamentally important for planetary defense objectives.

Planetary Defense Rapid Mission Response Strategies

The first goal of PD is to develop the means to detect and characterize a hazardous NEO with enough warning time to implement an effective response. The combined results from current and upcoming NEO survey systems will likely provide adequate warning for a majority of potentially hazardous NEOs well in advance of an impact. However, assuming that a newly discovered object among the NEO population could pose an impact risk, it would be prudent to consider scenarios where warning times may be relatively short (~5–7 years), and thus create challenging situations for characterization and/or mitigation. Current spacecraft development processes, hardware integration practices, and launch vehicle infrastructure are not optimized for such situations, and often require long lead times (e.g., more than 4 years) for successful mission launch and reliable systems operation. Additional time is required for spacecraft transit and implementation/assessment of necessary activities at the NEO. Hence a comprehensive PD response requires the capability to rapidly assess a potentially hazardous object and take appropriate action in the case of short impact warning times.

There are several rapid mission strategies that could be evaluated in development of PD demonstration missions. These strategies have been discussed in the survey's RMA study (NASA 2021c) and include such options as the following:

- Rolling Phase A/B Design, which would enable an advanced starting point for development and construction of a mission;
- Build on Demand, which designates that a spacecraft be built as fast as possible via a streamlined process;
- Repurposed/Commandeered, wherein a mission could repurpose/commandeer parts, components, or possibly entire spacecraft;
- Build to Inventory, wherein entire spacecraft can be held ready in time of need, or modular components compatible for rapid assembly are placed on standby status; and,
- Store in Space, which places assets on station, ready to be deployed on very short notice (e.g., GOES weather satellites).

All these proposed rapid response strategies have advantages and disadvantages in addressing key PD mission aspects. Of particular importance for evaluating each of the rapid response strategies is consideration of the specific NEO hazard, the spacecraft size and complexity (e.g., SmallSats to medium-class missions), the available time for integration and testing, the required mission response/deployment time, and the resulting overall mission cost.

NASA has previous experience with science missions to NEOs (e.g., NEAR-Shoemaker and OSIRIS-REx), and thus has some knowledge of PD characterization requirements, but currently has no detailed experience in conducting mitigation missions (e.g., DART will not impact Dimorphos until the Fall of 2022). Therefore, concepts for characterization missions are likely more mature than mitigation mission designs. In addition, the challenges of rapid reconnaissance (flyby or rendezvous) are better understood, and so tests of these rapid-response strategies may be implemented sooner in comparison to doing so for mitigation demonstration missions.

Finding: Current practices and procedures for spacecraft development and deployment are not optimized to address planetary defense mission needs if the available warning time is short. However, several rapid response strategies exist which could be tested on planetary defense demonstration missions.

Finding: A study of specific rapid-response strategies as part of planetary defense demonstration missions would help assess what preparations and resources would enable a launch 1, 2, 3, or 4 years from time of alert.

Special focus on examining the feasibility of a standardized rapid reconnaissance design, applicable to the greatest number of short-warning impact scenarios, would be beneficial in developing rapid-response planetary defense capabilities.

INTERNATIONAL COOPERATION ON NEO PREPARATION

With this decadal survey effort focused on NASA and NSF activities and support, this section provides a high-level overview of those NASA and NSF activities that will foster—and benefit from—greater international cooperation. The risk of a NEO impact is of worldwide concern, and thus international cooperation on PD efforts will pay dividends in both warning and response.

NASA continues its efforts to raise international awareness of the NEO hazard through a vigorous presence at international meetings (e.g., International Academy of Astronautics Planetary Defense Conference) and taking an active leadership role at the Scientific and Technical Subcommittee of the United Nations Committee on the Peaceful Uses of Outer Space (UNCOPUOS). NASA accomplishes this by supporting NEO response planning at the UNCOPUOS, by promoting education and outreach on NEO impact effects among international disaster management organizations, and by engaging with nongovernmental organizations. By cooperating on planetary defense demonstrations with international spacefaring partners, NASA will increase the global ability to respond to a future NEO threat.

NASA is an active participant in two United Nations-endorsed groups addressing planetary defense. The first is the International Asteroid Warning Network (IAWN), a coordinated system of asteroid detection and warning activities that shares NEO discovery, orbit determination, and impact prediction information (IAWN 2021). IAWN's 35 signatories share their validated NEO findings, hazard analyses, and impact predictions with UN member states through the UNCOPUOS. IAWN also coordinates NEO observation campaigns to help refine early warning protocols and expand PD characterization efforts.

NASA also is active in the UN-endorsed Space Mission Planning Advisory Group (SMPAG), comprised of several space agencies and international organizations (ESA 2022). SMPAG coordinates international efforts in NEO impact mitigation and response. SMPAG's information exchange and NEO mitigation planning increases its members' ability to respond to a NEO threat. Through SMPAG, NASA can propose and pursue significant, joint NEO technology efforts with spacefaring partners. International participation in U.S.-led technology R&D programs, and in-space demonstrations of PD techniques will avoid duplication and get maximum return from limited global resources.

Finding: NASA leadership and participation in the International Asteroid Warning Network and the Space Mission Planning Advisory Group has produced significant progress toward development of international planetary defense capabilities.

International collaboration expressed through NASA's DART and ESA's Hera missions will generate insights surpassing what either would produce on its own. Cooperation on these missions may be followed by further joint demonstration missions through the decade and beyond.

Finding: Knowledge obtained by planetary defense demonstration missions, such as DART and Hera, will advance understanding of NEO mitigation techniques and further international collaboration in planetary defense efforts.

NEO IMPACT EMERGENCY PROCEDURES AND ACTION PROTOCOLS

Devastating NEO impacts are low-probability, high-consequence events that may result in extensive loss of life, and warrant appropriate levels of preparedness. As in the case of hurricanes and other natural disasters, damage prevention depends on developing and exercising response protocols to support reliable communications, sound decision-making, and employment of effective mitigation measures. These efforts necessarily span the responsibilities of many U.S. government and international entities. Because the scope of this decadal survey effort is focused on NASA and NSF planetary defense activities, this section outlines key activities that support NEO impact emergency procedures and action protocols.

NASA has developed a sound process for the collection, dissemination, and communication of information regarding specific NEO impact threats. If through data supplied by MPC an object is identified as a potential impact threat by CNEOS's high-precision orbit determination, NASA will inform the relevant U.S. government entities (e.g., National Security Council, Office of Science Technology Policy, DHS, and Federal Emergency Management Agency [FEMA]) (NSTC 2021). Depending on the specific impact location, impact severity, and warning time, these entities will then pursue appropriate steps to assess the risk, prepare detailed communications, and implement necessary PD mitigation efforts.

> **Finding:** The Minor Planet Center and the Center for Near-Earth Object Studies provide crucial data for identifying NEOs and evaluating their impact probabilities, and therefore, are vital components for an effective planetary defense response.

Specific actions related to Goal 5 are addressed in the *Report on NEO Impact Threat Emergency Protocols* (NSTC 2021). NASA coordinates NEO end-to-end observations campaigns, performs table-top PD exercises, and constructs hypothetical impact scenarios to exercise and refine NEO impact protocols.

As more capable surveys increase our knowledge of the NEO population, PD planners may take advantage of future close approaches by coordinating opportunistic observation campaigns as part of exercises simulating potential impact events. These campaigns may enlist space- and ground-based facilities for orbit determination and characterization, as has been done with previous IAWN observation campaigns (e.g., 1999 KW_4 and Apophis) (IAWN 2021).

Another method to improve PD readiness is through continued collaborations between U.S. government agencies and international partners. This decade's expected improvement in knowledge of the NEO population and characterization of relevant NEOs will add fidelity to NEO impact response exercises and increase national and international readiness for a potential impact. NASA, FEMA, DoD, and other U.S. government agencies have held several joint PD table-top exercises presenting realistic impact scenarios, and similar exercises involving hypothetical potentially hazardous NEOs have been conducted during biennial International Academy of Astronautics Planetary Defense Conferences; these provided useful information regarding impact consequences and disaster response preparations at the local, state, national, and international levels.

> **Finding:** NASA's continued commitment to propose, plan, and participate in intra- and intergovernmental planetary defense table-top response exercises and international observation campaigns, will enable NASA's Planetary Defense Coordination Office to broaden and solidify connections to relevant U.S. and international agencies. These activities aid planetary defense preparation and strengthen global NEO impact emergency protocols.

CONCLUSIONS

NASA, NSF, and other government agencies play a leading role in developing the capacity to understand the NEO hazard and build a long-term ability to counter a potential impact threat. Society now possesses sufficiently mature telescope and space operations technologies to provide two of the three elements necessary to prevent a NEO impact.

First, NASA is expanding its NEO search abilities and, with its U.S. government and international partners, can issue warning information for any asteroid discovered on a threatening trajectory. New ground- and space-based search systems will increase the ability to provide impact warning for more numerous smaller asteroids.

Second, NASA and its partners possess spaceflight technology that makes NEO impact prevention practical. NEO deflection demonstrations, like DART, are essential to provide the technology building blocks of impact mitigation capability.

The third element for NEO impact prevention is the readiness and determination to respond to a future Earth impact. However, focused in-space efforts over the next decade and beyond are necessary to develop a suite of proven, practical technologies for safely deflecting or disrupting a threatening NEO.

Without the development and testing of in-space mitigation technologies, the only possible response to a threatening impact would be evacuation of the impact area and subsequent disaster response. A robust program of activities in the coming decade will enable the U.S. planetary defense community to forge detection, warning, and mitigation capabilities that will stand as a global example of how to shield society from a destructive yet preventable natural disaster.

REFERENCES

Adamo, D. 2020. "Toward Greater Preparedness and Resilience in Planetary Defense." White paper #055 submitted to the Planetary Science and Astrobiology Decadal Survey 2022–2032. *Bulletin of the AAS* 53(4). https://baas.aas.org/pub/2021n4i055/release/1?readingCollection=7272e5bb.

Arakawa, M., T. Saiki, K. Wada, K. Ogawa, T. Kadono, K. Shirai, H. Sawada, et al. 2020. "An Artificial Impact on Asteroid (162173) Ryugu Formed a Crater in the Gravity-Dominated Regime." *Science* 368:67–71.

Barbee, B.W., M. Bruck Syal, D. Dearborn, G. Gisler, K. Greenaugh, K.M. Howley, R. Leung, et al. 2017. "Options and Uncertainties in Planetary Defense: Mission Planning and Vehicle Design for Flexible Response." *Acta Astronautica* 143: 37–61.

Binzel, R., B.W. Barbee, O.S. Barnouin, J.F. Bell, M. Birlan, A. Boley, W. Bottke, et al. 2020. "Apophis 2029: Decadal Opportunity for the Science of Planetary Defense." White paper #045 submitted to the Planetary Science and Astrobiology Decadal Survey 2023–2032. *Bulletin of the AAS* 53(4).

Bonsall, A., G. Watts, J. Lazio, P. Taylor, E. Rivera-Valentin, E. Howell, F. Ghigo, et al. 2020. "GBT Planetary Radar System." *Bulletin of the AAS* 51(7). https://baas.aas.org/pub/2020n7i208/release/1.

Boslough, M.B.E., M. Aftosmis, M.J. Berger, S.M. Ezzedine, G. Gisler, B. Jennings, R.J. LeVeque, et al. 2016. "Asteroid-Generated Tsunami and Impact Risk." *American Geophysical Union Fall Meeting Abstracts* 2016:NH13A-1763.

Boslough, M.B.E., and D.A. Crawford. 2008. "Low-Altitude Airbursts and the Impact Threat." *International Journal of Impact Engineering* 35(12):1441–1448.

Brown, P.G., A.R. Hilderbrand, M.E. Zolensky, M. Grady, R.N. Clayton, T.K. Mayeda, E. Tagliaferri, et al. 2000. "The Fall, Recovery, Orbit, and Composition of the Tagish Lake Meteorite: A New Type of Carbonaceous Chondrite." *Science* 290:320–325. https://doi.org/10.1126/science.290.5490.320.

Brown, P., R.E. Spalding, D.O. ReVelle, E. Tagliaferri, and S.P. Worden. 2002. "The Flux of Small Near-Earth Objects Colliding with the Earth." *Nature* 420(6913):294–296.

Brown, P., P. Wiegert, D. Clark, and E. Tagliaferri. 2016. "Orbital and Physical Characteristics from Meter-Scale Impactors from Airburst Observations." *Icarus* 216:96–111.

Brozović, M., L.A.M. Benner, J.G. McMichael, J.D. Giorgini, P. Pravec, P. Scheirich, C. Magri, et al. 2018. "Goldstone and Arecibo Radar Observations of (99942) Apophis in 2012–2013." *Icarus* 300:115–128.

Cheng, A., R.T. Daly, O.S. Barnouin, J.B. Plescia, R.P. Binzel, D.C. Richardson, J.V. DeMartini, et al. 2020. "Apophis 2029 Planetary Defense Mission Options." White paper #070 submitted to the Planetary Science and Astrobiology Decadal Survey 2023–2032. *Bulletin of the AAS* 53(4).

Dearborn, D.S.P., and P.L. Miller. 2014. "Defending Against Asteroids and Comets." Pp. 733–754 in *Handbook of Cosmic Hazards and Planetary* Defense, J. Pelton and F. Allahdadi, eds. Cham, Switzerland: Springer.

Dearborn, D.S.P., M. Bruck Syal, B.W. Barbee, G. Gisler, K. Greenaugh, K.M. Howley, R. Leung, et al. 2020. "Options and Uncertainties in Planetary Defense: Impulse-Dependent Response and the Physical Properties of Asteroids." *Acta Astronautica* 166:290–305.

Dellaguistina, D., M. Nolan, A. Polit, D. Golish, D. Lauretta, and OSIRIS-REx Extended Mission Team. 2021. "An OSIRIS-REx Extended Mission to Apophis." AAS Division of Planetary Science meeting #53, id. 412.02. *Bulletin of the AAS* 53(7):2021n7i412p02.

ESA (European Space Agency). 2022. "Space Mission Planning Advisory Group." https://www.cosmos.esa.int/web/smpag.

Ezzedine, S., and P.L. Miller. 2014. "Water Impact Modeling." Pp. 1–19 in *Handbook of Cosmic Hazards and Planetary Defense*. Cham, Switzerland: Springer.

Giorgini, J.D., L.A.M. Benner, S.J. Ostro, M.C. Nolan, and M.W. Busch. 2008. "Predicting the Earth Encounters of (99942) Apophis." *Icarus* 193(1):1–19.

Harrigan, G., A. Valinia, N. Trepal, P. Babuska, and V. Goyal. 2021. "Arecibo Observatory Auxiliary M4N Socket Termination Failure Investigation." NASA/TM–20210017934, NES-RP-20-01585. https://ntrs.nasa.gov/citations/20210017934.

Harris, A.W., and P.W. Chodas. 2021. "The Population of Near-Earth Asteroids Revisited and Updated." *Icarus* 365:114452.

Hirabayashi, M., Y. Kim, and M. Brozvić. 2021. "Finite Element Modeling to Characterize the Stress Evolution in Asteroid (99942) Apophis During the 2029 Earth Encounter." *Icarus* 365:114493.

IAWN (International Asteroid Warning Network). 2021. "Homepage." https://iawn.net.

Jenniskens, P., M.H. Shaddad, D. Numan, S. Elsir, A.M. Kudoda, M.E. Zolensky, L. Le, et al. 2009. "The Impact and Recovery of Asteroid 2008 TC$_3$." *Nature* 458:485–488. https://doi.org/10.1038/nature07920.

Johnson, C. 2019. "How Americans See the Future of Space Exploration." https://www.pewresearch.org/fact-tank/2019/07/17/how-americans-see-the-future-of-space-exploration-50-years-after-the-first-moon-landing.

Jones, R.L., Vera C. Rubin Observatory LSST Solar System Science Collaboration, M.T. Bannister, B.T. Bolin, C.O. Chandler, S.R. Chesley, S. Eggl, et al. 2020. "The Scientific Impact of the Vera C. Rubin Observatory's Legacy Survey of Space and Time (LSST) for Solar System Science." https://doi.org/10.48550/arXiv.2009.07653.

King, P.K., M. Bruck Syal, D.S.P. Dearborn, R. Managan, J.M. Owen, and C. Raskin. 2021. "Late-Time Small Body Disruptions for Planetary Defense." *Acta Astronautica* 188:367–386.

Lauretta, D.S., D.N. DellaGiustina, C.A. Bennett, D.R. Golish, K.J. Becker, S.S. Balram-Knutson, O.S. Barnouin, et al. 2019. "The Unexpected Surface of Asteroid (101955) Bennu." *Nature* 568:55–60.

Mainzer, A., P. Abell, M.T. Bannister, B. Barbee, J. Barnes, J.F. Bell, III, L. Benner, et al. 2020. "The Future of Planetary Defense in the Era of Advanced Surveys." White paper #259 submitted to the Planetary Science and Astrobiology Decadal Survey 2023–2032. *Bulletin of the AAS* 53(4). https://doi.org/10.3847/25c2cfeb.ba7af878.

Mathias, D.L., L.F. Wheeler, and J.L. Dotson. 2017. "A Probabilistic Asteroid Impact Risk Model." *Icarus* 289:106–119.

Morrison, D. 2018. "Tunguska Workshop: Applying Modern Tools to Understand the 1908 Tunguska Impact." NASA/Technical Memorandum (NASA/TM—220174). https://ntrs.nasa.gov/api/citations/20190002302/downloads/20190002302.pdf.

Naidu, S.P., L.A.M. Benner, J.-L. Margot, M.W. Busch, and P.A. Taylor. 2016. "Capabilities of Earth-Based Radar Facilities for Near-Earth Asteroid Observations." *Astronomical Journal* 152(4):99. https://iopscience.iop.org/article/10.3847/0004-6256/152/4/99/meta.

NASEM (National Academies of Science, Engineering, and Medicine). 2019. *Finding Hazardous Asteroids Using Infrared and Visible Wavelength Telescopes.* Washington, DC: The National Academies Press. https://doi.org/10.17226/25476.

NASA (National Aeronautics and Space Administration). 2003. "Study to Determine the Feasibility of Extending the Search for Near-Earth Objects to Smaller Limiting Diameters." Report of the Near-Earth Object Science Definition Team. Washington, DC: NASA Office of Space Science Solar System Exploration Division. https://cneos.jpl.nasa.gov/doc/neoreport030825.pdf.

NASA. 2017. "Update to Determine the Feasibility of Enhancing the Search and Characterization of NEOs." Report of the Near-Earth Object Science Definition Team. Washington, DC: NASA Science Mission Directorate. https://cneos.jpl.nasa.gov/doc/2017_neo_sdt_final_e-version.pdf.

NASA. 2021a. "NASA Analysis: Earth Is Safe from Asteroid Apophis for 100-Plus Years." NASA Jet Propulsion Laboratory. https://www.jpl.nasa.gov/news/nasa-analysis-earth-is-safe-from-asteroid-apophis-for-100-plus-years.

NASA. 2021b. "Near-Earth Object Observations Program." Planetary Defense Coordination Office. https://www.nasa.gov/planetarydefense/neoo.

NASA. 2021c. Planetary Defense Missions: Rapid Mission Architecture Study. Mission Concept Study Report for the Planetary Science and Astrobiology Decadal Survey 2023–2032. Pasadena, CA: Jet Propulsion Laboratory, California Institute of Technology. https://tinyurl.com/2p88fx4f.

Nolan, M.C., M.M. Al Asad, O.S. Barnouin, L.A.M. Benner, M.G. Daly, C. Drouet d'Aubigny, R.W. Gaskell, et al. 2019. "Comparing the radar shape model of (101955) Bennu with ground truth from OSIRIS-REx." 50th Luna and Planetary Science Conference 2019 (LPI Contrib. No. 2132), Abstract #2162. https://www.hou.usra.edu/meetings/lpsc2019/pdf/2162.pdf.

NRC (National Research Council). 2010. *Defending Planet Earth: Near-Earth-Object Surveys and Hazard Mitigation Strategies.* Washington, DC: The National Academies Press.

NSTC (National Science and Technology Council). 2018. *National Near-Earth Object Preparedness Strategy and Action Plan.* Washington, DC: Executive Office of the President. https://www.nasa.gov/sites/default/files/atoms/files/ostp-neo-strategy-action-plan-jun18.pdf.

NSTC. 2021. *Report on NEO Impact Threat Emergency Protocols.* Washington, DC: Executive Office of the President. https://trumpwhitehouse.archives.gov/wp-content/uploads/2021/01/NEO-Impact-Threat-Protocols-Jan2021.pdf.

Popova, O.P., J. Borovička, W.K. Hartmann, P. Spurný, E. Gnos, I. Nemtchinov, and J.M. Trigo-Rodríguez. 2011. "Very Low Strengths of Interplanetary Meteoroids and Small Asteroids." *Meteoritics and Planetary Science* 46:1525–1550. https://doi.org/10.1111/j.1945-5100.2011.01247.x.

Popova, O.P., P. Jenniskens, V. Emel'yanenko, A. Kartashova, E. Biryukov, S. Khaibrakhmanov, V. Shuvalov, et al. 2013. "Chelyabinsk Airburst, Damage Assessment, Meteorite Recovery, and Characterization." *Science* 342:1069–1073. https://doi.org/10.1126/science.1242642.

Pravec, P., P. Scheirich, J. Ďurech, J. Pollock, P. Kušnirák, K. Hornoch, A. Galád, et al. 2014. "The Tumbling Spin State of (99942) Apophis." *Icarus* 233:48–60.

Pravec, P., P. Scheirich, P. Kušnirák, L. Šarounová, S. Mottola, G. Hahn, P. Brown, et al. 2006. "Photometric Survey of Binary Near-Earth Asteroids." *Icarus* 181:63-93.

Reinhardt, J., X. Chen, W. Liu, P. Manchev, and M.E. Paté-Cornell. 2016. "Asteroid Risk Assessment: A Probabilistic Approach." *Risk Analysis* 36(2):244–261.

Scheeres, D.J. 2019. "Stationkeeping About Apophis Through Its 2029 Earth Flyby," IAA-AAS-SciTech2019-034—AAS 19-953. *Proceedings of the IAA–AAS SciTech Forum 2019 on Space Flight Mechanics and Space Structures and Materials* 174. San Diego, CA: American Astronautical Society.

Scheeres, D.J., L.A.M. Benner, S.J. Ostro, A. Rossi, F. Marzari, and P. Washabaugh. 2005. "Abrupt Alteration of Asteroid 2004 MN4's Spin State During Its 2029 Earth Flyby." *Icarus* 178(1):281–283.

Seaman, R., J. Bauer, M. Brucker, E. Christensen, T. Grav, L. Jones, A. Mainzer, J. Masiero, and T. Spahr. 2020. "NEO Surveys and Ground-Based Follow-Up." White paper #241 submitted to the Planetary Science and Astrobiology Decadal Survey 2023–2032. *Bulletin of the AAS* 53(4). https://doi.org/10.3847/25c2cfeb.9eb9da4e.

Stickle, A., B. Barbee, P. Chodas, T. Daly, M. DeCoster, J. Dotson, R. Klima, et al. 2020. "The Need for a Well-Defined Modeling Pipeline for Planetary Defense." White paper #468 submitted to the Planetary Science and Astrobiology Decadal Survey 2023–2032.

Walker, C., J. Hall, L. Allen, R. Green, P. Seitzer, A. Tyson, A. Bauer, et al. 2021. *Impact of Satellite Constellations on Optical Astronomy and Recommendations Toward Mitigations.* Tucson, AZ: NOIRLab. https://noirlab.edu/public/products/techdocs/techdoc003.

Watanabe, S., M. Hirabayashi, N. Hirata, N.A. Hirata, R. Noguchi, Y. Shimaki, H. Ikeda, et al. 2019. "Hayabusa2 Arrives at the Carbonaceous Asteroid 162173 Ryugu—A Spinning Top-Shaped Rubble Pile." *Science* 364(6437):268–272.

19

Human Exploration

THE PIVOTAL ROLE OF SCIENCE IN HUMAN EXPLORATION

Human exploration of space inspires our nation and the world while simultaneously benefiting our technology development, economic standing, and scientific knowledge.[1] Human and robotic exploration of the solar system over the next decade and beyond will benefit from a logical, sustained, and science-focused approach. In this chapter, the committee addresses the opportunities for science within the context of current human exploration plans and priorities, as well as areas of planetary science that can support human flight activities. Although humans may eventually travel to the far reaches of the solar system, the committee focuses on the surfaces of the Moon and Mars as the most likely near-term, science-rich destinations, based on stated NASA and commercial exploration plans.

There are many important motivations for human exploration of the Moon and Mars. The committee's discussion and recommendations reflect an overarching premise: that a robust science program—that is, one capable of addressing decadal-level science[2]—is a required element to ensure the maximum value and longevity of human exploration programs such as Artemis for the Moon and planned exploration of Mars. The promise of exciting discovery is a core element of ambitious and enduring space programs, as evidenced by the success of sustained robotic programs such as the Hubble Space Telescope and the Mars Exploration Program. Merging human exploration with scientific discovery benefits our nation's investments in its trailblazing planetary exploration program.

> **Finding:** Human exploration is an aspirational and inspirational endeavor, and NASA's Moon-to-Mars exploration plans hold the promise of broad benefits to the nation and the world. Human exploration can potentially enable breakthrough science at the Moon and Mars. Communicating the process and importance of scientific discovery, as enhanced by human explorers, will inspire the next generation of STEM professionals (NASA 2015).

> **Recommendation: Conducting decadal-level science should be a central requirement of the human exploration program.**

---

[1] A glossary of acronyms and technical terms can be found in Appendix F.

[2] Decadal-level science is that which results in significant, unambiguous progress in addressing at least one of the survey's 12 priority science questions.

SCIENCE ENABLED BY HUMAN EXPLORERS

Planetary science and astrobiology field studies benefit from an astronaut's ability to observe sites in striking detail, recognize unexpected observations, analyze critically in real-time to create and refine conceptual models, and react to changing conditions, hypotheses, and interpretations while in the field (McPhee and Charles 2020). Humans can efficiently make targeted in situ measurements and conduct sampling activities that require careful but relatively rapid decisions based on local geological context. Even as robotic exploration capabilities have grown, human explorers can conduct scientific operations much more rapidly than robotic assets (Bartels 2018) and are particularly adept at installing and operating complex infrastructure and scientific assets, especially when unforeseen issues or difficulties require decision-making and on-the-spot innovation (Slakey and Spudis 2008).

These points are reflected in one of the greatest legacies of NASA's human exploration program to date: the scientific bonanza afforded by the Apollo in situ measurements and sample collections, which continue even today to yield breakthrough discoveries about the Earth-Moon system, for example, as new analytical approaches are applied to lunar samples in terrestrial laboratories. Science benefited tremendously from the Apollo program, which helped to create the field of planetary science by inspiring a generation of scientists and engineers from chemistry, physics, and the geosciences to turn their attention to the Moon; many of these individuals have become the leaders of the space program. Major investments in analytical instrumentation, the enlisting of a substantial cohort of students and postdoctoral scientists, an explosion of the planetary literature, and the launch of one of the premier annual meetings in planetary science can all be traced to Apollo.

Certain key science objectives at the Moon and Mars (Table 19-1) can be strongly enabled by future human missions. Of primary importance is the ability of human missions to return carefully chosen samples with increased quality, diversity, and volume. Similarly, in situ investigations of surface and subsurface ice samples on the Moon and Mars, for example, can be enabled by the ability of humans to manipulate complex sampling tools as well as the potential to return intact samples and/or ice cores at cryogenic temperatures. Other measurements that could be facilitated by humans include those requiring deployment of geophysical/subsurface investigations and modern atmosphere/exosphere measurement packages.

Astronauts are most effective when they are well trained not only in the engineering and operations of vehicle and hardware components but also in field geology (and astrobiology, where appropriate) and scientific research techniques. The remarkable science derived from the Apollo "J" missions (Apollo 15, 16, and 17) was in part owing to the geologic field training of the astronauts prior to the flights (Phinney 2015). While this basic geological training of the Apollo astronauts was sufficient for short lunar sorties, future astronauts exploring the Moon and Mars will require more comprehensive scientific training (Hodges and Schmitt 2019). Training activities are most effective when classroom and laboratory learning are complimented by intense field training. Sustained field experiences are critical to master the iterative observational and interpretive skills needed to translate field observables into scientific hypotheses by the astronaut onsite, thereby capitalizing on the powerful human ability to understand large amounts of interrelated data (Compton 1985; Lofgren et al. 2011). The quality and utility of these scientific interpretations, however, are highly dependent on the field scientist's experience (Schmitt et al. 2011). The complex mental processes of developing and refining working hypotheses directly influences tactical decision-making for determining priorities, tasks, and handling contingencies while conducting fieldwork. Therefore, astronauts need to be exquisitely trained to develop the science knowledge base and mental thought processes to optimize scientific measurements, sample collection and high grading, and operational activities while conducting fieldwork on the lunar and/or martian surface. Involvement of some professionally trained planetary scientists in astronaut teams would be ideal, in keeping with the successful record of Dr. Harrison Schmitt's involvement in the scientifically impactful Apollo 17 mission.

Finding: A crucial driver of sustained human exploration is the ability of human explorers—with appropriate training and mission planning—to conduct and enable the highest quality, decadal-level science that expands humankind's understanding of Earth, the solar system, and the universe.

Recommendation: NASA should engage with the science community to (1) define scientific goals for its human exploration programs at the early stages of program planning; and (2) ensure scientific expertise in field geology, planetary science, and astrobiology in its astronaut teams.

TABLE 19-1 Science Objectives (Non-Exhaustive List) Enabled or Facilitated by Humans at the Moon and/or Mars

Human Expertise	Science Objective	Priority Science Questions
Astronauts can be well-equipped to conduct sorties, and sample and return intact cores deeper (>1 m) than is easily accomplished by robotic missions	Determine the origin, composition, and history of ice deposits	4.3, 5.5, 6.1, 10.3, 10.4
	Establish internal heat flow and determine near-surface stratigraphy using geophysical probes and cores	5.2, 5.5
Astronauts can collect more and better geologic samples than static robotic missions by virtue of their ability to more rapidly assess geologic context to select the optimal samples, conduct traverses to allow for increased sample diversity, and to return larger sample quantities. Astronauts could also retrieve samples robotically cached. On Mars, astronauts could deploy more widespread and sophisticated in situ monitoring to track gas fluxes and conduct sophisticated life detection investigations.	Establish the impact flux through time in the inner solar system, the nature of impactors, and whether there was a late heavy bombardment	2.4, 3.1, 3.2, 4.1, 4.2, 9.1, 10.2
	Probe of volcanic, tectonic, and magmatic processes, including the formation of planetary dichotomy/asymmetry	3.5, 4.3, 5.2, 5.3, 5.6, 8.2, 8.3
	Determine the timing and characteristics of the giant impact that produced the Earth-Moon system	3.3, 4.3
	Determine changes in the ancient atmosphere, climate, and habitable environments with liquid water	3.6, 4.3, 5.3, 5.4, 6.1, 6.2, 10.1, 10.2, 10.3, 10.5, 10.7
	Determine whether there is/was life	11.1, 11.3, 11.4
Astronauts can efficiently deploy stations over a wide area to make measurements of modern properties, can conduct in situ tests to determine optimal layouts predeployment, and can conduct tests using an initial layout and redeploy as necessary post-testing.	Measure interactions of atmospheres and exospheres with the space environment	4.1, 6.5, 10.2
	Determine interior structure and history of the magnetic field	3.3, 4.4, 5.1, 5.2, 8.2
	Determine if liquid water currently exists in subsurface aquifers	10.1, 10.3

NEAR-TERM HUMAN EXPLORATION PLANS, RELATIONSHIP TO SCIENCE, AND IN SITU RESOURCE UTILIZATION

Multiple entities have ambitious plans for exploration of the Moon and Mars in the coming decades. NASA's Artemis plan (NASA 2020a) calls for landing humans near the south polar region of the Moon within the 2020s in pursuit of the development of a basecamp designed for longer stays. Artemis consists of several elements including the Space Launch System (SLS) rocket, Orion spacecraft, Human Landing System (HLS), and the Gateway (an orbiting outpost in a highly elliptical rectilinear halo lunar orbit). Artemis plans for Orion to be launched on an SLS, rendezvous with a crew transfer to the HLS, and then for the HLS vehicle (SpaceX Starship) to deliver astronauts to the lunar surface and back to Orion for return to Earth. Early Artemis missions are to focus on proving various architecture elements through short duration missions to the lunar surface. Artemis missions are planned to enable longer surface expeditions with enhanced capabilities. NASA plans for an Artemis Base Camp (Figure 19-1) that will include advanced elements such as an unpressurized lunar terrain vehicle, a habitable mobility platform (a pressurized rover), habitat module, power systems, and in situ resource utilization (ISRU) capabilities (NASA 2020a). Building on the Artemis experiences on the Moon, NASA intends to fly humans to Mars perhaps as soon as the late 2030s. In addition to NASA, U.S. commercial entities and non-U.S. space agencies

FIGURE 19-1 Artist rendition of a planned Artemis Base Camp. SOURCE: Courtesy of NASA.

and entities have expressed interest in and plans for sending humans to the Moon and Mars, as discussed further in the final sections of this chapter.

There are destinations and measurements that will help both robotic and human exploration to achieve their common scientific goals in this decade (NASEM 2019). Precursor missions can enable monitoring measurements that contribute to both science and to optimizing crew safety and planetary protection compliance. Examples include characterization of radiation environments, dust properties, thermophysical environments, and, on Mars, properties of chemical, mineral, and trace organic constituents in the regolith (e.g., form and abundance) that would inevitably be brought into human habitats. Additionally, high spatial resolution landing site reconnaissance for landing site safety, efficient traverse planning, and planning for infrastructure placement (e.g., landforms, roughness, geotechnical properties, and compositional variation) enables well-planned human activities while simultaneously advancing science objectives by the acquisition of datasets that meet science observation requirements and improve our ability to deftly target the most scientifically meaningful landing sites.

A particular issue relevant to both exploration and science is the substantial overlap between studies to characterize the potential for ISRU and scientific measurements desired to quantify the state and evolution of near-surface volatile reservoirs (e.g., depth, distribution, and composition of water ice) at the Moon and Mars. Measurement and characterization of lunar polar volatiles is central to scientific questions pertaining to the age, origin, and evolution of lunar and solar system volatiles (e.g., the INSPIRE mission concept study report [NASA 2021a]). Similar measurements are critical for informing the design and development of ISRU pilot demonstration systems as well as full-scale ISRU processing plants envisioned to enable long-term human presence on the Moon (and eventually Mars). While in situ resource *prospecting* goals for science and exploration share commonalities, these goals and synergies will begin to diverge when *extraction* activities commence.

The Artemis Accords (see final section of this chapter) call for the sustainable utilization of resources from the Moon (NASA 2020e). A reasonable interpretation in the context of the Accords is that resources are utilized in a manner that appropriately balances consumptive uses (e.g., using lunar polar water ice to manufacture rocket fuel or for human uses) with scientific uses (e.g., using lunar polar water ice to study the lunar volatile cycle).

BOX 19-1
Sustainability

NASA has used the word "sustainable" to describe one goal for human lunar exploration through Artemis. As "sustainable" has not yet been defined in this context, we provide our working definition of "sustainable" as meaning that there are widely accepted reasons to continue human lunar exploration that justify the continued investment, commitment, and risk beyond a few missions. The reasons to continue the lunar program include ongoing scientific discoveries; investing in potential commercial development; technology development; educating the next generation of science, technology, engineering, and mathematics (STEM) professionals and a scientifically literate public; and inspiring the public about our individual and collective opportunities and future.

Such resource allocations will ultimately require identifying the nature and occurrence of a resource, determining the extent of resource recoverability, understanding variability that may make some deposits more scientifically important than others, and development of means to ensure oversight and equitable use. Therefore, to ensure the availability of resources, especially water ice, for scientific use, the capability for NASA to map and understand the resources is critical. Such an understanding would be used as input into allocation for different purposes. An oversight structure could be employed to appropriately allocate resources, again to ensure their availability for both science and utilization by humans; such an oversight structure would have to be international, to account for the international participation in the Artemis program and to ensure equitable allocation of resources within a sustainable manner (Box 19-1). In parallel, NASA would benefit from convening a team of experts to review the ethics of planetary ISRU and determine optimal plans and processes to ensure sustainable and responsible resource utilization. Humans are the trustees of our planetary environments for future generations.

Finding: With a renewed national human spaceflight program for destinations beyond Earth, as well as commercial entrants with interests in establishing space-based economic activities, there is ample opportunity for decadal science objectives to infuse, and ideally drive, choices of human destinations and activities on the Moon and Mars.

Finding: A strategic plan is needed to identify measurements most critical to informing ISRU architecture options, ensuring sustainable exploration, and the connection to addressing decadal-level science questions.

INTEGRATING SCIENCE INTO HUMAN EXPLORATION

The decadal survey *Vision and Voyages for Planetary Science in the Decade 2013–2022* emphasized the importance of budgetary firewalls between human and robotic spaceflight, reduction of "turmoil" caused by incorporation of human exploration requirements in robotic science mission post-selection, and the importance of carefully crafted collaboration (NRC 2011). For this decade with a near-term plan for human exploration of the Moon and preparatory activities at Mars, the committee emphasizes the importance of carefully crafted collaboration. A program of scientific exploration can be constructed this decade whereby science enables human exploration and human exploration enables science.

The National Research Council's Committee on Human Exploration (NRC 1997) included among its principles for management of science in programs of human exploration the need for an integrated science program, specifying that the "scientific study of specific planetary bodies, such as the Moon and Mars, should be treated as an integral part of an overall solar system science program and not separated out simply because there may be concurrent interest in human exploration of these bodies." Integrated science programs at the Moon and Mars are critical to the successful integration of science in plans for human exploration.

Integrated programs of science for the Moon and Mars are only one component of ensuring the success of science in human exploration endeavors. Of critical importance is how the identified science goals and objectives for the Moon and Mars are incorporated into systems-level requirements, for example, for Artemis. As described in the following section, the organizational structure currently in place for determining science requirements for human exploration diminishes the great potential of Artemis to accomplish transformational science.

NASA PROGRAMMATIC CONSIDERATIONS FOR ARTEMIS AND BEYOND: CHALLENGES OF INTEGRATING SCIENCE AND HUMAN EXPLORATION

The United States embarked on a new era of human space exploration with the advent of Space Policy Directive-1 (SPD-1) in 2017. SPD-1 set a new national space policy direction and stated that "the United States will lead the return of humans to the Moon for long-term exploration and utilization, followed by human missions to Mars and other destinations." In response, the NASA Planetary Science Division (PSD) created the Lunar Discovery and Exploration Program (LDEP) in 2019. A 2019 Committee on Astrobiology and Planetary Science (NASEM 2019) report concluded that "PSD has taken early measures to ensure participation of the lunar science community and that decadal lunar science priorities are or will be addressed in its Lunar Discovery and Exploration Program." This early commitment in Science Mission Directorate (SMD) to supporting human exploration activities was promising, and here the committee evaluates subsequent SMD activities as well as interactions with the human exploration directorates within NASA.

NASA has a framework for governance and strategic management as well as defined roles for each of its directorates. In this agency structure, SMD is responsible for planning and executing the science priorities established by decadal surveys for the five scientific disciplines assigned to SMD while also considering factors such as cost, technical readiness, and programmatic balance. From the outset of Artemis planning, Human Exploration and Operations Mission Directorate (HEOMD) was to be responsible for planning, developing, and operating human exploration systems (NASA 2020a). In September 2021, HEOMD was divided into two separate directorates, a Space Operations Mission Directorate (SOMD) to focus on launch and space operations, and an Exploration Systems Development Mission Directorate (ESDMD), which will define and manage development of the Artemis program and NASA's Moon-to-Mars exploration approach. This high-level structure implies that it would be the role of SMD to provide the science priorities and requirements to the human exploration directorates (particularly ESDMD), which then would implement these within the context of overall human exploration plans.

Implementing Lunar Scientific Exploration in Artemis: Misaligned Responsibilities, Accountability, and Authority Across NASA's Science and Human Exploration Directorates

Maximizing the science that can be accomplished by Artemis requires the scientific enterprise to be fully integrated into human exploration planning, thus necessitating cross-directorate collaboration among SMD, ESDMD, SOMD, and the Space Technology Mission Directorate (STMD) that reflects the roles and responsibilities assigned to these directorates. Barriers to such collaboration can be bureaucratic, cultural, or owing to political factors (such as the cancelation of programs and/or changing program capabilities that can leave science wary of relying on human exploration to accomplish its high-priority goals). The challenge of integrating science requirements into NASA human spaceflight planning and programs is not new, dating back to the beginning of the Agency as early as 1962 (Beattie 2001). With science planned on decadal timescales, human exploration is encouraged to have similarly long-term continuity of purpose in order to be successfully integrated into science planning and to maximize the collaborative scientific output between robotic and human exploration.

SMD's PSD has the responsibility to accomplish the goals and priorities for lunar science identified by the decadal survey, including development of missions, instruments and technologies needed by such missions, and research and analysis to identify and answer priority science questions. PSD relies on the decadal survey for long-term prioritization of lunar science and on independent science assessment groups (e.g., the Lunar Exploration Analysis Group, LEAG) to provide community input to assist with planning the achievement of decadal priorities. The PSD director is responsible for the scientific exploration of the Moon. However, as detailed below, currently neither the PSD director nor the SMD associate administrator have the authority to implement science objectives within the Artemis program. This basic structural conflict compromises the Agency's ability to achieve decadal-level science through human exploration and undermines the optimal synergies between science and human exploration programs.

NASA's Organizational Structure for Incorporating Science into Artemis

NASA's Artemis activities to date have been primarily conducted within HEOMD, and these are the focus of discussion here. How the Fall 2021 reorganization of HEOMD into ESDMD and SOMD will affect Artemis is unclear at this time. However, the committee notes that this reorganization presents an opportunity for NASA to rectify issues in how science is being incorporated into the human exploration program.

As HEOMD has been the primary entity driving the planning and architecture for Artemis, the organizational structure of HEOMD in relation to SMD is of key interest in terms of understanding how planetary science and astrobiology objectives would be incorporated into Artemis. Figure 19-2 shows a detailed NASA organizational chart with the key directorates contributing to Artemis (HEOMD, SMD, STMD). The HEOMD deputy associate administrator (DAA) for systems integration and engineering has managed five offices, including the Science and Technology Utilization Office (S&TU). There is not yet an Artemis science team, but a Multicenter Support Team (sometimes referred to as a "Multicenter Science Team") was formed from select NASA centers, reporting to the S&TU.

According to a presentation given to the committee, the role of the S&TU Office has been to "integrate science and technology goals from mission directorates and international partners to develop HEO utilization goals, objectives, and requirements for Artemis missions, and the cross-platform research strategy to prepare for human missions to Mars" (NASA 2021b). Thus, it would presumably be through this office that SMD goals and objectives would be translated into requirements for Artemis, with S&TU representatives providing a "strategic view of competing priorities to optimize the advancement of knowledge from human spaceflight missions" (NASA 2021b).

Given this information, science requirements were to be developed for Artemis by the Utilization, Coordination, and Integration Group (UCIG) formed by the S&TU. UCIG has three co-chairs—one each from HEOMD, STMD, and the Exploration Surface Strategy Integration Office (ESSIO) within SMD—and additional members that represent HEOMD and the seven Directorates within the Agency that have interest in utilizing Artemis assets.[3] UCIG leadership does not include PSD. This group has been working to create the HEO-006 Utilization Plan,[4] which would define SMD's utilization goals and objectives for Artemis (as well as those for HEOMD and STMD) and include three relevant annexes that define the plan and requirements for both individual Artemis missions and the long-term program (10-year timeframe).[5]

The most visible effort of SMD to provide input into Artemis was through the Artemis III Science Definition Team (SDT), established by the SMD Associate Administrator. The SDT was composed of eleven NASA civil servants and three consultants from outside NASA.[6] The SDT sought additional input by soliciting white papers and then community comments on the draft report. The SDT operated under a short timeline: white papers were solicited on 20 August 2020 and were due on 8 September, and a draft report was released on 16 October 2020 (NASA 2020c). Input on landing site selection was not solicited; the landing site was specified as to be within

---

[3]Seven mission directorate representatives that fund utilization: SMD/ESSIO, SMD/BPSD, SMD/DAA Programs, STMD, HEOMD/HRP, HEOMD/AES Enabling Capabilities, Office of Planetary Protection.

[4]"HEO Double 0 Documents" (about eight in total) represent the top-level HEOMD technical policy. The HEO-006 Utilization Plan will include plans for all HEOMD platforms, including ISS, commercial LEO, and Artemis.

[5]Annex 1, titled "Cornerstone Utilization Capabilities that Enable Multiple Objectives" includes the highest-level priorities (e.g., traverse approaches, sampling strategies, extended duration surface missions, robotic utilization of HEOMD assets, instruments, PSR operations, science team coordination). Annex 2 defines the "Ten-Year Utilization Phasing Plan" for how to build capabilities over time to accomplish the highest-priority science goals. Annex 4 is where the utilization objectives and requirements are described for each individual Artemis mission.

[6]The three Artemis III SDT consultants included the LEAG chair, the SSERVI chief scientist, and the CAPTEM Lunar Sample Subcommittee chair.

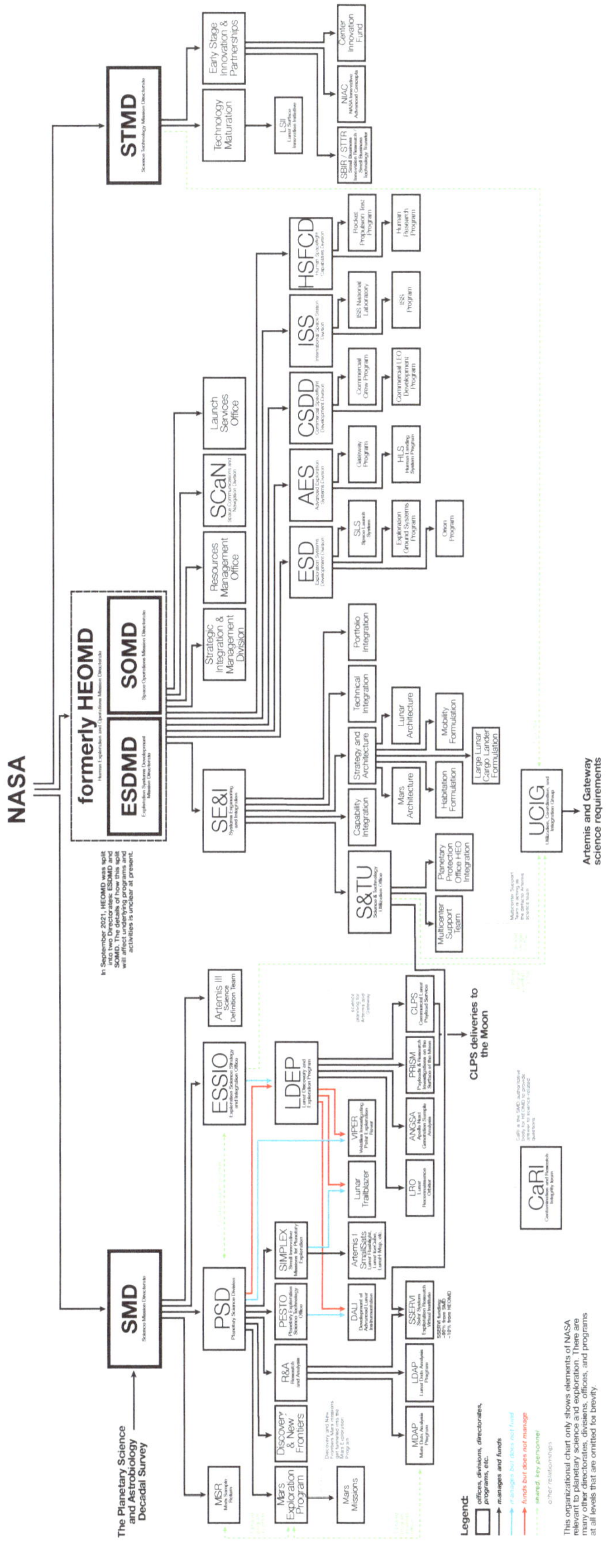

FIGURE 19-2 NASA organization structure highlighting the lunar-relevant entities within SMD, human exploration, and STMD.

6° of the south pole and consequently the SDT prioritized science that can be accomplished at a polar landing site. However, whether and how the SDT recommendations will be translated into requirements is currently at the discretion of the human exploration directorates.

The PSD director is accountable for the scientific exploration of the Moon without the proper authority to accomplish this responsibility. SMD planning for near- and long-term utilization of Artemis is being led by ESSIO, which is not part of PSD and therefore limits PSD participation in this process (despite having one PSD scientist part-time in ESSIO). ESSIO does not have demonstrated authority to levy requirements within the human exploration directorates, where current ownership of Artemis science utilization and requirements resides.

> **Finding:** The separation of roles and responsibilities across multiple divisions and offices therein is not conducive to the development or implementation of a cohesive lunar science and exploration program. ESSIO has also not demonstrated the existence of a process for determining lunar science requirements for Artemis nor has ESSIO shown any plan or prioritization of specific science goals for Artemis. This situation is exacerbated by the lack of an LDEP director and/or "chief scientist" role for lunar science with the required authority to ensure that lunar science in PSD is optimized through a unified program that uses human exploration as one implementation option to achieve decadal-level science.

As a result of this organizational structure (Figure 19-2), SMD has no authoritative mechanism to influence the Artemis design to accomplish lunar science objectives. The S&TU allowed for collaboration in terms of how SMD will "utilize" HEOMD assets, but not the required capabilities of those assets. Science requirements are thus not a priori considerations for Artemis activities, and instead SMD presently retrofits science objectives to be compatible with the capabilities provided by the human exploration program. The responsibilities, authority, and accountability for executing the planetary science and astrobiology decadal survey are misaligned within NASA, and the prioritization of human exploration capabilities have not been consistent with the current lunar science priorities. This situation makes it unlikely that Artemis will address priority lunar science goals outlined by the decadal survey, absent reform.

> **Finding:** The systems engineering approach necessary to incorporate science objectives and requirements needs to occur in the early stages of human mission planning and hardware development. The later such integration occurs, the greater the risk of prohibitive expense associated with scientific requirements and/or the exclusion of priority science altogether.

> **Finding:** The organizational structure for lunar science and exploration within NASA at the outset of Artemis has compromised the Agency's ability to ensure that the highest-priority scientific goals will be accomplished at the Moon because science requirements have not been adequately provided to or incorporated into the human exploration architecture. As a result, the scientific return from the first human landings in the Artemis program will not be as significant as it could be and may be minimal. Although Artemis leaders at NASA state that science requirements will be added onto subsequent missions, after the HLS hardware has proven itself initially, NASA has not demonstrated that actions are being taken now that will result in the effective and timely integration of science into the program.

> **Finding:** Absent an integrated, coherent strategy, it is unlikely decadal-level planetary science will be accomplished at the Moon through Artemis. Human exploration can be used to achieve lower-level science goals until Artemis incorporates top-level requirements capable of achieving decadal science priorities.

Recommendation: PSD should develop a strategic lunar program that includes human exploration as an additional option to robotic missions to achieve decadal-level science goals at the Moon.

Recommendation: NASA should adopt an organizational approach in which SMD has the responsibility and authority for the development of Artemis lunar science requirements that are integrated with human exploration capabilities. NASA should consider establishing a joint program office at the Associate Administrator level for the purpose of developing Artemis program-level requirements across SMD, ESDMD, SOMD, and other directorates as appropriate.

Human Enabled Decadal-Level Science at the Moon

As human exploration advances into the solar system, it is essential that science is a driving motivation and that the science and human directorates within NASA work together efficiently and effectively. This desired synergy is exemplified by the Endurance-A rover mission (see Chapter 22 and Appendix C).[7] Endurance-A would traverse diverse terrains and robotically collect a large (~100 kg) suite of carefully selected samples from scientifically important locations across the SPA basin and deliver them to a human landing site for retrieval by astronauts and return to Earth. This mission would conduct the highest priority lunar science to revolutionize our understanding of the Moon and the early history of the solar system recorded in its most ancient impact basin.

SCIENTIFIC AND HUMAN EXPLORATION OF MARS

Past Cooperation and Plans for the Next Decade and Beyond

As noted in the report of the Mars Architecture Strategy Working Group (MASWG 2020), although "human Mars exploration has been limited to a series of architecture studies, current national policy explicitly calls for eventual human missions to Mars." Scientific robotic exploration of Mars, coordinated by the Mars Exploration Program, has benefited the human Mars program by assisting in risk mitigation for crewed systems. For example, dust storm solar illumination data collected by rovers has been instrumental in sizing human surface power system concepts to allow tighter contingency margins, reducing mass and cost (Rucker et al. 2016). Future weather satellites and landed measurements are likely to be needed for landing/return launch operations. The Radiation Assessment Detector instrument on the Mars Science Laboratory measured radiation during cruise and landing to understand doses likely to be experienced by astronauts. The Mars Oxygen In Situ Resource Utilization Experiment on the Perseverance rover has manufactured oxygen from carbon dioxide, demonstrating in situ utilization of martian resources. Landing and return technologies are also a key area where benefit is likely. Present approaches require upscaling to meet human crew needs but may then benefit scientific payloads.

As discussed above in the context of the Artemis lunar program, human Mars missions will require coordination and advanced planning within and between SMD and the human exploration directorates. Human missions to Mars are being planned by NASA to fly in the late 2030s; thus, internal NASA interactions and conceptual development of missions to support both scientific and programmatic objectives need to appear in long-term planning now. Robotic missions often have lead times of 4 to 7 years and development needs to begin early enough for scientific payloads to be available for precursor flight missions and for any robotic precursor missions to inform the human mission designs. Astronaut extravehicular activities can benefit from scientific field tools now common in terrestrial geology (e.g., hand-held compositional sensors, scene multispectral or hyperspectral imagers), well beyond the simple mechanical tools that were state of the art for Apollo. Such spaceflight qualification would need to begin now. Keys to successful implementation of science-driven human exploration at Mars include:

1. Integrating science and human exploration planning and expertise to ensure that adequate, accurate, and appropriate Mars-specific knowledge and experience are provided in support of human missions, and that scientific progress will be sustained and advanced by human missions to Mars.
2. Identification of robotic science missions with overlap with human exploration precursor measurement needs, human landing site-specific knowledge, or infrastructure capability, including delineating "required" versus "desired" data.
3. Providing opportunities within PI-led science mission competitions for additional funding if the mission can also serve the human exploration needs in (2).
4. Funding that transparently follows from the entity setting requirements to the entity implementing the project or mission such that both funding and management responsibility reside within the same office or organization.

---

[7]The full Endurance mission study report is available at https://tinyurl.com/2p88fx4f.

Community engagement and selection of a science planning team to work with the mission-architecture team would be central to accomplishing many of these goals.

Recommendation: PSD should have the authority and responsibility for integrating science priorities into the human exploration plans for Mars.

Key issues to be addressed for Mars include defining the science goals and requirements for human missions and identifying areas where scientific expertise and/or robotic mission data are needed prior to the development of the mission architecture and hardware.

Science and Preparation for Human Missions to Mars

Human missions to the Moon offer opportunities to reduce risk for future human missions to Mars, for example, learning how to operate for long periods of time in a harsh planetary environment while being only 3 days away from Earth in case of a problem or emergency and developing hardware for use on a planetary surface that might be utilized on Mars as well. However, there are other risk-reduction areas for human missions to Mars for which scientific research and analysis is needed but is not being carried out or planned at a sufficient level of detail. These risk areas include, for example, mitigating the effects on humans of solar and galactic radiation and microgravity effects in space, dust on the surface and in the atmosphere of Mars, the potential physical/chemical effects on humans of perchlorates and other oxidizing agents in the regolith, and the effects on humans of trace amounts of liquid water in the regolith or dust associated with deliquescent minerals that are present. The MEPAG goals document includes a comprehensive list of topics where precursor measurements at Mars can be used to reduce risk to human missions (MEPAG 2020). Confirming prioritization of these areas of investigation and mapping these to required precursor flight missions early would provide sufficient time for robotic Mars missions that might be able to address specific concerns prior to human exploration.

Human missions to Mars also need to consider the issues of planetary protection (PP) and mitigate against both forward from Earth and back contamination from Mars (Coustenis et al. 2021; NASEM 2021). PP is critical both for maintaining the integrity of scientific investigations that seek to determine if life exists or existed on Mars, to protect the crew from any harmful indigenous martian life forms, and to protect the terrestrial biosphere from a potentially harmful release of martian life on Earth (Spry et al. 2020).

NASA has plans to send humans to Mars in the late 2030s while the commercial sector is describing plans for human flights to the Red Planet as early as the mid-2020s. It would be prudent for NASA to consider the earliest plausible human Mars mission scenario and develop a planetary protection protocol in accordance with this timeline. The general approach to planetary protection is to minimize the risk of both forward and back contamination. Characterizing organic constituents, understanding the natural transport of contamination on Mars (e.g., atmospheric and/or subsurface transport), and understanding microbial survivability in martian environments are identified areas of planetary protection knowledge gaps to be addressed in advance of human exploration of Mars, particularly at the human landing sites (NRC 2002; McKay et al. 2020; NASA 2020b; Spry et al. 2020; NASEM 2021). Many of these goals have measurement similarities to the Mars Life Explorer medium-class mission prioritized by the decadal survey for Mars (see "Mars Exploration Program" in Chapter 22), highlighting an opportunity for science and human exploration collaboration.[8]

Finding: NASA has not yet developed a planetary protection plan and related research activities specifically tailored to understand and mitigate the risk of forward and back contamination from human missions to Mars on timescales consistent with the earliest plausible human missions.

---

[8]The full Mars Life Explorer mission study report is available at https://tinyurl.com/2p88fx4f.

A TALE OF TWO ORBITERS: LRO AND IMIM

A key tenet of enabling high-priority planetary science and astrobiology investigations to be accomplished through human exploration is the importance of carefully crafted collaboration between SMD and the human exploration directorates. NASA has utilized different approaches to this cross-directorate collaboration in various capacities, and here the committee considers a case study by comparing two mission examples—Lunar Reconnaissance Orbiter (LRO) and the international Mars Ice Mapper (iMIM)—with varying degrees of success in the coupling of science-human exploration programmatics.

Lunar Reconnaissance Orbiter: A Case Study of Well-Coupled Science-Human Programmatics

During the Vision for Space Exploration: Moon, Mars, and Beyond era (2004), science was a prominent driver for future human exploration activity. It was decided by NASA leadership that before lunar exploration could be well planned, a thorough reconnaissance of the surface was to be performed from orbit to map and characterize the Moon in unprecedented detail. To accomplish this, a multi-directorate "Objectives Requirements Definition Team" was assembled following the model of a science definition team traditionally performed in SMD. This coupled SMD-HEOMD team, composed of NASA and outside experts across both science and human spaceflight disciplines, successfully defined the highest priority measurements, from which instruments were competed in the scientific community. The mission became the Lunar Reconnaissance Orbiter, funded by NASA's human exploration program but executed according to SMD project management fundamentals at NASA's Goddard Space Flight Center. After exploration objectives were achieved, LRO was transferred to PSD where it has continued to yield important scientific data.

Results from LRO are essential for the planning of future human and robotic exploration missions. Instruments aboard LRO have enabled scientists to produce the most detailed 3-D map of the lunar surface, map the mineralogy of the Moon, characterize the radiation environment and its biological impacts, and enable the search for lunar volatiles, including water ice.

Finding: LRO is perhaps the most successful example of cooperation and mission performance in a joint SMD and human exploration project and represents a template for how to initiate and manage joint collaborations between science and human exploration directorates at NASA in the future.

International Mars Ice Mapper: A Case Study of the Need for Better Coupled Science-Human Programmatics

Mapping near-surface ice on Mars is an area of high potential synergy between the measurement requirements of the robotic science and human exploration programs (MEPAG 2015; MASWG 2020; MEPAG 2020). Systematic mapping of near surface ice would help us to understand Mars's climate and recent climate change (see Q5, Q6, and Q10) as well as determine the potential of ice for martian ISRU. The international Mars Ice Mapper is in pre-Phase A at the time of this writing. iMIM has been promulgated by NASA as a mission joint between NASA, CSA (Canada), JAXA (Japan), and ASI (Italy) with a primary goal of mapping the distribution of the depth to mid-latitude ground ice. iMIM was presented to the decadal survey as having originated in priorities for human exploration (planning for resource access and science goals) and assigned to SMD to implement because of PSD/MEP expertise in executing robotic missions (NASA 2020d). However, unlike LRO, the initial concept was developed without significant input from the Mars science community on the objectives or on the measurement requirements. As currently articulated, the radar chosen for iMIM would determine the depth to the upper surface of ground ice in select locales but would not have the ability to determine properties relevant to the Mars climate science goals, such as the distribution of ice below its upper surface, the ice purity, or the degree of pore filling. Furthermore, there is concern that the measurements (as presented 23 November 2020, Mars Panel and 21 June 2021, MEPAG) would not achieve the mission objectives for radar penetration depth owing to the scattering properties of Mars's near-surface at the proposed radar wavelength, which is significantly shorter (higher frequency) than that chosen by multiple recent predecadal mission concept studies (MORIE, Calvin et al. 2021; MOSAIC, Lillis et al. 2021).

At the time of this writing, iMIM's relationship to either the robotic Mars science program or the human Mars program is unclear. At the most recent public discussion (NASA 2021c), NASA presented iMIM as an agency priority assigned to SMD to implement as a science-based mission, and that, while of interest to human exploration, it was not a mission that HEOMD was supporting.

Despite a rocky start, the mission concept has potential to follow a more LRO-like path where the scientific and human exploration communities work to define priority measurements. NASA is in the process of convening a Measurement Definition Team (MDT) that could reevaluate the measurement objectives and requirements and formulate a mission that meets both the scientific and human precursor exploration goals. For example, the MDT may consider changing the radar frequency and/or adding a second, longer wavelength band to the sounder mode meet the science objectives. iMIM is a case study for how stronger programmatic coupling and organizational structures are needed between the science and human exploration communities to ensure that precursor and eventual human missions to Mars collect meaningful scientific measurements while simultaneously supporting human exploration precursor measurement needs.

Finding: iMIM measurements presently articulated do not or only minimally address the prioritized science goals and measurement requirements for Mars ice mapping as defined in the MEPAG goals document, the Mars planetary mission concept studies prepared for the decadal survey (MORIE, MOSAIC), and this decadal survey.

Finding: With engagement of the scientific community in measurement definition, iMIM has the potential to be a pathfinding example of how Mars human exploration objectives can simultaneously advance high priority science questions related to Mars climate and how scientific expertise can help successfully realize human exploration objectives for ISRU.

Additional discussion and two recommendations regarding iMIM are given in the MEP section of the Recommended Program chapter (Chapter 22).

RESEARCH PROGRAMS TO ENABLE AND OPTIMIZE HUMAN EXPLORATION

With NASA's goal to send Artemis astronauts to the lunar surface within this decade, multiple research and technology development investments are required now to enable and optimize high-priority lunar science activities with humans at the Moon. Table 19-2 is a nonexhaustive list of example high-priority areas that require dedicated funding from SMD to enable high priority lunar science research via Artemis. A significant portion of this work will also be extensible to human exploration of Mars.

To adequately include science requirements in lunar human exploration plans, an Artemis Science Team is necessary to identify and advocate for the highest-priority science questions to be addressed for Artemis. SMD/PSD need to define both near and long-term science goals that are keyed to specific Artemis design decisions (e.g., landing site selection, landed capabilities, and upmass/downmass requirements). NASA's human exploration directorates are forging ahead with architecture and capability decisions, and SMD does not have an Artemis Science Team in place to provide timely inputs to influence Artemis capabilities. SMD/PSD needs to identify long-lead activities and fund work to close knowledge gaps and develop key hardware and capabilities to enable high priority science within Artemis missions (e.g., develop astronaut tools and hand-held instruments for use on the Moon, long-lived deployable instrument packages, next generation sample collection systems for volatiles, and cryogenic sample collection and curation capabilities). The absence of these capabilities will undoubtedly inhibit the ability to conduct decadal-level science by humans on the Moon.

Finding: SMD has not formulated an Artemis Science Team nor developed a plan for creating the science capabilities required for achieving high priority lunar science through human exploration. SMD has the potential to conduct higher priority science through Artemis by expeditiously identifying outstanding issues that need to be addressed in order to optimize Artemis science return and developing a funded program to conduct this work.

TABLE 19-2 Representative SMD Research and Development Activities to Enable and Optimize High-Priority Science from Artemis

Activity	Description
Cold sampling and curation	Laboratory studies to determine how to sample, transport, and curate volatile samples at cold temperatures to optimize sample integrity and science return. Identify and develop facilities required to handle, store, and analyze volatile samples from the lunar pole, including cryogenic transport and curation.
Sample science	Determine sampling details for all sample types (sample masses, types, containers, collection operations, contamination knowledge, curation, lab analysis, etc.).
In situ measurements	Identify required measurements to, for example, characterize samples (including volatiles) prior to collection, in situ science measurements not associated with sample collection (e.g., identify specific in situ instrumentation required, path to flight qualification of handheld science instruments, and concepts of operations [ConOps] for data acquisition), and develop long-lived instrument packages.
Science requirements	Develop science requirements for Artemis missions (near-term and long-term strategic); feed forward to required surface capabilities, landing site selection.
Landing site selection	Provide assessments and inputs regarding landing site selection to optimize science return from human missions.
Science accommodation trades	Analyze science return and high grading for decisions on what assets to deliver to lunar surface (e.g., trading sample sizes vs tools vs in situ instruments), develop a process for decision-making within SMD for science trades, assess options for upmass/downmass to surface (e.g., CLPS for Moon, commercial options for Mars, and HLS).
Planetary protection	Develop plans to mitigate forward and back contamination, identify and fly required precursor missions.
Imaging and situational awareness	Low-angle light and limited illumination present significant challenges to surface operations near the lunar south pole. Requires assessment for Artemis planning (e.g., required cameras, lighting, and a priori mapping) for both science and safe surface operations.
Science concept of operations	Detailed science planning to enable near-real time science support from Earth-based science team, optimization of science ConOps, crew time/scheduling/activities, mobility requirements.
Crew training	Develop crew training curriculum (classroom, fieldwork, and lab work); comprehensive science/geology/astrobiology training required.

ROLE OF COMMERCIAL SPACE AND HUMAN-SCALE VEHICLE CAPABILITIES

Multiple commercial companies are developing vehicles for human flights to the surface of the Moon and Mars. These human-scale vehicles have the potential to provide unprecedented payload capacity and the distinct possibility of lowering the cost of surface access. Commercial vehicles designed for human planetary exploration thus provide tremendous scientific research potential, but NASA currently lacks a mechanism for the planetary science and astrobiology community to develop or fly payloads on these platforms to either the Moon or Mars.

To capitalize on such opportunities, NASA could develop a funded program aligned with the development approach of the commercial vendor pool, including a rapid development schedule, relatively high-risk tolerance compared to traditional planetary science missions, and ultimately a high ratio of potential science value for the dollars spent if successful. Such a program would enable the planetary community to participate in sending science and exploration payloads to these planetary destinations, which will advance science objectives outlined in this decadal survey, the NASA strategic plan, and similar guiding documents.

While NASA's Payloads and Research Investigations on the Surface of the Moon (PRISM) and Commercial Lunar Payload Services (CLPS) programs are pathways for some robotic lunar mission payloads to reach the Moon via commercial lander services, NASA does not have a similar on-ramp for the planetary community to develop or fly payloads that could fully take advantage of human flights to the Moon (or Mars). As an example, the committee considers the SpaceX Starship that has been selected by NASA for the first HLS contract, although the same logic applies to any vendor providing human surface access to the Moon and/or Mars. Starships can accommodate payloads that are significantly larger and heavier than traditional NASA planetary payloads, significantly reducing the need for the costly reductions in size and mass required for traditional NASA payloads. Starships can fly multiple payloads and instruments on individual flights to reduce overall risk, and significantly more power can be available for the payload. Additionally, SpaceX's intended mission cadence provides more frequent flight opportunities, which has several advantages for enabling increased science and broadening community participation in planetary science and astrobiology activities. Along with their role in HLS, SpaceX also has a stated goal of sending humans to Mars in the mid-late 2020s, with cargo flights of Starship to Mars followed by crewed missions to the martian surface (Figure 19-3). At the time of this writing, Starship development is in the early stages and planned capabilities have not yet been demonstrated in orbit or in space, but both cargo and crew flights to Mars offer significant potential science opportunities.

To take advantage of such capabilities, a funding mechanism within NASA is needed to provide the opportunity for members of the community to fly robotic payloads on these high-capacity human flights. This program has to be consistent with the mission timelines for rapid flights planned by SpaceX (and other vendors as appropriate), especially if commercial entities send test or human flights to Mars prior to NASA. Traditional cost modeling and estimating practices based on historical data and traditional approaches would need to be revised given this new paradigm of planetary flight opportunity. To be most effective, planning within NASA needs to begin immediately to prepare for payloads on the first uncrewed Starship flights, likely first to the Moon and then for Mars. Planetary

FIGURE 19-3 SpaceX Starship landers on Mars. SOURCE: Courtesy of SpaceX. CC BY 2.0.

science and astrobiology payloads sent to the Moon and Mars onboard human-scale vehicles (test flights as well as crewed missions) would bring much increased payload capacities that could provide uniquely science-enabling capabilities. The types of payloads that might be used to achieve SMD, human exploration, and STMD objectives could be much different than those designed for traditional NASA flight opportunities with their stringent mass and volume constraints.

Finding: Commercial human spaceflight missions to the lunar and martian surfaces will provide unprecedented payload capacity and potentially offer tremendous opportunities for planetary science and astrobiology. These vehicles may lower the cost of surface access, which can enable a new paradigm for planetary science and astrobiology investigations, technology development and testing, and human exploration of space.

Recommendation: NASA should develop a strategy to utilize opportunities to fly science payloads on commercial test flights and crewed missions to the Moon and Mars as such opportunities arise.

EXTERNAL COOPERATION

Multiple entities including, but not limited to, NASA (via Artemis), private-sector organizations, international entities such as non-U.S. space agencies and interest groups, have expressed interest in human exploration of the Moon, Mars, and beyond. To this end, NASA has established the Artemis Accords (NASA 2020e), which describe "a shared vision for principles, grounded in the Outer Space Treaty of 1967, to create a safe and transparent environment which facilitates exploration, science, and commercial activities for all of humanity to enjoy." International cooperation will be a key component of human exploration within the solar system, and an important aspect of ensuring a common set of principles to guide these exploration activities. In addition, NASA benefits by working with other government agencies where appropriate to take advantage of government-wide expertise as applicable to support human exploration activities, such as the U.S. Geological Survey for resource prospecting and astronaut field training, and the National Science Foundation for field analog science and operations research in Antarctica.

Finding: International participation in human programs has the benefit of (1) spreading the cost out over a larger number of participating entities and making it more affordable to each, (2) providing wider participation of scientists, engineers, and the public from different countries and cultures, and (3) enhancing international cooperation in peaceful endeavors.

Finding: International participation carries with it enhanced risk, in terms of coordination and management of schedules, potential for increased cost, mismatch or miscommunication of requirements, and potential for withdrawal of partners at inopportune times.

Finding: NASA's continued encouragement of international participation in human missions in the solar system (the Moon, Mars, near-Earth objects, other potential targets) is beneficial as a way of enhancing the scientific return from the missions and of providing a forum for constructive and peaceful interactions between countries.

REFERENCES

Bartels, M. 2018. "Why We Can't Depend on Robots to Find Life on Mars." *Space*, August 22. https://www.space.com/41551-finding-mars-life-robots-versus-humans.html.

Beattie, D.A. 2001. *Taking Science to the Moon*. Baltimore, MD: Johns Hopkins University Press.

Calvin, W.M., N.E. Putzig, C.M. Dundas, A.M. Bramson, B.H. Horgan, K.D. Seelos, H.G. Sizemore, et al. 2021. "The Mars Orbiter for Resources, Ices, and Environments (MORIE) Science Goals and Instrument Trades in Radar, Imaging, and Spectroscopy." *Planetary Science Journal* 2(2):76.

Compton, R.R. 1985. "Geology in the Field." New York: Wiley.

Coustenis, A., G. Kminek, and N. Hedman. 2021. "The COSPAR Panel on Planetary Protection." P. 2232 in *43rd COSPAR Scientific Assembly Abstracts*. January 28–February 4.

Heldmann, J.L., A.M. Bramson, S. Byrne, R. Beyer, P. Carrato, N. Cummings, M. Golombek, et al. "Accelerating Martian and Lunar Science Through SpaceX Starship Missions." White paper #518 submitted to the Planetary Science and Astrobiology Decadal Survey 2023–2032.

Hodges, K.V., and H.H. Schmitt. 2019. "Imagining a New Era of Planetary Field Geology." *Science Advances* 5(9). https://doi.org/10.1126/sciadv.aaz2484.

Lillis, R.J., D. Mitchell, L. Montabone, N. Heavens, T. Harrison, C. Stuurman, S. Guzewich, S. England, P. Withers, and M. Chaffin. 2021. "MOSAIC: A Satellite Constellation to Enable Groundbreaking Mars Climate System Science and Prepare for Human Exploration." *Planetary Science Journal* 2:211.

Lofgren, G.E., F. Horz, and D. Eppler. 2011. "Geologic Field Training of the Apollo Astronauts and Implications for Future Manned Exploration." *Geological Society of America Special Papers* 483:33–48. https://doi.org/10.1130/2011.2483(03).

MASWG (Mars Architecture Strategy Working Group), B.M. Jakosky, R. Zurek, S. Byrne, W. Calvin, S. Curry, B. Ehlmann, et al. 2020. *Mars, the Nearest Habitable World—A Comprehensive Program for Future Mars Exploration*. Washington, DC: National Aeronautics and Space Administration.

MEPAG (Mars Exploration Program Analysis Group). 2020. "Mars Science Goals, Objectives, Investigations, and Priorities: 2020 Version," D. Banfield, ed. White paper submitted to the Planetary Science and Astrobiology Decadal Survey 2023–2032. https://mepag.jpl.nasa.gov/reports.cfm.

MEPAG. 2015. "Report from the Next Orbiter Science Analysis Group (NEX-SAG)." December 14. https://mepag.jpl.nasa.gov/reports/NEX-SAG_draft_v29_FINAL.pdf.

McKay, C., A. Davila, J. Eigenbrode, C. Lorentson, R. Gold, J. Canham, A. Dazzo, et al. 2020. "Contamination Control Technology Study for Achieving the Science Objectives of Life-Detection Missions." White paper #340 submitted to the Planetary Science and Astrobiology Decadal Survey 2023–2032. *Bulletin of the AAS* 53(4).

McPhee, J.C., and J.B. Charles. 2020 "Human Planetary and Astrobiology Exploration: How Will Radiation, Low Gravity, and Isolated and Confined Conditions Affect Our Health?" White paper #027 submitted to the Planetary Science and Astrobiology Decadal Survey 2023–2032. *Bulletin of the AAS* 53(4).

Musk, E. 2018. "Making Life Multi-Planetary." *New Space* 6(1):2–11. https://doi.org/10.1089/space.2018.29013.emu.

NASA (National Aeronautics and Space Administration). 2015. *Historical Studies in the Societal Impact of Human Spaceflight*. S.J. Dick, ed. NASA SP-2015-4803. Washington, DC: Office of Communications, NASA History Program Office, p. 530.

NASA. 2020a. *Artemis Plan: NASA's Lunar Exploration Program Overview*. Washington, DC. https://www.nasa.gov/sites/default/files/atoms/files/artemis_plan-20200921.pdf.

NASA. 2020b. "Biological Planetary Protection for Human Missions to Mars." NID 8715.129. Washington, DC: Office of Safety and Mission Assurance.

NASA. 2020c. *Artemis III Science Definition Team Report: A Bold New Era of Human Discovery*. NASA/SP-20205009602. Washington, DC: Science Mission Directorate. https://www.nasa.gov/sites/default/files/atoms/files/artemis-iii-science-definition-report-12042020c.pdf.

NASA. 2020d. "Presentation from R. Davis, J. Garvin, T. Haltigin to Panel on Mars." November 23. Washington, DC: National Academies of Sciences, Engineering, and Medicine. https://webapp.nationalacademies.org/napar/projectview.aspx?key=52188#MeetingId34067.

NASA. 2020e. "The Artemis Accords: Principles for Cooperation in the Civil Exploration and Use of the Moon, Mars, Comets, and Asteroids for Peaceful Purposes." Washington, DC. https://www.nasa.gov/specials/artemis-accords/img/Artemis-Accords-signed-13Oct2020.pdf.

NASA. 2021a. *INSPIRE: IN Situ Solar System Polar Ice Roving Explorer*. Mission Concept Study Report for the Planetary Science and Astrobiology Decadal Survey 2023–2032. Pasadena, CA: Jet Propulsion Laboratory, California Institute of Technology. https://tinyurl.com/2p88fx4f.

NASA. 2021b. Presentation from J. Robinson to Panel on Mercury and the Moon. April 9. Washington, DC: National Academies of Sciences, Engineering, and Medicine. https://webapp.nationalacademies.org/napar/projectview.aspx?key=52183.

NASA. 2021c. Presentation from E. Ianson and M. Meyer to Mars Exploration Program Analysis Group. September 27. https://mepag.jpl.nasa.gov/meetings.cfm.

NASEM (National Academies of Sciences, Engineering, and Medicine). 2019. *Report Series: Committee on Astrobiology and Planetary Science: Review of the Planetary Science Aspects of NASA SMD's Lunar Science and Exploration Initiative*. Washington, DC: The National Academies Press. https://doi.org/10.17226/25373.

NASEM. 2021. *Report Series: Committee on Planetary Protection: Evaluation of Bioburden Requirements for Mars Missions*. Washington, DC: The National Academies Press. https://doi.org/10.17226/26336.

NRC (National Research Council). 1997. *The Human Exploration of Space*. Washington, DC: National Academy Press.

NRC. 2002. *Safe on Mars: Precursor Measurements Necessary to Support Human Operations on the Martian Surface*. Washington, DC: National Academy Press. https://doi.org/10.17226/10360.

NRC. 2011. *Vision and Voyages for Planetary Science in the Decade 2013–2022*. Washington, DC: The National Academies Press. https://doi.org/10.17226/13117.

Phinney, W.C. 2015. *Science Training History of the Apollo Astronauts.* NASA SP-2015-626. Washington, DC: National Aeronautics and Space Administration.

Rucker, M.A., S.R. Oleson, P. George, G. Landis, J. Fincannon, A. Bogner, J. McNatt, et al. 2016. "Solar vs. Fission Surface Power for Mars." P. 5452 in *AIAA SPACE* 2016.

Schmitt, H.H., A.W. Snoke, M.A. Helper, J.M. Hurtado, K.V. Hodges, and J.W. Rice, Jr. 2011. "Motives, Methods, and Essential Preparation for Planetary Field Geology on the Moon and Mars." *Geological Society of America Special Papers* 483:1–15. https://doi.org/10.1130/2011.2483(01).

Slakey, F., and P.D. Spudis. 2008. "Robots vs. Humans: Who Should Explore Space?" *Scientific American* February 1. https://www.scientificamerican.com/article/robots-vs-humans-who-should-explore.

Spry, J. A., B. Siegel, G. Kminek, C. Bakermans, J. N. Benardini, E. Beltran, R. Bonaccorsi, et al. 2020. "Planetary Protection Knowledge Gaps and Enabling Science for Human Mars Missions." Whitepaper #205 submitted to the Planetary Science and Astrobiology Decadal Survey 2023–2032. *Bulletin of the AAS* 53(4).

20

Infrastructure for Planetary Science and Exploration

Planetary science is highly dependent on infrastructure—that is, the equipment, instrumentation, and facilities that enable advanced study of planetary bodies, through simulations, experiments, remote observations, and space-craft exploration.[1] In response to the survey's statement of task for the development of a comprehensive research strategy, this chapter provides background, findings, and recommendations on key infrastructure elements that support the priority activities in planetary science, astrobiology, and planetary defense identified in other parts of the report. Infrastructure, as defined here, includes the facilities, services, and organizational relationships needed to advance the mission of NASA's Planetary Science Division that are not directly supported by individual program elements. The critical infrastructure for these disciplines is housed at NASA, the National Science Foundation (NSF), and other government facilities, but also in individual institutions throughout the world. Owing to the sheer volume of facilities, the committee necessarily describes only a subset here. This discussion is not an exhaustive and complete list of every possible facility, but instead provides an overview of key facility types, noting specific needs and areas of concern when applicable. Any omissions are unintentional and are not meant to imply that a specific facility is unimportant. Development and improvement of infrastructure are long-term investments and need to be viewed with a longer time horizon than a decade; thus, recommendations made that would result in investments in the coming decade need to consider requirements beyond that time horizon.

NASA INFRASTRUCTURE

NASA directly funds many facilities relevant to planetary science, astrobiology, and planetary defense missions, ranging from telescopes to spacecraft communication to test facilities. Not all of these facilities fall solely under the aegis of the Planetary Science Division, but a subset of those of critical importance to our community is included.

NASA Test and Environmental Simulation Facilities

NASA's field centers are home to many test facilities used for technology, instrument, or spacecraft integration and testing. The centers have ISO 9001-certified cleanrooms, vibration, acoustic, and electromagnetic interference test capabilities, and thermal vacuum chambers; several have limited radiation testing facilities,

---

[1] A glossary of acronyms and technical terms can be found in Appendix F.

as well. These facilities are available for use by the community through partnership agreements and the committee supports this effort.

In addition, NASA centers host a variety of facilities capable of replicating other planetary environments, including high pressure planetary interiors, very low-pressure atmospheres, extremely hot surface temperatures, and atmospheres and surfaces so cold that methane and other organic compounds form condensates and ices. These facilities permit instrument and technology development and testing, but also enable scientific experiments to be performed under conditions like those encountered on other planetary bodies. The size and capabilities of the different simulation facilities vary, and this variety is needed to accommodate the different needs of the planetary science community. One example is the NASA Glenn Extreme Environments Rig (GEER), which can simulate many different extreme environments, but has thus far focused much of its effort on replicating conditions at the surface of Venus (Lukco et al. 2018). Another example is the Planetary Aeolian Laboratory (PAL) at the NASA Ames Research Center (ARC), which over the past few decades has been used to study aeolian processes at pressures representative of the surface of Venus, Mars, and Titan (e.g., Greeley and Iversen 1985; Burr et al. 2015; Swann et al. 2020). Environmental simulations may also be needed to understand instrument measurements at extreme conditions. For example, the high surface temperatures on Venus have been shown to affect mineral spectra in the visible and NIR wavelengths, and this type of spectroscopy is one of the only available tools for examining variations in surface composition from orbit (Helbert et al. 2021). It is thus critical to have laboratory facilities capable of replicating extreme environments to develop spectral libraries for interpretation of mission measurements, or calibration of mission instruments. Scientific experiments cover many areas of research, including material properties, chemistry (of geologic materials, gases, organic/prebiotic), aeolian processes, the microphysics of nucleation, condensation, and ice formation, and physicochemical properties of planetary materials.

However, NASA Centers are not immune from management, staffing, and funding issues. A review of SMD planetary facilities in 2015 (Mackwell et al. 2016) noted that a lack of on-site personnel with an active interest in using and improving the NASA ARC PAL wind tunnels, coupled with a lack of engagement of management in facility operation or experiments, had likely resulted in limited community use of the PAL facility and a growing need for modernization of equipment. This is despite PAL's Mars Surface Wind Tunnel being the only facility in the United States capable of simulating Mars surface pressures over a 13 m active wind tunnel into which sand and dust may be introduced.

Finding: NASA's diverse planetary environment experimental facilities would benefit from regular review by an external review board to determine if their capabilities and operations are meeting community needs. For those that are underutilized, an evaluation of both their technological and administrative aspects, including a feasibility study and cost-benefit analysis of whether refurbishment, rebuilding, or retirement is appropriate. Retained facilities need adequate funding for both on-site management and staff, as well as future modernization and expansion. These facilities could also be more effectively utilized by the planetary science community by increasing their discoverability via a database of available systems and their capabilities.

NASA Telescope Facilities

Observatories (on the ground and in space) provide both unique discoveries and essential support for planetary missions as well as the continuing search for and characterization of exoplanets by providing spatial, temporal, and spectral context for observations from spacecraft. Both ground and space-based facilities support the major subsets of the decadal survey's 12 key science questions by providing essential monitoring of dynamic or unique solar system phenomena, including atmospheres (Q7, Chapter 10), comets (Q1, Q2, and Q3, Chapters 4–6), cryo-volcanic/plume activity (Q5, Q6, and Q8, Chapters 8–9 and 11), occultations (Q4, Chapter 7), and many more, all varying on timescales of hours to multiple decades. Changes over long timescales are challenging for a visiting spacecraft, so telescope observations fill the gaps between missions. There are excellent synergies between planetary missions and ground/space-based observatories, such as Earth-based support campaigns, which encompass both professional facilities (multi-spectral from radio to X-ray) and amateur observers (in visible and near-infrared).

Ground-based facilities have the benefit of longevity, and their capabilities can increase over time as science instruments can be upgraded, repaired, or replaced. These instruments have relaxed mass and size constraints compared to spacecraft instrumentation, delivering capabilities such as high spectral resolution, spectral multiplexing, and the strong light-gathering power of large apertures. Currently, NASA's Infrared Telescope Facility devotes 50 percent of its time to solar system studies, but all other facilities rely on competitive proposals each semester, with limited NASA/ESA guidance on priorities for spacecraft mission science support. Observatories may also consider offering the possibility of long-term status for monitoring programs, extending over multiple cycles with mutually agreed renewal procedures. Extremely large telescopes coming online late in the next decade (see NSF section below) will deliver advanced capabilities with angular resolution comparable to Voyager approach data (Wong et al. 2020). NASA investment in these observatories would enable new capabilities for spacecraft mission support, potentially even affecting mission payload decisions if designed to operate in conjunction with dedicated support programs.

Finding: Planetary science at NASA-funded observatory facilities benefits from a proposal mechanism for spacecraft mission support, for example, the Keck 2022A call for mission support proposals, as well as multicycle programs.

Space-based facilities can access spectral regions invisible from the ground and are not subject to blurring owing to atmospheric turbulence at visible wavelengths, or high sky backgrounds at infrared wavelengths. Suborbital and airborne telescopes partially escape these atmospheric effects without the high cost of spacecraft development and launch. Specific spectral ranges are only accessible from space-based facilities including X-ray, the ultraviolet, and atmosphere-opaque regions of the infrared. Observing and instrument conditions (e.g., thermal and photometric) can also be stabilized on space platforms, enabling high-precision measurements and long time series. Thus, it is important that astrophysical instruments and observatory capabilities are designed with planetary observations in mind. An excellent approach for enabling solar system observations is the purposeful inclusion of solar system scientists on science working groups, development teams, and instrument and operations staff which has proven to be highly successful for JWST (Hammel and Milam 2020). It is also desirable to allow flexible scheduling for unanticipated phenomena, through mechanisms such as director's discretionary programs, and a variety of observation cadences, from short-term to long-term monitoring of objects, potentially over multiple years with long term programs.

Finding: To enable planetary observations, astrophysical telescope assets need to continue to include tracking non-sidereal rapidly moving objects, with dynamic range accommodations for both bright and faint targets.

The Deep Space Network

The Deep Space Network (DSN) is a critical element of NASA's solar system exploration program. It is the only asset available for communications with deep space missions. The DSN is currently composed of three stations located in Goldstone, California; Madrid, Spain; and Canberra, Australia; along with operations control and other services in the United States. Each station has one 70-meter antenna, with three 34-meter beam waveguide antennas at Goldstone and Canberra and four at Madrid. NASA has plans to add at least one more 34-meter antenna at each complex in the near future (Lazio et al. 2020). These antennas support more than three dozen missions with downlink and uplink capabilities in S-, X-, and Ka-band (limited). Collectively, these stations provide nearly continuous full-sky coverage. The 70-meter dishes are in high demand, particularly during critical events, because of their downlink capability, sensitivity, and ability to satisfy other mission requirements. As such, they are heavily oversubscribed, and current deep-space missions are limited mostly by downlink rather than onboard storage capacity (Johnston 2020). The DSN also contends with aging infrastructure, particularly the 70 m antennas that were constructed in the 1960s; a 2020 NASA OIG audit (NASA 2020b) raised concerns about the inability to properly maintain these systems.

Nonetheless, the DSN continues to perform extraordinarily well, returning data with a very low drop-out rate and achieving command and telemetry availability of better than 95 percent to most operating missions. In the coming decade, with the launch of JWST and expansion of human exploration, demands for DSN support will increase dramatically, projected at another factor of 10 by the early 2030s (Figure 20-1). Future capabilities afforded by optical communication, transponder advances, advanced onboard compression and data processing software,

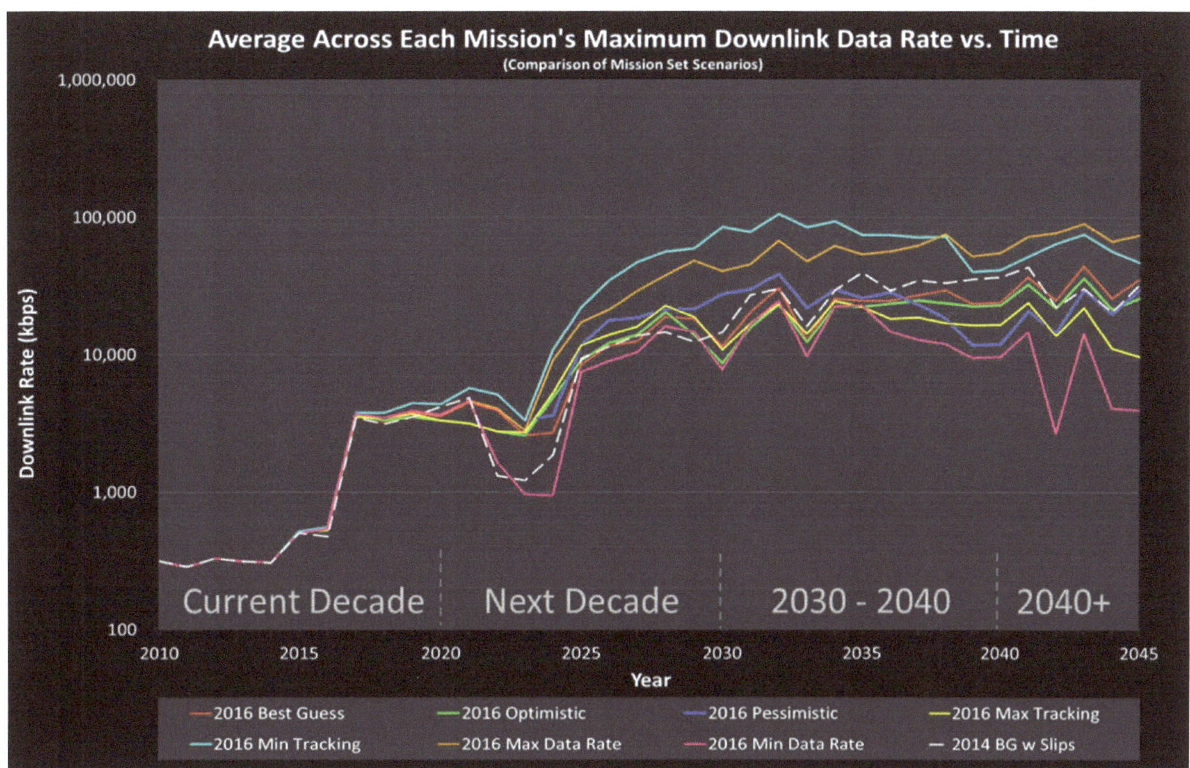

FIGURE 20-1 Past and predicted loading of the Deep Space Network for 2010 to 2045, based on a past notional mission manifest. Although the planned missions have changed, the challenge remains that a factor of 10 increase in data downlink is expected by 2030 owing to mission complexity and instrument advances. SOURCE: D.S. Abraham, B. MacNeal, D. Heckman, Y. Chen, J. Wu, K. Tran, A. Kwok, and C.-A. Lee, 2018, "Recommendations Emerging from an Analysis of NASA's Deep Space Communications Capacity," Paper AIAA 2018-2528 presented at 2018 SpaceOps Conference, https://doi.org/10.2514/6.2018-2528, used with permission from the California Institute of Technology.

and other means may provide future increases in returned data volumes and will be important to meeting mission demands (Lichten 2021), but the DSN infrastructure needs to keep pace.

Owing to differences in atmospheric opacity, planetary missions require downlink capability in multiple frequency bands. For example, three-band telemetry during outer planet atmospheric occultations allows sounding of different pressure depths within the atmosphere. In addition, S-band capacity is required for communications from Venus during probe, balloon, lander, and orbit insertion operations because communications in other bands cannot penetrate the atmosphere. X-band capability is required for communication through the atmosphere of Titan, and also for emergency spacecraft communications. Although Ka-band downlink has a clear capacity advantage, there is a need to maintain multiple-band downlink capability, because Ka-band imposes additional power and pointing demands on spacecraft with limited resources that may be difficult to meet, especially in the event of spacecraft safe mode entry. Last, the DSN is crucial for precision spacecraft ranging and navigation, and this capability needs to be maintained.

Finding: S-, Ka-, and X-bands are critical to planetary mission communication, requiring careful management of these key frequencies. X-band also remains crucial for science data downlink, recognizing that Ka-band pointing- and spacecraft-stability requirements pose challenges for resource-limited missions, particularly in safe mode conditions.

Outer solar system exploration also requires either 70-meter antennas or equivalent arrays to achieve the data rates needed for distant missions, such as to Uranus, Neptune, and the Kuiper belt. To support multiple missions, the DSN needs to be able to receive data from more than one mission at one station simultaneously. If new arrays can only mimic the ability of one 70-meter station and nothing more, missions will remain downlink-constrained and will have to compete against one another for limited downlink resources.

> **Recommendation: NASA should expand uplink and downlink capacities as necessary to meet the navigation and communication requirements of the missions recommended by this decadal survey, with adequate margins, while also balancing the demands from other projects, including JWST, Roman Space Telescope, Artemis, and others.**

The Goldstone Solar System Radar (GSSR) is a key facility for ground-based planetary radar observations, specifically, the 70 m DSS-14 and 34 m DSS-13 elements of the DSN. GSSR is a fully steerable radar facility that can transmit at X-band (8560 MHz, 3.5 cm) at a continuous power of 500 kW from DSS-14; the 80 kW, C-band (7190 MHz, 4.2 cm) DSS-13 is primarily used for asteroid and lunar studies. After the loss of the Arecibo Telescope in December 2020, which at the time was the largest, most sensitive, and most powerful planetary radar facility, GSSR is now the primary facility for radar observations of near-Earth objects (NEOs) and other planetary bodies. Radar mapping of the venusian surface is most feasible at L- and S-band; the current GSSR infrastructure is unable to support radar observations of the venusian surface. Additionally, GSSR is some 15 times less sensitive than Arecibo was, which results in limited observing capabilities in monostatic configuration.

> **Finding:** GSSR is the remaining key facility for ground-based planetary radar observations and can provide critical planetary defense and other planetary science observations.

Planetary Sample Curation and Associated Laboratory Facilities

Just as geological samples from Earth record the natural history of our planet, astromaterials hold the natural history of the solar system and beyond. Astromaterials include samples of extraterrestrial origin collected on Earth (e.g., meteorites and cosmic dust) and samples returned from space as part of a sample return mission (e.g., Apollo, Stardust, and Genesis). With a large increase in sample return missions (e.g., Hayabusa, Hayabusa-2, OSIRIS-REx, and the Mars Sample Return campaign), curation facilities are at a critical juncture; these samples are stored in curation facilities designed to maintain the integrity of the samples, and these facilities represent a key infrastructure that supports planetary sample science. Moreover, the curatorial facilities are managed by a workforce of curators who have the knowledge to responsibly conserve the samples to maximize their science value in perpetuity. Additionally, they are responsible for allocating astromaterials samples to the scientific community. NASA's astromaterials represent an invaluable resource for the planetary science community, and NASA allocates >1,000 samples to scientists across the globe each year. Sample analysis work is supported by numerous NASA R&A programs.

> **Finding:** Continued funding is needed to support and maintain the curatorial facilities that host NASA's astromaterials collections for past, present, and future sample return efforts. The Mars Sample Receiving Facility costs are best handled within the existing budget for MSR (see Chapter 22).

The Mars Sample Receiving Facility (SRF) is a critical part of the Mars Sample Return (MSR) Program. The SRF will conduct, under strict containment, preliminary examination and analysis of unsterilized samples collected and returned to Earth. The SRF will be required to implement planetary protection Category V (Restricted Earth Return) requirements and contamination control requirements. Evaluation of martian samples in the SRF requires an ISO-certified positive-pressure cleanroom and biosafety level-4 (BSL-4) negative-pressure containment in a cabinet or suit laboratory; existing BSL-4 laboratories are not designed to host cleanrooms, and there are many approvals needed for certifying the BSL-4 status of a new facility. For Class V restricted missions, public health officials and other regulatory agencies need to be involved in planning and implementation. Current MSR schedules suggest that the SRF will need to be ready in about 10 years (early 2030s), which is approximately equal to the amount of time it would take to build and commission a new BSL-4 laboratory, if this is the implementation approach needed. Furthermore, the SRF will need to have a clear workflow toward release of samples to long-term curation and to the community for planned analyses that take advantage of leading geochemical and microanalytical facilities around the world (Carrier et al. 2022; see also Chapter 22). Development of a plan for the design and construction of the SRF is still in early stages and may already be behind schedule for sample return in the early 2030s timeframe.

> **Recommendation: NASA, in partnership with ESA and community stakeholders, should develop the plan for the end-to-end processing of samples returned from Mars. This plan should include the definition, design, and construction of the Mars Sample Receiving Facility to ensure that it is ready to receive the samples by 2031. The plan should also outline the approach for expeditiously distributing the samples to the scientific community for analysis and to a long-term curation facility.**

In addition to sample return missions, astromaterials collected on Earth as meteorites and cosmic dust represent a critical feedstock for planetary sample science. Many sample collection efforts are supported by federal funding and rely on infrastructure. For example, cosmic dust is captured in collectors mounted on high-altitude aircraft such as WB-57 (Ellington Airfield at Johnson Space Center) and ER-2 (Armstrong Flight Research Center), which are both maintained and supported by NASA. The United States also supports the Antarctic Search for Meteorites (ANSMET) through a three-agency agreement between NASA, NSF, and the Smithsonian. Nearly half of all astromaterials sample requests made to NASA each year request samples from the ANSMET collection. Furthermore, this collection is composed of samples from the Moon, Mars, undifferentiated and differentiated asteroids, further highlighting the breadth of the sample science community that uses the ANSMET collection as a critical resource.

> **Finding:** Continued efforts are needed to support and maintain the infrastructure that enables the collection of astromaterials on Earth, including Antarctic meteorites and cosmic dust.

As technological advancements and new ideas expand the variety and scope of scientific questions that can be asked with astromaterials samples, so expands the need for better storage, processing, and sample handling capabilities of curation laboratories that house and process astromaterials samples. Over the coming decade, we will need to improve our ability to curate and process under "cold" conditions. The ever-expanding plans for the return of samples from volatile-rich solar system targets and/or targets of astrobiological significance (e.g., volatile-rich samples from the lunar poles, comet nucleus samples, and future ocean world samples of biological importance) necessitate the development of curation at temperatures below that of typical curation facilities (20°C). Temperature requirements depend primarily on which volatiles are expected within the returned sample, which in turn relate to the conditions under which the material formed and has since been preserved. The curatorial temperatures for terrestrial materials, including tissue samples and ice cores, include: $\leq -20°C$ (the temperature of typical walk-in freezers in which physical processing and documentation takes place); $\leq -40°C$ for archival storage (e.g., of ice cores); and -80 to $-196°C$ (liquid nitrogen) for biological samples (e.g., Anchordoquy and Molina 2007; Rissanen et al. 2010). Thus, except for biological tissue storage, the field of Earth materials curation has not yet entered the realm of cryogenics, although cold curation and processing of astromaterials are in development (Herd et al. 2016). Most importantly, the infrastructure requirements for cold curation vary substantially depending on the temperature requirements for the samples. For example, without specific temperature requirements for Artemis samples, detailed planning for an Artemis cold curation facility cannot proceed. Even after a decision is made about sample temperature requirements, R&D work will be needed to optimize infrastructure requirements for the curation and processing of geological materials under cold conditions.

> **Finding:** Further work is needed to define the sample temperature requirements for curating and processing future cold or volatile samples and for developing the appropriate facilities.

Analytical laboratories also are a critical piece of infrastructure for planetary science and astrobiology, providing a key role in the analysis and understanding of returned samples, meteorites, and mission data. As stated in the last decadal survey, The most important instruments for any sample return mission are the ones in the laboratories on Earth (NRC 2011). Maintenance of sample analysis capability is critical for planned sample return efforts, as is development of new techniques and instrumentation. As an example, the Apollo Next Generation Sample Analysis Program has demonstrated the usefulness of the technique of X-ray computed tomography for documenting materials in previously unopened lunar samples from Apollo missions without disturbing the sample (Zeigler et al. 2020). Along with returned samples, instrument capability is critical for analysis of meteorites and laboratory experimental products. Additionally, sample return efforts are being sent to worlds with the potential of returning materials beyond typical geologic materials (e.g., ice, organic molecules, gases) that require special care while analyzing or entirely new techniques to measure. NASA provides funding critical support, maintain, and expand the analytical and experimental capabilities of U.S.-based laboratory facilities to conduct planetary science and astrobiology; this includes equipment investment, as well as adequate technical support staff. Long-term support of PI-led laboratories is essential for maintaining national leadership in specialized microanalytical capabilities crucial to planetary sample analysis (e.g., metal isotopes, age dating work with less common isotope systems, zircon analysis, FIB and TEM work). The Department of Energy (DOE) and NSF support synchrotron capabilities play a key role in nanometer-scale ancient earth fossil analysis and will be important to the next decade's astrobiological samples. The committee endorses the finding of the National Academies report *Strategic Investments in Instrumentation and Facilities for Extraterrestrial Sample Curation and Analysis* (NASEM 2019) that longer-term funding support for technical staff is desirable to ensure laboratories maintain this expertise and capability better than current short term soft money funding programs allow.

Plutonium-238 Production

Radioisotope power systems (RPS) are essential to the exploration of the solar system. The *Vision and Voyages* (*V&V*) decadal survey (NRC 2011) recognized that plutonium-238 (^{238}Pu) fuel is essential for comprehensive exploration of the solar system and made a recommendation to restart production with a goal of producing 1.5 kg/year. NASA's RPS Program Office and DOE have begun such a program and are now successfully producing fuel (Sutliff et al. 2020), on track to accomplish a production rate of 1.5 kg/year in 2026 (Dudzinksi 2021; Zakrajsek 2021). Spacecraft power systems use ^{238}Pu formed into fuel pellets of 0.15 kg of PuO_2 each and then encased in an iridium clad. Depending on the RPS system required, the number of clads per system vary from 32 (4.8 kg of ^{238}Pu) for one multi-mission radioisotope thermoelectric generator (MMRTG) system to 72 (10.8 kg) for one general-purpose heat source (GPHS) RPS system. The RPS Program Office (with DOE) has supported a one-time production of up to 22 clads in 1 year to support Mars 2020, but the currently planned production rates are 10 to 15 clads/year.

The transformation of source material into $^{238}PuO_2$ pellets, and subsequent fabrication of clads, requires significant infrastructure to develop and represents a major investment by NASA. The current remaining inventory from the DOE stockpile is ~30 kg of PuO_2, which can be mixed with newly produced material to meet the power specification required for RPS units. Production is expected to increase to 1.5 kg/year by 2026. With a rate of $^{238}PuO_2$ production of 1.5 kg/year, clads can be manufactured steadily at 10 clads/year; clad production can be increased by depleting the stockpile of older $^{238}PuO_2$. The committee commends the NASA/DOE approach to steadily produce $^{238}PuO_2$ clads independent of mission new starts or selections. However, the currently planned production rate remains a significant limiting factor in NASA's ability to develop new deep space missions. Based on the Planetary Mission Concept Studies, as well as the studies commissioned by this decadal survey, RPS demand far outweighs availability in the upcoming decade. With a regular cadence of New Frontiers and flagship missions launched, and assuming the use of next gen mod 1 radioisotope thermoelectric generator (RTGs) (requiring 64 $^{238}PuO_2$ clads) (Zakrajsek 2021), the committee derives an example scenario in Figure 20-2; this modest power scenario ignores any demands from the Discovery, Lunar, or Mars programs, human exploration needs or other NASA programmatic needs, and assumes 20 kg of ^{238}Pu is available now. Even with a steady 1.5 kg/year production, the type and number of missions that can be flown will be constrained by the number of available clads, far less than the desired program of missions, until the current PuO_2 inventory is exhausted. After that point, production of ^{238}Pu will be the rate-limiting step.

> **Recommendation: NASA should evaluate plutonium-238 production capacity against the mission portfolio recommended in this report and against other NASA and national needs, and increase it, as necessary, to ensure a sufficient supply to enable a robust exploration program at the recommended launch cadence.**

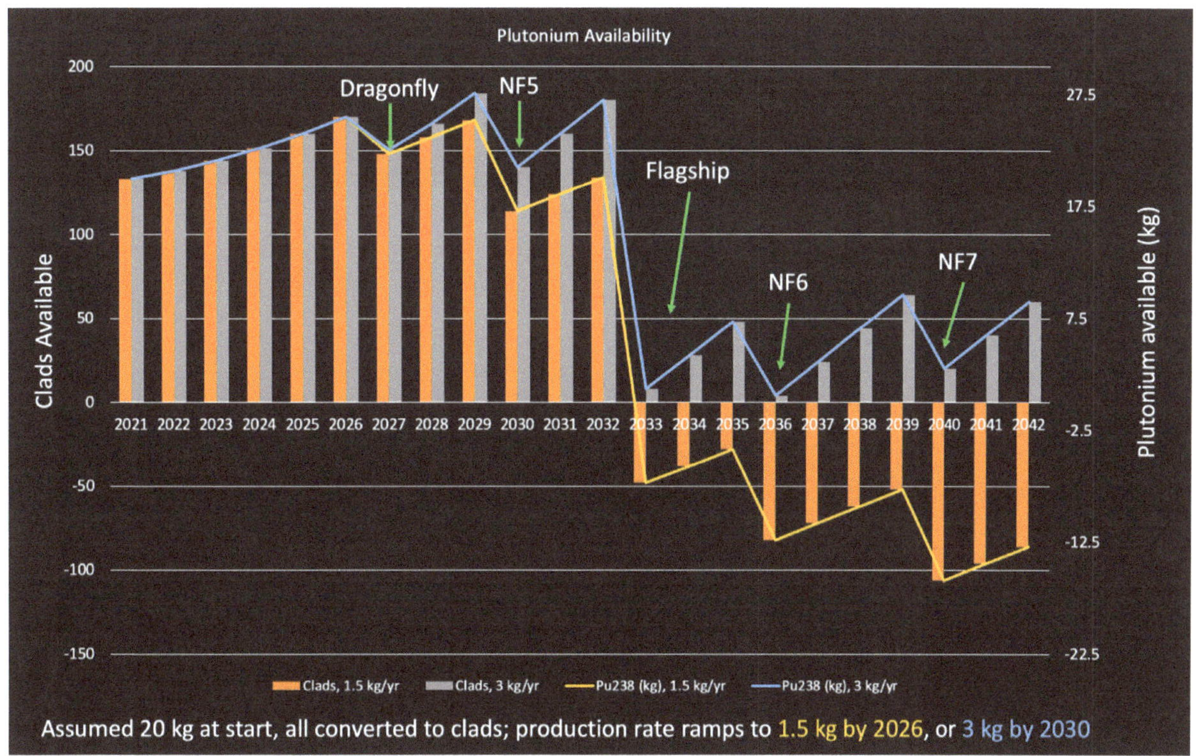

FIGURE 20-2 Example illustrating the need to increase production of $^{238}\text{PuO}_2$ above 1.5 kg/year. With the cadence assumed here (Dragonfly with one multi-mission radioisotope thermoelectric generator (MMRTG), New Frontiers (NF) opportunities with one next gen mod 1 radioisotope thermoelectric generator (RTG) each, and a flagship with three next gen mod 1 RTGs), a steady 1.5 kg/year production is insufficient for even a modest program of missions.

New technology investments may also help to mitigate the dearth of the ^{238}Pu supplies in certain scenarios. For example, the RPS Program Office is developing a dynamic RPS system. Such systems could increase the thermal to electric efficiency by a factor of ~4 over current GPHS RPS technology. Such units are not likely to be available for long duration missions endorsed by the current decadal study, but a demonstration of a dynamic RPS for a mission of shorter duration could pave the way for future missions in later decades with a significantly lower demand for ^{238}Pu.

Recommendation: NASA should continue to invest in maturing higher efficiency radioisotope power system technology to best manage its supply of plutonium-238 fuel.

Launch Services

Launch vehicle availability and capability continue to pose a challenge to NASA's program of planetary exploration. The workhorse Atlas V and Delta IV vehicles are expected to be retired and will no longer be available for this decadal survey's prioritized missions. The primary launch vehicles likely to be available include the existing Falcon 9 series, both Recoverable and Expendable versions, and the Vulcan Centaur; NASA will need to be agile in selecting appropriate options for planetary exploration to enable missions both small and large, near and far. The currently available vehicles can support missions to the Moon, the inner solar system, and Mars, but continued exploration of the outer solar system requires multiple gravity assists, and often long cruise durations, or high-performance propulsion systems, in-orbit assembly, or use of high-performance launch vehicles (high C_3, i.e., the square of the hyperbolic excess velocity; Figure 20-3). Studies of outer solar system missions show that the use of singular Jupiter gravity assists,

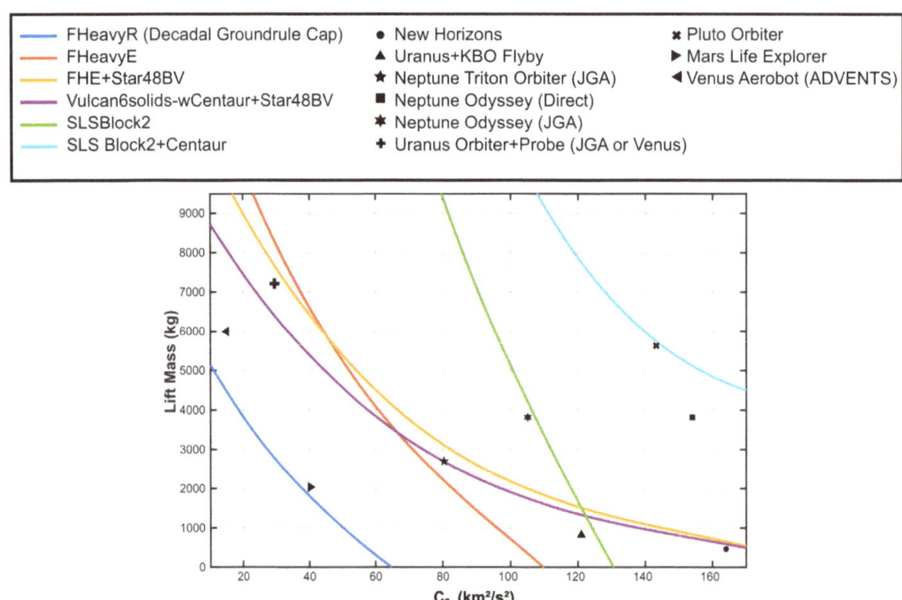

FIGURE 20-3 Performance comparison curves for launch vehicles proposed to be available during the time period covered by this decadal survey; the heavy lift options (SLS, Vulcan) are not yet ready for flight, constraining distant and/or large mission launches. The curves show how each vehicle's payload lift mass varies with the C_3 parameter, and it is necessary that the mission fit below the desired vehicle's curve, with margin. Also shown are the required C_3 values for several mission concepts studied under PMCS and as part of this decadal survey. NOTE: FHeavy, Falcon Heavy; JGA, Jupiter Gravity Assist; KBO, Kuiper belt object; SLS, Space Launch System.

which enables larger spacecraft and/or shorter flight times, tightly constrains launch windows for large missions and is not an optimal solution. Commercial ride shares, emerging small launch providers, and commercial delivery services (e.g., the Commercial Lunar Payload Services) may also provide other opportunities for some destinations.

> **Recommendation: NASA should develop a strategy to focus and accelerate development of high energy launch capability, or its equivalent, and in-space propulsion to enable robust exploration of all parts of the solar system. Any new systems that are developed should also build the pedigree required to permit the launch of nuclear materials.**

Data Archiving and Distribution

In 2020, NASA convened an independent review board (IRB) to examine the state of the Planetary Data Ecosystem (PDE) which includes all aspects of data collection and archiving. Rather than repeat that effort, the committee refers to key recommendations from the IRB's report (NASA 2021) where appropriate.

NASA missions and most R&A programs require a data management plan (DMP) that includes the timely archiving of all raw and calibrated data, mission and instrument documentation, and calibration procedures. The production and archiving of higher-level science products have been encouraged, but these activities are often completed only after the mission ends. DMPs should ensure that ground-based datasets are archived in a manner compliant with findability, accessibility, interoperability, and reusability criteria (Wilkinson et al. 2016), in high level (reduced) formats useful for planetary researchers. This is particularly important for observations of time-variable targets, where future discoveries can generate the need to access much older data for comparison.

> **Finding:** An index covering all planetary science, astrobiology, and field sample databases, would assist in data sharing and analysis. This index could include links to relevant spectral databases (e.g., USGS SpecLib, RELAB, GEISA, HIITRAN), other dedicated data archives (e.g., MAST, ASTROMAT), and sample archives that are available for outside users and/or collaboration.

Planetary Data System

The Planetary Data System (PDS) has been a pioneering repository and archive of planetary mission data for many decades. Development of the PDS4 Information Model to handle the ever-expanding amounts of mission data, to make it more user accessible, and to provide advanced tools for database access and analysis should be continued. The International Planetary Data Alliance has endorsed the PDS4 standard, which is being utilized by many agencies to ensure interoperability of archives and data sets. The PDS now accepts a range of nonmission datasets, including ground-based observations and planetary analog laboratory and field measurements. PDS does offer some archiving training, including a data users workshop every 2 years, but additional training opportunities with wide community reach would aid nonmission data providers with little archiving experience.

Finding: As recommended in the IRB PDE report, "regular, accessible, and effective training programs for researchers" is needed for "data producers, mission specialists, and others who need to archive with the PDS."

Sharing of Software, Algorithms, and Model Output

NASA DMPs increasingly require that most software, algorithms, and model output generated under NASA funding be made publicly available for wider use. The potential benefits are profound: enabling new users to work without needing to redevelop existing tools; allowing users to develop more advanced codes rather than start from scratch; enabling existing model output to be used for a whole range of new purposes. Examples include retrieval algorithms that extract temperature profiles from radiance data; or taking output generated by climate models to study weather patterns and instead using it as input to chemistry models, to assess entry-descent-landing risks, or for outreach purposes. Such repurposing of existing tools and output would avoid duplication of effort and save NASA scientists, engineers, and EPO officers significant time and money.

Finding: As recommended in the IRB PDE report, a plan is needed "for the preservation of models and model output, beginning with requirements for how these should be preserved and linked to other Ecosystem elements."

However, there are concerns over how to provide access to multiple iterations of software that are continuously evolving and improving, and how to preserve and distribute the huge datasets (often Terabytes in size) generated by model simulations, when new and improved simulations may render earlier datasets obsolete. Planetary research, particularly concerning time-varying phenomena or hard-to-observe bodies, continues to use data more than 100 years old. By contrast, the longevity of software code may be limited by the compilers, hardware, and operating systems needed to run it, and/or superseded by more sophisticated code. Similarly, model output may be superseded by that from more sophisticated modeling systems.

Finding: Community awareness of and access to the best software and model output datasets is needed, not necessarily preservation of every code, simulation, or output version indefinitely.

Preserving Model Data

Model input and output data generated by NASA funding is not currently accepted by the PDS. Examples of this type of data include N-body simulations, hydrocode impact simulations, atmospheric circulation modeling, and magnetosphere modeling. Such data are valuable for a diverse range of NASA-supported activities, often different from those for which the modeling was originally performed. These include assessing atmospheric and radiation risks for human exploration of Mars, testing theories of planetary formation, exploring the capabilities and requirements of proposed planetary missions or instruments, and advanced visualization. Hosting model output data in community-recognized repositories would allow them to be easily located. Making these repositories also discipline-specific, with mandated data and metadata formats, would allow them to be supplemented by generalized tools designed to enable easier access to and sampling of these often-huge datasets (Newman et al. 2020).

Repositories of model output can also be designed to contain only the most recent complete outputs from a particular model, with earlier datasets simply documented for posterity.

Finding: A clear plan for the preservation and sharing of planetary model input and output data, involving both the PDS and planetary modeling communities, is needed to develop a network of discipline-specific, community-recognized repositories, rather than via rigorous, costly, and often unnecessary archiving.

Sharing Software and Algorithms

Descriptions of methods and simple algorithms may be documented with publications, but more complex algorithms and software require more elaborate documentation and repositories. Minimum requirements ensure results from ad hoc analysis code can be replicated or at least understood, and that specific NASA-funded software development efforts (including mission data processing pipelines) comply with NASA open-source software policies, for example using repositories such as the NASA-managed GitHub; exceptions are made for code containing proprietary or ITAR/EAR restricted material. However, software developers span a range of effort levels, from those satisfying the minimum DMP requirements to those who plan to provide training and ongoing support to all users wishing to use their codes, including regular updates to the code base, user's guides, and tutorials.

Finding: Grant programs that offer the ability to fund community-oriented software projects need to also consider code maintenance, documentation, and user support.

Other Specialized Services

The Navigation and Ancillary Information Facility (NAIF), which operates under the PDS, is critical for maintaining spacecraft mission SPICE kernels that contain all the ancillary information relating to how data from a particular spacecraft instrument was collected. NASA planetary missions are expected to adopt SPICE as a standard during mission planning, operations, and archiving for consistency and usability with existing tools. The committee also encourages the availability of ephemeris and visualization tools, such as JPL Horizons and the PDS Rings Node tools, for more casual use. The availability of higher-level pointing and ancillary information in the mission data itself—either in the PDS4 files, or in higher level search tools, such as PDS OPUS, PDS Imaging Atlas, and other community tools—is a valuable resource for the community. Lastly, The IAU Minor Planet Center's (MPC's) role as the worldwide repository for positional measurements of small bodies and responsibility for their initial orbit computation is crucial for planetary defense efforts to identify and track NEOs. The Center for NEO Studies utilizes the MPC data to compute high-precision orbits, produces comprehensive assessments of Earth impact probabilities, and hosts the results publicly, maintenance of which are vital to inform global planetary defense efforts (see Chapter 18, Planetary Defense).

Supercomputing Facilities

Many research tasks require large computational resources including circulation modeling, N-body simulations of dynamical processes and solar system formation, as well as other high-resolution modeling and computations. With advances in neural net processing and machine learning techniques, a desire to run models at higher spatial and/or temporal resolution to delve deeper into processes and phenomena, as well as the collection of increasingly large quantities of data, computational resource demands are higher than ever. In addition to individual and institutional resources, NASA operates multiple advanced supercomputer clusters, some of which are available for science community use. Access to these systems is primarily granted in conjunction with R&A proposal requests, and the committee supports their use and continued modernization and expansion of capabilities in the future.

SUPPORTING NSF INFRASTRUCTURE

NSF has supported and continues to support a wide variety of current and future astronomical facilities as well as laboratory and other facilities. NASA and NSF recently reaffirmed their commitment to partnering on space research activities in a new 2020 memorandum of understanding (NASA-NSF 2020). This includes the use of NSF-managed stations and facilities in the Antarctic, as well as collaborative science and technology programs. Here the committee highlights several prominent facilities, but the list is not all-inclusive.

Existing Ground-Based Astronomical Facilities

The NSF's National Optical-Infrared Astronomy Research Laboratory (NOIRLab) provides the mainstay of ground-based telescopic science for astronomy. NOIRLab has five programs, some of which had previously operated under a different management structure: The Cerro Tololo Inter-American Observatory (CTIO) in Chile; the Community Science and Data Center (CSDC), International Gemini Observatory (operating in Hawaii and Chile); the Kitt Peak National Observatory (KPNO) near Tucson, Arizona; and the Vera C. Rubin Observatory which is under construction and is expected to become operational in coming years.

The NOIRLab southern-hemisphere assets, including telescopes at CTIO and Gemini South, provide a unique opportunity for U.S. researchers to perform ground-breaking research in the southern hemisphere on objects out of the ecliptic plane (e.g., comets). In the coming decade, it is expected that the CSDC will provide an increasing role in supporting the future of astronomy and planetary science as the amount of astronomical data, both PI-led and public survey data, has and continues to increase tremendously. NSF support for independent PI-led investigations in big data science will be important to satisfy growing needs over the next decade for development and implementation of new workhorse methodologies and algorithms.

The NSF-supported ecosystem of both 8–10 m class (e.g., Gemini) and 4 m class (e.g., KPNO) telescopes will continue to be important in the era of extremely large ground-based telescopes (see below). Smaller telescopes enable high-risk/high-reward observations, proof-of-concept studies for programs at larger telescopes, follow up on transient discoveries, provide testbeds for new instrumentation, enable longer-term observing campaigns, and provide training opportunities for students and early career researchers (Chanover et al. 2020). Flexible access options (e.g., fast-turnaround programs at Gemini) are geared toward publishable results on short timescales. These facilities are enabling for strategic research that happens on timescales faster than the mission development cycle, such as impact monitoring (Q4, Q7, Chapters 7 and 10), small body mutual events (Q1, Q4, Chapters 4 and 7), dynamic atmospheric events (Q6, Q7, Q8, Chapters 9–11), and others.

The Atacama Large (Sub)Millimeter Array (ALMA) is capable of high-resolution thermal and atmospheric observations of planetary objects such as the Galilean satellites and Titan, enabling important science, particularly in combination with and/or as a complement to space mission data. For example, global and long-term perspectives of Titan provided by ALMA—and ultimately by the next-generation Very Large Array (VLA; ngVLA, see next section)—will complement local studies to be performed by Dragonfly. ALMA and ngVLA wavelength sensitivities are well-suited for detection of complex organics in atmospheres and plumes, and thermal surface/subsurface anomalies important for understanding endogenic heat flow in outer planetary satellites. More broadly, ALMA observations of protoplanetary disks around other stars reveal important constraints on protoplanetary disk evolution and the initial stages of planet accretion (Q1 and Q12, Chapters 4 and 15) relevant to solar system studies.

Future Facilities

The Vera C. Rubin Observatory is currently under construction and will revolutionize planetary astronomy with an unprecedented inventory of solar system objects as well as time-domain astronomical investigations. The NSF has invested heavily in the construction and preparation of this facility. To provide the maximum benefit to the scientific community, the NSF needs to provide the greatest possible support to this facility (while balancing with other valuable NSF outlays) including, at the earliest opportunity, funding access for both Rubin consortium guided science as well as innovative PI-led science so the maximum benefit of the observatory can be achieved. This ought to include ample support for software infrastructure that can benefit all researchers on a rapid timescale, as well as PI-led science, to allow for new innovations as the many discoveries by the observatory sculpt and change our view of the solar system in the next decade.

The next generation of extremely large (larger than 20 m effective diameter) optical telescopes (ELTs) will see first light in the next decade. These include the European Extremely Large Telescope and the Giant Magellan Telescope (GMT) in Chile, and the Thirty Meter Telescope (TMT) in the northern hemisphere. Access to TMT and GMT for the full U.S. community will rely on NSF's US-ELT program. These observatories will offer unprecedented angular resolution, supported by robust adaptive optics instrumentation programs. Without NSF investment in guest observing programs at these next generation telescopes, the dramatically new science research that will be performed will only benefit from the skills, knowledge, and innovative perspectives from the subset of U.S. researchers who are members of TMT and GMT consortia. In particular, the US-ELT program could support key planetary science programs involving long-term time-domain science across all key questions, transient phenomena (Q5, Q6, Q7, Q8, Chapters 8–11), and small body characterization surveys (Q1, Q2, Q3, Q4, Chapters 4–7) (Wong et al. 2020).

> **Finding:** As stated in the *Pathways to Discovery in Astronomy and Astrophysics for the 2020s* decadal survey, the committee expects the ELT facilities to provide transformational planetary science, but only if the observing programs are adequately funded.

Construction in ngVLA is planned to start in the next few years, with eventual baselines as large as almost 9,000 km delivering spatial resolution at Neptune of ~130 km (de Pater et al. 2020). This facility promises advances in planetary science, particularly for the giant planets, building on prior radio science results from Juno, Cassini, the VLA, and ALMA. As with space astrophysical observatories, groundbreaking planetary research at future facilities will be aided by the purposeful inclusion of planetary astronomers within development teams and observatory staff (Hammel and Milam 2020).

> **Recommendation: NSF-supported, ground-based telescopic observations provide critical data to address important planetary science questions. The NSF should continue (and if possible expand) funding to support existing and future observatories (e.g., NOIRLab, ALMA, TMT, GMT, ngVLA) and related PI-led and guest observer programs. Planetary astronomers should be included in future observatory plans and development to maximize the science return from solar system observations.**

Laboratory and Other Facilities (including Antarctic)

The 2020 NASA and NSF MoU state their commitment to "continue working together to advance NASA and NSF—sponsored science programs . . . with special emphasis on those that make use of the NSF-managed facilities, including those in the Antarctic." A continued NASA-NSF Antarctic commitment, based on mutual understanding of needs and resources, is needed to maximize the use of facilities that can advance the scientific goals of both agencies. For example, NASA could consider having an NSF program manager or U.S. Antarctic Program representative participate in the review of Antarctic proposals to NASA programs.

> **Finding:** Joint NASA-NSF reviews of Antarctic research proposals and facilities would strengthen the science programs at both agencies.

Ground-Based Planetary Radar

Ground-based planetary radar observations have been primarily conducted at shared-use facilities, at the NSF facilities of Arecibo Observatory in Puerto Rico and Green Bank Telescope (GBT) in West Virginia. Previously, the planetary radar project at the Arecibo Observatory was supported by NASA through the Near-Earth Object Observations Program and recently through the Solar System Observations Program, tasked through the Planetary Defense Coordination Office, while the facility was NSF managed under a cooperative agreement. NASA supported all planetary radar observations at Arecibo, as well as partially supported facility maintenance and infrastructure. Although the Arecibo Telescope was lost, the GBT can continue to serve as a receiving antenna for GSSR. Additionally, GBT is currently considering the addition of a radar transmitter (Bonsall et al. 2019), providing much needed complementary observations to GSSR (see Planetary Defense chapter). To date, NASA and NSF have informally cooperated at Arecibo and GBT by leveraging their grantees and contractors as intermediaries. As the remaining facilities for ground-based planetary radar observations, GSSR and GBT are critical for planetary defense and science observations.

> **Recommendation: NASA and NSF should review the current radar infrastructure to determine how best to meet the community's needs, including expanded capabilities at existing facilities, to replace those lost with Arecibo.**

INTRA-AGENCY, INTERAGENCY, AND INTERNATIONAL COLLABORATIONS

Planetary science and astrobiology are multi-disciplinary, with associated fields including hydrology, geology, meteorology, microbiology, oceanography, heliophysics, astrophysics, and many more. Most of these areas of expertise have their own organizations, agencies, and research programs in every country across the world. Collaboration and sharing of infrastructure and ideas is therefore an important way to maximize efficiency and results, as recognized in NASA's *Explore: Science 2020–2024* (NASA 2020a). Here, the committee highlights a few key infrastructure partnerships and discusses areas for future collaboration.

Intra-Agency Facilities

Planetary scientists currently collaborate across all of NASA divisions, for example, with exoplanets science in both Planetary and Astrophysics divisions; space weathering, solar wind, magnetospheres interests in both the Planetary and Heliophysics divisions; and many others (Mandt et al. 2020). Another example is the overlap between resources used for human exploration and scientific uses: wind tunnels and environmental chambers may be used for spacesuit testing, as well as for instrumentation development and for scientific studies. Other examples of cross-divisional infrastructure include NASA's ground and space-based telescope facilities, supercomputers, and other assets.

> **Finding:** The committee encourages the continued involvement of the Planetary Science Division in developing, enabling, and supporting cross-divisional facilities with input from the community on the design, capabilities, and resource allocations from these facilities.

Interagency Facilities

In emerging disciplines such as ocean worlds and astrobiology, as well as in more established fields such as atmospheric and geologic processes, cross-agency interactions can provide additional insights, data, tools, and facilities to advance our knowledge. As a first example, the National Oceanographic Partnership Program facilitates partnerships between 16 federal agencies (including NASA), plus academia and industry, to advance ocean science research and education. As a second example, NASA Goddard's highly successful Heliophysics Data Portal, which facilitates the sharing and use of heliophysics data from both NASA and international missions, could serve as a template for repositories of other types of planetary output. Similarly, the Community Coordinated Modeling Center, which supports and performs space weather modeling, is a partnership between eight agencies (including NASA, NOAA, and NSF) and numerous modeling groups across different agencies and institutions; this paradigm can provide great insight to planetary modelers aiming to set up something similar.

> **Finding:** Already established, and newly emerging, mechanisms for facility and data collaborations across other federal science agencies can serve as a good model for future NASA collaborations. Such partnerships ought to span from theoretical modeling and simulations to data ecosystems to data analysis, laboratory experiments, and field investigations across multiple entities.

International Facilities

In mid-2021, the European Space Agency (ESA) released its *Voyage 2050* exploration themes for the period 2035–2050 (ESA 2021). One of the three large class mission themes is focused on exploration of the moons of the giant planets, in particular a goal of detecting biosignatures at an ocean world. In the medium-class mission category, recommendations included exploration of magnetospheres as a complex system, Venus geology and geophysics, as well as possible contributions to a U.S.-led ice giants or solar system origins mission. Continued ESA partnerships are vital to the planetary science community, as there is a long history of successful international cooperation on mission and instrument development, testing, and operations, for example Cassini-Huygens, the Hubble Space Telescope, and the upcoming ExoMars rover and Mars Sample Return, among many others; some of these missions used ESA-led test facilities, launch vehicles, or ground communications.

> **Finding:** The committee encourages NASA to continue the history of successful international cooperation to advance the goals of both NASA and ESA. Further new and emerging partnerships, with JAXA, CSA, ISRO, and others, for example, are also encouraged and would enable even broader planetary science opportunities.

REFERENCES

Abraham, D.S., B. MacNeal, D. Heckman, Y. Chen, J. Wu, K. Tran, A. Kwok, and C.-A. Lee. 2018. "Recommendations Emerging from an Analysis of NASA's Deep Space Communications Capacity." Paper AIAA 2018-2528, presented at SpaceOps Conference. https://doi.org/10.2514/6.2018-2528.

Anchordoquy, T.J., and M.C. Molina. 2007. "Preservation of DNA." *Cell Preservation Technology* 5:180–188.

Bonsall, A., G. Watts, J. Lazio, P. Taylor, E. Rivera-Valentin, E. Howell, F. Ghigo, et al. 2019. "GBT Planetary Radar System." *Bulletin of the AAS* 51(7). https://assets.pubpub.org/mwnvuyv4/31598545550747.pdf.

Burr, D.M., N.T. Bridges, J.K. Smith, J.R. Marshall, B.R. White, and D.A. Williams. 2015. "The Titan Wind Tunnel: A New Tool for Investigating Extraterrestrial Aeolian Environments." *Aeolian Research* 18:205. https://doi.org/10.1016/j.aeolia.2015.07.008.

Carrier, B.L., D.W. Beaty, A. Hutzler, A.L. Smith, G. Kminek, M.A. Meyer, T Haltigin, et al. 2022. "Science and Curation Considerations for the Design of a Mars Sample Return (MSR) Sample Receiving Facility (SRF)." *Astrobiology* 22(S1): S-217–S-237. https://doi.org/10.1089/ast.2021.0110.

Chanover, N., C. Schmidt, and D. DeColibus. 2020. "The Continued Relevance of 4m Class Telescopes to Planetary Science in the 2020s." White paper #497 submitted to the Planetary Science and Astrobiology Decadal Survey 2023–2032. *Bulletin of the AAS* 53(4). https://doi.org/10.3847/25c2cfeb.752e4fa4.

de Pater, I., C. Moeckel, J. Tollefson, B. Butler, K. de Kleer, L. Fletcher, M.A. Gurwell, et al. 2020. "Prospects to Study the Ice Giants with the ngVLA." White paper #063 submitted to the Planetary Science and Astrobiology Decadal Survey 2023–2032. *Bulletin of the AAS* 53(4). https://doi.org/10.3847/25c2cfeb.029d5009.

Dudzinksi, L. 2021. "RPS Program Technology Updates." Presentation to the Committee on Planetary Science and Astrobiology Decadal Survey. August. Washington, DC: National Academies of Sciences, Engineering, and Medicine.

ESA (European Space Agency). 2021. *Voyage 2050: Final Recommendations from the Voyage 2050 Senior Committee.* Paris, France. https://www.cosmos.esa.int/documents/1866264/1866292/Voyage2050-Senior-Committee-report-public.pdf/e2b2631e-5348-5d2d-60c1-437225981b6b?t=1623427287109.

Greeley, R., and J.D. Iversen. 1985. *Wind as a Geological Process on Earth, Mars, Venus, and Titan.* Cambridge, UK: Cambridge University Press.

Hammel, H., and S. Milam. 2020. "A Lesson from the James Webb Space Telescope: Early Engagement with Future Astrophysics Great Observatories Maximizes Their Solar System Science." White paper #214 submitted to the Planetary Science and Astrobiology Decadal Survey 2023–2032. *Bulletin of the AAS* 53(4). https://doi.org/10.3847/25c2cfeb.c23241af.

Helbert, J., A. Maturilli, M.D. Dyar, and G. Alemanno. 2021. "Deriving Iron Contents from Past and Future Venus Surface Spectra with New High-Temperature Laboratory Emissivity Data." *Science Advances* 7(3):eaba9428.

Herd, C.D.K., R.W. Hilts, A.W. Skelhorne, and D.N. Simkus. 2016. "Cold Curation of Pristine Astromaterials: Insights from the Tagish Lake Meteorite." *Meteoritics & Planetary Science* 51:499–519.

Johnston, M.D. 2020. "Scheduling NASA's Deep Space Network: Priorities, Preferences, and Optimization ICAPS 2020." https://icaps20subpages.icaps-conference.org/wp-content/uploads/2020/10/SPARK-2020_paper_1.pdf.

Lazio, J., B. Arnold, B. Giovanelli, M. Levesque, J. Berner, and A. Smith. 2020. "The Deep Space Network Status and Future." Presentation to Giant Planet Systems Panel. November. Washington, DC: National Academies of Sciences, Engineering, and Medicine.

Lichten, S.M. 2021. "NASA Deep Space Network Resource Loading in the 2020s—Update 23—Feb- 2021." Presentation to Decadal Survey Ocean Worlds Dwarf Planets Panel. February. Washington, DC: National Academies of Sciences, Engineering, and Medicine.

Lukco, D., D.J. Spry, R.P. Harvey, G.C.C. Costa, R.S. Okojie, A. Avishai, L. M. Nakley, P.G. Neudeck, and G.W. Hunter. 2018. "Chemical Analysis of Materials Exposed to Venus Temperature and Surface Atmosphere." *Earth and Space Science* 5:270–284. https://doi.org/10.1029/2017EA000355.

Mackwell, S.J., et al. 2016. "Review of Currently Funded SMD Planetary Facilities." https://www.lpi.usra.edu/psd-facilities.

Mandt, K., A. Rymer, J. Kalirai, R. Allen, A. Cocoros, K. Stevenson, D. Hurley, et al. 2020. "Advancing Space Science Requires NASA Support for Coordination Between the Science Mission Directorate Communities." White paper #414 submitted to the Planetary Science and Astrobiology Decadal Survey 2023–2032. *Bulletin of the AAS* 53(4). https://doi.org/10.3847/25c2cfeb.53e7ca7e.

NASA (National Aeronautics and Space Administration). 2020a. *Explore: Science 2020–2024: A Vision for Science Excellence.* Washington, DC. https://science.nasa.gov/science-pink/s3fs-public/atoms/files/2020-2024_Science.pdf.

NASA. 2020b. "NASA's Planetary Science Portfolio." IG-20-023. Washington, DC: NASA Office of Inspector General Office of Audits. September 16.

NASA. 2021. "Final Report of the Planetary Data Ecosystem Independent Review Board." https://science.nasa.gov/science-pink/s3fs-public/atoms/files/PDE%20IRB%20Final%20Report.pdf.

NASA and NSF (National Science Foundation). 2020. "Memorandum of Understanding Between the National Aeronautics and Space Administration and the National Science Foundation Regarding Achievement of Mutual Research Activities Advancing Space, Earth, and Biological Sciences." https://www.nasa.gov/sites/default/files/atoms/files/2020_nasa-nsf_mou.pdf.

NASEM (National Academies of Sciences, Engineering, and Medicine). 2019. *Strategic Investments in Instrumentation and Facilities for Extraterrestrial Sample Curation and Analysis.* Washington, DC: The National Academies Press. https://doi.org/10.17226/25312.

Newman, C., V. Airapetian, M. Battalio, S. Bougher, A. Brown, S.D. Domagal-Goldman, S. Fan, et al. 2020. "An Urgently Needed Repository for Planetary Atmospheric Model Output." White paper #505 submitted to the Planetary Science and Astrobiology Decadal Survey 2023–2032. *Bulletin of the AAS* 53(4). https://doi.org/10.3847/25c2cfeb.6974fd2e.

NRC (National Research Council). 2011. *Vision and Voyages for Planetary Science in the Decade 2013–2022.* Washington, DC: The National Academies Press. https://doi.org/10.17226/13117.

Rissanen, A.J., E. Kurhela, T. Aho, T. Oittinen, and M. Tiirola. 2010. "Storage of Environmental Samples for Guaranteeing Nucleic Acid Yields for Molecular Microbiological Studies." *Applied Microbiology and Biotechnology* 88:977–984.

Sutliff, T., P.W. McCallum, and S.G. Johnson. 2020. "Establishing a Supply of Pu-238 and Associated Radioisotope Power Systems Capabilities and Policy Improvements—A Multi-Part Success Story." White paper #455 submitted to the Planetary Science and Astrobiology Decadal Survey 2023–2032. *Bulletin of the AAS* 53(4). https://doi.org/10.3847/25c2cfeb.735c2861.

Swann, C., D.J. Sherman, and R. Ewing. 2020. "Experimentally-Derived Thresholds for Windblown Sand on Mars." *Geophysical Research Letters* 47. https://doi.org/10.1029/2019GL084484.

Wilkinson, M.D., M. Dumontier, I.J. Aalbersberg, G. Appleton, M. Axton, A. Baak, N. Blomberg, et al. 2016. "The FAIR Guiding Principles for Scientific Data Management and Stewardship." *Scientific Data* 3:160018. https://doi.org/10.1038/sdata.2016.18.

Wong, M., K. Meech, M. Dickinson, T. Greathouse, R.J. Cartwright, N. Chanover, and M.S. Tiscareno. 2020. "Transformative Planetary Science with the US ELT Program." White paper #490 submitted to the Planetary Science and Astrobiology Decadal Survey 2023–2032. *Bulletin of the AAS* 53(4). https://doi.org/10.3847/25c2cfeb.6e93d41f.

Zakrajsek, J. 2021. "RPS Program Technology Updates." Presentation to the Ocean Worlds Dwarf Planets Panel. March. Washington, DC: National Academies of Sciences, Engineering, and Medicine.

Zeigler, R.A., D. Edey, R. Hanna, S.A. Eckley, R.A. Ketcham, J. Gross, and F.M. McCubbin. 2020. "Using X-Ray Computed Tomography to Image Apollo Drive Tube 73002." American Geophysical Union Fall Meeting 2020, abstract #V017-03.

21

Technology

Technology is the foundation of scientific exploration. Two Voyager spacecraft were launched over four decades ago, now far beyond any human made object, and are still returning valuable scientific data about the interstellar medium. This achievement was enabled by new technologies such as the radioisotope thermoelectric generators, instruments, and high reliability electronics. The term "technology" has many meanings in current literature, but for the purposes of this decadal survey, technology means "the systematic application of scientific or other organized knowledge to practical tasks" (Galbraith 1978) In this way, technology applies to everything it takes to develop and deliver science missions into the solar system to perform science investigations. Today, our current technological capabilities are considered "state of the art." However, there are state of the art technologies that exist today (e.g., flight processors) that have not yet been adopted to improve the science return of planetary missions. As we progress, we are constantly developing new or improved technologies (any set of productive techniques which offer a significant improvement in terms of higher resolution, unique techniques, survival in extreme environments, reliability, or lower costs over the state of the art) to answer science questions that were previously out of reach.

New science findings create new questions. Answering these questions has made science investigations more demanding, requiring the advancement of technology at a pace challenging our experience. Planetary science presents unique challenges in terms of instrumentation, power generation, mass control, miniaturization, propulsion, precision landing, hazard avoidance, communication, and extreme environments to name a few. Enabling planetary science missions requires early and substantial technology investments.

The *Vision and Voyages for Planetary Science in the Decade 2013–2022 (V&V)* decadal survey recommended that "a substantial program of planetary exploration technology development should be reconstituted" and "consistently funded at approximately 6 to 8 percent of the total NASA Planetary Science Division (PSD) budget" (NRC 2011). The NASA PSD has been generally successful in implementing these recommendations and has made good strides in developing technologies for this coming decade. Terrain Relative Navigation (TRN), precision landing, organics detection, sample collection and preservation, imaging and target characterization, hazard detection and avoidance (HD&A), are perfect examples.

This decadal survey offers observations about the progress made in technology management and development over the past decade, and identifies areas where improvements can further enable the exciting science missions identified in this decade and those that follow. Just as the science landscape changes, so does the technology landscape. Technological advances are being made both within NASA as well as in other government, academia, and industry. Such advances can be better leveraged if the technology roadmaps are assessed and adjusted to take advantage of them as they emerge. A technology development program that holistically oversees this work can ensure the highest return on investment.

TECHNOLOGY DEVELOPMENT IN NASA

The approach to technology development within NASA has changed multiple times through the decades from centralized control, to distributed management, to the present-day hybrid approach. At the beginning of the *V&V* decadal survey, technology development within SMD was uncoordinated and lacked focus, and the *V&V* decadal survey recommended that the PSD reestablish a cohesive technology program. At that time the Agency also did not have a focused mission directorate for technology. During the *V&V* period, PSD implemented the *V&V* recommendations and established several new programs, while NASA established the Space Technology Mission Directorate (STMD). The *V&V* midterm review found that NASA had made substantial progress in meeting the technology goals of *V&V* (NASEM 2018).

As *V&V* was being developed, NASA had taken steps to improve the way it approached technology development across the full scope of work in the mission directorates. In the fiscal year 2011 budget request, NASA requested separate funding for space technology, and its budget has grown considerably since then. STMD develops crosscutting, pioneering new technologies and capabilities necessary for NASA to achieve current and future missions. These high-payoff, revolutionary technologies are rapidly developed, demonstrated, and integrated through collaborative partnerships. STMD invests in and matures broadly applicable, disruptive technology not currently available in industry, to meet the needs of future NASA missions in science and exploration, as well as to lower the cost for other government agencies and commercial space initiatives. Today, STMD is largely focused on advancing technologies and testing new capabilities at the Moon that will be critical for crewed missions to Mars: The Moon is essentially serving as a technology testbed and proving ground for Mars.

In 2017 PSD established the Planetary Exploration Science Technology Office (PESTO) at Glenn Research Center (GRC) to (1) manage planetary technology investments that are not yet specific to a mission in development; (2) coordinate planetary-relevant technology investments across PSD and other NASA organizations; and (3) meet the goal of maximizing infusion into specific missions. This office recommends annual technology investment strategies (considering investments being made by STMD and other NASA and government organizations) and updates the roadmap to achieve the strategic goals, based primarily on future mission needs. PESTO manages the development of technologies that have not yet been adopted by missions (except nuclear systems) and fosters a coordinated technology investment portfolio across NASA by collating and tracking existing investments across the Agency to communicate needs, identify gaps and promote cross-directorate collaboration. Last, PESTO promotes technology infusion by communicating technology development results to mission planners, facilitating technical exchanges between mission engineers and technologists throughout the development life cycle, hosting technology reviews, and tracking infusion success stories.

To assess the efficacy of NASA's technology efforts, the committee researched best practices for technology strategies and approaches and developed a flow diagram, Figure 21-1, to use in assessing how well NASA's technology efforts reflect best practices. In addition, best practices in technology development programs reflect principles (see Box 21-1) that guide the program's execution and facilitate the assessment.

A comparison of NASA's technology efforts with best practices revealed areas where NASA can improve. These are shown as red and blue text in Figure 21-1 and are described in findings and recommendations below. The technology principles in Box 21-1 provide additional clarity on these best practices in managing an effective technology development portfolio that will enable additional and more advanced science investigations in the future.

Planetary Science Division Technology Development

Whereas the *V&V* midterm review (NASEM 2018) found that NASA's PSD efforts over the first half of the decade had generally met *V&V* technology goals, NASA has not been able to financially sustain the same percentage of funding over the past 5 years of the decade. During the second half of the decade, based on

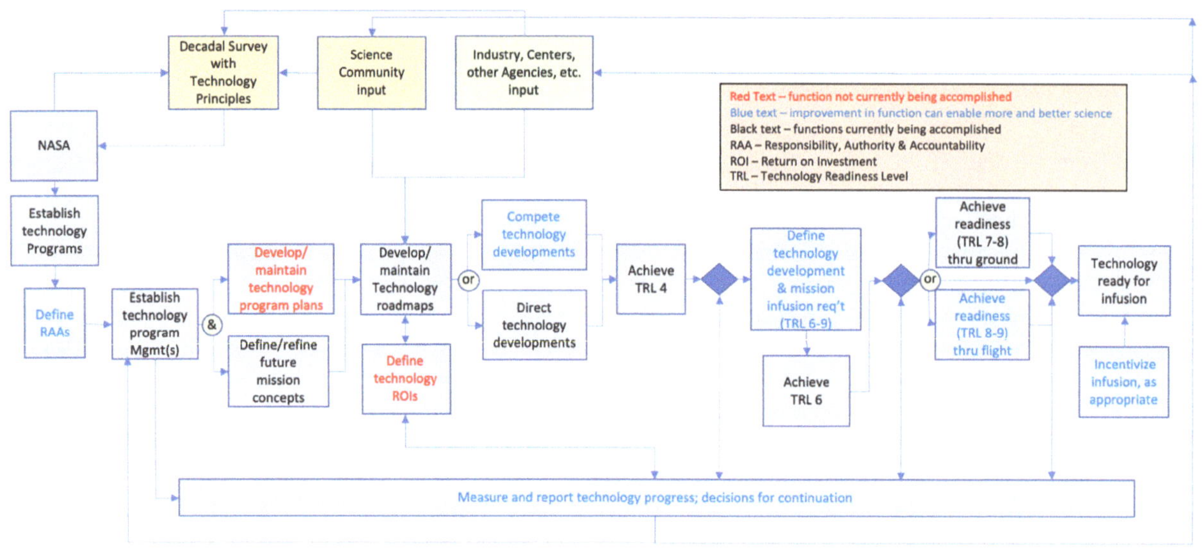

FIGURE 21-1 Technology management reflecting best practices.

BOX 21-1
Science Enterprise Technology Development Principles

- The technology strategy considers both short-term and long-term development efforts that encompass the upcoming decade and at least two decades after.
- Technology development covers the scope of the planetary science and astrobiology decadal survey, including planetary defense and planetary protection.
- Technology development investments enable accomplishment of priority decadal science goals and objectives, as defined in the strategic research (SR) objectives, when the current state of the art cannot satisfy those goals and objectives.
- Technology development investments improve the science enterprise return on investment (RoI)[a] by, for example, reducing mass, power, volume, cost, development time, increasing capability performance in both instrumentation and infrastructure, and/or enabling novel mission concepts.
- Technology development investments are balanced across the science disciplines consistent with the balance of decadal science goals, objectives and priorities and the Science Enterprise strategy.
- Technology development investments are prioritized by forecasted RoI.
- Technology development investments provide the ways and means to mature and transition evolving technologies into flight projects that benefit the Science Enterprise holistically.
- Emerging technologies with high RoI to the Science Enterprise that are near ready (Technology Readiness Level (TRL) 5–6) are incentivized to be incorporated in upcoming missions.
- Technology development allows the broadest participation possible to ensure the highest probability of success by engaging all organizations involved in planetary science and astrobiology through some form of competitive process that invites broad participation and ideas.
- Technology development may not always come to fruition and that is acceptable, but knowledge gained is documented.
- The measures of success of technology development are based on both qualitative and quantitative metrics.
- Technology plans and progress are shared transparently with all interested parties.

---

[a] RoI is based on factors such as: science return; cross-cutting improvements in sensor, instrument or infrastructure capabilities; technologies for single applications; or a combination of them.

NASA data, the level of funding fell short of the recommended 6–8 percent level with it declining to about 4 percent. The *V&V* recommended that it was critical that technology funding not be used to cover over-run costs of missions stating, "Reallocating technology funds to cover tactical exigencies is tantamount to "eating the seed corn." The National Academy of Sciences, National Academy of Engineering, and Institute of Medicine report *Rising Above the Gathering Storm: Energizing and Employing America for a Brighter Economic Future* stated: "At least 8 percent of the budgets of federal research agencies should be set aside for discretionary funding managed by technical program managers in those agencies to catalyze high-risk, high-payoff research" (NAS et al. 2007). This finding supports a similar level of PSD funding for the significant technology advancements that will be needed to accomplish strategic research and missions prioritized in this report (see Chapter 22).

Finding: NASA has not sustained the recommended level of planetary technology funding, 6–8 percent of the PSD budget, with the level declining to about 4 percent over the past 5 years. This is now significantly below the level of investment recommended in *V&V*.

Recommendation: NASA PSD should strive to consistently fund technology advancement at an average of 6 to 8 percent of the PSD budget.

Technology funding within PSD is distributed across numerous programs including competed programs within R&A (e.g., PICASSO, MatISSE, and DALI; see Table 17-1), as well as for example, the Mars Exploration Program, the Lunar Discovery and Exploration Program, and the icy satellites surface technology program. As a result, the recommendations in this report for 10 percent of the PSD budget to be devoted to R&A (Chapter 17) and 6–8 percent to be invested in technology development contain significant overlap.

Progress toward the recommended funding level is difficult owing to the lack of a centralized accounting of technology funding throughout PSD. Currently, there is no single focal point in PSD that oversees and approves all technology development activities. The establishment of PESTO as a lead organization has been a good first step, but PESTO does not have responsibility for coordination of all technology work in PSD. The committee was unable to identify who in PSD has the authority to make technology investment decisions. This approach has led to a technology program that is piecemeal; that is, it is managed at individual technology program level as opposed to one with an integrated strategy for the advancement of planetary and astrobiology science. In fact, the committee learned that there was no technology plan in PSD that described, for example, how technology was managed, who was involved, how those organizations worked together, and who made decisions. As a result, there appears to be no way for PSD to effectively consider how the various technology efforts can be integrated holistically such that the combination of certain technologies would provide even higher returns on investment. In addition, there appears to be no document with STMD that describes how PSD and STMD work together to coordinate and exercise their respective roles and responsibilities. This lack of documented agreement, particularly between PSD and STMD, inhibits the execution of a cohesive technology development program that ensures technology investments improve the ability of all missions, regardless of destinations, to meet their science goals with improved capabilities.

The lack of documented structure and agreements also inhibits the science community and supporting communities from understanding the technology development strategy for planetary and astrobiology future endeavors. A transparent approach, with all program level documentation available to all interested parties, would help the technology program to take advantage of the breadth and depth of creativity available in the science and supporting communities, particularly because the proposal selection rates are so low. As it stands, creativity is stymied by a piecemeal solicitation approach, and new ideas are difficult to bring forward. Coupled with this, there appears to be no obvious source for understanding the status of important technology developments. PESTO does publish some information on its website, but it does not appear to be complete.

Last, the relationship between the science community and the NASA technology program could be strengthened if the technology program regularly reported progress at both major conferences and the discipline analysis/assessment groups (e.g., Mercury, Venus, Lunar, Mars, Small Bodies, and Outer Planets). Reporting of technology progress metrics, such as the return on investment (RoI) metric, suggested below, would contribute substantially to a transparent organizational approach, which valued the input and feedback of its stakeholders.

Finding: The committee found it difficult to uncover what technology activities were currently active and how much funding was being allocated to technology development, an issue that was also identified in the *V&V* midterm review (NASEM 2018). Transparency is important to the science community as they plan for and develop approaches to accomplishing the next set of science objectives so that their implementation approaches can take advantage of the technology work being pursued by PSD and STMD.

Finding: The charter for PESTO includes the responsibility to "Work with partnering organizations to develop partnership agreements and approaches," but the committee could not find any evidence that there were documented agreements with its major partner, STMD, where STMD committed to provide PSD with its fair share of the STMD resources and how those resources were to be expended in pursuit of PSD technology needs. At present, PSD is dependent on STMD to provide technology flight opportunities for those technologies important to the pursuit of planetary science and astrobiology. However, STMD's current focus is on enabling the human exploration of the Moon, and there is no mechanism or agreement with PSD that ensures STMD resources are allocated to PSD's needs consistent with PSD's percentage of the agency's budget.

Finding: The committee could not find evidence that PSD or PESTO had documented how the PSD technology program is managed and executed, with the only evidence being the PESTO charter and many individual technology efforts documented in solicitations like ROSES. As a result, the science community and the supporting organizations, such as industry, other government agencies, and so on have only limited visibility into PSD's technology program, mostly through personal connections and advertisements for technology efforts (e.g., ROSES) thus making this program less than fully transparent.

Finding: As the single focal point for technology program management within PSD, PESTO needs to be cognizant of all technology efforts important to accomplishing the science priorities in the decadal survey. Based on the charter, it appears that several important technology areas are missing from their purview, such as planetary defense and planetary protection.

Recommendation: The PSD technology program should create a PSD Technology Program Plan that provides the details on what the program goals are, how the program operates, who is involved, and how the science community and supporting organizations can play a role. This plan should include how plans, funding levels, solicitation approaches, including selection rates, and results are communicated to the community at large. This plan should be prominent on the PSD PESTO website and updated annually. Based on PESTO's charter, this office should be cognizant of all technology efforts related to planetary science, astrobiology and planetary defense and could serve as the single organization responsible for all technology development or as a minimum for integrating all technology development.

Recommendation: PSD should establish a standard mechanism for the science community and other relevant organizations to provide input into PSD on technology needs, including new and creative approaches to technology, similar to how the science community provides input through the various science analysis/assessment groups (AGs). Two possible examples could be a PSD Technology AG, similar to the science AGs, or a collaboration among existing AG technology leads. A mechanism of this sort would be an effective way to increase transparency in the technology program.

Technologies in development typically fall into 3 categories:

- Enabling—those technologies that currently do not exist but are necessary to accomplish the science priorities;
- Enhancing—those technologies which can improve the science RoI by reducing resources required or improving performance so that more science can be accomplished for the same amount of dollars; and
- Dormant—those technologies that may be sufficiently advanced to use in missions but are not yet accepted by the implementers of science missions.

Enabling technologies, by their nature, are the highest priority for investment but the remaining technologies, enhancing and dormant, reduce future costs and increase future mission performance. However, there does not currently appear to be a way to know how significant these technologies could be from a science return on investment. This lack of information impairs PSD's ability to increase the amount of science that is ultimately pursued.

Recommendation: PSD should develop a set of return on investment metrics that guide the investment and encourage incorporation of technologies. These metrics should be transparent to the planetary science and astrobiology community.

The maturation of technologies in the aerospace community from the early concept through actual use in a space mission have been defined in nine technology readiness levels (TRLs). NASA requires that any new technologies be at TRL level 6 before a mission's preliminary design review to ensure the successful incorporation of that technology. *V&V* identified the transition of technologies from TRL-4 to TRL-6 as a "valley of death," where there was no mechanism to bring these technologies to a level of maturity needed for insertion in flight projects. PSD embraced this recommendation and created the MatISSE program to help solve this obstacle for instrumentation and worked with STMD to include planetary technologies in their flight project technology lines. These changes have been very beneficial. Now, NASA's technology development efforts are geared to bringing new technologies to TRL-6 with the expectation that flight projects will bring those technologies from TRL-6 to flight readiness status (TRL-8) and fly them. While this strategy works for many technology developments, in some cases, when a TRL-6 technology is evaluated for insertion by a flight project during its early phases (e.g., pre-Phase A), the technology might be deemed too programmatically and/or technically risky to be included in the mission. There are several reasons that might lead to this situation. Among the most important are (1) despite NASA best efforts, TRL definitions still have a certain degree of ambiguity that might result in a premature conclusion of a technology development task, leaving too much scope for a flight project to accomplish within its resources; and (2) not all technologies at TRL-6 are created equal. Some take more resources and risks to mature them than flight projects can afford. This has created a second obstacle where technologies judged to be insertable at TRL-6 are not being used (e.g., aerocapture).

PSD and SMD have established some specific programs, such as the Mars Exploration Program, that can and do consider integration of important technologies into flight projects, and these programs have been successful in incorporating the technologies. Mars 2020 is a good example; however, other science missions without a parent program do not have these opportunities. NASA has instituted a process in its competed mission lines, where some technologies can be added to the mission without penalty by stipulating in the Announcement of Opportunity (AO) specific technologies. A great example is the Psyche Discovery mission, which is flying an optical communication experiment as part of its payload to bring that technology to a TRL of 9. This approach is creative, and the committee encourages NASA to continue this approach and even consider expanding it. Unfortunately, the list of technologies to be included has been limited, leaving many technologies without a mechanism for integration in flight projects.

Finding: There are several important technologies that could improve PSD's science return on investment that are not being integrated into flight projects because they are deemed too risky by the flight projects.

Recommendation: This second obstacle (technology at TRL-6 deemed too risky) should be addressed by PSD, and a solution implemented that considers the long-term return on investment of all technologies under development.

Solutions could include:

- Directing some technologies to be used or providing incentives for using technologies in this category, such as increasing the number of technologies offered in AOs; allowing technology demonstration mission in SIMPLEx AOs; or similar approaches in any new programs;

- Allowing missions to include technologies with high RoI for future missions by allocating additional reserves over and above any cost caps to cover unknowns;
- Creating a separate technology line similar to the former New Millennium Program where multiple technologies could be demonstrated in small flight missions; and
- Adopting a systematic way of bounding the risks, the cost, and the schedule of technologies at TRL-6 by requiring additional information at TRL-6 such as defining work required to complete the space qualification of all components necessary to achieve flight status and documenting the attendant list of technical and programmatic risks.

Space Technology Mission Directorate Technology Development

Collaboration between SMD and STMD has enabled technology development for several significant planetary spaceflight exploration technologies. As an example, STMD-supported space technologies used for Mars 2020 include Mars Environmental Dynamics Analyzer (MEDA), Mars Oxygen In Situ Resource Utilization Experiment (MOXIE), Mars Entry, Descent and Landing Instrument 2 (MEDLI2), and Terrain Relative Navigation (TRN). STMD further supports technology development for launch and landing systems and thrusters, some of which aim for future lunar explorations, among others.

An analysis of STMD spending over the past 5 years shows that it has invested approximately 10.6 percent of its budget on planetary science technologies. Given that STMD is chartered to support all NASA efforts, a budget allocation to PSD similar to its share of the Agency budget is appropriate. SMD's budget is about 30 percent of the NASA budget, with PSD representing about 11 percent of the Agency's budget, so STMD's investment has been about right. However, over the past several years, STMD's priorities seem to have been shifted to more commercial and human exploration technology developments putting some pressure on plans for robotic technologies needed for future PSD science missions. As noted earlier, there does not seem to be a documented agreement between STMD and PSD that ensures STMD's continued investment in planetary and astrobiology technologies.

Finding: During the past decade, SMD/PSD and STMD have worked together on developing high risk technologies important to the future of planetary science and astrobiology missions.

Finding: STMD investment in PSD technology needs can be reprioritized by other parts of the Agency when other Agency needs are deemed greater.

Recommendation: STMD should ensure that its level of investment in SMD mission technologies is balanced at approximately 30 percent of its overall budget with the PSD portion at no less than 10 percent.

TECHNOLOGIES FOR THIS DECADE AND BEYOND

Many of the strategic research (SR) objectives in each of the science chapters of this survey require new and improved technologies to accomplish them. The science missions pursuing these SRs or planetary defense (PD) also require new or improved capabilities to reach their destination, operate there, and accomplish the necessary measurements. Seventeen technology areas were identified for this decade and beyond that need advancement based on an assessment of the SR objectives, priority missions, and the current technological state of the art. Besides benefiting these critical science investigations, technology advancements can benefit all classes of missions, large and small, and potentially enable missions in higher cost categories to be accomplished in the next class below. Lucy and Psyche both use solar arrays for power, at distances from the Sun that were not feasible in the early days of the Discovery Program.

Table 21-1 summarizes the technology priorities for the coming decadal, the key science questions that benefit from these technologies, the destinations where the strategic research activities associated with particular key questions would be conducted, and gives a summary of the rationale for the technology improvements. Each technology area in Table 21-1 is described in more detail in the following sections.

TABLE 21-1 Technologies Identified to Be Advanced in This Decade and Beyond

Technology Area	Rationale	Key Science Questions/ PD Chapter	Applicable Destinations
Instrumentation			
General in situ instruments	Instruments to perform in situ measurements require technology developments to improve sensitivity and dynamic range, mitigate noise sources, and reduce mass, power, and volume requirements.	1, 2, 3, 4, 5, 6, 7, 8, 9, 10, 11, 12	Venus, Moon, Mars, small bodies, ocean worlds, gas giants, ice giants
General remote sensing instruments	Remote sensing instruments require improved sensitivity and dynamic range, wavelength coverage, spectral, and/or spatial resolution, as well as reduced mass, power, and volume. Active systems such as EM sounding, radar tomography, and atmospheric lidar for wind or composition benefit from continued development.	All	All
Instruments to search for evidence of life	While instruments capable of meeting or exceeding astrobiology mission measurement requirements already exist, more work is needed to improve performance and add robustness to the life detection technology portfolio. Integrated instrument suites, improved front-end sample handling and mitigation of contamination sources are key areas of development.	9, 10, 11	Mars, ocean worlds, Venus (Moon and small bodies for false positives)
In situ sample handling, preprocessing, and analysis	Priority missions need this technology area this decade. Sample collection without modifying/destroying sample physical/chemical properties, robust material separation (sample handling) and high accuracy and precision detectors (sample analysis) need to mature.	1, 2, 3, 4, 5, 6, 7, 8, 9, 10, 11, 12	Venus, Moon, Mars, small bodies, ocean worlds, giant planets
General Technology Areas (alphabetically)			
Autonomy	Autonomy advancements are required at a system level to integrate and harmonize subsystems to make decisions and execute planned operations on remote, complex and potentially unknown planetary bodies.	All	All
Challenging environments	Priority missions and SR objectives need technologies for overcoming extreme temperatures, pressures, radiation, and dust accumulation this decade.	1, 3, 4, 5, 6, 8, 9, 10, 11, 12	Earth, Moon, Venus, Mars, small bodies, ocean worlds
Cold/cryogenic sample return	Maintaining cold/cryogenic samples is the next step in sample return, and cold/cryogenic sample return missions are being considered as soon as early next decade.	1, 3, 4, 5, 6, 9, 10, 11, 12	Moon, Mars, Venus, small bodies, ocean worlds
Communication systems	As missions to achieve SR objectives become more complex, current radio capabilities will be inadequate in the future. New optical communication will be capable of meeting future communication demands. However, further improvements in current capabilities will enable a higher science RoI.	All	All

continues

TABLE 21-1 Continued

Technology Area	Rationale	Key Science Questions/ PD Chapter	Applicable Destinations
Entry/deorbit, descent, and landing systems	Further advancements in TRN and HD&A are required to enable recommended missions and SR objectives in this decade. Other technologies include anchoring and sampling on low-gravity bodies, TPS, aero-decelerators and high-ISP throttleable descent engines.	3, 4, 5, 6, 10, 11, 12	Venus, Moon, Mars, small bodies, ocean worlds
In situ mobility (aerial/surface)	Improved in situ mobility is required for priority missions later this decade. Aerial mobility benefits from further advances in rotor vehicles and balloon platforms, while surface mobility needs autonomy (see Autonomy) and higher mechanical endurance.	1, 3, 4, 5, 6, 8, 10, 11, 12	Venus, Moon, Mars, small bodies, ocean worlds
Launch, cruise, and encounter optimization	Aerocapture is considered ready for infusion and can enhance/enable a large set of missions. Additional technology improvements in SEP, trajectory design and propellants can provide increased performance and higher RoI.	All	All
Planetary defense	Advancement of planetary defense technologies to address characterization and mitigation objectives require development in the coming decade.	PD	Small bodies
Planetary protection and contamination control	New up-to-date NASA Standard Assay techniques are needed, along with better contamination control processes. Other technologies like terminal sterilization are also needed.	1, 2, 3, 4, 5, 6, 8, 9, 10, 11, 12	Moon, Mars, small bodies, ocean worlds
Radioisotope thermoelectric generator/Radio-isotope power systems	Multiple missions require NGRTG-based power and are planned to start in this decade, some using Mod 1 and some Mod 2. Dynamic conversion technologies will improve overall RTG efficiency and reduce plutonium-238. Improvements are also applicable to smaller class missions.	1, 2, 3, 4, 5, 6, 7, 8, 10, 11, 12	All
Solar array and batteries	Priority missions need advanced technologies for highly efficient, flexible solar arrays and energy storage systems this decade.	All	All
Subsurface access	Priority future missions targeting surface/subsurface exploration require access to pristine/unmodified materials. Technologies include drills, melt probes, tethers, submersibles, emplaced communication nodes, telemetry from the probe/drill tip, and materials capable of meeting stringent planetary protection requirements.	1, 3, 4, 5, 6, 8, 9, 10, 11, 12	Earth, Moon, Mars, small bodies, ocean worlds
Technology system engineering and integration	Many technology areas are best advanced when integrated with other technology areas, particularly for automated landing, sampling, mobility, and surface operations.	All	All

INSTRUMENTATION

General In Situ Instruments

Instruments to perform in situ measurements need to continue to be advanced by technology developments to improve sensitivity and dynamic range, mitigate noise sources, and reduce mass, power and volume requirements. Many of these developments are enabling for the strategic research objectives within this survey, including for priority missions. For example, the continued maturation of in situ geochronology techniques will enable direct measurement of the timing of key geologic events on terrestrial planet surfaces, now only possible by sample return. Similarly, improvements in the range, resolution, and size of mass spectrometers have broad applicability, from improved measurement precision of noble gas isotopes in atmospheres, to improved discrimination of both organic and inorganic samples on planetary surfaces. Current instrument development programs such as PICASSO (TRL 1–4), MatISSE (TRL 4–6), ICEE (TRL 5–6), and targeted programs for extreme environments (i.e., HOTTech, COLDTech) can continue to make advancements. Future flight instrument development can also take advantage of terrestrial field environments for demonstration (i.e., via PSTAR). This approach fills the gap between in situ instruments and advanced laboratory equipment.

General Remote Sensing Instruments

Instruments for electro-optical remote sensing also benefit from continued technology developments to improve sensitivity and dynamic range, wavelength coverage, spectral and/or spatial resolution, and the constant push to reduce mass, power, and volume. Active instruments in particular provide new measurement capabilities for priority missions: improved radar and EM sounding of planetary subsurfaces and interiors, lidar techniques for interrogation of composition, wind and dust profiles, as well as measurement of surface seismic motion, and sub-millimeter sounding of atmospheres. Operation in extreme environments also defines a set of technology needs, such as sensors for the surfaces of Europa or Venus (also see Challenging Environments). Beyond this decade, sensing techniques such as muon tomography and ultra-precise gravimetry offer exciting new capabilities to interrogate planetary interiors remotely. Last, improvements in both space- and ground-based telescopic instruments will enable remote characterization, including biosignature searches, of exoplanet atmospheres (NASEM 2019a); also covered in Chapter 20, Infrastructure.

Instruments to Search for Evidence of Life

The sensitivity of in situ instruments focused on the search for life continues to improve, driving down limits of detection. Current detectors capable of meeting or exceeding mission measurement requirements already exist, as evidenced by the maturation of several mission concepts to search for evidence of life on Europa and Enceladus. However, more work is needed. For example, development of new technologies for microscale and nanoscale analysis (e.g., optical microscopy, Raman and infrared spectroscopy, and laser-induced breakdown spectroscopy) and implementation of commercial compact, low-power RNA and DNA sequencing devices could add robustness to the life detection technology portfolio (NASEM 2019a). The performance of these and other detectors can be further improved with front-end sample handling and sample preparation systems (including chromatography and other separation techniques to distinguish biomolecules from a complex abiotic organic background), as well as careful mitigation of contamination sources. Current instrument development programs need to continue; the ICEE-2 program, for example, is supporting development of a range of organic analyzers as well as instruments for vibrational spectrometry and microscopy. For astrobiology instruments and other payloads that would benefit from integrated instrument techniques/suites with multiple independent biosignature tests, candidate subsystem solutions for life detection (e.g., sampling tools, sample processing tools, and sensitive detectors) could then be selected for integration under targeted programs such as COLDTech.

Finding: Instrument development (in situ and remote sensing) continues through existing programs such as PICASSO, MatISSE, and targeted programs such as COLDTech and HOTTech. Instruments, in particular those focused on the search for biosignatures, would benefit from early integration (with other instruments as well as sample acquisition, handling, and preprocessing systems) to enable multiple analytical techniques to be applied to the same collected sample.

In Situ Sample Acquisition, Handling, and Preprocessing

Planetary missions featuring in situ sample analysis need to meet requirements for sample acquisition, handling, and processing. Significant technological development has already been achieved for active and passive sample acquisition. Successful examples of active acquisition include the scoops used on the Viking and Phoenix missions to Mars, the TAGSAM system on the OSIRIS-REx mission, and the MSL drill and sample processing system. Regarding passive sample acquisition systems, the Cassini mission successfully sampled gas and icy materials from the Enceladus plume, the Stardust mission collected dust samples from the comet Wild 2, and a large funnel system is in development for a future Enceladus plume flythrough mission under the COLDTech Program. However, more work is needed for specific sample acquisition cases (e.g., high velocity >10 km/s plume sampling at Triton), and future missions will place increasingly challenging requirements on the handling and preprocessing of acquired samples, particularly for sensitive life detection and geochemical isotopic measurements. These requirements arise both from advanced measurement capabilities of one or more instruments that depend on precise, intensive sample preparation (NASEM 2019b), and from the need to control potential forward contamination from Earth (McKay et al. 2020) (more in Planetary Protection and Contamination Control). It is critical that science requirements drive sample handling technologies, including nondestructive acquisition and preparation, rather than off-the-shelf engineering solutions or ease of implementation (NASEM 2019b).

Sample handling encompasses the mostly physical manipulation of an acquired sample to prepare it for subsequent analysis steps. Examples may include ingestion of material (in any physical state), metering, shaping (powdering, leveling, sectioning), physical subsampling/separation, and fine positioning. Ingestion may include containment of the acquired sample or downstream steps such as preparation of a liquid sample for injection into a MEMS-scale processor. Some preliminary measurements, such as verification sensors, mass determination, or inspection imaging, may be folded into the handling steps.

Sample preprocessing encompasses the mostly chemical treatment and modification of an acquired sample to prepare it for analysis. Examples may include controlled mixing with reagents and labels, filtering, concentration, extraction, derivatization, with some preliminary verification measurements potentially folded in, prior to analysis by one, or ideally multiple, analytical techniques.

Sample handling and preprocessing can enable measurements of chemical or morphological biosignatures that are at trace concentration (e.g., nanomolar levels and below), are isolated to certain minor phases of a heterogeneous sample or are otherwise difficult to detect owing to analytical interferences that occur in a particular technique. These benefits also apply to high-priority geochemical objectives, such as precision isotopic analysis (e.g., oxygen isotope systematics) in individual minerals or compounds, for which sample preparation is needed to improve quantitative capability and lower potential ambiguity of results. Focused development in sample handling and preprocessing additionally would enable multiple instruments to share a common front-end, which can lead not only to savings of mass and complexity, but also to improved science return owing to alignment of analyses to a common sample.

Finding: Sample acquisition has benefited from significant technology development, although work is still needed for specific cases. Sample analysis requires significant handling and preprocessing of acquired samples prior to sensor analysis. Sample handling and preprocessing technology need urgent attention to extract target materials accurately and efficiently from acquired samples, and these implementations need to be science-requirements-driven.

GENERAL TECHNOLOGY AREAS

Autonomy

The development of more capable and robust autonomous mission and spacecraft systems is a strongly growing need in planetary science, astrobiology, and planetary defense. Autonomy is implemented as a set of computer algorithms that permit missions and spacecraft to operate and achieve their objectives with some degree of independence from human decision makers. As such, autonomy can provide truly mission-enabling functionality through rapid response to spacecraft conditions including anomalies, processing and evaluation of large raw data sets (e.g., optical imaging), and eventually achievement of science objectives through goal-seeking and optimization processes. Autonomy is primarily associated with onboard decision making, and thereby bounded by computing resources on a spacecraft. It can apply as well to algorithms implemented in a ground segment (e.g., "autonomous mission control") to increase the efficiency and effectiveness of Earth-based decision-making procedures under interplanetary communication time constraints.

Onboard autonomy development has been driven by the need for real-time planetary spacecraft control and management, including fault detection and recovery; navigation, orbit insertion, and landing; and in situ mobility (e.g., autonomous roving) subsystems. Based on a record of success to date, these applications can be developed further and implemented more broadly, with increasing degrees of independence, on priority missions in the coming decade. Future missions, particularly those under severe communications or environmental constraints (e.g., ocean world missions such as the Europa Lander concept, Venus surface, and others), will depend additionally on autonomy realized at the *system level*, to include evaluation of multiple, complex data inputs to make balanced decisions based on higher-level mission priorities, even when the precise conditions and inputs may be unpredictable a priori (Amini et al. 2020). This leads to the need for greater application of artificial intelligence, particularly utilizing machine learning tools that can be developed and tested to a high standard of robustness (NASEM 2019a). Selected topics that highlight particularly urgent areas of autonomy development in the coming decade follow here.

Development and infusion of system-level autonomy: Whereas component- or subsystem-level autonomy addresses decision making within a specific domain, such as instrument target detection/prioritization or mobility planning, its application on future missions may ultimately be limited by the need for repeated human interaction to resolve competing drivers at the mission level (e.g., optimizing target selection and mobility planning together). System-level autonomy targets the entire system to harmonize every system component and reduce redundancy and incompatibility between subsystems. Traditionally development and infusion of significant system-level autonomous systems have been limited by factors related to the distributed "ownership" of system-level autonomy requirements and to the perception of risk associated with removal of humans from decision-making loops (Amini et al. 2020). In the past, such advancements may have often been seen as enhancements to mission capability, but not necessarily critical to core mission objectives. Going forward, missions highlighted by this survey and beyond, particularly those to remote planetary environments and generating vastly greater volumes of raw data (Theiling et al. 2020), system-level autonomy, enabled by new advancements in artificial intelligence and machine learning, will be seen as a critical, mission-enabling technology.

Finding: Autonomy needs to evolve at a systems level to integrate and harmonize subsystems to make decisions and execute planned operations on remote yet complex planetary science and astrobiology missions. Machine learning/artificial intelligence can support the implementation of autonomy in such environments.

Extended in situ mobility: Surface rovers, aerial vehicles, or other mobile elements continue to be high-priority systems for autonomy. Autonomy enables long-traverse rover missions for this decade, combining multiple sensor fusion for real-time hazard detection and path optimization to yield order-of-magnitude increases in range and operational efficiency (Amini et al. 2020; Matthies et al. 2020). For a lunar rover, autonomous driving is critical in darkness, and even strongly enhancing during sunlit traverses, where even slight communication delays can limit traverse speed. On more distant bodies, autonomous mobility similarly enables more efficient access to more terrain or riskier yet high-priority terrain, such as caves or ravines, by minimizing ground-in-the-loop control.

Finding: Long-traverse rover and other extended mobility missions are enabled by higher-speed, hazard-avoiding autonomous mobility over longer durations, particularly where human interactions are limited or impossible. Future remote missions with, for example, rovers and aerial vehicles, will increasingly rely on mobility autonomy to access a greater range of surface regions and features.

Science acquisition and analysis: Achieving autonomous science analysis (including target selection/sample acquisition, instrument operation, data analysis/interpretation, and follow up by optimization or redirection) has the potential to vastly increase the effectiveness and reduce the cost of planetary missions. Large data volumes generated by modern, high-resolution instruments are not efficiently transmitted from remote environments and analyzed on Earth within reasonable timeframes to address complex scientific analyses (Theiling et al. 2020). Furthermore, some complex science objectives, such as life detection or identification of surface features from orbit, are not amenable to prescriptive algorithms, necessitating iterative analyses and synthesis of data from multiple instruments. Such requirements imply that future missions with severe resource limitations are critically enabled by science autonomy, again at a system level guiding activities and instruments to achieve broad goals.

> **Finding:** Autonomy applied to all science acquisition and analysis activities can greatly benefit science return from remote missions. Where science objectives require onboard data prioritization and iterative analysis, science autonomy will be critically enabling.

Challenging Environments

Extreme environment: Technologies for resisting extreme temperatures have advanced in the past decade. Solar power generation in extremely high/low temperatures has been demonstrated both near the Sun (at 0.046 AU by the Parker Solar Probe) and far from it (5 AU by the Juno mission). Advances have been made in silicon carbide (SiC) integrated circuit (IC) electronics through the Long-Lived In Situ Solar System Explorer (LLISSE) that bring the technology to the level of 1970–1980 silicon-based electronics used in Viking and Voyager (Hunter et al. 2020). While identifying progress in the past decade, technologies for protecting spacecraft from extreme environments continue to remain enabling as strategic research expands to new and more hostile environments. Future missions under consideration plan to explore Mercury's surface, Venus's atmosphere and surface, lunar polar regions, Mars's polar regions, and ocean worlds. Extremely cold/high temperatures impact hardware, constraining mission operational periods, and high pressure, such as on Venus or ocean world interiors, is another strong constraint on mission operations. Some targets, such as Venus and potentially Europa, host corrosive chemicals such as sulfuric acid. Systems and materials isolating and resisting extreme conditions enable longer mission durations in extreme environments. Technologies necessary to be advanced include power storage and generation systems, materials, mechanical actuators, and electronics, including memory, among others.

> **Finding:** Protecting spacecraft from extreme environments (e.g., temperature/pressure/chemical corrosion) needs to be advanced to enable in situ priority missions. Technologies needing further advances include power generation and storage, materials, actuators, and electronics, including memory, among others.

Dust mitigation: The mission of NASA's Opportunity rover ended in mid-2018 because of the lack of solar power generation caused by dust covering owing to a dust storm. NASA's InSight lander also suffered from dust covering its solar panels, and attempted to remove the dust by using the spacecraft's arm had limited success. Similar issues can happen on other planets and moons. Therefore, effective dust removal/mitigation methods for spacecraft systems are considered an enabling technology for in situ priority missions planning long-term operations in dusty environments. Unfortunately, while various approaches such as mechanical, fluid, and electric methods have been proposed, current technology does not offer effective dust mitigation processes.

> **Finding:** Dust deposition may not be considered extreme but may be fatal to missions under some circumstances. Proper mitigation technologies warrant further advances to enable in situ long-term priority missions to rocky bodies.

Radiation: Radiation continues to be a major risk to spacecraft that requires extensive testing, analysis, and parts screening to prevent failures of avionics that can cause the premature termination of science missions. New approaches and technologies to mitigating or eliminating radiation damage have the potential to reduce costs and increase spacecraft capabilities by reducing the parts processes and/or eliminating the extensive shielding required for such destinations as Jupiter. Ultimately, advances in radiation protection could enable the use of commercial grade products (see Game-changing Trends). In addition, higher specific energy primary batteries based on lithium carbon monofluoride will play key roles in future lander missions to icy moons, but they are susceptible to significant capacity loss when exposed to high radiation. While studies have shown initial tests to quantify radiation effects on the batteries, further technologies for protecting batteries from radiation are essential to enhance such missions.

Finding: Approaches and technologies to protecting spacecraft electronics and power storage systems from radiation have been developed, but further technical advances will enhance future missions targeting icy moons around giant planets. Such advances will also reduce costs and increase capabilities of future spacecraft, potentially enabling the use of commercial grade parts.

Cold/Cryogenic Sample Return

Cold/cryogenic sample return is a key technology area that enables returning cold/cryogenic volatile samples from planetary bodies to curation and laboratory facilities on Earth without compromising sample integrity, and as such, requires significant development for upcoming mission opportunities in the next decade.

This technology area has been identified as crucial for addressing several strategic research activities that can only be accomplished by future sample return missions. In contrast to traditional sample return technologies that only deal with uncooled materials, cold/cryogenic sample return technologies require systems that acquire, contain, and preserve volatile samples in cold/cryogenic environments under ambient conditions. Volatile sampling, handling, and containment technologies require strict temperature and pressure controls to avoid phase changes and chemical alterations over a long-term period (Milam et al. 2020). Spacecraft systems and instruments required for sample acquisition and handling have to be thermally isolated and kept at cryogenic conditions during sampling operations. Cryocoolers need to be maintained at constant cryogenic temperatures and pressures during potentially long Earth-return cruise phases requiring increased thermal containment capability and reliability. One of the most challenging operational phases occurs during reentry and recovery, when return capsules contact and interact with Earth's atmosphere, exposing the sample containment systems to adverse conditions (e.g., high heat). All of these enabling technologies have to be capable of maintaining the samples' integrity throughout every mission phase so that they can be successfully delivered to planetary sample curation and laboratory analysis facilities (Milam et al. 2020).

Finding: Cold/cryogenic sample return requires significant development of technologies to enable the acquisition, containment, and preservation of cold/cryogenic volatile materials at ambient sampling conditions. Such technologies are needed to be employed during all phases of the mission in order to preserve and maintain the scientific integrity of the samples.

Communication Systems

As planetary science and astrobiology missions become more complex and the science data volume continues to increase, maintaining high-data-rate communication is essential to return the science data collected. While this technology area has constantly been advancing to meet the mission requirements in the past decade, it needs to keep evolving to achieve higher data communication rates over the limitations of radio frequency (RF) communications, including bandwidth, spectrum, and overall size of frequency packages and power used. The priority mission concepts considered by this decadal all used these current, state of the art data communication techniques and equipment. However, a review of the science strategic research objectives recognized that current capabilities will be inadequate in the future and that the emerging optical communication capabilities will be capable of meeting future communication demands.

NASA is developing the Laser Communications Relay Demonstration (LCRD), which was launched in late 2021. LCRD will be NASA's first two-way optical communications relay satellite that will demonstrate the benefits of optical technologies, such as higher data transmission and less size, weight, and power requirements. Two other demonstrations are planned in 2022 including the Deep Space Optical Communications demonstration on the Psyche spacecraft. Despite the significant improvements in data rates, optical communication will require improvements in antenna/spacecraft pointing for missions beyond 5 AU.

Finding: Despite the continuing advances in radio frequency communications, future science missions will have data volumes that surpass the technology's capabilities. Optical communication technologies currently in development within NASA will be able to achieve much higher-data-rate communication. Continued investment in this technology is necessary to achieve the required capabilities.

In the meantime, the following technology advances can continue improve data transfer efficiency: higher radio frequencies and channel coding and modulations can avoid spectrum congestion while using limited power and spectrum; better compression processes can reduce data sizes, leading to efficient communication; and large, deployable mesh reflectors dramatically improve communication rates (Hamkins 2020).

Finding: Further advances in higher frequency transponders and antennas, efficient data processing, and large, deployable reflectors can enhance the communication capability.

Entry/Deorbit, Descent and Landing Systems

Several missions under consideration involve scientific exploration with landed platforms at several destinations (e.g., Mars, Europa, and Venus). While some of these destinations have been previously visited and are well surveyed, other destinations are much less explored and information about their environment and surface characteristics is lacking. That information is needed to design landing systems that deliver the science payload to the surface with acceptable risk. Even for those places that have been visited before, there is a need to access ever more challenging landing sites driven by the desire to maximize science value. To respond to these challenges, continued development of enabling landing technologies is required.

NASA has been investing heavily in TRN and HD&A technologies as part of its Mars Exploration Program, its Safe and Precise Landing Integrated Capabilities Evolution (SPLICE) project and predecessors (ALHAT), and its Europa Lander mission concept technology investments. These investments led to the highly successful development and use of TRN by the Perseverance rover on Mars, which allowed the project to land in Jezero Crater, a site of great scientific value but with unprecedented landing risk: a great example of how proper technology investments can lead to superior science investigations.

As we look into the future, more TRN technology developments are required to improve its robustness to landing in poor illumination conditions, including the development of active techniques (e.g., LIDAR), which can have great benefit for landing on places like the poorly illuminated Moon's south pole or to relax the stringent landing site selection constraints in hard to access destinations such as Europa's surface. These advances in TRN and HD&A will require high performance computing to perform the complex computations needed. In addition to TRN and HD&A, soft landing requires robust sensing of the lander's altitude and velocity with respect to the landing surface. Past soft landers used heavy, bulky, and costly radio frequency sensors tailor made for each mission, which are difficult, or in some cases impossible, to afford for smaller landers and nonflagship missions. NASA, through its SPLICE project, has been addressing this problem with the development of the Navigation Doppler LIDAR landing sensor but despite many years of development, landing missions still struggle to find landing altimeters and velocimeters that can be afforded in developing time/cost and spacecraft resources.

Finding: As more difficult terrains are envisioned for future missions, continued technology investments in TRN/HD&A can enable spacecraft to safely land in ever more challenging and constrained landing situations.

Missions contacting the surface of small bodies require interactions with the body's surface—e.g., sampling and drilling—that involve reaction forces that can overcome the spacecraft gravitational forces. Missions like NASA's OSIRIS-REx side-stepped this problem by using the Touch and Go (TAG) sampling technique, which sampled surface regolith, but generally could not reach the deeper depths for more primordial or unmodified material. Potential solutions to this include the use of anchoring devices or propulsive reaction forces.

Finding: Efforts have been limited to develop technologies to enable landers to acquire deep samples; for example, 10s of cm to 1m, or other interactions that require large reactive forces in low-gravity regimes. Investment in such technologies would enable access to primordial/unmodified subsurface materials of small bodies.

NASA has invested considerable resources in the development of thermal protection systems (TPS) like the Heatshield for Extreme Entry Environment Technology (HEEET), and the Phenolic-Impregnated Carbon Ablator (PICA). These TPS technologies are currently capable of operating over a wide range of entry conditions and are crucial for the landing of larger payloads on Mars and for enabling atmospheric probes on Venus, Saturn, Titan, Uranus, and Neptune. HEEET (currently at TRL-6 for certain conditions) was developed in the past decade, as the heritage carbon phenolic used for the Galileo entry probe is no longer available (Ellerby et al. 2020).

Finding: NASA's investments on TPS technologies have enabled several landing missions and atmospheric probes in the past and, together with current developments like HEEET, stand to enable many future missions to multiple destinations.

In addition to these enabling technologies, there are also enhancing technologies and engineering developments that can also benefit from investments prior to a project start.

NASA has invested considerable resources in the development of deployable aero-decelerators (e.g., HIAD, SIAD, and ADEPT) that have the potential to dramatically increase landed mass on future missions. These technologies, however, seem to fall into the "second valley of death"—that is, are too risky—and do not yet seem to be under consideration for future missions.

Finding: NASA and the science community would benefit from studying how the maturing aero-decelerator technologies can be integrated into future missions to increase science value.

An area where NASA technology development has fallen short is in the redevelopment of high specific impulse (ISP) throttleable descent engines to enable the landing of heavier payloads with higher lander precision. The Mars Sample Retrieval Lander (SRL), currently under development, requires a pinpoint landing capability that demands large and fuel-consuming position corrections during powered flight, meaning that the low efficiency throttleable mono-prop engines developed first by Viking and later resurrected for Curiosity and Perseverance are insufficient.

Finding: The Mars SRL as well as current and future missions to the Moon and other destinations could benefit greatly from the availability and use of high ISP (~300 sec) bi-prop throttleable engines.

In Situ Mobility (Aerial/Surface)

Aerial mobility: As in Earth aviation, aerial mobility can provide a vantage for rapid, precise surface analysis over regional scales, in situ studies of atmospheric properties, and unique access to hazardous terrain (Bapst et al. 2020; Cutts et al. 2020; Mittelholz et al. 2020). This capability comes in many forms, such as rotorcraft, balloons, and airplanes, among others. Rotorcraft such as the Mars Ingenuity technology demonstration and the Dragonfly mission have provided and are expected to provide important measurements over multiple terrain types. Balloon platform technology can address SR objectives but needs advances this decade to meet the requirements of in situ atmospheric explorations on Venus and other planetary atmospheres. This technology requires the capability to inflate after storage in the parent spacecraft while remaining ultralight and resisting damage during deployment and controlling altitude during long-term operations. (Cutts et al. 2020; Matthies et al. 2020).

Finding: Balloon platform technology has not yet achieved the maturity of rotorcraft and airplanes and is enabling for rapid, precise surface analysis and in situ studies of atmospheric properties on Venus and other planetary atmospheres. The technology requires the capability of inflation, given ultralight materials and structures, without damage and for controlling altitude during science operations.

Surface mobility: A number of long duration, long-traverse rover missions were identified as necessary to address SRs at the Moon this decade. Rover technology has made major strides over the past decades. NASA's Perseverance rover is currently demonstrating advanced surface mobility capabilities, following the earlier Mars rover missions. However, priority missions are now planning traverse operations as far as 2,000 km, far longer than ever completed previously. Autonomous mobility capabilities of traversing over rough and steep terrains can optimize the path selection process. Autonomy required for these rover capabilities are covered in the Autonomy section, but other technology advances are also required. At present, the titanium-based shape memory alloy tire, developed by NASA Glenn Research Center, is one example of a technological advance that improves a rover's performance, compared to Apollo's LRV tires. Also, onboard actuators consume electric power to warm up and keep their performance. NASA has currently made efforts in heatless mobility actuators that can reduce electric power consumption.

> **Finding:** Rover surface mobility on terrestrial bodies has evolved over decades. Now long-traverse surface mobility is identified as an enabling technology that allows smooth traversing regardless of large rocks and steep slopes at traverse rates much greater than current technology. Such capabilities need to mature this decade.

Some locations identified to be scientifically valuable, such as deep pits, caves, crevasses, and rough terrains, prohibit access by regular rover-type vehicles having wheels. To access these regions, different technology capabilities, such as crawling, hopping, and flying, can enable challenging in situ investigations, including remote sensing and sample acquisition. For example, robotic probes detachable from their parent rovers or landers that can crawl over steep slopes can enable accessing the bottom of cliffs, fractured terrains, or crevasses where high-priority astrobiology targets (e.g., permanently shadowed regions and plume sources) may be located. Such technologies (e.g., JPL's Axel) are currently under development, although further work is essential.

> **Finding:** Strategic research has identified scientifically valuable regions that traditional rovers and landers cannot easily access, such as caves, craters, crevasses, and other rough or fractured terrain. Technologies for accessing such challenging regions are still immature and need advancement.

Launch, Cruise, and Encounter Optimization

Several missions of interest to this decadal survey involve travel to far destinations, complex orbital tours of multiple targets, sample returns from small and distant bodies, and landers on hard-to-reach places. These mission types are characterized by long transit times that drive up Phase-E costs, large ΔVs (i.e., velocity changes) that reduce the science payload delivered to the destination, the need for large launch vehicles that drive costs beyond Discovery or New Frontiers mission class opportunities, and complex trajectories with constrained launch windows and multiple gravity assists that extend transit times and might require special spacecraft accommodations. There are potential technologies that can minimize these burdens.

Aerocapture is an orbital insertion technique which utilizes a single pass through a planetary atmosphere to dissipate enough orbital energy for planetary capture (Dutta 2020). It can deliver large orbit insertion ΔVs with minimum fuel, resulting in significant reductions in transit time, and/or increases in science payload mass. Aerocapture can also enable planet orbit insertion of smallsats, launched as secondary payloads, on targets like Venus and Mars (Austin 2020a,b). Advances in atmospheric entry guidance and control techniques, as demonstrated successfully by Curiosity and Perseverance on Mars, and advances in autonomous optical navigation, as demonstrated by the Deep Impact mission to comet Tempel 1, combined with the development of new TPS, like PICA used also by Curiosity and Perseverance and HEEET that will be used by the Mars Sample Return's Earth Entry Vehicle, make aerocapture a technology ready for mission infusion. Because aerocapture is not being proposed for use in missions, it is considered a "dormant" technology that is perceived as high risk in a mission competitive environment.

> **Finding:** Aerocapture is a technology that is ready for infusion and that can enhance/enable a large set of missions, but that will require special incentives by NASA to be proposed and used in a science mission.

Solar-electric propulsion (SEP) has been successfully flown in several science missions (e.g., Deep Space 1, Dawn, Hayabusa, BepiColombo) and will be flown in new missions currently under development (Psyche, Mars Sample Return Earth Return Orbiter). Continued improvements on SEP technologies can benefit a variety of future missions and destinations (Polk et al. 2020). These improvements include increases in ISP, lower power thrusters for small spacecraft, and advanced power processing units.

> **Finding:** SEP has been successfully demonstrated in several missions and further technology improvements can enable SEP in small spacecraft and enhance a large set of science missions.

There have also been recent advances in trajectory design and optimization that combine gravity assists, low-thrust high-ISP propulsion, aero-assist, and other tools available to the trajectory and mission designer to increase delivery mass and shorten transit times for a variety of mission types (landers, orbiters, flyby). An example of these efforts is NASA's Astrodynamics in Support of Icy Worlds Missions program that supports the maturity of astrodynamics tools in support of the exploration of icy moons orbiting gas/icy giants.

Finding: Astrodynamics and mission design tools can have a large impact on the design of innovative mission concepts that maximize science value and reduce mission costs.

New technology developments on cold propellant propulsion can result in a significant reduction on heater power needs that lead to large mass reductions on solar arrays and spacecraft mass (Casillas et al. 2020). This is of particular benefit to solar-electric powered missions to the outer planets, comets, and asteroids. NASA has been investing in this technology through the Advancement for Low-Temperature Operation in Space (TALOS) program for the Astrobotic Moon lander Peregrine, and JPL has been partnering with MSFC to extend this technology for deep space missions.

Finding: The cold propellant propulsion technology currently being developed by NASA has the potential to achieve important mass reductions in solar-electric powered missions to the outer planets, comets, and asteroids.

Planetary Defense

Planetary defense (PD) focuses on protecting Earth from devastating near-Earth asteroid and comet impacts. This technology area is organized around the stages required for effective PD implementation for this decade: near-Earth object (NEO) discovery, follow-up, and tracking; in situ characterization; and hazard mitigation.

NEO discovery, follow-up, and tracking: New survey systems coming on-line in the mid-2020s will significantly increase the number of known NEOs, which will require additional follow-up and enhanced tracking observations. Special focus on capabilities to improve the sensitivity, reliability, and range of radars are particularly useful for PD interests (e.g., phased array radar and solid-state amplifier technologies) (Lazio et al. 2020). Advancement in radar capabilities would enhance tracking and physical characterization of previously detected NEOs at further ranges, which will increase orbit refinement and provide additional physical characterization data vital for determining impact risk and aid the development of impact mitigation strategies.

Finding: Development of advanced radar technologies to improve NEO follow-up, tracking, and characterization capabilities would enhance planetary defense preparation.

In situ characterization: Spacecraft reconnaissance fly-by or rendezvous missions have been successfully demonstrated by small body missions over multiple decades. However, further advances are needed to determine key physical characteristics (e.g., mass, composition, and internal structure) of NEOs required for subsequent mitigation efforts (Abell et al. 2020). It is difficult to accurately determine these characteristics from fast flyby missions targeting small bodies of ~50 m and larger in diameter with mass being one of the most critical to measure for effective mitigation planning. Technologies that need improvement/development are terminal guidance navigation and control (GNC) algorithms and systems for targeting small NEOs during hypervelocity flybys, spacecraft systems/instruments to track such NEOs during these rapid flybys, and instruments/systems to determine the mass of NEOs during high-speed encounters (Barbee et al. 2020). Technological development of instruments that could be deployed or landed in either flyby or rendezvous missions to determine additional NEO properties (e.g., internal structure and strength) would also be useful for PD characterization objectives.

Finding: Development of technologies to obtain critical characterization information of NEOs during reconnaissance missions, particularly fast flybys, would inform planetary defense mitigation efforts.

ΔΔ*Hazard mitigation:* NEO impact scenarios vary depending on the impactor's physical characteristics and potential warning times. Hence, there is no mitigation technique that is appropriate for every circumstance, because the NEO may need to be deflected or disrupted by impulsive means, or gradually moved off course via slow-push methods. Impulsive techniques involve kinetic impactors or deployment of nuclear explosive devices (NEDs), whereas slow-push methods include ion beam deflection (IBD) or gravity tractor (GT) concepts. For kinetic impactors, improvement in targeting and imaging systems is required to precisely impact NEOs over a variety of encounter geometries during hypervelocity intercepts. NEDs would benefit from sensor technologies that would increase the trigger timing accuracy to ensure detonation at the optimum distance during high encounter velocities. For IBD, improvements in beam density and energy increase the effectiveness of this technique, as does decreasing the plume dispersion from the ion source or thruster. Improvements in spacecraft autonomy for long duration complex proximity operations are also needed for IBD or GT mitigation techniques (Barbee et al. 2020).

> **Finding:** Development of multiple technologies for both impulsive and slow-push mitigation techniques is prudent given the variety of potential impact scenarios. Further demonstration of these technologies would enable planetary defense mitigation missions.

Planetary Protection and Contamination Control

Planetary protection involves protecting solar system bodies from forward contamination by terrestrial organisms and protecting Earth from possible extraterrestrial organisms that may be returned from other solar system bodies. This requires spacecraft developers to design and build spacecraft and procedures that meet both the planetary protection bioburden requirements for the particular body, for example, Mars or an ocean world like Europa or Enceladus, and the contamination control requirements necessary to ensure the science measurements are not compromised. Thus, spacecraft bioburden and the presence of specific molecular contaminants on (or outgassed from) payload surfaces need to be adequately understood, controlled, and documented.

Planetary protection and contamination control advancements involve a systemic view of both the science of microorganisms and contaminants that might overwhelm or confound measurements, and the technologies to minimize these on spacecraft surfaces. Considerable scientific work is under way to conduct more accurate and precise microbial diversity assessments in cleanrooms and on spacecraft, understand the probability of biological contamination, and develop planetary protection conventions and contamination control requirements for future missions (for, in particular, those focused on astrobiology investigations); this is outside of the scope of technology addressed here. Here the committee focuses on specific technology developments that address planetary protection and contamination control for spacecraft.

A hybrid technology that combines both culture-based and multiple next-generation sequencing methods needs to be explored as the standard methodology for spacecraft bioburden assessments. It is a high priority in the upcoming decade to mature a nucleic acid-based approach for bioburden assessments, adapting it from industry and academia approaches. This methodology needs to be tailored for key areas including, but not limited to (1) sampling and sample processing (swab to sequencing) capability from low-biomass spacecraft surfaces; (2) viable organism enumeration; (3) ability to rapidly identify PP-relevant organisms; (4) bioinformatic pipeline and database standardization; and (5) phylogenetic identification assessment of the broadest spectrum of organisms on the surface.

> **Finding:** The current NASA Standard Assay (NSA) is insufficient and inefficient to accurately estimate the number, diversity, and functional capabilities of spore-formers and other organisms associated with spacecraft surfaces.

Attention needs to be given to controlling the possible introduction of contaminants over and above those used for planetary protection bioburden, especially chemical species of astrobiological interest such as amino acids, nucleic acids, carboxylic acids or lipids and molecules that may confound measurements of these (e.g., isobaric species), at concentrations that might interfere with the scientific exploration of planetary bodies. Investigation of new technologies that can eliminate and/or reduce this contamination is extremely important to life detection measurements. Future life detection missions may need to use new materials that do not contain or produce these contaminants and that can withstand the cleaning and sterilization processes needed. Further technology work in this area would be extremely beneficial.

> **Finding:** Life detection investigations are beginning in earnest, and contamination control in these planetary missions is becoming even more critical to enable the very precise and difficult astrobiology measurements that are susceptible to contamination.

Concepts for terminal sterilization—that is, the complete elimination of all biological contamination at the landing site following the completion of all scientific investigation—are in the formulation phase as part of the Europa Lander mission concept technology efforts. At bodies where the timescales of surface-subsurface transport exceed the 1,000-year period of biological exploration, missions might not require such extreme measures, providing significant cost savings. For bodies where surface-subsurface transport is less than 1,000 years, further technology development of terminal sterilization, in concert with planetary protection requirements tailored for the specific body and mission science requirements, would provide a robust strategy to minimize the risk of contamination while maximizing unambiguous science return.

Finding: Terminal sterilization has been identified as one possible technology that could prevent the possible contamination of an icy world's ocean within the 1,000-year timeframe.

Radioisotope Power Systems

Radioisotope power systems (RPSs) convert heat generated by radioactive decay, traditionally from plutonium-238 (^{238}Pu), into electric power. Radioisotope thermoelectric generators (RTGs), which generate an electric current across thermoelectric couples when subjected to a temperature difference, have been the workhorse power source for missions to locations with limited solar irradiance, or where the use of solar arrays is impractical. Recent and upcoming missions such as Curiosity, Perseverance, and Dragonfly are using multi-mission RTGs (MMRTGs); however, the applicability of this design is limited by an end-of-design-life (EODL, typically 17 years) power of 62 W.

Most of the priority missions for this decade and many beyond require RTGs for power, at levels several times greater than currently available, and thus would benefit from improvements in specific power (power per unit mass). In addition, this high demand for RTGs as a power source—which drives ^{238}Pu production—motivates the need for more efficient conversion technologies (see Chapter 20). The NASA RPS Program has anticipated these needs and has been working on improved RTG designs that can provide up to EODL power of 210 W. Efforts have started with a refurbished, legacy general-purpose heat source (GPHS)-RTG expected to be available in 2024 (Overy et al. 2020). The RPS Program's goal is to have two of these improved RTGs ready for fueling by 2026 with availability for 2030 missions. Further, the RPS Program is planning a second modification early in the next decade to further improve both conversion efficiency and power output.

Continued improvements in available power and conversion efficiency will enable future missions beyond this decade. Dynamic RPS systems, using Stirling or Brayton conversion technologies, are currently under development and could increase conversion efficiency by a factor of ~4 over the current GPHS technology. Such units are not likely to be available for missions endorsed by the current decadal study, but a flight demonstration in this decade could pave the way for infusion in later decades, resulting in a significantly lower demand for ^{238}Pu.

Finding: It is critical that the planned development and delivery of improved RTGs with higher power output stay on schedule, as multiple missions planned for the upcoming decade depend on them as a power source. Further advancements in higher efficiency RPS technology will alleviate the demand for plutonium-238.

Solar Arrays and Energy Storage

Many future planetary science missions will require advanced solar power generation and storage technologies, often in challenging environments. While these technologies have been steadily advancing, several priority missions in this decadal point to the need for development in specific areas, usually driven by extreme environments. Inner solar system missions to Mercury and Venus push operational temperatures, for example, ~465C for the Venus surface. While missions like Parker Solar Probe and BepiColombo have demonstrated thermal management of solar arrays under high irradiance, improved array performance at higher temperatures than were experienced by MESSENGER is still needed for extended operations. In the lower Venus atmosphere, low irradiance, high temperature (LIHT) conditions and corrosive gases at high pressure present a significant challenge for solar array technology. Similarly, the Mercury surface presents sustained high irradiance, high temperature (HIHT) conditions, compromising cell performance and lifetime.

For the outer solar system, solar array performance under low irradiance, low temperature (LILT) conditions continues to improve, with Juno operating successfully at 5AU, and recent advanced array developments that make solar power at Saturn (~10 AU) competitive with RTGs (Schwartz 2020), a needed development as radioisotope power is a limited resource. Flexible blanket arrays continue to grow in area and specific power, pushing the boundaries of feasible solar powered missions.

Solar arrays for landers and rovers present unique challenges. Power needs for these flight systems are generally high enough to need large arrays that cannot be readily accommodated, so smaller arrays are used, and operations are often limited by power cycling. But improvements in the accommodation and reliability of retractable arrays can make larger arrays a viable option for future missions.

> **Finding:** Recent NASA and industry investments in solar array technologies have resulted in improvements in photovoltaic efficiency, specific power, and array size, pushing solar as a viable power option for the outer solar system. Additional array improvements are needed for high temperature environments in the inner solar system.

Batteries are the main energy storage component of spacecraft power systems, functioning either as the primary power source, used for periods of hours or days (for applications such as atmospheric probes), or functioning as a secondary, rechargeable power source, used together with solar arrays or RTGs.

Battery technology for planetary applications has been steadily improving, as measured by specific energy (energy per unit mass) and energy density. Li-Ion is the standard chemistry for rechargeable, secondary batteries, routinely providing a specific energy of 200 Wh/kg at the battery level. However, cell-level specific energy is typically 25 percent higher; improvements in cell packaging are expected to improve this value over the next decade. In addition, new chemistries—many based on solid electrolytes with a Li anode—show promise in specific energy increases above 300 Wh/kg. Primary battery technology is also improving, driven largely by investments for targeted flight projects. For example, the Europa Lander concept has prompted the development of Li-CF$_x$ chemistry for flight, ~2× improvement in specific energy from the conventional Li-SOCl$_2$ batteries, a critical improvement when battery mass is driving the system design (Bugga 2020).

In addition to specific energy, planetary missions push the need for longer battery lifetimes, particularly missions to the outer solar system with long cruise times. Improvements in calendar, operational and cycle lives are needed for outer planetary missions, as they currently limit planned mission lifetimes. In addition, improved resilience to either high temperatures on Venus and Mercury, or low temperatures on Mars and ocean worlds, is needed for future missions. These applications for specific, extreme environments often require unique electrolytes that are suited to the expected temperature conditions.

> **Finding:** Battery capability (specific energy, energy density, lifetime) require steady improvements to keep up with future planetary mission needs. Extending the operational temperature range for extreme environments often requires special electrolytes and materials that may be a unique application.

Subsurface Access

Development of technologies enabling access to pristine/unmodified materials, ocean materials, and subsurface zones is necessary for priority future missions targeting surface/subsurface exploration on planetary bodies in the inner solar system and ocean worlds, particularly those searching for evidence of life (NASEM 2019a). Possible technologies may include drills, melt probes, tethers, submersibles, emplaced communication nodes, telemetry from the probe/drill tip, and materials capable of meeting stringent planetary protection requirements (Dachwald et al. 2020). While maturation of some of these technologies is under way through programs such as SESAME and COLDTech, significant additional work is needed. This technology area further requires rigorous validation processes at dedicated laboratory facilities for simulating surface/subsurface conditions not found on Earth, as well as extensive field testing (Howell et al. 2020; Schmidt 2020; Stone et al. 2020).

Enabling technology for missions in this decade includes developments to provide robust and pristine access to depths of at least 2 meters. Achieving a depth of 2–10 meters will provide access to samples protected from significant gamma radiation exposure (which is dependent on impact gardening erosion, and other processes), and also allows penetration through a range of overburden/soil, sediment, and possibly seasonal ice (though again, this is body-dependent). Drilling technology capable of reaching 2 meters has been developed in Europe and will be flown on the ExoMars rover. The United States has also been developing technology for these depths but additional advancement is necessary to meet capabilities required for a priority mission in this decade. A subsequent challenge would be to deploy a multi-string sampling drill that can penetrate hard materials in the 2–10 meter range and incorporate autonomy features (to sense and react to changes in, for example, density and subsurface obstacles).

Finding: While 1–2 m drill technology is maturing and planned for lunar missions, 2–10 m drill technology is critical but not mature enough to robustly sample pristine materials from subsurface layers of the widest variety of rock and ice materials on Mars, the Moon, and other bodies.

Additional key subsurface access technologies that would enable missions in future decades include:

- Deep drilling in the 10–100 m range (following decade) and km range (longer term) in rock or ice by investing in technologies such as hybrid wire-line-type drills with the potential to access subsurface reservoirs (ice on Mars or melt lenses/brine pockets on ocean worlds) and subsurface oceans.
- Alternative subsurface access probes (e.g., melt probes) for icy/ocean worlds, combining vertical access and probe mobility in solid materials, while tackling environmental stability, communication, and planetary protection/contamination control challenges.
- Submersibles for interior oceans of ocean worlds. These may include either tethered or free submersibles and would likely implement autonomy for navigation as well as sample collection and analysis.

Finding: Technology development to reach beyond 10 meters and access subsurface reservoirs and oceans would revolutionize our understanding of the interiors of terrestrial and icy/ocean worlds, and enable unprecedented astrobiology investigations in the coming decades.

Technology System Engineering and Integration

As planetary science, astrobiology and planetary defense missions have become more and more complex and sophisticated over the past five decades, the need to integrate technology advancements across multiple technology areas has become necessary to accomplish the new strategic research advocated by this decadal. In several of the following technology areas, there will be discussions where this is particularly important. As noted in earlier in the chapter, a capability to examine technology advancements from a system level is missing in the management of technology development.

Finding: Many technology areas are best advanced when integrated with other technology areas, particularly for automated landing, sampling, mobility, and surface operations.

Recommendation: NASA should initiate or continue activities that pursue the technologies identified in this decadal survey, with particular emphasis on those technologies that enable the recommended science (missions and strategic research), those enhancing technologies that will improve the overall science return on investment, and those dormant technologies that have achieved TRL-6 but are not yet deemed sufficiently mature for inclusion in flight missions.

DISRUPTIVE AND GAME-CHANGING TRENDS IN TECHNOLOGIES

Technology advancement is accelerating at an exponential pace driven by forces in the government for needs like defense and space exploration, in the commercial sector for improved products and market share, in academia through both support to government and commercial enterprises and for basic research, and by other nations seeking to improve their status on the global stage. Many of these organizations are pushing the technological boundaries and inventing new ways of thinking about difficult problems along with new techniques for how to solve them. The emergence of self-driving cars is just one example.

As these technologies become available, the aerospace community can take advantage of them to improve the way we build and operate our space missions as they continue to explore the solar system and beyond. The following technology trends have the potential to have game-changing impacts in future SMD missions. These are technology areas that are being driven mostly by forces outside NASA's SMD, or even space exploration, but need to be monitored and explored through conferences, workshops, and other means, and invested in, with forethought to take advantage of the capabilities that they offer.

New Commercial Launch Systems

Ongoing launch vehicle developments associated with the human exploration of Moon and Mars, if successful, could result in SLS class launch vehicle capability at orders-of-magnitude lower costs. Launch system concepts like the emerging super-heavy lift launch vehicles, which involve full vehicle reusability and on-orbit refueling, have the potential to dramatically reduce launch costs and increase launch mass to the point that mass will no longer be a driver for spacecraft design. In this scenario, spacecraft miniaturization and optimization for launch mass reduction will be replaced by a brute force approach that reduces cost by using the ample mass and volume resources made available by these new launch systems. If this scenario comes to pass, NASA and the space community need to be ready to adapt its culture to make maximum use of these new launch vehicle capabilities, not only to reduce costs but to also formulate new mission concepts that are currently beyond our imagination.

Advanced Materials and Manufacturing

Significant advances in materials sciences such as carbon nanotubes and graphene have the potential for game changing improvements in spacecraft design by enabling lighter and stronger structures, higher capacity batteries, higher efficiency and lower mass solar cells, thermal management systems, electronics, and sensors. Similarly, smart structures such as origami structures and flexible structures are at a new phase of innovation to replace traditional structural designs.

New manufacturing capabilities, such as additive manufacturing (Sacco 2019), have the potential to enable the creation of new materials and parts that may one day be made in microgravity only and may only function there. This would permit a shift in the logistic and planning of space missions given this in-space construction would be a reality, ultimately allowing construction of space mission components directly in space. A recent example relevant to planetary exploration are bulk metallic glasses that allow for actuator systems that don't require lubrication and integrated thermal management systems. Implementation of new capabilities would have the potential to lower costs and manufacture times enormously.

Quantum Computing and Artificial Intelligence/Machine Learning

While quantum computing is still in the early phases, there have already been many innovations and break-throughs. It now appears that some of the most prominent and widely used AI and ML algorithms can be sped-up significantly if run on quantum computers, which does not mean performing a task faster, but rather taking a previously impossible task and making it possible, or even easy. As it pertains to AI/ML, there is the potential for classical and quantum machines to work together leveraging the elastic nature of the cloud and the powerful, specific problem-solving capabilities of quantum computing. Over time, both computing formats will continue to advance, but the ability to accelerate workloads on traditional graphics processing units and application-specific integrated circuits while also leveraging the power of quantum computing could be a recipe for faster, more robust computational capabilities, which we can expect to see as quantum computing becomes more widely accessible. The possibilities for space are significant with potential applications including fully autonomous science operations in challenging environments (roving, sampling, sample processing); sensor processing and interpretation; and in situ mobility with both major assets such as rovers and aircraft as well as probes. In addition, quantum sensors are also making great progress and have the potential to beat the performance of our traditional approaches. Quantum sensors have applications in a wide variety of fields including microscopy, positioning systems, communications, electric and magnetic field sensors, and seismology.

Small Fission Reactors for Power and Propulsion

Fission power systems (FPS) can offer a distinct advantage over other systems for higher power requirements and can offer new possibilities for more capable missions and access to the farthest reaches of the solar system and beyond. Nuclear-thermal propulsion and nuclear-electric propulsion (NEP), as an example, have been studied for decades but are not yet accessible because of their mass and large size. These technologies can significantly lower costs while enabling heavier science payloads, and more frequent missions (Polzin et al. 2020) and, more importantly, open exciting new science mission concepts. NEP technology, for example, can enable electric propulsion in the outer solar system where solar-electric propulsion is not feasible (Casani et al. 2020). Studies by NASA and the Department of Energy have shown that electric propulsion driven by a 10 kW space fission power reactor can enable a large number of exciting science missions to the outer solar system.

Commercial Exploration of Space

The current trend of the commercialization of the exploration of space, like the Commercial Lunar Payload Services and Human Landing System programs, has the potential of bringing new industry players with new ideas that go beyond the initial targets. The capability also expands the infrastructure of space components and services (e.g., communications, spacecraft platforms) that can be leveraged by future SMD missions to achieve great improvements in capability at lower costs.

Automotive Electronics

The past decade has seen great progress in the development of driverless car technology motivated by great advances in sensors, high performance computers, and navigation and control algorithms. It is worth noting that the electronics and software used in this application not only have to perform very complex autonomous functions that were undreamed-of a few years ago, but accomplished in the harsh demanding automotive environment while achieving the reliability demanded by the presence of human lives. All these properties are very relevant to the design and implementation of space missions. The technologies and commercial resources being developed to meet the needs of the autonomous car industry can have a profound impact in the development of future NASA's science missions.

Pulsar Navigation

X-ray pulsar-based navigation (XNAV) is a technique that uses X-ray signals emitted by pulsars and sensed by a vehicle's on-board X-ray sensors to determine its position in space. Its main advantage is that the spacecraft can autonomously and accurately determine its position without requiring Earth resources like the Deep Space Network. This capability could reduce our reliance on ground-based navigation capabilities and/or reduce the operations costs associated with these capabilities.

In Situ Resource Utilization

The Moon, Mars, and asteroids are key places for in situ resource utilization (ISRU) in the near-term. Human exploration of the Moon and Mars are currently driving the development of ISRU technologies. If successful, the resulting infrastructure and technology could be leveraged by robotic science missions at great benefit. Other solar system bodies such as Pluto, Charon, or other Kuiper belt objects could be further explored with a single mission to the surface. Given the possibility of ISRU, additive manufacturing and robotic assembly of spacecraft components could enable science mission extensions, such as the manufacturing of mobile probes, rovers, and aircraft, that could expand the observational capability of a single landed mission.

Finding: Emerging technologies in many different sectors offer game-changing opportunities to increase capabilities of our science investigations, while reducing the development burden and associated costs.

Recommendation: NASA should maintain cognizance of emerging new technologies and encourage the science and engineering communities to explore new ways that these technologies can enable greater science while reducing development and operations costs.

REFERENCES

Abell, P.A., C. Raymond, T. Daly, D.R. Adamo, B.W. Barbee, M. Bruck Syal, K. Carte, et al. 2020. "Near-Earth Object Characterization Priorities and Considerations for Planetary Defense." White paper #270 submitted to the Planetary Science and Astrobiology Decadal Survey 2023–2032. *Bulletin of the AAS* 53(4).

Amini, R., A. Azari, S. Bhaskaran, P. Beauchamp, J. Castillo-Rogez, R. Castano, S. Chung, et al. 2020. "Advancing the Scientific Frontier with Increasingly Autonomous Systems." White paper #513 submitted to the Planetary Science and Astrobiology Decadal Survey 2023–2032. *Bulletin of the AAS* 53(4).

Austin, A., G. Afonso, S. Albert, H. Ali, A. Alunni, J. Arnold, G. Bailey, et al. 2020a. "Aerocapture as an Enhancing Option for Ice Giants Missions." White paper #046 submitted to the Planetary Science and Astrobiology Decadal Survey 2023–2032. *Bulletin of the AAS* 53(4).

Austin, A., G. Afonso, S. Albert, H. Ali, A. Alunni, J. Arnold, G. Bailey, et al. 2020b. "Enabling and Enhancing Science Exploration Across the Solar System: Aerocapture Technology for SmallSat to Flagship Missions." White paper #057 submitted to the Planetary Science and Astrobiology Decadal Survey 2023–2032. *Bulletin of the AAS* 53(4).

Bapst, J., T.J. Parker, J. Balaram, T. Tzanetos, L.H. Matthies, C.D. Edwards, A. Freeman, et al. 2020. "Mars Science Helicopter Compelling Science Enabled by an Aerial Platform." White paper #361 submitted to the Planetary Science and Astrobiology Decadal Survey 2023–2032. *Bulletin of the AAS* 53(4).

Barbee, B.W., P.A. Abell, J. Atichson, O. Barnouin, S. Bhaskaran, J. Cahill, P. Chodas, et al. 2020. "Technology Development for Planetary Defense In Situ Spacecraft Missions to Near-Earth Objects." White paper #113 submitted to the Planetary Science and Astrobiology Decadal Survey 2023–2032. *Bulletin of the AAS* 53(4).

Bugga, R., E. Brandon, E. Darcy, R. Ewell, P. Faguay, B. Futz, R. Gitzendanner, et al. 2020. "Energy Storage Technologies for Planetary Science and Astrobiology Missions." White paper #201 submitted to the Planetary Science and Astrobiology Decadal Survey 2023–2032. *Bulletin of the AAS* 53(4).

Casani, J.R., M.A. Gibson, D.I. Poston, N.J. Strange, J.O. Elliott, R.L. McNutt, S.L. McCarty, et al. 2020. "Enabling a New Generation of Outer Solar System Missions: Engineering Design Studies for Nuclear Electric Propulsion." White paper #011 submitted to the Planetary Science and Astrobiology Decadal Survey 2023–2032. *Bulletin of the AAS* 53(4).

Casillas, A.R., G. Barnett, C. Engelbrecht, C.S. Guernsey, J. McKinnon, M. Preudhomme, J.R. Reh, et al. 2020. "Affordability of Outer-Planet Exploration: A Pragmatic Rationale for Implementing a Cold-Propulsion Based Energy-Efficient Spacecraft Infrastructure." White paper #308 submitted to the Planetary Science and Astrobiology Decadal Survey 2023–2032. *Bulletin of the AAS* 53(4).

Cutts, J.A., S. Aslam, S. Atreya, K. Baines, P. Beauchamp, J. Bellan, D.C. Bowman, et al. 2020. "Scientific Exploration of Venus with Aerial Platforms." White paper #194 submitted to the Planetary Science and Astrobiology Decadal Survey 2023–2032. *Bulletin of the AAS* 53(4).

Dachwald, B., S. Ulamec, F. Postberg, F. Sohl, J.-P. de Vera, C. Waldmann, R.D. Lorenz, et al. 2020. "Key Technologies and Instrumentation for Subsurface Exploration of Ocean Worlds." *Space Science Reviews* 216(83):1–45. https://doi.org/10.1007/s11214-020-00707-5.

Dutta, S., G. Afonso, S.W. Albert, H.K. Ali, G.A. Allen, A.I. Alunni, J.O. Arnold, et al. 2020. "Aerocapture as an Enhancing Option for Ice Giants Missions." White paper #046 submitted to the Planetary Science and Astrobiology Decadal Survey 2023–2032. *Bulletin of the AAS* 53(4).

Ellerby, D., H. Hwang, M. Gasch, R. Beck, and T. White. 2020. "TPS and Entry Technologies for Future Outer Planet Exploration." White paper #393 submitted to the Planetary Science and Astrobiology Decadal Survey 2023–2032. *Bulletin of the AAS* 53(4).

Galbraith, J.K. 1978. *The New Industrial State*. Boston, MA: Houghton Mifflin.

Hamkins, J., D. Antsos, J. Border, G. Davis, L. Deutsch, J. Lazio, and J. Velazco. 2020. "Communication and Navigation Technologies." White paper #085 submitted to the Planetary Science and Astrobiology Decadal Survey 2023–2032. *Bulletin of the AAS* 53(4).

Howell, S., W.C. Stone, K. Craft, C. Manager, A. Murray, and A. Rhoden. 2020. "Ocean Worlds Exploration and the Search for Life." White paper #191 submitted to the Planetary Science and Astrobiology Decadal Survey 2023–2032. *Bulletin of the AAS* 53(4).

Hunter, G., T. Kremic, and P. Neudeck. 2020. "High Temperature Electronics for Venus Surface Applications: A Summary of Recent Technical Advances." White paper #399 submitted to the Planetary Science and Astrobiology Decadal Survey 2023–2032. *Bulletin of the AAS* 53(4).

Lazio, T.J.W., A.K. Virkki, N. Pinilla-Alonso, L.A.M. Benner, M. Brozovic, B.J. Butler, B.A. Campbell, et al. 2020. "The Next-Generation Ground-Based Planetary Radar." White paper #434 submitted to the Planetary Science and Astrobiology Decadal Survey 2023–2032. *Bulletin of the AAS* 53(4).

Matthies, L.H., P.G. Backes, J.L. Hall, B.A. Kennedy, S.J. Moreland, H.D. Nayar, I.A. Nesnas, J. Sauder, and K. Zacny. 2020. "Robotics Technology for In Situ Mobility and Sampling." White paper #353 submitted to the Planetary Science and Astrobiology Decadal Survey 2023–2032. *Bulletin of the AAS* 53(4).

McKay, C., A. Davila, J. Eigenbrode, C. Lorenston, J. Canham, A. Dazzo, T. Errigo, et al. 2020. "Contamination Control Technology Study for Achieving the Science Objectives of Life-Detection Missions." White paper #340 submitted to the Planetary Science and Astrobiology Decadal Survey 2023–2032. *Bulletin of the AAS* 53(4).

Milam, S.N., J.P. Dworkin, J.E. Elsila, D.P. Glavin, P.A. Gerakines, J.L. Mitchell, K. Nakamura-Messenger, et al. 2020. "Volatile Sample Return in the Solar System." White paper #049 submitted to the Planetary Science and Astrobiology Decadal Survey 2023–2032. *Bulletin of the AAS* 53(4).

Mittelholz, A., J. Epsley, J. Connerney, R. Fu, C.L. Johnson, B. Langlais, R.J. Lillis, et al. 2020. "Mars' Ancient Dynamo and Crustal Remanent Magnetism." White paper #006 submitted to the Planetary Science and Astrobiology Decadal Survey 2023–2032. *Bulletin of the AAS* 53(4).

NAS, NAE, and IOM (National Academy of Sciences, National Academy of Engineering, and Institute of Medicine). 2007. *Rising Above the Gathering Storm: Energizing and Employing America for a Brighter Economic Future.* Washington, DC: The National Academies Press. https://doi.org/10.17226/11463.

NASEM (National Academies of Sciences, Engineering, and Medicine). 2018. *Visions into Voyages for Planetary Science in the Decade 2013–2022: A Midterm Review.* Washington, DC: The National Academies Press.

NASEM. 2019a. *An Astrobiology Strategy for the Search for Life in the Universe.* Washington, DC: The National Academies Press. https://doi.org/10.17226/25252.

NASEM. 2019b. *Strategic Investments in Instrumentation and Facilities for Extraterrestrial Sample Curation and Analysis.* Washington, DC: The National Academies Press. https://doi.org/10.17226/25312.

NRC (National Research Council). 2011. *Vision and Voyages for Planetary Science in the Decade 2013–2022.* Washington, DC: The National Academies Press. https://doi.org/10.17226/13117.

Overy, R.D., G.G. Sadler, J.P. Fleurial, and D.G. Hall. 2021. "Radioisotope Power for Scientific Exploration." 52nd Lunar and Planetary Science Conference 2021 (LPI Contrib. No. 2548).

Polk, J.E., G. Soulas, and A. Hoskins. 2020. "Electric Propulsion: A Key Enabling Technology for Planetary Exploration." White paper #480 submitted to the Planetary Science and Astrobiology Decadal Survey 2023–2032. *Bulletin of the AAS* 53(4).

Polzin, K.A., C.R. Joyner, T. Kokan, S. Edwards, A. Irvine, M. Rodriguez, and M. Houts. 2020. "Enabling Deep Space Science Missions with Nuclear Thermal Propulsion." White paper #093 submitted to the Planetary Science and Astrobiology Decadal Survey 2023–2032. *Bulletin of the AAS* 53(4).

Sacco, E., and S.K. Moon. 2019. "Additive Manufacturing for Space: Status and Promises." *International Journal of Advanced Manufacturing Technology* 105(10):4123–4146. https://doi.org/10.1007/s00170-019-03786-z.

Schmidt, B. 2020. "A Sustainable Partnership Between NASA and NSF for Planetary Science and Astrobiology Research in Antarctica." White paper #508 submitted to the Planetary Science and Astrobiology Decadal Survey 2023–2032. *Bulletin of the AAS* 53(4).

Schwartz, J.A., R. Ewell, N. Haegel, S. Liu, J. Mcnatt, E. Plichta, and S. Surampudi. 2020. "Solar Array Technologies for Planetary Science and Astrobiology Missions." White paper #080 submitted to the Planetary Science and Astrobiology Decadal Survey 2023–2032. *Bulletin of the AAS* 53(4).

Stone, W.C., S.M. Howell, N. Bramall, C. German, A. Murray, and V. Siegel. 2020. "National Ocean Worlds Analog Test Facility and Field Station." White paper #416 submitted to the Planetary Science and Astrobiology Decadal Survey 2023–2032. *Bulletin of the AAS* 53(4).

Theiling, B.P., W. Brinckerhoff, J. Castillo-Rogez, L. Chou, V. Da Poian, H. Graham, S.S. Hosseini, et al. 2020. "Non-Robotic Science Autonomy Development." White paper #048 submitted to the Planetary Science and Astrobiology Decadal Survey 2023–2032. *Bulletin of the AAS* 53(4).

22

Recommended Program: 2023–2032

This chapter outlines a prioritized program of research activities to advance the frontiers of planetary science, astrobiology, and planetary defense in the 2023–2032 decade. The recommended program, traced directly from the priority science questions, defines an integrated portfolio of flight projects, high-priority research activities, and technology development that will produce transformative advances in our knowledge and understanding. The program is shown to be achievable within realistic funding profiles, and decision rules are identified to accommodate budgetary changes or new scientific or technical developments. The recommended program is balanced across activities of varied scale and scientific focus and includes key areas for cooperation with NASA's human exploration program and U.S. agency, industry, and international partners. This chapter addresses items three through nine[1] in the Statement of Task (see Appendix A).

The chapter begins with background on approaches and definitions and an assessment of key elements of NASA's ongoing missions and existing programs and activities. This discussion is followed by recommendations and prioritizations of future missions, and a detailed description of the recommended overall program for representative budgetary profiles, including decision rules. The final sections provide recommendations on major aspects of NASA's activities that have broad importance for both the recommended program and the continued success of the nation's planetary science and astrobiology efforts, informed by key findings and recommendations detailed in Chapters 16–21.

SCIENTIFIC THEMES AND PRIORITY SCIENCE QUESTIONS

This decadal report is the first to be organized according to significant, overarching science questions. The committee first identified three high-level scientific themes[2] as intellectual drivers for the pursuit of planetary science and astrobiology:

Origins: How did the solar system and Earth originate, and are systems like ours common or rare in the universe?
Worlds and Processes: How did planetary bodies evolve from their primordial states to the diverse objects seen today?
Life and Habitability: What conditions led to habitable environments and the emergence of life on Earth, and did life form elsewhere?

---

[1] Note: this chapter only includes summary discussions of and key recommendations for the state of the profession (Chapter 16), research and analysis (Chapter 17), planetary defense (Chapter 18), infrastructure (Chapter 20), and technology (Chapter 21). Readers interested in more details on any and all of these topics should consult the relevant chapter.

[2] It is from these themes that the report's title—*Origins, Worlds, and Life*—is derived.

Across these themes, the committee defined 12 priority science questions, shown in Table 22-1, each comprised of a single overarching topic and a brief description of the question's scope. Chapters 3 through 15 describe these priority questions, as well as specific sub-questions and strategic research activities for each that would provide substantial advances in understanding over the next decade. The strategic research activities span basic theory and modeling, laboratory and experimental work, ground-based observations, and spacecraft measurements and investigations.

TABLE 22-1 The Twelve Priority Science Question

Scientific Themes	Priority Science Question Topics and Descriptions
(A) Origins	Q1. *Evolution of the protoplanetary disk.* What were the initial conditions in the solar system? What processes led to the production of planetary building blocks, and what was the nature and evolution of these materials?
	Q2. *Accretion in the outer solar system.* How and when did the giant planets and their satellite systems originate, and did their orbits migrate early in their history? How and when did dwarf planets and cometary bodies orbiting beyond the giant planets form, and how were they affected by the early evolution of the solar system?
	Q3. *Origin of Earth and inner solar system bodies.* How and when did the terrestrial planets, their moons, and the asteroids accrete, and what processes determined their initial properties? To what extent were outer solar system materials incorporated?
(B) Worlds and Processes	Q4. *Impacts and dynamics.* How has the population of solar system bodies changed through time, and how has bombardment varied across the solar system? How have collisions affected the evolution of planetary bodies?
	Q5. *Solid body interiors and surfaces.* How do the interiors of solid bodies evolve, and how is this evolution recorded in a body's physical and chemical properties? How are solid surfaces shaped by subsurface, surface, and external processes?
	Q6. *Solid body atmospheres, exospheres, magnetospheres, and climate evolution.* What establishes the properties and dynamics of solid body atmospheres and exospheres, and what governs material loss to space and exchange between the atmosphere and the surface and interior? Why did planetary climates evolve to their current varied states?
	Q7. *Giant planet structure and evolution.* What processes influence the structure, evolution, and dynamics of giant planet interiors, atmospheres, and magnetospheres?
	Q8. *Circumplanetary systems.* What processes and interactions establish the diverse properties of satellite and ring systems, and how do these systems interact with the host planet and the external environment?
(C) Life and Habitability	Q9. *Insights from terrestrial life.* What conditions and processes led to the emergence and evolution of life on Earth; what is the range of possible metabolisms in the surface, subsurface, and/or atmosphere; and how can this inform our understanding of the likelihood of life elsewhere?
	Q10. *Dynamic habitability.* Where in the solar system do potentially habitable environments exist, what processes led to their formation, and how do planetary environments and habitable conditions co-evolve over time?
	Q11. *Search for life elsewhere.* Is there evidence of past or present life in the solar system beyond Earth, and how do we detect it?
Cross-cutting A–C linkage	Q12. *Exoplanets.* What does our planetary system and its circumplanetary systems of satellites and rings reveal about exoplanetary systems, and what can circumstellar disks and exoplanetary systems teach us about the solar system?

Topics that appear frequently in the science question chapters include:

- The central role of sample return and in situ analyses for providing breakthrough science and ground-truth constraints;
- The dearth of knowledge of the ice giant[3] systems, which may represent the most common class of exoplanets, and the importance of ice versus gas giant comparative studies;
- Effects of nebular processes on compositional mixing and the formation of planetary building blocks and primitive bodies, and the need for further constraints on the solar system's dynamical evolution, from primordial planet migration to ongoing bombardment;
- The complex interplay of internal and external processes—many still ongoing—that affect planets, moons, rings, and small bodies, and the factors responsible for the varied initial states and divergent evolutionary paths of our terrestrial planets;
- The conditions that led to the emergence of life on Earth, and the compelling rationale for understanding habitability beyond Earth, particularly at Mars and icy ocean worlds; and
- A strong desire in the coming decade to make substantive progress in understanding whether life existed (or exists) elsewhere in the solar system.

These topics figured prominently in the committee's discussions and provided a backdrop for the committee's work in developing a balanced portfolio of recommended research activities and missions.

Program Evaluation Approach

The committee developed its recommended program by evaluating potential research activities against several general criteria. First and foremost was the capacity to deliver breakthrough science return as a function of activity cost level and technical readiness. Science return was judged with respect to the priority scientific questions[4] (Q1 through Q12), while cost and technical readiness were assessed through an independent evaluation process, described below. A second criterion was programmatic balance across the priority science questions and target destinations, together with an appropriate mix of small, medium, and large activities. Other criteria included the potential for cooperation with planned human exploration efforts and other key partners, as well as the availability of trajectory opportunities within the period 2023–2032. There was not a fixed weighting for each of these criteria: for example, for a mission that uniquely addresses a crucial science question, science was considered a proportionally larger factor, while for a mission without a viable trajectory, technical feasibility was proportionally more important. Committee decisions considered the state of science, technology, and mission studies at the time of this report's writing.

Definition of Mission Classes and Mission Lines

The statement of task calls for the evaluation of NASA's planetary missions in three separate cost classes—small, medium, and large. The Discovery program supports small missions, while the New Frontiers program supports medium-class missions. Large missions are commonly referred to as Flagship missions. Any mission class may also be supported through destination-specific programs within NASA's Planetary Science Directorate (PSD), namely the Mars Exploration and Lunar Discovery and Exploration programs. Small- and medium-class missions may either be led by a principal investigator (PI) or be strategically directed by NASA; large missions are strategically directed by NASA and are not PI-led. Mission classes are differentiated not only by their costs but also by their timescale of execution, span of technology, and involvement of the scientific community.

The Discovery program supports PI-led missions that address focused science objectives, with each mission selected through a competitive process. Rapid mission development (3–5 years) is feasible and desirable, and

---

[3]While the mass of a gas giant (e.g., Jupiter or Saturn) is dominated by hydrogen and helium, the mass of an ice giant (e.g., Uranus or Neptune) is dominated by heavier elements (e.g., carbon, nitrogen, and oxygen), some portion of which is in the form of ices.

[4]Priority science questions, detailed in Chapters 4 through 15, are referenced by question number as Q1 to Q12, respectively.

emphasis is placed on maintaining a high cadence of launches on average every 24 months. The Discovery program does not specify mission objectives or destinations, and proposers may address any science that can be accommodated within a specified cost cap. The overall program structure allows Discovery missions to rapidly respond to new discoveries and nimbly address changing scientific priorities, while encouraging creative new ideas and approaches to maximizing science return per dollar. The program has been remarkably successful, with 13 launches since its initiation, and remains a vibrant and highly valuable component of NASA's planetary mission portfolio.

New Frontiers (NF) missions address a broader suite of science goals than those that can be implemented within Discovery but are still more scientifically focused than Flagship missions. NF missions can be executed on timescales of less than a decade, and while they are complex and challenging, they typically take advantage of technological developments from prior missions. NF missions, like those in Discovery, are PI-led and are selected via a competitive process, a model proven effective in maximizing innovation and community involvement. In contrast to the Discovery program, NF solicitations have been more strategic, restricting proposals to a small number of specific mission themes identified primarily through decadal surveys.

Flagship missions (e.g., Viking, Voyager, Galileo, Cassini, Mars Science Laboratory, Mars 2020, Europa Clipper, and subsequent missions in the Mars Sample Return campaign) have an approximately 10-year development cycle. These missions address a wide range of important scientific objectives at high priority targets and utilize sophisticated instrument payloads that involve some degree of new technology development. Flagship missions typically require very capable launch vehicles, large teams of investigators, and a complex organization of supporting institutions, often with multi-agency and international involvement. Because of their scientific breadth, high cost, technical complexity, and high strategic importance to NASA and the nation, flagship missions are directed missions, typically with PI-led individual instruments selected through open competition. Although large in cost, Flagship missions have consistently proven to have a high science return per dollar and have engaged a large fraction of the planetary science community (NASEM 2017b).

Program Balance Considerations

The prior decadal survey, *Vision and Voyages* (NRC 2011), provided a strong rationale for the importance of balance across mission cost classes when defining a mission portfolio to maximize overall scientific return. Its logic remains pertinent today. A program consisting of a single Flagship per decade would be unable to respond to ongoing scientific developments and result in long stretches of time with relatively little new data, whereas a portfolio of only Discovery-class missions could not address the most complex science questions that require an integrated suite of sophisticated instruments, complex mission designs, long duration investigations, and/or travel to distant targets. The former could lead to stagnant science and engineering communities, while the latter could fail to yield transformative scientific advances.

The past 30 years of planetary robotic exploration has generally followed a progression of mission types—from reconnaissance flybys, to orbital investigations, to in situ exploration, to sample return—at a growing number of target destinations. Each step along this progression allows us to address more sophisticated and challenging scientific questions, with a commensurate increase in mission and instrumentation complexity and cost. Different objects in the solar system are currently at different stages in this exploration continuum, and as such a balanced portfolio will naturally contain a range of mission classes. For example, as described below, sample return from Mars is planned for the same overall timeframe as the initial flyby reconnaissance of the distant Trojan asteroids.

Vision and Voyages provided the following criteria by which a flight program may be assessed:

- Capacity to make steady progress—Does the proposed program make reasonable progress toward the scientific goals set forth in the decadal survey? Are the cadence of missions and the planning process such that new scientific discoveries can be followed up rapidly with new missions, such as small missions in the Discovery program? Does the program smoothly match and complement programs initiated by prior decadal surveys?
- Stability—Can one construct an orderly sequence of missions, meeting overarching scientific goals, developing advanced technology, sizing and nurturing the research and technical community, and providing

for appropriate interactions with the international community? Is the program stable under the inevitable budgetary perturbations as well as the occasional mission failures?

- Balance—Is the program structured to contain a mix of small, medium, and large missions that together make the maximum progress toward the scientific goals envisioned by this decadal survey? Can some of the scientific objectives be reached or approached via missions of opportunity and by means of piggyback or secondary flights of experiments on other NASA missions?
- Robustness—Is the program robust in that it provides opportunities for the training and development of the next generation of planetary scientists? Is it robust in that it lays the technological foundation for a period longer than the present decade?

The criteria cited above are not orthogonal. "Balance" in various guises permeates the other three criteria. For example, a balanced portfolio of missions enhances overall program stability, and provides better assurance of a continuing stream of new results. A balanced portfolio also helps prevent large excursions in workforce demands and cost, therefore fitting more easily into the relatively smooth year-to-year NASA budget.

> **Recommendation: NASA's suite of planetary missions should continue to consist of a balanced mix of small, medium, and large missions, enabling both a steady stream of new discoveries and the capability to make major scientific and technical advances, as well as the needed training of future generations of planetary scientists.**

ONGOING MISSIONS AND EXISTING PROGRAMS

The overall goal of this report is to recommend an integrated portfolio of flight projects, technology development, and supporting research activities to maximize the advancement of planetary science and astrobiology, as well as planetary defense, over the next decade. Recommended future activities need to be considered within the context of the existing planetary exploration program. In this section, the committee discusses and provides recommendations related to existing activities, including (1) operating missions; (2) missions and international contributions in development; (3) Mars Sample Return; and (4) the existing Mars Exploration, Lunar Discovery and Exploration, Research and Analysis, Astrobiology, and Planetary Defense programs.

Operating Missions

NASA has been remarkably successful in launching and operating planetary missions in the past decade. Currently operating PSD spacecraft include the ongoing Mars orbiter missions, the Curiosity and Perseverance Mars rovers, the Lunar Reconnaissance Orbiter, the InSight and Lucy Discovery missions, and the New Horizons, Juno, and OSIRIS-REx New Frontiers missions. These missions are at varied stages of operation, ranging from newly launched to those in extended mission phases. Extended missions can provide important new data and scientific insights with high science value per additional dollar expended (NASEM 2016). Another important aspect of extended missions is their role in providing mission experience and leadership for the development of the next generation of scientists and engineers who will lead the missions of the future. To ensure appropriate and high-value scientific return, NASA evaluates missions proposing to enter into or operating within an extended mission phase through the Senior Review process.

> **Finding:** The committee endorses the Senior Review process for evaluation of the merit of extended missions and supports continued operation of missions prioritized by the Senior Review as representing high scientific return and/ or programmatic importance.

New Frontiers, Discovery, and SIMPLEx Missions in Development

Vision and Voyages recommended a total of seven New Frontiers mission themes: five for the New Frontiers 4 (NF-4) call, and two to be added to the New Frontiers 5 (NF-5) call. NASA subsequently added an ocean worlds mission theme to NF-4. NASA selected the Dragonfly mission to Titan in NF-4, which is currently in development.

In development Discovery missions include the Psyche mission to the asteroid Psyche, and the DAVINCI and VERITAS missions to Venus. Four small satellite missions under the recently introduced Small Innovative Missions for Planetary Exploration (SIMPLEx) line are also in development.

Finding: The committee strongly endorses the continued development of the Dragonfly, Psyche, DAVINCI, VERITAS, and small satellite missions. The committee finds the projected costs of these missions to be commensurate with their expected scientific return.

Contributions to International Missions in Development

NASA is providing hardware and other contributions to a variety of international space missions, including, for example, the ESA-led JUICE mission to Ganymede and the Jupiter system, the ESA-led ENVISION mission to Venus, and the JAXA-led MMX mission to the small moons of Mars.

Finding: The committee strongly endorses NASA's partnerships with international space agencies and recognizes the benefits these bring to the U.S. space science community.

Europa Clipper

The second highest priority Flagship mission identified in *Vision and Voyages* was the Jupiter Europa Orbiter (JEO). NASA followed the decadal survey's recommendations to descope the mission, which will now perform multiple flybys rather than orbit Europa, and to use solar rather than radioisotope thermoelectric power. The resulting Europa Clipper Flagship mission is in development, with a planned launch in October 2024. This mission will provide a critical foundation for the exploration of ocean worlds through its focused exploration of an important target of high astrobiological interest. As a large Flagship class mission, its budgetary growth has the potential to undercut other parts of the planetary program if not closely monitored and thoughtfully controlled (NASEM 2018b).

Recommendation: NASA should continue the development of the Europa Clipper mission and closely monitor the mission's cost.

Mars Sample Return

The prior decadal survey recommended a Mars sample caching rover as its top priority Flagship mission, and as the first step in a campaign to return martian samples to Earth. *Vision and Voyages* envisioned Mars Sample Return as being "broken into three separate missions that can be spaced out over two or even three decades, reducing the per-year costs, and thus making it easier for programmatic balance to be maintained." Following the recommendations of *Vision and Voyages*, the Mars Astrobiology Explorer-Cacher (MAX-C) mission was redesigned to include a single rover, allowing the reuse of existing Mars landing technologies. NASA implemented this recommendation with the successful landing of the Perseverance rover on Mars on February 18, 2021, which has begun its primary mission goal of acquiring and caching high-quality samples for return to Earth.

In 2017, NASA SMD introduced a concept to commence Mars Sample Return (MSR) through a "focused and rapid" campaign with essential participation from ESA; planning and implementation of this concept has proceeded subsequently. NASA asked the committee to assess whether NASA's MSR plans play an appropriate role in the research strategy for the next decade (see "Considerations for Planetary Defense Flight Programs" in Appendix A), and, as appropriate, to provide recommendations related to the plans themselves.

While many science questions remain to be explored from orbit and the surface (e.g., MASWG 2020), key scientific objectives for Mars and, more broadly, for planetary and astrobiological science, can only be achieved via study of carefully selected martian samples in terrestrial laboratories. The Perseverance rover is collecting samples from Jezero crater, a former lake basin with a feeding channel system that was carved into Noachian (>3.7 Ga) stratigraphy. Distinct types of sedimentary, igneous, water-altered, and impact-formed rocks accessible in this region will provide a geological record of a time interval particularly important for understanding Mars's environmental evolution and, potentially, its biology. Sample return will provide geologic materials that are not represented among martian meteorites (e.g., sediments and water-altered rocks), and whose volatile, organic, and secondary mineral components have not been altered by impact. Sample context—location and stratigraphic position—will enable a chronology and age-dating of key episodes in Mars's evolution and calibrate timings inferred from crater size frequency distributions that constrain the early impactor flux in the inner solar system.

The scientific impact of sample return is broad, both because of the types of samples to be collected and the measurements possible in laboratories on Earth (iMOST 2019). Certain types of measurements (e.g., phase-specific stable and radiogenic isotopes, trace elements, nanometer-scale composition and texture, and precise organics characterization) cannot be done remotely because they require sample preparations and analytical precisions only possible in specialized laboratories. Key science investigations that will be conducted on the returned samples include:

- Coordinated analyses in the search for life, including analyzing potential biosignatures and inventories of prebiotic organic molecules; measuring compound-specific isotopic compositions and chirality; and establishing that such features are indigenous through preprocessing, nanometer-scale textural interrogation and follow-up observations with multiple methods.
- Constraining the nature and longevity of environments with liquid water and assessing habitability and volatile inventory. Samples of ancient atmosphere (trapped in inclusions in minerals/glasses) will allow measurements of elements and rare isotopes (e.g., noble gases) that cannot be determined remotely (iMOST 2019). Samples of modern atmosphere will allow noble and trace gas measurement (Jakosky et al. 2020).
- Determining radiometric ages, critical trace element abundances, nucleosynthetic isotope anomalies, and paleomagnetic properties to quantify and further understand Mars's origin, differentiation, and geologic evolution, as well as compositional heterogeneity and dynamical mixing in the early solar system.

In addition, sample return will allow for future analyses by instruments and techniques not yet developed. As has been the case with the Apollo samples from the Moon, future analyses are expected to yield profound results for many decades after sample return.

Finding: The committee reaffirms the broad and fundamental scientific importance of Mars Sample Return recognized in *Vision and Voyages*, the 2018 Decadal Midterm Review (NASEM 2018a), and the 2020 Independent Review Board Report (NASA 2020b). MSR will enable investigations to address many fundamental issues, including crucial elements of Q3 through Q6, Q10, and Q11. MSR will also provide an invaluable sample collection to the benefit of future generations. As such, the committee finds that the aspirational, groundbreaking MSR campaign plays an appropriately central role in the research strategy for planetary science and astrobiology in the next decade.

NASA's plans for MSR were extensively reviewed by an Independent Review Board (IRB) in December 2020. This technically demanding endeavor includes three interconnected missions: Perseverance, an Earth Return Orbiter (ERO), and a Sample Retrieval Lander (SRL). The SRL will carry a Sample Fetch Rover and Sample Transfer Arm to be supplied by the European Space Agency (ESA), as well as a Mars Ascent Vehicle to be developed and provided by NASA. The ERO is being developed by ESA. While the IRB concluded that decades of Mars exploration and years of preliminary planning have prepared NASA and ESA to undertake MSR, they found the plan they reviewed to be high-risk. The IRB identified several options to reduce both technical and programmatic risks, including a replan for SRL and ERO launches in 2028 (a shift from a baseline 2026 launch plan), a budgetary increase, further exploration of mission architectural and vehicle options, and simplification/consolidation of MSR organizational and management structures. The overall IRB recommendation was that MSR proceed, owing to its extraordinary scientific value and potential for world-changing discoveries. The committee endorses the findings and recommendations of the December 2020 IRB report on the MSR Program.

Recommendation: The highest scientific priority of NASA's robotic exploration efforts this decade should be completion of Mars Sample Return as soon as is practicably possible with no increase or decrease in its current scope.

The IRB described MSR as "arguably the most technically difficult and operationally demanding robotic space mission NASA and ESA have ever undertaken," and asserted that, given its ambitious goals and complexity, it is essential that MSR be conducted with mission success as its top priority. In such a situation, there is clearly potential for mission cost growth, particularly if the 2026–2028 launch window is not achieved. Given the financial scale of MSR, cost growth has the potential to impact other NASA planetary science programs severely.

Mars exploration has historically figured prominently in NASA's planetary program, and annual funding for Mars exploration has varied from ~25 to ~35 percent of the PSD budget over the past three decades. The representative PSD programs presented later in this chapter (Table 22-2) adopt a total MSR cost of $5.3 billion (this includes NASA's contribution to a Sample Receiving Facility),[5] and restrict funding for the Mars Exploration Program (MEP) to a relatively low level until the end of the decade when MSR's annual costs decrease. The resulting decade-long total for Mars exploration in these plans, including both MSR and MEP, is ~20 percent of the total PSD budget over the decade, and ≤30 percent in any single year, which is lower than over the past several decades. Even with up to an additional 20 percent growth in MSR's total cost, the next decade's funding for Mars exploration would remain at or below previous percentage levels. However, cost growth of MSR at this level would, without an associated budget augmentation, undermine the programmatic balance across the priority scientific questions, mission classes, and target destinations, damaging the health of NASA's overall planetary science program. Delaying the completion of MSR would undoubtably increase its total cost, which would also negatively impact the long-term health of the planetary program. Therefore, the completion of MSR as rapidly and efficiently as possible is highly desirable. The committee also notes that further reductions in the MEP program beyond that in the Level Program outlined here (Table 22-2) could not offset significant MSR cost growth without severely impacting the long-term health of the nation's strategic program at Mars.

> **Recommendation: Mars Sample Return is of fundamental strategic importance to NASA, U.S. leadership in planetary science, and international cooperation, and should be completed as rapidly as possible. However, its cost should not be allowed to undermine the long-term programmatic balance of the planetary portfolio. If the cost of MSR increases substantially (≥20 percent) beyond the $5.3 billion level adopted here, or goes above ~35 percent of the PSD budget in any given year, NASA should work with the Administration and Congress to secure a budget augmentation to ensure the success of this strategic mission.**

Key Existing Programs and Activities

In this section, the committee discusses prominent programs and activities within NASA's PSD, including the Mars Exploration and Lunar Discovery and Exploration programs, the Research and Analysis and Astrobiology programs, and the Planetary Defense Coordination Office.

Throughout the history of planetary exploration NASA has used a variety of approaches for mission selection and development, from typically decade-long, or longer, flagship missions that address a broad suite of fundamental science questions to smaller, higher cadence Discovery missions that address focused questions at diverse targets. In addition, NASA has created programs to define and coordinate its efforts for scientific exploration of particular objects that have broad importance across a combination of scientific and programmatic priorities, including coordination with NASA's human exploration plans.

The Mars Exploration Program (MEP) is a scientific success story, providing stability across decades to support long-term, strategic science planning; coordination across multiple missions and assets (often operating simultaneously and in support of each other); development of a multi-generational science community that defines the goals and priorities of the program; international coordination and involvement; infrastructure and technology development; and development of scientific connections to human exploration plans. The relatively short travel times to Mars have allowed for a structured, multi-mission approach, with each mission building on its predecessor and addressing science questions of increasing sophistication. The 2019 formation of the Lunar Discovery and Exploration Program (LDEP) has the potential to emulate the successful MEP model by supporting activities focused on the Moon such as those listed above for Mars. The Moon's greater proximity and accessibility affords

---

[5] All dollar amounts are in real-year dollars unless specified otherwise.

tremendous opportunities for frequent missions with increasingly sophisticated payloads, growing commercial and international partnerships, and advancement of decadal-level science[6] through coordination with near-term human exploration activities planned for the next decade.

> **Finding:** Mars and the Moon are currently unique in the breadth of activities focused on them. Further, they each provide the opportunity to investigate a wide range of priority science questions at destinations that are relatively easy to reach. These aspects justify the existence of MEP and LDEP as dedicated programs and their associated administrative costs.

Other objects or classes of objects in the solar system, while currently not subject to the same breadth of activity as Mars and the Moon, also merit coordinated, strategic scientific planning. As the number of NASA and international missions to specific destinations or classes of objects increases, such as has recently occurred in the selection of three missions to Venus and increased interest in ocean world missions, the need for coordination and the opportunities for collaboration also increase. Scientific exploration strategies may consider, for example, (1) coordination within NASA to support key research topics encompassing remote-sensing, laboratory, theoretical, and ground- and space-based telescopic investigations focused on upcoming missions; (2) a technology development plan to enable future missions; and (3) collaboration on possible future activities between U.S. and international and commercial partners to maximize NASA's investments, aid in the selection of an optimal suite of missions, and enhance the exchange of scientific knowledge and data.

> **Recommendation: NASA should develop scientific exploration strategies, as it has for Mars, in areas of broad scientific importance, for example, Venus and ocean worlds that have an increasing number of U.S. missions and international collaboration opportunities.**

Mars Exploration Program

Over the past three decades, MEP has maintained a portfolio of small to large Mars science missions. As detailed in the MASWG 2020 report, three attributes drive the programmatic approach at Mars:

- Scientific: Mars is a key destination in the search for past and present extraterrestrial life. It provides the opportunity to search for extant life and, thanks to its 4.5-billion-year-old rock record that includes a relatively pristine record of the first billion years of solar system history, it offers a unique opportunity to investigate the full range of interacting processes on habitable terrestrial planets (e.g., geoscience, climate/ atmosphere, space weather, and potentially biology) under different conditions from Earth (see priority science questions 5, 6, 10, and 11).
- Programmatic: Mars is accessible, allowing multiple mission classes to explore different components of the martian environment and their interactions, including access to the surface and near subsurface.
- Exploration: Mars is NASA's stated long-term destination for human exploration. Mars is also a destination of rapidly increasing interest for multiple international space agencies (including those of Europe, China, Japan, India, UAE, and Russia), as well as commercial entities, with whom our scientific understanding will benefit from long-term coordination.

The committee reaffirms findings of the *Vision and Voyages* midterm review (NASEM 2018a) regarding the importance of coordinating Mars exploration and managing the MEP as a program, rather than just as a series of missions, to optimize science at the architectural level.

> **Finding:** MEP has a record of success, and it has advanced our understanding of Mars and the evolutionary paths of terrestrial planets while also fostering, for example, technological developments, identification of opportunities for joint mission implementation, and public enthusiasm for planetary science.

---

[6]Decadal-level science is that which results in significant, unambiguous progress in addressing at least one of the survey's 12 priority science questions.

The committee recommends above that completion of MSR be the top priority for NASA's scientific exploration of Mars in the next decade. MSR is managed as a separate program outside of MEP, a logical approach given the size and complexity of this international endeavor.

Finding: Maintaining strong scientific, programmatic, and strategic connections between the MEP and MSR will provide the scientific leadership for MSR and ensure the feed forward from the sample-return campaign to future Mars science activities.

There remain many fundamental science questions at Mars beyond those addressed by MSR. Many of these require a series of measurements that build on prior data, and/or measurements at varied locations, times, or parts of the Mars system. Thanks to Mars's relative accessibility, international partners eager to pursue partnerships with NASA, and increasing capabilities of small spacecraft, effective coordination by MEP this decade can support a mission cadence to enable ongoing discovery along multiple arcs of priority science goals (see MASWG 2020). New, rapid, and low-cost exploration techniques using proven technology advancements—such as innovative landing methods, small satellites, and aerial vehicles—can be part of the MEP strategy to advance scientific and human exploration goals. MEP may also have opportunities this decade to prioritize science on future human exploration missions (see Chapter 19).

Recommendation: NASA should maintain the Mars Exploration Program, managed within the PSD, that is focused on the scientific exploration of Mars. The program should develop and execute a comprehensive architecture of missions, partnerships, and technology development to enable continued scientific discovery at Mars.

As highlighted in community documents (MEPAG 2020), certain high-priority science objectives at Mars—including subsurface ice composition, detailed organics characterization to search for modern biosignatures, and in situ stable and radiogenic isotopic measurements of rocks—likely require medium-class mission implementations. The committee considered a range of medium-class Mars mission concepts and whether such missions are best performed under the auspices of MEP or included in the New Frontiers program.

Finding: Retaining medium-class Mars missions within MEP affords key advantages for coordination across multiple missions and international partnerships, and for fostering development and utilization of common enabling technological aspects.

The committee considered four medium-class Mars missions (two studied by NASA prior to the survey and two identified and studied during the survey) and prioritized two of these for TRACE: the Mars Life Explorer (MLE) and In Situ Mars Geochronology (see Appendixes C and D, respectively). The committee ranked MLE as the highest priority medium-class mission for the MEP. "Are we alone?" is one of the most profound questions that can be addressed by solar system exploration (Q11). Ancient biosignatures are a focus of Perseverance and MSR, and multiple types of past habitable environments with liquid water and organic matter have been discovered on Mars (Q10). Mars has subsequently undergone profound climate change (Q6), and key questions are whether any habitable environments persist to the present and whether they are inhabited. The focus of Mars Life Explorer is to seek extant life and assess modern habitability.[7] The notional mission concept examines Mars's lowest latitude ice deposits that preserve a record of recent climate change and may provide a recent habitat for life. MLE would land and drill into the ice to characterize and quantify organics, trace gases, and isotopes at a fidelity suitable for biosignature detection. It would also assess ice habitability and the question of modern liquid water via analysis of elemental chemistry, salts, conductivity, and ice thermophysical properties. Long-term atmospheric measurements over a martian year would determine the current stability versus instability of the ice deposits.

Recommendation: Subsequent to the peak-spending phase of MSR, the next priority medium-class mission for MEP should be Mars Life Explorer.

---

[7]The full Mars Life Explorer mission study report is available at https://tinyurl.com/2p88fx4f.

The complex interplay among Mars science, human exploration, and international partnerships requires carefully crafted collaborations. High potential synergy of objectives exists between the science of martian climate change (Q5, Q6, and Q10) and human exploration efforts relating to in situ resource utilization (ISRU), including the desire to map near-surface ice on Mars. The international Mars Ice Mapper (iMIM), in pre-Phase A at the time of this writing, was proposed by NASA as a mission with priorities for human exploration and multiple international partners. However, as presently articulated, iMIM measurements only minimally address the science goals and measurement requirements for Mars ice mapping defined in community documents and by planetary mission concept studies prepared for this decadal survey (see Appendixes C and D). NASA is in the process of convening a Measurement Definition Team (MDT); however, this activity postdates the basic international agreements, including a choice of a specific radar instrument. The MDT needs to include stakeholders from both the Mars science and human exploration communities. With incorporation of measurement requirements based on current state-of-the-art in understanding Mars's crustal properties and Mars science, iMIM could address priority science questions related to Mars climate while also realizing human exploration objectives (see Chapter 19 for further discussion).

> **Finding:** While leveraging international partnerships can potentially lower NASA mission costs, coordination is necessary to achieve priority science. Stronger programmatic coupling is needed between science and human exploration communities to ensure that precursor missions, such as iMIM and eventual human missions to Mars, achieve decadal-level science.

> **Recommendation: The development of the goals and measurement requirements for missions addressing both science and human exploration interests should be developed to meet the objectives of both communities.**

> **Recommendation: NASA should consider implementation of an ice mapping mission that prepares for ISRU by humans *and* addresses the priority climate science questions at Mars related to near-surface ice.**

Lunar Discovery and Exploration Program

The 2017 Space Policy Directive-1 (SPD-1) instructed NASA to explore the Moon with commercial and international partners and to return humans to the Moon for long-term exploration and resource utilization to achieve sustainable human presence. The Lunar Discovery and Exploration Program (LDEP), begun in 2019 in response to this directive, is intended to support commercial partnerships and innovative approaches to accomplish lunar exploration and science goals. LDEP is executed through the Exploration Science Strategy and Integration Office (ESSIO) and integrates and coordinates the Artemis science efforts across the SMD divisions, across NASA directorates, and with other U.S. and international partners.

DEP has implemented new research and technology developments; established the Commercial Lunar Payload Services (CLPS) program for lunar landing services; and supported the development of lunar science instruments, lunar CubeSats and SmallSats, and the development of lunar rovers. There are currently two LDEP flight projects managed by PSD. The Volatiles Investigating Polar Exploration Rover (VIPER) is a lunar volatiles detection and measurement mission that will be launched as a payload on a CLPS to the lunar south pole, characterizing the distribution and physical state of lunar polar water and other volatiles in lunar cold traps and studying the potential for in situ resource utilization from the Moon's polar regions. Lunar Trailblazer, a SmallSat orbiter selected in the SIMPLEx program, will characterize the form, abundance, distribution, and time variability of H_2O/OH in sunlit terrains and ice in permanently shadowed regions.

Commercial Lunar Payload Services Program

The goal of the CLPS program is to enable reliable and affordable access to the lunar surface by helping to establish a viable commercial lunar sector. Like the Commercial Cargo and Crew programs, NASA pays CLPS providers for services rather than spacecraft. These services include delivery to the lunar surface from Earth (launch vehicle and spacecraft), and ground systems. Thus far, NASA has contracted eight CLPS deliveries to the lunar surface, the largest being for the VIPER rover, although no landings have yet occurred. CLPS capabilities are currently limited but have the potential to meet additional NASA needs such as lunar night survival, mobility, delivery of larger payloads, and sample return. The CLPS model for purchasing delivery services to the Moon, once landing systems have proven to be reliable, is a promising and innovative procurement model within SMD that can benefit planetary science, if used as one component of a PSD-integrated program to accomplish important science objectives. Competition and firm fixed-price contracts incentivize the commercial sector to keep costs low, allowing for a higher mission cadence and a larger acceptance of risk. The higher cadence will provide opportunities to test and advance new technology, address focused science questions, and train new scientists and engineers.

Recommendation: NASA should continue to support commercial innovation in lunar exploration. Following demonstrated success in reaching the lunar surface, NASA should develop a plan to maximize science return from CLPS by, for example, allowing investigators to propose instrument suites coupled to specific landing sites. NASA should evaluate the future prospects for commercial delivery systems within other mission programs and consider extending approaches and lessons learned from CLPS to other destinations, for example, Mars and asteroids.

Structuring LDEP to Achieve Decadal-Level Science

LDEP is funded within the PSD budget, but the responsibility for its budget is split between PSD and ESSIO. ESSIO is focused primarily on inter-division activity coordination and commercial partnerships, whereas PSD is responsible for accomplishing lunar science goals. LDEP funds many (but not all) lunar programs in PSD but does not currently manage or coordinate them. In the current LDEP organizational structure, no single organizational chain has authority for executing lunar science missions and accomplishing lunar science. Further, there is as of yet no overall strategy for lunar scientific exploration or a program director or chief scientist to lead such a plan. As a result, despite substantial investment and tremendous potential for innovative lunar exploration, LDEP activities are not well coordinated or optimized to accomplish high-priority planetary science goals at the Moon.

Recommendation: PSD should execute a strategic program to accomplish planetary science objectives for the Moon, with an organizational structure that aligns responsibility, authority, and accountability.

The 2007 report, *The Scientific Context for the Exploration of the Moon*, provided a set of lunar science concepts, goals, and recommendations that have informed subsequent studies and lunar community activities (NRC 2007). More recently, the Lunar Exploration Analysis Group (LEAG) charged the Advancing Science of the Moon Specific Action Team (ASM-SAT) with evaluating progress made in accomplishing lunar science goals and identifying key directions for future research. The resulting 2018 ASM-SAT report (LEAG 2018) highlighted three additional science concepts beyond those in the 2007 report. Given that the National Academies has not addressed the most compelling science goals to be addressed at the Moon since 2007, Box 22-1 outlines current priority lunar science themes identified by the committee.

In recognition of the wide range of current and planned lunar exploration activities, it is now essential that a prioritized set of objectives and measurements to address decadal-level science questions at the Moon be developed. The scientific success of MEP provides an example of the utility and impact of having well-defined goals, prioritized science questions, specific measurements to address those questions, and a long-term plan. The MEPAG "goals document" was begun in the 1990s and has continuously evolved via sustained and well-organized inputs from the Mars science community through workshops and committees.

Finding: A structured approach to setting science goals and measurement objectives at the Moon, led by the lunar science community in a manner similar to that led by the Mars community, would allow for scientific prioritization and coordination of lunar missions, instrumentation, landing site selections, and other activities performed within LDEP.

Lunar science can benefit enormously from available and planned robotic, commercial, and human implementation options. Managing these different assets through a well-coordinated Science Mission Directorate/Exploration Systems Development Mission Directorate program would maximize synergy among existing and future domestic

BOX 22-1
Science Themes for Lunar Exploration

The central goal of a science-driven program of lunar discovery and exploration is to reveal the history of major events and processes that have shaped the Earth–Moon system and the solar system. The committee prioritizes three overarching science themes that address (1) Solar System History, (2) Geologic Processes, and (3) Water and Volatiles.

Science Theme 1: Uncover the lunar record of solar system origin and early history. The Moon's composition, structure, and ancient surface preserve a record of early events, from the giant impact that produced the Earth–Moon system to ongoing bombardment as life on Earth emerged and evolved.

The Moon's composition and physical properties constrain how the Earth–Moon system formed, while its ancient surface preserves the primordial history of asteroid and comet impacts and other processes. This record holds clues to environmental conditions on Earth as life emerged that have been all but erased on the Earth. Analyses of returned lunar samples also provide the basis for a chronology used to understand the sequence and timing of events across the solar system. Key questions remain about the nature of planet formation in the inner solar system, early bombardment, and related hypotheses of giant planet orbital migration. While exploration of other worlds can contribute to addressing these questions, the Moon will remain unique in its ability to provide accessible ground-truth constraints on terrestrial planet assembly and solar system bombardment history and chronology.

Science Theme 2: Understand the geologic processes that shaped the early Earth, and that are best preserved on the Moon. The Moon retains a record of processes that established the evolutionary paths of rocky worlds, including volcanism, magnetism, tectonism, and impacts.

The Moon is amongst the rocky worlds of the solar system that best preserve ancient geologic processes. Like the terrestrial planets, including Earth and Mars, the Moon started as a partially (or completely) molten world. As the magma ocean cooled, it produced the Moon's current differentiated structure, including a crust, mantle, and partially molten core. While differentiation is a ubiquitous process, most other worlds have a poorly preserved record of these early epochs—making the Moon one of the best locations to understand these processes. Even as the Moon's geologic activity waned, volcanism of varied styles and chemistries persisted, impacts and tectonism reshaped the crust, and radiation processes weathered its surface—all of which are relevant to worlds across the solar system.

Science Theme 3: Reveal inner solar system volatile origin and delivery processes. The Moon hosts water and other volatiles in its interior, across its surface, and in ice deposits at its poles, providing a record that may help constrain theories of the origins of Earth's oceans and the building blocks for life, as well as ongoing volatile delivery processes.

The Moon holds a record of volatiles obtained throughout its history. Primordial volatiles contained in the Moon's interior suggest that water was retained even through the energetic Moon-forming impact. At the surface, volatiles migrate in an active cycle and are trapped in permanently shadowed regions near the poles, processes that occur at bodies with tenuous atmospheres across the solar system. The origin, composition, concentration, and distribution of the Moon's volatiles remain uncertain. Determining the source(s) of the Moon's water and other volatiles may shed light on the source(s) of Earth's water and on mechanisms that act as ongoing sources of volatiles in the present day. Lunar volatile reservoirs also have implications for in situ resource utilization by human explorers.

and international missions, execute a healthy and comprehensive technology pipeline at the architectural (versus individual mission) level, and ensure sustenance of foundational infrastructure.

NASA's Artemis Plan (NASA 2020a) calls for landing humans near the Moon's south pole within the 2020s and ultimately developing a basecamp to support sustainable exploration and longer stays. NASA's historic investment in the Artemis program merits a substantive scientific component capable of transforming our understanding of the Earth-Moon and the early solar system. The successful integration of science into programs of human exploration has historically been a challenge and remains so for Artemis (Beattie 2001; see Chapter 19). Currently, science requirements do not drive the Artemis capabilities. However, in the committee's view, *it is imperative that Artemis support breakthrough, decadal-level science, as outlined in Box 22-1*. To not do so would be a missed opportunity for NASA and the nation that would undermine the value of the envisioned long-term Artemis program.

> **Recommendation: The advancement of high priority lunar science objectives, as defined by PSD based on inputs from this report and groups representing the scientific community, should be a key requirement of the Artemis human exploration program. Design and implementation of an integrated plan responsive to both NASA's human exploration and science directorates, with separately appropriated funding lines, presents management challenges; however, overcoming these is strongly justified by the value of human-scientific and human-robotic partnerships to the agency and the nation.**

Transformative Science Enabled by a Synergistic Robotic-Human Partnership

While the Discovery, current CLPS, and SIMPLEx programs provide opportunities to address focused science objectives at the Moon, broader science goals require more ambitious and complex missions. A fundamental science goal is to utilize the Moon to investigate the early dynamical and impact history of the solar system (see Box 22-1). Impacting comets and asteroids are thought to have substantially influenced the origin and early evolution of life on Earth through, for example, the delivery of water, key elements, and organics; effects on geologic and climate evolution; and impact-driven mixing and energy deposition. An impact on the Moon created its oldest and largest known basin, the roughly 2,500 km South Pole–Aitken (SPA) basin, and excavated the Moon's interior, potentially exposing lunar mantle materials. There are no known mantle samples in Apollo or lunar meteorite collections; analysis of such material would provide crucial new understanding of the Moon's bulk composition and interior structure, which in turn are key constraints on Earth-Moon system origin and the Moon's primordial evolution (Moriarty et al. 2021). The record of early impacts preserved on the Moon can be used to test hypotheses of early solar system evolution, including models of the migration of the giant planets that sent asteroids and comets on collision courses with Earth and other planets. Understanding early bombardment and its implications has been the longstanding highest priority for planetary science and astrobiology at the Moon. Furthermore, the lunar geologic record provides a basis for understanding how large terrestrial bodies evolve through time forming crusts, mantle, and cores.

The committee studied two mission concepts designed to address this highest priority lunar science, which would revolutionize our understanding of the Moon and the early history of the solar system recorded in its most ancient impact basin. Both concepts, Endurance-R and Endurance-A, utilize a CLPS-delivered rover to traverse nearly 2,000 km of diverse terrains within the SPA basin.[8] The rover conducts in situ measurements and collects samples for return to Earth for detailed analyses in existing and future laboratories. The first concept, Endurance-R, requires two medium-class missions to robotically collect and return about ~2 kg of total sample mass. This mission was viewed by the committee as too costly given the small returned sample mass.

The second concept, Endurance-A, involves a single medium-class mission to robotically collect 100 kg of samples, which are delivered to a location where they can be collected by astronauts for return to Earth. Retrieval and return of substantial Endurance-A samples by Artemis astronauts would be the ideal synergy between NASA's human and scientific exploration of the Moon, producing flagship-level science at a fraction of the cost to PSD

---

[8] The full report of the Endurance mission concept studies is available at https://tinyurl.com/2p88fx4f.

through strategic coordination with Artemis exploration efforts. This would be a forward-looking, inspirational partnership to deliver ground-breaking science not possible through the local collection of limited samples.

Science objectives of Endurance-A include determining:

- The age of the largest and oldest impact structure on the Moon, SPA basin, to anchor the earliest impact history of the solar system;
- When post-SPA farside basins formed to test the giant planet migration and terminal cataclysm hypotheses, and to better constrain the inner solar system impact chronology used to date the surfaces of other planetary bodies;
- The age and mineralogical and geochemical composition of deep and crustal materials exposed in SPA to understand the bulk composition of the Moon, its primordial differentiation and geologic evolution, and the significance of chronologic measurements completed on nearside samples for timing lunar solidification;
- The age and nature of volcanic features and compositional anomalies on the lunar farside to characterize the thermochemical evolution of terrestrial worlds and constrain the origin of the Moon's nearside-farside asymmetry; and
- The geologic diversity of the SPA Terrane to provide geologic context for returned samples, ground truth for orbital measurements, and characterize the surface processes that shape planetary bodies.

Recommendation: Endurance-A should be implemented as a strategic medium-class mission as the highest priority of the Lunar Discovery and Exploration Program. Endurance-A would utilize CLPS to deliver the rover to the Moon, a long-range traverse to collect a substantial mass of high-value samples, and astronauts to return them to Earth.

If timelines or plans for Artemis render this partnership infeasible, NASA, with guidance from CAPS,[9] could evaluate options for a robotic return of the minimum set of samples needed to accomplish the core science objectives, leveraging international partnerships and commercial capabilities as appropriate, while maintaining life cycle costs to NASA commensurate with a medium class mission.

Research and Analysis

Research and analysis (R&A) activities are the foundation for advancing the scientific knowledge from NASA's missions and projects, training and maintaining a diverse science and engineering workforce to meet NASA's needs, and preparing for the Agency's future activities. The fraction of the NASA PSD budget devoted to R&A decreased from 14 percent in 2010, to 8 percent in 2019, to a projected level of 7.7 percent by FY 2023 (see Figure 17-1 of the Research and Analysis chapter). Negative impacts on the community, and the portion of time spent writing and reviewing proposals instead of doing science, have greatly increased of late, as evidenced by multiple white papers and presentations to the committee that characterize the current situation as a crisis. Reversing this trend of decreasing *proportional* investment in R&A is essential to maintaining the health of the nation's planetary science efforts (see Chapter 17 for detailed discussion).

Finding: Robotic exploration of the solar system is driven by the desire to increase scientific knowledge. Strong, steady investment in R&A is needed to ensure that the scientific return from past and ongoing missions is maximized; that new data drives the development of improved understanding and novel, testable hypotheses; and that these advances feed into the development of innovative techniques and future mission concepts that will deliver breakthrough, high-impact scientific results.

PSD R&A efforts include a variety of programs. Among these, it is the openly competed R&A programs—to which any PI may propose and whose funds are awarded through a highly competitive, peer-review process—that

---

[9] The Committee on Astrobiology and Planetary Sciences, https://www.nationalacademies.org/our-work/committee-on-astrobiology-and-planetary-sciences.

are best suited for driving innovation, responding rapidly to new developments, identifying the most meritorious ideas worthy of support, and attracting new and increasingly diverse individuals into the field. These programs are identified in Table 17-1 of the Research and Analysis chapter.

Finding: Rigorous peer-review and open, equitable competitions are fundamental to NASA's success and to maximizing excellence in its activities and its workforce. The openly competed R&A programs drive innovation, direct funding to the most meritorious ideas, and encourage broad access and participation.

The committee recommends that PSD's investment in R&A activities be tied to the scale of the PSD program, a logic similar to that used in industry to establish investment levels for supportive research and development activities. The past decade has been enormously successful for PSD, accompanied by an approximate doubling of its annual budget. This reflects broad-based expansion across most of the primary PSD budgetary categories, with flagship (MSR and Europa Clipper) and the Discovery/New Frontiers mission lines approximately doubling from 2018 to 2023, in addition to proportionally even larger growth in planetary defense and LDEP during this period. However, proportional investments in R&A have not kept up with this growth in PSD activities. Recent levels in the 8 to 9 percent range threaten the continued scientific payoff from NASA's mission investments. Returning the annual PSD percentage investment in R&A activities to ≥10 percent through an increase in support to the openly competed programs would represent an approximately 40 percent increase in those programs relative to their recent levels. This illustrates that a relatively small increase in fractional PSD investment into R&A would strongly (and disproportionately) enhance the value delivered by R&A to its flight programs. Adjustments in R&A funding are ideally managed over a multi-year period to smooth out annual fluctuations, either upward are downward, with the goal of maintaining the stability of the workforce, associated national technical capabilities, and the pool of high-quality R&A proposals. The committee repeats here the following recommendation from Chapter 17:

Recommendation: PSD should increase its investment in R&A activities to achieve a minimum annual funding level of 10 percent of the PSD total annual budget. This increase should be achieved through a progressive ramp-up in funding allocated to the openly competed R&A programs, as defined in this decadal survey. Mid-decade, NASA should work with an appropriately constituted independent group to assess progress in achieving this recommended funding level.

Chapter 17 also presents recommendations that relate to specific large programs within PSD R&A, and to R&A proposal submission and evaluation processes. While planetary science and astrobiology R&A is funded primarily through NASA's PSD, NSF also provides related support, primarily within its Division of Astronomical Sciences (AST). NSF-AST supports ground-based astronomy as well as basic planetary science research. Ground-based observations are an important complement to data returned by planetary missions (see Telescopic Observations section later in this chapter). Other relevant activities at NSF, such as support of and access to field site for analog studies in extreme Earth environments, are detailed in Chapter 1.

Recommendation: NASA and NSF would realize greater return on their R&A investments by working together to streamline the mechanisms by which researchers can propose and conduct science that is of benefit to both agencies.

Astrobiology Program

Astrobiology encompasses a diverse set of topics, ranging from the study of prebiotic chemistry, terrestrial life's origin and evolution, the co-evolution of planets and life, and the search for habitability and for life elsewhere in the solar system and beyond to the exoplanets (see Q9 through Q11, Chapters 12, 13, and 14, respectively). Astrobiology is thus an inherently interdisciplinary field involving integrative study of physical, chemical, biological, geologic, planetary, and astrophysical systems. Over the past 25 years, astrobiology has become a crucial element of NASA's planetary program and is the central scientific motivation for large strategic missions currently in development (MSR and Europa Clipper), and a compelling component of many New Frontiers and Discovery mission themes. Significant advances have occurred in astrobiology over the past decade, as discussed in the *NASA Astrobiology Strategy* (NASA 2015) and the *Astrobiology Strategy for the Search for Life in the Universe* (NASEM 2019a). Three ideas advanced in these reports are particularly relevant to the coming decade of exploration of the solar system:

- Dynamic habitability and the co-evolution of planets and life (see Q9, Chapter 12) are key concepts to be addressed in the next decade, requiring integration across the planetary science and astrobiology communities because they involve the study of terrestrial environments, solar system bodies, and exoplanets.
- Understanding life in extreme environments on Earth is highly relevant to the next decade of solar system exploration. First, life in isolated refugia and ephemeral environments on Earth (e.g., in terrestrial deserts) implies that habitability is a continuum defined over varying time and spatial scales. Second, awareness of the habitability of saline and hypersaline terrestrial environments, together with the discovery of potential brines on Mars, has led to a resurgence of the idea of life adapted to saline fluids. Third, discovery of life in the ocean floor subsurface and continental lithosphere provides new models for rock-hosted, chemosynthetic life that might exist elsewhere in the solar system, for example, in ocean worlds.
- The search for life hinges on the ability to validate potential biosignatures, building on what has been discovered to date regarding life on Earth. Issues related to record bias, preservational bias, false negatives, and false positives all play a role in biosignature detectability and interpretation. However, the planning, implementation, and operations of planetary exploration missions with astrobiological objectives have tended to be more strongly defined by geological perspectives than by astrobiology-focused strategies.

As a result of these factors, the committee specifically endorses the following three recommendations from the 2019 National Academies' report:

Recommendation: NASA and other relevant agencies should catalyze research focused on emerging systems-level thinking about dynamic habitability and the coevolution of planets and life, with a focus on problems and not disciplines—that is, using and expanding successful programmatic mechanisms that foster interdisciplinary and cross-divisional collaboration.

Recommendation: NASA's programs and missions should reflect a dedicated focus on research and exploration of subsurface habitability in light of recent advances demonstrating the breadth and diversity of life in Earth's subsurface, the history and nature of subsurface fluids on Mars, and potential habitats for life on ocean worlds.

Recommendation: To advance the search for life in the universe, NASA should accelerate the development and validation, in relevant environments, of mission-ready, life detection technologies. In addition, it should integrate astrobiological expertise in all mission stages—from inception and conceptualization to planning, development, and operations.

Ongoing NASA and NSF research activities on Earth are also essential to progress in astrobiology (see Chapter 17).

Planetary Defense Program

NASA's Planetary Defense Program coordinates and supports activities to protect our world from asteroid and comet impacts. Its charter, per congressional mandate, is to detect and track all near-Earth objects and assess the threat and consequences of Earth impact. As awareness of the hazard posed to life and property by Earth-approaching asteroids and comets has grown, the U.S. Congress and presidential administrations have directed NASA, NSF, and other government agencies to pursue activities in support of planetary defense (NSTC 2018). The establishment of NASA's Planetary Defense Coordination Office (PDCO) in 2016 brought leadership and strategic direction to national planetary defense efforts. The placement of PDCO within the Planetary Science Directorate reflects a synergistic relationship, wherein scientific knowledge about the characteristics and population of small bodies, as well as experience in successful implementation of small body space missions, is applied to developing methods to assess the risk, and protect the planet from, hazardous objects. While not primarily science-driven missions, NASA's recently launched Double Asteroid Redirection Test (DART) kinetic-impactor demonstration—in conjunction with ESA's Hera follow-up assessment mission, and the Near-Earth Object Surveyor Mission (NEO Surveyor)— exemplify this synergy (e.g., NASEM 2019b).

Chapter 18 provides a detailed discussion of planetary defense for the next decade and includes 38 findings and 11 recommendations. Here the committee presents several of the key elements and recommendations from that chapter.

Finding: A dedicated space-based mid-infrared survey is the most effective architecture to accomplish congressionally directed NEO survey goals. NEO Surveyor, currently pending confirmation, will conduct that survey and provide real-time information on object diameter, critical for rapid impact hazard assessment.

Recommendation: NASA should fully support the development, timely launch, and subsequent operation of NEO Surveyor to achieve the highest priority planetary defense NEO survey goals.

A critical next step is to develop a flexible implementation approach to quickly characterize threatening objects via reconnaissance missions in order to plan for mitigation if needed. After NEO Surveyor, the next priority planetary defense mission is a rapid-response, flyby reconnaissance of an object representative of the most hazardous class of objects (~50 to 100 m diameter; see Chapter 18). A rapid response capability may also provide a template for responding to newly identified, high-value science targets such as interstellar objects or dynamically new comets.

Recommendation: In the coming decade, NASA should develop an approach for a rapid-response, flyby characterization of emerging, short-warning-time (<3 years) threats and science opportunities.

DISCOVERY, NEW FRONTIERS, AND FLAGSHIP RECOMMENDATIONS FOR THE DECADE 2023–2032

In this section, the committee first addresses issues related to the Discovery and New Frontiers programs, including cost structure, cadence, and, in the case of New Frontiers, whether mission themes will continue to be specified. It then describes the process by which candidate medium and large missions were studied and assessed, followed by a prioritization of such missions.

In its guidelines to the committee, NASA provided cost definitions for three mission classes, each of which was exclusive of the launch vehicle and mission operations (i.e., Phase E-F) costs: less than ~$500 million for small missions, between $500 million and $900 million for medium missions, and more than $900 million for large, strategic missions. The committee carefully evaluated these cost classes and recommends revisions to the cost structures for the Discovery and New Frontiers programs, as specified below, based on several considerations: (1) growing trajectory, instrumentation, and technology requirements needed to address decadal-level science questions, as evidenced both by recent mission selections and results of independent mission concept analyses; (2) the different roles played by each program within NASA's planetary mission portfolio; (3) a desire to more clearly anticipate mission life cycle cost, both for community awareness and for NASA budgetary planning; and (4) effects of inflation.

Discovery and SIMPLEx Missions

The Discovery program was initiated in 1990 as a means to ensure frequent access to space for planetary science investigations through competed PI-led missions (Table 22-1). The low cost and short development times of Discovery missions provide flexibility to address new scientific discoveries on a timescale significantly less than 10 years. Identification of specific Discovery missions is outside the scope of decadal recommendations, but the overall program warrants evaluation and recommendations.

TABLE 22-2 Discovery Program Mission Selections to Date

Year of AO	Mission Selected	Launch Date	Description
n/a	Near-Earth Asteroid Rendezvous	February 17, 1996	Asteroid orbiter and rendezvous
n/a	Mars Pathfinder	December 4, 1996	Mars lander and Sojourner rover
1994	Lunar Prospector	January 6, 1998	Lunar orbiter
1994	Stardust	February 7, 1999	Comet particle sample return
1996	Genesis	August 8, 2001	Solar wind sample return.
1996	CONTOUR	July 3, 2002	Comet nuclei flybys (lost contact 6 weeks after launch)
1998	MESSENGER	August 3, 2004	Mercury orbiter
1998	Deep Impact	January 12, 2005	Comet impactor and flyby
2000	Dawn	September 27, 2007	Orbit of asteroid Vesta and dwarf planet Ceres
2000	Kepler	March 6, 2009	Extrasolar planets telescope
2004	*No Selection*		
2006	GRAIL	September 9, 2011	Lunar orbiters for gravity mapping
2010	InSight	May 5, 2018	Terrestrial planet seismicity
2014	Lucy	October 16, 2021	Trojan asteroid tour
2014	Psyche	2023	Psyche asteroid rendezvous
2018	DAVINCI	2028–2030	Venus entry probe
2018	VERITAS	2028–2030	Venus orbiter

The strategic research activities given in Chapters 4–15 demonstrate that many science questions can be addressed at multiple destinations and provide examples of the rich array of science that can be addressed with future Discovery missions. Primitive body investigations are ideally suited for Discovery missions. The vast number and diversity of asteroids and comets provide opportunities to benefit from frequent launches. The proximity of some targets allows missions that can be implemented within the context of the Discovery program. Near the limit of the Discovery cost cap, it may be possible to collect and return samples from nearby objects. The diversity of targets means that proven technologies may be reflown to new targets, reducing mission risk and cost. And the population of scientifically compelling targets is not static but is continually increasing because of discoveries in the supporting research and analysis programs. Opportunities for lunar Discovery missions could build on the success of the CLPS program and incorporate more sophisticated instrumentation and/or longer-lived missions. Missions to Mars can benefit from collaboration with international and commercial partners and innovative approaches and technologies that could provide more affordable access to the martian surface and expand the geologic diversity of sites that have been visited. Even the outer solar system can be accessible in the Discovery program, as evidenced by the 2021 launch of Lucy to the Trojan asteroids and the selection of the Io Volcano Observer and the Trident mission to Triton for Phase A studies following the most recent call for Discovery missions. Thus, there is still much compelling science that can be addressed by Discovery missions.

Recommendation: The Discovery program has made important and fundamental contributions to planetary exploration and should continue to be supported in the coming decade.

The SIMPLEx program of very small, low-cost missions that is managed within the Discovery program takes advantage of recent advances in small spacecraft technologies and/or ride-share opportunities on other launches. Inclusion of this innovative program promises to increase launch cadence overall within the Discovery program and provides flexibility to achieve a balanced portfolio across targets and maintain a continuous stream of new data to the planetary science community. The combination of higher risk tolerance and shorter mission lifetimes complements the Discovery mission line by embracing rapidly improving commercial deep space SmallSats capabilities and allowing infusion of new technologies that will ultimately benefit larger missions. In addition, SIMPLEx represents an entry-level mission that could allow early career scientists and nontraditional institutions to participate and gain experience in the NASA mission process.

Finding: SIMPLEx plays a unique role within the PSD mission portfolio that capitalizes on new technology and innovative launch strategies at modest cost. It is well-placed within the Discovery management structure, where it can be flexibly accommodated as budgets and ride-share opportunities allow.

However, the challenging budgets, schedules, and coordination of rideshares or small launch procurement for SIMPLEx missions requires strong NASA stakeholder engagement with the mission team. Five SIMPLEx missions have been selected to date, with the latest SIMPLEx round having a mission cost cap of $55 million. In development SIMPLEx missions are the LunaHMap mission to map hydrogen at the lunar south pole, the Janus mission to send twin spacecraft to explore two binary asteroid systems, the Lunar Trailblazer mission to provide maps of water abundance on the Moon, and the ESCAPADE mission to study the martian upper atmosphere. These missions provide high science value, but may represent the "low-hanging fruit" at the current cost cap. A higher cost cap would enhance the potential of the program to achieve decadal-level science and drive innovation across more destinations.

Recommendation: NASA should provide a ~50 percent increase in the SIMPLEx cost cap for future calls to expand the range of possible destinations and increase the scientific return from this program.

Discovery Program Cost Structure and Cadence

The Discovery program was designed to support a high cadence of missions within a specified cost cap but without science or target constraints. This structure encourages innovation to maximize science return per dollar. The first generations of Discovery missions targeted objects and material in the inner solar system (see Table 22-2). Recently, Discovery has supported missions to targets in the outer solar system, including both proposals selected for Phase A study (e.g., the Io Volcano Observer and the Trident mission to Triton) and for flight (Lucy, which will study Jupiter's Trojan asteroid swarms). This evolution reflects a natural and beneficial progression of scientific discovery. It is highly desirable that the Discovery mission class retain the ability to access objects in the outer solar system, while continuing to prioritize launch cadence and maximal science return per dollar—core attributes of this program.

Vision and Voyages recommended a Phase A–F Discovery cost cap, exclusive of the launch vehicle, that is equivalent to ~$700 million when inflated to FY 2025 dollars. The most recent Discovery competitions had a cost cap of ~$500 million for the hardware development phases (Phase A–D), with the mission operations phase (Phase E) and launch vehicle costs excluded from the cost cap. By excluding Phase E from the cap, this cost structure allowed for missions to outer solar system targets—which generally have longer trajectories and higher Phase E costs—to be competitive within Discovery, a very favorable outcome. However, estimated total mission costs for selections made with this cost structure (Psyche and Lucy) proved to be roughly twice the original cost cap once Phase E and launch vehicle costs were included. Such a large difference between the original cost cap and the actual life cycle cost is problematic: it makes budgetary forecasting challenging, while also introducing a potential mismatch between community expectations for launch cadence and realities of program budgetary constraints. Launch vehicle costs are largely outside of a proposing team's control (and are relatively predictable by NASA), and the committee reaffirms *Vision and Voyages* arguments that launch vehicles be excluded from the Discovery cost cap. However, Phase E costs can vary substantially across different missions. Excluding Phase E costs from the cost cap makes it challenging to assess the science return per true total mission cost and also undermines budgetary forecasting needed to maintain a predictable cadence of frequent selections and launches.

The committee carefully weighed the cost cap for Discovery going forward, considering several factors. Addressing many of the priority science questions identified in this survey will require higher levels of instrumentation and/or mission complexity, and potentially longer missions (including notably to outer solar system objects), than were required in the past. Indeed, as Discovery missions accomplish many of the high-priority science objectives achievable at the current cost, a larger cost cap will be needed to ensure the breadth of breakthrough science that has driven the success of the Discovery program.

> **Finding:** A single Discovery cost cap that covers Phase A–F activities will allow each proposing team to allocate their costs between hardware development and operations in a manner that best suits the specific mission and maximizes the science that is achieved.

> **Recommendation: The Discovery Phase A through F cost cap should be $800 million in FY 2025 dollars, exclusive of the launch vehicle, and periodically adjusted throughout the decade to account for inflation. This cap will enable the Discovery Program to continue to support missions that address high-priority science objectives, including those that can reach the outer solar system.**

A key element for a vibrant planetary exploration program is the regular acquisition of new information to test hypotheses, and mission cadence is an essential element of the Discovery Program. Discovery Announcements of Opportunity (AOs) have been released in 1994 (two selected), 1996 (two selected), 1998 (two selected), 2000 (two selected), 2004 (no selection), 2006 (one selected), 2010 (one selected), 2014 (two selected), and 2018 (two selected) (see Table 22-2). The preparation and review of Discovery proposals is time and resource intensive, and many high-quality proposals are received in each competition. In each of the past two Discovery competitions, NASA selected two missions. Although this increased the length of time between Discovery AOs, an overall high cadence of launches has generally been maintained.

> **Finding:** The committee commends NASA for maintaining a high cadence of Discovery mission opportunities and finds that the selection of two Discovery missions per announcement of opportunity is a good approach to maximizing return on the substantial time and costs associated with proposal preparation and reviews.

> **Finding:** The interval from release of the announcement of opportunity to launch has recently increased from 4 to 9 years. This increase appears to be owing to a combination of factors, including delayed selections, funding availability, unforeseen schedule delays, and growing technical complexity. The trend toward a longer implementation time is cause for concern because it undercuts the scientific responsiveness of the Discovery program.

New Frontiers Program and Cost Structure

NASA asked the committee to consider whether specific flight investigations in New Frontiers will ideally continue to be specified or whether this mission class be open like the Discovery program (see "Considerations" in Appendix A). New Frontiers missions address broader and/or more technically challenging scientific questions than Discovery-class missions. They are complex, may have a large suite of instruments and/or complex mission operations, and can be managed by only one of three centers (JPL, APL, and GSFC). Extensive time and resources go into planning and proposing NF missions, and only a relatively small number of concepts can be developed by each of the centers. The higher cost of NF missions also means that a single mission can consume a significant fraction of the planetary science budget and that they occur infrequently. It is therefore essential that NF missions be strategically designed to address the most important questions put forward by the science community. Decadal surveys provide the opportunity for a large, diverse group of scientists that represent the broad science community to devote significant time and resources to the evaluation and prioritization of candidate mission themes to best address priority science questions.

> **Recommendation: Mission themes for the NF-6 and NF-7 calls should continue to be specified by the decadal survey. Additional concepts that may arise mid-decade owing to new discoveries should be evaluated by an appropriately constituted group representing the scientific community and considered for addition to NF-7.**

Mission life cycle costs are the primary factor in determining launch cadence for a cost-bounded program like New Frontiers. In evaluating the NF cost structure, the committee prioritized enabling access to all targets across the solar system at the potential expense of launch cadence. Missions to the outer (and innermost) solar system can have very long cruise phases to reach their target(s); sample return missions can also require long cruise phases to return their samples to Earth. Even in a quiet cruise operational mode (defined below), these durations can result in significantly higher costs than for missions to nearby objects with comparable cost instrument suites. Further, while the committee's recommended cost structure for Discovery can support targeted outer solar system science, accomplishing much of this science requires a medium-class mission. As with Discovery, the committee carefully weighed the option of separating the NF Phase A–D and Phase E–F costs, and again concluded that leaving the development and operational phases together in a single cost cap provides the greatest flexibility for mission teams to maximize scientific return.

> **Recommendation: New Frontiers should have a single cost cap that includes both Phase A–D and the primary mission Phase E–F costs, with a separate, additional cost cap allocation for a mission's quiet cruise phase. This approach will enable the NF Program to optimize mission science, independent of cruise duration.**

Vision and Voyages recommended a New Frontiers Phase A-F cost cap (excluding launch vehicle) equivalent to $1.34 billion in FY 2025 dollars. This recommendation merits reassessment based on actual and projected NF mission costs. The initial reconnaissance of the solar system has demonstrated its remarkable diversity and complexity, raising questions in planetary and astrobiological science of increasing sophistication that require increasingly advanced instrumentation and/or mission design. The exciting and aspirational Dragonfly mission selected in NF-4—involving a drone that lands on Titan and then performs multiple flights to explore varied regions and perform in situ analyses— is an example case. While the NF-4 competition had a Phase A-D cost cap equivalent to $1.14 billion in FY 2025 dollars, the NASA prelaunch budget for Dragonfly through FY 2026 is about $1.7 billion (launch is planned for 2027), suggesting that the total life cycle mission cost will likely be significantly higher than the original cost cap. As indicated earlier in this chapter, the committee endorses the Dragonfly mission at this budgetary level. Indeed, its costs are not too dissimilar from those of the scientifically compelling NF mission concepts considered by the committee, which had independently estimated (see next section) Phase A–D costs in the $1.2 billion to $2 billion range (FY 2025 dollars). These missions are representative of the nature and breadth of the science that will optimally be accomplished in the NF program in the coming decade, and a Phase A–D cost of some $1.5 billion is thus representative of the associated hardware costs. A nominal 2-year NF primary mission with a cost of $80 million per year would yield a primary mission Phase E cost of $160 million. Examination of recent Discovery and NF cruise costs indicates that a representative quiet cruise phase costs approximately $30 million per year.

> **Recommendation: The NF Phase A-F cost cap, exclusive of quiet cruise phase and launch vehicle costs, should be increased to $1.65 billion in FY 2025 dollars. A quiet cruise allocation of $30 million per year should be added to this cap, with quiet cruise to include normal cruise instrument checkout and simple flyby measurements, outbound and inbound trajectories for sample return missions, and long transit times between objects for multiple-target missions.**

Mission Study Process and Technical Evaluation

The program portfolio recommended below was designed to achieve an appropriate balance among mission classes. To maintain this balance, it is crucial that all missions be initiated with a reasonable understanding of their probable costs. This decadal survey, like its predecessor, has placed considerable emphasis on cost realism, and the technical and cost evaluation process used in this decadal survey was specifically designed to provide a realistic assessment of mission costs. The committee has relied on detailed mission studies and cost estimates derived using a methodology designed to quantify the technical, schedule, and cost risks that are inherent in concepts with modest degrees of technical maturity (see Appendix C).

Prior to the start of this decadal survey, NASA undertook the selection and study of a suite of candidate missions through the Planetary Mission Concept Study (PMCS) program.[10] This program completed 11 Center-led studies (Appendix C) whose final reports were made available to the science community. The committee examined the PMCS studies, together with missions recommended by *Vision and Voyages* and existing concepts explored

---

[10] The final report of each of the PMCS concepts is available at https://science.nasa.gov/solar-system/documents.

by science definition teams and identified additional mission concepts needed to address the full breadth of the priority science questions, including input from white papers submitted by the scientific community (Appendix B). The committee prioritized these gap-filling candidate missions and commissioned 11 additional mission studies (Appendix C).[11] Each study was overseen by one or more committee members, selected based upon their expertise, who acted as "science advocates" for the missions during the study. The studies were conducted by the Jet Propulsion Laboratory (JPL), the Applied Physics Laboratory (APL), or the Goddard Space Flight Center (GSFC) and were funded by NASA and delivered to the agency, which then delivered them to the decadal survey. Although NASA was aware of the contents of the studies, the Agency was not involved in their prioritization.

Seventeen of the available mission studies were prioritized by the committee for further Technical, Risk, and Cost Evaluation (TRACE) using criteria described at the beginning of this chapter. This independent evaluation was performed by The Aerospace Corporation, a contractor to the National Academies. The TRACE process is designed to provide an independent assessment of the technical feasibility of the mission candidates, as well as to produce a rough estimate of their costs. The process considers many factors when evaluating a mission's potential costs, including the actual costs of analogous previous missions. It therefore reflects cost impacts that may be beyond the control of project managers and principal investigators. It includes a probabilistic model of cost growth owing to technical and schedule risks, and hence projects cost growth resulting from insufficient technical maturity. Appendix C discusses the TRACE process in more detail.

The TRACE process typically resulted in cost estimates that were higher than the estimates produced by the study teams, driven in large part by inclusion of costs related to probable threats to technical implementation and schedule. Independent cost estimates based on analogue missions, and independent risk assessment attempt to remove biases inherent in advocate estimation processes. Only the independently generated cost estimates were used in evaluation of the candidate missions by the committee in formulating the final recommendations.

It is stressed that the studies carried out were of specific "point designs" for the mission concepts. They provide a proof-of-concept demonstration that a mission is feasible and provide a basis for developing a cost estimate for the purpose of this decadal survey and are only intended to be representative of a potential implementation approach.

Prioritized New Large Strategic Missions

The decadal survey considered six candidate flagship missions for the decade 2023–2032 that were judged to have exceptional scientific merit, based on their ability to address priority science questions Q1 through Q12. In alphabetical order, these are the Enceladus Orbilander, the Europa Lander, the Mercury Lander, the Neptune-Triton Odyssey Flagship, the Uranus Orbiter and Probe, and the Venus Flagship.[12] TRACE analyses were performed on all six, which were found to have medium-low to medium technical risk, except for the Venus Flagship, which has numerous system elements that increased its technical risk to medium-high (Appendix C). Three of the missions proposed—Europa Lander, the Enceladus Orbilander, and the Neptune Odyssey Flagship—require the SLS launch vehicle, but options are available for launch on a heavy lift vehicle with the inclusion of a solar-electric propulsion stage and/or a Jupiter gravity assist (if available). The TRACE cost estimates in FY 2025 dollars for these missions were $4.9 billion for Enceladus Orbilander, $5.8 billion for Europa Lander, $2.8 billion for Mercury Lander, $5.2 billion for Neptune-Triton Odyssey, $4.2 billion for the Uranus Orbiter and Probe, and $7.8 billion for the Venus Flagship.

Uranus and Neptune, the so-called "ice giants"—although whether they are predominantly ice versus rock remains uncertain (see Q2 and Q7)—are the only planets that have never been studied with a dedicated orbital tour. An ice giant system mission was judged to be the top priority flagship for the next decade, primarily for its ability to produce transformative, breakthrough science across a broad range of topics and key science questions. A secondary consideration was that the system-oriented, multi-target emphasis of such a mission is programmatically complementary to the flagships currently under way (MSR and Europa Clipper) that focus on

---

[11]The full mission study reports examining the concepts identified by the decadal survey are available at https://tinyurl.com/2p88fx4f.

[12]The mission study reports for Enceladus Orbilander, Mercury Lander, Neptune-Triton Odyssey, and Venus Flagship are available at https://science.nasa.gov/solar-system/documents.

single targets. As part of the prioritization process, the committee carefully considered flagship missions to perform system science at either Uranus or Neptune. While missions to both systems have outstanding scientific merit, the committee concluded that a Uranus mission is favored because an end-to-end mission concept exists that can be implemented in the 2023–2032 decade on currently available launch vehicles. Neptune Odyssey does not have demonstrated viable trajectories for a launch within the decade covered by this survey on currently available launch vehicle configurations, and there is a potential need for modifications to existing fairings. In addition, there are uncertainties in power requirements and the possible need for solar electric propulsion to reach Neptune if neither the SLS nor a Jupiter gravity assist are available. The Uranus mission has flexible trajectory opportunities, with the ability to be initiated at times throughout the decade including a launch as early as 2031.

The highest priority new Flagship mission for the decade 2023–2032 is the Uranus Orbiter and Probe mission. The Uranus Orbiter and Probe (UOP) will deliver an in situ atmospheric probe and conduct a multi-year orbital tour that would transform our knowledge of ice giants in general and the uranian system in particular. Uranus itself is one of the most intriguing bodies in the solar system: an extreme axial tilt; low internal energy; high speed winds and active atmospheric dynamics; and complex magnetic field all present major puzzles. It is unclear when and where Uranus formed, or if it swapped positions with Neptune during early solar system migration. It has been proposed that an early catastrophic impact caused the planet's extreme tilt and odd characteristics, possibly forming its rings and satellites, but this has yet to be validated. Uranus's large ice-rock satellites represent potential ocean worlds that could have astrobiological importance. Some of these moons display surprising degrees of geological activity or evidence of past internal heat release, particularly Ariel and Miranda, and yet they are part of the least explored regular satellite system. Detailed study of an ice giant system will provide vital ground-truth to exoplanetary science, given that exoplanets with similar masses are perhaps the most abundant class of exoplanet, and an inherently different class of planet than gas-rich Jupiter and Saturn. The committee's prioritization of the UOP mission reaffirms its identification in *Vision and Voyages* as the next highest priority flagship after MAX-C and the Jupiter Europa Orbiter, whose derivative missions (Perseverance and Europa Clipper) are already in operation or development. The TRACE found UOP to have medium-low technical risk, the only flagship concept to receive this rating among those studied.

Key science questions for the Uranus Orbiter and Probe are:

1. Origin, Interior, and Atmosphere (Q1, Q2, Q7, Q12):
   - How does atmospheric circulation function, from interior to thermosphere, in an ice giant?
   - What is the 3D atmospheric structure in the weather layer?
   - When, where, and how did Uranus form, how did it evolve both thermally and spatially, including migration, and how did it acquire its retrograde obliquity?
   - What is Uranus's bulk composition and its depth dependence?
   - Does Uranus have discrete layers or a dilute core, and can this be tied to its formation and tilt?
   - What is the true rotation rate of Uranus, does it rotate uniformly, and how deep are the winds?
2. Magnetosphere (Q7):
   - What dynamo process produces Uranus's complex magnetic field?
   - What are the plasma sources and dynamics of Uranus's magnetosphere and how does it interact with the solar wind, Uranus's upper atmosphere, and satellite surfaces?
3. Satellites and rings (Q2, Q4, Q5, Q8, Q10)
   - What are the internal structures and rock-to-ice ratios of the large uranian moons and which moons possess substantial internal heat sources or possible oceans?
   - How do the compositions and properties of the uranian moons constrain their formation and evolution?
   - What geological history and processes do the surfaces record and how can they inform outer solar system impactor populations? What evidence of exogenic interactions do the surfaces display?
   - What are the compositions, origins and history of the uranian rings and inner small moons, and what processes sculpted them into their current configuration?

The Uranus Orbiter and Probe mission will deliver an in situ probe into Uranus's atmosphere, then complete a multi-year orbital tour of all aspects of the uranian system including the atmosphere, interior, magnetosphere, rings, and satellites. The orbital science on the study payload includes visible and thermal imaging, visible/near-infrared imaging spectroscopy, fields and particles science, and radio/gravity science. The probe instruments measure atmospheric composition and isotopic ratio profiles, provide critical ground truth for the hydrogen ortho-para fraction and the vertical temperature profile, and determine abundances of the noble gases and their isotopes, as well as the vertical wind profile, parameters inaccessible to remote sensing. For more detail, refer to the complete science traceability matrix in the full mission study report.[13] UOP can be launched on an existing heavy lift expendable rocket, with or without a Jupiter gravity assist. The primary and secondary launch opportunities occur in June 2031 and April 2032, and both benefit from a Jupiter gravity assist available at those times to place ~5,000 kg in orbit at Uranus after a ~13-year cruise. These optimal launch and cruise times could be achieved with a FY 2024 start of the UOP mission, as is included in the committee's Recommended Flight Program (see Table 22-3 and Figure 22-1 below). Other launch opportunities from 2032 through 2038 (and beyond) utilize multiple inner solar system gravity assists (including a Venus flyby) to place up to 5,900 kg in orbit with an increased ~15-year cruise time. These diverse launch opportunities provide significant schedule flexibility and were considered by the committee to be a major strength of the Uranus mission concept. Further, international interest in an ice giant mission offers the opportunity for partnership, in analogy to the highly successful Cassini/Huygens partnership between NASA and ESA. Indeed, the 2021 report of the Voyage 2050 Senior Committee (ESA 2021) recommends that ESA pursue a substantial, medium class contribution to an ice giant orbiter mission led by an international partner.

The second highest priority new Flagship mission for the decade 2023–2032 is the Enceladus Orbilander. Enceladus is a small, active ice world in which gas and particles from its subsurface ocean are being jetted into space. Conditions at Enceladus thus allow for direct investigation of the habitability of an ocean world and assessment of whether it is inhabited. This addresses one of the most fundamental questions in solar system science: is there life beyond Earth and if not, why not?

Direct in situ sampling of plume materials by Cassini showed evidence of water vapor, carbon dioxide, methane, ammonia, complex organic molecules, and various salts, and ongoing hydrothermal activity in Enceladus's rocky core is inferred. However, Cassini flyby velocities were high, leading to fragmentation of large molecules, and ambiguity as to the precise identity of the parent organic compounds.

Enceladus Orbilander will sample an extant subsurface ocean through study of freshly ejected plume material originating from a well-characterized location.[14] Orbilander will execute a 1.5-year set of orbits of Enceladus, collecting plume samples from orbit, prior to a 2-year landed mission when more voluminous plume material is acquired in both passive and active (i.e., scooping) modes. Approximately ~300 µl of sample can be passively collected in ~10 days on the surface or in a single scoop. There are two main science objectives: (1) to search the plume materials for evidence of life (e.g., via multiple complementary approaches including the detection of amino acids, lipids, polyelectrolyte, and cell-like morphologies) at the level of fidelity necessary for biosignature detection (Q10, Q11, Q12); and (2) to obtain geochemical and geophysical context for life detection experiments (e.g., conditions in the ocean, structure/dynamics of the interior, and the structure of the jet vents; Q5, Q8). In addition to life detection, landed science includes a seismometer and radio science. Orbital science includes laser altimetry, radar sounding, gravity/radio science, thermal and visible imaging, and landing site reconnaissance. Viable launch opportunities on existing heavy-lift launch vehicles occur in 2037, 2038, and during the 2040s; these lead to Enceladus landing during favorable south pole illumination and Earth-communication conditions that begin in the early 2050s. The committee's *Recommended Program* starts Orbilander in FY 2029 (see Table 22-3 and Figure 22-1), in support of these launch times.

The remaining four Flagship missions are described next in unranked, alphabetical order.

- *Europa Lander*—This mission would characterize the biological potential of Europa's ocean through direct study of any chemical, geological, and possibly biological, signatures as expressed at the surface of Europa. The search for signs of life on Europa's surface would incorporate an analytical payload to perform quantitative organic compositional, microscopic, and spectroscopic analysis on five samples acquired from

---

[13]The full mission study report is available at https://tinyurl.com/2p88fx4f.

[14]The Enceladus Orbilander mission study report is available at https://science.nasa.gov/solar-system/documents.

centimeters beneath the icy surface, with supporting context imaging observations. This mission would significantly advance our understanding of Europa as an ocean world, even in the absence of any definitive signs of life, and would provide the foundation for the future robotic exploration of Europa. However, it was considered to be a lower priority than Enceladus Orbilander because of challenges associated with a short surface mission lifetime, effects of extensive radiolytic processes, and a lack of known continuously active, large volume plumes to supply fresh material from the underlying ocean. The committee notes that NASA's investments in Europa Lander studies have advanced the concepts and technology for landing and conducting surface operations on an icy body. They have also aided the maturation of instruments and tools for surface science on icy worlds. Continuing work in these areas will support a potential future landed mission to Europa as well as to other ocean worlds.

- *Mercury Lander*—This mission, as proposed, would deliver a lander with a suite of instruments to the surface of the innermost planet to gain insight into the original distribution of elements in the earliest stages of solar system development and to learn how planets and exoplanets form and evolve near to their host stars. The Mercury Lander would investigate: (1) the chemistry and mineralogy of Mercury's extremely reduced and unexpectedly volatile-rich surface, (2) Mercury's interior structure and magnetic field, (3) the active processes that produce Mercury's exosphere and alter its regolith, and (4) the geologic processes that have shaped its evolution. It was ranked lower in priority because of the narrower scientific scope of the mission as proposed compared to an ice giant system mission, and the high priority placed on the transformative science possible with the astrobiologically focused Enceladus Orbilander. The Mercury Lander concept would benefit from development work to enable enhanced spacecraft thermal control and high-temperature subsystems that would allow for longer duration surface operations and cost-effective circular and low-altitude orbits. Further, mission concept development to broaden the science goals, for example, to enable characterization of isotopic composition, would be valuable.
- *Neptune-Triton Odyssey*—This mission would deliver an orbiter and atmospheric probe to the Neptune-Triton system to study an ice giant planet, its rings, small satellites, space environment, and its large irregular moon, Triton, using a single launch of the SLS or, for example, heavy lift with a Jupiter gravity assist or a solar-electric propulsion kick stage. The mission would address: (1) how the interiors and atmospheres of ice giant (exo)planets form and evolve; (2) what causes Neptune's strange magnetic field, and how its magnetosphere and aurora work; (3) whether Triton is an ocean world, what causes its plumes, and the nature of its atmosphere; (4) how Triton's geophysics and composition can expand our knowledge of dwarf planets like Pluto; and (5) the connections between Neptune's rings, its small inner satellites, and Triton's orbital evolution. The Neptune-Triton flagship was ranked lower than the Uranus Orbiter and Probe primarily owing to the lack of demonstrated viable trajectories to Neptune for a launch within the decade covered by this survey on available launch vehicles, uncertainties in heavy lift launch vehicle accommodation, and the potential need for solar-electric propulsion to reach Neptune if neither the SLS nor a Jupiter gravity assist is available. The concept would benefit from a new mission study in which a Jupiter gravity assist is not assumed, and existing launch vehicles and power systems are utilized. Such a study will likely require consideration of the technological feasibility of solar electric propulsion, identification of any technology developments required, and prioritization of potential science descopes while retaining Flagship-level science.
- *Venus Flagship*—This mission would deliver an orbiter, lander, variable-altitude aerobot, and two small satellites on a single launch that will use multiple instruments to probe and measure the exosphere, atmosphere, and surface at multiple scales with high precision. The science goals would be to (1) understand the history of volatiles and liquid water on Venus and determine if Venus has ever been habitable, (2) understand the composition and climatological history of the surface of Venus and the present-day couplings between the surface and atmosphere, and (3) understand the geologic history of Venus and whether Venus is active today. West Ovda Regio tessera would be the nominal landing site to examine rocks considered to be among the most likely to have formed in a habitable climate regime. The Venus Flagship included multiple elements and was intended to operate in a challenging environment. The committee's assessment was that critical, flagship-class science advances could be made from the lander and a descoped

orbiter/lander mission was evaluated. However, the TRACE cost of the descoped mission is estimated to be $5.7 billion (see Box C-2), which is significantly higher than the other prioritized flagship missions. Its relatively high cost, together with the technical risk of landing on a challenging surface environment and sample handling at the Venus surface, led to the Venus Flagship being ranked lower than the top two prioritized flagship missions. It is possible that its science objectives could be met by landing in the plains, at a lower risk and cost. The concept would benefit from a study to assess the scientific rationale and the technical and cost feasibility of such an approach.

Prioritized New Frontiers Missions

New Frontiers (NF) Announcement of Opportunities (AOs) have included a list of mission themes that each specify destination and primary science objectives, with proposals required to address one of the themes. In general, the NF mission themes have been specified by decadal surveys, although subsequent to *Vision and Voyages* NASA added an Ocean Worlds (Titan and/or Enceladus) theme to the NF-4 call. Per above, the committee recommends that the New Frontiers Program continue to specify mission themes in future calls, and in this section the committee describes its related recommendations.

New Frontiers 5

In a November 5, 2020, Community Announcement, NASA indicated an October 2022 (target) release date for the NF-5 AO, and that the NF-5 mission themes would be (parenthetical words are NASA's):

- Comet Surface Sample Return (CSSR)
- Lunar South Pole–Aitken Basin (SPA) Sample Return (pending Artemis landing site selection(s) and science objectives)
- Ocean Worlds (only Enceladus)
- Saturn Probe
- Venus In Situ Explorer
- Io Observer (pending flight selection(s) for the Discovery program)
- Lunar Geophysical Network (LGN)

Midway through the decadal process, on May 12, 2021, NASA issued a Community Announcement that the NF-5 announcement of opportunity (AO) was to be delayed until a target release of October 2024. That announcement indicated that NASA intended to use the results of this survey to guide the NF-5 AO. However, when the National Academies initiated this decadal survey, it was with the understanding that the NF-5 mission themes would not be determined by the survey committee. Therefore, committee membership was not designed nor vetted to provide impartial findings and recommendations on NF-5. On May 25, 2021, the survey chairs released a letter notifying the community that the committee would not adjust the mission themes for NF-5 and would retain those listed above.

New Frontiers 6 and 7

The committee prioritized 13 (potentially) medium-class missions for TRACE (see Appendix C): Calypso Uranus and KBO flyby, Triton Ocean World Surveyor, Enceladus Multiple Flyby (EMF), Intrepid lunar rover, Endurance lunar rover, INSPIRE lunar polar volatiles rover, Titan Orbiter, Titan Orbiter and Probe, Ceres Sample Return, Mars Life Explorer (MLE), Centaur Orbiter and Lander (CORAL), Mercury Lander, and Mars In Situ Geochronology. An additional 6 medium-class missions were considered that had undergone independent cost and technical evaluation as part of *Vision and Voyages*: Venus in situ explorer, Lunar SPA Sample Return, LGN, CSSR, Io Observer, and Saturn probe. As indicated above, the committee recommends that Mars Life Explorer and Endurance-A be supported through MEP and LDEP, respectively, and the Mercury Lander was ultimately considered as a flagship mission, leaving a total of 16 mission concepts to be considered for inclusion in New Frontiers.

The panels prioritized NF mission themes based on how well each would address the priority science questions. Using the panel scientific prioritizations as key input, the steering group then prioritized themes based on a combination of science merit, programmatic balance across different science questions and destination class, cost, and technical readiness. Careful consideration was given to the number of NF mission themes to be prioritized. NF-4 and NF-5 calls had, or will have, six and seven mission themes, respectively. As emphasized by *Vision and Voyages*, "Because preparation and evaluation of New Frontiers proposals places a substantial burden on the community and NASA, it is important to restrict each New Frontiers solicitation to a manageable number of candidate missions." Indeed, with only three major design centers (APL, JPL, GSFC) that can manage NF missions and proposals, a restricted list is needed so they can appropriately allocate resources. On the other hand, after the NF-5 selection, six mission themes from the prior decade will remain unselected. Adding themes based on concepts studied in this survey, which is desirable to ensure that the NF list continues to address the currently highest priority science, then requires increasing the number of mission themes and/or removing some of the prior themes. In consideration of this balance, the committee decided to recommend eight NF themes per call.

In nonprioritized order, the mission themes recommended for New Frontiers 6 are:

- Centaur Orbiter and Lander (CORAL)
- Ceres Sample Return
- Comet Surface Sample Return (CSSR)
- Enceladus Multiple Flyby (EMF)
- Lunar Geophysical Network (LGN)
- Saturn Probe
- Titan Orbiter
- Venus In Situ Explorer (VISE)

If one of the above mission themes is selected in NF-5, it would be removed from the NF-6 list.

The mission themes recommended for New Frontiers 7 include all nonselected mission themes from the NF-6 list above, with the addition of:

- Triton Ocean World Surveyor

The following provides descriptions for the NF-6 and NF-7 mission themes in alphabetical order, as well as additional contextual information in some cases. The listed science objectives reflect the judgment of the committee and are not necessarily identical to those identified in the mission concept studies.

CORAL investigates a Centaur from orbit and in situ, exploring one of a population of dynamically evolved but compositionally primitive small icy bodies from the Kuiper belt that currently reside between Jupiter and Neptune. The proximity of Centaurs provides an opportunity to conduct a comprehensive study of the geochemical and physical properties of primordial ice-rich planetesimals, which trace the composition of nebular volatiles such as H_2O, CO_2, CO, and NH_3, revealing the nature of early solar system compositional reservoirs. The mission will map the surface and measure the ices and organics in situ.[15]

Science Objectives:
- Determine the chemical and physical properties of a Centaur to understand the nature of primitive planetesimals.

---

[15]The full mission study report is available at https://tinyurl.com/2p88fx4f.

- Perform in situ elemental, isotopic, and organic analyses of a Centaur to develop a comprehensive understanding of the composition and initial conditions of the protoplanetary disk.
- Determine the shape, topography, geological landforms, and density of a Centaur to understand the evolutionary history of this population of objects.
- Determine degree of aqueous alteration on a Centaur to investigate the biologic potential of icy planetesimals and potential brine reservoirs.

The mission shall address all four objectives.

Ceres Sample Return focuses on quantifying Ceres's current habitability potential and its origin, which is important for understanding habitability of mid-sized planetary bodies. Habitability is addressed through orbital and in situ investigation of the surface and subsurface environment around a hypothesized brine extrusion zone and via detailed compositional investigations in Earth labs of a returned samples. These samples will be collected from young carbonate salt deposits, typified by those identified by the Dawn mission at Occator crater, as well as some of Ceres's typical dark materials. A sample of adequate mass to achieve the science objectives, acquired in pristine condition and returned at a temperature of lower than $-20°C$ to prevent alteration, would allow investigations of the origin and evolution of Ceres's organic matter, its brine chemistry and sources, and its accretional environment.

Science Objectives:
- Characterize the depth and extent of potential deep brine layer(s) to determine whether liquid exists beneath Ceres today near hypothesized brine extrusion zones.
- Characterize the nature of Ceres's brines from salt deposits to determine the chemistry of waters and their potential habitability.
- Determine the composition, structure, and isotopic composition of Ceres's organics to understand processes of abiotic organic synthesis and evolution.
- Determine the elemental abundances and isotopic ratios of Ceres's materials via measurements on returned samples to determine its accretional environment.

The mission shall address all four objectives.

Comet Surface Sample Return seeks to understand the nature of cometary formation and mixing of materials in the protosolar nebula; compositional reservoirs present in the early solar system; the role of comets in the delivery of water and organic molecules to the early Earth, terrestrial planets and satellites; and evolutionary processes spanning from the protoplanetary disk to current cometary activity. The mission will map the nucleus of a Jupiter family comet, select an optimal sampling site, and acquire a sample from the surface for return to Earth for laboratory analysis. The sample will be acquired and transported in a manner that preserves organics and prevents aqueous alteration of the sample. Volatile material will be characterized via onboard analysis and/or by capture and return at noncryogenic temperatures.

Science Objectives:
- Determine the elemental, isotopic, and structural composition of the organic and inorganic components of a comet nucleus to understand early compositional reservoirs.
- Sample, preserve, and analyze cometary organic material to determine how complex organic molecules form and evolve in interstellar, nebular, and planetary environments.
- Determine the isotopic composition of cometary water to address the role of comets in delivering volatiles to Earth's atmosphere and interior.
- Determine if cometary organic matter contributed significantly to prebiotic chemistry and homochirality of life on Earth.

The mission shall address all four objectives.

Enceladus Multiple Flyby seeks to characterize Enceladus's habitability and look for evidence of life via multiple flybys and analysis of plume material. Enceladus, an active icy moon with a subsurface ocean in a relatively benign radiation environment, provides the best opportunity to directly sample a potential habitable subsurface ocean. Prior Cassini observations demonstrate the presence of alkali and carbonate salts and complex organic molecules in plume icy grains; gas-phase nitrogen- and oxygen-bearing as well as aliphatic and aromatic organic molecules; redox couples (e.g., H_2 and CO_2), habitable temperature, salinity, and pH; alkaline hydrothermal activity; and water–rock reactions. However, Cassini flyby velocities were high, leading to fragmentation of large compounds, and ambiguity as to the precise identity of the parent organic molecules.[16]

Science Objectives:
- Search for and identify complex organic molecules in Enceladus plume materials, with velocities <4 km/s and sample volume >1 μl with appropriate contamination control to enable life-detection investigations.
- Determine the composition, energy sources, and physicochemical conditions of Enceladus's ocean to assess its habitability.
- Characterize Enceladus's cryovolcanic activity to determine spatial and compositional variations in plume activity and the processes causing ocean material ejection and modification.

The mission shall address all three objectives.

The question of whether Enceladus harbors life merits study and would be best addressed by the Orbilander flagship mission (above), which would provide high fidelity in situ analyses for life detection. However, budgetary constraints might not allow for a second new flagship in this decade. Given this, the committee includes the Enceladus Multiple Flyby mission theme in NF, providing an alternative pathway for progress this decade on this crucial question, albeit with smaller sample volume and relatively high sample collection velocity. There are two principal differences between Orbilander and the NF Enceladus Multiple Flyby mission. First, EMF would likely be restricted to a limited number of passes at high velocity (~4 km/s), thus greatly restricting the sample volume available and possibly increasing sample degradation. Conversely, Orbilander in its landed and active scooping mission phase can acquire sample volumes 10^2–10^4 times larger. Second, EMF would have a much smaller instrument component, which reduces the number of orthogonal or cross-checking life-detection techniques available, as well as the geological and geophysical context that can be provided. If the Enceladus Orbilander flagship can be initiated this decade as recommended, then EMF would be removed from the NF-6 and NF-7 lists.

Lunar Geophysical Network examines the physical properties of the present-day Moon by deploying a global, long-lived ($\geq$6 years) network of geophysical instruments on its surface. Although all large terrestrial bodies are thought to form cores, mantles, and primordial crusts through solidification of magma oceans, the Moon retains the most faithful record of the nature of this process. LGN will reveal the nature and evolution of the lunar interior and facilitate understanding of the initial solidification and primordial geologic processes that have shaped all terrestrial bodies. These measurements (e.g., seismic, heat flow, laser ranging, and magnetic-field/electromagnetic sounding) will allow the bulk composition of the Moon to be calculated, elucidate the dynamical processes that are active during the early history of terrestrial planets, provide new constraints on the collision process that generated our unique Earth-Moon system, and illuminate processes currently active on the Moon.

Science objectives:
- Determine the internal structure and size of the crust, mantle, and core to constrain the composition, mineralogy, and lithologic variability of the Moon.
- Determine the distribution and origin of lunar seismic activity in order to better understand the origin of moonquakes and provide insights into the current dynamics of the lunar interior and the interplay with external phenomena such as tidal interactions with Earth.

---

[16]The full mission study report is available at https://tinyurl.com/2p88fx4f.

- Determine the global heat-flow budget for the Moon in order to constrain more precisely the distribution of heat-producing elements in the crust and mantle, the origin and nature of the Moon's asymmetry, its thermal evolution, and the extent it was initially melted.

The mission shall address all three objectives.

Saturn Probe obtains in situ measurements of the atmosphere from an entry probe. Understanding the initial conditions in the protosolar nebula requires measurements of each of the giant planets' elemental and isotopic compositions. Constraining giant planet formation mechanisms is particularly dependent on knowing when and where Saturn formed, over how long, and if its orbit has migrated over time to stop Jupiter's inward movement. Noble gas abundances are also crucial for determining if helium rain has prolonged Saturn's thermal evolution. Additionally, comparisons of what governs the diversity of giant planet climates, circulation, and meteorology require constraints on the vertical temperature and wind profiles, as well as vertical circulation. Although some measurements may be obtained via remote sensing, many of the science objectives require in situ sampling.

Science Objectives:
- Determine the in situ noble gas, elemental, and isotopic abundances to understand conditions in the protosolar nebula, as well as constrain Saturn's formation, evolution, and migration.
- Determine the in situ tropospheric temperature-pressure profile to quantify Saturn's heat transport and convective stability.
- Determine the in situ vertical wind shear to characterize Saturn's tropospheric circulation and meteorology.
- Constrain vertical mixing in Saturn's troposphere to bound transport from the deeper interior

The mission shall address all four objectives.

Titan Orbiter globally characterizes Titan's dense N_2 atmosphere that harbors prebiotic molecules, its Earth-like methane hydrological cycle and seas, and its subsurface liquid water ocean, including how they evolve over time, in order to assess Titan's potential habitability. Cassini flybys revealed complex organic chemistry, methane-ethane lakes and seas, and meteorology on Titan; however, these processes could not be thoroughly studied owing to instrumentation and flyby coverage limitations. Titan orbiter will investigate how the organic chemical factory on Titan works, both in the atmosphere and on the surface, providing important context for data from Dragonfly and complementary global measurements.[17]

Science Objectives:
- Determine Titan's internal structure, the depth and thickness of the ice shell and subsurface ocean, and whether the former is convecting; and determine rates of interior-surface solid or gas interchange.
- Characterize Titan's global geology and its landscape-shaping processes.
- Characterize Titan's global methane hydrological and sedimentological system, including surface transport/flow rates and cloud distributions.
- Quantify the production, transport and fate of organic molecules in Titan's upper atmosphere and atmospheric and climate evolution in general.

The mission shall address all four objectives.

The CAPS 2020 report on NF-5 (NASEM 2020) stated "With the selection of Dragonfly in the NF-4 competition and now under development, reconsideration by NASA of including a Titan mission in the NF-5 call under the Ocean Worlds mission theme is warranted on programmatic grounds and removing Titan from the list of potential targets would be appropriate. The next steps for Titan exploration are best evaluated and prioritized by the upcoming planetary science and astrobiology decadal survey." The committee evaluated these issues and concluded that Titan Orbiter provides important and complementary science to Dragonfly, including global-scale monitoring of the atmosphere, seasons, and surface geologic processes, versus the several hundreds of kilometers

---

[17]The full mission study report is available at https://tinyurl.com/2p88fx4f.

to be studied by Dragonfly along its flightpath. Cassini was not a Titan orbiter and did not achieve global coverage, and Titan's upper atmospheric composition is not known and will not be measured by Dragonfly. *Vision and Voyages* evaluated the Titan Saturn System Mission (TSSM), which addressed a wide range of science goals, and Titan Orbiter in combination with Dragonfly would cover most of the science envisaged for TSSM, which was broader in scope than the NF-4 Ocean Worlds Titan mission theme.

Triton Ocean World Surveyor orbits Neptune and performs multiple flybys of its largest and retrograde orbiting satellite, Triton. Triton is likely a captured KBO and a candidate ocean world with a geologically young surface and active geysers. It has a hazy atmosphere like Pluto's and a uniquely strong ionosphere.

Proposed Science Objectives:
- Determine whether Triton is an ocean world, ascertain its interior structure, and decide whether Triton's ice shell is in hydrostatic equilibrium and decoupled from the interior.
- Characterize Triton's surface composition and geology, and look for changes, including plumes and their composition.
- Determine the nature of the moon–magnetosphere interaction at Triton.
- Determine the composition, density, temperature, pressure, and spatial/temporal variability of Triton's atmosphere.

The mission shall address all four objectives.

The recommendation that Triton Ocean World Surveyor be delayed until NF-7 took into consideration launch trajectories, which benefit from a Jupiter gravity assist likely available in the NF-7 timeframe.

Venus In Situ Explorer (VISE) investigates the processes and properties of Venus that cannot be characterized from orbit or from a single descent profile. These include (1) complex atmospheric cycles (e.g., radiative balance; chemical cycles, atmospheric dynamics, variations of trace gases, light stable isotopes, and noble gas isotopes, and the couplings between these processes); (2) surface–atmosphere interactions (e.g., physical and chemical weathering at the surface, near-surface atmospheric dynamics, and effects upon the atmosphere by any ongoing geological activity); and (3) surface properties (e.g., elemental and mineralogical composition of surface materials, heat flow, seismic activity, and any magnetization). VISE will provide breakthrough information on the origin of the terrestrial planets, the evolution of their interiors and surfaces, atmospheric evolution and climate, and critical insights into the nature and habitability of exoplanets.

Science objectives:
- Characterize past or present large-scale spatial and temporal (global, longitudinal and/or diurnal) processes within Venus's atmosphere.
- Investigate past or present surface–atmosphere interactions at Venus.
- Establish past or present physical and chemical properties of the Venus surface and/or interior.

The mission shall address at least two of these three objectives.

Two missions on the NF-5 list of mission themes do not appear on the above lists for NF-6 and NF-7: Io Observer and SPA Sample Return. The committee carefully considered the Io Observer NF theme in light of the success of the IVO Discovery mission in reaching Phase A. CAPS 2020 stated the following: "If NASA's exploration of Io proceeds via the selection of the IVO Discovery mission, then based on the IVO Step 1 proposal, inclusion of Io Observer would be redundant scientifically and its inclusion in NF5 would strongly warrant reconsideration." The committee reaffirms the importance of Io as a unique body. Not only is it important to understanding tidal dissipation and resulting active volcanic, tectonic, and plasma processes, but also, for example, to providing an important analog to young terrestrial planets and tidally heated exoplanets. The committee anticipates that Io Observer will have an opportunity to compete in NF-5. The selection of IVO for Phase A study demonstrates that fundamental Io science can also be achieved via the Discovery program, and this may be increasingly true with time as power systems and launch vehicles continue to evolve. These factors placed this theme at lower priority for NF-6 and NF-7 than other themes that clearly require a medium-class mission to complete their core science.

The SPA Sample Return mission addresses the highest priority lunar science. However, achieving the top science objectives with a fixed lander, as has been typically envisioned, is challenging. The committee concluded that the Endurance-A rover mission is a superior approach for acquiring abundant samples across diverse terrains to address multiple top-level science questions for the Moon and the solar system. The committee recommends that NASA pursue Endurance-A as a strategic medium-class mission within LDEP (see LDEP section above).

REPRESENTATIVE FLIGHT PROGRAMS FOR THE DECADE

Following the recommendations, priorities, and decision rules outlined in this report, the committee developed two representative program portfolios—*Recommended* and *Level*—for solar system exploration in the decade 2023–2032. The *Level Program* is designed to fit within the decadal funding projected for PSD, calculated by inflating the FY 2023 NASA planetary science budget by 2 percent per year through the remainder of the decade. The *Recommended Program* describes a vision to address the priority questions identified in this report and can be met with a total decade budget that is ~17.5 percent higher than the *Level Program*.

Both programs:

- Continue support for missions in operation and in development.
- Continue the Mars Sample Return campaign as currently planned.
- Initiate the Uranus Orbiter and Probe Flagship mission.
- Increase R&A funding to 10 percent or more of the annual PSD budget by mid-decade.
- Continue support of the Mars (MEP) and Lunar (LDEP) Programs.
- Return the MEP to its pre-MSR funding levels as MSR costs decrease, reaching a level of $500 million in FY 2032.
- Incorporate the cost cap recommendations for Discovery and New Frontiers.
- Incorporate cost realism from the TRACE studies.
- Assume life cycle costs of a representative Discovery mission to be $900 million.
- Assume a representative NF mission has an 11-year cruise and a life cycle cost of $2 billion in the first half of the decade and $2.2 billion in the second half.
- Maintain support for Planetary Defense, enabling a new start, rapid-response flyby reconnaissance mission.
- Sustain and increase plutonium production.

Table 22-3 summarizes the total decade funding levels recommended for each of the major programmatic elements in both the *Recommended* and *Level Programs*. Where the two programs differ most is in the level of support for new initiatives to address the science questions discussed in Chapters 4–15 (Tables 22-3 and 22-4). The *Recommended Program* captures the highest priorities of the community as outlined in this report and is both aspirational and inspirational. This program enables the robust training and development of diverse planetary science and engineering communities, drives technology development and implementation, and maintains U.S. leadership in exploration across the solar system. This program begins the highest-priority Flagship mission, the Uranus Orbiter and Probe, in FY 2024 with a launch in the early 2030s to take advantage of a Jupiter gravity assist that is available through 2032 to minimize cruise length and complexity. This UOP start date maintains the strategic mission funding as the MSR costs decrease (Figure 22-1). The *Recommended Program* also begins the second prioritized flagship, the Enceladus Orbilander in the latter part of the decade, allowing for state-of-the-art life detection techniques and sample volume that cannot be achieved in a medium-class mission. Orbilander provides an outstanding opportunity to explore the astrobiological conditions of ocean worlds and will revolutionize our understanding of these worlds. The *Recommended Program* restores the strong recommendation from *Vision and Voyages*, endorsed by this committee, for a cadence of two NF missions per decade. The NF-5 selection, which was to be the second NF mission from the decade of *Vision and Voyages*, would be completed early in the decade and would be followed by the selection of two NF missions in a mid-decade NF-6 opportunity. The *Recommended Program* increases the total funding for R&A across the decade by $1.25 billion relative to the current planned FY 2023 level adjusted

TABLE 22-3 Representative PSD Programs for the Decade

Program Element	Recommended Program ($ million)	Level Program ($ million)
R&A[a]	3,870	3,350
Europa Clipper	1,700	1,700
Mars Sample Return	5,300	5,300
Discovery	5,250	5,250
New Frontiers	7,300	5,100
Mars Exploration[b]	2,850	2,650
Lunar Exploration[c]	4,760	4,760
Planetary Defense	1,700	1,700
Radioisotope power	1,750	1,750
Planetary Other[d]	2,150	2,150
New Flagship #1	3,450	1,280
New Flagship #2	1,040	—
Total	**41,120**	**34,990**

NOTE: Costs are in real year dollars assuming 2 percent annual inflation.

[a] R&A budget levels reflect a 10 percent per year funding increase for openly competed R&A programs sufficient to bring the annual investment in R&A to ≥10 percent of the annual PSD budget by mid-decade; nonopenly competed R&A activities are increased by inflation at 2 percent/year; see "Recommended Funding for NASA Planetary R&A" section of Chapter 17 for details.

[b] MEP would support extended mission science and directed activities (e.g., small spacecraft missions, international collaborations, technology development, and/or execution of a science-enhanced iMIM) during the main MSR development phase, followed by a gradual ramp-up of the MEP yearly budget to its pre-MSR level by the decade's end, and, in the Recommended Program, supporting initiation of a medium-class Mars mission.

[c] LDEP would support ongoing activities and Endurance-A.

[d] Other includes non-R&A elements in the Planetary Science Research line (e.g., planetary data system and astromaterial curation); management in the Discovery, MEP, and LDEP lines; international mission contributions in the Discovery and Outer Planets and Ocean Worlds lines; and icy satellites surface technology. It is increased by 2 percent/year relative to current levels to account for inflation. Technology development is included in most elements (e.g., R&A, Flagships, Mars Exploration, Lunar Exploration, and Planetary Other) and does not have a separate line in this table.

forward with inflation (see Chapter 17). Figure 22-1 shows the funding profile for the *Recommended Program* and Table 22-4 summarizes the primary differences between the two programs.

The *Level Program* can be conducted assuming the currently projected NASA budget (Table 22-3). The *Level Program* increases the total funding for R&A across the decade by $730 million relative to the current level including inflation. However, the *Level Program* budget profile (Figure 22-2) reveals several substantial disadvantages compared with the *Recommended Program*. This budget scenario does not include Enceladus Orbilander or MLE. The Uranus Orbiter and Probe Flagship mission could be initiated, but it would not begin until late in the decade, delaying its launch until 2038 or later, well into the decade beyond the horizon of this survey, and more than 14 years after the planned launch of Europa Clipper. Regular launch opportunities for UOP are available in this period but would have 2–3 years longer cruise durations and require Venus as well as Earth flybys. The program also loses the continuity of development begun in the Europa Clipper and MSR missions, with a drop in strategic mission funding from FY 2027 through FY 2030 raising the potential for loss of optimized workflow and/or critical expertise. Indeed, delay of UOP until the end of the decade would focus interim new developments

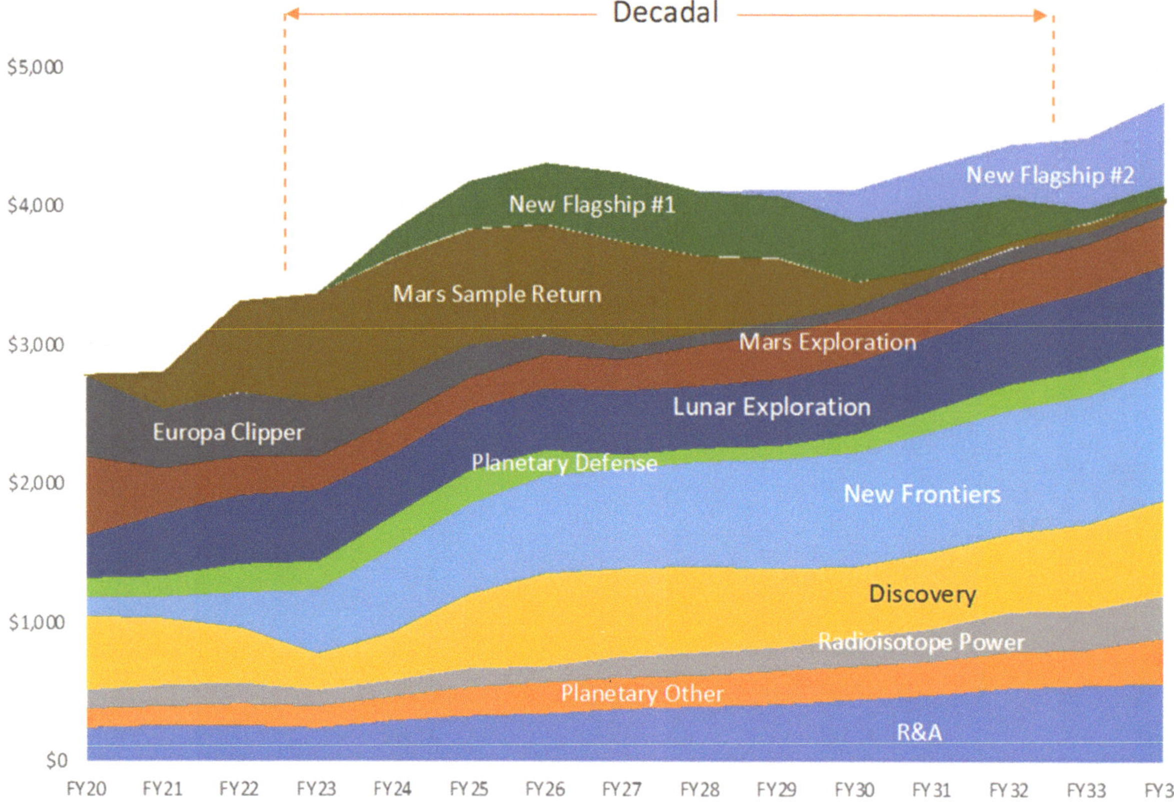

FIGURE 22-1 Recommended Program for the 2023 to 2032 decade. Costs are in real year dollars assuming 2 percent annual inflation. "Planetary Other" includes non-R&A elements in Planetary Science Research, management in Discovery, MEP, and LDEP; international mission contributions, and icy satellites surface technology.

TABLE 22-4 Comparison of Representative Programs

Recommended Program	Level Program
Continue Mars Sample Return	Continue Mars Sample Return
Five new Discovery selections at recommended cost cap	Five new Discovery selections at recommended cost cap
Support LDEP with mid-decade start of Endurance-A	Support LDEP with mid-decade start of Endurance-A
R&A increased by $1.25 billion	R&A increased by $730 million
Continue Planetary Defense Program with NEO Surveyor and a follow-on NEO characterization mission	Continue Planetary Defense Program with NEO Surveyor and a follow-on NEO characterization mission
Gradually restore MEP to pre-MSR level with late decade start of Mars Life Explorer	Gradually restore MEP to pre-MSR level in late decade with no new start for Mars Life Explorer
New Frontiers 5 (1 selection)	New Frontiers 5 (1 selection)
New Frontiers 6 (2 selections)	New Frontiers 6 (late, or not included)
Begin Uranus Orbiter and Probe in FY 2024	Begin Uranus Orbiter and Probe in FY 2028
Begin Enceladus Orbilander in FY 2029	No new start for Enceladus Orbilander this decade

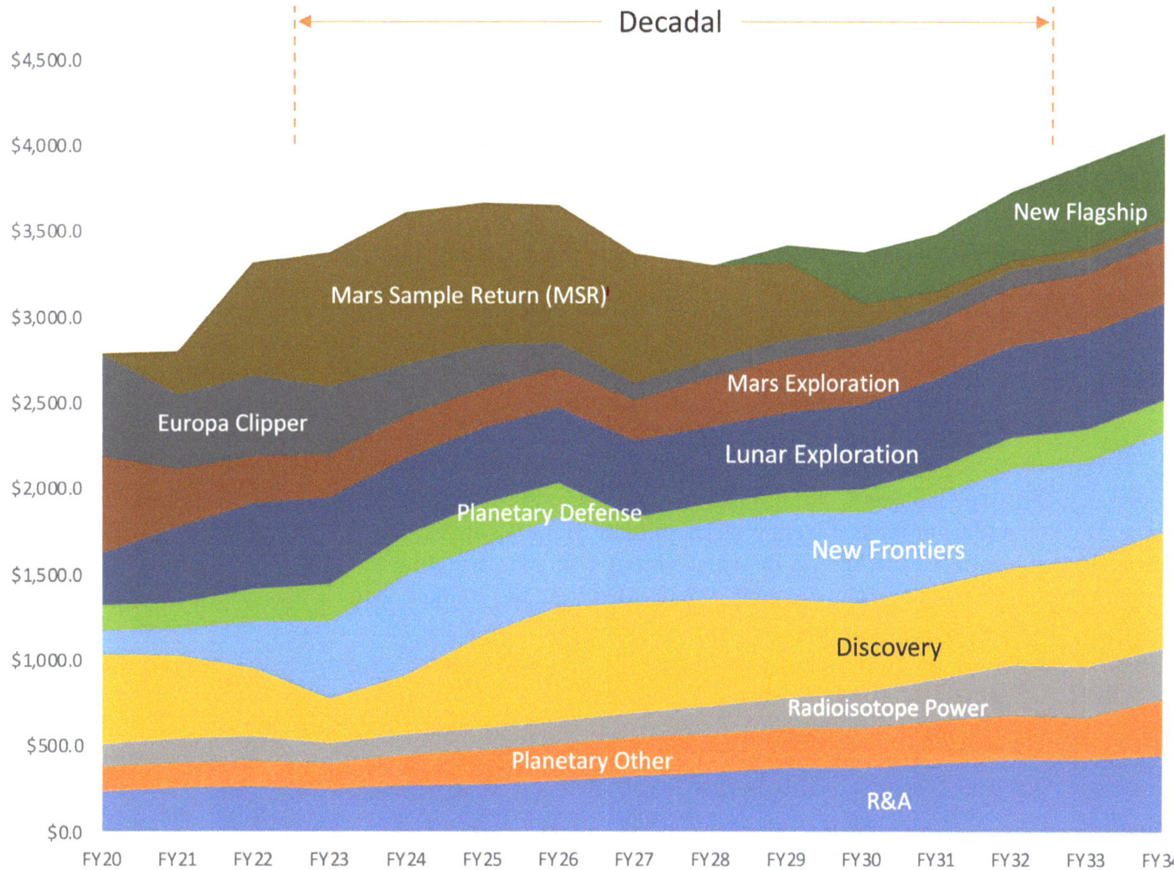

FIGURE 22-2 Level Program for the 2023 to 2032 Decade. Costs are in real year dollars assuming 2 percent annual inflation. "Planetary Other" includes non-R&A elements in Planetary Science Research, management in Discovery, MEP, and LDEP; international mission contributions, and icy satellites surface technology.

on smaller programmatic elements and maintaining the vitality of the whole program and the community in these circumstances may be very challenging. The first opportunity to select from the highly regarded new NF mission concepts studied in this report—including CORAL, Ceres Sample Return, and Titan Orbiter—will be in NF-6. In the *Level Program*, this selection would likely not occur until very late in the decade or in the next because of the delayed NF-5 selection and the funding required to complete Dragonfly.

This undermines the potential for high-impact science, including notably that involving outer solar system targets and/or sample return. In summary, the reductions associated with the *Level Program* would result in a less balanced portfolio with a significantly lower science return compared to the *Recommended Program*.

Budgetary Decision Rules

The committee strongly endorses the aspirational *Recommended Program*. The *Level Program* has been designed to fit within the projected PSD budget for the coming decade. However, if the budget is reduced below the *Level Program* budget then the committee has developed the following prioritized programmatic reductions to new initiatives to accommodate those changes. In priority order, with the first program element the first to be reduced, the committee recommends:

1. Delay the start of the next Flagship mission;
2. Reduce the number of new Discovery missions to four;

3. Reduce the funding level for Planetary Defense by removing the new-start mission after NEO Surveyor;
4. Reduce the cadence of New Frontiers in the coming decade;
5. Reduce the funding level for LDEP with a late-decade start of Endurance-A;
6. Reduce the funding level for MEP below the Level program;
7. Reduce the number of new Discovery missions to three; and
8. Reduce R&A funding.

Science Traceability of Prioritized Large and Medium-Class Missions

The large- and medium-class strategic and PI-led missions prioritized and recommended in the preceding sections of this report were selected based on their ability to address the priority science questions, as well as programmatic balance, technical risk and readiness, and cost. After these missions had been selected, the committee evaluated this portfolio of new missions to assess how well they covered the breadth of the priority science questions (Q1–Q12) discussed in Chapters 4–15. The committee considered whether each mission would likely contribute to a "substantial," "breakthrough," or "transformative" advance for each of the sub-questions in Q1 through Q12. The tabulated and normalized results are displayed in a mission portfolio assessment matrix (Table 22-5) on a scale of modest (yellow) to high (dark green) contribution. This matrix illustrates that the collective suite of prioritized missions does an excellent job of addressing the full breadth of the priority planetary science questions and does so at a diverse set of destinations. The committee notes that Q9 focuses on terrestrial

TABLE 22-5 Mission Portfolio Assessment Matrix

Mission Name	Priority Science Questions											
	1	2	3	4	5	6	7	8	9	10	11	12
Mars Sample Return												
Uranus Orbiter and Probe												
Enceladus Orbilander												
Endurance-A												
Mars Life Explorer												
Centaur Orbiter and Lander												
Ceres Sample Return												
Comet Surface Sample Return												
Enceladus Multiple Flyby												
Lunar Geophysical Network												
Saturn Probe												
Titan Orbiter												
Triton Ocean World Surveyor												
Venus In Situ Explorer												

NOTES: Assessment of the science questions addressed by MSR and each of the other large- and medium-class missions prioritized in this report. The top rows include MSR and the two new large strategic missions prioritized here. Endurance-A and Mars Life Explorer are highly ranked medium-class missions recommended for the LDEP and MEP programs, respectively. The remaining rows are the prioritized New Frontiers mission themes in alphabetical order. Yellow represents a modest contribution—typically a "substantial" advance in addressing one to a few of a priority science sub-questions—whereas the increasing intensity of green indicates increasing levels of "breakthrough" or "transformative" advances—that is, addressing an increasing number of sub-questions. Note that Q9 focuses on terrestrial life and is therefore not the primary focus of most planetary missions, but rather is supported through astrobiology research programs.

life and is therefore not the primary focus of most planetary spacecraft missions, but rather is supported through astrobiology research programs. The committee further emphasizes that the table is not intended to, and should not be used to, prioritize between the missions; for example, a mission that definitively answers a single question can be as impactful as one that makes progress on several questions.

The *Level Program* would not include the Enceladus Orbilander or MLE, the Uranus Orbiter and Probe mission would be delayed, the sixth New Frontiers competition may not occur, and there would be fewer PI-led missions overall. These reductions would result in a less balanced portfolio with a significantly lower science return compared to the *Recommended Program*.

STATE OF THE PROFESSION

The committee's Statement of Task explicitly requested an assessment of the state of the planetary science and astrobiology communities, and Chapter 16 is devoted to this topic. The state of the profession (SoP), including issues of diversity, equity, inclusivity, and accessibility (DEIA), is central to the success of the planetary science enterprise. Its inclusion here, *for the first time in a planetary science decadal survey*, reflects its importance and urgency.

Ensuring broad access and participation is essential to maximizing excellence in an environment of fierce competition for limited human resources, and to ensuring continued American leadership in planetary science and astrobiology (PS&AB). A strong system of equity and accountability is required to recruit, retain, and nurture the best talent into the PS&AB community. The committee applauds the hard-earned progress that has been made—most notably with respect to the entry and prominence of women in the field—as well as the exemplary goals and intentions of NASA science leadership with respect to DEIA. However, much work remains to be done, to address persistent and troubling issues of basic representation by race/ethnicity.

The committee's eight SoP recommendations (see Chapter 16) address four primary topical areas:

1. *An evidence gathering imperative.* Equity and accountability require accurate and complete data about the SoP. There is an urgent need for data concerning the size, identity, and demographics of the PS&AB community; and workplace climate. Without such data, it cannot be known if the best available talent is being utilized, nor how involvement may be undermined by adverse experiences.
2. *Education of individuals about the costs of bias and improvement of institutional procedures, practices, and policies.* The committee recommends that the PSD adopt the view that bias can be both unintentional and pervasive, and provides actionable steps to assist NASA in identifying where bias exists and in removing it from its processes.
3. *Broadening opportunities to advance the SoP.* Engaging underrepresented communities at secondary and college levels to encourage and retain them along PS&AB career pathways is essential to creating and sustaining a diverse community.
4. *Creating an inclusive and inviting community free of hostility and harassment.* Ensuring that all community members are treated with respect, developing and enforcing codes of conduct, and providing ombudsperson support to address issues is important for maintaining healthy and productive work environments.

Together, the SoP findings and recommendations aim to assist NASA's PSD in boldly addressing issues that concern its most important resource: the people who propel its planetary science and exploration missions. The reader is referred to Chapter 16 for detailed discussion of these important issues.

OTHER KEY PROGRAMMATIC RECOMMENDATIONS

Detailed rationales, findings, and recommendations for infrastructure, human exploration, and technology are given in their respective chapters. Key topics discussed in these three chapters and those recommendations having broad programmatic and budgetary implications are summarized below.

Infrastructure

Chapter 20 provides a detailed discussion of PSD's infrastructure needs for the next decade and includes 17 findings and 7 recommendations. Here the committee presents several of the key elements and recommendations from that chapter.

Mars Sample Receiving Facility

The processing and analysis of samples returned from Mars will occur in three separate stages: (1) initial receiving and characterization to verify that the samples can be safely distributed; (2) distribution to the science community for detailed analysis; and (3) long-term curation. An end-to-end plan is needed for all three of these elements, and this planning needs to include early engagement with the sample science community, government stakeholders, and the public. The samples will be initially delivered to the Mars Sample Receiving Facility (SRF), which will be the first facility since Apollo to implement planetary protection Category V, Restricted Earth Return requirements. These requirements include life detection, biohazard assessment, and, if necessary, sterilization procedures. Samples are anticipated from Mars in 2031 and the required biosafety level facility (BSL-4) can take up to 10 years to build (see Infrastructure chapter). Therefore, the finalization of requirements and the implementation of the SRF needs to begin immediately. Other government agencies have relevant BSL-4 facilities, and NASA can leverage this infrastructure and expertise to construct the SRF while minimizing cost. MSR/MEP and international participation and contribution in SRF requirements definition and funding is essential. Because the driving aim of MSR is analysis of samples by the science community, the SRF analysis capability needs to focus on the tools necessary to verify sample safety and does not need the full range of specialized microanalytical instrumentation available at other labs. The long-term curation of the Mars samples is best overseen by a NASA/ESA curatorial team. While MSR is driving the development schedule, it is prudent for NASA to consider needs beyond the MSR program, as the acquisition, curation, and analysis of samples is a many-decades-spanning investment and handling astrobiologically relevant samples from other destinations is an anticipated future need (see Infrastructure chapter).

> **Recommendation: NASA, in partnership with ESA and community stakeholders, should develop the plan for the end-to-end processing of samples returned from Mars. This plan should include the definition, design, and construction of the Mars Sample Receiving Facility to ensure that it is ready to receive the samples by 2031. The plan should also outline the approach for expeditiously distributing the samples to the scientific community for analysis and to a long-term curation facility.**

Plutonium-238 Production

Plutonium-238 (^{238}Pu) and Radioisotope Power Systems (RPS) are essential to the exploration of the solar system. *Vision and Voyages* recommended that NASA restart ^{238}Pu production with a goal of producing 1.5 kg/year and the program is on track to reach this goal in 2026.

> **Recommendation: NASA should evaluate plutonium-238 production capacity against the mission portfolio recommended in this report and other NASA and national needs, and increase it, as necessary, to ensure a sufficient supply to enable a robust exploration program at the recommended launch cadence.**

> **Recommendation: NASA should continue to invest in maturing higher efficiency radioisotope power system technology to best manage its supply of plutonium-238 fuel.**

Launch Services

The capability, cost, and availability of launch vehicles are fundamental components of the timing and implementation of NASA's planetary mission portfolio. The workhorse Atlas V and Delta IV vehicles are expected to be retired and will no longer be available for use by this survey's prioritized missions. At the same time, new launch vehicle entrants—e.g., SpaceX Starship and Blue Origin New Glenn—are emerging from a variety of companies, both established aerospace manufacturers and "new-space" enterprises. Some of these vehicles may have performance metrics suitable for planetary science missions on all scales. In particular, there may be opportunities for innovative solutions to sending missions on trajectories to the outer solar system. Experimental flight testing of new launch vehicles may also provide a mechanism for PSD technology flight validation missions.

> **Recommendation: NASA should develop a strategy to focus and accelerate development of high energy launch capability, or its equivalent, and in-space propulsion to enable robust exploration of all parts of the solar system. Any new systems that are developed should also build the pedigree to permit the launch of nuclear materials.**

Telescopic Observations

Telescopic observations from space- and ground-based observatories provide essential support for planetary science and astrobiology through synergies with data returned from flight missions (see Chapter 20). Ground-based facilities have the benefit of longevity, and their capabilities can increase over time as science instruments can be upgraded, repaired, or replaced. Observations are utilized to monitor dynamic solar system phenomena, including planet and satellite atmospheres, comets, interstellar objects, and cryovolcanic/plume activity that can vary on timescales of hours to multiple decades. Occultations provide data important to understanding distant small body populations in general, and targets of planned space missions in specific. Such investigations are called out among the strategic research activities in priority science questions Q2 through Q8 (Chapters 5–11). Ground- and space-based observations of protoplanetary disks and exoplanetary systems provide crucial constraints to understanding how the solar nebula formed and ultimately evolved into our planetary system (see Q1–Q3 and Q12, Chapters 4–6 and 15, respectively).

Extremely large ground-based telescopes coming on-line late in the next decade will deliver advanced capabilities with angular resolution comparable to Voyager approach data. NASA investment in these observatories would enable new science and capabilities for spacecraft mission support. The NSF's National Optical-Infrared Astronomy Research Laboratory (NOIRLab) provides the mainstay of ground-based telescopic science for astronomy. The Vera C. Rubin Observatory is currently being commissioned and will revolutionize planetary astronomy with an unprecedented inventory of solar system objects and time-domain observations. The next generation of extremely large optical telescopes will see first light in the next decade. Without NSF investment in guest observing programs at these next generation telescopes, the dramatically new science research that will be performed will only benefit from the skills, knowledge, and innovative perspectives from the subset of U.S. researchers. Construction of the next-generation VLA, planned to start in the next few years, promises major advances in planetary science, particularly for the giant planets and their largest moons (see Chapter 20). Planetary research at these future facilities will be aided by the inclusion of planetary astronomers within development teams and observatory staff.

> **Recommendation: NSF-supported, ground-based telescopic observations provide critical data needed to address important planetary science questions. The NSF should continue and, if possible, expand the funding to support the existing and future observatories, provide guest observer programs, and include planetary astronomers in future observatory development in order to maximize the science return from solar system observations**

Planetary Radar

The loss of the Arecibo Observatory planetary radar facility has resulted in a significant gap in solar system observations, particularly in support of planetary defense. Radar observations are the most precise method for NEO astrometry, and provide important data needed to constrain small body size and spin states, improving our knowledge of the NEO population (see Chapters 7 and 18).

> **Recommendation: NASA and NSF should review the current radar infrastructure to determine how best to meet the community's needs, including expanded capabilities to replace those lost at Arecibo Observatory.**

Human Exploration

Chapter 19 provides a detailed discussion of human exploration activities and includes 16 findings and 6 recommendations. Here the committee presents several of the key elements and recommendations from that chapter.

Human exploration is an aspirational and inspirational endeavor, and NASA's Moon-to-Mars exploration plans hold the promise of broad benefits to the nation and the world. Human explorers can conduct cutting edge scientific investigations and collect carefully chosen samples, significantly enhancing quality of the science that can be achieved (see LDEP section above). In turn, a robust science program provides the motivating rationale for a sustainable human exploration program that will achieve the maximum return from programs such as Artemis at the Moon. The Human Exploration chapter provides a detailed discussion of the importance of science in human exploration and the need for an effective NASA organizational structure to maximize the scientific return from its human exploration endeavors. Here the committee presents two key recommendations from that chapter.

Recommendation: Conducting decadal-level science should be a central requirement of the overall human exploration program.

Recommendation: NASA should adopt an organizational approach in which SMD has the responsibility and authority for the development of Artemis lunar science requirements that are integrated with human exploration capabilities. NASA should consider establishing a joint program office at the associate administrator level for the purpose of developing Artemis program-level requirements across SMD, ESDMD, SOMD, and other directorates as appropriate.

Technology

Technology is the foundation of scientific exploration and significant technology investment by NASA is needed to ensure that the priority missions recommended by this survey can be accomplished. Within PSD, technology funding is included in the R&A program, Mars and lunar exploration, large mission development, the icy satellite surface technology program and elsewhere. Chapter 21 provides a detailed discussion of technology development needs and includes 44 findings and 7 recommendations to guide the necessary investments for the future. Here the committee presents the key budgetary recommendation from that chapter.

Recommendation: NASA PSD should strive to consistently fund technology advancement at an average of 6 to 8 percent of the PSD budget.

REFERENCES

Beattie, D. 2001. *Taking Science to the Moon*. Baltimore, MD: Johns Hopkins University Press.

ESA (European Space Agency). 2021. *Voyage 2050: Final Recommendations from the Voyage 2050 Senior Committee*. Paris, France. https://www.cosmos.esa.int/documents/1866264/1866292/Voyage2050-Senior-Committee-report-public.pdf/e2b2631e-5348-5d2d-60c1-437225981b6b?t=1623427287109.

iMOST (International MSR Objectives and Samples Team) D.W. Beaty, M.M. Grady, H.Y. McSween, E. Sefton-Nash, B.L. Carrier, F. Altieri, et al. 2019. "The Potential Science and Engineering Value of Samples Delivered to Earth by Mars Sample Return." *Meteoritics & Planetary Science* 54(3):667–671. https://onlinelibrary.wiley.com/doi/full/10.1111/maps.13232.

Jakosky, B., M. Amato, S. Atreya, D. Des Marais, P. Mahaffy, M. Mumma, M. Tolbert, B. Toon, C, Webster, and R. Zurek. 2020. "Scientific Value of Returning an Atmospheric Sample from Mars." White paper 159 submitted to Planetary Science and Astrobiology Decadal Survey 2023–2032. *Bulletin of the AAS* 53(4). https://baas.aas.org/pub/2021n4i159/release/1.

LEAG (Lunar Exploration Analysis Group). 2018. *Advancing Science of the Moon: Report of the Lunar Exploration Analysis Group Special Action Team*. https://www.lpi.usra.edu/leag/reports/ASM-SAT-Report-final.pdf.

MASWG (Mars Architecture Strategy Working Group). 2020. B.M. Jakosky, R. Zurek, S. Byrne, W. Calvin, S. Curry, B. Ehlmann, et al. 2020. *Mars, the Nearest Habitable World—A Comprehensive Program for Future Mars Exploration*. Washington, DC: National Aeronautics and Space Administration.

MEPAG (Mars Exploration Program Analysis Group). 2020. "Mars Science Goals, Objectives, Investigations, and Priorities: 2020 Version." https://mepag.jpl.nasa.gov/reports/MEPAGGoals_2020_MainText_Final.pdf.

Moriarty III, D.P., N. Dygert, S.N. Valencia, R.N. Watkins, and N.E. Petro. 2021. "The Search for Lunar Mantle Rocks Exposed on the Surface of the Moon." *Nature Communications* 12(4659). https://www.nature.com/articles/s41467-021-24626-3.pdf.

NASA (National Aeronautics and Space Administration). 2015. *NASA Astrobiology Strategy 2015*. Washington, DC: NASA Astrobiology Program. https://astrobiology.nasa.gov/nai/media/medialibrary/2015/10/NASA_Astrobiology_Strategy_2015_151008.pdf.

NASA. 2020a. *Artemis Plan: NASA's Lunar Exploration Program Overview.* Washington, DC. https://www.nasa.gov/sites/default/files/atoms/files/artemis_plan-20200921.pdf.

NASA. 2020b. *Summary of NASA Response to Mars Sample Return Independent Review Board Recommendations*. Washington, DC: Science Mission Directorate. https://www.nasa.gov/sites/default/files/atoms/files/nasa_esa_mars_sample_return_irb_report.pdf.

NASEM (National Academies of Sciences, Engineering, and Medicine). 2015. *The Space Science Decadal Surveys: Lessons Learned and Best Practices*. Washington, DC: The National Academies Press.

NASEM. 2016. *Extending Science: NASA's Space Science Mission Extensions and the Senior Review Process*. Washington, DC: The National Academies Press.

NASEM. 2017b. *Powering Science: NASA's Large Strategic Science Missions*. Washington, DC: The National Academies Press.

NASEM. 2017c. *Report Series: Committee on Astrobiology and Planetary Science: Getting Ready for the Next Planetary Science Decadal Survey*. Washington, DC: The National Academies Press.

NASEM. 2018a. *Visions into Voyages for Planetary Science in the Decade 2013–2022: A Midterm Review*. Washington, DC: The National Academies Press.

NASEM. 2019a. An *Astrobiology Strategy for the Search for Life in the Universe*. Washington, DC: The National Academies Press. https://doi.org/10.17226/25252.

NASEM. 2019b. *Finding Hazardous Asteroids Using Infrared and Visible Wavelength Telescopes*. Washington, DC: The National Academies Press. https://doi.org/10.17226/25476.

NASEM. 2019d. *Report Series: Committee on Astrobiology and Planetary Science: Review of the Planetary Science Aspects of NASA SMD's Lunar Science and Exploration Initiative*. Washington, DC: The National Academies Press.

NASEM. 2019e. *Report Series: Committee on Astrobiology and Planetary Science: Review of the Commercial Aspects of NASA SMD's Lunar Science and Exploration Initiative*. Washington, DC: The National Academies Press.

NASEM. 2020. *Report Series: Committee on Astrobiology and Planetary Science: Options for the Fifth New Frontiers Announcement of Opportunity*. Washington, DC: The National Academies Press.

NRC (National Research Council). 2007. *The Scientific Context for the Exploration of the Moon*. Washington, DC: The National Academies Press.

NRC. 2008. *Grading NASA's Solar System Exploration Program: A Midterm Report*. Washington, DC: The National Academies Press.

NRC. 2011. *Vision and Voyages for Planetary Science in the Decade 2013–2022*. Washington, DC: The National Academies Press, p. 314.

NSTC (National Science and Technology Council) Interagency Working Group for Detecting and Mitigating the Impact of Earth-Bound Near-Earth Objects. 2018. *National Near-Earth Object Preparedness Strategy and Action Plan*. Washington, DC: Executive Office of the President. https://www.nasa.gov/sites/default/files/atoms/files/ostp-neo-strategy-action-plan-jun18.pdf.

23

The Future

A decadal survey, by definition, looks 10 years ahead.[1] Predicting the future is fraught with challenges. Moreover, many of the concepts and recommendations contained in the preceding chapters may not be implemented or their consequences known until multiple decades from now.

The coming years will witness events that were unforeseen when this report was drafted. New scientific discoveries, technical developments, and budgetary booms or busts are inevitable and all will impact the implementation of the recommendations in this report. Thus, the survey report and the policy ecosystem within which it exists is most effective if it can accommodate the inevitability of change. Therefore, the survey committee has endeavored, via the two representative programs and the decision rules in Chapter 22, to provide some guidance as to how budgetary changes in the near-to-mid-term might be accommodated. Longer term resilience is beyond the purview of the survey committee but is provided by via two mechanisms, continuing oversight of the implementation of decadal surveys and the midterm review process.

CONTINUING OVERSIGHT

Vision and Voyages noted the following: *A decadal survey should not be blindly followed if external circumstances dictate that a change in strategy is needed. But, who decides if change warrants a deviation from a decadal plan? . . . A group specifically tasked to monitor and assess progress toward decadal goals is essential. Such a group should be able to provide the necessary strategic guidance needed to achieve the decadal science goals in a timely manner and consistent with the survey recommendations* (NRC 2011).

In response to this call for strategic guidance, the National Academies established the Committee on Astrobiology and Planetary Science (CAPS) in 2012 and gave it the specific tasks to provide continuing oversight of the implementation of the decadal survey and to act as the organizing committee for future surveys and for other relevant reports. In 2016, CAPS was re-chartered to give it the ability to draft short reports on topical issues of relevance to the implementation of the prior decadal survey and related reports. At NASA's request, CAPS completed four such reports in the period 2017–2020 on topics as diverse as how the agency could best prepare for this decadal survey (NASEM 2017), the Lunar Exploration and Development Program (NASEM 2019a,b), and options for the next New Frontiers announcement of opportunity (NASEM 2020).

---

[1] A glossary of acronyms and technical terms can be found in Appendix F.

Finding: The Committee on Astrobiology and Planetary Science is succeeding in its role of monitoring the implementation of the recommendations in the most recent relevant decadal survey as evidenced by the steady stream of requests from NASA for the committee's input.

THE MIDTERM REVIEW

Section 301(a) of the NASA Authorization Act of 2005 directed NASA to have "[t]he performance of each division in the Science directorate . . . reviewed and assessed by the National Academy of Sciences at 5-year intervals." The primary reason cited in the Act for undertaking such reviews was to evaluate the progress or lack of progress of the agency at meeting the goals of the decadal surveys. Such information could then be used by legislators to improve NASA's responsiveness to the survey recommendations. *Vision and Voyages* noted that in addition to assessing NASA's progress toward implementing decadal goals, such midterm reviews could also be used to gauge the degree to which scientific understanding has advanced since release of the last survey. Indeed, the most recent midterm review committee was tasked to describe "the most significant scientific discoveries, technical advances, and relevant programmatic changes in planetary sciences over the years since the publication of the planetary decadal survey (*Vision and Voyages*)" (NASEM 2018).

A question frequently asked is whether a 10-year planning horizon is too long or too short. Arguments for a decade-plus horizon often note that the timescale for planning, building, and launching even small-class spacecraft missions is approaching a decade. Large-class missions often require multi-decade timescales from inception to first results, and even longer if necessary technology development activities are considered. However, given the rate of scientific and technical development and the vagaries of the annual budget cycle, any attempt to plan for several decades hence seems doomed to failure. Moreover, given the substantial effort required of a significant fraction of the community to undertake a decadal survey, repeating the exercise on a sub-decade timescale would meet with many obstacles. Therefore, efforts to turn the midterm review into a semi-decadal survey are misguided. Indeed, one of the best practices identified in the 2015 review of the decadal survey process (NASEM 2015) was that, "Midterm assessment reports are most useful when they engage and inform the broad community by providing a progress report on implementing the decadal program, together with sufficient context to understand the rationale behind the program's current implementation strategy."

Finding: The survey committee echoes the sentiment expressed in the *Vision and Voyages* that a 10-year assessment and planning timescale and a 5-yearly review cycle is appropriate at the present time. Moreover, the midterm review is most effective if it reinforces the prior decadal survey and is not used to reprioritize or otherwise make significant modifications to the latter's key recommendations.

PREPARING FOR THE NEXT PLANETARY DECADAL SURVEY

All three planetary science decadal surveys conducted to date have been asked to identify and prioritize the highest priority large- and medium-class spacecraft missions to be initiated in the next 10 years. The first planetary decadal was forced to rely on prior mission studies of variable fidelity and to devise its own concepts. The lack of technical realism embodied in some of its priority mission recommendations became apparent soon after the report's release.

The second decadal survey fared better because it could draw upon several detailed mission studies undertaken by NASA and was also able to commission some medium-fidelity mission studies. The latter were performed over a relatively short timescale while the survey was under way. Formulating and studying 25 mission concepts in parallel with the drafting of the decadal survey caused considerable stress to the survey's schedule and on the workload of those members participating in the studies. This less than ideal situation prompted *Vision and Voyages* to comment that a "more effective method would be for NASA to sponsor studies for potential flagship and New Frontiers missions that capture the broadest possible science questions as well as reduce the time pressure on the decadal survey itself."

As a result, that survey recommended that "NASA sponsor community-driven, peer-reviewed mission studies in the years leading up to the next decadal survey, using a common template for the study reports." This call was reiterated by CAPS and the midterm review (NASEM 2017, 2018), and NASA eventually initiated the predecadal planetary mission concept study (PMCS) process. Eleven concepts were studied via the PMCS mechanism (see Appendix C) and were completed in 2020.

The availability of the PMCS reports greatly helped the survey committee formulate this report. However, the 11 PMCS concepts did not adequately represent the full scope of scientifically interesting and technically feasible projects that might be initiated in the coming decade. Fortunately, NASA was able to provide sufficient resources for the survey committee to commission 10 additional studies (nine addressing science and one addressing planetary defense goals).

Finding: The predecadal mission concept study process represented an important innovation in NASA's support of the decadal process and greatly assisted the work of the survey committee in the formulation of this report.

While helpful, the PMCS process was not without its faults. If the process or something similar is used to prepare for future decadal surveys, the following changes are suggested:

- Consider forming an appropriately constituted group representing the scientific community to recommend a strategic set of mission studies needed to address priority science questions prior to an open call for such studies. This list would not be restrictive but could help ensure that needed studies are performed;
- Increase the number of concepts studied beyond 11 to relieve the decadal survey of some of the burden of conducting its own studies;
- Discourage the formulation of overelaborate concepts incorporating comprehensive instrument complements and/or multiple interacting spacecraft;
- Encourage the inclusion of multiple descope options;
- Disincentivize reliance on new launch and/or power systems whose availability might not be consistent with a realistic mission timeline;
- Avoid reliance on a single potential launch vehicle; and
- Consider multiple launch opportunities in the decade(s) beyond that covered by the survey.

REFERENCES

NASEM (National Academies of Sciences, Engineering, and Medicine). 2015. *The Space Science Decadal Surveys: Lessons Learned and Best Practices*. Washington, DC: The National Academies Press.

NASEM. 2017. *Report Series: Committee on Astrobiology and Planetary Science: Getting Ready for the Next Planetary Science Decadal Survey*. Washington, DC: The National Academies Press.

NASEM. 2018. *Visions into Voyages for Planetary Science in the Decade 2013–2022: A Midterm Review*. Washington, DC: The National Academies Press.

NASEM. 2019a. *Report Series: Committee on Astrobiology and Planetary Science: Review of the Planetary Science Aspects of NASA SMD's Lunar Science and Exploration Initiative*. Washington, DC: The National Academies Press.

NASEM. 2019b. *Report Series: Committee on Astrobiology and Planetary Science: Review of the Commercial Aspects of NASA SMD's Lunar Science and Exploration Initiative*. Washington, DC: The National Academies Press.

NASEM. 2020. *Report Series: Committee on Astrobiology and Planetary Science: Options for the Fifth New Frontiers Announcement of Opportunity*. Washington, DC: The National Academies Press.

NRC (National Research Council). 2008. *Grading NASA's Solar System Exploration Program: A Midterm Report*. Washington, DC: The National Academies Press.

NRC. 2011. *Vision and Voyages for Planetary Science in the Decade 2013–2022*. Washington, DC: The National Academies Press, p. 314.

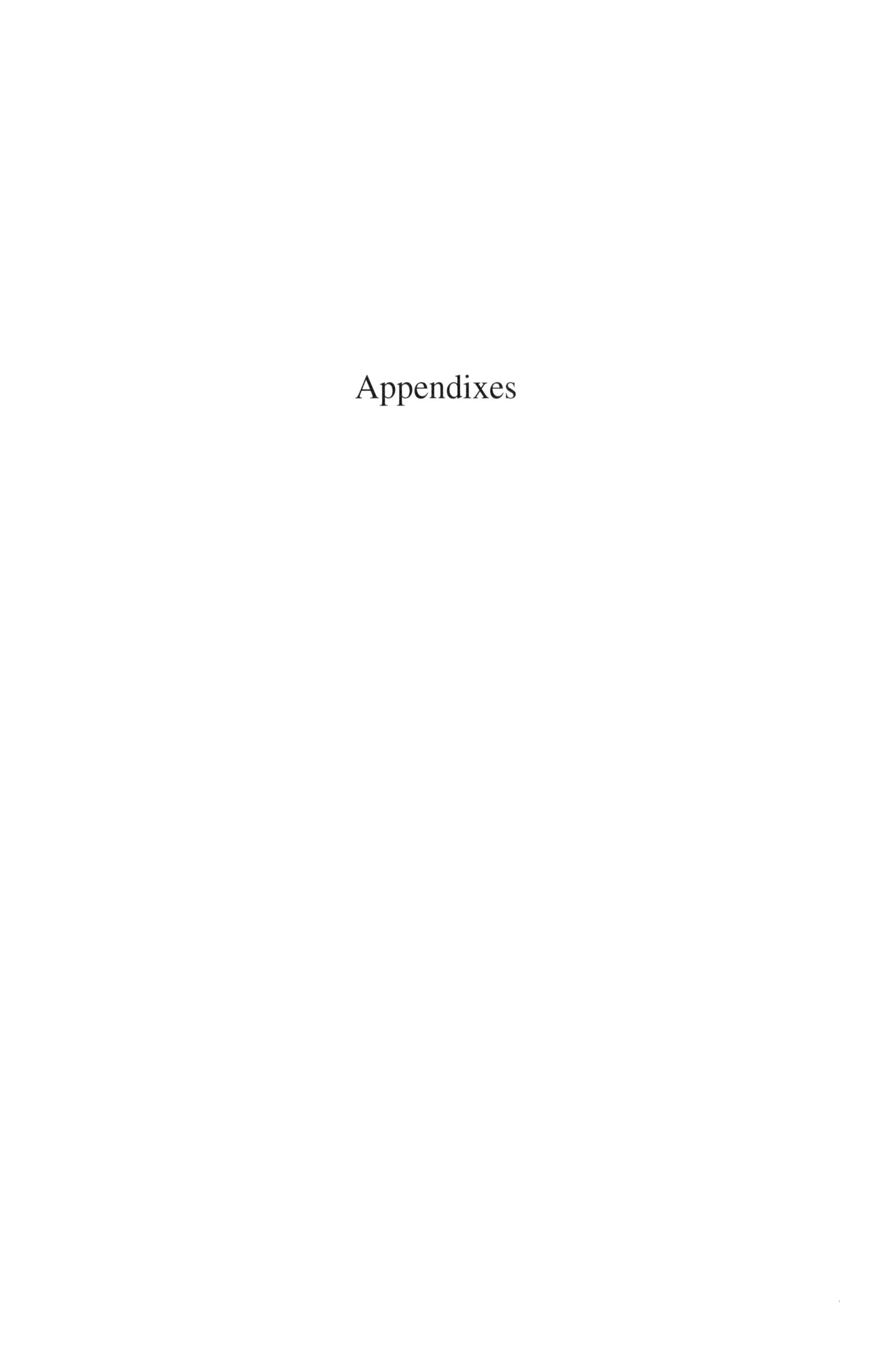

Appendixes

Appendix A

Letter of Request, Statement of Task, and Other Guidance

LETTER OF REQUEST

National Aeronautics and Space Administration
Headquarters
Washington, DC 20546-0001

October 24, 2019

SMD/Planetary Science Division

Dr. Margaret Kivelson Chair, Space
Studies Board
The National Academies of Sciences, Engineering, and Medicine
500 5th Street NW Washington, DC
20001

Dear Dr. Kivelson:

The National Research Council (NRC) decadal survey concept was born with a comprehensive survey for ground-based astronomy, published in 1964. Since that time, the practice of canvassing the scientific community for program planning guidance has been extended to other fields of study. The first true Planetary Science decadal survey, entitled "New Frontiers in the Solar System: An Integrated Exploration Strategy," was issued in 2003. The second decadal survey, entitled "Vision and Voyages for Planetary Science in the Decade 2013-2022," was issued in 20111. Impressive progress has been made in many areas of planetary science since then, including, for example, enhanced understanding of small and primitive bodies, of the geology and history of Mars, and of the diverse character of satellites of the gas giant planets. As a result of this progress and a number of important programmatic developments affecting NASA' s science programs and in cognizance of the National Foundation's (NSF) continuing interest in and support of planetary science research, NASA's Science Missions Directorate (SMD) would like to initiate a follow-on survey effort at this time. The Enclosed Statement of Task outlines the scope and assumptions to be used in the development of this new survey.

I would like to request that the National Academies of Science, Engineering, and Mathematics (NASEM) submit a plan to NASA for the development of the decadal study defined by the Statement of Task. Because we would like to use the findings of the survey to inform development of the fiscal year 2024 budget, the results of the scope should be available to us by March 31, 2022. Once agreement with NASEM on the scope and cost for the proposed study has been achieved, the NASA contracting officer for this effort will issue a task order for implementation. Ms. Doris Daou will be the technical point of contact for this effort and may be reached at (XXX) XXX-XXXX or Doris.Daou@nasa.gov.

Sincerely,
Lori S. Glaze, Ph.D.
Director,
Planetary Science Division

Concurrance:
ThomasH.Zurbuchen, Ph.D.
Associate Administrator,
Science Mission Directorate

Enclosure

CC: Space Studies Board/C. Hartman
D. Smith
D. Day
Science Mission Directorate/M. New
D. Daou

Planetary Science and Astrobiology Decadal Survey 2023–2032

The guiding document for the decadal survey is the statement of task. The items described in the sections on scope, considerations, approach, and so on are additional, nonbinding counsel for the survey committee and its staff while they carry out their work.

STATEMENT OF TASK

The Space Studies Board shall establish a survey committee (the "committee") to develop a comprehensive science and mission strategy for planetary science that updates and extends the board's current solar system exploration decadal survey, *Vision and Voyages for Planetary Science in the Decade 2013–2022* (2011).

The new decadal survey shall broadly canvas the field of space- and ground-based planetary science to determine the current state of knowledge and to identify the most important scientific questions to be addressed during the interval 2023–2032.

For the first time, this decadal survey will also study aspects of planetary defense, now that this activity is fully incorporated as an element of NASA's planetary science endeavors. The survey will also take into account planned human space exploration activities.

In addition, the survey and report shall address relevant programmatic and implementation issues of interest to NASA and the National Science Foundation (NSF). Because the content and structure of the program portfolios of the two agencies are distinct from one another, implementation and investment recommendations specific to each agency should be elaborated in separate sections of the final report. This will ensure that the report's investment guidance will be clearly addressed to the appropriate agency.

It is critically important that the recommendations of the committee be achievable within the boundaries of anticipated funding. NASA and NSF will provide an up-to-date understanding of these limitations to the committee at the time of survey initiation.

The report should provide a clear exposition of the following:

1. An overview of planetary science, astrobiology, and planetary defense—what they are, why they are compelling undertakings, and the relationship between space- and ground-based research.
2. A broad survey of the current state of knowledge of the solar system.
3. The most compelling science questions, goals, and challenges which should motivate future strategy in planetary science, astrobiology, and planetary defense.
4. A coherent and consistent traceability of recommended research and missions to objectives and goals.
5. A comprehensive research strategy to advance the frontiers of planetary science, astrobiology, and planetary defense during the period 2023–2032 that will include identifying, recommending, and ranking the highest priority research activities (research activities include any project, facility, experiment, mission, or research program of sufficient scope to be identified separately in the final report). For each activity, consideration should be given to the scientific case, international and private landscape, timing, cost category and cost risk, as well as technical readiness, technical risk, lifetime, and opportunities for partnerships. The strategy should be balanced, by considering large, medium, and small research activities for both ground and space.
6. Recommendations for decision rules, where appropriate, for the comprehensive research strategy that can accommodate significant but reasonable deviations in the projected budget or changes in urgency precipitated by new discoveries or technological developments.
7. An awareness of the science and space mission plans and priorities of NASA human space exploration programs and potential foreign and U.S. agency partners reflected in the comprehensive research strategy and identification of opportunities for cooperation, as appropriate.
8. The opportunities for collaborative research that are relevant to science priorities between SMD's four science divisions (e.g., comparative planetology approaches to exoplanet or astrobiology research); between NASA SMD and the other NASA mission directorates; between NASA and the NSF; between NASA and other U.S. government entities; between NASA and private sector organizations; between NASA and its international partners.

9. The state of the profession, including issues of diversity, inclusion, equity, and accessibility; the creation of safe workspaces; and recommended policies and practices to improve the state of the profession. Where possible, provide specific, actionable, and practical recommendations to the agencies and community to address these areas.

SCOPE

In order to ensure that the committee provides actionable advice and to ensure consistency with other advice developed by the National Academies, guidelines for the scientific scope of the survey are as follows:

1. The report should address and be organized according to the significant, overarching questions in planetary science, astrobiology, and planetary defense.
2. Basic or supporting ground- and space-based, laboratory, field, and theoretical research in astrobiology is within scope. Any findings and recommendations in the area of astrobiology should take into consideration the National Academies' report *An Astrobiology Strategy for the Search for Life in the Universe* (2018).
3. Interactions between solar and heliospheric phenomena and the atmospheres, magnetospheres, and surfaces of solar system bodies are within scope. Reassessment of recommendations treated in the National Academies' report *Solar and Space Physics: A Science for a Technological Society* (2012) is out of scope.
4. Excluding analog studies, focused study of the Earth system—including its atmosphere, magnetosphere, surface, and interior—is out of scope; these topics are treated in the National Academies' report *Thriving on Our Changing Planet: A Decadal Strategy for Earth Observation from Space* (2017).
5. Studies of meteorites and other extraterrestrial materials in terrestrial laboratories that further planetary science goals are in scope, but findings and recommendations in this area should take into consideration the National Academies' report *Strategic Investments in Instruments and Facilities for Extraterrestrial Sample Curation and Analysis* (2018).
6. Recommendations for ground- and space-based investigations to detect exoplanets are out of scope. (These topics are being addressed by *ASTRO2020: Decadal Survey on Astronomy and Astrophysics* currently in progress.) However, the identification of scientific issues and questions related to the study of exoplanets, including the comparative planetology and potential habitability of solar and extrasolar planets, is in scope.
7. Scientific investigations of near-Earth objects, both for the impact hazard presented to Earth and the future exploration and resource opportunities, are within scope. Findings and recommendations in this area should take into consideration the National Academies' report *Defending Planet Earth: Near-Earth-Object Surveys and Hazard Mitigation Strategies* (2010) as well as more recent National Academies' and community studies related to this area such as *Near-Earth Object Observations in the Infrared and Visible Wavelengths* (2018).
8. Findings and recommendations concerning planetary protection policies are out of scope. But the identification of planetary protection considerations for recommended missions—as recommended in the National Academies' report *Review and Assessment of the Planetary Protection Policy Development Processes* (2018)—and research or technology development to mitigate concerns about biological contamination are in scope.
9. Recommendations regarding construction of major new ground-based observatories are out of scope; these are addressed within the scope of the ASTRO2020 decadal survey currently in progress. The role that current and contemplated new ground-based facilities can play in advancing planetary science is in scope. How the facilities under consideration in the ASTRO2020 survey (when available) could benefit planetary science is within scope.
10. The scientific identification and initial validation of technosignatures is in scope but the application of such signatures in survey studies is out of scope. Recommendations in this area should take into account the summary of the Technosignatures Workshop found in the meeting report *NASA and the Search for Technosignatures* (2018) as well as the National Academies' reports *An Astrobiology Strategy for the Search for Life in the Universe* (2019) and *Exoplanet Science Strategy* (NASEM 2018).

CONSIDERATIONS

National Science Foundation Recommendations

For NSF, the survey will be most effective if it is aspirational, inspirational, and transformative. The decadal survey should assess how the current NSF portfolio of facilities and individual investigator grants address these priorities, as well as how currently planned and new facilities under consideration in the ASTRO2020 survey could benefit the planetary science priorities. The study may recommend changes to NSF's portfolio of facilities, including initiating divestment actions, as it deems necessary to advance the science and to optimize the value of current and future facilities.

The decadal survey steering committee is encouraged to comment on NSF opportunities for expanding partnerships, whether private, interagency, or international.

National Aeronautics and Space Administration Recommendations

The report should reflect NASA's statutory responsibility for flight mission investigations. The committee is strongly encouraged to adhere to the following guidelines as they draft the principal components of the NASA implementation portion of the report:

1. Recommendations for individual flight investigations for initiation between 2023 and 2032 as follows (note that dollar values given below do not include launch vehicle or Phase E costs; full life cycle costs for PI-led missions may be as much as double these values):

    a. Flight investigations believed executable for less than approximately $500 million (candidates for the Discovery or SIMPLEx programs) should not be identified or prioritized. They will be proposed by community investigators to address the science goals and challenges called for in the statement of task.
    b. The report should consider whether specific flight investigations with costs in the approximate range $500 million to $900 million (New Frontiers class) should continue to be specified or whether this mission class should be open in a manner similar to the Discovery program. If specific flight investigations are recommended, the report should provide a candidate list of objectives for each mission.
    c. The report should identify specific destinations and science goals for "large strategic missions" with costs projected to exceed $900 million.
    d. The prioritization of flight investigations for Mars and the Moon should be integrated with flight investigation priorities for other solar system objects into a single prioritized list of all recommended missions.
    e. The findings and recommendations contained in *Visions and Voyages for Planetary Science in the Decade 2013–2022*—together with other recent National Academies reports on topics relevant to planetary science, astrobiology, and planetary defense—should be used as input to the decadal efforts. Missions identified in these reports that have not yet been confirmed for implementation must be reprioritized.
    f. The study will assess whether NASA's plans for Mars Sample Return (MSR) play an appropriate role in the research strategy for the next decade. The study may include findings and recommendations regarding those plans, as appropriate, including substantive changes in NASA's plans. Recommendations may include, but are not limited to, actions ranging from increased investments (upscopes) to reduced investments (descopes) and termination. It is not necessary to rank MSR among other recommended activities for space.
    g. It is understood that initiation of missions on these lists will depend on actual resource availability.

2. Recommendations for NASA-funded supporting research required to maximize the science return from the flight mission investigations and to provide the context and impetus for future flight mission investigations.

3. A discussion of strategic technology development needs and opportunities relevant to NASA planetary science programs.
4. A discussion of the following:

   a. How planned and potential human space flight activities will provide new opportunities for planetary science.
   b. What areas of planetary science will provide information needed to support human space flight activities.

CONSIDERATIONS FOR PLANETARY DEFENSE FLIGHT PROGRAMS

1. The flight missions in this line are focused on NEO search and/or characterization, or in-space mitigation technology development. Currently, these are strategic, directed investigations as opposed to competed. As such, the report could identify specific prioritized planetary defense goals for "strategic missions," even if the anticipated costs are below the current $500 million competed mission threshold.
2. The panel should look for opportunities with Planetary Science flight missions where Planetary Defense objectives could be achieved with augmentation by Planetary Defense funding to add capability or enhance operations of otherwise purely Planetary Science missions.
3. Planetary Defense flight investigations believed executable for less than approximately $500 million should be identified and prioritized. They could be either directed to a specific purpose or, if for a more broadly identified objective (e.g., Apophis encounter), proposed by community investigators through an AO process to address the Planetary Defense goals and challenges identified.
4. It is not foreseen that the projected budget for Planetary Defense would allow flight projects in excess of $500 million development costs. However, if the panel finds that specific flight investigations are needed with life cycle costs in the approximate range $500 million to $1 billion, the report should provide the candidate objectives to be achieved for each mission and they should be prioritized.

APPROACH

The organization of the study is sized based upon prior planetary decadal surveys. The committee will consist of a steering group—approximately 15–20 members, responsible for the overall organization and execution of the study, and the production of a final consensus report that will undergo the usual National Academies' review processes—and five or six supporting panels—approximately 10–12 members each—responsible for providing the scientific and technical breadth to span the diverse suite of scientific topics and potential solar system destinations.

The scheme used to allocate the domain of study among the panels should support delivery of a report organized according to the significant, overarching questions in planetary science, astrobiology, and planetary defense. Individual panels may span multiple solar system target bodies, with specific panel structure determined by the National Academies and the committee's chair(s). An important role of the panels will be to evaluate input from the research community about issues of scientific and programmatic priorities in the field. In keeping with prior planetary science decadal surveys, the work of the study panels will be integrated and incorporated as chapters in the final survey report.

One representative from each of the panels shall serve on the steering group. The composition of the steering group and panels will take full advantage of the diversity of the planetary science, astrobiology, and planetary defense communities in factors such as gender, race, ethnicity, career stage, types and sizes of institutions, geographic distribution, and so on. It is imperative that some early career researchers be invited to serve on panels.

In assembling the committee and panels, calls for nominations will be sent to the planetary science, astrobiology, and planetary defense communities and sponsors. National Academies' staff will nominate a candidate for chair after consultation with the Space Studies Board, the Committee on Astrobiology and Planetary Science, and other relevant stakeholders. The chair will work with committee staff and others to develop the structure for the study and a slate of nominees for the balance of the committee's membership.

In assembling the slate of nominees for the steering group and panels, committee staff will follow National Academies' procedures for reducing and balancing biases, and for ensuring that the steering group and panels have the needed expertise across disciplines and diversity among their members, including gender, career stage, underrepresented groups, types and sizes of institutions, and geographic distribution.

In designing and pricing the study, the National Academies' should include resources for independent and expert cost analysis support to ensure that all flight mission cost estimates can be meaningfully intercompared and are as accurate as possible, given the varying maturity of project concepts and other recognized uncertainties. The prioritized list of science missions should be developed with the anticipated resources in mind.

The final report must represent a comprehensive and authoritative analysis of the subject domain and represent the community stakeholders. The study activity will include town hall meetings, sessions at geographically dispersed professional meetings, and aggressive use of electronic communications for soliciting and aggregating inputs from across the community and country. It is anticipated that a call for white papers will be issued prior to the commencement of the study itself. The committee may also convene focused workshops on special topics of interest. Other input-gathering methods will be explored and used, including a prestudy event to inform early-career researchers about the scope of, and their potential role in, the decadal survey.

PRODUCTS

It is suggested that the committee produce three products: a complete, integrative report of the findings and recommendations of the study, incorporating the reports of the supporting panels; an abbreviated high-level presentation of the main findings and recommendations suitable for distribution to the general public; and a web-based archive of report-relevant documents, including all community white papers and mission studies.

Appendix B

White Papers Received

Broad community participation is one of the defining features of a decadal survey. One of the most effective means to engage the community is solicitation of input from any and all interested parties via the submission of white papers. The decadal survey solicited such input via the standard means used in past surveys—for example, through community newsletters and solicitations during town hall meetings held in the months prior to the initiation of the survey. An innovation for this survey was a series of webinars designed to introduce early-career researchers to the decadal survey. One of these webinars was devoted to white papers: why they are important, how to write them, and how to submit one.

To facilitate document management, several submission guidelines were imposed. These included imposition of a seven-page limit, a prescribed format, submission to the survey committee via a specific individual (the first or submitting author) through a special website set up by the National Academies, and a series of cascading subject-specific deadlines—science-related papers by 15 July 2020, mission-related papers by 15 August, and those concerning the state of the profession by 15 September—to ensure that all contributions were available for consideration and discussion prior to the first meeting of the survey committee. Everyone in the planetary science community was encouraged to author white papers.

In total, 527 unique white papers were received (compared to 24 and 199 for the first and second planetary decadal surveys, respectively) and are listed below alphabetically by their respective submitting author. The full text of each white paper listed below can be found in the 18 March 2021 issue of the *Bulletin of the American Astronomical Society (AAS)*,[1] except for six whose authors opted out of publication (their contributions are indicated below by a terminal *).[2]

The committee expresses its sincere thanks to the AAS for its continuing support of the space science decadal surveys. These thanks are extended to every individual who contributed to the 527 white papers received. The time, energy, and financial resources devoted by the planetary science and astrobiology communities writ large to support the decadal survey process is clear evidence of its desire to openly discuss and set priorities to guide our future activities as we explore planetary environments both near and far.

---

[1]American Astronomical Society, "Whitepapers for the Planetary Science and Astrobiology Decadal Survey 2023–2032," *Bulletin of the AAS* 53(4), https://doi.org/10.3847/25c2cfeb.7272e5bb.

[2]The full text of all the white papers received is also available in a single searchable file at https://tinyurl.com/2p88fx4f.

EACH UNIQUE WHITE PAPER RECEIVED

Abell, P., C. Raymond, T. Daly, and D.R. Adamo. 2020. Near-Earth Object Characterization Priorities and Considerations for Planetary Defense.

Adamo, D. 2020. Toward Greater Preparedness and Resilience in Planetary Defense.

Adamo, D., P.A. Abell, R.C. Anderson, B.W. Barbee, J.B. Hopkins, T.D. Jones, R.R. Landis, et al. 2020. Exploration Leading to Low-Latency Telepresence on Mars from Deimos.

Adamo, D., P.A. Abell, R.C. Anderson, B.W. Barbee, T.D. Jones, D.D. Mazanek, and G. Podnar. 2020. Near-Earth Object Exploration Leading to Human Visits.

Ahrens, C. 2020. Advancing Experimental Research and Instrumentation in the Laboratory in Understanding the Moon.

Ahrens, C. 2020. Geologic Science Research Priorities for the Moons of Uranus.

Ahrens, C. 2020. Spectral Science Priorities for an Infrared Spectrometer with Interstellar Probe.

Ahrens, C. 2020. The Need for a Facility to Understand Volatile Ice Rheology.

Airapetian, V., G. Gronoff, K. Kobayashi, D.N. Mvondo, A. Zestos, J. Kasting, B. Hayworth, et al. 2020. The Impact of Extreme Space Weather on the Rise of Biomolecules: Early Earth and Exo-Earth Environments.

Aloisi, A., and N. Reid. 2020. (Un)conscious Bias in the Astronomical Profession: Universal Recommendations to Improve Fairness, Inclusiveness, and Representation.

Amini, R., A. Azari, S. Bhaskaran, P. Beauchamp, J. Castillo-Rogez, R. Castano, S. Chung, et al. 2020. Advancing the Scientific Frontier with Increasingly Autonomous Systems.

Arney, G., N. Izenberg, S.R. Kane, K.E. Mandt, V.S. Meadows, A.M. Rymer, L.C. Quick, et al. 2020. Exoplanets in Our Backyard: A Report from an Interdisciplinary Community Workshop and a Call to Combined Action.

Arnold, J., T.R. Spilker, D.M. Cornelius, G.A. Allen, A.M. Brandis, D.A. Saunders, M. Qu, et al. 2020. Heatshields for Aerogravity Assist Vehicles Whose Deceleration at Titan Saves Mass for Future Flagship Class Exploration of Enceladus.

Ashley, J., C. Schröder, A.W. Tait, A.G. Tomkins, P.J. Boston, R.C. Wiens, D.F. Wellington, et al. 2020. Continued Use of Exogenic Materials Found on Mars as Planetary Research Tools.

Asmar, S., R.A. Preston, P. Vergados, D.H. Atkinson, T. Andert, H. Ando, C.O. Ao, et al. 2020. Solar System Interiors, Atmospheres, and Surfaces Investigations via Radio Links: Goals for the Next Decade.

Austin, A., G. Afonso, S. Albert, H. Ali, A. Alunni, J. Arnold, G. Bailet, et al. 2020. Enabling and Enhancing Science Exploration Across the Solar System: Aerocapture Technology for SmallSat to Flagship Missions.

Aye, M., C. Million, and A. Annex. 2020. The Role of (Open Source) Software in Planetary Science.

Azari, A., J.B. Biersteker, R.M. Dewey, E.J. Forsberg, C.D.K. Harris, H.R. Kerner, K.A. Skinner, et al. 2020. Integrating Machine Learning for Planetary Science: Perspectives for the Next Decade.

Bailey, S. 2020. Will Key Lunar Decadal Objectives Be Missed in the Lunar Land Rush?

Bailey, S., and D. Paige. 2020. Armada: A Flagship Multi-Mission Architecture.

Baines, K., A. Akins, D.H. Atkinson, S. Atreya, M. Bullock, K. Cheung, J.A. Cutts, et al. 2020. New-Frontiers (NF) Class In-Situ Exploration of Venus: The Venus Climate and Geophysics Mission Concept.

Balcerski, J., K. Jessup, G. Hunter, M. Zborowski, A. Colozza, and D. Makel. 2020. Exploration of Venus' Atmosphere by Low-Cost Distributed Sensing Architecture.

Balint, T., D. Atkinson, A. Babuscia, J. Baker, C. Bradford, C. Elder, R. Conversano, et al. 2020. Uranus System Exploration Under the New Frontiers Mission Class (A Novel Perspective).

Balint, T., Y.H. Lee, S.M. Howell, S.M. Perl, K. Craft, T.A. Hurford, M.L. Cable, et al. 2020. Enabling the Next Frontiers in Astrobiology—Ocean and Ice Worlds Explorations with a Radioisotope Power System Inside a Pressure Vessel.

Banfield, D., J. Stern, A. Davila, S.S. Johnson, D. Brain, R. Wordsworth, B. Horgan, et al. 2020. Summary of the Mars Science Goals, Objectives, Investigations, and Priorities.

Bapst, J., T.J. Parker, J. Balaram, T. Tzanetos, L.H. Matthies, C.D. Edwards, A. Freeman, et al. 2020. Mars Science Helicopter: Compelling Science Enabled by an Aerial Platform.

Barba, N., A. Austin, D. Banfield, A.B. Chmielewski, P. Clark, W. Coogan, R. Conversano, et al. 2020. High Science Value Return of Small Spacecraft at Mars.

Barbee, B., P.A. Abell, D.R. Adamo, J. Atchison, O. Barnouin, J. Cahill, P. Chodas, et al. 2020. Research and Analysis for Planetary Defense In-Situ Spacecraft Missions to Near-Earth Objects.

Barbee, B., P.A. Abell, J. Atchison, O. Barnouin, S. Bhaskaran, J. Brophy, J. Cahill, et al. 2020. Technology Development for Planetary Defense In Situ Spacecraft Missions to Near-Earth Objects.

Barbee, B., P.A. Abell, R.P. Binzel, M. Brozovic, J. Cahill, P. Chodas, T. Daly, et al. 2020. Future Spacecraft Missions for Planetary Defense Preparation.

Barge, L.M., L. Rodriguez, J.M. Weber, and B. Theiling. 2020. Beyond "Biosignatures": Importance of Applying Abiotic/Prebiotic Chemistry to the Search for Extraterrestrial Life.

Barnes, J.W., A.G. Hayes, J.M. Soderblom, S.M. MacKenzie, J.D. Hofgartner, R.D. Lorenz, E.P. Turtle, et al. 2020. New Frontiers Titan Orbiter.

Bauer, J., S. Milam, G. Bjoraker, S. Carey, D. Daou, L. Fletcher, W. Harris, et al. 2020. Planetary Science with Astrophysical Assets: Defining the Core Capabilities of Platforms.

Beauchamp, P., J. Moore, L. Spilker, A. Hendrix, J. Bowman, M. Cable, S. Edgington, et al. 2020. Technologies for Ocean Worlds.

Becerra, P., A. Bramson, A. Brown, S. Byrne, S. Diniega, B. Horgan, M. Sori, et al. 2020. The Importance of the Climate Record in the Martian Polar Layered Deposits.

Beck, R., M. Mahzari, H. Hwang, M. Stackpoole, E. Venkatapathy, D. Ellerby, T. White, et al. 2020. TPS and Entry System Technologies for Future Mars and Titan Exploration.

Beddingfield, C., C. Li, S. Atreya, P. Beauchamp, I. Cohen, J. Fortney, H. Hammel, et al. 2020. Exploration of the Ice Giant Systems.

Benardini, J., N. Singh, and K. Venkateswaran. 2020. Molecular Biology Modernizing Planetary Protection Engineering to Enable Science for Biologically Sensitive Robotic and Human Missions.

Benner, S., E. Biondia, H.-J. Kima, and J. Špaček. 2020. Chemical Guidance in the Search for Past and Extant Life on Mars.

Bennett, K., M. McAdam, M. Milazzo, P. Garcia, J. Shelton, P. Gardiner, S. Diniega, et al. 2020. The Preventing Harassment in Science Workshop: Summary and Best Practices for Planetary Science and Astrobiology.

Bering, E.A., A. Parker, M. Giambusso, M. Carter, J. Squire, F.C. Díaz, and S. M. Hörst. 2020. Solar and Hybrid Electric Propulsion to the Kuiper Belt and Beyond.

Bering, E.A., L. Andersson, M. Moldwin, and P. Withers. 2020. MARSCat: Imaging of the Martian Ionosphere Using a CubeSat Constellation.

Binzel, R., B.W. Barbee, O.S. Barnouin, J.F. Bell, M. Birlan, A. Boley, W. Bottke, et al. 2020. Apophis 2029: Decadal Opportunity for the Science of Planetary Defense.

Blake, D., K. Zacny, T. Bristow, S. Morrison, P. Sarrazin, E. Rampe, V. Tu, et al. 2020. MER-Class Rover Investigations of Mars in the Coming Decades.

Blake, D., R. Hazen, S. Morrison, T. Bristow, K. Zacny, E. Rampe, R. Downs, et al. 2020. In-Situ Crystallographic Investigations of Solar Systems in the Next Decade.

Blake, D., R. Walroth, T. Bristow, P. Sarrazin, M. Gailhanou, R. Downs, and K. Thompson. 2020. MapX: An In-situ X-ray μ-Mapper for Habitability and Biosignature Studies.

Blake, D., T. Bristow, P. Sarrazin, and K. Zacny. 2020. In-Situ Mineralogical Analysis of the Venus Surface Using X-ray Diffraction.

Blanc, M., K. Mandt, O. Mousis, N. André, D.H. Atkinson, S. Atreya, T. Balint, et al. 2020. Science Goals and Mission Objectives for the Future Exploration of Ice Giants Systems: A Horizon 2061 Perspective—Part I.

Blanc, M., O. Prieto-Ballesteros, N. André, J. Gomez-Elvira, G. Jones, V. Sterken, et al. 2020. Joint Europa Mission (JEM) a Multiscale, Multi-Platform Mission to Characterize Europa's Habitability and Search for Extant Life.

Blanc, M., K. Mandt, O. Mousis, N. André, D.H. Atkinson, S. Atreya, T. Balint, et al. 2020. Science Goals and Mission Objectives for the Future Exploration of Ice Giants Systems: A Horizon 2061 Perspective—Part II: Mission Scenarios.

Blank, J.G., A. Agha-mohammadi, E.R. Bell, D.A. Crown, B. Morrell, C.J. Patterson, K. Uckert, et al. 2020. Volcanic Caves as Priority Sites for Astrobiology Science.

Blewett, D., J. Halekas, D. Waller, J. Cahill, A. Deutsch, T.D. Glotch, L. Regoli, et al. 2020. Science Case for a Lander or Rover Mission to a Lunar Magnetic Anomaly and Swirl.

Bottke, W. 2020. Exploring the Bombardment History of the Moon.

Bottke, W. 2020. The Evolution of Small Body Populations: From Planet Migration to Thermal Drift Forces.

Bottke, W., and H.F. Levison. 2020. Tenets of an Effective and Efficient Research and Analysis Program for NASA.

Brain, D., R. Lillis, Y. Ma, and R. Ramstad. 2020. The Influence of Planetary Magnetic Fields on Atmospheric Retention and Habitability.

Bramson, A., C. Andres, J. Bapst, P. Becerra, S.W. Courville, C.M. Dundas, et al. 2020. Mid-Latitude Ice on Mars: A Science Target for Planetary Climate Histories and an Exploration Target for In Situ Resources.

Brecht, A., S. Brecht, J. Luhmann, J. Bellan, K. Jessup, T. Navarro, S. Lebonnois, et al. 2020. Closing the Gap Between Theory and Observations of Venus Atmospheric Dynamics with New Measurements.

Brisset, J., E. Fernandez-Valenzuela, A. Sickafoose, F. Venditti, A. Whizin, E. Beltran, A.F.B. de Almeida Prado, et al. 2020. Understanding the Formation and Evolution of the Kuiper Belt by Exploring the Haumea System.

Bristow, T., B. Lafuente, N. Stone, M. Parenteau, S.R. Wolfe, S.P. Rojo, K. Boydstun, et al. 2020. A Strategy for Managing NASA's Long Tail of Planetary Research Data.

Brooks, S., and T.M. Becker. 2020. Frontiers in Planetary Rings Science.

Brown, A., S. Byrne, E. Zubko, S. Diniega, N. Heavens, M. Mishchenko, J. Schlegel, et al. 2020. The Case for a Multi-Channel Polarization Sensitive Lidar for Investigation of Insolation-Driven Ices and Atmospheres.

Bugga, R., E. Brandon, E. Darcy, R. Ewell, P. Faguay, B. Futz, R. Gitzendanner, et al. 2020. Energy Storage Technologies for Planetary Science and Astrobiology Missions.

Buie, M. 2020. Occultation Studies of Primitive and Small Bodies.

Buratti, B., and M. Sykes. 2020. Small Bodies Assessment Group Community Decadal Survey Summary.

Buratti, B., E. Asphaug, J. Bauer, J. Bellerose, D. Blewett, W. Bottke, D. Britt, et al. 2020. The Small Satellites of the Solar System: Priorities for the Decadal Study.

Burr, D., D. Banfield, M. Day, S. Diniega, L. Fenton, M. Lapôtre, L. Neakrase, et al. 2020. NASA Planetary Wind Tunnel Facilities.

Byrne, P.K., A.J. Dombard, C. Elder, S.A. Hauck, M.M. Daswani, P.V. Regensburger, and S.D. Vance. 2020. Towards a Fuller Understanding of Icy Satellite Seafloors, Interiors, and Habitability.

Byrne, P.K., D.T. Blewett, N.L. Chabot, S.A. Hauck, E. Mazarico, K.E.V. Kaaden, and R.J. Vervack. 2020. One the Case for Landed Mercury Science.

Byrne, P., C.R. Richey, J.C. Castillo-Rogez, and M.V. Sykes. 2020. White Paper on Improvements to the NASA Research and Analysis Proposal and Review System.

Cable, M., S.E. Waller, R. Hodyss, A.E. Hofmann, M.J. Malaska, R.E. Continetti, A. Jaramillo-Botero, et al. 2020. Plume Grain Sampling at Hypervelocity: Implications for Astrobiology Investigations.

Cable, M., S. MacKenzie, M. Neveu, T.M. Hoehler, A.R. Hendrix, J. Eigenbrode, F. Postberg, et al. 2020. The Case for a Return to Enceladus.

Cabrol, N., J. Bishop, S.L. Cady, C. Demergasso, N. Hinman, M. Hoffman, I. Kanik, et al. 2020. Addressing Strategic Knowledge Gaps in the Search for Biosignatures on Mars.

Cabrol, N., L.K. Fenton, K. Warren-Rhodes, J. Hines, N. Hinman, J. Moersch, P. Sobron, et al. 2020. BIOMARS: A Foundational High-Resolution Environmental Sensor Array.

Cahill, J., E.J. Speyerer, D.H. Needham, R. Weber, I. Daubar, E. Costello, D. Moriarty, et al. 2020. Assessing the Present-Day Impact Flux to the Lunar Surface via Impact Flash Monitoring and Its Implications for Sustained Lunar Exploration.

Carpenter, K., M.L. Cable, M. Choukroun, S. Iacoponi, S. Moreland, M. Heverly, H. Nayar, et al. 2020. The Science Case for Surface Mobility on Icy Worlds.

Carpenter, K., M.L. Cable, M.N. Choukroun, H. Ono, R.A. Thakker, M.D. Ingham, P. Mcgarey, et al. 2020. Venture Deep, the Path of Least Resistance: Crevasse-Based Ocean Access Without the Need to Dig or Drill.

Carr, C., M. Samnani, J. Tani, J. McKaig, E. Hammons, D.J. Newman, K. Ho, et al. 2020. Space Drones: An Opportunity to Include, Engage, Accelerate, and Advance.

Carr, G., E. Brandon, R. Ewell, R. Surampudi, A. Barchowsky, C. Iannello, A. Hernandez-Pelle, et al. 2020. Power Electronic Technologies for Planetary Science and Astrobiology Missions.

Carson, J., M. Munk, R. Sostaric, A. Johnson, F. Amzajerdian, J.B. Blair, A.D. Cianciolo, et al. 2020. Precise and Safe Landing Navigation Technologies for Solar System Exploration.

Cartwright, R., C. Beddingfield, T. Nordheim, C. Elder, W. Grundy, B. Buratti, A. Bramson, et al. 2020. The Science Case for Spacecraft Exploration of the Uranian Satellites.

Casani, J.R., M.A. Gibson, D.I. Poston, N.J. Strange, J.O. Elliott, R.L. McNutt, S.L. McCarty, et al. 2020. Enabling a New Generation of Outer Solar System Missions: Engineering Design Studies for Nuclear Electric Propulsion.

Casillas, A., G. Barnett, C. Engelbrecht, C.S. Guernsey, J. McKinnon, M. Preudhomme, J.R. Reh, et al. 2020. Affordability of Outer-Planet Exploration: A Pragmatic Rationale for Implementing a Cold-Propulsion Based Energy-Efficient Spacecraft Infrastructure.

Cassell, A., N. Cheatwood, S. Hughes, C. Kazemba, and G. Swanson. 2020. Deployable Entry Vehicles for Future Science and Exploration Missions.

Castillo-Rogez, J., C. Richey, R. Pappalardo, S. Brooks, I.J. Daubar, E. Turtle, et al. 2020. Mission Roles: Status, Issues, and Recommendations for the Planetary Science and Astrobiology Decadal Committee Consideration.

Castillo-Rogez, J., J.D. Hofgartner, K. Singer, C. Cockell, B.J. Holler, M. Neveu, M. Bose, et al. 2020. Habitability of Small Bodies—State of Knowledge and Motivations for Exploration in the Next Decade.

Castillo-Rogez, J., J. Scully, M. Neveu, D. Wyrick, G. Thangjam, A. Rivkin, M. Sori, et al. 2020. Science Motivations for the Future Exploration of Ceres.

Castillo-Rogez, J., B. Donitz, I. Nesnas, T. Swindle, J. O'Rourke, M. Villarreal, et al. 2020. Smallsats for Small Body Exploration and Technology Infusion.

Castillo-Rogez, J., M. Sykes, A. Hendrix, J. Rathbun, C. Richey, and P.K. Byrne. 2020. NASA Research and Analysis: Status, Issues, and Recommendations for the Planetary Science and Astrobiology Decadal Survey Committee.

Chanover, N., C. Schmidt, and D. DeColibus. 2020. The Continued Relevance of 4m Class Telescopes to Planetary Science in the 2020s.

Cheng, A., R.T. Daly, O.S. Barnouin, J.B. Plescia, R.P. Binzel, D.C. Richardson, J.V. DeMartini, et al. 2020. Apophis 2029 Planetary Defense Mission Options.

Chou, L., N. Grefenstette, S.S. Johnson, H. Graham, P. Mahaffy, C. Kempes, J.E. Elsila, et al. 2020. Towards a More Universal Life Detection Strategy.

Choukroun, M., P. Backes, M.L. Cable, R. Hodyss, M. Badescu, J.L. Molaro, S. Moreland, et al. 2020. Sampling Ocean Materials, Traces of Life or Biosignatures in Plume Deposits on Enceladus's Surface.

Clark, K. 2020. Classical and Quantum Cell-Cell Signaling by Microbial Life on Earth and Possible Other Livable Worlds.

Clark, K. 2020. Classical and Quantum Information Processing in Aneural to Neural Cellular Decision Making on Earth and Perhaps Beyond.

Clark, K. 2020. Eco-Evolutionary Origins, Nature, and Impact of Paired Reproduction in Earth and Possible Extraterrestial Microbiota.

Clark, K. 2020. Ultrsociality, Goods Theory, and Primitive Agriculture in Cosmopolitan Earth and Putative Extraterrestial Microbial Symbionts.

Clement, B.G., J.N. Benardini, R.C. Hendrickson, E.F. Klonicki, and A. L. Smith. 2020. Planetary Protection Conventions for Landed Missions to Ocean Worlds.

Cohen, B., G.F. Herzog, S.J. Jaret, J.I. Simon, T.D. Swindle, S.E. Suarez, M.M. Tremblay, et al. 2020. Geochronology as a Framework for Inner Solar System History and Evolution.

Cohen, B., S. Lawrence, B. Denevi, T. Glotch, D. Hurley, C.R. Neal, M. Robinson, et al. 2020. Lunar Missions for the Decade 2023–2033.

Cohen, I., C. Beddingfield, R. Chancia, G. DiBraccio, M. Hedman, S. MacKenzie, B. Mauk, et al. 2020. New Frontiers-Class Uranus Orbiter: Exploring the Feasibility of Achieving Multidisciplinary Science with a Mid-Scale Mission.

Collinson, G., M. Benna, and R. Pfaff. 2020. Exploring the Martian Atmosphere's "Dynamo Region" to Better Understand Phenomena That Impact Radio Communication at Earth.

Collinson, G., S. Boucher, S. Curry, G. DiBraccio, C. Dong, C. Fowler, Y. Futaana, et al. 2020. Space Physics at Venus: Exploring Atmospheric Dynamics, Escape, and Evolution.

Costello, E., R.W.K. Potter, D.M.H. Baker, R.R. Ghent, J. Gillis-Davis, J. Grier, S.P.S. Gulick, et al. 2020. Investigating Impact Processes at All Scales: The Moon as a Laboratory.

Courville, S., N.E. Putzig, M.R. Perry, G.W. Patterson, G.A. Morgan, A.J. Gemer, P.C. Sava, et al. 2020. Developing Active Source Seismology for Planetary Science.

Crary, F., G.B. Clark, P. Delamere, C. Dong, R.W. Ebert, C. Harris, G. Hospodarsky, et al. 2020. The Magnetosphere of Jupiter: Moving from Discoveries Towards Understanding.

Crichton, D., N.J. Chanover, L.R. Gaddis, M.K. Gordon, M. Marley, L. Mayorga, L.M. Prockter, et al. 2020. On the Use of Planetary Science Data for Studying Extrasolar Planets.

Crum, R. 2020. Advanced Technology Developments for Europa Lander and Oher In-Situ Ocean World Missions.

Cuk, M., A.K. Virkki, T. Kohout, E. Lellouch, and J.J. Lissauer. 2020. Pathways to Sustainable Planetary Science.

Curry, S., J. Luhmann, C. Gray, and G. Collinson. 2020. Venus and Solar Storms: Solar Energetic Particles, Stream Interaction Regions and Coronal Mass Ejections.

Cutts, J., K. Baines, P. Beauchamp, C. Bower, A. Davis, L. Dorsky, D. Dyar, et al. 2020. Venus Corona and Tessera Explorer (VeCaTEx).

Cutts, J., S. Aslam, S. Atreya, K. Baines, P. Beauchamp, J. Bellan, D.C. Bowman, et al. 2020. Scientific Exploration of Venus with Aerial Platforms.

Czaja, A., A. Corpolongo, A. Gangidine, B. Horgan, L. Kah, J. Osterhout, S. Ruff, et al. 2020. Mars as a Compelling Target in the Continuing Search for Signs of Ancient Extraterrestrial Life.

Dahl, E., S. Brueshaber, R. Cosentino, C. Palotai, N. Rowe-Gurney, R. Sankar, K. Sayanagi, et al. 2020. Ice Giant Atmospheric Science.

Daubar, I., R.A. Beyer, V. Hamilton, A. McEwen, N. Bardabelias, S.M. Brooks, P.K. Byrne, et al. 2020. Extended Missions in Planetary Science: Impacts to Science and the Workforce.

Davidsson, B., J. Brisset, R.T. Daly, T. Denk, A. Ermakov, L. Feaga, M. Gritsevich, et al. 2020. What Do Small Bodies Tell Us About the Formation of the Solar System and the Conditions in the Early Solar Nebula?

Davila, A. 2020. Astrobiology on Mars: Organic Chemical Evolution on an Earth-Like Planet.

de Kleer, K., B. Butler, M. Cordiner, I. de Pater, M. Gurwell, J. Lazio, S. Milam, et al. 2020. Mapping Satellite Surfaces and Atmospheres with Ground-Based Radio Interferometry.

de Pater, I., C. Moeckel, J. Tollefson, B. Butler, K. de Kleer, L. Fletcher, M.A. Gurwell, et al. 2020. Prospects to Study the Ice Giants with the ngVLA.

Deutsch, A., A. Maiti, A. Luspay-Kuti, A. Kereszturi, A. Lucchetti, A. Colaprete, A. Vorburger, et al. 2020. Science Opportunities Offered by Mercury's Ice-Bearing Polar Deposits.

Dhaliwal, J. 2020. Meteoritical Constraints on Our Solar System and Beyond.

Diniega, S., A. Bramson, P. Buhler, B. Buratti, D. Burr, M. Chojnacki, S. Conway, et al. 2020. Mars as a "Natural Laboratory" for Studying Surface Activity on a Range of Planetary Bodies.

Diniega, S., D. Burr, C.M. Dundas, B. Jackson, M. Mischna, S. Rafkin, I. Smith, et al. 2020. A Critical Gap: In Situ Measurements of Planetary Surface-Atmosphere Interactions Beyond Earth.

Diniega, S., J. Castillo-Rogez, I. Daubar, J. Filiberto, T. Goudge, K. Lynch, A. Rutledge, et al. 2020. Ensuring a Safe and Equitable Workspace: The Importance and Feasibility of a Code of Conduct, Along with Clear Policies Regarding Authorship and Team Membership.

Diniega, S., T. Putzig, S. Byrne, W. Calvin, C. Dundas, L. Fenton, P. Hayne, et al. 2020. White Paper Summary of the Final Report from the Ice and Climate Evolution Science Analysis Group (ICE-SAG).

Ditzler, M., A.C. Rios, M. Nuevo, M. Popovic, R. Mancinelli, J.T. Broddrick, D. Summers, et al. 2020. Beyond Targeted Searches: The Need for System-Level Approaches to Understanding the Connection Between Astrochemistry and the Emergence of Life.

Domagal-Goldman, S., G. Arney, and E.D. Lopez. 2020. Astrobiology as a NASA Grand Challenge.

Donitz, B., K.J. Meech, J. Castillo-Rogez, K.M. Moore, S.W. Courville, S. Ferguson, et al. 2020. New Frontiers Mission Concept Study to Explore Oort Cloud Comets.

Dorsey, J., W. Doggett, G. McGlothin, N. Alexandrov, B.D. Allen, M. Chandarana, J. Cooper, et al. 2020. State of the Profession: NASA Langley Research Center Capabilities/Technologies for Autonomous In-Space Assembly and Modular Persistent Assets.

Dove, J. 2020. Cracks Emerging in NASA's Discovery Program.

Dreier, C., and B. Nye. 2020. Increasing the Scope of Planetary Defense Activities: Programs, Strategies, and Relevance in a Post-COVID-19 World.

Dreier, C., and B. Nye. 2020. The Search for Life as a Guidepost to Scientific Revolution.

D'Souza, S., and A. Alunni. 2020. Maximizing Planetary Science Return by Advancing Non-Propulsive Control Systems.*

Dundas, C., S. Byrne, M. Chojnacki, S. Diniega, I. Daubar, C. Hamilton, C. Hansen, et al. 2020. Current Activity on the Martian Surface: A Key Subject for Future Exploration.

Dutra, F. 2020. Astrobiological Humanities Group for Interplanetary Protection and Defense Policies.

Dutra, F., and A. Dodsworth. 2020. Astrobiological Group of Social Issues for the Planetary Science and Astrobiology Decadal Survey 2023–2032.

Dutta, S., M. Perez-Ayucar, A. Fedele, R. Gardi, G.D. Calabuig, S. Schuster, J. Lebreton, et al. 2020. Aerocapture as an Enhancing Option for Ice Giants Missions.

Dyar, M., N.R. Izenberg, G. Arney, J. Balcerski, P. Byrne, L. Carter, C. Gray, et al. 2020. Revision of New Frontiers Goals for a Venus Mission.

Edwards, C., M.J. Amato, J.D. Baker, N J. Barba, J. Balaram, E.J. Brandon, P. Briggs, et al. 2020. Emerging Capabilities for Mars Exploration.

Ellerby, D., H. Hwang, M. Gasch, R. Beck, and T. White. 2020. TPS and Entry Technologies for Future Outer Planet Exploration.

Elston, J., M.A. Bullock, M.Z. Stachura, S. Lebonnois, S.S. Limaye, D.H. Grinspoon, M. Pauken, et al. 2020. In Situ Exploration of Venus' Clouds by Dynamic Soaring.

Engelhart, A., J.G. Blank, C. Carr, H.J. Cleaves, and K. Lynch. 2020. Astrobiology on Habitable Worlds: The Case for Considering Prebiotic Chemistry in Mission Design.

Ermakov, A., J.C. Castillo-Rogez, R.S. Park, C. Sotin, J. Lazio, S.M. Howell, et al. 2020. A Recipe for Geophysical Exploration of Enceladus.

Eubanks, T., J. Schneider, A.M. Hein, A. Hibberd, and R. Kennedy. 2020. Exobodies in Our Back Yard: Science from Missions to Nearby Interstellar Objects.

Fackrell, L., R. Rotz, H. Demir, and P. Schroeder. 2020. A Critical Zone Network Approach to the Study of Mars.*

Farrell, W., E. Beltran, P. Prem, M. Poston, A. Deutsch, D. Hurley, K. Retherford, et al. 2020. Lunar Polar Volatile Resources: Obtaining Their Origin Prior to Extraction.

Farrell, W., O. Tucker, M. Neveu, D. Bower, L. Quick, J. Renaud, R. Tyler, et al. 2020. The Need for a Large-Scale, Integrated Approach to Ocean World Modeling.

Fayolle, E., L. Barge, M. Cable, B. Drouin, J.P. Dworkin, J. Hanley, B. Henderson, et al. 2020. Critical Laboratory Studies to Advance Planetary Science and Support Missions.

Fortney, J., M. Marley, L. Mayorga, and A. Rymer. 2020. Synergy Between Ice Giant and Exoplanet Exploration: The Solar System's Planets "As Exoplanets."

Fournier, G., J.P. Gogarten, A.D. Goldman, A.S. Petrov, L. Rothschild, D. Segrè, E. Smith, et al. 2020. Understanding the Early Major Transitions in Evolutionary History Part 2: Ancient Evolution of Biological Systems and the Biosphere.

Fowler, C., R. Ramstad, M. Chaffin, S. Xu, R. Jarvinen, G. Collinson, M. Fillingim, et al. 2020. Unravelling the Drivers of Venusian Ionospheric Structure, Energy Balance and Evolution, Through In-Situ Plasma Measurements at Venus.

Frank, E., C. Drier, B. Clark, and C.R. Neal. 2020. Why and How to Leverage the Commercial Space Sector for the Benefit of Planetary Science and Its Community.

Frank, E., M.D. Dyer, S. Solomon, S. Curry, J. Helbert, L. Jozwiak, A. Komjathy, et al. 2020. The Thalassa Venus Mission Concept.

Frank, E., P.K. Byrne, S.Z. Weider, and L. Elkins-Tanton. 2020. Normalizing Non-Academic Career Paths in Planetary Science.

Freeman, A., M. Mischna, C.R. Richey, J. Lazio, C. Raymond, and M. Cable. 2020. Re-Imagining NASA Planetary Mission Science Opportunities for the Next Two Decades.

French, R., R. Hunter, M. Loucks, J. Currie, D. Sinclair, E. Mosleh, and P. Beck. 2020. Photon-Enabled Planetary Small Spacecraft Missions for Planetary Science.

Fries, M., J. Ashley, L. Beegle, R. Bhartia, P. Bland, A. Burton, A.L. Butterworth, et al. 2020. The Scientific Need for a Dedicated Interplanetary Dust Instrument at Mars.

Gangidine, A., D.G. Willingham, and E.E. Groopman. 2020. Surmounting Return Sample Science Barriers with NRL's NAUTILUS: The Next Generation of Secondary Ion Mass Spectrometry.

Gardner-Vandy, K., D. Scalice, J.C. Chavez, D.M. David-Chavez, K.J. Daniel, E. Gonzales, A. Lee, et al. 2020. Relationships First and Always: A Guide to Collaborations with Indigenous Communities.

Garrick-Bethell, I., D.A. Paige, M. Burton, J. Boland, J.N.H. Abrahams, C. van der Bogert, Y.-J. Choi, et al. 2020. NanoSWARM: NanoSatellites for Space Weathering, Surface Water, Solar Wind, and Remanent Magnetism.

Garvin, J., G. Arney, S. Atreya, M. Gilmore, S. Getty, D. Grinspoon, N. Johnson, et al. 2020. Deep Atmosphere of Venus Probe as a Mission Priority for the Upcoming Decade.

Garvin, J., J.S. Jones, J.W. Head, A. McAdam, N. Petro, K. Young, R.A. Kent, et al. 2020. Lunar Exploration in 3D: Volumetric Imaging of Lunar Materials and Manufactured Materials via X-ray Computer Tomography.

Garvin, Z., E. Boyd, M. Floyd, R.L. Harris, H. Kalucha, P. Mahaffy, J.E. Moores, et al. 2020. Mars Trace Gas Fluxes: Critical Strategies and Implications for the Upcoming Decade.

Gasch, M., H. Hwang, D. Ellerby, M. Stackpoole, E. Venkatapathy, A. Cassell, J. Feldman, et al. 2020. Technologies for Future Venus Exploration.

Gertsch, L. 2020. Space Resources Science.

Ghent, R., N.E.B. Zellner, I. Daubar, C.I. Fassett, M. Kirchoff, S. Marchi, S.J. Robbins, et al. 2020. Assessing the Recent Impact Flux in the Inner Solar System: 1 Ga to Present.

Gladden, R., C.H. Lee, C.D. Edwards, M.A. Viotti, and R.M. Davis. 2020. Enabling Robotic and Human Exploration: A Relay Network for the Future of Mars Exploration.

Glotch, T., L. Carter, P. Clark, B. Denevi, B. Greenhagen, W. Patterson, N. Petro, et al. 2020. A Next Generation Lunar Orbiter Mission.

Goel, A., R. Anderson, S. Bandyopadhyay, E. Brandon, J.V. Hook, M. Mischna, and F. Rossi. 2020. Distributed Instruments for Planetary Surface Science.

Goldman, A., G. Fournier, D. Segrè, and L. Williams. 2020. Understanding the Early Major Transitions in Evolutionary History Part 1: Stages in the Emergence of Complex Life.

Grace, A. 2020. Promoting the A in SPACE: Arts Run the Places STEM Takes Us.

Graham, H., and A. Murray. 2020. Collaborative Partnerships for Improved Astrobiology Science Outcomes.

Graham, H., K.H. Freeman, L. Chou, and M.J. Pasterski. 2020. Appeal for Improved Sample Selection, Preparation and Interpretation Standards for Organic Biosignature Experiments Performed by Flight Instruments.*

Grandidier, J., P. Jaffe, W.T. Roberts, M.W. Wright, A.A. Fraeman, C.A. Raymond, A. Austin, et al. 2020. Laser Power Beaming for Lunar Night and Permanently Shadowed Regions.

Grau Galofre, A., C.N. Andres, P. Becerra, A. Bhardwaj, A. Bramson, F. Butcher, P.R. Christensen, et al. 2020. A Comparative View of Glacial and Periglacial Landforms on Earth and Mars.

Gray, C., P.K. Byrne, S. Curry, J.G. O'Rourke, and E. Royer. 2020. Science on the Fly! The Importance of Venus Flyby Observations.

Gray, C., E. Young,, T. Kouyama, Y.J. Lee, A. Mahieux, E. Marcq, K. McGouldrick, et al. 2020. The Need for Ground-Based Observations of Venus.

Green, J., V. Airapetian, R. Bamford, N. Call, C. Dong, J. Hollingsworth, S. Hubbard, et al. 2020. Interdisciplinary Research in Terraforming Mars: State of the Profession and Programmatics.

Guillot, T., J. Fortney, E. Rauscher, M.S. Marley, V. Parmentier, M. Line, H. Wakeford, et al. 2020. Keys of a Mission to Uranus or Neptune, the Closest Ice Giants.

Guzewich, S., J. Abshire, L. Carter, D. Cremons, L. Hanson, D.H. Baker, D. Kao, et al. 2020. The Mars Atmospheric and Polar Science Mission.

Guzewich, S., J. Abshire, M. Baker, J. Battalio, T. Bertrand, A. Brown, A. Colaprete, et al. 2020. Measuring Mars Atmospheric Winds from Orbit.

Hallinan, G., J. Burns, J. Lux, A. Romero-Wolf, L. Teitelbaum, T. Chang, J. Kocz, et al. 2020. FARSIDE: A Low Radio Frequency Interferometric Array on the Lunar Farside.

Hamkins, J., D. Antsos, J. Border, G. Davis, L. Deutsch, J. Lazio, and J. Velazco. 2020. Communications and Navigation Technologies.

Hammel, H., and S.N. Milam. 2020. A Lesson from the James Webb Space Telescope: Early Engagement with Future Astrophysics Great Observatories Maximizes Their Solar System Science.

Hand, K., A.E. Murray, J.B. Garvin, W.B. Brinckerhoff, B. Christner, K.E. Edgett, B. Ehlmann, et al. 2020. Science of the Europa Lander Mission Concept.

Hand, K., C.B. Phillips, C.F. Chyba, B. Toner, K. Katija, V. Orphan, J. Huber, et al. 2020. On the Past, Present, and Future Role of Biology in NASA's Exploration of Our Solar System.

Hansen, C. 2020. Triton: Fascinating Moon, Likely Ocean World, Compelling Destination!

Hardin, G. 2020. Categorizing Potential Planetary Mission Types.

Harman, C., N.R. Izenberg, K.B. Stevenson, M. Zemcov, C.M. Lisse, G. Arney, S. Redfield, et al. 2020. Looking Back Is Looking Forward: The Need for Retrospective Solar System Observations in Advance of Exoplanet Retrievals.

Harris, W., Y.R. Fernandez, G. Sarid, J.K. Steckloff, K. Volk, M. Womack, and L.M. Woodney. 2020. Active Primordial Bodies: Exploration of the Primordial Composition of Ice-Rich Planetesimals and Early-Stage Evolution in the Outer Solar System.

Hauck, S., D. Blewett, P.K. Byrne, N.L. Chabot, C.M. Ernst, C.L. Johnson, E. Mazarico, et al. 2020. Fundamental and Interdisciplinary Questions Drive the Scientific Exploration of Mercury.

Hayne, P., D. Paige, and A. Ingersoll. 2020. New Approaches to Lunar Ice Detection and Mapping.

Haynes, M., A. Virkki, F. Venditti, D. Hickson, N. Pinilla-Alonso, J. Brisset, L. Benner, et al. 2020. Asteroids Inside Out: Radar Tomography.

Hein, A., T.M. Eubanks, A. Hibberd, D. Fries, J. Schneider, M. Lingam, R.G. Kennedy, et al. 2020. Interstellar Now! Missions to and Sample Returns from Nearby Interstellar Objects.

Helbert, J., M.D. Dyar, D. Kappel, A. Maturilli, and N. Mueller. 2020. Importance of Orbital Spectroscopy on Venus.

Heldmann, J., A. Bramson, S. Byrne, R. Beyer, P. Carrato, N. Cummings, M. Golombek, et al. 2020. Accelerating Martian and Lunar Science Through Space X Starship Missions.*

Henderson, R. 2020. From Files to Vials: A Framework for a Tissue-on-Chips Facility Aboard the International Space Station.

Hendrix, A., and J. Rathbun. 2020. Results of the 2020 Planetary Science Workforce Survey Conducted by the AAS-DPS.

Hendrix, A., and T.A. Hurford. 2020. Ocean Worlds: A Roadmap for Science and Exploration.

Hendrix, A., and T. Hurford. 2020. Ocean Worlds: Science Goals for the Next Decade.

Hendrix, A., T. Becker, D. Bodewits, T. Bradley, S. Brooks, B. Byron, J. Cahill, et al. 2020. Ultraviolet-Based Science in the Solar System: Advances and Next Steps.

Hendrix, A.R., T. Hurford, G.W. Patterson, K.E. Mandt, C.B. Beddingfield, R.J. Cartwright, et al. 2020. Potential Ocean Worlds.

Hibbitts, C., R.N. Clark, M.S. Gudipati, M.T. Mellon, U. Raut, T.M. Becker, K. Retherford, et al. 2020. Laboratory Architecture as an Infrastructural Capability to Increase the Science Returned by Ocean World Missions.

Hoehler, T., L. Bebout, W. Brinckerhoff, J. Broddrick, C. Dateo, A. Davila, D.D. Marais, et al. 2020. Groundwork for Life Detection.

Holler, B., M.T. Bannister, K.N. Singer, S.A. Stern, S.D. Benecchi, C.M.D. Ore, L.N. Fletcher, et al. 2020. Prospects for Future Exploration of the Trans-Neptunian Region.

Holler, B., S.N. Milam, J.M. Bauer, J.W. Kruk, C. Alcock, M.T. Bannister, G.L. Bjoraker, et al. 2020. Minor Body Science with the Nancy Grace Roman Space Telescope.

Holt, T., B. Buratti, J. Castillo-Rogez, B.J.R. Davidsson, T. Denk, B.J. Holler, J. Horner, et al. 2020. Captured Small Solar System Bodies in the Ice Giant Region.

Hong, J., S. Romaine, L. Nittler, M. Elvis, I. Crawford, G. Branduardi-Raymont, et al. 2020. X-ray Studies of Planetary Systems: A 2020 Decadal Survey White Paper.

Hong, J., S. Romaine, L. Nittler, M. Elvis, I. Crawford, G. Branduardi-Raymont, et al. 2020. Lunar X-ray Imaging Spectrometer (LuXIS).

Horanyi, M., N. Turner, T. Balint, C. Alexander, N. Altobelli, J. Castillo-Rogez, B. Draine, et al. 2020. Interplanetary and Interstellar Dust as Windows into Solar System Origins and Evolution.

Horgan, B., J.L. Bishop, A. Brown, W. Calvin, C. Edwards, T. Goudge, L.C. Kah, et al. 2020. The Evolution of Habitable Environments on Terrestrial Planets: Insights and Knowledge Gaps from Studying the Geologic Record of Mars.

Horzempa, P. 2020. Calypso Venus Scout.

Horzempa, P. 2020. Europa Exploration Philosophy.

Horzempa, P. 2020. Ice Giant Exploration Philosophy: Simple, Affordable.

Howell, S., W.C. Stone, K. Craft, C. Manager, A. Murray, and A. Rhoden. 2020. Ocean Worlds Exploration and the Search for Life.

Hsu, H., A. Sulaiman, H. Cao, M.M. Hedman, O. Agiwal, N. Altobelli, K. Baillie, et al. 2020. Ice Giants—The Return of the Rings.

Hsu, H., F. Crary, J. Parker, I. de Pater, G. Holsclaw, J. Pitman, K. Sayanagi, et al. 2020. Jupiter System Observatory at Sun-Jupiter Lagrangian Point One.

Hudson, R., B. Theiling, D. Bower, H. Graham, M. Trainer, C. Nixon, and S. Milam. 2020. Laboratory Studies in Support of the Exploration of Ocean Worlds and NASA Missions.

Hunter, G., T. Kremic, and P.G. Neudeck. 2020. High Temperature Electronics for Venus Surface Applications: A Summary of Recent Technical Advances.

Hurley, D., D.T. Blewett, J. Cahill, N. Chabot, B. Greenhagen, C. Hibbitts, R. Klima, et al. 2020. Mission to Characterize Volatiles in Old, Cold, Permanently Shadowed Regions on the Moon.

Hurley, D., E. Adams, S. Arnold, G. Ho, J. Kalirai, M. Paul, and D. Srinivasan. 2020. Perspectives on NASA Planetary Programs from a Mission Implementation Center.

Iacovino, K., N.G. Lunning, G.M. Moore, K.V. Kaaden, F.M. McCubbin, K. Righter, K.B. Prissel, et al. 2020. Making Planets on Earth: How Experimental Petrology Is Essential to Planetary Exploration.

Ishii, H., C.M. Corrigan, M. Bose, J. Davidson, M. Fries, J. Gross, J. Karner, et al. 2020. Terrestrial Recovery of Extraterrestrial Materials: Providing Continued, Long-Term Sample Analysis Opportunities for Research and Mission Support.

Izenberg, N. 2020. Planetary and Astrobiology Blank Papers: Science White Papers Cancelled or Downscaled Due to Direct Impact of COVID-19 and National-Scale Civil Action.

Izenberg, N., D.M. Gentry, D.J. Smith, M.S. Gilmore, D. Grinspoon, M.A. Bullock, P.J. Boston, et al. 2020. The Venus Life Equation.

Izenberg, N., G.A. Landis, S.R. Oleson, P. Abel, M. Bur, A. Colozza, B. Faller, et al. 2020. Hopper Missions to Triton and Pluto Using a Vehicle with In-Situ Refueling.

Izenberg, N., L.W. Esposito, T.K.P. Gregg, and P.K. Byrne. 2020. Venus Exploration Targets: Update of 2014 Venus Exploration Targets Workshop Tables to 2019 Venus GOI, and Findings from the 2019 Venera-D Landing Site Workshop.

Izenberg, N., M.D. Dyar, A. Treiman, J. O'Rourke, J. Cutts, G. Hunter, M. Amato, et al. 2020. The Venus Strategic Plan.

Izenberg, N., R.L. McNutt, K.D. Runyon, P.K. Byrne, and A. MacDonald. 2020. Human Assisted Science at Venus: Venus Exploration in the New Human Spaceflight Age.

Jacobson, S., D. Bodewits, M. Bose, M. Fries, D. Jha, P. Mane, L. Nittler, et al. 2020. Small Bodies Tell the Story of the Solar System: A Rationale for a Small Body Sample Return Program Including Laboratory Analysis of Returned Samples.

Jakosky, B., M. Amato, S. Atreya, D.D. Marais, P. Mahaffy, M. Mumma, M. Tolbert, et al. 2020. Scientific Value of Returning an Atmospheric Sample from Mars.

Jakosky, B., S. Byrne, W. Calvin, S. Curry, B. Ehlmann, J. Eigenbrode, T. Hoehler, et al. 2020. Mars, the Nearest Habitable World—A Comprehensive Program for Future Mars Exploration.

Jawin, E. 2020. Planetary Science Priorities for the Moon in the Decade 2023–2032: Lunar Science Is Planetary Science.

Jawin, E., T. Glotch, R. Watkins, L. Jozwiak, S. Valencia, H. Meyer, R.A. Yingst, et al. 2020. Exploring End-Member Volcanism on the Moon at the Aristarchus Plateau.

Jhoti, E., D. Paige, T. Horvath, and T. Powell. 2020. An Ultra-Low Altitude Lunar Orbiter.

Jolliff, B., M. Robinson, and S. Ravi. 2020. Origin and Evolution of the Moon's Procellarum KREEP Terrane.

Jolliff, B., N. Petro, D. Moriarty, R. Watkins, J. Head, and R. Potter. 2020. Sample Return from the Moon's South Pole–Aitken Basin.

Juanola-Parramon, R., H.B. Hammel, G. Arney, A. Roberge, G. Villanueva, W. Harris, B. Schmidt, et al. 2020. Solar System Science with Space Telescopes.

Kaçar, B., A. Anbar, A. Garcia, L. Seefeldt, Z. Adam, and K. Konhauser. 2020. Between a Rock and a Living Place: Natural Selection of Elements and the Search for Life in the Universe.

Kaluna, H., C.K. Baybayan, and B. Kamai. 2020. Creating Spaces for Indigenous Voices Within Planetary Science—Part 1.

Kamai, B., C.K. Baybayan, and H. Kaluna. 2020. Creating Spaces for Indigenous Voices Within Planetary Science—Part 2.

Kane, S., G. Arney, P. Byrne, D. Crisp, S. Domagal-Goldman, C. Goldblatt, D. Grinspoon, et al. 2020. Venus as a Nearby Exoplanetary Laboratory.

Karunatillake, S., A. Bramson, K. Zacny, C. Dundas, L. Ojha, O. Aharonson, E. Vos, et al. 2020. GANGOTRI Mission Concept on the Glacial Key to the Amazonian Climate of Mars.

Katalenich, J., and J. Sholtis. 2020. Microsphere Plutonium-238 Oxide Fuel to Revolutionize New Radioisotope Power Systems and Heat Sources for Planetary Exploration.

Kaufman, J., D. Banfield, J.S. Boland, M.A. Eby, P. Hayne, J.A. Lang, M.A. Mischna, et al. 2020. Expanding Mars Science Return in the MSR Era: The Need for, Capabilities of, and Challenges Associated with Small Mars Science Missions.

Keane, J., A.A. Ahern, F. Bagenal, A.C.B. Mlinar, K. Basu, P. Becerra, T. Bertrand, et al. 2020. The Science Case for Io Exploration.

Keane, J., A.A. Ahern, F. Bagenal, A.C.B. Mlinar, K. Basu, P. Becerra, T. Bertrand, et al. 2020. Recommendations for Addressing Priority Io Science in the Next Decade.

Kelley, M., H.H. Hsieh, C.O. Chandler, S. Eggl, T.R. Holt, L. Jones, M. Jurić, et al. 2020. Community Challenges in the Era of Petabyte-Scale Sky Surveys.

Kerber, L., P.K. Byrne, A.G. Davies, C.W. Hamilton, D. Jha, J.T. Keane, L. Keszthelyi, et al. 2020. The Importance of Planetary Volcanism and Key Investigations for the Next Decade.

Khan, Z., E. Nesvold, R. Smith, S. Swiersz, and L.M. Walkowicz. 2020. Military Work by Space Exploration Organizations: A Barrier to Inclusion and Safe Workspaces for Marginalized Communities.

Kim, T., D.B. Reisenfeld, P.A. Fernandes, C.A. Maldonado, H.T. Smith, M.F. Thomsen, and A.D. Vira. 2020. Composition and Dynamical Processes of Ions in Giant Magnetospheres: The Importance of In Situ Measurements for Advancing the Current Knowledge.

Kleinboehl, A., D.M. Kass, G.S. Orton, M.D. Hofstadter, C. Leung, M.A. Mischna, S.J. Greybush, et al. 2020. Extended Climatological Observations of Solar System Atmospheres.*

Klima, R., C. Ernst, N. Chabot, K.V. Kaaden, S. Besse, and M. Fries. 2020. Mercury's Low Reflectance Material—Evidence for Graphite Flotation in a Magma Ocean?

Klimczak, C., C.B. Beddingfield, P.K. Byrne, H.C.J. Cheng, K.T. Crane, and A. Annex. 2020. Opportunities and Challenges for Structural Geology and Tectonics in the Planetary Sciences.

Kofman, V., C. Mockel, G. Orton, F. Venditti, A. Migliorini, S. Faggi, M. Cordiner, et al. 2020. Synergies Between Ground-Based and Space-Based Observations in the Solar System and Beyond.

Kohler, E., C. He, S.E. Moran, S.D. Shim, K.K. Brugman, A.C. Johnson, P.C. Vergeli, et al. 2020. The Importance of Prioritizing Exoplanet Experimental Facilities.

Kollmann, P., F. Allegini, R.C. Allen, N. André, A.R. Azari, F. Bagenal, et al. 2020. Magnetospheric Studies: A Requirement for Addressing Interdisciplinary Mysteries in the Ice Giant Systems.

Kopparapu, R. 2020. Strange New Worlds: Comparative Planetology of Exoplanets and the Solar System.

Kramer, G., J. Deca, S. Shukla, T. Kohout, X. Wang, and R. Watkins. 2020. The Plethora of Science Afforded by a Lunar Swirl.

Kramer, G., S. Bailey, J.M. Hurtado, M.J. Laine, C.F. Radley, S. Shukla, and R. Watkins. 2020. The Mutuality Between Science and Commercial Exploration of the Moon.

Kremic, T., and G.W. Hunter. 2020. Long-Lived In-Situ Solar System Explorer (LLISSE): Potential Contributions to Solar System Exploration.

Kremic, T., M. Amato, J. Balcerski, M. Gilmore, G. Hunter, W. Kiefer, N. Izenberg, et al. 2020. Venus Surface Platforms.

Kremic, T., N. Chanover, A. Cheng, T. Hurford, M. Hoffmann, and L. Paganini. 2020. Stratospheric Balloon Platforms for Planetary Science.

Krishnamoorthy, S., A. Komjathy, J.A. Cutts, P. Lognonne, R. Garcia, M.P. Panning, P.K. Byrne, et al. 2020. Seismology on Venus with Infrasound Observations from Balloon and Orbit.

Kurtz, M., A. Accomazzi, and E. Henneken. 2020. Enabling Synergy: Improving the Information Infrastructure for Planetary Science.

Landis, G., and S.R. Oleson. 2020. Sample Return from Titan.

Landis, M., J.C. Castillo-Rogez, P.O. Hayne, H.H. Hsieh, K.H.G. Hughson, K.E. Miller, D. Kubitschek, et al. 2020. Why We Should Study the Themis Asteroid Family in the 2023–2032 Decade.

Lauer, T., and J. Blakeslee. 2020. NSF's National Optical Infrared Astronomy Research Laboratory and Planetary Science.

Lazio, J., A.K. Virkki, N. Pinilla-Alonso, L.A.M. Benner, M. Brozovic, B.J. Butler, B.A. Campbell, et al. 2020. The Next-Generation Ground-Based Planetary Radar.

Lazio, J., B. Arnold, M. Levesque, J. Berner, A. Smith, S. Lichten, S. Townes, et al. 2020. The Deep Space Network: Enabling Richer Data Sets for Future Planetary Science Missions.

Lazio, J., S. Asmar, F. Cordova, E.J. Murphy, D.H. Atkinson, N. Chabot, I.J. Cohen, et al. 2020. Collaborative Actions to Enable Richer and More Complex Planetary Science Mission Data.

Lee, J., P.J. Boston, D. Buckner, R.C. Everroad, S.M. Ledford, J.E. Moores, G. Reitz, et al. 2020. SOTERIA: Searching for Organisms Through Equipment Recovery at Impact Areas.

Leonard, E., C. Elder, T. Nordheim, R. Cartwright, D.A. Patthoff, C. Beddingfield, M. Tiscareno, et al. 2020. A New Frontiers Class Mission for the Uranian System That Focuses on Moon, Magnetosphere, and Ring Science.

Lewicki, C., A. Graps, M. Elvis, P. Metzger, and A. Rivkin. 2020. Furthering Asteroid Resource Utilization in the Next Decade Through Technology Leadership.

Li, L., R. West, M. Kenyon, C. Nixon, P. Fry, D. Wenkert, M. Hofstadter, et al. 2020. Radiant Energy Budgets and Internal Heat of Planets and Moons.

Ligterink, N., A. Riedo, M. Tulej, R. Lukmanov, V. Grimaudo, C. de Koning, P. Wurz, et al. 2020. Detecting the Elemental and Molecular Signatures of Life: Laser-Based Mass Spectrometry Technologies.

Limaye, S., J.W. Head, M.A. Bullock, L. Zasova, I.D. Kovalenko, M. Nakamura, J. Cutts, et al. 2020. Future Exploration of Venus: International Coordination and Collaborations.

Limaye, S., K.H. Baines, M.A. Bullock, C. Cockell, J.A. Cordova, J.A. Cutts, D. Grinspoon, et al. 2020. Venus, an Astrobiology Target.

Limaye, S., N.M. Abedin, C.O. Ao, T. Bocanegra, M.A. Bullock, J.P. Carrico, V. Cottini, et al. 2020. Venus Observing System.

Lin, Y., F. Zhong, B.L. Henderson, V. Abrahamsson, I. Kanik, J. Gross, L. Newlin, et al. 2020. MASEX—A Dedicated Life Detection Mission on Mars.

Lindensmith, C., J. Nadeau, E. Serabyn, P. Willis, and P. Boston. 2020. Microscopy for Detection of Extant Life.

Lis, D., S. Milam, E.A. Bergin, D.B. Morvan, P.F. Goldsmith, N. Biver, G.A. Blake, et al. 2020. Isotopic Ratios in Water and the Origin of Earth's Oceans.

Lisse, C.M., M. Zemcov, K. Mandt, K. Runyon, C. Ahrens, C. Beichman, J. Bock, et al. 2020. Instrumentation for Producing Groundbreaking Planetary and Astrophysical Science on an Interstellar Probe Mission.

Lister, T., M.S.P. Kelley, G. Gyuk, Q. Ye, and J. Li. 2020. Rapid Response and Robotic Telescopes for Understanding Small Body Transient Science.

Lucey, P.G., P. Prem, M.L. Cable, C. Hibbitts, C.I. Honniball, C.M. Pieters, et al. 2020. Lunar Volatiles Orbiters.

Lyons, T., K. Rogers, R. Krishnamurthy, L. Williams, S. Marchi, E. Schwieterman, D. Trail, et al. 2020. Constraining Prebiotic Chemistry Through a Better Understanding of Earth's Earliest Environments.

MacDonald, M., J. Balboni, C. Cornelison, J. Hartman, M. Haw, E. Fretter, B. Cruden, et al. 2020. NASA Ames Thermophysics Ground Test Facilities Supporting Future Planetary Atmospheric Entry.

MacKenzie, S., S. Birch, S. Hörst, C. Sotin, E. Barth, J. Lora, M. G. Trainer, et al. 2020. Titan: Earth-Like on the Outside, Ocean World on the Inside.

Mahabal, A., T. Hare, V. Fox, and G. Hallinan. 2020. In-Space Data Fusion for More Productive Missions.

Mahaffy, P., M. Weng, R. Arevalo, M. Benna, R. Summons, W. Brinckerhoff, M. Cable, et al. 2020. Agnostic Biosignature Exploration at Europa Through Plume Sampling.

Mainzer, A., P. Abell, M.T. Bannister, B. Barbee, J. Barnes, J. Bell III, L. Brenner, et al. 2020. The Future of Planetary Defense in the Era of Advanced Surveys.

Mandt, K. 2020. Advancing Space Science Requires NASA Support for Coordination Between the Science Mission Directorate Communities.

Margot, J. 2020. Venus: Moment of Inertia and Length-of-Day Variations.

Marley, M., C. Harman, H.B. Hammel, P.K. Byrne, J. Fortney, A. Accomazzi, S.E. Moran, et al. 2020. Enabling Effective Exoplanet/Planetary Collaborative Science.

Mathies, R., and A. Butterworth. 2020. The Case for an Orbital Mission to Characterize the Organic Content of the Enceladus Plumes.

Matthies, L., P. Backes, J. Hall, B. Kennedy, S. Moreland, H. Nayar, I. Nesnas, et al. 2020. Robotics Technology for In Situ Mobility and Sampling.

Mayyasi, M., B. Sanchez-Cano, K. Peter, M. Benna, R. Lillis, R. Ramstad, X. Fang, et al. 2020. How Closely-Knit Is the Martian Atmosphere System?

McAdam, M., A.S. Rivkin, L.F. Lim, J. Castillo-Rogez, F. Marchis, and T.M. Becker. 2020. Main Belt Asteroid Science in the Decade 2023–2032: Fundamental Science Questions and Recommendations on Behalf of the Small Bodies Assessment Group.

McCubbin, F., J.H. Allton, J.J. Barnes, M.J. Calaway, C.M. Corrigan, J. Filiberto, M.D. Fries, et al. 2020. Advanced Curation of Astromaterials for Planetary Science Over the Next Decade.

McEnerney, B., D.R.P. Dillon, J. Vickers, D.S.D. Jolly, P.A. Rollett, and K.M. Taminger. 2020. Enabling New Science Mission Capabilities with Additive Manufacturing Technologies.

McGouldrick, K., G. Arney, A. Brecht, A. Colaprete, S. Curry, J. Deighan, T. Fukuhara, et al. 2020. Venus Orbital Mission Concept: Kythiran Eolian dYnamics from the Surface to the Thermosphere from an Orbital NEtwork (KEYSTONE).

McGouldrick, K., N. Izenberg, J. Whitten, C. Tsang, E. Young, J. Balcerski, S. Kane, et al. 2020. The Atmospheric eXploration and Investigative Synergy (AXIS) Group: Proposal for a New Interdisciplinary NASA Assessment/Analysis Group (AG).

McGovern, P. 2020. Venus: A Natural Volcanological Laboratory.

McKay, C., A. Davila, J. Eigenbrode, C. Lorentson, R. Gold, J. Canham, A. Dazzo, et al. 2020. Contamination Control Technology Study for Achieving the Science Objectives of Life-Detection Missions.

McNutt, R., W.J. Adams, J.L. Green, and G.F. Squibb. 2020. Planetary Science Mission Operations for the Next Decade and Beyond.

McPhee, J., and J. Charles. 2020. Human Planetary and Astrobiology Exploration: How Will Radiation, Low Gravity, and Isolated and Confined Conditions Affect Our Health?

McSween, H., M. Grady, K. McKeegan, D. Beaty, and B. Carrier. 2020. Why Mars Sample Return Is a Mission Campaign of Compelling Importance to Planetary Science and Exploration.

Meech, K., J. Castillo-Rogez, E. Bufanda, M. Buie, O. Hainaut, H. Ishii, J.V. Keane, et al. 2020. In-Situ Exploration of Objects on Oort Cloud Comet Orbits: OCCs, Manxes and ISOs.

Meech, K., J. Castillo-Rogez, M. Choukroun, G. Filacchione, O. Hainaut, H. Hsieh, T. Bergin, et al. 2020. Main Belt Comets as Clues to the Distribution of Water in the Early Solar System.

Mendez, A., E.G. Rivera-Valentín, D. Schulze-Makuch, J. Filiberto, R. Ramírez, T.E. Wood, A. Dávila, et al. 2020. Habitability Models for Planetary Sciences.

Mesick, K., P. Gasda, T. Gabriel, C. Hardgrove, and B. Feldman. 2020. Nuclear Spectroscopy for the Exploration of Mars and Beyond.

Messenger, K., A.G. Hayes, S. Sandford, C. Raymond, S.W. Squyres, L.R. Nittler, S. Birch, et al. 2020. The Case for Non-Cryogenic Comet Nucleus Sample Return.

Michaud, E., A.P.V. Siemion, J. Drew, and S.P. Worden. 2020. Lunar Opportunities for SETI.

Milam, S., H. Hammel, J. Bauer, M. Brozovic, T. Grav, B. Holler, C. Lisse, et al. 2020. Combined Emerging Capabilities for Near-Earth Objects (NEOs).

Milam, S., J.P. Dworkin, J.E. Elsila, D.P. Glavin, P.A. Gerakines, J.L. Mitchell, K. Nakamura-Messenger, et al. 2020. Volatile Sample Return in the Solar System.

Milazzo, M., C. Richey, J. Piatek, A. Vaughan, and A. Venkatesan. 2020. The Growing Digital Divide and Its Negative Impacts on NASA's Future Workforce.

Milazzo, M., C. Richey, J. Rathbun, E. Rivera-Valentín, S. Diniega, J. Piatek, B. Schmidt, et al. 2020. DEIA White Papers for Planetary 2023 Supported by the Cross-AG EDI Working Group.

Miller, I. 2020. Constraints on Biogenesis.

Miller, K., B. Theiling, A.E. Hofmann, J. Castillo-Rogez, M. Neveu, S.S. Hosseini, J. Barnes, et al. 2020. The Value of CHONS Isotopic Measurements of Major Compounds as Probes of Planetary Origin, Evolution, and Habitability.

Mills, F., K.L. Jessup, and A. Brecht. 2020. Atmospheric Chemistry on Venus—New Observations and Laboratory Studies to Progress Significant Unresolved Issues.

Mittelholz, A., J. Espley, J. Connerney, R. Fu, C. L. Johnson, B. Langlais, R.J. Lillis, et al. 2020. Mars' Ancient Dynamo and Crustal Remanent Magnetism.

Montabone, L., N. Heavens, J.L. Alvarellos, M. Aye, A. Babuscia, N. Barba, J.M. Battalio, et al. 2020. Observing Mars from Areostationary Orbit: Benefits and Applications.

Moore, J., L. Spilker, J. Bowman, M. Cable, S. Edgington, A. Hendrix, et al. 2020. Exploration Strategy for the Outer Planets 2023–2032: Goals and Priorities (Updated).

Moore, K., J. Castillo-Rogez, K.J. Meech, S.W. Courville, B.P. Donitz, S. Ferguson, K. Llera, et al. 2020. Rapid Response Missions to Explore Fast, High-Value Targets Such as Interstellar Objects and Long Period Comets.

Moores, J., H.M. Sapers, D. Oehler, C. Newman, and L. Whyte. 2020. High-Frequency Near-Surface Gas Measurement: An Opportunity to Solve Puzzles in Planetary Atmospheric Processes in Martian Methane and Beyond.

Moriarty, D., C. Neal, N. Dame, and S. Lawrence. 2020. The Moon Is a Special Place.

Moriarty, D., N.E. Petro, S.N. Valencia, S. Bailey, T. Morton, and C.R. Neal. 2020. Lunar Sample Return from Multiple Locations Is a Critical Capability for Addressing High-Priority Planetary Science Goals.

Mouginis-Mark, P., D. Burr, P. Byrne, K. Coles, D.A. Crown, A. Patthoff, M.S. Phillips, et al. 2020. Planetary Geologic Mapping.

Muirhead, B., A.K. Nicholas, C. Edwards, J. Umland, S. Vijendran, and R. Zurek. 2020. Mars Sample Return Campaign Concept Architecture.

Murchie, S., R.E. Arvidson, J.L. Bishop, W.M. Calvin, J. Carter, J. Christian, R.N. Clark, et al. 2020. Maximizing the Science and Resource Mapping Potential of Orbital VSWIR Spectral Measurements of Mars.

Mustafi, S., C.A. Nixon, N. Petro, X. Li, L. R. Purves, A. Douglawi, S.P. Simpson, et al. 2020. Cryogenic Hydrogen Oxygen Propulsion System (CHOPS) for Planetary Science Missions.

Neal, C., C. Pieters, A. Abbud-Madrid, J. Burns, K.D. Hanna, N. Dygert, B. Ehlmann, et al. 2020. Long-Term Commitment to Explore and Sustain Our Earth-Moon Environment.

Neal, C., S.P. Gulick, B. Baker, S. D'Hondt, N. Eguchi, T. Gregg, et al. 2020. Forging Partnerships with Other Federal Programs: NASA and the National Science Foundation (NSF) Through Scientific Ocean Drilling.

Nenon, Q., G. Clark, P. Kollmann, B. Mauk, T.A. Nordheim, A.R. Poppe, E. Roussos, et al. 2020. Open Science Questions and Missing Measurements in the Radiation Belts of Jupiter.

Neveu, M., A. Anbar, A. Davila, D.P. Glavin, S.M. MacKenzie, C. Phillips-Lander, B. Sherwood, et al. 2020. Returning Samples from Enceladus for Life Detection.

Neveu, M., and S. MacKenzie. 2020. Responsiveness to Discovery.

Neveu, M., J.C. Aponte, J. Castillo-Rogez, B.L. Ehlmann, H.B. Franz, C.H. House, E. Mazarico, et al. 2020. Exploring Solar System Organic Chemistry Evolution Through the Surfaces of Ceres and Large Asteroids.

Newman, C., T. Bertrand, J. Battalio, M. Day, M. De La Torre Juárez, M.K. Elrod, F. Esposito, et al. 2020. Toward More Realistic Simulation and Prediction of Dust Storms on Mars.

Newman, C., V. Airapetian, M. Battalio, S. Bougher, A. Brown, S.D. Domagal-Goldman, S. Fan, et al. 2020. An Urgently Needed Repository for Planetary Atmospheric Model Output.

Niles, P., N. Barba, S.H. Bertsch, J. Bishop, M. Day, M. Evans, K. Fisher, et al. 2020. Mars Commercial Rover Payload Services.

Nittler, L., and R. Stroud. 2020. On the Importance of Presolar Grains for Planetary Science.

Nixon, C., J. Abshire, A. Ashton, J.W. Barnes, N. Carrasco, M. Choukroun, A. Coustenis, et al. 2020. The Science Case for a Titan Flagship-Class Orbiter with Probes.

Nolan, M., P. Michel, F.C.F. Venditti, N. Samarasinha, S. Sugita, M.E. Perry, T. Daly, et al. 2020. Near-Earth Objects.

Nunn, C., S.B. Calcutt, P.E. Clark, T.M. Eubanks, S. Kedar, M.P. Panning, W.T. Pike, et al. 2020. Theiling. 2020. MoonShake: A Future Lunar Seismic Network Delivered by Penetrators.

O'Rourke, J., J. Buz, R.R. Fu, R.J. Lillis, and C. Dong. 2020. Searching for Crustal Remanent Magnetism on Venus Is a High-Impact Science Objective.*

Orton, G., D. Atkinson, T. Balint, M. Hofstadter, O. Mousis, K. Sayanagi, and T. Spilker. 2020. Science Return from In Situ Probes in the Atmospheres of the Ice Giants.

Orton, G., D. Atkinson, T. Balint, M. Hofstadter, O. Mousis, K. Sayanagi, and T. Spilker. 2020. In Situ Probes in the Atmospheres of the Ice Giants.

Paganelli, F., B.A. Archinal, C.H. Acton, A. Conrad, T.C. Duxbury, D. Hestroffer, J.L. Hilton, et al. 2020. The Need for Recommendations in Support of Planetary Bodies Cartographic Coordinates and Rotational Elements Standards.

Paige, D. 2020. Seeking Out Life on Mars in the Next Decade.

Paige, D., C. Ahrens, S. Bailey, M. Eubanks, P. Hayne, K. Mesick, S. Price, et al. 2020. Small Lunar Penetrators.

Pappalardo, R., T. Becker, D. Blaney, D. Blankenship, J. Burch, P. Christensen, K. Craft, et al. 2020. The Europa Clipper Mission: Understanding Icy World Habitability and Blazing a Path for Future Exploration.

Paranicas, C., L. Regoli, N. Ligier, T. Nordheim, and K. Hibbitts. 2020. Exogenic Versus Endogenic Features of the Planetary Satellites.

Parenteau, N., S. Domagal-Goldman, N.Y. Kiang, E. Schwieterman, V. Meadows, C.T. Reinhard, H.E. Hartnett, et al. 2020. Synergies Between Exoplanet and Solar System Life Detection Efforts: Encouraging Collaboration to Enhance Science Return.

Perl, S., S. Adeli, C. Basu, B.K. Baxter, J. Bowman, E. Boyd, M. Cable, et al. 2020. Salty Environments: The Importance of Evaporites and Brine Environments as Habitats and Preservers of Biosignatures.

Phillips, C., K. Hand, J. Scully, J. Pitesky, K. Craft, M. Cameron, T. Nordheim, et al. 2020. An Exploration Strategy for Europa.

Phillips-Lander, C., A. Agha-mohamamdi, J. Wynne, T. Titus, N. Chanover, C. Demirel-Floyd, K. Uckert, et al. 2020. Mars Astrobiological Cave and Internal Habitability Explorer (MACIE): A New Frontiers Mission Concept.

Piatek, J., K.E.V. Kaaden, T.A. Goudge, J.L. Molaro, and M.P. Milazzo. 2020. Breaking Down Barriers: Accessibility in Planetary Science.

Pital, A. 2020. Scientific Consensus Is Dead, Long Live Scientific Consensus.

Polk, J., G. Soulas, and A. Hoskins. 2020. Electric Propulsion: A Key Enabling Technology for Planetary Exploration.

Polzin, K., C.R. Joyner, T. Kokan, S. Edwards, A. Irvine, M. Rodriguez, and M. Houts. 2020. Enabling Deep Space Science Missions with Nuclear Thermal Propulsion.

Poppe, A., J.R. Szalay, C.M. Lisse, M. Horányi, M. Zemcov, S. Hsu, P. Pokorny, et al. 2020. Dust Measurements from Interstellar Probe: Exploring the Zodiacal Cloud and the Interstellar Dust Environment.

Prem, P., Á. Kereszturi, A.N. Deutsch, C.A. Hibbitts, C.A. Schmidt, C. Grava, C.I. Honniball, et al. 2020. Lunar Volatiles and Solar System Science.

Prockter, L., M.R. Wheeler, K. Aye, K.H. Baines, M. Bland, D. Blewett, S. Diniega, et al. 2020. The Value of Participating Scientist Programs to NASA's Planetary Science Division.

Prockter, L., M.S. Tiscareno, E.J. Grayzeck, C.H. Acton, R.E. Arvidson, J.M. Bauer, R. Beebe, et al. 2020. The Planetary Data System: A Vital Component in NASA's Science Exploration Program.

Radebaugh, J. 2020. The Value of a Dual Anonymous System for Reducing Bias in Reviews of Planetary Research and Analysis Proposals and Scientific Papers.

Radebaugh, J., B. Thomson, B. Archinal, R. Beyer, D. DellaGiustina, C. Fassett, L. Gaddis, et al. 2020. Maximizing the Value of Solar System Data Through Planetary Spatial Data Infrastructures.

Rampe, E., D. Blake, P. Sarrazin, T. Bristow, M. Gailhanou, V. Tu, B. Lafuente, et al. 2020. CheMinX: A Next Generation XRD/XRF for Quantitative Mineralogy and Geochemistry on Mars.

Rapin, W., A. Fraeman, B. Ehlmann, A. Mittelholz, B. Langlais, R. Lillis, V. Sautter, et al. 2020. Critical Knowledge Gaps in the Martian Geological Record: A Rationale for Regional-Scale in Situ Exploration by Rotorcraft Mid-Air Deployment.

Rask, J., P. Boston, and E. Goolish. 2020. Strategy for Integration of Life Sciences in Space.

Rathbun, J., C.R. Richey, E.G. Rivera-Valentín, J.H. Roberts, R.N. Watkins, N. Zellner, and M. Kirven. 2020. Enabling the Planetary Workforce to Do the Best Science by Funding Work That Is a Service to the Profession.

Rathbun, J., C. Richey, B.A. Cohen, J.L. Piatek, J.H. Roberts, I.J. Daubar, S. Diniega, et al. 2020. Ensuring Inclusivity in the 2023 Planetary Science and Astrobiology Decadal Survey.

Rathbun, J., E.G. Rivera-Valentín, J.T. Keane, K. Lynch, S. Diniega, L.C. Quick, C. Richey, J. Vertesi, O.J. Tucker, and S.M. Brooks. 2020. Who Is Missing in Planetary Science?: Strategic Recommendations to Improve the Diversity of the Field.

Raymond, C., J. Castillo-Rogez, K. Krohn, D. Buczkowski, K. Singer, B. Bottke, A. Rivkin, et al. 2020. What Are the Main Geological Processes That Determined the Evolution and Current State of Small Bodies and Are They Similar to Those on Larger Bodies?

Rayner, J., and V. Reddy. 2020. The NASA Infrared Telescope Facility (IRTF) 2023–2032.

Reale, O., T. Fauchez, S. Teinturier, S. Guzewich, S. Greybush, and J. Wilson. 2020. Building a Standardized Observing System Simulation Experiment (OSSE) Framework for Mars.

Redwing, E., C.P. McKay, and I. de Pater. 2020. Fluorescence Spectroscopy to Detect Complex Biomolecules in Plumes of the Outer Solar System: A Fast, Low-Cost, Flyby Mission to Search for Biomarkers.

Reeves, G., B.A. Kennedy, G.H. Tan-Wang, P.G. Backes, S.A. Chien, V. Verma, K.P. Hand, et al. 2020. Development of Autonomous Actions to Enable the Next Decade of Ocean World Exploration.

Reinecke, D., P.C. Brandt, G.H. Fountain, A.M. Rymer, and J.A. Vertesi. 2020. The Team Science Challenges of Very Long Duration Spaceflight Missions.

Retherford, K. 2020. Boldly Increase the Planetary Science Research and Analysis Budget by 50% Immediately.

Rhoden, A., J. Englander, E. Morse, G. Benavides, K. Walsh, and L.F. Lim. 2020. Delivering Critical Science by Enabling Small Interplanetary Missions Beyond Mars.

Richardson, J., N. Whelley, P. Whelley, M. Milazzo, C. Knudson, R. Romo, and S.K. Nawotniak. 2020. Building Safer and More Inclusive Field Experiences in Support of Planetary Science.

Richey, C., A. Venkatesan, J. Grier, K. Sheth, M. McAdam, M.B. Wilhelm, M.S. Tiscareno, et al. 2020. A Call to Planetary2023 Panels to Implement Actionable Recommendations from Recent National IDEA Studies.

Richey, C., K.B. Clancy, K.M.N. Lee, and E.M. Rodgers. 2020. Recommendations from the CSWA Survey on Workplace Climate.

Richey, C., M. S. Tiscareno, P. Prem, A. Venkatesan, J. Rathbun, and M. McAdam. 2020. Lessons Learned on IDEA from the Astro2020 Decadal Survey.

Richter, L., J. Flahaut, J. Hamilton, S. Jacquet, P. Lognonne, U. Mall, B. Ordoubadian, et al. 2020. "Land and Fly" Methods for Effective, Future Lunar Exploration.

Rivera-Valentín, E., G. Martínez, J. Filiberto, K. Lynch, V.F. Chevrier, R.V. Gough, M. Tolbert, et al. 2020. Resolving the Water Cycle on a Salty Mars: Planetary Science and Astrobiology Exploration Strategies for the Next Decade.

Rivera-Valentín, E., J. Rathbun, J.T. Keane, K. Lynch, C. Richey, S. Diniega, and J. Vertesi. 2020. Who Is Missing in Planetary Science: A Demographic Study of the Planetary Science Workforce.

Rivera-Valentín, E., P.A. Taylor, C.R. Sanchez-Vahamonde, D. Hickson, C. Neish, M. Brozović, J. Whitten, et al. 2020. The Importance of Ground-Based Radar Observations for Planetary Exploration.

Rivkin, A., M. Milazzo, A. Venkatesan, E. Frank, M.R. Vidaurri, P. Metzger, and C. Lewicki. 2020. Asteroid Resource Utilization: Ethical Concerns and Progress.

Rivkin, A., T. Daly, J. Atchison, B. Barbee, P. Abell, A. Stickle, T.S. Sotirelis, et al. 2020. The Case for a Planetary Defense-Optimized NEO Characterization Tour.

Robbins, S., A. Stern, R. Binzel, W. Grundy, D. Hamilton, R. Lopes, B. McKinnon, et al. 2020. Pluto System Follow-On Missions: Background, Rationale, and New Mission Recommendations.

Robbins, S., M.R. Kirchoff, N.G. Barlow, C.R. Chapman, S. Marchi, E.A. Silber, and A.M. Stickle. 2020. The Importance of Continuing Solar System-Wide Impact Crater Studies.

Roberge, A., D. Fischer, B. Peterson, J. Bean, D. Calzetti, R. Dawson, C. Dressing, et al. 2020. The Large UV/Optical/Infrared Surveyor (LUVOIR): Telling the Story of Life in the Universe.

Rogers, K., U. Pedreira-Segade, P. Fox, J.T. Shelley, A. Steele, and D. Trail. 2020. Reimagining Origins of Life Research: Innovation and Synthesis via Experimentation, Instrumentation, and Data Analytics.

Roser, J., E. Sciamma-O'Brien, R. Cartwright, C. Beddingfield, M. Nuevo, D. Cruikshank, Y. Pendleton, et al. 2020. Optical Constants of Outer Solar System Materials and Radiative Transfer Modeling.

Roth, N. 2020. Volatiles in the Next Decade (2023–2032).

Royer, E., C. Gray, A. Brecht, D. Gorinov, and S. Bougher. 2020. Importance of Airglow and Auroral Emissions as Tracers of Venus' Upper Atmosphere Dynamics and Evolution.

Rufu, R., J. Salmon, K. Pahlevan, C. Visscher, M. Nakajima, and K. Righter. 2020. The Origin of the Earth-Moon System as Revealed by the Moon.

Runyon, K., C.J. Ahrens, C.B. Beddingfield, J.T. Cahill, R. Cartwright, I.J. Cohen, B. Holler, et al. 2020. Comparative Planetology of Kuiper Belt Dwarf Planets Enabled by the Near-Term Interstellar Probe.

Rymer, A., K. Runyon, J. Vertisi, K. Hansen, K. Soderlund, K. Sayanagi, A. Stern, et al. 2020. Neptune and Triton: A Flagship for Everyone.

Sanchez-Cano, B., M. Mayyasi, K. Peter, C. Bertucci, X. Fang, C.M. Fowler, Z.R. Girazian, et al. 2020. Coordinated Multi-Spacecraft Observations of the Martian Plasma Environment.

Sanmartin, J.R. 2020. Bare-Tether Missions Paradigm for Exploration of Ocean Worlds in Plumes of Icy Moons Enceladus, Europa, and Triton.

Santos, A., J. Balcerski, D.M. Burr, J. Helbert, G. Hunter, N. Izenberg, N. Johnson, et al. 2020. The Importance of Venus Experimental Facilities.

Santos, A., J. Filiberto, I. Ganesh, M. Gilmore, J.A. Lewis, and A.H. Treiman. 2020. Venus Petrology: The Need for New Data.

Santos, J., K. Edquist, H. Hwang, S. Berry, A. Brandis, F.M. Cheatwood, B. Cruden, et al. 2020. Entry, Descent, and Landing Instrumentation.

Sapers, H., J.E. Moores, D. Banfield, D.Z. Oehler, M.G. Daly, C.F. Lange, T. Onstott, et al. 2020. The Martian Atmospheric Gas Evolution (MAGE) Experiment: Off-Axis Integrated Cavity-Enhanced Output Spectrometer (OA-ICOS).

Sarrazin, P., R. Obbard, N. Vo, K. Zacny, N. Hinman, B. Lafuente, J. Bishop, et al. 2020. In Situ MicroCT for Planetary Exploration.

Sayanagi, K., C.L. Young, B. Holler, J.T. Clarke, N. Schneider, B. Meinke, B. J. Naasz, et al. 2020. Architectures and Technologies for a Space Telescope for Solar System Science.

Sayanagi, K., M. Wong, T. Becker, S. Brooks, S.R. Brueshaber, E.K. Dahl, I. de Pater, et al. 2020. Priority Questions for Jupiter System Science in the 2020s and Opportunities for Europa Clipper.

Scalice, D., M. Kirven-Brooks, A. Gronstal, M. Milazzo, L. Seyler, J. Foster, N. Cabrol, et al. 2020. Power and Responsibility.

Schenk, P., E.G. Rivera-Valentín, M. Kirchoff, S. Robbins, B. Cohen, L. Dones, K. Singer, et al. 2020. The Chronology Problem in the Outer Solar System: Constraining the "WHEN" of Major Dynamical and Geological Events.

Schilffarth, A., S.A. Stern, and D. Lauretta. 2020. Planetary CHoPS.

Schmidt, B. 2020. A Sustainable Partnership Between NASA and NSF for Planetary Science and Astrobiology Research in Antarctica.

Schmidt, B., K. Craft, T. Cwik, K. Zacny, M. Smith, V. Singh, B. Stone, et al. 2020. Dive, Dive, Dive: Accessing the Subsurface of Ocean Worlds.

Schmidt, B., S.S. Johnson, T. Hoehler, H. Graham, J. Bowman, S. Som, L. Barge, et al. 2020. Enabling Progress Towards Life Detection on NASA Missions.

Schmidt, B., S. Som, E. Quartini, J. Buffo, C. Chivers, K. Soderlund, C.E. Carr, et al. 2020. Diversity in Action: Solutions for a More Diverse and Inclusive Decade of Planetary Science and Astrobiology.

Schmidt, G., and K. Gibbs. 2020. Value of Virtual Institutes and the Synergy of Science and Exploration.

Schubert, P. 2020. Natural Gamma Transmutation Studies.

Schwartz, J., and C. Dreier. 2020. Our Ethical Obligation to Planetary Science in the Age of Competitive Space Exploration.

Schwartz, J., R. Ewell, N. Haegel, S. Liu, J. Mcnatt, E. Plichta, and S. Surampudi. 2020. Solar Array Technologies for Planetary Science and Astrobiology Missions.

Seaman, R., J. Bauer, M. Brucker, E. Christensen, T. Grav, L. Jones, A. Mainzer, et al. 2020. NEO Surveys and Ground-Based Follow-Up.

Sheikh, S., A. Berea, R. Davis, G.G. De la Torre, J. DeMarines, T. Fisher, S. Foote, et al. 2020. Technosignatures as a Priority in Planetary Science.

Sibille, L., R.W. Moses, R.P. Mueller, M.A. Viotti, M.M. Munk, P.J. van Susante, K. Zacny, et al. 2020. Mars Reconnaissance: Civil Engineering Advances for Human Exploration.

Siegler, M., and J. Feng. 2020. Science Case for Microwave Wavelength Measurements.

Singer, K., C.C. Walker, K.L. Craft, M. Neveu, R.A. Beyer, M. Sori, C.A. Nixon, et al. 2020. The Importance of Further Studies and Missions to Understand Cryovolcanism.

Skok, J.R., S. Karunatillake, K. Zacny, E.B. Hughes, J. Blank, J. Gaskin, A. Williams, et al. 2020. SPRING Mission: Exploring the Past and Enabling the Future of Mars.

Smith, D., F. Nimmo, K. Khurana, C. Johnson, M. Wieczorek, M. Zuber, C. Paty, et al. 2020. Callisto: A Guide to the Origin of the Jupiter System.

Smith, I.B., W.M. Calvin, D.E. Smith, C. Hansen, S. Diniega, A. McEwen, et al. 2020. Solar-System-Wide Significance of Mars Polar Science.

Smith, I.B., P.O. Hayne, S. Byrne, P. Becerra, M. Kahre, W. Calvin, C. Hvidberg, et al. 2020. Unlocking the Climate Record Stored Within Mars' Polar Layered Deposits.

Smrekar, S., J. Andrews-Hanna, D. Breuer, P. Byrne, D. Buczkowski, B. Campbell, A. Davaille, et al. 2020. Habitability, Geodynamics, and the Case for Venus.

Soderlund, K., M. Bethkenhagen, I. de Pater, J. Fortney, S. Hamel, R. Helled, Y. Kim, et al. 2020. The Underexplored Frontier of Ice Giant Interiors and Dynamos.

Som, S., T.D. Robinson, and A.D. Perez. 2020. Pale Blue Dot Explorer: A Case for Adding Earth to the Planetary Sciences List of Targets.

Sori, M., A. Ermakov, J. Keane, B. Bills, A. Bramson, A. Evans, D. Hemingway, et al. 2020. Transformative Science Unlocked by Future Geodetic Data at Mars, Venus, and Ocean Worlds.

Spilker, T., and P.M. Beauchamp. 2020. Technologies for the Scientific Exploration of the Outer Planets.

Spry, J.A., B. Siegel, G. Kminek, C. Bakermans, J.N. Benardini, E. Beltran, R. Bonaccorsi, et al. 2020. Planetary Protection Knowledge Gaps and Enabling Science for Human Mars Missions.

Stallard, T., A. Rymer, I. Cohen, S. Miller, H. Melin, J. O'Donoghue, L. Moore, et al. 2020. The Importance of Exploring Neptune's Aurora and Ionosphere.

Stamenkovic, V. 2020. Deep Trek: Science of Subsurface Habitability and Life on Mars.

Stamenkovic, V., K. Lynch, J. Tarnas, P. Boston, C.D. Edwards, B. Sherwood-Lollar, et al. 2020. Deep Trek: Mission Concepts for Exploring Subsurface Habitability and Life on Mars—A Window into Subsurface Life in the Solar System.

Steinbruegge, G., A. Romero-Wolf, S. Peters, D. Schroeder, L. Carrer, C. Hamilton, et al. 2020. PRIME—A Passive Radar Sounding Concept for Io.

Stern, J., M. Weng, H. Graham, J. Bowman, S. Hooker, M. Neveu, L. Quick, et al. 2020. Building Consensus, Collaboration, and Capability for Ocean Worlds Field Science.

Stickle, A., B. Barbee, P. Chodas, T. Daly, M. DeCoster, J. Dotson, R. Klima, et al. 2020. The Need for a Well-Defined Modeling Pipeline for Planetary Defense.

Stoker, C., J.G. Blank, P. Boston, L. Chou, S. DasSarma, J. Eigenbrode, N. Grefenstette, et al. 2020. We Should Search for Extant Life on Mars in This Decade.

Stone, W., S. Howell, N. Bramall, C. German, A. Murray, and V. Siegel. 2020. National Ocean Worlds Analog Test Facility and Field Station.

Stough, R., D.K.F. Robinson, J.B. Holt, D.A. Smith, W.D. Hitt, and B.A. Perry. 2020. NASA's Space Launch System: Capabilities for Ultra-High C_3 Missions.

Strauss, B., S.R. Borges, T. Faridani, J.A. Grier, A. Kiihne, E.R. Maier, C. Olsen, et al. 2020. Nonbinary Systems: Looking Towards the Future of Gender Equity in Planetary Science.

Streetman, B. 2020. Enabling Technology Development Recommendations for the Next Decade and Beyond.

Stroud, R., J. Barnes, L. Nittler, J. Gross, J. Davidson, C. Corrigan, H. Ishii, et al. 2020. Strategic Investment in Laboratory Analysis of Planetary Materials as Ground Truth for Solar System Exploration.

Sutliff, T., P.W. McCallum, and S.G. Johnson. 2020. Establishing a Supply of Plutonium-238 and Associated Radioisotope Power Systems Capabilities and Policy Improvements—A Multi-Part Success Story.

Sykes, M., A. Pathare, G. Kramer, N. Samarasinha, and R. Watkins. 2020. NASA Planetary Research and Analysis: Strategy for Reorganization.

Sykes, M., J. Castillo-Rogez, C. Richey, and P.K. Byrne. 2020. NASA Planetary Research and Analysis: What Is R&A?

Tackett, B., S. Dutta, R. Powell, R. Lugo, and D. Engel. 2020. Guidance and Control Approaches That Enable Titan Aerogravity Assist for an Enceladus Mission.

Tamppari, L., A. Brecht, K. Baines, B. Drouin, L. Esposito, B.S. Guzewich, R. Hofer, et al. 2020. Terrestrial Planets Comparative Climatology (TPCC) Mission Concept.

Tavares, F., D. Buckner, D. Burton, J. McKaig, P. Prem, E. Ravanis, N. Trevino, et al. 2020. Ethical Exploration and the Role of Planetary Protection in Disrupting Colonial Practices.

Taylor, P., E. Rivera-Valentín, and A. Bonsall. 2020. Ground-Based Radar for Planetary Science and Planetary Defense.

Thangavelautham, J., H. Kalita, A.Q. Retis, A. Wissing, B. Haugh, C. Angie, G. Nail, et al. 2020. Mission Concept: Cave and Lave Tube Exploration on Moon, Mars and Icy Moons for Eventual Settlement.

Theiling, B., W. Brinckerhoff, J. Castillo-Rogez, L. Chou, V.D. Poian, H. Graham, S.S. Hosseini, et al. 2020. Non-Robotic Science Autonomy Development.

Thompson, M., J. Barnes, D. Blewett, J. Cahill, B. Denevi, K.D. Hanna, J. Gillis-Davis, et al. 2020. Space Weathering Across the Solar System: Lessons from the Moon and Outstanding Questions.

Thronson, H., B. Thomas, L. Barbier, and A. Buonomo. 2020. Using Artificial Intelligence to Support Science Prioritization by the Decadal Surveys.

Tiscareno, M.S., D.M. Scalice, M.L. Thompson, J.L. Noviello, V. White, A. Venkatesan, et al. 2020. Planetary Nomenclature and Indigenous Communities.

Tiscareno, M.S., M. Vaquero, M.M. Hedman, H. Cao, P.R. Estrada, A.P. Ingersoll, K.E. Miller, et al. 2020. The Saturn Ring Skimmer Mission Concept: The Next Step to Explore Saturn's Rings, Atmosphere, Interior and Inner Magnetosphere.

Titus, T., D. Robertson, and J.B. Sankey. 2020. Planetary Defense Preparedness: Identifying the Potential for Post-Asteroid Impact Time Delayed and Geographically Displaced Hazards.

Titus, T., S. Diniega, L. Fenton, L. Neakrase, J. Nienhuis, J. Radebaugh, K.E. Williams, et al. 2020. Aeolian Processes and Landforms Across the Solar System: Science and Technology Requirements for the Next Decade.

Titus, T.N., J.J. Wynne, P. Boston, P. de Leon, C. Demirel, H. Jones, et al. 2020. Science and Technology Requirements to Explore Caves in Our Solar System.

Traore, D. 2020. Cis-Lunar Space Neutrino Subterranean Asset.

Treiman, A., J. Filiberto, and M. McCanta. 2020. Laboratory Studies in Support of Venus Exploration: Surface and Near-Surface.

Trubl, G., K. Stedman, K. Bywaters, P.J. Boston, J. Kaelber, S. Roux, J.B. Emerson, et al. 2020. Astrovirology: Expanding the Search for Life.

Turyshev, S., H. Helvajian, L.D. Friedman, T. Heinsheimer, D. Garber, and A. Davoyan, and V.T. Toth. 2020. Exploring the Outer Solar System with Solar Sailing SmallSats on Fast-Transit Trajectories and In-Flight Autonomous Assembly of Advanced Science Payloads.

Udry, A., J. Filiberto, J. Gross, M. Schmidt, D. Rogers, L. Hausrath, R. Wiens, et al. 2020. The Importance of the Study of Igneous Rocks and Compositions to Constrain the Martian Planetary Evolution.

Unterborn, C., P.K. Byrne, A.D. Anbar, G. Arney, D. Brain, S.J. Desch, B.J. Foley, et al. 2020. Exogeoscience and Its Role in Characterizing Exoplanet Habitability and the Detectability of Life.

Valencia, S., J.A. Richardson, T. Glotch, E. Jawin, S. Ravi, B.L. Jolliff, and R.N. Watkins. 2020. End-Member Volcanism in the Absence of Plate Tectonics: Silicic Volcanism on the Moon.

Valencia, S., N. Curran, J. Flahaut, J. Gross, C.M. Mercer, D.P. Moriarty, C.R. Neal, et al. 2020. High Priority Returned Lunar Samples.

van Belle, G., J. Kugler, N. Moskovitz, and J. Piness. 2020. LightBeam: Flyby-Like Imaging Without the Flyby.

Van Kranendonk, M., R. Baumgartner, S. Cady, K. Campbell, B. Damer, D. Deamer, T. Djokic, et al. 2020. Terrestrial Hydrothermal Fields and the Search for Life in the Solar System.

Vance, S., C. Elder, A. Hofmann, S. Howell, M. Milazzo, R.T. Pappalardo, J. Noviello, et al. 2020. Addressing Mental Health in Planetary Science.

Vance, S., M. Behounkova, B.G. Bills, P. Byrne, O. Cadek, J. Castillo-Rogez, G. Choblet, et al. 2020. Distributed Geophysical Exploration of Enceladus and Other Ocean Worlds.

Vance, S., D.N. DellaGiustina, K. Hughson, T. Hurford, S. Kedar, A.G. Marusiak, et al. 2020. Planetary Seismology: The Solar System's Ocean Worlds.

Vander Kaaden, K., C. Ryan, E.G. Rivera-Valentín, C.B. Phillips, J. Haber, J. Filiberto, and A. Denton. 2020. Creating Inclusive, Supportive, and Safe Environments in Planetary Science for Members of the LGBTQ+ Community.

Vander Kaaden, K., M.S. Thompson, F.M. McCubbin, C.L. Johnson, C.M. Ernst, N.L. Chabot, and P.K. Byrne. 2020. Mercury Sample Return to Revolutionize Our Understanding of the Solar System.

Vander Kaaden, K., R.J. Vervack, E. Rampe, F.M. McCubbin, C. Klimczak, C.J. Cline, P.K. Byrne, et al. 2020. Recommended Laboratory and Field Studies Ahead of Future Mercury Exploration.

Vander Kaaden, K., S.A. Hauck, C.M. Ernst, R.J. Vervack, C. Klimczak, C.L. Johnson, G. DiBraccio, et al. 2020. The Next Decade of Funding Opportunities for Mercury-Related Science and Mission Support.

Varatharajan, I., D. Angerhausen, E. Antoniadou, V. Bickel, M. D'Amore, M. Faragalli, I. López-Francos, et al. 2020. Artificial Intelligence for the Advancement of Lunar and Planetary Science and Exploration.

Venkatapathy, E., D. Prabhu, G. Allen, and M. Gasch. 2020. Thermal Protection System to Enable Ice Giant Aerocapture Mission for Delivering Both an Orbiter and an In Situ Probe.

Venkatapathy, E., J. Feldman, D.S. Adams, R. Beck, D. Ellerby, M. Gasch, H. Hwang, et al. 2020. Sustaining Mature Thermal Protection Systems Crucial for Future In-Situ Planetary Missions.

Vera C. Rubin Observatory LSST Solar System Science Collaboration, R.L. Jones, M.T. Bannister, B.T. Bolin, C.O. Chandler, S.R. Chesley, S. Eggl, et al. 2020. The Scientific Impact of the Vera C. Rubin Observatory's Legacy Survey of Space and Time (LSST) for Solar System Science.

Vertesi, J., and A. Rymer. 2020. Human Factors for Long Duration Space Missions.

Vidaurri, M., A. Wofford, G. Black-Planas, J. Brande, and S. Domagal-Goldman. 2020. Absolute Prioritization of Planetary Protection, Ethics, and Avoiding Imperialism in All Future Science Missions: A Policy Perspective.

Vidaurri, M., and A. Gilbert. 2020. Environmental Considerations in the Age of Space Exploration: The Conservation and Protection of Non-Earth Environments.

Villanueva, G., C. Nixon, L. Paganini, M. Cordiner, S. Milam, and G. Chin. 2020. The Present and Future of Observational Studies of Ocean Worlds.

Villarreal, M., R. Lillis, J. Luhmann, C. Lee, J. O'Rourke, R. Oran, K. Moore, et al. 2020. The Importance of Plasma and Magnetic Investigations in Small Body Missions.

Virkki, A., P.A. Taylor, M.W. Busch, E.S. Howell, E.G. Rivera-Valentín, Y. Fernandez, L.A. Benner, et al. 2020. Near-Earth Object Characterization Using Ground-Based Radar Systems.

Viswanathan, V., E. Mazarico, S. Merkowitz, J.G. Williams, S.G. Turyshev, D.G. Currie, A.I. Ermakov, et al. 2020. Extending Science from Lunar Laser Ranging.

Wang, X., D.T. Blewett, G. Kramer, D. Barker, C. Hartzell, D. Han, M. Horányi, et al. 2020. Electrostatic Dust Transport Effects on Shaping the Surface Properties of the Moon and Airless Bodies Across the Solar System.

Watkins, R., N. Zellner, M. McAdam, N. Whelley, I. Daubar, C. Hartzell, and K. Gardner-Vandy. 2020. Professional Development in the Next Decade: Supporting Opportunities in All Career Paths and Life Events.

Watkins, R., P.T. Metzger, M. Mehta, D. Han, P. Prem, L. Sibille, A. Dove, et al. 2020. Understanding and Mitigating Plume Effects During Powered Descents on the Moon and Mars.

Weber, R., C.R. Neal, R. Grimm, M. Grott, N. Schmerr, M. Wieczorek, J. Williams, et al. 2020. The Scientific Rationale for Deployment of a Long-Lived Geophysical Network on the Moon.

Wellington, D. 2020. Planetary Data Architecture: Lessons from Terrestrial Remote Sensing.

Westphal, A., and K.S.J. Pister. 2020. Rapid Cometary Sample Return Enabled by Low-Cost, 10-Gram Interplanetary Spacecraft.

Westphal, A., L.R. Nittler, R. Stroud, M.E. Zolensky, N.L. Chabot, N.D. Russo, J.E. Elsila, et al. 2020. Cryogenic Comet Sample Return.

Whelley, P., C.N. Achilles, A.M. Baldridge, M.E. Banks, E. Bell, H. Bernhardt, J. Bishop, et al. 2020. The Importance of Field Studies for Closing Key Knowledge Gaps in Planetary Science.

White, T., H. Hwang, D. Ellerby, R. Beck, M. Gasch, J.V. Kam, and E. Venkatapathy. 2020. Thermal Protection System Materials for Sample Return Missions.

Whitehead, J. 2020. REVISED: Mars Ascent Vehicle Needs a Sustained Development Effort, Regardless of Sample Return Mission Timelines.

Whitehead, J. 2020. REVISED: Technology Development Can Lead to Smaller Mars Ascent Vehicles, for Multiple Affordable Sample Returns.

Whitehead, J. 2020. REVISED: The Challenge of Launching Geology Samples Off of Mars Is Easily Underestimated, Due to Tempting Misconceptions.

Whitten, J., M.S. Gilmore, J. Brossier, P.K. Byrne, J.J. Knicely, and S.E. Smrekar. 2020. Venus Tesserae: The Importance of Venus Tesserae and Remaining Open Questions.

Wilhelm, M.B., A. Ricco, D. Oehler, D. Buckner, A. Rodriguez, P. Mahaffy, J. Eigenbrode, et al. 2020. Abzu: Uncovering the Origin of Ancient Organics on Mars.

Williams, D., R.C. Anderson, S. Byrne, F. Costard, A. Hayes, R. Jaumann, P. Mouginis-Mark, et al. 2020. RPIFs to PDUCs: New Planetary Data Utilization Centers to Support NASA's Planetary Data Ecosystem.

Willis, P., M.F. Mora, A. Noell, J. Creamer, F. Kehl, K. Zamuruyev, E. Jaramillo, et al. 2020. How to Search for Chemical Biosignatures on Ocean Worlds.

Woerner, D., S. Johnson, J. Fleurial, S. Howell, B. Bairstow, and M. Smith. 2020. Radioisotope Heat Sources and Power Systems Enabling Ocean Worlds Subsurface and Ocean Access Missions.

Wong, M.H., H.M. Fahad, J. Li, K. Sayanagi, P. Steffes, and S. Atreya. 2020. Advancing Chip-Based Gas Composition Sensors for Efficient Planetary Exploration.

Wong, M.H., K.J. Meech, M. Dickinson, T. Greathouse, R.J. Cartwright, N. Chanover, and M.S. Tiscareno. 2020. Transformative Planetary Science with the US ELT Program.

Wong, M.H., S. Luszcz-Cook, K. Sayanagi, L. Moore, T. Koskinen, J.I. Moses, I. de Pater, et al. 2020. Gas Giant and Ice Giant Atmospheres: Focused Questions for 2023–2032.

Woodney, L., A.S. Rivkin, W. Harris, B.A. Cohen, G. Sarid, M. Womack, O. Barnouin, et al. 2020. Strength in Diversity: Small Bodies as the Most Important Objects in Planetary Sciences.

Yang, C.P., L.F. Zambrano-Marin, K. Carte, A.K. Davis, D.M. Graninger, K. Greenaugh, C. McCarthy, et al. 2020. Critical Data Identification, Information Communication, and Readily Available Knowledge Base for Effectively Mitigating Impact of Near-Earth Objects.

Yingst, R.A., and MEPAG. 2020. Mars as a Competitive Candidate for Inclusion in the New Frontiers Mission List: MEPAG Community Perspectives.

Yingst, R.A., and MEPAG. 2020. Mars System Science: Why Mars Remains a Compelling Target for Solar System Science.

Yingst, R.A., and MEPAG Steering Committee. 2020. MEPAG Steering Committee Diversity, Equity, Inclusion and Accessibility White Paper Statement.

Young, C., M. Wong, K. Sayanagi, S. Curry, K. Jessup, T. Becker, et al. 2020. The Science Enabled by a Dedicated Solar System Space Telescope.

Young, E., and W.D. Leavitt. 2020. Developing Methane Isotopologues as Interplanetary Biosignatures.

Young, E., M. Beasley, M. Bullock, C. Cantrall, N. Chanover, J. Fox, G. Holsclaw, et al. 2020. Stratospheric Venus Observatory.

Young, K., J.M. Hurtado, T.G. Graff, D.P. Moriarty, S. Valencia, R. Watkins, N. Petro, et al. 2020. The Importance of Human Exploration in Accomplishing High Priority Lunar Science Objectives.

Zhu, F., L. French, N. Schorghofer, A. Blachowicz, S. Li, L. Schurmeier, and M. Paton. 2020. Robot Technology Advancements for In-Situ Exploration of Subsurface Environments.

Zurek, R., B. Campbell, S. Byrne, W. Calvin, L.M. Carter, R.T. Clancy, B. Ehlmann, et al. 2020. Mars Next Orbiter Science Analysis Group (NEX-SAG): White Paper Report to the 2023–2032 Planetary Sciences and Astrobiology Decadal Survey.

Appendix C

Technical Risk and Cost Evaluation of Priority Missions

BACKGROUND

The survey's statement of task (see Preface and Appendix A) calls for "identifying, recommending, and ranking the highest priority research activities" in planetary science, astrobiology, and planetary defense. However, the identification and ranking of priority activities alone is insufficient. The statement of task mandates that consideration be given to the technical readiness, technical risk, and likely cost of the major activities identified in the survey report. Concern about readiness and risk are not new. Alarm about the accuracy of the mission cost estimates used in past decadal studies was raised in a 2006 report by the National Academies. The latter noted that "major missions in space and Earth science are being executed at costs well in excess of the costs estimated at the time when the missions were recommended . . . in decadal surveys for their disciplines. Consequently, the orderly planning process that has served the space and Earth science communities well has been disrupted, and the balance among large, medium, and small missions has been difficult to maintain" (NRC 2006). As a result, the 2006 report recommended that "NASA should undertake independent, systematic, and comprehensive evaluations of the cost-to-complete of each of its space and Earth science missions that are under development, for the purpose of determining the adequacy of budget and schedule" (NRC 2006, p. 33)

The technical readiness of activities prioritized in decadal surveys and the associated cost estimates for candidate missions were discussed extensively during a lessons-learned workshop convened by the National Academies (NRC 2007a). Workshop participants commented that decadal surveys would benefit greatly if they conducted their own assessments of the technical risk and cost associated with priority missions, rather than relying on NASA's own estimates. Moreover, the adoption a uniform risk- and cost-estimating methodology would enable cross comparison between competing activities within a given survey (NRC 2007a, pp. 21–30).

The first occasion on which this advice was put into practice was when NASA included a call for an independent evaluation of cost and technology readiness in the statements of task for a review of the agency's Beyond Einstein program (NRC 2007b). Soon thereafter, Congress recognized the benefit of such evaluations and mandated that the National Academies "include independent estimates of the life cycle costs and technical readiness of missions assessed in the decadal survey wherever possible" (U.S. Congress 2008). This requirement was first implemented during the 2010 decadal survey of astronomy and astrophysics via the use of the so-called cost and technical evaluation (CATE) process (NRC 2010). The CATE methodology was developed by the Aerospace Corporation and is particularly suitable for comparing costs and risks associated with a population of low-maturity mission concepts. Subsequently, the CATE methodology was successfully implemented by three additional decadal surveys, including *Vision and Voyages* (NRC 2011).

A comprehensive review of the then most recent round of space-science decadal surveys was conducted during a workshop held in 2012 (NASEM 2013) and again in a 2015 consensus study (NASEM 2015). The CATE process was one of the topics examined in detail during both activities. One of the principal lessons learned about the CATE process, as identified in the 2015 report, was that "it is most useful as a reasonableness check on what is being recommended" (NASEM 2015, p. 52). In other words, how does the technical feasibility of one concept under consideration rank relative to its peers? Moreover, the 2015 report found that the details "used to support the CATE analysis are not necessarily indicative of how a mission will ultimately be implemented" (NASEM 2015, p. 52) Therefore, more emphasis needed to be placed on the technical readiness and risk aspects of the evaluation rather than the cost estimation. To reflect this important change in emphasis, the CATE process was revised somewhat and its name was changed to technical risk and cost evaluation (TRACE). The first decadal to implement the TRACE methodology was the 2020 astronomy and astrophysics survey (NASEM 2023).

THE CHALLENGE OF TECHNICAL RISK, COST, AND SCHEDULE EVALUATIONS

The concepts assessed by a decadal survey are typically in preliminary stages of development—that is, pre-Phase A or Phase A concepts. However, the cost of a mission is typically not well understood until it has gone through its preliminary design review (PDR). Even after PDR, unanticipated increases in mass, cost, and schedule can occur. Another challenge to accurate evaluation is the fact that not all pre-Phase A concepts are equal. Some may be more mature than others because more resources have been available during their formulation. Accordingly, ensuring that a mission evaluation is fair and equitable requires that the relative maturity of concepts be considered.

Several different varieties of technical risk/cost/schedule/evaluations are used when discussing mission concepts. The best known are the so-called ICE (independent cost estimates) and NASA's TMC (technical, management, and cost). A third is the TRACE process adopted by the most recent round of decadal surveys. Each has its own strengths and weaknesses (Table C-1).

The objective of the TRACE process is to perform a technical risk and cost evaluation for a set of concepts that may have a broad range of maturity, and to ensure that the evaluations are consistent, fair, and informed by historical data. Typically, concepts evaluated via the TRACE process are early in their lifecycle, and therefore are likely to undergo significant subsequent design changes. Historically, such changes have resulted in cost growth. ICEs, on the other hand, are usually done later in the lifecycle of a project after it has matured—that is, typically at

TABLE C-1 Similarities and Differences Among Three Different Approaches to Assessing the Technical, Cost, and Risk Characteristics of Spacecraft Missions

	TMC	ICE	TRACE
Used consistently to compare several concepts	Yes	No	Yes
Concept cost is evaluated with respect to	Cost Cap	Project Budget	NASA Budget
Maturity of concept	Phases A–B	Phases B–D	Pre-Phase A
Evaluation Process Includes			
Quantified schedule growth cost threat	No	Typically	Yes
Quantified design growth cost threat	No	No	Yes
Cost threat for increase in launch vehicle capability	No	No	Yes
Independent estimates for non-U.S. contributions	No	No	Yes
Reconciliation performed with project team	No	Yes	No
Technical and cost risk rating (low, medium, high)	Yes	No	Yes

NASA's key decision points KDP-B and KDP-C and design reviews such as PDR and the Critical Design Review (CDR). ICEs often do not consider certain aspects of cost growth associated with design evolution in the earliest phases of a project. Therefore, a robust process is required that fairly treats a concept of low maturity relative to one that has undergone several iterations and review. TRACE evaluations consider several components of risk assessment (see Table C-1). However, an essential prerequisite for a TRACE evaluation is a particular mission design specified in sufficient detail that it has a concept maturity level (CML) of four or five.

FORMULATION AND IDENTIFICATION OF MISSIONS FOR TRACE

NASA prepared for the planetary science and astrobiology decadal survey via a twofold approach. First, the agency commissioned science definition teams (SDT) and related groups to study various mission concepts.[1] The best known of the latter was that for the Europa Lander (NASA 2017b). Second, NASA worked with the planetary science and astrobiology communities via a competitive process to identify, fund, study, and create the appropriate documentation for 11 predecadal planetary mission concept studies (PMCS).

The survey's six panels reviewed the SDT and PMCS mission study reports and assessed those concepts proposed by the community in white papers and prior proposals (e.g., *Vision and Voyages* mission study reports). Then the panels identified 18 additional large- and medium-class mission concepts (including a pair jointly developed by two panels and one proposed by a cross-panel group) that could address key scientific questions within their respective purviews. Following documentation by their originating groups, the steering group assessed each one and selected nine for additional study.

To ensure that all mission concepts are mature enough for subsequent evaluation by the TRACE team, the nine concepts were sent for detailed technical studies: three each at the Jet Propulsion Laboratory, the Goddard Space Flight Center, and the Applied Physics Laboratory. One or more "science champions" drawn from the ranks of the panels was attached to the mission design teams at each of the centers to ensure that the concepts remained true to the scientific and measurement objectives of their originating panel.[2]

A tenth study was later initiated by JPL to create a series of low fidelity (CML-2) concepts addressing planetary defense objectives (see Chapter 18 for details). None of the resulting small-class mission concepts were submitted to the TRACE team because the survey's statement of task did not call for the prioritization of such likely low-cost missions.[3]

Once the nine additional studies were completed, the scientific and technical feasibility of the resulting concepts was assessed by their originating panels. Also assessed by the panels were those PMCS and SDT concepts of relevance, and two additional concepts that had been contenders in the competition for the fourth New Frontiers opportunity.[4] Following deliberations and prioritization by their respective panel, the most promising mission concepts were forwarded to the steering group for additional discussion and an independent, multistep ranking process. The outcome of that process was that the 17 most promising concepts (one with two variants) were identified and submitted to the TRACE team for evaluation.

In summary, from an initial group of 33 concepts, the survey's panels and steering group identified 17 missions, each of which had been studied to a sufficient fidelity to undergo the TRACE process. A full list of all of the missions considered by the decadal survey, their origins, and ultimate disposition can be found in Table C-2.[5]

---

[1] It is important to note that not all such activities resulted in concepts of sufficient technical maturity—e.g., NEX-SAG (NASA 2015), Ice Giant (NASA 2017a), and ICE-SAG (NASA 2019)—to be submitted for TRACE analysis without additional work. As such, the work of these groups was used as input to the survey's mission formulation deliberations.

[2] The full mission study reports for the nine concepts identified by the decadal survey are available at https://tinyurl.com/2p88fx4f.

[3] The complete final report of the planetary defense rapid mission architecture study is available at https://tinyurl.com/2p88fx4f.

[4] Required to ensure that representative concepts responsive to the New Frontiers 4 and 5 Ocean Worlds mission theme undergo technical risk and cost evaluation and thus remain potentially viable candidates for future New Frontiers opportunities. Full study reports for these two missions are available at https://tinyurl.com/2p88fx4f.

[5] A notable omission from Table C-2 is the Venus In Situ Explorer. This concept did not undergo TRACE analysis because the survey committee concurred with *Vision and Voyages'* decision that the phase-A study—conducted when a concept responsive to VISE was in the step-two competition for the third New Frontiers launch opportunity—was equivalent to or better than a CATE. "The committee assumes that the ongoing NASA evaluation . . . has validated [its] ability to be performed at a cost appropriate for New Frontiers" (NRC 2011, p. 15).

TABLE C-2 Mission Concepts Considered by the Decadal Survey for Technical Risk and Cost Evaluation

Name	Origin	Disposition	Additional Information
Mercury Lander	PMCS	Selected for TRACE	NASA 2020g and Appendix C
Venus Flagship	PMCS	Selected for TRACE	NASA 2020k and Appendix C
Venera D	SDT	Not selected for TRACE	Venera-D 2019 and Appendix D
VISAN: Venus In Situ Seismic and Atmospheric Network	PV	Not selected for study or TRACE	Appendix E
VSCA: Venus Sub-Cloud Aerobot	PV	Not selected for study or TRACE	Appendix E
VLP: Venus Life Potential	PV	Not selected for study or TRACE	Appendix E
VIDEO: Venus Investigation of Dynamics from an Equatorial Orbit	PV	Not selected for study or TRACE	Appendix E
ADVENTS: Assessment and Discovery of Venus's Past Evolution and Near-Term Climatic and Geophysical State	PV	Selected for study; not selected for TRACE	NASA 2021a and Appendix D
Lunar Geophysical Network	PMCS and *V&V*	Prioritized following CATE by *V&V*; no further action needed	NASA. 2020d and Appendix C of *V&V*
Intrepid: Lunar Long-Range Rover Traverse	PMCS	Selected for TRACE	NASA 2020c and Appendix C
Endurance: Lunar South Pole Aitken–Basin Sample Collecting Rover	PMM	Selected for study and two variants selected for TRACE	NASA 2021f and Appendix C
INSPIRE: In Situ Solar System Polar Ice Roving Explorer	PMM	Selected for study and selected for TRACE	NASA 2021g and Appendix C
MOSAIC: Mars Orbiter for Surface-Atmospheric-Ionospheric Connections	PMCS	Not selected for TRACE	NASA 2020f and Appendix D
MORIE: Mars Orbiter for Resources, Ices, and Environments	PMCS	Not selected for TRACE	NASA 2020e and Appendix D
Mars Life Explorer	PM	Selected for study and selected for TRACE	NASA 2021h and Appendix C
Mars In Situ Geochronology	PMCS	Selected for TRACE	NASA 2020b, additional detail from originating team, and Appendix C
Mars Deep Time Rover	PM	Not selected for study; not selected for TRACE	Appendix E
Mars Polar Ice, Climate and Organics	PM	Not selected for study; not selected for TRACE	Appendix E
Ceres Sample Return	PMCS	Selected for TRACE	NASA 2020h and Appendix C
Cryogenic Comet Nucleus Sample Return	PSSSB	Selected for study. Not selected for TRACE	NASA 2021d and Appendix D
Europa Lander	SDT	Selected for TRACE	NASA 2017b, additional detail from originating team, and Appendix C

continued

TABLE C-2 Continued

Enceladus Orbilander	PMCS	Selected for TRACE	NASA 2020a, additional details from originating team, and Appendix C
Enceladus Multiple Flyby	Prior study documentation made available to survey	Selected for TRACE	NASA 2021e and Appendix C
Titan Orbiter and Probe	Prior study documentation made available to survey	Version without the sea probe selected for TRACE	NASA 2021i and Appendix C
Saturn Ring Skimmer	PGPS	Not selected for study or TRACE	Appendix E
Centaur Orbiter and Lander	PSSSB	Selected for study and selected for TRACE	NASA 2021c, and Appendix C
Uranus Orbiter and Probe	PGPS and POWDP	Selected for study and selected for TRACE	NASA 2021j and Appendix C
Calypso: Uranus Moon and KBO Flyby	PSSSB and POWDP	Selected for study and selected for TRACE	NASA 2021b and Appendix C
Odyssey: Neptune Orbiter and Probe	PMCS	Selected for TRACE	NASA 2020i, additional details from originating team, and Appendix C
Triton Ocean World Surveyor	POWDP	Selected for study and selected for TRACE	NASA 2021k and Appendix C
Persephone: Pluto System Orbiter and Kuiper Belt Explorer	PMCS	Not selected for TRACE	NASA 2020j and Appendix D
Interstellar Object Rapid Response Mission	PSSSB	Not selected for study or for TRACE	Appendix E
Solar System Space Telescope	Cross-panel group	Not selected for study or for TRACE	Appendix E

NOTES: Green, orange, and pink shading, respectively, indicate that a concept was studied and selected for TRACE, was studied but not selected for TRACE, and was suggested but not studied. PGPS, Panel on Giant Planet Systems; PM, Panel on Mars; PMM, Panel on Mercury and Moon; POWDP, Panel on Ocean Worlds and Dwarf Planets; PSSSB, Panel on Small Solar System Bodies; PV, Panel on Venus; PMCS, Predecadal Mission Concept Study; SDT, Science Definition Teams; *V&V, Vision and Voyages* decadal survey (NRC 2011).

OVERVIEW OF THE TRACE PROCESS

The National Academies engaged the services of the Aerospace Corporation to perform independent TRACE evaluations of mission concepts identified by the committee's steering group during this survey. Aerospace's TRACE team consists of experts in the evaluation of technical, cost, and schedule risks.

The TRACE began when the survey committee forwarded the full mission study reports for the 17 concepts selected for detailed examination to the Aerospace Corporation team. If the supplied documentation was insufficient or more details were required, the TRACE team requested additional information from the decadal survey. In three cases, the decadal survey was able to work with the relevant originating team to obtain the information required to complete the TRACE. The members of the TRACE team worked interactively to determine an initial assessment of technical risk and cost and schedule estimates for each of 17 missions analyzed. The TRACE team was diligent and worked, to the extent feasible, to treat all 17 concepts in a consistent and evenhanded manner.

Following an initial internal review within Aerospace to ensure that the 17 assessments were mutually consistent, the results were presented to the survey's steering group. The latter provided feedback to the TRACE team who, in turn, incorporated this feedback into revised technical, cost, and schedule risk evaluations.

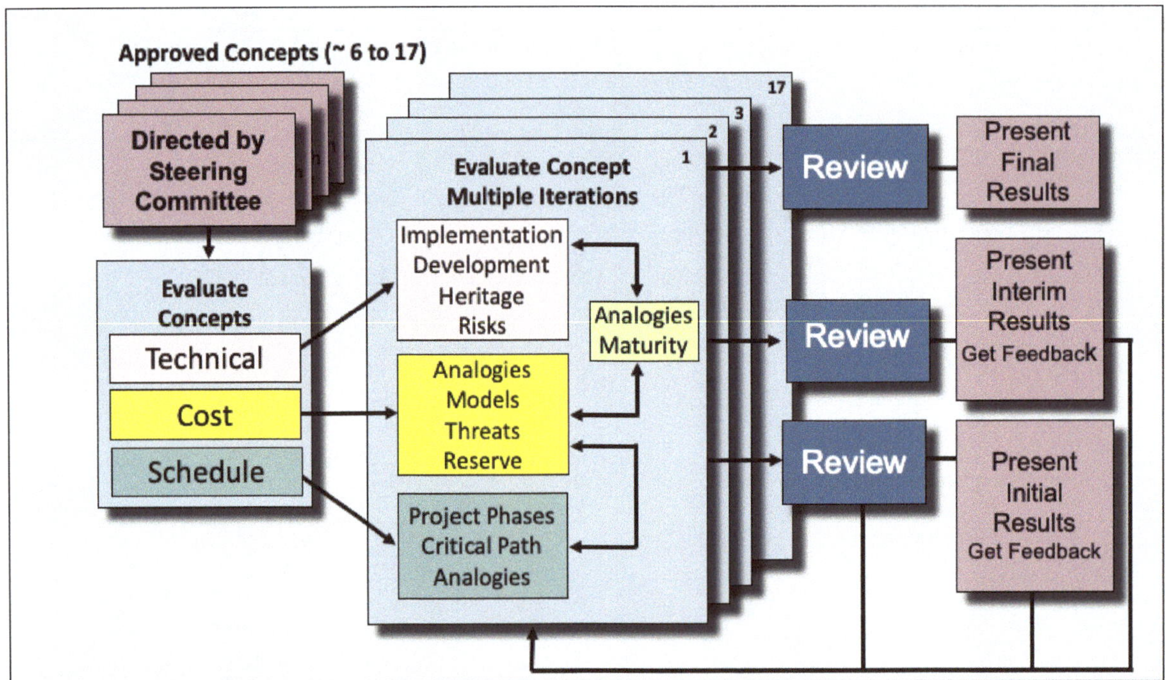

FIGURE C-1 Schematic illustration of the flow of the Aerospace Corporation's technical risk and cost evaluation (TRACE) process. The blocks in purple indicate interaction by the TRACE team with the committee.

The TRACE team's approach (Figure C-1) is based on the following principles:

- Use multiple methods and databases containing the pertinent details of past space systems so that no one model or dataset biases the results. Specifically, the TRACE team used proprietary Aerospace models (e.g., Small Satellite Cost Model, Missions Operations Cost Estimating Tool) and space-industry standards (e.g., NASA's Project Cost Estimating Capability).
- Use analogy-based estimating; tie costs and schedule estimates to NASA systems that have already been built with known cost and schedule.
- Use both system-level estimates as well as a build-up-to-system level by appropriately summing subsystem data so as not to underestimate system cost and complexity.
- Use verification tools to check cost and schedule estimates for internal consistency and risk assessment.
- Quantify the total threats to costs from schedule growth, the costs of maturing technology, additional scope, and the threat to costs owing to mass growth resulting in the need for a larger, more costly launch vehicle in an integrated fashion.

In summary, an analogy-based methodology estimates the costs of future systems by comparison to the known cost of systems that have previously been built. The TRACE process provides an independent estimate of cost and complexity of pre-Phase A, proof of mission concepts anchored with respect to previously built hardware informed by the NASA historical record of concept design growth. Thus, the TRACE process is designed to inform the decadal survey about the potential future impact of the concepts evaluated to best enable the prioritization of a balanced program within a defined future budget scenario. The use of multiple methods—for example, analogies and standard cost models—ensures that no one model or database biases the estimate. The use of system-level estimates and arriving at total estimated costs by statistically summing the costs of all individual work breakdown structure (WBS) elements ensures that elements are not omitted and that the system-level complexity is properly represented in the cost estimate.

The evaluation of technology, cost, and schedule are inextricably intertwined. However, it is easier to describe each element of the overall analysis (e.g., technical, schedule, and cost) separately, noting in each instance the linkages to the overall TRACE evaluation.

Technical Evaluation

The evaluation of technology readiness, risk, and maturity in the TRACE process focuses on identifying the most important technical threats to achieving the necessary mission performance and stated scientific goals. The assessment is limited to a consideration of the top-level technical maturity and risk. Deviations from the current state of the art, system and operational complexity, and integration concerns associated with the use of heritage components are identified. Technical maturity and the need for specific additional development are evaluated by the TRACE team by assessing the readiness levels of key technologies and hardware. Technical risk assessment also included available resource margins for the reference design, resilience of the program architecture, testing challenges, and operational requirements. During the assessment of technology risks and concept maturity, the technical, cost, and schedule teams interact so that technological threats can be translated into schedule and cost risks.

The technical evaluation phase of the TRACE process is limited to the identification of high-level technical risks that could potentially impact schedule and cost. The TRACE process places no cost cap on mission concepts and hence risk as a function of cost is not considered. Concept maturity and technical risk are evaluated by considering the ability of a concept to meet a specified performance with adequate mass, power, and performance margins, given the proposed launch date.

TRACE evaluations also assess proposed mass and power contingencies with respect to technical maturity using AIAA guidelines to achieve a consistent evaluation. If the TRACE technical team concludes that the proposed contingencies are insufficient, they are increased in accordance with historical data on mass and power growth as summarized in the AIAA guidelines. In some cases, growth in mass and power requirements necessitate the selection of a larger launch vehicles to execute the proposed mission. In addition, the need for a more capable launch vehicle—for example, to accommodate potential growth in mass and power requirements—are passed on to the TRACE cost and schedule teams for incorporation into their estimates.

Schedule Evaluation

To aid in the assessment of concept risk, independent schedule estimates are incorporated as part of the TRACE cost estimate. This is especially true for assessment of risk with respect to proposed mission development and execution timelines. Like the TRACE assessment of cost risk, schedule risk is also derived from analogies in the historical NASA record. Historical data from past analogous NASA missions, properly adjusted, are used to gauge the realism of the proposed durations of the development phases. Similarly, the time to critical mission reviews (e.g., PDR and CDR) and the time required for integration and testing are evaluated for each concept and contrasted with appropriate historical experience.

A statistical approach is used to create a schedule probability "S-curve"—that is, a curve of the probability that the development time will exceed some specific value as a function of that value. The overall schedule, as proposed, is then adjusted with the historical data in mind. If the proposed schedule for a particular mission cannot be met, the next available launch window is selected. Additional costs incurred because the proposed schedule cannot be met are then added to the total cost of the mission. The committee requested that the TRACE team use the 70th percentile value in its schedule estimate—that is, there is a 70 percent probability that the schedule will be shorter than indicated and a 30 percent probability that it will be longer.

Cost Evaluation

The primary goal of the TRACE cost evaluation is to provide independent estimates (in fiscal year 2025 dollars) that can be used to prioritize various concepts within the context of the expected NASA budgetary constraints for the coming decade (see Chapter 22). The TRACE team developed high-level cost estimates based on the information provided by the various mission study teams with a focus on treating all projects equally. To be consistent for all concepts, the TRACE cost process allows an increase in cost resulting from increased contingency mass and power, increased schedule, increased required launch vehicle capability, and other cost threats depending on the concept maturity and specific risk assessment of a particular concept.

All cost assessments for the TRACE process are probabilistic in nature and are based on the NASA historical record and documented project lifecycle growth studies. Traditional S-curves of cost probability versus cost are provided for each concept with both the project estimate and the TRACE estimate at the 70th percentile indicated.

The focus of the TRACE costing process is to estimate the cost of conceptual hardware—for example, instruments, spacecraft bus, landers—using multiple analogies and cost models based on historical data (see Figure C-2). A probabilistic cost-risk analysis is employed to estimate appropriate cost reserves. Ensuring consistency across the range of concepts—from those that are immature to those that are significantly more mature—the cost estimates are updated and adjusted with information from the technical TRACE team with respect to mass and power contingencies, technical maturity and development risk, and potentially required additional launch vehicle capability. Using independent schedule estimates, costs are adjusted using appropriate "burn rates" (cost per month) to properly reflect the impact of schedule changes.

Last, the results are integrated, cross-checked, and verified for consistency before being presented to the survey committee.

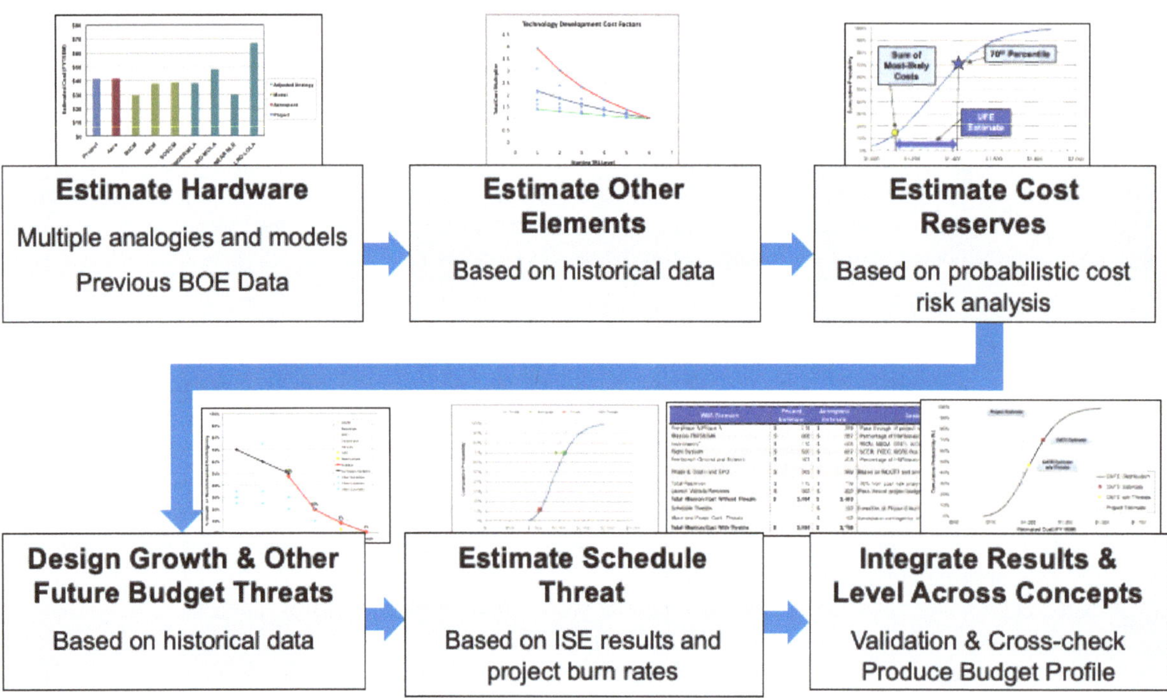

FIGURE C-2 Schematic illustration of the TRACE cost process. The top row represents the basic elements of the cost-evaluation process. Typical of all cost estimates, the TRACE evaluation begins with the hardware costs as defined in WBS 5 (payload), WBS 6 (flight systems), and WBS 10 (integration and test) using multiple analogies and different cost models. Typical cost wraps or percentages of the hardware (WBS 1, 2, 3, 4, 7, and 9) are based on the historical record. Operations cost is estimated with a combination of parametric models and the monthly operations cost of historical analogies. Then each of the WBS cost elements are probabilistically evaluated to produce cost reserves in the form of a typical S-curve. The 70th percentile represents the initial cost estimate without consideration of future cost threats. The lower row represents the evaluation of the additional cost threats, typically not evaluated by the NASA concept study teams, associated with design and schedule growth. Last, all analysis is integrated and leveled against other concept evaluations.

TRACE RESULTS FOR PRIORITY MISSIONS

Summaries of the results of the TRACE evaluations of the 17 priority missions identified by the decadal survey because of their potential to address the 12 key science goals (see Chapters 4 to 15) and potential technical viability presented in Boxes C-1 through C-18. These missions are as follows (not in priority order):

- Mercury Lander (Box C-1);
- Venus Flagship (Box C-2);
- Intrepid: lunar long-range rover traverse (Box C-3);
- Endurance: South Pole–Aitken Basin sample collecting rover (A-variant Box C-4, R-variant Box C-5);
- INSPIRE: In Situ Solar System Polar Ice Roving Explorer (Box C-6);
- Mars In Situ Geochronology (Box C-7);
- Mars Life Explorer (Box C-8);
- Ceres Sample Return (Box C-9);
- Europa Lander (Box C-10);
- Enceladus Orbilander (Box C-11);
- Enceladus Multiple Flyby (Box C-12);
- Titan Orbiter (Sea Probe Descoped) (Box C-13);
- Centaur Orbiter and Lander (Box C-14);
- Uranus Orbiter and Probe (Box C-15);
- Calypso: Uranus Moon and KBO Flyby (Box C-16);
- Odyssey: Neptune Orbiter and Probe (Box C-17); and
- Triton Ocean World Surveyor (Box C-18).

SUMMARY

The TRACE evaluation is a forward-looking technical readiness and budget (cost and schedule) evaluation process typically used to assess pre-Phase A mission concepts. Linked technical, cost, and schedule evaluations were developed for each of the priority mission concepts selected by the committee. The use of historical databases and evaluation of the technical risk, cost, and schedule histories of analogous space systems that have already flown provide a high degree of confidence that the resulting assessments are realistic and credible.

The TRACE-process-derived mission costs are typically higher than the cost estimates provided by mission advocates and design center study teams. The reason for this is that project-derived cost estimates are typically done via a bottoms-up or "grass roots" approach, and beyond standard contingencies, they do not include probabilities of risk incurred by necessary redesigns, schedule slips, and other unforeseen required adjustments. In other words, these estimates typically do not account for the "unpleasant surprises" that historically happen in nearly all space mission developments.

TRACE evaluations include a probabilistic assessment of required reserves, assuming that the concept achieves the mass and power as allocated or constrained by the respective stated project contingencies within the schedule as stated by the project. In addition to these reserves, additional cost threats are also included that quantify potential cost growth based on design maturity (mass and power growth) and schedule growth. Potential cost threats for larger required launch vehicle capability are also included, if required. It is the combination of these reserves and cost threats that are often the main reason for the large differences between the TRACE evaluation and the project estimate, when they occur. Differences in the estimates for hardware costs (instruments and flight systems) can also be a contributing factor.

Cost increases and schedule slippage have plagued spacecraft missions since the dawn of the Space Age. As such, fiscal uncertainties are a significant threat to long-term planning and budget management. Even with the most careful evaluation, the ultimate cost of a spacecraft is poorly constrained until relatively late in its development cycle. As a result, a decadal survey assessing concepts in their earliest phases of development faces a quandary: Throw caution to the wind and accept the assurance of mission advocates that their concept is doable within a specific cost and schedule, or adopt a specific approach to program evaluation. While not a panacea, the TRACE process's use of the history record of "unpleasant surprises" provides the best tool currently available to add a degree of realism to long-term program planning.

BOX C-1
Mercury Lander

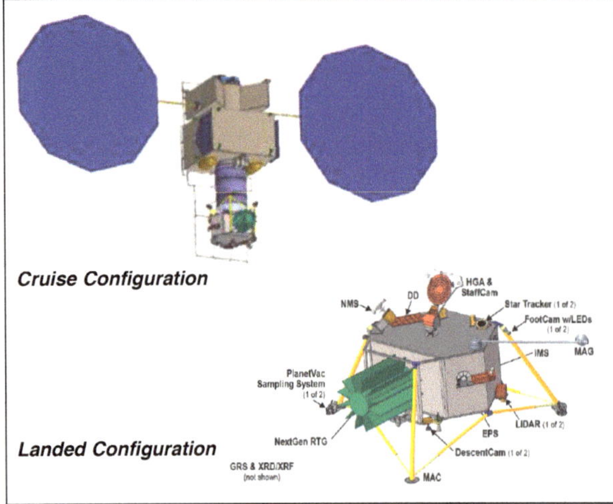

Cruise Configuration

Landed Configuration

Scientific Objectives as Studied

Geochemistry: Investigate the mineralogy and chemistry of Mercury's surface

Geophysics: Characterize Mercury's interior structure and magnetic field

Space Environment: Determine the active processes that produce Mercury's exosphere and alter its regolith

Geology: Characterize the landing site at a variety of scales and provide context for landed measurements

Key Features

Lander (10.5-yr life, 10-yr cruise, 0.2-yr orbital mission, 0.3-yr landed mission)

  Payload: (11 instruments) Gamma-Ray Spectrometer (GRS), X-Ray Diffractometer/X-Ray Fluorescence Spectrometer (XRD/XRF), Magnetometer (MAG), Accelerometer/Short Period Seismometer, Neutral Mass Spectrometer (NMS), Ion Mass Spectrometer (IMS), Energetic Particle Spectrometer (EPS), Dust Detector (DD), Regolith Imagers, Panoramic Imager (StaffCam); Sampling System (PlanetVac)

  Flight System: Next Generation RTG, 4.5 A-hr battery, redundant avionics, X-Band and Ka-Band communications, liquid bipropellant system, sun shield

Descent System: STAR 48 GXV solid rocket motor

Orbital Stage: Liquid bipropellant system with OMAC and CCTCap thrusters, GN&C, 3 m^2 solar array, 60 A-hr battery, sun shield

Cruise Stage: Solar Electric Propulsion with 2× NEXT-C engines, 60 m^2 Ultraflex solar array

Launch: Falcon Heavy Expendable, *nominal launch: March 2035*

Orbit: 100 × 6,000 km orbit at Mercury followed by landing

Key Challenges

Maintaining safe thermal control at close solar distance

Conducting surface operations with limited communications

Providing sufficient power for lander surface operations

System mass multipliers with lander growth

Technical Risk Rating

Medium: Medium new development, adequate to optimistic margins, and/or medium risk of achieving major mission objectives as proposed

Cost-Element Comparison

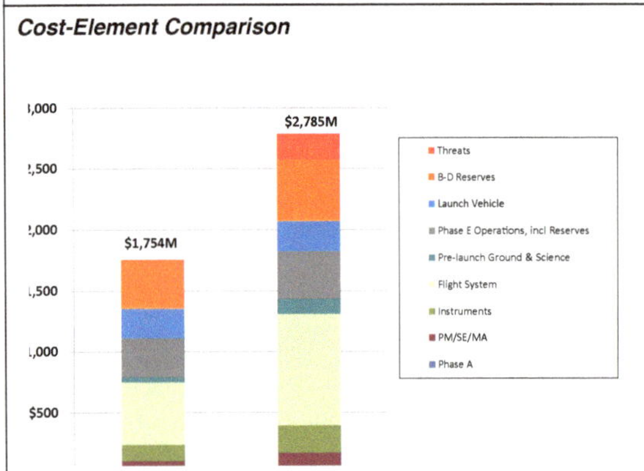

Cost-Risk Analysis S-Curve

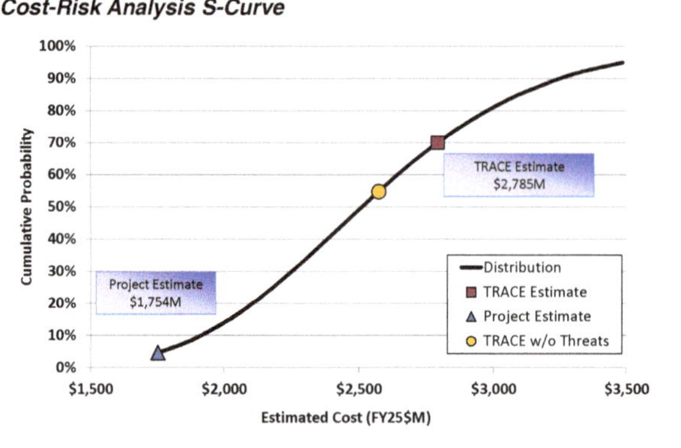

SOURCE: Images in upper-left box from NASA, 2020g, *Mercury Lander: Transformative Science from the Surface of the Innermost Planet*, Mission Concept Study Report for the Planetary Science and Astrobiology Decadal Survey 2023–2032, Columbia, MD: Johns Hopkins University Applied Physics Laboratory, https://science.nasa.gov/solar-system/documents.

BOX C-2
Venus Flagship

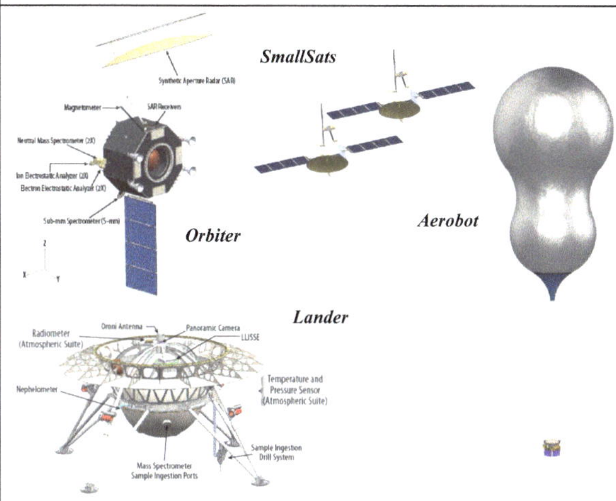

Scientific Objectives as Studied

Habitability: Understand the history of volatiles and liquid water on Venus and determine if Venus has ever been habitable

Determine if Venus once hosted liquid water at the surface
Identify and characterize the origins and reservoirs of Venus's volatiles today
Place constraints on whether there are habitable environments on Venus today and search for organic materials and biosignatures

Climate: Understand the composition and climatological history of the surface of Venus and the present-day couplings between the surface and atmosphere

Constrain surface composition and chemical markers of past and present climate

Geology: Understand the geologic history of Venus and whether Venus is active today

Determine if Venus shows evidence of a current or past plate tectonic regime
Determine whether Venus is tectonically and volcanically active today

Key Features

Orbiter (12.4-yr life)

Payload (7 instruments): SAR, NIR-I, S-mm, Mag, NMS, ESA-I, ESA-e
Flight System: Chemical propellant system, redundant avionics, solar powered

SmallSats ×2 (6-yr life)

Payload (7 instruments): LP, ESA-I, ESA-e, SEPD, Mag, EUV, E-Fd
Flight System: Electric propulsion, solar powered, commercially procured

Lander (6–8-hr mission life Lander, 60-day for LLISSE)

Payload (11 instruments): NMS, TLS, AS, NI, GRS, Nephelometer, XRD, XFS, PC, Raman-LIBS, LLISSE
Flight System: 4-leg lander, 4-fan divert system, drill, and sample handling system

Aerobot (60-day mission life)

Payload (5 instruments): AMS-N, FM, Mag, MET, VI
Flight System: Variable altitude (52–62 km) helium balloon system, solar powered

Launch

Falcon Heavy, *nominal launch: 2031*

Key Challenges

Lander design for hazards in challenging landing environment

Lander sample collection and handling at Venus surface with mission time constraints

Advanced small satellite development for multi-year Venus mission

Mission complexity for technology development and coordinated operations

Technical Risk Rating

Medium-High: Significant new development, optimistic to negative margins, and/or significant risk of achieving major mission objectives

Descope Options

Deleting the Aerobot and the two SmallSats reduces the estimated cost to ~$5,700 million

Deleting the Lander and the two SmallSats reduces the estimated cost to ~$3,800 million

Cost-Element Comparison

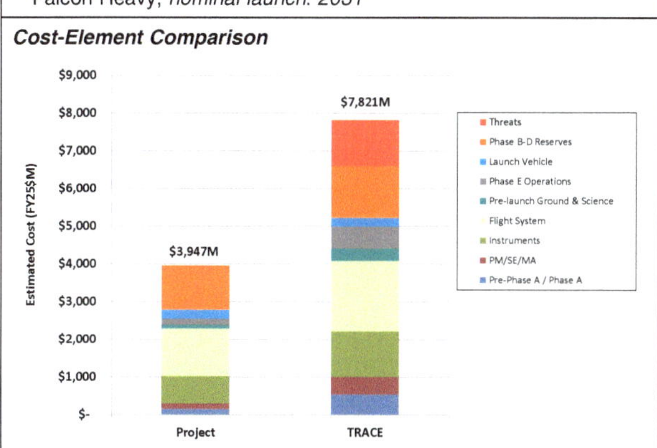

Cost-Risk Analysis S-Curve

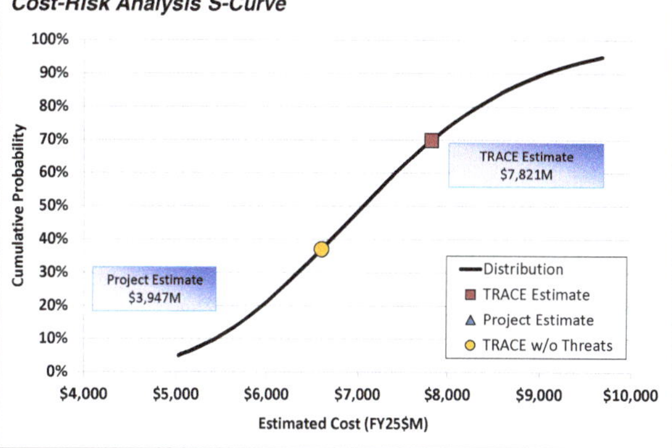

SOURCE: Images in upper-left box from NASA, 2020k, *2020 Venus Flagship Mission Study: A Mission to Explore the Habitability of Venus and the Origins of Earth-Sized Planets Both Near and Far*, Mission Concept Study Report for the Planetary Science and Astrobiology Decadal Survey 2023–2032, Greenbelt, MD: NASA Goddard Space Flight Center, https://science.nasa.gov/solar-system/documents.

BOX C-3
Intrepid—Long-Range Lunar Rover

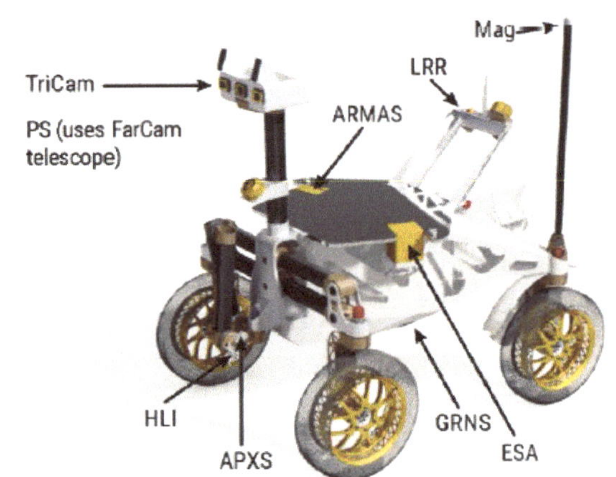

Scientific Objectives as Studied

Evolution of the Lunar Interior and Nature of the Procellarum KREEP Terrane:

Determine the cause of extended volcanism in the Procellarum region
Determine the cause of the lunar crustal asymmetry
Test hypotheses for the origin of non-basaltic volcanism
Determine the composition of deep mantle pyroclastic deposits
Determine decline of core dynamo and magnetic field over time

Diversity of Styles of Magmatism:

Characterize flood basalt emplacement, rilles, flows, and vents
Determine origin and composition of cones, domes, and shields
Characterize pyroclastic volcanism processes: composition and physical state
Determine the relationship between intrusive and effusive materials

Post-Emplacement Modification of Magmatic Materials:

Test hypotheses of impact crater formation, ballistic sedimentation, ray formation
Determine target material influence on impact crater formation
Determine the causes of magnetic anomalies, swirls, and space weathering

Key Features

Rover (4-yr life)

Payload: (8 instruments + retroreflector) TriCam (pair of color stereo imagers and single monochromatic narrow angle camera), Point Spectrometer (PS), Hand Lens Imager (HLI), Alpha Particle X-ray Spectrometer (APXS), Magnetometer (MAG), Gamma Ray and Neutron Spectrometer (GRNS), Automated Radiation Measurements for Aerospace Safety (ARMAS), Electrostatic Analyzer (ESA), Laser Retroreflector (LRR)

Robotic Arm: 5-DOF robotic arm for placement of APXS and HLI

Rover: 4-wheeled rover with passive suspension and dual-sided rocker, system powered by Mod 2 NGRTG (12 GPHS) and battery, Direct to Earth communications in S-band, Warm Electronics Box for avionics, combination of aluminum and carbon fiber structure

Launch Delivered via: NASA Commercial Lunar Payload Services (launch and medium-class lander), *nominal launch: April 2030*

Destination: Land near Lunar Reiner Gamma formation and traverse to Irregular Mare Patch near Aristarchus Crater

Key Challenges

Availability of NGRTG Mod 2 in time for mission

Long-range traverse with high degree of autonomy

Impact of dust on rover performance over time

Technical Risk Rating

Medium-Low: Moderate new development, adequate margins, and/or moderate risk to achieve major mission objectives as proposed

Cost-Element Comparison

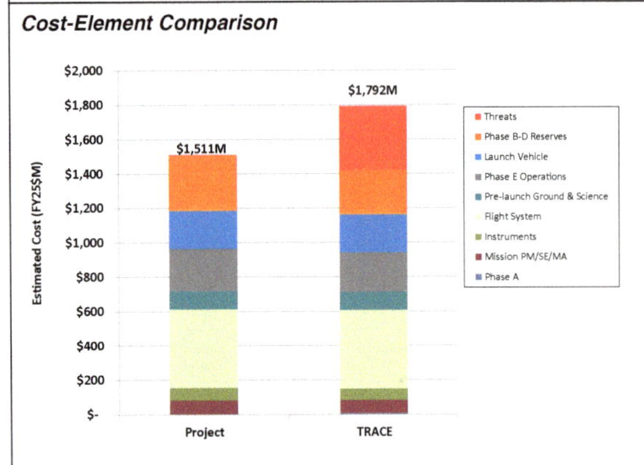

Cost-Risk Analysis S-Curve

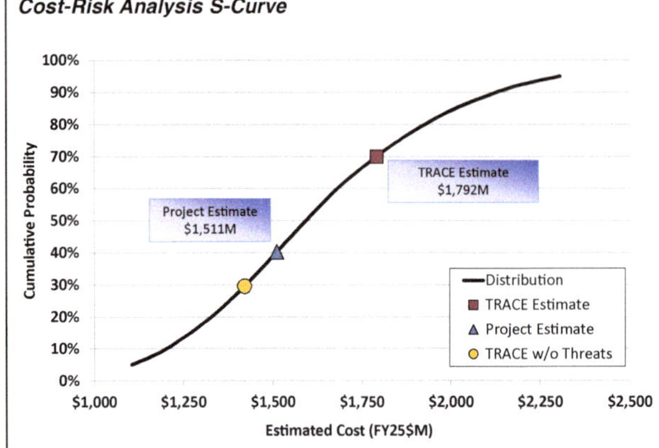

SOURCE: Image in upper-left box from NASA, 2020c, *Intrepid Planetary Mission Concept Study Report*, Mission Concept Study Report for the Planetary Science and Astrobiology Decadal Survey 2023–2032, Pasadena, CA: Jet Propulsion Laboratory, California Institute of Technology, https://science.nasa.gov/solar-system/documents.

BOX C-4
Endurance-A, South Pole–Aitken Basin Sample Collecting Rover (Astronaut Return)

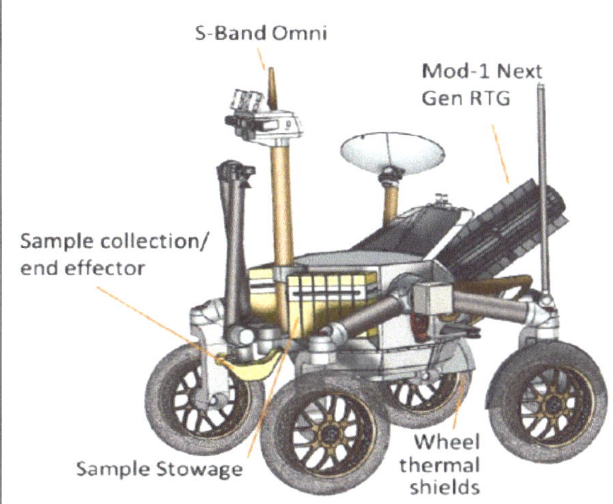

Scientific Objectives as Studied

Solar System Chronology:

Anchor the earliest impact history of the solar system by determining the age of the largest and oldest impact basin on the Moon, the South Pole–Aitken basin

Test the giant planet instability, impact cataclysm, and late heavy bombardment hypothesis by determining when large far side lunar impact basins formed

Anchor the "middle ages" of solar system chronology (between 1 and 4 billion years ago) by determining the absolute age of a cratered, far side lunar mare basalt

Planetary Evolution:

Test the magma ocean paradigm and characterize the thermo-chemical evolution of terrestrial planets by determining the age and nature of volcanic features and compositional anomalies on the far side of the Moon

Explore a giant impact basin from floor to rim by characterizing the geologic diversity across the South Pole–Aitken basin

---

Key Features

Rover: Four-wheeled rover with passive suspension and dual-sided rocker, system powered by Mod 1 NGRTG and battery, comm uplink/downlink to/from commercial relay in S-band, Warm Electronics Box for avionics, combination of aluminum and carbon fiber structure. (4-yr life)

Payload: (8 instruments + retroreflector) TriCam (pair of color stereo imagers and single monochromatic narrow angle camera), Point Spectrometer (PS), Hand Lens Imager (HLI), Alpha Particle X-ray Spectrometer (APXS), Magnetometer (MAG), Gamma Ray and Neutron Spectrometer (GRNS), Automated Radiation Measurements for Aerospace Safety (ARMAS), Electrostatic Analyzer (ESA), Laser Retroreflector (LRR)

Sampling System: 5-DOF robotic arm with scoop and sieve end effector, side-mounted cache capable of storing 100 kg of samples in 12 astronaut-removable containers

Launch: Delivered via NASA Commercial Lunar Payload Services (launch and medium-class lander), *nominal launch: April 2030*

Destination: Lunar South Pole–Aitken Basin

Key Challenges

Long-range traverse with high degree of autonomy

Limit of cargo mass on commercial lunar landers

Impact of dust on rover performance over time

Reliance on uncertain commercial lunar communications relay service

Availability of Mod 1 NGRTG in time for mission

Technical Risk Rating

Medium: Medium new development, adequate to optimistic margins, and/or medium risk of achieving major mission objectives as proposed

---

Cost-Element Comparison

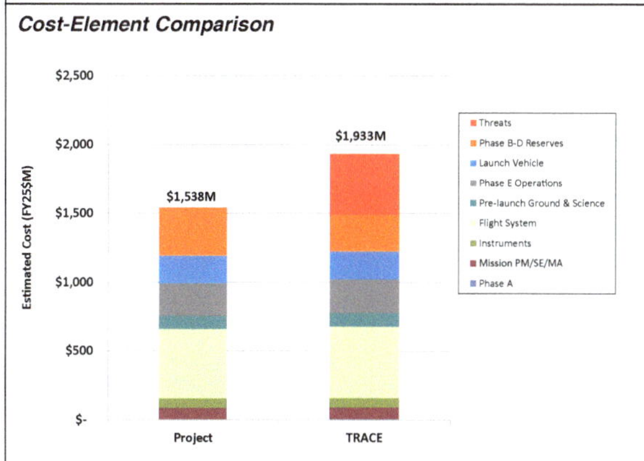

Cost-Risk Analysis S-Curve

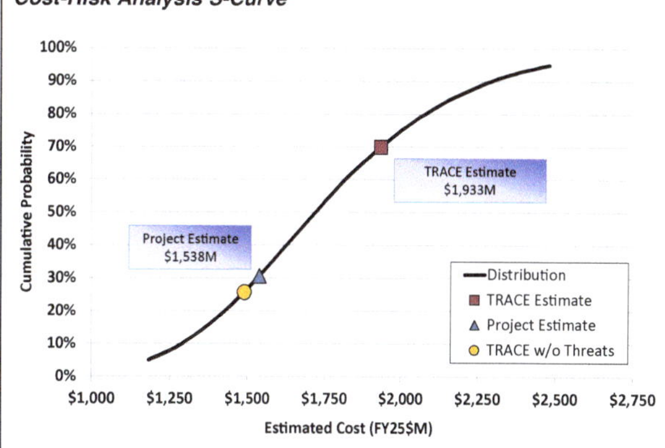

SOURCE: Image in upper-left box from NASA, 2021f, *Endurance: Lunar South Pole–Aitken Basin Traverse and Sample Return Rover*, Mission Concept Study Report for the Planetary Science and Astrobiology Decadal Survey 2023–2032, Pasadena, CA: Jet Propulsion Laboratory, California Institute of Technology, https://tinyurl.com/2p88fx4f.

BOX C-5
Endurance-R—South Pole–Aitken Basin Sample Collecting Rover (Robotic Return)

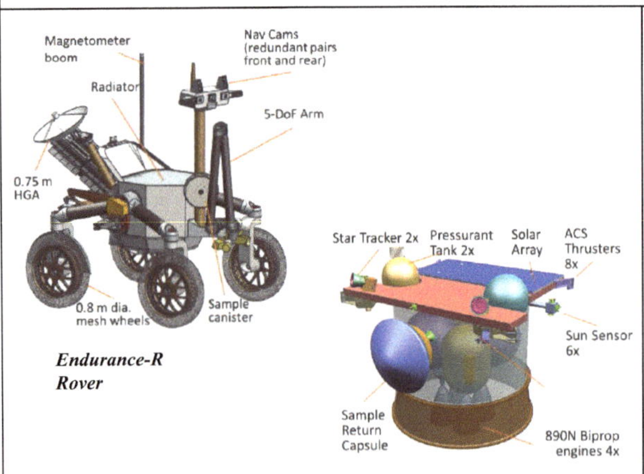

Scientific Objectives as Studied

Solar System Chronology:

Anchor the earliest impact history of the solar system by determining the age of the largest and oldest impact structure on the Moon: South Pole–Aitken basin

Test the giant planet instability, impact cataclysm, and late heavy bombardment hypotheses by determining when large far side lunar impact basins formed

Anchor the "middle ages" of solar system chronology (between 1 and 4 billion years ago) by determining the absolute age of a cratered, far side lunar mare basalt

Planetary Evolution:

Test the magma ocean paradigm and characterize the thermochemical evolution of terrestrial planets by determining the age and nature of volcanic features and compositional anomalies on the far side of the Moon

Explore a giant impact basin from floor to rim by characterizing the geologic diversity across the South Pole–Aitken basin

Key Features

Rover (4-yr life)

Payload: (8 instruments + retroreflector) TriCam (pair of color stereo imagers and single monochromatic narrow angle camera), Point Spectrometer (PS), Hand Lens Imager (HLI), Alpha Particle X-ray Spectrometer (APXS), Magnetometer (MAG), Gamma Ray and Neutron Spectrometer (GRNS), Automated Radiation Measurements for Aerospace Safety (ARMAS), Electrostatic Analyzer (ESA), Laser Retroreflector (LRR)

Sampling System: 5-DOF robotic arm with scoop and sieve end effector, sample canister with 12 separate chambers with build-in lids for holding 2.2 kg of samples

Rover: 4-wheeled rover with passive suspension and dual-sided rocker, system powered by Mod 1 NGRTG and battery, comm uplink/downlink to/from commercial relay in S-band, Warm Electronics Box for avionics, combination of aluminum and carbon fiber structure

Earth-Return Vehicle: (7-mo life) Sample Return Capsule (SRC), sample transfer arm, 2536 m/s delta V biprop, structure, GNC, avionics, 1.55 m² array and Li-Ion battery

Launch: Delivered via 2 × NASA Commercial Lunar Payload Services (launch and medium-to large class landers), *nominal launch: April 2030* (Rover), *2034* (ERV)

Destination: South Pole–Aitken Basin

Key Challenges

Limit of cargo mass on available commercial lunar landers

Long-range traverse with high degree of autonomy

Impact of dust on rover performance over time

Reliance on uncertain commercial lunar communications relay service

Availability of NGRTG in time for mission

Technical Risk Rating

Medium: Medium new development, adequate to optimistic margins, and/or medium risk of achieving major mission objectives as proposed

Cost-Element Comparison

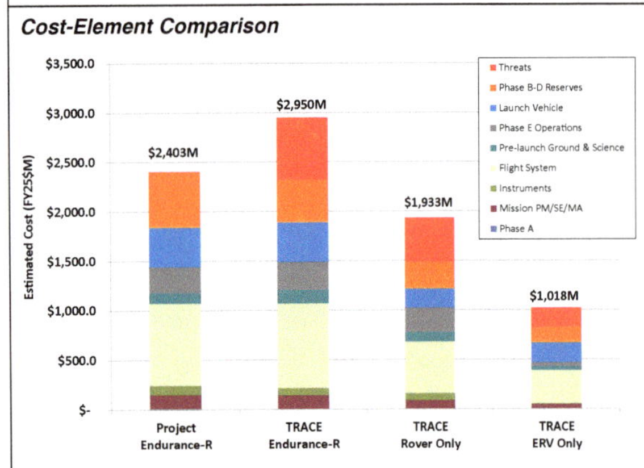

Cost-Risk Analysis S-Curve

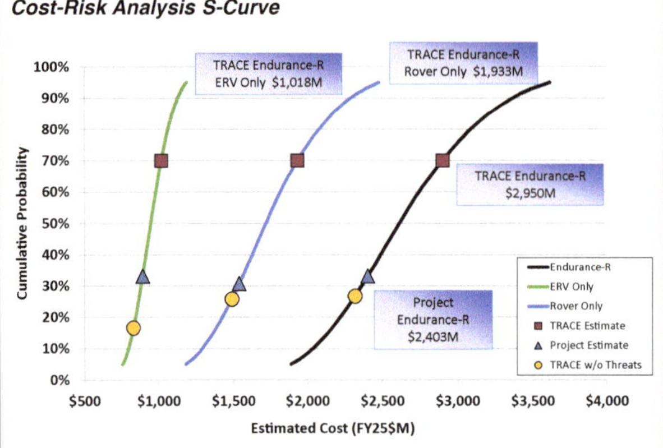

SOURCE: Images in upper-left box from NASA, 2021f, *Endurance: Lunar South Pole–Aitken Basin Traverse and Sample Return Rover*, Mission Concept Study Report for the Planetary Science and Astrobiology Decadal Survey 2023–2032, Pasadena, CA: Jet Propulsion Laboratory, California Institute of Technology, https://tinyurl.com/2p88fx4f.

BOX C-6
INSPIRE, Lunar Polar Volatiles Rover

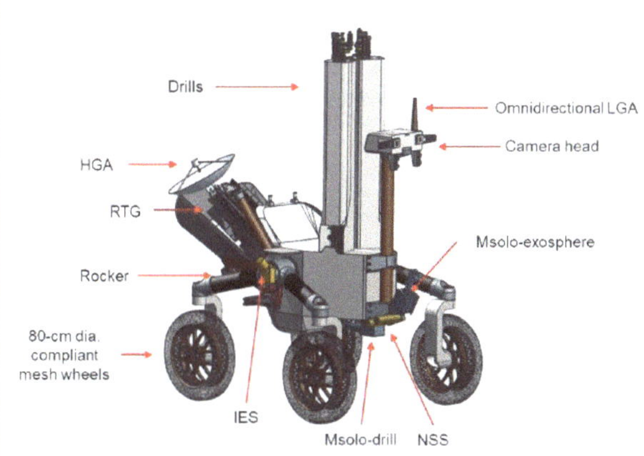

Scientific Objectives as Studied

Origins: Determine the origins of volatiles in the inner solar system

Determine the abundance and distribution of lunar volatiles including water by measuring sulfur-bearing molecules, isotopic ratios, carbon-based molecules, and D/H ratios

Use volatile distributions (lateral and vertical) and the physical form of volatiles to distinguish between sources including early outgassing, asteroid impacts, comet impacts, solar wind–regolith interactions, and ongoing small meteoroid bombardment

Ages: Evaluate the timescales of volatile delivery in the inner solar system

Determine the age of lunar volatiles by measuring the form and distribution and comparing with past and present-day environmental conditions

Evolution: Assess how volatiles evolve on solar system airless bodies

Determine how lunar volatiles have evolved over time by measuring the distribution of volatiles and correlating with environmental factors and geological context

Key Features

Rover (3-yr life)

Payload: (7 instrument types) Near IR Spectrometer (NIRVSS), Neutron Spectrometer (NSS), Mass Spectrometer (MSolo, Qty 2), Ion Electron Sensor (IES), Ground Penetrating Radar (Mini-GPR), Thermal IR Spectrometer (TIRS), Cameras

Drill System: 2 m TRIDENT drill, including control electronics

Rover: 4-wheeled rover with passive suspension and dual-sided rocker, system powered by Mod 1 NGRTG and battery, comm uplink/downlink to/from commercial relay in S-band, Warm Electronics Box for avionics, combination of aluminum and carbon fiber structure

Launch: Delivered via NASA Commercial Lunar Payload Services (launch and medium-class lander), *nominal launch: April 2030*

Destination: Land at Cabeus crater and explore five permanently shadowed regions

Key Challenges

Long-range traverse with high degree of autonomy

Limit of cargo mass on commercial lunar landers

Impact of dust on rover performance over time

Reliance on uncertain commercial lunar communications relay service

Availability of Mod 1 NGRTG in time for mission

Technical Risk Rating

Medium: Medium new development, adequate to optimistic margins, and/or medium risk of achieving major mission objectives as proposed

Cost-Element Comparison

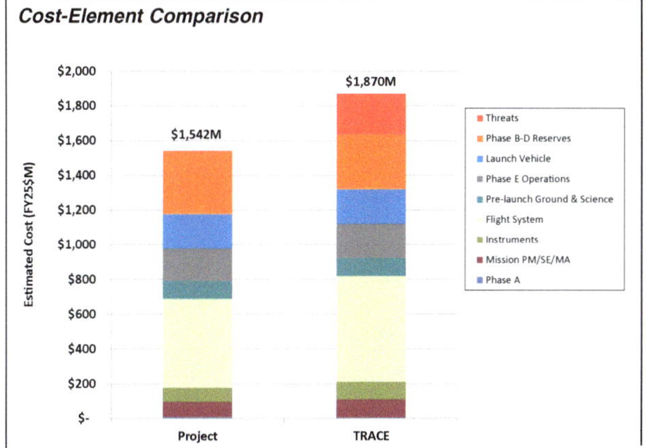

Cost-Risk Analysis S-Curve

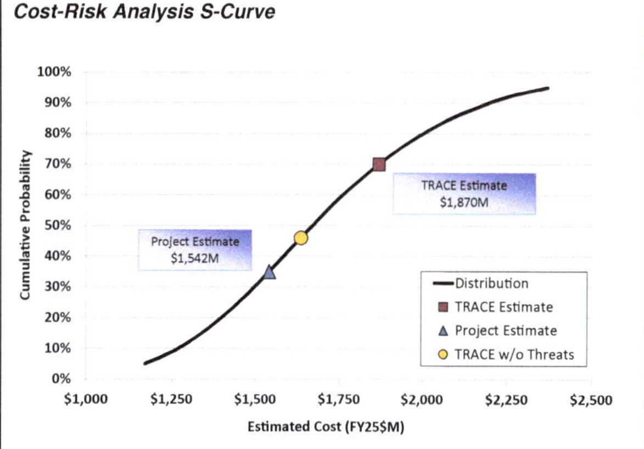

SOURCE: Image in upper-left box from NASA, 2021g, *INSPIRE: IN Situ Solar System Polar Ice Roving Explorer*, Mission Concept Study Report for the Planetary Science and Astrobiology Decadal Survey 2023–2032, Pasadena, CA: Jet Propulsion Laboratory, California Institute of Technology, https://tinyurl.com/2p88fx4f.

BOX C-7
Mars In Situ Geochronology

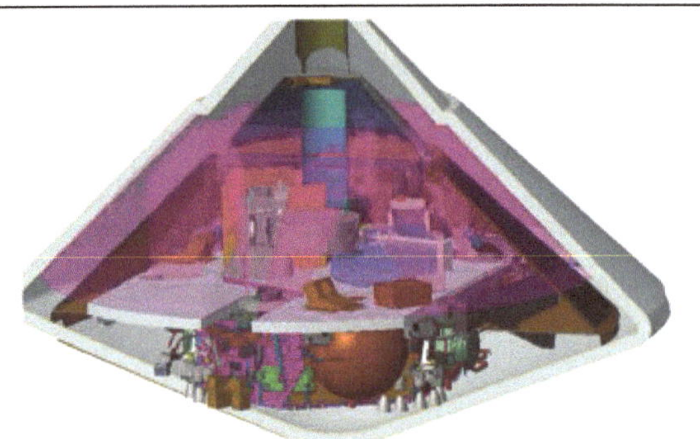

Mars In-Situ Geochronology Lander in its Aeroshell

Scientific Objectives as Studied

Determine the chronology of basin-forming impacts to constrain the time period of heavy bombardment in the inner solar system and thus address fundamental questions related to inner solar system impact processes and chronology

Reduce the uncertainty for inner solar system chronology in the "middle ages" (1–3 Ga) to improve models for planetary evolution, including volcanism, volatiles, and habitability

Establish the history of habitability across the solar system

Calibrate the body-specific chronology for Mars

Key Features

Lander (40-mo life: 27.5-mo cruise, 11-mo data collection, 1.5-mo data downlink)

Science Payload:
CDEX—Chemistry and Dating Experiment
KArLE—Potassium-Argon Laser Experiment
ICPMS—Inductively Coupled Plasma Mass Spectrometer
UCIS—Ultra-Compact Imaging Spectrometer
Panoramic Imagers (×2)
MicroImager

Sampling/Engineering:
PlanetVac sampling system and triage station for sample

Lander Bus: 5.1 m^2 Ultraflex array, 2×25 Ahr Li Ion batteries, Hydrazine propulsion, redundant avionics, UHF and X-Band communications, GNC

Entry System: Enlarged aeroshell and backshell based on InSight and Viking parachute

Cruise Stage: X-Band communications, 5 m^2 fixed solar array, star trackers

Launch: Atlas V 411, *nominal launch: July 2030*

Destination: Mars Surface, 21 deg N latitude

Key Challenges

Lander mass increase impact on design of the entry, descent, and landing system

Lander energy margin toward end of science operations

Uncertainty in development instrumentation for this mission application

Technical Risk Rating

Medium-Low: Moderate new development, adequate margins, and/or moderate risk to achieve major mission objectives as proposed

Cost-Element Comparison

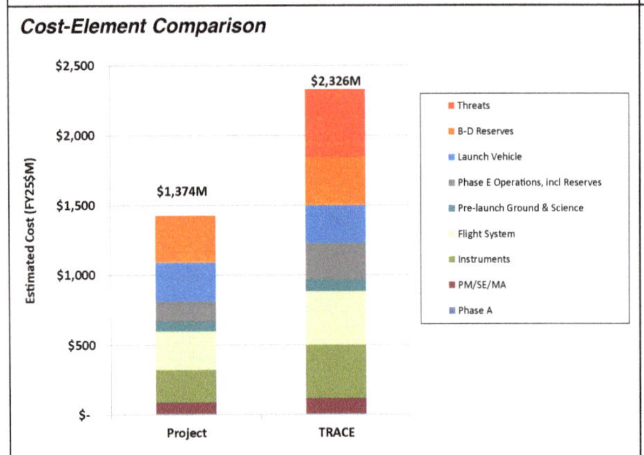

Cost-Risk Analysis S-Curve

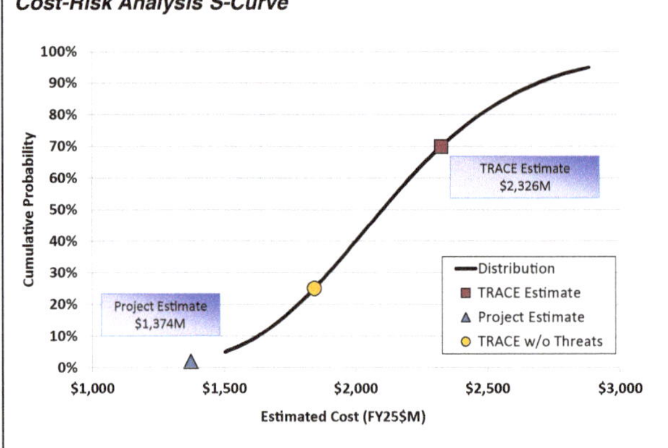

SOURCE: Image in upper-left box from NASA, 2020b, *In Situ Geochronology for the Next Decade*, Final Report Submitted in Response to NNH18ZDA001N-PMCS: Planetary Mission Concept Studies, Greenbelt, MD: NASA Goddard Space Flight Center, https://science.nasa.gov/solar-system/documents.

BOX C-8
Mars Life Explorer

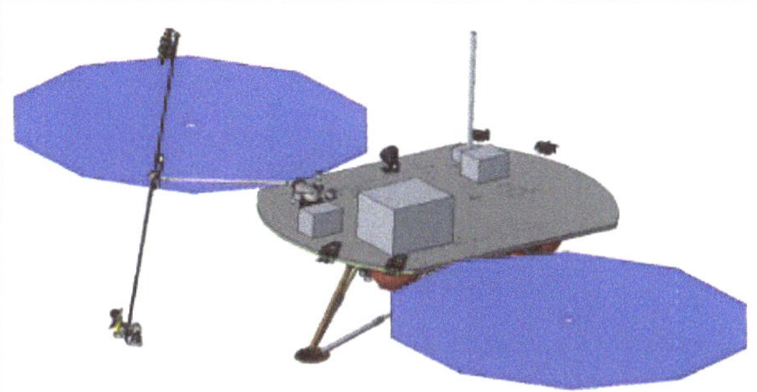

Mars Life Explorer with Drill Deployed

Scientific Objectives as Studied

Search for organic molecules, non-equilibrium gases, and isotopes associated with ice and regolith, and evaluate their possible biological origin

Assess the habitability of the near-subsurface environment with respect to required elements to support microbial life, microbial energy sources, and compounds toxic to microbes

Quantify the down borehole thermophysical properties of the ice/ice-cemented regolith and any role for liquid water in its creation or modification

Determine the processes that preserve/modify/destroy these ice deposits in the modern climate

Key Features

Lander (34-mo life: 10-mo cruise, 3-mo sample, 21-mo meteorology only)

Science Payload:
Biosignature Detection Suite (DRaMS w/EGA + Mini-TLS)
Chemistry, Mineral, Conductivity Suite (CheMinX + MECA + EGA)
Wind, Water, Radiative Flux Suite (Anemometer + TLS + MEDA temp/pres/TIRS)
Downhole Sensor & Imager

Sampling/Engineering:
Single-segment 2m Drill, 3 DOF Drill Boom, Drill Avionics Box, Biobarrier
Gas Sample Transfer System, Context Camera

Lander Bus: 2×5.8 m^2 Ultraflex array, 80 Ahr Li Ion battery, Hydrazine propulsion, redundant avionics, UHF and X-Band communications, GNC

Entry System: 3.65 m diameter aeroshell, backshell

Cruise Stage: X-Band communications, 5 m^2 fixed solar array, star trackers

Launch: Falcon Heavy Recoverable, *nominal launch: May 2035*

Destination: Martian surface, 40–50° north latitude

Key Challenges

Challenge accessing near-surface ice with lander accuracy and drill mobility

Lander mass increase beyond design limits of the entry, descent, and landing system

Design variation due to science payload suite uncertainty

Impact to power subsystem sizing from uncertainty in payload and concept of operations

Technical Risk Rating

Medium-Low: Moderate new development, adequate margins, and/or moderate risk to achieve major mission objectives as proposed

Cost-Element Comparison

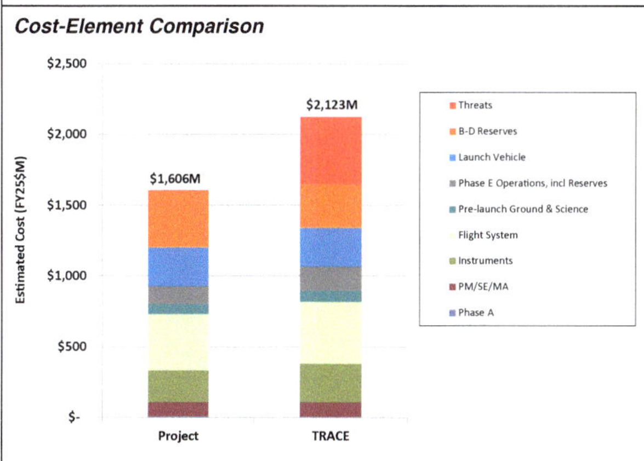

Cost-Risk Analysis S-Curve

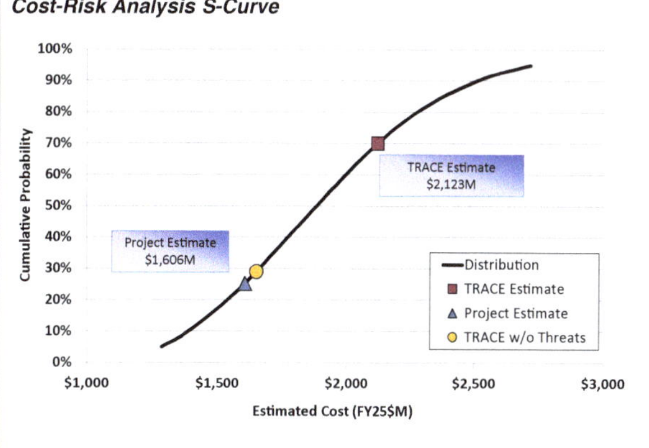

SOURCE: Image in upper-left box from NASA, 2021h, *Mars Life Explorer*, Mission Concept Study Report for the Planetary Science and Astrobiology Decadal Survey 2023–2032, Pasadena, CA: Jet Propulsion Laboratory, California Institute of Technology, https://tinyurl.com/2p88fx4f.

BOX C-9
Ceres Sample Return

Ceres Orbiter/Lander – Landed Configuration

Scientific Objectives as Studied

Test if extrusion from a brine-rich mantle occurred during Ceres's recent history

Test if endogenic activity is ongoing at Occator crater

Determine the depth of liquid water below Occator crater

Characterize Ceres's deep brine environment at Occator crater

Characterize the evolution of organic matter in long-lived brines

Determine Ceres's accretional environment

Key Features

Hybrid Orbiter/Lander (13.5-yr life, 6.3-yr outbound cruise, 1.4-yr orbital mission, 3-wk landed mission, 5.75-yr inbound cruise)

Payload:
Narrow Angle Camera: Imaging at pixel scale of 1 m from 100 km altitude; includes image compression
Magnetotelluric Sounder: Determines depth-dependent electrical conductivity of the subsurface from frequency-dependent magnetic and electric fields
Infrared Point Spectrometer: Miniature spectrometer covering 2–4 micron range
Sampling System: PlanetVac sample collector, sample transfer system

Flight System: 3× NEXT thruster electric propulsion, Hydrazine propellant system, powered by 95 m^2 roll-out solar array (ROSA) (two-axis gimballed) and 126 A-hr Li Ion battery, X-band communications, GNC with Terrain Relative Navigation, redundant avionics, Sample Return Capsule (SRC)

Launch: Falcon Heavy Recoverable, *nominal launch: Dec 2030*

Orbit: 275 km and 28 km orbit at Ceres followed by landing, ascent, and return to Earth

Key Challenges

New use of large ROSA for hybrid orbiter/lander mission

Lack of definition for sample handling and preservation

Uncertainty in sample collection requirements and approach

Lack of definition for surface energy generation and usage

Technical Risk Rating

Medium: Medium new development, adequate to optimistic margins, and/or medium risk of achieving major mission objectives as proposed

Cost-Element Comparison

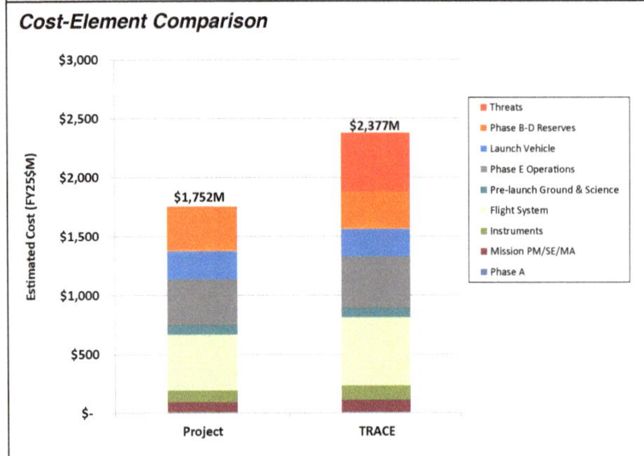

Cost-Risk Analysis S-Curve

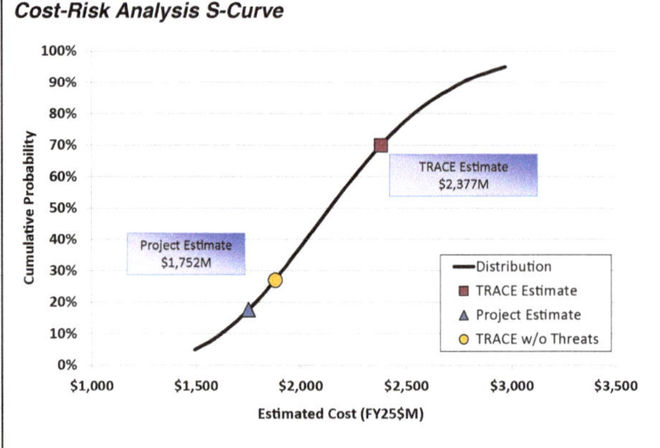

SOURCE: Image in upper left hand box from NASA, 2020h, *Mission Concept Study: Ceres: Exploration of Ceres' Habitability.* Mission Concept Study Report for the Planetary Science and Astrobiology Decadal Survey 2023–2032. Pasadena, CA: Jet Propulsion Laboratory, California Institute of Technology. https://science.nasa.gov/solar-system/documents.

BOX C-10
Europa Lander

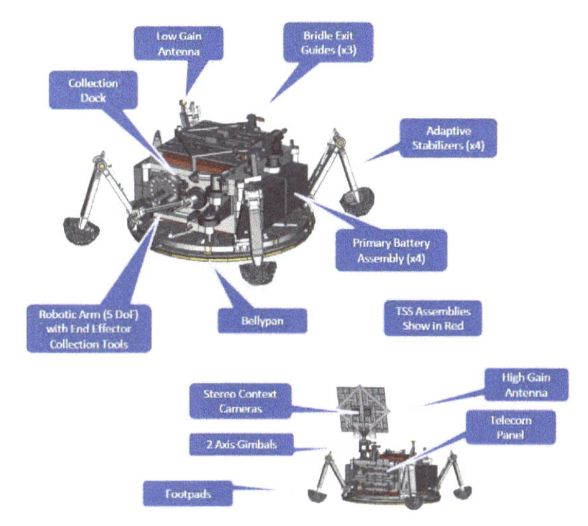

Scientific Objectives as Studied

Biosignatures: Search for evidence of biosignatures on Europa

Identify potential biosignatures through a minimum of 9 lines of evidence: organic abundance, organic patterns, chirality, isotopes, microscale structures, macroscale structures, cellular properties, and biominerals

Habitability: Assess the habitability of Europa via in situ techniques

Characterize non-ice composition of Europa's near-surface material to determine whether there are environmental factors essential for life

Determine the proximity to liquid water and recently erupted materials at the lander's location

Surface Properties and Dynamics: Characterize the surface and subsurface of Europa

Observe properties of surface materials and connect local properties with those seen from remote sensing

Characterize dynamic processes of Europa's surface and ice shell over the mission duration

Key Features

Carrier Stage (7-yr life)

Flight System: Chemical propellant system, 127 m2 solar array, X-band communications, radiation vaults to protect avionics, radiation TID 1.5 Mrad, Bio Barrier for Lander

Deorbit Stage

Flight System: Solid rocket motor with 5,000–5,700 kN-s total impulse, reference design based on Star-48 motor, system includes thermal and separation hardware

Descent Stage:

Flight System: Hydrazine monopropellant system with 410 kg propellant, redundant avionics and GNC with Terrain Relative Navigation (TRN) including LIDAR

Europa Lander: (22-day surface mission)

Payload (5 instruments): Organic Composition Analyzer, Vibrational Spectrometer, Microscope for Life Detection, Context Remote Sensing Imager, Geophysical Sounding System

Flight System: Sample collection system, sterilization subsystem, Direct to Earth communications, primary battery, vault-enclosed avionics, radiation TID 2.0 Mrad

Launch: SLS Block 1B, *nominal launch: Nov 2026*

Key Challenges

Lander science operation within energy and thermal constraints

Landing safely on uneven and uncertain terrain

Long-range direct to Earth downlink of science data from lander

Lander and instrument contamination control for biosignatures science

System mass multipliers with hardware growth

Technical Risk Rating

Medium: Medium new development, adequate to optimistic margins, and/or medium risk of achieving major mission objectives as proposed

Cost-Element Comparison

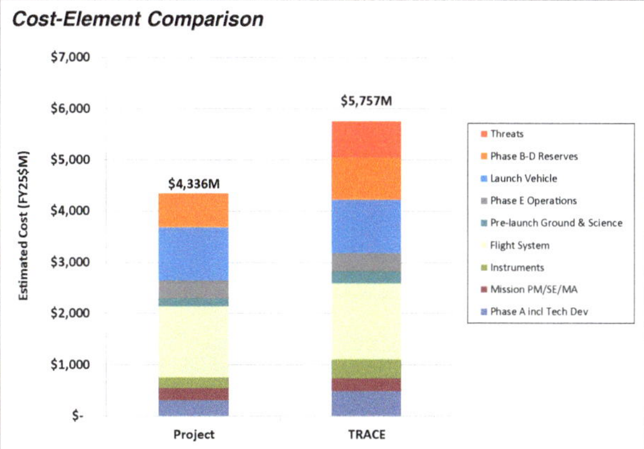

Cost-Risk Analysis S-Curve

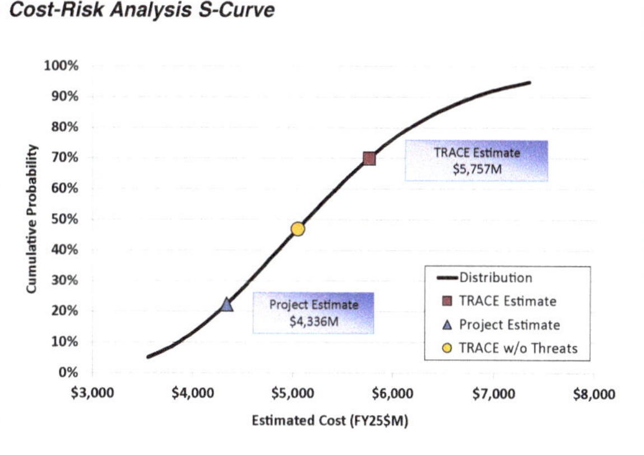

SOURCE: Images in upper-left box from NASA, 2017b, *Report of the Europa Lander Science Definition Team*, Washington, DC, https://europa.nasa.gov/resources/58/europa-lander-study-2016-report.

BOX C-11
Enceladus Orbilander

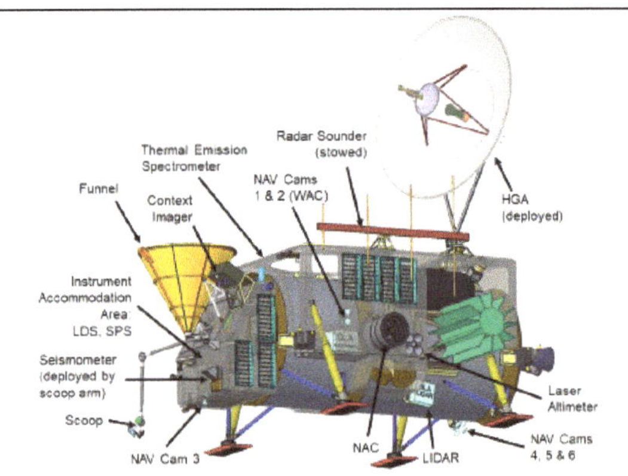

Deployed Orbilander

Scientific Objectives as Studied

Determine If Enceladus Is Inhabited: Search for biosignatures
Characterize amino acids and lipids
Search for a polyelectrolyte and any cell-like morphologies

Assess to What Extent Enceladus's Ocean Is Able to Sustain Life and Why: Provide geochemical and geophysical context for life detection
Quantify physical and chemical environment
Determine internal structure, vent structures

Find Locations to Land and Actively Sample: Balance science, safety, and planetary protection
Find a scientifically compelling landing site with a sufficiently high plume fallout rate
Perform reconnaissance for both safe landing and active sampling
Find sites with adeqVuate surface temperature to avoid melt through ice crust

Key Features

Orbilander (10.5-yr life, 7-yr cruise, 1.5-yr orbital mission, 2-yr landed mission)

Payload:
Life Detection: High Resolution Mass Spectrometer (HRMS), Separation Mass Spectrometer (SMS), Electrochemical Sensor Array (ESA), Microcapillary Electrophoresis-Laser Induced Fluorescence (mCE-LIF), Microscope, Nanopore Sequencer,
Remote Sensing: Radar Sounder, Thermal Emission Spectrometer (TES), Laser Altimeter, Narrow Angle Camera, Wide Angle Camera,
In Situ: Seismometer, Context Imager
Sampling System: Passive Collector, Active Collector, Sample Transfer and Processing Unit

Flight System: Chemical bi-propellant system, powered by 2x NGRTGs with battery, X-band/Ka-band Direct to Earth communications, GNC with Terrain Relative Navigation (TRN), redundant avionics

Launch:
SLS Block 2B with Castor 30B upper stage, nominal launch: Oct 2038
Falcon Heavy Expendable (FHE), nominal launch: September 2037, backup December 2038 (adds roughly 2 years to cruise phase for alternate trajectory)

Orbit: 12-hr halo orbit at Enceladus

Key Challenges

Complexity of TRN + pitch-over landing strategy

Lander/Instrument contamination control for life detection

Orbilander sample collection and handling

Lander/Instrument concept growth in mass and complexity

Technical Risk Rating

Medium: Medium new development, adequate to optimistic margins, and/or medium risk of achieving major mission objectives as proposed

Cost-Element Comparison

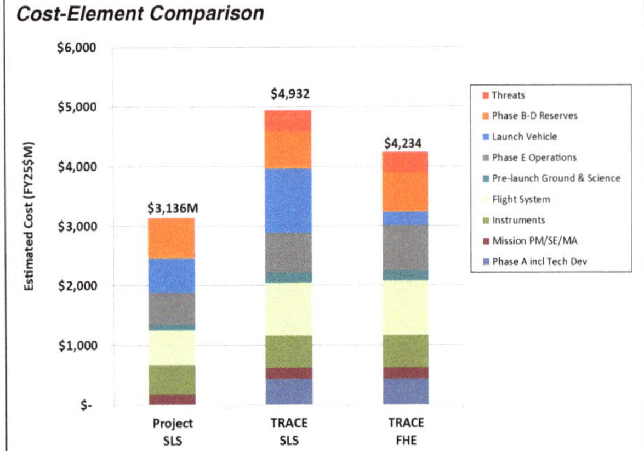

Cost-Risk Analysis S-Curve

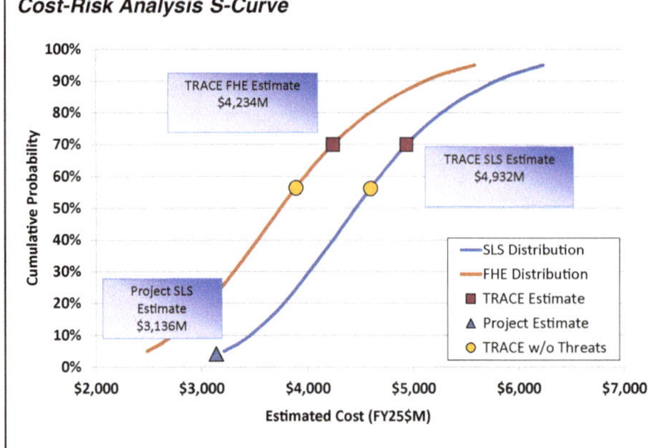

SOURCE: Image in upper-left box from NASA, 2020a, *Enceladus Orbilander: A Flagship Mission Concept for Astrobiology,* Mission Concept Study Report for the Planetary Science and Astrobiology Decadal Survey 2023–2032, Columbia, MD: Johns Hopkins University Applied Physics Laboratory, https://science.nasa.gov/solar-system/documents.

BOX C-12
Enceladus Multiple Flyby

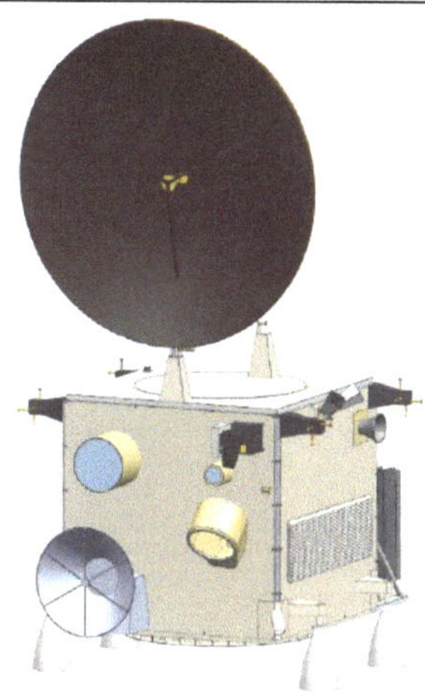

Scientific Objectives as Studied

Search for Signs of Life in Enceladus Plume Materials:

Search for multiple features of life (biosignatures)

Multiple, independent measurements of molecular qualities in organic compounds

Assess the Habitability of the Enceladus Ocean:

Quantitative measurements of key habitability parameters including sources of essential elements and micronutrients, sources of chemical energy, and key physiochemical parameters

Characterize Enceladus's Cryovolcanic Activity:

Observe details of the structure of Enceladus plume and how it varies in space and time, including a more precise estimation of the relative contributions of jets and curtains to the overall plume

---

Key Features

Flyby Vehicle (12-yr life, 9-yr cruise, 3-yr repeat flyby mission)

Payload:
Organic Chemical Analyzer (OCA): Two instruments
Inorganic Chemical Analyzer (ICA)
Narrow Angle Camera
Sampling Handling System: Collects icy grains from Enceladus plume and delivers to OCA and ICA instruments

Flight System: Chemical bi-propellant system, powered by 2x NGRTGs with battery, Ka-band communications, 3-axis controlled GNC, redundant avionics

Launch: Falcon Heavy Expendable, *nominal launch: Oct 2038*

Orbit: Multiple flybys at >50 km and <50 km altitude at Enceladus

Key Challenges

NGRTG performance at required power output for mission

Flyby plume sample collection and handling system development

Orbilander sample collection and handling

Instrument contamination control for biosignature detection

Technical Risk Rating

Medium-Low: Moderate new development, adequate margins, and/or moderate risk to achieve major mission objectives as proposed

---

Cost-Element Comparison

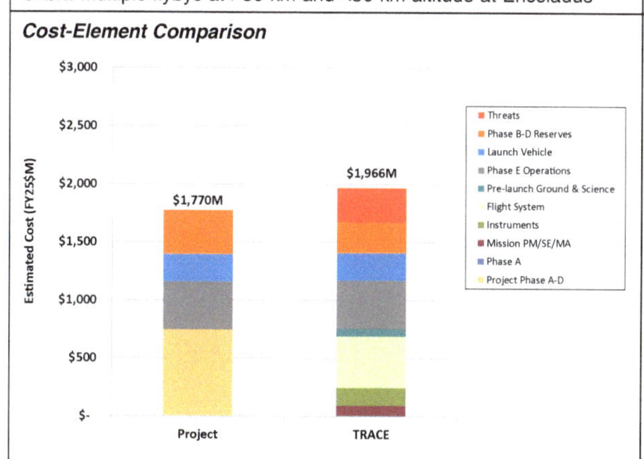

Cost-Risk Analysis S-Curve

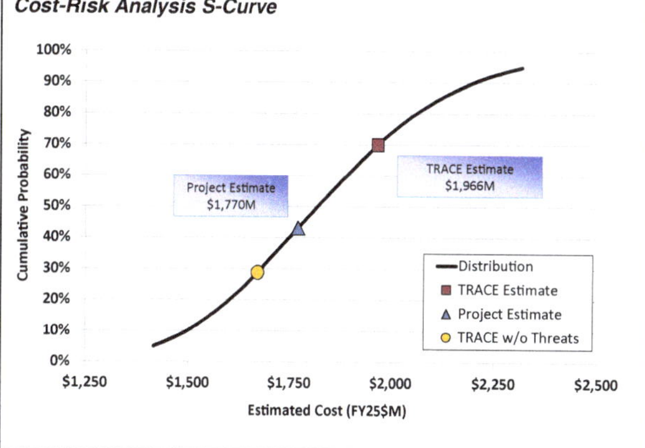

SOURCE: Image in upper-left box from NASA, 2021e, *Enceladus Multiple Flybys: Is There Life Beyond Earth?* Mission Concept Study Report for the Planetary Science and Astrobiology Decadal Survey 2023–2032. Greenbelt, MD: NASA Goddard Space Flight Center. https://tinyurl.com/2p88fx4f.

BOX C-13
Titan Orbiter (Sea Probe Descoped)

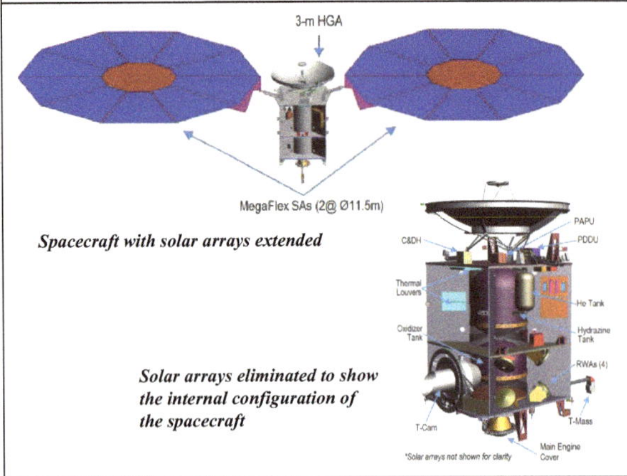

Scientific Objectives as Studied

Geology: Understand the processes actively shaping Titan's surface

Geophysics: Understand Titan's interior structure and surface–interior exchange processes

Astrobiology/Chemistry: Understand Titan's organic chemistry and path to prebiotic molecules

Atmosphere: Understand Titan's climate as a source of surface modification

Key Features

Orbiter (14-yr life: 10-yr cruise phase, 2-yr Saturn tour phase, 2-yr Titan orbit phase)

Science Payload: Mass Spectrometer (T-MASS) for profiling Titan's atmospheric constituents, Infrared Imager (T-CAM), Radar Altimeter (T-ALT)

Orbiter Bus: 2 x 11.5 m diameter Megaflex array, 78 Ahr Li Ion battery, dual mode chemical propulsion, redundant avionics, X-Band, Ka-Band, and UHF communications, GNC

Launch: Atlas V 551 or Falcon Heavy Recoverable, *nominal launch: 2031 to 2039*

Orbit: 2-yr Saturn tour with 20 flybys of Titan followed by 2 years at 1,500 km near polar orbit around Titan

Key Challenges

Titan atmosphere uncertainty impact on flight system for aerosampling passes and possibly aerobraking

Development and qualification of T-MASS hypervelocity aerosampling at Titan

Development of Megaflex arrays for Titan mission application

Technical Risk Rating

Medium-Low: Moderate new development, adequate margins, and/or moderate risk to achieve major mission objectives as proposed

Cost-Element Comparison

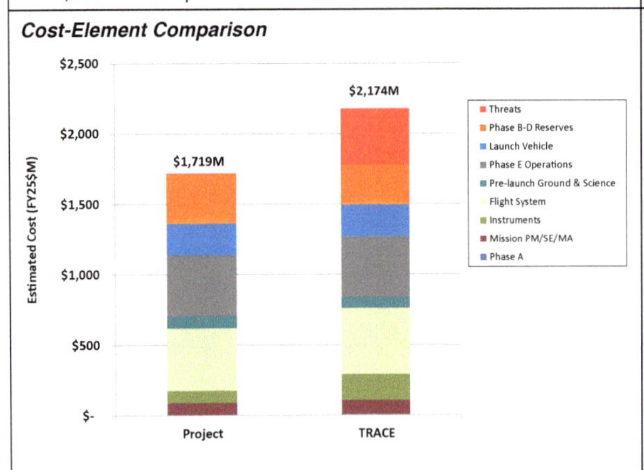

Cost-Risk Analysis S-Curve

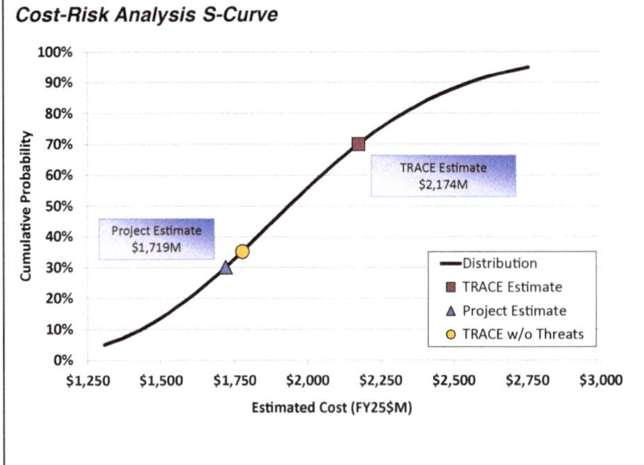

SOURCE: Images in upper-left box from NASA, 2021i, *Titan Orbiter and Probe*, Mission Concept Study Report for the Planetary Science and Astrobiology Decadal Survey 2023–2032, Pasadena, CA: Jet Propulsion Laboratory, California Institute of Technology, https://tinyurl.com/2p88fx4f.

BOX C-14
Centaur Orbiter and Lander

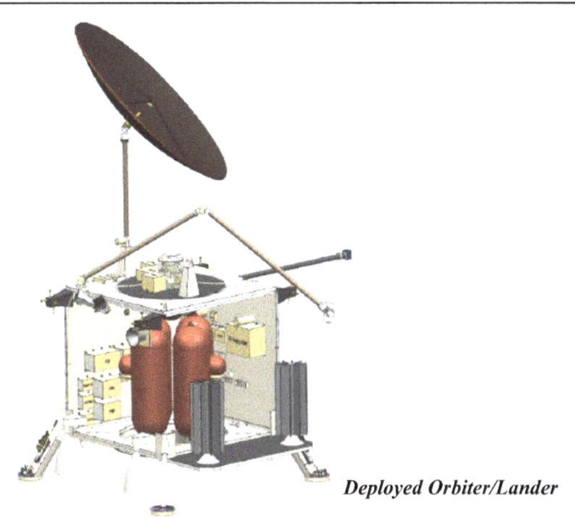

Deployed Orbiter/Lander

Scientific Objectives as Studied

Understand Early Solar System Compositional Reservoirs: Determine isotopic composition, large-scale mineralogical make-up, grain-scale composition, and interior volatile composition

Understand the Accretion and Dynamical Evolution of Primordial Icy Planetesimals: Determine impact history and relative ages, physical characteristics of the body, internal mass distribution, and magnetism present during formation and accretion

Determine the Geological and Evolutionary Processes That Have Influenced Icy Planetesimals: Determine landforms and any evidence of changes over the mission; icy regolith characteristics, surface weathering, source and cause of activity (if present), and characteristics of ring systems

Investigate the Biologic Potential of Icy Planetesimals and Potential Brine Reservoirs: Determine the thermal history by looking for alteration minerals; determine the composition, form, and distribution of organic material.

Key Features

Orbiter/Lander: (14-yr life, 9-yr cruise, 4-yr orbital mission, 2-mo nominal landed mission)

Payload: (8 instrument types) Gas Chromatograph Mass Spectrometer (GCMS), Xray Fluorescence (XRF), Raman/UV Spectrometer, IR Spectral Imager, UV Spectral Imager, Narrow Angle Camera (x2), Wide Angle Camera, Magnetometer

Sample Acquisition and Handling: PlanetVac-type sample collection (x2) with drill and pneumatic sample collection with carousel for handling
Navigation Instruments: Navigation cameras, panoramic camera, LIDAR
Flight System: Chemical bi-propellant system, system powered by NGRTG and Lion battery, Ka-band Direct to Earth communications, GNC with Terrain Relative Navigation (TRN), redundant avionics, landing anchor system

Launch: Falcon Heavy Expendable, *nominal launch: January 2040*

Orbit: 50 km altitude polar mapping orbit at 2015 BQ311 Centaur

Key Challenges

Low power margins during orbital phase and drilling

Potential changes to vehicle configuration impacting design

Impact of surface property unknowns on sampling and anchoring

Technical Risk Rating

Medium-Low: Moderate new development, adequate margins, and/or moderate risk to achieve major mission objectives as proposed

Cost-Element Comparison

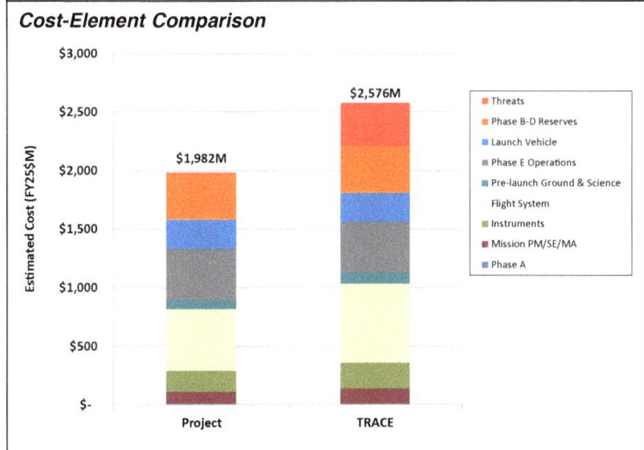

Cost-Risk Analysis S-Curve

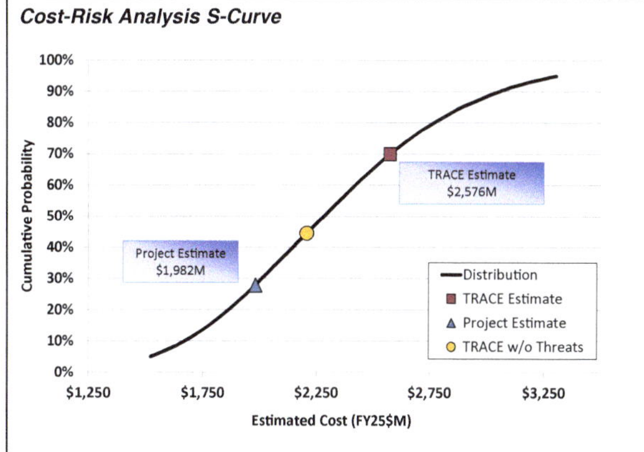

SOURCE: Image in upper-left box from NASA, 2021c, *CORAL: Centaur Orbiter and Lander*, Mission Concept Study Report for the Planetary Science and Astrobiology Decadal Survey 2023–2032, Greenbelt, MD: NASA Goddard Space Flight Center, https://tinyurl.com/2p88fx4f.

BOX C-15
Uranus Orbiter and Probe

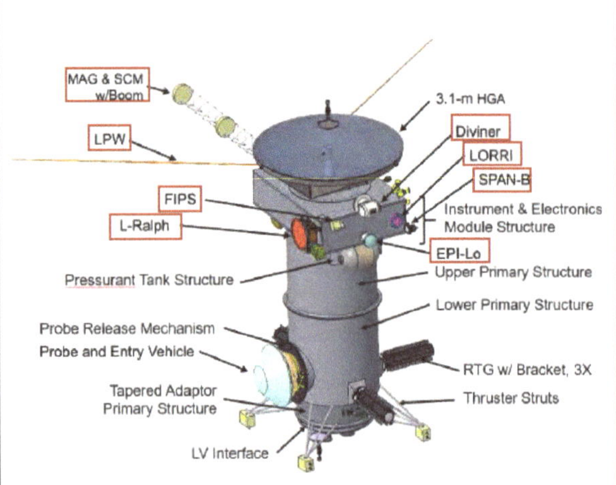

Scientific Objectives

Origins:
When and where did Uranus form in the protosolar nebula?
Did Uranus and Neptune migrate or swap positions?
Did a giant impact tilt Uranus, rearrange its interior, and form satellites?

Processes:
What mechanisms are transporting heat/energy in the planet and satellites?
How do all the components in the uranian system interact with each other?
What external factors are altering the planet, satellites, and ring compositions?
What interior structure produces Uranus' complex magnetosphere?

Habitability:
Did any of Uranus's moons have oceans in the past?
Are any of the moons presently ocean worlds?

Interconnections:
How do ice giant planets form and evolve in exoplanetary systems?
How does solar wind couple to a magnetosphere with a rapidly changing orientation?

Key Features

Orbiter

Payload: (9 instruments) Magnetometer (MAG), Narrow-Angle Camera, Thermal IR Camera, Langmuir Probe & Waves (LPW), Search Coil Magnetometer (SCM), Fast Imaging Plasma Spectrometer (FIPS), Solar Wind Electrons Alphas and Protons (SWEAP), Solar Probe Analyzer-B (SPAN-B), Energetic Particle Instrument (EPI-Lo), Vis/NIR Imaging Spectrometer & Wide-Angle Camera

Flight System: Chemical propellant system, redundant avionics, powered by 3 x Next Generation Radioisotope Thermal Generators (NGRTGs) Mod 1 with secondary battery

Probe (60-day mission)

Payload: (3 instruments) Mass Spectrometer, Atmospheric Structure Instrument, Ortho-Para H2 Detector

Flight System: HEEET TPS, primary battery, avionics, comm

Launch: Falcon Heavy Expendable, *nominal launch: 2031*

Orbit: Uranus orbit with moon tour of Uranus system

Key Challenges

Power margin sensitivity to mission trajectory/lifetime alternatives

Flight system lifetime for 18.5-year mission

Payload requirements growth leading to more complex mission

Technical Risk Rating

Medium-Low: Moderate new development, adequate margins, and/or moderate risk to achieve major mission objectives as proposed

Cost-Element Comparison

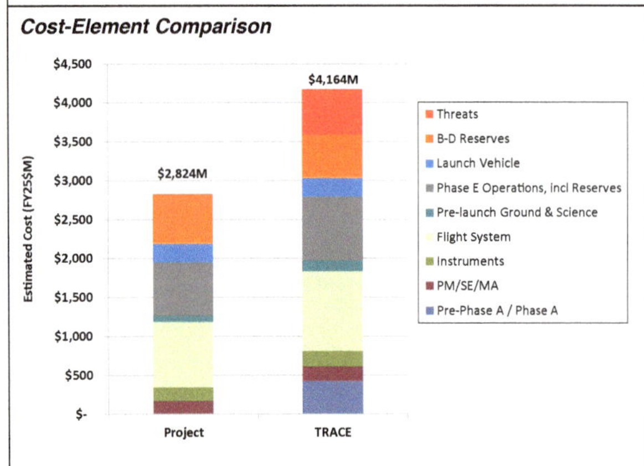

Cost-Risk Analysis S-Curve

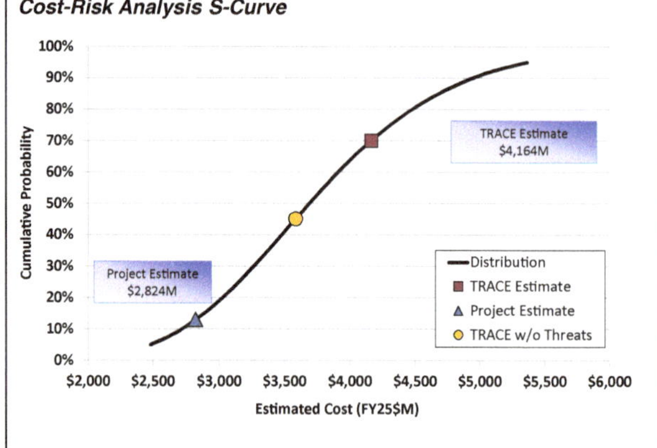

SOURCE: Image in upper-left box from NASA, 2021j, *Uranus Orbiter and Probe: Journey to an Ice Giant System*, Mission Concept Study Report for the Planetary Science and Astrobiology Decadal Survey 2023–2032, Columbia, MD: Johns Hopkins University Applied Physics Laboratory, https://tinyurl.com/2p88fx4f.

BOX C-16
Calypso: Ariel and KBO Flyby

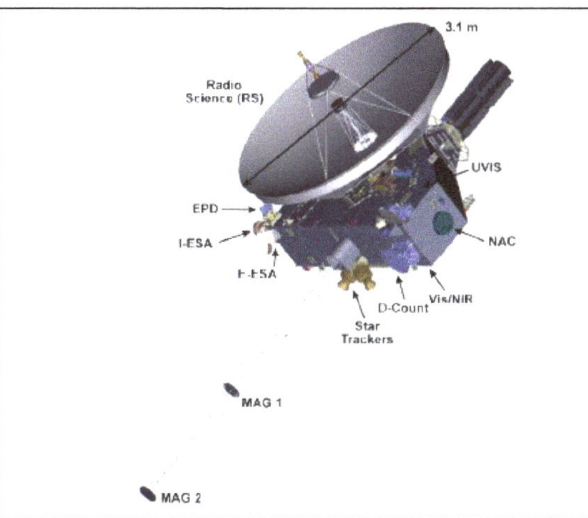

Scientific Objectives as Studied

Explore the Uranus System and Its Impact-Generated Moons: (potential ocean worlds)

Search for Ariel subsurface ocean and internal structure

Characterize Ariel crater populations

Uranian System Science: Uranus, rings, ring-moons, and satellites

Study Large Kuiper Belt Objects (KBOs):

Characterize the global geology and morphology of KBOs and their rotational and physical properties

Map G!kúnll'hòmdímà surface composition and volatile distribution

Characterize G!kúnll'hòmdímà crater populations

Key Features

Spacecraft: (17.1-yr life: 6-yr cruise to Uranian moon, 11-yr cruise to KBO)

Science Payload: (8 instrument types + radio science) Vector Magnetometer (x2) (MAG), Narrow Angle Camera (NAC), Vis/NIR Imaging Spectrometer (Vis/NIR), Radio Science Experiment (RS), Ultraviolet Imaging Spectrometer (UVIS), Ion Electrostatic Analyzer (I-ESA), Electron Electrostatic Analyzer (E-ESA), Energetic Particle Detector (EPD), Dust Counter (D-Count)

Spacecraft Bus: Powered by NGRTG Mod 0, 78 Ahr Li Ion battery, hydrazine propulsion, redundant avionics, X-Band and Ka-Band communications including 3.1 m high gain antenna, GNC

Launch: Vulcan or Falcon Heavy, both with STAR 48BV kick stage, *nominal launch: 2035*

Orbit: Flyby of Uranian moon (nominally Ariel) followed by flyby of KBO 11 years later (nominally G!kúnll'hòmdímà)

Key Challenges

Low power margin, particularly if NGRTG Mod 0 is not available

Mission destination bodies are uncertain and dependent on launch year

Critical closest approach operations robustness is unclear for flyby

Technical Risk Rating

Medium-Low: Moderate new development, adequate margins, and/or moderate risk to achieve major mission objectives as proposed

Cost-Element Comparison

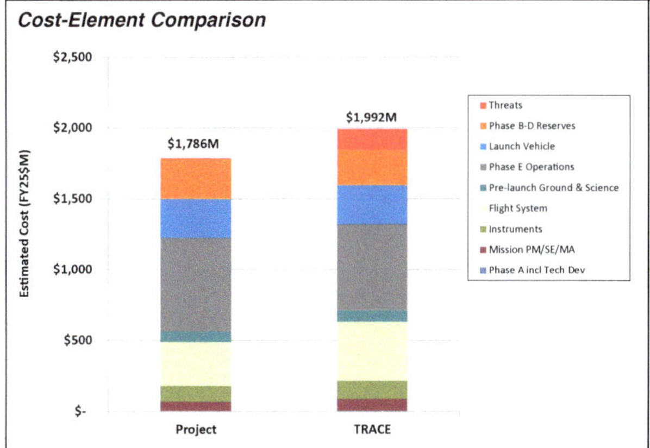

Cost-Risk Analysis S-Curve

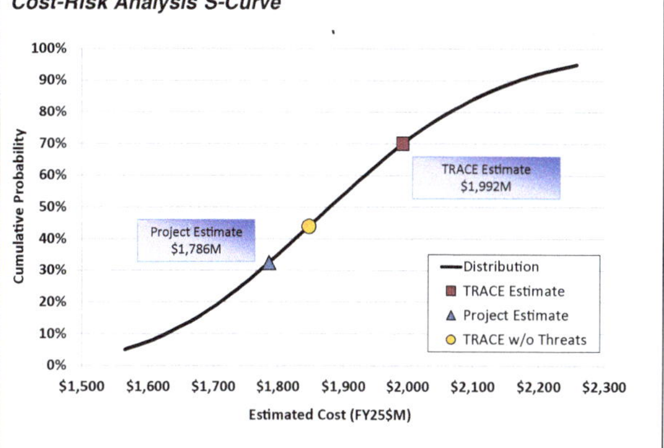

SOURCE: Image in upper-left box from NASA, 2021b, *Calypso: In Search of Ocean Worlds in the Uranian System and Kuiper Belt*, Mission Concept Study Report for the Planetary Science and Astrobiology Decadal Survey 2023–2032, Columbia, MD: Johns Hopkins University Applied Physics Laboratory, https://tinyurl.com/2p88fx4f.

BOX C-17
Neptune Odyssey, Neptune–Triton Orbiter and Probe

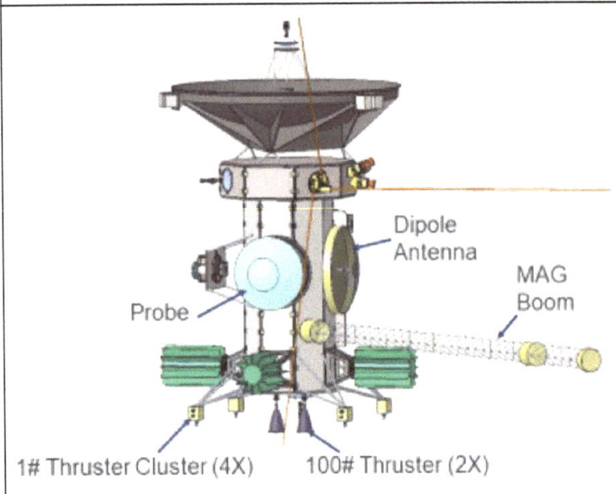

Scientific Objectives as Studied

Origins: Understand Neptune's origin and how it evolved, placing it in context with other planetary types

Particles and Fields: Study Neptune's aurora and magnetic field to understand processes critical to the Neptune system, including exploring Triton's interactions and whether Triton contains a subsurface ocean by observing auroral activity and magnetic induction

Ocean Worlds: Study Triton's surface, interior, atmosphere, and magnetic interaction with Neptune to determine if it is an ocean world, understand plume activity, and how Triton's ionosphere is coupled with Neptune's magnetosphere

Comparative Planetology: Compare and contrast attributes of Triton with other Kuiper Belt objects for a better understanding of dwarf planets

Satellite and Ring Systems: Understand Neptune's ring-moon system

Key Features

Orbiter (20-yr life)

Payload (14 instruments): Magnetometer, Color Narrow-Angle Camera, UV Imaging Spectrograph, Ion and Neutral Mass Spectrometer, Laser Altimeter, Vis-NIR Imaging Spectrometer, Radio and Plasma Wave Detector, Thermal Infrared Imager, Microwave Radiometer, Thermal Plasma Spectrometer, Energetic Charged Particle Detector, Energetic Neutral Atom Imager, ToF Dust Spectrometer, EPO Camera

Flight System: Chemical propellant system, redundant avionics, powered by 3 x Next Generation Radioisotope Thermal Generators (NGRTGs)

Probe (30-day mission)

Payload: (7 instruments) Mass Spectrometer, Atmospheric Structure Instrument, He Abundance Detector, Ortho-Para H_2 Detector, Nephelometer, Net Flux Radiometer, EPO Camera

Flight System: HEEET TPS, primary battery, avionics, communications system

Launch: SLS Block 2 with Centaur Upper Stage, *nominal launch: 2033*

Orbit: Neptune orbit with repeat flybys of Triton

Key Challenges

Availability of SLS Block 2 with Centaur upper stage

NGRTG performance at required power output for mission

HEEET TPS performance at mission peak pressure and heat flux

Probe power growth and lifetime of primary batteries with long storage

Complexity of large suite of instruments on an orbiter an a probe

Technical Risk Rating

Medium: Medium new development, adequate to optimistic margins, and/or medium risk of achieving major mission objectives as proposed

Cost-Element Comparison

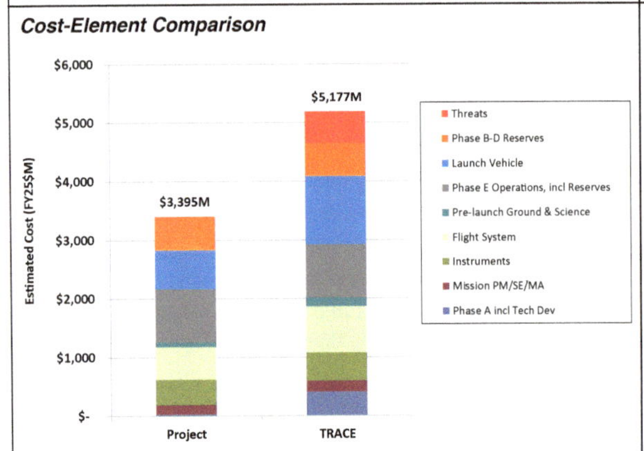

Cost-Risk Analysis S-Curve

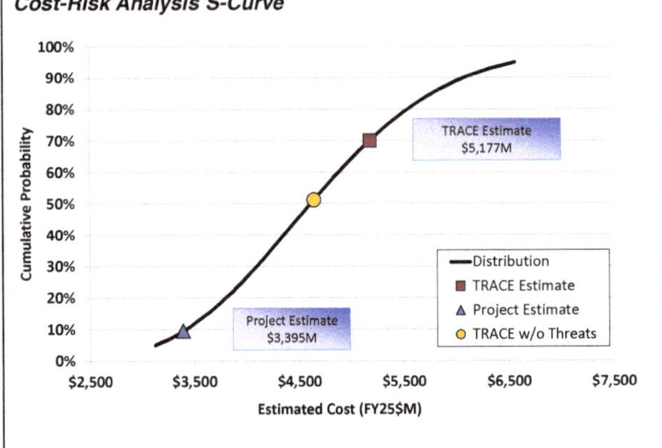

SOURCE: Image in upper-left box from NASA, 2020i , *Neptune Odyssey: Mission to the Neptune-Triton System*, Mission Concept Study Report for the Planetary Science and Astrobiology Decadal Survey 2023–2032, Columbia, MD: Johns Hopkins University Applied Physics Laboratory, https://science.nasa.gov/solar-system/documents.

BOX C-18
Triton Ocean World Surveyor

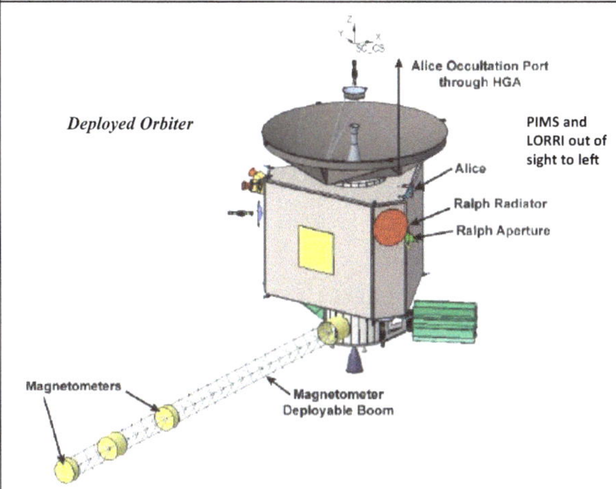

Scientific Objectives as Studied

Determine whether Triton is an ocean world, ascertain its interior structure, and decide whether Triton's ice shell is in hydrostatic equilibrium and de-coupled from the interior

Characterize Triton's surface composition and geology, and look for changes, including plumes and their composition

Determine the nature of the moon–magnetosphere interaction at Triton

Determine the composition, density, temperature, pressure, and spatial/temporal variability of Triton's atmosphere

---

Key Features

Orbiter/Lander (20-yr life, 16-yr cruise, 4-yr orbital mission)

Payload: (6 instruments)
Ultraviolet Imaging Spectrograph (Alice), Long Range Imager (LORRI), Multispectral Visible Imaging Camera Linear Etalon Imaging Spectral Array (Ralph), Plasma for Magnetic Sounding (PIMS), Ion and Neutral Mass Spectrometer, Fluxgate Magnetometer with boom

Flight System: Chemical bi-propellant system, system powered by NGRTG and battery, Ka-band and X-Band communications, redundant GNC, redundant avionics

Launch:
Falcon Heavy Expendable with STAR 48BV, *nominal launch: February 2031, backup launches in March 2032 and in 2040s*

Orbit:
Highly inclined orbit around Neptune with 35–45 flybys for Triton

Key Challenges

Lifetime reliability and power issues for long mission duration

Availability of NGRTG

Unplanned growth of instruments from model payload suite

Technical Risk Rating

Medium-Low: Moderate new development, adequate margins, and/or moderate risk to achieve major mission objectives as proposed

---

Cost-Element Comparison

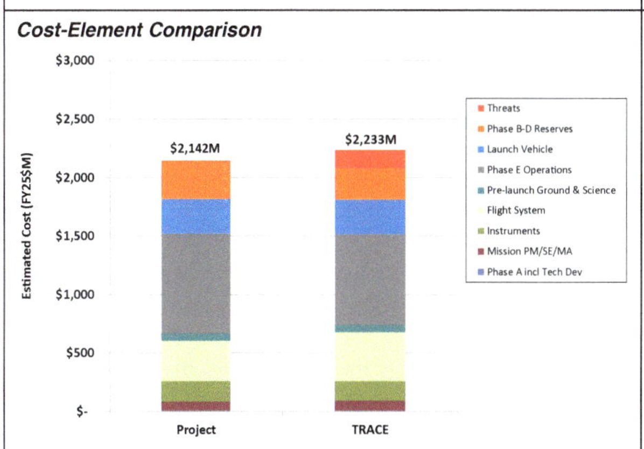

Cost-Risk Analysis S-Curve

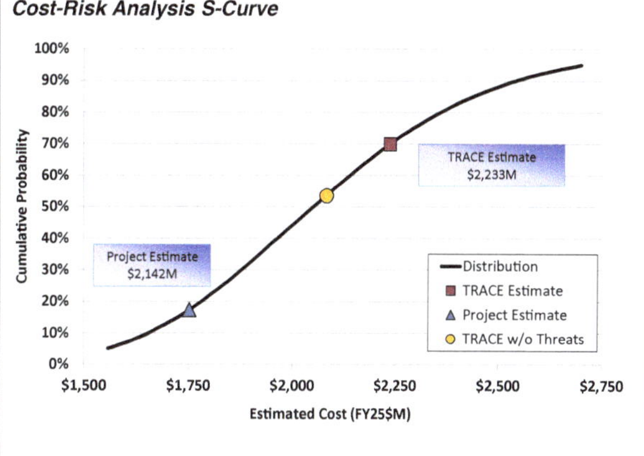

SOURCE: Image in upper-left box from NASA, 2021k, *Triton Ocean World Surveyor: A Neptune Orbiter and Triton Explorer.* Mission Concept Study Report for the Planetary Science and Astrobiology Decadal Survey 2023–2032. Columbia, MD: Johns Hopkins University Applied Physics Laboratory. https://tinyurl.com/2p88fx4f.

REFERENCES

NASA (National Aeronautics and Space Administration). 2015. *Report from the Next Orbiter Science Analysis Group (NEX-SAG)*. MEPAG NEX-SAG Final Report. Pasadena, CA: MEPAG (Mars Exploration Program Analysis Group) NEX-SAG (Next Orbiter Science Analysis Group).

NASA. 2017a. *Ice Giants Pre-Decadal Study Final Report*. JPL D-100520. Pasadena, CA: Jet Propulsion Laboratory.

NASA. 2017b. *Report of the Europa Lander Science Definition Team*. Washington, DC. https://europa.nasa.gov/resources/58/europa-lander-study-2016-report.

NASA. 2019. *MEPAG ICE-SAG Final Report (2019)*. Report from the Ice and Climate Evolution Science Analysis Group (ICE-SAG). Pasadena, CA: MEPAG (Mars Exploration Program Analysis Group).

NASA. 2020a. *Enceladus Orbilander: A Flagship Mission Concept for Astrobiology*. Mission Concept Study Report for the Planetary Science and Astrobiology Decadal Survey 2023–2032. Columbia, MD: Johns Hopkins University Applied Physics Laboratory. https://science.nasa.gov/solar-system/documents.

NASA. 2020b. *In Situ Geochronology for the Next Decade*. Final Report Submitted in Response to NNH18ZDA001N-PMCS: Planetary Mission Concept Studies. Greenbelt, MD: NASA Goddard Space Flight Center. https://science.nasa.gov/solar-system/documents.

NASA. 2020c. *Intrepid Planetary Mission Concept Study Report*. Mission Concept Study Report for the Planetary Science and Astrobiology Decadal Survey 2023–2032. Pasadena, CA: Jet Propulsion Laboratory, California Institute of Technology https://science.nasa.gov/solar-system/documents.

NASA. 2020d. *The Lunar Geophysical Network*. Final Report Submitted in Response to NNH18ZDA001N-PMCS: Planetary Mission Concept Studies. Greenbelt, MD: NASA Goddard Space Flight Center. https://science.nasa.gov/solar-system/documents.

NASA. 2020e. *MORIE: Mars Orbiter for Resources, Ices, and Environments*. Mission Concept Study Report for the Planetary Science and Astrobiology Decadal Survey 2023–2032. Pasadena, CA: Jet Propulsion Laboratory, California Institute of Technology. https://science.nasa.gov/solar-system/documents.

NASA. 2020f. *MOSAIC: Mars Orbiters for Surface-Atmosphere-Ionosphere Connections*. Mission Concept Study Report for the Planetary Science and Astrobiology Decadal Survey 2023–2032. Pasadena, CA: Jet Propulsion Laboratory, California Institute of Technology https://science.nasa.gov/solar-system/documents.

NASA. 2020g. *Mercury Lander: Transformative Science from the Surface of the Innermost Planet*. Mission Concept Study Report for the Planetary Science and Astrobiology Decadal Survey 2023–2032. Columbia, MD: Johns Hopkins University Applied Physics Laboratory. https://science.nasa.gov/solar-system/documents.

NASA. 2020h. *Mission Concept Study: Ceres: Exploration of Ceres' Habitability*. Mission Concept Study Report for the Planetary Science and Astrobiology Decadal Survey 2023–2032. Pasadena, CA: Jet Propulsion Laboratory, California Institute of Technology. https://science.nasa.gov/solar-system/documents.

NASA. 2020i. *Neptune Odyssey: Mission to the Neptune-Triton System*. Mission Concept Study Report for the Planetary Science and Astrobiology Decadal Survey 2023–2032. Columbia, MD: Johns Hopkins University Applied Physics Laboratory. https://science.nasa.gov/solar-system/documents.

NASA. 2020j. *Persephone: A Pluto-System Orbiter and Kuiper Belt Explorer*. Mission Concept Study Report for the Planetary Science and Astrobiology Decadal Survey 2023–2032. Columbia, MD: Johns Hopkins University Applied Physics Laboratory. https://science.nasa.gov/solar-system/documents.

NASA. 2020k. *2020 Venus Flagship Mission Study: A Mission to Explore the Habitability of Venus and the Origins of Earth-Sized Planets Both Near and Far*. Mission Concept Study Report for the Planetary Science and Astrobiology Decadal Survey 2023–2032. Greenbelt, MD: NASA Goddard Space Flight Center. https://science.nasa.gov/solar-system/documents.

NASA. 2021a. *ADVENTS: Assessment and Discovery of Venus' Past Evolution and Near-Term Climatic and Geophysical State*. Mission Concept Study Report for the Planetary Science and Astrobiology Decadal Survey 2023–2032. Greenbelt, MD: NASA Goddard Space Flight Center. https://tinyurl.com/2p88fx4f.

NASA. 2021b. *Calypso: In Search of Ocean Worlds in the Uranian System and Kuiper Belt*. Mission Concept Study Report for the Planetary Science and Astrobiology Decadal Survey 2023–2032. Columbia, MD: Johns Hopkins University Applied Physics Laboratory. https://tinyurl.com/2p88fx4f.

NASA. 2021c. *CORAL: Centaur Orbiter and Lander*. Mission Concept Study Report for the Planetary Science and Astrobiology Decadal Survey 2023–2032. Greenbelt, MD: NASA Goddard Space Flight Center. https://tinyurl.com/2p88fx4f.

NASA. 2021d. *CROCODILE: Cryogenic Return of Cometary Organics, Dust, and Ice for Laboratory Exploration*. Mission Concept Study to Report to the Planetary Science and Astrobiology Decadal Survey 2023–2032. Greenbelt, MD: NASA Goddard Space Flight Center. https://tinyurl.com/2p88fx4f.

NASA. 2021e. *Enceladus Multiple Flybys: Is There Life Beyond Earth?* Mission Concept Study Report for the Planetary Science and Astrobiology Decadal Survey 2023–2032. Greenbelt, MD: NASA Goddard Space Flight Center. https://tinyurl.com/2p88fx4f.

NASA. 2021f. *Endurance: Lunar South Pole-Aitken Basin Traverse and Sample Return Rover.* Mission Concept Study Report for the Planetary Science and Astrobiology Decadal Survey 2023–2032. Pasadena, CA: Jet Propulsion Laboratory, California Institute of Technology. https://tinyurl.com/2p88fx4f.

NASA. 2021g. *INSPIRE: In Situ Solar System Polar Ice Roving Explorer.* Mission Concept Study Report for the Planetary Science and Astrobiology Decadal Survey 2023–2032. Pasadena, CA: Jet Propulsion Laboratory, California Institute of Technology. https://tinyurl.com/2p88fx4f.

NASA. 2021h. *Mars Life Explorer.* Mission Concept Study Report for the Planetary Science and Astrobiology Decadal Survey 2023–2032. Pasadena, CA: Jet Propulsion Laboratory. https://tinyurl.com/2p88fx4f.

NASA. 2021i. *Titan Orbiter and Probe.* Mission Concept Study Report for the Planetary Science and Astrobiology Decadal Survey 2023–2032. Pasadena, CA: Jet Propulsion Laboratory, California Institute of Technology. https://tinyurl.com/2p88fx4f.

NASA. 2021j. *Uranus Orbiter and Probe: Journey to an Ice Giant System.* Mission Concept Study Report for the Planetary Science and Astrobiology Decadal Survey 2023–2032. Columbia, MD: Johns Hopkins Applied Physics Laboratory. https://tinyurl.com/2p88fx4f.

NASA. 2021k. *Triton Ocean World Surveyor: A Neptune Orbiter and Triton Explorer.* Mission Concept Study Report for the Planetary Science and Astrobiology Decadal Survey 2023–2032. Columbia, MD: Johns Hopkins University Applied Physics Laboratory. https://tinyurl.com/2p88fx4f.

NASEM (National Academies of Sciences, Engineering, and Medicine). 2013. *Lessons Learned in Decadal Planning in Space Science: Summary of a Workshop.* Washington, DC: The National Academies Press.

NASEM. 2015. *The Space Science Decadal Surveys: Lessons Learned and Best Practices.* Washington, DC: The National Academies Press

NASEM. 2023. *Pathways to Discovery in Astronomy and Astrophysics for the 2020s.* Washington, DC: The National Academies Press.

NRC (National Research Council). 2006. *An Assessment of Balance in NASA's Science Programs.* Washington, DC: The National Academies Press, p. 32.

NRC. 2007a. *Decadal Science Strategy Surveys: Report of a Workshop.* Washington, DC: The National Academies Press, pp. 21–30.

NRC. 2007b. *NASA's Beyond Einstein Program: An Architecture for Implementation.* Washington, DC: The National Academies Press, pp. 66–114.

NRC. 2010. *New Worlds, New Horizons in Astronomy and Astrophysics.* Washington, DC: The National Academies Press, Appendix C.

NRC. 2011. *Vision and Voyages for Planetary Science in the Decade 2013–2022.* Washington, DC: The National Academies Press, pp. 331–353.

NRC. 2013. *Lessons Learned in Decadal Planning in Space Science: Summary of a Workshop.* Washington, DC: The National Academies Press.

U.S. Congress, House. 2008. National Aeronautics and Space Administration Authorization Act of 2008. Public Law 110-422. H.R. 6063. 110th Congress, Section 1104b.

Venera-D Joint Science Definition Team. 2019. *Venera-D: Expanding Our Horizon of Terrestrial Planet Climate and Geology through the Comprehensive Exploration of Venus.* Phase II Final Report of The Venera-D Joint Science Definition Team. Washington, DC: NASA Headquarters and Moscow, Russia: Space Research Institute.

Appendix D

Missions Studied But Not Sent for TRACE

As explained in Appendix C, the science mission concepts considered in this report came from three main sources: (1) missions studied by NASA science definition teams (SDTs); (2) missions selected and studied via NASA's predecadal mission concept study (PMCS) process, and (3) missions identified by the decadal survey's panels and prioritized by the steering group. The latter group of nine concepts were studied at the committee's request by leading design centers (including the Jet Propulsion Laboratory, the Goddard Space Flight Center, and the Johns Hopkins University Applied Physics Laboratory). The complete set of SDTs, PMCS, and decadal mission studies were further prioritized by the steering group and 17 of them were selected for technical risk and cost evaluation (TRACE) by the Aerospace Corporation.

This appendix describes the eight missions that were fully studied to Concept Maturity Level (CML) ~4 to 5 via one of the three mechanisms outlined above but were not selected for TRACE. These missions are the following:

- Venera-D;
- ADVENTS—Assessment and Discovery of Venus's Past Evolution and Near-Term Climatic and Geophysical State;
- Lunar In Situ Geochronology;
- Vesta In Situ Geochronology;
- MORIE—Mars Orbiter for Resources, Ices, and Environment;
- MOSAIC—Mars Orbiter for Surface-Atmosphere-Ionosphere Characterization;
- CROCODILE—Cryogenic Return of Cometary Organics, Dust, and Ice for Laboratory Exploration; and
- Persephone—Pluto System Orbiter and KBO Flyby.

These missions were not subjected to TRACE analysis because the committee considered them to have lower science merit and/or to be less technically ready than the missions discussed in Appendix C. Each of the above eight missions is described in more detail in the sections below.

VENERA-D

Origin and Rationale

This mission concept study was performed by the Venera-D Joint Science Definition Team at the request of the Russian Space Agency, the Institute for Space Research Institute of the Russian Academy of Sciences, and NASA (Venera-D 2019). The purpose of this study was to define a comprehensive mission to investigate the atmosphere and surface of Venus using an orbiter, a lander, and the NASA-contributed Long-Lived In Situ Solar System Explorer (LLISSE).

Goals

Science goals of Venera-D address key outstanding questions related to Venus's atmosphere and surface. The goals of Venera-D's orbiter are, in priority order, to:

- Study the dynamics and nature of superrotation, radiative balance, and nature of the greenhouse effect;
- Characterize the thermal structure of the atmosphere, winds, thermal tides, and solar-locked structures;
- Measure the composition of the atmosphere; study the clouds, their structure, composition, microphysics, and chemistry;
- Study the composition of the low atmosphere and low clouds, study surface emissivity, and search for volcanic events on the night side; and
- Investigate the upper atmosphere, ionosphere, electrical activity, magnetosphere, the atmospheric escape rate, and solar wind interaction.

The goals of the Venera-D's lander and LLISSE are to:

- Measure elemental and mineralogical abundances of the surface materials and near subsurface (~a few cm), including radiogenic elements;
- Study the interaction between the surface and atmosphere;
- Investigate the structure and chemical composition of the atmosphere down to the surface, including abundances and isotopic ratios of the trace and noble gases;
- Perform direct chemical analysis of cloud aerosols;
- Characterize the geology of local landforms at different scales;
- Study variation of near-surface wind speed and direction, temperatures, and pressure (LLISSE).
- Measure incident and reflected solar radiation (LLISSE);
- Measure near-surface atmospheric chemical composition (LLISSE); and
- Detect seismic activity, volcanic activity, and volcanic lightning.

Implementation

The mission consists of an orbiter and a 1980s-vintage Soviet VEGA-type lander equipped with modern instrumentation and LLISSE, an independent NASA-provided meteorological, radiation, and compositional instrument package designed to survive on the surface for approximately 3 months. Possible augmentations include subsatellites, additional small long-lived landers, and various types of airborne platforms.

The Venera-D spacecraft (orbiter and landers) is designed for launch on an Angara-5 rocket. After entering a 24-hour, near-polar orbit around Venus and deploying the landers, the orbiter conducts its own investigations and acts as a communications relay with Earth. The operating lifetime of the orbiter is about 3 years.

Following release from the orbiter, the lander will sample the atmosphere and image the surface during descent to a landing site in a high-latitude region of the northern hemisphere. The lander includes a thermally insulated titanium pressure vessel with thermal storage batteries and is designed to operate for about 3 hours after landing.

Mission Challenges

This mission has not yet received full support from the Russian Space Agency to begin development. The comprehensive nature of the mission and the multiple, as yet undefined, enhancements and international contributions will require substantial coordination.

Conclusions

Venera-D is designed to meet the highest-priority science objectives for Venus. This mission goes far beyond the currently selected VERITAS, DAVINCI, and Envision missions by landing and conducting in situ science on the Venus plains. Continuing support for its definition and development is required if this joint U.S.–Russian activity is to be implemented.

ADVENTS

Origin and Rationale

This concept was proposed by the decadal survey's Panel on Venus, and the study (CML ~4–5) was conducted at NASA's Goddard Space Flight Center in the early months of 2021 (NASA 2021a). This mission explores whether a medium-class mission addressing aspects of the Venus Sub-Cloud Aerobot, Venus Life Potential, and Venus Investigations of Dynamics from an Equatorial Orbit proposals (see Appendix E) could be formulated by descoping the large lander, radar, and SmallSats from the Venus Flagship mission (see Appendix C).

Goals

The goals of this concept are to:

- Understand how Venus formed and evolved for comparison to other rocky planets and exoplanets;
- Study the potential past habitability of the Venus surface; and
- Determine the composition, dynamics, and potential habitability of the present-day atmosphere of Venus.

Implementation

ADVENTS will deliver an orbiter and a variable-altitude aerobot to Venus. Synergistic instruments on both platforms collect in situ and remote measurements of the atmosphere, along with remote sensing of the surface, for at least 60 Earth-days. A single dropsonde is also deployed by the aerobot to sample the chemistry of the atmosphere from the clouds to the surface. The orbiter is placed in a 12-hour period equatorial orbit and conducts its own observations and acts as a communications relay between the aerobot and Earth. Instruments on the orbiter include near-infrared surface and cloud imagers, a magnetometer, an extreme ultraviolet monitor, and a radio occultation package. Instruments on the aerobot include an aerosol mass spectrometer with nephelometer, a tunable laser spectrometer, a magnetometer, a meteorological package, and the dropsonde.

Mission Challenges

The ADVENTS mission is complex, and faces mass and cost challenges associated with delivering to Venus and deploying an aerobot and a dropsonde within the medium-class mission cost cap. The need for a dedicated orbital relay element also represents a considerable mission cost.

Conclusions

ADVENTS can address many of the high-priority science objectives for Venus not covered by the VERITAS, DAVINCI, or Envision missions. However, ADVENTS overlaps with some objectives of the VISE mission, resulting in it being given a low priority for a TRACE study. Continued development of numerous Venus-relevant technologies—including atmospheric entry, balloon altitude control, and instrument miniaturization—would help enable the ADVENTS concept to meet the constraints of a medium-class mission.

LUNAR IN SITU GEOCHRONOLOGY

Origin and Rationale

This concept was proposed in response to the PMCS announcement of opportunity, and the study (CML ~4–5) was conducted at NASA's Goddard Space Flight Center (NASA 2020a). The purpose of the study was to determine the absolute age of a selected location on the Moon. This concept was developed as a part of a larger effort to study the absolute dating of features on various planetary bodies including Mars (see Appendix C), Vesta (see below), and the Moon. The relative dating of features on a planetary body can be determined by applying the basic principle that younger features overlay those that are older. Similarly, a more heavily cratered surface is likely to be older than an adjacent surface with fewer craters. Turning these relative ages into absolute dates requires the isotopic analysis in terrestrial laboratories of carefully selected samples from key geological features. While the Apollo and former Soviet Union's Luna missions returned samples to Earth, as will planned human exploration missions, the number of scientifically significant terrains far exceed any currently conceivable ability to collect and return samples to terrestrial laboratories for geochronological analysis. This concept explores the idea that technology may have advanced sufficiently to perform geochronological analysis in situ using a robotic spacecraft.

Goals

The goals of the lunar version of the in situ geochronology mission concept are to:

- Establish the chronology of basin-forming impacts by measuring the radiometric age of samples directly sourced from the impact melt sheet of a pre-Imbrian lunar basin. In situ dating of an impact-melt sheet of a lunar basin thought to be significantly older than the Imbrium basin would place it either within the canonical cataclysm (3.9 Ga) or as part of a declining bombardment in which most impacts are 4.2 Ga or older.
- Establish the age of a very young lunar basalt to correlate crater size–frequency distributions with crystallization ages. In situ dating would reduce the uncertainty in absolute model ages derived from crater size–frequency distribution measurements to no more than 20 percent of the current uncertainty associated with different lunar chronology functions.

Implementation

This proposed medium-class mission relies on a static lunar lander designed to launch on a Falcon 9 Heavy Recoverable. Once placed in lunar transfer trajectory, the lander would enter polar lunar orbit, overfly the selected landing site, deorbit, and land. The lander relies on solar power, limiting its operations to daytime and relying on batteries to survive the lunar night. The design life of the mission is 12 months. Two different on-board instruments—one examining rubidium and strontium isotopes and the other potassium and argon isotopes—are used to determine independent age estimates. Sampling is accomplished by a pneumatic acquisition system and associated sample-handling and sample-preparation subsystems. Additional instruments carried include an imaging spectrometer and cameras to document the geological context of the landing site and to characterize the samples themselves and a trace-element analyzer to augment sample contextualization.

Mission Challenges

A lack of mobility in this mission concept leads to uncertainty in the ability to obtain the needed samples to provide accurate dating of key events, particularly in geologically complex terrains. Top science objectives associated with dating ancient impact basins may be particularly challenging. Lunar rock samples may display a complex collection of mineral phases, which even in the laboratory require careful separation before being analyzed for their isotopic composition. As such, it is not clear that the in situ capabilities for sample selection and phase separation would produce accurate dates for basin-forming impacts, and the return of samples via the Endurance-A mission was considered a scientifically superior strategy to address this highest-priority science. Dating young lunar basaltic terrains may be more favorable for in situ analysis, but this was judged to have less scientific importance than other medium-class missions.

Conclusions

In situ geochronology is a promising but still relatively new technique, and demonstration of its ability to produce meaningful ages from a potentially limited choice of geologically complex lunar samples is needed. A static lander exacerbates such challenges.

VESTA IN SITU GEOCHRONOLOGY

Origin and Rationale

This concept was proposed in response to the PMCS announcement of opportunity, and the study (CML ~4–5) was conducted at NASA's Goddard Space Flight Center. The purpose of the study was to determine the absolute age of two selected locations on the main belt asteroid Vesta. This concept was developed as a part of a larger effort to study the absolute dating of features on various planetary bodies including Mars (see Appendix C), the Moon (see above), and Vesta (NASA 2020a). While the relative dating of features on a planetary body can be determined by techniques such as crater counting, determining the actual age of specific features requires radiometric studies of samples of known geological context. While the so-called HED meteorites originated on Vesta, their context is unknown and thus, of no use to dating specific features on this large asteroid. Unfortunately, there are no current plans to return samples from Vesta via robotic or other means.

Goals

The goal of the vestan version of the in situ geochronology mission concept is to determine the radiometric ages of vestan samples from locations whose geological context is known. In situ dating would constrain Vesta's geologic timescale by dating key stratigraphic craters and adjacent geologic terrains. Given the large disagreement in ages derived by various indirect methods, a few absolute dates would not only reveal the ages of key basins but would also set firm constraints on impactor flux estimates, used throughout the asteroid belt to establish relative dates via the crater counting technique.

Implementation

This proposed medium-class mission is conceptually similar to, and carries the same instrumentation, as the lunar in situ geochronology concept (see above). The principal difference is that here the lander is not static: Vesta's low gravity enables the spacecraft to hop once so that a second landing site can be examined.

Following launch on a Falcon 9 Heavy Expendable rocket, and a 49-month journey, the lander enters a 250 km altitude circular orbit. After 6 months of orbital mapping, the lander descends to the first of its two landing sites. Following 142 days of surface operations, the lander retracts its large solar arrays and hops to a second site several hundred km distant. Once at the new site, the lander redeploys its solar panels and begins a second set of surface operations for another 142 days.

Similar to the lunar dating mission (see above), two different on-board instruments—one examining rubidium and strontium isotopes and the other potassium and argon isotopes—are used to determine independent age estimates. Sampling is accomplished by a pneumatic acquisition system and associated sample-handling and sample-preparation subsystems. Additional instruments carried include an imaging spectrometer and cameras to document the geological context of landing sites and to characterize the samples themselves, and a trace-element analyzer to augment sample contextualization.

Mission Challenges

The PMCS report identified four major challenges. First, mission-enabling payload and lander technologies have not yet reached TRL 6. Second, the best highest-resolution images of Vesta are too coarse to confirm that a suitable landing site exists prior to the spacecraft's arrival. Challenges associated with the addition of landing algorithms, terrain-relative navigation, and other related capabilities could drive the mission concept design and approach. Third, landing and operating at multiple sites is a complexity that could drive the design and operations. Fourth, a reliable sample acquisition and distribution system is necessary to achieve the science goals and drives the mission design. In addition, there remain uncertainties associated with whether a set of in situ isotopic measurements of a limited suite of samples can be confidently interpreted to represent an age.

Conclusions

The concept is notable for being a versatile medium-class mission that, with technology maturation, would advance in situ instrumentation. However, the science addressed by returning to Vesta was deemed to be narrower and more focused in comparison to other missions considered that would visit unexplored targets or return samples. Thus, other medium-class missions under consideration were evaluated as higher science priorities for the limited TRACE opportunities.

MORIE

Origin and Rationale

This concept was proposed in response to the PMCS announcement of opportunity, and the study (CML ~4–5) was conducted at the Jet Propulsion Laboratory. The purpose of this mission is to address when, where, and how water has modified the martian surface through time.

Goals

The goals of the mission are to (NASA 2020b):

- Determine when elements of the cryosphere formed and how ice deposits are linked to the planet's ancient, recent, and current climate;
- Explore the evolution of surface environments and their transition through time; and
- Prospect for in situ resources necessary to support future human activities on the surface.

Implementation

This medium-class mission is designed to conduct the first synthetic aperture radar imaging from Mars orbit, fine-scale radar sounding of the depth to buried ice, compositional mapping of outcrops, monitoring ice and dust aerosols, and surface mapping at 1 m resolution. The orbiter would use solar-electric propulsion (SEP) to enable a 2-year journey to Mars and allow it to enter a entered a Sun-synchronous orbit crossing the equator at 3:00 p.m. local solar time. After one martian year, the orbit's inclination would be changed from 92.7 degrees to 90 degrees to enable radar sounding of previously unobserved portions of Mars's polar caps.

The spacecraft's instrument complement includes the following: full polarization, ultra-high frequency synthetic aperture radar; a dual-band ice-sounding radar; a high-resolution, multiband imager; shortwave and thermal infrared imaging spectrometers, dual stereo cameras, and a wide-angle multispectral imager.

Mission Challenges

The mission had no novel risks or technical challenges; accommodation of the large radar imager and sounder and their power and data needs drove the SEP implementation choice. The full mission design achieves the optimal science at a cost (per the PMCS estimate) commensurate with implementation as a medium-class mission.

Conclusions

MORIE science was deemed high priority. Given the initiation of iMIM by NASA as a directed mission (with its objective of ice mapping for in situ resources), which was announced after the PMCS was completed, MORIE may be better positioned as a follow-on to iMIM or elements of MORIE may be incorporated into iMIM to achieve Mars cryospheric science objectives.

MOSAIC

Origin and Rationale

This concept was proposed in response to the PMCS announcement of opportunity, and the study (CML ~4–5) was conducted at the Jet Propulsion Laboratory (NASA 2020c). The purpose of this mission is twofold: first, to understand the processes determining Mars's contemporary climate and, in particular, to delineate the interconnections between Mars's surface, upper and lower atmospheres, and ionized environment; and second, to identify hazards, characterize resources, and demonstrate technologies that might enable future human exploration activities on Mars.

Goals

The scientific questions this mission is designed to address are as follows:

- How do volatiles move between the subsurface, surface, and atmosphere?
- How does the martian lower-middle atmosphere respond diurnally, on meso- and global scales, to the seasonal cycle of insolation?
- How does coupling from the lower atmosphere combine with the influence of space weather (i.e., solar wind, solar energetic particles and solar extreme ultraviolet) to control the upper atmospheric system and drive atmospheric escape?

In addition, this concept is designed to address the following questions relating to the human exploration of Mars:

- How, where, and when can future astronauts access extractable water ice resources?
- With what degree of accuracy can martian weather be forecast, for operational purposes?
- How will mesospheric and thermospheric winds affect aerobraking spacecraft?
- How will space weather effects on the martian ionosphere affect surface–surface and surface–orbit communications?
- How will energetic particle radiation affect astronauts in Mars orbit?
- Can reliable high-bandwidth Earth–Mars communication be maintained?

Implementation

This large-class mission is designed to conduct simultaneous and systematic observations of the martian climate via eight science investigations carried out by 22 unique science instruments, hosted on 10 individual spacecraft. All are launched on a single Falcon Heavy Recoverable launch vehicle. The 10 spacecraft are deployed into three different orbits about Mars: low, near-polar Sun-synchronous; inclined elliptical; and areostationary. The Sun-synchronous orbits enable vertical profiling of wind, aerosols, water, and temperature, as well as mapping of surface and subsurface ice. The elliptical orbits sampling all of Mars's plasma regions enable multi-point in situ measurements necessary to understand mass/energy transport and ion-driven escape. The areostationary orbits enable synoptic views of the lower atmosphere, global views of the hydrogen and oxygen exospheres, and upstream measurements of space weather conditions.

Mission Challenges

MOSAIC is very clearly a large-class mission addressing multiple questions with a complex array of instruments deployed on 10 orbiters. Portions that might be flown as descoped constellations or stand-alone missions still may be too costly for consideration as medium-class missions. Moreover, such descope would lose the advantage of simultaneous, multi-point measurements at/from what was prioritized by the concept study team.

Conclusions

Given that the Mars Sample Return campaign is now under way, programmatic balance considerations argue against prioritization of another large-class Mars mission at this time. Considerable portions of the science may also be achievable within lower cost implementation choices, albeit with loss of synchronicity.

CROCODILE

Origin

This concept was proposed by the decadal survey's Panel on Small Solar System Bodies, and the study (CML ~4–5) was conducted at NASA's Goddard Space Flight Center in the early months of 2021 (NASA 2021b). The rational for the mission is to determine if the return of cryogenic samples of cometary material is feasible within the scope of a medium-class mission in the next decade. The concept explores the possibility of rendezvousing with a Jupiter-family comet, mapping its nucleus, selecting an optimal sampling site, sampling the nucleus below the surface, and returning cryogenically preserved material to Earth for laboratory analysis.

Goals

The overall science goal of the mission is to assess the elemental, isotopic, and structural composition of the volatile, organic, and inorganic components of a comet nucleus to address the following issues:

- The compositional reservoirs present in the early solar system;
- The role of comets in the delivery of water and organic molecules to the early Earth, terrestrial planets, and satellites; and
- The evolutionary processes spanning from the protoplanetary disk to current cometary activity.

Implementation

The study focused on sampling Comet 67P/Churyumov-Gerasimenko because of its suitable distance from Earth and its previous characterization by ESA's Rosetta spacecraft. Following launch on a Falcon Heavy Recoverable, the baseline mission envisages a solar-electric powered spacecraft undertaking a 5.7-year cruise to 67P.

Following arrival, 4.1 years is devoted to the selection of sampling sites and sample-collection campaign. The return cruise to Earth takes another 5.7 years.

The spacecraft's payload consists of a suite of infrared imaging spectrometers and cameras, a sample collection system, cryogenic sample Dewars, and the Earth Entry Vehicle. The Dewars containing solid argon are maintained at the desired temperatures by two-stage Stirling cryocoolers. A sample return temperature of 120 K was selected to preserve amorphous water ice and entrained volatiles, as well as water-soluble organics and salt. Two ~1–100 g samples are collected from ~25 cm below the surface of 67P's nucleus via the so-called shoot-and-go technique using a harpoon while the spacecraft remains 5–10 m above the surface.

Mission Challenges

The mission study report identified four major technological challenges: (1) operation of the cryocooler over the full 15.5-year mission duration; (2) confirmation of the successful collection of a small sample, possibly only ~1 g; (3) multiple aspects of the cryogenic sample acquisition and storage systems: for example, heat transfer associated with sampling and retrieval systems, the design and testing of long-term cryogenic storage assemblies and mechanisms, and long-life automated cryogenic seals; and (4) the high power requirements of the cryocooler and the solar-electric propulsion systems.

Conclusions

While the concept can address key science questions, the mission concept study suggested that overcoming the identified technical challenges pushed the estimated cost of the spacecraft beyond that for a medium-class mission.

PERSEPHONE

Origin and Rationale

This mission concept was proposed in response to the PMCS announcement of opportunity, and its study (CML ~4–5) was conducted at the Applied Physics Laboratory (NASA 2020d). Persephone was designed to address several key questions arising from the results returned by the New Horizons mission to Pluto and the Kuiper belt.

Goals

Persephone is designed to address the following primary questions:

- What are the internal structures of Pluto and Charon, and what is the evidence for a subsurface ocean on Pluto?
- How have surfaces and atmospheres in the Pluto system evolved?
- How has the KBO population evolved?

In addition, the mission could address the following secondary questions:

- What is Pluto's internal heat budget?
- What is Charon's magnetic field environment?
- How do KBOs and the heliosphere interact?

Implementation

This large-class mission is designed to launch on an SLS Block 2 with a high-energy upper stage. A combination of a Jupiter gravity assist (in the early 2030s or early 2040s) and a radioisotope electric propulsion system enables Persephone to put itself on a 28-year-long trajectory to the Pluto system. Some 19 years after launch, the spacecraft will encounter a 50–100 km diameter KBO. Once at Pluto, the spacecraft conducts a 3-year tour of the dwarf planet and its five known satellites. An optional extended mission, following the completion of the Pluto-system tour, could enable Persephone to encounter another 50–100 km diameter KBO some 36 years after launch. The spacecraft is equipped with 11 instruments including the following: plasma and spectrometers; panchromatic and color narrow-angle, panchromatic wide-angle, and thermal infrared cameras; near-infrared and ultraviolet spectrometers; a subsurface radar sounder; a laser altimeter; and a radio science system.

Mission Challenges

The 30+-year prime mission duration and the requirement for five next-generation radioisotope power systems strongly suggest that this concept's requirements are beyond the likely acceptable parameter range for implementation in the near to mid-term. Reliance on an SLS Block 2 also added significant risk. Moreover, detection of an internal ocean at Pluto is not straightforward because of the absence of a strong time-varying background magnetic field that would be certain to produce an inductive field in a conductive ocean, and the lack of time-varying tidal forcing owing to Pluto and Charon's dual synchronous state.

Conclusions

A Pluto orbiter mission is technically much more challenging than a flyby. Although determining whether Pluto has an ocean would be of great geophysical and astrobiological interest, the associated technical challenges and extremely long mission lifetime led to it being rated less highly than other large-class missions considered.

REFERENCES

NASA. 2020a. *In Situ Geochronology for the Next Decade*. Final Report Submitted in Response to NNH18ZDA001N-PMCS: Planetary Mission Concept Studies. Greenbelt, MD: NASA Goddard Space Flight Center. https://science.nasa.gov/solar-system/documents.

NASA. 2020b. *MORIE: Mars Orbiter for Resources, Ices, and Environments*. Mission Concept Study Report for the Planetary Science and Astrobiology Decadal Survey 2023–2032. Pasadena, CA: Jet Propulsion Laboratory, California Institute of Technology. https://science.nasa.gov/solar-system/documents.

NASA. 2020c. *MOSAIC: Mars Orbiters for Surface-Atmosphere-Ionosphere Connections*. Mission Concept Study Report for the Planetary Science and Astrobiology Decadal Survey 2023–2032. Pasadena, CA: Jet Propulsion Laboratory, California Institute of Technology. https://science.nasa.gov/solar-system/documents.

NASA. 2020d. *Persephone: A Pluto-System Orbiter and Kuiper Belt Explorer*. Mission Concept Study Report for the Planetary Science and Astrobiology Decadal Survey 2023–2032. Columbia, MD: Johns Hopkins University Applied Physics Laboratory. https://science.nasa.gov/solar-system/documents.

NASA. 2021a. *ADVENTS: Assessment and Discovery of Venus' Past Evolution and Near-Term Climatic and Geophysical State*. Mission Concept Study Report for the Planetary Science and Astrobiology Decadal Survey 2023–2032. Greenbelt, MD: NASA Goddard Space Flight Center. https://tinyurl.com/2p88fx4f.

NASA. 2021b. *CROCODILE: Cryogenic Return of Cometary Organics, Dust, and Ice for Laboratory Exploration*. Mission Concept Study to Report to the Planetary Science and Astrobiology Decadal Survey 2023–2032. Greenbelt, MD: NASA Goddard Space Flight Center. https://tinyurl.com/2p88fx4f.

Venera-D Joint Science Definition Team. 2019. *Venera-D: Expanding Our Horizon of Terrestrial Planet Climate and Geology Through the Comprehensive Exploration of Venus*. Phase II Final Report of the Venera-D Joint Science Definition Team. Washington, DC: NASA Headquarters and Moscow: Space Research Institute. https://www.lpi.usra.edu/vexag/meetings/meetings-of-interest/Venera-D-Report.pdf.

Appendix E

Panel Missions Not Selected for Additional Study

INTRODUCTION

As described in Appendix C, each of the survey's six panels reviewed the SDT and PMCS reports and assessed those concepts proposed by the community in white papers and prior proposals. Then the panels identified 18 additional large- and medium-class mission concepts that could address key scientific questions within their respective purviews. Following documentation by their originating groups, the steering group assessed each of these concepts and selected nine for additional study. This appendix describes the nine concepts that were not selected for additional study. These concepts are as follows, in no specific order:

- VISAN—Venus In Situ Seismic and Atmospheric Network;
- VSCA—Venus Sub-Cloud Aerobot;
- VLP—Venus Life Potential;
- VIDEO—Venus Investigation of Dynamics from an Equatorial Orbit;
- Mars Deep Time Rover;
- Mars Polar Ice, Climate, and Organics;
- Saturn Ring Skimmer;
- Interstellar Object Rapid Response Mission; and
- Solar System Space Telescope.

The following sections describes these nine concepts

VENUS IN SITU SEISMIC AND ATMOSPHERIC NETWORK

Origin and Goals

This concept was proposed by the Panel on Venus to address the following goals:

- Determine if Venus is seismically and/or volcanically active;
- Characterize the interior structure and thermal state of Venus;
- Establish meteorological conditions at the Venus surface; and
- Determine the morphology of the Venus surface at various spatial scales.

Implementation

This concept was proposed as a potential medium-class mission to be implemented by five long-long landers and a supporting orbiter. The landers touch down in a preplanned array within 500 ± 200 km of one another. The orbiter acts as communications relay and conducts a search for seismic activity via airglow emissions.

Nature of Study Requested

The panel requested that a rapid mission architecture (RMA) study (i.e., Concept Maturity Level [CML] ~1–3) be undertaken to assess the feasibility of precisely landing multiple assets simultaneously or in series, and sustaining communication.

VENUS SUB-CLOUD AEROBOT

Origin and Goals

This concept was proposed by the Panel on Venus to address the following goals:

- Establish the early evolution of Venus and determine if liquid water was ever present;
- Measure atmospheric dynamics and composition on Venus; and
- Characterize the geologic history preserved on the surface of Venus.

Implementation

This concept was proposed as a potential medium- or large-class mission to be implemented by a variable-altitude aerial platform and a supporting orbiter. The aerobot would conduct near-infrared imaging and spectroscopy from an altitude of 47 km (compared with 56–61 km for the aerobot component of the Venus Flagship concept). The concept envisaged that either entire platform descends to the requisite altitude or that a tethered instrument platform is lowered from an aerobot at a higher altitude.

Nature of Study Request

The panel requested a low-fidelity (CML ~1–3) study to descope the PMCS Venus Flagship mission and redesign it to focus on the feasibility of using an aerobot to conduct sub-cloud investigations.

VENUS LIFE POTENTIAL

Origin and Goals

This concept was proposed by the Panel on Venus to address the following goals:

- Characterize Venus's water history based on geologic and atmospheric markers;
- Characterize Venus's ultraviolet absorption component and its interdependency on atmospheric conditions; and
- Identify Venus's atmosphere and surface chemistry and the potential for extant cloud life.

Implementation

This concept was proposed as a large-class mission to be implemented via an orbiter, a large lander, and a fixed-altitude aerial platform. The orbiter is in a high, equatorial, retrograde orbit (cf. versus the polar orbit for radar-equipped VFM). The lander is designed to target a representative radar-smooth plains unit at low latitudes (cf. versus the tesserae for the Venus Flagship mission).

Nature of Study Requested

The panel requested a full study (i.e., CML ~4–5) using the Venus Flagship as its starting point to investigate an alternative to concept emphasizing a focus on the current potential for life in the Venus clouds.

VENUS INVESTIGATION OF DYNAMICS FROM AN EQUATORIAL ORBIT

Origin and Goals

This concept was proposed by the Panel on Venus to address the following goals:

- Characterize the dynamics within the Venus atmosphere;
- Measure the chemistry of the Venus atmosphere;
- Characterize the Venus ionosphere, and types and rates of atmospheric escape;
- Determine if Venus is seismically and/or volcanically active; and
- Establish the composition of major geological units on Venus.

Implementation

This concept was proposed as a large-class mission to be implemented via an orbiter, SmallSats, a fixed-altitude aerial platform, a small lander, and dropsondes. The orbiter is placed in a high, equatorial orbit (cf. versus the polar orbit of the Venus Flagship) so that it can continuously monitor the entire planetary disk, and thus increase communication duration with lander and aerial platform.

Nature of Study Requested

The panel requested a full study (i.e., CML ~4–5) using the Venus Flagship as its starting point to create a new concept emphasizing the study of atmospheric dynamics and the planet's super-rotation.

MARS DEEP TIME ROVER

Origin and Goals

This concept was proposed by the Panel on Mars to investigate Mars's Noachian/pre-Noachian terrains, the solar system's only preserved geologic record of the emergence of a habitable world (>3.5 Ga). The widespread presence of clay minerals in these terrains indicates long-duration water–rock interaction in environments whose nature is indeterminate from orbit. Examination of such terrains addresses the following goals relating to fundamental processes occurring during the first billion years in the evolution of habitable worlds, in that it would:

- Determine the composition and history of volcanic, sedimentary, and groundwater/hydrothermal pre-Noachian and Noachian martian units;
- Determine the suites of habitable conditions present on early Mars and what controls observed variability over space and time;
- Evaluate the surface inventory of water and other volatiles, including change with time by isotopic and mineralogical/chemical analysis; and
- Measure, if feasible, the ages of rock units and surfaces.

Implementation

This concept was proposed as a medium-class mission to be implemented via a rover using an upgraded airbag landing system and having greater operating autonomy than its MER predecessors. The rover would carry mast and navigational cameras, an infrared spectrometer, a magnetometer, a mass spectrometer, and a meteorological package, as well as an arm-mounted X-ray and Mössbauer spectrometers, and a microscope. The arm would carry a rock-powder acquisition system.

Nature of Study Request

The panel requested a full mission study (i.e., CML ~4–5) to assess if the PMCS Intrepid lunar rover concept (see Appendix C) could be adapted for use on Mars. Among the issues to be assessed during the study were the following: Can the landing ellipse be reduced to <25 km? Can the rover drive 300 m per day? Is it feasible to incorporate a sample-handling system and a mass spectrometer on an MER-class rover? Is a solar power system sufficient?

MARS POLAR ICE, CLIMATE, AND ORGANICS

Origin and Goals

This concept was proposed by the Panel on Mars to understand the processes controlling the modern-day evolution of the martian climate by studying the climatic record encoded in the northern polar layered deposits (NPLD). Such studies are needed to connect orbital measurements of NPLD with modeling of climate change induced by variations in the planet's orbital elements over the past 10 million years. The specific goals of the proposed mission are as follows:

- Measure the composition, structure, and sub-mm layering of ice and dust contained in the top 2 m of the NPLD;
- Quantify the evolution of the NPLD surface, including the net accumulation of dust and ice over 1 martian year;
- Measure the present-day surface energy balance and the fluxes of heat, momentum, dust, and volatiles between the atmosphere and surface over 1 martian year; and
- Measure the polar atmospheric circulation and the transport of water and dust in the planetary boundary layer over 1 martian year.

Implementation

This concept was proposed as a medium-class lander equipped with a drill capable of penetrating at least 50 cm (ideally 2 m) into ice layers for in situ measurement. Extraction of samples and their delivery instruments mounted on the spacecraft's deck is also required. The spacecraft lands on ice-covered NPLD shortly after the loss of the north polar seasonal ice cap. Drilling continues through the summer and is followed by meteorological and flux measurements for the remainder of 1 martian year. Instruments include a meteorology and flux package; a borehole microscopic imager, and a thermoelectric conductivity probe, as well as a suite of sample analysis instruments on the lander's deck for compositional measurements.

Nature of Study Request

The panel requested a full study (CML ~4–5) to investigate the adaptation of the ICE-SAG (NASA 2019) and *Vision and Voyages* polar lander (NRC 2011, p. 362) concepts to one full Mars year of operations and the use of a radioisotope power source. The study would also assess how to optimize the drill and on-board instrumentation for the search for organic material in the ice.

SATURN RING SKIMMER

Origin and Goals

This concept was proposed by the Panel on Giant Planet Systems to address the following goals:

- Quantify the processes operating in particle-rich astrophysical disks, including particle aggregation and fragmentation; and
- Probe the origins and history of Saturn's rings.

Implementation

This concept was proposed as a medium-class Saturn orbiter in a low-inclination orbit (pericenter 90,000–130,000 km to avoid both the F and main rings). Repeated passes over the rings at altitudes <1,000 km (i.e., 100 times lower than Cassini) would enable extremely high-resolution imaging and sampling of the material flowing between the rings and the planet. To maintain pointing at one location (not predetermined) in the rings for at least 1,000 seconds requires that the spacecraft incorporate a scan platform or be able to slew. The in situ instruments are pointed in the appropriate ram directions to obtain sufficient fluxes to measure composition or be slewed to constrain particle and molecular velocity distributions.

Instruments include the following: a high-resolution camera (to image five regions in the rings at ≤1 m spatial resolutions, taking at least four images of the same region separate in time by at least ~250 seconds); a dust detector (capable of sensing particles down to 10 nm in radius, determining particle masses and velocities, and sensitive enough to measure dust fluxes from the rings between 10^{-15} and 10^{-13} kg/m^2/s and densities between 1 and 1,000/cm^3 over each of the main rings); and an ion and neutral mass spectrometer (capable of sensing between 1 and 100 amu and determining molecular velocities, and sensitive enough to measure average ion/molecule fluxes from the rings between 10^{-15} and 10^{-13} kg/m^2/s and densities between 0.1 and 10^5/cm^3 over each of the main rings). The instruments are also required to measure the mass fractions of the material flowing out of the rings in the form of water, organics, and silicates to an accuracy of at least 10 percent and determine isotope ratios for selected CHO materials.

Nature of Study Request

The panel requested a full mission study (i.e., CML ~4–5).

INTERSTELLAR OBJECT RAPID RESPONSE MISSION

Origin and Goals

This concept was proposed by the Panel on Small Solar System Bodies to investigate the acquisition of time-critical spacecraft observations of novel solar system phenomena. The discovery of the first interstellar objects—1I/'Oumuamua in 2017 and 2I/Borisov in 2019—transiting the solar system was one of the key scientific results of the past decade. When the National Science Foundation's Vera C. Rubin Observatory becomes operational later this decade, the rate at which interstellar objects are discovered is expected to rise to one per year. The ability to design, develop, and deploy a spacecraft mission on a prompt timescale is also of interest for addressing other classes of opportunistic objects (e.g., Oort cloud comets) and for planetary defense purposes (see Chapters 18 and 22). However, the current structure of mission opportunities within NASA's Planetary Science Division does not lend itself to opportunistic, rapid response missions to newly identified targets of high scientific value. Therefore, the goal of the study is to investigate novel programmatic approaches to the rapid procurement, development, and targeting of objects of intrinsic scientific or societal interest.

Implementation

This concept was proposed as an examination of novel mission architectures to enable small-class spacecraft to undertake rapid response missions in the next decade. Three principal architectures are to be considered: first, build the spacecraft and store on ground until needed; second, build the spacecraft and store in space until discovery of the object of interest (e.g., ESA Comet Interceptor); and third, build the spacecraft after the discovery of the object of interest.

Nature of Study Request

The panel requested that an RMA study (i.e., CML ~1–3) be undertaken to assess the feasibility, resources, and cost implications of the three programmatic architectures described above. Also to be examined were the benefits and consequences of a redundant multi-spacecraft architecture.

SOLAR SYSTEM SPACE TELESCOPE

Origin and Goals

This concept was proposed by a cross-panel group to promote new innovative science and to support and exploit discoveries made by past/current/future missions to specific targets. A space telescope dedicated to observations of solar system bodies would benefit the planetary science community by enabling capabilities beyond those of current astrophysics missions and provide more opportunities for high-priority solar system science studies. The goals of the telescope are as follows:

- Provide significant improvement in capability and mission timeline compared to current observatories to explore dynamic processes and systems via long baseline time-domain measurements; and
- Expand spectroscopic mapping and improves angular resolution of small body populations, including KBOs.

Implementation

This concept was proposed as a medium-class, 2–10 m dedicated telescope with ultraviolet/visible/near-infrared imaging and spectroscopic capabilities, optimized for cadence and survey observations of transient, evolving, and interacting processes in the solar system. Desirable characteristics include the following: ability to operate at solar elongations $\geq 30°$; a tracking capability of $\geq 216''$/hour; instrument field of view of $60''$; and minimum lifetime of 7 years.

Nature of Study Request

The cross-panel group requested that an RMA study (i.e., CML ~1–3) be undertaken to investigate the benefits of a dedicated solar system space telescope, to perform trade studies to define an overall optimal architecture and payload, and to demonstrate feasibility within the constraints of the New Frontiers program.

REFERENCES

NASA. 2019. *MEPAG ICE-SAG Final Report (2019)*. Report from the Ice and Climate Evolution Science Analysis Group (ICE-SAG). Pasadena, CA: MEPAG (Mars Exploration Program Analysis Group).

NRC (National Research Council). 2011. *Vision and Voyages for Planetary Science in the Decade 2013–2022*. Washington, DC: The National Academies Press.

Appendix F

Glossary and Acronyms

AAAC: Astronomy and Astrophysics Advisory Committee.

Abiosignature: A characteristic that mimics a biosignature but is not associated with life.

Abiotic: Of or relating to nonliving things; independent of life or living organisms.

Accretion: The growth of a massive object by gravitationally attracting more matter, typically gaseous matter, in an orbiting accretion disk, causing the object to grow larger, hotter, and more luminous. A related term is used in meteorology for the process of accumulation of frozen water as precipitation over time as it descends through the atmosphere—that is, the basis of cloud formation.

Additive manufacturing: Set of manufacturing techniques based on adding layer-upon-layer of material. The most well known of such methods is 3D printing.

Aero-assist: Use of a body's atmosphere to modify the trajectory of a spacecraft. Aerocapture is an example of such a maneuver.

Aerocapture: Deliberate use of an aerodynamic drag in an atmosphere to facilitate orbital capture of a spacecraft without the use of thrusters.

AIAA: American Institute of Aeronautics and Astronautics.

Albedo: The fraction of light that is reflected from the surface of a planetary body.

ALHAT (Autonomous Landing Hazard Avoidance Technology): Safe landing technology development program, predecessor to SPLICE.

AO: Announcement of Opportunity.

Archean: The second geologic eon on Earth, occurring after the Hadean and lasting from 4 billion to 2.5 billion years ago.

Artemis: NASA's current program to return humans to the Moon later this decade.

ASI: Agenzia Spaziale Italiana, the Italian space agency.

ASTROMAT: Astromaterials data system, NASA's repository for extraterrestrial materials and laboratory data related to them.

AU (astronomical unit): The mean distance from the Earth to the Sun.

Autotroph: An organism that utilizes inorganic sources as nutrition and energy sources.

AVGR: NASA's Ames Vertical Gun Range.

Axel: Family of rover platforms developed at the Jet Propulsion Laboratory, providing versatile mobility on planetary surfaces.

Biogenesis: The production of substances through biological processes.

Biosignature: A characteristic that can be interpreted as evidence of life.

Biosphere: The layer of a planet where life exists; the sum total of a planet's ecosystems.

Biotic: Relating to life or living organisms.

Bistatic: A radar technique where the transmitter and receiver are at different locations.

BOE: Basis of estimate.

Bolide: An extremely bright meteor caused when an extraterrestrial body enters into and explodes in Earth's atmosphere. Portions of the object's body—meteorites—may survive to reach Earth's surface.

Bouguer gravity: A gravitational anomaly that requires the incorporation of terrain height for correction.

Breccias: A type of sedimentary rock that comprises various minerals bound in a matrix that holds them together.

CAPTEM: Curation and Analysis Planning Team for Extraterrestrial Materials.

CDR (Critical Design Review): During phase C, evaluation of full maturity of a mission or technology that it is ready to complete design and fabrication.

Channel coding: Process of detecting and correcting bit errors in digital communication systems.

Chemoautotropic: An organism, typically bacteria or archaea, that can derive energy from chemical reactions involving inorganic molecules.

Chemolithoautotroph: A chemoautotropic organism whose metabolism is supported via inorganic compounds derived from minerals.

Chirality: The right- or left-handedness of an asymmetric molecule.

Chondrite: A stony meteorite, unaltered from its parent body.

Chondrule: Round grains that make up a fraction of chondrites, formed from molten or partially molten droplets of minerals.

Circumstellar disk: A broad ring of material orbiting around a star.

Clathrate: A chemical substance consisting of a lattice of one type of molecule (e.g., water) trapping and containing a second type of molecule (e.g., methane).

CLPS: Commercial Lunar Payload Services.

CML (Concept Maturity Level): Index of mission concept maturity ranging from 1 (rudimentary sketch) to 9 (critical design review stage).

Co-I: Co-investigator.

COLDTech: Concepts for Ocean Worlds Life Detection Technology.

Corona: The outermost layer of a star.

CRISM: Compact Reconnaissance Imaging Spectrometer for Mars.

Cryogenics: The branch of physics dealing with the behavior of matter at very low temperatures.

Cryovolcanism: The eruption of water and other volatile materials onto the surface of a planet or moon owing to internal heating.

CSA: Canadian Space Agency.

D/H ratio: Deuterium to hydrogen ratio.

DAPR: Dual-anonymous peer review.

DART: Double Asteroid Redirection Test.

Delta-V (ΔV): Impulse per unit of spacecraft mass (i.e., velocity change) that is needed to perform a given maneuver.

Derivatization: Modifying a chemical compound to give it properties more amenable to a particular analytical method.

Diagenesis: The physical and chemical changes occurring in sedimentary materials during and after rock formation at temperatures and pressures less than what is required for the formation of metamorphic rocks but that excludes surface alteration.

Dilute core: An extended region of enriched elements heavier than hydrogen or helium in the deep interior of the giant planets.

DNA (deoxyribonucleic acid): The genetic biopolymer found in most if not all organisms on Earth. Its helical structure is maintained by the presence of four nucleobases: adenine, guanine, cytosine, and thymine.

DOE: Department of Energy.

Dragonfly: A NASA rotorcraft mission scheduled to be sent to Titan in the late 2020s.

Dynamo: An electromagnetic process in which the movement of conductive material gives rise to a magnetic field.

EAR: Export Administration Regulations.

Eccentricity: A measurement of the degree to which an elliptical orbit deviates from a circular orbit. An ellipse of zero eccentricity is a circle.

Egalitarianism: The idea that all people deserve equal opportunities.

EIL: NASA Johnson Space Center's Experimental Impact Laboratory.

Electromagnetic sounding: Determination of variations in electrical conductivity with depth.

EM: Electromagnetic.

Endogenic: Relating to a process of internal origin.

Energy density: The amount of energy per unit volume.

EPO: Education and Public Outreach.

ESA: European Space Agency.

ESDMD: NASA's Exploration Systems Development Mission Directorate.

ESSIO: NASA's Exploration Science Strategy and Integration Office.

Europa Clipper: A large NASA multiflyby mission scheduled for launch to Europa in the mid-2020s.

EVA: Extravehicular activity.

ExMAG: Extraterrestrial Materials Assessment Group.

Exogenic: Relating to a process of external origin.

Exoplanets: Planets formed around stars other than the Sun.

Extremophile: An organism that is capable of living in extreme physical or chemical conditions, including high or low temperatures, high salinity, and intense radiation.

False negative: Results that wrongly indicate a negative result.

False positive: Results that wrongly indicate a positive result.

Felsic: Igneous rocks enriched in lighter elements (e.g., silicon, oxygen, aluminum, sodium, and potassium) that form feldspar and quartz.

FIB: Focused ion beam.

Flux: A measure of the energy or number of particles passing through unit area in unit time.

Flyby: Operation in which a spacecraft passes in close proximity to another body but is not in orbit about it.

Forward contamination: Contamination of extraterrestrial bodies with terrestrial life forms (see planetary protection).

FPS: Fission power system.

Ga: A billion years ago.

GBT: Green Bank Telescope.

GEISA: A spectroscopic database maintained by NASA.

GEO: Geosynchronous Earth orbit.

Geodetic: Relating to investigations of the shape of a planetary body.

Geosphere: The solid layers of a planet, including its core.

GitHub: A repository for software.

GNC: Guidance, navigation, and control.

GOES: Geostationary Operational Environmental Satellites.

GPHS: General-purpose heat source.

GPU (graphics processing unit): Specialized electronic circuit designed to rapidly manipulate and alter memory to accelerate the creation of images.

GRAIL (Gravity Recovery and Interior Laboratory): A NASA orbiter that investigated the Moon's interior structure.

Graphene: Allotrope of carbon with desirable physical and electrical properties.

Gravity assist: Use of the relative movement and gravity of an astronomical object to alter the path and speed of a spacecraft.

GSFC: Goddard Space Flight Center.

GSSR: NASA's Goldstone Solar System Radar.

Habitability, habitable: The ability of a planet to harbor life at some specific time, but it does not necessarily do so.

Habitable zone: The zone surrounding a star at which liquid water could exist on the surface of a planetary body with a sufficiently dense atmosphere.

Hadean: The first geologic eon on Earth, lasting from the planet's formation 4.54 billion years ago to 4 billion years ago.

Hadley: A type of convective circulation pattern seen in planetary atmospheres.

Hayabusa 2: The JAXA sample-return mission to asteroid Ryugu.

HD&A: Hazard detection and avoidance.

HEEET: Heatshield for Extreme Entry Environment Technology.

HEOMD: Human Exploration and Operations Mission Directorate.

HIAD: Hypersonic Inflatable Aerodynamic Decelerator.

HIHT: High Irradiance, High Temperature.

HIITRAN: The high-resolution transmission molecular absorption database.

HLS: Human Landing System.

HOTTech (High Operating Temperature Technology): A NASA technology development program.

Hydrocode: Complex computer simulations of dynamic fluid events such as the formation of an impact crater.

Hydrosphere: All bodies of water on a planet, as distinguished from the lithosphere and the atmosphere.

Hydrothermal: Relating to the action of hot liquid or gas within or on the surface of a planet.

IAA: International Academy of Astronautics.

IAWN: International Asteroid Warning Network.

IBD: Ion Beam Deflection.

IMIM: International Mars Ice Mapper, a proposed radar mapping mission.

Infrared spectroscopy: Measurement of the interaction of infrared radiation with matter by absorption, emission, or reflection.

Inorganic: Compounds not associated with life and/or not containing carbon.

Interstellar medium: Space between the stars, made of gas (primarily helium and hydrogen).

Ionosphere: The region of a planet's atmosphere that is kept partially ionized by solar ultraviolet and X-ray irradiation.

IRTF: NASA's Infrared Telescope Facility.

ISE: Initial Schedule Estimate.

ISFM: Internal Scientist Funding Model.

Isotope: One of two or more atoms of the same element that have the same number of protons in the nucleus but a different number of neutrons.

Isotopic: Species of the same element that have different atomic weights.

ISP: Specific impulse of a propulsion device. A measure of how efficiently a reaction system creates thrust.

ISRO: Indian Space Research Organization.

ISRU: In Situ Resource Utilization.

ITAR: International Traffic in Arms Regulations.

JAMSTEC: Japan Agency for Marine-Earth Science and Technology

JAXA: Japan Aerospace Exploration Agency.

Jovian: Pertaining to the planet Jupiter or a similar-size exoplanet.

JPL: Jet Propulsion Laboratory.

JUICE: Jupiter Icy Moons Explorer. An ESA Ganymede orbiter mission scheduled for launch in the mid-2020s.

JWST (James Webb Space Telescope): A joint NASA-ESA 6.5-meter space telescope launched in December 2021.

KBO: Kuiper Belt object.

KDP (key decision point): Events at which the decision authority determines the readiness of a program/project to progress to the next phase of the life cycle.

Keck: Keck Observatory.

Kepler mission: NASA missions that searched for exoplanets using the transit technique.

Kerogen: Naturally occurring insoluble matter found in sedimentary rock.

Kuiper Belt: A region of the outer solar system containing icy planetesimals distributed in a roughly circular disk some 40 to 100 AU from the Sun.

LADEE: Lunar Atmosphere, Dust, and Environment Explorer.

LCRD: Laser Communications Relay Demonstration. NASA mission to showcase the use of optical communication in space.

LEAG: Lunar Exploration Analysis Group.

LEO: Low Earth orbit.

Lidar: Light detection and ranging.

LIHT: Low irradiance, high temperature.

Lithology: The study of rocks' physical characteristics and formation.

Lithosphere: The outermost shell of a rocky planet. On Earth, the lithosphere is the crust and the relatively elastic portion of the upper mantle.

LLNL: Lawrence Livermore National Laboratory.

Lobate scarp: Curvilinear structure found on the surfaces of some planetary bodies.

LRO: Lunar Reconnaissance Orbiter.

LRV: Lunar Roving Vehicle.

M dwarf: Sometimes called "red dwarfs," these stars are the smallest type on the main sequence.

Ma: A million years ago.

Macromolecule, macromolecular: Of or pertaining to a molecule—for example, a nucleic acid, protein, or synthetic polymer—containing a very large number of atoms.

Magnetosphere: The region of space in which a planet's magnetic field dominates that of the solar wind.

Magnetotail: The portion of a planetary magnetosphere pulled downstream by the solar wind.

Mantle: The part of a planet between its crust and core, composed of relatively dense materials.

MAPSIT: Mapping and Planetary Spatial Infrastructure Team.

Mars 2020: NASA's Perseverance sample-collecting rover mission.

Mascons: A large positive gravitational anomaly located on an area of a planetary body's surface.

Mass spectrometry: Analytical technique used to measure the mass-to-charge ratio of ions.

MAST (Mikulski Archive for Space Telescopes): A NASA repository for data from space astronomy missions.

MASWG: Mars Architecture Strategy Working Group.

MatISSE: Maturation of Instruments for Solar System Exploration. NASA program that supports the development of spacecraft-based instruments, covering TRL-4 to TRL-6.

MAVEN: Mars Atmosphere and Volatile Evolution mission.

MEDA (Mars Environmental Dynamics Analyzer): Instrument that makes weather measurements including wind speed and direction, temperature and humidity, and also measures the amount and size of dust particles in the martian atmosphere.

MEDLI (Mars Entry, Descent and Landing Instrument): Used to monitor the performance of a spacecraft's heatshield during atmospheric entry.

MEMS: Micro-electromechanical systems.

MEP: Mars Exploration Program.

MEPAG: Mars Exploration Program Analysis Group.

MESSENGER: NASA's Mercury Surface, Space Environment, Geochemistry, and Ranging mission.

Metabolism: A set of chemical reactions that change nutrients into the energy needed by a living organism to maintain structure, grow, and replicate.

MExAG: Mercury Exploration Assessment Group.

MMRTG: Multi-Mission Radioisotope Thermoelectric Generator converts the heat from the decay of radioactive plutonium-238 to electrical power.

MMX (Martian Moons Exploration): A Japanese mission to return samples from Phobos.

Monomer: A subunit of a polymer, when bonded to other identical subunits.

MOXIE (Mars Oxygen In Situ Resource Utilization Experiment): An oxygen-producing ISRU experiment on the Perseverance rover.

MSFC: NASA's Marshall Space Flight Center.

MSR: Mars sample return.

Muon tomography: Use of cosmic ray muons to construct three-dimensional models of the density distribution in an object.

MWG: Modeling Working Group.

Nanomolar level: Measure of concentration of a given chemical component.

NASA: National Aeronautics and Space Administration.

Nascent: Emerging and coming into existence.

N-body simulations: A computer simulation of a dynamic behavior of a system consisting of a large number of particles as they interact in response to gravitational and/or other forces.

Nebula: A cloud of gas and dust in space.

NED: Nuclear explosive device.

NEO: Near-Earth object.

NEOCam (Near-Earth Object Camera): The former name of NASA's NEO Surveyor mission.

NEOWISE: The NEO-search phase of NASA's Wide-field Infrared Survey Explorer mission.

NEP: Nuclear-electric propulsion.

New Frontiers mission: NASA missions of intermediate size and cost between Discovery and large-class (flagship) missions.

New Millennium: A now-discontinued technology demonstration program initiated by NASA in 1995.

NEX-SAG: Next Orbiter Science Analysis Group.

NGRTG: New Generation Radioisotope Thermoelectric Generator (see RTG).

NLSI: NASA Lunar Science Institute.

NNSA: National Nuclear Security Administration.

NOAA: National Oceanic and Atmospheric Administration.

Noachian: The oldest of three time periods into which the geologic history of Mars has been divided, spanning from about 4.1 billion to about 3.5 billion years ago.

NoDD: No Due Date, a type of research proposal that has no specific submission deadline.

Nonsidereal: The motion of a celestial body independent of that which appears to be caused by a planet's rotation.

NSA (NASA Standard Assay): Procedure to verify spacecraft cleanliness for missions with bioburden requirements.

NSF: National Science Foundation.

NSTC: National Science and Technology Council.

NTP: Nuclear-thermal propulsion.

Nucleobase (or Base): A class of compounds with the ability to bond pairwise with each other and stack vertically. These properties lead to the helical structure of DNA and RNA. The five primary nucleobases are guanine, cytosine, adenine, thymine, and uracil. Some viruses make use of aminoadenine in place of adenine.

Occultation: An event that occurs when one object is hidden by another object that passes between it and the observer. Contrast with a transit, when a smaller object passes in front of a larger one.

Ocean world: A planetary body having substantial amounts of liquid water on its surface or as a subsurface ocean. In the solar system, established examples include Earth, Callisto, Enceladus, Europa, Ganymede, and Titan.

Oort cloud: A spherical distribution of comets having semi-major axes between 1,000 and 50,000 AU, typically with low orbital eccentricity.

OPAG: Outer Planets Assessment Group.

Opportunity: Second of the two rovers launched in 2003 to land on Mars and search for signs of ancient water.

OPUS (Outer Planets Unified Search): A NASA database containing images and spectra of objects in the outer solar system.

Organic molecule: A chemical that contains carbon and hydrogen bonds.

ORR (operational readiness review): During Phase D, the state of the built system and procedures are examined to ensure that expectation match reality for the system and support hardware, software, personnel, procedures, and documentation.

OSIRIS-REx: Origins, Spectral Interpretation, Resource Identification, Security, Regolith Explorer. A NASA mission currently returning a sample from the asteroid Bennu to Earth.

Outgassing: Venting of gasses from the crust of a planetary body.

Oxygenic photosynthesis: Photosynthesis during which oxygen is produced and released.

PAH: Polycyclic aromatic hydrocarbons

Paleoclimate: The climate of some former period of geologic time.

Paleomagnetic: Area of study investigating the history of a planetary body's magnetic field through rock and sediment.

Pan STARRS: Panoramic Survey Telescope and Rapid Response System.

PD (Planetary Defense): Measures designed to protect Earth from the hazards posed by the impact by asteroids and comets.

PDCO: NASA's Planetary Defense Coordination Office.

PDR (Preliminary Design Review): A technical review held during Phase B of a project designed to demonstrate that a system's preliminary design meets all requirements with acceptable risk and within cost and schedule constraints.

PDS (data): NASA's Planetary Data System.

PDS (star): An astronomical object catalogued during a survey conducted at the Pico dos Dias Observatory in Brazil.

Perchlorate: A salt containing the ClO_4^- ion.

PESTO: Planetary Exploration Science Technology Office.

Petrologic: Area of geology that studies the conditions under which rocks form.

Phase A: Concept and technology development.

Phase B: Preliminary design and technology completion.

Phase C: Final design and fabrication.

Phase D: System assembly, integration and test, and launch and checkout.

Phase E: Operations and sustainment.

Phase F: Closeout.

Photodissociation/Photolysis: The breakup of molecules through exposure to light.

Phylogenetic: Relating to the study of the evolutionary diversification of organisms over time.

PI: Principal investigator.

PICA (Phenolic-Impregnated Carbon Ablator): A thermal protection technology developed by NASA for spacecraft heatshields.

PICASSO (Planetary Instrument Concepts for the Advancement of Solar System Observations): NASA program supporting the initial development of spacecraft-based instrument systems, covering technology readiness level 1 to 3.

Plagioclase: A group of minerals under the feldspar classification.

Planetary protection: Measures designed to protect Earth and other planetary bodies from cross-contamination by biological materials carried on spacecraft.

Planetesimal: A rocky and/or icy body a few kilometers to several tens of kilometers in size, which was formed in the protoplanetary nebula.

Plasma: A highly ionized gas, consisting of almost equal numbers of free electrons and positive ions.

Porosity: The percentage of the total volume of a body that is made up of open spaces.

Prebiotic: Relating to the chemical reactions that naturally occur in planetary environments that are important for the creation of biologically relevant compounds (e.g., amino acids and sugars). Such reactions may be relevant to the origins of life.

Pre-Phase A: Concept studies.

PRISM: Payloads and Research Investigations on the Surface of the Moon.

Protoplanet: A planet in the process of accreating material in a protoplanetary disk.

Protoplanetary disk: A circumstellar disk of matter, including gas and dust from which planets may eventually form or be in the process of forming.

Protoplanetary gas nebula: The phase in the evolution of a protoplanetary disk prior to the dispersal of its gaseous component.

PSD: NASA's Planetary Science Division.

PSTAR: Planetary Science and Technology from Analog Research. NASA program addressing the need for field experiments to prepare for human and robotic missions.

Psyche: A NASA spacecraft scheduled to orbit the asteroid Psyche to investigate the origin of planetary cores.

Plutonium-238 (^{238}Pu): A radioactive isotope of plutonium decaying via the emission of an alpha particles and used as a source of heat and electrical power.

Pyroclastic: Materials that are composed solely or primarily of fragments of volcanic rocks.

Radiative balance: Accounting for all sources of ingoing and outgoing sources of radiation in a system.

Radioisotope thermoelectric generator: Nuclear power source using the heat produced by radioactive decay (typically **Phase A:** Plutonium-238) to generate electricity.

Radiolysis: The dissociation of molecules by ionizing radiation.

Radionuclides: A radioactive nuclide.

Redox reaction: Any of a large class of chemical reactions characterized by the reduction of one reactant and the simultaneous oxidation of another.

Regolith: The layer of dust and fragmented rocky debris that forms the uppermost surface on many planets, satellites, and asteroids. It is formed by a variety of processes, including meteoritic impact.

RELAB: Reflectance Experiment Laboratory.

Rendezvous: Set of orbital maneuvers during which two spacecraft arrive at the same orbit and approach to a very close distance.

Retrograde: Orbital motion in the opposite sense to the rotation of Earth and most of the other planetary bodies in the solar system.

RF: Radio frequency.

RNA (ribonucleic acid): A biopolymer, similar to DNA, whose single- or double-stranded helical structure is maintained by four nucleobases: adenine, uracil, cytosine, or guanine. Its principal role is to act as a messenger carrying instructions, encoded in DNA's nucleobase pairs, for the synthesis of proteins. In some viruses, genetic information is encoded in RNA rather than DNA.

RoI: Return on investment.

ROSES (Research Opportunities in Space and Earth Sciences): NASA's omnibus program for funding external research.

Rosetta: An ESA spacecraft rendezvous with Comet 67P/Churyumov-Gerasimenko in 2014.

RPS: Radioisotope power system.

RTG: Radioisotope Thermoelectric Generator.

S&TU: Science and technology utilization.

Saltation: A type of particle transport, occurring when loose materials are removed from a surface and carried by a fluid.

SAM: Sample Analysis at Mars instrument on the Curiosity rover.

SBAG: Small Bodies Assessment Group.

SDT: Science Definition Team.

Secondary cratering: Cratering created from the projectiles during larger crater's creation.

SEP: Solar Electric Propulsion.

SESAME: Scientific Exploration Subsurface Access Mechanism for Europa.

SHARAD: Mars Shallow Radar sounder. An instrument on Mars Reconnaissance Orbiter identifying water and ice deposits under the Mars surface.

SIAD: Supersonic Inflatable Aerodynamic Decelerator.

Siderophile: Those elements that tend to sink into a planets core rather than form part of its silicate crust.

SIMPLEx: Small Innovative Missions for Planetary Exploration. NASA's class of very small, low-cost, high-risk missions.

SLS: Space Launch System.

SMD: NASA's Science Mission Directorate.

SOFIA: Stratospheric Observatory for Infrared Astronomy.

Solar nebula: The cloud of gas and dust from which the Sun, the planets, and other bodies in the solar system were formed.

SOMD: NASA's Space Operations Mission Directorate.

SoP: State of the profession.

South Pole–Aitken Basin: The largest, deepest, and oldest impact basin on the Moon.

Space weathering: Alteration of an atmosphereless planetary body's surface materials by exposure to the space environment.

Specific energy: Energy per unit mass.

SpecLib: Spectral Library.

Spectral resolution: A measurement of the ability to resolve the features of an electromagnetic spectrum.

Spectroscopy: The process of dissecting electromagnetic radiation from an object into its component wavelengths to determine its chemical composition.

Spectrum congestion: Radio signal integrity issues created when multiple sources transmit simultaneously on similar frequencies.

SPHERE (Spectro-Polarimetric High-Contrast Exoplanet Research): Instrument on the European Southern Observatory's Very Large Telescope (VLT) providing images and characterization of exoplanets.

SPICE (Spacecraft, Planet, Instrument, Camera Matrix, Events): A series of files (known as kernels) containing ancillary data relating to the time, location, orientation, and other parameters describing the circumstances under which a specified instrument made specific observations.

SPLICE: Safe and Precise Landing Integrated Capabilities Evolution.

SR: Strategic research.

SRL: Sample Return Lander.

SRR (System Requirements Review): During Phase A, confirms that program requirements are well posed and align with agency strategic objectives.

SSERVI: Solar System Exploration Research Virtual Institute.

Stardust: NASA mission to return samples of a comet's coma to Earth.

STMD: NASA's Space Technology Mission Directorate.

Stratigraphy: An area of geology that investigates rock layers and their interpretation.

Subduction: Tectonic movement where one plate will move downward underneath another plate for recycling.

Sublimation: The act of changing a substance directly from a solid to a gas without it passing through a liquid stage.

Surface morphology: The structure and form of a particular surface.

Synestia: A rapidly rotating toroidal cloud of vaporized rock resulting from an impact between two planetary bodies.

TAGSAM (Touch and Go Sample Acquisition Mechanism): Typically used to collect material from planetary bodies with very weak gravitational fields.

TALOS (Thruster Advancement for Low-temperature Operation in Space): Development program for small and light thrusters operating at low temperatures.

Taphonomy, taphonomic: Of or relating to the study of processes that result in the formation, preservation, alteration, and destruction of biosignatures.

Tectonism: The processes of faulting, folding, or other deformation of the lithosphere of a planetary body, often resulting from large-scale movements below the lithosphere.

TEM: Transmission electron microscopy.

Terrain relative navigation: Autonomous localization and navigation based on the detection of specific terrain features with optical camera, radar or lidar systems.

Terrestrial: Relating to the planet Earth.

TESS: Transiting Exoplanet Survey Satellite.

Tesserae: Unique geological features found on some of the plateau highlands on Venus and characterized by their extremely rugged topography. They are also known as complex ridged terrain and are believed to be formed when crustal stresses cause the surface to fold, buckle, and break.

TPS: Thermal protection system.

Trans-Neptunian Region: The region of space beyond Neptune's orbits.

TRAPPIST: Transiting Planets and Planetesimals Small Telescope. TRAPPIST consists of a pair of robotic optical telescopes. One is located at La Silla Observatory in Chile and the other is at Oukaïmeden Observatory in Morocco.

TRL (technology readiness level): Method of estimating the maturity of a technology. NASA defines nine readiness levels, ranging from consistent with basic physical principles (TRL-1) to fully ready "flight proven" systems (TRL-9).

TRN (terrain relative navigation): Autonomous localization and navigation based on the detection of specific terrain features with optical camera, radar, or lidar systems.

Trojan: Very small body located near the two stable gravitational points 60 degrees preceding and following a planet in its orbit.

UCIG: Utilization, Coordination, and Integration Group.

UNCOPUOS: United Nations Committee on the Peaceful Uses of Outer Space.

URC: Under Represented Communities.

USGS: U.S. Geological Survey.

UV: Ultraviolet.

VEXAG: Venus Exploration Analysis Group.

Vibrational spectrometry: See infrared spectroscopy.

Volatile: A substance that vaporizes at a relatively low temperature.

Volcanic resurfacing: Significant changes to a planetary object's surface owing to the effects of volcanic activity.

Voyager 1 and 2: NASA probes launched in the late 1970s to study the outer planets.

V&V: Vision and Voyages for Planetary Science in the Decade 2013–2022, the second planetary science decadal survey.

WISE (Wide-Field Infrared Survey Explorer): A NASA infrared space telescope launched in 2009.

X-ray computed tomography: An analytical technique used to determine the internal structure of an object via the convolution of X-ray images obtained from multiple vantage points.

YORP (Yarkovsky-O'Keefe-Radzievskii-Paddack): This effect occurs when thermal emissions from and the solar radiation reflected off a surface change the rotation state of an object in space.

YORPD: Yearly Opportunities for Research in Planetary Defense.

Appendix G

Biographical Information for Committee Members

STEERING GROUP

ROBIN M. CANUP (*Co-Chair*) is assistant vice president at Southwest Research Institute, where she leads the Planetary Sciences Directorate in Boulder, Colorado. Dr. Canup is a theoretician, utilizing numerical simulations and analytical methods to study the formation and early evolution of planets and their moons. She has modeled many aspects of the formation of the Moon, including hydrodynamical simulations of lunar-forming giant impacts, the accumulation of the Moon, and its initial composition and orbital evolution. Dr. Canup has also developed models for an impact origin of the satellites of Pluto and Mars. Another major area of her work has addressed the origin of the systems of rings and satellites around the outer giant planets, including models of circumplanetary disk formation during late gas accretion, satellite accretion and migration, giant impacts and early dynamical evolution, and the compositional and interior properties of outer rings and moons. She was the recipient of the 2003 Urey Prize of the Division of Planetary Sciences and the 2004 Macelwane Medal of the American Geophysical Union and is a fellow of the American Geophysical Union. Dr. Canup was elected to the National Academy of Sciences in 2012 and served on the National Academies 2015 and 2018 J. Lawrence Smith Medal Selection Committees, and the 2014 and 2017–2019 NAS Class I Membership Committees. She was elected to the American Academy of Arts and Sciences in 2017. She earned her PhD and MS in astrophysics and planetary sciences from the University of Colorado Boulder.

PHILIP R. CHRISTENSEN (*Co-Chair*) is a Regents Professor and the Ed and Helen Korrick Professor in the School of Earth and Space Exploration at Arizona State University. His research interests focus on the composition, processes, and physical properties of Mars, Earth, asteroids, Europa, and other planetary surfaces. Dr. Christensen uses spectroscopy, radiometry, field observations, and numerical modeling to study the geology and history of planets and moons. A major facet of his research is the development of spacecraft instruments, and he has built eight science instruments that have flown on NASA's Mars Observer, Mars Global Surveyor, Mars Odyssey, Mars Exploration Rovers, OSIRIS-Rex, and Lucy missions, as well as an infrared spectrometer for the UAE's Hope Mars mission. Dr. Christensen's group is currently developing an infrared camera for NASA's Europa Clipper mission. Over the past 25 years, he has developed an extensive K–12 education and outreach program to bring the excitement of science and exploration into the classroom. Dr. Christensen is a fellow of the American Geophysical Union and the Geological Society of America and received the AGU's Whipple Award in 2018, the GSA's G.K. Gilbert Award in 2008, NASA's Public Service Medal in 2005, and NASA's Exceptional Scientific Achievement Medal in 2003. He received his PhD in geophysics and space physics from the University of California, Los Angeles.

Dr. Christensen has previously served as a member of the NRC's Committee on Planetary and Lunar Exploration, served as chair of the Mars Panel of the NRC planetary science decadal survey in 2010–2011, and was co-chair of the NRC's Committee on Astrobiology and Planetary Science from 2012–2015.

MAHZARIN R. BANAJI is the Richard Clarke Cabot Professor of Social Ethics in the Department of Psychology at Harvard University. Dr. Banaji is an elected fellow of the American Academy of Arts and Sciences, the British Academy, and the National Academy of Sciences; Herbert Simon Fellow of the Academy of Political and Social Science; and a fellow of the American Philosophical Society. She has received awards for teaching excellence from both Yale and Harvard. Dr. Banaji also received a Guggenheim Fellowship, APA's Award for Distinguished Scientific Contribution, APS's William James Fellow Award for "a lifetime of significant intellectual contributions to the basic science of psychology," and the Cattell Fellow Award for "a lifetime of applied psychological research." She is the recipient of five honorary degrees, the Golden Goose Award from the U.S. Congress, and the Atkinson Prize in Psychological and Cognitive Science from the National Academy of Sciences. Dr. Banaji is a co-author of *Blindspot: Hidden Biases of Good People.*

STEVEN J. BATTEL is president of Battel Engineering, providing engineering, technology development, and review services to NASA, the Department of Defense, as well as university and industry clients. Mr. Battel is also an adjunct clinical professor of engineering at the University of Michigan. Prior to starting Battel Engineering, he worked as an engineer, researcher, and manager at the University of Michigan, the Lockheed Palo Alto Research Laboratory, the University of California, Berkeley, and the University of Arizona's Lunar and Planetary Laboratory. Over the past 45 years, he has developed scientific instruments and electronic systems for more than 30 NASA and ESA missions, including Gravity Probe-B, Mars-Curiosity, Mars-Phoenix, Cassini, Huygens, HST, LADEE, MAVEN, ExoMars, and Mars 2020. Mr. Battel is a fellow of the AIAA and AAAS, a senior member of IEEE, and a member of Sigma Xi. He is a member of the National Academy of Engineering and a current member of the Committee on Solar and Space Physics and a former member of the Space Studies Board, the Aeronautics and Space Engineering Board, the AURA Space Telescope Institute Council, and the AURA Solar Observatory Council. Mr. Battel has served on multiple previous National Academies' committees, including the NASA Astrophysics Performance Assessment, the Committee on Assessment of Options for Extending the Life of the Hubble Space Telescope, and three decadal survey committees: Astronomy and Astrophysics (2010), Solar and Space Physics (Heliophysics, 2012), and Earth Science and Applications from Space (2017). He also chaired the recently released 2022 report *Leveraging Commercial Space for Earth and Ocean Remote Sensing.* Mr. Battel received his BS in engineering from the University of Michigan.

LARS E. BORG is Cosmochemistry group lead within the Nuclear and Chemical Sciences Division of the Lawrence Livermore National Laboratory. His research interests involve isotopic, geochemical, and petrologic analysis of terrestrial bodies, including the Moon, Earth, Mars, and asteroids, as well as materials from the nuclear fuel cycle. One area of Dr. Borg's research involves using the measured ages of planetary samples, and suites of samples, to constrain processes occurring in the protoplanetary disk and in differentiated bodies during the earliest stages of solar system history. Another area of research is focused on understanding the origin of interdicted nuclear materials and development of isotopic techniques to understand the behavior of historical nuclear tests. Work in planetary science has defined age relations of various suites of lunar crustal and mantle rocks, as well as martian basaltic meteorites in order to understand the timing and style of global-scale differentiation on these bodies. Dr. Borg has also completed both radiogenic and stable isotope analyses on lunar samples to define the timing and extent of volatile element depletion in the inner solar system. He has served on numerous NASA subcommittees, including the NASA Advisory Council's Planetary Sciences Subcommittee, the Curation and Analysis Planning Team for Extraterrestrial Materials, the Lunar Sample Subcommittee (chair), and the Mars Exploration Program Analysis Group. Dr. Borg received his PhD in isotope geochemistry from The University of Texas at Austin.

ATHENA COUSTENIS is director of research with the National Centre for Scientific Research and is based at Paris Observatory in Meudon, France. She earned her PhD in astrophysics and space techniques and her

Habilitation to Direct Research from the University of Paris. Dr. Coustenis works in the field of planetology and her research focuses on the use of ground- and space-based observatories and space missions to study solar system bodies. Her current interests include planetary atmospheres and surfaces, with particular emphasis on the satellites of the giant planets. Dr. Coustenis is also interested in the characterization of the atmospheres of extrasolar planets. In recent years, she has been leading efforts to define and select future space missions to be undertaken by the European Space Agency (ESA) and its international partners. She is currently the chair of the ESA Human Spaceflight and Exploration Science Advisory Committee, chair of the COSPAR Panel on Planetary Protection, and chair of the Science Advisory Committee of the French National Center for Space Studies (CNES). She has also chaired and served on numerous ESA and NASA advisory groups and international associations such as AAS/DPS, IUGG/IAMAS, and EGU. Dr. Coustenis previously served as an ex officio member on the National Academies' Space Studies Board and as a member of the National Academies' Committee on Survey of Surveys: Lessons Learned from the Decadal Survey Process.

JAMES H. CROCKER is vice president and general manager, retired, of Space Systems at Lockheed Martin Corporation. The focus of Mr. Crocker's career has been the design, construction, and management of very large, complex systems and instruments for astrophysics and space exploration both in the United States and internationally. These include space missions both human and robotic such as Apollo 17, Skylab, Orion; missions to Mars, Jupiter, Saturn, asteroids, the Moon, comets; and the Hubble Space Telescope, the Spitzer Space Telescope, and the James Webb Space Telescope. In ground-based astronomy, Mr. Crocker was program manager for the Sloan Digital Sky Survey and head of the Program Office for the European VLT, an array of optically phased 8-meter telescopes in the Atacama Desert in Chile. He is a current board member of the Association of Universities for Research in Astronomy. Mr. Crocker is a past board chair of the Denver Museum of Nature and Science and the Universities Space Research Association. He is a fellow of the AIAA and the AAS, a full member of the International Academy of Astronautics, and a member of the National Academy of Engineering. Crocker earned a BEE from the Georgia Institute of Technology, an MS in engineering from University of Alabama in Huntsville, and an MS in engineering management from the Johns Hopkins University.

BRETT W. DENEVI is a principal staff scientist at the Johns Hopkins University Applied Physics Laboratory. Dr. Denevi's research focuses on the origin and evolution of planetary surfaces, particularly the history of volcanism, the effects of impact cratering, and space weathering. She has been a member of numerous NASA mission teams, including as deputy principal investigator of the Lunar Reconnaissance Orbiter Camera, deputy instrument scientist for the Mercury Dual Imaging System on board the MESSENGER spacecraft at Mercury, participating scientist on the Dawn mission at asteroid Vesta, co-investigator and instrument lead on the Lunar Vertex investigation, and co-investigator on NASA's ShadowCam instrument on the Korean Pathfinder Lunar Orbiter. Dr. Denevi is the science chair of NASA's Lunar Exploration Analysis Group and is the recipient of the 2015 Maryland Academy of Science Outstanding Young Scientist Award, a NASA Early Career Fellowship, and seven NASA group achievement awards. Asteroid 9026 Denevi was named in her honor. She received her PhD in geology and geophysics from the University of Hawaii.

BETHANY L. EHLMANN is a professor of planetary science at the California Institute of Technology. Dr. Ehlmann's research interests include planetary surface processes, infrared spectroscopy, the evolution of Mars, and water–rock interactions throughout the solar system. Previously, she was a European Union Marie Curie Fellow and a collaborator on the Mars Exploration Rovers during their primary and first extended missions and an affiliate of the Dawn science team for its Ceres phase. Dr. Ehlmann is co-investigator and a deputy principal investigator for the Compact Reconnaissance Imaging Spectrometer for Mars, participating scientist on the Mars Science Laboratory mission, co-investigator for the Mars-2020 rover's Mastcam-Z and SHERLOC instruments, and principal investigator of Lunar Trailblazer. She is a recipient of the Division for Planetary Sciences Urey Prize, the American Geophysical Union's Macelwane medal, the Committee on Space Research's Zeldovich medal, National Geographic's Emerging Explorer award, and the Mineralogical Society of America Distinguished Lecturer award, as well as NASA Group Achievement awards. Dr. Ehlmann earned a PhD in geological sciences from Brown University. She has served on the National Academies' Committee on Astrobiology and Planetary Science.

LARRY W. ESPOSITO is a professor at the University of Colorado Boulder and at the Laboratory for Atmospheric and Space Physics. He is the principal investigator of the Ultraviolet Imaging Spectrograph experiment on the Cassini space mission to Saturn. Dr. Esposito was chair of the Voyager Rings Working Group and, as a member of the Pioneer Saturn Imaging Team, he discovered Saturn's F ring. His research focuses on the nature and history of planetary rings. Dr. Esposito has been a participant in numerous U.S., Russian, and European space missions and used the Hubble Space Telescope for its first observations of Venus. Dr. Esposito was awarded the Harold C. Urey Prize from the American Astronomical Society, the Medal for Exceptional Scientific Achievement from NASA, and the Richtmyer Lecture Award from the American Association of Physics Teachers and the American Physical Society. He received his PhD in astronomy from the University of Massachusetts Amherst. Dr. Esposito has extensive experience with National Academies' activities, including chairing the Task Group on the Forward Contamination of Europa and the Committee on Planetary and Lunar Exploration.

ORLANDO FIGUEROA is the president of Orlando Leadership Enterprise, LLC, which focuses on providing expert assessment and advice in space mission systems and technology, organization and enterprise/program management, strategic planning, and team and leadership development. Prior to starting his current role, Mr. Figueroa retired from NASA as a senior executive with 33 years of experience in the management, planning, and development of scientific space programs, missions, and related technologies. He is versed in interacting with national and international government and nongovernment organizations. Mr. Figueroa is the recipient of numerous awards, including the 2016 National Space Society Pioneer Award, the 2010 NASA Distinguished Service Medal, the Senior Executive Service Presidential Rank awards, the 2008 Smithsonian Latino Center Legacy award for contributions to American culture in science, and the 2005 Service to America Medal Federal Employee of the Year. He received his BS in mechanical engineering from the University of Puerto Rico.

JOHN M. GRUNSFELD is the president and chief executive officer of Endless Frontier Associates, LLC. Dr. Grunsfeld has more than 30 years of experience in program management and research. Prior to starting his current role, he was the associate administrator for the Science Mission Directorate at NASA, where he managed the portfolio of the agency's space and Earth science programs and joint agency programs. In addition, he served in numerous positions, including NASA chief scientist, deputy director of the Space Telescope Science Institute, professor at the Johns Hopkins University, NASA astronaut, and senior research fellow at the California Institute of Technology. Dr. Grunsfeld is a veteran of five Space Shuttle missions, including STS 67, STS-81, STS-103, STS-109, and STS-125. He is the recipient of multiple awards, including the NASA Space Flight Medals, the NASA Exceptional Service Medals, the NASA Distinguished Service Medal, and a NASA Constellation Award. Dr. Grunsfeld earned his PhD in physics from the University of Chicago.

JULIE HUBER is a senior scientist at the Woods Hole Oceanographic Institution. Dr. Huber is an oceanographer who studies microbial life in deep-sea crustal ocean habitats. These microbes provide critical ecosystem services such as primary production to sustain deep-sea food webs, nutrient and element recycling, carbon sequestration, and symbiotic relationships with diverse animals. In her research, she unravels the mystery of what microbes live in oceanic crustal environments, how they harness energy from the fluids and rocks that surround them, their activities, evolutionary trajectories, and interactions, and how they contribute to carbon and element cycling that influences the broader ocean system. Dr. Huber has led and participated in numerous deep-sea expeditions using human-occupied, remotely operated, and autonomous vehicles to study microbial life on and beneath the seafloor around the world. She previously served as a member of the National Academies' Committee on Scientific Ocean Drilling. A 2007 NASA Astrobiology Institute postdoctoral fellow, Dr. Huber also received the L'Oréal USA for Women in Science Fellowship. She received her PhD in oceanography from the University of Washington.

KRISHAN KHURANA is a senior research geophysicist at the Institute for Geophysics and Planetary Physics and the Department of Earth and Space Sciences at the University of California, Los Angeles. He has worked on many theoretical and empirical investigations relating to the magnetospheres of Venus, Earth, Jupiter, and Saturn, and is currently a co-investigator on the magnetometer experiments onboard the Cluster, THEMIS, JUICE, and Europa Clipper. Dr. Khurana's recent research has covered studies of subsurface oceans in Europa, Ganymede, and Callisto; ULF waves in outer magnetospheres; the structure and composition of the jovian plasma sheet; and the maintenance of corotation in the jovian magnetosphere. He was elected as a fellow to the American Geophysical Union in 2011. Dr. Khurana received his PhD in the field of magnetohydrodynamic waves in rotating fluids from the University of Durham. He previously served on the National Academies' Committee on Planetary and Lunar Exploration, the Committee on Solar and Space Physics, and the Committee on Heliophysics Performance Assessment.

WILLIAM B. McKINNON is professor of Earth and planetary sciences at Washington University in Saint Louis. He also serves as a fellow of the McDonnell Center for the Space Sciences. Dr. McKinnon's research interests include the structure, origin, evolution, geology, and bombardment history of outer planet satellites and dwarf planets and impact mechanics on rocky and icy bodies. He is a co-investigator on NASA's New Horizons and Europa Clipper missions and on the European Space Agency's Jupiter Icy Moons Explorer, and serves as an editor of *Earth and Planetary Science Letters.* Dr. McKinnon has received three group achievement awards from NASA, has asteroid (9526) Billmckinnon named after him, and in 2014 received the G.K. Gilbert Award from the Planetary Geology Division of the Geological Society of America. He earned his PhD in planetary science and geophysics from the California Institute of Technology. Dr. McKinnon has served as co-chair of the National Academies' Committee on Astrobiology and Planetary Science, and has been a member of the Committee on Planetary and Lunar Exploration and the Committee on Priorities for Space Science Enabled by Nuclear Power and Propulsion: A Vision for Beyond 2015.

FRANCIS NIMMO is a professor in the Department of Earth and Planetary Sciences at the University of California, Santa Cruz. His research interests cover Mars, Venus, Europa, Ganymede, Mercury, the Moon, and Pluto (as well as other icy satellites). Dr. Nimmo's research accomplishments include showing that a giant impact could have generated the martian hemispheric dichotomy; identifying shear-heating as an important process on Enceladus, Europa, and Triton; proposing true polar wander as an important process on Enceladus and Pluto; and explaining the link between plate tectonics and dynamo activity on Mars and Venus. He is the recipient of the 2007 Macelwane medal and Urey prize, the 2018 Farinella Prize and the 2019 Jeffreys lectureship. He received his PhD in volcanism and tectonics on Venus from Cambridge University, United Kingdom. Dr. Nimmo previously served on the National Academies' Committee for the Review of the Next Decadal Mars Architecture, the Satellites Panel for the *Vision and Voyages* decadal survey, and the Committee on Planetary and Lunar Exploration. He was elected to the National Academy of Sciences in 2020.

CAROL RAYMOND is a senior research and principal scientist at the Jet Propulsion Laboratory (JPL) and serves as the program scientist for both the Mission Formulation and the Small Bodies and Planetary Defense Offices within JPL's Planetary Science Directorate. Dr. Raymond's current research focuses on the geophysical evolution of small solar system bodies and icy moons, including Vesta, Ceres, Psyche, and Europa, and what they reveal about the early evolution of the solar system. In addition, planetary magnetic fields have been a long-term research interest. She led the NASA Dawn Mission as deputy principal investigator and assumed the principal investigator role in the extended mission phase. Dr. Raymond has held various positions at JPL since 1990 and was a visiting associate in the Division of Geological and Planetary Sciences at the California Institute of Technology and an adjunct associate research scientist at the Lamont-Doherty Earth Observatory of Columbia University. She is the recipient of three NASA Exceptional Achievement Medals, the Antarctic Service Medal, the Shoemaker Award of the American Geophysical Union, and is a fellow of the Geological Society of America. She received her PhD in geological sciences from Columbia University.

BARBARA SHERWOOD LOLLAR, Companion of the Order of Canada, FRS, FRSC, FRCGS, is a university professor and Dr. Norman Keevil Chair in Earth Sciences at the University of Toronto. She is past-president of the Geochemical Society and co-director of the CIFAR program Earth 4D—Subsurface Science and Exploration. In 2015, Dr. Sherwood Lollar was named a fellow of the American Geophysical Union and, in 2019, a fellow of the Geochemical Society and European Association of Geochemistry. She is a member of the National Academy of Sciences and the National Academy of Engineering. She has published on stable isotope geochemistry and hydrogeology, the fate of carbon-bearing fluids and gases such as CO_2, CH_4, and H_2 in ancient fracture waters in the Earth's crust, deep subsurface microbiology, and the remediation of surface drinking water supplies. Dr. Sherwood Lollar has been a recipient of academic awards, including the 2012 Eni Award for Protection of the Environment, the 2012 Geological Society of America Geomicrobiology and Geobiology Prize, the 2014 International Helmholtz Fellowship, the 2016 NSERC John Polanyi Award, the 2016 Bancroft Award for the Royal Society of Canada, the 2018 Logan Medal of the Geological Association of Canada, the 2019 Herzberg Gold Medal for Canada, the 2019 C.C. Patterson Award in environmental geochemistry, and the Canada Council for the Arts 2020 Killam Prize in Natural Sciences. She has served as a member on many National Academies' committees, including the Committee on Astrobiology and Planetary Sciences, the Committee on the Origin and Evolution of Life, and on the National Academies' Space Studies Board. In addition, Dr. Sherwood Lollar chaired the committee that authored the 2019 National Academies' *Astrobiology Strategy for the Search for Life in the Universe*.

AMY SIMON is the senior scientist for Planetary Atmospheres Research in the Solar System Exploration Division at the NASA Goddard Space Flight Center. Dr. Simon's scientific research involves the study of the composition, dynamics, and cloud structure in jovian planet atmospheres, primarily from spacecraft observations. She is also involved in multiple robotic flight missions, as well as future mission concept development. Dr. Simon was a co-investigator on the Cassini Composite Infrared Spectrometer and is the deputy instrument scientist for the OSIRIS-REx Visible and near-IR Spectrometer as well as the Landsat 9 TIRS2 instrument. She is the Lucy L'Ralph instrument deputy principal investigator and the principal investigator of the Hubble Outer Planet Atmospheres Legacy Program. Dr. Simon earned her PhD in astronomy from New Mexico State University, Las Cruces. She was a member of the *Visions and Voyages* decadal survey steering group and vice chair of that survey's Giant Planets Panel.

PANEL ON GIANT PLANET SYSTEMS

JONATHAN I. LUNINE (*Chair*) is the David C. Duncan Professor in the Physical Sciences at Cornell University, where he is also chair of the Department of Astronomy. Dr. Lunine's research focuses on how planets form and evolve, what processes maintain and establish habitability, and what are the limits of environments capable of sustaining life. He was an interdisciplinary scientist on the Cassini-Huygens mission to Saturn, and is currently co-investigator on the Juno mission in orbit around Jupiter and an interdisciplinary scientist for the James Webb Space Telescope. Dr. Lunine has chaired or served on a number of advisory and strategic planning committees for NASA and the National Science Foundation. He is the recipient of numerous awards, including the Harold C. Urey Prize of the Division for Planetary Sciences of the American Astronomical Society, the Macelwane Medal of the American Geophysical Union (AGU), the Zeldovich Prize awarded jointly by COSPAR and the Russian Academy of Sciences, the Basic Science Award of the International Academy of Astronautics, and the Jean-Dominique Cassini medal of the European Geosciences Union. Dr. Lunine is a fellow of the AGU and American Association for the Advancement of Science and a member of the National Academy of Sciences. He received a PhD in planetary science from the California Institute of Technology. Dr. Lunine has served on several National Academies' committees, including the Committee on the Origins and Evolution of Life (co-chair), the Committee for a Review of Programs to Determine the Extent of Life in the Universe (co-chair), the Committee on Decadal Survey on Astronomy and Astrophysics 2010 (member), and the Committee on Human Spaceflight (co-chair).

AMY SIMON (*Vice Chair*), see Steering Group entry.

FRANCES BAGENAL is a senior research scientist at the Laboratory for Space and Atmospheric Physics at the University of Colorado Boulder and leads its Magnetospheres of the Outer Planets Group. Prior to that, she was a professor of Astrophysical and Planetary Sciences at the same institution. Her research interests focus primarily on the physics of gas giant planets, specifically understanding the magnetospheres by combining data analysis and theoretical models. Dr. Bagenal has been co-investigator on several highly successful NASA missions, including Voyager, Galileo, New Horizons, and Juno, and has chaired or led working groups within those mission teams. She is a member of the National Academy of Sciences. In her National Academies' work, she chaired the Committee on the Assessment of the Role of Solar and Space Physics in NASA's Space Exploration Initiative and has been a member of the Space Studies Board, the Committee on Planetary and Lunar Exploration, and the Committee on International Space Programs. Dr. Bagenal is the recipient of multiple awards, including the James Van Allen Lecture Award from the American Geophysical Union, and the Boulder Faculty Assembly's Excellence in Research Award. Dr. Bagenal earned her PhD in Earth and planetary sciences from the Massachusetts Institute of Technology.

RICHARD W. DISSLY is a senior manager for civil space at Ball Aerospace, where he is responsible for program management, instrument development, and mission formulation, primarily in support of NASA science missions. He has led or supported a wide range of proposal teams covering Earth, planetary, heliophysics, and astrophysics science opportunities. Dr. Dissly recently served as the Ball project manager on IXPE, an astrophysics SMEX mission. Prior to joining Ball, he was a research scientist at the NOAA Aeronomy Laboratory, where he developed instrumentation for the measurement of trace atmospheric gases from aircraft platforms. Dr. Dissly received his PhD in planetary science from the California Institute of Technology.

LEIGH N. FLETCHER is a professor of planetary science at the University of Leicester. His research interests span planetary atmospheres (dynamics, chemistry, and climate), planet formation, and planetary satellites. In particular, his research focuses on the exploration of giant planet systems, including the circulation of the gas and ice giant atmospheres, via a combination of planetary missions and ground- and space-based planetary astronomy. Dr. Fletcher is a co-investigator on the Cassini mission to Saturn, a participating scientist on the Juno mission to Jupiter, and interdisciplinary scientist for ESA's forthcoming Jupiter Icy Moons Explorer. Prior to joining the University of Leicester, he was a NASA fellow at the Jet Propulsion Laboratory and a Royal Society Research Fellow at the University of Oxford. Dr. Fletcher is the recipient of numerous awards, including the Harold C. Urey Prize of the Division for Planetary Sciences of the American Astronomical Society. He received his doctorate in planetary physics from the University of Oxford.

TRISTAN GUILLOT is CNRS (National Center for Scientific Research) director of research at the Nice Observatory in France. His research interests include studying the interiors of Jupiter, Saturn, Uranus, and Neptune; the formation of planetesimals and planets; the evolution of giant exoplanets; and the spectra and detectability of giant planets and brown dwarfs. Dr. Guillot was formerly the associate editor of the publication *Astronomy & Astrophysics*, a research fellow at the University of Reading, and a research associate at the Lunar and Planetary Laboratory of the University of Arizona. He is the recipient of numerous awards, including NASA's Group Achievement Award for the Juno Science Team, the Harold C. Urey Prize of the Division for Planetary Sciences of the American Astronomy Society, and the Zeldovich Medal from COSPAR and the Russian Academy of Sciences. Dr. Guillot received a PhD in astrophysics from the Observatoire de la Côte d'Azur/Université Paris.

MATTHEW HEDMAN is an associate professor at the University of Idaho. Dr. Hedman's research focuses on orbital dynamics, planetary rings, Enceladus's plume, the Moon's dust exosphere, and infrared spectroscopy. Prior to joining his current institution, he was a research associate at Cornell University and a research fellow at the University of Chicago. Currently, Dr. Hedman is a co-investigator on the Europa Clipper mission. Dr. Hedman has served on numerous review panels for NASA, the Ice-Giant predecadal study, and two planetary mission concept studies on Enceladus and Neptune. He received his PhD in physics from Princeton University.

RAVIT HELLED is a professor of theoretical astrophysics at the University of Zurich. Her research interests include planetary structure and evolution, planet formation, solar system exploration, and exoplanets. Previously, Dr. Helled was a postdoctoral fellow and researcher at the University of California, Los Angeles, and a professor at Tel Aviv University. Current mission involvement includes co-investigator of the Juno NASA mission; co-investigator of the ESA's JUICE mission; science team member of the ESA's PLATO mission; and working group leader and consortium member of the ESA's ARIEL mission. She is a member of the European Space Sciences Committee's Astronomy and Fundamental Physics Panel and the National Research Council of the Swiss National Science Foundation and leads the Academic Platform of the NCCR PlanetS. Dr. Helled serves the community intensively by frequently being an invited speaker and is involved in panels, committees, scientific forums, and review panels. She is the recipient of various awards and was selected as one of the 50 most influential women of Forbes Israel in 2015. Dr. Helled received her PhD in planetary science from Tel Aviv University.

KATHLEEN E. MANDT is the chief scientist for exoplanets at the Johns Hopkins University Applied Physics Laboratory. Dr. Mandt's research interests focus on the dynamics, chemistry, and evolution of planetary atmospheres, and she is particularly interested in advancing both solar system and exoplanet understanding of planetary system formation and evolution by forging new connections between the planetary science and exoplanet communities. Previously, Dr. Mandt was an adjoint professor in the Department of Physics and Astronomy at The University of Texas at San Antonio and a senior research scientist at Southwest Research Institute. She serves in several community and NASA mission leadership roles, including membership of the steering committee of the Outer Planets Assessment Group. Dr. Mandt previously served as a member of the Division for Planetary Science Professional Culture and Climate Subcommittee and as the volatiles theme lead for the Lunar Reconnaissance Orbiter (LRO) mission. She is the project scientist for the LRO Lyman Alpha Mapping Project instrument and for the Io Volcano Observer phase A study, the deputy project scientist for the Heliophysics Division–funded Interstellar Probe predecadal mission study, and a science team member on the Europa Clipper Plasma Instrument for Magnetic Sounding teams. She earned her PhD in environmental science and engineering from The University of Texas at San Antonio. Dr. Mandt previously served on the Astro2020 Panel on Exoplanets, Astrobiology, and the Solar System.

ALYSSA RHODEN is a principal scientist within the Planetary Science Directorate at the Southwest Research Institute. Her expertise is in the geophysics of icy satellites, particularly those containing oceans, and the evolution of giant planet satellites systems through time. Dr. Rhoden utilizes a novel combination of numerical modeling, statistical analysis, and photogeology to interpret the surfaces of icy ocean worlds to assess their interiors, orbits, and habitability through time. She is co-leader of NASA's Network for Ocean Worlds Research Coordination Network. Prior to joining her current institution, Dr. Rhoden was an assistant professor of planetary science at Arizona State University in the School of Earth and Space Exploration. She participated in planning NASA's multi-flyby mission to Europa and a future mission to land on Europa's surface. Dr. Rhoden is the recipient of numerous awards, including the Kavli Institute Frontiers of Science Fellowship and an Early Career Fellowship by NASA. She received her PhD in Earth and planetary science from the University of California, Berkeley. Dr. Rhoden was a member of the National Academies' Committee on Astrobiology and Planetary Science.

PAUL M. SCHENK is a staff scientist at the Lunar and Planetary Institute. His research over the years has focused on using Voyager, Galileo, Cassini, Dawn, and New Horizons stereo- and monoscopic images to map the topography and geology of the icy bodies in the outer solar system from Ceres to Arrokoth. Dr. Schenk also studies impact cratering on small bodies and plume deposition processes on Enceladus. He was the lead editor of *Enceladus and the Icy Moons of Saturn* and a participating scientist on the Cassini project and the Dawn mission to Vesta and Ceres. In addition, Dr. Schenk is currently a co-investigator for the New Horizons mission, the scientific editor of the *Lunar and Planetary Information Bulletin*, and a recipient of the Fred Whipple Award from the American Geophysical Union. He received his PhD in geology from the Washington University in St. Louis.

MICHAEL H. WONG is a research scientist at the SETI Institute. Dr. Wong also has appointments at the University of California, Berkeley, and at the University of Michigan. His research interests focus on planetary atmospheres, where he leads Hubble/WFC3 and Gemini North/NIRI imaging programs in support of the NASA Juno mission to Jupiter. Dr. Wong serves on the Hubble Space Telescope Users Committee and is a founding member of the OPAL program that conducts annual imaging of the giant planets with HST. Prior to his current roles as a participating scientist on the Juno mission, collaborator on the Mars Science Laboratory mission, and science team member for TMT/IRIS and Keck/Liger instruments, he was affiliated with science and instrument teams for Cassini/CIRS and the Galileo Probe Mass Spectrometer. Dr. Wong was previously a National Research Council research associate at NASA Goddard Space Flight Center, after earning his PhD in atmospheric and space sciences from the University of Michigan, Ann Arbor.

PANEL ON MARS

VICTORIA E. HAMILTON (*Chair*) is an institute scientist in the Department of Space Studies at Southwest Research Institute in Boulder, Colorado. Dr. Hamilton is a geologist with extensive experience in laboratory spectroscopy and remote sensing data analysis of Mars and asteroids. She was an affiliate of the Mars Global Surveyor TES science team, was a participating scientist on the Mars Science Laboratory mission, and is deputy principal investigator for the THEMIS instrument on the 2001 Mars Odyssey mission. Dr. Hamilton is a mission spectroscopy scientist, a science team co-investigator, and the OTES deputy instrument scientist on the OSIRIS-REx and OSIRIS-APEX asteroid sample return missions. She is also a co-investigator and deputy instrument principal investigator (for L'TES) on the Lucy Trojan asteroid survey mission. Dr. Hamilton has published on laboratory mineral and meteorite spectroscopy, numerical modeling of infrared spectra, martian surface and atmospheric aerosol composition, martian thermophysical properties, and the composition of asteroid Bennu. She has built, operated, and managed a spectroscopy laboratory equipped with spectrometers for measuring visible, near infrared, and thermal infrared properties of rocks, minerals, and meteorites in reflectance and emission. Dr. Hamilton has received the NASA Group Achievement Award for the OSIRIS-REx Thermal Emission Spectrometer Team, Mars Science Laboratory Science Office Development and Operations Team, 2001 Mars Odyssey Thermal Emission Imaging System Team, and Mars Global Surveyor Thermal Emission Spectrometer Team. She received her PhD in geology from Arizona State University. Dr. Hamilton was a member of the National Academies' Committee on Cost Growth in NASA Earth and Space Science Missions, co-chair of the Committee on NASA Science Mission Extensions, member of the Committee on NASA Large Strategic Missions: Science Value and Role in a Balanced Portfolio, and vice chair of the Committee to Review the NASA Science Mission Directorate Science Plan (2019).

BETHANY L. EHLMANN (*Vice Chair*), see Steering Group entry.

WILLIAM B. BRINCKERHOFF is a senior scientist at NASA's Goddard Space Flight Center in the Solar System Exploration Division. His research interests include small body geochemistry, planetary mission design, and astrobiology. Dr. Brinckerhoff currently serves as project scientist for the Mars Organic Molecule Analyzer instrument on the European Space Agency's ExoMars rover mission, deputy lead for the Dragonfly Mass Spectrometer, and co-investigator for the Mars Science Laboratory's Sample Analysis at Mars investigation. He has served in numerous positions at NASA GSFC, including chief of the Planetary Environments Laboratory, and research space associate. In addition, Dr. Brinckerhoff was a senior professional staff member of the Johns Hopkins University Applied Physics Laboratory. He received a PhD in experimental condensed matter physics from the Ohio State University.

TRACY K.P. GREGG is an associate professor at the University of Buffalo in the Department of Geology. Her research focuses on the behavior of volcanic deposits, particularly lava flows, and how emplacement processes are affected by different ambient conditions. Previously, she was an NSF Ridge postdoctoral fellow and an assistant research scientist at the Woods Hole Oceanographic Institution. Dr. Gregg served as chair for NASA's Geologic Mapping Standards Committee (2002–2007). Most recently, Dr. Gregg served as the U.S. co-chair for the NASA/Roscosmos Venera-D Joint Science Definition Team to generate the Venera-D Phase II report. She received a PhD in geological sciences from Arizona State University.

JASPER S. HALEKAS is an associate professor at the University of Iowa in the Department of Physics and Astronomy. His research focuses on the interaction between the solar wind and the planets and moons in the solar system. Dr. Halekas designs and builds spaceflight instruments to make high-fidelity measurements of charged particles and uses them to understand the plasma physics that occurs in the interplanetary medium and the environments near planetary bodies. As such, his research spans the intersection between planetary science and space plasma physics, touching on planetary geology and atmospheres, magnetic reconnection, shocks, plasma sheaths, and plasma waves and turbulence. He is involved in multiple spacecraft missions, including Mars Express, Mars Atmosphere and Volatile Evolution, Parker Solar Probe, and ARTEMIS. Dr. Halekas received a PhD in physics from the University of California, Berkeley.

JOHN "JACK" W. HOLT is a professor at the University of Arizona in the Lunar and Planetary Laboratory and the Department of Geosciences. He is a co-investigator on the SHARAD radar instrument on the Mars Reconnaissance Orbiter, and co-investigator on NASA's Operation IceBridge program. In addition, Dr. Holt conducts field studies of Alaskan glaciers and Mars analog debris-covered glaciers using airborne and surface-based geophysical methods and has conducted seven field campaigns in Antarctica. He has developed clutter mitigation techniques for airborne and orbital radar sounding and airborne radar hardware for terrestrial deployments. Previously, Dr. Holt was research professor at the University of Texas, the Herbert J. Reich Professor of Natural Sciences at Deep Spring College, and both a postdoctoral scholar and a technical staff member at the Jet Propulsion Laboratory. He received a BS in electrical engineering from Rice University, and an MS and a PhD in geology from the California Institute of Technology.

JOEL HUROWITZ is an associate professor at Stony Brook University in the Department of Geosciences. His research projects include in situ exploration of the surface of Mars, studying the geochemical and mineralogical composition of clastic sediments through analog field studies, analyzing the reactivity and toxicity of planetary regolith, and using experimental approaches to aqueous geochemistry. Previously, Dr. Hurowitz was a research scientist at Stony Brook University, a research scientist at the Jet Propulsion Laboratory, and a hydrogeologist at Leggette, Brashears & Graham, Inc. He received a PhD in geosciences from Stony Brook University.

BRUCE M. JAKOSKY is a professor and associate director for science in the Laboratory for Atmospheric and Space Physics and the Department of Geological Sciences at the University of Colorado Boulder. His research interests are in the geology of planetary surfaces, the evolution of the martian atmosphere and climate, the potential for life on Mars and elsewhere, and the philosophical and societal issues in astrobiology. Over the years, Dr. Jakosky has been involved with the Viking, Solar Mesosphere Explorer, Clementine, Mars Observer, Mars Global Surveyor, Mars Odyssey, Mars Science Laboratory, and Lunar Reconnaissance Orbiter spacecraft missions. He is the principal investigator of the Mars Atmosphere and Volatile Evolution mission to Mars. Dr. Jakosky received his PhD in planetary science and geophysics from the California Institute of Technology. He chaired the National Academies' Committee on Origins and Evolution of Life and the Committee on Astrobiology Strategy for the Exploration of Mars.

MICHAEL MANGA is a professor in and chair of the Department of Earth and Planetary Science at the University of California, Berkeley. His research interests include studying geological processes involving fluids, including problems in physical volcanology, geodynamics, hydrogeology, and geomorphology. Previously, he was a Miller Research Fellow at the University of California, Berkeley, and an assistant professor at the University of Oregon. Dr. Manga is the recipient of numerous awards, including a 2005 MacArthur Fellowship, the Robert Wilhelm Bunsen Medal from the European Geoscience Union, and the Donath Medal from the Geological Society of America. He is a member of the National Academy of Sciences. He received a PhD in Earth and planetary sciences from Harvard University. Dr. Manga chaired the National Academies' Committee on Improving Understanding of Volcanic Eruptions.

HARRY Y. McSWEEN is the Chancellor's Professor Emeritus at the University of Tennessee. His research focuses on meteorites. Previously, Dr. McSween was the head of the Department of Earth and Planetary Sciences and interim dean of the College of Arts and Sciences at the University of Tennessee. He was a member of the science teams for the Mars Pathfinder and the Mars Global Surveyor spacecraft missions, co-investigator for the Mars Exploration Rovers and the Dawn spacecraft missions, and is currently a co-investigator for the THEMIS instrument on Mars Odyssey. Dr. McSween is the recipient of numerous awards, including the Leonard Medal of the Meteoritical Society, the J. Lawrence Smith Medal of the National Academy of Sciences, and the Whipple Award of the American Geophysical Union. He was president of the Geological Society of America and is a member of the National Academy of Sciences and a fellow of the American Academy of Arts and Sciences, the American Geophysical Union, and the Geological Society of America. Dr. McSween received a PhD in geological sciences from Harvard University. He has previously served on numerous National Academies' committees, including the Committee on Planetary and Lunar Exploration and the Committee on Planetary Science Decadal Survey: 2013–2022.

CLAIRE E. NEWMAN is a research scientist and co-owner of Aeolis Research, a small private research company. Dr. Newman's research focuses on the atmospheres of Mars and Titan, with a special interest in understanding the connections between the surface and atmosphere, from measuring weather in the lowest few meters of the atmosphere, to studying aeolian processes on multiple bodies, to simulating martian dust lifting and dust storms using numerical weather prediction models. She is a team member on the Mars Science Laboratory, InSight, Mars 2020, and Dragonfly missions, and currently co-leads the Mars 2020 atmospheres working group and is a member of InSight's solar array cleaning team. Dr. Newman has received NASA Group Achievement Awards for her work on the Mars Science Laboratory mission, including observations of the 2018 global dust storm. She was previously a research scientist at Ashima Research, a staff scientist and postdoctoral research scholar at the California Institute of Technology, and a postdoctoral researcher at the University of Oxford, where she received her doctorate in atmospheric physics. Dr. Newman was an associate editor at the *Journal of Geophysical Research: Planets* from 2017 through 2021, and is convenor of the session "Aeolian Processes on Earth and Other Planetary Bodies" at the Fall meeting of the American Geophysical Union.

ALEJANDRO M. SAN MARTIN is the chief engineer for the Guidance, Navigation, and Control Section at the Jet Propulsion Laboratory (JPL). His research interests include spacecraft guidance, navigation, and control, with a specialization in the problems surrounding landing spacecraft on planetary bodies. In addition, Mr. San Martin has many years of experience with designing end-to-end guidance, navigation, and control systems. Previously, he has served in many different roles at JPL, including chief engineer for articulation and attitude control system of the Mars Pathfinder mission, chief engineer of the guidance and control system of the Mars Exploration Rovers, and chief engineer of the guidance, navigation, and control system of the Mars Science Laboratory. Mr. San Martin is the recipient of numerous awards, including the NASA Exceptional Achievement in Engineering Medal of Honor, and was named a fellow at the Jet Propulsion Laboratory in 2013. He is a member of the National Academy of Engineering. He received an MS in aeronautical and astronautical engineering from the Massachusetts Institute of Technology.

KIRSTEN L. SIEBACH is an assistant professor at Rice University in the Department of Earth, Environmental, and Planetary Sciences. She is a participating scientist on the Mars 2020 mission and a member of the Science and Operations Team for the Mars Science Laboratory. Dr. Siebach researches source-to-sink sedimentary processes on Mars and early Earth to interpret the history of water and surface environments early in the solar system. She is also actively engaged in promoting education and outreach related to Earth and planetary sciences. Dr. Siebach has received several NASA Group Achievement Awards for her work on the Mars Science Laboratory, Mars Exploration Rovers, and Phoenix missions. Prior to joining Rice University, she completed her PhD in geology at the California Institute of Technology and then did postdoctoral research at Stony Brook University studying the geochemistry of martian sediments. Dr. Siebach was a member of the National Academies' Committee to Review the Report of the NASA Planetary Protection Independent Review Board.

AMY WILLIAMS is an assistant professor of geology at the University of Florida. Her research interests include the formation and preservation of physical and molecular biosignatures in terrestrial environments as an analog for putative biosignature formation on Mars. She has been a member of the NASA Curiosity rover science team since 2009, and currently works with the Sample Analysis at Mars instrument team to explore the distribution of organic molecules on Mars's surface. Dr. Williams is a participating scientist on the NASA Perseverance and Curiosity rover missions. She received a nomination in 2017 for the Maryland Academy of Sciences Outstanding Young Scientist Award and the 2022 University of Florida Excellence Award for Assistant Professors. Dr. Williams holds a PhD in geology from the University of California, Davis.

ROBIN D. WORDSWORTH is a professor of environmental science and engineering at Harvard University, with a joint appointment in Earth and planetary sciences. Dr. Wordsworth leads the Wordsworth Planetary Climate and Atmospheric Evolution Research Group at Harvard University, which focuses on the boundary between solar system and exoplanet atmospheres and climates. Past work of the Wordsworth group has included studies of Mars's early climate during warm and wet episodes, snowball and hothouse climates in Earth's distant past, the link between ultraviolet radiation and life, and the detectability of water and life on exoplanets. Wordsworth received an NSF CAREER award in 2018 and previously served as a committee member on NASA's Mars Architecture Strategy Working Group. He received his doctorate in physics from the University of Oxford.

PANEL ON MERCURY AND THE MOON

TIMOTHY L. GROVE (*Chair*) is the Robert R. Shrock Professor of Earth and Planetary Sciences in the Department of Earth, Atmospheric, and Planetary Sciences at the Massachusetts Institute of Technology. Dr. Grove is interested in the processes that have led to the chemical evolution of the Earth and other planets, including the Moon, Mars, Mercury, and meteorite parent bodies. His approach to understanding planetary differentiation is to combine field, petrologic, and geochemical studies of igneous rocks with high-pressure, high-temperature experimental petrology. Dr. Grove is a past president of the American Geophysical Union (AGU) and the recipient of multiple awards, including the Harry H. Hess Medal of the AGU, the V.M. Goldschmidt Award of the Geochemical Society, the Bowen Award of the AGU, and the Pick Award of the Rocky Mountain Association of Geologists. He is a member of the National Academy of Sciences. Asteroid (9276) Timgrove is named in his honor. He is the recipient of honorary degrees from the University of Liege and the Université de Lausanne and earned his PhD in geology at Harvard University.

BRETT W. DENEVI (*Vice Chair*), see Steering Group entry.

JAMES DAY is a professor in the Geosciences Research Division at the University of California, San Diego. He is also director of the Scripps Isotope Geochemistry Laboratory. His research interests include isotope geochemistry, cosmochemistry, petrogenesis of igneous and metamorphic rocks, planetary dynamics and geodynamics, and planet formation and accretion. In addition, Dr. Day is a geologist and geochemist whose research focuses on volcanism and what the mineralogy and composition of rocks can tell about how the planets formed and evolved to their present-day states. He studies asteroids and products formed in the mantle of Mars, the Earth, and the Moon. Dr. Day also studies terrestrial basaltic volcanism to further understand crust formation processes and the role of volcanism on Earth system cycles. He is the recipient of the Nier Prize from the Meteoritical Society, the Houtermans Award from the European Association of Geochemistry, and the Antarctic Service Medal. Dr. Day received his PhD in geochemistry from the University of Durham. He has not previously served on a National Academies' committee.

ALEXANDER J. EVANS is an assistant professor of Earth, environmental, and planetary sciences at Brown University. His research interests include understanding the evolutionary, tectonic, geodynamic, and geophysical processes of solid planets. Dr. Evans's work includes analyses of altimetry, gravity, geomorphology, and tectonics to determine the structure, surface, and internal evolution of solid planets. Prior to joining Brown University, he held postdoctoral research positions at the Lunar and Planetary Laboratory at the University of Arizona, the Southwest Research Institute, the Colorado School of Mines, and Columbia University. Dr. Evans is the recipient of numerous awards, including the National Association of Graduate-Professional Students' Lifetime Achievement Award and the Massachusetts Institute of Technology's Presidential Fellow Award. He received his PhD in planetary geophysics from the Massachusetts Institute of Technology. Dr. Evans has not previously served on a National Academies' committee.

SARAH FAGENTS is a researcher at the University of Hawaii at Manoa with the Hawaii Institute of Geophysics and Planetology. Her research interests include planetary volcanism, volcanic fluid dynamics, and icy satellite geology. Dr. Fagents focuses on understanding the mechanisms of formation of volcanic features on the Earth and other planetary bodies, including the physics of eruptive processes, and the influence that the planetary environment has on the style of eruptions and resulting landforms. Dr. Fagents is currently a co-investigator on the Mars 2020 Rover Mission Mastcam-Z, and on the JPL-NAI Titan project Habitability of Hydrocarbon Worlds: Titan and Beyond. She received her PhD in planetary science from Lancaster University. Dr. Fagents has not previously served on a National Academies' committee.

WILLIAM M. FARRELL is a plasma physicist at NASA's Goddard Space Flight Center in the Solar System Exploration Division. His research interests include the study of lightning storms on Earth and the planets, the dusty plasma environment at planetary moons and asteroids, the space environment of the Moon, and planetary auroral and magnetospheric processes. Dr. Farrell was a co-investigator on the Cassini mission to Saturn and is currently a co-investigator on the Parker Solar Probe and Wind spacecraft. From 2014 to 2019, he was the principal investigator of the DREAM2 center for space environments leading a team of more than 30 investigators on the space weather effects at the Moon and other airless bodies. Dr. Farrell is the recipient of numerous awards, including the NASA Exceptional Scientific Achievement Medal, the Robert H. Goddard Award for Exceptional Achievement in Science, and the NASA/Goddard Divisional Peer Award. He received his PhD in physics from the University of Iowa.

CALEB I. FASSETT is a planetary scientist at NASA's Marshall Space Flight Center. His research focuses on using a combination of remote sensing, geologic mapping, and numerical modeling to understand planetary surfaces and geomorphological processes. In addition, Dr. Fassett's research interests include how observations of impact crater populations can be used to infer the chronology and geologic history of planetary bodies. Prior to joining NASA's Marshall Space Flight Center, he was a visiting assistant professor at Mount Holyoke College and a postdoctoral research associate at Brown University. Dr. Fassett currently supports many activities at Marshall, including the Human Landing System and the Advanced Concept Office. He received a PhD in geosciences from Brown University. Dr. Fassett has not previously served on a National Academies' committee.

JENNIFER L. HELDMANN is a research scientist at NASA's Ames Research Center. Her research interests focus on the studies of the Moon and Mars, which includes improving our understanding of lunar volatile deposits and studies of recent water on Mars through analysis of spacecraft data, numerical modeling, and terrestrial analog fieldwork. Dr. Heldmann is the principal investigator (PI) of the Field Investigations to Enable Solar System Science and Exploration (FINESSE) and Resource Exploration and Science of OUR Cosmic Environment (RESOURCE) projects through NASA's Solar System Exploration Research Virtual Institute (SSERVI), where she manages more than 100 team members focused on science, technology, and mission operations research. She is a founding member of the NASA/SSERVI Analogs Focus Group. Dr. Heldmann is the recipient of numerous awards. Dr. Heldmann also received the Association of Women Geoscientists Professional Excellence Award in 2021. She served on the Science and Payload teams and as the Observation Campaign Coordinator for NASA's Lunar Crater Observation and Sensing Satellite and is a team member for NASA's Volatiles Investigating Polar Exploration Rover. She is interested in the intersection between robotic and human exploration of space and served as the Planetary Science Division Lead for NASA's Optimizing Science and Exploration Working Group, has served on multiple Mars Exploration Program Analysis Group committees, and was a member of the NASA Artemis III Science Definition Team focused on future scientific investigations to be conducted by astronauts on the Moon. Dr. Heldmann is also a member of the Space Camp Hall of Fame at the U.S. Space and Rocket Center in Huntsville, Alabama. She received her PhD in planetary science from the University of Colorado Boulder. Dr. Heldmann has not previously served on a National Academies' committee.

MASATOSHI HIRABAYASHI is an assistant professor in aerospace engineering, is an adjunct professor in geosciences at Auburn University, and leads Auburn's Space Technology and Applications Research Laboratory. Prior to joining his current institution, he was a visiting scholar at the Imperial College, London, a postdoctoral associate at Purdue University, and a research associate at the University of Colorado Boulder. Dr. Hirabayashi's research interests focus on the Moon, Mercury, and small bodies. This includes the study of geophysical characterizations of surface processes on the lunar surface, investigations of links between surface processes on Mercury and dust environments around this planet, geophysical characterizations of small bodies, and assessing the feasibility of proximity operations around Mercury, the Moon, and small solar system bodies. He received his PhD in aerospace engineering from the University of Colorado Boulder.

JAMES TUTTLE KEANE is a scientist at the Jet Propulsion Laboratory. His research focuses on studying the interactions between orbital dynamics, rotational dynamics, and geologic processes on rocky and icy worlds across the solar system. Prior to joining JPL, he was a postdoctoral fellow in the Joint Center for Planetary Astronomy under the Division of Geological and Planetary Sciences at the California Institute of Technology, and a graduate research associate at the University of Arizona. Dr. Keane has extensive experience with NASA missions, including the GRAIL lunar orbiter, the New Horizons mission to Pluto and the Kuiper belt, and the Juno mission to Jupiter. He is currently an affiliate of the Keck Institute for Space Studies, a member of the International Association of Astronomical Artists, the American Geophysical Union, the American Astronomical Society, the AAS Division of Dynamical Astronomy, and the AAS Division for Planetary Science. Dr. Keane is the recipient of numerous awards, including JPL's Voyager Award, the Editor's Citation for Excellence in Refereeing Award from *Geophysical Research Letters*, and the Pellas-Ryder Award of the Geological Society of America. Asteroid (36773) Tuttlekeane is named in his honor. He received his PhD in planetary science from the University of Arizona. Dr. Keane has not previously served on a National Academies' committee.

FRANCIS McCUBBIN is the astromaterials curator at NASA's Johnson Space Center within the Astromaterials Research and Exploration Science Division. As head curator, he is responsible for protecting the scientific integrity of NASA's priceless astromaterials collections and distributing select samples to the global community for further scientific examination. His research focuses on understanding the abundance, distribution, and origin of water in the inner solar system, as well as deciphering the thermal and magmatic evolution of the terrestrial planets, moons, and asteroids. Prior to joining his current organization, Dr. McCubbin was a research scientist at the Institute of Meteoritics at the University of New Mexico and a postdoctoral fellow at the Geophysical Laboratory of the Carnegie Institution for Science. He received his PhD in geosciences from Stony Brook University.

MIKI NAKAJIMA is an assistant professor in the Department of Earth and Environmental Sciences and in the Department of Physics and Astronomy at the University of Rochester. Dr. Nakajima's research interests include the origins of Earth and the Moon, the martian moons, the early Earth and lunar environments, and the impact history in the solar system. In particular, she investigates various impact processes and considers their geological, geophysical, and geochemical implications based on theoretical and numerical modeling. Dr. Nakajima is a member of the science team for the Japan Aerospace Exploration Agency's Martian Moons Exploration sample return mission. Prior to moving to Rochester in 2018, Dr. Nakajima was a postdoctoral fellow at the Carnegie Institution for Science. She received her PhD in planetary science at the California Institute of Technology.

MARK P. SAUNDERS is an independent consultant. Since retiring from NASA in December 2008, he has been consulting for various NASA offices, providing program/project management and systems engineering expertise. This has included support to the Office of Chief Engineer, the Office of Independent Program and Cost Evaluation, the Mars Program, and the Science Office for Mission Assessments (at NASA's Langley Research Center). Mr. Saunders has participated in the rewriting of NASA's policy on program/project management, advised and supported the agency's independent program/project review process, and supported the review of various programs and projects. Mr. Saunders served as director of the independent program assessment office at NASA headquarters, where he was responsible for enabling the independent review of the agency's programs and projects at life-cycle milestones to ensure the highest probability of mission success, and as program manager for the Discovery Program at the Office of Space Science. He received the Presidential Meritorious Rank Award in 2008, the Outstanding Performance awards in 1982, 1994 and 2008, and the NASA Outstanding Leadership Medals in 1998, 2004, and 2006. Mr. Saunders earned his BA in industrial engineering at the Georgia Institute of Technology. He has served on several National Academies' committees, including the Space Studies Board and the Committee on Astrobiology and Planetary Science.

SONIA M. TIKOO-SCHANTZ is an assistant professor at Stanford University in the Departments of Geophysics and, by courtesy, Geological Sciences. Dr. Tikoo-Schantz's research focuses on the use of paleomagnetism and fundamental rock magnetism as tools to investigate problems in the planetary sciences such as dynamo evolution, the origins of magnetic anomalies within planetary crusts, and impact cratering processes. Prior to joining Stanford University, she was an assistant professor at Rutgers University and a postdoctoral research associate at the University of California, Berkeley. Dr. Tikoo-Schantz received her PhD in planetary sciences from the Massachusetts Institute of Technology. She has not previously served on a National Academies' committee.

PANEL ON OCEAN WORLDS AND DWARF PLANETS

ALEXANDER G. HAYES (*Chair*) is an associate professor in the Department of Astronomy at Cornell University and director of the Cornell Center for Astrophysics and Planetary Science. Dr. Hayes has more than 15 years of experience in utilizing spacecraft-based platforms to study the properties of planetary surfaces. His NASA flight project experience includes Cassini, MER, MSL, Mars2020, Europa Clipper, and Dragonfly. He has also worked on instrument design and characterization for several Missile Defense Agency Programs, including THAAD, SM3, and EKV. Dr. Hayes's research program focuses on planetary surface processes and solar system exploration, with a special interest in the ocean worlds of the outer solar system, Mars, and comets. He is a recipient of the Zeldovich Medal from COSPAR and the Russian Academy of Sciences, the Ronald Greeley Early Career Award from the American Geophysical Union, the Sigma Xi Young Scholar Procter Prize, and a NASA Early Career Fellowship. Dr. Hayes received his PhD in planetary science from the California Institute of Technology. He is currently a member of the National Academies' Committee on Astrobiology and Planetary Science and the COSPAR Panel on Planetary Protection.

FRANCIS NIMMO (*Vice Chair*), see Steering Group entry.

MORGAN L. CABLE is a research scientist at the Jet Propulsion Laboratory. Her research interests focus on organic and biomarker detection through both in situ and remote sensing techniques. Dr. Cable also conducts fieldwork in extreme environments on Earth, searching for life in places such as the Atacama Desert, the summit of Mount Kilimanjaro, and the lava fields in Iceland. She has held numerous positions at the Jet Propulsion Laboratory including co-investigator for the Dragonfly mission, project staff scientist for the Europa Lander project, and supervisor of the Astrobiology and Ocean Worlds group. Dr. Cable is the recipient of numerous awards, including the NASA Early Career Public Achievement Medal for leadership roles in advancing NASA's missions, the Voyager Award for exceptional work on outer worlds projects, and the Bruce Murray Award for Excellence in Education and Public Engagement. She was also named by the American Chemical Society as one of the "Talented Twelve" rising stars in chemistry. Dr. Cable received her PhD in inorganic chemistry from the California Institute of Technology. She has not previously served on a National Academies' committee.

ALFONSO DAVILA is a research scientist at NASA's Ames Research Center. His research interests include the search for a second genesis of life in the solar system using terrestrial analog environments to assess the potential habitability of other planetary bodies and developing strategies for life detection based on first principles in biology and biochemistry. Dr. Davila's work includes conducting field investigations in the most extreme deserts on Earth, providing scientific advice for the maturation of instruments for space exploration, and developing mission concepts to search for evidence of life on Mars, Europa, and Enceladus. Prior to joining his current institution, Dr. Davila was a research scientist at the SETI Institute and a postdoctoral researcher at NASA's Ames Research Center. He received his PhD in geophysics from the University of Munich. Davila has not previously served on a National Academies' committee.

GLEN FOUNTAIN is a program manager (retired) at the Johns Hopkins University Applied Physics Laboratory (APL). He led the engineering team for the New Horizons mission to Pluto during the initial concept work and was the mission's project manager from 2004–2015. Although Mr. Fountain retired in 2016, he remains working in a part-time capacity, continuing to support the New Horizons mission and other activities of APL's Space Exploration Sector. At APL, Mr. Fountain was responsible for developing the attitude control system and sensors for the Small Astronomy Satellite series. As part of the system implementation for the MAGSAT satellite in the 1970s (for which he was assistant project scientist), Mr. Fountain led the team that developed one of the first microprocessor-based attitude control systems ever flown. He also led a group of engineers developing instruments flown on various missions, including Galileo, Ulysses, Delta 180, Delta 181, and Delta 183. Mr. Fountain oversaw hardware development to support missions including the Advance Composition Explorer, NEAR, MSX, CONTOUR, MESSENGER, and the New Horizons mission to Pluto. He received his MS in electrical engineering from Kansas State University. Mr. Fountain was a member of the Satellites Panel of the National Academies' *Vision and Voyages* decadal survey.

CHRISTOPHER R. GERMAN is a senior scientist at the Woods Hole Oceanographic Institution. Dr. German's research interests include the geological controls on seafloor fluid flow and its impacts on ocean budgets, chemosynthetic ecosystems, and astrobiology. He is an enthusiastic user of new technologies for deep ocean exploration. He is the co-lead for NASA's Network for Ocean Worlds, which advances comparative studies to characterize Earth and other ocean worlds across their interiors, oceans, and cryospheres; search for biosignatures; and understand life in relevant ocean world analogues and beyond. Dr. German is the recipient of numerous awards, including an Alexander Von Humboldt Foundation's Research Prize (Germany, 2014), the MBE medal (Queen Elizabeth II, UK, 2002), the Edward A. Flinn Award (International Lithosphere Panel, 2000), and the John Murray Award of the Royal Society (UK, 1988). He received his PhD in Earth sciences from the University of Cambridge. Dr. German has not previously served on a National Academies' committee.

CHRISTOPHER R. GLEIN is a lead scientist at the Southwest Research Institute. His research interests and projects include leading investigations of geochemical processes that involve water, other volatiles, and organic compounds on worlds in the outer solar system. Dr. Glein's work has contributed to the understanding of the origin and evolution of ocean worlds and their potential habitability. He was a member of the Cassini Ion and Neutral Mass Spectrometer team, and is currently on the Europa Clipper Mass Spectrometer for Planetary Exploration team. Prior to joining his current institution, Dr. Glein was a Deane postdoctoral fellow at the University of Toronto and a McClintock postdoctoral fellow at the Carnegie Institution for Science. He received his PhD in geological sciences from the Arizona State University. Dr. Glein has not previously served on a National Academies' committee.

CANDICE HANSEN is a senior scientist at the Planetary Science Institute. Her research interests include the seasonal behavior of volatiles, outer planet satellites' tenuous atmospheres, and Enceladus's plumes. In addition, Dr. Hansen's mission specialties include remote sensing, particularly visible and ultraviolet imagers, and operation of scientific instruments on spacecraft. She is a Juno co-investigator responsible for JunoCam. Dr. Hansen has held numerous positions at the Jet Propulsion Laboratory, including deputy principal investigator for the High Resolution Imaging Science Experiment, co-investigator on the Cassini Ultraviolet Imaging Spectrograph, and principal investigator for the New Frontiers mission study concept "Argo." She is a member of the American Geophysical Union and the American Astronomical Society. Dr. Hansen is the recipient of numerous awards, including the 2018 NASA Outstanding Public Leadership Medal for her work on JunoCam, the 2009 NASA Exceptional Scientific Achievement Medal, the 2005 NASA Exceptional Service Medal, and the 2002 Jet Propulsion Laboratory Exceptional Leadership Award. She received her PhD in geophysics and space physics from the University of California, Los Angeles. Dr. Hansen has not previously served on a National Academies' committee.

EMILY S. MARTIN is a research physical scientist for the Center for Earth and Planetary Studies at the Smithsonian Institution's National Air and Space Museum. Dr. Martin's research interests include planetary surface processes and tectonic deformation across the solar system, especially the icy bodies of the outer solar system. In addition, she is interested in using preserved geologic histories to better understand the development of ocean worlds. Previously, Dr. Martin was a postdoctoral research fellow and visiting researcher at the National Air and Space Museum. She received her PhD in geological sciences from the University of Idaho. Dr. Martin has not previously served on a National Academies' committee.

MARC NEVEU is an assistant research scientist at the University of Maryland, College Park. Dr. Neveu's research interests include planetary exploration and astrobiology and the search for the origins, evolution, and distribution of life in the universe. In particular, he seeks to understand whether oceans on other worlds harbor life by simulating their physics and chemistry, as well as through laboratory and field studies of similar places on Earth where microbial life thrives. Previously, Dr. Neveu was an astrobiology research assistant at NASA Headquarters and a postdoctoral researcher at Arizona State University. He is the recipient of numerous awards, including a Scialog Fellowship from the Research Corporation for Science Advancement, a NASA Postdoctoral Management Program Fellowship, and a NASA Early Career Fellowship. Dr. Neveu received his PhD in astrophysics from the Arizona State University. He has not previously served on a National Academies' committee.

CAROL S. PATY is a professor at the University of Oregon. Prior to joining her current organization, she was an associate professor in the School of Earth and Atmospheric Sciences at the Georgia Institute of Technology. Dr. Paty also worked at the Southwest Research Institute in the Space Sciences and Engineering Division. Her research is focused on understanding planetary magnetospheric dynamics and moon-magnetosphere interactions using a combination of computational simulations and data collected by various space-based instruments. Dr. Paty pioneered the application of multifluid plasma dynamic simulations to icy moons and outer planet magnetospheres, and the inclusion of plasma-neutral interactions in global simulations. She was a participating scientist on the Cassini mission to Saturn and is currently a co-investigator on the Plasma Environment Package for the European mission to Ganymede (JUICE). Dr. Paty is also a co-investigator on both the Plasma Instrument for Magnetic Sounding and the Radar for Europa Assessment and Sounding: Ocean to Near-Surface instruments for NASA's Europa Clipper mission. She earned her PhD in geophysics and space physics from the University of Washington. Dr. Paty previously served on the National Academies' Committee on Strategic NASA Science Missions.

LYNNAE C. QUICK is a research scientist at NASA's Goddard Space Flight Center. Her research focuses on geophysical processes on the ocean worlds in the solar system and in extrasolar planetary systems. She applies principles of fluid dynamics and heat transfer to model cryovolcanism and ocean crystallization on the icy moons of the giant planets and on dwarf planet Ceres. As a member of the Europa Clipper science team, Dr. Quick is a co-investigator on the Europa Imaging System, and a science team member and lead for the student and early career investigator program science enhancement option on the Dragonfly mission. She also served as a science collaborator on the Dawn mission. Dr. Quick is a member of the AAS Division for Planetary Sciences, the National Society of Black Physicists, the American Geophysical Union, and the American Astronomical Society. She received her PhD in Earth and planetary sciences from the Johns Hopkins University. Dr. Quick has not previously served on a National Academies' committee.

JASON M. SODERBLOM is a research scientist in the Department of Earth, Atmospheric, and Planetary Sciences at the Massachusetts Institute of Technology (MIT). His research interests include exploring the composition, operative geologic processes, and evolutionary history of planetary surfaces through analysis of visible and near-infrared images and spectra. Dr. Soderblom is a co-investigator on the Mapping Imaging Spectrometer for Europa and Europa Imaging System instruments selected for the NASA Europa Clipper mission and on the recently selected Dragonfly mission to Titan. Prior to joining MIT, he was a postdoctoral researcher at the University of Arizona and a visiting scientist at Cornell University. Dr. Soderblom is the recipient of several honors, including a NASA Early Career Fellowship, a New York–NASA Space Grant Fellowship, and several NASA Group Achievement Awards. He received his PhD and MS in astronomy from Cornell University. Dr. Soderblom has not previously served on a National Academies' committee.

KRISTA M. SODERLUND is a research scientist at The University of Texas at Austin, Institute for Geophysics (UTIG). Dr. Soderlund's research interests include planetary and satellite interiors, fluid dynamics, and dynamos across the solar system. Recent projects have focused on ocean circulation and ice-ocean coupling within the Galilean and saturnian satellites as well as magnetic field generation in Mercury, the Moon, and the ice giants. She is a co-investigator on the Radar for Europa Assessment and Sounding: Ocean to Near-Surface Instrument on NASA's Europa Clipper Mission, was a co-investigator on the Neptune-Triton Planetary Mission Concept

Study, and served on the Ice Giant Predecadal Mission Concept Study science definition team. Dr. Soderlund is the recipient of numerous awards, including the UTIG Outstanding Young Researcher Award and a NASA Early Career Fellowship. She received her PhD in geophysics and space physics from the University of California, Los Angeles. Dr. Soderlund has not previously served on a National Academies' committee.

PANEL ON SMALL SOLAR SYSTEM BODIES

NANCY L. CHABOT (*Chair*) is the planetary chief scientist at the Johns Hopkins University Applied Physics Laboratory. Previously, she was a senior research associate at the Case Western Reserve University and an NRC postdoctoral research scientist at NASA's Johnson Space Center. Dr. Chabot served as the instrument scientist for the Mercury Dual Imaging System on NASA's MESSENGER mission and was chair of the Geology Discipline Group. Currently, she is the coordination lead on NASA's Double Asteroid Redirection Test mission, the deputy principal investigator for the Mars-moon Exploration with Gamma rays and Neutrons instrument on the JAXA's Martian Moons Exploration mission, and an interdisciplinary scientist on the joint ESA-JAXA BepiColombo mission. Dr. Chabot received a PhD in planetary sciences from the University of Arizona. She has not previously served on a National Academies' committee.

CAROL RAYMOND (*Vice Chair*), see Steering Group entry.

PAUL A. ABELL is the chief scientist for Small Body Exploration at NASA's Johnson Space Center in the Astro-materials Research and Exploration Science Division. He has been a visiting astronomer at the NASA Infrared Telescope Facility at Mauna Kea Observatory and a team member of JAXA's Hayabusa and Hayabusa2 sample return missions. Dr. Abell's research interests include the physical characterization of near-Earth objects (NEOs) via ground-based and spacecraft observations, examination of NEOs for future robotic and human exploration, and identification of potential resources within the NEO population for future in situ utilization. He is a member of an internal NASA team that is examining the possibility of sending astronauts to NEOs for human missions, a science team member of the Vera C. Rubin Observatory Solar System Collaboration, and an investigation team member on both NASA's Double Asteroid Redirection Test and Near-Earth Object Surveyor planetary defense missions. Dr. Abell received a PhD in geology from the Rensselaer Polytechnic Institute.

WILLIAM F. BOTTKE is the director of the Department of Space Studies at Southwest Research Institute. He uses numerical simulations to study the formation, evolution, and bombardment history of the solar system's planets, satellites, and small bodies. In his work, Dr. Bottke has created collisional and dynamical evolution models of the main asteroid belt, the primordial Kuiper belt, Trojans, irregular satellites, and near-Earth objects (NEOs). He has also explored how small body populations were affected by giant planet migration and nongravitational forces. Dr. Bottke has been the first author of 10 articles for *Nature, Science,* and *Nature Geosciences* and 2 articles for *Annual Reviews of Earth and Planetary Science,* and was an editor for the books *Asteroids III* and *IV.* He has also led two teams within NASA's Lunar Science Institute and NASA's Solar System Exploration Research Virtual Institute. Dr. Bottke is currently a science team member on four NASA missions: OSIRIS-REx, Lucy, Psyche, and NEO Surveyor. He has been the recipient of numerous prizes, including the first Paolo Farinella Prize and a fellowship in the Meteoritical Society, and has also been a Shoemaker Lecturer for the American Geophysical Union as well as a Kavli Lecturer for the American Astronomical Society. Dr. Bottke received his PhD in planetary sciences from the University of Arizona. He was a Texaco Prize Postdoctoral Fellow at Caltech and a research associate at Cornell University.

HAROLD C. CONNOLLY, JR., is the founding chair and a professor within the Department of Geology at Rowan University. Prior to joining his current university, he was a professor at the City University of New York. Dr. Connolly has held visiting positions at the Lunar and Planetary Laboratory of the University of Arizona, Hokkaido University, and the University of Tokyo. He is a geologist who specializes in problems founded in petrology and geochemistry through the study of primitive planetary meteorites known as chondrites. His research aims to constrain the processes and timing of the formation and evolution of rock bodies within the solar system with an emphasis on asteroids. Dr. Connolly is mission sample scientist and co-investigator of the OSIRIS-REx asteroid sample return mission. He is also a co-investigator of the Japan Aerospace Exploration Agency's asteroid sample return mission, Hayabusa2. Dr. Connolly is a founding member of the Hayabusa2 and OSIRIS-REx Joint Comparative Asteroid Working Group. He is the recipient of the U.S. Antarctic Service Medal from the Department of the Navy, and a fellow of the Meteoritical Society. Dr. Connolly received a PhD in geological sciences from Rutgers University and held a postdoctoral position at the California Institute of Technology.

THOMAS D. JONES is a senior research scientist at the Institute for Human and Machine Cognition, a not-for-profit research institute of the Florida State University System. He is a planetary scientist and consultant to NASA and the aerospace community. As an astronaut with NASA, Dr. Jones logged 53 days in space. He conducted science operations for Space Radar Laboratory (SRL-1) on STS-59, was the payload commander on STS-68's SRL-2, deployed and recovered science satellites on STS-80, and, with the STS-98 crew, delivered the U.S. Lab Destiny to the International Space Station. Previous employers include Science Applications International Corporation, the Central Intelligence Agency, and the U.S. Air Force's Strategic Air Command (as a B-52 pilot). Previously, Dr. Jones served on the NASA Advisory Council and was a board member of the Association of Space Explorers and the Astronauts Memorial Foundation. He is the recipient of the NASA Distinguished Service Medal, four NASA Space Flight Medals, the NASA Exceptional Service award, the NASA Outstanding Leadership Medal, and the Air Force Commendation Medal. In 2018, Dr. Jones was inducted into the U.S. Astronaut Hall of Fame. He graduated from the Air Force Academy and received a PhD in planetary sciences from the University of Arizona.

STEFANIE N. MILAM is the James Webb Space Telescope deputy project scientist for planetary science at NASA's Goddard Space Flight Center. Dr. Milam's research interests include measurements of the volatile isotopic enrichments from nearby evolved stars and chemical processes that are found in the interstellar medium, star-forming regions, and planetary systems. She also conducts remote observations of comets and interstellar objects and has a laboratory dedicated to studying chemical processing that occurs on these bodies. Previously, Dr. Milam was a principal investigator at the SETI Institute and at NASA's Ames Research Center. She is a member of the American Astronomical Society (AAS), the American Chemical Society (ACS), the AAS Division for Planetary Sciences, the AAS Laboratory Astrophysics Division, the ACS Astrochemistry Subdivision, and the International Astronomical Union. Dr. Milam received her PhD in chemistry from the University of Arizona.

EDGARD G. RIVERA-VALENTÍN is a senior scientist with the Universities Space Research Association at the Lunar and Planetary Institute. Their research interests are in applied planetary science, particularly in the study of processes the help inform planetary defense, planetary protection, and planetary surface exploration. Previously, Dr. Rivera-Valentín was a planetary radar astronomer at the Arecibo Observatory, a postdoctoral research associate at Brown University, and a senior graduate assistant at the University of Arkansas. They are the recipient of the NASA Planetary Science Division's Early Career Fellow Award, a doctoral academy fellow at the University of Arkansas, the Diversity Leadership Award and the Metzger Award in Astronomy at Alfred University. Asteroid 2010 ER87 is now named (389478) Rivera-Valentín in their honor. Dr. Rivera-Valentín received a PhD in space and planetary science from the University of Arkansas.

DANIEL J. SCHEERES is the A. Richard Seebass Endowed Chair and distinguished professor at the University of Colorado Boulder in the Smead Department of Aerospace Engineering Sciences. Dr. Scheeres has studied the dynamics of the asteroid environment from a scientific, engineering, and navigation perspective since 1992 and has been involved with NASA's NEAR mission to asteroid Eros and the Japanese Hayabusa missions to asteroids Itokawa and Ryugu. He is currently a co-investigator on NASA's OSIRIS-REx mission to asteroid Bennu, leading the Radio Science team of that mission, and is the principal investigator of Janus, a NASA SIMPLEx mission currently scheduled to launch in August 2022. Dr. Scheeres has published a Springer Praxis book on orbital mechanics about small bodies titled *Orbital Motion in Strongly Perturbed Environments: Applications to Asteroid, Comet and Planetary Satellite Orbiters.* Asteroid 8887 is named "Scheeres" in recognition of his contributions to the scientific understanding of the dynamical environment about asteroids. He is a fellow of both the American

Institute of Aeronautics and Astronautics and the American Astronautical Society and is a member of the International Academy of Astronautics and the National Academy of Engineering. Dr. Scheeres has been awarded the Dirk Brouwer Award from the American Astronautical Society. He earned his PhD for aerospace engineering from the University of Michigan. Dr. Scheeres has served on the National Academies' Committee on Assessment of the U.S. Air Force's Astrodynamic Standards and the NEO Mitigation Panel.

RHONDA STROUD is the head of the Nanoscale Materials Section at the Naval Research Laboratory. Dr. Stroud is a materials physicist and planetary scientist focusing on nanostructures, including quasicrystals and aerogel, and on the materials that make up comets and cosmic dust. In addition, she pioneered the use of the focused ion beam technology in the study of meteorites. Dr. Stroud joined the Naval Research Laboratory as a postdoctoral researcher after completing her doctorate. She has served as an external reviewer for the Materials Division at Argonne National Laboratory, on the Department of Energy's external review committee for the electron microscopy user facilities at Oak Ridge National Laboratory, and the National Center for Electron Microscopy at Lawrence Berkeley National Laboratory. Dr. Stroud is a fellow of the American Physical Society, the Meteoritical Society, and the Microscopy Society of America. She received a PhD in physics from the Washington University in Saint Louis.

MEGAN BRUCK SYAL is a physicist at Lawrence Livermore National Laboratory (LLNL) in the Design Physics Division. Dr. Syal specializes in experimental and numerical simulation of planetary impacts, including hypervelocity impact experiments (with an emphasis on porous and volatile-rich materials) and modeling of impact events in a variety of shock physics codes. Additionally, she is very active in the field of planetary defense, supporting a variety of activities including the following: NASA's DART mission with simulations of the planned 2022 spacecraft impact at Dimorphos, NASA–FEMA Asteroid Impact Tabletop Exercises, and a NASA–NNSA interagency collaboration on hazardous asteroid mitigation case studies. Previously, Dr. Syal was a postdoctoral researcher at LLNL, and a data specialist at the Smithsonian Astrophysical Observatory's Chandra X-Ray Center. She is a recipient of a NASA Earth and Space Science Fellowship, a NASA Group Achievement Award (for her involvement in the Deep Impact—EPOXI mission Science Team), and a Brown University Graduate Fellowship. Dr. Syal obtained her PhD in planetary geosciences at Brown University.

MYRIAM TELUS is an assistant professor at the University of California, Santa Cruz. Her research interests include studying the constraining timescales and conditions of planetesimal formation and evolution via elemental, isotopic and petrographic analyses of meteorites. Prior to joining the University of California, Santa Cruz, Dr. Telus was a postdoctoral fellow at the Carnegie Institute of Washington. She is the recipient of numerous awards, including the 2019 NASA Planetary Science Early Career Award, the 2011 Watumull Scholarship for excellence in graduate research from the University of Hawaii at Manoa, and the Gates Millennium Scholarship. Dr. Telus received a PhD in geology and geophysics from the University of Hawaii at Manoa. She has not previously served on a National Academies' committee.

AUDREY THIROUIN is a senior research scientist at Lowell Observatory. Dr. Thirouin's research interests include physical and dynamical properties of the small bodies in the solar system such as asteroids, comets, Centaurs, Trojans, and trans-neptunian objects. She is particularly interested in the rotational properties of these bodies and what can learned from them. Previously, Dr. Thirouin was a research assistant at the Instituto de Astrofisica de Andalucia, the Centre de Spectrométrie Nucléaire et de Spectrométrie de Masse, and the Institut de Mécanique Céleste et Calcul d'Éphémérides. She is a member of the American Astronomical Society's Division for Planetary Sciences and was a member of the Spanish Society of Astronomy. Dr. Thirouin received her PhD in astronomy and planetary science from the University of Granada, Spain. She has not previously served on a National Academies' committee.

CHAD TRUJILLO is an associate professor at Northern Arizona University. His research interests focus on the Kuiper belt, inner Oort cloud, outer solar system, planet formation, Titan, and active asteroids. Previously, Dr. Trujillo was the head of adaptive optics, an astronomer, and a science fellow at the International Gemini

Observatory. In addition, he was a postdoctoral scholar at the California Institute of Technology and a research assistant at the University of Hawaii. Dr. Trujillo is the principal investigator on three previous NASA programs: Exploring the Inner Oort Cloud, Beyond the Kuiper Belt Edge, and Primordial Solar System Ices. He is the recipient of numerous awards, including the 2015 AURA Service Award for Science and the *Science Spectrum Magazine* 2005 Trailblazer award. Dr. Trujillo received a PhD in astronomy from the University of Hawaii. He has not previously served on a National Academies' committee.

BENJAMIN P. WEISS is a professor of planetary sciences at the Massachusetts Institute of Technology (MIT). He also serves as chair of the Program in Planetary Sciences and director of the MIT Paleomagnetism Laboratory. His research interests include the study of the formation, evolution, and history of planetary bodies, with a focus on paleomagnetism and geomagnetism, geophysics, meteoritics, and habitability. Previously, Dr. Weiss was a visiting Miller Professor at the University of California, Berkeley; a visiting professor at the Institut de Physique du Globe de Paris; a Victor O. Starr assistant professor at MIT; and a scientist at the California Institute of Technology. He is the recipient of the James B. Macelwane Medal from the American Geophysical Union, the Charles E. Reed Faculty Initiatives Award, and the Milton and Francis Clauser Doctoral Prize. Dr. Weiss received a PhD in planetary science and geology from the California Institute of Technology. He has not previously served on a National Academies' committee.

PANEL ON VENUS

PAUL K. BYRNE (*Chair*) is an associate professor of Earth and planetary sciences in the Department of Earth and Planetary Sciences at Washington University in St. Louis. His research focuses on comparative planetary geology—comparing and contrasting the surfaces and interiors of planetary bodies, including Earth, to understand geological phenomena at the systems level. Dr. Byrne's research projects span the solar system from Mercury to Pluto and, increasingly, extend to the study of extrasolar planets. Prior to joining Washington University, he was a postdoctoral research fellow, first at the Earth and Planets Laboratory of the Carnegie Institution for Science in Washington, DC, and then at the Lunar and Planetary Institute in Houston, Texas, followed by becoming an assistant and then associate professor at North Carolina State University. Dr. Byrne is the president-elect of the American Geophysical Union's Planetary Sciences section, serves on the steering committee of the Venus Exploration Analysis Group, and is a recipient of numerous awards, including the NASA Group Achievement Award as part of the MESSENGER mission team in 2018. He was named a NASA Early Career Fellow in 2019. Dr. Byrne earned his PhD in planetary geology from Trinity College in Dublin, Ireland. He has not previously served on a National Academies' committee.

LARRY W. ESPOSITO (*Vice Chair*), see Steering Group entry.

GIADA N. ARNEY is a research space scientist in the Planetary Systems Laboratory at NASA's Goddard Space Flight Center. Her research interests include solar system–exoplanet synergies, astrobiology, and terrestrial solar system worlds (particularly Venus and Earth). Dr. Arney currently serves as one of two deputy principal investigators of the DAVINCI mission to Venus. Previously, she led the Science Support Analysis Team for the LUVOIR astrophysics flagship mission concept study. Dr. Arney was a co-editor of a recent title in the Space Science Series, *Planetary Astrobiology*. She received the NASA Early Career Achievement Award in 2018 and was a recipient of the Presidential Early Career Award for Scientists and Engineers in 2019. Dr. Arney received her PhD in astronomy and astrobiology from the University of Washington. She has not previously served on a National Academies' committee.

AMANDA S. BRECHT is a space scientist in the Planetary Systems Branch of NASA's Ames Research Center. Her research interests include planetary science, planetary atmospheres, general circulation models, and comparative planetology. Prior to joining NASA Ames in her current capacity, she was a research scientist at the Bay Area Environmental Research Institute, a NASA postdoctoral program fellow at NASA Ames, and a graduate research assistant at the University of Michigan. Dr. Brecht is the recipient of numerous awards, including the Rackham Engineering Award and a NASA Postdoctoral Fellowship. She received her PhD in atmospheric and space science from the University of Michigan, Ann Arbor. Dr. Brecht has not previously served on a National Academies' committee.

THOMAS E. CRAVENS is a professor at the University of Kansas in the Department of Physics and Astronomy. His research interests focus on planetary ionospheres; the interaction of the solar wind with planets, including Earth, Venus, Mars, and comets; and solar system X-ray emission. He was a team member on the Cassini Ion and Neutral Mass Spectrometer and is an interdisciplinary scientist on NASA's Mars Atmosphere and Volatile Evolution mission. Dr. Cravens was a co-investigator on the Ion and Electron Sensor for the Rosetta Plasma Consortium and was a co-investigator on the TUNDE and PLASMAG instruments on the VEGA-1 and 2 missions to comet Halley. He is a fellow of the American Association for the Advancement of Science and the American Geophysical Union. Dr. Cravens received his PhD in astronomy from Harvard University.

KANDIS-LEA JESSUP is a mid-career lead scientist working in the Department of Space Studies at the Southwest Research Institute and an affiliate professor in the Department of Physics and Astronomy at Regis College. She specializes in the observation and modeling of planetary atmospheres across the solar system (e.g., Venus, Io, Titan, and Pluto). Her research ranges in breadth from spectroscopic observation and the related radiative transfer modeling, photochemical modeling, and microphysical modeling needed to interpret these data to broader planet-systems focused topics such as atmospheric evolution and climate. Additionally, Dr. Jessup is actively involved in the development of instrumentation designed to enable remote and/or in situ investigation of solar system atmosphere targets. She is the co-author of two academic book chapters: "Io's Atmosphere" by Lellouch et al. 2007 (*Io After the Galileo Era*) and "Photochemistry and Haze Formation," by Mandt et al. 2020 (*The Pluto System*). Dr. Jessup has substantial experience working internationally as an interdisciplinary scholar. She served on the joint NASA-Roscosmos Venera-D Mission Science Definition Team from 2015 to 2019 and is currently assigned as a guest investigator on ESA's BepiColombo mission to Mercury. In this latter role, Dr. Jessup actively supported collaborations between the BepiColombo Science Team and the Joint NASA–JAXA Akatsuki Mission Science Working Team. Dr. Jessup is also one of the key leaders of the series of Cloud Habitability Workshops sponsored by NASA and IKI in 2019 and 2021. She is a member of the American Geophysical Union and the American Astronomical Society. Dr. Jessup received her PhD in atmospheric and space science from the University of Michigan in 2002. She has not previously served on a National Academies' committee.

JAMES F. KASTING is an Evan Pugh Professor at Pennsylvania State University (PSU). His research interests include atmospheric evolution, planetary atmospheres, and paleoclimates. Before joining PSU, he spent 2 years at the National Center for Atmospheric Research in Boulder, Colorado, and 7 years in the Space Science Division at NASA's Ames Research Center. Dr. Kasting is a fellow of the American Association for the Advancement of Science, the International Society for the Study of the Origin of Life, and the American Geophysical Union and is a member of the National Academy of Sciences. He is the recipient of the Stanley Miller Medal, also known as the NAS Award in Early Earth and Life Sciences. Dr. Kasting earned his PhD in atmospheric science from the University of Michigan. His service on National Academies' committees includes membership on the Committee for US–USSR Workshop on Planetary Sciences, the Panel to Review Terrestrial Planet Finder Science Goals, and chair of the organizing committee for Searching for Life Across Space and Time: A Workshop.

SCOTT D. KING is a professor at the Virginia Polytechnic Institute and State University. Dr. King's research focuses on the study of thermomechanical processes that operate within planets using numerical models constrained by spacecraft data. Currently, he is a guest investigator of the Dawn at Ceres mission and a team member of the InSight mission. Dr. King has participated in multiple NASA R&A review panels, including as group chief. He is a fellow of the Geological Society of America and is the recipient of numerous awards, including a senior Alexander von Humboldt research prize. Dr. King was named a University Faculty Scholar at Purdue University.

He received his PhD in geophysics from the California Institute of Technology. Dr. King has not previously served on a National Academies' committee.

BERNARD MARTY is a professor at the Université de Lorraine in France. Prior to joining his current institution, Dr. Marty was a postdoctoral researcher at the Geophysical Institute of the University of Tokyo and a CNRS research scientist at the Université Pierre et Marie Curie. He was selected as a senior member of the Institut Universitaire de France, a knight in 2012, and promoted to officer in 2018 under the Ordre des Palmes académiques of the Republic of France. Dr. Marty is the recipient of the Grand Prix Dolomieu of the French Academy of Sciences, the Bowen Award of the American Geophysical Union, and the Goldschmidt Medal of the Geochemical Society. Currently, he is involved in many space missions, including principal investigator for the analysis of noble gases and nitrogen in Apollo samples and member of the Science Team for NASA's Genesis mission, the Preliminary Examination Team for NASA's Stardust mission and the JAXA Hayabusa2 and NASA OSIRIS-REx analysis teams. Dr. Marty received his PhD in physics from Université Pierre et Marie Curie.

THOMAS NAVARRO is a postdoctoral fellow at the McGill Space Institute at McGill University and an assistant researcher in the Department of Earth, Planetary, and Space Sciences at the University of California, Los Angeles. His research and projects focus on the development and utilization of state-of-the-art three-dimensional general circulation models, along with their validation and comparison to satellite observations. Dr. Navarro is a NASA participating scientist on JAXA's Akatsuki mission to Venus and was involved in the proposed HOVER Venus mission. Prior to starting his current role, he was a postdoctoral scholar at UCLA, a graduate research assistant at the Sorbonne University, and a research engineer at CNRS in Paris. Dr. Navarro received his PhD in planetary science from the Sorbonne University. He has not previously served on a National Academies' committee.

JOSEPH G. O'ROURKE is an assistant professor at Arizona State University in the School of Earth and Space Exploration. His research focuses on planetary interiors—how processes that occur deep within planets can govern conditions on their surfaces over geologic time. Dr. O'Rourke has served on the steering committee of the Venus Exploration Analysis Group. Prior to assuming his current position, he was a postdoctoral scholar at Arizona State University. Dr. O'Rourke is the recipient of numerous awards, including a National Science Foundation graduate research fellowship and awards from the American Geophysical Union for outstanding graduate research and presentations. He received his BS in astronomy and physics and geology and geophysics from Yale University and his PhD in planetary science from the California Institute of Technology.

JENNIFER M. ROCCA is a principal project systems engineer at the Jet Propulsion Laboratory (JPL), where she is the Engineering Technical Authority for the SPHEREx near-infrared all-sky spectral survey mission. She has been with JPL for more than 20 years, serving in numerous flight and project system design, development, and operations roles on GRACE, Deep Impact, SIM, Dawn, Juno, and NISAR. Ms. Rocca has proposal and concept study experience in systems engineering and capture-lead roles for proposals in Earth sciences, planetary sciences, and astrophysics. She is the recipient of numerous awards, including the 2019 JPL Explorer Award for SPHEREx Capture Leadership and NASA Exceptional Achievement Medals for her work on Deep Impact and Juno. Ms. Rocca received her MS in aeronautical and astronautical engineering from Stanford University. She has not previously served on a National Academies' committee.

ALISON R. SANTOS is a postdoctoral research associate at Wesleyan University in the Earth and Environmental Sciences Department. Her research interests include igneous petrology and geochemistry, volatiles in planetary systems, and planetary evolution, and she has conducted research in these areas on Mars, Venus, and the angrite parent body. Previously, Dr. Santos was a postdoctoral fellow with the Universities Space Research Association through NASA's Postdoctoral Program based at NASA's Glenn Research Center, and a research scientist with the Institute of Meteoritics at the University of New Mexico. She received her PhD in Earth and planetary sciences from the University of New Mexico and was a New Mexico Space Grant awardee while there. Dr. Santos has not previously served on a National Academies' committee.

JENNIFER L. WHITTEN is an assistant professor of planetary science at Tulane University. Her research interests are focused on geologic processes that generate and modify terrestrial planetary crusts, including volcanic and impact-related resurfacing processes. She is currently leading a study of the radiophysical properties of Venus surface units using Magellan data and Arecibo telescopic observations. Before joining Tulane, Dr. Whitten was a postdoctoral researcher at the National Air and Space Museum's Center for Earth and Planetary Studies. She is the associate principal investigator of the VERITAS Discovery mission proposal team and serves as a member of the Venus Exploration and Analysis Group Steering Committee. Dr. Whitten has received several awards, including being named a NASA Early Career Fellow. She received her PhD in planetary geology from Brown University.